D1325500
52 795 272 6

He
cor
Cor

PHILIP'S

ASTRONOMY ENCYCLOPEDIA

PHILIP'S

ASTRONOMY ENCYCLOPEDIA

Foreword by Leif J. Robinson
Editor Emeritus, *Sky & Telescope* magazine

Star Maps created by Wil Tirion

General Editor Sir Patrick Moore

PHILIP'S ASTRONOMY ENCYCLOPEDIA
First published in Great Britain in 1987 by Mitchell Beazley under the title The Astronomy Encyclopedia (General Editor Patrick Moore)

This fully revised and expanded edition first published in 2002 by Philip's, an imprint of Octopus Publishing Group
2–4 Heron Quays
London E14 4JP

ISBN 0–540–07863–8

Managing Editor *Caroline Rayner*
Technical Project Editor *John Woodruff*
Commissioning Editor *Frances Adlington*
Consultant Editor *Neil Bone*
Executive Art Editor *Mike Brown*
Designer *Alison Todd*
Picture Researcher *Cathy Lowne*
Production Controller *Sally Banner*

THE GREEK ALPHABET

α Α alpha	η Η eta	ν Ν nu	τ Τ tau
β Β beta	θ Θ theta	ξ Ξ xi	υ Υ upsilon
γ Γ gamma	ι Ι iota	ο Ο omicron	φ Φ phi
δ Δ delta	κ Κ kappa	π Π pi	χ Χ chi
ε Ε epsilon	λ Λ lambda	ρ Ρ rho	ψ Ψ psi
ζ Ζ zeta	μ Μ mu	σ Σ sigma	ω Ω omega

MULTIPLES AND SUBMULTIPLES USED WITH SI UNITS

Multiple	Prefix	Symbol	Submultiple	Prefix	Symbol
10^3	kilo-	k	10^3	milli-	m
10^6	mega-	M	10^6	micro-	m
10^9	giga-	G	10^9	nano-	n
10^{12}	tera-	T	10^{12}	pico-	p
10^{15}	peta-	P	10^{15}	femto-	f
10^{18}	exa-	E	10^{18}	atto-	a

HOW TO USE THE ENCYCLOPEDIA

Alphabetical order

'Mc' is treated as if it were spelled 'Mac', and certain shortened forms as if spelled out in full (e.g. 'St' is treated as 'Saint'). Entries that have more than one word in the heading are alphabetized as if there were no space between the words. Entries that share the same main heading are in the order of people, places and things. Entries beginning with numerals are treated as if the numerals were spelled out (e.g. **3C** follows **three-body problem** and precedes **3C 273**). An exception is made for **HI region** and **HII region**, which appear together immediately after **Hirayama family**. Biographies are alphabetized by surname, with first names following the comma. (Forenames are placed in parentheses if the one by which a person is commonly known is not the first.) Certain lunar and planetary features appear under the main element of names (e.g. **Imbrium, Mare** rather than Mare Imbrium).

Cross-references

SMALL CAPITALS in an article indicate a separate entry that defines and explains the word or subject capitalized. '*See also*' at the end of an article directs the reader to entries that contain additional relevant information.

Measurements

Measurements are given in metric (usually SI) units, with an imperial conversion (to an appropriate accuracy) following in parentheses where appropriate. In historical contexts this convention is reversed so that, for example, the diameter of an early telescope is given first in inches. Densities, given in grams per cubic centimetre, are not converted, and neither are kilograms or tonnes. Large astronomical distances are usually given in light-years, but parsecs are sometimes used in a cosmological context. Particularly in tables, large numbers may be given in exponential form. Thus 10^3 is a thousand, 2×10^6 is two million, and so on. 'Billion' always means a thousand million, or 10^9. As is customary in astronomy, dates are expressed in the order year, month, day. Details of units of measurement, conversion factors and the principal abbreviations used in the book will be found in the tables on this page.

Stellar data

In almost all cases, data for stars are taken from the HIPPARCOS CATALOGUE. The very few exceptions are for instances where the catalogue contains an error of which the editors have been aware. In tables of constellations and elsewhere, the combined magnitude is given for double stars, and the average magnitude for variable stars.

Star Maps pages 447–55

Acknowledgements page 456

FRONTMATTER IMAGES

Endpapers: **Andromeda Galaxy** The largest member of the Local Group, this galaxy is the farthest object that can be seen with the naked eye.

Half-title: **Crab Nebula** This nebula is a remnant of a supernova that exploded in the constellation of Taurus in 1054.

Opposite title: **M83** Blue young stars and red HII emission nebulae clearly mark out regions of star formation in this face-on spiral galaxy in Hydra.

Opposite Foreword: **NGC 4945** This classic disk galaxy is at a distance of 13 million l.y. Its stars are mainly confined to a flat, thin, circular region surrounding the nucleus.

Opposite page 1: **Earth** This photograph was obtained by the Apollo 17 crew en route to the Moon in 1972 December.

SYMBOLS FOR UNITS, CONSTANTS AND QUANTITIES

a	semimajor axis	*L*	luminosity	*t*	time
Å	angstrom unit	L_n	Lagrangian points (n = 1 to 5)	*T*	temperature (absolute), epoch (time of perihelion passage)
AU	astronomical unit	l.y.	light-year	T_{eff}	effective temperature
c	speed of light	m	metre, minute	*v*	velocity
d	distance	*m*	apparent magnitude, mass	W	watt
e	eccentricity	m_{bol}	bolometric magnitude	y	year
E	energy	m_{pg}	photographic magnitude	*z*	redshift
eV	electron-volt	m_{pv}	photovisual magnitude	α	constant of aberration, right ascension
f	following	m_V	visual magnitude	δ	declination
F	focal length, force	*M*	absolute magnitude, mass (stellar)	λ	wavelength
g	acceleration due to gravity	N	newton	μ	proper motion
G	gauss	*p*	preceding	ν	frequency
G	gravitational constant	*P*	orbital period	π	parallax
h	hour	pc	parsec	ϖ	longitude of perihelion
h	Planck constant	*q*	perihelion distance	Ω	observed/critical density ratio, longitude of ascending node
H_0	Hubble constant	q_0	deceleration parameter	°	degree
Hz	hertz	*Q*	aphelion distance	′	arcminute
i	inclination	*r*	radius, distance	″	arcsecond
IC	*Index Catalogue*	*R*	Roche limit		
Jy	jansky	s	second		
k	Boltzmann constant				
K	degrees kelvin				

CONVERSION FACTORS

Distances
1 nm = 10 Å
1 inch = 25.4 mm
1 mm = 0.03937 inch
1ft = 0.3048 m
1 m = 39.37 inches = 3.2808 ft
1 mile = 1.6093 km
1 km = 0.6214 mile
1 km/s = 2237 mile/h
1 pc = 3.0857×10^{13} km = 3.2616 l.y. = 206,265 AU
1 l.y. = 9.4607×10^{12} km = 0.3066 pc = 63,240 AU

Temperatures (to the nearest degree)	
°C to °F:	×1.8, +32
°C to K:	+273
°F to °C:	−32, ÷1.8
°F to K:	÷1.8, +255
K to °C:	−273
K to °F:	×1.8, −460

Note: To convert temperature *differences*, rather than points on the temperature scale, ignore the additive or subtractive figure and just multiply or divide.

CONTRIBUTORS

Philip's would like to thank the following contributors for their valuable assistance in updating and supplying new material for this edition:

Alexander T. Basilevsky, Vernadsky Institute of Geochemistry and Analytical Chemistry, Moscow, Russia
Richard Baum, UK
Peter R. Bond, FRAS, FBIS, Space Science Advisor for the Royal Astronomical Society, UK
Neil Bone, Director of the BAA Meteor Section and University of Sussex, UK
Dr Allan Chapman, Wadham College, University of Oxford, UK
Storm Dunlop, FRAS, FRMetS, UK
Tim Furniss, UK
Peter B. J. Gill, FRAS, UK
Dr Ian S. Glass, South African Astronomical Observatory, South Africa
Dr Monica M. Grady, The Natural History Museum, London, UK
Dr Andrew J. Hollis, BAA, UK
James B. Kaler, Department of Astronomy, University of Illinois, USA
William C. Keel, Department of Physics and Astronomy, University of Alabama, USA
Professor Chris Kitchin, FRAS, University of Hertfordshire, UK
Professor Kenneth R. Lang, Tufts University, USA
Dr Richard McKim, Director of the BAA Mars Section, UK
Mathew A. Marulla, USA
Steve Massey, ASA, Australia
Sir Patrick Moore, CBE, FRAS, UK
Dr François Ochsenbein, Astronomer at Observatoire Astronomique de Strasbourg, France
Dr Christopher J. Owen, PPARC Advanced Fellow, Mullard Space Science Laboratory, University College London, UK
Chas Parker, BA, UK
Neil M. Parker, FRAS, Managing Director, Greenwich Observatory Ltd, UK
Martin Ratcliffe, FRAS, President of the International Planetarium Society (2001–2002), USA
Ian Ridpath, FRAS, Editor *Norton's Star Atlas*, UK
Leif J. Robinson, Editor Emeritus *Sky & Telescope* magazine, USA
Dr David A. Rothery, Department of Earth Sciences, The Open University, UK
Robin Scagell, FRAS, Vice President of the Society for Popular Astronomy, UK
Jean Schneider, Observatoire de Paris, France
Dr Keith Shortridge, Anglo-Australian Observatory, Australia
Dr Andrew T. Sinclair, former Royal Greenwich Observatory, UK
Pam Spence, MSc, FRAS, UK
Dr Duncan Steel, Joule Physics Laboratory, University of Salford, UK
Nik Szymanek, University of Hertfordshire, UK
Richard L. S. Taylor, British Interplanetary Society, UK
Wil Tirion, The Netherlands
Dr Helen J. Walker, CCLRC Rutherford Appleton Laboratory, UK
Professor Fred Watson, Astronomer-in-Charge, Anglo-Australian Observatory, Australia
Dr James R. Webb, Florida International University and the SARA Observatory, USA
Dr Stuart Weidenschilling, Senior Scientist, Planetary Science Institute, USA
Professor Peter Wlasuk, Florida International University, USA
John Woodruff, FRAS, UK

Contributors to the 1987 edition also include:
Dr D. J. Adams, University of Leicester, UK
Dr David A. Allen, Anglo-Australian Observatory, Australia
Dr A. D. Andrews, Armagh Observatory, N. Ireland
R. W. Arbour, Pennell Observatory, UK
R. W. Argyle, Royal Greenwich Observatory (Canary Islands)
H. J. P. Arnold, Space Frontiers Ltd, UK
Professor W. I. Axford, Max-Planck-Institut für Aeronomie, Germany
Professor V. Barocas, Past President of the BAA, UK
Dr F. M. Bateson, Astronomical Research Ltd, New Zealand
Dr Reta Beebe, New Mexico State University, USA
Dr S. J. Bell Burnell, Royal Observatory, Edinburgh, UK
D. P. Bian, Beijing Planetarium, China
Dr D. L. Block, University of Witwatersrand, South Africa
G. L. Blow, Carter Observatory, New Zealand
Professor A. Boksenberg, Royal Greenwich Observatory (Sussex, UK)
Dr E. Bowell, Lowell Observatory, USA
Dr E. Budding, Carter Observatory, New Zealand
Dr P. J. Cattermole, Sheffield University, UK
Von Del Chamberlain, Past President of the International Planetarium Society
Dr David H. Clark, Science & Engineering Research Council, UK
Dr M. Cohen, University of California, USA
P. G. E. Corvan, Armagh Observatory, N. Ireland
Dr Dale P. Cruikshank, University of Hawaii, USA
Professor J. L. Culhane, Mullard Space Science Laboratory, UK
Dr J. K. Davies, University of Birmingham, UK
M. E. Davies, The Rand Corporation, California, USA
Professor R. Davis, Jr, University of Pennsylvania, USA
D. W. Dewhirst, Institute of Astronomy, Cambridge, UK
Professor Audouin Dollfus, Observatoire de Paris, France
Commander L. M. Dougherty, UK
Dr J. P. Emerson, Queen Mary College, London, UK
Professor M. W. Feast, South African Astronomical Observatory, South Africa
Dr G. Fielder, University of Lancaster, USA
Norman Fisher, UK
K. W. Gatland, UK
A. C. Gilmore, Mt John Observatory, University of Canterbury, New Zealand
Professor Owen Gingerich, Harvard-Smithsonian Center for Astrophysics, USA
Dr Mart de Groot, Armagh Observatory, N. Ireland
Professor R. H. Garstang, University of Colorado, USA
L. Helander, Sweden
Michael J. Hendrie, Director of the Comet Section of the BAA, UK
Dr A. A. Hoag, Lowell Observatory, USA
Dr M. A. Hoskin, Churchill College, Cambridge, UK
Commander H. D. Howse, UK
Professor Sir F. Hoyle, UK
Dr D. W. Hughes, University of Sheffield, UK
Dr G. E. Hunt, UK
Dr R. Hutchison, British Museum (Natural History), London, UK
Dr R. J. Jameson, University of Leicester, UK
R. M. Jenkins, Space Communications Division, Bristol, UK
Dr P. van de Kamp, Universiteit van Amsterdam, The Netherlands
Professor W. J. Kaufmann, III, San Diego State University, USA
Dr M. R. Kidger, Universidad de La Laguna, Tenerife, Canary Islands
Dr A. R. King, University of Leicester, UK
Dr Y. Kozai, Tokyo Astronomical Observatory, University of Tokyo, Japan
R. J. Livesey, Director of the Aurora Section of the BAA, UK
Sir Bernard Lovell, Nuffield Radio Astronomy Laboratories, Jodrell Bank, UK
Professor Dr S. McKenna-Lawlor, St Patrick's College, Co. Kildare, Ireland
Dr Ron Maddison, University of Keele, UK
David Malin, Anglo-Australian Observatory, Australia
J. C. D. Marsh, Hatfield Polytechnic Observatory, UK
Dr J. Mason, UK
Professor A. J. Meadows, University of Leicester, UK
Howard Miles, Director of the Artificial Satellite Section of the BAA, UK
L. V. Morrison, Royal Greenwich Observatory (Sussex, UK)
T. J. C. A. Moseley, N. Ireland
Dr P. G. Murdin, Royal Greenwich Observatory (Sussex, UK)
C. A. Murray, Royal Greenwich Observatory (Sussex, UK)
I. K. Nicolson, MSc, Hatfield Polytechnic, UK
J. E. Oberg, USA
Dr Wayne Orchiston, Victoria College, Australia
Dr M. V. Penston, Royal Greenwich Observatory (Sussex, UK)
J. L. Perdrix, Australia
Dr J. D. H. Pilkington, Royal Greenwich Observatory (Sussex, UK)
Dr D. J. Raine, University of Leicester, UK
Dr R. Reinhard, European Space Agency, The Netherlands
H. B. Ridley, UK
C. A. Ronan, East Asian History of Science Trust, Cambridge, UK
Professor S. K. Runcorn, University of Newcastle upon Tyne, UK
Dr S. Saito, Kwasan & Hida Observatories, University of Kyoto, Japan
Dr R. W. Smith, The Johns Hopkins University, USA
Dr F. R. Stephenson, University of Durham, UK
E. H. Strach, UK
Professor Clyde W. Tombaugh, New Mexico State University, USA
R. F. Turner, UK
Dr J. V. Wall, Royal Greenwich Observatory (Sussex, UK)
E. N. Walker, Royal Greenwich Observatory (Sussex, UK)
Professor B. Warner, University of Cape Town, South Africa
Professor P. A. Wayman, Dunsink Observatory, Dublin, Ireland
Dr G. Welin, Uppsala University, Sweden
A. E. Wells, UK
E. A. Whitaker, University of Arizona, USA
Dr A. P. Willmore, University of Birmingham, UK
Dr Lionel Wilson, University of Lancaster, UK
Professor A. W. Wolfendale, University of Durham, UK
Dr Sidney C. Wolff, Kitt Peak National Observatory, USA
K. Wood, Queen Mary College, London, UK
Les Woolliscroft, University of Sheffield, UK
Dr A. E. Wright, Australian National Radio Astronomy Observatory, Australia

FOREWORD

The progress of astronomy – or, more precisely, astrophysics – over the past century, and particularly the past generation, is not easily pigeon-holed.

On the one hand, profound truths have tumbled abundantly from the sky. Here are four diverse examples:

1. Our universe began some 14 billion years ago in a single cataclysmic event called the Big Bang.
2. Galaxies reside mainly in huge weblike ensembles.
3. Our neighbouring planets and their satellites come in a bewildering variety.
4. Earth itself is threatened (at least within politicians' horizons) by impacts from mean-spirited asteroids or comets.

On the other hand, ordinary citizens may well feel that astronomers are a confused lot and that they are farther away than ever from understanding how the universe is put together and how it works. For example, 'yesterday' we were told the universe is expanding as a consequence of the Big Bang; 'today' we are told it is accelerating due to some mysterious and possibly unrelated force. It doesn't help that the media dine exclusively on 'gee-whiz' results, many of them contradictory and too often reported without historical context. I can't help but savour the pre-1960s era, before quasars and pulsars were discovered, when we naïvely envisioned a simple, orderly universe understandable in everyday terms.

Of course, all the new revelations cry out for insightful interpretation. And that's why I'm delighted to introduce this brand-new edition. So much has been discovered since it first appeared in 1987 . . . so much more needs to be explained!

It's sobering to catalogue some of the objects and phenomena that were unknown, or at least weren't much on astronomers' minds, only a generation or two ago.

The one that strikes me most is that 90% (maybe 99%!) of all the matter in the universe is invisible and therefore unknown. We're sure it exists and is pervasive throughout intergalactic space (which was once thought to be a vacuum) because we can detect its gravitational influence on the stuff we can see, such as galaxies. But no one has a cogent clue as to what this so-called dark matter might be.

Masers, first created in laboratories in 1953, were found in space only 12 years later. These intense emitters of coherent microwave radiation have enabled astronomers to vastly improve distance determinations to giant molecular clouds and, especially, to the centre of our Galaxy.

A scientific 'war' was fought in the 1960s as to whether clusters of galaxies themselves clustered. Now even the biggest of these so-called superclusters are known to be but bricks in gigantic walls stretching across hundreds of millions of light-years. These walls contain most of the universe's visible matter and are separated from each other by empty regions called voids.

The discovery of quasars in 1963 moved highly condensed matter on to astronomy's centre stage. To explain their enormous and rapidly varying energy output, a tiny source was needed, and only a black hole having a feeding frenzy could fill the bill. Thus too was born the whole subdiscipline of relativistic astrophysics, which continues to thrive. Quasars are now regarded as having the highest energies in a diverse class called active galaxies.

Gamma-ray bursts, the most powerful outpourings of energy known in the universe, only came under intense scrutiny by astronomers in the 1990s (they had been detected by secret military satellites since the 1960s). The mechanism that leads to this prodigious output is still speculative, though a young, very massive star collapsing to form a black hole seems favoured.

A decades-long quest for extrasolar planets and closely related brown-dwarf (failed) stars came to an abrupt end in 1995 when the first secure examples of both entities were found. (By a somewhat arbitrary convention, planets are regarded as having masses up to several times that of Jupiter; brown dwarfs range from about 10 to 80 Jupiters.) Improved search strategies and techniques are now discovering so many of both objects that ordinary new ones hardly make news.

One of the greatest successes of astrophysics in the last century was the identification of how chemical elements are born. Hydrogen, helium, and traces of others originated in the Big Bang; heavier elements through iron derive from the cores of stars; and still heavier elements are blasted into space by the explosions of very massive stars.

The discovery of pulsars in 1967 confirmed that neutron stars exist. Born in supernova explosions, these bodies are only about 10 kilometres across and spin around as rapidly as 100 times a second. Whenever a pulsar's radiation beam, 'focused' by some of the strongest magnetic fields known, sweeps over the Earth, we see the pulse. In addition to being almost perfect clocks, pulsars have allowed studies as diverse as the interstellar medium and relativistic effects. Finally, unlike any other astronomical object, pulsars have yielded three Nobel Prizes!

Tantalizing though inconclusive evidence for extraterrestrial life accumulates impressively: possible fossil evidence in the famous Martian meteorite ALH 84001, the prospect of clement oceans under the icy crust of Jupiter's satellite Europa, and the organic-compound rich atmosphere of Saturn's moon Titan. And then there is the burgeoning catalogue of planets around other stars and the detection of terrestrial life forms in ever more hostile environments. All this suggests that we may not be alone. On a higher plane, despite many efforts to find extraterrestrial intelligence since Frank Drake's famous Ozma experiment in 1960, we haven't picked up E.T.'s phone call yet. But the search has barely begun.

The flowering of astrophysics stems from the development of ever larger, ever more capable telescopes on the ground and in space. All the electromagnetic spectrum – from the highest-energy gamma rays to the lowest-energy radio waves – is now available for robust scrutiny, not just visible light and long-wavelength radio emission as was the case in as recently as the 1950s.

Equally impressive has been the development of detectors to capture celestial radiation more efficiently. In the case of the CCD (charge-coupled device), trickle-down technology has allowed small amateur telescopes to act as though they were four or five times larger. Augmented by effective software, CCDs have caused a revolution among hobbyists, who, after nearly a century-long hiatus, can once again contribute to mainstream astrophysical research.

Increasingly, astronomers are no longer limited to gathering electromagnetic radiation. Beginning late in the last century, they started to routinely sample neutrinos, elementary particles that allow us to peek at such inaccessible things as the earliest times in the life of the universe and the innards of exploding stars. And the gravitational-wave detectors being commissioned at the time of this writing should allow glimpses of the fabric of spacetime itself.

Astronomy has involved extensive international collaborations for well over a century. The cross-disciplinary nature of contemporary research makes such collaborations even more compelling in the future. Furthermore, efforts to build the next generation of instruments on the ground and especially in space are so expensive that their funding will demand international participation.

Where do astronomers go from here? 'Towards the unknown' may seem like a cliché, but it isn't. With so much of the universe invisible or unsampled, there simply have to be many enormous surprises awaiting!

When it comes to the Big Questions, I don't know whether we are children unable to frame our thoughts, or teenagers at sea, or adults awash in obfuscating information. Researchers find the plethora of new discoveries – despite myriad loose ends and conundrums – to be very exciting, for it attests to the vibrancy and maturation of the science. Yet, as we enter the 21st century, astronomers are still a very long way from answering the two most common and profound questions people ask: what kind of universe do we live in, and is life pervasive?

Leif J. Robinson
Editor Emeritus, *Sky & Telescope* magazine

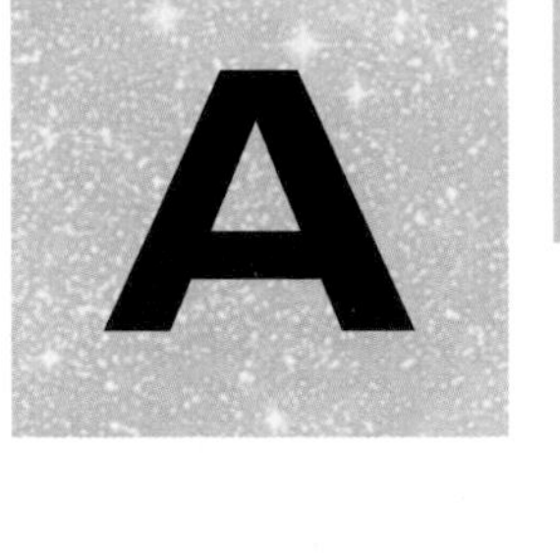

AAO Abbreviation of ANGLO-AUSTRALIAN OBSERVATORY

AAT Abbreviation of ANGLO-AUSTRALIAN TELESCOPE

AAVSO Abbreviation of AMERICAN ASSOCIATION OF VARIABLE STAR OBSERVERS

Abbot, Charles Greeley (1872–1961) American astronomer who specialized in solar radiation and its effects on the Earth's climate. He was director of the Smithsonian Astrophysical Observatory from 1907. Abbot made a very accurate determination of the solar constant, compiled the first accurate map of the Sun's infrared spectrum and studied the heating effect of the solar corona. He helped to design Mount Wilson Solar Observatory's 63-ft (19-m) vertical solar telescope.

Abell, George Ogden (1927–83) American astronomer who studied galaxies and clusters of galaxies. He is best known for his catalogue of 2712 'rich' clusters of galaxies (1958), drawn largely from his work on the PALOMAR OBSERVATORY SKY SURVEY. The Abell clusters, some of which are 3 billion l.y. distant, are important because they define the Universe's large-scale structure. Abell successfully calculated the size and mass of many of these clusters, finding that at least 90% of the mass necessary to keep them from flying apart must be invisible.

aberration (1) (aberration of starlight) Apparent displacement of the observed position of a star from its true position in the sky, caused by a combination of the Earth's motion through space and the finite velocity of the light arriving from the star. The effect was discovered by James BRADLEY in 1728 while he was attempting to measure the PARALLAX of nearby stars. His observations revealed that the apparent position of all objects shifted back and forth annually by up to 20″ in a way that was not connected to the expected parallax effect.

The Earth's movement in space comprises two parts: its orbital motion around the Sun at an average speed of 29.8 km/s (18.5 mi/s), which causes **annual aberration**, and its daily rotation, which is responsible for the smaller of the two components, **diurnal aberration**. The former causes a star's apparent position to describe an ELLIPSE over the course of a year. For any star on the ECLIPTIC, this ellipse is flattened into a straight line, whereas a star at the pole of the ecliptic describes a circle. The angular displacement of the star, α, is calculated from the formula $\tan \alpha = v/c$, where v is the Earth's orbital velocity and c is the speed of light.

Diurnal aberration is dependent on the observer's position on the surface of the Earth. Its effect is maximized at the equator, where it produces a displacement of a stellar position of 0″.32 to the east, but drops to zero for an observer at the poles.

Bradley's observations demonstrated both the motion of the Earth in space and the finite speed of light; they have influenced arguments in cosmology to the present day.

aberration (2) Defect in an image produced by a LENS or MIRROR. Six primary forms of aberration affect the quality of image produced by an optical system. One of these, CHROMATIC ABERRATION, is due to the different amount of refraction experienced by different wavelengths of light when passing through the boundary between two transparent materials; the other five are independent of colour and arise from the limitations of the geometry of the optical surfaces. They are sometimes referred to as Seidel aberrations after Ludwig von Seidel (1821–96), the mathematician who investigated them in detail.

The five Seidel aberrations are SPHERICAL ABERRATION, COMA, ASTIGMATISM, curvature and distortion. All but spherical aberration are caused when light passes through the optics at an angle to the optical axis. Optical designers strive to reduce or eliminate aberrations and combine lenses of different glass types, thickness and shape to produce a 'corrected lens'. Examples are the composite OBJECTIVES in astronomical refractors and composite EYEPIECES.

Curvature produces images that are not flat. When projected on to a flat surface, such as a photographic film, the image may be in focus in the centre or at the edges, but not at both at the same time. Astronomers using CCD cameras on telescopes can use a field flattener to produce a well-focused image across the whole field of view. Often this is combined with a focal reducer to provide a wider field of view.

Distortion occurs where the shape of the resulting image is changed. Common types of distortion are pin-cushion and barrel distortion, which describe the effects seen when an image of a rectangle is formed. Some binocular manufacturers deliberately introduce a small amount of pin-cushion distortion as they claim it produces a more natural experience when the binoculars are panned across a scene. Measuring the distortion in a telescope is extremely important for ASTROMETRY as it affects the precise position measurements being undertaken. Astrometric telescopes once calibrated are maintained in as stable a condition as possible to avoid changing the distortion.

Abetti, Giorgio (1882–1982) Italian solar physicist, director of ARCETRI ASTROPHYSICAL OBSERVATORY (1921–52). As a young postgraduate he worked at Mount Wilson Observatory, where pioneering solar astronomer George Ellery HALE became his mentor. Abetti designed and constructed the Arcetri solar tower, at the time the best solar telescope in Europe, and used it to investigate the structure of the chromosphere and the motion of sunspot penumbras (the Evershed–Abetti effect).

ablation Process by which the surface layers of an object entering the atmosphere (for example a spacecraft or a METEOROID) are removed through the rapid intense heating caused by frictional contact with the air. The heat shields of space vehicles have outer layers that ablate, preventing overheating of the spacecraft's interior.

absolute magnitude (M) Visual magnitude that a star would have at a standard distance of 10 PARSECS. If m = apparent magnitude and r = distance in parsecs:

$$M = m + 5 - 5 \log r$$

For a minor planet this term is used to describe the brightness it would have at a distance of 1 AU from the Sun, 1 AU from the Earth and at zero PHASE ANGLE (the Sun–Asteroid–Earth angle, which is a physical impossibility).

absolute temperature Temperature measured using the absolute temperature scale; the units (obsolete) are °A. This scale is effectively the same as the modern thermodynamic temperature scale, wherein temperature

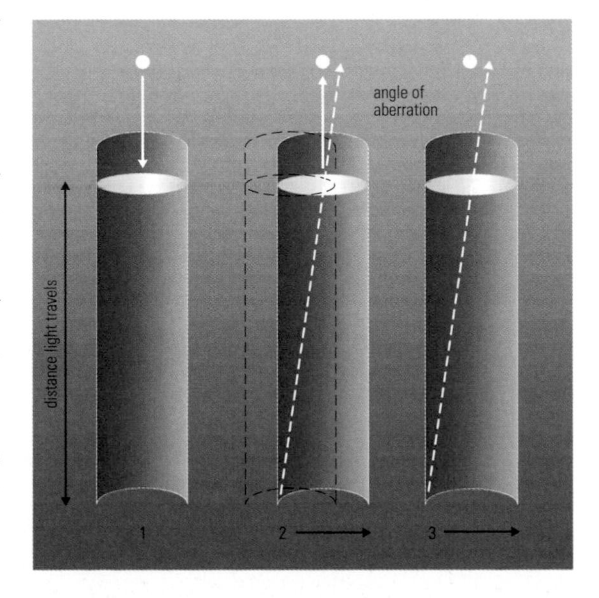

◀ **aberration** Aberration can cause displacement in the position of a star relative to its true position as viewed in a telescope. Bending of the light path away from the optical axis produces coma, drawing the star image into a 'tail' (3). Offset of the star's position (2) can reduce the effectiveness of the telescope for astrometry.

is defined via the properties of the Carnot cycle. The zero point of the scale is ABSOLUTE ZERO, and the freezing and boiling points of water are 273.15°A and 373.15°A, respectively. 1°A is equivalent to 1 K and kelvins are now the accepted SI unit. *See also* CELSIUS SCALE

absolute zero Lowest theoretically attainable temperature; it is equivalent to −273.15°C or 0 K. Absolute zero is the temperature at which the motion of atoms and molecules is the minimum possible, although that motion never ceases completely because of the operation of the Heisenberg uncertainty principle. (This principle states that an object does not have a measurable position and momentum at the same time, because the act of measuring disturbs the system.) Absolute zero can never be achieved in practice, but temperatures down to 0.001 K or less can be reached in the laboratory. The COSMIC MICROWAVE BACKGROUND means that 2.7 K is the minimum temperature found naturally in the Universe.

absorption As a beam of light, or other ELECTROMAGNETIC RADIATION, travels through any material medium, the intensity of the beam in the direction of travel gradually diminishes. This is partly due to scattering by particles of the medium and partly due to absorption within the medium. Energy that is absorbed in this way may subsequently be re-radiated at the same or longer wavelength and may cause a rise in temperature of the medium.

The absorption process may be general or selective in the way that it affects different wavelengths. Examples can be seen in the colours of various substances. Lamp black, or amorphous carbon, absorbs all wavelengths equally and reflects very little, whereas paints and pigments absorb all but the few wavelengths that give them their characteristic colours.

Spectral analysis of starlight reveals the selective absorption processes that tell us so much about the chemical and physical conditions involved. The core of a star is a hot, incandescent, high-pressure gas, which produces a CONTINUOUS SPECTRUM. The atoms of stellar material are excited by this high-temperature environment and are so close together that their electrons move easily from atom to atom, emitting energy and then being re-excited and so on. This gives rise to energy changes of all possible levels releasing all possible colours in the continuous SPECTRUM.

The cooler, low-pressure material that comprises the atmospheres of both star and Earth, and the interstellar medium that lies between them, can be excited by constituents of this continuous radiation from the star core, thus absorbing some of the radiation. Such selective absorption produces the dark ABSORPTION LINES that are so typical of stellar spectra. These lines are not totally black; they are merely fainter than the continuum because only a fraction of the absorbed energy is re-radiated in the original direction. *See also* FRAUNHOFER LINES; MORGAN–KEENAN CLASSIFICATION

absorption line Break or depression in an otherwise CONTINUOUS SPECTRUM. An ABSORPTION line is caused by the absorption of photons within a specific (usually narrow) band of wavelengths by some species of atom, ion or molecule, any of which has its own characteristic set of absorptions. Absorption lines are produced when electrons associated with the various species absorb incoming radiation and jump to higher energy levels. The analysis of absorption lines allows the determination of stellar parameters such as temperatures, densities, surface gravities, velocities and chemical compositions (*see* SPECTRAL CLASSIFICATION).

absorption nebula *See* DARK NEBULA

Abu'l-Wafā' al Būzjānī, Muhammad (940–997/8) Persian astronomer and mathematician. His *Kitāb al-kamil* ('Complete Book' [on Astronomy]) and his astronomical tables were used by many later astronomers, and he was the first to prove that the sine theorem is valid for spherical surfaces (for example, the celestial sphere). Abu'l-Wafā' discovered irregularities in the Moon's motion which were explained only by advanced theories of celestial mechanics developed centuries later.

Acamar The star θ Eridani, visual mag. 2.88, distance 161 l.y. Through small telescopes it is seen to be double. The components are of mags. 3.2 and 4.3, with spectral types A5 IV and A1 Va. The name comes from the Arabic *akhir al-nahr*, meaning 'river's end', for in ancient times it marked the southernmost end of Eridanus, before the constellation was extended farther south to ACHERNAR.

acapulcoite–lodranite Association of two groups of ACHONDRITE meteorites. They show a range of properties that grade into each other, with similar oxygen isotopic compositions. Acapulcoite–lodranites are thought to be partial melts of chondritic precursors. They have been described as primitive achondrites, suggesting that they are a bridge between CHONDRITES and achondrites.

acceleration of free fall Acceleration experienced by a body falling freely in a gravitational field. A body in free fall follows a path determined only by the combination of its velocity and gravitational acceleration. This path may be a straight line, circle, ellipse, parabola or hyperbola. A freely falling body experiences no sensation of weight, hence the 'weightlessness' of astronauts, since the spacecraft is continuously free falling towards the Earth while its transverse motion ensures that it gets no closer. The free fall acceleration is 9.807 m/s^2 at the Earth's surface. It varies as the inverse square of the distance from the Earth's centre.

accretion Process by which bodies gain mass; the term is applied both to the growth of solid objects by collisions that result in sticking and to the capture of gas by the gravity of a massive body. Both types of accretion are involved in the formation of a planetary system from a disk-shaped nebula surrounding a PROTOSTAR. When newly formed, such a disk consists mostly of gas, with a small fraction (*c.*1%) of solid material in small dust particles, with original sizes of the order of a micrometre. Grains settle through the gas towards the central plane of the disk; they drift inwards toward the protostar at rates that vary with their sizes and densities, resulting in collisions at low velocities. Particles may stick together by several mechanisms, depending on their compositions and local conditions, including surface forces (van der Waals bonding), electrical or magnetic effects, adsorbed layers of organic molecules forming a 'glue', and partial melting of ices. This sticking produces irregularly shaped fluffy aggregates, which can grow further by mechanical interlocking.

When bodies reach sizes of the order of a kilometre or larger, gravity becomes the dominant bonding mechanism. Such bodies are called PLANETESIMALS. Mutual perturbations cause their orbits to deviate from circularity, allowing them to cross, which results in further collisions. The impact velocity is always at least as large as the escape velocity from the larger body. If the energy density exceeds the mechanical strength of the bodies, they are shattered. However, a large fraction of the impact energy is converted into heat, and most fragments move at relatively low velocities, less than the impact velocity. These fragments will fall back to the common centre of gravity, resulting in a net gain of mass unless the impact velocity greatly exceeds the escape velocity. The fragments form a rubble pile held together by their mutual gravity. As this process is repeated, bodies grow ever larger. At sizes greater than a few hundred kilometres gravitational binding exceeds the strength of geological materials, and the mechanical properties of the planetesimals become unimportant.

Accretion is stochastic, that is, the number of collisions experienced by a body of a given size in an interval of time is a matter of chance. This process produces a distribution of bodies of various sizes. Often, the size distribution can be described by a power law, with an index *s*, defined so that the number of objects more massive than a specific mass, *m*, is proportional to m^{-s}. If *s* is less than 1, most of the mass is in the larger objects; for larger values of *s*, the smaller bodies comprise most of the mass. Power law size distributions may be produced by either accretion or fragmentation, with the latter tending to have somewhat larger *s* values. However, accretion of planetesimals subject to gravitational forces can produce another type of distribution. If relative velocities are low compared with a body's escape velocity, its gravity can deflect the trajectories of other objects, causing impacts for encounters that would otherwise be near misses. This 'gravitational focusing' is more effective for more massive bodies, and it increases their rate of mass gain by accretion, allowing the largest bodies to grow still more rapidly. In numerical simulations of accretion, the first body to reach a size such that its escape velocity exceeds the mean relative velocity experiences 'runaway growth', quickly becoming much larger than the mean size. Its own perturbations then stir up velocities of the smaller bodies near its orbit, preventing them from growing in the same manner. At greater distances, its effects are weaker, and the process can repeat in another location. The result is a series of PROTOPLANETS in separated orbits; these can grow further by sweeping up the residual population of small planetesimals.

In the inner Solar System, the final stage of accretion probably involved collisions between protoplanets. Impacts of this magnitude would have involved enough energy to melt the planets; such an event is theorized to be responsible for the origin of the Moon. If a protoplanet attained sufficient mass before dissipation of the SOLAR NEBULA, then its gravitation could overcome the pressure of the nebular gas, and it could accrete gas from the nebula. This process can begin at a critical mass that depends on a number of factors, including the density and temperature of the gas, and the protoplanet's distance from the Sun. The rate at which gas is accreted is limited by the escape of energy, which must be radiated away by the gas as it cools. The original protoplanet would then become the CORE of a planet that consists mostly of hydrogen. Plausible estimates imply that the critical mass is at least a few times Earth's mass, but protoplanets of this size could have accreted in the outer Solar System. Jupiter and Saturn probably formed by accretion of gas. *See also* COSMOGONY

accretion disk Disk of matter that surrounds an astronomical object and through which material is transferred to that object. In many circumstances, material does not transfer directly from one astronomical object to another. Instead, the material is pulled into equatorial orbit about the object before accreting. Such material transfer systems are known as ACCRETION disks. Accretion disks occur in protostellar clouds, close BINARY STAR systems and at the centre of galaxies.

Accretion disks are difficult to observe directly because of their small size or large distance from Earth. The disks that appear the largest (because they are nearest) are PROTOPLANETARY DISKS, around 100 AU in size, some of which have been imaged by the Hubble Space Telescope. Accretion disks in CLOSE BINARIES range in diameter from a few tenths to a few solar radii. Details about size, thickness and temperature of accretion disks can be provided by observing eclipses occurring between the disk and the secondary star from which the material is being pulled.

The energy output of the material being accreted depends on the mass and radius of the accreting object. The more massive and the smaller the accreting object, the higher the speed of material arriving, and the greater the amount of energy released on impact. Energy continues to be radiated as the material loses energy by slowing down within the disk. If the accreting star in a binary system is a WHITE DWARF, as in a CATACLYSMIC VARIABLE, the inner part of the disk will radiate in the ultraviolet, while the outer part radiates mostly in the visible. The MASS TRANSFER in such systems is often unstable, causing DWARF NOVAE outbursts.

In an X-RAY BINARY the accreting star is a NEUTRON STAR or stellar-mass BLACK HOLE and the inner disk radiates in X-rays. Unstable mass transfer across these disks produces soft X-RAY TRANSIENTS and sometimes relativistic JETS. The greatest amounts of energy are released when matter accretes on to a SUPERMASSIVE BLACK HOLE at the centre of a galaxy. This is the power source of an ACTIVE GALACTIC NUCLEUS, the central region of which radiates in ultraviolet and X-ray and can produce relativistic jets.

ACE Acronym for ADVANCED COMPOSITION EXPLORER

Achernar The star α Eridani, visual mag. 0.46, distance 144 l.y. Achernar is the ninth-brightest star in the sky and has a luminosity over a thousand times that of the Sun. Its spectral type is B3 V with additional features that suggest it is a SHELL STAR. The name, which comes from the Arabic meaning 'river's end' (the same origin as ACAMAR), was given to this star in Renaissance times when the constellation Eridanus was extended southwards.

Achilles First TROJAN ASTEROID to be recognized, by Max WOLF in 1906; number 588. It is *c.*147 km (*c.*91 mi) in size.

achondrite STONY METEORITE that formed from melted parts of its parent body. Achondrites usually have differentiated compositions. They generally do not contain CHONDRULES, and they have very low metal contents. There are many different groups of achondrites, some of which can be linked to form associations allied with specific parents. The separate associations have little, if any, genetic relationship to each other. *See also* ACAPULCOITE–LODRANITE; ANGRITE; AUBRITE; BRACHINITE; HOWARDITE–EUCRITE–DIOGENITE ASSOCIATION; LUNAR METEORITE; MARTIAN METEORITE; UREILITE; WINONAITE

achromat (achromatic lens) Composite LENS designed to reduce CHROMATIC ABERRATION. The false colour introduced into an image by a lens can be reduced by combining the action of two or more lenses with different characteristics. In an achromat, two lenses of different materials are used together.

The most common example is the OBJECTIVE of a good quality but inexpensive astronomical REFRACTOR. This is usually made of a crown glass lens and a flint glass lens that have different refractive indices and introduce different levels of dispersion. By making one lens diverging and the other converging the optical designer can produce a converging composite lens that brings light of two differ-

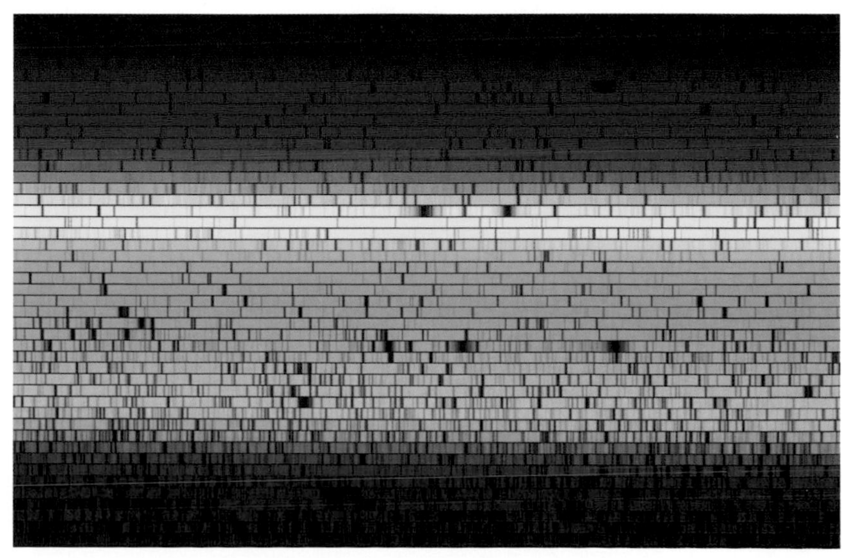

▼ **absorption line** The visible light spectrum of the cool giant star Arcturus (α Boötis) is shown here. The dark vertical lines in the spectrum are caused by atoms in the star's atmosphere absorbing radiation. Because each element absorbs radiation at characteristic wavelengths, the spectrum of a star can be used to determine which elements are present.

A

▲ **Acidalia Planitia** When this area of Mars was originally imaged by Viking some astronomers interpreted the linear feature as an ancient shoreline. Mars Global Surveyor images showed that it is actually layered rock.

ent wavelengths to a focus at the same point. This reduces considerably the false colour that would be produced by a single lens, but it does not eliminate it altogether: bright objects observed against a dark background, such as the Moon at night, will have a coloured edge. There is also a reduction in overall contrast compared with a completely colour-corrected optical system such as an APOCHROMAT.

Acidalia Planitia Main dark area in the northern hemisphere of MARS (47°.0N 22°.0W).

Acrux The star α Crucis (of which 'Acrux' is a contraction), visual mag. 0.77, distance 321 l.y. Small telescopes split it into two components of mags. 1.3 and 1.7. Their spectral types are B0.5 IV and B1 V, so both appear blue-white. Acrux is the southernmost first-magnitude star, declination −63°.1.

actinometer (pyrheliometer) Instrument used for measuring at any instant the direct heating power of the Sun's radiation. Sir William HERSCHEL first noted, in 1800, that the heating effect of the Sun's rays was greatest beyond the red end of the spectrum. This INFRARED RADIATION was further investigated by his son Sir John HERSCHEL, who invented the actinometer around 1825.

active galactic nucleus (AGN) Central energy-producing region in some GALAXIES. AGNs are distinct in having substantial portions of their energy output coming from processes that are not associated with normal stars and their evolution. The observed guises of this energy release define the various types of active nuclei.

At the lowest power levels are LINERS (Low Ionization Nuclear Emission-line Regions), generally recognized only by the ratios of fairly weak EMISSION LINES; not all LINERS are genuine active nuclei. Activity characterized by strong, broad emission lines occurs in SEYFERT GALAXIES, most of which are spirals; Seyfert types 1 and 2 have different patterns of line width. Seyfert galaxies also show strong X-ray emission and, often, far-infrared radiation. RADIO GALAXIES are most notable for their strong radio emission, usually from a pair of lobes symmetrically placed about the galaxy, often accompanied by JETS and radio emission from the nucleus itself. This activity may have little or no trace in the optical region, although some radio galaxies do have spectacular optical emission lines similar to those of both types of Seyfert galaxy. Higher-luminosity objects are QUASISTELLAR OBJECTS (QSOs), which are known as QUASARS (quasi-stellar radio sources) if they exhibit strong radio emission. These objects are so bright that the surrounding galaxy can be lost for ordinary observations in the light of the nucleus. Members of another class, BL LACERTAE OBJECTS, show featureless spectra and rapid variability, suggesting that they are radio galaxies or quasars seen along the direction of a relativistic jet, the radiation of which is strongly beamed along its motion. These categories share features of strong X-ray emission, large velocities for the gas seen in emission lines, and a very small emitting region as seen from variability. Many show variation in the ultraviolet and X-ray bands on scales of hours to days, implying that the radiation is emitted in a region with light-crossing time no longer than these times.

The most popular model for energy production in all these kinds of active galactic nuclei involves material around a supermassive BLACK HOLE (of millions to a few thousand million solar masses). The power is released during ACCRETION, likely in an ACCRETION DISK, while jets may be a natural by-product of the disk geometry and magnetic fields.

active optics Technique for controlling the shape and alignment of the primary MIRROR of a large reflecting TELESCOPE. As a telescope tilts to track the path of a celestial object across the sky, its mirror is subject to changes in the forces acting upon it, as well as temperature variations and buffeting from the wind, which can cause it to flex, giving rise to SPHERICAL ABERRATION or ASTIGMATISM of the image.

Active optics compensate for these effects, through the use of a number of computer-controlled motorized mirror supports, known as actuators, which continually monitor the shape of the mirror and adjust it into its correct form. These adjustments are typically only about 1/10,000 the thickness of a human hair but are enough to keep light from a star or galaxy precisely focused.

For many years it was considered impossible to build telescopes of the order of eight metres in diameter using a single mirror because it would have had to be so thick and heavy in order to maintain its correct shape as to make it impractical. The development of active optics technology has meant that relatively thin primary mirrors can now be built that are lighter and cheaper and are able to hold their precise shape, thereby optimizing image quality. The system also compensates for any imperfections in the surface of the mirror caused by minor manufacturing errors. *See also* ADAPTIVE OPTICS; SEGMENTED MIRROR

active region Region of enhanced magnetic activity on the Sun often, but not always, associated with SUNSPOTS and extending from the solar PHOTOSPHERE to the CORONA. Where sunspots occur, they are connected by strong magnetic fields that loop through the CHROMOSPHERE into the low corona (coronal loops). Radio, ultraviolet and X-ray radiation from active regions is enhanced relative to neighbouring regions of the chromosphere and corona. Active regions may last from several hours to a few months. They are the sites of intense explosions, FLARES, which last from a few minutes to hours. NOAA (National Oceanic and Atmospheric Administration), which monitors solar activity, assigns numbers to active regions (for example, AR 9693) in order of their visibility or appearance. The occurrence and location of active regions varies in step with the approximately 11-year SOLAR CYCLE. Loops of gas seen as FILAMENTS or PROMINENCES are often suspended in magnetic fields above active regions.

Adams, John Couch (1819–92) English mathematician and astronomer who played a part in the discovery of Neptune. In 1844, while at St John's College, Cambridge, he began to investigate the orbital irregularities of Uranus, which he concluded could be accounted for by gravitational perturbations by an undiscovered planet. He calculated an orbit for this planet, and identified a small region of sky where it might be found. He approached James CHALLIS, director of the Cambridge Observatory, and George Biddell AIRY, the Astronomer Royal. However, communications

► **active optics** The active optics actuators on the reverse of the primary mirror of the WIYN telescope at Kitt Peak National Observatory allow the mirror to be flexed continually to compensate for the effects of gravity as the telescope moves. This system means that far thinner mirrors can be used without risking distortion of the images and data obtained.

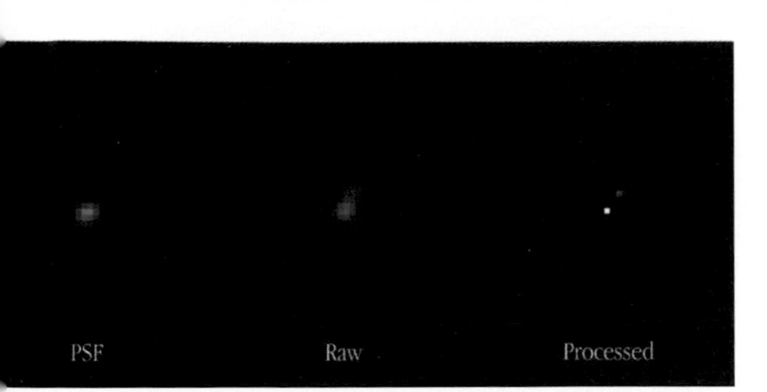

◄ **adaptive optics** Telescopes using adaptive optics, such as the Very Large Telescope (VLT), have far better resolving power than earlier ground-based telescopes. Here the light from a close binary pair with a separation of only 0″.03 has been reflected from the primary mirror on to a subsidiary mirror, which is continually adjusted to compensate for variations in the Earth's atmosphere; it is then computer processed.

between Adams and Airy did not run smoothly, and no search was mounted from Britain. When the Uranus-disturbing planet, Neptune, was located on 1846 September 23, it was by Johann GALLE and Henrich D'ARREST, observing from Berlin and guided by a position calculated independently by Urbain LE VERRIER.

Adams was a brilliant scientist, but shy and rather retiring, and he refused a knighthood in 1847. He returned to Cambridge as Lowndean Professor in 1858, becoming director of the Cambridge Observatory in 1860. Adams' subsequent researches on the lunar parallax and other small motions, and the celestial mechanics of meteor streams following the 1866 Leonid storm, won him numerous honours. In spite of the Neptune affair, which led to arguments over the conduct of British science and a souring of Anglo-French scientific relations, Adams enjoyed the friendship of Airy, Challis and Le Verrier.

Adams, Walter Sydney, Jr (1876–1956) American astronomer, born in Syria to missionary parents, who succeeded George Ellery HALE as director of Mount Wilson Observatory (1923–46). At Yerkes Observatory (1898–1904), Adams became an expert at using spectroscopic techniques to determine stellar radial velocities. He followed Hale to Mount Wilson, where a great new observatory specializing in solar astronomy was being built. Adams and Hale obtained solar spectra showing that sunspots were cooler than the rest of the Sun's surface, and by measuring Doppler shifts in solar spectra he was able accurately to measure our star's differential rotation, which varies with latitude. In 1914 he began studying the intensity of spectral lines of stars beyond the Sun, which could be used to calculate the stars' absolute magnitudes; during his Mount Wilson years, Adams computed and catalogued the radial velocities of 7000 stars, determining the absolute magnitudes of another 6000.

Adams discovered that the intensities differed for main-sequence, giant and dwarf stars, and used this knowledge to identify Sirius B as the first example (1915) of a white dwarf. His calculations showed that Sirius B is an extremely hot, compact star containing 80% of the Sun's mass packed into a volume roughly equal to that of the Earth. Ten years later he was able to measure a Doppler shift of 21 km/s (13 mi/s) for Sirius B, a result predicted by Arthur EDDINGTON's model of white dwarfs, which, because they are very dense, produce powerful localized gravitational fields manifested in just such a spectral redshift. Adams' discovery was therefore regarded as an astrophysical confirmation of Albert EINSTEIN's theory of general relativity. In 1932 Adams found that the atmosphere of Venus is largely composed of CO_2; he also discovered that the interstellar medium contained the molecules CN and CH. The climax of Adams' career was his role in the design and building of Mount Palomar's 200-inch (5-m) Hale Telescope.

adaptive optics Technique that compensates for distortion caused in astronomical images by the effects of atmospheric turbulence, or poor SEEING.

Adaptive optic technology uses a very thin, deformable mirror to correct for the distorting effects of atmospheric turbulence. It operates by sampling the light using an instrument called a wavefront sensor. This takes a 'snapshot' of the image from a star or galaxy many times a second and sends a signal back to the deformable mirror, which is placed just in front of the focus of the telescope. The mirror is very thin and can be flexed in a controlled fashion hundreds of times a second, compensating for the varying distortion and producing an image almost as sharp as if the telescope were in space. The control signals must be sent from the wavefront detector to the mirror fast enough so that the turbulence has not changed significantly between sensing and correction. *See also* ACTIVE OPTICS; SPECKLE INTERFEROMETRY

ADC Abbreviation of ASTRONOMICAL DATA CENTER

Adhara The star ε Canis Majoris, visual mag. 1.50, distance 431 l.y., spectral type B2 II. It has a 7th-magnitude COMPANION, which is difficult to see in very small telescopes as it is drowned by Adhara's light. The name comes from an Arabic phrase meaning 'the virgins', given to an asterism of four of five stars of which Adhara was the brightest.

adiabatic Process in thermodynamics in which a change in a system occurs without transfer of heat to or from the environment. Material within the convective regions of stars moves sufficiently rapidly that there is little exchange of energy except at the top and bottom of the region. The material therefore undergoes adiabatic changes, and this leads to a simple pressure law of the form:

$$P = k\rho^{5/3}$$

where P is the pressure, ρ the density, and k a constant. Such a pressure law is called polytropic, and it enables the region to be modelled very simply.

Adonis Second APOLLO ASTEROID to be discovered, in 1936. It was lost but became numbered as 2101 after its recovery in 1977. Because of its low inclination orbit, Adonis makes frequent close approaches to the Earth. It has been suggested to be the parent of a minor METEOR SHOWER, the Capricornid–Sagittariids, and may, therefore, have originated as a cometary nucleus. *See* table at NEAR-EARTH ASTEROID

Adrastea One of the inner moons of JUPITER, discovered in 1979 by David Jewitt (1958–) and Edward Danielson in images obtained by the VOYAGER project. It is irregular in shape, measuring about 25 × 20 × 15 km (16 × 12 × 9 mi). It orbits near the outer edge of Jupiter's main ring, 129,000 km (80,000 mi) from the planet's centre, taking 0.298 days to complete one of its near-circular equatorial orbits. *See also* METIS

Advanced Composition Explorer (ACE) NASA spacecraft launched in 1997 August. It is equipped with nine instruments to determine and compare the isotopic and elemental composition of several distinct samples of matter, including the solar CORONA, interplanetary medium, interstellar medium and galactic matter. The craft was placed into the Earth–Sun Lagrangian point, or L_1, 1.5 million km (940,000 mi) from Earth, where it remains in a relatively constant position with respect to the Earth and the Sun.

aerolite Obsolete name for STONY METEORITE

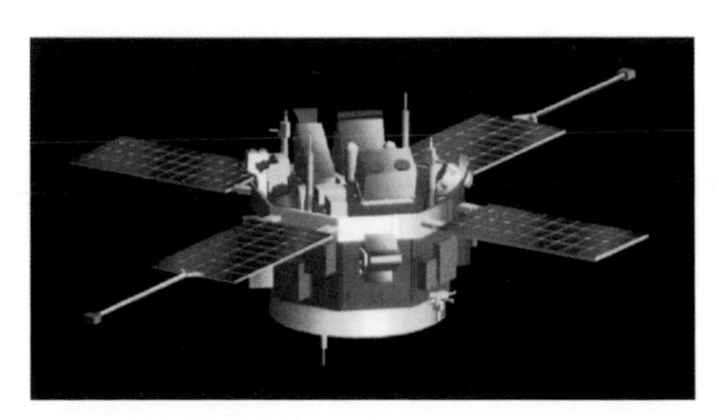

◄ **Advanced Composition Explorer** ACE's nine instruments sample a wide range of accelerated particles from the Sun and interstellar and galactic sources. One of their main functions is to give warning of geomagnetic storms that might endanger astronauts and disrupt power supplies and communications on Earth.

A

aeronomy Study of the physics and chemistry of the upper ATMOSPHERE of the Earth and other planets. On Earth, this region is rather inaccessible, being generally above the height that meteorological balloons can reach, so research techniques rely heavily on the use of rockets and satellites together with remote sensing by radio waves and optical techniques.

The primary source of energy for the processes investigated is incident solar energy absorbed before it reaches the surface of the planet. This energy may ionize the upper atmosphere to form ionospheric plasma or may cause chemical changes, such as the photodissociation of molecules to form atoms or the production of exotic molecules such as ozone and nitrous oxide. Some minor constituents have an important catalytic role in the chemistry of the upper atmosphere, hence, for example, the significant influence of chlorine compounds on the ozone concentration in the STRATOSPHERE.

Consideration of the atmosphere as a fluid leads to an understanding of the various winds and circulation patterns. Fluid oscillations include atmospheric tides, internal gravity waves and disturbances that propagate because of buoyancy forces. The tides are predominantly caused by solar heating, rather than by gravitational forces, and, on Earth, are the principal component driving the wind system at an altitude of about 100 km (about 60 mi). Under certain conditions the upper atmosphere may become turbulent, which leads to mixing and enhanced heat transport.

Optical phenomena include AIRGLOW, in which photoemission may be caused by a range of physical and chemical processes, and AURORAE, where the visible emissions are produced by charged particles from the MAGNETOSPHERE. Associated with aurorae are electric current systems, which create perturbations in the magnetic field. There are other currents in the upper atmosphere caused by tidally driven dynamos.

aether All-pervasive fluid through which electromagnetic waves were originally thought to propagate. Electromagnetic theory showed that light needed no such medium to propagate and experimental tests such as the MICHELSON–MORLEY EXPERIMENT failed to detect signs of such a medium, so the idea of aether was dropped from physical theory.

Agena Alternative name for the star β Centauri. Also known as HADAR

Agena One of the most successful US rockets. It was used extensively for rendezvous and docking manoeuvres in the manned GEMINI PROJECT, launching satellites and as a second stage for US lunar and planetary missions.

▼ **Agena** The Agena target vehicle is seen here from Gemini 8 during the rendezvous. Testing docking procedures was vital to the success of the Apollo missions.

Aglaonike Ancient Greek, the first woman named in the recorded history of astronomy. She was said to have predicted eclipses, and some of her contemporaries regarded her as a 'sorceress' who could 'make the Moon disappear at will'.

AGN Abbreviation for ACTIVE GALACTIC NUCLEUS

airglow Ever-present faint, diffuse background of light in the night sky resulting from re-emission of energy by atmospheric atoms and molecules following excitation during daylight by solar radiation. Airglow emissions, which occur in the upper ATMOSPHERE, mean that Earth's night sky is never completely dark.

Prominent among airglow emissions is green light from excited oxygen, at 557.7 nm wavelength, which is found mainly in a roughly 10 km (6 mi) deep layer at around 100 km (60 mi) altitude. Red oxygen emissions at 630.0 nm and 636.4 nm occur higher in the atmosphere; together with those from sodium, these emissions become more prominent in the twilight airglow. The night-time airglow varies in brightness, probably in response to changing geomagnetic activity. The day-time airglow is about a thousand times more intense than that seen at night but is, of course, a great deal more difficult to study because of the bright sky background.

Airy, George Biddell (1801–92) English astronomer, the seventh ASTRONOMER ROYAL. The son of an excise officer, he grew up in Suffolk and won a scholarship to Trinity College, Cambridge. Airy became a professor at the age of 26, and was offered the post of Astronomer Royal in 1835, having already refused a knighthood on the grounds of his relative poverty. (He turned down two further offers, before finally accepting a knighthood in 1872.) Academic astronomy in Airy's day was dominated by celestial mechanics. Astronomers across Europe, especially in Germany, were making meticulous observations of the meridional positions of the stars and planets for the construction of accurate tables. These tables provided the basis for all sorts of investigations in celestial mechanics to be able to take place. As a Cambridge astronomy professor and then as Astronomer Royal at Greenwich, Airy was to be involved in such research for 60 years.

In addition to such mathematical investigations, Airy was a very practical scientist, who used his mathematical knowledge to improve astronomical instrument design, data analysis, and civil and mechanical engineering. Upon assuming office as Astronomer Royal he began a fundamental reorganization of the Royal Observatory, Greenwich. He did little actual observing himself, but developed a highly organized staff to do the routine business, leaving him free for analytical, navigational and government scientific work. Airy was quick to seize the potential of new science-based techniques such as electric telegraphy, and by 1854, for instance, the Observatory was transmitting time signals over the expanding railway telegraph network.

It is sad that in the popular mind Airy is perhaps best remembered as the man who failed to enable John Couch ADAMS to secure priority in the discovery of Neptune in 1846. Yet this stemmed in no small degree from Adams' own failure to communicate with Airy and to answer Airy's technical questions. Airy made no single great discovery, but he showed his generation how astronomy could be made to serve the public good.

Airy disk Central spot in the DIFFRACTION pattern of the image of a star at the focus of a telescope. In theory 84% of the star's light is concentrated into this disk, the remainder being distributed into the set of concentric circles around it. The size of the Airy disk is determined by the APERTURE of the telescope. It limits the RESOLUTION that can be achieved. The larger the aperture, the smaller the Airy disk and the higher the resolution that is possible.

◀ **Airy, George Biddell** Portrait in ink of controversial 19th-century Astronomer Royal George Biddell Airy.

Aitken, Robert Grant (1864–1951) Leading American double-star observer and director of LICK OBSERVATORY (1930–35). His principal work was the *New General Catalogue of Double Stars* (1932), based largely on data he gathered at Lick beginning in 1895. It contains magnitudes and separations for more than 17,000 double stars, including many true binary systems. Aitken discovered more than 3000 doubles, and computed orbits for hundreds of binaries.

AI Velorum star Pulsating VARIABLE STAR, similar to the DELTA SCUTI type, with period shorter than 0.25 days and amplitude of 0.3–1.2 mag. AI Velorum stars belong to the disk population and are not found in star clusters. They are sometimes known as dwarf Cepheids.

AL Abbreviation of ASTRONOMICAL LEAGUE

Alba Patera Low-profile shield volcano on MARS (40°.5N 109°.9W). It is only 3 km (1.9 mi) high but some 1500 km (930 mi) across.

Albategnius Latinized name of AL-BATTĀNĪ

Albategnius Lunar walled plain (12°S 4°E), 129 km (80 mi) in diameter. Its walls are fairly high, 3000–4250 m (10,000–14,000 ft), and terraced; they are broken in the south-west by a large (32 km/20 mi) crater, Klein. Albategnius is an ancient impact site, and its eroded rims display landslips and valleys. The terrain surrounding this crater is cut by numerous valleys and deep trenches, evidence of the Mare IMBRIUM impact event. A massive pyramid-shaped mountain and many bowl craters mark the central floor.

al-Battānī, Abu'Abdullah Muhammad ibn Jābir (Latinized as **Albategnius**) (*c*.858–929) Arab observational astronomer (born in what is now modern Turkey) who demonstrated that the Sun's distance from the Earth, and therefore its apparent angular size, varies, which explains why both total and annular solar eclipses are possible. He made the first truly accurate calculations of the solar (tropical) year (365.24056 days), the ecliptic's inclination to the celestial equator (23° 35′) and the precession of the equinoxes (54″.5 per year).

albedo Measure of the reflecting power of the surface of a non-luminous body. Defined as the ratio of the amount of light reflected by a body to the total amount falling on it, albedo values range from 0 for a perfectly absorbing black surface, to 1 for a perfect reflector or white surface. Albedo is commonly used in astronomy to describe the fraction of sunlight reflected by planets, satellites and asteroids: rocky bodies have low values whereas those covered with clouds or comprising a high percentage of water-ice have high values. The average albedo of the MOON, for example, is just 0.07 whereas VENUS, which is covered in dense clouds, has a value of 0.76, the highest in the Solar System. The albedo of an object can provide valuable information about the composition and structure of its surface, while the combination of an object's albedo, size and distance determines its overall brightness.

Albert One of the AMOR ASTEROIDS; number 719. It is *c*.3 km (*c*.2 mi) in size. Albert was an anomaly for many decades in that it was numbered and named after its discovery in 1911 but subsequently lost. Despite many attempts to recover it, Albert escaped repeated detection until the year 2000.

Albireo The star β Cygni, visual mag. 3.05, distance 386 l.y. Albireo is a beautiful double star of contrasting colours. It comprises an orange giant (the brighter component, spectral type K3 II) twinned with a companion of mag. 5.1 and spectral type B9.5 V which appears greenish-blue. The two are so widely spaced, by about 34″, that they can be seen separately through the smallest of telescopes, and even with good binoculars (if firmly mounted). The name Albireo is a medieval corruption, and is meaningless.

al-Būzjānī *See* ABU'L-WAFĀ' AL BŪZJĀNĪ, MUHAMMAD

Alcor The star 80 Ursae Majoris, visual mag. 3.99, distance 81 l.y., spectral type A5 V. Alcor is a spectroscopic binary, though no accurate data are known. The name may comes from an Arabic word meaning 'rider'. Alcor forms a naked-eye double with MIZAR; the two are not a genuine binary, but Alcor is part of the URSA MAJOR MOVING CLUSTER.

Alcyone The star η Tauri, distance 368 l.y., spectral type B7 III. At visual mag. 2.85, it is the brightest member of the PLEIADES star cluster. In Greek mythology, Alcyone was one of the seven daughters of Atlas and Pleione.

Aldebaran The star α Tauri, visual mag. 0.87 (but slightly variable), distance 65 l.y. It is an orange-coloured giant of spectral type K5 III. It marks the eye of Taurus, the bull. Its true luminosity is about 150 times that of the Sun. Although Aldebaran appears to be a member of the V-shaped Hyades cluster, it is a foreground object at about half the distance, superimposed by chance. The name comes from the

▼ **albedo** This albedo map of Mars was produced by NASA's Mars Global Surveyor. Red areas are bright and show where there is dust while blue areas show where the underlying, darker rocks have been exposed.

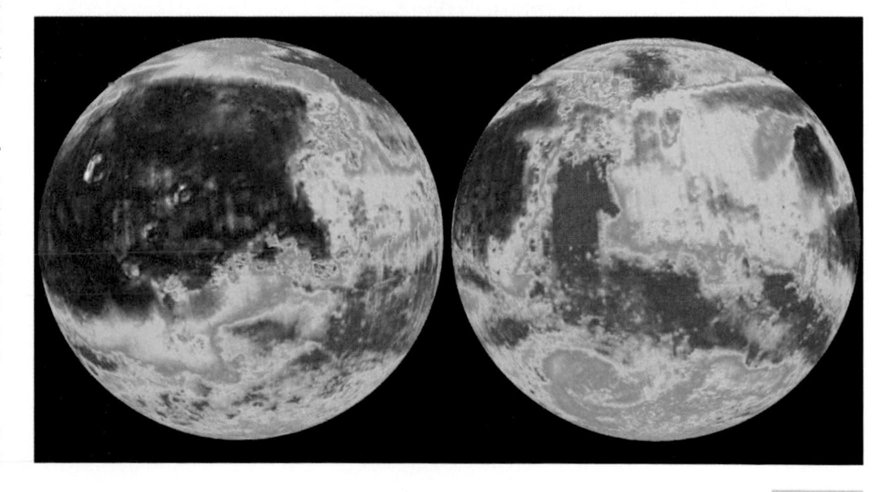

A

Arabic meaning 'the follower' – Aldebaran appears to follow the Pleiades cluster across the sky.

Alderamin The star α Cephei, visual mag. 2.45, distance 49 l.y., spectral type A7 V. Its name comes from an Arabic expression referring to a forearm.

Aldrin, Edwin Eugene ('Buzz'), Jr (1930–) American astronaut. After setting a record for space-walking during the Gemini 12 mission in 1966, Aldrin was assigned to Apollo 11 as Lunar Module pilot, and on 1969 July 20 he became the second man to walk on the Moon, after Neil ARMSTRONG.

Alfvén, Hannes Olof Gösta (1908–95) Swedish physicist who developed much of the theory of MAGNETOHYDRODYNAMICS, for which he was awarded the 1970 Nobel Prize for Physics. In 1942 he predicted the existence of what are now called ALFVÉN WAVES, which propagate through a plasma, and in 1950 he identified synchrotron radiation from cosmic sources, helping to establish radio astronomy.

Alfvén waves Transverse MAGNETOHYDRODYNAMIC waves that can occur in a region containing plasma and a magnetic field. The electrically conducting plasma is linked to and moves with the magnetic field. Sometimes this phenomenon is referred to as the plasma and magnetic field being 'frozen-in' to each other. The plasma follows the oscillations of the magnetic field and modifies those oscillations. Alfvén waves may transfer energy out to the solar CORONA, and they are also found in the SOLAR WIND and the Earth's MAGNETOSPHERE. They are named after Hannes ALFVÉN.

Algenib The star γ Pegasi, visual mag. 2.83, spectral type B2 IV, distance 333 l.y. It is a BETA CEPHEI STAR – a pulsating variable that fluctuates by 0.1 mag. with a period of 3.6 hours. The name Algenib comes from the Arabic meaning 'the side'; it is also an alternative name for the star α Persei (*see* MIRPHAK).

Algieba The star γ Leonis, visual mag. 2.01, distance 126 l.y. Small telescopes show it to be a beautiful double star with golden-yellow components of mags. 2.6 and 3.5, spectral types K1 III and G7 III. The pair form a genuine binary with an orbital period of nearly 620 years. The name may come from the Arabic *al-jabha*, meaning 'the forehead', referring to its position in a much larger figure of a lion visualized by Arab astronomers in this region.

▼ **Aldrin, Edwin Eugene ('Buzz'), Jr** As well as piloting the lunar module of Apollo 11, 'Buzz' Aldrin also deployed and monitored experiments on the Moon's surface. Here, he is seen with the Solar Wind Composition experiment.

Algol Prototype of the ALGOL STARS, a subtype (EA) of ECLIPSING BINARY stars. It is now known, however, that Algol is somewhat atypical of its eponymous subtype.

The first recorded observation of Algol was made by Geminiano MONTANARI in 1669. In 1782 John GOODRICKE established that Algol's variability was periodic, with a sudden fade occurring every 2.867 days. Gradually, the concept of an eclipsing companion became accepted and was finally confirmed when, in 1889, Hermann VOGEL showed that the radial velocity of Algol varied with the same period as that of the eclipses.

By this time, Algol was known to be a triple system. In 1855 F.W.A. ARGELANDER had observed that the period between primary minima had shortened by six seconds since Goodricke's observations. Fourteen years later he noted that the period between the times of minima varied in a regular fashion with a period of about 680 days. This was attributed to the variation in the distance that the light from the system had to travel because of orbital motion around the common centre of gravity with a third star (Algol C).

In 1906 the Russian astronomer Aristarkh Belopolski (1854–1934) confirmed the existence of Algol C by showing that radial-velocity variations in the spectral lines of Algol also had a period of 1.862 years superimposed on the period of 2.867 days for Algol AB. Several years later, Joel STEBBINS, a pioneer in stellar photometry, found that there was a secondary minimum of much smaller amplitude occurring exactly halfway between the primary eclipses. This showed for the first time that the companion was not dark at all, but merely much fainter than A. Photoelectric observations showed the depth of the secondary minimum to be 0.06 mag. and that the light from star A increased as the secondary minimum approached. This was interpreted as a reflection of light from the body of star B.

There are two stars, of spectral types B8 and G, rotating about each other. The B8 star is a dwarf and is the visible component. The fainter star (whose spectrum was only observed directly for the first time in 1978) is a subgiant. The orbit is inclined to the line of sight by 82°, which results in mutual eclipses corresponding to a drop in light of 1.3 mag. when A is eclipsed by B. Hipparcos data give a distance to Algol of 93 l.y. This corresponds to luminosities of 100 and 3 for A and B respectively. From the length and depth of the eclipses, sizes of 2.89 and 3.53 solar radii have been derived for A and B respectively. The corresponding masses are 3.6 and 2.89 solar masses, and this apparent anomaly gives rise to what is known as the 'Algol paradox'.

In current theories of STELLAR EVOLUTION, stars advance in spectral type as they evolve, and the rate at which they do so is a function of their initial masses. Thus, if two stars form together from interstellar material, the more massive of the two should evolve more quickly. In Algol the more evolved star is the less massive and the cause seems to be MASS TRANSFER from B to A. A stream of material between the two stars has been detected in the radio observations, and the current transfer rate is thought to be at least 10^{-7} solar masses per year. Optical spectra have shown very faint lines, which are thought to be emitted by a faint ring of material surrounding star A.

Algol C was first resolved by speckle interferometry in 1974 and on several occasions since. Its angular separation has never exceeded 0″.1, which explains why the star has never been seen by visual observers.

The real nature of the Algol system is still far from clear. Even after 200 years of continuous observation it still evokes considerable interest from astronomers.

Algol stars (EA) One of the three main subtypes of ECLIPSING BINARY. The light-curves of Algol stars exhibit distinct, well-separated primary minima. Secondary minima may be detectable, depending on the characteristics of the system. Outside eclipses, the light-curve is essentially flat, although it may exhibit a small, gradual increase and decrease around secondary minimum, which is caused by the reflection effect, where light from the bright (MAIN SEQUENCE) primary irradiates the surface of the cooler secondary, thus raising its temperature and luminosity. The components of the binary system may be DETACHED or, as in ALGOL itself, one component may be SEMIDETACHED. Systems in which the semidetached component is transferring mass to the non-evolved component are sometimes described as being 'Algol-type' or 'Algol-like' binaries.

In the Algol-type binaries, the detached component is a main-sequence star and its less-massive companion is a red SUBGIANT that fills its ROCHE LOBE. Such systems are differentiated physically from the closely related BETA LYRAE STARS by the fact that an ACCRETION DISK is never present around the detached component.

Some systems that begin as Algol stars (with detached components) may evolve into BETA LYRAE systems, with a high rate of MASS TRANSFER and massive accretion disks. When the mass-transfer rate drops, the accretion disks disappear to reveal the unevolved stars, and the systems display all the characteristics of an Algol-like binary.

The eventual fate of an Algol system depends on many factors, most notably on the stars' masses, which determine their rates of evolution. If the main-sequence star in an Algol system is comparatively massive, it will evolve rapidly and expand to fill its Roche lobe while the companion star is still filling its own Roche lobe. The result is a CONTACT BINARY in which both stars share the same photosphere. These binaries are often called W URSAE MAJORIS STARS after the prototype for this subtype.

If the main-sequence star in an Algol system evolves slowly, then its companion may become a WHITE DWARF before the primary swells to fill its Roche lobe. When the primary finally does expand to become a RED GIANT, gas flows across the inner Lagrangian point and goes into orbit about the white dwarf, forming an accretion disk. Such systems, called U GEMINORUM STARS or DWARF NOVAE, exhibit rapid irregular flickering from the turbulent hot spot where the mass-transfer stream strikes the accretion disk. However, most of these systems are not eclipsing pairs. It is important to note that eclipsing variables only appear to fluctuate in light because of the angle from which they are observed.

Algonquin Radio Observatory (ARO) One of Canada's principal RADIO ASTRONOMY facilities, operated by the National Research Council and situated in Ontario's Algonquin Provincial Park, well away from local radio interference. Instruments include a 32-element ARRAY of 3-m (10-ft) dishes for solar observations and a 46-m (150-ft) fully steerable radio dish for studies of stars and galaxies. The ARO began work in 1959, and the 46-m dish was built in 1966.

ALH 84001 Abbreviation of ALLAN HILLS 84001

Alhena The star γ Geminorum, visual mag. 1.93, distance 105 l.y., spectral type A1 IV. The name comes from an Arabic term that is thought to refer to the neck of a camel, from a former constellation in this area.

Alioth The star ε Ursae Majoris, visual mag. 1.76, distance 81 l.y. It is one of the so-called peculiar A stars, of spectral type A0p with prominent lines of chromium. It is, by a few hundredths of a magnitude, the brightest star in the PLOUGH (Big Dipper). Its name may be a corruption of the Arabic for 'tail'.

Alkaid (Benetnasch) The star η Ursae Majoris, visual mag. 1.85, distance 101 l.y., spectral type B3 V. The name comes from an Arabic word meaning 'the leader'. Its alternative name, Benetnasch, is derived from an Arabic phrase referring to a group of mourners accompanying a coffin formed by the quadrilateral of stars which is now known as the bowl of the PLOUGH (commonly referred to in the US as the Big Dipper).

Allan Hills 84001 (ALH 84001) METEORITE that was found in Antarctica in 1984 and identified as a MARTIAN METEORITE in 1994. It has a mass of *c.*1.93 kg. A complex igneous rock, it has suffered both thermal and shock processes. In composition, it is an orthopyroxenite rich in carbonates, which form patches up to *c.*0.5 mm across. Few hydrated minerals have been identified amongst the alteration products in ALH 84001, so it has been proposed that the carbonates were produced at the surface of Mars in a region of restricted water flow, such as an evaporating pool of brine. Tiny structures (*c.*200 nm in size) within the carbonates have been interpreted by some as fossilized Martian bacteria; however the claim is controversial, and it is subject to continued investigation.

Allegheny Observatory Observatory of the University of Pittsburgh, located 6 km (4 mi) north of Pittsburgh. The observatory, which dates from 1859, became part of the university in 1867. During the 1890s its director was James E. KEELER, who used a 13-inch (330-mm) refractor to discover the particulate nature of Saturn's rings. Later, Allegheny was equipped with the 30-inch (76-cm) Thaw telescope (the third-largest refractor in the USA) and the 31-inch (0.79-m) Keeler reflector. The observatory now specializes in astrometric searches for EXTRASOLAR PLANETS.

Allende METEORITE that fell as a shower of stones in the state of Chihuahua, Mexico, on 1969 February 8. More than 2 tonnes of material is believed to have fallen. Allende is classified as a CV3 CARBONACEOUS CHONDRITE. Studies of components, such as CAIs and CHONDRULES, within Allende have been instrumental in understanding the structure, chemistry and chronology of the pre-solar nebula. The first INTERSTELLAR GRAINS (nanometre-sized diamonds) to be identified in meteorites were isolated from Allende.

Allen Telescope Array (ATA) Large-area radio telescope – formerly called the **One-hectare Telescope** (1hT) – that will consist of 350 steerable parabolic antennae 6.1 m (20 ft) in diameter. The ATA is a joint undertaking of the SETI Institute and the University of California at Berkeley. When completed in 2005, it will permit the continuous scanning of up to 1 million nearby stars for SETI purposes, and will serve as a prototype for the planned SQUARE KILOMETRE ARRAY.

ALMA Abbreviation of ATACAMA LARGE MILLIMETRE ARRAY

Almaak The star γ Andromedae, visual mag. 2.10, distance 355 l.y. It is a multiple star, the two brightest components of which, mags. 2.3 and 4.8, are divisible by small telescopes, forming a beautiful orange and blue pairing, spectral types K3 II and B9 V. The fainter star has a close 6th-magnitude blue companion that orbits it every 61 years. Its name comes from the Arabic referring to a caracal, a wild desert cat, and is also spelled *Almach* and *Alamak*.

Almagest Astronomical treatise composed in *c.*AD 140 by PTOLEMY. It summarizes the astronomy of the Graeco-Roman world and contains a star catalogue and rules for calculating future positions of the Moon and planets according to the PTOLEMAIC SYSTEM. The catalogue draws from that of HIPPARCHUS, though to what extent is a matter of controversy. In its various forms the *Almagest* was a standard astronomical textbook from late antiquity until the Renaissance. Its original name was *Syntaxis* ('[Mathematical] Collection'), but it became known as *Megiste*, meaning 'Greatest [Treatise]'. Around AD 700–800 it was translated into Arabic, acquiring the prefix *Al-* (meaning 'the'). It was subsequently lost to the West but was treasured in the Islamic world; it was reintroduced to European scholars via Moorish Spain in the form of a translation of the Arabic version into Latin completed in 1175 by Gerard of Cremona (*c.*1114–87). It remained of great importance until the end of the 16th century, when its ideas were supplanted by those of Nicholas COPERNICUS, Tycho Brahe and Johannes Kepler.

almanac, astronomical Yearbook containing information such as times of sunrise and sunset, dates for phases of the Moon, predicted positions for Solar System objects and details of other celestial phenomena such as eclipses. For astronomical and navigational purposes the leading publication is *The ASTRONOMICAL ALMANAC*.

Alnair The star α Gruis, visual mag. 1.73, distance 101 l.y., spectral type B7 V. Its name means 'bright one', from an Arabic expression meaning 'bright one from the fish's tail', given by an unknown Arab astronomer who visualized the tail of the southern fish, Piscis Austrinus, as extending into this area.

Alnath The star β Tauri, visual mag. 1.65, distance 131 l.y., spectral type B7 III. The name, which is also spelled *Elnath*, comes from the Arabic meaning 'the butting one' – it marks the tip of one of the horns of Taurus, the bull.

Alnilam The star ϵ Orionis, visual mag. 1.69, distance about 1300 l.y. A blue-white supergiant, spectral type B0 Ia, it is the central star of the three that make up the belt of Orion and is marginally the brightest of them. Its name comes from an Arabic phrase meaning 'string of pearls', referring to the belt.

Alnitak The star ζ Orionis, visual mag. 1.74, distance 820 l.y. It is a binary, with individual components of mags. 1.9 and 4.0, spectral types O9.5 Ib and B0 III, and an orbital period of around 1500 years. A telescope of 75-mm (3-in.) aperture or more should show both stars. Alnitak is a member of the belt of Orion, and its name comes from the Arabic meaning 'belt'.

Alpes (Montes Alps) Cross-faulted lunar mountains that rise 1–3 km (3600–9800 ft) above the north-east margins of Mare IMBRIUM. The Alps are 290 km (180 mi) long, and are traversed by the ALPES VALLIS. The majority of the bright peaks have altitudes between 2000 and 2500 m (6000–8000 ft), but some are significantly higher: Mount Blanc, one of the Moon's greatest mountains, is nearly 3500 m (11,500 ft) tall.

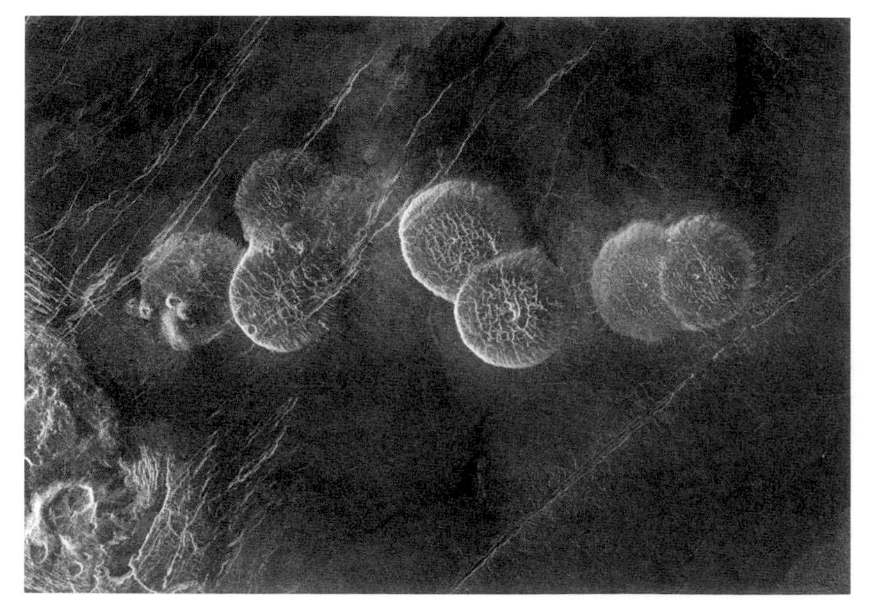

▼ **Alpha Regio** This Magellan radar image shows multiple volcanic domes in Alpha Regio on Venus.

Alpes Vallis (Alpine Valley) Darkened gap 200 km (120 mi) long that cuts through the lunar mountain range known as the Montes ALPES. The Alpine Valley is a GRABEN that developed as a result of Mare IMBRIUM'S tectonic adjustment. Varying in width from 7 to 18 km (4–11 mi), it irregularly tapers away from Mare Imbrium. Two delicate faults cut at right angles across the valley's floor, which has otherwise been smoothed by the lavas that have filled it. Running down the middle of the valley is a SINUOUS RILLE, which seems to originate in a vent crater, which may be a volcanic feature, probably a collapsed lava tube.

Alpha Unofficial name for the INTERNATIONAL SPACE STATION

Alpha² Canum Venaticorum star (ACV) Type of main-sequence VARIABLE STAR that exhibits photometric, magnetic and spectral fluctuations, primarily as a result of stellar rotation. Periods range from 0.5 to 150 days; amplitudes from 0.01 to 0.1 mag.; and spectra from B8p to A7p. The subtype ACV0 exhibits additional low-amplitude (*c.*0.1 mag.) non-radial pulsations, with periods of 0.003 to 0.1 days. *See also* SPECTRUM VARIABLE

Alpha Capricornids Minor METEOR SHOWER, active from mid-July until mid-August and best seen from lower latitudes. Peak activity occurs around August 2, from a RADIANT a few degrees north-east of α Capricorni. Rates are low, about six meteors/hr at most, but the shower produces a high proportion of bright, flaring meteors with long paths. The meteor stream may be associated with the short-period (5.27 years) comet 45P/Honda–Mrkos–Pajdusaková; it has a low-inclination orbit close to the ecliptic. Spreading of stream METEOROIDS by planetary perturbations means that the radiant is rather diffuse.

Alpha Centauri (Rigil Kentaurus, Toliman) Closest naked-eye star to the Sun, 4.4 l.y. away, with a visual mag. of −0.28, making it the third-brightest star in the sky. Small telescopes reveal that it is a triple system. The two brightest components are of solar type, mags. −0.01 and 1.35, spectral types G2 V and K1 V, forming a binary with an orbital period of 79.9 years. The third member of the system is the red dwarf PROXIMA CENTAURI, which is the closest star of all to the Sun. Alpha Centauri is also known as *Rigil Kentaurus* (Rigil Kent for short), from the Arabic meaning 'centaur's foot'. An alternative name, *Toliman*, is derived from an Arabic term meaning 'ostriches', the figure visualized by Arab astronomers in the stars of this region.

Alpha Monocerotids Normally very minor METEOR SHOWER, active around November 21–22. The shower produced outbursts of more substantial activity in 1925, 1935, 1985 and 1995, suggesting a ten-year periodicity with several stronger displays having been missed. In 1995 rates of one or two meteors per minute were sustained for only a short interval. The shower is apparently associated with comet C/1943 W1 van Gent-Peltier-Daimaca.

alpha particle Helium nucleus, consisting of two protons and two neutrons, positively charged. Helium is the second-most abundant element (after Hydrogen), so alpha particles are found in most regions of PLASMA, such as inside stars, in diffuse gas around hot stars and in cosmic rays. Alpha particles are also produced by the radioactive decay of some elements. In the PROTON–PROTON chain of nuclear fusion reactions inside stars, four protons (hydrogen nuclei) are converted to one alpha particle (helium nucleus) with release of fusion energy, which powers stars. In the

TRIPLE-α PROCESS, which is the dominant energy source in red giant stars, three alpha particles fuse to form a carbon nucleus with release of energy.

Alphard The star α Hydrae, visual mag. 1.99, distance 177 l.y., spectral type K3 II or III. Its name comes from an Arabic word meaning 'the solitary one', a reference to its position in an area of sky in which there are no other bright stars.

Alpha Regio Isolated highland massif on VENUS (25°.5S 0°.3E), showing complex structure; it is best described as a plateau encircled by groups of high volcanic domes. The circular central area has a mean elevation of 0.5 km (0.3 mi).

Alphekka (Gemma) The star α Coronae Borealis, visual mag. 2.22, distance 75 l.y., spectral type A0 IV. It is an ALGOL STAR; its brightness drops by 0.1 mag. every 17.4 days as one star eclipses the other. Its name, which is also spelled *Alphecca*, comes from the name *al-fakka*, meaning 'coins', by which Arab astronomers knew the constellation Corona Borealis. More recently, the star has also become known as *Gemma*, since it shines like a jewel in the northern crown.

Alpheratz The star α Andromedae, visual mag. 2.07, distance 97 l.y. It has a peculiar SPECTRUM, classified as B9p, which has prominent lines of mercury and magnesium. Its name is derived from the Arabic *al-faras*, meaning 'the horse', since it used to be regarded as being shared with neighbouring Pegasus (and was also designated δ Pegasi); indeed, it still marks one corner of the SQUARE OF PEGASUS. Its alternative name, *Sirrah*, is derived from the Arabic *surrat al-faras*, meaning 'horse's navel'.

Alphonso X (1221–84) King of Léon and Castile (part of modern Spain), known as Alphonso the Wise, a patron of learning and especially of astronomy. He commissioned a new edition of the highly successful *Toledan Tables* of the motions of the Sun, Moon and five naked-eye planets, prepared originally by AL-ZARQĀLĪ in Toledo a century before. The new *Alphonsine Tables*, incorporating ten years of revised observations and completed in 1272, were not superseded for almost 400 years.

Alphonsus Lunar crater (13°.5S 3°W), 117 km (72 mi) across. Its fault-dissected walls rise to over 3000 m (10,000 ft) above the floor. Running nearly north–south across the floor is a ridge system, which is 15 km (9 mi) wide and, at the point where it forms a prominent central peak, about 1000 m (3000 ft) high. Within Alphonsus are a series of kilometre-sized elliptical features with haloes of dark material; they are oriented roughly parallel to the central ridge system and are considered by many planetary geologists to be of volcanic origin. In 1958 Soviet astrophysicist Nikolai Kozyrev (1908–83) obtained a spectrum showing blue emission lines, which he interpreted as proof of a gaseous emission from the crater's central peak, but these results have never been duplicated. The north wall of Alphonsus overlaps the south wall of PTOLEMAEUS, indicating that Alphonsus formed following the Ptolemaeus impact event.

ALPO Abbreviation of ASSOCIATION OF LUNAR AND PLANETARY OBSERVERS

Alrescha The star α Piscium, visual mag. 3.82, distance 139 l.y. It is a close binary with a calculated orbital period of around 930 years. The brighter component, of mag. 4.2, is a peculiar A star of spectral type A0p with strong lines of silicon and strontium; the fainter companion, mag. 5.2, is a metallic-line A star, type A3m. The name Alrescha, sometimes also spelled *Alrisha*, comes from an Arabic word meaning 'the cord'.

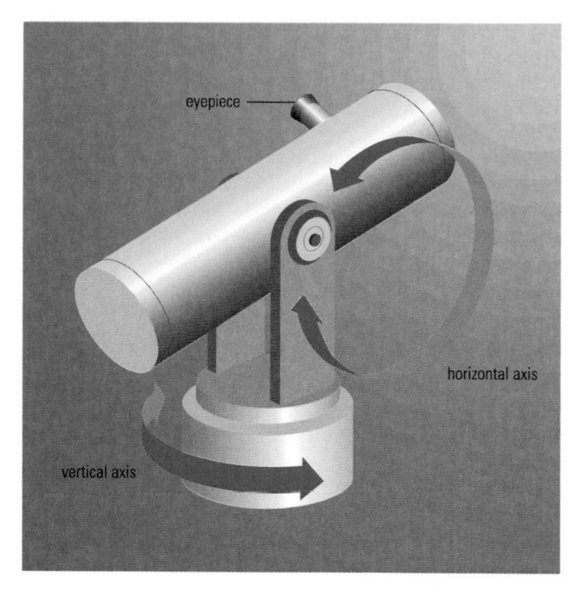

◄ altazimuth mounting
This simple form of telescope mount allows free movement in both horizontal and vertical axes, but is not suitable for use with motordrives, unless they are computer controlled.

al-Sūfī, Abu'l-Husain (Latinized as **Azophi**) (903–986) Arab astronomer (born in modern Iran) famous for his *Kitāb suwar al-kawākib al-thābita* ('Book on the Constellations of the Fixed Stars'), a detailed revision, based upon his own observations, of PTOLEMY's star catalogue. In this work he identified the stars of each constellation by their Arab names, providing a table of revised magnitudes and positions as well as drawings of each constellation. Al-Sūfī was the first to describe the two brightest galaxies visible to the naked eye: the Andromeda Galaxy and the Large Magellanic Cloud, which he called the White Bull.

Altair The star α Aquilae, visual mag. 0.76, distance 16.8 l.y. It is a white main-sequence star of spectral type A7 V, with a luminosity 10 times that of the Sun. Altair is the 12th-brightest star and forms one corner of the SUMMER TRIANGLE. Its name comes from an Arabic expression meaning 'flying eagle'.

Altai Rupes Range of lunar mountains (25°S 22°E) cut by four deep cross-faults. The Altais curve 505 km (315 mi) from the west wall of Piccolomini to the west side of the large formation CATHARINA. They rise very steeply from the east to an average altitude of 1800 m (6000 ft), with highest peaks at 3500–4000 m (11,000–13,000 ft). The scarp is roughly concentric with the south-west margins of Mare NECTARIS. It may be the sole remnant of an outer ring of a much larger, multi-ring impact BASIN.

altazimuth mounting Telescope mounting that has one axis (altitude) perpendicular to the horizon, and the other (azimuth) parallel to the horizon. An altazimuth (short for 'altitude–azimuth') mounting is much lighter, cheaper and easier to construct than an EQUATORIAL MOUNTING for the same size telescope, but is generally not capable of tracking the apparent motion of celestial objects caused by the Earth's rotation. Many amateur instruments with altazimuth mountings can therefore be used for general viewing, but are not suitable for long-exposure photography.

Historically, large professional telescopes were invariably built with massive equatorial mountings, which often dwarfed the instrument they held. The lightweight and simple nature of altazimuth mountings, combined with high-speed computers, has led to almost all modern instruments being built with altazimuth mountings. On these telescopes, computers are used to control the complex three-axis motions needed for an altazimuth mount to track the stars. Both the altitude and azimuth axes are driven at continuously varying rates but, in addition, the field of view will rotate during a long photographic exposure, requiring an additional drive on the optical axis to counter FIELD ROTATION. Some amateur instruments, especially DOBSONIAN TELESCOPES, are now being equipped with these three-axis drive systems, controlled by personal computers.

altitude Angular distance above an observer's horizon of a celestial body. The altitude of a particular object depends both on the location of the observer and the time the observation is made. It is measured vertically from 0° at the horizon, along the great circle passing through the object, to a maximum of 90° at the ZENITH. Any object below the observer's horizon is deemed to have a negative altitude. *See also* AZIMUTH; CELESTIAL COORDINATES

al-Tūsī, Nasīr al-Dīn (Latinized as **Nasireddin** or **Nasiruddin**) (1201–74) Arab astronomer and mathematician from Khurāsān (in modern Iran) who designed and built a well-equipped observatory at Marāgha (in modern Iraq) in 1262. The observatory used several quadrants for measuring planet and star positions, the largest of which was 3.6 m (12 ft) in diameter. Twelve years of observations with these instruments allowed him to compile a table of precise planetary and stellar positions, titled *Zīj-i ilkhānī*. Al-Tūsī's careful measurement of planetary positions convinced him that the Ptolemaic Earth-centred model of the Solar System was incorrect. His work may have influenced COPERNICUS.

aluminizing Process of coating the optics of a reflecting telescope with a thin, highly reflecting layer of aluminium. The optical component to be aluminized is first thoroughly cleaned and placed in a vacuum chamber, together with pure aluminium wire, which is attached to tungsten heating elements. After removing the air from the chamber, the heating elements are switched on, vaporizing the aluminium, which then condenses on to the clean surface of the optical component. This forms an evenly distributed coating, usually just a few micrometres thick.

Alvan Clark & Sons American firm of opticians and telescope-makers whose 19th-century refracting TELESCOPES include the largest in the world. After a career as a portrait painter and engraver, **Alvan Clark** (1804–87) started an optical workshop in 1846 under the family name with his sons, **George Bassett Clark** (1827–91) and **Alvan Graham Clark** (1832–97), the latter joining the firm in the 1850s. Alvan Graham Clark discovered over a dozen new double stars, including, in 1862, the 8th-magnitude Sirius B.

During the second half of the 19th century, Alvan Clark & Sons crafted the fine objective lenses for the largest refracting telescopes in the world, including the UNITED STATES NAVAL OBSERVATORY's 26-inch (0.66-m) (1873), PULKOVO OBSERVATORY's 30-inch (0.76-m) (1878), Leander McCormick (Charlottesville, Virginia) Observatory's 28-inch (0.7-m) (1883), LICK OBSERVATORY's 36-inch (0.9-m) (1888), LOWELL OBSERVATORY's 24-inch (0.6-m) (1896) and YERKES OBSERVATORY's 40-inch (1-m) (1897). In addition to these large professional instruments, the firm made numerous smaller refractors, 4–6 inch (100–150 mm) in aperture, which are prized by today's collectors of antique telescopes.

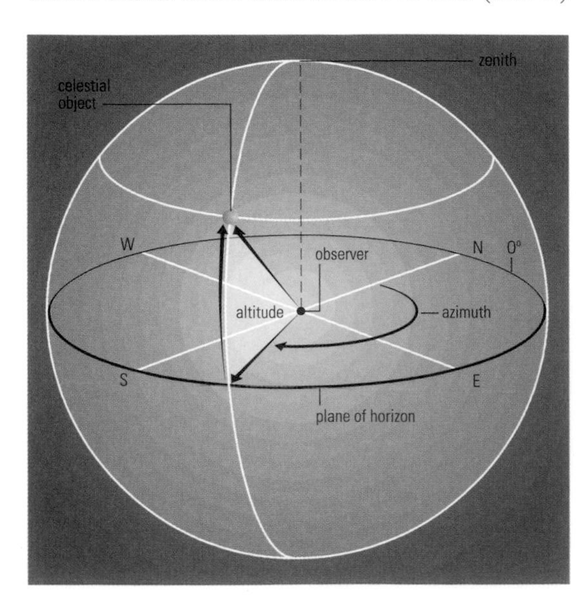

▶ **altitude** The altitude of a celestial object relative to an observer is measured on a scale of 0–90° from the observer's horizon to the zenith – the point directly overhead.

Alvarez, Luis Walter (1911–88) American physicist who first identified the layer of clay enriched by the element iridium that appears in the strata separating the Cretaceous and Tertiary geological periods, known as the K/T boundary. Since meteorites contain much higher amounts of iridium than do terrestrial rocks and soil, Alvarez' discovery supported the hypothesis that a giant meteorite impact (*see* CHICXULUB) may have caused a mass extinction event on our planet 65 million years ago.

al-Zarqāli, Abū Ishaq Ibrahim ibn Yahya (Latinized as **Arzachel**, and other variants) (1028–87) Arab astronomer who worked in Toledo, Spain, and prepared the famous *Toledan Tables* of planetary positions, which corrected and updated the work of Ptolemy and Muhammad ibn Mūsā al-Khwārizmī (c.780–c.850). Al-Zarqālī also accurately determined the annual rate of apparent motion of the Earth's aphelion relative to the stars as 12″, remarkably close to the correct value of 11″.8.

Amalthea Largest of JUPITER'S inner satellites. Amalthea was the fifth Jovian moon to be found, in 1892 by E.E. BARNARD, and the first since the four much larger GALILEAN SATELLITES were discovered in 1610. The discovery was made visually, the last such discovery for a planetary satellite. Amalthea is irregular in shape, measuring about 270 × 165 × 150 km (168 × 103 × 93 mi). Amalthea orbits Jupiter at a distance of only 181,400 km (112,700 mi), under SYNCHRONOUS ROTATION with a period of 0.498 days such that it always keeps the same blunt end towards the planet. Its orbit is near-circular, inclined to the Jovian equator by only 0°.4. Amalthea is notable as being the reddest object in the Solar System, possibly because of the accumulation of a surface covering of sulphur derived from the EJECTA of IO's volcanoes. Amalthea has considerable surface relief, with two large craters, called Pan and Gaea, and two mountains, named Mons Ida and Mons Lyctos. Some sloping regions appear very bright and green, the cause of this phenomenon being unknown.

amateur astronomy, history of From at least as early as the 17th century until around 1890, astronomical research in Britain was invariably undertaken by those who worked for love and considered themselves 'amateurs' (from Latin *amat*, 'he loves'). The reasons were political and economic, as successive governments operated low-taxation, low-state-spending policies that encouraged private rather than public initiatives. Amateur astronomy, while it existed on Continental Europe, was less innovative, largely because the governments of France, Germany and Russia taxed more heavily and invested in professional science as an expression of state power. The United States had a mixed astronomical research tradition, with outstanding amateurs, such as the spectroscopist Henry DRAPER, engaged in front-rank research, and major professional observatories financed by millionaire benefactors.

Although the British astronomical tradition was predominantly amateur, its leading figures were 'Grand Amateurs' in so far as fundamental research was their dominant concern. In the Victorian age, wealthy gentleman scientists were willing to spend huge sums of money to pursue new lines of research and commission ground-breaking technologies, such as big reflecting telescopes. The quality of Grand Amateur research enjoyed peer recognition from European and American

professionals, while its own esprit de corps was expressed through membership of the ROYAL ASTRONOMICAL SOCIETY and the Royal Society in London, academic honours and a clearly defined social network. This was, indeed, professional-quality research paid for by private individuals. Grand Amateurs pioneered work on the gravitation of double star systems, cosmology, planetary studies, selenography, photography and spectroscopy, and included between 1820 and 1900 such figures as John HERSCHEL, William DAWES, Lord ROSSE, Admiral William SMYTH, William LASSELL, William HUGGINS, Norman LOCKYER and the master-builder and astrophotographer Isaac Roberts (1829–1904). The results of their researches transformed our understanding of the Universe.

Yet Victorian Britain also saw a fascination with astronomy spreading to the less well-off middle and even working classes. School teachers, modest lawyers, clergymen and even artisans took up astronomy; the self-educated telescope-maker John Jones worked for a few shillings per week as a labourer on Bangor docks, Wales. People with modest and often home-made instruments (especially after the silvered glass mirror replaced speculum in the 1860s) did not expect, like the Grand Amateurs, to change the course of astronomy, but enjoyed practical observation as a serious and instructive hobby. The Reverend Thomas WEBB's celebrated *Celestial Objects for Common Telescopes* (1859) became the 'bible' for these serious amateurs.

The big-city astronomical societies of Leeds (1859, 1892), Liverpool (1881), Cardiff (1894), Belfast (*c.*1895) and others became the foci for these observing amateurs, with their lectures, meetings and journals. In 1890 the BRITISH ASTRONOMICAL ASSOCIATION (BAA) became the national organizing body for British amateurs, with branches in Manchester (1892, 1903) and elsewhere, many of which later became independent societies. Unlike the Royal Astronomical Society, they all admitted women as members. These societies, which dominated amateur astronomy well into the 20th century, remained predominantly middle-class, and it was not until the major social and economic changes in Britain following 1945 that the demographic base of British amateur astronomical societies began to widen significantly. The BAA established a system whereby amateurs would send their observations to a central clearing house where they would be synthesized by an expert and the collective results published. The great majority of amateurs who contribute observations on behalf of science continue to operate within such systems (*see also* THE ASTRONOMER).

Amateur astronomy changed dramatically after World War II, and much of the emphasis moved across the Atlantic. Before the war, in the 1920s, the ranks of active amateur observers were swelled with the founding of the amateur telescope-making (ATM) movement by Russell PORTER and Albert G. Ingalls (1888–1958). Now that inexpensive war-surplus optical equipment was widely available, it was no longer essential for an amateur astronomer to build a telescope from the ground up as a rite of passage. By the mid-1950s, a wide variety of commercial instruments had entered the marketplace; later designs, such as the SCHMIDT–CASSEGRAIN and DOBSONIAN TELESCOPES, owed much to amateur observers and remain extremely popular. The numbers of amateur observers grew rapidly, particularly in the United States. It is no coincidence that the ASTRONOMICAL LEAGUE (1946) and the ASSOCIATION OF LUNAR AND PLANETARY OBSERVERS (1947) were formed at this time. A watershed for professional–amateur collaboration came in 1956 with the establishment of the Moonwatch programme, in anticipation of satellite launches for the International Geophysical Year (1957–59). Energized by the Soviet Union's launch of Sputnik 1, and guided by astronomers at the Smithsonian Astrophysical Observatory, Moonwatch galvanized amateurs around the world in a unique and grand pro–am effort.

The appearance of affordable charge-coupled devices (CCDs) in the final decade of the 20th century had an even greater impact on amateurs than had the war-surplus items of two generations before. Digital data, exponentially increasing computing power, and ever more sophisticated commercial SOFTWARE together created a revolution. They allowed amateurs to become competitive with ground-based professionals in the quality of data obtained in such areas as astrometry, photometry and the imaging of Solar System objects.

New organizations with new ideas sprang up. The INTERNATIONAL AMATEUR–PROFESSIONAL PHOTOELECTRIC PHOTOMETRY group, founded in 1980, is the prototype organization representing this new era. It encourages joint amateur–professional authorship of technical papers. Similar, though focused on campaigns to study cataclysmic variable stars, is the Center for Backyard Astrophysics. One of the latest groups to form is The Amateur Sky Survey, a bold venture to develop the hardware and software needed to patrol automatically the sky in search of objects that change in brightness or move. Other groups, such as the INTERNATIONAL OCCULTATION TIMING ASSOCIATION, have graduated from visual observations of lunar events to video recordings that determine the profiles of asteroids. The INTERNATIONAL DARK-SKY ASSOCIATION campaigns on an issue of concern to professionals and amateurs alike.

As the present era of mammoth all-sky surveys from Earth and space culminates, the need for follow-up observations – particularly continuous monitoring of selected objects – will grow dramatically. In a traditional sense, because of their numbers and worldwide distribution, sophisticated amateurs are ideally suited for such tasks, not as minions but as true partners with professionals. And, in the era of the Internet, amateurs should be able to plumb online sky-survey DATABASES just as readily as professionals can. The challenge facing the entire astronomical community today is to educate both camps about rewarding possibilities through mutual cooperation.

Ambartsumian, Viktor Amazaspovich (1908–96) Armenian astronomer who became an expert on stellar evolution and founded Byurakan Astrophysical Observatory. His development of the theory of radiative

◄ **Alvarez, Luis Walter** Many years after Alvarez first proposed that an anomaly in the iridium levels at the boundary between the Cretacious and Tertiary geological periods might have been caused by a meteor impact, geologists looking for oil found evidence of a massive impact centred near Chicxulub on Mexico's Yucatán Peninsula. Shown here is a radar image of the impact site.

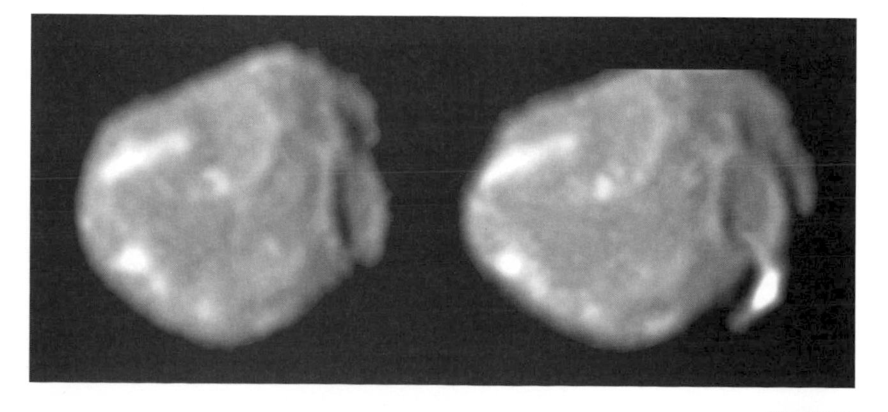

▼ **Amalthea** The bright streak to the left on Amalthea's surface is about 50 km (30 mi) long. It is not clear whether this feature (called Ida) is the crest of a ridge or material ejected from the crater to its right.

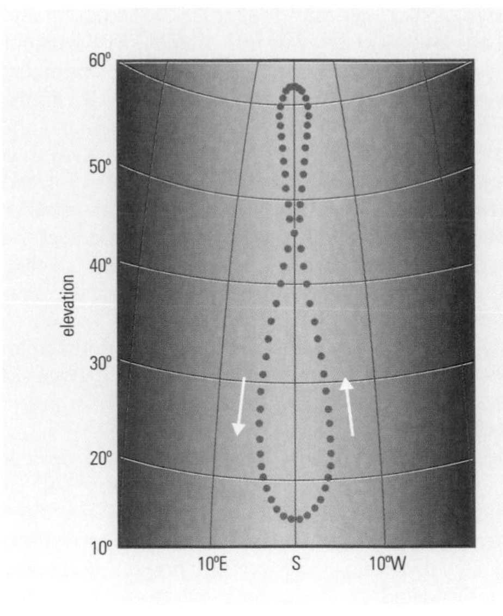

► **analemma** A plot of the Sun's apparent position from 52°N, looking south at midday, at 5-day intervals throughout the course of a year. The Sun is at the top of the figure 8 at the summer solstice and at the bottom at the winter solstice.

transfer allowed him to show that T Tauri stars are extremely young. He greatly advanced the understanding of the dynamically unstable stellar associations and extended principles of stellar evolution to the galaxies, where he found much evidence of violent processes in active galactic nuclei.

AM Canum Venaticorum Unique blue VARIABLE STAR with fluctuating period of about two minutes. It has primary and secondary minima, the latter sometimes disappearing. It is probably a SEMIDETACHED BINARY of two white dwarfs, an ACCRETION DISK and a hot spot.

American Association of Variable Star Observers (AAVSO) Organization of amateur and professional astronomers, based in the USA but with an international membership. Founded in 1911, it originally collected mainly visual estimates of the changing brightnesses of mainly long-period variable stars, but its programme now encompasses all manner of variable objects, from pulsating RR LYRAE STARS and ECLIPSING BINARIES to exotic GAMMA-RAY BURSTERS. The AAVSO continues to provide timely data to researchers, including those using instruments on board spacecraft such as HIPPARCOS and High-Energy Transient Explorer 2. By 2001 the AAVSO International Database contained more than 9 million observations.

Ames Research Center NATIONAL AERONAUTICS AND SPACE ADMINISTRATION (NASA) research institute located at Moffett Field, California, in the heart of 'Silicon Valley'. It is NASA's centre of excellence for information technology and its lead centre in Aeronautics for Aviation Operations Systems. Ames also develops science and technology requirements for current and future flight missions relevant to astrobiology. Moffett Field has been a government airfield since 1933, but was closed as a military base in 1994. It is now a shared facility known as Moffett Federal Airfield.

AM Herculis star (AM) Binary system, with period in the range 1 to 3 hours, that shows strongly variable linear and circular polarization and also eclipses. AM Herculis stars are strongly variable X-ray sources and their light-curves change from orbit to orbit. They also show changes in brightness and in variability with time scales of decades. The total range of light variations may reach 4–5 magnitude V. AM Herculis stars seem to be related to DWARF NOVAE, in that one component is a K-M-type dwarf and the other a compact object, but they differ in that the magnetic field of the compact component is sufficiently strong to dominate the mass flow and thus cause the effects observed. Am Herculis stars are also known as Polars. *See also* CATACLYSMIC VARIABLE

AMiBA Abbreviation of ARRAY FOR MICROWAVE BACKGROUND ANISOTROPY

Amor asteroid Any member of the class of ASTEROIDS that approach, but do not cross, the orbit of the Earth; their perihelion distances range from the terrestrial aphelion at 1.0167 AU to an arbitrary cut-off at 1.3 AU. Like the other MARS-CROSSING ASTEROIDS, Amor asteroids have limited lifetimes because of the chance of a collision with that planet. Over extended periods of time many Amors will evolve to become APOLLO ASTEROIDS, reducing their lifetimes further because of the greater chance of an impact on the Earth, or one of the other terrestrial planets. The disparity of compositional types observed indicates that Amors derive from various sources, including extinct cometary nuclei, the KIRKWOOD GAPS (through jovian perturbations) and the inner MAIN BELT (through perturbations imposed by Mars).

The first Amor-type asteroid to be discovered was EROS, in 1898, but the archetype giving them their collective name is (1221) Amor. That object was found in 1932, the same year as the first Apollo asteroid. Over 760 Amors had been discovered by late 2001. Notable examples listed in the NEAR-EARTH ASTEROID table include (719) ALBERT, (887) Alinda, (1036) GANYMED, (1580) Betulia, (1627) Ivar, (1915) Quetzalcoatl, (3552) Don Quixote and (4954) Eric.

amplitude In the study of VARIABLE STARS, the overall range in magnitude of a variable, from maximum to minimum. This definition is in contrast to the normal usage in physics, where the term is applied to half of the peak-to-peak value assumed by any parameter.

Am star Metallic-line class A STAR with high abundances of particular metals. These class 1 CHEMICALLY PECULIAR STARS (CP1) extend to class Fm. Am stars are enriched by factors of 10 or so in copper, zinc, strontium, zirconium, barium and the rare earths, but are depleted in calcium and scandium. As slow rotators that lack outer convection layers, these stars are apparently braked by the gravitational effects of close companions. In the quiet atmospheres, some atoms fall under the action of gravity, while others rise by means of radiation pressure. Sirius is an Am star.

amu Abbreviation of ATOMIC MASS UNIT

analemma Long, thin figure-of-eight shape obtained by plotting (or photographing) the position of the Sun on the sky at the same time of day at regular intervals throughout the year. The elongated north–south variation is due to the INCLINATION of the Earth's equator to its orbit, and the much shorter east–west variation is due to the ECCENTRICITY of the Earth's orbit. A considerable degree of patience and technical skill is required to record the analemma photographically.

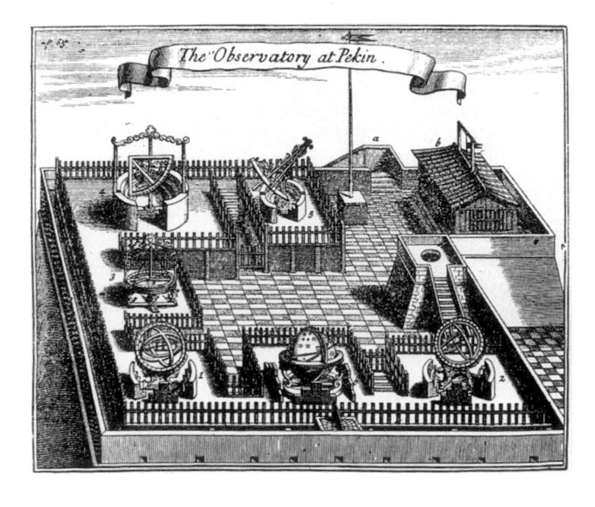

► **Ancient Beijing Observatory** This engraving shows the Imperial Astronomical Observatory at Beijing in the late 17th century. The instruments were used for mapping the skies extremely accurately.

Ananke One of JUPITER's outer moons, *c.*30 km (*c.*20 mi) in size. All members of this group, which includes Carme, Pasiphae and Sinope, are in RETROGRADE MOTION (Ananke's inclination is 149°). They are thought to be fragments of a captured asteroid that subsequently broke apart. Ananke was discovered in 1951 by Seth Nicholson. It takes 631 days to orbit Jupiter at an average distance of 21.28 million km (13.22 million mi) in an orbit of eccentricity 0.244. The population of known outer satellites of Jupiter is increasing rapidly, with eleven more having been discovered since 1999.

anastigmat Compound lens designed to be free of ASTIGMATISM. In practice the astigmatism will only be eliminated in some areas of the lens but other ABERRATIONS will be sufficiently well corrected to give excellent definition across the whole field of view.

Anaxagoras Comparatively young crater (75°N 10°W), 51 km (32 mi) in diameter, near the Moon's north pole. Like other freshly formed impact sites, Anaxagoras is the centre of a bright system of rays and steep, finely terraced walls. Its rays extend south to Plato. The rims rise to a height of 3000 m (10,000 ft) above the floor. Anaxagoras has a very bright, 300-m (1000-ft) high central peak, which is part of a larger range that crosses the crater's floor. To the east, Anaxagoras overlaps Goldschmidt, a degraded ring, 80 km (50 mi) in diameter .

Anaxagoras of Clazomenae (*c.*499–428 BC) Greek philosopher (born in what is now modern Turkey) whose theory of the origin and evolution of the Solar System is, in terms of today's 'standard model', correct in its basic premise. He believed it originated as a disk whose rotation caused the matter in it to separate according to its density, the densest materials settling at the centre and the more rarefied materials spreading out towards the periphery. He was imprisoned for teaching that the Sun was not a deity but a red-hot stone, and that the Moon, the phases of which he correctly explained, shone by reflected sunlight.

Anaximander of Miletus (*c.*611–547 BC) Greek philosopher (born in what is now modern Turkey) who believed the Earth to be one of many existing worlds, and the Sun and Moon rings of fire. He taught that Earth moves freely in space – not fixed upon anything solid. In his cosmogony, the Universe came into existence from an 'eternal reservoir', rotation having spread fire (the stars) to outer regions, leaving heavy matter (Earth) at the centre. He was said to have discovered the equinoxes and the obliquity of the ecliptic, but there is little evidence for this.

Ancient Beijing Observatory Astronomical observatory founded in 1442, situated in central Beijing on an elevated platform 14 m (46 ft) above street level. In about 1670, the Flemish Jesuit missionary Ferdinand Verbiest (1623–88) began re-equipping the observatory, and six of the eight large bronze instruments remaining on the site date from 1673. The other two were built in 1715 and 1744. It is not known why Verbiest based his instruments on outmoded designs by Tycho BRAHE well into the era of telescopic astronomy.

Anderson, John August (1876–1959) American astronomer who, with Francis PEASE, used the Michelson stellar interferometer at the prime focus of Mount Wilson Observatory's 100-inch (2.5-m) Hooker Telescope to measure the diameter of the red giant star Betelgeuse. Using this arrangement, Anderson was also able to separate very close double stars. He supervised the grinding and polishing of the primary mirror for Mount Palomar Observatory's 200-inch (5-m) Hale Telescope.

Andromeda *See* feature article

Andromeda Galaxy (M31, NGC 224) One of the two giant spiral galaxies in the LOCAL GROUP of galaxies, the other being our Galaxy, the Milky Way. M31 is the nearest spiral to the Milky Way, some 2.4 million l.y. away. Its proximity has led to intensive studies by astronomers, yielding fundamental advances in such diverse fields as star formation, stellar evolution and nucleosynthesis, dark matter, and the distance scale and evolution of the Universe.

The Andromeda spiral is visible to the naked eye. Found close to the 4th-magnitude star ν Andromedae, M31 (RA $00^h\ 42^m.7$ dec. +41°16′) appears as a faint patch of light, best seen on a transparent, moonless night from a dark site. It was recorded by the 10th-century Persian astronomer AL SŪFĪ as a 'little cloud'. Binoculars and small telescopes show the central regions as an elongated haze; long-exposure imaging with large instruments is required to show the galaxy's spiral structure.

The Andromeda Galaxy played an important role in the 'GREAT DEBATE' among astronomers in the 1920s regarding the nature of the spiral nebulae: were these 'island universes' – complete star systems outside our own – as proposed by the 18th-century philosopher Immanuel KANT, or were they gas clouds within the Milky Way collapsing to form stars? Photographs taken in 1888 by Isaac Roberts (1829–1904) using a 20-inch (0.5-m) telescope revealed M31's spiral nature, but it was not until the 1920s that the most important clues were uncovered by Edwin HUBBLE. In 1923–24, using the 100-inch (2.5-m) Hooker Telescope at Mount Wilson, California, Hubble was able to image individual CEPHEID VARIABLES in the Andromeda spiral. Applying the PERIOD–LUMINOSITY RULE to the derived light-curves showed that the spiral was a galaxy in its own right beyond our own.

The next important stage in the study of M31 came between 1940 and 1955, with the painstaking observations of Walter BAADE from Mount Wilson during the wartime blackout, and later with the 200-inch (5-m) Hale

ANDROMEDA

Constellation of the northern sky between the Square of Pegasus and the 'W' of Cassiopeia. In mythology, Andromeda, the daughter of King Cepheus and Queen Cassiopeia, was chained to a rock as a sacrifice to the sea monster Cetus and was rescued by Perseus. ALPHERATZ (or Sirrah), its brightest star, lies at the north-eastern corner of the Square of Pegasus and was once also known as δ Pegasi. ALMAAK is a fine double, with orange and bluish-white components, mag. 2.3 and 4.8. υ And is orbited by three planets (*see* EXTRASOLAR PLANET). The most famous deep-sky object is the ANDROMEDA GALAXY (M31, NGC 224), which is just visible to the unaided eye as a faint misty patch; the first extragalactic supernova, S ANDROMEDAE, was first observed here in 1885. NGC 572 is an open cluster of several dozen stars fainter than mag. 8; NGC 7662 is a 9th-magnitude planetary nebula.

BRIGHTEST STARS

	Name	RA h m	dec. ° ′	Visual mag.	Absolute mag.	Spectral type	Distance (l.y.)
α	Alpheratz	00 08	+29 05	2.07	−3.0	B9	97
β	Mirach	01 10	+36 37	2.07	−1.9	M0	199
γ	Almaak	02 04	+42 20	2.10	−3.1	K3 + B9	355
δ		00 39	+30 52	3.27	0.8	K3	101

ANDROMEDA GALAXY

Other designations	Andromeda Nebula, M31, NGC 224
Apparent size	3°.1 × 1°.25
Apparent (integrated) magnitude m_v	3.4 mag
Absolute magnitude M_v	−21.1 mag
Type (G de Vaucouleurs)	SA(s)b
Angle between plane of galaxy and line-of-sight	13°
Distance	740 Kpc, 2.4 million l.y.
Number of stars	4×10^{11}
Total mass	$3.2. \times 10^{11}\ M_\odot$
Diameter (optical)	50 Kpc
Dimensions of optical nucleus	5 × 8 Kpc
Satellite galaxies	M32, NGC 147, NGC 185, NGC 205, IC 10, LGS 3, And I, II, III, V, VI

Reflector at the Palomar Observatory, California. Baade succeeded in resolving stars in the Andromeda Galaxy's central bulge; they appeared to be mainly old and red, substantially fainter than the bright blue stars of the outer regions, and apparently similar to those in globular clusters. Baade referred to the bulge stars as Population II, labelling the hot disk stars as Population I (*see* POPULATIONS, STELLAR). This distinction remains in current use and is an essential feature of accepted theories of star formation, and stellar and galaxy evolution.

The discovery of the two stellar populations led in turn to a crucial finding for cosmology. The Cepheid variables turned out to be of two subsets, one belonging to each population, obeying different period-luminosity rules. Since the Cepheids observed by Hubble were of Population I, the derived distance of M31 had to be revised upwards by a factor of two – as, were all other distances to galaxies, which had used the Andromeda Galaxy as a 'stepping stone'.

The neutral hydrogen (HI) distribution in M31 has been extensively studied by radio astronomers, observing the TWENTY-ONE CENTIMETRE EMISSION LINE. Neutral hydrogen is a constituent of the galaxy's gas and is distributed like other Population I components. The gas shows a ZONE-OF-AVOIDANCE near the galactic centre, which is where Population II stars dominate. The gas is distributed in a torus, the innermost parts of which seem to be falling towards the nucleus.

Radial velocities of hot gas clouds across the galaxy have been mapped. Together with HI observations, these measurements allow a rotation curve to be constructed as a function of galactic radius. HI measurements, particularly, suggest that the outer regions of M31 contain substantial amounts of unseen additional mass. Such halos of DARK MATTER are crucial to current theories of galaxy formation and clustering, and cosmology.

Observations with the HUBBLE SPACE TELESCOPE in 1993 showed the nucleus of M31 to be double, with its components separated by about 5 l.y. This may be the result of a comparatively recent merger between the Andromeda Galaxy and a dwarf companion. Several small satellite galaxies surround M31, the most prominent being M32 (NGC 221) and M110 (NGC 205).

The disk of M31 shows a number of star clouds, the most obvious being NGC 206, which covers an area of 2900 × 1400 l.y. About 30 novae can be detected in M31 each year by large telescopes. M31 was the site, in 1885 August, of the first SUPERNOVA to be observed beyond the Milky Way: it was designated S Andromedae and reached a peak apparent magnitude +6.

It might be expected that the proximity of M31 would mean that it could make a substantial contribution to theories for the development of spiral structure. Instead, it has contributed controversy, partly because the galaxy is so close to edge-on that details of the spiral structure are hard to delineate. Indeed, it is not even known how many spiral arms there are. Halton ARP has proposed two trailing spiral arms, one of these disturbed by the gravitational pull of M32. A. Kalnajs proposes instead a single leading spiral arm, set up via gravitational resonance with M32. The dust clouds do not help in deciding between these two models. Resolution of the debate will ultimately advance our understanding of the mechanism generating spiral structure (*see* DENSITY WAVE THEORY).

M31 is surrounded by a halo of globular clusters, which is some three times more extensive than the halo around our Galaxy. The stars in these clusters show a generally higher metallicity than is found in our own Galaxy's globulars. The great spread in element abundances in the M31 globular clusters suggests slower and more irregular evolution than has occurred in the Milky Way.

Nearly every galaxy in the Universe shows a REDSHIFT, indicative of recession from the Milky Way. The spectrum of M31, however, shows it to be approaching at a velocity of about 35 km/s (22 mi/s). In some 3 billion years, M31 and the Milky Way will collide and merge eventually to form a giant elliptical galaxy.

M31 is our sister galaxy, the nearest spiral galaxy that is similar in most attributes to the Milky Way. Much of our home Galaxy is hidden from our perspective by massive dust clouds; we rely on the Andromeda Galaxy for an understanding of our own Galaxy, as well as of the rest of the Universe.

▼ **Andromeda Galaxy** The Andromeda Galaxy, M31, is the largest member of the Local Group and is the farthest object that can be seen with the naked eye. Many of the star-like points in this image are in fact globular clusters within its galactic halo.

Andromedids (Bielids) METEOR SHOWER associated with comet 3D/BIELA. The parent comet split into two fragments in 1845 and has not been definitely seen since 1852; it is now considered defunct. Swarms of METEOROIDS released from the comet have given rise to spectacular meteor showers. Its name derives from its RADIANT position, near γ Andromedae.

The shower's first recorded appearance was in 1741, when modest activity was observed. Further displays were seen in 1798, 1830, 1838 and 1847, in each case during the first week of December. The 1798 and 1838 displays produced rates of over 100 meteors/hr. When seen in 1867 the Andromedids appeared on the last day of November. The NODE, where the orbit of the meteoroid swarm and the orbit of the Earth intersect, is subject to change as a result of gravitational perturbations by the planets. The Andromedid node is moved earlier (regresses) by two or three weeks per century.

In 1867 the association between a meteor shower and a comet was demonstrated by Giovanni SCHIAPARELLI in the case of the PERSEIDS; other such connections were sought. It was known that the orbit of Biela's comet approached that of the Earth very closely, so that its debris could conceivably give rise to a meteor shower, and when the radiant was calculated it was found to agree closely with that of the meteor showers previously seen to emanate from Andromeda.

Biela's comet, if it still existed, would have been in the vicinity of the Earth in 1867, and since the meteoroid swarm would not be far displaced from its progenitor, a display could be expected. A good, though not spectacular, Andromedid shower was seen on November 30, confirming the prediction. Since the orbital period was about 6.5 years, Edmund Weiss (1837–1917), Heinrich D'ARREST and Johann GALLE, who had made the first calculations, predicted another display for 1872 November 28.

Soon after sunset on 1872 November 27, a day earlier than expected, western European observers were treated to an awesome spectacle; meteors rained from the sky at the rate of 6000 per hour. The event caused less alarm and terror among the general population than it might have done, there having occurred only six years previously an equally dramatic LEONID display, which had caused no harm. Although the Andromedids were about as numerous as the 1866 Leonids, they were less brilliant due to their lower atmospheric velocity, at 19 km/s (12 mi/s) compared with the Leonids' 70 km/s (43 mi/s).

The shower of 1872 led, incidentally, to a mysterious episode in astronomical history. E.F. Wilhelm Klinkerfues (1827–84), at Göttingen Observatory, reasoning that it should presumably be ahead of the meteoroid swarm, suggested that the comet should be visible in the opposite part of the sky direction to the Andromedid radiant. He accordingly cabled to Norman POGSON, an astronomer at Madras (the comet would not be visible from high northern latitudes): 'Biela touched Earth 27 November – search near Theta Centauri.' On December 2, Pogson's search found a comet near the indicated position.

The object was observed again the following night, but then clouds intervened and when a clearance finally came, there was no sign of it. The observations were inadequate for the calculation of an orbit and the prediction of future positions, so the comet, if such it was, was lost. If Biela's comet still existed and was pursuing its original orbit, it would have passed the position indicated by Klinkerfues some months previously. We must conclude that if Pogson, who was an experienced observer, did see a comet, it was not 3D/Biela, and was either a fragment of that object or another comet that just happened to be there at the time. Both alternatives are hard to believe, and the question remains open.

The next encounter was badly timed, and no shower was seen. In 1885, two revolutions after the 1872 event, European observers were delighted and thrilled by an even greater meteor storm on November 27, during which Andromedid rates were estimated (counting was virtually impossible) at 75,000 per hour. This rate compares with that of 140,000 per hour estimated for the 1966 Leonid peak. The most intense activity in the 1885 Andromedids was over in about six hours (such meteor storms are invariably short-lived), though lower rates were detected for a few days to either side. This suggests that the core filament in the Andromedid meteor stream in 1885 had a width of about 160,000 km (100,000 mi).

The fall of an iron METEORITE at Mazapil, Mexico, during the 1885 Andromedid display can be dismissed as coincidence. Meteoroids shed by comets are generally small, have a dusty composition, and never survive ABLATION in Earth's atmosphere.

Since 1885 the Andromedids have been quite undistinguished, and the shower is now to all intents and purposes defunct. Planetary perturbations have shifted the orbit of the meteoroid swarm so that it does not at present meet that of the Earth. A fairly strong display occurred on 1892 November 23, while on 1899 November 24 about 200 Andromedids per hour were seen. W.F. DENNING recorded 20 Andromedids per hour in 1904, and visual observations in 1940 yielded rates of about 30 per hour on November 15. A few individual shower members were caught by the Harvard Super-Schmidt meteor cameras in the USA in 1952 and 1953. A computer model of the Andromedid stream by David Hughes (1941–), of the University of Sheffield, UK, suggests that further planetary gravitational perturbations will bring the shower back to encounter the Earth around 2120.

Angara New fleet of Russian satellite launch vehicles to be operational in about 2003 to replace the PROTON launch vehicle. The Angara will be based on a first stage, which forms the basic vehicle for flights to low Earth orbit. This stage can be clustered together with two types of upper stage, forming a more powerful first stage, for flights to geostationary transfer orbit (GTO). The largest Angara, with five core first stages and a high-energy KVRD upper stage, will be able to place 6.8 tonnes into GTO.

Anglo-Australian Observatory (AAO) Australia's principal government-sponsored organization for optical astronomy, funded jointly by Australia and the UK. The AAO operates the 3.9-m (153-in.) ANGLO-AUSTRALIAN TELESCOPE (AAT) and the 1.2-m (48-in.) UNITED KINGDOM SCHMIDT TELESCOPE (UKST) at SIDING SPRING OBSERVATORY near Coonabarabran in New South Wales. A laboratory in the Sydney suburb of Epping, 350 km (200 mi) from the telescopes, houses the AAO's scientific, technical and administrative staff; while the operations staff are based at Coonabarabran. Of increasing importance within the AAO is its External Projects Division, which contracts to build large-scale instrumentation for, among others, the GEMINI TELESCOPES, the SUBARU TELESCOPE and the VERY LARGE TELESCOPE.

Anglo-Australian Telescope (AAT) Optical telescope funded jointly by the British and Australian governments, located at SIDING SPRING OBSERVATORY in New South Wales. It is operated by the ANGLO-AUSTRALIAN OBSERVATORY. Inaugurated in 1974, the 3.9-m (153-in.) AAT remains the largest optical telescope in Australia, although both partner countries now have access to larger southern-hemisphere instruments. Its IRIS2 infrared imager and 2dF (two-degree field) system, which allows the spectra of 400 target objects to be obtained simultaneously, provide unique facilities for wide-field observations.

angrite Subgroup of the ACHONDRITE meteorites. Angrites are medium- to coarse-grained basaltic igneous rocks. Although the angrites have similar oxygen isotopic compositions to the HOWARDITE–EUCRITE–DIOGENITE ASSOCIATION (HEDs) meteorites, they are unrelated.

Ångström, Anders Jonas (1814–74) Swedish physicist who helped prepare the ground for the application of spectroscopy in astronomy. He used diffraction gratings to make high-precision measurements of the Sun's spectral lines, and in 1868 he published an atlas of the solar spectrum. It contained measurements of over a thousand lines, expressed in units of 10^{-7} mm, a quantity later named the ANGSTROM UNIT in his honour.

angstrom unit (symbol Å) Unit of length, formerly used to express the wavelength of light, particularly in spectroscopy; it is equal to 10^{-10} m. It is named after Anders Jonas ÅNGSTRÖM, who first used it in his atlas of the solar spectrum in 1868. The angstrom has now been replaced by the SI measurement the NANOMETRE (nm).

angular measure Measure of the apparent diameter of a celestial object, or the distance between two objects, expressed as an angle, usually in degrees, arcminutes or arcseconds. The angle subtended by an object is determined by its true diameter and its distance from the observer; if the distance to an object is known, its true diameter may be calculated by measuring its apparent diameter.

angular momentum Property of rotating or orbiting bodies that is the rotational equivalent of the momentum of an object moving in a straight line. It is the product of the angular velocity and the moment of inertia (I). The moment of inertia is the rotational equivalent of mass, and for a small particle it is given by the mass of that particle multiplied by the square of its distance from the rotational axis. For large objects the overall moment of inertia must be found by adding together the individual moments of inertia of its constituent particles. A planet orbiting the Sun may be regarded as a small particle and so has a moment of inertia of $I = M_p R_o^2$, where M_p is the mass of the planet and R_o the distance of the planet from the Sun. A spherical, uniform, rotating planet will have $I = 0.4\ M_p R_p^2$ where R_p is the radius of the planet. Angular momentum is always conserved (that is, its total value for the system remains constant) during any changes.

The rotational kinetic energy is given by $I\omega^2/2$, where ω is the angular velocity (compare this with the kinetic energy of an object moving in a straight line: $Mv^2/2$).

For a planet in an elliptical orbit, R_o is smaller at PERIHELION than at APHELION; conservation of angular momentum therefore means that the orbital angular velocity, ω_o, must decrease from perihelion to aphelion in order that the product, $R_o\omega_o$, remains constant. Kepler's second law of planetary motion is thus the result of conservation of angular momentum. Similarly conservation of angular momentum results in ACCRETION DISKS forming in CLOSE BINARY stars where mass is being exchanged, and so is linked to many types of VARIABLE STAR including NOVAE and Type I SUPERNOVAE.

Ann Arbor Observatory (University of Michigan Detroit Observatory) Historic institution in Ann Arbor, 58 km (36 mi) west of Detroit. It was founded (*c.*1854) as part of the university's bid for pre-eminence in scientific teaching and research. It is equipped with a 12½-inch (320-mm) refractor dating from 1857. Today, the observatory is used largely for public outreach, having been restored in 1999 as a centre for 19th-century science, technology and culture.

annual parallax (heliocentric parallax) Difference in the apparent position of a star that would be measured by hypothetical observations made from the centre of the Earth and the centre of the Sun.

Due to the Earth's movement in its orbit around the Sun, nearby stars appear to shift their position relative to more distant, background stars, over the course of a year, describing a path known as the parallactic ellipse. When observed six months apart from opposite sides of the Earth's orbit, the position of a nearby star will appear to have shifted by an angle $\Delta\theta$. Half of this angle, π, gives the annual PARALLAX, which is equal to the angle subtended at the observed star by the semimajor axis of the Earth's orbit. The reciprocal of annual parallax in arcseconds gives the distance to the star in PARSECS. *See also* DIURNAL PARALLAX; TRIGONOMETRIC PARALLAX

annular eclipse Special instance of a SOLAR ECLIPSE, occurring when the Moon is close to APOGEE, such that the Sun has a significantly greater apparent diameter than the Moon. As a result, the Moon's central passage across the solar disk leaves a ring – or *annulus* – of bright sunlight visible at mid-eclipse. Such events, while interesting in their own right, are not generally considered as dramatic as a total solar eclipse, since they do not allow the solar corona and prominences to become visible, or cause a noticeable darkening even at their maximum extent. Under favourable circumstances, the annular phase, with the dark body of the Moon silhouetted against the Sun's brilliant photosphere, can last up to $12^m 30^s$.

▼ **annular eclipse** The bright photosphere of the Sun can be seen surrounding the Moon during an annular eclipse. This eclipse was photographed on 1994 May 10 from Mexico.

anomalistic month Time taken for the Moon to complete a single orbit around the Earth, measured from PERIGEE to perigee. An anomalistic month is shorter than the more commonly used SYNODIC MONTH, being equivalent to 27.55455 days of MEAN SOLAR TIME. *See also* DRACONIC MONTH; MONTH; SIDEREAL MONTH; TROPICAL MONTH

anomalistic year Period of a single orbit of the Earth around the Sun, measured from PERIHELION to perihelion. Equivalent to 365.25964 days of MEAN SOLAR TIME, an anomalistic year is about $4^m 43^s.5$ longer than a SIDEREAL YEAR because of the gradual eastward movement of the point of perihelion. *See also* TROPICAL YEAR

anomaly Angular measurement used for describing the position of a body in an elliptical orbit, measured around the orbit in the direction of motion from the pericentre. It can be defined as a true, eccentric or mean anomaly. In the diagram the point X represents the position of the body, and the angle PSX is the true anomaly. The mean anomaly is measured similarly, but to the position of an imaginary body that orbits at constant angular speed with the same period as the real body. It cannot be indicated by a simple geometrical construction. The eccentric anomaly is the angle PCX′, where X′ is the position of the body projected on to a circumscribing circle. The eccentric and mean anomalies are intermediate angles used in the calculation of the position of the object in its orbit at any time. The difference between the true and mean anomaly is the equation of the centre. *See also* ELLIPSE

anorthosite Type of basaltic rock found in the lunar highland crust. Highland basalts are richer in aluminium and calcium, and poorer in iron, magnesium and titanium, than basalts found in the lunar maria (low-lying plains).

ansae Term applied by early telescopic observers to define the opposite extremities of Saturn's ring system, which, when viewed foreshortened from Earth, resemble handles to the planet.

antapex Point on the CELESTIAL SPHERE from which the Sun and the entire Solar System appears to be moving away, at a velocity of around 19–20 km/s (*c.*12 mi/s), relative to nearby stars. The antapex lies in the direction of the constellation Columba at around RA 6^h dec. −30°. It is diametrically opposite on the celestial sphere to the APEX, the direction towards which the Sun and Solar System appear to be moving.

Antarctic astronomy Collective term for the astronomical activities conducted in the special conditions of the Earth's south polar continent. The Amundsen–Scott South Pole Station sits atop a 2800-m (9200-ft) cap of ice – the coldest, driest place on Earth. This venue is ideal for astronomers who want to work at infrared, submillimetre and millimetre wavelengths. These portions of the electromagnetic spectrum are compromised by water vapour, which is pervasive at most other locations on Earth and makes observations at these long wavelengths difficult or impossible. But at the pole, the vapour is frozen out, leading to a dark, transparent sky ideal for investigating star-formation processes in molecular clouds and the evolution of protostars and other young objects. Similarly, distant, primeval galaxies can be effectively studied because their visible light has been redshifted to long wavelengths. The pole is also a premier site for assaying variations in the so-called COSMIC MICROWAVE BACKGROUND (CMB), subtle differences that reflect the large-scale distribution of matter and energy in the very early Universe.

In addition, of course, a polar site features a totally dark sky 24 hours a day during midwinter, which allows continual monitoring of targets. And during midsummer the Sun can also be watched continuously, which was crucial in the 1980s, during the early days of HELIOSEISMOLOGY, when we got the first glimpse of the interior of our star.

Antarctic astronomy was born around 1980, spurred

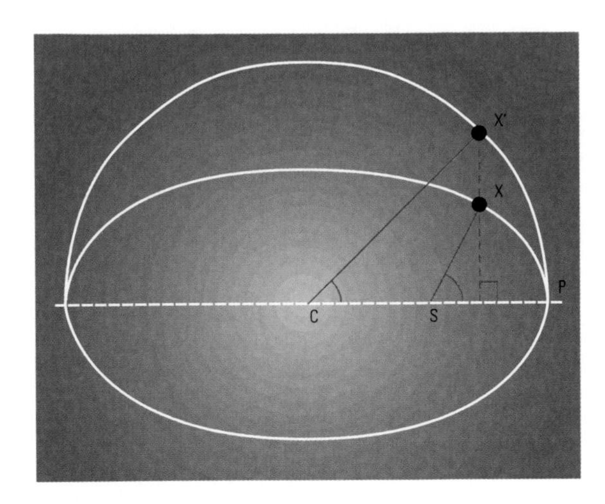

◀ **anomaly** The position of a body (X) in an elliptical orbit can be described in terms of its angular distance in the direction of motion from the pericentre. The true anomaly is described by the angle PSX in the diagram here. The eccentric anomaly, PCX', is given by projecting the body's position on to a circumscribing circle, and can be used as a matter of convenience in orbital calculations.

by Martin Pomerantz (1916–) of the Bartol Research Institute. The principal organization is the Center for Astrophysical Research in Antarctica. Among other facilities, it supports research with the Antarctic Submillimeter Telescope/Remote Observatory (AST/RO), a 1.7-m (67-in.) instrument that focuses on atomic and molecular gas in our Galaxy and others nearby, and the 13-element Degree Angular Scale Interferometer (DASI) for studies of the CMB.

One 'telescope' at the south pole looks down rather than up – the Antarctic Muon and Neutrino Detector Array (AMANDA). Its mission is to count ultra-high-energy neutrinos from our Galaxy and beyond that have passed through the Earth, which screens out 'noise'. AMANDA detects the light that is emitted when a neutrino interacts with an atom in the Antarctic icecap. Sources of extremely high-energy neutrinos include active galactic nuclei, black holes, supernovae remnants and neutron stars.

There is much more to Antarctic astronomy than what happens at the pole. This continent provides one of our best 'laboratories' to prepare for the search for exotic life forms elsewhere in the Solar System. On the polar plateau in East Antarctica lies Lake Vostok. The size of Lake Ontario in North America, it has been buried under thousands of metres of ice for millions of years. Although unexplored, Lake Vostok might hold extremophiles that could help us determine how organisms might survive in the putative ocean under the ice mantle of Jupiter's satellite Europa.

Off the plateau and within a helicopter flight of McMurdo Station, the 'capital' of Antarctica, are the Dry Valleys, frozen landscapes that, though more benign, nevertheless mimic those of Mars. Yet life goes on in these valleys, sometimes hidden inside rocks that protect cyanobacteria from desiccation, thermal extremes and overexposure to ultraviolet radiation. Perhaps this strategy also operates, or once did, on Mars and elsewhere in the Solar System.

Antarctica is also conducive for launching experiments aboard long-duration balloons that ascend 40 km (25 mi) into the upper stratosphere. The prevailing westerly winds carry such balloons around the continent and return them to near the launch site at McMurdo Station in about 10 days. Gamma-rays, X-rays and CMB radiation have all been sampled during such balloon flights.

One of the most surprising discoveries in Antarctica took place in 1969, when nine meteorites were found near the Yamato Mountains. Since then concerted searches have uncovered some 20,000 rocks from space – more than half of all known meteoritic specimens. Antarctica is not favoured by falling meteorites. The abundance occurs because when meteorites land they become embedded in flowing ice – as if on a conveyor belt – and carried to sites where they become exposed by wind erosion. A few have been identified as pieces of the Moon and Mars. An Antarctic meteorite known as ALH 84001 became famous in 1996 when scientists announced that it contains fossil evidence for life on Mars. *See also* BALLOON ASTRONOMY; LIFE IN THE UNIVERSE

Antares The star α Scorpii, marking the heart of Scorpius, the scorpion, distance 604 l.y. An irregular variable, it fluctuates between visual mags. 0.9 and 1.2 and is the brightest member of the Scorpius–Centaurus Association, the nearest OB ASSOCIATION. Antares is a red supergiant of spectral type M1 Ib, about 400 times the Sun's diameter and more than 10,000 times as luminous. It has a much smaller and hotter companion, mag. 5.4 and spectral type B2.5 V, which orbits it in about 900 years; this companion can be seen through telescopes with apertures of 75 mm (3 in.) or more. The name Antares reflects its pronounced red colour: it comes from a Greek expression that can be translated either as 'like Mars' or 'rival of Mars'.

antenna Metal wire, rod, dish or other structure used to transmit and receive RADIO WAVES, whereas a normal RADIO TELESCOPE only receives radio waves. RADAR ASTRONOMY uses antennae (or aerials) to make either continuous wave or pulsed transmissions to bodies in the Solar System, either from Earth or from space satellites. Amateur astronomers can study meteors by using low-power transmitters in the 10–20 MHz band and observing the reflections off the ionized trails made by the meteors in the upper atmosphere. Other work requires powerful transmitters and high-gain antenna systems. Transmissions have been made to the Moon, planets and comets with large radio telescopes receiving the reflections, for example JODRELL BANK and ARECIBO. Arecibo has acted as an antenna transmitting signals for Project Ozma. The Haystack antenna (of Lincoln Laboratory) was used to test the theory of GENERAL RELATIVITY in 1967. For a radar pulse passing near the Sun, and reflected from a planet, there should be a small time delay of 2×10^{-4} s, equivalent to 60 km (40 mi) difference in the distance of the planet. Mars, Mercury and Venus were used by Haystack, and Arecibo used Mercury and Venus.

Antennae (NGC 4038 & 4039) Pair of interacting galaxies – a spiral and a lenticular – in the constellation Corvus (RA $12^h\ 01^m.9$ dec. −18°52′). Each has an apparent mag. +10.5 and maximum diameter 5′. Two long curved tails, comprising stars and gas ejected by the collision, extend away for an apparent distance of 20′. The pair lie at a distance of 60 million l.y. They are sometimes known as the Ring-Tail Galaxy.

anthropic principle Idea that the existence of the Universe is intimately related to the presence of life. The principle exists in two distinct forms, known as the weak and strong versions.

The **weak anthropic principle** (WAP) arises from the notion that any observations made by astronomers will be biased by selection effects that arise from their own existence. Characteristics of the Universe that appear to be quite improbable may merely arise from the fact that certain properties are necessary for life to exist. Cosmologist John D. Barrow (1952–) has given this definition of the WAP: the observed values of all physical and cosmological quantities are not equally probable, but take on values restricted by the requirements that (1) there exist sites where carbon-based life can evolve and (2) the Universe be old enough for it to have already done so.

The **strong anthropic principle** (SAP) goes further, stating that the Universe must have fundamental properties that allow life to develop within it at some stage in its history. This implies that the constants and laws of nature must be such that life can exist. A number of quite distinct interpretations of the SAP are possible, including the suggestion that there exists only one possible Universe 'designed' with the goal of generating and sustaining 'observers' – life. The American mathematician John Archibald Wheeler (1911–) has pointed out that this argument can be interpreted as implying that observers are necessary to bring the Universe into being, an idea he calls the 'participatory anthropic principle' (PAP). A third

▼ **Antarctic astronomy** A plot of a neutrino 'event' at the AMANDA-B detectors. The impacting neutrino decayed to a muon (the turquoise line is the latter's projected track), which then decayed into other subatomic particles. The purple circles indicate which detectors were hit by radiation and the number beside each shows the time elapsed in nanoseconds from the first detector being hit.

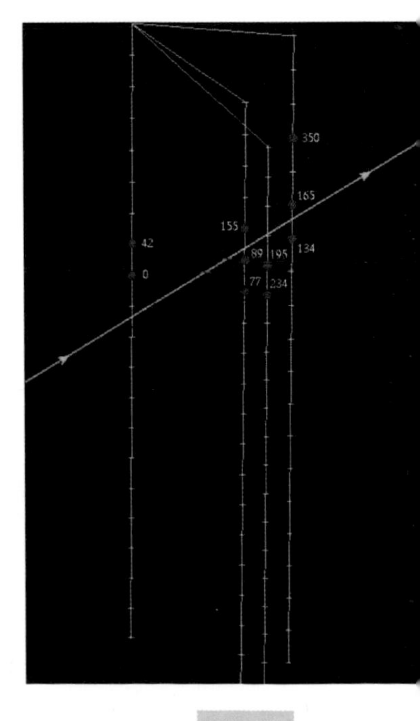

A

possible interpretation of the SAP is that our Universe is just one of an ensemble of many different universes, and that by chance its properties are optimized for the existence of life. This idea is consistent with the 'many-worlds' or 'sum-over-histories' approach of quantum cosmology (*see* QUANTUM GRAVITY), which requires the existence of many possible real 'other universes'.

antimatter Matter consisting of particles with opposite quantum numbers to normal particles. The marriage of the QUANTUM THEORY with SPECIAL RELATIVITY led Paul Dirac (1902–84) in 1929 to propose the existence of electrons with positive charge and spin opposite that of normal electrons. This anti-electron, commonly known as a positron, was the first hint that antimatter existed. The tracks characteristic of positrons were subsequently found in COSMIC RAY cloud chamber experiments by Carl Anderson (1905–91) in 1929, proving the existence of these particles and of antimatter in general. Consequently, quantum theorists now associate every known particle with a complementary antiparticle and most have been seen in high-energy physics experiments. If the particle and its antimatter partner interact, they mutually annihilate and the rest mass is converted into electromagnetic energy via Einstein's formula $E = mc^2$. Antimatter has played important roles in cosmology and extragalactic astronomy. The question of why the Universe consists primarily of ordinary matter and very little antimatter has defied explanation. If entire galaxies of antimatter existed, the effects of these galaxies should be seen. These effects have not been observed.

It was proposed that quasars might actually be galaxies of ordinary matter and antimatter galaxies colliding, but this model quickly gave way to the more common idea of massive black holes in the centres of protogalaxies. Antimatter is definitely present in astrophysical objects, but not in significant amounts. For instance, pair production is the creation of an electron–positron pair from a gamma ray scattering off a baryon. This is thought to be a prominent process in both pulsars and quasars. The opposite process, pair annihilation, forms a gamma-ray photon, which has energy of 0.511 MeV (million electron volts); this spectral line is seen in the spectra of ACTIVE GALACTIC NUCLEI (AGN).

Antiope MAIN-BELT ASTEROID, number 90, which in 2000 was found to be binary in form. The two components of Antiope are each *c*.80 km (*c*.50 mi) in size, and they orbit their mutual centre of gravity with a separation of *c*.160 km (*c*.100 mi).

antitail COMET tail that appears to point towards the Sun. Solar radiation pressure drives the dust tails of comets away from the Sun. Under certain circumstances, however, active comets may be observed to have a sunward-pointing 'spike' or fan. Such antitails result from ejection of dust from the comet nucleus in thin sheets, which, when viewed in the plane of the comet's orbit, can appear to point towards the Sun as a result of perspective. Perhaps the best-known example is the antitail sported by Comet AREND–ROLAND in 1957; antitails were also shown by comets KOHOUTEK and HALE–BOPP.

Antlia *See* feature article

Antoniadi, Eugène Michael (1870–1944) French astronomer, born in Turkey of Greek parents, who became an expert observer of planetary features. As a young man, Antoniadi made many fine drawings of sunspots and the planets Mars, Jupiter and Saturn using 3- and 4½-inch (75- and 110-mm) telescopes. This work attracted the attention of the French astronomer and writer Camille FLAMMARION, who had built a fine private observatory at Juvisy, France. From 1894 to 1902 Antoniadi used Juvisy's 9½-inch (240-mm) Bardou refracting telescope to make detailed studies of the planets, especially Mars. From 1896 to 1917 he directed the British Astronomical Association's Mars Section, publishing ten *Memoirs* that collected and discussed the observations of the section's members; many of the observations were by Antoniadi himself, made with the Juvisy telescope and the great 33-inch (0.83-m) refractor at Meudon, France.

At a time when the 'canals' supposedly seen on Mars by Giovanni SCHIAPARELLI and Percival LOWELL were generally accepted as real features by many astronomers, Antoniadi interpreted these linear markings as optical illusions produced by the tendency of human vision to transform disconnected features into continuous lines. His observations of Mars during the 1909 opposition convinced him that the canals were natural features similar to Earth's valleys; two years later he correctly described the clouds obscuring the planet's surface as due to windblown sand and dust from the Martian deserts. In 1924 he made one of the first sightings of the Tharsis volcanoes. Antoniadi summarized both his own and historical observations of Mars in the classic book *La Planète Mars* (1930). Four years later he wrote another classic work, *La Planète Mercure*, which remained a standard for decades. Antoniadi was also an expert on ancient languages, a skill he used to research and write *L'Astronomie Egyptienne* (1940).

Antoniadi scale Standard scale of SEEING conditions devised in the early 1900s by Eugène ANTONIADI to describe the conditions under which lunar and planetary observations are made. The five gradations on this scale are: (I) perfect seeing, without a quiver; (II) slight undulations, with periods of calm lasting several seconds; (III) moderate seeing, with larger tremors; (IV) poor seeing, with constant troublesome undulations; and (V) very bad seeing, scarcely allowing a rough sketch to be made.

Apache Point Observatory (APO) Major optical/infrared observatory owned and operated by the Astrophysical Research Consortium (ARC) of several prominent US universities. Located at an elevation of 2790 m (9150 ft) in the Sacramento Mountains of New Mexico approximately 225 km (140 mi) south-east of Albuquerque, this world-class observatory site is managed by the New Mexico State University on the ARC's behalf. Its principal instrument is a 3.5-m (138-in.) reflector, a general-purpose telescope with a lightweight spun-cast mirror. The APO is also home to the 2.5-m (98-in.) telescope of the SLOAN DIGITAL SKY SURVEY.

ap-, apo- Prefixes referring to the farthest point of an orbit from the primary body, as in APASTRON, APHELION, APOGEE, apocentre, apoapse. *See also* APSIDES

apastron Point in an orbit around a star that is farthest from that star. The term is usually used to describe the positions of the two components in a binary star system, moving around their common centre of mass in an elliptical orbit, when they are farthest away from one another. *See also* APSIDES

Apennines, Montes (Apennine Mountains) Most impressive lunar mountain range, 1400 km (870 mi) long; it rises to 4 km (2.5 mi) above the lava-filled southeast 'shores' of Mare IMBRIUM. The Apennines' highest peak is Mount Huygens (at 5600 m/18,400 ft), followed by Mount Bradley (at 5000 m/16,400 ft), Mount Hadley (4500 m/14,800 ft) and Mount Wolff (3500 m/11,500 ft). The numerous valleys and gorges

ANTLIA (GEN. ANTLIAE, ABBR. ANT)

Small, faint southern constellation representing an air pump; it lies between Hydra and Vela. Antlia was introduced by Lacaille in the 18th century. Its brightest star, α Ant, is mag. 4.3; ζ Ant is a wide double with components of mags. 5.9 and 6.2. Antlia's brightest deep-sky object is NGC 2997, a 10th-magnitude spiral galaxy.

A

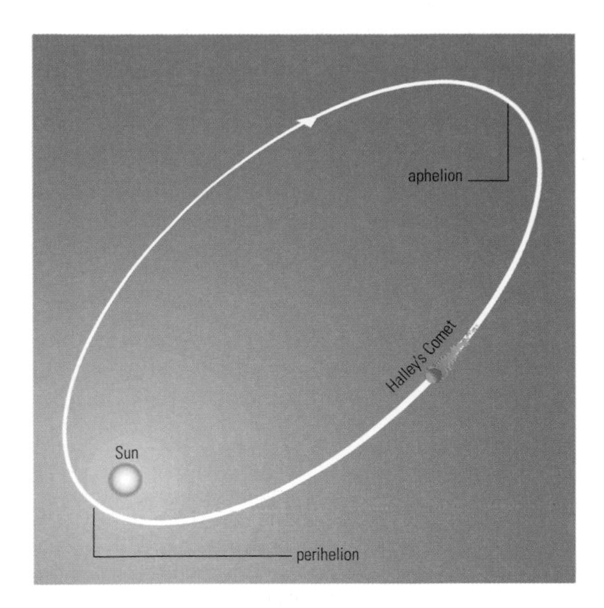

◀ **aphelion** Halley's Comet has a highly eccentric orbit: its aphelion – its greatest distance from the Sun – is 35.295 AU (roughly halfway between the mean orbits of Uranus and Pluto), while its perihelion – its closest approach to the Sun – is just 0.587 AU, well within Venus' orbit.

that cut through the range are probably shock fractures from the impact that created Mare Imbrium.

aperture Clear diameter of the primary lens or mirror of a TELESCOPE. The aperture is the single most important factor in determining both the RESOLVING POWER and light-gathering ability of the instrument.

aperture synthesis ARRAY of RADIO TELESCOPES that work together to observe the same astronomical object and to produce an image with higher resolution than the telescopes individually could achieve. Resolution approaching that of Earth-bound optical telescopes – say, one arcsecond – would require a single radio dish several kilometres in diameter. To construct such a radio telescope, and make it steerable to observe anywhere in the sky, has been impossible in the past, so aperture synthesis has been a most valuable tool for mapping in detail small-scale structure. The individual radio telescopes play the role of different parts of the surface of a hypothetical 'super-telescope', which is synthesized by correctly combining their outputs. The technique is further enhanced by using Earth's rotation to fill in missing parts of the super-telescope via the movement of the telescopes relative to the astronomical target. This technique of using Earth's rotation was developed by Martin RYLE. Usually some parts of the super-telescope surface are missing, that is, it is not 'fully filled', causing defects in the map such as bands or rings of spurious emission. The problem can be overcome to some extent either by complementing the map with an observation made by a single large radio telescope, or by computing techniques such as maximum-entropy methods which clean up the defects. The Cambridge One-mile telescope and the VERY LARGE ARRAY are examples of aperture synthesis telescopes.

apex (solar apex) Point on the CELESTIAL SPHERE towards which the Sun and the entire Solar System appear to be moving, at a velocity of around 19–20 km/s (*c.*12 mi/s) relative to nearby stars. It lies in the direction of the constellation Hercules at around RA 18^h dec. +30°. It is diametrically opposite on the celestial sphere to the ANTAPEX, the direction from which the Sun and Solar System appear to be moving.

The apex is also the point on the celestial sphere towards which the Earth appears to be moving at any given time, due to its orbital motion around the Sun.

aphelion Point at which a Solar System body, such as a planet, asteroid or comet, moving in an elliptical orbit, is at its greatest distance from the Sun. For the Earth, this occurs around July 4, when it lies 152 million km (94 million mi) from the Sun. *See also* APSIDES; PERIHELION

Aphrodite Terra Largest and most extensive of the continent-like highland regions on VENUS. About the size of Africa, Aphrodite Terra exhibits considerable variation in physiography. It extends for 10,000 km (6200 mi) from 45° to 210°E in a roughly east–west direction along and south of Venus' equator. The western and central mountainous regions, known respectively as Ovda Regio and Thetis Regio, rise to between 3 and 4 km (1.9–2.5 mi) above the mean planetary radius. East of the central massif is a complex regional fracture system with east-north-east-trending ridges and linear troughs, the most prominent of which are Dali and Diana Chasmata, 2077 km (1291 mi) and 938 km (583 mi) in length respectively. Each trough or rift valley is at least 1 km (0.6 mi) deep. Diana Chasma may be tectonic in origin. To the south is the semicircular Artemis Chasma, thought to be a massive corona-type structure (a concentric arrangement of ridges and grooves). In the east the broad upland dome of Atla Regio, with its fissured surface and volcanic centres, is somewhat higher than the western and central mountainous districts, but is itself overshadowed by the imposing volcanic peaks Ozza Mons (6 km/4 mi high and 500 km/310 mi across) and Maat Mons (9 km/6 mi high and 395 km/245 mi across). The absence of craters suggests the whole region is of comparatively recent origin.

apochromat Composite LENS designed to be free of CHROMATIC ABERRATION. Modern materials and design techniques make it possible to produce lenses that introduce no discernible false colour into the image. Apochromatic refractors used in astronomy usually have an OBJECTIVE made of three separate pieces of glass, each of different materials with different characteristics. Some or all of the materials may be highly specialized and expensive, but the overall effect is to produce some of the best optical systems that are currently available to amateur astronomers.

Telescopes incorporating apochromatic objectives are sometimes referred to as 'fluorites' or 'ED' telescopes, the names being derived from the materials used.

Apochromats may also be incorporated into eyepieces and BARLOW LENSES. The smaller sized lenses make these more common and affordable than apochromatic objectives. They are particularly useful with reflecting telescopes that are inherently free of chromatic aberration. *See also* ACHROMAT

apogee Point in the orbit of the Moon or an artificial satellite that is farthest from the Earth. *See also* APSIDES

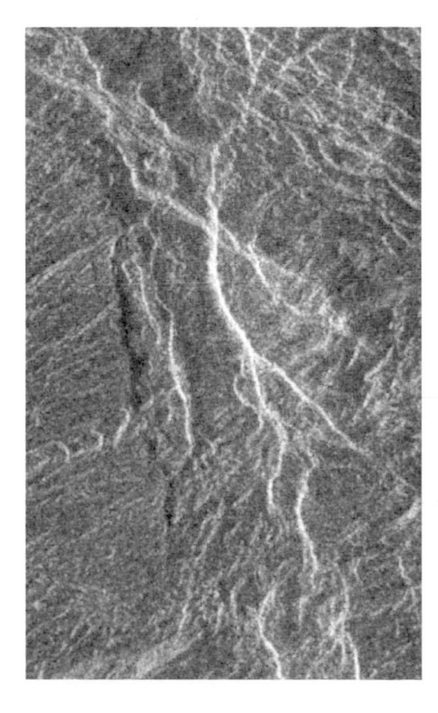

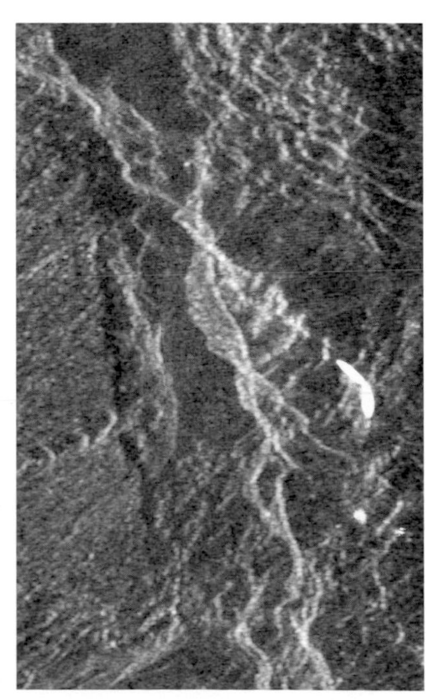

▼ **Aphrodite Terra** The surface of this upland region of Venus is extremely varied, containing deep troughs and high ridges as well as volcanoes. The area to the left of the ridge shown in this Magellan radar image was first interpreted as a landslide but is now thought to be simply an effect of the image-processing methods used.

▲ **Apollo programme** The Saturn V rocket, carrying the Apollo 11, is launched. The first manned mission to the Moon, it was launched on 1969 July 16.

Apollo asteroid Any member of the class of ASTEROIDS with orbits that cross that of the Earth (requiring a perihelion distance less than 1.0167 AU and an aphelion distance greater than 0.9833 AU), and with orbital periods longer than one year. The archetype, (1862) Apollo, was discovered in 1932 as the first Earth-crossing asteroid. By late 2001 more than 740 Apollo asteroids of varying sizes had been discovered, mostly since 1990. Because these asteroids collide with the Earth or other terrestrial planets on time scales of order ten million years, or may be lost through other dynamical paths, they cannot have been in place since the formation of the Solar System. They must, therefore, be delivered to their present orbits on a continual basis. At one time it was thought that they might mostly be extinct cometary nuclei, but it has been shown that they may also be derived from MAIN-BELT ASTEROIDS pumped by Jupiter's gravity out of the KIRKWOOD GAPS and into high-eccentricity orbits in the inner Solar System.

Some characteristics of the following noteworthy Apollo asteroids are shown in the NEAR-EARTH ASTEROID table: (1566) Icarus, (1620) Geographos, (1685) Toro, (1862) Apollo, (1863) Antinous, (1866) Sisyphus, (2063) Bacchus, (2101) Adonis, (2102) Tantalus, (2201) Oljato, (2212) Hephaistos, (3200) Phaethon, (4015) Wilson-Harrington, (4179) Toutatis, (4183) Cuno, (4581) Asclepius, (4660) Nereus, (4789) Castalia, (6489) Golevka and 1937 UB Hermes. *See also* ATEN ASTEROID; AMOR ASTEROID

Apollonius of Perga (262–190 BC) Greek mathematician and astronomer, known as the Great Geometer. His most famous published work was the eight-volume *Conics*, which deals with the geometry of conic sections – the circle, ellipse, parabola and hyperbola, all of which are important in describing the motions of celestial bodies. Although Apollonius' theory of planetary motions was based upon the erroneous geometrical model of epicycles and eccentrics promoted by PTOLEMY, he correctly observed and described the retrograde motion of Mars and other planets.

Apollo programme Name given to the US programme to land men on the Moon and return them safely to Earth. The cost of the NATIONAL AERONAUTIC AND SPACE ADMINISTRATION's (NASA's) triumph in the Moon race with the Soviet Union was US$25 billion. At its peak, the programme employed some 500,000 people.

The first public announcement of the intention of the USA to achieve the first manned landing on another world was made by President Kennedy in 1961. However, NASA was already preparing three series of robotic missions, namely RANGER, SURVEYOR and LUNAR ORBITER, to investigate the feasibility of manned missions and to search for suitable landing sites.

The Apollo spacecraft consisted of four units in two pairs: the Command and Service Module (CSM), together with the descent and ascent stages of the LUNAR MODULE (LM). The Command Module (CM) housed the astronauts during the journeys to the Moon and back. It remained attached to the Service Module (SM), which contained the rocket engines, fuel and electrical supply, until shortly before re-entering the Earth's atmosphere.

After achieving lunar orbit, the LM, containing two astronauts, undocked from the CSM, decelerated, and landed at the pre-selected location on the Moon's surface. The third astronaut remained orbiting the Moon in the CSM. At the completion of extra-vehicular activities (EVAs) on the surface, the descent stage served as a launch platform for the ascent stage, which then redocked with the orbiting CSM. Following the transfer of the two astronauts, film cassettes and Moon rocks from the ascent stage to the CM, the former was jettisoned. It was usually targeted to impact the lunar surface, thereby providing a seismic signal of known energy to calibrate seismometers deployed on the surface.

Apollo 1 should have been the first Earth orbital test flight of the CM, but a fire in the CM during ground testing on 1967 January 27 killed astronauts Roger Chaffee (b.1935), Virgil Grissom (b.1926) and Edward White (b.1930) and prompted numerous design changes. This was followed by Apollo 4, 5 and 6, all unmanned flights to test the SATURN ROCKET.

The manned flights with three astronauts aboard – the Commander (CDR), the Lunar Module Pilot (LMP) and the Command Module Pilot (CMP) – began with Apollo 7. Originally 14 manned Apollo flights were planned but funding cuts later reduced this to 11. As confidence was gained, each mission gathered more scientific data than its predecessors. As durations of EVAs and distances covered increased, more comprehensive selections of rock and soil samples were obtained. More

► **Apollo programme** The Lunar Roving Vehicles allowed the astronauts on the final three Apollo lunar missions to range farther from the landers and so collect more varied samples from the Moon's surface.

experiments were also deployed or conducted on the lunar surface, and the later orbiting CSMs obtained hundreds of high-resolution stereophotographs and other measurements of the surface.

Within the constraints dictated by spacecraft design and landing safety, the landing sites were chosen to give a reasonably representative selection of different surface types, as determined from earlier ground-based or spacecraft studies. In the case of Apollo 11, a near-equatorial site provided the simplest landing and redocking conditions, with a relatively smooth and level descent path and landing point. A location in Mare Tranquillitatis was chosen, not too far from the impact point of Ranger 8 and the landing point of Surveyor 5. Neil ARMSTRONG and 'Buzz' ALDRIN became the first men to step on the Moon. Apollo 12 was again targeted for an equatorial site on Oceanus Procellarum, very close to the earlier landing site of Surveyor 3. This enabled the astronauts to inspect and return spacecraft samples that had been exposed to the lunar environment for two and a half years.

The ill-fated Apollo 13 was targeted for Fra Mauro, a near-equatorial site in the lunar highlands that was thought to be covered by ejecta from the impact event that produced the Imbrium Basin. Unfortunately, an oxygen tank in the SM exploded halfway to the Moon, disabling most of the spacecraft systems. Despite many hardships, the astronauts returned safely. Apollo 14 was targeted for the same landing location and was completely successful. The exploration by astronauts Alan SHEPARD and Stuart Roosa (1933–) was aided by a wheeled trolley known as the Modular Equipment Transporter.

The three remaining missions touched down well away from the equator and had the benefit of a battery-powered roving vehicle that widened the field of exploration. The Apollo 15 landing site combined a mare region (Palus Putredinis), a nearby sinuous valley (Hadley Rille) and very high mountains (the Apennines). Apollo 16 landed in a highlands area, on the so-called Cayley Formation – the loose material that fills in craters and covers the lower slopes here – but very close to the Descartes Formation, which was thought to represent possible highlands volcanism.

The choice of the Taurus-Littrow site for the final mission resulted largely from the Apollo 15 observations of small, dark-haloed craters, identified as possible volcanic vents, in the area. Additionally, the valley floor was one of the darkest mare surfaces on the Moon. Geologist-astronaut Harrison Schmitt (1935–) was the only scientist to set foot on the Moon.

The quantity of data resulting from the Apollo programme was overwhelming. The 380 kg of rock and soil samples have attracted particular attention, having been subjected to every kind of test and analysis imaginable. However, many of the stored samples have yet to be examined.

Another major archive is the collection of many thousands of photographs taken during the transit and orbital phases of the missions. In particular, the metric and panoramic cameras used on the last three missions produced high-resolution, wide-angle, stereoscopic coverage of about 20% of the lunar surface. Of the 16 or so other experiments conducted from orbit, ten were concerned with 'remote sensing', allowing a comparison with data from the landing sites.

About 25 different types of experiment were carried out or deployed on the surface. These included seismometers, magnetometers and heat-flow experiments to investigate the subsurface structure and properties. Laser reflectors helped to refine the Moon–Earth distance.

apparent magnitude Apparent brightness of a star (a measure of the light received at Earth) measured by the stellar magnitude system. Because stars are at different distances from us and the interstellar medium has a variable absorption, apparent magnitude is not a reliable key to a star's real (intrinsic) LUMINOSITY.

APOLLO MISSIONS

dates of mission () = landing	astronauts (CDR, LMP, CMP)	landing site and coordinates	revs around Moon	days on Moon	hours of EVA*	wt (kg) of samples	remarks
Apollo 7 1968 Oct. 11–22	Schirra Eisele Cunningham	Earth orbit only	–	–	–	–	only CSM put into Earth orbit; 163 revolutions
Apollo 8 1968 Dec. 21–27	Borman Lovell Anders	orbited Moon, no landing	10	–	–	–	only CSM used; first manned circumnavigation of Moon
Apollo 9 1969 Mar. 3–13	McDivitt Scott Schweickart	Earth orbit only	–	–	–	–	both LM and CSM put into Earth orbit; undocking and docking; 151 revolutions
Apollo 10 1969 May 18–26	Stafford Young Cernan	orbited Moon, no landing	31	–	–	–	first full test of complete spacecraft (except LM landing and lift-off); LM 4 solo orbits
Apollo 11 1969 July 16 (20)–24	Armstrong Aldrin Collins	south-west part of Mare Tranquillitatis 0°.8N 23°.46E	30	0.9	2.5	21	first lunar landing, in the actual spot now officially named Statio Tranquillitatis
Apollo 12 1969 Nov. 14 (19)–24	Conrad Bean Gordon	south-east part of Oceanus Procellarum 3°.04S 23°.42W	45	1.3	7.7	34	landed close to Surveyor 3, parts of which were removed and returned to Earth
Apollo 13 1970 Apr. 11–17	Lovell Swigert Haise	no landing made	0.5	0	0	0	mission aborted owing to oxygen tank explosion in SM
Apollo 14 1971 Jan. 31 –Feb. (5)–9	Shepard Mitchell Roosa	north of Fra Mauro (highlands area) 3°.65S 17°.48W	34	1.4	9.4	43	first landing in lunar highlands; first use of 'push-cart' to carry equipment
Apollo 15 1971 July 26 (30)–Aug. 7	Scott Irwin Worden	Hadley Rille and Apennine Mountains 26°.08N 3°.66E	74	2.8	18.5	77	first use of lunar rover, and of semi-automatic metric and panoramic cameras in SM
Apollo 16 1972 Apr. 16 (21)–27	Young Duke Mattingly	north of Descartes, Cayley Plains 8°.97S 15°.51E	64	3.0	20.3	94	generally similar to previous mission
Apollo 17 1972 Dec. 7 (11)–19	Cernan Schmitt Evans	Taurus-Littrow, east Mare Serenitatis 20°.17N 30°.77E	75	3.1	22.1	110	first mission with fully trained geologist (Schmitt)

*EVA = extra-vehicular activity (on lunar surface)

apparent solar time Local time, based on the position of the Sun in the sky, and as shown by a sundial. Apparent noon occurs when the Sun crosses the local MERIDIAN at its maximum altitude. The length of a solar day measured this way is not uniform. It varies throughout the year by as much as 16 minutes because of the elliptical orbit of the Earth and the fact that the Sun appears to move across the sky along the ecliptic, rather than the celestial equator. Clock time is therefore based on the uniform movement of a fictitious mean Sun. The difference between apparent and MEAN SOLAR TIME is called the EQUATION OF TIME.

apparition Period of time during which it is possible to observe a celestial body that is only visible periodically. The term is usually used to describe a particular appearance of a comet, such as the 1985–86 apparition of Comet HALLEY.

appulse Close approach in apparent position in the sky between two celestial bodies. Unlike an OCCULTATION, when one body passes directly in front of another, such as a planet occulting a star, an appulse occurs when their directions of motion on the celestial sphere converge, the two just appearing to touch. The impression is caused by a line-of-sight effect, the two bodies actually lying at greatly differing distances from the observer. *See also* CONJUNCTION

apsides The two points in an elliptical orbit that are nearest to and farthest away from the primary body. The line joining these two points is the line of apsides. For an object orbiting the Sun the nearest apse is termed the PERIHELION and the farthest apse is the APHELION. For the Moon or an artificial satellite orbiting the Earth the apsides are the PERIGEE and APOGEE. The components of a binary star system are at PERIASTRON

APUS (GEN. APODIS, ABBR. APS)

Small, faint southern constellation, representing a bird of paradise, between Triangulum Australis and the south celestial pole. It was introduced by Keyser and de Houtman at the end of the 16th century. The brightest star, α Aps, is mag. 3.8; δ Aps is a wide double, with red and orange components, mags. 4.7 and 5.3.

AQUARIUS (GEN. AQUARII, ABBR. AQR)

Tenth-largest constellation and one of the signs of the zodiac. It represents a water-bearer (possibly Ganymede in Greek mythology) and lies between Pegasus and Piscis Austrinus. Aquarius is a rather inconspicuous constellation; its brightest stars, α Aqr (Sadalmelik) and β Aqr (Sadalsuud), are both mag. 2.9. ζ Aqr is a visual binary with pale yellow components, mags. 4.4 and 4.6, separation 1″.9, period 856 years. The constellation's deep-sky objects include M2 (NGC 7089), a mag. 7 globular cluster; and two planetary nebulae: the HELIX NEBULA (NGC 7293) and the SATURN NEBULA (NGC 7009), respectively mag. 7 and 8. The ETA AQUARID, DELTA AQUARID and IOTA AQUARID meteor showers radiate from this constellation.

BRIGHTEST STARS

	Name	RA h m	dec. ° ′	Visual mag.	Absolute mag.	Spectral type	Distance (l.y.)
α	Sadalmelik	22 06	−00 19	2.95	−3.9	G2	759
β	Sasalsuud	21 32	−05 34	2.90	−3.5	G0	612
δ	Skat	22 55	−15 49	3.27	−0.2	A3	160

when they are closest together and at APASTRON when farthest apart. For objects orbiting, say, the Moon or Jupiter, one sometimes sees terms such as periselenium or perijove. Such terms are rather cumbersome, however, and it has become more usual to use the general terms pericentre and apocentre (or less commonly periapse and apoapse) for all cases where the primary is not the Sun, the Earth or a star, or when discussing elliptic motion in general.

Ap star Class A STAR with peculiar chemical composition. Ap stars extend to types Fp and Bp. In these class 2 CHEMICALLY PECULIAR STARS (CP2), elements such as silicon, chromium, strontium and the rare earths are enhanced (europium by a factor of 1000 or more). The odd compositions derive from diffusion, in which some atoms sink in the quiet atmospheres of slowly rotating stars, whereas others rise. The strong, tilted magnetic fields of Ap stars concentrate element enhancements near the magnetic poles. As the stars rotate, both the fields and spectra vary.

Apus *See* feature article

AQUILA (GEN. AQUILAE, ABBR. AQL)

Equatorial constellation, representing an eagle (possibly that belonging to Zeus in Greek mythology), between Sagitta and Sagittarius. ALTAIR, its brightest star, marks one corner of the SUMMER TRIANGLE and is flanked by two slightly fainter stars, β (Tarazed) and γ Aql, forming an unmistakable configuration. η Aql is one of the brightest Cepheid variables (range 3.5–4.4, period 7.18 days). Van Biesbroeck's Star, a mag. 18 red dwarf, is one of the intrinsically faintest stars known, absolute mag. 19.3. NGC 6709 is an open cluster of about 40 stars fainter than mag. 9.

BRIGHTEST STARS

	Name	RA h m	dec. ° ′	Visual mag.	Absolute mag.	Spectral type	Distance (l.y.)
α	Altair	19 51	+08 52	0.76	2.2	A7	17
γ	Tarazed	19 46	+10 37	2.72	−3.0	K3	460
ζ		19 05	+13 52	2.99	1.0	A0	83
θ		20 11	−00 49	3.24	−1.5	B9.5	287
δ		19 25	+03 07	3.36	2.4	F0	50
λ		19 06	−04 53	3.43	0.5	B9	125

Aquarius *See* feature article

Aquila *See* feature article

Ara *See* feature article

Arab astronomy *See* ISLAMIC ASTRONOMY

Arago, (Dominique) François (Jean) (1786–1853) French scientist and statesman, director of Paris Observatory from 1830, who pioneered the application of the photometer and polarimeter to astrophysics. He discovered that two beams of light polarized at right angles do not interfere with each other, allowing him to develop the transverse theory of light waves. Arago supervised the construction in 1845 of Paris Observatory's 380-mm (15-in.) refractor, at the time one of the world's largest telescopes.

Aratus of Soli (315–245 BC) Ancient Greek poet (born in what is now modern Turkey) whose *Phaenomena* elaborated upon the descriptions of the 'classical' constellations given a century before by EUDOXUS. Aratus' patron, King Antigonus of Macedonia, was so inspired by the *Phaenomena* that he had the constellation figures described in the epic poem painted on the concave ceilings of the royal palace, thus creating what was probably the earliest celestial atlas.

Arcetri Astrophysical Observatory Observatory of the University of Florence, situated at Arcetri, south of the city. The present observatory was built by Giovanni DONATI in 1872, replacing a much older one in the city centre. A solar tower was added in 1924. Arcetri is now a major centre of Italian astrophysics, and its scientists use both national (GALILEO NATIONAL TELESCOPE) and international (mainly EUROPEAN SOUTHERN OBSERVATORY) facilities. It has a major role in the development of the LARGE BINOCULAR TELESCOPE.

archaeoastronomy Study of ancient, essentially prehistoric, astronomical theories and practices through evidence provided by archaeology, the interpretation of ancient artefacts, the written records (if available) and ethnological sources such as tribal legends (sometimes called ethnoastronomy).

Many ancient peoples – for example the Stone Age populations of the European countries and the Eastern Mediterranean – are known to have been fully aware of celestial phenomena, as manifested in the alignments of their megalithic monuments such as Stonehenge, Newgrange and the dolmens of north-west France. Measurements of the Egyptian pyramids show that they are orientated within a small fraction of a degree in the north–south and east–west directions, and this must have required considerable practical observational skill (*see* EGYPTIAN ASTRONOMY).

Ancient astronomical observations were not limited to the Old World: for example, Meso-American peoples such as the Maya (*see* NATIVE AMERICAN ASTRONOMY) and the Aztec have left codices and other evidence which reveal knowledge of celestial events and the determination of celestial cycles.

In some ancient civilizations, observation of the HELIACAL RISING of Sirius provided a reference point that enabled their luni-solar calendar to be corrected to match the true solar year. This was essential to the civilisation if the calendar was to be used to decide the proper time for planting crops.

What may be called **applied historical astronomy** makes use of ancient records to derive data of value to modern astronomical research. For example, ancient observations of eclipses from Babylon, China, Europe and the Islamic world can be used to determine the variation in the Earth's rotational period over the past 2700 years. In addition, ancient records of the extremely rare outbursts of supernovae, often recorded as 'guest' or 'new' stars, are invaluable in the study of the development and present-day remnants of these events.

Archimedes Lunar crater (30°N 4°W), 82 km (51 mi) in diameter, the largest in Mare IMBRIUM. Archimedes' terraced walls average only 1200 m (4000 ft) above its almost featureless floor: it is filled almost to the brim by Imbrium's lava flows. A few peaks are as high as 2250 m (7400 ft). Scattered bright streaks crossing the crater's floor are probably part of the Autolycus RAY system. Archimedes' inner walls are steeper than its outer ramparts, which are about 11 km (7 mi) wide; these ramparts are highlighted by bright EJECTA. To the south and south-east of Archimedes lies a group of rilles; these rilles are probably faults caused by the cumulative weight of Imbrium's massive lava flows. North of Archimedes are a very bright group of mountains and a sinuous wrinkle ridge. The Spitzbergen Mountains lie to the north-west.

archive Collection of records. The archives resulting from the space instruments like the Hubble Space Telescope or IUE are organized as large databases: they can be queried to get the observations matching various criteria, and the observed images can frequently be downloaded on the Internet. The large ground-based observatories are also installing similar large archives.

arcminute, arcsecond Small units of angular measure. An arcminute (symbol ′) is 1/60 of a degree and an arcsecond (symbol ″) 1/60 of an arcminute, or 1/360 of a degree. The units are widely used in astronomy, particularly as a measure of angular separation or diameter of celestial bodies. The RESOLVING POWER of a telescope is also usually expressed in arcseconds.

Arcturus The star α Boötis, visual mag. −0.05 (but variable by a few hundredths of a magnitude), distance 37 l.y. Arcturus, the fourth-brightest star, is an orange-coloured giant, spectral type K2 III, estimated to be 25 times the Sun's diameter and 100 times as luminous. Its name comes from the Greek and means 'bear watcher' or 'bear guardian', from its proximity to Ursa Major, the great bear.

Arecibo Observatory Site of the world's largest single-dish radio telescope, and one of the most important facilities for radio astronomy, planetary radar and terrestrial atmospheric studies. The observatory is located in the Guarionex Mountains of north-western Puerto Rico about 12 km (8 mi) south of the city of Arecibo. A natural depression provided a bowl in which the 300-m (1000-ft) spherical reflector was built by Cornell University, beginning in 1960. The telescope was dedicated in 1963 November as the Arecibo Ionospheric Observatory, with radar studies of the Earth's atmosphere and the planets as its primary mission. In 1971, with radio astronomy investigations increasing, it became part of the National Astronomy and Ionospheric Centre (NAIC) operated by Cornell, and in 1974 the mirror's surface was upgraded to allow centimetre-wavelength observations to be made. This event was celebrated by the transmission of the coded **Arecibo message** towards the globular cluster M13, some 25,000 l.y. away, containing information about terrestrial civilization. The telescope was further upgraded in 1997.

The fixed, upward-pointing dish is 51 m (167 ft) deep, covers 8 hectares (20 acres) and is surfaced with almost 40,000 perforated aluminium panels supported by a network of steel cables underneath. Suspended 137 m (450 ft) above the dish is a 915-tonne platform in which the telescope's highly sensitive radio receivers are located. They can be moved along a rotating bow-shaped azimuth arm 100 m (330 ft) long to allow the telescope to point up to 20° from the vertical. The near-equatorial location of Arecibo (latitude 18°N) means that all the planets are accessible, together with many important galactic and extragalactic objects.

ARA (GEN. ARAE, ABBR. ARA)

Small, rather inconspicuous southern constellation representing an altar (possibly that upon which the gods swore allegiance before their battle against the Titans in Greek mythology), between Scorpius and Apus. Its brightest stars, α Ara and β Ara, are both mag. 2.8. The constellation's deep-sky objects include NGC 6193, a 5th-magnitude open cluster, and NGC 6397, one of the closest globular clusters, just visible to the naked eye.

BRIGHTEST STARS

	RA h	RA m	dec. °	dec. ′	Visual mag.	Absolute mag.	Spectral type	Distance (l.y.)
α	17	32	−49	53	2.84	−1.5	B2	242
β	17	25	−55	32	2.84	−3.5	K3	603
ζ	16	59	−55	59	3.12	−3.1	K5	574
γ	17	25	−56	23	3.31	−4.4	B1	1136

Arend–Roland, Comet (C/1956 R1) Comet discovered by Silvain Arend (1902–92) and Georges Roland (1922–91), Uccle Observatory, Belgium, on photographic plates obtained on 1956 November 8. Comet Arend–Roland reached perihelion, 0.32 AU from the Sun, on 1957 April 8. After perihelion, the comet came closest to Earth (0.57 AU) on April 21, becoming a prominent naked-eye object in northern hemisphere skies in the latter parts of the month. Peak brightness approached mag. −1.0, and the tail reached a length of 25–30°. A prominent, spiked ANTITAIL 15° long developed in late April to early May. The comet's orbit is hyperbolic.

areo- Prefix pertaining to the planet MARS, such as in areography, the mapping of Martian surface features.

Ares Vallis Valley on MARS (9°.7N 23°.4W). MARS PATHFINDER landed here in 1997, not far from the VIKING 1 lander site CHRYSE PLANITIA. Ares Vallis exhibits the characteristics of an ancient flood-plain.

Argelander, Friedrich Wilhelm August (1799–1875) German astronomer, born in Memel (in modern Lithuania), noted for his compilation of fundamental star catalogues and atlases. From 1817 to 1822 Argelander worked at the observatory of the University of Königsberg, under its director Friedrich Wilhelm BESSEL. There he revised the positions of 2848 bright stars catalogued by John FLAMSTEED, amongst the first compiled after the invention of the telescope. In 1823 Argelander became director of the Finnish observatory at Turku and, later, Helsinki, where he remained until his appointment as director of the Bonn Observatory in 1836. In 1843 he published *Uranometria nova*, a catalogue of positions and magnitudes for stars visible to the naked eye; for this project, Argelander invented his 'stepwise method' of estimating a star's visual magnitude by comparing it to the relative brightness of neighbouring stars.

Argelander's greatest cartographic achievement was the *BONNER DURCHMUSTERUNG* (BD). Employing Bonn Observatory's 75-mm (3-in.) Fraunhofer 'comet seeker'

◄ **Arecibo Observatory** As well as its more famous role in the SETI project, the radio telescope is a powerful radar transmitter and receiver, and is used to map small bodies such as asteroids and to examine the Earth's upper atmosphere.

A

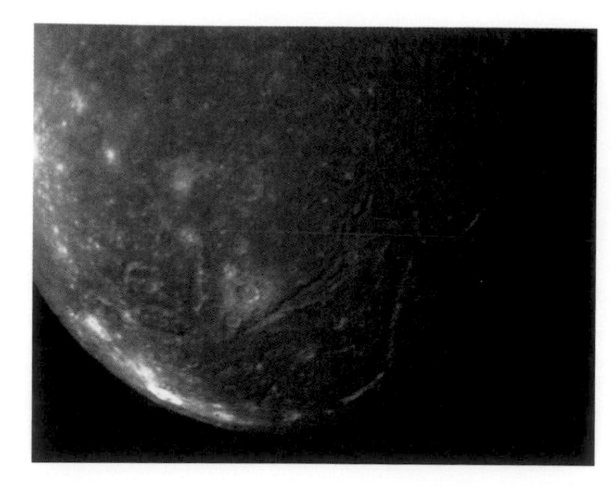

► **Ariel** Voyager 2 captured this image of Uranus' satellite Ariel in 1986. Ancient faults and valleys overlain by abundant impact craters indicate that the moon has not been geologically active for billions of years.

refractor, which they equipped with a micrometer eyepiece, from 1852 to 1863 he and his assistants Eduard Schönfeld (1828–91) and Adalbert Krüger (1832–96) mapped 324,198 stars to the 9th magnitude between the north celestial pole and −2° declination. In 1886 Schönfeld extended the survey to include another 133,659 stars to the declination zone of −23°, resulting in the *Southern Bonner Durchmusterung*. In 1863 Argelander and Wilhelm Förster (1832–1921) co-founded the Astronomische Gesellschaft (Astronomical Society). This organization published the first 'AGK' catalogue of 200,000 fundamental stars in 1887.

Argelander step method Visual method of estimating the magnitude of a VARIABLE STAR, based upon assessing the ease with which the variable may be distinguished from individual comparison stars. The arbitrary steps thus derived (known as 'grades') may subsequently be used to obtain the variable's magnitude. The method has the advantage of being relatively objective, in that the magnitudes of any comparisons are unknown at the time of making the estimate.

Argo Navis Former southern constellation, one of Ptolemy's original 48, representing the ship of the Argonauts. Argo Navis was huge, covering a quarter as much sky again as the largest present-day constellation, Hydra. In the 18th century, Lacaille divided it into the new constellations CARINA, the ship's keel, PUPPIS, the deck, and VELA, the sails.

argument of perihelion (ω) Angle measured along the orbital plane of a planet from the ASCENDING NODE to the PERIHELION. *See also* ORBITAL ELEMENTS

Argyre Planitia Classical circular feature on MARS (50°.0S 44°.0W). The Mariner spacecraft identified it as a basin 900 km (560 mi) in diameter, with a featureless floor covered by dust deposits. The floor is a light ochre colour when it is not covered by ice-fogs or frosts. The basin was formed by a meteoroid impact which created its 3-km-high (1.9-mi-high) mountainous rims. It is encircled by Nereidum Montes, on the north, and Charitum Montes, on the south.

Ariane Fleet of European satellite launchers developed by the EUROPEAN SPACE AGENCY and operated by Arianespace, a commercial satellite launching company. Arianespace became operational in 1981, two years after the maiden flight of Ariane 1. Further Ariane 2 and 3 models were developed and the present fleet consists of six types of Ariane 4 and an initial version of Ariane 5, four more models of which are under development. The Ariane 4 fleet will be retired in 2002–2003, leaving the Ariane 5 models in service, complemented by a fleet of smaller Vega boosters and Eurockot and Soyuz vehicles, operated by associated companies, for launches of lighter payloads into low Earth and other orbits. The major market for the Ariane 4 and 5 fleets is geostationary transfer orbit (GTO), the staging post for flights to equatorial geostationary orbit, using spacecraft on-board propulsion units. Launches take place from Kourou, French Guiana, which being located close the equator provides cost- and fuel-effective launches.

Communications satellites weighing up to 6 tonnes are being launched to GTO. On some flights, two smaller satellites can be launched together, reducing the launch cost for the customer. An Ariane launch costs in the region of US$100 million. The six Ariane 4 models provide a capability of placing 2.1–4.4 tonnes into GTO, with one model with no strap-on boosters and five with two to four solid or liquid boosters or combinations of these. Ariane 5 can place 6 tonnes into GTO. Four new Ariane 5 models, the 5E/S, 5E/SV, 5/ECA and 5/ECB, will improve GTO performance to between 7.1 and 12 tonnes.

Ariel Icy SATELLITE of URANUS, discovered in 1851 by William LASSELL. Ariel has a highly fractured surface, in marked contrast to its similar-size neighbour, UMBRIEL. Most fractures are paired to produce valleys with down-dropped floors, known as GRABENS. These fractures may have been caused by stresses induced by TIDES. Alternatively, if the interior had frozen after originally been molten, they could have been produced by the stretching of Ariel's LITHOSPHERE. Many areas, particularly valley floors, can be seen to have been flooded by some kind of viscous fluid during an episode of CRYOVOLCANISM. This flooding may have been a result of TIDAL HEATING, although Ariel no longer shares orbital RESONANCE with any other satellite of Uranus. Past heating events may have been sufficient to allow a small rocky CORE to settle out at Ariel's centre by DIFFERENTIATION.

The ice of which most of Ariel's volume is composed is believed to be rich in ammonia, although this has not been detected by spectroscopy. Ammonia-water ices have the remarkable property of being able to liberate small percentages of melt at temperatures as low as 176 K. As this is only about 100 K warmer than Ariel's present surface temperature, it goes some way to explaining the former molten nature of the interior and the apparent ease with which cryovolcanic melts have been able to reach and spread over the surface. However, even the youngest areas are scarred by abundant impact craters, indicating that Ariel was last geologically active a long time, maybe 2 billion years, ago. *See* data at URANUS

Ariel Series of six UK scientific satellites launched between 1962 and 1979. The first five were a collaboration with the USA. Ariel 1–4 studied the ionosphere and galactic radio sources; Ariel 5 and 6 carried X-ray and cosmic ray detectors.

Aries *See* feature article

A ring Outermost of the rings of SATURN visible from Earth, spanning a width of 14,600 km (9100 mi) to a maximum radius of 136,800 km (85,000 mi) from the planet's centre.

ARIES (GEN. ARIETIS, ABBR. ARI)

Rather inconspicuous northern constellation and sign of the zodiac, representing the ram whose golden fleece was sought by Jason and the Argonauts in Greek mythology. It lies between Taurus and Pisces, and used (*c*.150 BC) to contain the VERNAL EQUINOX, which is still sometimes called the first point of Aries. HAMAL, its brightest star, is mag. 2.0; γ Ari (Mesarthim) is a fine visual binary with bluish-white components, both mag. 4.6, separation 7″.6. The brightest deep-sky object is NGC 772, a 10th-magnitude spiral galaxy.

BRIGHTEST STARS

	Name	RA h m	dec. ° ′	Visual mag.	Absolute mag.	Spectral type	Distance (l.y.)
α	Hamal	02 07	+23 28	2.01	0.5	K2	66
β	Sheratan	01 55	+20 48	2.64	1.3	A5	60

◀ **A ring** This view of Saturn's A ring was obtained on August 23, when Voyager 2 was about 2.8 million km (1.7 million mi) from the planet. The Cassini Division is at bottom right, the Encke Division at top left and one of the F ring's shepherding satellites near the top.

Aristarchus Relatively young lunar crater (23°.6N 47°.4W), 40 km (25 mi) in diameter; it is the brightest object on the Moon's surface. Aristarchus reflects 20% of the sunlight falling upon it, which gives it an ALBEDO about twice that of typical lunar features. Its outer ramparts are highlighted by a white blanket of EJECTA, and its brilliant rays radiate south and south-east. It is possible to see Aristarchus by EARTHSHINE. The crater walls are terraced through a drop of 3000 m (10,000 ft) to the lava-flooded floor, which is only half the diameter of the crater, supporting a small central peak. The dark vertical bands visible on the inner walls of Aristarchus are likely an effect of offsetting from the landslips that formed its terraces. With neighbouring HERODOTUS and Vallis SCHRÖTERI, this impact crater has long been the focus of intensive study. Many LUNAR TRANSIENT PHENOMENA have been observed here. There is evidence of volcanic activity, including plentiful volcanic domes, rilles and areas of fire fountaining.

Aristarchus of Samos (*c*.310–*c*.230 BC) Greek astronomer and mathematician who measured the distances of the Sun and Moon, and who is also cited as the first astronomer to propound a heliocentric theory. Aristarchus reckoned the Sun's and Moon's distances by measuring the apparent angle between the Earth and Sun during the first quarter Moon, when the Moon–Sun angle is 90°. Though his method was mathematically valid, the exact moment at which the Moon was 50% illuminated proved very difficult to measure in practice, and any slight error in its value would introduce gross errors in the final result. Aristarchus observed the Sun–Earth–Moon angle as 87°, determining that the Sun was twenty times as far from the Earth as the Moon, and hence was twenty times the actual size of our natural satellite. Because the actual angle at first quarter is 89° 50′, the Sun is really about 400 times as distant, but Aristarchus' result, announced in his work *On the Sizes and Distances of the Sun and Moon*, nevertheless demonstrated that the Sun is a much more distant celestial object than the nearby Moon.

The idea of a moving Earth seems to have originated with the followers of Pythagoras (*c*.580–500 BC), though they had it orbiting a central fire, not the Sun. But according to Archimedes (*c*.287–212 BC) and Plutarch (AD 46–120), it was Aristarchus who put the Sun at the centre of a very large Universe where the stars were 'fixed' and only appeared to move because of the Earth's rotation on its axis. Aristarchus' hypothesis called for a circular Earth orbit, but that could not account for the unequal lengths of the seasons and other irregularities, while a moving Earth was contrary to ancient Greek cosmology; his heliocentric model was therefore not accepted in his own day.

Aristotle (384–322 BC) Greek philosopher and polymath who developed a cosmology based upon a 'perfect' Earth-centred Universe. In 367 BC he enrolled at Plato's Academy, where he studied under EUDOXUS; in 335 BC he founded his own school at Athens, called the Lyceum. It was there that Aristotle developed a cosmology incorporating the 'four elements' of earth, air, fire and water, and the four 'fundamental qualities' – hot, cold, dry and wet. These ideas led him to reject the idea of a vacuum and, therefore, any atomic theory, because this demanded the presence of particles existing in empty space. Aristotle also considered and rejected any idea of a moving Earth. Aristotle taught that the Universe was spherical, with the Sun, Moon and planets carried round on concentric spheres nesting inside one another. The Earth itself was a sphere, Aristotle reasoned, because of the shape of the shadow cast on the Moon in a lunar eclipse. All these celestial bodies were eternal and unchanging, and he thought of them being composed of a fifth element or essence. The outermost sphere carried the 'fixed stars'; it controlled the other spheres, and was itself controlled by a supernatural 'prime mover'.

In Aristotle's Universe everything had its natural place. 'Earthy' materials fell downwards because they sought their natural place at the centre of the spherical Earth, which was itself at the centre of the Universe. Water lay in a sphere covering the Earth, hence its tendency to 'find its own level'. Outside the sphere of water lay that of the air, and beyond this was the sphere of fire; flames always burned upwards because they were seeking their natural place above the air. Change was confined to the terrestrial world, the 'sublunary' region lying inside the Moon's sphere. As a result meteors, comets and other transitory events were thought to occur in the upper air; they were classified as meteorological phenomena.

The Aristotelian world-view, as modified by Ptolemy, would hold sway for nearly two millennia, thanks in part to its incorporation into the doctrines of the Catholic Church. *See also* GEOCENTRIC THEORY

Arizona Crater *See* METEOR CRATER

Armagh Observatory Astronomical research institution close to the centre of Armagh, Northern Ireland. The observatory was founded in 1790 and equipped with a $2\frac{1}{2}$-inch (60-mm) telescope, which is still in existence. A 10-inch (250-mm) refractor was added in 1885. During the 19th century, the observatory excelled in positional astronomy, its most important contribution being the

▼ **Aristarchus** This false-colour image of the Aristarchus plateau on the Moon shows the composition of the surface. Red indicates pyroclastic flows from the region's many volcanic craters, while the bright blue area is the 42-km-wide (26-mi) Aristarchus crater, surrounded by ejecta from the impact that caused it.

▶ Armstrong, Neil Alden Neil Armstrong is pictured here with an X-15 rocket-powered aircraft in 1960.

NEW GENERAL CATALOGUE compiled by J.L.E. Dreyer. Today's research topics include stellar physics and Solar System dynamics.

armillary sphere Oldest known type of astronomical instrument, used for both observing the heavens and teaching. It is a skeletal celestial sphere, the centre of which represents the Earth-bound observer. A series of rings represent great circles such as the equator and the ecliptic, while other rings represent the observer's horizon and meridian. The armillary sphere can be rotated to show the heavens at a particular time of day. Instruments of this kind were used from the time of the early Greek astronomers to that of Tycho Brahe.

Armstrong, Neil Alden (1930–) American astronaut who, on 1969 July 20, became the first human to walk on the Moon. After serving as a fighter pilot, he became a test pilot for NASA's experimental X-15 and other rocket planes. He commanded the first docking mission (Gemini 8, 1966) and was back-up commander for Apollo 8 (1968); he commanded the Apollo 11 lunar landing mission, with 'Buzz' ALDRIN and Michael Collins (1930–) completing the crew. Armstrong served on the 1986 panel that investigated the space shuttle *Challenger* tragedy.

Arneb The star α Leporis, visual mag. 2.58, distance about 1300 l.y., spectral type F0 Ib. Its name comes from the Arabic meaning 'hare'.

Arp, Halton Christian (1927–) American astronomer known for his controversial theories of redshifts, which he developed from his studies of peculiar galaxies. In 1956 he determined the relation between the absolute magnitude of a nova at maximum brightness and the rate at which this brightness decreases after the outburst. From his studies of galaxies, galaxy clusters and active galactic nuclei (AGNs), Arp proposed that AGNs, which often display very high redshifts, are ejected from the cores of nearby active galaxies with lower redshifts, the higher apparent redshifts of the AGNs being the result of the violent ejection process.

array Arrangement of RADIO TELESCOPES or ANTENNAE working together, most commonly in RADIO ASTRONOMY, to improve the resolution of the image created. The common shapes for an array are linear, Y-shaped or circular. An array is used in APERTURE SYNTHESIS and in RADIO INTERFEROMETRY. *See also* VERY LARGE ARRAY

Array for Microwave Background Anisotropy (AMiBA) Compact array telescope being built on Mauna Loa, Hawaii, by the Academia Sinica Institute of Astronomy and Astrophysics (Taiwan) to investigate structure in the cosmic microwave background radiation. The instrument consists of 19 small dishes 1.2 and 0.3 m (48 and 12 in.) in diameter on a single mounting of unusual form; it should begin operation in 2004.

Arrhenius, Svante August (1859–1927) Swedish chemist known for his explanation of why liquid solutions of ions conduct electricity. In astronomy he proposed the PANSPERMIA THEORY, according to which life was brought to Earth by way of spore-containing meteorites.

Arsia Mons Shield volcano (8°.4S 121°.1W), 800 km (500 mi) in diameter, situated in the THARSIS region of MARS. It rises to 9 km (6 mi) in height, and has a large summit caldera and radiating volcanic flows.

artificial satellite Man-made object that is placed into orbit around the Earth, Sun or other astronomical body. The first artificial satellite, SPUTNIK 1, was launched by the Soviet Union on 1957 October 4. By 2002 there had been almost 5000 successful launches, with an annual

▶ Arp, Halton Christian The interacting galaxy system Arp 102 consists of a spiral and a fainter elliptical Seyfert galaxy that has an active nucleus. It was his observations of the peculiarities of such interacting pairs that led Arp to his controversial theory about the nature of active galactic nuclei.

launch rate of around 90. Most of these carried single satellites, but it is increasingly common for two or more to be carried by one launch vehicle. At the same time, other objects such as the upper rocket stage and protective nose cone may also be released, creating a cloud of artificial debris – an undesirable class of satellites – around Earth.

All satellites move in elliptical orbits governed by NEWTON'S LAWS OF MOTION. However, the orbit is modified continuously by external forces, such as friction with the upper atmosphere and variations in the gravitational pull of the Earth. The rates of change depend on the height of the satellite and the inclination of the orbital plane to the equator.

Changes in the density of the atmosphere, due principally to solar activity, cause variable drag on a satellite. The overall effect is to change the ECCENTRICITY, with the orbit becoming more and more circular. Eventually, the satellite will spiral inwards and experience increasing drag until it finally plunges into the denser regions of the atmosphere and burns up.

Other factors affecting a satellite's orbit include atmospheric tides and winds, whether the PERIGEE of the orbit occurs in the northern or southern hemisphere and the local times at which this occurs. Solar radiation pressure has a major effect on satellites with a large area/mass ratio, such as those with large solar arrays. The overall result is that no satellite can have a stable orbit below a height of about 160 km (100 mi). At this height the orbital period is about 88 minutes.

Satellites are generally classified as military or civil, although there is considerable overlap. About 25% of current launches are for military purposes, such as photoreconnaissance, communications, electronic listening and navigation. Categorizing US military satellites is fairly straightforward, but Russian or Chinese sources rarely divulge the purposes of their military satellites. For example, the 2386 Russian Cosmos satellites launched by the end of 2001 included all types of experimental, scientific and military spacecraft.

Common names are used extensively instead of technical designations. Such names as Spot, Landsat, Meteosat and Mir are well known, but confusion may arise when two organizations use the same or similar names. For example, the US GEOS was used for geodetic studies, while the EUROPEAN SPACE AGENCY used the same name for a satellite that investigated charged particles in the upper atmosphere. Acronyms have been used widely, adding to the confusion.

Fortunately, COSPAR (International Committee for Space Research) has provided a standard system for identifying all satellites and related fragments, based on the year of launch and the chronological order of successful launches in that year. For example, the MIR space station, launched on 1986 February 19, was known as 1986-17A, since it was the 17th launch in 1986. The A referred to the satellite. The rocket that put it into orbit was designated 1986-17B. If several satellites are launched by a single rocket, they are usually given the letters A, B, C, and so on. If more than 24 fragments are created by an explosion, the letter Z is followed by AA, AB, ..., BA, BB, ..., and so on. Before 1963 a system involving Greek letters was used. For example, Sputnik 3 was designated 1958 δ 2 whilst the associated rocket was known as 1958 δ 1.

Photographic, visual and laser techniques are all used to track satellites. However, most satellites are tracked through their telemetry, by measuring the phase difference between a signal transmitted from the ground and a return signal from the satellite. This method can determine an Earth orbiting satellite's distance to within 10 m (33 ft).

If the sky is clear, satellites can be seen after dusk or just before dawn, when the sky is dark but the satellite is still illuminated by the Sun. The length of these visibility periods depends on the time of the year, the latitude of the observer and the orbit of the satellite. The satellite's brightness depends on the nature and curvature of the reflecting surface, the PHASE ANGLE (the angle between the Sun and the observer as seen from the satellite) and its distance and altitude from the observer. Some objects, such as the INTERNATIONAL SPACE STATION, rival the brightest planets, but others can only be seen through medium-sized telescopes.

The rate at which satellites cross the sky depends on their height above the Earth. Low satellites may cross the sky in about 2 minutes, but those at heights of about 2000 km (1200 mi) may take half an hour. Satellites at heights of about 36,000 km (22,400 mi) take 24 hours to complete one revolution. If the orbital inclination is 0°, the satellite appears motionless in the sky. Many meteorological and communication satellites have been launched into these GEOSTATIONARY ORBITS.

Roughly two-thirds of all the satellites launched have been from the former Soviet Union and most of these have been from the northern launch complex at Plesetsk, near Archangel. The other main launch site in the former Soviet Union is at BAIKONUR (Tyuratam), north-east of the Aral Sea. A relatively small number of satellites have been launched from Kapustin Yar, near Volgograd, and a new Russian launch site is being developed at Svobodny in the Russian Far East.

The Americans have three main launch sites: Cape Canaveral in Florida, Vandenberg Air Force Base in California and Wallops Island in Virginia. A number of commercial launch sites are also being developed for small rockets.

In recent years, as other nations have begun to develop their space industries, launch sites have sprung up in South America and Asia. European ARIANE rockets lift off from Kourou in French Guiana. Japan uses two southern sites, Kagoshima and Tanegashima. India's main launch centre is on Sriharikota Island, northeast of Madras. China has three main launch centres, at Jiuquan, Taiyuan and Xichang, in the less populated areas of north, west and southwest China respectively. The most recent innovation is the development of an ocean-going launch platform for the US–Ukraine–Norway SEA LAUNCH programme.

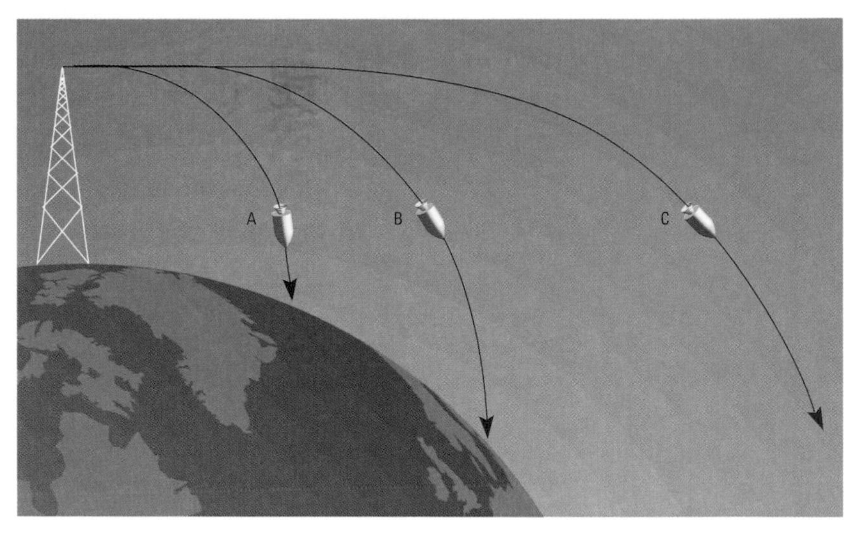

▲ **artificial satellite** If a high enough tower could be built, satellites could be fired from it directly into orbit. The type of orbit would depend on the initial velocity. If the starting velocity were low (A or B), the satellite would soon hit the ground. But at 8 km/s (5 mi/s) it would stay in its path (C), falling all the time, but never coming any closer to the ground.

Āryabhata (AD 476–*c*.550) Indian mathematician and astronomer who wrote three astronomical treatises, only one of which, the *Āryabhatīya* (AD 499), survives. In it he attributed the apparent movements of the planets in the sky to both their periods of revolution about the Sun and their orbital radii, presaging Copernican theory. Āryabhata also believed that the Earth rotated while the stars remain stationary.

Arzachel Latinized name of AL-ZĀRQALĪ

Arzachel Lunar walled plain (18°S 2°W), 97 km (60 mi) in diameter. Arzachel is the youngest of the three great walled plains that border Mare Nubium's east side. Its lofty (3000–4100 m/9800–13,500 ft), sharply defined walls show many terraces, especially on the west side. Its rugged, 1500-m-high (4900-ft-high) central peak confirms the crater's youth. A broad and lengthy canyon divides the south-east wall. A prominent rille runs from north–south across the east side of Arzachel's floor.

ASCA Abbreviation of Advanced Satellite for Cosmology and Astrophysics. *See* ASUKA

ascending node (Ω) Point at which an orbit crosses from south to north of the reference plane used for the orbit. For the planets the reference plane is the ECLIPTIC. For satellites it is usually the equator of the planet. *See also* INCLINATION; ORBITAL ELEMENTS

Asclepius APOLLO ASTEROID, number 4581. On its discovery as 1989 FC it caused a media furore because of its close approach to the Earth, to within 685,000 km (426,000 mi), which is less than twice the lunar distance. Its discovery stimulated political action in the United States aimed at taking steps to determine whether any possible asteroid impact on the Earth might be foreseen and perhaps obviated. *See* table at NEAR-EARTH ASTEROID

Ascraeus Mons Shield volcano situated in the THARSIS region of MARS (11°.9N 104°.5W). It is 18 km (11 mi) high, with complex caldera.

Aselli Latin name meaning 'the asses' applied to the stars γ and δ Cancri, which lie north and south, respectively, of the star cluster PRAESEPE, 'the manger'. γ Cancri (known as *Asellus Borealis*, the northern ass) is of visual mag. 4.66, 158 l.y. away, and spectral type A1 V. δ Cancri (*Asellus Australis*, the southern ass) is visual mag. 3.94, 136 l.y. away and spectral type K0 III.

ashen light (Fr. *lumière cendre*, ash-coloured light) Expression sometimes given to EARTHSHINE on the Moon, but more precisely given to the faint coppery glow infrequently seen on the dark side of VENUS, where the planet is visible as a thin crescent near INFERIOR CONJUNCTION. Sometimes the entire night hemisphere is affected, at other times the glow is patchy and localized, but it has no preferred position. In appearance the phenomenon is analogous to earthshine, although it is produced in a very different way. The ashen light was apparently seen by Giovanni Battista RICCIOLI in 1643, but it was first accurately described in 1715 by the English cleric William Derham, Canon of Windsor. It is possibly the oldest unsolved mystery in the observational history of the Solar System.

The ashen light phenomenon is rare, fugitive and suspect. It is only seen, and then with extreme difficulty, if Venus is observed on a dark sky and has its bright part hidden by an occulting device located inside the eyepiece of the telescope. Even this precaution is insufficient to dispel thoughts of illusion, and in the absence of photographic confirmation there is a natural tendency towards scepticism. Still, too many experienced observers have claimed sightings of ashen light to completely discount the phenomenon.

Volcanic activity, phosphorescence of the surface, self-luminosity, accidental combustion and other illuminating processes, including lightning, have all been proposed by way of explanation. The only conceivable physical mechanism that would be dependent on the phase or position of Venus as viewed from Earth would in fact be earthlight, but calculation clearly demonstrates that theoretical earthlight falls well below the threshold of visibility. Possibly the cause is to be found in an electrical phenomenon in Venus' upper atmosphere. Another possibility emerged in 1983 when David A. Allen (1946–94) and John W. Crawford imaged the planet in infrared and found cloud patterns on its night side. These clouds, with a retrograde rotation period of 5.4 ± 0.1 days, are thought to be at a lower level than the ultraviolet features and may be sufficiently lit by radiation scattered from the day side to become visible occasionally in integrated light.

ASP Abbreviation of ASTRONOMICAL SOCIETY OF THE PACIFIC

aspect Position of the Moon or a planet, relative to the Sun, as viewed from Earth. The angle on the celestial sphere between the Sun and another Solar System body is known as the ELONGATION of that body. When the elongation is 0°, that body is said to be at CONJUNCTION. When it is 180°, and the two are opposite one another in the sky, it is at OPPOSITION. When the elongation is 90°, the body is at QUADRATURE.

association, stellar Group of stars that have formed together but that are more loosely linked than in a star cluster. Stellar associations lie in the spiral arms of the Galaxy and help to define the shape of the arms, being highly luminous. OB ASSOCIATIONS are groups of massive O and B stars. R associations illuminate REFLECTION NEBULAE and are similar stars but of slightly lower mass (3 to 10 solar masses). T ASSOCIATIONS are groups of T TAURI STARS.

Association of Lunar and Planetary Observers (ALPO) Organization of amateur observers that collects reports of observations of all Solar System objects (Sun, Moon, planets, comets, meteors and meteorites), founded in 1947. These observations, especially of Mars, Jupiter and Saturn, are summarized annually in the organization's *Journal*, which provides synopses of notable happenings on these dynamic bodies.

Association of Universities for Research in Astronomy (AURA) Consortium of US universities and other educational and non-profit organizations; it was founded in 1957. It operates world-class astronomical observatories. Current members include 29 US institutions and six international affiliates. The facilities operated by AURA include the GEMINI OBSERVATORY, the NATIONAL OPTICAL ASTRONOMY OBSERVATORY and the SPACE TELESCOPE SCIENCE INSTITUTE. In addition, AURA has a New Initiatives Office formally established in January 2001 to work towards the goal of a 30-m (98-ft) GIANT SEGMENTED-MIRROR TELESCOPE.

A star Member of a class of white stars, the spectra of which are characterized by strong hydrogen ABSORPTION LINES. Ionized metals are also present. MAIN-SEQUENCE dwarf A stars (five times more common than B stars) range from 7200 K at A9 to 9500 K at A0. Their zero-age masses range from 1.6 to 2.2 times that of the Sun, and their zero-age luminosities from 6 to 20 times that of the Sun. Their lifetimes range from 2.5 billion to 900 million years. Evolved A DWARF STARS can have masses up to three times that of the Sun. Class A GIANT STARS and SUPERGIANTS are distinguished by the narrowing of the hydrogen lines with increasing luminosity.

Though A dwarfs can rotate rapidly, averaging from 100 km/s (60 mi/s) at A9 to 180 km/s (110 mi/s) at A0, their envelopes are not in a state of convection, and they lack solar-type chromospheres. A dwarfs exhibit a variety of spectral anomalies, especially among the slower rotators. Metallic-line AM STARS are depleted in calcium and scandium, while enriched in copper, zinc and the rare earths. The AP STARS (A-peculiar) have strong magnetic fields that range to 30,000 gauss. They exhibit enhancements of silicon, chromium, strontium and the rare earths (europium is enriched by over 1000 times), which are concentrated into magnetic starspots. These odd compositions result principally from diffusion, in which some chemical elements gravitationally settle, while others are lofted upwards by radiation. Rare LAMBDA BOÖTIS stars have weak metal lines.

DELTA SCUTI STARS (Population IA dwarfs and subgiants) pulsate in under a day with multiple periods and amplitudes of a few hundredths of a magnitude (the large-amplitude variety are known as dwarf Cepheids). Population II RR LYRAE STARS (A and F horizontal branch) have similar pulsation periods, but amplitudes of a few tenths of a magnitude.

Bright A stars include Sirius A1m V, Vega A0 V, Altair A7 V and Deneb A2 Ia.

asterism Distinctive pattern formed by a group of stars belonging to one or more constellations. Perhaps the most famous asterism is the PLOUGH (Big Dipper), a shape formed by seven stars in Ursa Major. Other famous asterisms within individual constellations include the SICKLE OF LEO and the TEAPOT in Sagittarius. The SQUARE OF PEGASUS is an example of an asterism that is composed of stars from two constellations (in this case, Andromeda and Pegasus); another is the SUMMER TRIANGLE. The term is also used for smaller groupings of stars visible in binoculars or telescopes, some of which have been given descriptive names such as the COATHANGER (in Vulpecula) and Kemble's Cascade (in CAMELOPARDALIS).

asteroid (minor planet) Rocky, metallic body, smaller in size than the major planets, found throughout the Solar System. The majority of the known asteroids orbit the Sun in a band between Mars and Jupiter known as the MAIN BELT, but asteroids are also found elsewhere. There is a population known as NEAR-EARTH ASTEROIDS, which approach the orbit of our planet, making impacts possible. Other distinct asteroid classes include the TROJAN ASTEROIDS, which have the same orbital period as Jupiter but avoid close approaches to that planet. Farther out in the Solar System are two further categories of body that are at present classed as asteroids, although in physical nature they may have more in common with comets in that they seem to have largely icy compositions. These are the CENTAURS and the TRANS-NEPTUNIAN OBJECTS, members of the EDGEWORTH–KUIPER BELT.

Although it is likely that there might be larger bodies in the Edgeworth–Kuiper belt awaiting discovery (*see* VARUNA), the largest asteroid in the inner Solar System is (1) CERES, which is *c.*933 km (*c.*580 mi) in diameter. It was discovered in 1801. Through a telescope, asteroids generally appear as pin-pricks of light, which led to the coining of the term 'asteroid', meaning 'star-like', by William HERSCHEL. Broadly spherical shapes are attained by asteroids if their self-gravity is sufficient to overcome the tensile strength of the materials of which they are composed, setting a lower limit for sphericity of about 200 km (120 mi); there are about 25 main-belt asteroids larger than this. All other asteroids are expected to be irregular in shape (*see* CASTALIA, DEIMOS, EROS, GASPRA, GEOGRAPHOS, GOLEVKA, IDA, KLEOPATRA, MATHILDE, PHOBOS, TOUTATIS, VESTA). Use of the HUBBLE SPACE TELESCOPE, sophisticated techniques employing ground-based telescopes and the application of radar imaging have recently led to the ability to resolve the shapes of some asteroids. OCCULTATIONS also allow asteroid sizes and shapes to be determined observationally: on rare occasions when asteroids pass in front of brighter stars, observers on the ground may see a fading or disappearance of the star, the duration of which (usually a few seconds) defines a cord across the asteroid profile. For many asteroids, sizes have been calculated on the basis of a comparison of their brightness both in scattered visible light from the Sun and also in the thermal infrared radiation emitted, which balances the solar flux they absorb. Such measurements allow the ALBEDO to be derived. In most cases, however, sizes are estimated simply on the basis of the observed absolute magnitude and an assumed value for the albedo.

The lower limit on size at which a solid body might be considered to be an asteroid is a matter of contention. Before the application of photography, starting in the 1890s, asteroids were discovered visually. Long photographic exposures on wide-field instruments, such as astrographic cameras and Schmidt telescopes, led to many thousands of asteroids being detected thereafter. However, the limited sensitivity of photographic emulsions makes it impossible to detect asteroids smaller than a few hundred metres in size, even if they are in the vicinity of the Earth. In the late 1980s the introduction by the SPACEWATCH project of charge-coupled devices for asteroid searching made it feasible to detect objects only a handful of metres in size during passages through cis-lunar space. A convenient size at which to draw a line is 10m (30ft): larger solid objects may be regarded as being asteroids, while smaller ones may be classed as METEOROIDS.

The total mass of all the asteroids in the inner Solar System (interior to Jupiter's orbit) is about 4×10^{21} kg, which is about 5% of the mass of the Moon. A large fraction of that total is held in the three largest asteroids, Ceres, PALLAS and Vesta. A decreasing fraction of the overall mass is represented by the smaller asteroids: although there may be over a million main-belt asteroids each about 1 km (0.6 mi) in diameter, their total mass is less than that of a single asteroid of diameter 200 km (120 mi). As a rule of thumb the number of asteroids increases by about a factor of one hundred for each tenfold decrease in size. Because the mass depends upon the cube of the linear dimension, the smaller asteroids thus represent a decreasing proportion of the total mass of the population.

When an asteroid is found its position is reported to the International Astronomical Union's Minor Planet Center, and it is allotted a preliminary designation. This is of the form nnnn pq, where nnnn is the year and pq represents two upper-case letters. The first letter (p) denotes the half-month in which the discovery was made, January 1–15 being labelled A, January 16–31 as B, and so on. The letter I is not used, so that December 16–31 is labelled Y in this convention, and the letter Z is also not used. The second letter (q) provides an identifier for the particular asteroid; in this case Z is used but not I, making 25 letters/asteroids in all. Thus the third asteroid reported in the second half of March 1989 was labelled as 1989 FC. Until the last couple of decades this designation system was adequate because no more than 25 asteroids were being discovered in any half-month, but now thousands are reported and so an additional identifier is required. This takes the form of a subscripted numeral. Thus 1999 LD_{31} was the 779th asteroid reported in the first half of June 1999. Many asteroids are reported

▼ **asteroid** Plots of the larger known asteroids: most of them lie in the main belt between the orbits of Mars and Jupiter, while the Trojans move in the same orbit as Jupiter, 60° in front of or behind it.

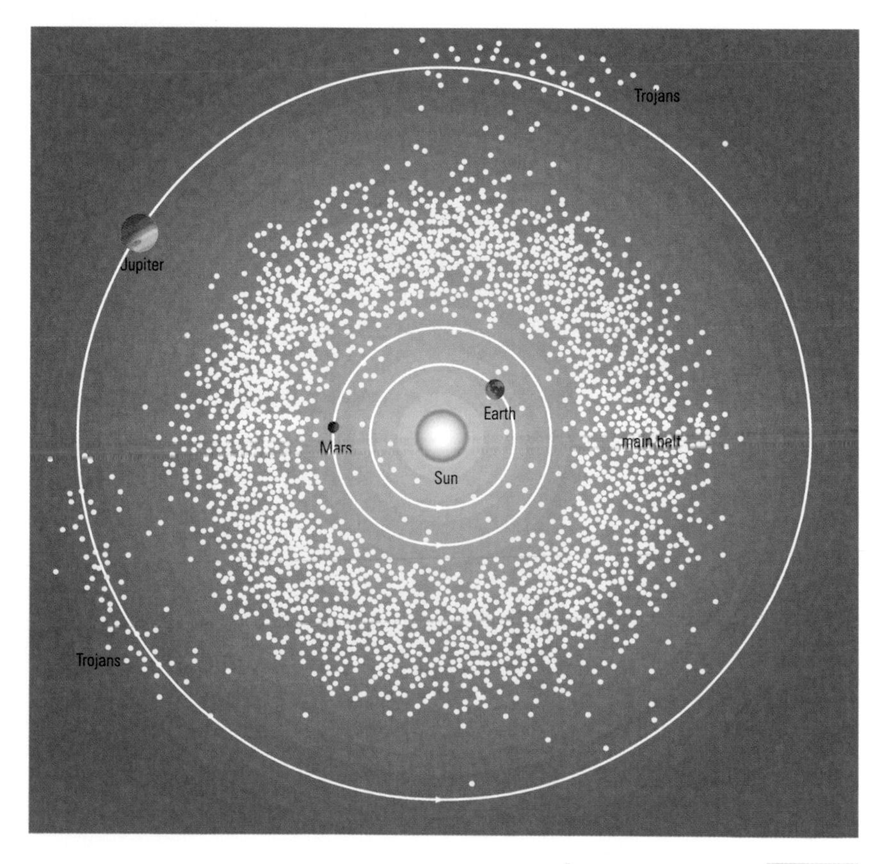

several times at different epochs, and so receive multiple designations of this form.

After an asteroid has been observed for a sufficient length of time, such that our determination of its orbit is secure (this generally requires three oppositions), it is allocated a number in the master list, which begins with (1) Ceres and continues with (2) Pallas, (3) Juno, (4) Vesta and so on. This list exceeded 25,000 during the year 2001. Upon numbering, the discoverers of each asteroid have the right to suggest a name for it to the naming committee of the International Astronomical Union, and in most cases that suggestion is adopted. Certain conventions apply to distinct categories of asteroid, such as the Trojans and the Centaurs. In the past, before strict rules were adopted for numbering and naming, a few exceptions to these procedures occurred, such as with ALBERT and HERMES.

Although some asteroids may be cometary nuclei that have become dormant or extinct, most of these bodies are thought to have originated in the region of the present main belt. They seem to be the remnant debris from a large complex of PLANETESIMALS that failed to form a major planet due to the gravitational stirring effect of Jupiter. Soon after their formation, some of the larger proto-asteroids underwent sufficient internal heating, through radioactive decay, to melt to some degree, acquiring metallic cores and layered mantles. This chemical DIFFERENTIATION is represented in the variety of known METEORITE classes – these being in effect asteroid fragments.

A consequence of this differentiation is that asteroids present a variety of spectroscopic classes, representing different surface (and, assumedly, bulk) compositions. The most abundant group is called C-type (for 'carbonaceous') asteroids, these being especially prevalent as one moves outwards through the main belt. C-type asteroids reflect more light at the red end of the spectrum, but their low albedos (below 0.05) make them very dark, so they may be thought of as being black-brown. They are thought to contain the same materials as CARBONACEOUS CHONDRITE meteorites. The inner part of the main belt is dominated by S-type (stony or 'silicaceous') asteroids. These have moderate albedos (0.15–0.25) and are thought to be analogous to the metal-bearing stony meteorites known as ordinary CHONDRITES. The third most populous class are the M-type ('metallic') asteroids. These also have moderate albedos and are thought to be derived, like nickel-iron meteorites, from the metal-rich cores of large differentiated parent bodies that have become exposed by collisional break-up. Members of a less common class, the E-types ('enstatite'), have elevated albedos of 0.40 or more, as has the V-class Vesta. There are many other subdivisions of each of these classes, and other rare categories that have been recognized as being distinct from the common groupings. The distributions of the various asteroid types in the main belt – the preponderance of S-types at the inner edge but with more C-types at greater distances – suggest that the volatile-rich asteroids formed farther from the Sun, as a result of the presumed temperature gradient in the SOLAR NEBULA. Similarly, the homogeneous composition of the HIRAYAMA FAMILIES suggests that most asteroids remain in orbits similar to those where they formed in that nebula.

► **asteroid** Because they have low albedos, asteroids are very difficult to spot and it is often only their proper motion relative to the background stars that gives away their presence.

Another important clue to the nature of asteroids comes from their spin rates. The rotation period of an asteroid may be measured from observations of its repeating trends in brightness variation, generally displaying two maxima and two minima in each revolution. This pattern indicates that the changes in brightness are dominated by the effect of the asteroid's non-spherical shape, rather than any albedo distribution across its surface. Typical amplitudes are only up to 0.2 mag., or 20% in the brightness, although some near-Earth asteroids, such as Geographos, vary by a much greater amount, as a result of their elongated shapes. Typically, asteroids have rotation periods in the order of ten hours, although a few are much longer (some weeks), possibly due to the damping effect of a companion satellite (as with Ida, EUGENIA and PULCOVA). At the other end of the scale, short rotation periods are of interest because self-gravitation cannot hold together an asteroid having a rotation period of less than about two hours. Until recently all asteroids for which light curves were available displayed periods in excess of this amount, meaning that they could be 'rubble piles' held together by gravity. (Another piece of evidence pointing to such a structure is the presence of voids within asteroids, as indicated by the low average densities determined for some of them, for example Ceres and Mathilde.) The first asteroid to be measured to have a shorter period (about 97 minutes) was 1995 HM, which thus appears to be a monolith. Since then several other asteroids have been shown to have even higher spin rates, with periods as short as ten minutes. These must similarly be single rocks rather than rubble piles, being held together by their tensile strength.

The meteoroids that produce meteorites, and indeed some complete asteroids, leave the main belt through a variety of mechanisms. Inter-asteroid collisions may grossly change their orbits (evidence of past collisions derives from the Hirayama families), and such collisions also change the spin rates discussed above. A more significant avenue through which asteroids can escape the main belt is the rapid dynamical evolution that occurs if an object acquires an orbital period that is a simple fraction of that of Jupiter. This leads to depleted regions of orbital space known as the KIRKWOOD GAPS. Asteroids leaving those regions tend to acquire high-eccentricity orbits, which can lead to them either being ejected from the Solar System by Jupiter, or else joining the population of MARS-CROSSING ASTEROIDS or near-Earth asteroids.

Viewed in three dimensions, the main belt is a somewhat wedge-shaped torus, increasing in thickness from its inner to its outer edge. The median inclination increases from about 5° to about 9°, whereas the median orbital eccentricity is around 0.15 throughout the main belt. Most asteroids, therefore, follow heliocentric orbits that are only a little more inclined than those of the planets, and with modestly greater eccentricities. It follows that the majority of main-belt asteroids move among the stars, as seen from the Earth, in a similar way to the planets: they stay close to the ecliptic, and retrograde through opposition. There are, however, exceptions among various dynamical classes such as the near-Earth asteroids, the Trojans, and various specific main-belt classes that are grouped together in terms of common dynamical behaviour rather than composition or origin.

asteroid belt General term for the region between the orbits of Mars and Jupiter in which the orbits of most asteroids lie. *See* MAIN-BELT ASTEROID

asteroseismology Study of the internal structure of stars using observations of the frequencies and strengths of global oscillations detected at the surface by their Doppler shifts. The very low amplitude of stellar oscillations has limited its current application to pulsating stars. *See also* HELIOSEISMOLOGY

asthenosphere Weak, uppermost layer of a planet's MANTLE, in which solid-state creep first plays a predominant role; it lies immediately below the relatively rigid LITHOSPHERE. At the asthenosphere the rise in temperature with depth reaches the threshold at which plastic flow may occur in response to small stress differences. In the case of the Earth, seismological evidence – the attenuation of S waves and a decrease of P wave velocities between depths of about 100–250 km (62–155 mi) – is taken to define the asthenosphere. The viscosity of the asthenosphere is largely similar to that of the bulk of the underlying mantle, but it is far less than that of the lithosphere, enabling plate-tectonic motions to take place. At subduction zones the asthenosphere appears to penetrate down to approximately 700 km (430 mi), which is comparable with the transition zone between the upper and lower mantle. Other planets such as Venus and Mars are assumed to have asthenospheres, as are some of the larger satellites. The Moon is known to have an asthenosphere at a depth of approximately 1000 km (620 mi), and recent seismological evidence suggests that it may be even more fluid at a greater depth. *See also* SEISMOLOGY; TECTONICS

astigmatism ABERRATION in lenses and mirrors that prevents star images away from the centre of the field of view from being focused into sharp points. When astigmatism is present point objects such as stars usually appear elongated, either into a line or an ellipse. Attempts to reduce the elongated image to a point by refocusing will cause the orientation of the elongation to change through 90°. The best or least distorted result occurs as the orientation changes and the image becomes round; this is known as the circle of least confusion. *See also* ANASTIGMAT

Astraea MAIN-BELT ASTEROID; number 5. It is about 117 km (73 mi) across, and was found in 1845, almost four decades after the discovery of (4) VESTA.

Astro Fleet of Japanese ARTIFICIAL SATELLITES. When launched, they are given individual names.

astrochemistry Chemistry in stars. The chemical composition of most stars is dominated by hydrogen, with helium in second place and the remaining elements a long way behind. The relative proportions (or abundances) of the elements are quantified either by the number of atoms or the mass involved. In terms of mass, average material from outer layers of the Sun and the rest of the Solar System contain 70.7% hydrogen, 27.4% helium and 1.6% of all the other elements. These three quantities are called mass fractions and are conventionally labelled X, Y and Z. In terms of the number of atoms, hydrogen dominates even more, with 92.0% hydrogen, 7.8% helium, and all the rest just 0.12%. Elements other than hydrogen and helium are often termed the HEAVY ELEMENTS, and their relative abundances are shown in the accompanying table.

These figures, which are representative compositions throughout the Universe, are changed by NUCLEOSYNTHESIS reactions inside stars. Thus at the centre of the Sun, the abundance of hydrogen has currently fallen to 38% by mass, and that of helium has increased to about 60%. Stars that have completed their MAIN-SEQUENCE lives, will have cores of almost pure helium. Solar mass stars will go on to synthesize carbon and oxygen, and higher mass stars will create the elements all the way to iron. The remaining elements are produced during SUPERNOVA explosions.

Nucleosynthesis implies that hydrogen is continually being converted into heavier elements, and so abundances change slowly. But only a few per cent of helium by mass can have been produced this way during the lifetime of the Universe. Most helium, therefore, was produced during the early stages of the Universe (*see* COSMOCHEMISTRY). The remaining elements have been produced inside stars, and then ejected to mix with the interstellar medium through stellar winds, supernova explosions and so on. The mass fraction, Z, of the heavy elements is therefore increasing with time. The oldest stellar POPULATIONS, known as Population II stars, found in the galactic nucleus and halo and in globular clusters, have values of Z around 0.5%. Stars in the disk of the Galaxy are younger, have a higher proportion of heavy elements and are called Population I stars. Thus the Sun, which formed about 4.5 billion years ago, has $Z = 1.6\%$, while stars forming today have $Z \approx 4\%$.

PERCENTAGE ABUNDANCE OF HEAVY ELEMENTS IN STARS

	By mass	By number
carbon	0.292	0.0305
nitrogen	0.095	0.0084
oxygen	0.769	0.0608
neon	0.122	0.0076
sodium	0.003	0.0002
magnesium	0.048	0.0024
aluminium	0.004	0.0002
silicon	0.069	0.0030
sulphur	0.038	0.0015
argon	0.018	0.0006
calcium	0.006	0.0002
iron	0.165	0.0037
nickel	0.009	0.0002
remainder	0.007	0.0002

Some stars are exceptions to the norms, the most important being the WHITE DWARFS. These stars, which are at the ends of their lives, have lost their outer layers and so have their cores exposed. The cores contain the products of nucleosynthesis, and thus the composition of white dwarfs ranges from helium-rich through relatively pure carbon to calcium. WOLF–RAYET stars subdivide into the WN stars, formed of helium and nitrogen, and the WC stars, within which helium, carbon and oxygen predominate. T TAURI stars have an over-abundance of lithium. AP STARS are over-abundant in elements such as mercury, strontium, silicon, europium, holmium and chromium. It is thought that these latter peculiarities occur just in a thin surface layer and are perhaps brought about by diffusion. Beneath that thin layer the remaining parts of the star are of normal composition. CARBON STARS have over-abundances of lithium and carbon, while S-type stars additionally have over-abundances of zirconium and yttrium. Technetium has been detected in both carbon and S stars, and since all isotopes of technetium are radioactive with short half-lives, it must have been produced inside the stars. Nucleosynthesis products are thus being brought to the surface in these stars. There are other anomalies to be found, and such stars are termed CHEMICALLY PECULIAR STARS (CP).

In the outer layers of cooler stars, simple molecules such as CH, TiO, CN and C_2 may be observed, but stars are too hot for most molecules to form. Molecules are therefore to be found in cooler regions such as planets (*see* GEOCHEMISTRY) and GIANT MOLECULAR CLOUDS (GMCs). Some 120 molecule species have been identified so far in GMCs, including water (H_2O), ammonia (NH_3), salt (NaCl) and ethanol (CH_3CH_2OH). Much of the material between the stars is molecular hydrogen (H_2).

Astro E Japanese X-ray observation satellite that has not yet been scheduled for launch. The satellite will also carry a US X-ray instrument, which was to have flown originally on the CHANDRA X-RAY OBSERVATORY. Astro E will observe in the soft X-ray region, with its main observational subjects being hot plasmas, spectroscopy of black hole candidates and pulsars.

astrograph Telescope specially designed for taking wide-angle photographs of the sky to measure star positions. Traditionally, astrographs are refracting telescopes characterized by their OBJECTIVES, which have relatively fast FOCAL RATIOS (sometimes as fast as *f*/4) and fields of view up to 6° wide – large by astronomical standards. The objectives may have three or four component lenses to achieve this performance, and the biggest astrographs have lenses 50 cm (20 in.) or so in diameter. Astrographs are always placed on EQUATORIAL MOUNTINGS, which are motor-driven to allow the telescope to follow the apparent motion of the sky.

The first recognizable astrograph was built in 1886 at the Paris Observatory. It had an aperture of 330 mm (13 in.) and its objective was designed to give the best images in blue light, early photographic emulsions being insensitive to other colours. The success of this telescope led, in 1887, to a major international conference of astronomers agreeing to embark on a photographic survey of the whole sky – the *CARTE DU CIEL*. The Paris astrograph was adopted as a standard model for this ambitious project, and several similar telescopes were built with focal lengths of 3.4 m (134 in.) and objectives corrected for colour and COMA, including one at Greenwich in 1890.

Astrographic photographs were always taken on glass plates rather than film, which lacks the stability of glass. The instruments were principally used for ASTROMETRY, the accurate measurement of star positions, but have now been superseded by the SCHMIDT CAMERA.

Astrographic Catalogue *See CARTE DU CIEL*

astrolabe Early disk-shaped astronomical instrument, for measuring positions on the celestial sphere, equippped with sights for observing celestial objects. The classical form, known as the **planispheric astrolabe**, originated in ancient Greece, attained its greatest refinement in the hands of medieval Arab astronomers (*see* ISLAMIC ASTRONOMY) and reached the Christian West in about the 10th century AD. The basic form consists of two flat disks, one of which (the *mater*) is fixed and represents the observer on the Earth. The other (the *rete*) is movable and represents the celestial sphere. The altitudes and azimuths of celestial bodies can be read. A pointer or alidade can be used for measuring altitudes when the instrument is suspended by a string.

Given the latitude, the date and the time, the observer can read off the altitude and azimuth of the Sun, the bright stars and the planets, and measure the altitude of a body and find the time. Its function of representing the night sky for a certain latitude at various times is reproduced in the modern PLANISPHERE. The astrolabe can be used as an analogue computer for many problems in spherical trigonometry, and it was used even for terrestrial surveying work such as determining the altitude of a tower.

The **mariner's astrolabe** is a simplified instrument designed for observing the altitudes of the Sun and stars while at sea, as an aid to navigation. It was developed by the Portuguese and was used mainly in the 17th century, being supplanted by the SEXTANT.

The **precision** (or **prismatic**) **astrolabe** is a 19th-century instrument for determining local time and latitude by accurately measuring when a star reaches a certain altitude. It was further developed as the **impersonal astrolabe** (so called because it eliminates the observer's personal error) by André DANJON in 1938.

► **astrolabe** Astronomy was a vital aid to navigation for centuries, and astrolabes, like this medieval Islamic brass example, were important instruments.

astrology In its modern form, a pseudoscience that claims to be able to assess personality traits and predict future events from the positions of celestial objects, particularly the visible planets. Today, more newspaper space is devoted to horoscopes than to astronomy, and if anything the influence of astrology is probably increasing. Yet there is no scientific justification for the claims made by astrologers.

Astrology has a long history and was of great importance to primitive and superstitious peoples. The positions of the planets and the constellations at the time of a person's birth were believed to have a great influence on their subsequent life and career.

Unusual celestial events such as eclipses of the Sun and Moon, and the appearance of comets and 'guest stars', were significant in times when most people believed that the heavens could influence their fates, and there emerged a class of astrologers who made their living from the gullibility of their clients. Kings often employed astrologers whose job was to keep up the morale of their subjects by assuring them that the dates of particular events such as royal births were auspicious ones. The efforts of astrologers and priests to predict such apparently ominous events as eclipses helped to establish astronomy as a legitimate subject of study.

astrometric binary BINARY STAR system in which one component is too faint to be observed directly. Its binary nature is detected by perturbations in the visible component's PROPER MOTION, caused by the orbital motion of the unseen companion. Some astrometric binaries, just beyond the limit resolution, appear as a single elongated image, with the orientation of the long axis changing as a result of orbital rotation.

astrometry Branch of astronomy concerned with the precise measurement of the apparent positions of celestial objects through the creation of a fundamental non-rotating reference frame on the CELESTIAL SPHERE. This frame is used not only for precisely recording the relative positions of celestial objects, but also for measurements of stellar PARALLAX and PROPER MOTION, and as a reference for relating optical observations to those in other regions of the electromagnetic spectrum. The frame has the Sun at its centre and is based on the plane of the Earth's equator (from which DECLINATION is measured) and the FIRST POINT OF ARIES (the zero point for measuring RIGHT ASCENSION).

The position of the equator and the First Point of Aries for a given date (called the EPOCH) has traditionally been defined by making observations of the Sun and planets using a meridian or TRANSIT CIRCLE. This instrument is mounted on a fixed east–west horizontal axis and can only rotate in a north–south vertical plane. The declination of a celestial object is found from the inclination of the telescope when the object is observed crossing the MERIDIAN; RIGHT ASCENSION is reckoned by timing the exact moment this occurs.

Once the positions of the celestial equator and the First

Point of Aries are established, the positions of stars can be measured relative to these by further observations. Measuring stellar positions in this way provides fundamental observations of the celestial bodies since they are measured with respect to an inertial reference frame rather than to other 'fixed' stars. The initial, raw observations from the transit circle first need to be corrected for the Earth's movement through space, that is, referred to the centre of the Sun (heliocentric coordinates), and also corrected for the individual motions of the stars themselves. The components of these motions across the line of sight are called proper motions and were first discovered in 1718 by Edmond HALLEY.

The observations are first referred to the centre of the Earth by allowing for the effect of REFRACTION, diurnal parallax and diurnal ABERRATION. By making further corrections for PRECESSION, NUTATION and polar drift, the coordinates are then referred to the fundamental plane of the equator. Heliocentric coordinates are obtained by correcting for the orbital motion of the Earth by allowing for the effects of ANNUAL PARALLAX and aberration. It is then possible to go one step further and make a correction for the motion of the Sun itself, which appears 'reflected' in the proper motions of nearby stars. The ultimate accuracy of our measurement of stellar positions therefore depends on our knowledge of the astronomical constants of precession, aberration, nutation and the stellar proper motions.

In the past, much of our knowledge of positions, parallaxes and proper motions was achieved through photographic astrometry, undertaken over many years using long-focus telescopes to obtain photographic plates of individual areas of the sky. The stellar images on these plates were then measured very accurately on dedicated measuring machines to determine their relative positions.

Nowadays our knowledge comes from instruments such as the Carlsberg Meridian Telescope (CMT) at the Roque de los Muchachos Observatory on the Canary Island of La Palma and the astrometric satellite HIPPARCOS. The Carlsberg Meridian Telescope is dedicated to carrying out high-precision optical astrometry and is currently being used to map the northern sky using a CCD detector. This will provide accurate positions of stars, allowing a reliable link to be made between the bright stars measured by the Hipparcos satellite and fainter stars seen on photographic plates. So far, the CMT has made measurements of some 180,000 positions, proper motions and magnitudes for stars down to magnitude 15, and over 25,000 positions and magnitudes of 180 Solar System objects, ranging from the outer planets including Pluto to some of the many asteroids that orbit the Sun between Earth and Mars.

Hipparcos was the first space experiment designed specifically for astrometry. The satellite, which operated between 1989 and 1993, extended the range of accurate parallax distances by roughly 10 times, at the same time increasing the number of stars with good parallaxes by a much greater factor. Of the 118,218 stars in the Hipparcos catalogue, the distances of 22,396 are now known to better than 10% accuracy. Prior to Hipparcos, this number was less than 1000. The companion Tycho database, from another instrument on the satellite, provides lower accuracy for 1,058,332 stars. This includes nearly all stars to magnitude 10.0 and many to 11.0.

Information on star positions, parallaxes and proper motion is published in FUNDAMENTAL CATALOGUES and between 1750 and 1762 James BRADLEY compiled a catalogue that was to form the basis of the first fundamental one ever made, by Friedrich BESSEL, in 1830. By comparing Bradley's work with that of Giuseppe PIAZZI in Palermo in about 1800, Bessel was able to calculate values of proper motions for certain bright stars and also to derive the constants of precession for the epochs involved. Bessel's name is also inextricably linked with two other aspects of astrometry. He was the first, in 1838, to measure the TRIGONOMETRIC PARALLAX of a star and derived a distance of 9.3 l.y. (the modern value is 11.3 l.y.) for the nearby binary star SIXTY-ONE CYGNI. From his work on the fundamental catalogue he also found that the proper motion of SIRIUS was not constant and he correctly attributed this 'wobble' to the presence of an unseen companion pulling Sirius from its path. The companion was discovered in 1862, making Sirius the first example of an ASTROMETRIC BINARY.

Further fundamental catalogues followed with those of Simon NEWCOMB (1872, 1898) and Arthur AUWERS (1879) being especially important. The latter was the first in the *Fundamental Katalog* series compiled in Germany, the latest of which is the *FK5 Part II.* This catalogue provides mean positions and proper motions at equinox and epoch J2000.0 for 3117 new FUNDAMENTAL STARS.

The European Space Agency also has plans for a next-generation astrometric satellite, called GAIA. Instead of orbiting the Earth, GAIA will be operated at the L_1 Sun–Earth Lagrangian point, located 1.5 million km (0.9 million mi) from Earth in the direction away from the Sun. Here it will have a stable thermal environment and will be free from eclipses and occultations by the Earth. The purpose of the satellite will be repeatedly to measure positions of more than a billion stars to an accuracy of a few microarcseconds, the goal being to achieve 10-microarcsecond precision for stars as faint as 15th magnitude, with potentially four or five microarcseconds achievable for stars brighter than 10th magnitude. This will be more than a hundred times more accurate than the observations from Hipparcos.

astronautics Science of space flight. *See also* APOLLO PROGRAMME; ROCKET ASTRONOMY

Astronomer Royal Honorific title conferred on a leading British astronomer. Until 1971 the Astronomer Royal was also the director of the ROYAL GREENWICH OBSERVATORY. Those who have held the position are listed in the accompanying table.

The post of **Astronomer Royal for Scotland** was created in 1834 initially to provide a director for the Royal Observatory, Edinburgh; the present holder, appointed in 1995, is John Campbell Brown (1947–). The last **Astronomer at the Cape**, working originally at the Royal Observatory at the Cape of Good Hope (now in South Africa), was Richard Hugh Stoy (1910–94); the post was abolished in 1968. The post of **Royal Astronomer for Ireland** was created to provide a director for the DUNSINK OBSERVATORY. It lapsed in 1921; the last incumbent was Henry Crozier Plummer (1875–1946).

Astronomical Almanac, The Publication containing fundamental astronomical reference data for each calendar year. It includes positions of the Sun, Moon and planets, data for physical observations, positions of planetary satellites, sunrise and sunset times, phases of the Moon, eclipses, locations of observatories and astronomical constants. It is published jointly by the US Naval Observatory and Her Majesty's Nautical Almanac Office.

ASTRONOMERS ROYAL

Name	Held office
John FLAMSTEED (1646–1719)	1675–1719
Edmond HALLEY (1656–1742)	1720–42
James BRADLEY (1693–1762)	1742–62
Nathaniel BLISS (1700–1764)	1762–64
Nevil MASKELYNE (1732–1811)	1765–1811
John POND (1767–1836)	1811–35
George Biddell AIRY (1801–92)	1835–81
William CHRISTIE (1845–1922)	1881–1910
Frank DYSON (1868–1939)	1910–33
Harold Spencer JONES (1890–1960)	1933–55
Richard WOOLLEY (1906–86)	1956–71
Martin RYLE (1918–84)	1972–81
Francis GRAHAM-SMITH (1923–)	1982–90
Arnold Wolfendale (1927–)	1991–94
Martin REES (1942–)	1995–

Astronomical Data Center (ADC) Part of the Space Sciences Directorate at NASA's GODDARD SPACE FLIGHT CENTER, located in Greenbelt, Maryland. The ADC specializes in archiving and distributing astronomical datasets, most of which are in the form of machine-readable catalogues rather than images. The Center's data collection is accessible via the World Wide Web.

Astronomical Journal Major US journal for the publication of astronomical results. Founded in 1849 by Benjamin GOULD, the journal is published monthly by the University of Chicago Press, with a bias towards observational rather than theoretical papers. Historically, the '*AJ*' emphasized traditional fields of astronomy, such as galactic structure and dynamics, astrometry, variable and binary stars, and Solar System studies. It has now broadened its coverage to all aspects of modern astronomy.

Astronomical League (AL) Blanket organization for amateur astronomers in the United States. It is the largest astronomical organization in the world, with nearly 20,000 members from some 250 local societies. Its goal is to promote amateur astronomy through educational and observational activities. Founded in 1946, the AL holds a nationwide meeting every year and publishes a quarterly newsletter, *The Reflector*, which describes amateur activities throughout the country.

Astronomical Society of Australia Australian organization for professional astronomers, founded in 1966. The Society publishes a peer-reviewed journal, PASA (*Publications of the Astronomical Society of Australia*). In 1997 it introduced an electronic counterpart (called el-PASA) to improve international access to Australian astronomical research.

Astronomical Society of the Pacific (ASP) Claimed to be the largest general astronomy society in the world, founded in 1889 by Californian amateur and professional astronomers. It retains its function as a bridge between the amateur and professional worlds, and it has developed into a leading force in astronomy education. It publishes a monthly magazine, *Mercury*, and original research is reported in the *Publications of the ASP*. The Society's headquarters are in San Francisco.

Astronomical Technology Centre (ATC) UK's national centre for the design and production of state-of-the-art astronomical technology. It is located on the site of the ROYAL OBSERVATORY EDINBURGH (ROE). It was created in 1998 to replace the instrument-building functions of both ROE and the ROYAL GREENWICH OBSERVATORY. The ATC provides expertise from within its own staff and encourages collaborations involving universities and other institutions. Customers include the GEMINI OBSERVATORY (for which ATC was a prime contractor for the GMOS multi-object spectrometers), the Herschel Space Observatory, the UK INFRARED TELESCOPE and the JAMES CLERK MAXWELL TELESCOPE.

astronomical twilight *See* TWILIGHT

astronomical unit (AU) Mean Sun–Earth distance, or half the semimajor axis of the Earth's orbit, used as a unit of length for distances, particularly on the scale of the Solar System. Its best current value is 149,597,870.66 km.

The determination of the astronomical unit was one of the most important problems of astronomy until comparatively recent times. From the ancient Greeks up to the early 18th century, the Earth–Sun distance was known very imperfectly because it depended on very fine angular measurements that were beyond the capabilities of the available instruments. Johannes Kepler's best estimate, for example, was only one-tenth of the actual value. In the 18th and 19th centuries attempts to measure the AU were made by making accurate observations of Mars or Venus from two widely separated places on the Earth's surface, and calculating the planet's distance by triangulation. Kepler's third law was then used to compute the Sun–Earth distance. The first such attempt was made by Giovanni Domenico CASSINI, who got within 7% of the true value. Some improvement was achieved by measuring the distances of asteroids, especially close-approaching ones, such as 433 Eros. Such refinements were achieved in the late 19th century by David GILL and by Harold Spencer JONES in the 20th. The current technique makes use of radar measurements, and the probable error has been reduced by several orders of magnitude compared to the best visible-light values.

astronomy, history of Astronomy is the oldest of the sciences and predates the written word. ARCHAEOASTRONOMY has shown that early peoples made use of astronomical observations in the alignment of their megalithic monuments.

Written records of astronomical observations in Egypt, Babylon and China (*see* EGYPTIAN ASTRONOMY; BABYLONIAN ASTRONOMY; CHINESE ASTRONOMY) and Greece date from pre-Christian times and it is to the Greeks that we owe the beginnings of scientific theorizing and systematic efforts to understand the Universe. The main legacy of Greek astronomy was a system of thought dominated by the work of ARISTOTLE, though many other Greek scientists of great originality contributed to early astronomical knowledge.

Greek ideas survived the medieval period, largely through the *ALMAGEST* of Ptolemy, a Graeco-Roman compendium of astronomical knowledge preserved by scholars in the Islamic world and the Byzantine Empire (*see* MEDIEVAL EUROPEAN ASTRONOMY; ISLAMIC ASTRONOMY).

'Modern' astronomy can be said to have begun with the rejection by Nicholas COPERNICUS of Aristotle's complex picture of planetary orbits, which assumed that the Sun was the centre of the Universe. Although Copernicus put forward a simpler and more logical view, which placed the planets at distances from the Sun that increased with their periods, its details were very complex and unconvincing to many of his contemporaries (*see* COPERNICAN SYSTEM). Only after the meticulous observations of Tycho BRAHE and their interpretation by Johannes KEPLER did the new ideas begin to gain general acceptance (*see* RENAISSANCE ASTRONOMY).

The invention of the telescope and its exploitation by GALILEO began an era of astronomical discovery that completely destroyed the picture formulated by the ancients. Building on the mathematical and physical discoveries of the 17th century, Isaac NEWTON produced a synthesis of ideas which explained earthly mechanics and planetary motions, based on his three laws of motion and his inverse-square law of gravitational attraction.

The last few centuries show an ever-increasing number of astronomical discoveries. As early as the 17th century, Ole RÖMER showed that the speed of light is finite. Investigations of the heavens beyond the Milky Way began in the 1700s and are principally associated with William HERSCHEL. He also began the trend towards building telescopes of greater and greater power, with which the Universe could be explored to ever-increasing distances.

The 19th century saw the development of astrophysics as the centre of astronomical interest: astronomers became interested in the physics of celestial phenomena and the nature of celestial bodies, rather than just describing their positions and motions. The development of stellar SPECTROSCOPY and its use by William HUGGINS and others led the way to understanding the nature of the many nebulous objects in the sky. ASTROPHOTOGRAPHY emerged as a sensitive and reliable technique that allowed sky maps to be prepared in an objective and repeatable way. Sky SURVEYS began to be made that showed up spiral nebulae and variable phenomena that had been beyond the reach of purely visual techniques.

At the start of the 20th century were two astounding discoveries that revolutionized physics and astronomy:

Max PLANCK's realization that energy is quantized (1901), and Albert EINSTEIN's theory of special relativity (1905). They were the heralds of what was to come. Observational astronomy has grown exponentially ever since. The giant telescopes in excellent climates promoted by George Ellery HALE soon led to the discovery by Edwin HUBBLE that the spiral nebulae are external galaxies and that the Universe is expanding. The morphology of our own Galaxy began to be understood.

The great outburst of technological development associated with World War II led to the development of sensitive radio and infrared detectors that allowed the Universe to be investigated at wavelengths where new phenomena such as pulsars and quasars were discovered. In the late 1950s space probes had been developed that could take X-ray and gamma-ray detectors above the Earth's atmosphere and open yet further vistas to astrophysics. Even visible-light observations have been improved by the development of CCD detectors many times more sensitive than photographic plates.

The early 21st century is characterized by a new generation of giant telescopes and a vigorous investigation of conditions near the time when the Universe was formed. Massive data-handling techniques are allowing surveys of unprecedented detail to be made, searching for new phenomena hitherto unobserved or unobservable because of their faintness or rarity. *See also* AMATEUR ASTRONOMY, HISTORY OF; INFINITE UNIVERSE, IDEA OF

astrophotography Recording images of celestial objects on traditional photographic emulsion (film), as opposed to electronic means. From the late 19th century until the last two decades of the 20th, almost all professional astronomical observations were made using photography. From its introduction in 1839, it was 50 years before photography became accepted as a legitimate research tool by professional astronomers, and another quarter of a century before improvements to equipment and plates had made it vital to the progress of astronomical research.

What the human eye cannot see in a fraction of a second it will never see, whereas with both photography and electronic imaging light is accumulated over a period of time, revealing objects that are invisible even through the largest telescope, and creating a permanent record that is not affected by the observer's imagination or accuracy. However, early astrophotography suffered from insufficiently accurate TELESCOPE DRIVES, lack of sensitivity ('speed') in film, and film's inability to record light equally well across the whole visible spectrum – for many years it was 'blind' to red.

Despite these drawbacks, the new technology was put to immediate use, mainly by enthusiastic amateurs. In this first form of practicable photography, a silvered copper plate sensitized with iodine vapour was exposed, developed with mercury vapour and fixed with sodium hyposulphite to produce a Daguerrotype, named after Louis-Jacques-Mandé Daguerre (1789–1851), who perfected the process. The Moon was an obvious (because bright) target, and the first successful Daguerreotype of it is attributed to John DRAPER (1840 March); the exposure time was 20 minutes, and the image showed only the most basic detail. Hippolyte FIZEAU and Léon FOUCAULT obtained a successful Daguerreotype of the Sun in 1845 April. In 1850 W.C. BOND obtained excellent images of the Moon and the first star images, of Vega and Castor.

With the replacement of Daguerrotypes by the faster and more practical wet plate collodion technique in the early 1850s came a more systematic approach to astrophotography. John HERSCHEL proposed using photography on a daily basis to record sunspot activity, and this 'solar patrol' was taken up and directed by Warren DE LA RUE. This trail-blazing work, which continued until 1872, was continued and developed by Walter MAUNDER and Jules JANSSEN. De La Rue subsequently obtained excellent images of the 1860 July 18 total solar eclipse, which proved that the prominences were solar and not lunar phenomena. In the United States, G.P. BOND pursued stellar photography with improved equipment and considerable success.

An improvement came in the 1870s with the introduction of the more convenient dry photographic plates, and talented individuals such as Lewis RUTHERFORD and Henry DRAPER in the United States, and subsequently Andrew COMMON and Isaac Roberts (1829–1904) in Britain, pushed forward the frontiers of astrophotography. In 1880 Draper succeeded in taking a photograph of the Orion Nebula (M42), and a few years later Common and Roberts obtained excellent photographs of M42. There were also significant advances in photographic spectroscopy, by Draper, William HUGGINS, E.C. PICKERING and others.

The 1880s marked photography's breakthrough in astronomy. In 1882 David GILL obtained the first completely successful image of a comet. He was impressed by the huge numbers of stars on the images, accumulated in exposures lasting up to 100 minutes. With Jacobus KAPTEYN, he set out to construct a photographic star map of the southern sky; the resulting catalogue contained the magnitudes and positions of over 450,000 stars. Equally enthusiastic were the HENRY brothers at Paris Observatory, who helped to instigate the international project to prepare a major photographic sky chart that became known as the *CARTE DU CIEL*.

With the new century came a desire to discern the true nature of astronomical bodies by photographing their spectra. Increasing telescope apertures enabled greater detail to be resolved in the spiral nebulae by, for example, James KEELER and George RITCHEY. Edwin HUBBLE and others, using the 100-inch (2.5-m) Hooker Telescope at Mount Wilson Observatory, showed these objects to be other galaxies far beyond the Milky Way, and spectroscopy showed them to be receding from Earth, in an expanding Universe.

As the 20th century progressed, photographic materials became increasingly sensitive and pushed the boundaries of the known Universe ever outward. But by the end of the century the technique that had contributed so much to astronomy had been dropped far more rapidly than it was ever accepted, in favour of electronic imaging using CCDs. Although astrophotography was well established and initially cheaper, their greater light sensitivity, linear response to light and digital output saw CCDs rapidly sweep the board. As CCDs have become larger and cheaper, electronic imaging has almost completely replaced astrophotography

Astrophotography at major observatories may be a thing of the past, but many amateur astronomers still use conventional film, despite the availability of CCDs and digital cameras, partly because of its simplicity but also because it continues to yield results of superlative quality at low cost. Some advanced amateurs use film for the exposures themselves, but then scan the images for manipulation on a computer, borrowing one of the main advantages of digital photography.

For astrophotography and CCD imaging alike, the telescope or camera must be driven so as to follow objects as they are carried across the sky by the Earth's rotation. The long exposure times required by photographic film, often over an hour, necessitate a means of monitoring the accuracy of tracking and making corrections to the drive rate: either a separate GUIDE TELESCOPE or an OFF-AXIS GUIDING system. In simple systems the astrophotographer looks through an eyepiece equipped with cross-wires and adjusts the drive rate manually, while more advanced systems use an AUTOGUIDER, which makes automatic corrections if the chosen guide star starts to drift out of the field of view. Autoguiders on amateur instruments incorporate a CCD whose purpose is to monitor the guide star – the imaging task may be left to the superior data acquisition of photography.

Virtually every type of film has its use in amateur astrophotography. An important characteristic of film is its graininess: the surface is covered with minute light-sensitive grains, which are visible under high magnification after exposure. Fine-grained films provide excellent sharpness and contrast, but are relatively insensitive to light. More sensitive films can record images with shorter exposure times, or fainter objects in a given exposure time, but have the disadvantage of coarse grain. Negative films generally record a wider range of brightness than transparency films, but require more effort afterwards to get a satisfactory print. This drawback is now largely overcome with the ready availability of film scanners, computer manipulation and high-quality colour printers. HYPERSENSITIZATION is an advanced technique to increase the sensitivity and speed of film for long exposures.

As CCDs become cheaper and capable of recording images of ever higher resolution, even amateur astrophotography is declining in importance. It is rare to find advanced amateurs using film for planetary imaging or for supernova patrolling, for example. It remains ideal for aurora and total eclipse photography, and for general constellation shots. Many deep-sky photographs, particularly of wide fields, remain firmly in the province of film. It is not hard to foresee the day when even ordinary film is hard to find, but until then, photography will still have a place in amateur astronomy.

Astrophysical Journal Foremost research journal for theoretical and observational developments in astronomy and astrophysics. Dating from 1895, when it was launched by George Ellery HALE and James KEELER, the '*ApJ*' reported many of the classic discoveries of the 20th century. In 1953 *The Astrophysical Journal Supplement Series* was introduced to allow the publication of extensive astronomical data papers in support of the main journal.

astrophysics Branch of astronomy that uses the laws of physics to understand astronomical objects and the processes occurring within them. The terms 'astronomy' and 'astrophysics' are however often used as synonyms. Astrophysics effectively dates from the first applications of the SPECTROSCOPE to study astronomical objects. Sir William HUGGINS' identification of some of the chemical elements present in the Sun and stars in the 1860s was the first major result of astrophysics. Today, the use of computers for data processing, analysis and for modelling objects and processes is an integral part of almost all aspects of astrophysics.

SPECTROSCOPY is the most powerful technique available in astrophysics. Individual atoms, ions or molecules emit or absorb light at characteristic frequencies (or 'lines', from their appearance in a spectroscope) and can thus be identified and their quantity assessed. Information on the physical and chemical state of an object can be obtained from an analysis of the way its SPECTRUM is modified by the collective environmental conditions within the object. Line-of-sight velocities of the whole object or regions within it are indicated by the DOPPLER SHIFT. The most intensively studied spectral region is the ground-accessible 'optical' region (the NEAR-ULTRAVIOLET to the NEAR-INFRARED), but much information now comes from observations in the RADIO, MICROWAVE, FAR-ULTRAVIOLET, X-RAY and GAMMA-RAY regions.

Stellar spectra characteristically show a continuous bright background that by its colour distribution approximately indicates the 'surface' temperature (but is made up of radiation contributed in some measure from throughout the whole of the stellar atmosphere). Several different processes produce these continua. In cooler stars (like the Sun) the interaction chiefly involves electrons and hydrogen atoms; in hotter stars, electrons and protons; and in the hottest stars, electrons alone. The dark spectral lines superimposed on the continuum, which are due to atoms, ions and molecules, strictly do not arise from absorption by a cooler outer layer as a simple visualization suggests, but in a complex depth-dependent manner in which radiation dominantly escapes from outer and therefore cooler (darker) regions of the atmosphere near the centre of an absorption line, or deeper, hotter (brighter) regions for the continuum. The fact that the lines appear dark indicates that the outer layers are cooler, and indeed the temperature generally is expected to decrease outwards from the centre of any star. For the Sun and certain other stars some bright lines are also seen, particularly in the far-ultraviolet region, indicating that in these the temperature is rising again in extended, extremely tenuous layers of the outer atmosphere. The strengths, widths and polarization of spectral lines give information on the chemical composition, ionization state, temperature structure, density structure, surface gravity, magnetic field strength, amount of turbulent motion within the atmosphere of the star and stellar rotation.

A striking result of astronomical spectroscopy is the uniformity of chemical composition throughout the Universe (*see* ASTROCHEMISTRY), extending even to the most distant galaxies. Although the atomic constituents of the Universe are everywhere the same, the proportions of these elements are not identical in all astronomical objects. Measurements of these differences give information on the nuclear processes responsible for the formation of the elements and the evolutionary processes at work. A star is kept in a steady state for a long time by the balance of the inward pull of gravity and the opposing force arising from the increase of pressure with depth that is maintained by a very high central temperature. For a star like the Sun (mass of 2×10^{30} kg, radius 700,000 km/435,000 mi) the central temperature is about 15,600,000 K. The energy from such hot matter works its way tortuously outwards through the star until it is radiated away from the atmospheric region near its surface. This process yields a relationship between the masses and luminosities of stars, with the luminosity increasing very dramatically with mass. Thus the luminosities of stars within the relatively small range of masses between 0.1 and 10 times that of the Sun differ by a factor of over a million. The high rate of loss of energy requires a powerful energy-producing process within the star, identified as thermonuclear fusion of light atomic nuclei (mainly hydrogen nuclei) into heavier ones (mainly helium nuclei). The transformation of hydrogen into helium in the case of the Sun – a small star – makes enough energy available to enable it to shine as it does today for a total of about 10^{10} years. In contrast, the lifetime of massive, very luminous stars, which burn up their hydrogen very rapidly, can be as short as a million years. During this burning process the core contracts, and it eventually becomes hot and dense enough for helium itself to ignite and to form carbon and oxygen. Although the contraction of the core is thus halted, the star then has a rather complicated structure, with two nuclear burning zones, and may become unstable. Later stages in the evolutionary cycle present many possibilities depending principally on the initial mass of the star. It may become a WHITE DWARF, such as the companion of Sirius, which is a very dense, hot star with about the mass of the Sun but a radius nearer that of the Earth. The immense pressures built up in such stars result in the matter forming them becoming DEGENERATE.

More massive stars than the Sun progress to advanced stages of NUCLEOSYNTHESIS, experiencing ignition of successively heavier nuclei at their cores. In some rare cases much of the outer layers of the star may be violently ejected in a SUPERNOVA outburst in which heavy elements (some synthesized in the explosion) are scattered widely into the general interstellar medium. Out of this enriched material later generations of stars are formed. The residues of these events may be NEUTRON STARS, which can be observed as PULSARS, and represent a still higher state of compression than white dwarfs, with electrons and protons crushed together to form a degenerate neutron gas. On the other hand, a star may go on contracting until it becomes a BLACK HOLE, whose violent accretion of nearby material makes its influence observable by strong emission in the X-ray region.

A great deal of theoretical work has gone into explaining how the distribution of radiation over the spectrum of a star depends on the structure of its outermost layers, and, more broadly, the processes occurring during the life cycles of stars. The latter includes the stars' formation from collapsed regions of the interstellar medium, their active burning phases, and their deaths, which can range from being an event in which a star mildly fades away to being a violent explosion that at its maximum brightness can appear as brilliant as a whole galaxy.

Such applications of physics to astronomical problems in general are progressing strongly on most fronts and much success has been achieved. Currently a great deal of attention is being directed towards the processes that occurred during the earliest phases of the formation of the Universe, very close to the time of the Big Bang. But here astronomers and physicists are venturing out hand in hand, for not only is our knowledge of the behaviour of the Universe incomplete, but so too is our knowledge of physics itself. New physical theories unifying all forces of nature are needed before we can begin to understand the most fundamental properties of the evolving Universe and, indeed, the origin of matter itself.

Asuka Japan's fourth cosmic X-ray astronomy satellite. It is called the Advanced Satellite for Cosmology and Astrophysics (ASCA), also known as Astro D. It was launched in 1993 February, equipped with four large-area telescopes. ASCA was the first X-ray astronomy mission to combine imaging capability with a broad-wavelength bandwidth, good spectral resolution and large-area coverage, as well as the first to carry charge-coupled devices (CCDs) for X-RAY ASTRONOMY.

Atacama Large Millimetre Array (ALMA) Millimetre-wavelength telescope planned by a major international consortium that includes the NATIONAL RADIO ASTRONOMY OBSERVATORY (USA), the PARTICLE PHYSICS AND ASTRONOMY RESEARCH COUNCIL (UK) and the EUROPEAN SOUTHERN OBSERVATORY, planned for completion in 2010. The instrument, at an elevation of 5000 m (16,400 ft) at Llano de Chajnantor in the Chilean Andes, will consist of at least 64 antennae 12 m (39 ft) in diameter, allowing resolution to a scale of 0″.01. The high altitude will give ALMA access to wavebands between 350 μm and 10 mm, otherwise accessible only from space.

ataxite *See* IRON METEORITE

ATC Abbreviation of ASTRONOMICAL TECHNOLOGY CENTRE

Aten asteroid Any member of a class of asteroids that, like the APOLLO ASTEROIDS, cross the orbit of the Earth, but that are distinguished by having orbital periods of less than one year. In consequence they spend most of the time on the sunward side of our planet. The archetype is (2062) Aten, discovered in 1976, although an asteroid of this type was also discovered in 1954 and subsequently lost. By late 2001 over 120 Aten-type asteroids were catalogued. Characteristics of the following are listed in in the NEAR-EARTH ASTEROID table: 1954 XA, (2062) Aten, (2100) Ra-Shalom and (2340) Hathor. *See also* INTRATERRESTRIAL ASTEROID

Athena Small US satellite launch vehicle powered by solid propellants. It was originally called the Lockheed Martin Launch Vehicle (LMLV). There are two Athena models, with two and three stages, respectively, capable of placing satellites weighing a maximum of 2 tonnes into low Earth orbit. The booster has flown six times since 1995, with four successful launches. The launcher has now been discontinued.

Atlantis One of NASA's SPACE SHUTTLE orbiters. *Atlantis* first flew in 1985 October.

Atlas Lunar crater (47°N 44°E), 88 km (55 mi) in diameter, located to the east of Mare Frigoris. It forms a notable pair with the smaller crater Hercules to the west. Atlas' walls vary in altitude between 2750 and 3500 m (9000–11,500 ft). The floor of Atlas is uneven and has been observed to 'glitter'. It contains a group of central hills, which resemble a ruined crater ring, many craterlets and a very noticeable cleft on the east side. To the north is O'Kell, a large, heavily eroded, ancient ring.

Atlas One of the small inner satellites of SATURN, discovered in 1980 by Richard Terrile during the VOYAGER missions. It is spheroidal in shape, measuring about 38 × 34 × 28 km (24 × 21 × 17 mi). Atlas has a near-circular equatorial orbit just outside Saturn's A RING, at a distance of 137,700 km (85,600 mi) from the planet's centre, where its orbital period is 0.602 days. It appears to act as a SHEPHERD MOON to the outer rim of the A ring.

Atlas One of the USA's primary satellite launchers. It began its career as the country's first intercontinental ballistic missile in 1957. The Atlas was equipped with a Centaur upper stage in 1962 and, since that failed maiden flight, has operated in a number of Atlas Centaur configurations, making over 100 flights. This configuration later became known as the Atlas 2, 2A and 2AS, which by 2001 had together made almost 50 flights. The new Atlas 3, which first flew in 1999, and the Atlas 5 fleets (there is no Atlas 4) will use a Russian first-stage engine. The Atlas 2 fleets will be retired in 2002, leaving the Atlas 3 and 5 offering launches to geostationary transfer orbit within the 4- to 8.6-tonne range. The Atlas 5 has been built primarily as a US Air Force launcher, under the Evolved Expendable Launch Vehicle programme.

atlas, star Collection of charts on which are plotted the positions of stars and, usually, numbers of deep-sky objects, over the whole of the sky or between certain limits of declination, down to a certain LIMITING MAGNITUDE. The classic star atlases of the 17th and 18th centuries in particular are now appreciated for their beauty. The 19th century saw a division appear between atlases produced for professional and for amateur astronomers, and the emergence of the photographic atlases. Today's professionals have little

▼ **Atlas** An Atlas 2 is launched in 1966. The rocket is carrying an Agena target vehicle towards a rendezvous with Gemini 11.

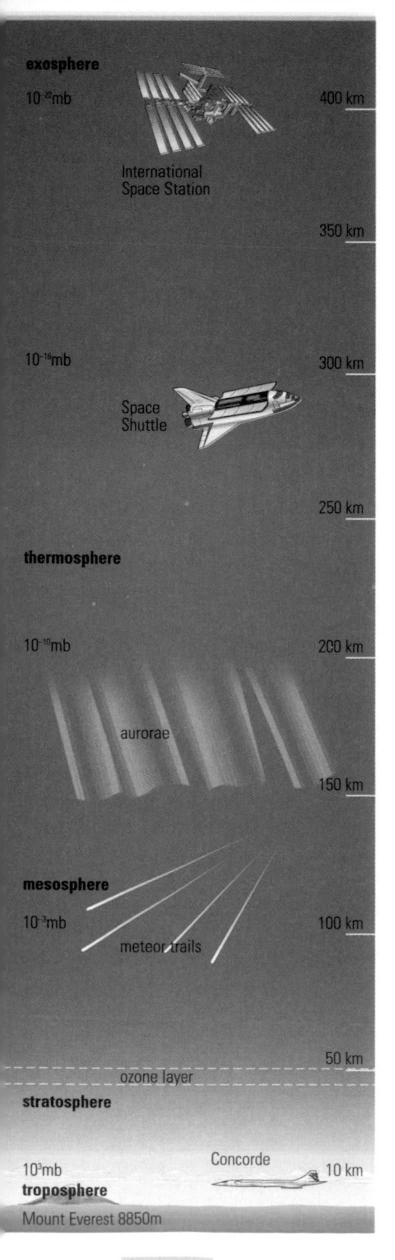

▼ **atmosphere** The Earth's atmosphere consists of the troposphere, extending from ground level to a height of between 8 and 17 km (5–10 mi); the stratosphere extends up to about 50 km (30 mi); the mesosphere, between 50 and about 80 km (50 mi); the thermosphere from about 80 up to 200 km (125 mi); beyond this lies the exosphere.

use for printed atlases, and amateurs are beginning to rely more on virtual atlases than on printed ones.

Until the 17th century, accuracy of position in atlases was often the victim of artistic licence, stars being placed where they looked best on intricately drawn constellation figures. The first truly modern atlas was Johann BAYER's *Uranometria* of 1603, which plotted accurate positions from Tycho BRAHE's star catalogue, distinguished between different magnitudes, and introduced the practice of labelling stars with Greek letters (*see* BAYER LETTERS); the allegorical constellation figures were depicted faintly to the stars. Other notable atlases, based on observations by their originators, were those of Johannes HEVELIUS and John FLAMSTEED. The last great general all-sky 'artistic' star atlas was Johann BODE's *Uranographia* (1801), which showed over 17,000 stars and nebulae on huge pages with beautifully engraved constellation figures.

The 19th and 20th centuries saw the introduction of photographic atlases for the professional, such as the *CARTE DU CIEL* and the PALOMAR OBSERVATORY SKY SURVEY, and the appearance of atlases for amateur observers. The most venerable is perhaps *Norton's Star Atlas*, by Arthur Philip Norton (1876–1955), first published in 1919 and with its 20th edition planned for 2002. The lavish two-volume *Millennium Star Atlas* (R.W. Sinnott and M. Perryman, 1997), showing stars to magnitude 11, may be the last such large printed star atlas for the amateur observer, now that virtual atlases are readily available (*see* SOFTWARE).

atmosphere Gaseous envelope that surrounds a planet, satellite or star. The characteristics of a body that determine its ability to maintain an atmosphere are the temperature of the outer layers and the ESCAPE VELOCITY, which is dependent on the body's mass. Small bodies, such as the Moon, Mercury and the satellites of the planets in our Solar System (apart from TITAN and TRITON), do not have any appreciable permanent atmosphere. The escape velocity for these bodies is sufficiently low for it to be easily exceeded by gas molecules travelling with the appropriate thermal speeds for their masses and temperature. The speed of a gas molecule increases with temperature and decreases with the molecular weight. Consequently, the lighter molecules, such as hydrogen, helium, methane and ammonia, escape more readily to space than the heavier species, such as nitrogen, oxygen and carbon dioxide. MERCURY has a transient, tenuous atmosphere, consisting of material that it captures from the SOLAR WIND, which it retains for a short period. In the cold outer reaches of the Solar System, PLUTO and Triton have potential atmospheric gases frozen out on their surfaces. When Pluto is close to perihelion (as in the late 20th century), the gases sublime to produce a tenuous atmosphere of methane and ammonia. It is estimated that these gases will refreeze on to the surface by about the year 2020.

The primary atmospheres of the bodies in the Solar System originated from the gaseous material in the SOLAR NEBULA. The lighter gases were lost from many of these objects, particularly the terrestrial planets VENUS, EARTH and MARS, which now have secondary atmospheres formed from internal processes such as volcanic eruptions. The main constituent of the atmospheres of Venus and Mars is carbon dioxide (CO_2); the main constituents of the Earth's atmosphere are nitrogen and oxygen (approximately 78% and 21% by volume, respectively). Free oxygen in the Earth's atmosphere is uniquely abundant and is believed to have accumulated as a result of photosynthesis in algae and plants. Earth's is thus regarded by some authorities as a tertiary atmosphere. The only other significant nitrogen atmosphere found in the Solar System is on Titan, the largest satellite of Saturn, although nitrogen gas produces the active geyser-like plumes observed on Triton. Only the giant outer planets, JUPITER, SATURN, URANUS and NEPTUNE, still retain their primordial atmospheres of mainly hydrogen and helium. However, these major planets are also surrounded by envelopes of escaped hydrogen. Titan, too, is associated with a torus of neutral hydrogen atoms, which have escaped from the satellite's upper atmosphere and populate the satellite's orbit. It is also possible for a body to gain a temporary atmosphere through a collision with an object containing frozen gases, such as a cometary nucleus.

The gaseous layers of an atmosphere are divided into a series of regions organized on the basis of the variation of temperature with altitude. For worlds with definite solid surfaces, such as the Earth, Venus, Mars and Titan, the levels start from the ground and relate to existing observations. The outer planets are huge gaseous bodies with no observable solid surface, and with atmospheres that have been investigated to just beneath the cloud tops. Consequently, the characteristics of the lowest layers are based upon theoretical models and the levels usually start from a reference level at which the pressure is 1 bar. The names given to the various structured layers of the terrestrial atmosphere are also applied to all other planetary atmospheres. The precise variation of the atmospheric temperature with altitude depends on the composition of the atmosphere and the subsequent solar (and internal) heating and long-wave cooling at the various levels.

The lowest layer is the TROPOSPHERE, which on the Earth extends from sea level, where the average pressure is 1 bar, to the TROPOPAUSE, at an altitude of between 5–8 km (3–5 mi) at the poles and 14–18 km (9–11 mi) at the equator. This region contains three-quarters of the mass of the atmosphere; it is the meteorological layer, containing the cloud and weather systems. The troposphere is heated from the ground and the temperature decreases with height to the tropopause, where the temperature reaches a minimum of approximately 218 K at the poles and 193 K over the equator. Above the tropopause, the temperature increases with height through the STRATOSPHERE because the heating from ozone (0_3) absorption dominates the long-wave cooling by CO_2. The temperature increases to about 273 K at an altitude of 50 km (31 mi), which marks the stratopause. In the Earth's atmosphere, the region between 15 and 50 km (9 and 31 mi), where the ozone is situated, is often called the ozonosphere. Beyond the stratopause the temperature again decreases with altitude through the MESOSPHERE to the atmospheric minimum of 110–173 K at the mesopause, which lies at an altitude of 86 or 100 km (53 or 62 mi) at different seasons. The atmosphere at these levels is very tenuous and is therefore sensitive to the heating from the Sun. Consequently, this region and the further layers of the atmosphere have a significant diurnal temperature cycle. Beyond the mesopause, we enter the THERMOSPHERE, where the temperature increases with height until it meets interplanetary space. The heating that produces this layered structure is primarily created by the absorption of far-ultraviolet solar radiation by oxygen and nitrogen in the atmosphere. Solar X-rays also penetrate through to the mesosphere and the upper atmospheric layers, which are therefore sensitive to changes in the solar radiation and atmospheric chemistry. The solar ultraviolet radiation photo-ionizes the atmospheric constituents in these outer layers to produce ionized atoms and molecules; this results in the formation of the IONOSPHERE at an altitude of about 60–500 km (40–310 mi). This region is also the domain of METEORS and AURORAE, which regularly occur in the upper atmosphere. Above 200–700 km (120–430 mi) is the EXOSPHERE, from which the atmospheric molecules escape into space.

The individual planctary atmospheres each possess some or all of these basic structures but with specific variations as a consequence of their size, distance from the Sun, and atmospheric chemistry. The massive CO_2 atmosphere of Venus has a deep troposphere, which extends from the surface to an altitude of about 100 km (62 mi). About 90% of the volume of the atmosphere is contained in the region between the surface and 28 km

(17 mi). Venus has no stratosphere or mesosphere but does have a thermosphere. Its planetary atmosphere is, therefore, quite different from that of the Earth.

By terrestrial standards, the Martian atmosphere is very thin, since the surface pressure is only 6.2 millibars. The atmosphere does have some similarity in the vertical layering, however, with the presence of a troposphere, stratosphere and thermosphere. During the global dust storms, the structure of the Martian troposphere may change dramatically to an isothermal layer (that is, one with a constant temperature throughout) or may even display an inversion. However, there is much less ozone in the Martian atmosphere than in that of Earth, so there is no reversal in the temperature gradient as there is in the terrestrial stratosphere. Lower temperatures are found in the Martian thermosphere as a consequence of the CO_2 cooling effect.

The basic structure of the atmospheres of the outer planets resembles that of the Earth. There is a troposphere, which extends to unknown depths beneath the clouds, and a well-defined stratosphere, created by the heating due to absorption of solar radiation by methane (CH_4). Knowledge of the higher levels of the atmospheres is still incomplete. Titan has a well-defined atmospheric structure, comprising a troposphere, stratosphere and thermosphere, created more by the complex aerosol chemistry than by gaseous composition alone.

These atmospheric structures vary on a local scale as a consequence of the planetary weather systems. The atmosphere of the Earth receives its energy from the Sun. The atmosphere (including clouds) reflects 37% of the incident radiation back to space and absorbs the remainder, which becomes redistributed so that 48% of the incident radiation actually reaches the Earth's surface, which then reflects 5% of the total directly back to space. The surface reradiates most of its absorbed energy back into the atmosphere as long-wave radiation, with a small amount (8%) passing directly to space. The heated atmosphere reradiates some of the long-wave radiation to the surface while a portion (50%) is ultimately lost to space. The atmosphere behaves like a giant heat engine, trying to balance the absorbed solar energy with the long-wave radiation emitted to space: the small imbalances that are always present give rise to the weather, and any longer-term variations produce changes in climate. In the Earth's atmosphere, clouds, which cover approximately 50% of the surface, are a key factor in weather and climate because a small change in their geographical and vertical distribution will have a profound effect on the environment.

Exploration of the Solar System has now provided a unique opportunity to examine the weather systems of all the planetary atmospheres. They have very different properties: Venus' atmosphere is slowly rotating and totally covered with cloud; Mars' thin atmosphere is strongly affected by the local topography and annual global dust storms; and the outer planets Jupiter, Saturn and Neptune have huge, rapidly rotating gaseous envelopes and meteorologies that are largely driven by the planets' internal sources of energy. These planetary atmospheres of the Solar System are natural laboratories for investigating geophysical fluid dynamics.

atmospheric extinction Reduction in the brightness of light from astronomical objects when it passes through the Earth's atmosphere. The atmosphere attenuates the light from these objects so that they appear fainter when seen from the surface than they would outside the atmosphere. This is caused by absorption and scattering by gas molecules, dust and water vapour in the atmosphere. The amount of extinction is variable and can be determined by measuring the brightness of standard stars photometrically. Atmospheric extinction is most pronounced close to the horizon, where astronomical bodies are observed through a greater volume of air.

atmospheric pressure Force exerted by the gas forming an atmosphere on a unit area. The units are pascals (Pa), with 1 Pa = 1 N/m^2. Commonly encountered non-SI units are the bar or millibar (1 bar = 10^5 Pa) and the atmosphere (1 Atm = 101,325 Pa). Sea-level pressure on the Earth is around 10^5 Pa, surface pressures on Venus and Mars are 9.2×10^6 Pa and 620 Pa respectively. Atmospheric pressure represents the weight of a vertical column of the atmosphere whose cross-section is 1 m^2.

ATMOSPHERIC REFRACTION

Altitude (a) in degrees above horizon	Refraction (R) in arcminutes
15°	3.69
20°	2.79
25°	2.19
30°	1.79
35°	1.49
40°	1.29
45°	1.09
50°	0.89
55°	0.79
60°	0.69
65°	0.59
70°	0.49
80°	0.29
90°	0.0

Atmospheric refraction at a pressure of one atmosphere (1013.25 millibars) and a temperature of 283 K

atmospheric refraction Small increase in the apparent altitude of a celestial object, as viewed by an observer on the surface of the Earth, caused by light from the object changing direction as it passes through the Earth's atmosphere. When light passes from one medium to another it is bent, or refracted, and the same is true when light from a star passes from the vacuum of space into the Earth's atmosphere. The result is to make the object appear at a higher altitude than is really the case, and the nearer it is to the horizon, the more pronounced the effect, even causing objects below the horizon to appear visible. The degree by which the light is refracted is also dependent on atmospheric pressure and temperature. All astronomical observations have to be corrected for atmospheric refraction to obtain true, as opposed to apparent, positions.

ATNF Abbreviation of AUSTRALIA TELESCOPE NATIONAL FACILITY

atom Smallest part of an ELEMENT. Atoms have a NUCLEUS containing PROTONS and NEUTRONS, together with a surrounding cloud of ELECTRONS (*see* ATOMIC STRUCTURE). The nucleus is positively charged, and contains almost all the mass of the atom. The mass of the atom is called the atomic mass, symbol A, and is given approximately in ATOMIC MASS UNITS by the total number of protons and neutrons in the nucleus. The number of protons in the nucleus is the atomic number, symbol Z, and it determines the element to which the atom belongs. In the normal atom there are equal numbers of protons and electrons, so that the atom as a whole is electrically neutral. If an atom loses or gains one or more electrons it becomes an ION. The number of neutrons in the nucleus may vary and results in the different ISOTOPES of the element.

There are a number of variations in the nomenclature for atoms. Amongst the commonest are the chemical symbol plus atomic number and mass as superscripts and subscripts, for example: $Fe^{26}_{55.847}$, $^{26}Fe_{55.847}$ or $_{26}Fe^{55.847}$. Here, the atomic mass is the average for the element as it is found on Earth with the isotopes in their natural abundances. When an individual isotope is symbolized, the atomic mass will be much closer to a whole number. Thus the naturally occurring isotopes of iron on Earth are $Fe^{26}_{53.9396}$ (5.8%), $Fe^{26}_{55.9349}$ (91.7%) and

$Fe^{26}_{56.9354}$ (2.2%). An isotope may also be written down as the element name or symbol followed by the number of nuclear particles, for example iron-56 or Fe-56 or in the previous notation Fe^{26}_{56}.

atomic mass unit (amu) Unit used for the masses of ATOMS. It is defined as one-twelfth of the mass of the carbon-12 isotope. It is equal to 1.66033×10^{-27} kg.

atomic structure The gross structure of an ATOM comprises a nucleus containing PROTONS and NEUTRONS, together with a surrounding cloud of ELECTRONS. The nucleus has a density of about 2.3×10^{17} kg/m^3 and is a few times 10^{-15} m across. The atom as a whole has a size a few times 10^{-11} m (note that neither the size of the nucleus nor of the atom can be precisely defined). Although there is structure inside the nucleus, the term 'atomic structure' is usually taken to mean that of the electrons.

Quantum mechanics assigns a probability of existence to each electron at various points around the nucleus, resulting in electron orbitals. While the quantum mechanical description may be needed to describe some aspects of the behaviour of atoms, the conceptually simpler Bohr–Sommerfeld model is adequate for many purposes. This model was proposed in 1913 by the Danish physicist Niels Bohr (1885–1962). It treats the subatomic particles as though they are tiny billiard balls and the electrons move in circular orbits around the central nucleus. Different orbits correspond to different energies for the electrons, and only certain orbits and hence electron energies are permitted (that is, the orbits are quantized). Arnold Sommerfeld (1868–1951, born in East Prussia at Königsberg, now Kaliningrad) modified the theory in 1916 by adding elliptical orbits. When an electron changes from one orbit to another a PHOTON is emitted or absorbed, the wavelength of which corresponds to the energy difference between the two levels. Since the orbits are fixed, so too are the energy differences, and thus the photons are emitted or absorbed at only specific wavelengths, giving rise to the lines observed in a SPECTRUM.

The electron energies are determined by quantum numbers; the principal quantum number, n, the azimuthal quantum number, l, the magnetic quantum number, m_l, and the spin, m_s. These numbers can take values: $n = 1, 2, 3, 4, \ldots \infty$; $l = 0, 1, 2, \ldots n - 1$; $m_l = 0, \pm1, \pm2, \ldots \pm l$; $m_s = \pm\frac{1}{2}$. The value of n determines the shell occupied by the electron, with $n = 1$ corresponding to the K shell, $n = 2$ to the L shell, $n = 3$ to the M shell, and so on. The value of l determines the ellipticity of the orbit, $l = n - 1$ gives a circular orbit; smaller values of l correspond to increasingly more elliptical orbits.

The PAULI EXCLUSION PRINCIPLE requires that no two electrons within an atom can have the same set of quantum numbers, and its operation determines the electron structure of the atom and so the properties of the chemical elements. For $n = 1$, the other quantum numbers must be $l = 0$, $m_l = 0$, and $m_s = \pm\frac{1}{2}$. The K shell can thus contain at most two electrons, with quantum numbers, $n = 1$, $l = 0$, $m_l = 0$, $m_s = +\frac{1}{2}$ and $n = 1$, $l = 0$, $m_l = 0$, and $m_s = -\frac{1}{2}$. Thus we have helium with two electrons in the K shell and a nucleus with two protons and (normally) two neutrons. Hydrogen has just one electron in the K shell. But the third element, lithium, has to have one electron in the L shell as well as two in the K shell, and so on. The maximum number of electrons in each shell is $2n^2$, so the L shell is filled for the tenth element, neon, and the eleventh, sodium, has two electrons in the K shell, eight in the L shell, and one in the M shell. Atoms with completed shells are chemically very unreactive, giving the noble gases, helium, neon, argon, krypton, xenon and radon.

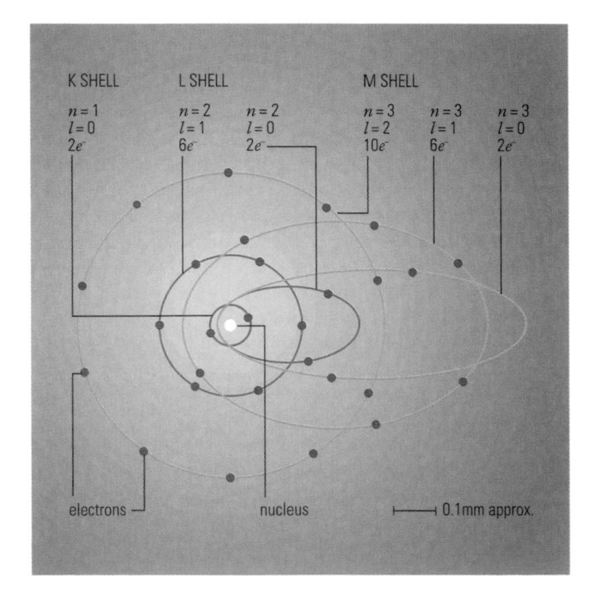

▶ **atomic structure** The Bohr–Sommerfeld model of atomic structure visualizes electrons orbiting around the nucleus in discrete shells. The first three shells are: K (blue), which can contain a maximum of two electrons; L (red), which can contain a maximum of eight electrons; and M (yellow), which has a maximum of eighteen.

atomic time System of accurately measuring intervals of time based on the transitions between energy levels of the caesium-133 atom. When an atom of caesium changes from a lower-energy level to a higher one, it absorbs radiation of a very precise frequency, namely 9,192,631,770 Hz. This varies by less than one part in 10 billion and has therefore been used to define the basic unit of time, the SI second, which is used in international timekeeping. In 1967 the SI second was defined as being 'the duration of 9,192,631,770 periods of the radiation corresponding to the transition between two hyperfine levels of the ground state of the caesium-133 atom'.

Prior to this, the second was defined in terms of the mean solar day and astronomical observations provided a definitive measure of time to which clocks were accordingly adjusted to keep them in step. The Earth, however, is not a good timekeeper; its rotation rate is irregular and slowing due to the effects of tidal braking. With the advent of more accurate quartz clocks, followed in 1955 by the caesium-beam atomic clock, it therefore became necessary to re-define the standard unit for measuring time. Today's atomic clocks are now accurate to one second within a few thousand years and are used to form INTERNATIONAL ATOMIC TIME (TAI). *See also* TIMEKEEPING

AU Abbreviation of ASTRONOMICAL UNIT

aubrite (enstatite achondrite) Subgroup of the ACHONDRITE meteorites. Aubrites are highly reduced meteorites with mineralogies and oxygen isotopic compositions similar to those of ENSTATITE CHONDRITES, leading to the suggestion that aubrites might have formed by partial melting of an enstatite chondrite precursor.

Aura Abbreviation of ASSOCIATION OF UNIVERSITIES FOR RESEARCH IN ASTRONOMY

Auriga *See* feature article

aurora Illumination of the night sky, known popularly as the northern and southern lights, produced when electrons accelerated in Earth's MAGNETOSPHERE collide with atomic oxygen and molecular and atomic nitrogen in the upper atmosphere, at altitudes in excess of 80 km (50 mi). These collisions produce excitation, and as the atmospheric particles return to their ground state, they re-emit the excess energy as light at discrete wavelengths; aurorae are often coloured green and red.

Geographically, the aurora is present more or less permanently in two oval regions, encircling either geomagnetic pole. The ovals show a marked offset, being on average 20° from the pole on the dayside, 30° on the nightside. Under normal conditions, the ovals remain comparatively narrow and are confined to high latitudes. Release of stress within the magnetosphere gives rise to occasional brightenings of the ovals, with a westward-travelling surge of increased activity during substorms, of which as many as five may occur each day even in quiescent conditions.

The rather more major disturbances brought about by GEOMAGNETIC STORMS, following the arrival of energetic, highly magnetized plasma thrown from the Sun during CORONAL MASS EJECTION or FLARE events, can dramatically alter the configuration of the auroral ovals. During the biggest disturbances, each oval brightens and broadens, particularly on the nightside, pushing auroral activity towards the equator. During such expansions, the aurora can become visible from lower latitudes, such as those of the British Isles and southern USA. Aurorae at these latitudes are most commonly seen a year or so ahead of sunspot maximum, with a secondary, less intense peak 12–18 months after sunspot maximum.

During a major display at lower latitudes, the aurora may first appear as a structureless glow over the polewards horizon. The glow may rise higher, taking on the form of an arc, with folding giving rise to a ribbon-like structure described as a band. Arcs and bands may be homogeneous, lacking internal structure, or may show long, vertical striations known as rays. Individual rays can appear like searchlight beams, stretching over the horizon. Where a rayed band has a high vertical extent, its movement gives rise to the commonly portrayed 'curtain' effect. If a display is particularly intense, the rays and other features may pass overhead and on into the equatorwards half of the sky. At this stage – which is comparatively rare in displays at lower latitudes – the aurora takes on the form of a corona, with rays appearing to converge on the observer's magnetic zenith as a result of perspective. Corona formation often marks the short-lived peak of an auroral display, following which activity again retreats polewards. In the most major low-latitude storms – perhaps three or four times in each roughly 11-year sunspot cycle – activity may go on all night, with several coronal episodes separated by quieter interludes.

As well as showing occasionally rapid movement, auroral features change in brightness. Some changes are slow (pulsing), others rapid (flaming – waves of brightening sweeping upwards from the horizon). No two displays are ever quite the same, and at lower latitudes the most an observer might see on many occasions are simply the upper parts of a display as a horizon glow towards the pole. It is from this appearance, the 'northern dawn' (a description first used in the 6th century by Gregory of Tours), that the aurora borealis takes its name.

Increased auroral activity is often found in the declining years of the sunspot cycle at times when Earth becomes immersed in persistent high-speed SOLAR WIND streams. These coronal hole-associated aurorae are much less dynamic and extensive than the events that follow coronal mass ejections, and they often recur at intervals of 27 days, roughly equivalent to the Sun's rotation period as seen from Earth. Activity from these displays usually comprises a quiet homogeneous arc, with only occasional rayed outbursts, and often lasts for several successive nights.

Auroral rayed bands may have a vertical extent of several hundred kilometres and a lateral extent covering tens of degrees in longitude. Their base extent in latitude, however, is only a few kilometres, reflecting the narrow field-aligned sheets along which electrons undergo their final acceleration in the near-Earth magnetosphere before impacting on the upper atmosphere.

The aurora shows an emission spectrum that is brightest at its red end. The human eye, however, is most sensitive to the green oxygen 557.7 nm emission, which predominates in the aurora at altitudes of about 100 km (62 mi). Higher up, from about 150–600 km (90–370 mi) altitude, red oxygen emissions are found; auroral rays often show a colour gradation from green at their base to red at the top. At extreme altitudes of 1000 km (600 mi), molecular nitrogen can produce purple emissions under further excitation by solar ultraviolet radiation. During intense displays, the lower border of bands may show red nitrogen emissions.

Ionization produced during auroral activity enhances the ionospheric E layer at 112 km (70 mi) altitude. Radio operators can use the aurora to scatter short-wave signals over longer-than-normal distances to make 'DX' contacts. Doppler shifts allow radar measurements of the particle motions in auroral arcs.

References to the aurora are found in pre-Christian, Greek, Chinese, Japanese and Korean texts. European medieval church records often mention 'battles' in the sky, equating the auroral red emissions with blood; such imagery also appears in Viking chronicles. Treated with due caution, such records provide useful information on solar-terrestrial activity from pre-telescopic times.

Scientific investigations of the aurora began in earnest in the 18th century, with the work of Edmond HALLEY, Jean-Jacques de Mairan (1678–1771) and others. The first modern account of the aurora australis was made by Captain Cook in 1773. The long-suspected connection between solar activity and the aurora was confirmed when a major solar flare was observed in 1859, followed a day or so later by a huge auroral storm. Theoretical work and experiments by Kristian Birkeland (1867–1917) and Carl Störmer (1874–1957) in the early 20th century began to clarify some of the mechanisms by which the aurora occurs. Störmer undertook an intensive programme of parallactic auroral photography from Norway beginning in 1911. Great progress in understanding the global distribution of auroral activity was made during

AURIGA (GEN. AURIGAE, ABBR. AUR)

Northern constellation, representing a charioteer (possibly Erichtonius, legendary king of Athens and inventor of the four-horse chariot), between Perseus and Gemini. It is easily recognized by virtue of CAPELLA, at mag. 0.1 the sixth-brightest star in the sky. EPSILON AURIGAE is a remarkable eclipsing binary, mag. range 3.7–4.0, period 27.1 years – the longest of any such known star; another eclipsing binary is ZETA AURIGAE. The constellation's deep-sky objects include the open clusters M36, M37 and M38, which each contain several dozen stars fainter than mag. 8–9.

BRIGHTEST STARS

	Name	RA h m	dec. ° ′	Visual mag.	Absolute mag.	Spectral type	Distance (l.y.)
α	Capella	05 17	+46 00	0.08	−0.5	G2 + G6	42
β	Menkalinan	06 00	+44 57	1.90	−0.1	A2	82
θ		06 00	+37 13	2.65	−1.0	A0	173
ι		04 57	+33 10	2.69	−3.3	K3	512
η		05 07	+41 14	3.18	−1.0	B3	219

◄ **aurora** The colours in aurorae are an indication of the height at which particles in the Earth's atmosphere are being ionized: the red areas are higher than the green.

PRINCIPAL AURORAL EMISSIONS

wavelength (nm)	relative intensity	emitting species	typical altitude (km)	visual colour
391.4	47.4	molecular N	1000	violet-purple
427.8	24.4	molecular N	1000	violet-purple
577.7	100.0	atomic O	90–150	green
630.0	10 to 600	atomic O	>150	red
636.4	3 to 200	atomic O	>150	red
661.1	–	atomic N	65–90	red
676.8	–	atomic N	65–90	red

the INTERNATIONAL GEOPHYSICAL YEAR of 1957–58. Current models owe much to the work of the Japanese researcher Syun-Ichi Akasofu (1930–).

Continuing investigations of the aurora, which is the most visible of several effects on near-Earth space resulting from the Sun's varying activity, are further aided by observations from spacecraft, which can directly sample particles in the solar wind and magnetosphere, and image the auroral ovals from above.

Australia Telescope Compact Array *See* AUSTRALIA TELESCOPE NATIONAL FACILITY

Australia Telescope Long Baseline Array *See* AUSTRALIA TELESCOPE NATIONAL FACILITY

Australia Telescope National Facility (ATNF) Set of eight radio telescopes that can be used either individually or together. Six of the telescopes, at Narrabri in New South Wales, constitute the *Australia Telescope Compact Array*. Each dish has an aperture of 22 m (72 ft). A seventh 22-m antenna, the Mopra Telescope, is located near SIDING SPRING OBSERVATORY, 100 km (60 mi) to the south-west. The final component is the PARKES RADIO TELESCOPE, a further 220 km (140 mi) away. When all eight instruments are used together they constitute the *Australia Telescope Long Baseline Array*. The ATNF is funded by the Commonwealth Scientific and Industrial Research Organisation (CSIRO).

autoguider Electronic device that ensures a TELESCOPE accurately tracks the apparent movement of a celestial object across the sky. Although a telescope is driven about its axes to compensate for the rotation of the Earth, it is still necessary to make minor corrections, particularly during the course of a long observation. An autoguider achieves this by using a photoelectric device, such as a quadrant photodiode or a CCD, to detect any drifting of the image. The light-sensitive surface of the detector is divided into four quarters, with light from a bright guide star in the field of view focused at the exact centre. If the image wanders from the centre into one of the four quadrants, an electrical current is produced. By knowing from which quadrant this emanated, it is possible to correct the telescope drive to bring the image back to the exact centre again.

autumnal equinox Moment at which the Sun's centre appears to cross the CELESTIAL EQUATOR from south to north, on or near September 23 each year at the FIRST POINT OF LIBRA. At the time of the EQUINOX, the Sun is directly overhead at the equator and rises and sets due east and due west respectively on that day, the hours of daylight and darkness being equal in length. The term is also used as an alternative name for the First Point of Libra, one of the two points where the celestial equator intersects the ECLIPTIC. The effects of PRECESSION cause these points to move westwards along the ecliptic at a rate of about one-seventh of an arcsecond per day. *See also* VERNAL EQUINOX

Auwers, (Georg Friedrich Julius) Arthur (1838–1915) German astronomer, a specialist in astrometry, and a founder of Potsdam Astrophysical Observatory. He compiled many important star catalogues, and computed the orbits of the binaries Sirius and Procyon before astronomers confirmed the existence of those bright stars' faint companions by direct observation.

Auzout, Adrien (1622–91) French astronomer who, independently of William GASCOIGNE, invented the micrometer. He developed the wire micrometer for measuring separations in the eyepiece, and was one of the first to fit graduated setting circles to telescopes to facilitate the measurement of coordinates. As a member of the 'Paris School' of scientists, which also included Giovanni Domenico Cassini, Jean Picard, Ole Römer and Christiaan Huygens, Auzout was instrumental in founding PARIS OBSERVATORY.

averted vision Technique for viewing faint objects through a telescope. The retina of the eye is equipped with two different types of photoreceptors: cone cells and rod cells. Cone cells are responsible for colour vision and respond only to high levels of light; rod cells are more sensitive to light, but do not provide colour vision. This is why colours cannot be seen under low levels of light. The fovea, a small area of the retina located directly behind the lens, contains only cones, so when one looks directly at an object, no rods are used. A dim object that cannot be seen when viewed directly will often come into view if one looks slightly to the side. This allows light from the object to fall on a region of the retina populated by rods, which are about 40 times (equivalent to a difference of 4 stellar magnitudes) as sensitive as cones. The best averted vision is achieved when the object is about 8° to 16° away from the fovea, towards the nose. The regions above and below the fovea are almost as sensitive, but the area towards the ear should be avoided since the blind spot (where the optic nerve enters the eye) is 13° to 18° in that direction.

Avior The star ε Carinae, visual mag. 1.86 (but slightly variable), distance 632 l.y., spectral type K3 III. The name Avior is of recent application and its origin is unknown.

AXAF Abbreviation for Advanced X-Ray Astrophysics Facility. It was renamed the CHANDRA X-RAY OBSERVATORY.

axion Elementary particle proposed to explain the lack of CP (Charge and Parity) violation in strong interactions. Axions have been proposed as a type of COLD DARK MATTER, the mass of which contributes to the gravitational potential in galaxy haloes, thus helping to explain the radial velocity curves of galaxies. Recent background radiation measurements and Hubble Space Telescope observations of brown dwarfs in the haloes of galaxies have limited the need for axions in cosmology.

axis Imaginary line through a celestial body, which joins its north and south poles, and about which it rotates or has rotational symmetry. The angle between this spin axis and the perpendicular to its orbital plane is called the axial inclination, which in the case of the Earth is 23°.45. In the cases of Venus, Uranus and Pluto, the values exceed 90°, because their axial spins are RETROGRADE. The gravitational pull of the Sun and Moon on the Earth's equatorial bulge causes its rotational axis to describe a circle over a period of 25,800 years, an effect known as PRECESSION. If a body possesses a magnetic field, its rotational and magnetic axes do not necessarily coincide. On the Earth the two are inclined at about 10°.8 to one another, while on Uranus the magnetic axis lies at 58°.6 to the rotational axis and is also offset from the centre of the planet.

azimuth Angular distance of an object measured westwards around the horizon from due north at 0°, through due east at 90°, due south at 180° and so on. An object's azimuth is determined by the vertical circle (meridian) running through it. Azimuth is one of the two coordinates in the horizontal (or horizon) coordinate system, the other being ALTITUDE. *See also* CELESTIAL COORDINATES

Azophi Latinized name of AL-SŪFĪ

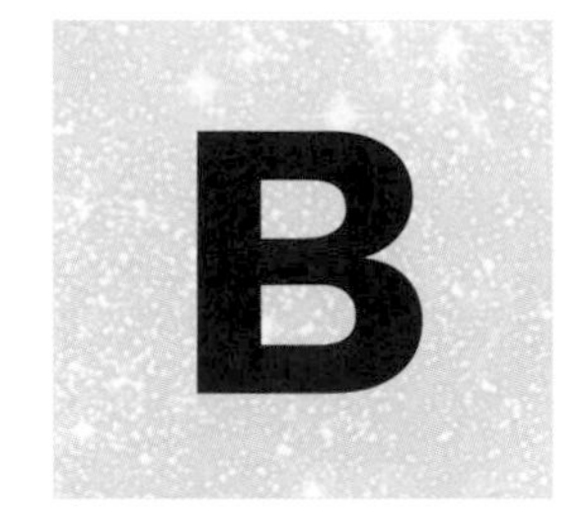

BAA Abbreviation of BRITISH ASTRONOMICAL ASSOCIATION

Baade, (Wilhelm Heinrich) Walter (1893–1960) German-American astronomer who used the large reflecting telescopes at Mount Wilson and Palomar Observatories to make many fundamental discoveries about the Milky Way's stellar POPULATIONS and the size of the observable Universe. From 1919 to 1931 Baade worked at Germany's Hamburg Observatory, where he made many observations of a wide variety of celestial objects – comets, asteroids, variable stars and star clusters, and galaxies. He discovered the unusual asteroids HIDALGO (1920) and ICARUS (1949). His skills as an observer were recognized by the Rockefeller Foundation, which awarded him a scholarship in 1929 to pursue his work at Mount Wilson Observatory. Two years later he became a staff member there, where he remained until 1959, using the 100-inch (2.5-m) Hooker Telescope and, later, Mount Palomar's 200-inch (5-m) Hale Telescope, to accumulate a large number of high-quality photographic plates of the Milky Way's stars, clusters and nebulae, as well as many galaxies.

In 1934 Baade proposed, with Fritz ZWICKY, that neutron stars were the end product of SUPERNOVAE and that many cosmic rays originated from these violent explosions of supermassive stars. With Zwicky and Rudolph MINKOWSKI he classified supernovae into two types that differed in their absolute magnitudes and spectral characteristics. With Minkowski, he identified the optical counterparts of the Cygnus A and Cassiopeia A radio sources. Observing extensively during the World War II blackout of Los Angeles, which darkened Mount Wilson's skies, Baade was able to resolve for the first time stars in the core of the ANDROMEDA GALAXY (M31), as well as in its companions, M32 and NGC 205. In 1944 he announced that the Milky Way consists of younger, metal-rich, highly luminous (usually blue) Population I stars found mostly in our Galaxy's spiral arms, and older, reddish Population II stars of the galactic nucleus and halo. Baade also discovered that there are two populations of CEPHEID VARIABLES. In 1952 and 1953 he used the new Hale Telescope to measure more accurately the brightness of Cepheid and RR LYRAE VARIABLES. He found no Cepheids in M31, but they should have been visible if the galaxy lay at the distance then currently believed, so he reasoned that M31 had to be at least twice as distant – a discovery that necessitated the doubling of the cosmic distance scale.

Baade's Window Best known of several small regions in the sky where there is a partial clearing of the dust clouds towards the centre of the Milky Way galaxy. Through these 'windows' it is possible visually to observe objects that are within and beyond the central bulge of the Galaxy. Baade's window is about three-quarters of a degree northwest of γ Sgr and surrounds the globular cluster NGC 6522. It is 4° east of the galactic centre and so lines-of-sight through the window pass within about 1800 l.y. of the centre. The region is named after Walter BAADE.

Babcock, Harold Delos (1882–1968) American astronomer who, as a long-serving staff member at Mount Wilson Observatory (1908–48), made many discoveries about solar and stellar magnetic fields. In 1908 he detected an extremely weak solar magnetic field. After several decades of studying the solar spectrum with a magnetograph, he was able accurately to measure the strength of the Sun's magnetic field and its variability. With his son Horace BABCOCK, he discovered that other stars possess strong magnetic fields.

Babcock, Horace Welcome (1912–) American astronomer at the Mount Wilson Observatory and the son of Harold BABCOCK, with whom he co-invented the solar magnetograph, an instrument used to study the Sun's magnetic field and its variation. He also designed and built many instruments for the telescopes of Mount Wilson and Palomar Observatories, including diffraction gratings, photometers, guiding electronics, polarizers and spectrographs. He was the first (1953) to propose the technique known as ADAPTIVE OPTICS.

Babylonian astronomy Astronomy as practised in Mesopotamia by various cultures from early in the 2nd millennium BC to about the 1st century AD. The city of Babylon was the dominant influence in this region for much of this period. It was established by King Hammurabi (1792–1750 BC) on the banks of the Euphrates, about 100 km (60 mi) south of Baghdad. Apart from periods of independence, it was successively controlled by Hittites, Kassites, Assyrians and Persians before its conquest in 311 BC by Alexander the Great. Thanks to the preservation of Babylonian written records on baked clay tablets from around 700 BC onwards, Babylonian mathematics and astronomical observations can be traced further back than those of other peoples. From Babylon came the sexagesimal system of numbers, based on 60 rather than our familiar 10. The subdivisions of the hour into 60 minutes and the minute into 60 seconds, and of the degree into 60 arcminutes and the arcminute into 60 arcseconds, are the legacy of this system.

Like many other ancient peoples, the Babylonians used a mixed luni-solar calendar, with extra months inserted when necessary to correct for the fact that there are between 12 and 13 lunar months in one year. They kept accurate records of celestial events and eventually (*c.*5th century BC) recognized the METONIC CYCLE (that a timespan of 235 lunar months is very close to 19 solar years). They set up a predictable calendar based on this relationship. By studying the motions of the planets they were able to identify cycles in their behaviour and could construct predictive almanacs (ephemerides). They also recognized that the annual apparent motion of the Sun appears to be uneven, and they developed a simple model to allow for this when predicting its position.

Although most of the surviving Babylonian records date from after 700 BC, a compilation of celestial omens that supposedly warned of impending disasters, called the Enūma Anu Enlil, contains references to events dating back to the 3rd millennium BC. This text was used to

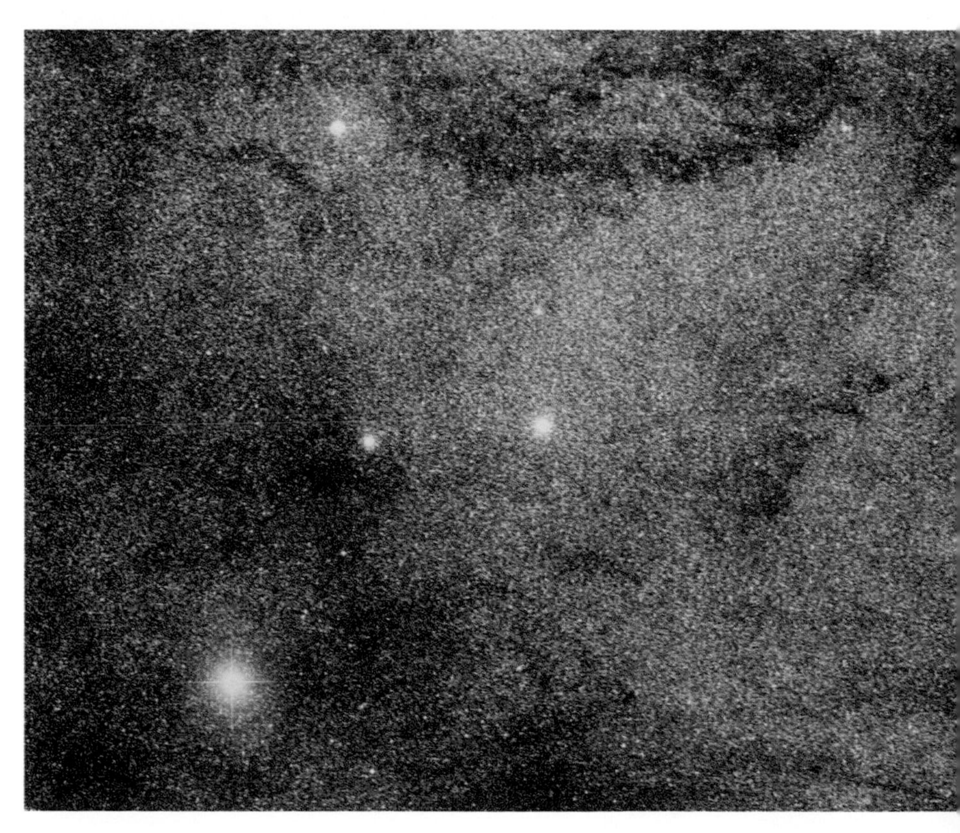

▼ **Baade's Window** The partial clearing of dust clouds known as Baade's Window allowed Walter Baade to observe stars near the centre of the Milky Way, including RR Lyrae variables. He used the period–luminosity relation to calculate their absolute magnitude and thus calculated the distance of the Solar System from the centre of our Galaxy.

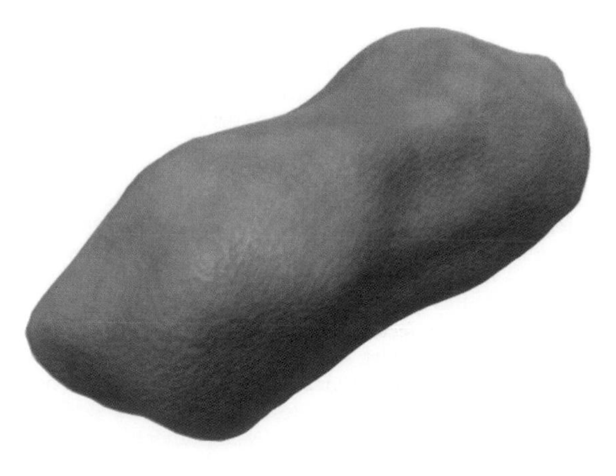

► **Bacchus** Radar images of asteroid (2063) Bacchus have shown it to be about 1.5 km (0.93 mi) long. It also appears to be smooth in texture in comparison with such asteroids as Eros. [S. Ostro (JPL/NASA)]

interpret new omens that appeared from time to time and their possible consequences, with a view to advising the king on how to avoid them by performing appropriate religious rituals. Out of the recording and interpretation of celestial omens came 'diaries' of the movements of the Sun, Moon and planets, the discovery of cycles in these movements, and eventually a mathematical model of these movements built around arithmetical sequences.

The precision of the Babylonian observations was complementary to the more speculative nature of Greek astronomy. After Alexander's conquest of Babylon drew their attention to Babylonian achievements, Greek astronomers such as HIPPARCHUS became interested in making precise observations and in developing geometrical models that fitted the data.

Bacchus APOLLO ASTEROID; number 2063. Radar images show Bacchus to have a smooth, elongated shape. *See* table at NEAR-EARTH ASTEROID

background noise Intrinsic noise in the detector or the sky background noise. For radio and infrared astronomers the main concern is usually detector noise. For optical astronomers, the night sky itself can be brighter than the astronomical object being observed. In all cases astronomers attempt to measure the background noise so that it can be removed. *See also* SIGNAL-TO-NOISE RATIO

backscattering Reflection of light back towards the direction of the light source by an angle greater than 90°. Light may be scattered from its direction of travel by fine particles of matter. For particles significantly larger than the wavelength of the incident light, reflection or backscattering of the light occurs. *See also* FORWARD SCATTERING

Baikonur Important launch centre situated in a semi-arid desert, north-east of the Aral Sea. Now leased from Kazakstan by Russia, it is the site from which all Soviet/Russian manned spaceflights have been launched. It is also known as Tyuratam.

Bailey, Solon Irving (1854–1931) American astronomer at Harvard College (1887–1931) who pioneered photographic surveys of the southern skies from Peru and South Africa. In 1889 Bailey made the first of several expeditions to South America to find a suitable locale for Harvard's Boyden Station, and eventually chose a site near Arequipa, Peru, at an altitude of 2500 m (8200 ft). From there, Bailey measured the brightness and positions of almost 8000 stars visible only from the southern hemisphere, and studied over 500 variable stars in the globular clusters ω Centauri, M3, M5 and M15. In 1908, from Cape Colony (modern South Africa), Bailey completed a photographic survey of the southern regions of the Milky Way.

Bailly Vast lunar enclosure (67°S 63°W), 298 km (185 mi) in diameter, described as a 'field of ruins'. Though technically the largest of the Moon's walled plains, Bailly is a difficult feature to make out because of its location at the south-west limb and its broken-down walls. Its rims rise to heights of 4250 m (14,000 ft) in some places, but in others are barely discernible. Bailly's floor lacks a central peak, but it does include two sizeable craters, known as Bailly A and B, and many smaller impact sites.

Baily, Francis (1774–1844) English stockbroker and amateur astronomer whose description of the phenomenon visible during annular and total solar eclipses now known as BAILY'S BEADS, published in 1838, was the first to attract attention. Baily also made the first highly accurate calculation of the Earth's mean density.

Baily's beads Phenomenon usually seen at SECOND CONTACT and THIRD CONTACT of a total SOLAR ECLIPSE, when several dazzling spots of the Sun's photosphere become visible through depressions in the Moon's irregular limb. Baily's beads may also be visible when the Moon at an ANNULAR ECLIPSE only just appears too small completely to cover the Sun. Indeed, it was at just such an eclipse, seen from Jedburgh in Scotland on 1836 May 15, that Francis BAILY first described the 'row of lucid points, like a string of bright beads' around the Moon's limb, from which the phenomenon takes its name.

Baker–Schmidt camera Form of SCHMIDT CAMERA used for photographing meteors; it incorporates design modifications by J.G. Baker. With a typical focal ratio of *f*/0.67, the Baker–Schmidt camera employs near-spheroidal primary and secondary mirrors, together with a correcting plate, which produce a wide, flat field of view and images that are largely free from CHROMATIC ABERRATION, ASTIGMATISM and distortion.

Ball, Robert Stawell (1840–1913) Irish astronomer, the fourth Royal Astronomer for Ireland (1873–92). He discovered several galaxies with the 'Leviathan of Parsonstown', Lord ROSSE's great 72-inch (1.8-m) reflecting telescope. Ball supported the nebular hypothesis of the Solar System's origin, according to which the Sun and planets condensed from a hot primordial cloud of gas. He believed that meteorites were produced by terrestrial volcanoes and expelled by violent eruptions, falling back to the ground after temporarily orbiting our planet.

balloon astronomy Astronomical research carried out using instruments flown on balloons. In the late 19th century, scientists began using balloons to lift telescopes and other astronomical instruments above the turbulent lower atmosphere. Even after the advent of ROCKET ASTRONOMY and ARTIFICIAL SATELLITES, the convenience and cheapness of balloons ensured their continued use for research and testing new space technology. The development of high-altitude balloons was an important spur for balloon astronomy. Recent applications include study of the cosmic microwave background (CMB); balloons have also been used in planetary exploration.

The first astronomical use of balloons was probably in 1874 when Jules JANSSEN sponsored two aeronauts in an unsuccessful attempt to record the solar spectrum with a small hand-held spectroscope. In 1899 July, an automatic photographic spectrograph developed by Aymar de la Baume-Pluvinel (1860–1938) demonstrated that water vapour in the Earth's atmosphere was showing up in the solar spectrum. Victor HESS discovered cosmic rays with a balloon-borne electroscope in 1912; further cosmic-ray studies were conducted in the 1930s by twin scientists Jean Piccard (1884–1963) and Auguste Piccard (1884–1962), who ascended in a balloon to above 20,000 m (65,000 ft).

The first use of a balloon-borne astronomical telescope was by Audouin Charles Dollfus (1924–) on 1954 May 30, during an attempt to detect water vapour in the Martian atmosphere. A few years later, Dollfus and the English astronomer Donald Eustace Blackwell (1921–) took the first astronomical photographs from the air; their high-resolution images of the Sun's photosphere proved that solar GRANULATION was the product of internal convection. Higher-flying unmanned balloons were pioneered by Martin SCHWARZSCHILD with Stratoscope I, whose 0.3-m (12-in.) automated Sun-pointing telescope recorded sharper images of the granulation and the photosphere from an altitude of 24,000 m (79,000 ft). Stratoscope II, which carried a 0.86-m (34-in.) telescope, obtained infrared spectra of the Martian atmosphere and red giant stars during its first flights in 1963.

Large, high-altitude, unmanned helium balloons are still used for both astronomical observations and for trials of new technologies and payloads for space missions. Observations at X-ray, gamma-ray and infrared wavelengths continue to be made from balloons. Some of the most significant balloon-borne campaigns of recent years have involved studies of the CMB.

In the Boomerang experiment (1998/1999), a balloon-borne 1.2-m (48-in.) telescope circumnavigated Antarctica, taking measurements at four frequencies to separate faint galactic emissions from the CMB. The resultant map covered approximately 2.5% of the sky with an angular resolution 35 times better than had been achieved by the COSMIC BACKGROUND EXPLORER satellite, revealing variations as small as 0.0001 K in the temperature of the CMB. Similar endeavours include MAXIMA (Millimetre Anisotropy Experiment Imaging Array), balloon-borne millimetre-wave telescopes designed to measure fluctuations in the CMB. MAXIMA 1 (1998 August) observed 124 square degrees of sky, while MAXIMA 2 (1999 June) observed roughly twice that area.

The CMB and interstellar dust in our Galaxy were also observed by submillimetre telescopes on the 3-t French PRONAOS (Programme National d'Astronomie Submillimétrique) stratospheric balloons in 1994, 1996 and 1999. The successor to PRONAOS is the Elisa international project to observe the interstellar medium. 2000 June saw the inaugural balloon flight of Claire, the first gamma-ray telescope to use a separate detector and collector. A follow-up flight with an improved version took place in 2001 June.

Extremely large balloons that can stay aloft for up to 100 days are now being designed. The 30-day maiden flight of NASA's Ultra Long Duration Balloon (ULDB), carrying the Trans-Iron Galactic Element Recorder (TIGER) experiment, began in Antarctica on 2001 December 21. TIGER is designed to measure the abundances of elements from iron to zirconium in galactic cosmic rays. The ULDB is a pressurized balloon made of extremely strong materials, such as polyester, inside which the helium gas is sealed. The balloon maintains a constant volume and so stays at the same altitude. The French pioneered the use of such balloons to explore other worlds. Two helium balloons were delivered by Soviet VEGA spacecraft into the atmosphere of Venus on 1985 June 11 and 15. Inflated about 54 km (34 mi) above the surface, each lasted for about 46½ hours, and travelled more than a third of the way around the planet.

Traditional 'Montgolfier' hot-air or 'zero-pressure' balloons are much lighter and more easily deployed than their helium counterparts, and have potential for flights above other planets. Infrared Montgolfiers could not operate at Mars during the very cold nights, but a novel hot-air venting system could allow repeated, precision soft landings. Although they would be limited to daytime flights – up to 10 hours at lower Martian latitudes, or perhaps months during the long polar summers – such balloons could drag an instrumented 'snake' over the surface or soft-land a large rover, re-ascend, and continue imaging before landing again, possibly with a 'nanorover', at dusk. Other Montgolfier balloon missions that could explore the atmospheres of Venus, Jupiter and Titan are under study.

Balmer lines Series of EMISSION or ABSORPTION LINES in the HYDROGEN SPECTRUM resulting from electron transitions down to or up from the second energy level of that atom. The Balmer lines are named with Greek letters: Hα (or Balmer α), which connects levels 2 and 3, falls at 656.3 nm; Hβ , which connects levels 2 and 4, falls at 486.1 nm; Hγ falls at 434.0 nm; Hδ at 410.1 nm, and so on. The series ends at the Balmer limit in the ultraviolet at 364.6 nm.

Bappu, (Manali Kallat) Vainu (1927–82) Indian astronomer and director of the Kodaikanal Observatory who, while at Mount Wilson and Palomar Observatories, discovered with Olin Chaddock Wilson (1909–94) the Wilson–Bappu effect, relating the luminosity of a late-type star to the strength of its calcium K line. He pioneered ultra-low dispersion spectroscopy of stars and galaxies, and made important studies of red stars in the Magellanic Clouds and of planetary ring systems.

bar Non-SI unit by which pressure is measured. It is particularly used in astronomy to describe the pressure in a planetary atmosphere, one bar being equivalent to the average atmospheric pressure on the Earth at sea level. Atmospheric pressure is often quoted in millibars, one bar being equal to 1000 millibars.

barium star Giant G STAR or K STAR with an excess of carbon, barium, strontium and other heavy elements (including those from slow neutron capture) in its atmosphere. Barium stars have white dwarf companions. They most likely result from MASS TRANSFER that took place when the current barium star was a dwarf and the white dwarf was an evolving, chemically enriched, mass-losing giant. Barium stars are loosely related to CH and dwarf CARBON STARS.

Barlow lens Concave (negative) LENS placed between a telescope objective and eyepiece to increase the magnification, usually by two or three times. The negative lens reduces the angle of convergence of the light cone, effectively making it appear to the eyepiece that the primary has a longer focal length. It was invented in the early 1800s by English physicist Peter Barlow (1776–1872).

Barnard, Edward Emerson (1857–1923) American astronomer noted for his discoveries of comets and Jupiter's satellite AMALTHEA, and for his use of photography to map the Milky Way. While still an amateur astronomer he came to notice for his discovery of eight comets, which allowed him to overcome poverty and gain

▼ **balloon astronomy** The Boomerang balloon is shown here just before launch, with Mount Erebus in the background. At a height of 35 km (22 mi), the balloon flew above 99% of the Earth's atmosphere for 10 days, which allowed it to make unprecedented measurements of the cosmic microwave background radiation.

B

an education at Vanderbilt University. His observing skills won him a position at the new Lick Observatory in 1887, where he became the first astronomer to discover a comet using photography. At Lick, Barnard used the 36-inch (0.9-m) refractor to prove the transparency of Saturn's C Ring (1889) and to discover Amalthea (1892), the last planetary satellite to be found without the aid of photography. Three years later, he moved to the Yerkes Observatory, where he remained for the rest of his life, making numerous detailed visual observations of the planets and comets with the 40-inch (1-m) refractor. He used Yerkes' 10-inch (250-mm) Bruce Photographic Telescope to take wide-field images of the Milky Way's star clouds, discovering DARK NEBULAE. This work, culminating in his classic *A Photographic Atlas of Selected Regions of the Milky Way* (1927), continued photographic surveys of the Galaxy that Barnard had started at Lick in the 1890s using the 6-inch (150-mm) Willard wide-field telescope. He also discovered (1916) the star with the largest proper motion, named BARNARD'S STAR in his honour.

Barnard's Loop (Sh2-276) Large, diffuse and very faint arc of nebulosity centred on the Orion OB ASSOCIATION, which is the group of hot stars south of Orion's Belt. The nebula is visible only on the eastern and southern sides of the constellation. Barnard's Loop was named after the pioneer astronomical photographer Edward Emerson BARNARD, who discovered it in 1894. If it were a complete ring, Barnard's Loop would almost completely fill the region between Betelgeuse and Rigel, covering almost 20° of sky. Barnard's Loop appears to be a structure resulting from clearance of dust and gas in the interstellar medium by radiation pressure from the Orion OB association stars.

Barnard's Star Closest star to the Sun after the α Centauri triple system, lying 5.9 l.y. away in the constellation Ophiuchus, with a visual magnitude of only 9.54, well below naked-eye visibility. It is a red dwarf, spectral type M4, with only 0.05% of the Sun's luminosity. Discovered in 1916 by Edward Emerson BARNARD, it has the largest PROPER MOTION of any star, 10″.4 per year – so fast that it moves across the sky a distance equivalent to the Moon's apparent diameter in under two centuries. In the 1960s and 1970s the Dutch-American astronomer Peter Van de Kamp (1901–95) reported an apparent wobble in the proper motion of the star, which he attributed to the presence of two orbiting planets similar in size to Jupiter and Saturn. This was not confirmed, and the star's apparent 'wobble' seems to have been due to instrumental error (*see also* EXTRASOLAR PLANET).

▼ **barred spiral galaxy** NGC 1365 in Fornax has a prominent bar and open arms. The blue areas in the spiral arms are regions of intense star formation, caused by the gravitational influence of the bar.

barn-door mount *See* SCOTCH MOUNT

barred spiral galaxy SPIRAL GALAXY in which the distribution of stars near the nucleus is elongated in the disk plane. Strong bars are reflected in the HUBBLE CLASSIFICATION. Weaker bars are common among spirals, with the de Vaucouleurs types indicating that about 60% of spirals have distinct bars. Infrared observations, less sensitive to dust absorption and the confusing influence of young stars, may indicate an even greater fraction of spirals with small or weak bars, among them our Galaxy. Stellar dynamics in a disk can produce a bar because of instabilities, and a bar in turn may dissolve into a ring structure over cosmic time. Indeed, many bars are accompanied by stellar rings. Bars can alter the chemical content of a galaxy's gas, since flow along the bar mixes gas originally located at a wide range of radii, which thus started with different chemical compositions.

Barringer Crater *See* METEOR CRATER

barycentre Centre of mass of two or more celestial bodies. For example, when computing the perturbing effect of the Earth and Moon on other planets it is usual to regard the Earth and the Moon as a single body of their combined mass, located at their barycentre. Similarly when computing the orbits of the five outer planets one would treat the Sun and the four inner planets as a single body of their combined mass located at their barycentre.

barycentric dynamical time (TDB) *See* DYNAMICAL TIME

baryon Elementary particle that participates in the strong interaction. Two common baryons are protons and neutrons. Other, more massive baryons, such as the lambda and sigma particles, are commonly called hyperons. They have spins of $\frac{1}{2}$, and a baryon number of +1, as well as even intrinsic parity.

basalt Volcanic rock (solidified lava) that consists mostly of the minerals pyroxene $(Mg,Fe,Ca)_2Si_2O_6$ and plagioclase $NaAlSi_3O_8$-$CaAl_2Si_2O_8$. Basalts are abundant on Earth, the Moon, Venus and Mars, forming vast plains and volcanic constructs. They are probably abundant on Mercury and on Jupiter's satellite Io, where they also form plains. Basalts make up the surfaces of some asteroids, such as VESTA. Basalt is formed by solidification of partially melted planet MANTLE material.

basin Extensive topographically depressed area. In planetary science the term basin is commonly used with the adjective multi-ring. Multi-ring basins are circular depressions of a few tens to about 4000 km (2500 mi) in diameter; they typically have an elevated rim and one or two concentric rings of elevated terrain inside. They are observed on Mercury, Venus and Mars, as well as on large satellites, including the Moon. A multi-ring basin is an impact feature created by a collision with a comet or an asteroid: it is a type of impact CRATER. The onset diameter at which an impact crater becomes sufficiently large to form a multi-ring basin depends on the character of the target material (for example rock or ice) and on the body's gravitational acceleration. The onset diameter is 40–70 km (20–40 mi) on Venus, 100–130 km

(60–80 mi) on Mercury, 130–200 km (80–120 mi) on Mars, and 200–300 km (120–190 mi) on the Moon. The largest such structure known is the 4000-km (2500-m) VALHALLA on Jupiter's satellite CALLISTO. A prominent basin on the Moon's far side, called the SOUTH POLE–AITKEN BASIN, is about 2500 km (1600 mi) in diameter and 12 km (7 mi) deep. Next in size is the HELLAS PLANITIA basin on Mars, which is about 1800 km (1100 mi) in diameter and 5 km (3 mi) deep. Impacts of such size happened in the early history of the Solar System and are not known in later times.

Bayer, Johann (1572–1625) German magistrate of Augsburg and amateur astronomer, who in 1603 published his *Uranometria* star ATLAS, which identified each constellation's main stars by Greek letters; this convention of so-called **Bayer letters** is still in use. *Uranometria*, which consisted of 49 constellation charts, with stellar data taken from the work of Tycho BRAHE, is notable as the first star atlas to depict constellations around the south celestial pole. These constellations – Apus, Chamaeleon, Dorado, Grus, Hydrus, Indus, Musca (which Bayer called Apis, the Bee), Pavo, Phoenix, Triangulum Australe, Tucana and Volans – had recently been defined by the Dutch navigator Pieter Dirkszoon Keyser (*c*.1540–96), based on earlier observations by Amerigo Vespucci and others.

BD Abbreviation of *BONNER DURCHMUSTERUNG*

Beagle 2 UK Mars lander, almost 50% funded by the EUROPEAN SPACE AGENCY (ESA), which will be launched in 2003 June, flying piggyback on ESA's MARS EXPRESS orbiter. The spacecraft will arrive at Mars in 2003 December. The 30-kg Beagle 2 will land at a site in the ISIDIS PLANITIA region at about 13°N latitude, which is the best site given the constraints for a safe landing in a smooth area and the scientific objectives of the mission. Beagle will take samples of Martian soil and analyse it for signs of water and bacterial activity, as well as investigating the chemical isotopes present. It will also measure methane in the atmosphere and send back images of the surface. Beagle is being part-sponsored by commercial companies. The project is being managed by the UK's Open University. It will make a bouncy landing encased in inflated balloons, which will be detached as it comes to rest. If Beagle 2 touches down successfully on Mars, it will become only the fourth spacecraft, after the US VIKINGS 1 and 2 in 1976 and MARS PATHFINDER in 1997, to do so.

Becklin–Neugebauer object (BN object) Strong infrared source seen in the sky within the Orion Nebula but actually located behind that nebula inside the KLEINMANN–LOW NEBULA. It is thought to be a young B-type star, the optical and ultraviolet radiation of which is absorbed by a dense expanding envelope of dust that surrounds the star and then re-emits the energy as infrared radiation. It was found in 1967 by the American astronomers Eric Becklin and Gerry Neugebauer.

Becrux (Mimosa) The star β Crucis (of which 'Becrux' is a contraction), visual mag. 1.25, distance 353 l.y., spectral type B0.5 III. It is a BETA CEPHEI STAR, a type of eclipsing binary, with a period of 0.2 days; the total range of variation is less than 0.1 mag., which is too slight to be noticeable to the eye.

Bečvář, Antonín (1901–65) Czech astronomer, meteorologist and celestial cartographer who founded and directed (1943–50) the Skalnaté Pleso Observatory, which became known for its solar astronomy and photography of meteors using a specially constructed battery of wide-field cameras. Bečvář is best known for his celestial ATLASES. *Atlas coeli* (1950) charts 35,000 objects to the visual magnitude limit of 7¾ and many star clusters, nebulae and galaxies. This atlas was the first to include the many extraterrestrial radio sources discovered after World War II.

▲ **Beagle 2** This artist's impression of the Beagle 2 lander shows it with its solar panels open and its sampling arm deployed. The experiments are intended to discover whether the conditions for life ever existed, or still exist, on Mars. [All Rights Reserved Beagle 2. http://www.beagle2.com]

Bede, 'The Venerable' (*c*.673–735) First English astronomer. His arguments for a spherical Earth overturn the myth that all medieval people believed the world to be flat. Much of Bede's astronomy was concerned with updating the calendar and calculating the date of Easter from the Moon's phases, and as such formed part of his wider historical and theological studies. His calendrical formulae came to be used across Europe, and he regularized the use of AD dating (from the birth of Christ).

Beer, Wilhelm (1797–1850) German banker and amateur astronomer who, with Johann Heinrich MÄDLER, compiled *Mappa selenographia* (1837), which surpassed all previous maps and catalogues of lunar features. It contained accurate measures for the heights of over 800 lunar mountains/crater rims and the diameters of almost 150 craters. The two German astronomers also produced the first map of Mars to show its albedo features (1830).

Beehive *See* PRAESEPE

Beijing Astronomical Observatory (BAO) Major research institute of the Chinese Academy of Science, founded in 1958. The observatory has five observing stations of which Xinglong is the main optical/infrared observing site. Situated about 150 km (95 mi) north-east of Beijing at an elevation of 960 m (3150 ft), the observatory hosts China's largest telescope, a 2.16-m (85-in.) reflector. There is also a 1.26-m (50-in.) infrared telescope. Xinglong will be home to the giant LARGE SKY AREA MULTI-OBJECT FIBER SPECTROSCOPIC TELESCOPE (LAMOST). At the BAO's radio astronomy site at Miyun is the Metre-Wave Aperture Synthesis Radio Telescope (MSRT) used for survey astronomy. It consists of 28 dishes 9 m (30 ft) in diameter. *See also* ANCIENT BEIJING OBSERVATORY

Belinda One of the small inner satellites of URANUS, discovered in 1986 by the VOYAGER 2 imaging team. Belinda is *c*.68 km (*c*.42 mi) in size. It takes 0.624 days to circuit the planet at a distance of 75,300 km (46,800 mi) from its centre. It has a near-circular, near-equatorial orbit.

Bellatrix The star γ Orionis, visual mag. 1.64, distance 243 l.y., spectral type B2 III. Its name is derived from a Latin term, meaning 'the female warrior', and was first applied by medieval astrologers.

Bell Burnell, (Susan) Jocelyn (1943–) British astronomer who discovered the first four pulsars. She made the discovery in 1967 with a radio telescope that she

▼ **Bell Burnell, (Susan) Jocelyn** The team at Cambridge were looking for perturbations in radio waves that might indicate the presence of quasars when Jocelyn Bell Burnell discovered the first pulsar. The discovery of these rapidly spinning, but massive, objects confirmed the existence of neutron stars.

built while she was still a graduate student of Antony HEWISH at Cambridge University, but she did not share the Nobel prize subsequently awarded to Hewish

Benetnasch Alternative name for the star η Ursae Majoris. *See* ALKAID

Bennett, Comet (C/1969 Y1) Bright long-period comet discovered by John Caister (Jack) Bennett, South Africa, on 1969 December 28. The comet reached perihelion, 0.54 AU from the Sun, on 1970 March 20, approaching Earth in the following week. Comet Bennett became a prominent object in northern hemisphere skies, reaching peak magnitude +0.5 during April. At this time, Comet Bennett developed a strong dust tail 20° long and a very active, rapidly changing ion tail.

Bepi Colombo EUROPEAN SPACE AGENCY (ESA) mission to be launched in 2009 in conjunction with Japan to explore the planet MERCURY. After a two-year journey, including a fly-by of Venus, Bepi Colombo should become the second spacecraft to orbit Mercury (after the US MESSENGER) and will deploy a lander and a small sub-satellite to study the magnetosphere. This surface penetrator will send back data on the structure of the planet. Bepi Colombo will be powered by a solar electric propulsion system and will be protected against temperatures exceeding 673 K.

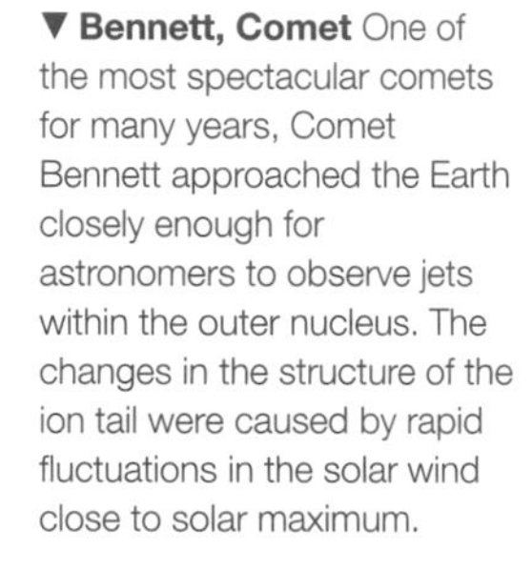

▼ **Bennett, Comet** One of the most spectacular comets for many years, Comet Bennett approached the Earth closely enough for astronomers to observe jets within the outer nucleus. The changes in the structure of the ion tail were caused by rapid fluctuations in the solar wind close to solar maximum.

BeppoSAX (Satellite for X-ray Astronomy) Italian–Dutch gamma-ray and X-RAY ASTRONOMY satellite. Soon after its launch in 1996 April, as the Satellite for X-ray astronomy, it was renamed in honour of Italian physicist Giuseppe Occhialini, whose nickname was 'Beppo'. It carries four spectroscopes and two wide-field cameras. In 1997 it played a major role in identification of GAMMA-RAY BURSTS by pinpointing X-ray emissions from these events, so enabling their positions and cosmological distances to be confirmed.

Berkeley Illinois Maryland Association (BIMA) Consortium consisting of the Radio Astronomy Laboratory of the University of California (Berkeley), the Laboratory for Astronomical Imaging at the University of Illinois (Urbana) and the Laboratory for Millimeter-Wave Astronomy at the University of Maryland, which operates the BIMA Millimetre Array at HAT CREEK OBSERVATORY.

Bessel, Friedrich Wilhelm (1784–1846) German astronomer and mathematician who was the first to measure stellar PARALLAX. In 1804, while employed as a shipping clerk, he calculated the orbit of Halley's Comet, which impressed Wilhelm OLBERS sufficiently to recommend that he be hired as assistant to Johann SCHRÖTER. At Schröter's private observatory in Lilienthal, Germany, Bessel observed Saturn and its rings and satellites, and comets. He continued to develop mathematical methods for celestial mechanics; it was while trying to solve the three-body problem of celestial mechanics (1817–24) that he developed the class of mathematical functions that bear his name (*see* BESSELIAN ELEMENTS). He reduced the positions for 3222 fundamental stars first observed by James BRADLEY at Greenwich Observatory between 1750 and 1762; his 1818 publication of over 63,000 star positions and proper motions, based on his and Bradley's observations, are considered the start of modern ASTROMETRY. In 1809 Friedrich Wilhelm III of Prussia appointed Bessel director of a new observatory at Königsberg.

Bessel's most famous achievement was his 1838 measurement of the parallax of the star 61 Cygni. He chose this relatively obscure star because of its high proper motion. Using Königsberg's Fraunhofer heliometer, Bessel determined that 61 Cygni has a parallax of 0″.31 (very close to the modern value of 0″.29), implying that the star was 10 l.y. from Earth. The Royal Astronomical Society awarded him its Gold Medal for this achievement, which marked the first step towards accurately measuring stellar distances. Bessel's 1840 paper on perturbations to Uranus' orbit suggested the existence of an eighth planet, six years before the discovery of Neptune by Johann GALLE. In 1841/1844 he predicted the existence of dark companions of Sirius and Procyon after discovering periodic variations in their proper motions – Sirius B was not confirmed until 1862.

Besselian elements These arise in a method devised by Friedrich BESSEL for calculating the circumstances of an ECLIPSE. For a solar eclipse a reference plane is used that is normal to the line passing through the centres of the Sun and Moon, and which passes through the centre of the Earth. The Earth's limb and the shadow of the Moon are projected on to this plane, and an eclipse will occur if the shadow intersects the limb. The Besselian elements define the location of the centre of the Moon relative to the centre of the Earth and the radii of the umbral and penumbral shadows within this plane, and the direction of projection.

Be star B STAR that shows a characteristic B-type spectrum with the addition of hydrogen emission lines. The emissions are commonly doubled, revealing a circumstellar disk with one side approaching the observer, the other receding. Be-SHELL STARS have thicker disks and metallic absorptions. Common among class B stars, Be stars are all rapid rotators with equatorial speeds that can exceed 300 km/s (190 mi/s). The origins are contentious, and include rotation, magnetic fields and pulsation.

Beta Centauri *See* HADAR

Beta Cephei star (Beta Canis Majoris star) Short-period pulsating VARIABLE STAR, of spectral type O8–B6, with light and radial-velocity periods of $0^d.1$ to $0^d.6$, and amplitudes of mags. 0.01 to 0.3V.

Beta Lyrae (Sheliak) Variable star that shows continuous variations in brightness; it is the prototype for one of the three subtypes of ECLIPSING BINARY. The Arabic name, Sheliak, comes from the Arabic for 'harp'. The star lies at a distance of 882 l.y.; its magnitude varies between 3.3 and 4.4. Small telescopes reveal a faint, mag. 7.2 companion. The variability and periodicity of Beta Lyrae were discovered in 1784 by John GOODRICKE. The first complete light-curve, obtained in 1859 by F.W.A. ARGELANDER, demonstrated that Beta Lyrae is an eclipsing binary.

All BETA LYRAE STARS show continuous magnitude variation between minima, resulting in a light-curve that has distinct primary and secondary minima. The double-humped light-curve of Beta Lyrae indicates that the stars are deformed into ellipsoids by their mutual gravitational interaction. Because the period is only 12.93854 days, the two stars must be so close to each other it becomes inevitable that significant tidal distortion in their shapes occurs. The light-curve can be explained by the changing total amount of stellar surface area presented to the Earth-based observer as the elongated stars revolve about each other.

Spectroscopic observations in the 1950s established that the primary star is a late-type B giant (B8.5II), but classification of the secondary was confounded by peculiar spectral features, some of which are caused by gas flowing between the stars and around the system as a whole. The secondary star is roughly four to five times more massive than the primary. According to the mass–luminosity relation, however, the more massive a star is, the brighter it is. Astronomers were puzzled because the massive secondary appeared significantly under-luminous, much dimmer than the primary.

Current theories about Beta Lyrae, supported by observations, suggest that the secondary star is enveloped in a thick ACCRETION DISK of gas captured from the primary. Material surrounding the secondary so severely subdues its luminosity that the star itself is impossible to observe. The primary star is overflowing its ROCHE LOBE, and gas streams across the inner LAGRANGIAN POINT on to the disk at the rate of 10^{-5} solar masses per year. Ultraviolet spectra from the OAO-3 (Copernicus) and Skylab orbiting observatories revealed clouds of gas at the L_4 and L_5 Lagrangian points. Gas is constantly escaping from the system.

Beta Lyrae star ECLIPSING BINARY (subtype EB) that shows a continuous variation in brightness throughout the orbital period, which is generally of one day or more in length. Both primary and secondary minima are always present. Although once thought to be CONTACT BINARIES (that is, systems that have tidally distorted components with a common atmospheric envelope, as in W URSAE MAJORIS STARS), the majority of Beta Lyrae stars appear to be semidetached systems in which gas escaping from a bloated primary star is falling on to an ACCRETION DISK surrounding the secondary star. Such a configuration is found in BETA LYRAE, which is the prototype for this type.

Astronomers now regard Beta Lyrae variables as members of a broad class of double stars known as SEMIDETACHED BINARIES, so called because only one of the two stars fills its ROCHE LOBE. They are probably binaries near the end of the initial rapid phase of MASS TRANSFER. Thus, some Beta Lyrae variables may evolve into Algol-type binaries (*see* ALGOL STAR). For example, as the primary star in a Beta Lyrae variable continues to evolve away from the main sequence, it will begin to resemble a red subgiant. As the rate of mass transfer subsides, the accretion disk shrouding the secondary may become transparent and a massive unevolved star will shine forth. The resulting system clearly would resemble ALGOL. Beta Lyrae stars, which account for about 20% of all eclipsing variables, are therefore an important but temporary stage in the early evolution of CLOSE BINARY systems. *See also* STELLAR EVOLUTION

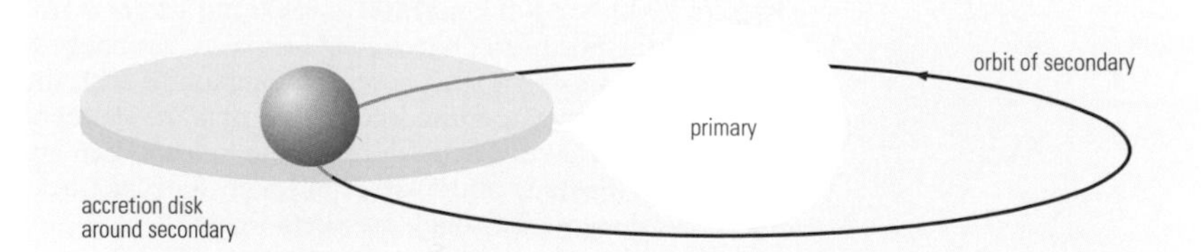

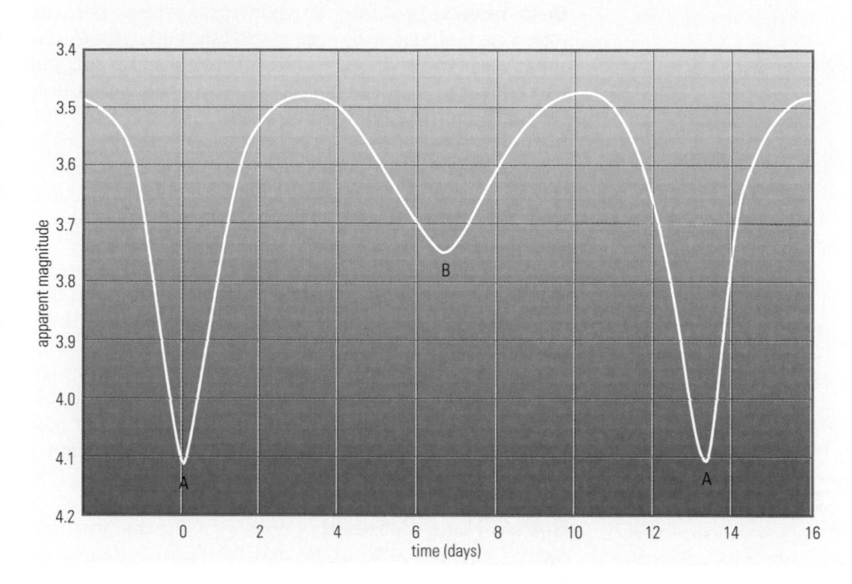

▲ **Beta Lyrae star** The continuous light-curve of Beta Lyrae stars is believed to be caused by an accretion disk of material which surrounds the secondary star and eclipses the primary. This theory also explains why the secondary star appears to be dimmer than the primary, despite being far more massive.

beta particle ELECTRON free of the atomic nucleus. It is a common by-product of nuclear reactions and decays.

Beta Pictoris MAIN-SEQUENCE A5 star in the constellation Pictor; it has visual magnitude 3.9 and is at a distance of 62 l.y. The star is surrounded by a PROTOPLANETARY DISK, which is edge-on as seen from Earth. One of the nearest planetary systems, it has been imaged by the Hubble Space Telescope and ground-based telescopes. The disk is about 400 AU in diameter, some four times the diameter of our Solar System. Both dust and hydrogen gas have been detected in the disk. Spectroscopic studies and dynamical modelling indicate that larger solid bodies (PLANETESIMALS) are present, and planets may have formed or be in the process of formation.

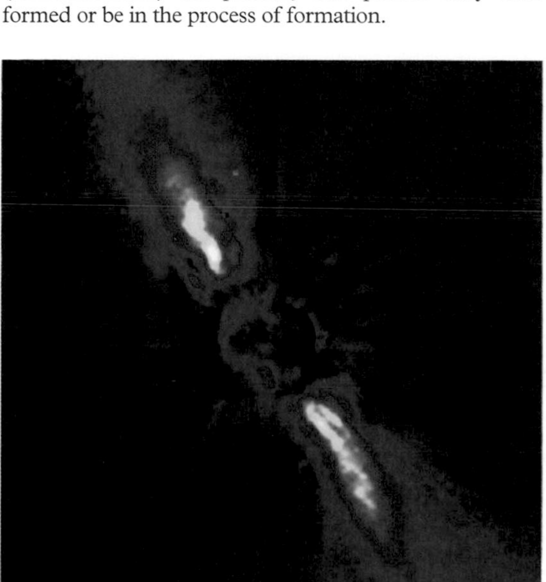

◄ **Beta Pictoris** This Very Large Telescope (VLT) false-colour image of the protoplanetary disk around Beta Pictoris shows that it is not even in composition. The visible kink in the disk is about 20 AU from the star, which corresponds roughly to the distance of Uranus from the Sun.

B

Beta Regio One of the three main uplands on VENUS, discovered by the PIONEER VENUS orbiter's radar. Beta Regio's 2000 × 3000 km (1200 × 1900 mi) extent is traversed by the north–south trough of Devana Chasma, generally inferred to be a fault complex comparable with the East African Rift. Superimposed on the rift at the southern end of Beta Regio is Theia Mons, a huge volcano 226 km (140 mi) in diameter and 5 km (3 mi) high. Satellite tracking shows that gravity is high over Beta Regio.

Beta Taurids DAYLIGHT METEOR STREAM active between June 5 and July 18, with peak around the end of June. The stream is associated with Comet 2P/ENCKE and is encountered again in October and November when it produces the night-time TAURIDS. The TUNGUSKA airburst of 1908 June 30 may have been produced by a particularly large piece of debris from the Beta Taurid stream.

Betelgeuse The star α Orionis, marking the right shoulder of Orion, distance 427 l.y. It is a red supergiant of spectral type M2 Ib, about 500 times the Sun's diameter and 10,000 times as luminous. Betelgeuse is an irregular variable with a considerable range, varying between about mags. 0.0 and 1.3 over a period of years. At its average value, magnitude 0.5, Betelgeuse is the 10th-brightest star. The origin of the name is often said to be from the Arabic *ibt al-jauzā'*, meaning 'armpit of the central one' but may come from *bait al-jauzā'*, 'house of the twins', a reference to neighbouring Gemini.

Bethe, Hans Albrecht (1906–) German-American atomic physicist who discovered the basic nuclear reactions that generate energy inside stars. After receiving an education in theoretical physics and holding teaching positions at the universities in Frankfurt, Stuttgart, Munich and Tübingen, Bethe emigrated to England, then to the USA, where he joined the faculty of Cornell University in 1935. During World War II he made significant contributions to the Los Alamos Science Laboratory's Manhattan Project, which developed the first atomic bomb, but this work convinced him of the importance of nuclear arms control, which he has since strongly advocated.

In the 1930s and 1940s, Bethe developed his models for stellar nuclear processes. He found that for stars of modest mass like the Sun, the most important nuclear reaction converts two protons (atomic hydrogen nuclei) into helium; Bethe also discovered that several protons can also combine with a carbon nucleus inside Sun-like stars to regenerate a carbon nucleus and helium 'ash'. He demonstrated that the powerful gravity of NEUTRON STARS, which are much more massive than the Sun, is sufficient to fuse protons and electrons to make neutrons. Bethe also modelled the carbon–nitrogen cycle of reactions, which powers many types of massive stars. In 1947 Bethe explained the Lamb shift observed in the spectrum of the hydrogen atom with a model that laid the foundations of a new branch of physics known as quantum electrodynamics.

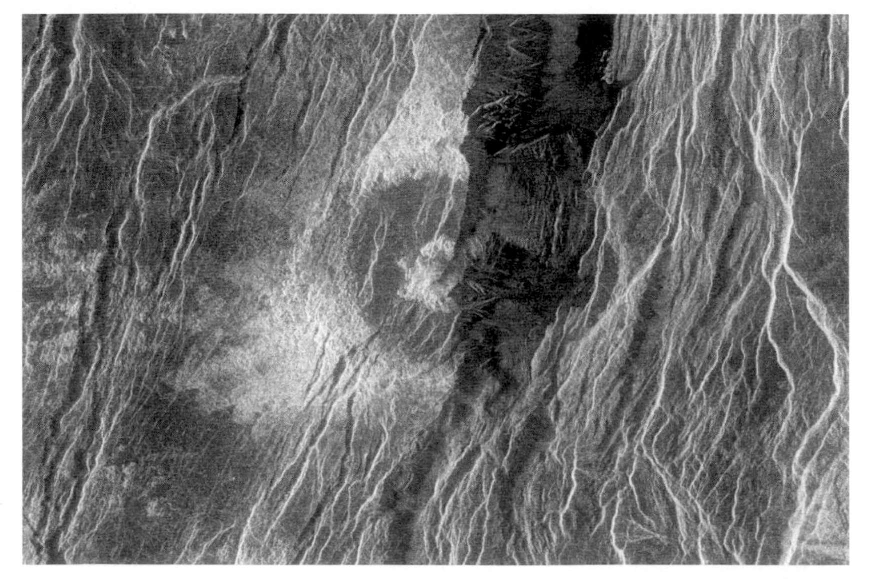

▼ **Beta Regio** The rift valley Devana Chasma in Beta Regio is as much as 20 km (12 mi) wide and here has cut through an earlier impact crater, which is itself 37 km (23 mi) across. The whole eastern side of the crater has collapsed and fallen into the rift.

Bianca One of the small inner satellites of URANUS, discovered in 1986 by the VOYAGER 2 imaging team. Belinda is about 44 km (27 mi) in size. It takes 0.435 days to circuit the planet at a distance of 59,200 km (36,800 mi) from its centre in a near-circular, near-equatorial orbit.

Biela, Comet 3D/ COMET discovered by Jacques Montaigne at Limoges, France, on 1772 March 8 and again by Jean Louis PONS on 1805 November 10. It was next noted by Wilhelm von Biela, an Austrian army officer and amateur astronomer, on 1826 February 27. He realized that it was the same as that which had already been observed in 1772 and 1805. It was subsequently named Biela's Comet and shown to be periodic, with a period of 6.6 years. It was seen again in 1832 and 1845–46.

On 1845 December 19 the comet appeared elongated and by the end of the year had split in two. By 1846 March 3 the two parts were over 240,000 km (150,000 mi) apart. At its next return, in 1852, the separation had increased to 2 million km (1.2 million mi). The comet was never seen again, but in November 1872 and 1885 tremendous showers of ANDROMEDID meteors – produced by debris released on the comet's disintegration – occurred on exactly the date when the Earth passed close to the comet's orbit.

Bielids Alternative name for the ANDROMEDIDS

Big Bang theory Theory concerning the explosive creation of the Universe from a single point. The first real strides in the study of cosmology occurred when Edwin HUBBLE and his collaborators noticed that all of the galaxies in the Universe, except for those in our local cluster, were receding from us. This fact was determined by measuring the REDSHIFT of the absorption lines in galaxy spectra. It was further noted that the more distant the galaxy, the faster the velocity of recession. If the galaxy velocities were extrapolated back into time, then all galaxies and stars apparently started at some time in the distant past at a single point. This cosmic singularity included not only all of the gas, dust, stars, galaxies and radiation, but all of SPACETIME as well.

Two types of models came out of these observations and theoretical arguments from general relativity. The STEADY-STATE THEORY pictured the Universe as expanding, with matter created between separating galaxies and clusters at precisely the rate needed to keep the average properties of the Universe the same, that is, homogeneous and isotropic. This idea suggested there was no beginning and will be no end, just constant expansion and filling in. The other major cosmological model, the Big Bang theory, held that the entire Universe emerged from a single point in an explosive event called the Big Bang. GENERAL RELATIVITY and the idea of a Big Bang are both consistent with the fact that more distant galaxies separate from each other more rapidly, following a mathematical relationship known as the HUBBLE LAW.

The Big Bang model predicted that the temperature of the very early universe was on the order of 10^9 K. At the start of the expansion, between $T = 0$ and $T = 10^{-43}$ s (the PLANCK TIME), the laws of physics are not well understood; in fact we would need a 'THEORY OF EVERYTHING' to be able to describe this era. The temperature continued to drop as the universe expanded. Between 10^{-43} and 10^{-35} s the universe had just two fundamental forces – gravity and the GUT force. After 10^{-35} s, the strong force separated from the electroweak force, but the universe was still far too hot for atoms to form. Finally, after about 10^{-12} s the weak and electromagnetic forces became distinct and the four fundamental forces looked as they do today. At 10^{-6} s the temperature had

◀ **Betelgeuse** The Hubble Space Telescope's first direct image of a star other than our Sun revealed that Betelgeuse (α Orionis) has a bright spot more than 2000 K hotter than the rest of the star's surface. The exact nature of the spot is not yet understood.

dropped sufficiently for the nuclei of atoms to form. Significant NUCLEOSYNTHESIS could begin and the COSMIC ABUNDANCE of elements was determined, primarily hydrogen and helium. The next significant event was at roughly T = 300,000 years after the Big Bang. At this time the universe became transparent to photons and the radiation decoupled from matter. As the universe continued to expand, matter evolved separately from the COSMIC MICROWAVE BACKGROUND (CMB) and atoms formed. Gravity caused small inhomogeneities in the matter distribution to collapse and form stars, galaxies and galaxy clusters.

The Big Bang model made a significant prediction: the Universe should have a temperature, and the temperature could be measured by looking at the CBR that permeates the entire Universe. Calculations within the framework of the Big Bang indicated that as the Universe expanded, the temperature of the CBR should constantly drop and the current temperature should be around 5 K. Two engineers from Bell Labs, Arno PENZIAS and Robert Wilson, were trying to determine why the Holmdel horn antenna, used to communicate with satellites, constantly picked up static at microwave frequencies. They failed to find a terrestrial or electrical explanation for the static and it was finally deduced that the noise was actually the CBR. This discovery, that the Universe has background radiation at a temperature very close to the theoretical predictions of the Big Bang model, was the most convincing piece of evidence that the Universe actually began as a condensed cosmic singularity many billions of years ago. Later observations by balloon-borne telescopes and the NASA satellite COBE precisely mapped this background radiation, even measuring anisotropies and our own motion relative to it.

Along with the successes of the Big Bang theory, there were also some problems. The horizon problem, the flatness problem, the lack of magnetic monopoles and the smoothness of the CBR all presented problems for the Big Bang model. Since the Universe is expanding rapidly in all directions, the two opposite horizons would be out of causal contact, and therefore would not have to show the same level of homogeneity as they do; this is the crux of the horizon problem. The flatness problem suggests that the Universe is so close to critical density that the initial conditions must have been fine tuned to extraordinary precision. The lack of magnetic monopoles present in the Universe was also considered a problem. After completely mapping the CBR, there seemed to be discrepancies between the amount of mass seen in the Universe and the critical density inferred from the smoothness of the CBR. The idea of an INFLATION epoch in the early universe solved both the horizon and flatness problems. Inflationary theory postulates that in the very early universe, just after $T = 10^{-34}$ s, the universe expanded rapidly and increased in size by a factor of 10^{50}. This rapid expansion was driven by Higgs field symmetry breaking. Because the universe actually expanded faster than the speed of light, the horizons were actually in causal contact in the early universe, thus disposing of the horizon problem. The flatness and magnetic monopole problems also go away since we see only a small part of the entire Universe. There remained a problem. The density of the Universe derived from the CBR should be equal to the critical density according to the Big Bang theory with inflation. Observations of the luminous matter in the Universe fell short of the critical density by a factor of four or more.

As observations from space platforms improved, scientists concentrated on trying to derive an accurate and consistent value for the HUBBLE CONSTANT. Determining this constant required the distances to distant galaxies to be measured independently of the redshift using bright standard candles like supernovae. Results from supernova studies indicated that the Universe was not expanding according to a constant value for the Hubble constant, but is accelerating. Far from invalidating the Big Bang theory, however, it actually explains the aforementioned problems with the Big Bang theory. To explain the acceleration, either the COSMOLOGICAL CONSTANT must be non-zero or there exists a different form of unobserved matter that has negative gravity, recently termed QUINTESSENCE. Either the non-zero cosmological constant or the presence of this exotic quintessence drives the expansion of the Universe.

Big Bear Solar Observatory (BBSO) Optical solar observatory at Big Bear Lake, California, built by the CALIFORNIA INSTITUTE OF TECHNOLOGY in 1969. Now managed by the New Jersey Institute of Technology, the observatory monitors the Sun with several optical telescopes, the largest of which has an aperture of 0.65 m (26 in.). Its location in the middle of the lake provides stable daytime observing conditions, eliminating the bad seeing caused by heat haze over land. BBSO participates in the GLOBAL OSCILLATION NETWORK GROUP.

Big Dipper Popular (chiefly US) name for the saucepan shape formed by seven stars in Ursa Major, also known as the PLOUGH.

BIMA Abbreviation of BERKELEY ILLINOIS MARYLAND ASSOCIATION

binary pulsar PULSAR orbiting another star, forming a

▼ **Big Bear Solar Observatory** The suite of telescopes in the dome of the Big Bear Solar Observatory observe the Sun in visible, near-infrared and ultraviolet light, thus giving astronomers detailed information about different parts of the solar surface and atmosphere. This information helps them to monitor and understand the way in which features on the Sun's surface form and evolve.

BINARY STAR system. The first binary pulsar to be discovered was PSR 1913+16, in 1974, for which its discoverers, Joseph Taylor and Russell HULSE, were awarded the 1993 Nobel Prize for physics. The pulsar, with a period of 0.059 seconds, orbits a neutron star with a period of almost 8 hours. Pulses from the companion neutron star have not been detected, but this might only be the result of an unfavourable viewing angle. The pulses from the pulsar arrive 3 seconds earlier at some times relative to others, showing that the pulsar's orbit is 3 light-seconds across, approximately the diameter of the Sun. Since this is a binary system, the masses of the two neutron stars can be determined, and they are each around 1–3 times the mass of the Sun. Observations have shown that the pulsar's orbit is gradually contracting, due to the emission of energy in the form of GRAVITATIONAL WAVES, as predicted by Einstein's theory of GENERAL RELATIVITY, causing the pulsar to reach PERIASTRON slightly early. Also, periastron advances 4° per year in longitude due to the gravitational field (thus the pulsar's periastron moves as far in a day as Mercury's moves in a century).

binary star Double star in which the two components are gravitationally bound to each other and orbit their common centre of mass. More than half of stars observed are binary or belong to multiple systems, with three or more components. In a CLOSE BINARY the stars interact directly with each other. Other double stars are optical doubles, which result when two stars appear to be close because they lie almost on the same line of sight as viewed from Earth, but in reality lie at a vast distance from each other. Binary systems are classified in a variety of ways, including the manner in which their binary nature is known.

Common proper motion binaries are seen as distinct objects that move across the sky together. No orbital motion is observed because they lie so far apart that their orbital period is very long.

VISUAL BINARIES are pairs that are resolved into separate components and for which orbital motion is observed.

ASTROMETRIC BINARIES are systems where only one component is seen. This star is observed to 'wobble' in a periodic manner as it moves across the sky, a movement caused by the gravitational pull of the unseen companion.

Spectrum binaries are systems where a SPECTROGRAM of an apparently single star reveals two sets of spectral lines characteristic of two different types of stars.

SPECTROSCOPIC BINARIES are systems where only one star is observed directly, but its spectrum shows one or two sets of spectral lines, which show a DOPPLER SHIFT, indicating orbital motion. These are known as single- and double-lined spectroscopic binaries respectively.

ECLIPSING BINARIES are systems aligned so that one or both members of the system are periodically observed to pass in front or behind the other, causing the total luminosity of the system to fluctuate in a periodic manner.

In general the nature of a binary system can be ascertained by the distance between the pair and thus the orbital period. Widely separated pairs that do not influence each other are known as double stars. Common proper motion pairs and visual binaries have the largest separation and longest orbital periods, which can be up to several million years. Eclipsing and spectroscopic binaries generally have the smallest separation, and their orbital periods can be as short as a few hours. Binaries with non-MAIN-SEQUENCE components, such as WHITE DWARFS, NEUTRON STARS or BLACK HOLES, can have orbital periods of as little as a few minutes, although these systems are short-lived, with lifetimes less than about 10 million years.

By convention, systems with orbital periods of less than 30 years are termed close binaries because the short orbital period implies separation of less than about 10 AU (the distance of Saturn from the Sun). Stars that remain within their own ROCHE LOBES are termed DETACHED BINARIES, while SEMIDETACHED BINARIES have one star (often a red giant) filling its Roche lobe. In semidetached binaries, material can spill from the lobe-filling star through the first LAGRANGIAN POINT on to its companion, often by transfer via an ACCRETION DISK. This MASS TRANSFER has considerable effect on the evolution of the two components. If both stars fill their Roche lobes, they are termed CONTACT BINARIES, and if material has escaped from the lobes to surround both stars, the system is known as a common envelope binary.

The first visual binary star was discovered by Joannes Baptista RICCIOLI in 1650, when he observed MIZAR and saw that the star was actually double, with a separation of 14″ (this was distinct from the known companion ALCOR). Small numbers of binaries were accidentally found by telescopic observers in the course of the next century, but it was not until 1767 that John MICHELL argued that stars appearing close together really were connected 'under the influence of some general law'. The observational proof of Michell's theory did not appear until 1804. William HERSCHEL had originally set out to determine the parallactic shift of the brighter (and, therefore, closer) member of an unequal pair of stars with respect to the fainter component. But for many stars, particularly CASTOR, the motion of the companion with respect to the primary star could be explained only if the two stars were regarded as physically connected, rotating around a common centre of gravity. Later observations showed that the apparent motion of the companion described an arc of an ellipse.

In 1827 the French astronomer Félix Savary (1797–1841) was the first to calculate an orbit that would predict the motion of the companion star with respect to the primary. An important consequence of this was that the masses of the two stars were directly obtainable from elements of the true orbit.

What is actually observed, the apparent orbit, is the projection of the true orbit on to the plane of the sky. The seven elements needed to define the size, shape and orientation of the true orbit are calculated from observations of POSITION ANGLE and separation (*see* ORBITAL ELEMENTS). The position angle is the angle that the line between the stars makes with the north point, and the separation is the angular distance between the two stars in arcseconds. By plotting position angle and separation from measurements made at different times the apparent orbit can be drawn. From the elements of the true orbit, the size of the semi-major axis (a) and the period of revolution (P years) will allow the sum of the masses of the two stars to be calculated, provided that the parallax of the binary is known. The relation connecting these quantities is (from Kepler's third law):

$$M_1 + M_2 = \frac{4\pi^2 a^3}{GP^2}$$

where M_1 and M_2 are in terms of the Sun's mass and G is the gravitational constant.

▼ **binary star** The two components of a binary system move in elliptical orbits around their common centre of gravity (G), which is not half way between them but nearer to the more massive component. In accordance with Kepler's second law of motion, they do not travel at uniform rates. In (2) they are greatly separated and moving slowly. In (6) they are closer and moving more rapidly.

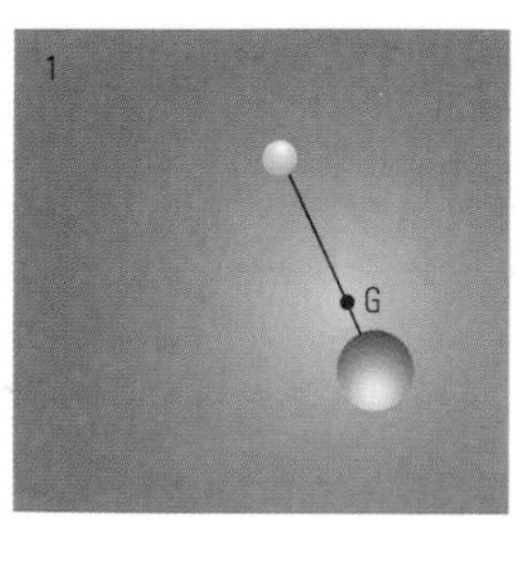

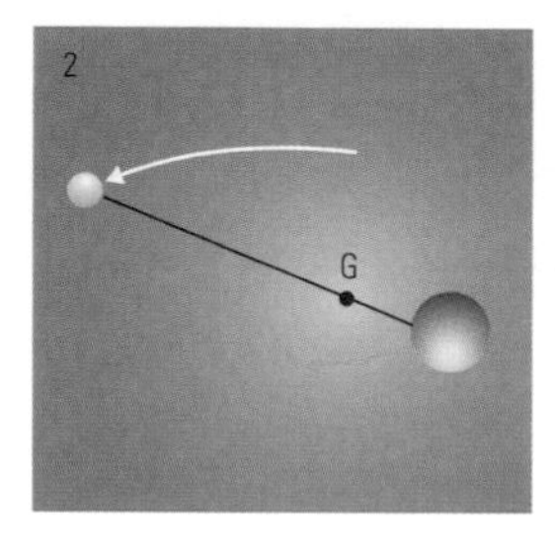

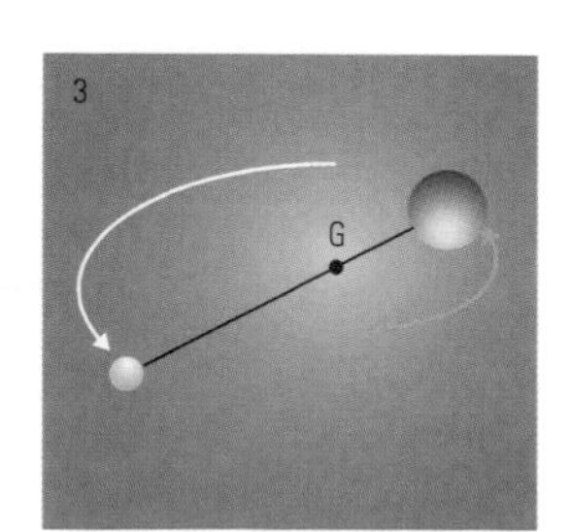

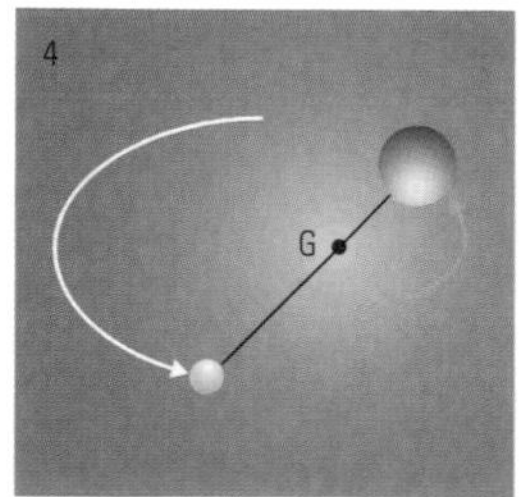

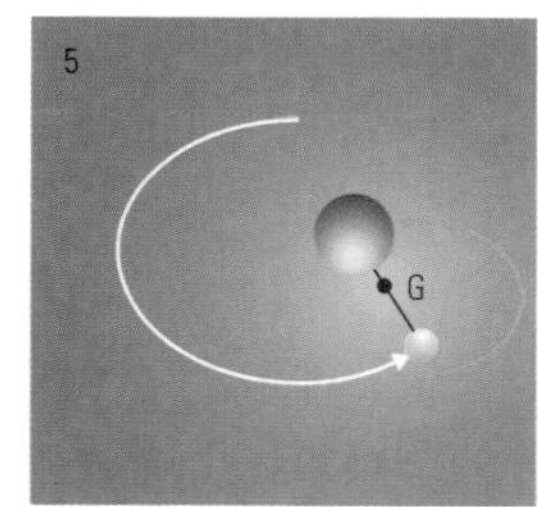

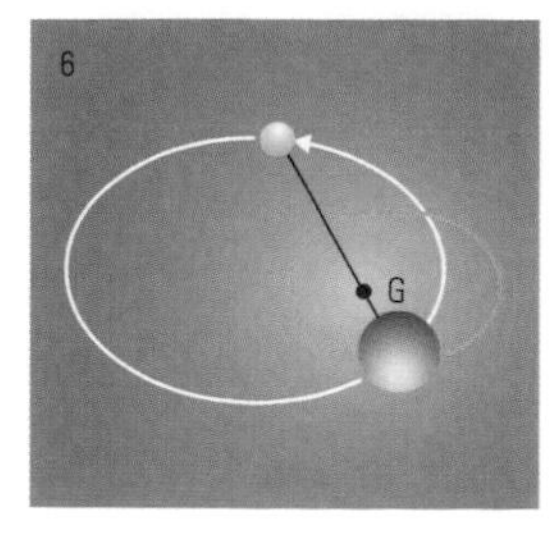

In order to derive individual masses, another relation between M_1 and M_2 is needed. This can be obtained by plotting the position of each star against the background of fainter, relatively fixed stars in the same field. The two stars will both appear to describe ellipses, which vary only in size depending on the mass of each star, thus:

$$\frac{M_2}{M_1} = \frac{a_2}{a_1}$$

where a_1 and a_2 are the semimajor axes of the apparent ellipses described by the two stars. This process is a very time-consuming job, requiring measurements of hundreds of images obtained over many years. It has only been carried out for a few systems.

The determination of masses from spectroscopic binaries is more difficult than for visual binaries because it is not possible to determine the inclination of the apparent orbit unless eclipses occur. The best that can be done is to assume that the inclination is 90° and to calculate from this the minimum values that the masses are likely to take. *See also* BLUE STRAGGLER; CATACLYSMIC VARIABLE; RECURRENT NOVA; X-RAY BINARY

binding energy Energy that is released when an atomic nucleus is formed; it is also called the mass defect because it is the difference in energy terms between the mass of the nucleus and the sum of the masses of its protons and neutrons. A star's energy is the binding energy that is released as hydrogen is built up to helium, carbon, oxygen, and so on. That process, however, stops at iron, for which the binding energy per nucleon is a maximum; building up heavier nuclei requires more energy than is released by the reaction. The consequent formation of an inert iron core in massive stars leads directly to type II SUPERNOVA explosions. *See also* FISSION; FUSION; NUCLEOSYNTHESIS

binoculars Two, usually small, TELESCOPES mounted side by side so that both eyes can be used simultaneously. Modern binoculars are usually refracting telescopes with fixed eyepieces and a set of built-in prisms to provide an upright, right-reading view (*see* PORRO PRISM). Binoculars are normally described by two numbers, for example, 7 × 50. The first number is the MAGNIFICATION, and the second number is the APERTURE in millimetres. Conveniently, the EXIT PUPIL in millimetres is calculated by dividing the second number by the first (about 7 mm in this example).

True binocular vision, where both eyes are used to form a stereoscopic view and gauge the distance to an object, is not effective when observing celestial objects. The distance from the objective lenses to the object is far too great compared with the distance between the lenses. However, apparent brightness and resolution are each improved when both eyes are used.

biosphere Shallow region over the surface of a planet that supports life. Earth's biosphere includes the land to about 8 km (5 mi) above mean sea level, the oceans to depths of at least 3.5 km (2 mi) and the atmosphere to a height of about 10 km (6 mi): in total, it is less than 20 km (12 mi) thick. Recently discovered microbial organisms called extremophiles (*see* LIFE IN THE UNIVERSE) occupy a secondary subsurface biosphere that is believed to extend to 4 km (2.5 mi) beneath the surface of the continental crust and 7 km (4.3 mi) below the seafloor. Specific habitats within a biosphere are known as **biomes**. Only Earth is known to possess a surface and a subsurface biosphere. Mars, Europa and Titan are targets for future space missions that will search for extraterrestrial subsurface biospheres, and in Titan's case, perhaps a surface biosphere as well.

Biot, Jean-Baptiste (1774–1862) French physicist who co-discovered the Biot–Savart law, which concerns the intensity of a magnetic field produced by a current flowing through a wire. He demonstrated the extraterrestrial origin of meteorites, after investigating the meteorite shower at L'Aigle, France, on 1803 April 26. The following year he accompanied the chemist Joseph-Louis Gay-Lussac on a balloon flight, to an altitude of 4000 m (13,000 ft), in order to collect scientific data about the upper atmosphere.

bipolar flow Non-spherical flow of material from a star. Occasionally stars lose matter at copious rates, for example at the end of their protostellar lives, and during their RED GIANT phases. If these flows are organized into two oppositely directed streams, they are called 'bipolar'. This phenomenon is responsible for the creation of the rapidly moving HERBIG-HARO OBJECTS, flying away from the precursors of T TAURI STARS. Very narrow, highly confined radio-emitting JETS characterize this youthful phase; these jets disturb the surrounding dark clouds and sweep up larger volumes of much slower molecular gas, also bipolar in pattern. For older stars the outflows may lead to bipolar PLANETARY NEBULAE.

Usually the stars that generate bipolar flows are surrounded by extensive, flattened, dusty envelopes – huge toroids – orientated perpendicular to the star's rotation axis. The interplay of stellar rotation and mass outflow is believed to create the bipolar outflows. Binarity of the driving star is implicated in at least some flows, as also may be magnetic fields.

Birr Castle astronomy Astronomy practised at Birr Castle, near Birr, County Offaly (formerly Parsonstown, King's County) in Ireland, with the 'Leviathan of Parsonstown', a 72-inch (1.8-m) telescope completed by William Parsons, Third Earl of ROSSE, in 1845. Using local workmen whom he trained, Rosse cast mirrors of speculum metal (a highly reflective copper–tin alloy) and constructed first a 36-inch (0.9-m) reflecting telescope. Later came the 72-inch, which remained the largest in the world until 1917, when Mount Wilson Observatory's 100-inch (2.5-m) Hooker Telescope went into service. The Leviathan's 72-inch mirror had a focal length of 54 ft (16.5 m), and the instrument was suspended between two walls 56 ft (17.1 m) high. It could not be effectively driven, and was unsuitable for photography.

Despite these drawbacks, the telescope's great size enabled Rosse to discern the spiral structure of certain nebulae which are now known to be galaxies, most notably the Whirlpool Galaxy (M51), which he sketched. Birr became an astronomical centre, and Rosse himself became a skilful observer, making his results freely available. After his death in 1867, his son Laurence Parsons, later the Fourth Earl, took over the observatory. J.L.E. DREYER was assistant at Birr between 1874 and 1878, during which period he accumulated much of the information that was to be later published in his NEW GENERAL CATALOGUE.

From 1900 activity gradually decreased; the Fourth Earl died in 1908, and the 72-inch was dismantled. In 1916 the last Birr astronomer, Otto Boeddicker (1853–1937), departed, and work ceased. In 1996–98 the 72-inch telescope was restored to working order with a new glass mirror.

BIS Abbreviation of BRITISH INTERPLANETARY SOCIETY

Blaauw, Adriaan (1914–) Dutch astronomer who specialized in the structure of the Milky Way galaxy. His research has covered the processes by which stars form, star clusters and associations, and measuring the cosmological distance scale. Blaauw has done much to advance European astronomy, playing a key role in founding the EUROPEAN SOUTHERN OBSERVATORY and establishing the leading journal *Astronomy & Astrophysics*. He helped to plan the HIPPARCOS astrometric satellite.

black body radiation Radiation emitted by an idealized perfect radiator. It has a CONTINUOUS SPECTRUM that depends only on the temperature of the source.

By KIRCHHOFF'S LAWS the efficiency of the emission by a heated object at a particular wavelength is proportional to the efficiency of its absorption at the same wavelength. Thus an object that absorbs with 100% efficiency over the whole spectrum, known as a black body, will also be the most efficient when it comes to emitting radiation. A good practical approximation to a black body is a small hole in the side of an otherwise closed box.

The emission from a black body, known as black body radiation, is a good fit to the emission from many astronomical objects, including stars and interstellar dust clouds (spectrum lines are mostly minor deviations from the overall emission). Black body radiation follows a bell-shaped distribution given by the PLANCK DISTRIBUTION. The peak of the distribution shifts to shorter wavelengths as the temperature increases (*see* WIEN'S LAW). This leads to the common experience that at moderate temperatures objects glow a dull red, then change colour successively through bright red, yellow, white to blue-white as the temperature is increased. The total emitted energy increases rapidly with temperature, leading to the STEFAN–BOLTZMANN LAW.

black drop Optical effect observed during the initial (ingress) and final (egress) stages of a TRANSIT of MERCURY or VENUS. Once the planet is fully projected on the Sun at ingress, but before its trailing edge breaks apparent contact with the solar limb, the expectation is of instantaneous separation. Instead, the planet seems to lengthen, with a dusky ligament appearing briefly to link it to the Sun's limb. This effect is the black drop, which may be likened to a drop of water before it falls from a tap. It is again seen at egress as the planet's leading edge approaches the opposite limb. The effect may be physiological in origin, but it has also been attributed to atmospheric turbulence and to instrument defects.

black dwarf Ultimate state of a WHITE DWARF star. A white dwarf does not have a means of maintaining its heat, and it therefore cools steadily. Given long enough, it cools to invisibility, and is then called a black dwarf. Such objects are hypothetical: none have been observed, and it is doubtful whether our Galaxy is old enough for any to have cooled sufficiently to enter this state. *See also* STELLAR EVOLUTION

Black Eye Galaxy (M64, NGC 4826) SPIRAL GALAXY in the constellation Coma Berenices (RA $12^h\ 56^m.7$ dec. $+21°41'$). It has a prominent lane of dark material close to its nucleus. The galaxy was originally discovered by J.E. BODE in 1779. It has apparent dimensions of $9'.2 \times 4'.6$ and magnitude +8.5. The Black Eye Galaxy's true diameter is 65,000 l.y., and it lies 24 million l.y. away.

black hole Object that is so dense and has a gravitational field so strong that not even light or any other kind of radiation can escape: its escape velocity exceeds the speed of light. Black holes are predicted by Einstein's theory of GENERAL RELATIVITY, which shows that if a quantity of matter is compressed within a critical radius, no signal can ever escape from it. Thus, although there are many black hole candidates, they cannot be observed directly. Candidates are inferred from the effects they have upon nearby matter. There are three classes of black hole: stellar, primordial (or mini) and supermassive.

A **stellar black hole** is a region of space into which a star (or collection of stars or other bodies) has collapsed. This can happen after a star massive enough to have a remnant core of more than 2.3 solar masses (the Landau–Oppenheimer–Volkov limit for NEUTRON STARS) reaches the end of its thermonuclear life. It collapses to a critical size, overcoming both electron and neutron degeneracy pressure, whereupon gravity overwhelms all other forces (*see* DEGENERATE MATTER).

Primordial black holes, proposed by Stephen HAWKING, could have been created at the time of the BIG BANG, when some regions might have got so compressed that they underwent gravitational collapse. With original masses comparable to that of Earth or less, these mini-black holes could be of the order of 1 cm (about half an inch) or smaller. In such small black holes, quantum effects become very important (*see* QUANTUM THEORY). It is possible to show that such a black hole is not completely black, but that radiation can 'tunnel out' of the event horizon at a steady rate; such radiation is known as HAWKING RADIATION. This then could lead to the evaporation of the hole. Primordial black holes could thus be very hot, and from the outside they could look like WHITE HOLES, the time-reversals of black holes.

It seems that **supermassive black holes** of the order of 100 million solar masses lie at the centres of ACTIVE GALACTIC NUCLEI, extreme examples of which are QUASARS. It is thought there may also be supermassive black holes at the centres of ordinary galaxies like the Milky Way.

The lifetime of a black hole can be shown to be proportional to the cube of its mass. For black holes of stellar mass, their potential lifetime is of the order of 10^{67} years. Many primordial black holes will have evaporated away completely by now.

There are various different models of black holes. The most straightforward is that of the Schwarzschild black hole, a non-rotating black hole that has no charge. In nature, however, it is expected that black holes do rotate but have little charge, so the model of a rotating Kerr black hole with no charge is probably the most applicable. A Kerr–Newman black hole is rotating and has a charge. A Reissner–Nordström black hole is a non-rotating black hole with a charge.

The radius of a non-rotating Schwarzschild black hole of mass M is given by $2GM/c^2$, where G is the gravitational constant and c is the speed of light. When a star becomes smaller than this SCHWARZSCHILD RADIUS, gravity completely dominates all other forces. The Schwarzschild radius determines the location of the surface of the black hole, called the EVENT HORIZON. Only the region on and outside the event horizon is relevant to the external observer; events inside the event horizon can never influence the exterior. There is no lower limit to the

▼ **Black Eye Galaxy** The prominent dust lane that gives the Sb-type spiral galaxy M64 its name is just detectable with large amateur instruments. Larger instruments or image processing are needed to resolve the outer spiral arms.

radius of a black hole. Some of the primordial black holes could be truly microscopic.

When a stellar black hole first forms, its event horizon may have a grotesque shape and be rapidly vibrating. Within a fraction of a second, however, the horizon settles down to a unique smooth shape. A Kerr black hole has an event horizon that is flattened at the poles rather than circular (just as rotation flattens the Earth at its poles). What happens to matter after it crosses the event horizon depends on whether or not the star is rotating. In the case of a collapsing but non-rotating star that is spherically symmetric, the matter is crushed to zero volume and infinite density at the SINGULARITY, which is located at the centre of the hole. Infinitely strong gravitational forces deform and squeeze matter out of existence at the singularity, which is a region where physical theory breaks down. In a rotating Kerr black hole, however, the singularity need not be encountered. Rotating black holes have fascinating implications for hypothetical space travel to other universes.

The density of matter in a star as it crosses the critical event horizon need not necessarily be very high: its density could even be less than that of water. This is because the density of any body is proportional to its mass divided by its radius cubed, and the radius of a black hole is, as we have seen, proportional to its mass. These two facts combined imply that the density at which a black hole is formed is inversely proportional to the square of the mass. Take a supermassive black hole with a mass of from 10,000 to 100 million solar masses – the mass of a black hole that might be found at the centre of certain active galaxies. Such a collapsing mass would reach the black hole stage when its average density was roughly that of water. If the mass of the collapsing sphere were that of an entire galaxy, the average density of matter crossing the event horizon would be less than that of air.

Attempts to discover stellar black holes must rely on the influence of their gravitational fields on nearby matter, and/or their influences on the propagation of radiation in the vicinity of the hole. Black holes within BINARY STAR systems are potentially the easiest to detect because of the influence on their companion. Material is pulled from the companion into the black hole via an ACCRETION DISK. The frictional heating within the disk leads to the emission of X-rays (*see* X-RAY BINARY). The first candidate where one companion in a binary system is thought to be a black hole is the X-ray source CYGNUS X-1. At the position of this X-ray source lies a SPECTROSCOPIC BINARY star HDE 226868, which has a period of 5.6 days. More recently, all-sky monitors on space-borne X-ray observatories have discovered soft X-ray transients (SXTs); objects that produce rare, dramatic X-ray outbursts (typically separated by decades). Around 75% of SXTs contain black hole candidates.

The existence of supermassive black holes in quasars and as the central sources in active galactic nuclei is generally accepted as the means of explaining the phenomena observed. Many ordinary galaxies like our own show enhanced brightening at their cores, along with anomalously high velocities of objects near the centre, suggesting the existence of a black hole.

There is also the MISSING MASS PROBLEM: the density of the observable matter in our Universe is much less than the theoretically computed value needed to 'close' the Universe (*see* CLOSED UNIVERSE), and it may be that at least some is in the form of black holes.

Blagg, Mary Adela (1858–1944) English astronomer who catalogued and mapped lunar features. She standardized the nomenclature of the Moon's topographic features (1907–13), collating and correcting thousands of names assigned by previous lunar cartographers. In 1920 the International Astronomical Union (IAU) appointed Blagg to its newly established Lunar Nomenclature Commission; twelve years of further research produced the authoritative *Named Lunar Formations* (1932, compiled with Karl Muller). With W.H. Wesley, she composed a *Map of the Moon* (1935), which remained the IAU's official lunar map until the 1960s.

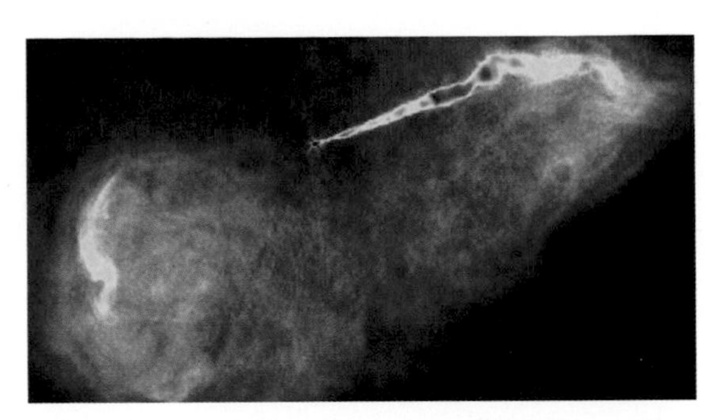

◀ **black hole** Radio jets, spewing from the centre of a galaxy, and lobes of radio emission are the signatures of a supermassive black hole. The giant elliptical galaxy M87 (NGC 4486) houses the radio source Virgo A (false colours indicate the strength of the radio emission).

blazar Term compounded from BL LACERTAE OBJECT and QUASAR; it refers to a specific kind of extragalactic object. The blazars are the most active of the galaxies with active nuclei, that is, the galaxies whose central regions are undergoing energetic processes that turn them into SEYFERT GALAXIES, BL Lacertac objects or quasars. The blazars show variable optical brightness, strong and variable optical polarization and strong radio emission. The variations in the optical region may be on timescales as short as days.

Much of the activity in ACTIVE GALACTIC NUCLEI is related to JETS of gas expelled from their central regions with relativistic velocities. The most probable explanation for the exceptional activity in blazars is that with these galaxies we are viewing jets directed straight towards us.

Blazhko effect Periodic change in the light-curves and periods of some RR LYRAE VARIABLE STARS; it was discovered by Sergei Nikolaevich Blazhko (1870–1956). The most likely cause is pulsation in two modes simultaneously, although certain stars appear to be 'oblique rotators', that is, stars in which the magnetic axis does not correspond to the rotational axis.

Blaze Star Popular name for the T Coronae Borealis, the brightest known RECURRENT NOVA. Normally around 11th magnitude, it has flared up to naked-eye brightness on two occasions, once in 1866 when it reached 2nd mag. and again in 1946 when it peaked at 3rd mag. T Coronae Borealis is a spectroscopic binary in which an M3 red giant orbits with a white dwarf every 227.5 days; gas falling from the red giant on to the companion causes the outbursts. A possible additional periodicity of 56.7 days may be caused by a third component.

blink comparator (blink microscope) Instrument that enables two photographs of the same area of sky, taken at different times, to be rapidly alternated to compare them. Any object that has changed position or brightness during the photograph intervals will show up.

The comparator has two optical paths so that the two photographs can be seen together in one viewing eyepiece. By careful adjustment, the separate images are brought into exact coincidence and then alternately illuminated, changing from one to the other about once a second. All features that are identical appear unchanged, but any object that is on only one of the photographs is seen to blink on and off. An object that has changed its position between the times the photographs were taken appears to jump to and fro and an object that has changed in brightness is seen to pulsate.

The eye is very efficient at detecting the few varying objects among what can be tens of thousands of star images. This simple technique makes possible the discovery of stars of large proper motion, minor planets, comets or variable stars, without the need individually to compare every star image on two photographs.

These days blink comparators are particularly used by hunters of novae and asteroids but past examples of their work include: the discovery of PLUTO by Clyde TOMBAUGH; the catalogue of over 100,000 stars brighter than magnitude 14.5 with detectable proper motion, produced by W.J. LUYTEN; and the majority of the nearly 30,000 known variable stars discovered at various observatories around the world.

Blinking Planetary (NGC 6826) PLANETARY NEBULA located in northern Cygnus (RA $19^h 44^m.8$ dec. $+50°31'$). The nebula has a compact 25″ diameter and overall magnitude +8.8. The central star is a relatively bright magnitude +10.6, which results in the interesting illusion that if the observer alternates between direct and AVERTED VISION, the nebula appears to blink in and out of view.

Bliss, Nathaniel (1700–1764) English astronomer, the fourth ASTRONOMER ROYAL (1762–64). Bliss was an able observer who successfully observed the 1761 transit of Venus. Bliss assisted James BRADLEY at Greenwich Observatory, succeeding him as Astronomer Royal, but made little impression on the observatory, dying only two years after his appointment.

BL Lacertae object Category of luminous ACTIVE GALACTIC NUCLEUS, the defining characteristics of which include very weak or unobservable emission lines and rapid variability. The prototype, BL Lacertae itself, was long listed as a variable star until its extragalactic nature became apparent from observations of the faint surrounding galaxy. When the objects' continuum light is faintest, weak emission lines similar to those of quasars may be detected, and their bright cores often have extensive haloes at radio wavelengths as well. These properties are well explained by a picture in which BL Lacertae objects are QUASISTELLAR OBJECTS (QSOs) or RADIO GALAXIES seen almost along the line of a relativistic JET, so that the radiation seen from the jet is Doppler boosted in frequency and intensity. This geometry also amplifies small changes in the velocity or direction of the jet, accounting for the strong variability of these objects, and fitting with the fact that many of them have small-scale radio jets showing superluminal motion. A few BL Lacertae objects have been detected at the highest photon energies, 1 TeV; this implies that their radiation is being beamed into a small angle (as in gamma-ray bursts), otherwise it would be lost to pair production within the source. BL Lacertae objects and the broadly similar optically violently variable (OVV) QSOs are often referred to collectively as BLAZARS.

blooming Natural film that formed on early uncoated lenses. It was noticed the bloom improved the transmission of the lenses and some opticians still use the term to refer to COATING. Blooming also refers to the loss of focus of a camera sensor because of excessive brightness and the term may occasionally have this meaning in astronomy.

blue moon Occasional blue colour of the Moon, due to effects in the Earth's atmosphere. It can be caused by dust particles, from volcanoes or forest fires, high in the upper atmosphere, which scatter light, making it appear blue. The expression sometimes refers to the occurrence of a second full moon in a calendar month, something which occurs about seven times every 19 years, and it is used in everyday speech to denote a rare event.

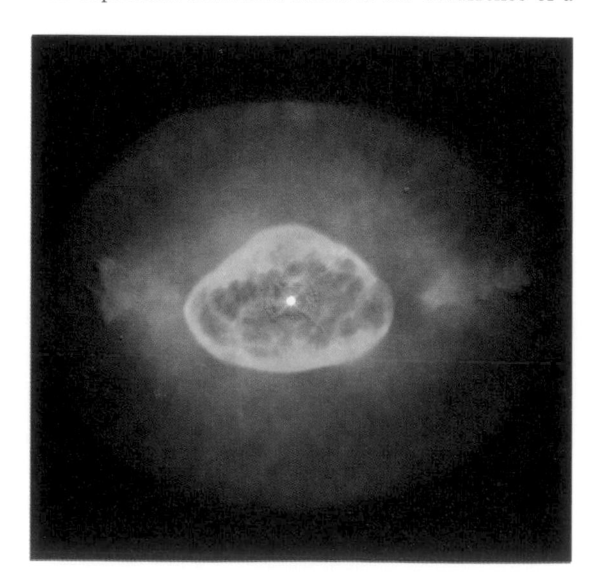

▶ **Blinking Planetary** The 'blink' of NGC 6826 is an optical effect rather than anything occurring within the planetary nebula itself.

blue straggler In a GLOBULAR CLUSTER, star that appears to be younger than the others because it still lies on the MAIN SEQUENCE after the majority of other cluster members have evolved off. It is observed to be on the main sequence beyond the TURNOFF POINT: it is bluer in spectral terms than the stars at the turnoff point. Being younger and more massive than the other stars in the cluster poses a problem as to the blue stragglers' origin. They are thought to be either stars in close binaries that have been rejuvenated by mass transfer from their companions, or stars that were produced by a stellar collision.

The Hubble Space Telescope has imaged many blue stragglers situated in the cores of globular clusters. They have also been found in OPEN CLUSTERS and in the DWARF GALAXY companions to the Milky Way. *See also* STELLAR EVOLUTION

BN object *See* BECKLIN–NEUGEBAUER OBJECT

Bode, Johann Elert (1747–1826) German astronomer, director of Berlin Observatory from 1772. He did much to popularize astronomy, founding the highly regarded *Astronomisches Jahrbuch* in 1774, which he edited for over half a century. In 1801 he published *Uranographia*, the most beautiful star ATLAS ever drawn, with a catalogue of over 17,000 stars and non-stellar objects. Bode's name is most famously associated with the empirical relationship between planetary distances known as BODE'S LAW, though he did not in fact discover it.

Bode's law (Titius–Bode law) Simple numerical relationship, first noticed by Johann Titius of Wittenberg, but popularized by Johann Elert BODE in 1772, which matches the distances of the then known planets from the Sun. The formula is produced by taking the numbers 0, 3, 6, 12, 24, 48, 96 and 192 (all, apart from the first two, being double their predecessor) and then adding four, giving the sequence 4, 7, 10, 16, 28, 52, 100, 196. If the distance of the Earth from the Sun is then taken to be 10, it is found that Mercury falls into place at 3.9, Venus at 7.2, Mars at 15.2, Jupiter at 52.0 and Saturn at 95.4. The discovery of Uranus at 191.8, by Sir William Herschel in 1781, initiated a hunt for the missing planet between Mars and Jupiter at 28. This led to the discovery in 1801 of the first minor planet, Ceres, and subsequently the ASTEROID BELT. Neptune, at 300.7, does not fit the sequence although Pluto, at 394.6, does. The relationship is now widely regarded simply as a mathematical curiosity, rather than indicating anything significant about the physical properties of the Solar System.

Bok, Bartholomeus Jan ('Bart') (1906–83) Dutch-American astronomer best known for his studies of the Milky Way and his discovery of BOK GLOBULES. At Harvard University (1929–57), Bok worked closely with his wife, **Priscilla Fairfield Bok** (1896–1975), to map the spiral arms of our Galaxy by 'star-counting' methods. In the course of this work, the two astronomers made highly detailed studies of the Carina region from Harvard's southern station in South Africa. During his tenure at Harvard, Bok was an early advocate of radio astronomy research, and he was instrumental in setting up Mexico's National Observatory at Tonantzintla. At Mount Stromlo (Australia) Observatory (1957–66), where he was director, they carried out similar surveys for the Magellanic Clouds. He was especially interested in the Galaxy's many different kinds of gas and dust clouds, and he was the first to identify the opaque, cool (10 K), tiny (0.6–2 l.y.) condensations of gas and dust now named after him that give rise to low-mass stars. The Boks wrote the classic book *The Milky Way*, which had run through five editions as of 1981. From 1966 to 1974, he directed the Steward Observatory of the University of Arizona;

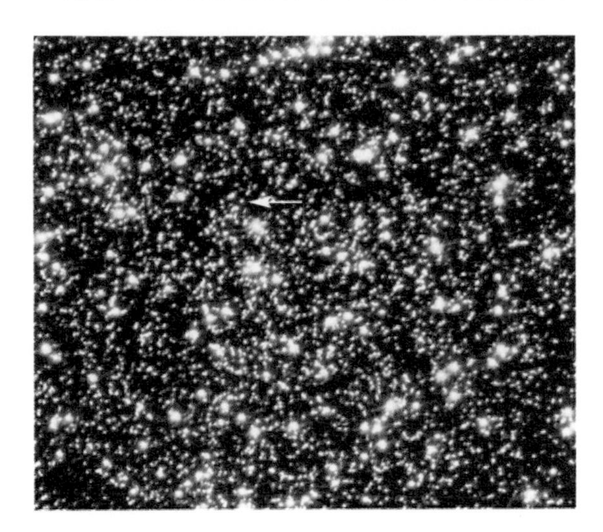

◄ **blue straggler** Surrounded by older yellow stars in the globular cluster 47 Tucanae is a massive young, blue star (arrowed). This blue straggler is thought to be the product of the slow merger of two stars in a double star system.

through his efforts Steward acquired a new 2.3-m (90-in.) reflector at Kitt Peak.

Bok globule *See* GLOBULE

bolide Term often used to describe a major FIREBALL that produces a sonic boom. Such events are frequently associated with the deposit of METEORITES.

bolometer Instrument to measure the total radiation received in a telescope from a celestial body by its effect upon the balance of an electrical circuit.

bolometric magnitude Measure of the total radiation of all wavelengths emitted by or received from a star expressed on the stellar magnitude scale.

Bolton, John Gatenby (1922–93) English scientist who spent much of his career in Australia as a pioneer of radio astronomy. Bolton used a radio interferometer to identify the extraterrestrial radio sources Taurus A (the Crab Nebula), Centaurus A (the galaxy NGC 5128) and Virgo A (the galaxy M87). His team was the first to confirm that the centre of the Milky Way galaxy was a strong source of radio emission, which they called Sagittarius A. Bolton founded CalTech's Owens Valley Radio Observatory in the 1950s and later directed Australia's Parkes Radio Observatory.

Boltzmann constant (symbol k) Constant defined as the universal gas constant (R) divided by Avogadro's constant. It is the gas constant per particle and has a value of 1.38066×10^{-23} J/K. It is often encountered in equations relating the properties of gases to temperature, for example the gas pressure law, in the form $P = nkT$, where P is the pressure, n the number of particles per cubic metre, and T the temperature. It also occurs in BOLTZMANN'S EQUATION.

Boltzmann's equation Equation, introduced by the Austrian physicist Ludwig Boltzmann (1844–1906), that gives the relative populations of atoms with electrons in different levels of excitation:

$$\frac{N_b}{N_a} = \frac{g_b}{g_a} e^{-(E_b-E_a)/kT}$$

where N_a and N_b are the number densities of atoms in excitation levels a and b, g_a and g_b are constants for the levels called statistical weights, E_a and E_b are the excitation energies of the levels, k is BOLTZMANN'S CONSTANT and T is the temperature. The equation is used in understanding stellar spectra and processes occurring in stellar atmospheres.

Bond, George Phillips (1825–65) American astronomer who in 1859 succeeded his father, William Cranch BOND, as director of the HARVARD COLLEGE OBSERVATORY. He made many contributions to the studies of double stars and stellar parallax. With his father, he discovered Saturn's Crepe Ring (the C Ring) and its eighth satellite, Hyperion, independently of William LASSELL. Bond argued against the then-popular theory that Saturn's rings were solid, having observed stars through the Crepe Ring. He used the new technique of photography (using 'wet' plates) to study the Moon and the planets and was the first to photograph a star, Vega, in 1850 and the first to image a double star, Mizar, in 1857.

Bond, William Cranch (1789–1859) American astronomer who founded the HARVARD COLLEGE OBSERVATORY and was its director from 1839, being succeeded by his son George Phillips BOND. A clock-maker by trade, he designed and built sophisticated chronometers for navigation and astronomy. He transferred his own private observatory to Harvard; Bond procured a 15-inch (380-mm) refracting telescope, then the largest in the world, for the observatory. With this, he and his son discovered Saturn's satellite Hyperion in 1848, and the planet's semitransparent Crepe Ring (the C Ring) in 1850. Bond also made detailed studies of sunspots and the Orion Nebula (M42).

Bondi, Hermann (1919–) British cosmologist and mathematician, born in Austria, who co-originated the STEADY-STATE THEORY of cosmology. After working for the British Admiralty during World War II, Bondi was appointed professor of applied mathematics at King's College, London in 1954. In 1948, with Fred HOYLE and Thomas GOLD, he constructed a new cosmological theory calling for the 'continuous creation' of matter at the rate of 10^{-10} nucleons per cubic metre per year. This process, explained Bondi, allowed the Universe to maintain a constant average density of matter that counterbalanced flat space expanding at a constant rate. The discovery in 1964 of the COSMIC MICROWAVE BACKGROUND cast serious doubt on the validity of the steady-state theory, which has since been abandoned in its original form. The work of Bondi and Hoyle on stellar structure, especially of the formation of heavier elements inside stars, was a valuable by-product of this theory. Bondi later demonstrated that gravitational waves are real, and not just theoretical, consequences of Einstein's general theory of relativity, and his work has defined the physical properties of these phenomena.

Bonner Durchmusterung (BD) Catalogue containing data on 324,000 stars down to magnitude 9.5, published in 1859–62 by Friedrich Argelander and extended southward in its coverage in 1886 by Eduard Schönfeld, adding a further 133,000 stars. The stars are numbered in declination zones from +90° to −23°, and are cited in the form 'BD +52° 1638'.

Boötes *See* feature article

BOÖTES (GEN. BOÖTIS, ABBR. BOO)

Large northern constellation, representing a huntsman, between Canes Venatici and Serpens Caput. It is easily recognized by virtue of ARCTURUS, mag. −0.1, which is the fourth-brightest star in the sky and lies on a continuation of the curved line through the stars ϵ, ζ and η Ursae Majoris. Two interesting binaries are ϵ Boo, IZAR (or Pulcherrima), which has orange and bluish-white components, mags. 2.6 and 4.8, separation 2″.9; and ξ Boo, which has yellow and orange components, mags. 4.8 and 7.1, separation 7″.1, period 152 years. There are no bright deep-sky objects in the constellation. The Quadrantid meteor shower radiates from the northern part of Boötes, which now incorporates the former constellation Quadrans Muralis, after which the shower is named.

BRIGHTEST STARS

	Name	RA h m	dec. ° ′	Visual mag.	Absolute mag.	Spectral type	Distance (l.y.)
α	Arcturus	14 16	+19 11	−0.05	−0.3	K2	37
ϵ	Izar	14 45	+27 04	2.35	−1.7	A0	210
η	Muphrid	13 55	+18 24	2.68	2.4	G0	37
γ	Seginus	14 32	+38 18	3.04	1.0	A7	85
δ		15 16	+33 19	3.46	0.7	G8	117
β	Nekkar	15 02	+40 23	3.49	−0.6	G8	219

Borrelly, Comet 19P/ SHORT-PERIOD COMET discovered on 1904 December 28 by Alphonse Borrelly of Marseilles. The comet is usually quite faint, reaching magnitude +8 at favourable returns. Its orbital period is 6.86 years, with the most recent return to perihelion coming on 2001 September 14. At this return, Comet Borrelly was visited by the DEEP SPACE 1 probe, becoming the second comet (after 1P/HALLEY) to have its nucleus imaged. Results from the spacecraft's September 22 flyby showed the nucleus to be a dark, elongated (6 × 3 km/3.7 × 1.9 mi) body with three emerging gas jets.

boson Elementary particle that obeys Bose–Einstein statistics. Bosons are symmetric particles, which have integral spins (0/1). Common bosons are the helium nucleus, the pi meson and the photon.

Boss, Lewis (1846–1912) American astronomer who in 1895 began an ambitious programme to measure star positions and magnitudes with unprecedented accuracy, culminating in two important star catalogues, the *Preliminary General Catalogue of 6,188 Stars* (1910) and the *General Catalogue of 33,342 Stars* (1937), the latter work completed decades after his death by his son **Benjamin Boss** (1880–1970). Boss observed the stars visible from Earth's northern hemisphere from New York's Dudley Observatory, where he became director in 1876, and the southern stars from Argentina.

Bowditch, Nathaniel (1773–1838) Self-educated American scientist, author and translator of Laplace's massive *Traité de mécanique céleste* (1829–39). Bowditch made many observations of meteors, comets and the Moon, which he wrote up between 1804 and 1820 – some of the first publications based upon original astronomical observations to appear in America.

Bowen, Edward George (1911–91) Welsh scientist who used radar and other equipment salvaged at the end of World War II to help found Australian radio astronomy. He played a key role in the design and construction of the Parkes Radio Observatory's 64-m (210-ft) radio telescope, which was used for several important surveys of radio sources. Bowen pioneered the use of radio telescopes to detect the radio-wavelength echoes produced by meteors.

Bowen, Ira Sprague (1898–1973) American astrophysicist who in 1927 showed that previously unidentified lines in the spectra of nebulae were due not to a new element, 'nebulium', but to so-called FORBIDDEN LINES of ionized oxygen and nitrogen. As long-time director of the Mount Wilson and Palomar Observatories (1946–64), Bowen oversaw the construction of the 200-inch (5-m) Hale Telescope and the 48-inch (1.2-m) Oschin Schmidt telescope used to make the PALOMAR OBSERVATORY SKY SURVEY.

bow shock Sharp boundary standing in the SOLAR WIND flow upstream of a planetary MAGNETOSPHERE or other obstacle, such as a cometary or planetary IONOSPHERE. The flow of the solar wind is supersonic and indeed faster than the various characteristic speeds associated with the magnetic field and plasmas of which it is constituted. As a consequence, information that the solar wind is approaching a planetary magnetosphere is unable to propagate upstream into the flow, and a curved standing shock wave is formed. At this shock, the solar wind plasma is rapidly decelerated and deflected around the magnetosphere; the upstream flow kinetic energy is converted into plasma heating. Thus a region of hot, slow-flowing, turbulent solar wind plasma and magnetic field appears downstream in the planet's magnetosheath (the region between the bow shock and the MAGNETOPAUSE).

Since the solar wind plasma is collisionless, it was originally assumed that a classical bow shock wave could not form. However, the magnetic field imparts a collective behaviour to the plasma, which thus acts in much the same way as do molecules in a classical gas. The magnetic field, therefore, has a strong influence on the structure of the bow shock. A quasi-perpendicular bow shock, in which the magnetic field points mostly at right angles to the shock surface normal, tends to be a very abrupt boundary. A quasi-parallel bow shock, in which the magnetic field is closely parallel to the shock surface normal, allows any very energetic particles to escape back into the upstream solar wind; it is thus a more diffuse boundary. Bow shocks and shock waves in general are efficient accelerators of particles, and they may be responsible for energization processes around planetary magnetospheres and indeed in many wider astronomical phenomena.

Boyden Observatory Optical observatory dating from 1887, situated 26 km (16 mi) east of Bloemfontein, South Africa. The observatory transferred from Peru to its present site in 1926 and was then equipped with a 1.5-m (60-in.) reflector, which remains the largest telescope on the site. Originally a southern station of HARVARD COLLEGE OBSERVATORY, Boyden has been operated since 1976 by the University of the Free State.

brachinite Subgroup of the ACHONDRITE meteorites. Brachinites are olivine-rich igneous rocks with approximately chondritic bulk composition. They have oxygen isotopic compositions similar to the HOWARDITE–EUCRITE–DIOGENITE ASSOCIATION (HEDs). Like the ACAPULCOITE–LODRANITE association, brachinites are considered to be primitive achondrites.

Brackett series Series of infrared EMISSION or ABSORPTION LINES in the HYDROGEN SPECTRUM resulting from electron transitions down to or up from the fourth energy level of that atom. The Brackett lines are named with Greek letters: Brackett α, which connects levels 4 and 5, lies at 4.0512 μm; Brackett β, which connects levels 4 and 6, lies at 2.6252 μm, and so on. The series ends at the Brackett limit at 1.4584 μm.

Bradford Robotic Telescope Autonomous 400-mm (18-in.) optical telescope located on the moors of West Yorkshire, England. It was the first instrument in the world to provide remote access via the Internet, principally for education.

Bradley, James (1693–1762) English astronomer and clergyman, Savilian professor of astronomy at Oxford and, later, at Greenwich Observatory. In 1742 he succeeded Edmond HALLEY, becoming the third ASTRONOMER ROYAL. While searching for stellar parallax, by making meticulous measurements of the star γ Draconis, Bradley discovered the ABERRATION of starlight in 1728; this was the first direct observational evidence of the Earth's orbital motion, confirming the Copernican model of the Solar System. His value for this constant was 20″.5, the modern value being taken as 20″.47. Bradley calculated the time it takes sunlight to reach the Earth as 8 minutes 12 seconds, just 7 seconds shorter than the currently accepted value. From observations covering a complete revolution of the nodes of the Moon's orbit (1727–48), he found a periodic nodding or NUTATION of the Earth's axis. At Greenwich, Bradley catalogued the positions of thousands of stars – because he took into account the effects of aberration and nutation, his catalogues contained more accurate stellar positions than those of his predecessors. He also made accurate measurements of Jupiter's diameter and carefully observed the eclipses and other phenomena of its satellites.

Brahe, Tycho (1546–1601) Danish astronomer who used pre-telescopic instruments of his own design to obtain planetary and star positions of unprecedented accuracy, later used by his assistant Johannes KEPLER to derive his three fundamental laws of planetary motions. Tycho was born into a noble family, his father Otto Brahe serving as a privy councillor and governor at Helsingborg, Denmark (now in Sweden). He was raised by his uncle, Jörgen Brahe, who left him a substantial inheritance. As a thirteen-year-old boy, Tycho was greatly impressed by astronomers' accurate prediction of the total solar eclipse of 1560 August 21. From 1559 to 1570, he attended the universities at Copenhagen, Leipzig, Wittenberg, Rostock and Basel, studying law and the humanities but maintaining a strong interest in astronomy. In 1566 Tycho duelled with a rival student, losing the tip of his nose, which he covered for the rest of his life with a metal prosthesis. During this period, his teachers helped him make astronomical globes and simple cross-staffs, forerunners of the more sophisticated instruments he would later use, and he read Ptolemy's *Almagest*, the only astronomy book then available.

In 1563 August, Tycho had observed a conjunction of Jupiter and Saturn, noticing that the Copernican tables were grossly in error (by several days) in predicting this event. He decided to devote the rest of his life to improving planetary and stellar positions, acquiring a large quadrant and building a private observatory at Skåne in 1571. The next year, Tycho independently discovered a supernova in Cassiopeia, and his report, *De stella nova* (1573), made him famous. The Danish king, Frederick II, gave him the Baltic island of Hveen in 1576, where Tycho built two observatories, URANIBORG and STJERNEBORG, carrying out 20 years' worth of very accurate observations with instruments such as the 6½-ft (2-m) mural quadrant. He described these instruments and his work at Hveen in two magnificent books, *Astronomiae instauratae mechanica* (1598) and *Astronomiae instauratae progymnasmata* (1602).

Whereas previous astronomers had contented themselves with observing the planets at opposition and quadrature, Tycho obtained positions at many intermediate points in their orbits, accurate to 30″ – the best previous planetary positions were accurate only to 15′. He was the first astronomer to take atmospheric refraction into account in correcting observed planetary and stellar positions. Tycho made the first truly scientific studies of comets, observing the position, magnitude, colour and orientation of the tail of the Great Comet of 1577. These observations led him to conclude that the comet's orbit, which he determined must have an elongated shape, lay beyond the Moon – a novel idea at the time. Tycho's careful measurements of the Sun's apparent movement allowed him to determine the length of a year to within 1 second, forcing 10 days to be dropped from the Julian calendar in 1582 after it was shown that the Julian year exceeded the 'true' year by this amount of time. He also compiled a catalogue of 777 stars.

Tycho was not a Copernican, but created a TYCHONIAN SYSTEM with the planets revolving around the Sun, and the Sun and Moon revolving around a fixed Earth. This model was widely accepted by many astronomers until the mid-17th century. His observations of the comet of 1577, which demonstrated that it moved among the planets, defeated the Aristotelian notion that the planets were contained within solid crystalline spheres through which a body like the comet could not possibly move. Tycho ended his days as Imperial Mathematician under the Holy Roman Emperor Rudolf II in Prague, where Kepler was one of his assistants. Kepler later reduced much of Tycho's data (which appeared in the *Rudolphine Tables* of 1627) to discover the three fundamental laws of planetary motion (1609–19).

Brahmagupta (598–*c*.670) Indian astronomer and author of the *Brāhmasphutasiddhānta* ('The Opening of the Universe'), a scholarly discussion of algebra, geometry and astronomy in verse form. Brahmagupta was an expert on the phases of the Moon, eclipses of the Sun and Moon, and the positions and movements of the planets.

Braille MARS-CROSSING ASTEROID (number 9969) visited by the DEEP SPACE 1 space probe in 1999 July. Braille is an irregularly shaped asteroid *c*.2.2 km (*c*.1.4 mi) long. Its perihelion distance is 1.32 AU and its orbital period 3.58 years.

Brans–Dicke theory Alternative theory of gravity to Einstein's general theory of relativity. It was proposed by Princeton physicists C.H. Brans and R.H. Dicke in 1961. Based on the ideas of Ernst Mach about reference frames being connected to the distribution of matter in the Universe, Brans and Dicke modified GENERAL RELATIVITY, introducing a scalar field, to include Mach's proposal. The strength of this field is governed by an arbitrary constant w, but experiments limited the Brans–Dicke parameter w to values closer and closer to one, at which the Brans–Dicke theory became indistinguishable from general relativity.

Brashear, John Alfred (1840–1920) American telescope-maker who founded a renowned optical company in Pittsburgh, Pennsylvania, named after himself, today known as Contraves Corp. Brashear made the 30-inch (0.76-m) Thaw refracting telescope for Pittsburgh's Allegheny Observatory. Besides telescopes, Brashear's firm made spectroscopes, prisms and gratings and other highly specialized astronomical equipment, including George Ellery HALE's spectroheliograph.

Braun, Wernher Magnus Maximilian von (1912–77) German-American rocket scientist who played a major role in the US space programme. His early career was spent at the rocket development and test centre at Peenemünde in north-eastern Germany, which produced a series of rockets culminating in the A-4, renamed the V-2 when it and the team were absorbed into Germany's war effort. At the close of World War II, von Braun and many other German rocket scientists were relocated first to White Sands (New Mexico) Proving Grounds, then to Huntsville, Alabama. At White Sands, the V-2s were put to peaceful use in high-altitude atmospheric and astronomical research. At Huntsville, von Braun led teams that built and launched the Jupiter and Redstone missiles and, on 1958 January 31, the USA's first artificial satellite, Explorer 1. He oversaw the development of the Saturn I, IB and V launch vehicles for the Apollo programme.

breccia Rock comprising angular fragments set in a fine-grained matrix. Breccias form as a result of rocks being crushed by meteoroid impacts, tectonic faulting,

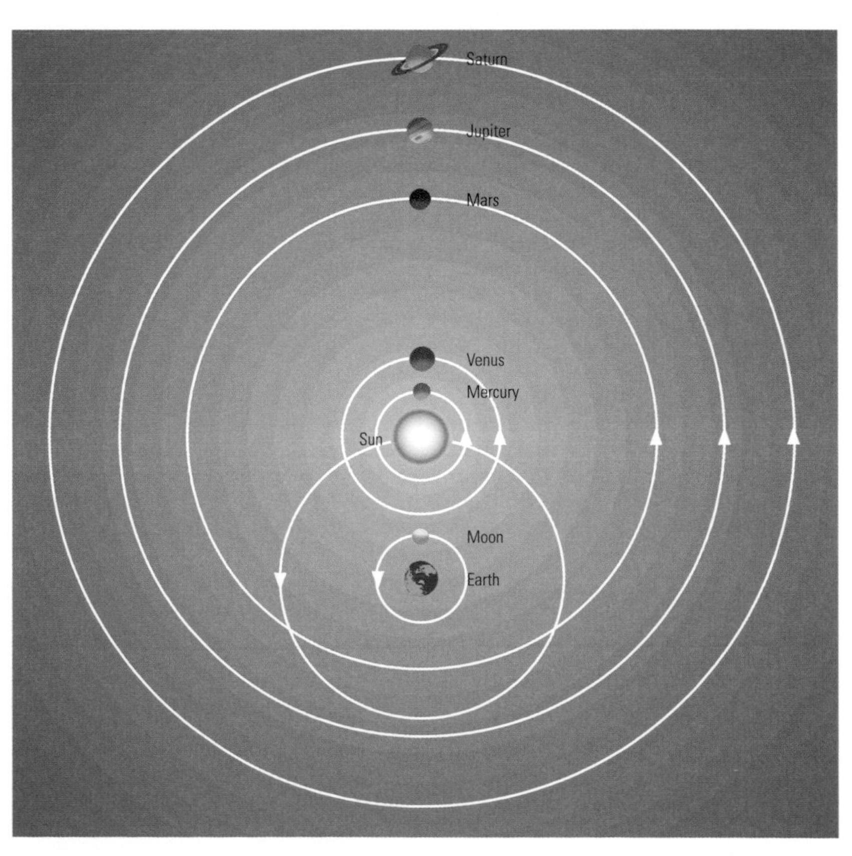

▼ **Brahe, Tycho** The Danish astronomer's model of the Solar System, with a static Earth at the centre. He thought that the Moon and Sun orbited the Earth and the other planets moved around the Sun.

▶ **B ring** Voyager 2's images of Saturn's rings revealed that they were made up of many thousands of narrow ringlets. The dark markings within the B ring are known as 'spokes' and are thought to be caused by the planet's magnetic field.

violent volcanic eruptions and certain other geological processes. Some METEORITES are breccias formed by collisions of their parent bodies. Impact-generated breccias, which may also contain glass, are the most abundant rocks in lunar highlands, and they were predominant among the rock samples returned by the Apollo astronauts.

bremsstrahlung *See* FREE–FREE TRANSITION

bright nebula Interstellar gas or dust cloud that may be seen in the visible by its own light, in contrast to DARK NEBULAE, which can only be seen when silhouetted against a bright background. Bright nebulae include all types of EMISSION NEBULAE plus REFLECTION NEBULAE.

Bright Star Catalogue (BS) Catalogue of all naked-eye stars, first compiled by Frank SCHLESINGER and published as the *Yale Bright Star Catalogue* by Yale University in 1930; it is now maintained by Dorrit HOFFLEIT. The current (5th preliminary, 1991) edition, in electronic form, lists 9110 stars brighter than m_v = 6.5 and gives their position, parallax, proper motion, magnitude, spectral type and other parameters.

B ring Second major ring of SATURN, lying inside the A RING at a distance between 82,000 km (51,000 mi) and 117,600 km (73,100 mi) from the planetary centre.

British Astronomical Association (BAA) Main organization co-ordinating the work of amateur astronomers in the UK, with a membership (2001) of around 2500. It was founded in 1890 as an egalitarian alternative to the Royal Astronomical Society. The BAA has a number of observing sections, each dedicated to a specific area of astronomy. Reports are published in a bimonthly *Journal*, and the annual *Handbook* provides ephemerides for astronomical phenomena.

British Interplanetary Society (BIS) World's longest-established organization that is devoted solely to supporting and promoting the exploration of space and astronautics; it was founded in 1933. Though its headquarters are in London, the society has a world-wide membership. Its monthly journal, *Spaceflight*, first appeared in 1956, the year before the launch of the first artificial satellite, and it is regarded as an authoritative source on international space programmes and the commercial exploration of space.

Brocchi's Cluster *See* COATHANGER

Brooks, Comet (C/1893 U1) COMET discovered in 1893, reaching peak magnitude +7.0. Its complex, rapidly changing ion tail was one of the first to have its activity photographed.

▼ **brown dwarf** The brown dwarf Gliese 229B (next to the cool red star Gliese 229), in images from the Hale Telescope (top) at Palomar and the Hubble Space Telescope (bottom). It is estimated to be 100,000 times dimmer than our Sun.

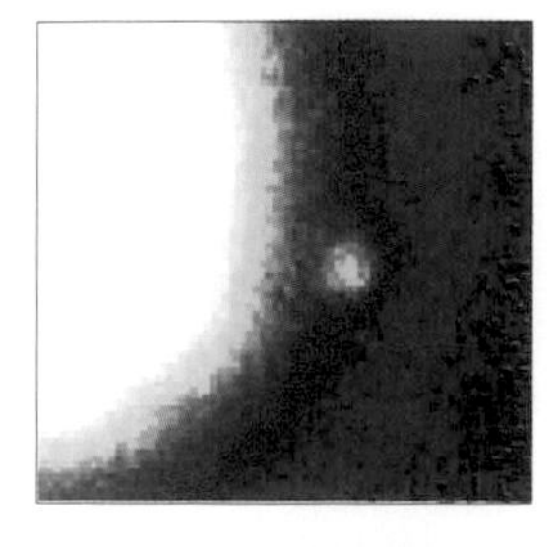

Brorsen, Comet 5D/ Faint short-period COMET discovered by Theodor Brorsen, Germany, on 1846 February 26. With an orbital period of 5.5 years, it was seen at five returns. At the last of these, in 1879, it underwent a dramatic fade and almost certainly disintegrated. Comet 5D/Brorsen has not been seen since.

Brorsen–Metcalf, Comet 23P/ Short-period COMET discovered by Theodor Brorsen, Germany, on 1847 July 20. The comet reached peak magnitude +6.5 in mid-August but was only observed over a short arc, leading to uncertainties as to its orbital period. The comet was rediscovered by Joel H. Metcalf on 1919 August 21, reaching magnitude +4.5 around perihelion on October 17. The comet next returned to perihelion, 0.48 AU from the Sun, on 1989 September 11, reaching magnitude +5, and proving a readily observable object during this apparition. Comet 23P/Brorsen–Metcalf appears to be subject to considerable NON-GRAVITATIONAL FORCE, which modifies the orbital parameters. The comet's current period is 70.54 years.

Brouwer, Dirk (1902–66) Dutch-born expert in celestial mechanics. He spent his career at Yale University, where he directed its observatory for a quarter-century and edited the *Astronomical Journal*. He was among the first to use computers to perform the complex and laborious calculations demanded by the methods of celestial mechanics that he developed for improving the orbital elements of planets, asteroids and comets. Brouwer extended these methods to calculate the orbits of the first artificial Earth satellites.

Brown, Ernest William (1866–1938) English mathematician, a specialist in celestial mechanics, who moved to America in 1891. Working at Yale University, Brown constructed tables of the Moon's motion accurate to 0″.01, based in part on a century and a half of observations made from England's Greenwich Observatory. His advances in lunar theory showed that the Earth's variable rotation rate explained much of the irregularity in the Moon's movements. Brown also made major contributions to our knowledge of the Trojan asteroids.

Brown, Robert Hanbury (1916–2002) English astronomer, born in India, who in the 1950s applied his expertise in radar developed during World War II to the new science of radio astronomy. His major accomplishment was the invention of the intensity interferometer, with which he was able to measure the diameters of stars, something impossible with optical telescopes. Brown played a major role in conducting radio surveys of the sky from Jodrell Bank, and spent much of his later career at the University of Sydney.

brown dwarf Star with mass greater than about 0.01 solar mass but less than 0.08 solar mass; its core temperature does not rise high enough to start thermonuclear reactions. It is luminous, however, because it slowly shrinks in size and radiates away its gravitational energy. As its surface temperature is below the 2500 K lower limit for RED DWARFS, it is known as a brown dwarf.

The first unambiguous detection and image of a brown dwarf, Gliese 229B (GL229B), was made in 1995. GL229B is a small companion to the cool red star Gliese 229, which is located 19 l.y. from Earth in the constellation of Lepus. It is estimated to be 20 to 50 times the mass of Jupiter. Infrared spectroscopic observations, made with the 200-inch (5-m) Hale telescope at Palomar, show that the brown dwarf has an abundance of methane, making it similar in composition to Jupiter and the other gas giant planets in the Solar System. More than 100 brown dwarf candidates have since been discovered in infrared surveys, many of which are warmer than GL229B but cooler than M dwarf stars. A new class of L dwarf *(see* L STAR) has thus been added to the SPECTRAL CLASSIFICATION system.

The Hubble Space Telescope (HST) has carried out a survey of brown dwarfs, showing that there are more low-mass brown dwarfs than high-mass ones. Brown dwarfs are far more abundant than previously thought, but they are probably not abundant enough to contribute significantly to the MISSING MASS. The HST survey also showed that many brown dwarfs are 'free-floating' isolated bodies rather than members of CLOSE BINARY systems, suggesting that they form in a manner similar to stars, not planets. However, some of the more massive EXTRASOLAR PLANETS may well be brown dwarfs.

Bruno, Giordano (1548–1600) Italian philosopher and monk. A sharp critic of the cosmology of Aristotle and an early supporter of Copernicus' heliocentric theory, he was burned at the stake in Rome for theological heresy and suspicion of magical practices. Bruno postulated a cosmology in which the universe was infinite, as elucidated in his 1584 work *De l'infinito universo e mondi.*

BS Abbreviation of BRIGHT STAR CATALOGUE

B star Member of a class of blue-white stars, the spectra of which are characterized by neutral helium and strong hydrogen lines. MAIN-SEQUENCE B dwarfs have a great range of stellar properties. Temperatures range from 10,000 K at B9.5 to 31,000 K at B0, zero-age masses from 2.3 to 14 solar masses, zero-age luminosities from 25 to 25,000 solar luminosities, and lifetimes from 800 million to 11 million years. B GIANTS and SUPERGIANTS have increasingly narrow hydrogen lines, from which the Morgan–Keenan class can be determined. Average dwarf rotation speeds maximize at B5 at over 250 km/s (160 mi/s). BE STARS are fast rotators surrounded by disks that radiate emission lines. Slower rotators can be CHEMICALLY PECULIAR STARS, appearing as MERCURY–MANGANESE STARS. AP STARS extend into the class as Bp.

Because of their youth, B stars are found in the galactic disk and in regions of star formation. Temperatures at B0 and B1 are high enough to produce diffuse nebulae. Light scattered by dust from cooler B stars creates reflection nebulae. Along with O STARS, B stars form loose, unbound OB ASSOCIATIONS. While rare (one for every 1000 or more stars), B stars are thousands of times more common than O stars, and help to outline the spiral arms of galaxies and to give them their bluish colours.

O dwarfs evolve into B supergiants, some of which turn into luminous blue variables, which can have luminosities up to two million times that of the Sun and mass loss rates up to a hundredth of a solar mass per year (Eta Carinae, for example). B dwarfs evolve into giants and may appear at some point as CEPHEID VARIABLES. Class B also contains the Beta Cephei variables, all giants or subgiants that pulsate with multiple periods of hours and amplitudes of a few hundredths of a magnitude. Bright examples of B stars include Achernar B5 IV, Regulus B7 V, Rigel B8Ia and Spica BI V.

Bullialdus Magnificent lunar crater (21°S 22°W), diameter 63 km (39 mi), in the middle of Mare Nubium; it has massive 2500-m-high (8000-ft) terraced walls. A broad, bright ejecta blanket flanks the ramparts of Bullialdus, highlighting a system of gullies flowing down the crater's outer slopes. The crater's squared-off interior walls contain many landslips. The floor features a complex group of central mountains, the highest reaching over 1000 m (3000 ft).

Burbidge, Geoffrey Ronald (1925–) and **Burbidge, (Eleanor) Margaret** (1919–) English husband-and-wife astrophysicists noted for their studies of the NUCLEOSYNTHESIS of elements inside stars, investigations into ACTIVE GALACTIC NUCLEI (AGNs) and research in COSMOLOGY. Working at Cambridge University, the California Institute of Technology, and the University of California at San Diego, Geoffrey Burbidge's research has involved the physics of non-thermal radiation processes, quasars and other AGNs, and alternative cosmological models, including the 'quasi-steady-state theory'. This model, unlike the original STEADY-STATE THEORY, explains the COSMIC MICROWAVE BACKGROUND. It proposes that matter is created by a series of 'mini Big Bangs' at a rate of 10^{16} solar masses on timescales of duration $c.1/H_0$ (the Hubble time). Burbidge's cosmology suggests that the isotope ^{4}He is produced, not by the traditional Big Bang, but by stellar hydrogen burning, as are other 'light' isotopes, up to ^{11}B. He has also directed Kitt Peak National Observatory.

Margaret Burbidge has worked at most of the same institutions as her husband, and has served as director of Royal Greenwich Observatory (1972–73). In much of her work she has applied spectroscopy to determine the masses and rotation rates of galaxies. In 1957 she, her husband, William FOWLER and Fred HOYLE published a seminal paper in which they set forth the basic processes that govern stellar nucleosynthesis. Their theory (known from the initials of its proponents as the B^{2}FH theory) explained that younger stars convert their abundant supplies of hydrogen into helium and light, and as they age, the helium is gradually burned into carbon, oxygen and other chemical elements. Heavier nuclei are formed when the carbon and oxygen atoms combine with protons or He nuclei – Mg, Si, S, Ar and Ca can be synthesized this way.

The Burbidges have made extensive investigations of quasars and AGNs, finding that the two often coincide. They concluded that AGNs emit enormous amounts of energy from relatively confined regions, and suggested that BLACK HOLES are the power source. Margaret Burbidge also helped develop the Faint Object Spectrograph for the HUBBLE SPACE TELESCOPE, which she has used to study the ultraviolet flux of AGNs.

Burnham, Sherburne Wesley (1838–1921) American observer and cataloguer of double stars who worked at Lick Observatory (1888–92) and Yerkes Observatory (1894–1914), discovering over a thousand new pairs. Burnham was able to use the largest refracting telescopes of the day, including Lick's 36-inch (0.9-m) and Yerkes' 40-inch (1-m). His life's work culminated in the publication of the massive two-volume *General Catalogue of Double Stars Within 121 Degrees of the North Pole*, which contained data for over 13,000 doubles.

Butterfly Cluster (M6, NGC 6405) Bright OPEN CLUSTER containing 80 stars of magnitude +6 and fainter together with associated nebulosity; it is located a few degrees north of the 'Sting' asterism marking the tail of Scorpius (RA 17^h 40^m.1 dec. −32°13′). With an integrated magnitude +4.2, the Butterfly Cluster is visible to the naked eye under good conditions. It was catalogued by

▼ Butterfly Cluster The open cluster M6 (NGC 6405) in Scorpius is predominantly composed of young, blue stars and is thought to be about 100 million years old. It is some 20 l.y. across.

B

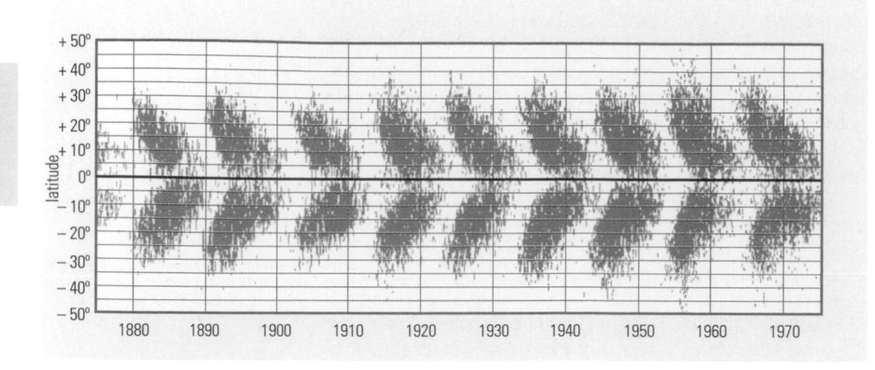

▲ **butterfly diagram** During the solar cycle, the latitude at which sunspots form changes. This is related to changes in the Sun's magnetic field during its cycle. Cycles usually overlap, with the first spots of a new cycle forming concurrently with the last spots of the old cycle.

PTOLEMY in the 2nd century AD as a 'little cloud'. The cluster lies 2000 l.y. away in the direction of the galactic centre. It takes its popular name from the distribution of its stars in two fan-shaped 'wings'. The Butterfly Cluster has an apparent diameter of 33′.

butterfly diagram (Maunder diagram) Plot of the distribution in solar latitude of SUNSPOTS over the course of the approximate 11-year SOLAR CYCLE. At the start of a cycle, sunspots tend to appear at solar latitudes of up to 40° north and south of the equator. Their numbers increase and their latitude distribution moves toward the equator as the cycle progresses. This graphical representation of sunspot latitudes against dates of observation was first plotted by the English astronomer Edward Walter MAUNDER in the early 20th century. The plot for each cycle resembles the wings of a butterfly, which gives the diagram its popular name. The diagram is a graphical representation of SPÖRER'S LAW.

Butterfly Nebula (M76, NGC 650 & 651) *See* LITTLE DUMBBELL

BY Draconis stars (BY) EXTRINSIC VARIABLE star, consisting of an emission-line dwarf star (with dKe–dMe spectra) that shows variations of 0.01–0.5 magnitude in luminosity. The changes arise from axial rotation of stars with markedly non-uniform surface luminosity, caused by large 'starspots' and chromospheric activity. The spots are larger than those on the Sun (covering up to 20% of the stellar surface) and have little or no penumbra. Periods range from a fraction of a day to about 120 days. A number (approximately 20) of these red dwarf stars also exhibit flare activity, in which case they are classed as FLARE STARS.

Byurakan Astrophysical Observatory Optical observatory situated some 40 km (25 mi) north-west of Yerevan, Armenia, at an elevation of more than 1500 m (4900 ft). The observatory was founded in 1946 by the Armenian Academy of Sciences, on the initiative of Viktor AMBARTSUMIAN. Its largest telescopes are a 1-m (40-in.) Schmidt and a 2.6-m (102-in.) reflector, opened in 1960 and 1976 respectively. The Schmidt is notable for the discovery of MARKARIAN GALAXIES.

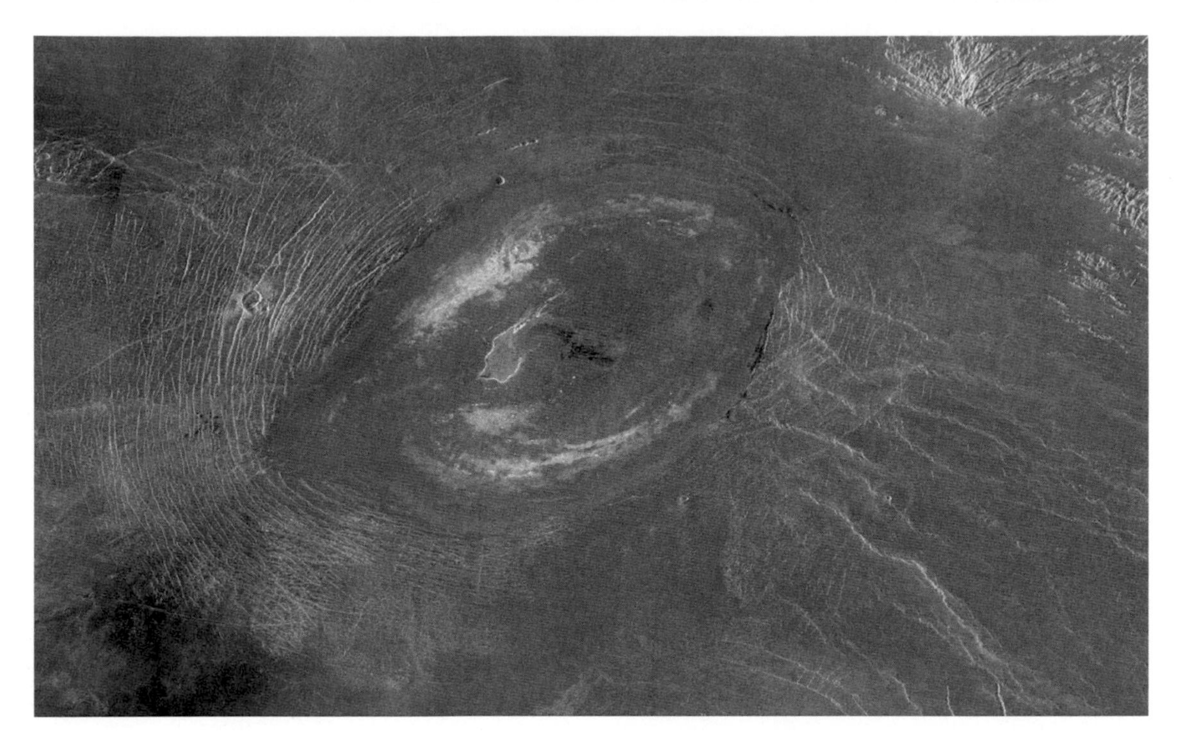

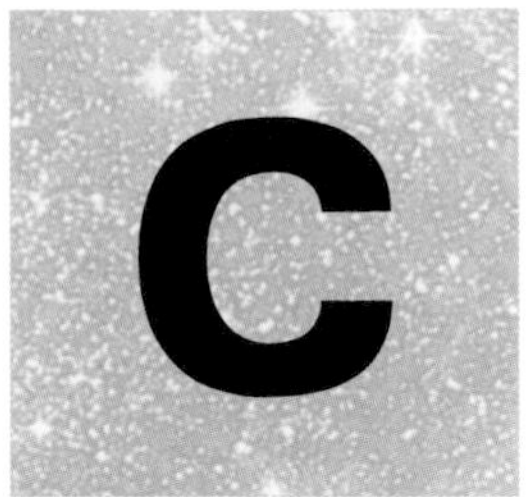

◄ **caldera** A Magellan radar image of Sacajawea Patera, in the western part of Ishtar Terra on Venus. With a depth of 1–2 km (0.6–1.2 mi), this caldera has a maximum diameter of 420 km (252 mi).

Caelum *See* feature article

CAI *See* CALCIUM–ALUMINIUM-RICH INCLUSION

Calar Alto Observatory German-Spanish optical observatory situated in the Sierra de los Filabres in Andalucía, southern Spain, at an elevation of 2170 m (7120 ft). Dating from 1973, it is operated by the MAX-PLANCK-INSTITUT FÜR ASTRONOMIE in Heidelberg; the Madrid Observatory also has a 1.5-m (60-in.) telescope there. Principal instruments of the German-Spanish Astronomical Centre are the 3.5-m (138-in.) and 2.2-m (87-in.) reflectors, dating from 1984 and 1979 respectively. There is also a 1.2-m (48-in.) telescope and a small Schmidt that was formerly at HAMBURG OBSERVATORY.

calcium–aluminium-rich inclusion (CAI) In a METEORITE, an irregular-shaped refractory inclusion of oxide and silicate minerals such as spinel, hibonite and melilite. CAIs can be up to *c.*1 cm (about half an inch) in size and frequently exhibit complex mineralogical zoning, both in their rims and in their cores. They contain the daughter products from now-extinct short-lived radionuclides, indicating rapid solidification after production of the nuclides. CAIs are the oldest solid objects from the presolar nebula.

caldera Large, roughly circular volcanic depression formed by collapse over an evacuated MAGMA chamber. Small calderas (less than 5 km/3 mi in diameter) are common at the crests of terrestrial basaltic and andesitic volcanoes. Calderas as large as 75 km (47 mi) across have formed on Earth during volcanic ash flow eruptions. The summit caldera of the Martian volcano OLYMPUS MONS, the largest volcano in the Solar System, is about 80 km (50 mi) across.

Caldwell Catalogue Listing of 109 deep-sky objects intended to complement those in the MESSIER CATALOGUE. Charles Messier omitted from his list several prominent objects, such as the Double Cluster in Perseus, and southern-hemisphere objects of which he was unaware. In 1993 English amateur astronomer Patrick MOORE, recognizing these omissions, gathered a further 109 deep-sky objects in a catalogue that takes its title from one of his middle names. The list enjoys some popularity among amateur observers who have observed all the Messier objects and are seeking further targets. *See* table, pages 66–67

calendar System for measuring longer intervals of time by dividing it into periods of days, weeks, months and years. The length of the day is based on the average rotation period of the Earth, while a year is based on the orbital period of the Earth around the Sun.

The MONTH was originally the period between successive full moons (29.5 days), giving a lunar year of 354 days, but in modern calendars it is now roughly one twelfth of a year. Because there are not a whole number of full moons in a year, the two cannot be simply reconciled into a common calendar, and while the modern civil calendar is based on the year, the dates of religious festivals (such as Easter) are still set by reference to the lunar month. The week is not based on any astronomical phenomena but instead derives from the Jewish and Christian tradition that every seventh day should be a day of rest.

The ancient Egyptians were the first to use a calendar based on a solar year, the Babylonians having used a lunar year of 12 months, adding extra months when it became necessary to keep the calendar in step with the seasons.

Our present calendar is based on that of the Romans, which originally had only ten months. Julius Caesar, on the advice of the Greek astronomer Sosigenes, introduced the Julian Calendar in 45 BC. This was a solar calendar based on a year of 365.25 days, but fixed at 365 days, with a leap year of 366 days every fourth year to compensate. The year commenced on March 25 and the Julian Calendar, which was completed by Caesar Augustus in 44 BC, also established the present-day names, lengths and order of the months.

The Julian year of 365.25 days is slightly different from the solar year of 365.2419 days (being 11^m14^s longer) but accumulates an error of almost eight days every 1000 years. By the 16th century this error had become noticeable, the VERNAL EQUINOX, for example, occurring 10 days early in 1582. To correct this error, Pope Gregory XIII removed 10 days from the calendar so in 1582, October 15 followed October 4, and instigated the use of a new system known as the Gregorian calendar in which only century years divisible by 400 (for

CAELUM (GEN. CAELI, ABBR. CAE)

Faint and obscure southern constellation, sandwiched between Columba and the southern end of Eridanus, introduced by Nicolas Lacaille in the 1750s. It represents a sculptor's chisel. Its brightest star, α Cae, is only mag. 4.4.

example, 1600, 2000) should be leap years. The start of the year was also changed to January 1.

By the time this 'New Style' calendar, as it was known, was adopted in Great Britain in 1752 the error had increased to 11 days, so that September 14 followed September 2. This is why the British financial year ends on April 5 – the old New Year's Day (March 25) plus the 11 days lost in 1752.

Today the Gregorian calendar, which repeats every 400 years since this equates to an exact number of weeks, is used throughout the Western world and parts of Asia. It is also known as the Christian calendar since it uses the birth of Christ as a starting point, subsequent dates being designated *anno domini* (in the year of our Lord) and preceding dates being BC (before Christ). The accumulated error between the Gregorian year and the true solar year now amounts to just three days in 10,000 years.

The two other major calendars in use today are the Jewish and Muslim calendars, both of which are based on a lunar year. The Jewish is derived from the ancient Hebrew calendar and is based on lunar months of 29 and 30 days alternately, with an intercalary month inserted every three years. The length of the Muslim year is also 12 months of alternate lengths 30 and 29 days, except for the 12th month which can have either 29 or 30 days.

Caliban One of the several small outer satellites of URANUS. It was discovered in 1997 by Brett Gladman and others. Caliban is about 60 km (40 mi) in size. It takes 579 days to circuit the planet at an average distance of 7.23 million km (4.49 million mi). It has a RETROGRADE orbit (inclination near 141°) with eccentricity 0.159. In origin Caliban is thought to be a

CALDWELL CATALOGUE

C	NGC/IC	Constellation	Type	RA (2000.0) h	min	Dec. °	′	Magnitude	Size (′)	Comment
1	188	Cepheus	open cluster	00	44.4	+85	20	8.1	14	
2	40	Cepheus	planetary nebula	00	13.0	+72	32	10.7	1.0 × 0.7	
3	4236	Draco	Sb galaxy	12	16.7	+69	27	9.6	23 × 8	
4	7023	Cepheus	reflection nebula	21	01.8	+68	10	—	18 × 18	
5	IC 342	Camelopardalis	SBc galaxy	03	46.8	+68	06	9.2	18 × 17	
6	6543	Draco	planetary nebula	17	58.6	+66	38	8.1	22 × 16	CAT'S EYE NEBULA
7	2403	Camelopardalis	Sc galaxy	07	36.9	+65	36	8.4	18 × 10	
8	559	Cassiopeia	open cluster	01	29.5	+63	18	9.5	4	
9	Sh2-155	Cepheus	bright nebula	22	56.8	+62	37	—	50 × 10	
10	663	Cassiopeia	open cluster	01	46.0	+61	15	7.1	16	
11	7635	Cassiopeia	bright nebula	23	20.7	+61	12	8	15 × 8	
12	6946	Cepheus	SAB galaxy	20	34.8	+60	09	8.9	11 × 10	
13	457	Cassiopeia	open cluster	01	19.1	+58	20	6.4	13	
14	869/884	Perseus	open clusters	02	20.0	+57	08	5.3 and 6.1	29 and 29	DOUBLE CLUSTER
15	6826	Cygnus	planetary nebula	19	44.8	+50	31	8.8	27 × 24	BLINKING PLANETARY
16	7243	Lacerta	open cluster	22	15.3	+49	53	6.4	21	
17	147	Cassiopeia	dE4 galaxy	00	33.2	+48	30	9.3	13 × 8	
18	185	Cassiopeia	dE0 galaxy	00	39.0	+48	20	9.2	12 × 10	
19	IC 5146	Cygnus	bright nebula	21	53.4	+47	16	—	10	COCOON NEBULA
20	7000	Cygnus	bright nebula	20	58.8	+44	20	—	120 × 100	NORTH AMERICA NEBULA
21	4449	Canes Venatici	Irregular galaxy	12	28.2	+44	06	9.4	6 × 5	
22	7662	Andromeda	planetary nebula	23	25.9	+42	33	8.3	17 × 14	
23	891	Andromeda	Sb galaxy	02	22.6	+42	21	9.9	14 × 3	
24	1275	Perseus	Seyfert galaxy	03	19.8	+41	31	11.6	3.5 × 2.5	
25	2419	Lynx	globular cluster	07	38.1	+38	53	10.4	4.1	
26	4244	Canes Venatici	Scd galaxy	12	17.5	+37	49	10.2	18 × 2	
27	6888	Cygnus	bright nebula	20	12.0	+38	20	—	20 × 10	
28	752	Andromeda	open cluster	01	57.8	+37	41	5.7	50	
29	5005	Canes Venatici	Sb galaxy	13	10.9	+37	03	9.5	6 × 3	
30	7331	Pegasus	Sb galaxy	22	37.1	+34	25	9.5	11 × 4	
31	IC 405	Auriga	bright nebula	05	16.2	+34	16	—	37 × 19	
32	4631	Canes Venatici	Sc galaxy	12	42.1	+32	32	9.3	17 × 3	
33	6992/5	Cygnus	SN remnant	20	56.8	+31	28	—	60 × 8	Eastern VEIL NEBULA
34	6960	Cygnus	SN remnant	20	45.7	+30	43	—	70 × 6	Western VEIL NEBULA
35	4889	Coma Berenices	E4 galaxy	13	00.1	+27	59	11.4	3 × 2	
36	4559	Coma Berenices	Sc galaxy	12	36.0	+27	58	9.8	13 × 5	
37	6885	Vulpecula	open cluster	20	12.0	+26	29	5.9	7	
38	4565	Coma Berenices	Sb galaxy	12	36.3	+25	59	9.6	16 × 2	
39	2392	Gemini	planetary nebula	07	29.2	+20	55	9.2	0.25	ESKIMO NEBULA
40	3626	Leo	Sb galaxy	11	20.1	+18	21	10.9	3 × 2	
41	Melotte 25	Taurus	open cluster	04	27	+16		0.5	330	HYADES
42	7006	Delphinus	globular cluster	21	01.5	+16	11	10.6	2.8	
43	7814	Pegasus	Sb galaxy	00	03.3	+16	09	10.3	6 × 3	
44	7479	Pegasus	SBb galaxy	23	04.9	+12	19	10.9	4.4 × 3.4	
45	5248	Boötes	Sc galaxy	13	37.5	+08	53	10.2	7 × 5	
46	2261	Monoceros	bright nebula	06	39.2	+08	44	—	3.5 × 1.5	HUBBLE'S VARIABLE NEBULA
47	6934	Delphinus	Globular cluster	20	34.2	+07	24	8.7	5.9	
48	2775	Cancer	Sa galaxy	09	10.3	+07	02	10.1	5 × 4	
49	2237-9	Monoceros	bright nebula	06	32.3	+05	03	—	80 × 60	ROSETTE NEBULA
50	2244	Monoceros	open cluster	06	32.4	+04	52	4.8	24	
51	IC 1613	Cetus	irregular galaxy	01	04.8	+02	07	9.2	12 × 11	
52	4697	Virgo	E4 galaxy	12	48.6	−05	48	9.3	6 × 4	
53	3115	Sextans	SO galaxy	10	05.2	−07	43	8.9	8 × 3	
54	2506	Monoceros	open cluster	08	00.2	−10	47	7.6	7	
55	7009	Aquarius	planetary nebula	21	04.2	−11	22	8.0	0.4 × 1.6	SATURN NEBULA

captured body, previously a CENTAUR. Other recently discovered outer Uranian satellites with a similar suspected origin to Caliban are PROSPERO, SETEBOS, STEPHANO and SYCORAX; an object discovered in VOYAGER 2 images in 1999 and provisionally designated as S/1986 U10 awaits confirmation.

California Extremely Large Telescope (CELT) Proposed 30-m (100-ft) optical/infrared telescope, sponsored jointly by the University of California and the CALIFORNIA INSTITUTE OF TECHNOLOGY. The instrument would build on the segmented-mirror technology used for the 10-m (33-ft) telescopes of the W.M. KECK OBSERVATORY, and would have a mirror composed of 1080 hexagonal segments. It would rely on ADAPTIVE OPTICS for high resolution, bringing the first generation of galaxies within its grasp when fully operational in 2010–15.

California Institute of Technology (Caltech) Small, independent university dedicated to exceptional instruction and research in engineering and science, located in Pasadena, California, USA. Its off-campus facilities include the JET PROPULSION LABORATORY, PALOMAR OBSERVATORY, OWENS VALLEY RADIO OBSERVATORY and the W.M. KECK OBSERVATORY.

C

California Nebula (NGC 1499) emission nebula in the eastern part of the constellation perseus (RA $04^h\ 00^m.7$ dec. +36°37′). The nebula is illuminated by the hot, young O-class star ξ Persei. Although visually faint, the nebula shows up well in long-exposure photographs taken on red-sensitive films. It takes its name from its resemblance, in outline, to the American state. The California Nebula covers an area of 160′ × 40′, elongated roughly north–south.

CALDWELL CATALOGUE (CONTINUED)

C	NGC/IC	Constellation	Type	RA (2000.0) h	min	Dec. °	′	Magnitude	Size (′)	Comment
56	246	Cetus	planetary nebula	00	47.0	−11	53	8.6	4 × 3	
57	6822	Sagittarius	irregular galaxy	19	44.9	−14	48	8.8	20 × 10	
58	2360	Canis Major	open cluster	07	17.8	−15	37	7.2	13	
59	3242	Hydra	planetary nebula	10	24.8	−18	38	7.8	0.27	GHOST OF JUPITER
60	4038	Corvus	Sc galaxy	12	01.9	−18	52	10.5	2.6 × 2	ANTENNAE
61	4039	Corvus	Sp galaxy	12	01.9	−18	53	10.5	2.6 × 2	ANTENNAE
62	247	Cetus	SAB galaxy	00	47.1	−20	46	9.1	20 × 7	
63	7293	Aquarius	planetary nebula	22	29.6	−20	48	7.3	13	HELIX NEBULA
64	2362	Canis Major	open cluster	07	18.8	−24	57	4.1	8	
65	253	Sculptor	Scp galaxy	00	47.6	−25	17	7.1	25 × 7	Silver Coin Galaxy
66	5694	Hydra	globular cluster	14	39.6	−26	32	10.2	3.6	
67	1097	Fornax	SBb galaxy	02	46.3	−30	16	9.2	9 × 7	
68	6729	Corona Australis	bright nebula	19	01.9	−36	58	—	1.0	
69	6302	Scorpius	planetary nebula	17	13.7	−37	06	9.6	2 × 1	
70	300	Sculptor	Sd galaxy	00	54.9	−37	41	8.1	20 × 15	
71	2477	Puppis	open cluster	07	52.3	−38	33	5.8	27	
72	55	Sculptor	SB galaxy	00	15.1	−39	13	7.9	25 × 4	
73	1851	Columba	globular cluster	05	14.1	−40	03	7.3	11	
74	3132	Vela	planetary nebula	10	07.7	−40	26	8.2	1.4 × 0.9	
75	6124	Scorpius	open cluster	16	25.6	−40	40	5.8	29	
76	6231	Scorpius	open cluster	16	54.0	−41	48	2.6	15	
77	5128	Centaurus	radio galaxy	13	25.5	−43	01	6.8	18 × 14	CENTAURUS A
78	6541	Corona Australis	globular cluster	18	08.0	−43	42	6.6	13	
79	3201	Vela	globular cluster	10	17.6	−46	25	6.7	18	
80	5139	Centaurus	globular cluster	13	26.8	−47	29	3.6	36	ω Centauri
81	6352	Ara	globular cluster	17	25.5	−48	25	8.1	7	
82	6193	Ara	open cluster	16	41.3	−48	46	5.2	15	
83	4945	Centaurus	SBc galaxy	13	05.4	−49	28	8.7	20 × 4	
84	5286	Centaurus	globular cluster	13	46.4	−51	22	7.6	9	
85	IC 2391	Vela	open cluster	08	40.2	−53	04	2.5	50	
86	6397	Ara	globular cluster	17	40.7	−53	40	5.7	26	
87	1261	Horologium	globular cluster	03	12.3	−55	13	8.4	7	
88	5823	Circinus	open cluster	15	05.7	−55	36	7.9	10	
89	6087	Norma	open cluster	16	18.9	−57	54	5.4	12	
90	2867	Carina	planetary nebula	09	21.4	−58	19	9.7	12	
91	3532	Carina	open cluster	11	06.4	−58	40	3.0	55	
92	3372	Carina	bright nebula	10	45.0	−59	50	—	120 × 120	η Carinae Nebula
93	6752	Pavo	globular cluster	19	10.9	−59	59	5.4	20	
94	4755	Crux	open cluster	12	53.6	−60	20	4.2	10	KAPPA CRUCIS CLUSTER
95	6025	Triangulum Australis	open cluster	16	03.7	−60	30	5.1	12	
96	2516	Carina	open cluster	07	58.3	−60	52	3.8	30	
97	3766	Centaurus	open cluster	11	36.1	−61	37	5.3	12	
98	4609	Crux	open cluster	12	42.3	−62	58	6.9	5	
99	—	Crux	dark nebula	12	53	−63		—	420 × 300	COALSACK
100	IC 2944	Centaurus	cluster	11	36.6	−63	02	4.5	60 × 40	
101	6744	Pavo	SBb galaxy	19	09.8	−63	51	8.3	16 × 10	
102	IC 2602	Carina	open cluster	10	43.2	−64	24	1.9	50	
103	2070	Dorado	bright nebula	05	38.7	−69	06	—	30 × 20	TARANTULA NEBULA
104	362	Tucana	globular cluster	01	03.2	−70	51	6.6	13	
105	4833	Musca	globular cluster	12	59.6	−70	53	7.3	14	
106	104	Tucana	globular cluster	00	24.1	−72	05	4.0	31	FORTY-SEVEN TUCANAE
107	6101	Apus	globular cluster	16	25.8	−72	12	9.3	11	
108	4372	Musca	globular cluster	12	25.8	−72	40	7.8	19	
109	3195	Chamaeleon	planetary nebula	10	09.5	−80	52	8.4	40 × 30	

▲ **Callisto** Asgard, a multi-ring impact basin on Callisto, imaged by the Galileo spacecraft. Callisto's icy surface is heavily scarred by impact cratering.

Callisto Outermost and darkest of the GALILEAN SATELLITES of JUPITER. It has an albedo of only 0.2. Callisto is heavily cratered and is the only member of the group to show no clear surface signs of current or past geological activity. Callisto's density of 1.9 g/cm^3 suggests that it is composed of rock and ice in a 40:60 mixture. Gravity measurements made by the Galileo orbiter during close flybys of Callisto suggest that its interior is only weakly differentiated (*see* DIFFERENTIATION), and that the rock is not all concentrated into a core, in contrast to Callisto's neighbour GANYMEDE. Perplexingly, measurements made by the Galileo orbiter hinted at a magnetic field, apparently generated at a shallow depth within Callisto as a result of its orbital passage through Jupiter's MAGNETOSPHERE. One way to explain this would be if Callisto has a salty ocean no more than 100 km (60 mi) below its icy surface, but this is hard to reconcile with the great age of the surface implied by the density of CRATERS.

Although Callisto's craters are clearly produced by meteoroidal or cometary impact, they are rather different from those found on the Moon. For example, Callisto has very few craters less than about 60 km (40 mi) in diameter, showing that the satellite was bombarded by a different population of impactors from that responsible for lunar craters. The largest craters have a subdued shape, as if Callisto's LITHOSPHERE has not been strong enough to support their topography. There are also a few enormous impact basins marked by concentric rings of fractures; the largest of these, named VALHALLA, is about 4000 km (2500 mi) across. About a dozen linear chains of craters have been identified on Callisto, each of which was probably produced by the serial impact of fragments of a comet broken up by tidal forces during a close passage of Jupiter (*see* SHOEMAKER–LEVY 9). *See* data at JUPITER

Caloris Planitia ('Basin of Heat') Largest single feature on MERCURY imaged by MARINER 10. It is situated near one of the planet's hot poles and centred on 30°.5N 189°.8W. The Caloris Planitia is an enormous multi-ring impact structure 1300 km (800 mi) in diameter – a quarter that of the planet. Imaged half-lit from the departing spacecraft, it is defined by a ring of discontinuous mountains, the Montes Caloris, roughly 2 km (1.2 mi) high. The BASIN floor consists of smooth plains with quasi-concentric and other ridges, transected by younger crack-like GRABEN. The ejecta blanket extends to a distance of approximately 700 km (430 mi) beyond the Montes Caloris; it comprises tracts of uneven hummocky plains and lineated terrain. The whole structure is undoubtedly the modified scar left by the impact of an asteroid-sized body, the floor being the end result of refilling of the crater by the crusting-over, semi-molten ASTHENOSPHERE. On the other side of Mercury, antipodal to Caloris, is a region of 'weird terrain', an extraordinary place of hills and valleys that break into other landforms; it probably formed as a result of shockwaves from the impact that created Caloris.

Caltech Submillimeter Observatory (CSO) Enclosed 10.4-m (34-ft) radio dish with a hexagonally segmented mirror, at MAUNA KEA OBSERVATORY, Hawaii. The telescope is operated by the CALIFORNIA INSTITUTE OF TECHNOLOGY under contract from the National Science Foundation, and has been in regular use since 1988. The instrument operates at wavelengths between 350 μm and 1.3 mm, and can be linked with the nearby JAMES CLERK MAXWELL TELESCOPE for short-baseline interferometry.

Calypso Small satellite of SATURN, discovered in 1980 by Dan Pascu (1938–) and others in images from the VOYAGER missions. It is irregular in shape, measuring about 30 × 16 × 16 km (19 × 10 × 10 mi). With a distance from the centre of the planet of 294,700 km (183,100 mi), it is co-orbital with TETHYS and TELESTO. Calypso and Telesto have circular, near-equatorial orbits, near the L_5 and L_4 LAGRANGIAN POINTS, respectively, of Tethys' orbit around Saturn, with a period of about 1.89 days.

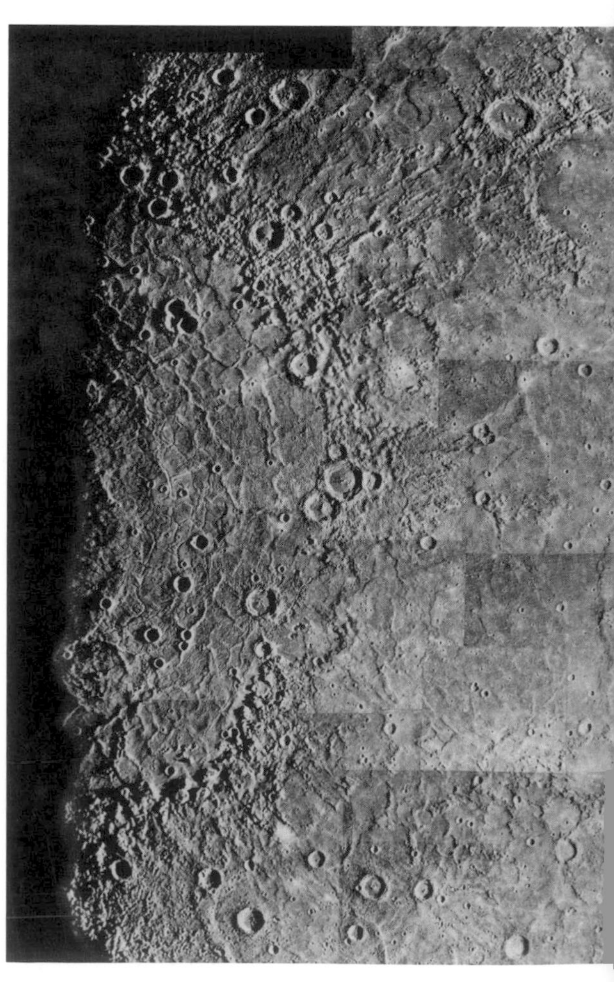

► **Caloris Planitia** This vast (1300 km/800 mi diameter) multi-ring impact basin on Mercury was imaged during the Mariner 10 mission in 1974. Shockwaves from the impact, early in Mercury's history, have produced unusual geological features on the opposite hemisphere of the planet.

Cambridge Low Frequency Synthesis Telescope (CLFST) East–west aperture synthesis radio telescope consisting of 60 trackable Yagi antennae, operated by the MULLARD RADIO ASTRONOMY OBSERVATORY and situated close to the RYLE TELESCOPE. The individual antennae are located on a 4.6-km (2.9-mi) baseline, and have a working frequency of 151 MHz

Cambridge Optical Aperture Synthesis Telescope (COAST) Instrument built by the MULLARD RADIO ASTRONOMY OBSERVATORY to extend the interferometric image-reconstruction techniques used in radio astronomy to optical and near-infrared wavelengths. It consists of an array of five 400-mm (16-in.) telescopes, of which up to four are in use at any one time. They can be used to synthesize a virtual telescope mirror 100 m (330 ft) in diameter, yielding images showing detail as fine as 0″.001 (1 milliarcsecond). COAST produced its first images in 1995.

Cambridge Radio Observatory *See* MULLARD RADIO ASTRONOMY OBSERVATORY

Camelopardalis *See* feature article

Camilla MAIN-BELT ASTEROID; number 107. It is notable because it is accompanied by a small moon.

Campbell, (William) Wallace (1862–1938) American astronomer and mathematician who measured a large number of radial velocities. As director of Lick Observatory (1900–1930), he founded the observatory's southern-hemisphere station in Chile and designed important accessory instruments for Lick's telescopes, including the Mills spectrographs. Campbell led an eclipse expedition in 1922, which confirmed Einstein's general theory of relativity by showing that the Sun's mass was sufficient to deflect light waves from other stars. With Heber CURTIS, he undertook a huge photographic survey of stellar spectra to determine the radial velocities of stars. This project had two important results: it allowed Campbell to map the local Milky Way and our Sun's motion relative to other nearby stars, and it led to the discovery of over a thousand spectroscopic binaries.

Canada–France–Hawaii telescope Optical/infrared telescope of 3.6-m (142-in.) aperture located at MAUNA KEA OBSERVATORY and operated jointly by the National Research Council of Canada, the Centre National de la Recherche Scientifique of France and the University of Hawaii. The instrument began operation in 1979, and its suite of instruments includes high-resolution wide-field imagers.

canals, Martian Elusive network of dark linear markings on the surface of Mars, reported by some observers from around 1870 until well into the 20th century. In 1877 Giovanni SCHIAPARELLI marked a number of very narrow features on his map of Mars which he referred to as *canali*, which in Italian means 'channels' or 'canals'. But when his findings were translated into English, it was the latter sense, with its implication of artificial construction, that found its way into reports. Controversy continued in the 1880s, some astronomers claiming they could see the 'canals', while others could not. In the 1890s, their existence was championed by Percival LOWELL, who founded Lowell Observatory in Flagstaff, Arizona, largely to study them. He became convinced that they were waterways constructed by intelligent beings to irrigate a desiccating planet. Only when close-up images were returned by the Mariner craft and later Mars probes was the existence of canals disproved with certainty.

Martian 'canals' make an intriguing episode in the annals of observational astronomy. No canal was ever convincingly photographed, yet highly detailed maps were prepared from sketches made at the eyepiece that showed canals in prodigious numbers. Their explanation is part psychological, part physiological. There is no denying the integrity of some who reported having observed canals, but equally there is no doubting that other observers were over-zealous in seeking evidence to fit a theory. The physiological factor was explained by Eugène ANTONIADI and others. When the eye is straining to discern detail at the very limits of visibility – and the disk of Mars is small, even at favourable oppositions – the brain can misinterpret what the eye beholds. In particular, in conditions of low contrast, it 'joins the dots' to form lines between discrete features, and conjures firm borders between areas of differing brightness.

Cancer *See* feature article

CAMELOPARDALIS (GEN. CAMELOPARDALIS, ABBR. CAM)

Fairly large but faint northern constellation, representing a giraffe; it extends from the northern borders of Perseus and Auriga towards the north celestial pole. Camelopardalis was introduced by Petrus Plancius in 1613, supposedly to commemorate the Biblical animal that carried Rebecca to Isaac. Its brightest star, β Cam, is a wide double, mags. 4.0 and 8.6. NGC 1502 is a small open cluster from which a chain of stars called Kemble's Cascade runs for $2\frac{1}{2}°$ towards neighbouring Cassiopeia.

CANCER (GEN. CANCRI, ABBR. CNC)

Faintest constellation of the zodiac, lying between Gemini and Leo. Mythologically, it represents the crab that was crushed underfoot by Hercules during his fight with the Hydra. The brightest star is β Cnc, a K4 giant of mag. 3.52, distance 290 l.y. ζ Cnc is a long-period binary divisible through small telescopes, mags. 5.0 and 6.2. Even easier to divide is ι Cnc, mags. 4.0 and 6.6. The constellation's most celebrated feature is M44, also known as PRAESEPE or the Manger, a large open star cluster. North and south of it lie γ and δ Cnc, known as the ASELLI ('asses'). In the south of the constellation, next to α Cnc, binoculars show M67, a smaller and fainter open cluster 2500 l.y. away.

CANES VENATICI (GEN. CANUM VENATICORUM, ABBR. CVN)

Northern constellation positioned beneath the tail of Ursa Major, representing a pair of hunting dogs, Asterion and Chara, held on a leash by neighbouring Boötes. Canes Venatici was introduced by Johannes Hevelius in 1687. α CVn, known as COR CAROLI, is an easy double, mags. 2.9 and 5.6. Y CVn is a red supergiant semiregular variable known as La Superba, with range 5.0–6.5 and period roughly 160 days. The constellation's most famous feature is the spiral galaxy M51, known as the WHIRLPOOL GALAXY. Other spirals visible with small instruments are M63 and M94. One of the best globular clusters in northern skies is M3, just on the naked-eye limit at 6th magnitude.

CANIS MAJOR (GEN. CANIS MAJORIS, ABBR. CMA)

Prominent constellation just south of the celestial equator, containing the brightest star in the sky, SIRIUS. Canis Major represents the larger of the two dogs of Orion and is one of the constellations recognized since the time of the ancient Greeks. ϵ (ADHARA) and μ CMa are difficult double stars with much fainter companions. UW CMa is an eclipsing binary, range 4.8–5.3, period 4.4 days. M41 is a naked-eye open cluster 4° south of Sirius, similar in apparent size to the full moon and containing some 80 stars of 7th magnitude and fainter. NGC 2362 is a small open cluster surrounding the mag. 4.4 blue supergiant τ CMa, its brightest member. *See also* MIRZAM; WEZEN

BRIGHTEST STARS

	Name	RA h	m	dec. ° ′	Visual mag.	Absolute mag.	Spectral type	Distance (l.y.)
α	Sirius	6	45	−16 43	−1.44	1.45	A0m	8.6
ϵ	Adhara	6	59	−28 58	1.50	−4.10	B2	431
δ	Wezen	7	08	−26 24	1.83	−6.87	F8	1800
β	Mirzam	6	23	−17 57	1.98	−3.95	B1	499
η	Aludra	7	24	−29 18	2.45	−7.51	B5	3200
ζ	Furud	6	20	−30 03	3.02	−2.05	B2.5	336
o^2		7	03	−23 50	3.02	−6.46	B3	2600

C

C

CANIS MINOR (GEN. CANIS MINORIS, ABBR. CMI)

Constellation on the celestial equator, representing the smaller of the two dogs following Orion, the other being Canis Major. Its leading star is PROCYON, the eighth-brightest in the sky. Apart from Procyon and β CMi, known as Gomeisa, there is little of note.

BRIGHTEST STARS

	Name	RA h m	dec. ° ′	Visual mag.	Absolute mag.	Spectral type	Distance (l.y.)
α	Procyon	7 39	+5 14	0.40	2.68	F5	11.4
β	Gomeisa	7 27	+8 18	2.89	−0.70	B8	170

Candy, Michael Philip (1928–94) English-born astronomer who worked at Royal Greenwich Observatory (1947–69) and Perth Observatory (1969–93). From Perth, he directed an observational programme that contributed greatly to the study of Halley's Comet at its 1986/87 apparition. Candy was an expert on the orbits of comets and asteroids, and has one of each type of object named after him.

Canes Venatici *See* feature article, page 69

Canis Major *See* feature article, page 69

Canis Minor *See* feature article

cannibalism Merging of a GALAXY into a much larger and more massive one, so that its content is incorporated without a major change in the structure of the larger galaxy. This process is thought to explain how CD GALAXIES at the centres of clusters have become so bright and massive. Cannibalism of dwarf satellites also seems to have played a role in the growth of the halos of SPIRAL GALAXIES, exemplified by the distinct streams of stars in the Milky Way and the ANDROMEDA GALAXY, which may be the assimilated remnants of former companions.

▼ **cannibalism** A Hubble Space Telescope image of the elliptical galaxy NGC 1316 in Fornax. The dark dust clouds and bluish star clusters are probably remnants of a collision 100 million years ago, during which NGC 1316 consumed a smaller galaxy.

Cannon, Annie Jump (1863–1941) American astronomer who, working at the Harvard College Observatory under the direction of Edward C. PICKERING, refined the system for classifying stellar spectra. In 1896, after teaching physics at Wellesley College, Cannon joined Harvard's staff of 'Pickering's women', a group of computing assistants hired mainly to work on the *HENRY DRAPER CATALOGUE* of stellar spectra.

Taking the alphabetical spectral classification begun by Williamina FLEMING and Antonia C. MAURY, Cannon dropped several categories and rearranged the sequence to give O, B, A, F, G, K, M, from the hottest stars to the coolest. White or blue stars were classified as type O, B or A, yellow stars as F or G, orange stars were designated K, and red stars as M. This scheme, which is the basis for the present-day HARVARD SYSTEM of spectral classification, was used by Cannon in the first ever catalogue of stellar spectra, for the 1122 brightest stars (1901). Later, she added types R, N and S, plus ten subcategories based upon finer spectral features.

In the course of her work, Cannon visually examined and classified hundreds of thousands of spectra: the *Henry Draper Catalogue*, filling nine volumes of the *Harvard Annals*, ultimately contained almost 400,000 stars sorted by spectral class. Cannon also discovered five novae, 300 new variable stars and published extensive catalogues of these objects in 1903 and 1907. She was the first woman to be elected an officer of the American Astronomical Society, but because of a reluctance of the scientific community to accept women in astronomy, she did not receive a regular appointment at Harvard until 1938, just two years before she retired.

Canopus The star α Carinae, the second-brightest star in the entire sky, visual mag. −0.62, distance 313 l.y. It is a white supergiant of spectral type F0 Ib, more than 10,000 times as luminous as the Sun. The Hipparcos satellite detected variations of about 0.1 mag., but the period (if any) and cause of the variation are not known. Canopus is named after the helmsman of the Greek King Menelaus.

Cape Canaveral *See* KENNEDY SPACE CENTER

Capella The star α Aurigae, at visual mag. 0.08 the sixth-brightest star in the sky, distance 42 l.y. It is a spectroscopic binary, consisting of two yellow giants with an orbital period of 104 days. Various spectral types have been given for this pair, but they are likely to be near G6 III and G2 III. The star's name comes from the Latin meaning 'she-goat'.

Cape Observatory *See* ROYAL OBSERVATORY, CAPE OF GOOD HOPE

Cape Photographic Durchmusterung (CPD) Catalogue compiled from the first large photographic survey of the southern sky, made at the Cape Observatory between 1885 and 1900 under the direction of David GILL. Data for the CPD were taken from Gill's photographic plates by Jacobus KAPTEYN. It contains 455,000 stars to 10th magnitude from dec. −18° to −90° and, achieved before the *CÓRDOBA DURCHMUSTERUNG*, complements the *BONNER DURCHMUSTERUNG*. It was later revised by Robert INNES.

Capricornus *See* feature article

captured rotation *See* SYNCHRONOUS ROTATION

Carafe Galaxy (Cannon's Carafe Galaxy) SEYFERT GALAXY in the southern constellation of Caelum (RA 04^h $28^m.0$ dec. −47°24′), part of a group with NGC 1595 and NGC 1598. Long-exposure images show a curved jet of emerging material, possibly the result of gravitational interaction with NGC 1595.

carbon Sixth element, chemical symbol C; it is fourth in order of COSMIC ABUNDANCE. Its properties include:

CAPRICORNUS (GEN. CAPRICORNI, ABBR. CAP)

Smallest constellation of the zodiac, lying in the southern celestial hemisphere between Sagittarius and Aquarius. In Greek mythology, it was said to represent the god Pan after he jumped into a river to escape from the monster Typhon. The constellation's brightest star is δ Cap, known as Deneb Algedi ('kid's tail'). It is a BETA LYRAE STAR, an eclipsing binary with a range of 0.2 mag. and a period of 1.02 days. α Cap (known as Algedi or Giedi, from the Arabic meaning 'kid') is of a wide unrelated pair of yellow stars, mags. 3.6 and 4.3, spectral types G9 and G3. β Cap is a wide double, mags. 3.1 and 6.1, colours golden yellow and blue-white.

BRIGHTEST STARS

	Name	RA h m	dec. ° ′	Visual mag.	Absolute mag.	Spectral type	Distance (l.y.)
δ	Deneb Algedi	21 47	−16 09	2.85	2.49	A5mF2	39
β	Dabih	20 21	−14 47	3.05	−2.07	K0+ A5	344

atomic number 6; atomic mass of the naturally occurring element 12.01115; melting point *c*.3820 K (it sublimes at 3640 K); boiling point 5100 K; valence 2, 3 or 4. Carbon has seven ISOTOPES, with atomic masses from 10 to 16, two of which are stable (carbon-12 and carbon-13). Carbon-12 is used to define the ATOMIC MASS UNIT, and so its atomic mass is exactly 12. The naturally occurring element contains carbon-12 (98.89%) and carbon-13 (1.11%) plus variable but small amounts of carbon-14. In its free state, carbon exists as amorphous carbon, graphite and diamond.

Carbon is unique in the vast number and variety of compounds that it can form. The study of its reactions forms the entire discipline of organic chemistry. Carbon's property of forming hexagonal rings and long-chain molecules and of linking with hydrogen, nitrogen and oxygen makes it the basis of life. In the form of carbon dioxide (CO_2) and methane (CH_4) it produces the two most significant GREENHOUSE gases.

Carbon-14 is radioactive, with a half-life of 5730 years, and forms the basis for carbon-dating archaeological remains. It is continually produced in the Earth's atmosphere from nitrogen-14 by COSMIC RAYS. Once produced, it is incorporated into living material. After the death of the organism, the decay of the carbon-14 slowly reduces its abundance relative to carbon-12, and so allows determination of the age.

Carbon plays an important role in the CARBON–NITROGEN–OXYGEN CYCLE, which is the major helium-producing nuclear reaction in massive stars. It is unusually abundant in some types of peculiar star (*see* ASTROCHEMISTRY), and may comprise the bulk of the material forming some white dwarfs.

carbonaceous chondrite Chondritic METEORITE with atomic magnesium to silicon ratio greater than 1.02. Carbonaceous chondrites are subdivided, on chemical or textural grounds, into seven groups, each (apart from the CH group) named after its type specimen. The CI (for Ivuna) group has six members, including TAGISH LAKE. These meteorites have a composition very close to that of the Sun, without the volatiles. They are rich in water (up to *c*.20% by weight) and carbon (up to *c*.7% by weight); carbon is present mainly as organic compounds, including amino acids. Members of the CM (for Mighei) group are similar to CI chondrites, but they contain CHONDRULES and slightly less carbon (up to *c*.3% by weight). This group includes MURCHISON. MICROMETEORITES show many similarities to CM chondrites. Meteorites in the CV (for Vigarano) group have large CHONDRULES and are rich in refractory elements, such as aluminium and iridium. CV chondrites contain centimetre-sized CALCIUM–ALUMINIUM-RICH INCLUSIONS (CAIs), but little carbon or water. ALLENDE is a member of this group. The CO (for Ornans) group has smaller chondrules and CAIs. It is poorer in refractories than CV chondrites. CO chondrites also contain little water or carbon. The CR (for Renazzo) group of chondrites is characterized by abundant, well-defined chondrules and high metal content. The CK (for Karoonda) group is mostly thermally metamorphosed and re-crystallized chondrites. Members of the CH group (H for high iron content) have very small chondrules and high metal contents.

carbon–nitrogen–oxygen cycle (CNO cycle, CN cycle, Bethe–Weizsäcker cycle) Cycle of nuclear reactions that accounts for the energy production inside MAIN-SEQUENCE stars of mass greater than the Sun. The cycle was first described, independently, by Hans BETHE and Carl von WEIZSÄCKER in 1938.

The reactions involve the fusion of four hydrogen nuclei (PROTONS) into one helium nucleus at temperatures in excess of 4 million K. The cycle is temperature dependent, and at temperatures greater than 20 million K it becomes the dominant energy-producing mechanism in stellar cores. The mass of one helium nucleus (4.0027 atomic mass units) is less than the total mass of four protons (4.0304 units) so some matter is converted into energy and emitted as gamma rays. Positrons and NEUTRINOS are also emitted during the cycle.

The presence of carbon as one of the reactants is essential, but it behaves like a catalyst. Carbon-12 is used in the first reaction, but after a series of reactions, during which four hydrogen nuclei are absorbed and a helium nucleus is formed, carbon-12 is reproduced. Isotopes of carbon, oxygen and nitrogen occur as transient intermediate products during the reactions.

carbon star (C star) Classically, a cool GIANT STAR, between about 5800 and 2000 K, that exhibits strong spectral absorptions of C_2, CN and CH and that has an atmosphere containing more carbon than oxygen. The original classes, R STARS (comparable in temperature with G STARS and K STARS) and N STARS (comparable with M STARS), are combined into class C. N stars are on the asymptotic giant branch (AGB) of the HERTZSPRUNG–RUSSELL DIAGRAM (HR diagram). As main-sequence stars evolve on to the upper AGB, those of intermediate mass can dredge up carbon made by the TRIPLE-α PROCESS into their atmospheres, turning first into carbon-rich S STARS and then into genuine carbon stars. When carbon exceeds oxygen, the two combine to make molecules (particularly CO), leaving no oxygen to make metallic oxides. The rest of the carbon then combines with itself and other atoms.

Most N-type carbon stars are irregular, semi-regular, or Mira variables. Carbon dust forming in strong stellar winds can surround, or even bury, the stars in molecule-rich shrouds, the stars being visible only in the infrared. Such carbon stars are also rich in elements created by the s process. Carbon stars are a major source of galactic dust and a significant source of interstellar carbon.

The warmer R stars appear to be core-helium-burning

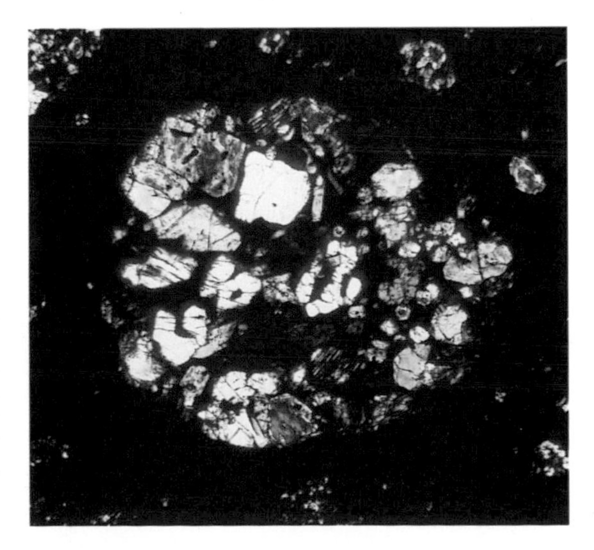

◄ **carbonaceous chondrite** A thin section of the Cold Bokkeveld meteorite, which fell in Cape Province, South Africa, in 1838. Taken with a petrological microscope, this photograph shows a near-circular chondrule about 1 mm in diameter.

C

CARINA (GEN. CARINAE, ABBR. CAR)

Prominent southern constellation, part of the old Greek figure of ARGO NAVIS, the ship of the Argonauts; Carina represents the ship's keel. Its leading star is CANOPUS, the second-brightest in the entire sky. Carina lies on the edge of a rich region of the Milky Way. ETA CARINAE is a violently variable star within the large and impressive ETA CARINAE NEBULA, NGC 3372. υ Car is an easy double for small telescopes, mags. 3.0 and 6.0. The variables R and S Car are MIRA STARS, reaching respectively 4th and 5th magnitude at maximum and with periods of 309 and 150 days, while l ('ell') Carinae is a bright Cepheid, range 3.3–4.2, period 35.5 days. IC 2602 is a prominent open cluster known as the Southern Pleiades, 480 l.y. away; its brightest member is θ Car, mag. 2.7. Other notable open clusters are NGC 2516, 3114 and 3532.

BRIGHTEST STARS

	Name	RA h m	dec. ° ′	Visual mag.	Absolute mag.	Spectral type	Distance (l.y.)
α	Canopus	06 24	−52 41	−0.62	−5.53	F0	313
β	Miaplacidus	09 13	−69 42	1.67	−0.99	A1	111
ε	Avior	08 23	−59 31	1.86	−4.58	K3 + B2	632
ι	Aspidiske	09 17	−59 16	2.21	−4.42	A7	692
θ		10 43	−64 23	2.74	−2.91	B0.5	439
υ		09 47	−65 04	2.92	−5.56	A6	1620
ω		10 14	−70 02	3.29	−1.99	B8	370
χ		07 57	−52 58	3.46	−1.91	B3	387

red giants that fall in a 'clump' in the HR diagram around absolute visual magnitude zero. Carbon-rich giant 'CH' stars (G and K stars with strong CH, CN and C_2 molecules), giant G and K BARIUM STARS, and dwarf carbon (dC) stars are binaries with white dwarf companions; they were probably contaminated by MASS TRANSFER when the white dwarf was a carbon-type giant. R Coronae Borealis and WC Wolf–Rayet stars are, respectively, carbon-rich helium giants and supergiants.

Carina *See* feature article

Carina arm Nearby spiral arm of the Milky Way GALAXY. It extends in the sky from Carina to Centaurus. It may be an extension of the SAGITTARIUS ARM, rather than a complete individual spiral arm. It may best be traced from its TWENTY-ONE CENTIMETRE radio emissions from hydrogen and by the presence of young blue stars.

Carme Outer moon of JUPITER, about 40 km (25 mi) in size. Carme was discovered in 1938 by Seth Nicholson. It takes 692 days to orbit Jupiter, at an average distance of 22.6 million km (14.0 million mi) in an orbit of eccentricity 0.253. It has a RETROGRADE path (inclination 165°), in common with other members of its group. *See also* ANANKE

Carnegie Observatories Observatories of the Carnegie Institution of Washington, founded in 1904. The main offices in Pasadena, California, house the observatories' scientific and technical staff. Until 1980 the Carnegie Observatories operated the MOUNT WILSON and PALOMAR Observatories under the name 'Hale Observatories' in partnership with the CALIFORNIA INSTITUTE OF TECHNOLOGY. Today, their main observing site is LAS CAMPANAS OBSERVATORY.

Carrington, Richard Christopher (1826–75) English brewer and amateur astronomer who built his own observatory at Redhill, Surrey. By making daily measurements of sunspot positions during the years 1853–61, he discovered, independently of Gustav SPÖRER, that the Sun's equatorial regions rotate faster than its more extreme latitudes. From a sunspot's heliographic latitude, he was able to predict how quickly the spot would move across the solar disk. In 1859 Carrington became the first person to observe a solar flare, one sufficiently energetic to be visible to the naked eye, and noted that it was followed by auroral displays. He also compiled an important catalogue (1857) of almost 4000 circumpolar stars.

Carrington rotation Mean period of 25.38 days introduced in the 19th century by the English astronomer Richard C. CARRINGTON from observations of SUNSPOTS as the mean length of time taken for the Sun to rotate on its axis at the equator with respect to the fixed stars. Since the Earth revolves about the Sun in the same direction that the Sun rotates, this SIDEREAL PERIOD is lengthened to a SYNODIC PERIOD of 27.28 days when observed from Earth. Beginning with that which commenced on 1853 November 9, the Sun's synodic rotations are numbered sequentially; Carrington defined the zero of solar longitude as the central solar meridian on this day. For example, Carrington Rotation 1985 commenced on 2002 January 7.

Carte du Ciel Programme initiated in 1887 by the brothers HENRY and others to map the whole sky in 22,000 photographs taken from 18 observatories, using identical 'astrograph' telescopes. One of the first large international astronomical projects, the *Carte du Ciel* took about 60 years to complete. The project also included the generation of the *Astrographic Catalogue*, listing several million stars down to 11th magnitude, which has since proved invaluable for the derivation of accurate proper motions. *See also HIPPARCOS CATALOGUE*

Carter Observatory National observatory of New Zealand, established in Wellington's Botanic Garden in 1941. It now functions as a centre for public astronomy, for which 230-mm (9-in.) and 150-mm (6-in.) refractors are used, together with a planetarium. The observatory also serves as a national repository for New Zealand's astronomical heritage.

Cartwheel Galaxy (MCG-06-02-022a) STARBURST GALAXY in the southern constellation Sculptor (RA 00^h $37^m.4$ dec. −33°44′). It lies about 500 million l.y. away. The galaxy's structure has been disrupted by the recent passage through it of a second, dwarf galaxy, which triggered an episode of star formation in the ring-shaped rim. The rim has a diameter of 150,000 l.y., and spokes of material extend towards it from the galaxy's centre.

cascade image tube Type of IMAGE INTENSIFIER used to amplify the brightness of faint optical images through a multistage process. In a simple image tube, the incident light beam falls on to a photocathode, liberating a stream of electrons. These electrons are accelerated by an electric field of around 40 kV and are focused by either a magnetic or electrostatic field on to a phosphor screen to form an

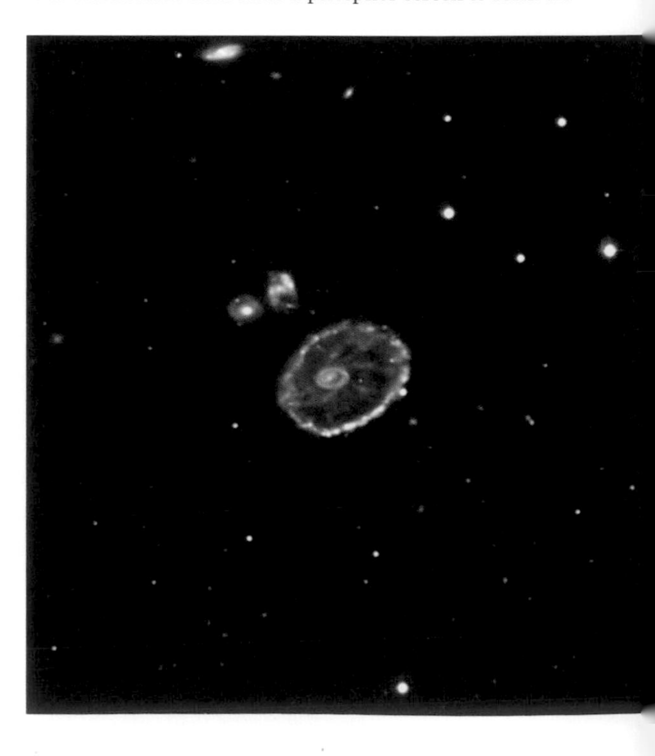

► **Cartwheel Galaxy** An Anglo-Australian Telescope photograph of this irregular galaxy, whose ring and spoke configuration is a result of a comparatively recent encounter with another galaxy.

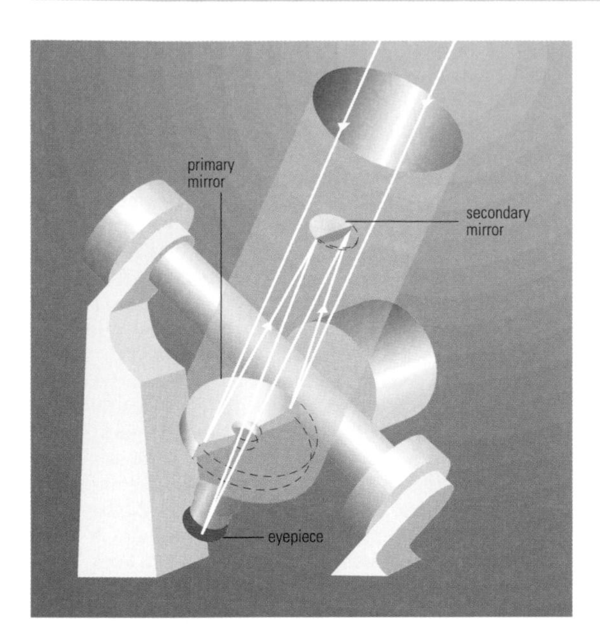

◄ **Cassegrain telescope** A cutaway showing the light path in a Cassegrain telescope. Light collected by the primary mirror is reflected from a hyperboloidal secondary to the eyepiece through a hole in the primary's centre.

image. A cascade image tube incorporates a number of photocathode stages; the output image from one section serves as the input for the next, resulting in increased amplification of the signal, and hence greater image brightness. The final detector may be a television camera or a CCD (charge-coupled device).

Cassegrain telescope REFLECTING TELESCOPE with a concave PARABOLOIDAL primary and a convex HYPERBOLOIDAL secondary. Light is gathered by the primary and reflected to the secondary, which is placed in the light path on the optical axis. The convex secondary increases the focal length by enough for the light to pass through a hole in the primary before coming to a focus behind it. This 'folding' of the light path results in an instrument that is much more compact than a refractor or Newtonian of the same focal length. The Cassegrain has no chromatic or spherical ABERRATION, but does suffer from slight astigmatism, moderate coma and strong field curvature.

The design is credited to the French priest Laurent Cassegrain (1629–93), but did not become popular until the late 19th century because of the difficulty of accurately figuring the convex hyperboloidal secondary mirror. However, this eventually became the commonest design for large professional telescopes, and popular for compact amateur instruments, leading to the development of many variations, such as the DALL–KIRKHAM TELESCOPE, SCHMIDT–CASSEGRAIN TELESCOPE and RITCHEY–CHRÉTIEN TELESCOPE.

Cassini NASA spacecraft launched to SATURN in 1997 November. In 2004 November, it will become the first spacecraft to orbit Saturn and will also deploy the European Space Agency craft HUYGENS, which is scheduled to land on the surface of the Saturnian moon TITAN. Cassini flew towards Saturn via two gravity assist flybys of Venus and one of the Earth, and in 2000–2001 it conducted, with the orbiting GALILEO, the first dual-spacecraft Jupiter science mission.

Cassini family French family of Italian origin that produced four astronomers and cartographers, all associated with PARIS OBSERVATORY, which was set up by Giovanni Domenico CASSINI (known to historians as Cassini I) at the invitation of Louis XIV of France. The elder Cassini's son, **Jacques** (or **Giacomo**) **Cassini** (1677–1756, Cassini II), assumed directorship of Paris Observatory in 1700. Jacques' major contributions were in geodesy and cartography: he established the definitive arc of the Paris Meridian, and conceived a method for finding longitudes by observing lunar occultations of stars and planets.

In 1771 Jacques' son, **César-François** (or **Cesare Francesco**) **Cassini de Thury** (1714–84, Cassini III), became the observatory's director. César-François was responsible for producing the first truly reliable map of France, which took half a century to complete. Despite this work, the fortunes of Paris Observatory declined during his directorship, due to lack of government support and competition from better-funded private observatories. After his death in 1784, his son **Jacques Dominique Cassini** (1748–1845, Cassini IV) took over the observatory and restored its former prestige.

Cassini, Giovanni Domenico (1625–1712) Italian astronomer and geodesist (also known as Jean Dominique after his move to France, and to historians as Cassini I) who founded PARIS OBSERVATORY and helped to establish the scale of the Solar System. As a result of his important studies of the Sun and Venus, Mars and Jupiter carried out at Bologna University, he was invited to plan the new Paris Observatory, built between 1667 and 1672 by order of Louis XIV. Cassini became the observatory's director and used its very long focal length, tubeless refracting telescopes to discover, in 1675, the division in Saturn's ring system that today bears his name. He also discovered four Saturnian moons (1671–84). He constructed a lunar map from eight years of original observations; this map served as the standard for the next 100 years.

Cassini accurately determined the rotation periods of Mars and Venus, and his studies of Jupiter's satellites enabled Ole RÖMER to calculate the velocity of light by observing the delay in satellite eclipse intervals caused by variations of Jupiter's distance from the Earth. Most importantly, Cassini established the scale of the Solar System. He derived a measure of Mars' parallax from his own Paris observations and those made by Jean RICHER from Cayenne, South America, that corresponded to a value of 140 million km (85 million mi) for the astronomical unit. Although 10 million km (6 million mi) short of the true figure, this was a huge improvement on previous values (for example, Johannes Kepler's estimate of 15 million km/10 million mi). After Cassini's health failed in 1700, his son Jacques assumed the directorship of Paris Observatory (*see* CASSINI FAMILY).

Cassini Division Main division in SATURN's ring system, separating the bright A RING and B RING at a distance of 117,600 km (73,100 mi) from the planet's centre. The Cassini Division is not empty: VOYAGER results show that it contains several narrow rings and that there are particles in the gaps between these rings.

▼ **Cassini** En route to Saturn, the Cassini spacecraft passed through the Jupiter system. This image of Jupiter's Great Red Spot and its environs, together with Io and its shadow on the cloud-tops, was obtained on 2000 December 12 from a distance of 19.5 million km (12.1 million mi).

C

CASSIOPEIA (GEN. CASSIOPEIAE, ABBR. CAS)

Prominent northern constellation, easily identified by the 'W' formed by its five brightest stars. Cassiopeia represents Queen of Ethiopia, wife of Cepheus, in the ANDROMEDA myth. γ Cas, which can unpredictably brighten or fade by half a magnitude or more, is the prototype GAMMA CASSIOPEIAE STAR. RHO CASSIOPEIAE is an immensely luminous pulsating supergiant of spectral type G2, estimated to lie more than 10,000 l.y. away. η Cas is a beautiful coloured double for small telescopes, mags. 3.5 and 7.5; it is a true binary with orbital period of 480 years. ι Cas is another double divisible in small telescopes (mags. 4.5 and 8.4). Closer and more difficult is σ Cas (mags. 5.0 and 7.3). Cassiopeia lies in the Milky Way and contains rich starfields. M52, NGC 457 and NGC 663 are good open clusters for binoculars. M103, smaller and elongated, is better seen through a telescope. TYCHO'S STAR appeared here in 1572.

BRIGHTEST STARS

	Name	RA h m	dec. ° ′	Visual mag.	Absolute mag.	Spectral type	Distance (l.y.)
γ		00 57	+60 43	2.15	−4.22	B0	613
α	Shedir	00 41	+56 32	2.24	−1.99	K0	229
β	Caph	00 09	+59 09	2.28	1.17	F2	55
δ	Ruchbah	01 26	+60 14	2.66	0.24	A5	99
ε		01 54	+63 40	3.35	−2.31	B3	442
η		00 49	+57 49	3.46	4.59	F9	19.4

Cassiopeia *See* feature article

Cassiopeia A Strongest radio source in the sky, apart from the Sun. It appears optically as a faint nebula. It is a SUPERNOVA REMNANT from an unrecorded supernova explosion in 1660.

Castalia APOLLO ASTEROID discovered in 1989; number 4769. Radar images have shown Castalia to have a twin-lobed shape. It is *c*.2.4 km (*c*.1.5 mi) long. *See* table at NEAR-EARTH ASTEROID

Castor The star α Geminorum, visual mag. 1.58, distance 52 l.y. Castor is a remarkable multiple star. Small telescopes show it as a double star, with blue-white components of mags. 2.0 and 2.9, spectral types A2 V and A5 V; these form a genuine binary with an orbital period of about 450 years, separation currently increasing. Each component is also a spectroscopic binary, with periods of 9.2 and 2.9 days respectively. There is also a third member of the Castor system, known as YY Geminorum, an ALGOL STAR eclipsing binary consisting of two red dwarfs, both of spectral type M1. During eclipses, which occur every 19.5 hours, their total visual magnitude drops from 9.3 to 9.8. Castor is named after one of the celestial twins commemorated by the constellation Gemini, the other being POLLUX.

▼ **Cassiopeia** The distinctive 'W' shape of this familiar northern constellation is marked out by five second-magnitude stars.

cataclysmic variable (CV) Term given to a diverse group of stars that undergo eruptions, irrespective of the cause of the outburst. The term may describe a SUPERNOVA, NOVA, RECURRENT NOVA, NOVA-LIKE VARIABLE, FLARE STAR, DWARF NOVA, some X-ray objects, and other erupting stars. CVs are very close binary systems, with outbursts caused by interaction between the two components. A typical system of this type has a low mass secondary which fills its ROCHE LOBE, so that material is transferred through its LAGRANGIAN POINT on to the primary, which is usually a WHITE DWARF. The transferred material has too much angular momentum to fall directly on to the primary; instead, it forms an ACCRETION DISK, on which a hot spot is formed where the infalling material impacts on its outer edge. For any particular star, outbursts occur at irregular intervals from about ten days to weeks, months or many years.

Most dwarf novae and recurrent novae show a relationship between maximum brightness and length of the mean cycle: the shorter the cycle, the fainter the maximum magnitude. Nova-like variables are a less homogeneous group and are also less well-studied. Some have bursts of limited amplitude; others have had no observed outburst but have spectra resembling old novae. There are many other objects that show some if not all of the characteristics of CVs, for example old novae, some X-ray objects, AM HERCULIS STARS and others.

Many models have been suggested to explain the observed outbursts. The two most probable theories are that they are caused either by variations in the rate of MASS TRANSFER or by instabilities in the accretion disk. Both models require transfer of mass from the red secondary. A few systems are so aligned that we see them undergoing eclipses, enabling the main light source to be studied in detail. Typical examples of such systems are Z Chamaeleontis and OY Carinae; these stars may be examined both at outburst and minimum. It appears that during outbursts the disk increases in brightness. The intervals between consecutive outbursts of the same type vary widely, as they do for most CVs. Z Chamaeleontis has a mean cycle for normal outbursts of 82 days; for super-outbursts the mean cycle is 287 days. For OY Carinae the respective values are about 50 and 318 days. Their semiperiodic oscillations, timed in seconds, are small.

The advantage of studying eclipsing systems is that both the primary and the hot spot may be eclipsed, indicating the probable sizes of the components. It is generally agreed that a variable mass-flow rate can account for many observations. For the mass-flow rate to vary, there must be instability in the red star, but no generally accepted theory has been advanced. The red star must relax after a burst until another surge again overflows, discharging another burst of gas, but exactly what causes this ebb and flow is a mystery. The angular momentum of the mass flow prevents it from falling directly on to the primary. Instead a disk is formed around it on which a hot spot is formed at the point of impact. Some of this matter must be carried away, and it is not known whether this material is lost to the system or whether it splashes back on to the disk. When an eclipsing system is at minimum, the main light source at primary eclipse comes from the red star. The disk is then in a steady state.

Theories stipulating that the cause of the outburst is instabilities in the disk, while accepting variable mass transfer as the origin of the subsequent observed phenomena, contend that it is what happens on the disk that gives rise to outbursts. A number of disk models have been proposed, but none appears to fit the observed facts. For example, the disk increases in brightness during outbursts, at least in the eclipsing systems, and presumably in other dwarf novae. Disk instability models differ on how this released energy spreads through the disk, which mainly radiates in the ultraviolet. This ultraviolet radiation does not behave in the same way as the visual radiation and appears to contradict these models. If the disk dumps energy on to the primary, nuclear burning would be expected, but evidence on this is unclear.

catadioptric system Telescope or other optical system that includes both mirrors and lenses. To be considered catadioptric, at least one lens and mirror must be optically active (it must converge, diverge or correct the optical path) – flat mirrors and optical 'windows' do not count. So, for instance, the SCHMIDT–CASSEGRAIN TELESCOPE and MAKSUTOV TELESCOPE are both catadioptric, the light passing through an optically figured corrector plate before striking the primary mirror, but a folded refractor or a Cassegrain telescope with a flat optical window is not. The catadioptric design is compact, and is often used for portable instruments.

catalogue, astronomical Tabular compilation of stars or other deep-sky objects, giving their CELESTIAL COORDINATES and other parameters, such as magnitude. The various types of catalogue are distinguished by their content (the kind of sources covered and the parameters given), their context (when, why and how they were created) and their scope (the LIMITING MAGNITUDE). Thousands of astronomical catalogues have been produced, from the one in Ptolemy's *ALMAGEST* (*c.*AD 150), which gave positions (to an accuracy of around 1°) and magnitudes of about 1000 stars, to modern catalogues – usually accessible only as electronic files – which may contain the positions (accurate to 0″.1 or better) and magnitudes of up to several hundred million sources. A list of catalogues is given in the *Dictionary of Nomenclature of Celestial Objects*, maintained under the auspices of the International Astronomical Union.

Before the modern era, cataloguers such as AL-SŪFĪ were concerned mainly with refining the positions and magnitudes given in the *Almagest*. The need for improved navigation that came with European expansion after the Renaissance provided the impetus for early modern catalogues such as John FLAMSTEED's *Historia coelestis brittanica*. As the power of telescopes increased, more and more stars were revealed for the cataloguers to measure. In the 19th and early 20th centuries, photographic methods supplanted visual observation as a means of gathering data, and catalogues swelled to contain hundreds of thousands of entries. Stars are known by their designations in catalogues, preceded by the catalogue's initials (*see* STELLAR NOMENCLATURE).

Three great *Durchmusterungen* ('surveys' in German) were compiled between 1855 and 1932 at Bonn, Córdoba and Cape Town, gathering together 1.5 million positions and magnitudes for about 1 million stars over the whole sky: the *BONNER DURCHMUSTERUNG* (BD), the *CÓRDOBA DURCHMUSTERUNG* (CD) and the *CAPE PHOTOGRAPHIC DURCHMUSTERUNG* (CPD). The first major catalogue of stellar spectra was the *HENRY DRAPER CATALOGUE* (HD), classifying 360,000 stars by spectral type, compiled by Annie CANNON and published between 1918 and 1936. The first major astrometric and photometric catalogues were produced by, respectively, Lewis and Benjamin BOSS and by Friedrich ZÖLLNER. Other notable star catalogues include the *Astrographic Catalogue* from the *CARTE DU CIEL* project; the *BRIGHT STAR CATALOGUE* (BS), with accurate values of the positions, parallaxes, proper motion, magnitudes, colours, spectral types and other parameters for the 9110 stars brighter than magnitude 6.5; the 1966 *SMITHSONIAN ASTROPHYSICAL OBSERVATORY STAR CATALOG* (SAO), the first large 'synthetic' catalogue created on a computer by combining data from several astrometric catalogues; the *GUIDE STAR CATALOGUE* (GSC), created for the Hubble Space Telescope in 1990; and the 1997 HIPPARCOS CATALOGUE (HIP) and Tycho Catalogue (TYC), the products of the Hipparcos astrometry mission.

Many other star catalogues have been compiled for specific categories of stars, such as the *Catalogue of Nearby Stars* (Gl, after its compiler, Wilhelm Gliese, 1915–93). Important early catalogues of double stars were prepared by Wilhelm STRUVE (Σ), Otto STRUVE (OΣ), Sherburne BURNHAM (BPS) and Robert AITKEN (ADS); the principal modern source is the 1996 *Washington Visual Double Star Catalog* (WDS) by C.E. Worley and G.C. Douglass. For variable stars the current reference is the *GENERAL CATALOGUE OF VARIABLE STARS*. Stars with high proper motion are listed in William LUYTEN's *Two-Tenths Catalog of Proper Motions* (1979–80). FUNDAMENTAL CATALOGUES give highly accurate positions and proper motions for selected stars against which the relative positions of other celestial objects may be measured.

There are also many catalogues of non-stellar objects. The earliest is the famous *MESSIER CATALOGUE* (M); there followed John HERSCHEL's *General Catalogue of Nebulae and Clusters*, and J.L.E. Dreyer's *NEW GENERAL CATALOGUE* (NGC) and *INDEX CATALOGUES* (IC). The NGC and the two ICs together represent the last attempt to include all known non-stellar objects in a single listing before the advent of today's powerful SURVEY capabilities. Subsequent catalogues dealing with specific objects included: Edward Emerson BARNARD's 1927 list of dark nebulae, *Catalogue of 349 Dark Objects in the Sky* (B); S. Sharpless' 1959 listing of emission nebulae, *A Catalogue of H II Regions* (Sh-2); and the *Catalog of Galactic Planetary Nebulae* by Luboš Perek (1919–) and Luboš Kohoutek (1935–) published in 1967 (PK). Catalogues of galaxies include the 1962–68 *Morphological Catalogue of Galaxies* (MCG or VV, from author Boris Alexandrovich Vorontsov-Vel'iaminov, 1904–94) and 1973 *Uppsala General Catalogue of Galaxies* (UGC) by Peter Nilson (1937–98). Peculiar galaxies have been catalogued by Halton ARP, and clusters of galaxies by George ABELL.

Catalogues have also been compiled for sources of radiation outside the optical region. The first Cambridge catalogue, of the 50 brightest radio sources, was published in 1950 (*see* THIRD CAMBRIDGE CATALOGUE). A recent radio catalogue, The NRAO VLA Sky Survey, nearing completion, has yielded a catalogue of over 1.7 million radio sources north of dec. −40°. At the other end of the electromagnetic spectrum, the Rosat catalogues (RX) are currently the most frequently used sources for X-ray emitters.

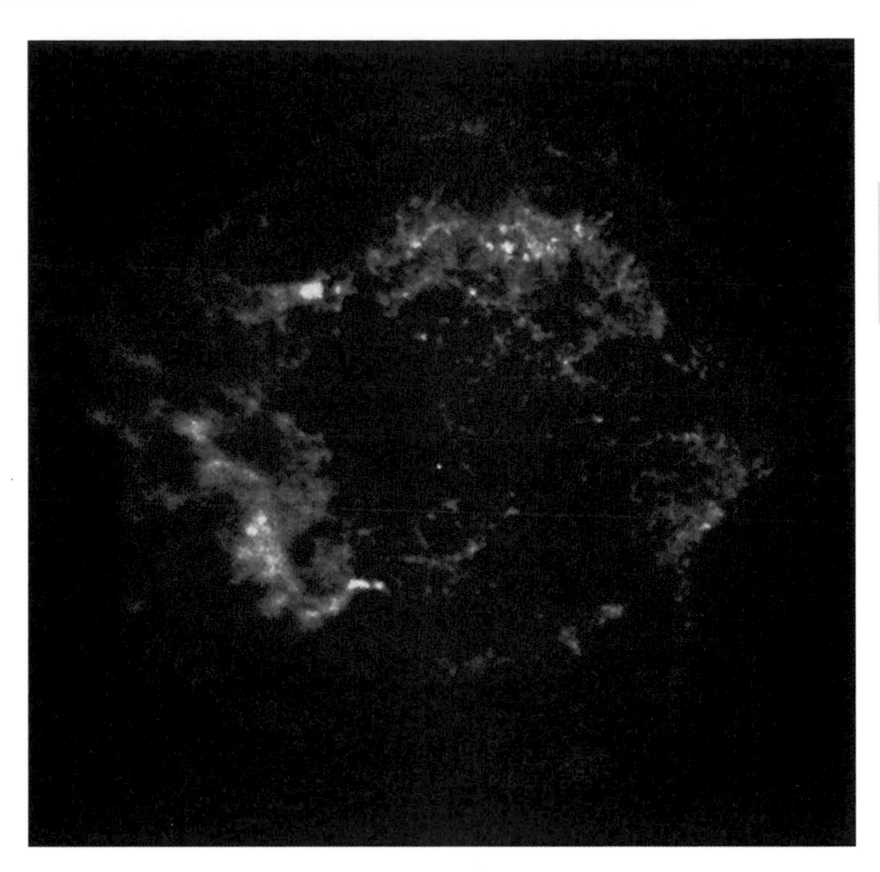

▲ **Cassiopeia A** Chandra X-ray image of this supernova remnant, which is extremely faint at visual wavelengths. The image reveals a shell of material 10 l.y. in diameter.

catena Chain of CRATERS. There are many named catenae on Earth's Moon, Mars, Jupiter's satellites Io, Ganymede and Callisto and on Neptune's satellite Triton.

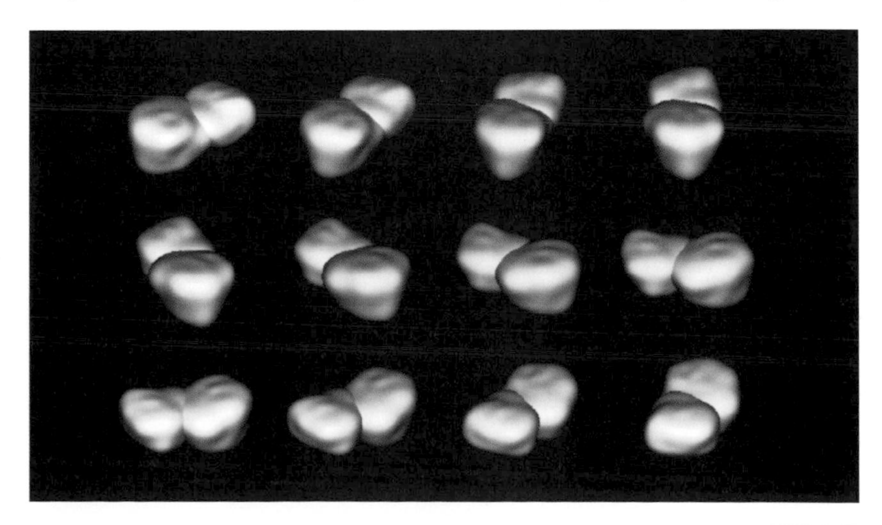

▼ **Castalia** This series of radar images of the near-Earth asteroid (4769) Castalia was obtained during its close approach in 1989 using the Arecibo telescope.
[S. Ostro (JPL/NASA)]

They are typically tens to hundreds of kilometres long and consist of craters of less than one kilometre to tens of kilometres in diameter. Catenae are most frequently chains of impact craters. Some catenae, such as those found on Io, may be chains of volcanic craters. Drainage of surface material into tectonic fault openings or collapsed lava tubes may also form small catenae.

Catharina Lunar crater (18°S 24°E), 88 km (55 mi) in diameter. With THEOPHILUS and CYRILLUS, it forms a trio of similar large craters. An older formation, Catharina shows much erosion, which gives its walls a highly irregular outline. Later impacts have almost completely destroyed its north-east wall. No terracing is visible on Catharina's inner slopes, and its floor is devoid of a central peak.

Cat's Eye Nebula (NGC 6543) Comparatively bright (magnitude +8.1) PLANETARY NEBULA; it shows a greenish tint when observed visually. Located in the northern constellation Draco, roughly midway between ζ and δ (RA $17^h\ 58^m.6$ dec. +66°38′), the Cat's Eye has an apparent diameter of 350″. It has an 11th-magnitude central star.

Caucasus Montes Lunar mountain range (36°N 8°E), dividing Mare IMBRIUM from Mare SERENITATIS. Some of its peaks rise to altitudes of 3650 m (12,000 ft). Material ejected by the Imbrium impact has scoured the Caucasus Mountains, which run roughly north–south. The most southerly part of the range consists mainly of isolated mountain peaks.

CCD (charge-coupled device) Small electronic imaging device, widely used in astronomy, which is highly efficient in its response to light and therefore able to detect very faint objects over a broad range of the spectrum.

CCDs consist of an oxide-covered silicon substrate with a two-dimensional rectangular array of light-sensitive electrodes on the surface. These electrodes form a matrix of pixels, or picture elements, each less than 0.03 nm in size and capable of storing electronic charges created by the absorption of light.

Photons imaged on to the surface of the CCD penetrate the electrode structure and enter the substrate, where electron–hole pairs are generated via the photoelectric effect, in numbers precisely proportional to the number of incident photons. The holes are conveniently lost by diffusion down into the depths of the substrate, while the electrons migrate rapidly to the nearest biased electrode, where they collect as a single charge packet in a 'potential well'.

Since it is impractical to wire the output from each individual electrode, the signal charge is transferred through the array, from one pixel to the next, by changing the voltages on each one. Pixels in adjacent rows are said to be 'charge-coupled' and the signal can move in parallel from row to row. As each row of signals reaches the end of the CCD it is read off into a serial register from which it can be stored in a computer, processed and the final image displayed.

Because the charge in each pixel is in proportion to the number of photons that have fallen on it, the output from a CCD is linear, which in turn means that the brightness of the final image produced is directly related to the brightness of object being observed, something which is not true for photographic emulsions.

In order to generate electron–hole pairs, photons have to pass through the electrode structure, and some are inevitably absorbed in this layer. Because of this, the quantum efficiency of front-illuminated CCDs is poor in the blue and UV regions of the electromagnetic spectrum. To overcome this, the silicon substrate can be thinned to around 15 μm and the CCD back-illuminated, thus eliminating the need for the photons to negotiate the electrode structure. This process can increase the sensitivity of a CCD by a factor of two and the use of anti-reflective coatings can also improve quantum efficiency.

One of the other problems that has to be overcome when using CCDs is the generation of 'dark current'. This is the unwanted charge that is created by the natural random generation and recombination of electron–hole pairs that occurs at temperatures above absolute zero. This thermally induced charge can mask the signal when observing very faint objects over a long period of time but can be effectively eliminated by cooling the CCD using liquid nitrogen. A low dark current makes it possible to store the signal charge for long periods of time, thus allowing exposures from a few tens of seconds to several hours to be achieved.

CCDs can thus be used to accumulate signals from very faint light sources. They are more sensitive than photographic emulsion, are linear, spatially uniform and stable with time. They are also sensitive over a broad range of the spectrum, have low noise levels and a large dynamic range so that they can be used to detect both bright and faint objects at the same time. They can be used for direct imaging applications, surveying large areas of sky, for photometry or as spectroscopic detectors. Because of their versatility, they have almost completely replaced photographic plates.

CD Abbreviation of *CÓRDOBA DURCHMUSTERUNG*

cD galaxy Largest, most luminous kind of normal GALAXY. Often found at the centres of rich clusters, they resemble giant ELLIPTICAL GALAXIES except for having a more extensive outer envelope of stars. Some cD galaxies can be traced over spans exceeding a million light-years. They may grow as a result of attracting stars that were stripped from surrounding galaxies by tidal encounters, either within the cluster as a whole or with the cD galaxy. This class was recognized by William W. MORGAN as part of the Yerkes system of galaxy classification, along with N GALAXIES.

CDS Abbreviation of CENTRE DE DONNÉES ASTRONOMIQUES DE STRASBOURG

celestial coordinates Reference system used to define the positions of points or celestial objects on the celestial sphere. A number of systems are in use, depending on the application.

EQUATORIAL COORDINATES are the most commonly used and are the equivalent of latitude and longitude on the Earth's surface. DECLINATION is a measure of an object's angular distance north or south of the CELESTIAL EQUATOR, values north being positive and those south negative. RIGHT ASCENSION, or RA, equates to longitude and is measured in hours, minutes and seconds eastwards from the FIRST POINT OF ARIES, the intersection of the celestial equator with the ECLIPTIC. HOUR ANGLE and POLAR DISTANCE can also be used as alternative measures.

The HORIZONTAL COORDINATE system uses the observer's horizon as the plane of reference, measuring the ALTITUDE (angular measure of an object above the horizon) and AZIMUTH (bearing measured westward around the horizon from north).

ECLIPTIC COORDINATES are based upon the plane of the ECLIPTIC and use the measures of CELESTIAL LATITUDE and CELESTIAL LONGITUDE (also known as ecliptic latitude and ecliptic longitude). Celestial latitude is measured in degrees north and south of the ecliptic, while celestial longitude is measured in degrees eastwards along the ecliptic from the First Point of Aries.

The GALACTIC COORDINATE system takes the plane of the Galaxy and the galactic centre (RA $17^h\ 46^m$ dec. −28°56′) as its reference points. Galactic latitude is measured from 0° at the galactic equator to 90° at the galactic pole, while galactic longitude is measured from 0° to 360° eastwards along the galactic equator.

celestial equator Projection on to the CELESTIAL SPHERE of the Earth's equatorial plane. Just as its terrestrial counterpart marks the boundary between the northern and southern hemispheres of the Earth and is the point from which latitude is measured, so the celestial equator delineates between the northern and southern hemispheres of the sky and is used as the zero point for measuring the celestial coordinate DECLINATION.

celestial latitude (symbol β, ecliptic latitude) Angular distance north or south of the ECLIPTIC and one coordinate of the ecliptic coordinate system. Designated by the Greek letter β, celestial latitude is measured in degrees from the ecliptic (0°) in a positive direction to the north ecliptic pole (+90°) and a negative direction to the south ecliptic pole (−90°). *See also* CELESTIAL LONGITUDE

celestial longitude (symbol λ, ecliptic longitude) Angular distance along the ecliptic from the FIRST POINT OF ARIES and one coordinate of the ecliptic coordinate system. Designated by the Greek letter λ, celestial longitude is measured from 0° to 360°, eastwards of the First Point of Aries. *See also* CELESTIAL LATITUDE

celestial mechanics (dynamical astronomy) Discipline concerned with using the laws of physics to explain and predict the orbits of the planets, satellites and other celestial bodies. For many years the main effort in the subject was the development of mathematical methods to generate the lengthy series of perturbation terms that are needed to calculate an accurate position of a planet. The advent of computers has eased this task considerably, with the result that the main emphasis of the subject has changed dramatically. Much effort is now devoted to the study of the origin, evolution and stability of various dynamical features of the Solar System, and in particular of the numerous intricate details of rings and satellite orbits revealed by the VOYAGER spacecraft.

The subject can be said to have started with the publication of Newton's *PRINCIPIA* in 1687, in which are stated his law of gravitation, which describes the forces acting on the bodies, and his three laws of motion, which describe how these forces cause accelerations of the motions of the bodies. Then the techniques of celestial mechanics are used to determine the orbits of the bodies resulting from these accelerations. One of the first results achieved by Newton was to give an explanation of KEPLER'S LAWS. These laws are descriptions deduced from observations of the motions of the planets as being elliptical orbits around the Sun, but until the work of Newton no satisfactory explanation of these empirical laws had been given. However, Kepler's laws are true only for an isolated system of two bodies; in the real Solar System the attractions of the other planets and satellites cause the orbits to depart significantly from elliptical motion and, as observational accuracy improved, these perturbations became apparent. The greatest mathematicians of the 18th and 19th centuries were involved in the effort of calculating and predicting the perturbations of the orbits, in order to match the ever-increasing accuracy and time span of the observations. The orbit of the Moon was the major problem, partly because the Moon is nearby, and so the accuracy of observation is high, but also because its orbit around the Earth is very highly perturbed by the Sun.

Various techniques were developed for calculating the perturbations of an orbit. An important technique is the method of the 'variation of arbitrary constants', developed by Leonhard EULER and Joseph LAGRANGE. An unperturbed orbit can be described by the six ORBITAL ELEMENTS, or arbitrary constants, of the ellipse. The effect of perturbations can be described by allowing these 'constants' to vary with time. Thus, for example, the eccentricity of the orbit may be described as a constant plus a number of periodic terms. The resulting expressions are called the 'theory of the motion of the body', and can be very lengthy in order to achieve the desired accuracy, perhaps hundreds of periodic terms. Around the middle of the 19th century an alternative method was developed in which the perturbations of the three coordinates of the body (for example, longitude, latitude and distance from the Sun) are calculated instead of the perturbations of the six elements. Variants of this method have been used ever since for lunar and planetary theories, but the variation of constants is still more suitable for many of the satellites. The latest theories used to calculate the positions of the Moon and planets in the almanacs used by navigators and astronomers are those derived by: Simon NEWCOMB for the five inner planets, Uranus and Neptune; by Ernest BROWN for the Moon; and by George HILL for Jupiter and Saturn.

Overall, Newton's four laws and the techniques of celestial mechanics have proved successful at explaining the motions of the planets and satellites. Problems with the orbit of the Moon were eventually resolved by improved theories, and anomalous perturbations of the orbit of Uranus led John Couch ADAMS and Urbain LE VERRIER to suspect the existence of a further planet, which resulted in the discovery of Neptune in 1846 close to the predicted position. Problems with the seemingly unaccountably rapid advance of Mercury's perihelion were eventually solved in 1915 with the publication of Einstein's general theory of relativity (*see* VULCAN). This is a more accurate representation of the laws of motion under the action of gravitation, but the differences from using Newton's laws are small, and only become noticeable in strong gravity fields.

By the early 20th century fairly good theories had been developed for the Moon and the planets, but these had just about reached the limit of what was possible with the means then available, of mechanical calculators, mathematical tables, and masses of hand algebra. There was some further effort on theories for other satellites and some asteroids, but much effort in the subject moved into other areas, particularly into theoretical studies such as the idealized THREE-BODY PROBLEM. With the advent of computers, renewed effort on the development of lunar and planetary theories became feasible. The initial idea was to use the same techniques as before, but to do the vast amounts of algebraic manipulation involved on a computer. Excellent theories of the Moon and some of the planets have been produced in this way. However it has proved to be more effective to use the less elegant but much simpler and more accurate method of integrating the equations of motion by numerical techniques. The positions of the Moon and planets in the almanacs are now all computed in this way, but the old algebraic methods are still the most effective for most of the natural satellites.

With the task of planetary theory development that had occupied celestial mechanics for 300 years now effectively solved, the main emphasis of the subject has changed dramatically, with many problems of origin, evolution and stability now being studied. The stability of the Solar System has long been of considerable interest, and recent numerical integrations suggest that the orbits of the major planets are stable for at least 845 million years (about 20% of the age of the Solar System), but the orbit of Pluto may be unstable over that time scale. Another problem that is now close to a full solution after more than a century of effort is to explain the origin of the KIRKWOOD GAPS in the ASTEROID BELT, which occur at certain distances from the Sun that correspond to commensurabilities of period with Jupiter. The explanation that was found for the gap at the 3:1 COMMENSURABILITY required the introduction of the concept of CHAOS into Solar System dynamics. Using a sophisticated combination of analytical and numerical methods, it was found that over a long period of time the orbital elements of an object at the commensurability could have occasional large departures (termed chaotic motion) from the small range of values expected from conventional analytical modelling of the motion. Subsequent work suggests that the same mechanism can explain the widest gap at the 2:1 commensurability. Many other instances of chaos have been identified, and with them is the realization that the Solar System is not such a deterministic dynamical system as had been supposed.

It is now realized that TIDAL EVOLUTION has been a major factor in adjusting the orbits of satellites, and modifying the spin rates of satellites and planets. There are many more occurrences of commensurabilities in the

C

► **celestial sphere** A useful concept for describing positions of astronomical bodies. As shown, key points of reference – the celestial pole and celestial equator – are projections on to the sphere of their terrestrial equivalents. An object's position can be defined in terms of right ascension and declination (equivalent, respectively, to longitude and latitude).

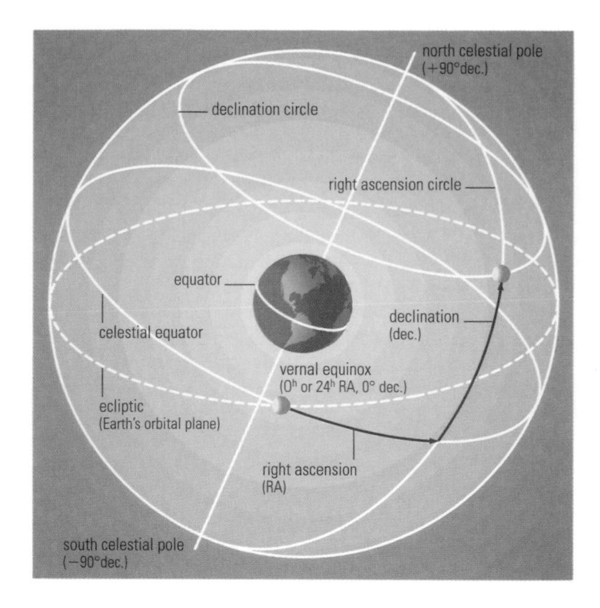

Solar System, particularly among the satellites, than would be expected by chance. It has been shown that some of these could have been caused by orbital evolution due to tidal action, which would continue until a commensurability was encountered, whereupon the satellites would become trapped.

A theory of shepherding satellites was proposed to explain the confinement of the narrow rings of Uranus, which were discovered by ground-based occultation observations. Subsequent images from the Voyager spacecraft have discovered small satellites close to one of the rings, and similar satellites close to one of Saturn's narrow rings, thus at least partially supporting the shepherding mechanism. Many other interesting problems of dynamics have arisen following the Voyager observations, such as the cause of the intricate structure of Saturn's rings, which consist of hundreds of individual ringlets. Some of the features have been explained by resonances with the satellites, which can cause various effects such as clumping around a ring, and radial variations of density. Other features are possibly caused by small unseen satellites orbiting within the rings, but there are many features still to be explained, and no doubt many new dynamical mechanisms still to be discovered.

CENTAURUS (GEN. CENTAURI, ABBR. CEN)

Large, prominent southern constellation, the ninth-largest in the sky, representing a centaur, a mythological beast with the legs and body of a horse and the upper torso of a man. This particular centaur was said to be Chiron, who taught the princes and heroes of Greek mythology. The constellation's brightest star is ALPHA CENTAURI, a triple system that includes the red dwarf PROXIMA CENTAURI, the closest star to the Sun. α and β Cen act as pointers to Crux, the Southern Cross. An easy double is 3 Cen, with components of 4.6 and 6.1 divisible with small apertures. R Cen is a variable MIRA STAR, ranging between 5th and 12th magnitude in about 18 months. ω Cen is the brightest globular cluster in the sky, so prominent that it was given a stellar designation by early observers. NGC 3766 and 5460 are open clusters for binoculars. NGC 5128 is a 7th-magnitude peculiar galaxy, also known as the radio source CENTAURUS A. NGC 3918 is a planetary nebula known as the Blue Planetary, resembling the planet Uranus in small telescopes.

BRIGHTEST STARS

	Name	RA h m	dec. ° ′	Visual mag.	Absolute mag.	Spectral type	Distance (l.y.)
α	Rigil Kent	14 40	−60 50	−0.28	4.07	G2 + K1	4.4
β	Hadar	14 04	−60 23	0.61	−5.42	B1	525
θ	Menkent	14 07	−36 22	2.06	0.70	K0	61
γ	Muhlifain	12 42	−48 58	2.20	−0.81	A1	130
ε		13 40	−53 28	2.29	−3.02	B1	367
η		14 36	−42 10	2.33	−2.55	B1	309
ζ		13 56	−47 17	2.55	−2.81	B2.5	385
δ		12 08	−50 44	2.58	−2.84	B2	395
ι		13 21	−36 43	2.75	1.48	A2	59
λ		11 36	−63 01	3.11	−2.39	B9	410
κ		14 59	−42 06	3.13	−2.96	B2	539
ν		13 50	−41 41	3.41	−2.41	B2	475
μ		13 50	42 29	3.47	−2.57	B2	527

celestial meridian Great circle on the CELESTIAL SPHERE that passes through the north and south celestial poles, together with the ZENITH and the NADIR. The term is usually used to refer to that part of the circle which is above the observer's horizon, intersecting it due north and due south.

celestial pole Either of the two points on the CELESTIAL SPHERE that intersect with a projection of the Earth's axis of rotation into space, and about which the sky appears to rotate. Because their position is dictated by the orientation of the Earth's axis, the celestial poles are subject to the effects of PRECESSION. This causes them to slowly drift, describing a circle in the sky of radius 23°.5 (the inclination of the Earth's axis) over a period of some 25,800 years, the effect only being noticeable over a few decades. At present, the north celestial pole is within 1° of the star Polaris, which is known as the POLE STAR, but in 3000 BC the north Pole Star was THUBAN, in Draco and by AD 10,000 it will be DENEB. For an observer, the ALTITUDE of the celestial pole is always equal to their latitude.

Celestial Police Name given themselves by the 24 astronomers who collaborated to search for a planet between the orbits of Mars and Jupiter, as predicted by BODE'S LAW. They were first convened in 1800 by Franz von ZACH, at Johann SCHRÖTER's observatory. Members of this international group, who also included Johann BODE, William HERSCHEL, Nevil MASKELYNE, Charles MESSIER and Wilhelm OLBERS, discovered three asteroids (Pallas, Juno and Vesta) between 1802 and 1807, after Giuseppe PIAZZI had found the first, Ceres, in 1801.

celestial sphere Inside of an imaginary sphere, with the Earth at its centre, upon which all celestial bodies are assumed to be projected.

The stars and planets are so far away that everything we see in the sky appears to be projected on to an enormous screen extending all around us, as if we were inside a gigantic planetarium. This is the illusion of the celestial sphere, half of which is always hidden from an observer on the Earth's surface, but upon which we base our charts of the sky and against which we make our measurements. The illusion is so strong that the early astronomers postulated the existence of a crystal sphere of very great radius to which the stars were fixed.

The most obvious behaviour of the celestial sphere is its apparent daily east to west rotation, due to the axial spin of the Earth. In the northern hemisphere we see some stars that never set – the CIRCUMPOLAR STARS; they appear to turn about the polar point near to the bright star Polaris. There is, of course, a similar polar point in the southern hemisphere marked by the fifth magnitude star, σ Octantis. The direction of the polar axis seems fixed in space, but it is in fact slowly drifting because of the effects of PRECESSION.

Having recognized one easily observed direction within the celestial sphere, it is possible to define another – the celestial equator. This is a projection of the Earth's equator on to the celestial sphere, dividing the sky into two hemispheres and enabling us to visualize a set of small circles similar to those of latitude on Earth, known as DECLINATION, which allow us to specify the position of an object in terms of angle above or below the celestial equator.

RIGHT ASCENSION, the celestial equivalent of longitude, is based upon the Earth's orbital motion around the Sun and is measured eastwards along the ECLIPTIC, the apparent path of the Sun through the constellations. For half of each year the Sun is in the northern hemisphere of the sky and for the other half it is in the southern hemisphere. On two dates each year, known as the EQUINOXES, it crosses the equator and it is the point at which it crosses into the northern hemisphere, at the spring or VERNAL EQUINOX, that is chosen as the zero of right ascension. This point on the celestial sphere, where the ecliptic and the celestial equator intersect, is called the FIRST POINT OF ARIES. This was an accurate description of its position thousands of years ago, but precession has now carried it into the constellation Pisces.

Right ascension can be measured in angular terms. However, it is more common to use units of time (hours, minutes, seconds or $^{h\ m\ s}$). Astronomers measure time by the rotation of the Earth relative to the stars, that is exactly 360° of axial rotation, rather than to the Sun. Their 'sidereal' timescale has a day of 24 sidereal hours, which in terms of civil time, is 23 hours, 56 minutes and 4 seconds long.

The beginning of the sidereal day is when the First Point of Aries lies on the MERIDIAN, the great circle linking the north and south points and passing directly overhead. After 24 sidereal hours it will be in that position again. The right ascension figures work like the face of a clock so that when the first point of Aries is on the meridian the sidereal time is 0^h. At 1^h sidereal time the sky will have rotated 15° and stars this angle east of the First Point of Aries will then lie on the meridian.

Apart from right ascension and declination, astronomers also use other systems of CELESTIAL COORDINATES to locate objects or points on the celestial sphere, dependent upon the particular application.

Although the stars seem relatively fixed in position, the Sun, Moon and planets (apart from Pluto) move across the celestial sphere in a band of sky about 8° either side of the ecliptic. This belt of sky is known as the ZODIAC and is divided into 12 signs named after the constellations they contained at the time of the ancient Greeks.

Celsius, Anders (1701–44) Swedish astronomer, best known for inventing the Centigrade temperature scale, now known as the CELSIUS scale. He took part in the 1736 expedition to Lapland organized by the French astronomer Pierre Louis Moreau de Maupertuis (1698–1759) to measure the length of a meridian arc. The results of this and a second expedition to Peru showed that the Earth is oblate. Celsius was a pioneer of stellar photometry, using a series of glass filters to measure the relative intensity of light from stars of different magnitudes, and he was one of the first to realize that the aurorae were related to Earth's magnetic field.

Celsius scale Temperature scale on which the freezing point of water is 0°C and the boiling point of water is 100°C. It is named after Anders CELSIUS. The magnitude of 1°C is the same as 1 K. It is also known as the Centigrade scale, although this name was officially abandoned in 1948. *See* ABSOLUTE TEMPERATURE

CELT Abbreviation of CALIFORNIA EXTREMELY LARGE TELESCOPE

Censorinus Small (5 km/3 mi) but brilliant lunar crater (0° 32°E); it is located on a bluff near the south-east border of Mare TRANQUILLITATIS.

Centaur Any of a group of planet-crossing objects in the outer planetary region that are classified as being ASTEROIDS, although it is likely that in nature they are actually large COMETS. Through to late 2001 more than 35 Centaurs had been discovered, taking the criterion for membership as a Neptune-crossing orbit (that is, perihelion distance less than 30 AU), implying that the Centaurs could not be classed as TRANS-NEPTUNIAN OBJECTS. At such great distances from the Sun, the temperature is extremely low and only the most volatile chemical constituents sublimate, making comet-like activity either totally absent or at least difficult to detect using Earth-based telescopes. Their faintness, coupled with their considerable helio- and geocentric distances, implies that all known Centaurs must be of substantial size (generally bigger than 50 km/30 mi). It is likely, however, that these are only the largest members of the overall population, with a much greater number of smaller objects awaiting identification.

The origin of the Centaurs is suspected to be as members of the EDGEWORTH–KUIPER BELT, having coalesced there when the rest of the Solar System formed over 4.5 billion years ago; they are thought to have been inserted into their present unstable orbits during just the last million years or so. The dynamical instability of the Centaurs derives from the fact that they will inevitably make close approaches to one or another of the giant planets, and the severe gravitational perturbations that result will divert the objects in question on to different orbits. Some Centaurs will be ejected from the Solar System on hyperbolic paths; others may fall into the inner planetary region and so become extremely bright active comets.

The fact that Centaurs show characteristics of both asteroids and comets leads to the convention for their naming. In Greek mythology the Centaurs were hybrid beasts, half-man and half-horse. The first Centaur object to be discovered was CHIRON, in 1977. The next was PHOLUS, in 1992. Others added since include (7066) Nessus, (8405) Asbolus, (10199) Chariklo and (10370) Hylonome. *See also* DAMOCLES; HIDALGO; JUPITER-CROSSING ASTEROID; LONG-PERIOD ASTEROID

Centaurus *See* feature article

Centaurus A (NGC 5128) Galaxy with powerful radio emission at a distance of around 3 Mpc (10 million l.y.). Optically it appears to be a normal elliptical galaxy, but it is crossed by a dark and very prominent dust lane. Very deep photographs reveal the galaxy to be more than 1° across, and the radio image is much bigger. This enigmatic object appears to be two galaxies in collision, a massive elliptical galaxy and a smaller dusty spiral (which can be seen in the infrared). Centaurus A is emitting a huge amount of energy at X-ray and optical wavelengths as well as at radio wavelengths.

Central Bureau for Astronomical Telegrams Bureau that rapidly disseminates information on transient astronomical events. It operates from the SMITHSONIAN ASTROPHYSICAL OBSERVATORY under the auspices of the INTERNATIONAL ASTRONOMICAL UNION. The bureau issues announcements concerning comets, asteroids, vari-

▼ **Centaurus A** Crossed by a prominent dark lane, this elliptical galaxy (also designated NGC 5128) is a powerful radio and X-ray source.

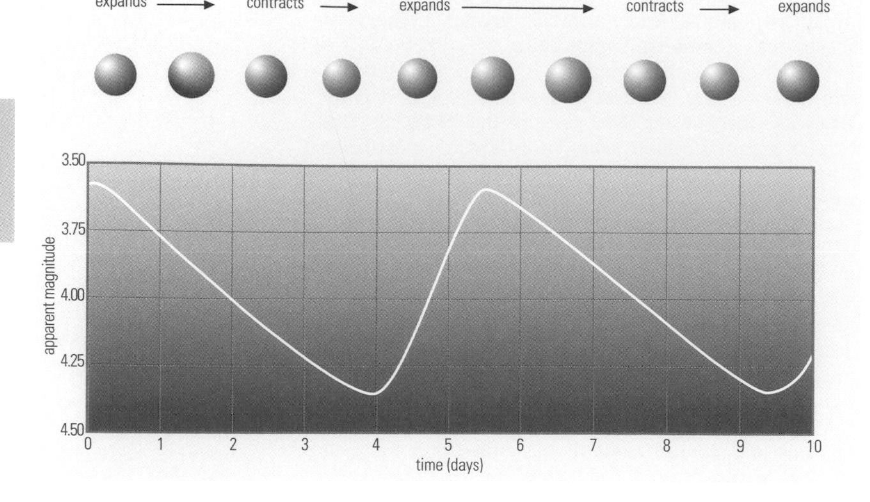

▲ **Cepheid variable** Their extremely regular light variations make Cepheids valuable standard candles: the longer the period from one maximum to the next, the greater the star's intrinsic luminosity. As shown, the typical Cepheid light-curve shows a rapid rise to peak brightness, followed by a slower decline to minimum.

able stars, novae and supernovae in the form of *IAU Circulars*, both electronically and in printed form.

central meridian Imaginary north–south line bisecting the disk of a planet, the Moon or the Sun. The central meridian passes through the poles of rotation of the object in question and is used as a reference point from which to determine the longitude of features on the disk as the body rotates. It is independent of any PHASE that may be present.

Centre de Données Astronomiques de Strasbourg (CDS) World's main astronomical DATA CENTRE dedicated to the collection and distribution of computerized astronomical data and related information from both ground- and space-based observatories. It hosts the SIMBAD (in full, Set of Identifications, Measurements and Bibliography for Astronomical Data) astronomical database, the world reference database for the identification of astronomical objects. The CDS is located at the STRASBOURG ASTRONOMICAL OBSERVATORY in France, and is a laboratory of the Institut National des Sciences de l'Univers.

The CDS was founded in 1972 as the Centre de Données Stellaires, its main aim being to cross-identify star designations in different catalogues – often the same object had a host of different catalogue identifications. Bibliographic references to objects were added, and SIMBAD was the result. By 2001 the database contained data for more than 2.25 million objects, and over 5 million references.

CEPHEUS (GEN. CEPHEI, ABBR. CEP)

Constellation of the northern polar region, between Cassiopeia and Draco. It represents a mythological king, husband of Cassiopeia and father of ANDROMEDA. Its most celebrated star is δ Cep, the prototype CEPHEID VARIABLE, a pulsating yellow supergiant varying between 3.5 and 4.4 with a period of 5.4 days; its variability was discovered in 1784 by John GOODRICKE. A wide bluish companion of mag. 6.3 makes it an attractive double for small telescopes or even binoculars. β Cep is another pulsating variable, though of much smaller amplitude (0.1 mag.) and far shorter period (4.6 hours); it is the prototype BETA CEPHEI STAR. μ Cep is a pulsating red supergiant known as the GARNET STAR, range 3.4–5.1, period about 2 years. Another variable red supergiant, VV Cep, is one of the largest stars known, with an estimated diameter about that of Jupiter's orbit. It varies semiregularly between mags. 4.8 and 5.4.

BRIGHTEST STARS

	Name	RA h m	dec. ° ′	Visual mag.	Absolute mag.	Spectral type	Distance (l.y.)
α	Alderamin	21 19	+62 35	2.45	1.58	A7	49
γ	Errai	23 39	+77 38	3.21	2.51	K1	45
β	Alfirk	21 29	+70 34	3.23	−3.08	B1	595
ζ		22 11	+58 12	3.39	−3.35	K1	726
η		20 45	+61 50	3.41	2.63	K0	47

centre of mass Position in space for an object or collection of objects at which the various masses involved act as though they were a single mass concentrated at that point. For an object in a uniform gravitational field it is the same as the centre of gravity. When two masses are linked by gravity, the centre of mass occurs on the line joining them at a distance from each object that is inversely proportional to the mass of that object. In this case the centre of mass is also called the BARYCENTRE.

centrifugal force Apparent force that appears when an object is forced to move along a circular or curved path. The force is actually the result of the inertia of the object attempting to keep the object moving in a straight line. It is the reaction to the CENTRIPETAL FORCE.

centripetal force Force acting on an object that causes it to move along a circular or curved path. It produces an acceleration towards the centre of curvature of the path, and the reaction to this acceleration is experienced as the CENTRIFUGAL FORCE. Gravity provides the centripetal force on an orbiting body, and the magnetic field on electrons producing SYNCHROTRON RADIATION.

Cepheid instability strip *See* INSTABILITY STRIP

Cepheid variable Yellow giant or supergiant pulsating VARIABLE STAR, so called because the first variable of the type to be discovered was DELTA CEPHEI. Cepheids pulsate in a particularly regular manner. These stars have left the MAIN SEQUENCE and occupy, on the HERTZSPRUNG–RUSSELL DIAGRAM, a position to the right of the upper main sequence and to the left of the red giants, termed the Cepheid INSTABILITY STRIP. Cepheids are passing through the first Instability Transition after leaving the main sequence.

During this brief period in their lives, these stars oscillate, alternately expanding and contracting so that in each cycle a star may change in size by as much as 30%. These regular, rhythmic changes in size are accompanied by changes in luminosity. The surface temperature also changes in the course of each cycle of variations in brightness, being at its lowest when the star is at minimum and at its highest when the star is brightest. This temperature change may equal 1500 K for a typical Cepheid. A change in temperature also means a change in SPECTRAL TYPE, so that the star may be F2 at maximum, becoming the later type, G2, at minimum, changing in a regular manner as the temperature falls or rises. A Cepheid may continue to pulsate in this manner for a million years, which is a comparatively short time compared to the life span of a star.

Most massive stars spend at least some time as Cepheid variables. Stars like Delta Cephei have amplitudes of around 0.5 mag. and periods usually not longer than 7 days; there are, however, Cepheids with larger amplitudes and longer periods, which form a separate subtype. This subtype includes the naked-eye stars 1 Carinae, β Doradus and κ Pavonis. The period of light changes is related to the average luminosity of the star. This means that the absolute magnitude of a Cepheid variable may be found by measuring the period of the light cycle. The apparent magnitude may be obtained directly. Once period, apparent and absolute magnitude are known, it becomes possible to determine the distance to the star.

Cepheids are visible in external galaxies, but their value as distance indicators is compounded by the fact that there are two types. Both follow a period–luminosity relationship, but their light-curves are different. First, there are the classical Cepheids, such as Delta Cephei itself, which are yellow supergiants of Population I. The second type, the W VIRGINIS STARS, are Population II stars found in globular clusters and in the centre of the Galaxy. In using Cepheids to determine distances it is necessary to know which type is being observed. At the time Cepheids were first used to determine distances it was not known that there were Cepheids with different period–luminosity values. This resulted in erroneously applying the value for type II to classical Cepheids. When this error was found, in 1952, the result was to double the size of the Universe. For periods of 3–10 days the light-curves of the two types closely resemble one another and classification is based on spectral differences. In particular, at certain phases, classical Cepheids exhibit calcium emission, whereas W Virginis stars show hydrogen emission.

The period–luminosity relationship means that the longer the period, the brighter the visual absolute magnitude. A comparison of the curves shows that classical Cepheids are about one magnitude brighter than type II Cepheids. The light-curves may be arranged in groups, according to their shapes, which progressively become more pronounced in each group as the period lengthens. Most Cepheid light-curves fall into one of about 15 such divisions, each with a longer average period. They all follow a period–luminosity relationship, which commences with the RR LYRAE stars of very short period and, after a break, is continued by the MIRA STARS. This regular progression – the longer the period, the later the spectral type – is called the Great Sequence. A typical Cepheid would have a surface temperature varying between 6000 and 7500 K and an absolute luminosity that is ten thousand times that of the Sun.

Since Cepheids are in a part of the Hertzsprung–Russell diagram where changes occur, observations are directed towards detecting changes in periods. Such changes are small but give information as to how stars progress through the instability strip; they can be detected by making series of observations separated by a few years.

Cepheus *See* feature article

Čerenkov radiation Electromagnetic radiation emitted when a charged particle passes through a transparent medium at a speed greater than the local speed of light in that medium (the speed of light in air or water is less than that in a vacuum). Radiation is emitted in a cone along the track of the particle. Cosmic rays ploughing into the Earth's atmosphere produce Čerenkov radiation, which can be detected at ground level. This type of radiation was discovered in 1934 by the Russian physicist Pavel Čerenkov (1904–90).

Ceres First ASTEROID to be discovered, on the opening day of the 19th century, hence it is numbered 1. Ceres, a MAIN-BELT ASTEROID, has a diameter of *c*.933 km (*c*.580 mi), although it is not precisely spherical; it is the largest asteroid. Ceres' mass, 8.7×10^{20} kg, represents about 30% of the bulk of the entire main belt, or about 1.2% of the mass of the Moon. Its average density, about 1.98 g/cm^3, is less than that of most meteorites. Ceres rotates in about nine hours, its brightness showing little variation, which is indicative of a fairly uniform surface, thought to be powdery in nature. It lies close to the middle of the main belt, at an average heliocentric distance of 2.77 AU.

The possibility of a planet between Mars and Jupiter had been suggested in the early 1600s by Johannes KEPLER, and in the late 18th century BODE'S LAW was interpreted as implying the likely existence of such a body. In 1800 a group of European astronomers formed the so-called CELESTIAL POLICE, having the aim of discovering this purported planet. Before they could begin their search, however, Ceres was discovered by Giuseppe PIAZZI from Palermo, Sicily. Piazzi was checking Nicolas LACAILLE's catalogue of zodiacal stars when he found an uncharted body that moved over the subsequent nights. Piazzi wanted to call the object Ceres Ferdinandea (Ceres is the goddess of fertility, the patron of Sicily, while Ferdinand was the name of the Italian king), but only the first part of that name was accepted by astronomers in other countries. Although he was prevented by illness from following it for an extended period, Piazzi's observations allowed Ceres to be recovered late in 1801. It soon became apparent that it was not large enough to be considered a major planet, and the rapid discovery of several other such objects in the following years, namely PALLAS, JUNO and VESTA, added to this view. William HERSCHEL coined the term asteroid for these new objects; they are also termed minor planets.

CETUS (GEN. CETI, ABBR. CET)

Fourth-largest constellation, lying on the celestial equator south of Aries and Pisces. It is not particularly prominent – its brightest star is β Cet, mag. 2.04, known as DENEB KAITOS; α Cet is known as MENKAR. Cetus represents the sea-monster from which ANDROMEDA was saved by Perseus. The constellation's most famous star is MIRA (ο Cet), the prototype long-period variable. τ Cet is the most Sun-like of all the nearby single stars. M77, a 9th-magnitude face-on spiral some 50 million l.y. away, is the brightest of all the Seyfert galaxies. Because Cetus lies close to the ecliptic, planets can sometimes be found within its boundaries.

BRIGHTEST STARS

	Name	RA h m	dec. ° ′	Visual mag.	Absolute mag.	Spectral type	Distance (l.y.)
β	Deneb Kaitos	00 44	−17 59	2.04	−0.30	G9	96
α	Menkar	03 02	+04 06	2.54	−1.61	M2	220
η		01 09	−10 11	3.46	0.67	K2	118
γ		02 43	+03 14	3.47	1.47	A2	82
τ		01 44	−15 56	3.49	5.68	G8	11.9

Cerro Tololo Inter-American Observatory (CTIO) Major optical observatory situated approximately 80 km (50 mi) east of La Serena, Chile, at an elevation of 2200 m (7200 ft). It is operated jointly by the ASSOCIATION OF UNIVERSITIES FOR RESEARCH IN ASTRONOMY and the National Science Foundation as part of the NATIONAL OPTICAL ASTRONOMY OBSERVATORIES; it is a sister observatory to KITT PEAK NATIONAL OBSERVATORY. The largest telescope on the site is the 4-m (158-in.) Victor M. Blanco Telescope, commissioned in 1974 to complement its northern-hemisphere twin, the MAYALL TELESCOPE, but now being developed to complement SOAR, the SOUTHERN ASTROPHYSICAL RESEARCH TELESCOPE, scheduled to begin operation in 2003. CTIO is also home to 1.5-m (60-in.), 1.3-m (51-in.) and 1.0-m (40-in.) reflectors.

Cetus *See* feature article

CfA *See* HARVARD–SMITHSONIAN CENTER FOR ASTROPHYSICS

CGRO Abbreviation of COMPTON GAMMA RAY OBSERVATORY

Challis, James (1803–82) English clergyman and astronomer, now chiefly remembered for his part in the search for Neptune. He succeeded George Biddell AIRY as Cambridge's Plumian Professor (1836), and served until 1861 as director of Cambridge Observatory. There in 1846 July, at the urging of John Couch ADAMS, who had calculated the theoretical position of a new planet from perturbations of Uranus, Challis initiated a rigorous search. He actually discovered Neptune on August 4, though he failed to recognize it as a planet before Johann GALLE and Heinrich D'ARREST of the Berlin Observatory announced the planet's discovery, on 1846 September 23.

Chamaeleon *See* feature article

CHAMAELEON (GEN. CHAMAELEONTIS, ABBR. CHA)

Small and unremarkable constellation near the south celestial pole, introduced at the end of the 16th century by Keyser and de Houtman, representing the colour-changing lizard. Its brightest star is α Cha, mag. 4.1. δ Cha is a binocular double, mags. 4.4 and 5.5.

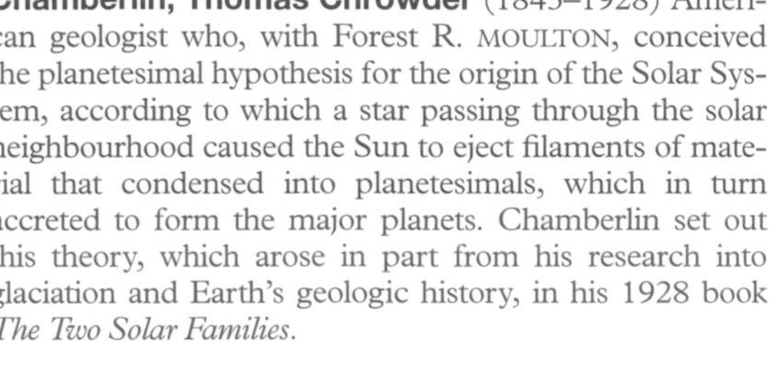

Chamberlin, Thomas Chrowder (1843–1928) American geologist who, with Forest R. MOULTON, conceived the planetesimal hypothesis for the origin of the Solar System, according to which a star passing through the solar neighbourhood caused the Sun to eject filaments of material that condensed into planetesimals, which in turn accreted to form the major planets. Chamberlin set out this theory, which arose in part from his research into glaciation and Earth's geologic history, in his 1928 book *The Two Solar Families*.

Chandler period 430-day period of variation in position of the Earth's axis of rotation. This small movement of the geographic poles is known as the 'Chandler wobble' after the American who first discovered it. It is thought to be caused by seasonal changes in the distribution of mass within the Earth.

Chandrasekhar, Subrahmanyan (1910–95) Indian-American astrophysicist who was awarded the 1983 Nobel Prize for Physics for his mathematical theory of black holes. At Cambridge University, he developed a theoretical model explaining the structure of white dwarf stars that took into account the relativistic variation of mass with the velocities of electrons that comprise their degenerate matter. He showed that the mass of a white dwarf could not exceed 1.44 times that of the Sun – the CHANDRASEKHAR LIMIT. Stars more massive than this must end their lives as a NEUTRON STAR or BLACK HOLE. He spent most of his career at the University of Chicago and its Yerkes Observatory, where he was on the faculty from 1937 to 1995, and he served as editor of the prestigious *Astrophysical Journal* (1952–71).

Chandrasekhar revised the models of stellar dynamics originated by Jan OORT and others by considering the effects of fluctuating gravitational fields within the Milky Way on stars rotating about the galactic centre. His solution to this complex dynamical problem involved a set of twenty partial differential equations, describing a new quantity he termed 'dynamical friction', which has the dual effects of decelerating the star and helping to stabilize clusters of stars. Chandrasekhar extended this analysis to the interstellar medium, showing that clouds of galactic gas and dust are distributed very unevenly. He also studied general relativity and black holes, which he modelled using two parameters – mass and angular momentum.

▼ **Chamaeleon** This complex of hot stars and nebulosity was photographed during testing of the Very Large Telescope (VLT). Designated Chamaeleon I, it lies close to the south celestial pole.

Chandrasekhar limit Maximum possible mass for a WHITE DWARF. It was first computed, in 1931, by the Indian astrophysicist Subrahmanyan CHANDRASEKHAR. The value computed by Chandrasekhar applies to a slowly rotating star composed primarily of helium nuclei and is about 1.44 solar masses.

A white dwarf is supported against its own gravitational attraction by electron degeneracy pressure *(see* DEGENERATE MATTER). The PAULI EXCLUSION PRINCIPLE states that no two electrons can occupy exactly the same state so that when all the low energy states have been filled, electrons are forced to take up higher energy states. With white dwarfs of progressively higher mass, as gravity attempts to squeeze the star into a smaller volume, so the electrons are forced into higher and higher energy states. They therefore move around with progressively higher speeds, exerting progressively higher pressures.

The greater the mass, the smaller the radius and the higher the density attained by a white dwarf before electron degeneracy pressure stabilizes it against gravity. As the mass approaches the Chandrasekhar limit, electrons are eventually forced to acquire velocities close to the speed of light (that is, they become 'relativistic'). As the limit is reached, the pressure exerted by relativistic electrons in a shrinking star cannot increase fast enough to counteract gravity. Gravity overwhelms electron degeneracy pressure, and a star that exceeds the Chandrasekhar limit collapses to a much denser state. Electrons combine with protons to form neutrons, and the collapse is eventually halted by neutron degeneracy pressure, by which time the star has become a NEUTRON STAR.

Chandrasekhar–Schönberg limit *See* SCHÖNBERG–CHANDRASEKHAR LIMIT

Chandra X-Ray Observatory (CXO) Third of NASA's GREAT OBSERVATORY series. The former Advanced X-Ray Astrophysics Facility, it was deployed into orbit by the Space Shuttle *Columbia* in 1999 July. The 5.2-tonne, 14-m-long (45-ft-long) spacecraft was renamed the Chandra X-Ray Observatory. It followed the HUBBLE SPACE TELESCOPE and the COMPTON GAMMA RAY OBSERVATORY in the Great Observatory series. Chandra is equipped with four science instruments – an imaging spectrometer, a high-resolution camera, and high- and low-energy spectrometers. With the European Space Agency's NEWTON X-Ray Telescope, which was also launched in 1999, the Chandra X-ray Observatory is providing astronomers with a wealth of data, including images showing a PULSAR inside a PLANETARY NEBULA and material that seems to be disappearing down a BLACK HOLE.

chaos Property of a mathematical model of a physical system that is akin to indeterminacy or instability, in which the final state of the system is very sensitively dependent on its initial state. The usual example quoted

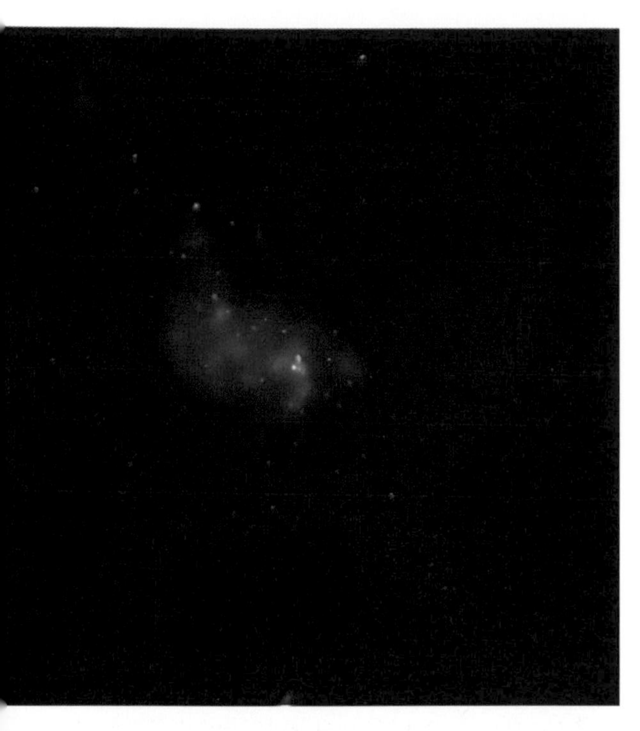

◄ **Chandra X-Ray Observatory** Among many valuable images obtained from Chandra has been this view showing a huge X-ray flare associated with the supermassive black hole at the centre of our Galaxy.

is for mathematical models of meteorology, where a small change of some apparently insignificant parameter of the system can cause a major change in the outcome of the model after running it for a time span of several days. It is important to note that the term chaos describes a property of the mathematical model, not of the outcome of the model, which in the meteorological case is still just normal weather.

In CELESTIAL MECHANICS the term has become very popular, and is used with various shades of meaning in different contexts. One of these is the problem of predicting the orbit of an object, such as an asteroid, that has a very close approach or repeated close approaches to a planet. The initial orbit has inevitable uncertainties due to limitations of the accuracy and coverage of positional observations, and these uncertainties are magnified greatly following a close approach to a planet, to the extent that the orbit after the close approach can be completely uncertain, or more likely that it is sufficiently uncertain that it cannot be used for predicting subsequent close approaches.

Another context in which the term arises is in the study of orbits at a COMMENSURABILITY. There are usually multiple resonant terms at a commensurability, associated with the eccentricity and inclination of each object. In analytical studies it is necessary to simplify the problem and to pick out just the single dominant term. However numerical integrations over long time spans show that in some cases the effects of the overlapping resonances can result in changes to the orbit that cannot be predicted by the analytical study. The nature of these changes is usually an increase in the eccentricity of the orbit, which could perhaps lead to a close approach to a planet causing the ejection of the body from the commensurability. This mechanism is a likely explanation of the KIRKWOOD GAPS at the 3:1 and 2:1 commensurabilities with Jupiter.

charge-coupled device *See* CCD

Charlier, Carl Vilhelm Ludvig (1862–1934) Swedish astronomer who worked at the University of Uppsala, Stockholm Observatory and the University of Lund, where he directed the observatory for thirty years. He made many detailed studies of the distribution and motions of stars and star clusters near the Sun, finding that the Milky Way was shaped like a disk, and rotated. Charlier proposed a hierarchical grouping of galaxies in an infinite universe.

Charon Only known SATELLITE of PLUTO. It was discovered on telescopic images by the American astronomer James Christy (1938–) in 1978. Charon is named after the ferryman who, in classical mythology, transported the ghosts of the dead across the river Styx into the underworld domain of the god Pluto. It is a mysterious body, never having been visited by any spacecraft.

The determination of Charon's orbit gave the first reliable measurement of Pluto's mass. Between 1985 and 1990, a fortunate series of mutual OCCULTATIONS between Pluto and Charon, when the plane of Charon's orbit lay in the line of sight from Earth, enabled the sizes of both bodies to be determined. Charon orbits exactly in Pluto's equatorial plane, and the rotations of the two bodies are mutually tidally locked so that they permanently keep the same faces toward each other.

Spectroscopic studies of Charon have revealed only water-ice, contaminated by rock or soot, with none of the more exotic ices found on Pluto. Charon's gravity is too weak to retain any kind of atmosphere, even in the cold outer reaches of the Solar System, where the surface temperature is only about 40 K. Competing tidal pulls on Charon from Pluto and the Sun could be responsible for sufficient TIDAL HEATING to have allowed Charon to differentiate (*see* DIFFERENTIATION), forming a rocky core, and even to maintain an ocean below the outer carapace of ice. *See also* EUROPA

Chassigny METEORITE that fell in Haute Marne, France, in 1815 October; approximately 4 kg of material was recovered. Chassigny is the sole member of its subgroup of MARTIAN METEORITES (SNCs). It is an unshocked, olivine-rich rock. It crystallized below the Martian surface *c.*1.3 billion years ago.

chemically peculiar stars Class A, B and F stars with odd chemical compositions caused by diffusion that enriches some chemical elements while depleting others. Chemically peculiar (CP) stars are subdivided into CP1 (metallic-line AM STARS), CP2 (magnetic AP STARS), CP3 (class B MERCURY–MANGANESE STARS) and CP4 (HELIUM-WEAK B STARS). The CP1 Am stars extend to class Fm, and the CP2 Ap stars to Fp and Bp. LAMBDA BOÖTIS STARS, not caused by diffusion, are sometimes included as well.

Chicxulub Impact site in the Yucatán Peninsula, Gulf of Mexico, where a huge METEORITE collided with the Earth. The crater is now buried, but geophysical surveys estimate its diameter to be between 180 and 320 km (110 and 200 mi). The collision is linked with a period of mass extinction, which marks the end of the Cretaceous Period, approximately 65 million years ago. There was a dramatic drop in the number of species present on the Earth: about 60% of all species suddenly disappeared.

Chi Cygni MIRA STAR of spectral type S7. It has the largest visual range of any known Mira star (at least 10 magnitudes) and is very red at minimum, corresponding to a temperature of *c.*2000 K. Its period is 407 days. It has

CHARON: DATA	
Globe	
Diameter	1250 km
Density	1.7 g/cm^3
Mass (Earth = 1)	0.0003
Sidereal period of axial rotation	6.3872 days
Escape velocity	0.58 km/s
Surface gravity (Earth = 1)	0.021
Albedo	0.36
Inclination of equator to orbit	0°
Surface temperature (average)	45 K
Orbit	
Semimajor axis	19,636 km
Eccentricity	0.01
Inclination to Pluto's equator	0°

C

a maximum observed magnitude of 3.3 and minimum 14.2; its distance is 228 l.y. There is strong infrared excess and circumstellar emission from molecules such as CO and SiO. The gas lost from the star's surface cools to form molecules and silicate dust grains in the circumstellar envelope. The dust then absorbs some of the star's radiation, is itself heated, and radiates in the infrared, producing the infrared excess.

Chinese astronomy Astronomy as practised in the Chinese empire from ancient times until the overthrow of the last imperial dynasty in 1911. Ancient Chinese astronomical records of eclipses and other unusual events in the sky date from as early as 1500 BC and formed part of the royal archives. Rulers could be made or unmade by portents from heaven, which explains their concern with celestial matters. Of particular concern were solar and lunar eclipses, and other transient phenomena such as novae and supernovae ('guest stars'), comets, auroras and conjunctions of the Moon and planets. Official astrologers or *tianwen* had to explain any unusual celestial event and, if possible, forewarn of its occurrence. Other officials, called *lifa*, dealt with the predictable parts of astronomy needed for calendrical studies. The latter made observations, kept records and developed mathematical models.

China experienced a violently unsettled period in the 3rd century BC and many early records were destroyed. However, by about 206 BC, the commencement of the Han Dynasty, a single calendar was in use everywhere. A luni-solar scheme was in use, with a correction of seven extra months in 19 years, according to the METONIC CYCLE, to keep pace with the seasons. Refinements were made from time to time, and the predictive powers of the almanacs were continuously improved. An interesting achievement of Han astronomers was to record in detail the passage of Halley's Comet through the constellations during its 12 BC apparition.

In contrast to early Western astronomy, the Chinese astronomers did not emphasize the zodiac. Instead, there were 28 unequally spaced 'lunar mansions' close to the celestial equator. Their constellations were small and very numerous, and almost completely different from the Western ones.

Observations making use of an instrument – the ARMILLARY SPHERE – were made as early as 52 BC. The imperial bureaucracy later encouraged the construction of observatories and the compilation of star catalogues. An engraved Song Dynasty (960–1277) star map survives from AD 1247 in a temple in Jiangsu province, containing 1440 stars with a typical positional accuracy of 1°.

In the early 17th century, Western influences began to arrive with Jesuit missionaries such as Matteo Ricci (1552–1610), and after the fall of the last imperial dynasty Chinese astronomy became aligned with that of the outside world. The ancient Chinese records have proved useful for, among other things, establishing the ages of supernovae that were not recorded in the West; the supernova of AD 1054 that led to the formation of the Crab Pulsar and the Crab Nebula is a notable example.

► **Chryse Planitia** Located near the east of Valles Marineris, Chryse Planitia was the landing site for the Viking 1 surface probe in 1976. The region is dominated by outflow channels created at a past epoch when liquid water flowed on Mars' surface.

Chiron An outer Solar System body, *c.*180 km (*c.*110 mi) in diameter, given dual ASTEROID and COMET designations as (2060) Chiron and 95P/Chiron respectively. Its orbit, with a period of just over 50 years, has perihelion near 8.4 AU and aphelion at 18.8 AU. Chiron, therefore, crosses the path of Saturn and also approaches Uranus; consequently it is dynamically unstable. Over a time scale of order a million years (that is, much less than the age of the Solar System) it is to be expected that Chiron's orbit will change radically. It was the first object to be discovered in a class of bodies known as the CENTAURS, these being characterized as large objects inhabiting the outer planetary region and having rapidly evolving orbits.

When found, in 1977 by Charles Kowal (1940–), Chiron appeared asteroidal in nature, but as its heliocentric distance decreased, passing perihelion in 1996 February, it began to develop characteristics associated with comets. In 1988 Chiron rapidly brightened, indicating the formation of some type of cometary coma; in the following years it developed a surrounding cloud of dust, and then the emission of cyanogen gas and other volatiles was detected. It is generally regarded as an escaped member of the EDGEWORTH–KUIPER BELT that has migrated inwards towards the Sun. *See also* ELST–PIZARRO; WILSON–HARRINGTON

Chladni, Ernst Florens Friedrich (1756–1827) German physicist known as the founder of acoustics. He was also a pioneer of meteoritics, the study of meteorites, which he was convinced were of extraterrestrial origin – a novel idea at the time. He assembled the first great collection of meteorites.

chondrite Most primitive and oldest of the METEORITES. Chondrites are STONY METEORITES that have not melted since their aggregation early in the history of the Solar System. They are mostly characterized by the presence of CHONDRULES, which are millimetre-sized spherules of rapidly cooled silicate melt droplets. On the basis of chemistry, chondrites are subdivided into three main groups, the CARBONACEOUS CHONDRITES (C), ORDINARY CHONDRITES (O) and ENSTATITE CHONDRITES (E). There are also two smaller classes, the Rumurutiites (R) and the Kakangari (K), each represented by only a single example. The groups have different oxygen isotope compositions, matrix, metal and chondrule contents and chondrule properties (such as size, type, and so on). The differences between the classes are primary, that is, they were established as the parent bodies accreted in different regions of the SOLAR NEBULA. In addition to these chemical classes, the chondrites are classified according to the processing that they have experienced, either thermal metamorphism or aqueous alteration. These secondary characteristics were established on the meteorites' parent bodies. A petrologic type from 3 to 6 indicates increasing thermal metamorphism. A petrologic type from 3 to 1 indicates increasing aqueous alteration. Type 3 chondrites are the least altered; they are further subdivided into 3.0 to 3.9, on the basis of silicate heterogeneity and thermoluminescence.

chondrule Submillimetre to millimetre-sized spherules of rapidly cooled silicate melt droplets found in METE-

ORITES. Chondrules normally consist of olivine and/or pyroxene, with a variety of textures, depending on the starting materials and the cooling regimen. Pyroxene-rich chondrules are often composed of crystallites radiating from a point offset from the centre of the sphere; olivine-rich chondrules frequently have a blocky or barred appearance. The term 'chondrule' comes from the Greek *chondros*, meaning grain or seed. The origin of chondrules is still not known with certainty. At one time, they were thought to be fused drops of 'fiery rain' from the Sun. Other theories include that they were formed by the cooling of droplets produced by collisions between asteroids, or by direct condensation from a gas. Alternatively, chondrules might have formed by melting and subsequent quenching of small aggregates of dust grains in the pre-solar nebula; the heat source for such melting might have been shock waves in the nebula or energetic outflow from the Sun.

Christie, William Henry Mahoney (1845–1922) English astronomer, the eighth ASTRONOMER ROYAL (1881–1910), who substantially improved the equipment at Greenwich Observatory, especially by acquiring the 28-inch (0.71-m) refractor. Christie was responsible for several important star catalogues produced by the Observatory, but he extended the observatory's work to include physical as well as positional astronomy, supervising programmes of photographic and spectroscopic stellar astronomy. With E. Walter MAUNDER he initiated daily sunspot observations which led to discoveries about solar activity.

chromatic aberration Introduction of false colour into images formed by a lens. When light passes through a lens, it is bent or refracted. The degree of bending depends on the colour or wavelength of the light, so different colours follow different paths. The consequence is that the different colours in any image formed by the lens come to a focus at different points. This is chromatic aberration. It was a serious drawback in the first refracting telescopes. In a telescope, chromatic aberration appears as coloured fringes around the edges of objects. Chromatic aberration can be reduced or eliminated by using an ACHROMAT or an APOCHROMAT.

chromosphere Layer or region of the solar atmosphere lying above the PHOTOSPHERE and beneath the CORONA. The name chromosphere comes from the Latin meaning 'sphere of colour'. The term is also used for the layer above the photosphere of a star. The Sun's temperature rises to about 10,000–20,000 K in the chromosphere. The chromosphere is normally invisible because of the glare of the photosphere, but it can often be seen near the beginning and end of a TOTAL SOLAR ECLIPSE when it is visible as a spiky pink or red rim around the Moon's disk at the limb or edge of the Sun. Today, the chromosphere can be observed at any time across the full solar disk with a SPECTROGRAPH or SPECTROHELIOGRAPH that isolates a single colour of the Sun's light – for example, the red light of the HYDROGEN-ALPHA LINE (H-alpha) at a wavelength of 656.3 nm, or the violet light of ionized calcium, Ca II, at wavelengths of 396.8 and 393.4 nm, known as the calcium H and K lines. The monochromatic image made with a spectrograph is known as a SPECTROGRAM or spectroheliogram. Spectrograms show features such as FIBRILS, FILAMENTS, FLOCCULI, PLAGES and PROMINENCES. A large cellular pattern, known as the **chromospheric network**, is also revealed in spectroheliograms. It appears at the boundaries of the photosphere SUPERGRANULATION, and contains magnetic fields that have been swept to the edges of these cells by the flow of material in the cell. A thin TRANSITION REGION separates the CHROMOSPHERE and the CORONA. SPICULES containing chromospheric material penetrate well into the corona (to heights of 10,000 km/6000 mi above the photosphere) at the edges of the cells.

chromospheric network *See* CHROMOSPHERE

CIRCINUS (GEN. CIRCINI, ABBR. CIR)

Fourth-smallest constellation, lying in the southern sky between Centaurus and Triangulum Australe. Representing a drawing compass, it was one of the constellations introduced by Lacaille. Its brightest star is α Cir, mag. 3.18, spectral type A7p, distance 53 l.y.; there is a companion of mag. 8.5.

chronometer *See* MARINE CHRONOMETER

Chryse Planitia Extensive plains region on MARS; it was the site of the VIKING 1 Lander probe. Centred near 27°.0N 40°.0W, it was shown by the Lander to consist of loose reddish material upon which were distributed large numbers of blocks of basaltic lava. Chryse occupies a large basin-like embayment into the cratered terrain of Mars, probably the infilled site of an ancient impact basin. A large number of prominent channels converge on the region. These channels have their origin in the eastern end of VALLES MARINERIS and are of presumed fluvial origin. Recent altimetry and imaging data from MARS GLOBAL SURVEYOR strongly suggest that the whole region from Chryse to the eastern end of Valles Marineris was once flooded.

Circinus *See* feature article

circular velocity Velocity of a body in a circular orbit around a massive primary. Its value is given by GM/R, where M is the mass of the primary, R is the radius of the orbit and G is the gravitational constant. For the Earth the circular velocity ranges from about 7.8 km/s (4.8 mi/s) for the lowest artificial satellites, to 3.1 km/s (1.9 mi/s) for satellites in geosynchronous orbit, and to 1.0 km/s (0.6 mi/s) for the Moon. *See also* ESCAPE VELOCITY

circumpolar star Star that never sets below the observer's horizon. For a star to be circumpolar at a given latitude its declination must be greater than 90° minus that latitude. For example, if the observer's latitude is 52°, by subtracting 52° from 90° we get 38°. Any star with a declination greater then 38° will therefore be circumpolar for that observer. At the equator, no stars are circumpolar whereas at the poles, all visible stars are circumpolar.

circumstellar matter Material in the form of gas, dust or larger solid particles in close proximity to a star. Such material can form through several different processes and at several different stages of a star's life. The angular momentum of the material normally causes it to form a disk or ring centred on and orbiting the star. However the material may still be falling in towards the star (*see* FU ORIONIS STAR) or be

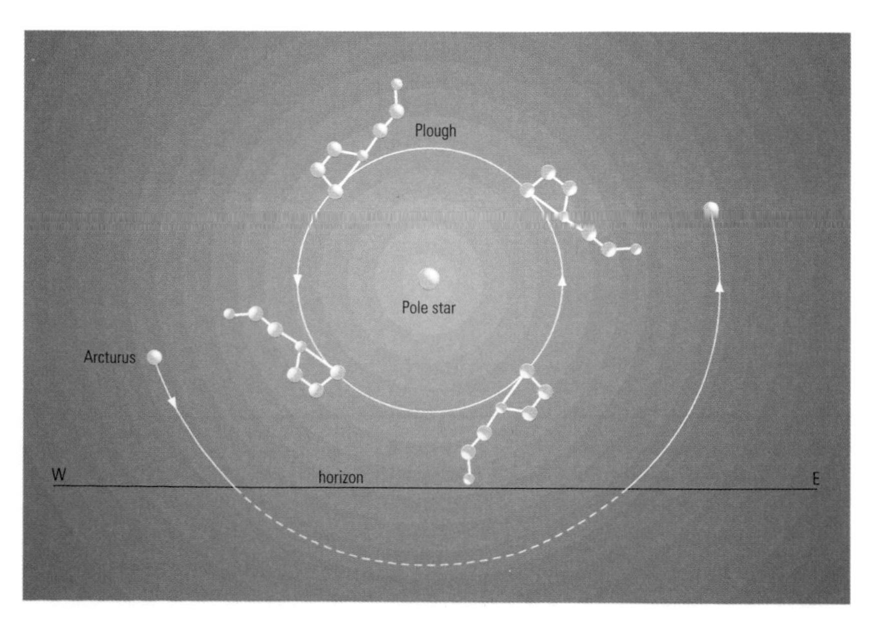

▼ circumpolar star Provided it lies sufficiently close to the celestial pole, a star may describe a complete circle once per sidereal day without disappearing below the observer's horizon. In the example shown, Alkaid, the end star on the Plough's 'handle', is such a circumpolar star, while Arcturus is not.

C

being ejected at velocities ranging from a few to many thousands of kilometres per second.

Almost all stars may be expected to be surrounded by circumstellar matter during and immediately after their formation. The collapse to a PROTOSTAR will leave material behind in the form of a flattened envelope surrounding the star and a more amorphous gas cloud further out. Condensation of the more refractory elements and compounds within the envelope may lead to the formation of dust particles (*see* BETA PICTORIS). These particles may then collide and stick together, building up to larger and larger sizes, and eventually perhaps forming PLANETS. The young star usually starts to expel material at velocities of a few hundred kilometres per second (*see* STELLAR WIND). Although the material is probably initially emitted isotropically, the surrounding equatorially concentrated envelope restricts the outflow to directions around the rotational poles of the protostar, resulting in BIPOLAR FLOWS and HERBIG–HARO OBJECTS. The stellar wind eventually evaporates the circumstellar envelope and brings a halt to planetary formation, if it is occurring. This process may explain why Uranus and Neptune are smaller than Jupiter and Saturn, since the protosun may have evaporated the remaining circumsolar material just as they were being formed. The EDGEWORTH–KUIPER BELT and the OORT CLOUD are the last remnants of this material that once surrounded the Sun. With the evaporation of the dense circumstellar material close to the star, the ultraviolet radiation from the more massive stars can then penetrate into the surrounding lower density nebula producing HII REGIONS (*see also* COSMOGONY).

Towards the end of the lives of solar-mass stars the outer layers of the star become unstable and are expelled at velocities of a few kilometres per second. The expelled material forms a cloud around the star up to a light-year across; it is heated by the ultraviolet radiation from the star. The resulting glowing material may then be seen as a PLANETARY NEBULA. Subsequent higher velocity winds from the central star may sweep the centre of the nebula clear of material to produce a spherical shell that can be seen projected on to the sky as a ring (*see* RING NEBULA). The rotation of the star, its magnetic fields or a binary central star may lead to bipolar and many other shapes for the planetary nebulae.

Circumstellar material also arises within CLOSE BINARY stars, when one star is losing mass to form an ACCRETION DISK around the other. It also arises as the stellar winds from MIRA VARIABLES, WOLF-RAYET and other hot stars, as dust particles around carbon stars and red giants, as the regions producing emission lines within the spectra of P CYG, Be (*see* GAMMA CASSIOPEIAE STAR), T TAURI STARS and so on. It also arises as the remnants of NOVA and SUPERNOVA explosions.

cislunar Term used to describe an object or an event that lies or occurs in the space between the Earth and the Moon or between the Earth and the Moon's orbit.

civil twilight *See* TWILIGHT

CK Vulpeculae Slow NOVA of 1670. Its magnitude varied from 2.7 to less than 17.0. There is a suspicion that it may be a RECURRENT NOVA.

Clairaut, Alexis-Claude (1713–65) French mathematician and physicist who applied his expertise to celestial mechanics, winning fame for predicting the 1759 return of Halley's Comet to within 1 month. His analysis of the comet's orbit, which took into account perturbations he had discovered during work on the three-body problem, suggested that the comet would arrive at perihelion on or near April 15; the actual date was March 13. His earlier work on lunar theory led to a refined orbit (1752) and tables of lunar motion (1754) that remained unsurpassed for over a century. This work also helped to confirm Newton's theory of gravitation.

Clark, Alvan *See* ALVAN CLARK & SONS

Clavius One of the largest lunar craters (58°S 14°W), diameter 225 km (140 mi); it is located in the southern lunar highlands. Clavius is an ancient impact site, the features of which have been largely obliterated by aeons of meteoric bombardment. It is often referred to as a walled depression because most of its rim is flush with the terrain outside its perimeter. Steep and rugged cliffs form the crater's inner walls, towering 3500 m (12,000 ft) above its convex floor.

▼ **Clementine** The Moon's south pole as imaged by the orbiting Clementine spacecraft in 1994. This picture has been compiled from some 1500 individual images. Results from Clementine support the possibility that ice deposits may be located in regions of permanent shadow at the lunar poles.

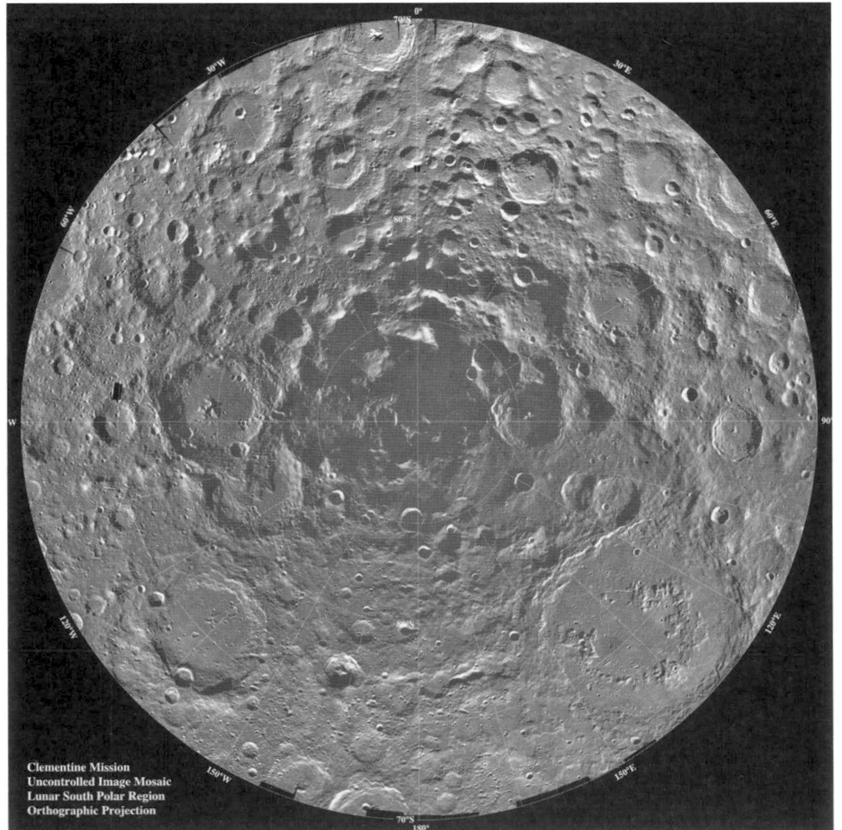

Clementine Joint project between the US Strategic Defense Initiative and the NATIONAL AERONAUTICS AND SPACE ADMINISTRATION (NASA). The objective of the mission was to test sensors and spacecraft components under extended exposure to the space environment and to make scientific observations of the MOON and the near-Earth asteroid 1620 GEOGRAPHOS. The observations included imaging at various wavelengths, including ultraviolet and infrared, laser ranging altimetry and charged-particle measurements.

Clementine was launched on 1994 January 25. After two Earth flybys, the spacecraft entered orbit around the Moon four weeks later. Lunar mapping, which took place over approximately two months, was divided into two phases. During the first month, Clementine was in a 5-hour elliptical polar orbit with a perilune (point at which it was closest to the Moon) of about 400 km (250 mi) at 28°S latitude. The orbit was then rotated to a perilune at 29°N, where it remained for one more month. This allowed the entire surface of the Moon to be mapped for the first time, as well as altimetry coverage from 60°S to 60°N. Near-infrared and ultraviolet measurements provided the best maps of surface composition and geology yet obtained. Clementine data also indicated the presence of water ice in deep craters close to the lunar poles.

After leaving lunar orbit, a malfunction in one of the on-board computers occurred on May 7, causing a thruster to fire until it had used up all of its fuel. This left the spacecraft spinning out of control and meant that the planned flyby of Geographos had to be abandoned. The spacecraft remained in geocentric orbit until July 20,

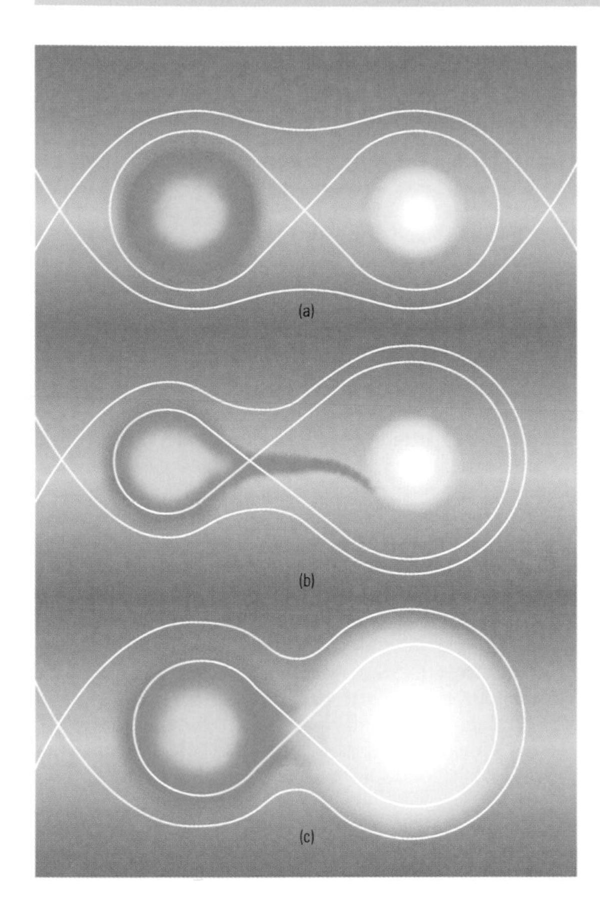

◀ **close binary** In close binary systems, the two stars may be completely detached (a). Semidetached systems have one star whose atmosphere fills its Roche lobe, leading to mass transfer to the other component (b). Where both stars fill their Roche lobes, the pair share a common atmosphere and the system becomes a contact binary (c).

when it made its last lunar flyby before going into orbit around the Sun.

Cleomedes Lunar walled plain (27°N 55°E), diameter 126 km (78 mi); it is located north of the Mare CRISIUM. Cleomedes is encircled by extremely broad mountains, which rise more than 2500 m (8500 ft) above its floor. Its walls are very uneven, with many shallow impact craters. A group of mountainous peaks arises just north-west of Cleomedes' centre, south of which is the crater Cleomedes B. A forked rille first appears north-east of these peaks, then runs south-east. A sizeable crater caps the south rim.

Clerke, Agnes Mary (1842–1907) Irish astronomy writer whose works won her widespread acclaim and election to the Royal Astronomical Society as an honorary fellow (1903), an honour previously accorded to only two other women. Her most famous book was the authoritative *A Popular History of Astronomy in the Nineteenth Century*, published in 1885 and in several revised editions until 1902.

clock *See* TIMEKEEPING

close binary BINARY STAR that has an orbital period of less than 30 years, implying that the two components are less than about 10 AU apart. Because of this proximity, most close binaries are SPECTROSCOPIC BINARIES and/or ECLIPSING BINARIES. MASS TRANSFER occurs at some stage in most close binaries, profoundly affecting the evolution of the component stars. If the two components in a close binary do not fill their ROCHE LOBES, the system is a detached binary. In a semidetached binary one star fills its Roche lobe and mass transfer occurs. In a CONTACT BINARY both stars fill their Roche lobes.

The evolution of close binaries depends on the initial masses of the two stars and their separation. The more massive star will evolve into a RED GIANT and fill its Roche lobe; material will spill through the inner LAGRANGIAN POINT on to its companion, thereby affecting its companion's evolution. The mass transfer can also alter the separation and orbital period of the binary star.

In binaries that are initially widely separated, material escaping from the Roche lobe of the evolved red giant immerses the system in material, creating a common envelope binary that contains the core of the red giant (a WHITE DWARF) and the companion star. Frictional forces cause the two components to approach, and thus the orbital period shortens. The common envelope is ejected and a CATACLYSMIC VARIABLE is left, in which the mass transfer from the companion on to the white dwarf causes the periodic outbursts seen in NOVAE, RECURRENT NOVAE, DWARF NOVAE and SYMBIOTIC STARS.

If one component of a close binary is massive enough, it may become a NEUTRON STAR or BLACK HOLE. Such binary systems are observed (*see* X-RAY BINARIES), but often a supernova explosion will blow the system apart into separate single stars.

closed universe Solution of Einstein's equations of GENERAL RELATIVITY in which the mass density of the Universe exceeds the CRITICAL DENSITY. This critical density is related to the HUBBLE CONSTANT and is estimated to be of the order of 9.2×10^{-27} kg/m^3. This implies that the Universe will eventually collapse back to a 'Big Crunch'. *See also* BIG BANG THEORY

Clown Face Nebula *See* ESKIMO NEBULA

Cluster Part of the European Space Agency's INTERNATIONAL SOLAR TERRESTRIAL PHYSICS MISSIONS. Cluster and the SOLAR HELIOSPHERIC OBSERVATORY (SOHO) are contributing to an international effort involving many spacecraft from Europe, the USA, Japan and other countries. The four Cluster satellites complement observations by SOHO, which was launched in 1995. The first four spacecraft were lost in the failure of the first Ariane 5 launch vehicle in 1996 June, but another four were put into orbit in 2000 on Starsem Soyuz boosters. The objective of these spacecraft is to study the three-dimensional extent and dynamic behaviour of Earth's plasma environment, observing how solar particles interact with Earth's magnetic field. The identical cylindrically shaped spacecraft are in high Earth orbits, flying in a tetrahedral formation passing in and out of the Earth's MAGNETOSPHERE, crossing related features such as the magnetopause, bow shock and magnetotail. The Cluster satellites are equipped with 11 instruments provided by France, Sweden, the USA, the UK and Germany.

cluster, star *See* STAR CLUSTER

cluster of galaxies *See* GALAXY CLUSTER

CME *See* CORONAL MASS EJECTION

▼ **Cluster** An artist's impression of the four Cluster spacecraft orbiting in formation in Earth's magnetosphere. Launched in 2000, the Cluster mission allows scientists to obtain three-dimensional measurements of particle densities and motions in near-Earth space.

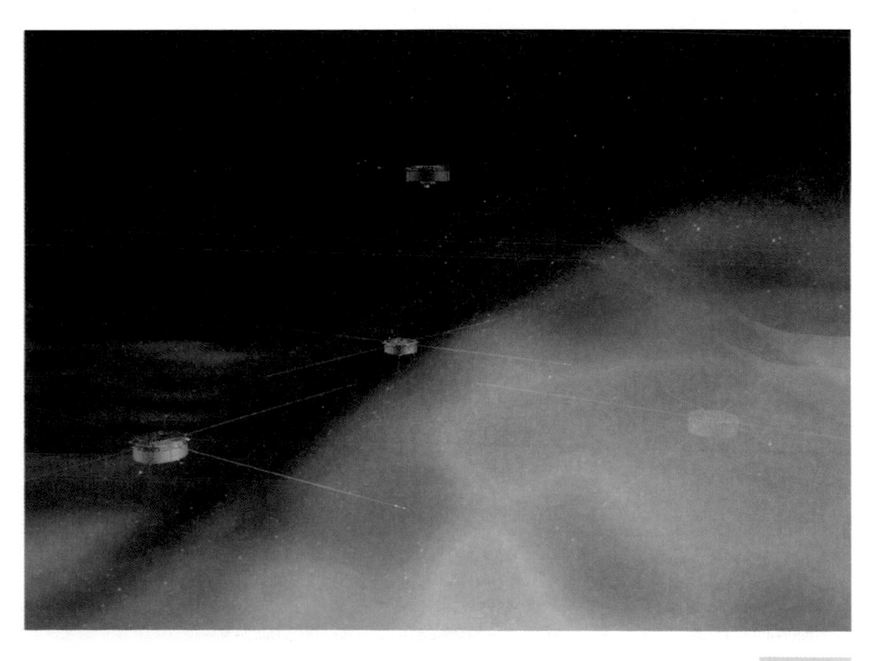

C

C

COLUMBA (GEN. COLUMBAE, ABBR. COL)

Southern constellation originated by Petrus Plancius in 1592 from stars between Canis Major and Eridanus not previously incorporated in any figure. It supposedly represents the Biblical dove that followed Noah's Ark. The name of its brightest star, Phact, comes from the Arabic meaning 'ring dove'. μ Col is a RUNAWAY STAR.

BRIGHTEST STARS

	Name	RA h m	dec. ° ′	Visual mag.	Absolute mag.	Spectral type	Distance (l.y.)
α	Phact	05 40	−34 04	2.65	−1.93	B7	268
β	Wazn	05 51	−35 46	3.12	1.02	K1.5	86

CNO cycle Abbreviation of CARBON–NITROGEN–OXYGEN CYCLE

Coalsack Dark, obscuring cloud, some 5° across, in the southern constellation CRUX (RA 12^h 50^m dec. −63°). The Coalsack, like all such clouds of dust-laden gas, is seen only in silhouette, because behind it lies the bright background of the Milky Way. On a dark night the Coalsack is very prominent: it appears to be the darkest spot in the entire sky, though that is purely an optical illusion. *See also* DARK NEBULA; GLOBULE

Coathanger (Cr 399) Asterism in the constellation Vulpecula, a few degrees north-west of the tail of Sagitta (RA 19^h $25^m.4$ dec. +20°11′). Visible to the naked eye as an almost-resolved patch, the Coathanger is shown by any small optical instrument to be a group of about 15 stars. It is dominated by six 5th-magnitude stars lying in an east–west line, with a 'hook' comprising a further four stars looping south from its middle. The Coathanger has an overall magnitude +3.6 and a diameter of 60′. It is not a true cluster: the stars that make up the pattern lie at greatly differing distances, ranging from 218 to 1140 l.y.

coating Thin layers of material applied to optical components to improve their reflectivity or transmission; also the process of applying these coatings. Thin metal coatings are applied to MIRROR surfaces to increase their reflectivity. Aluminium is the most common but gold, silver and other metals are used for special applications such as INFRARED ASTRONOMY.

Refractive-index-matching materials such as magnesium fluoride are often applied to lenses to reduce the light reflected from their surfaces, thus increasing the amount transmitted. This is sometimes called blooming because the slow accumulation of dirt on early, uncoated optics resembled the bloom on a plum or grape, and it was noticed that the bloom improved the transmission of the optics. Multi-layer coatings are often used to produce more effective anti-reflection coatings.

COBE *See* COSMIC BACKGROUND EXPLORER

Coblentz, William Weber (1873–1962) American physicist and astronomer who worked for the US National Bureau of Standards and Lowell Observatory. Coblentz studied the infrared spectra of stars, nebulae and planets, pioneering the science of infrared spectroscopy. He was the first to verify PLANCK'S LAW, through his studies of black-body radiation.

Cocoon Nebula (IC 5146) Faint EMISSION NEBULA surrounding a cluster of about 20 stars in the constellation CYGNUS (RA 21^h $53^m.4$ dec. +47°16′). The cluster has overall magnitude +7.2 and is at a distance of 3000 l.y. Star formation is probably still in progress in this rich Milky Way region. The Cocoon Nebula has an apparent diameter of 10′. It is crossed by obscuring lanes of dark material.

coelostat Two-mirror system that tracks motion of a celestial object, most usually the Sun, across the sky and allows light from it to be reflected into a fixed instrument.

A coelostat consists of a pair of plane mirrors, one of which is rotated by a motor east to west about a polar axis at half the Earth's rotation rate, thus counteracting the diurnal movement of the sky. Light from this mirror is reflected to a second, fixed mirror, which in turn directs the beam in a fixed direction, resulting in an image that is both stationary and non-rotating. Apparatus too heavy to be attached to a telescope, for example a SPECTROHELIOGRAPH, may then be positioned to receive these reflected rays. The primary characteristic of the coelostat, which distinguishes it from the similar heliostat, is that the image it produces is non-rotating.

Coelostats are often used in solar observatories, mounted at the top of a tower with instruments such as spectrographs placed at the bottom, or even underground. This arrangement allows long focal lengths to be achieved, enabling high-dispersion spectra of the Sun to be produced. *See also* COUDÉ FOCUS; SIDEROSTAT

Coggia, Comet (C/1874 H1) Bright long-period comet discovered on 1874 April 17 by Jérôme Eugène Coggia (1849–1919), Marseilles, France. The comet brightened rapidly during June as it approached Earth. Perihelion, 0.68 AU from the Sun, was reached on July 9. Around closest approach to Earth (0.25 AU) on July 18, the comet was of magnitude 0 and had a tail 60° long. Later in July, Comet Coggia faded rapidly as it headed southwards. The orbit is elliptical, with a period of 13,700 years.

cold dark matter Proposed as the missing mass component of galaxies. Flattened rotation curves of galaxies, and the velocities of stars at different distances from the centre of the galaxy, led astronomers to believe there was more mass present in the HALOES of galaxies than was being seen. Mass in the form of BARYONS, or AXIONS, was proposed as this cold dark component of galaxies. This matter might also provide enough mass to close the Universe (*see* CLOSED UNIVERSE). Recent measurements of the COSMIC MICROWAVE BACKGROUND and BROWN DWARFS in galactic haloes have reduced the need to invoke dark matter to account for missing mass. *See also* MISSING MASS PROBLEM

collapsar Obsolete name for a BLACK HOLE

collimation Process of aligning the optical elements of a telescope. Certain sealed instruments, such as refractors, are generally collimated at the factory and need never be adjusted, but most telescopes with mirrors do require occasional re-alignment, especially amateur Newtonian telescopes built with Serrurier trusses where the secondary and focuser assembly are routinely separated from the primary for transport. The procedure for collimating a Newtonian telescope is usually carried out in two steps, first aligning the primary to direct light to the centre of the secondary, then adjusting the secondary to direct the light cone down the centre of the focuser.

The term is also used to describe an optical arrangement of lenses or mirrors used to bring incoming light rays into a parallel beam before they enter an instrument such as a SPECTROSCOPE or X-RAY TELESCOPE.

▶ **coma, cometary**
Release of volatile materials due to solar radiation leads to rapid development of an extensive temporary 'atmosphere', the coma, around a cometary nucleus. These Hubble Space Telescope images show the evolution of the coma surrounding the nucleus of C/1995 O1 Hale–Bopp.

collisionless process Process occurring within a PLASMA on a timescale that is shorter than the average time for collisions between particles in that plasma. An example of a collisionless process is the formation of the BOW SHOCK between the SOLAR WIND and INTERSTELLAR MATTER.

colour index Difference in brightness of a star as measured at different wavelengths; it is used as a measure of the colour of the star. The different wavelengths are isolated by optical filters of coloured glass (for example blue and red) and light passing through each is expressed in magnitudes (B and R). The colour index B-R is zero for white stars (spectral type A0). It ranges from about −0.5 for the bluest stars to more than +2.0 for the reddest.

Colour index correlates well with the naked-eye perception of the colour of the brighter stars (the colour of fainter stars is hard to perceive for reasons to do with the physiology of the eye and the psychology of perception).

Colour index is principally a measurement of the temperature of stars. The bluest stars are hotter than 30,000 K, white stars have surfaces at temperatures of about 10,000 K. The reddest stars are very cool, say 3000 K. Cooler stars exist but emit so little light that they may best be studied by INFRARED techniques.

The colour index of light from a star may be changed as the light passes through interstellar space. Dust in space interacts more easily with the blue light from the star and disturbs the blue light from its straight path (scattering). Red light passes relatively freely around the dust and carries straight on to the observer's telescope. Thus the starlight is reddened (it would be more accurate to say 'de-blued'). The consequent increase of the star's colour index caused by the INTERSTELLAR DUST is called a colour excess due to INTERSTELLAR REDDENING.

The colour index of stars is also modified by the presence of atoms in their atmospheres. The light in different wavebands is affected to different degrees. It is possible to isolate the various effects by measuring the colour index between different pairs of wavelengths. A colour index formed with ultraviolet and with blue light (U-B) may be compared with (B-R) in a colour–colour plot. The position of a star in the plot gives clues about its chemical composition, temperature and interstellar reddening.

The colour index is also used for Solar System bodies and can give clues about their mineral compositions.

colour–magnitude diagram Plot of the MAGNITUDE of a collection of stars versus their COLOUR INDEX. It is used as a diagnostic tool to study star clusters. *See also* HERTZSPRUNG–RUSSELL DIAGRAM

Columba *See* feature article

Columbia Name of the first SPACE SHUTTLE orbiter. It flew in 1981.

Columbus Orbital Facility European module to be attached to the INTERNATIONAL SPACE STATION in 2004.

colure Great circle on the CELESTIAL SPHERE that passes through the two CELESTIAL POLES. The equinoctial colure passes through the celestial poles and the vernal and autumnal EQUINOXES. The solstitial colure passes through the celestial poles and the winter and summer SOLSTICES.

coma ABERRATION that makes off-axis star images grow small tails, giving them a comet-like appearance. Simple reflecting telescopes such as the NEWTONIAN TELESCOPE sometimes use a parabolic mirror to collect light and form an image. A perfect parabola will produce a perfect image in the centre of the field of view, that is, on its optical axis, and images close to the centre will also be very sharp. However, as the distance from the optical axis increases, the images start to grow tails pointing away from the centre. Eliminating these tails requires at least one extra optical component. Coma is not confined to mirrors but can appear in many optical systems.

COMA BERENICES (GEN. COMAE BERENICES, ABBR. COM)

Faint northern constellation, given permanent status in 1551 by Gerardus Mercator from stars previously regarded as belonging to Leo, though it had been referred to in ancient times. It supposedly represents the hair of Queen Berenice of Egypt, which she cut off in gratitude to the gods for the safe return of her husband from battle. Its brightest star is β Com, mag. 4.23. FS Com is a red giant that varies between 5th and 6th magnitudes every 2 months or so. 24 Com is an easy double for small telescopes, mags. 5.0 and 6.6, showing contrasting colours of orange and blue. Extending southwards for several degrees from γ Com is the Coma Star Cluster, a V-shaped scattering of faint stars, traceable in binoculars. M64 is the BLACK EYE GALAXY, and NGC 4565 is an edge-on spiral galaxy about 20 million l.y. away. Several members of the VIRGO CLUSTER of galaxies can be found in Coma, most notably the face-on spiral M100; it also contains the more distant COMA CLUSTER.

coma, cometary Teardrop-shaped cloud of gas and dust surrounding a comet's NUCLEUS. It is produced as a result of increased exposure to solar radiation once the COMET is sufficiently close to the Sun (usually within 3–4 AU).

Coma Berenices *See* feature article

Coma Cluster (Abell 1656) Nearest rich GALAXY CLUSTER, in the direction of COMA BERENICES. It lies at a distance of about 330 million l.y., and has over 450 member galaxies with known redshifts (and galaxy counts suggest several thousand members in all). The structure of the X-ray gas shows prominent clumps, suggesting that smaller groups have recently fallen into the cluster. These clumps also contain an excess of galaxies with recent (but not ongoing) star formation, indicating interaction between the cluster environment and the individual galaxies. The cluster mass as estimated from galaxy velocities is 2 thousand million million solar masses. The Coma cluster is part of the extensive Perseus–Pisces supercluster of galaxies, which stretches almost halfway around the sky from our vantage point.

comes Obsolete term for the fainter component of a BINARY STAR; it is now referred to as the COMPANION.

comet Small Solar System body, consisting of frozen volatiles and dust. Comets are believed to be icy planetesimals remaining from the time of the Solar System's formation 4.6 billion years ago. The word 'comet' derives from the Greek *kometes*, a long-haired star, which aptly describes brighter examples.

The main, central body of a comet is the NUCLEUS, typically only a few kilometres in diameter. The nucleus of Comet 1P/HALLEY has dimensions of about 15 × 8 km (9 × 5 mi), but most comet nuclei are smaller. At large distances from the Sun, a cometary nucleus is inactive, and indistinguishable from an ASTEROID. Comet nuclei are thought to have a dark outer crust, which, when heated by

◄ **Coma Cluster** This rich galaxy cluster, probably containing more than 1000 members, lies 350 million l.y. away. As in other such clusters, the most prominent members are giant elliptical galaxies.

▶ comet Radiation pressure and solar wind effects result in a comet's tail always pointing away from the Sun. An interesting consequence of this is that a comet therefore departs the inner Solar System tail-first.

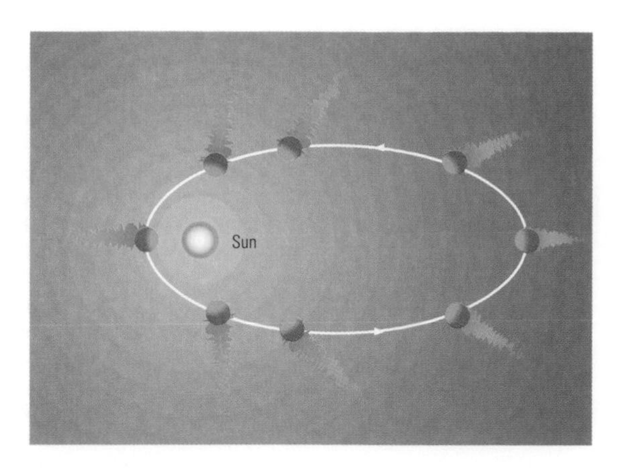

solar radiation, particularly at distances substantially less than that of Jupiter (5 AU) from the Sun, can crack to expose fresh volatile material below. Sublimation of frozen gases leads to development of a temporary atmosphere, or COMA, surrounding the nucleus. New comets, or known PERIODIC COMETS returning to PERIHELION, are usually found at this stage, as fluorescence in the coma causes them to brighten and appear as a teardrop-shaped fuzzy spot. As a comet approaches the Sun more closely, a TAIL or tails may form. Brighter comets often show a straight ion tail (type I) and a curved dust tail (type II). Enveloped in the coma, the nucleus is not directly visible to Earth-based telescopes, but a bright spot may mark its position, and jets of material can be seen emerging from it in the brightest and most active comets. The occurrence of such features indicates that much of the gas emission from a comet comes from persistent active regions, which may cover 15–20% of the nucleus. As the nucleus rotates, active regions are alternately 'turned on' by solar heating or abruptly shut down as they are carried into shadow. Such behaviour led to the distinctive spiral structure in the coma of C/1995 O1 HALE–BOPP. Intermittent gas and dust emission can produce other features, including hoods and shells in the coma.

The whole comet may be enveloped in a vast tenuous cloud of hydrogen, detectable at ultraviolet wavelengths from spacecraft. Gas jets emerging from the nucleus carry away dust particles, which depart on parabolic trajectories to form a curved dust tail. Dust tails appear yellowish, the colour of reflected sunlight as confirmed by spectroscopy. The dust particles appear to be silicate grains, some 10 micrometres in size. Bigger flakes of dusty material, perhaps a few millimetres in size, are also carried away; in large numbers, these METEOROIDS can end up pursuing a common orbit around the Sun as a METEOR STREAM. Comets' tails point away from the Sun. Under certain circumstances, however, thin sheets of ejected dusty material may appear to point towards the Sun from the coma, forming an ANTITAIL, as a result of perspective.

Gas emerging from the nucleus is rapidly ionized by solar ultraviolet radiation. Positively charged ions are picked up by the interplanetary magnetic field in the SOLAR WIND and dragged away from the coma to form an ion tail (also commonly described as a plasma tail). In contrast with the dust tail, a comet's ion tail appears relatively straight and may show a marked bluish colour, which results from emissions at 420 nm wavelength, characteristic of carbon monoxide (CO) excitation. The ion tail can exhibit knots and twists resulting from changes in the interplanetary magnetic field in the comet's vicinity; reversals of the field's polarity can result in complete shearing of the ion tail – a DISCONNECTION EVENT – after which a new, differently oriented ion tail may develop.

Comas and ion tails show emission spectra characteristic of a number of molecular species comprising combinations of hydrogen, carbon, nitrogen, oxygen and sulphur, such as water (H_2O), carbon monoxide (CO), carbon dioxide (CO_2) and radicals such as cyanogen (CN) and hydroxyl (OH). Methane (CH_4) and ammonia (NH_3) are certainly present but are difficult to detect. When a comet is very close to the Sun, metallic emissions (particularly from sodium) occur; observations of C/1995 O1 Hale–Bopp in 1997 revealed the presence of a third, distinct sodium tail.

Although comets can be ejected from the Solar System, never to return, on hyperbolic trajectories following planetary encounters, none has yet been shown to enter from interplanetary space: the comets that we observe are gravitationally bound to the Sun. A vast reservoir of cometary nuclei, the OORT CLOUD, surrounds the Solar System to a distance of 100,000 AU. Perturbations by passing stars or giant molecular clouds in the course of the Sun's orbit around the Galaxy can cause Oort cloud nuclei to fall inwards. There is strong evidence that such nuclei accumulate in a flattened disk – the EDGEWORTH–KUIPER BELT – at a distance of up to 1000 AU. From this region, further perturbations may lead nuclei to plunge inwards to perihelion as new long-period comets. Close encounter with one of the planets, particularly Jupiter, can dramatically alter the comet's orbit, changing some bodies from long- to short-period comets.

The division between short- and long-period comets is set purely arbitrarily at an orbital duration (perihelion to perihelion) of 200 years. Most comets have narrow elliptical orbits, which may be either direct or retrograde; 1P/Halley, for example, has a 76-year retrograde orbit. Comet 2P/ENCKE has the shortest period, of 3.3 years. In total, some 150 short-period comets are known.

Comets lose material permanently at each perihelion passage and must eventually become defunct; the lifetimes of short-period comets are probably of the order of 10,000 years. The ultimate fate of a short-period comet appears to vary. Some disperse entirely into a diffuse cloud of dust and gas. Comet nuclei are fragile and may break into smaller fragments close to perihelion. Others, depleted of volatile material, may simply become inert, leaving a dark asteroid-like core with no tail activity.

The brightness of a comet is expressed as the equivalent stellar magnitude, as in the case of nebulae. Most comets show fadings and outbursts caused mainly by varying nuclear jet activity and solar effects. Predicting the apparent brightness of comets is notoriously difficult. It does appear that proximity to the Sun is a more significant factor than closeness to Earth. Comets are usually at their most active and, therefore, brightest just after perihelion.

Comets are normally named after those who discover them, up to a maximum of three names. In some cases – increasingly common early in the 21st century – comet discovery by automated telescopes or spacecraft is reflected in their names, examples including the many named after LINEAR or SOHO. Some, notably 1P/Halley, 2P/Encke and 27P/CROMMELIN, are named after the analyst who first determined their orbit. Short-periodic comets are identified by the prefix P/ and a number indicating the

MAIN SHORT-PERIOD AND NOTABLE LONG-PERIOD COMETS

Designation	Name	Period (years)	Inclination (°)	Associated meteor stream
1P	Halley	76	162.2	Orionids, Eta Aquarids
2P	Encke	3.3	11.8	Taurids
3D	Biela	–	6.6	Andromedids
4P	Faye	7.34	9.1	
5D	Brorsen	–	29.4	
6P	D'Arrest	6.51	19.5	
7P	Pons–Winnecke	6.37	22.3	Pons-Winneckids
8P	Tuttle	13.51	54.7	Ursids
9P	Tempel-1	5.51	10.5	
10P	Tempel-2	5.47	12.0	
21P	Giacobini–Zinner	6.61	31.9	Giacobinids
55P	Tempel–Tuttle	33.22	162.5	Leonids
109P	Swift–Tuttle	135	113.4	Perseids
C/1956 R1	Arend–Roland	hyperbolic	119.9	
C/1965 S1	Ikeya–Seki	880	141.9	
C/1969 Y1	Bennett	1680	90.0	
C/1975 V1	West	500,000	43.1	
C/1996 B2	Hyakutake	14,000	124.9	
C/1995 O1	Hale–Bopp	2400	89.4	
C/2002 C1	Ikeya–Zhang	341	28.1	

order in which their orbit was defined. Defunct comets have the prefix D/, while long-period comets are denoted by C/. In addition to a name, long-period comets are identified by the year and date order of their discovery. The year is divided into 26 fortnightly intervals for this purpose, starting with A and B for January, and so on. A comet designated C/2002 B2 would be the second discovery made in the latter fortnight of 2002 January.

In most years, perhaps 25 comets become sufficiently bright to be observed with amateur telescopes. Spectacular naked-eye comets are rare and unpredictable: the brightest are usually new discoveries, making one of their first visits to the inner Solar System. The more predictable, short-period comets tend to be fainter, having already lost some of their volatile material. Among the brightest comets have been the KREUTZ SUNGRAZERS.

It has been speculated that the frequency of truly bright 'great' comets has been remarkably low in recent times compared with, say, the late 19th century. Statistically, however, the 20-year gap between C/1975 V1 WEST and C/1996 B2 HYAKUTAKE is not atypical. The next spectacular comet may appear at any time.

cometary globule Fan-shaped REFLECTION NEBULA that is usually closely associated with a pre-main-sequence star, such as a T TAURI STAR. The globule's appearance can superficially resemble that of a comet, but the two types of object are quite unrelated. There may be a bright rim to the 'head', and the 'tail' can be several light-years in length. Other recognized shapes for cometary globules include: an arc; a ring, sometimes with a star at the centre or on the rim; a biconical (hourglass) nebula with star at the 'waist'; and a linear wisp protruding from a star. Most cometaries shine by reflecting the light of their allied star, though some are ionized by the ultraviolet radiation of hot central stars (*see* HII REGION).

The heads of globules are denser regions within a larger nebula. The ultraviolet radiation and STELLAR WIND from the associated star ionizes the gases on the outer surface of the globule causing it to glow, so producing the bright rim around the head. The gas and radiation pressure also drives away surrounding material to leave the dark tail formed of nebula material sheltered by the head.

Recently, the biconical type of nebula has been generalized to include any bipolar system that consists of two separate nebulae with a star lying between. Enlarging the class to incorporate the bipolar nebulae makes cometary globules evolutionarily less homogeneous. The subclass of bipolar nebulae are not all indications of stellar youth: some RED GIANTS lose mass by a BIPOLAR FLOW, thereby generating bipolar nebulae.

The stars associated with cometary globules are often embedded within dusty equatorial disks, which are sometimes thick enough to render the stars optically invisible. Bright infrared sources are observed instead, which are the result of emission from the dust. The various cometary globule shapes may all be explained as viewing from different directions the basic model of a star plus a dust disk, with the disk constraining the star's light to shine into a bicone. *See also* HUBBLE'S VARIABLE NEBULA

commensurability There are many cases of pairs of planets or satellites whose orbital periods are close to the ratio of two small integers, and these are called 'commensurabilities'. They cause greatly increased PERTURBATIONS between the two bodies by a RESONANCE effect, and it is a major challenge of CELESTIAL MECHANICS to calculate these perturbations and the resulting effects on the orbits. The principal examples among the planets are the 5:2 commensurability between Jupiter and Saturn, 3:1 between Saturn and Uranus, 2:1 between Uranus and Neptune, and 3:2 between Neptune and Pluto. For Neptune and Pluto the commensurability is very close, and a special type of motion called a LIBRATION occurs. This prevents the two planets ever coming near to each other, even though their orbits cross, because they are never at that part of their orbits at the same time.

A similar effect occurs at the outer part of the ASTEROID BELT. The Hilda group of asteroids at the 3:2 commensurability with Jupiter, and the asteroid Thule at 4:3, can only remain in orbits so close to Jupiter because their commensurabilities protect them from close approaches. The TROJAN ASTEROIDS are an extreme example of this effect, as they move in the same orbit as Jupiter, but their 1:1 commensurabilities with Jupiter ensure that they remain far from Jupiter – close to 60° ahead or behind.

There are many more commensurabilities in the satellite systems of the major planets than would be expected by chance, and there has been much recent effort to explain their origin. In the Jupiter system there is a complex commensurability involving the three satellites Io, Europa and Ganymede, which is a combination of two 2:1 commensurabilities. It is likely that this was formed by TIDAL EVOLUTION of the orbits due to the tides raised in Jupiter by the satellites, combined with the effect of energy dissipation of the tide raised in Io by Jupiter. In the Saturn system, the 2:1 commensurability between Mimas and Tethys, and that between Enceladus and Dione, were also probably caused by tidal evolution. The likely explanation of the 4:3 commensurability between Titan and Hyperion is that Hyperion is the sole survivor from many objects originally in the region, and its resonance has protected it from close approaches to Titan. There are also many effects of commensurabilities in the ring systems of the planets.

Common, Andrew Ainslie (1841–1903) English engineer and amateur astronomer who designed and built large reflecting telescopes and took some of the finest astronomical photographs of his time. In the late 1870s he obtained a 36-inch (0.9-m) mirror from the English optician George Calver (1834–1927) and designed and supervised the construction of a telescope and a rotating observatory to house the instrument. In 1880 Common took his first photographs with this telescope, and for the next five years he continued to improve his photographic technology. His photographs of the Orion Nebula (M42), which showed stars that were invisible to visual observers, demonstrated the value of ASTROPHOTOGRAPHY.

Commonwealth Scientific and Industrial Research Organisation (CSIRO) One of the world's largest and most diverse scientific research organizations, with a history dating back to 1916. CSIRO is the umbrella organization for the AUSTRALIA TELESCOPE NATIONAL FACILITY.

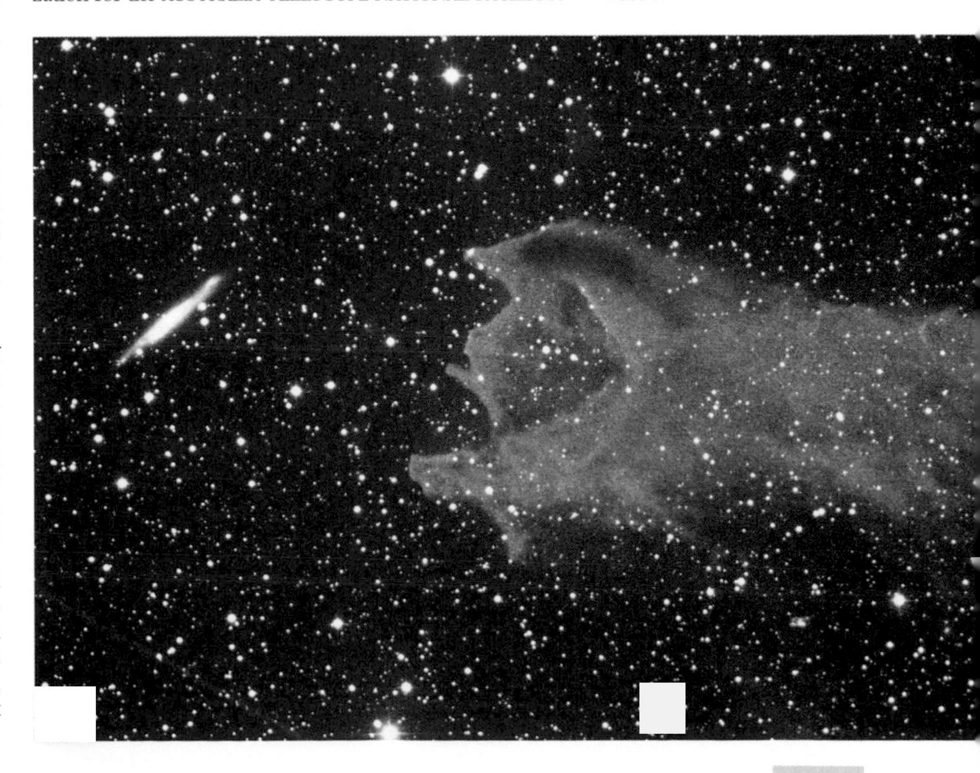

▼ **cometary globule** A strong outflowing stellar wind from a recently formed star draws this diffuse nebula into a comet-like shape. The young star's intense ultraviolet radiation ionizes gas in the nebula.

C

▲ **Cone Nebula** A striking region of bright and dark nebulosity associated with star formation in the Milky Way in Monoceros. The Cone Nebula is named for the tapering dark intrusion silhouetted against the bright emission nebulosity in the vicinity of the young Christmas Tree Cluster.

compact object Object, such as a WHITE DWARF or NEUTRON STAR, that is of high mass contained within a volume of space, indicating that it is formed of DEGENERATE MATTER. *See also* BLACK HOLE

companion Fainter of the two components of a BINARY STAR. Often this is the less massive component, lying farther from the centre of mass. It is sometimes called the secondary. *See also* PRIMARY

comparator Instrument that enables two photographs of the same area of sky, taken at different times, to be rapidly alternated in order to reveal objects that have changed position or brightness. The most common example is the BLINK COMPARATOR, in which discordant images will appear to blink on and off or pulsate. The STEREO COMPARATOR uses binocular vision to make them appear to stand out of the plane of the picture, while in another type of comparator, they appear a different colour from the unchanging stars. The instruments are particularly used by hunters of novae and asteroids.

Compton effect Loss of energy by, and consequent increase in wavelength of, a PHOTON that collides with an ELECTRON and imparts some of its energy to it. In the inverse Compton effect, the photon gains energy from the electron.

Compton Gamma Ray Observatory Second of NASA's GREAT OBSERVATORIES, launched from the Space Shuttle *Atlantis* on 1991 April 5. The observatory was named in honour of Arthur Holly Compton (1892–1962), who was awarded the Nobel Prize for physics for work on gamma-ray detection techniques.

Its four instruments provided unprecedented coverage of the electromagnetic spectrum, from 30 keV to 30 GeV. The Burst and Transient Source Experiment (BATSE) was designed to detect short outbursts. The Oriented Scintillation Spectrometer Experiment (OSSE) studied the spectrum of gamma-ray sources. The Imaging Compton Telescope (COMPTEL) mapped the gamma-ray sky at medium energies, and the Energetic Gamma Ray Experiment Telescope (EGRET) made an all-sky map of high-energy sources.

The observatory re-entered the Earth's atmosphere on 2000 June 4. During its lifetime, the telescope detected more than 400 gamma-ray sources, ten times more than were previously known. The origins of many of these sources remain unknown. Prior to Compton, 300 gamma-ray bursts had been detected; observations from the satellite recorded a further 2500.

One of BATSE's most important contributions was an all-sky map of gamma-ray burst positions, which confirmed that they originate well beyond our Galaxy. One of the major discoveries made by EGRET was a new class of quasars known as BLAZARS. The observatory also discovered a number of gamma-ray PULSARS, while observations of the galactic centre by OSSE revealed gamma radiation from the annihilation of positrons and electrons in the interstellar medium. *See also* GAMMA-RAY ASTRONOMY

Compton wavelength Length scale at which the wave-nature of a particle becomes significant. It is given by

$$\frac{h}{mc}$$

where h is the PLANCK CONSTANT, m is the rest mass of the particle and c is the velocity of light. For an electron its value is thus 2.4×10^{-12} m. The Compton wavelength is of significance for HAWKING RADIATION from BLACK HOLES and in COSMOLOGY, where it determines the earliest moment that can be understood using the laws of physics (the PLANCK TIME).

Cone Nebula (NGC 2264) Tapered region of dark nebulosity in the constellation MONOCEROS (RA $06^h\ 41^m.1$ dec. $+09°53'$); it obscures some of the extensive EMISSION NEBULA in the region of the Christmas Tree Cluster. The bright nebulosity into which the Cone Nebula intrudes covers an area about $20'$ across and lies at a distance of 2400 l.y. The Cone Nebula is difficult to observe visually: it is best revealed in long-exposure images.

conic section Curve that is obtained by taking a cross-section across a circular cone. The curve will be a circle, ellipse, parabola or hyperbola depending on the angle at which the cross-section is taken. The significance for astronomy is that these curves are also the possible paths of a body moving under the gravitational attraction of a primary body.

conjunction Alignment of two Solar System bodies with the Earth so that they appear in almost the same position in the sky as viewed from Earth. The INFERIOR PLANETS, Mercury and Venus, can align in this way either between the Sun and the Earth, when they said to be at INFERIOR CONJUNCTION, or when they lie on the opposite side of the Sun to the Earth, and are at SUPERIOR CONJUNCTION. The superior planets can only come to superior conjunction. When a planet is at conjunction its ELONGATION is 0°.

The strict definition of conjunction is when two bodies have the same celestial longitude as seen from Earth and because of the inclination of the various planetary orbits to the ecliptic, exact coincidence of position is rare. The term is also used to describe the apparent close approach in the sky of two or more planets, or of the Moon and one or more planets.

constellation Arbitrary grouping of stars, 88 of which are recognized by modern astronomers. Various constellation systems have been developed by civilizations over the ages; the one we follow is based on that of the ancient

Greeks, although it actually originated around 4000 years ago with the Sumerian people of the Middle East. The Chinese and Egyptian systems were completely different. In particular, the Chinese of the 3rd century AD had 283 constellations incorporating nearly 1500 stars. Many of the Chinese figures were small and faint, and some consisted merely of single stars.

A constellation pattern has no real significance: the stars are at very different distances from us and appear close together only because of a line-of-sight effect. This is well demonstrated by α and β Centauri, which lie side-by-side in the sky although in fact α is 4 l.y. away while β is over 100 times as distant.

PTOLEMY gave a list of 48 constellations in his *ALMAGEST*. He did not include the far southern sky, which was below the horizon from his observing site in Alexandria, and there were gaps between his constellations; but all the figures he listed are retained by modern astronomers, although with somewhat different boundaries. Many of those he named are drawn from ancient mythology; for example the legend of Perseus and Andromeda is well represented. The gaps left by Ptolemy were filled in by others, notably Johannes HEVELIUS.

Twelve new southern constellations, mostly representing exotic animals, including a bird of paradise and a flying fish, were formed in around 1600 by two Dutch navigators, Pieter Dirkszoon Keyser and Frederick de Houtman. A far more detailed survey of the southern sky was made in 1751–52 by the Frenchman LACAILLE, who catalogued nearly 10,000 stars from the Cape of Good Hope, modern South Africa. He introduced 14 new constellations to fill the gaps between Keyser and de Houtman's figures, mostly named after instruments of science and the arts, such as the Microscope, the Telescope and the Painter's Easel. Lacaille also divided up the large Greek constellation Argo Navis into three more manageable sections – Carina, Puppis and Vela.

After Lacaille there was a period when almost every astronomer who produced a star map felt obliged to introduce new constellations, often with cumbersome names such as Globus Aerostaticus, the hot-air balloon, Machina Electrica, the electrical machine, and Officina Typographica, the printing shop, all of which have been rejected. Another of these rejected constellations – Quadrans Muralis, the Mural Quadrant, invented by the Frenchman J.J. Lalande – is remembered because of the January meteor shower called the QUADRANTIDS, which radiates from the area it once occupied.

Finally, in 1922, the INTERNATIONAL ASTRONOMICAL UNION put matters on a more systematic footing. They fixed the accepted number of constellations at 88, and in 1928 adopted rigorous boundaries to the constellations based on circles of right ascension and declination, a system originally introduced for just the southern constellations in 1879 by Benjamin GOULD. The accompanying table gives the official list and their abbreviations.

It is clear that the constellations are very unequal in size and importance; they range from the vast HYDRA to the tiny but brilliant CRUX. It is also true that there are some constellations that seem to have no justification for a separate existence, HOROLOGIUM and LEO MINOR being good examples. Only three constellations have more than one first-magnitude star (taking the first-magnitude limit as the conventional 1.49) – CENTAURUS, CRUX and ORION. Note also that although Ophiuchus is not reckoned as a zodiacal constellation, it does cross the zodiac between SCORPIUS and SAGITTARIUS. *See also* individual constellations; table, pages 94–95

contact binary BINARY STAR in which both components fill their ROCHE LOBES. Common envelope binaries are systems in which material has escaped from the Roche lobes to surround both stars. The exchange of mass and energy within contact binaries is not well understood.

The best-known stable contact binaries are the W URSAE MAJORIS STARS, which consist of stars of spectral type G and K, with a typical mass ratio of 2:1. Material from the stars has escaped the Roche lobes and surrounds the components in a common envelope. Other observed stable contact binaries have massive early-type components or consist of cool SUPERGIANTS.

If a star in a contact binary is much more massive and larger than its companion, then it may undergo a catastrophic tidal instability, with the less massive star being pulled into its envelope. Resulting shock waves dissipate energy, expelling the outer envelope of the massive component and slowing the orbit of the binary system. PLANETARY NEBULAE with double cores may be the result of such a process.

continuous creation Idea that matter is constantly created in the gaps between galaxies so that the PERFECT COSMOLOGICAL PRINCIPLE applies in an expanding universe. This was a necessary feature in the STEADY-STATE THEORY of cosmology.

continuous spectrum (continuum) SPECTRUM containing photons with a smooth distribution of wavelengths; it has no breaks or gaps and no ABSORPTION LINES or EMISSION LINES, though these can be superimposed.

continuum Property that is seamless or has no 'smallest value'. A continuum can be subdivided into infinitely small pieces and is still the same thing, whereas a quantized medium has a 'smallest' quantity. Space in Isaac Newton's theory of mechanics is a 'continuum', and spacetime in relativity is a continuum.

Contour NASA DISCOVERY programme mission, to be launched in 2002 July, to fly past three comets, taking images, making comparative spectral maps of their nuclei and analysing the dust flowing from them. The spacecraft will fly past Comet 2P/ENCKE in 2003 November, Comet 73P/SCHWASSMANN–WACHMANN-3 in 2006 June and Comet 6P/D'ARREST in 2008 August.

convection Process of energy transfer in a gas or liquid as a result of the movement of matter from a hotter to a colder region and back again. For astronomy, convection

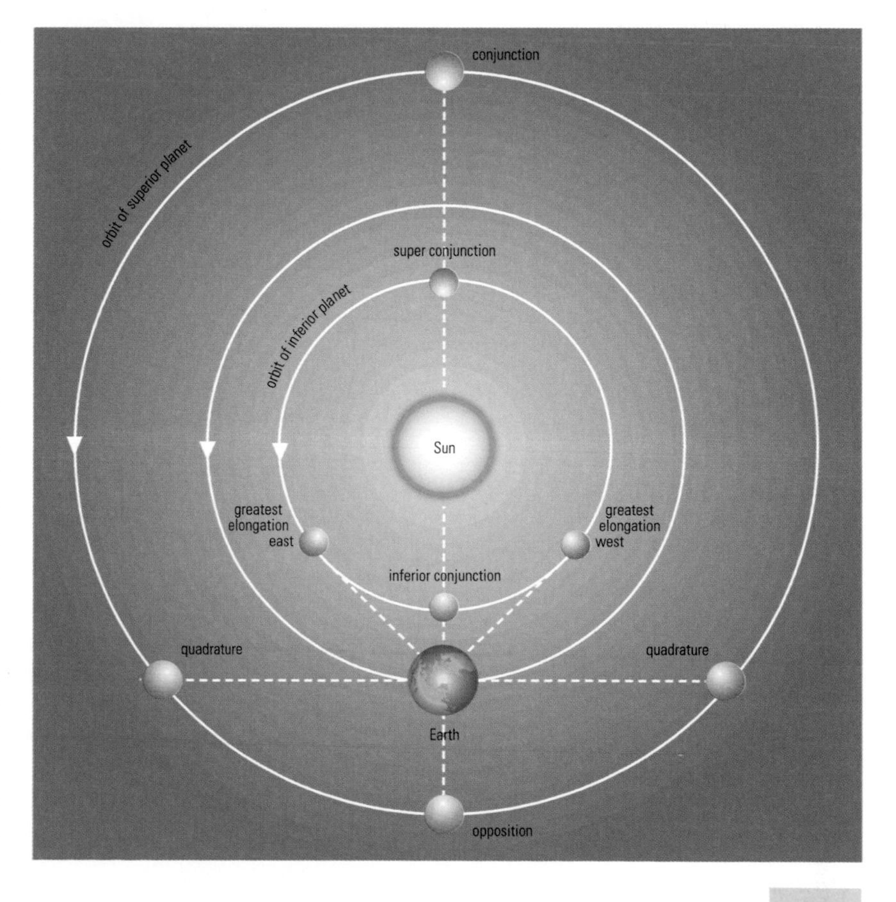

▼ **conjunction** Superior planets come to conjunction with the Sun when on the far side of it, as seen from Earth (and are therefore lost from view). The inferior planets, Mercury and Venus, can undergo conjunction at two stages in their orbit. At superior conjunction, they lie on the far side of the Sun from Earth, while at inferior conjunction they are between the Sun and the Earth. Under certain circumstance, Mercury and Venus can transit across the Sun's disk at inferior conjunction.

is mostly of significance in stellar interiors and in planetary atmospheres. In planetary atmospheres, convective circulation combined with Coriolis forces leads to the formation of HADLEY CELLS, where warm gas rises and moves away from the equator, eventually cooling and returning at lower levels. Convection in stars is still poorly understood, although numerical models are improving. The models are based upon mixing length theory. This assumes that the convective elements move a characteristic length, known as the mixing length, releasing all their excess energy only at the top of their movement, or absorbing their entire energy deficit only at the bottom. The transfer of energy thus occurs as a result of both the upward and downward phases of convective motion. Between the top and bottom of the mixing length the material changes ADIABATICALLY. For convection to occur, material moving upwards and changing adiabatically must be at a lower density than the surrounding material, so that it experiences a buoyancy force, and continues its upward motion. This requirement gives the SCHWARZSCHILD criterion for convection to occur:

$$\frac{d \log_e T}{d \log_e P} > K$$

where T is the temperature and P the pressure inside the star, and K a constant with the value of about 0.4 for completely ionized plasmas and about 0.1 for regions over which hydrogen is ionizing. In regions of stars where convection is occurring, the process transfers almost all the energy outwards. In other regions the energy is transferred by radiation. For the Sun, convection occurs over the outer third of its radius and although it stops just below the visible surface, its effects can still be seen as solar GRANULATION.

Cooke, Troughton & Simms English firm of telescope and scientific instrument manufacturers. **John Troughton, Sr** (*c.*1716–88) made high-precision sextants, quadrants and other scientific instruments from the mid-1750s. Around 1780, his nephew **John Troughton, Jr** (*c.*1739–1807) perfected a means of precisely dividing the circular scales for surveying and astronomical devices. John, Jr's brother, **Edward Troughton** (1756–1835), made the first mural circle, an innovative transit telescope, designed in 1806 and completed in 1812 for GREENWICH OBSERVATORY.

William Simms (1793–1860) improved methods for dividing transit circles. In 1824 he joined forces with the Troughtons, forming the partnership of Troughton & Simms. He was eventually succeeded by his son **William Simms, Jr** (1817–1907) and the latter's cousin, **James Simms** (1828–1915); James' two sons, **William Simms III** (1860–1938) and **James Simms, Jr** (1862–1939), also joined the family's optical business.

CONSTELLATIONS

Constellation	Name	Genitive	Abbr.	Area (sq deg)	First-magnitude stars
*Andromeda	Andromeda	Andromedae	And	722	
Antlia	The Air Pump	Antliae	Ant	239	
Apus	The Bird of Paradise	Apodis	Aps	206	
*Aquarius	The Water Carrier	Aquarii	Aqr	980 Z	
*Aquila	The Eagle	Aquilae	Aql	652	Altair
*Ara	The Altar	Arae	Ara	237	
*Aries	The Ram	Arietis	Ari	441 Z	
*Auriga	The Charioteer	Aurigae	Aur	657	Capella
*Boötes	The Herdsman	Botis	Boo	907	Arcturus
Caelum	The Chisel	Caeli	Cae	125	
Camelopardalis	The Giraffe	Camelopardalis	Cam	757	
*Cancer	The Crab	Cancri	Cnc	506 Z	
Canes Venatici	The Hunting Dogs	Canum Venaticorum	CVn	465	
*Canis Major	The Great Dog	Canis Majoris	CMa	380	Sirius
*Canis Minor	The Little Dog	Canis Minoris	CMi	183	Procyon
*Capricornus	The Sea Goat	Capricorni	Cap	414 Z	
Carina	The Keel	Carinae	Car	494	Canopus
*Cassiopeia	Cassiopeia	Cassiopeiae	Cas	598	
*Centaurus	The Centaur	Centauri	Cen	1060	Alpha, Beta Centauri
*Cepheus	Cepheus	Cephei	Cep	588	
*Cetus	The Whale	Ceti	Cet	1231	
Chamaeleon	The Chameleon	Chamaeleontis	Cha	132	
Circinus	The Compasses	Circini	Cir	93	
Columba	The Dove	Columbae	Col	270	
Coma Berenices	Berenice's Hair	Comae Berenices	Com	386	
*Corona Australis	The Southern Crown	Coronae Australis	CrA	128	
*Corona Borealis	The Northern Crown	Coronae Borealis	CrB	179	
*Corvus	The Crow	Corvi	Crv	184	
*Crater	The Cup	Crateris	Crt	282	
Crux	The Southern Cross	Crucis	Cru	68	Alpha, Beta Crucis
*Cygnus	The Swan	Cygni	Cyg	804	Deneb
*Delphinus	The Dolphin	Delphini	Del	189	
Dorado	The Goldfish	Doradus	Dor	179	
*Draco	The Dragon	Draconis	Dra	1083	
*Equuleus	The Foal	Equulei	Equ	72	
*Eridanus	The River	Eridani	Eri	1138	Achernar
Fornax	The Furnace	Fornacis	For	398	
*Gemini	The Twins	Geminorum	Gem	514 Z	Pollux
Grus	The Crane	Gruis	Gru	366	
*Hercules	Hercules	Herculis	Her	1225	
Horologium	The Pendulum Clock	Horologii	Hor	249	
*Hydra	The Water Snake	Hydrae	Hya	1303	
Hydrus	The Little Water Snake	Hydri	Hyi	243	
Indus	The Indian	Indi	Ind	294	

** Ptolemy's original 48 constellations*

The Englishman **Thomas Cooke** (1807–68), a maker of high-quality refracting telescope lenses, was joined by his sons **Charles Frederick Cooke** (1836–98) and **Thomas Cooke, Jr** (1839–1919). At their Buckingham Works in York, England, Thomas Cooke & Sons made many fine large refracting telescopes for observatories round the world. The telescope built in 1881 for Liège University, with a 10-inch (250-mm) lens, was widely regarded as one of their finest. In 1871 the firm completed what was then the world's largest refractor, a 25-inch (0.63-m) instrument for Robert Sterling Newall (1812–89). Cooke's master optician was Harold Dennis Taylor (1862–1943), who invented the 'Cooke photographic lens' in the 1880s. In 1892 he made the first three-element apochromatic lens, designed to virtually eliminate chromatic aberration and to have sufficient colour correction for use in visual studies and photography.

In 1922 Thomas Cooke & Sons bought out Troughton & Simms, forming by merger the firm of Cooke, Troughton & Simms, Ltd. The company continued to make telescopes and other scientific instruments until the late 1930s, when it sold its telescope-making operation to GRUBB, PARSONS & CO.

co-orbital satellite Natural satellite or moon of one of the major planets occupying the same orbital distance as a similar object. An example of such a pair is the Saturnian satellites EPIMETHEUS and JANUS. In some cases small satellites may occupy the LAGRANGIAN POINTS associated with one of the larger moons of the giant planets, for example, CALYPSO and TELESTO, the satellites of Saturn linked with TETHYS. Ring systems might also be regarded as being sets of tiny co-orbiting satellites. *See also* SHEPHERD MOON; TROJAN ASTEROID

Coordinated Universal Time (UTC) Tmescale derived from atomic clocks and based on civil time as kept at the Greenwich meridian. Popularly known as GREENWICH MEAN TIME (GMT), UTC is used as the basis for generating radio time signals. However, variations in the Earth's rotation mean that time as kept by atomic clocks (UTC) and that derived from observations of the stars (UT1, *see* UNIVERSAL TIME) gradually diverge. If this were allowed to go unchecked, the long term effect would be for time kept by clocks and that as shown by the Sun to become increasingly out of step. In order to avoid this, the two are kept within 0.9 second of one another through the periodic introduction of a LEAP SECOND into the UTC timescale, thus causing it to differ from INTERNATIONAL ATOMIC TIME (TAI) by an integral number of seconds.

Copenhagen Observatory One of the world's oldest astronomical institutions, established in 1642 to provide

CONSTELLATIONS (CONTINUED)

Constellation	Name	Genitive	Abbr.	Area (sq deg)	First-magnitude stars
Lacerta	The Lizard	Lacertae	Lac	201	
*Leo	The Lion	Leonis	Leo	947 Z	Regulus
Leo Minor	The Little Lion	Leonis Minoris	LMi	232	
*Lepus	The Hare	Leporis	Lep	290	
*Libra	The Balance	Librae	Lib	538 Z	
*Lupus	The Wolf	Lupi	Lup	334	
Lynx	The Lynx	Lyncis	Lyn	545	
*Lyra	The Lyre	Lyrae	Lyr	286	Vega
Mensa	Table Mountain	Mensae	Men	153	
Microscopium	The Microscope	Microscopii	Mic	210	
Monoceros	The Unicorn	Monocerotis	Mon	482	
Musca	The Fly	Muscae	Mus	138	
Norma	The Set Square	Normae	Nor	165	
Octans	The Octant	Octantis	Oct	291	
*Ophiuchus	The Serpent Bearer	Ophiuchi	Oph	948	
*Orion	Orion	Orionis	Ori	594	Rigel, Betelgeuse
Pavo	The Peacock	Pavonis	Pav	378	
*Pegasus	The Winged Horse	Pegasi	Peg	1121	
*Perseus	Perseus	Persei	Per	615	
Phoenix	The Phoenix	Phoenicis	Phe	469	
Pictor	The Painter	Pictoris	Pic	247	
*Pisces	The Fishes	Piscium	Psc	889 Z	
*Piscis Austrinus	*The Southern Fish*	Piscis Austrini	PsA	245	Fomalhaut
Puppis	The Stern	Puppis	Pup	673	
Pyxis	The Compass	Pyxidis	Pyx	221	
Reticulum	The Net	Reticuli	Ret	114	
*Sagitta	The Arrow	Sagittae	Sge	80	
*Sagittarius	The Archer	Sagittarii	Sgr	867 Z	
*Scorpius	The Scorpion	Scorpii	Sco	497 Z	Antares
Sculptor	The Sculptor	Sculptoris	Scl	475	
Scutum	The Shield	Scuti	Sct	109	
*Serpens	The Serpent	Serpentis	Ser	637	
Sextans	The Sextant	Sextantis	Sex	314	
*Taurus	The Bull	Tauri	Tau	797 Z	Aldebaran
Telescopium	The Telescope	Telescopii	Tel	252	
*Triangulum	The Triangle	Trianguli	Tri	132	
Triangulum Australe	The Southern Triangle	Trianguli Australis	TrA	110	
Tucana	The Toucan	Tucanae	Tuc	295	
*Ursa Major	The Great Bear	Ursae Majoris	UMa	1280	
*Ursa Minor	The Little Bear	Ursae Minoris	UMi	256	
Vela	The Sails	Velorum	Vel	500	
*Virgo	The Virgin	Virginis	Vir	1294 Z	Spica
Volans	The Flying Fish	Volantis	Vol	141	
Vulpecula	The Fox	Vulpeculae	Vul	268	

** Ptolemy's original 48 constellations*

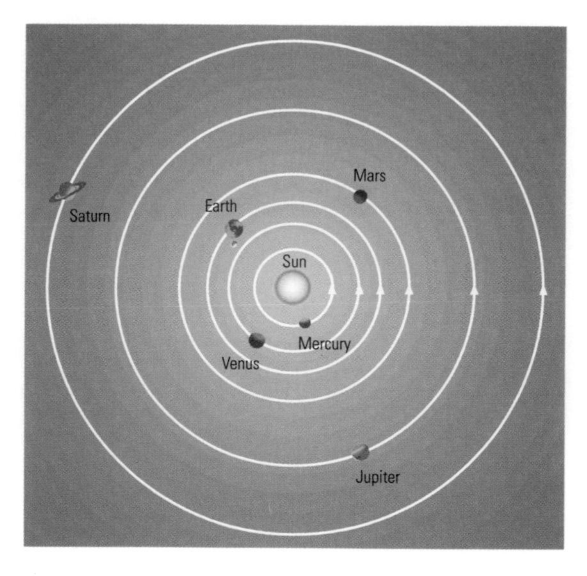

► Copernican system Copernicus' model of the Solar System put the Sun at the centre, with Earth and the other planets in orbit about it. Replacing the earlier geocentric system, this removed Earth from any special position in the Solar System.

accurate stellar positions for maritime navigation. Two hundred years later, it moved from its city centre site to the suburbs of Copenhagen. Today the observatory is operated by the University of Copenhagen as part of the Niels Bohr Institute for Astronomy, Physics and Geophysics. Its researchers use the telescopes of the EUROPEAN SOUTHERN OBSERVATORY as well as the Danish 1.54-m (60-in.) telescope, completed in 1979 at LA SILLA OBSERVATORY.

Copernican system HELIOCENTRIC THEORY of the Solar System advanced by Nicholas COPERNICUS, in which the Earth and the other planets revolve around the Sun, and only the Moon revolves around the Earth; the apparent daily motion of the sky is a consequence of the Earth's axial rotation, and the stars are otherwise motionless because they are extremely distant.

Shortly after 1500, Copernicus became interested in the problems inherent in the established PTOLEMAIC SYSTEM of an Earth-centred cosmology. Why, for instance, if the planets revolved around the Earth, did the retrograde, or reverse, motions of Mars, Jupiter and Saturn synchronize with the terrestrial year? The ancient geocentric cosmology required the operation of a host of eccentric circles and epicycles to explain such motions, whereas a heliocentric system could account for them simply as a consequence of the Earth's orbital motion.

By 1513 Copernicus had developed a workable heliocentric model that addressed various problems in planetary dynamics. It appears to incorporate elements drawn from the work of the Arab astronomers AL-TŪSĪ and Ibn al-Shātir (1304–75/6). The Copernican system seriously threatened the universally accepted physics of ARISTOTLE, and Copernicus feared ridicule from other scholars. However, he was finally prevailed upon to publish his *DE REVOLUTIONIBUS ORBIUM COELESTIUM* just before his death in 1543. The Copernican system was still based on circular rather than elliptical orbits (and was as a consequence worse at predicting planetary positions than the Ptolemaic system it sought to replace), and it placed the Sun at the centre of the cosmos, in a privileged position; but it did offer a logical sequence of planetary revolution periods that increased with distance from the Sun, an important step forward. More importantly, it marked the beginning of the end for geocentricism.

Copernicus, Nicholas (1473–1543) Polish churchman and astronomer who proposed that the planets revolve around a fixed Sun. ('Copernicus' is the Latinized form of his name, by which he is almost always known; the Polish form is Mikołaj Kopernik.) Through his study of planetary motions, he developed a heliocentric theory of the Universe in which the planets' motions in the sky were explained by having them orbit the Sun, as opposed to the geocentric (Earth-centred) model that had been favoured since the days of ARISTOTLE and PTOLEMY. The motion of the sky was now simply a result of the Earth's axial rotation, and, relative to the CELESTIAL SPHERE, the stars remained fixed as the Earth orbited the Sun because they were so distant. An account of his work, *DE REVOLUTIONIBUS ORBIUM COELESTIUM*, was published in 1543.

Copernicus studied canonical law and medicine at the universities of Cracow, Bologna and Padua. At Bologna (1496–1500), he learned astronomy and astrology from Domenico de Novara (1454–1504), using his observations of the 1497 March 9 lunar occultation of Aldebaran to calculate the Moon's diameter. Copernicus returned to Poland in 1503, as a canon in the cathedral chapter of Warmia; he also practised medicine but maintained a lively interest in astronomy, being invited (1514) to reform the JULIAN CALENDAR.

In Copernicus' day it was universally accepted that the Earth was solidly fixed at the centre of the cosmos. Ptolemy had explained planetary motions by having epicycles turn on larger circles around each planet's orbit (*see* PTOLEMAIC SYSTEM). Copernicus explored the consequences of fixing the Sun in the centre of the planetary system, with the Earth and the other planets orbiting it. Being a classical scholar rather than a self-conscious innovator, he began to examine the ancient Greek writers to see whether precedents for a heliocentric system existed, and found several, most notably in the writings of Heraclides of Pontus (388–315 BC) and ARISTARCHUS of Samos. Precisely what motivated Copernicus' radical departure from traditional astronomy is not known, but it is likely that he was also influenced by works that were critical of the Ptolemaic system, including the *Epitome of Ptolemy's Almagest* by REGIOMONTANUS and *Disputations Against Divinatory Astrology* by Giovanni della Mirandola (1463–94). While in Cracow, he may have encountered the writings of AL-TŪSĪ.

In the COPERNICAN SYSTEM, the epicycles of the superior planets (Mars, Jupiter and Saturn) could be discarded because their motions were explained by the effects of the Earth's orbit around a centralized Sun; for the inferior planets (Mercury and Venus) Copernicus centred their epicycles on the Sun instead of on separate carrying circles. Mercury, the swiftest planet, was closest to the Sun, and Saturn, the slowest, orbited at the outer bound of the Solar System, the other planets falling into place according to their periods of revolution.

In 1514 Copernicus first described his new model of the Solar System in a small tract, the *Commentariolus* ('Little Commentary'), which he distributed to only a few colleagues. The heliocentric theory was set forth in greater detail by his student RHAETICUS in the work *Narratio prima*, published in 1540/41. Copernicus' famous book *De revolutionibus orbium coelestium*, considered to be the definitive statement of his system of planetary motions, did not appear until the year of his death. It was Rhaeticus who had finally persuaded him to publish. Copernicus' reticence had had nothing to do with fear of persecution by the Catholic Church: as an ecclesiastical lawyer, he knew that the Church had no dogmatic rulings on scientific matters. What he had feared was academic ridicule in Europe's universities for seeming to contradict common sense.

In *De revolutionibus*, Copernicus refuted the ancient arguments for the immobility of the Earth, citing the advantages of the new Sun-centred model, which correctly ordered the planets by the rate at which they appeared to move through the heavens, and explaining the phenomenon of RETROGRADE MOTION. He correctly explained that the motions of the stars that would be produced by a moving Earth were not observable simply because the stars were so far away – 'so vast, without any question, is the divine handiwork of the Almighty Creator'.

Following the publication of *De revolutionibus*, most astronomers considered the Copernican system as merely a hypothetical scheme – a means of predicting planetary positions, lacking any basis in physical reality and impossible to confirm by astronomical observations. Most astronomers continued to follow Aristotle's physics, in which the Earth was viewed as a perfect, immovable body rather than a transient, moving entity, while the Ptolemaic system actually gave more accurate planetary positions than Copernicus' original scheme. Not until the discoveries of Johannes KEPLER and GALILEO did the Copernican system begin to make physical as well as geometrical sense. As a consequence of Galileo's writings, *De revolutionibus* was in 1616 placed on the *Index* of prohibited books by the Catholic Church. The book was not actually banned, however, but small alterations were introduced to make the Copernican system appear entirely hypothetical. But by then Copernicus' book was already being superseded by Kepler's *Astronomia nova* (1609) and the *Rudolphine Tables* (1627), which provided the basis for explaining the irregularities in the motions of the planets that had not been satisfactorily accounted for in Copernican system.

Copernicus Large lunar crater (10°N 20°W), known for its complex system of EJECTA and bright RAYS. Situated on the north shore of Mare Nubium, Copernicus dominates the Moon's north-west quadrant. The crater's lava-flooded floor, 92 km (57 mi) wide with multiple central peaks, lies nearly 4 km (2.5 mi) below its highly detailed walls. Dominating the inner ramparts of Copernicus are massive arc-shaped landslides, which formed by collapse and subsidence of the debris left over from the violent impact that created the main crater. A bright, broad (30 km/20 mi) blanket of ejecta surrounds the polygonal walls. Beyond this ring of bright material, Copernicus' majestic rays, best seen at full moon, radiate for hundreds of kilometres. These features testify to the relative 'youth' of Copernicus, which is 800 million years old. Numerous chains of craterlets curve outwards from the main crater in every direction.

Twin mountain ridges, running roughly east–west and separated by a spacious valley, divide the floor of Copernicus in half. The north group of mountains is composed of three major peaks of modest altitudes, the highest peak attaining 750 m (2400 ft). The south ridge, with a huge pyramidal mountain at its centre, is longer than the north ridge.

Copernicus (OAO-3) US astronomy satellite launched in 1972 August to study stellar ultraviolet and X-rays. It observed CYGNUS X-1, a BLACK HOLE candidate. It ceased operations in 1981.

Cor Caroli The star α Canum Venaticorum, visual mag. 2.89 (but slightly variable), distance 110 l.y. It is of spectral type A0p, with unusually prominent lines of silicon and europium. Cor Caroli is the prototype of the class of ALPHA^2 CANUM VENATICORUM STARS; like all stars in this class, it varies slightly as it rotates, due to surface patches of different composition brought about by strong localized magnetic fields. Small telescopes show it to be a double star, with an unrelated companion of mag. 5.61. Cor Caroli is Latin for 'Charles' heart', a name bestowed by British royalists to commemorate the beheaded King Charles I of England.

Cordelia One of the small inner satellites of URANUS, discovered in 1986 by the VOYAGER 2 imaging team. Cordelia is *c.*26 km (*c.*16 mi) in size. It takes 0.335 days to circuit the planet, at a distance of 49,800 km (30,900 mi) from its centre, in a near-circular, near-equatorial orbit. With OPHELIA it acts as a SHEPHERD MOON to Uranus' Epsilon Ring, Cordelia being just inside the orbit of that ring.

Córdoba Durchmusterung (CD) Catalogue of 614,000 southern stars to 10th magnitude, initiated by John Macon Thome (1843–1908) at Córdoba Observatory, Argentina, and covering the sky from dec. −22° to −90°. The four main volumes were completed by 1914, and a fifth was added in 1932 by Charles PERRINE. Together with the *BONNER DURCHMUSTERUNG* and *CAPE PHOTOGRAPHIC DURCHMUSTERUNG*, which it complements, the CD represents the last major survey made by visual observations, before photographic surveys became the norm.

Córdoba Observatory National observatory of Argentina, founded in 1871, some 650 km (400 mi) north-west of Buenos Aires; its most celebrated director was Benjamin GOULD, towards the end of the 19th century. The observatory is best known for its determinations of stellar positions published in the *CÓRDOBA DURCHMUSTERUNG* and the *Astrographic Catalogue* (*see* CARTE DU CIEL). It has a 1.5-m (60-in.) reflector, and its astronomers also use major multinational facilities such as the GEMINI OBSERVATORY.

core (planetary) Dense central region of a planet, having a composition distinct from the outer layers (MANTLE and CRUST). A planetary core forms when heavier components sink to the planet's centre during DIFFERENTIATION. Earth's core, which makes up nearly a third of the planet's mass, is composed mostly of iron alloyed with nickel and lighter elements, of which sulphur is probably the most abundant. The core is mostly liquid, but with a small solid inner core. Convective motions in the liquid iron generate Earth's magnetic field. Most of the core's properties are inferred by SEISMOLOGY. Mercury, Venus and Mars have iron cores as deduced from their densities and moments of inertia; the Moon is depleted in iron but may have a small core. The larger icy satellites of the outer planets probably possess rocky cores beneath icy mantles. The densities of the GAS GIANT planets imply that they possess cores of rock and icy material with masses of order 10 times Earth's mass; these cores probably formed by ACCRETION before the planets acquired their massive gaseous envelopes. *See also* COSMOGONY

core (stellar) Innermost region of a star; it is the region in which HYDROGEN BURNING takes place when the star is on the MAIN SEQUENCE. The Sun's core is believed to extend out to a quarter of the solar radius. Main-sequence stars of about 1.5 solar masses have radiative cores, in which the most important method of energy transport is by RADIATIVE DIFFUSION. Stars above this mass limit have convective cores, in which energy transport is mainly by CONVECTION. *See also* STELLAR INTERIOR

Coriolis effect To an Earthbound observer, anything that moves freely across the globe, such as an artillery shell or the wind, appears to curve slightly – to its right in the northern hemisphere and to its left in the southern hemisphere. This is the Coriolis effect. In 1835 Gaspard de

▼ **Copernicus** One of the most prominent rayed craters on the Moon, Copernicus has a diameter of 92 km (57 mi) and shows terraced walls and central peaks. The rays and ejecta blanket surrounding the crater are well shown in this Hubble Space Telescope image.

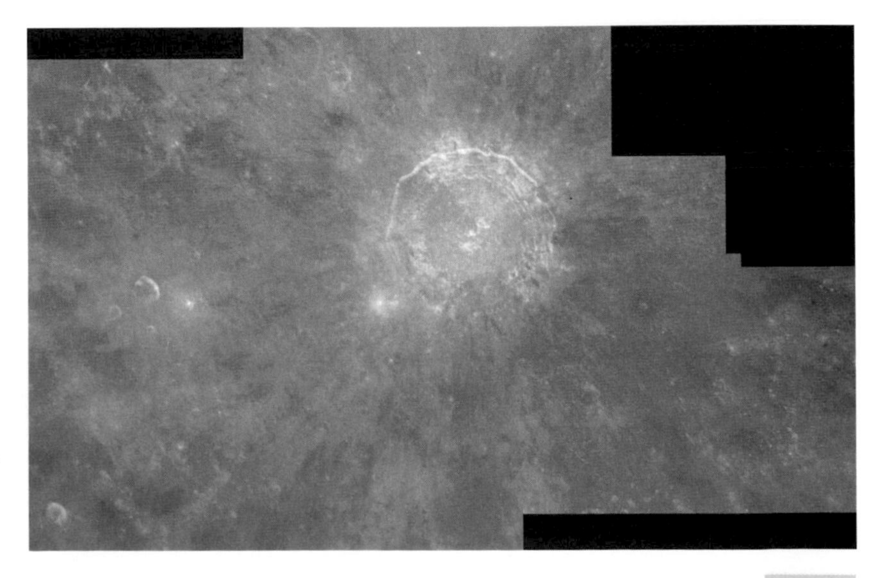

▶ **corona** (1) Loops in the Sun's inner corona imaged by the TRACE spacecraft in 1998. Like many other features in the solar corona, these are shaped by magnetic fields.

Coriolis (1792–1843) first explained that this apparent curvature was not caused by some mysterious force. It simply shows that the observer is on a rotating frame of reference, namely the spinning Earth. The Coriolis effect accounts for the direction of circulation of air around cyclones.

corona (1) Outermost region of the solar atmosphere, above the CHROMOSPHERE and TRANSITION REGION. The corona (from Latin 'crown') becomes visible as a white halo surrounding the Sun at a TOTAL SOLAR ECLIPSE, and can be observed at other times using a special instrument called a CORONAGRAPH. Coronal material is heated to temperatures of millions of Kelvin, and consequently emits energy at extreme-ultraviolet and X-ray wavelengths. Observations at these wavelengths show the corona over the whole of the Sun's face, with the cooler, underlying PHOTOSPHERE appearing dark.

The visible, white light corona may be divided into the inner K CORONA, within two or three solar radii of the photosphere, and the outer F CORONA. The K corona shines by visible sunlight that is weakly scattered by free electrons. It is about one-millionth as intense as the photosphere. Observations of the K corona indicate there are up to ten million billion (10^{12}) electrons per cubic metre at the base of the corona. The F corona is caused by sunlight scattered from solid dust particles. The shape of the inner K corona reflects constraint of electrons by MAGNETIC FIELDS. At the base of coronal streamers, the electrons are concentrated within magnetized loops rooted in the photosphere. Farther out, the streamers narrow into long, radial stalks that stretch tens of millions of kilometres into space; some may reach as far as halfway to Earth.

At times of reduced activity, near the minimum of the SOLAR CYCLE, coronal streamers are located mainly near the solar equator. The coronal streamers are the source of the low-speed SOLAR WIND. Near maximum in the activity cycle, the corona becomes crowded with streamers both near the equator and the Sun's poles.

The corona's high temperatures are capable of stripping iron atoms of 13 of their electrons, producing an ionized form whose emission in green light at wavelength 530.3 nm was first detected at the total eclipse of 1869 August 7. Initially taken – mistakenly – to be the signature of a new element ('coronium'), FeXIV emission provides a useful tracer for coronal structure. Further evidence for the corona's high temperatures comes from observation of its radio emission. Gas in the corona is completely ionized, and is therefore in the form of a PLASMA.

Recent observations of the corona from spacecraft suggest that at least some of the heat of the corona is related to the Sun's ever-changing magnetic fields. These observations have been made using a soft X-ray telescope on the YOHKOH spacecraft, and with ultraviolet and extreme-ultraviolet telescopes aboard SOHO and TRACE.

Images of the Sun at extreme-ultraviolet and X-ray wavelengths reveal dark regions, called **coronal holes**, and bright regions, known as **coronal loops**. Coronal holes are characterized by open magnetic fields that allow hot material to escape. At least some of the high-speed component of the solar wind flows out of the coronal holes nearly always present at the Sun's poles. Coronal holes extending towards the solar equator become common close to the minimum of the solar cycle.

The hottest and densest material in the low corona is located in thin, bright magnetized loops that shape, mould and constrain the million-degree plasma. These coronal loops have strong magnetic fields that link regions of opposite polarity in the underlying photosphere. Coronal heating is usually greatest where the magnetic fields are strongest. High-resolution TRACE images have demonstrated that the coronal loops are heated in their legs, suggesting the injection of hot material from somewhere near the loop footpoints in the photosphere below.

Coronal loops can rise from inside the Sun, sink back down into it, or expand out into space, constantly changing and causing the corona to vary in brightness and structure. These changing magnetic loops can heat the corona by coming together and releasing stored magnetic energy when they make contact. This method of coronal heating is termed MAGNETIC RECONNECTION. It can occur when newly emerging magnetic fields rise through the photosphere to encounter pre-existing ones in the corona, or when internal motions force existing coronal loops together.

There are about 50,000 small magnetic loops in the low corona; these come and go every 40 hours or so, forming an ever-changing magnetic carpet. SOHO observations indicate that energy is frequently released when these loops interact and reconnect, providing another possible source of coronal heating. Bursts of powerful energy released during magnetic reconnection can explain sudden, brief intense explosions on the Sun called FLARES. Numerous low-level flares, called microflares or nanoflares, might contribute to coronal heating. *See also* E CORONA; T CORONA

▼ **corona** (2) A radar image of Artemis Corona obtained by the Magellan orbiter. Artemis Corona is the largest such feature on Venus.

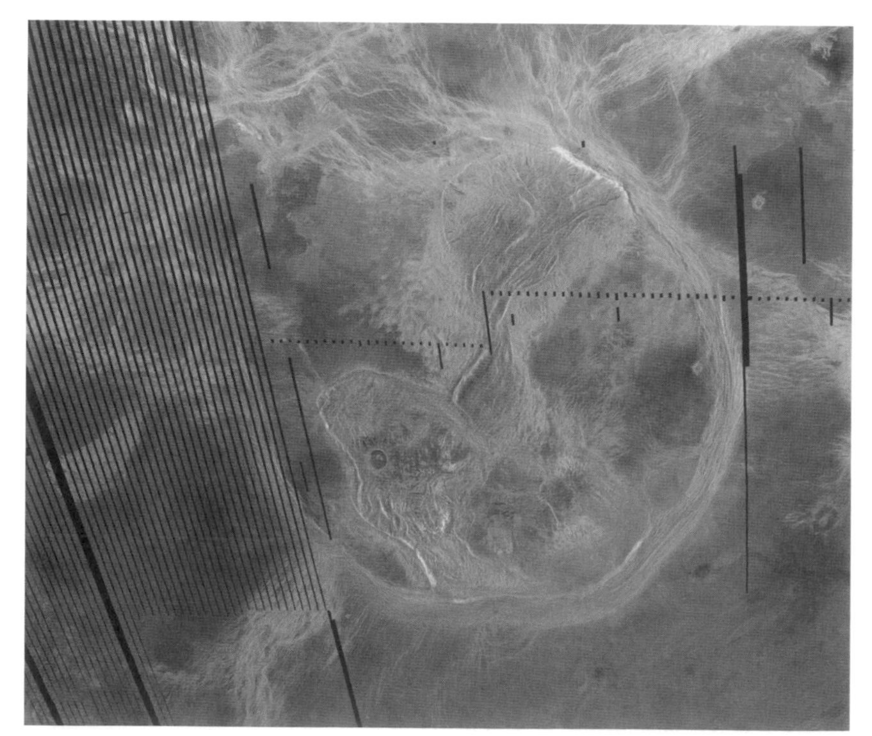

corona (2) Ovoid-shaped feature, of which several hundred are known on VENUS. Venusian coronae, typically 200 to 600 km (120–370 mi) across, are the foci of concentric and radial TECTONIC deformation and associated, probably basaltic, volcanism. They result from the uplift, followed by relaxation, of the LITHOSPHERE by thermally buoyant PLUMES. Uranus' satellite MIRANDA has three coronae, which are polygonal (rather than ovoid), concentrically banded features 200–300 km (120–190 mi) in diameter; this is comparable with Miranda's 235 km (146 mi) radius. Miranda's coronae are believed to be produced by a combination of tectonic disruption and CRYOVOLCANISM.

Corona Australis *See* feature article

Corona Borealis *See* feature article

coronagraph Instrument, used in conjunction with a telescope, that has an occulting disk to block out the brightness of the Sun's PHOTOSPHERE, providing an artificial eclipse, with additional precautions for removing all traces of stray light. It was invented by Bernard LYOT. A coronagraph is used to observe the faint solar CORONA in white light, or in all the colours combined, at any time. The best images, with the finest detail, are obtained from coronagraphs placed on satellites above the Earth's atmosphere, whose scattered light otherwise degrades the image. There are two coronagraph instruments aboard the SOHO spacecraft, for example.

coronal hole *See* CORONA

coronal mass ejection (CME) Transient ejection into interplanetary space of plasma and magnetic fields from the solar CORONA, seen in sequential images taken with a CORONAGRAPH. A coronal mass ejection expands away from the Sun at supersonic speeds up to 1200 km/s (750 mi/s), becoming larger than the Sun in a few hours and removing up to 50 million tonnes of material. CMEs are often associated with eruptive PROMINENCES in the chromosphere, and sometimes with solar FLARES in the lower corona.

Coronal mass ejections are more frequent at the maximum of the SOLAR CYCLE, when they occur at a rate of about 3.5 events per day. They can, however, occur at any time in the solar cycle, and unlike coronal streamers, they are not confined to equatorial latitudes at solar minimum.

CMEs are most readily seen when directed perpendicular to the line of sight, expanding outwards from the solar limb; Earth-directed events are seen as diffuse, expanding rings described as halo coronal mass ejections. Earth-directed CMEs can cause intense MAGNETIC STORMS, and trigger enhanced auroral activity. Coronal mass ejections produce intense shock waves in the SOLAR WIND, and accelerate vast quantities of energetic particles. A large CME may release as much as 10^{25} Joule of energy, comparable to that in a solar flare. Like flares, CMEs are believed to result from release of stored magnetic stress. In view of their influence on the near-Earth space environment, CMEs are monitored by national centres and defence agencies.

CORONA AUSTRALIS (GEN. CORONAE AUSTRALIS, ABBR. CRA)

Small southern constellation consisting of an arc of stars under the feet of Sagittarius, known since ancient Greek times when it was visualized as a crown or wreath. Its brightest stars, α and β CrA, are only of mag. 4.1. γ CrA is a tight 5th-magnitude binary requiring apertures over 100 mm (4 in.) to split.

CORONA BOREALIS (GEN. CORONAE BOREALIS, ABBR. CRB)

Distinctive northern constellation between Boötes and Hercules, representing the crown worn by Ariadne when she married Dionysus, who later threw it into the sky in celebration. Its brightest star is ALPHEKKA, mag. 2.22. ν CrB is an easy binocular double, mags. 5.2 and 5.4, while ζ CrB, mags. 5.0 and 6.0, requires a small telescope to divide. In the bowl of the crown is R CrB, a yellow supergiant (spectral type F8 or G0) that can suddenly fade by up to 9 magnitudes. when carbon accumulates in its atmosphere (*see* R CORONAE BOREALIS STAR). T CrB is a recurrent nova known as the BLAZE STAR.

coronal streamer *See* CORONA

coronium Chemical element initially hypothesized to explain an unidentified green emission line observed in the solar CORONA (as well as others found later). The coronium lines were eventually found to be FORBIDDEN LINES of highly ionized iron (Fe XIV) and other common elements.

correlation function Mathematical way of statistically quantifying the distribution of galaxies in cosmology. Two-point and three-point correlation functions are normally calculated using the observed galaxy distributions, and then compared with theoretical distributions derived from cosmological models. Since correlation functions are statistical in nature, if the models give the same correlation function as the actual data, then the model is successful, even though the exact locations of the individual galaxies and/or clusters are not precisely what we see. *See also* BIG BANG THEORY

Corvus *See* feature article, page 100

Cos B European GAMMA-RAY ASTRONOMY satellite; it was launched in 1975 August. It remained in operation until 1982 April, detecting 25 gamma-ray sources during its lifetime.

cosmic abundance Relative amounts of elements produced during the first three minutes of the BIG BANG. The Big Bang model predicted how the amounts of various elements were created under the conditions present in the early universe. Primarily hydrogen, helium and

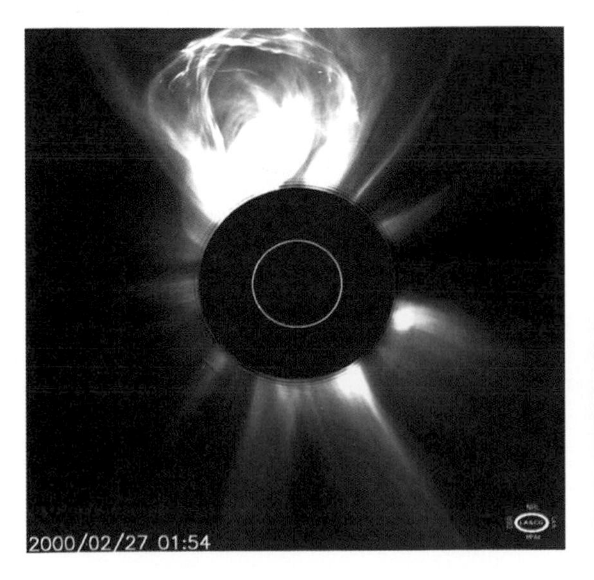

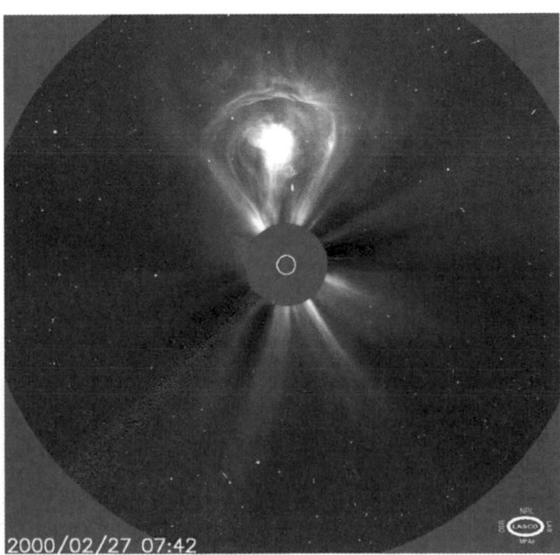

◄ **coronal mass ejection** SOHO images of a huge coronal mass ejection on 2000 February 27. The event was recorded using the LASCO C2 (left) and C3 (right) coronagraphs.

C

CORVUS (GEN. CORVI, ABBR. CRV)

Modest constellation south of Virgo whose most distinctive feature is a quadrilateral of stars similar to the Keystone of Hercules. It represents a crow, which in Greek legend was sent by Apollo to carry a cup (the neighbouring constellation Crater) to fetch water. δ Crv (Algorab, abbreviated from the Arabic meaning 'raven's wing') is a wide and unequal double for small telescopes, mags. 2.9 and 9.2. Corvus contains the ANTENNAE, a famous pair of colliding galaxies.

BRIGHTEST STARS

	Name	RA h m	dec. ° ′	Visual mag.	Absolute mag.	Spectral type	Distance (l.y.)
γ	Gienah	12 16	−17 33	2.58	−0.94	B8	165
β		12 34	−23 24	2.65	−0.51	G5	140
δ	Algorab	12 30	−16 31	2.94	0.79	B9.5	88
ε		12 10	−22 37	3.02	−1.82	K2	303

small amounts of lithium and beryllium were produced by NUCLEOSYNTHESIS reactions under high temperatures and densities in the early universe. After about 15 minutes, the temperature of the universe had dropped and no further nucleosynthesis could occur until stars formed. Thus all of the heavier elements must have been synthesized in stars and during SUPERNOVA explosions. Calculations indicate that only one ^{3}He or ^{4}He atom is formed for every 10,000 hydrogen atoms, and one lithium or beryllium atom for every hundred million hydrogen atoms. All of the remaining elements in the Universe were thus created in the centres of stars or during supernova explosions.

Cosmic Background Explorer (COBE) NASA satellite launched in 1989 November to study COSMIC MICROWAVE BACKGROUND radiation – the leftover radiation from the BIG BANG. COBE made precise measurements of the diffuse radiation between 1 μm and 1 cm over the whole celestial sphere. As well as measuring the spectrum of this radiation, COBE also studied fluctuations in its temperature distribution, and the spectrum and angular distribution of diffuse infrared background radiation.

The spacecraft's instruments were cooled by liquid helium and protected from solar heating by a conical sun shade. The satellite rotated at 1 rpm and performed a complete scan of the celestial sphere every six months.

COBE first determined that the cosmic microwave background has a black body temperature of 2.73 K. Its overall view of the sky showed that one-half of the sky is slightly warmer than the other half as the result of the Solar System's motion through space. Then, in 1992, scientists announced that COBE had found 'ripples' in the background radiation, regions that were no more than 30-millionths of a degree warmer or cooler than the average temperature. These minute variations were attributed to gaseous structures several hundred million light-years across that existed in the Universe around the time the first galaxies were forming.

COBE's scientific operations were terminated at the end of 1993. Five years later, astronomers released the first maps to show a background infrared glow across the sky. Based on data from the Diffuse Infrared Background Experiment aboard COBE, they revealed the 'fossil radiation' from dust that was heated by hidden stars in the very early Universe.

► **Cosmic Background Explorer** (COBE) This all-sky temperature map derived from COBE data shows radiation in the direction towards which Earth is moving to be blue-shifted and therefore hotter. Conversely, in the opposite direction, the radiation appears cooler due to redshift. The data suggest that our Local Group of galaxies moves at 600 km/s (375 mi/s) relative to the background radiation.

cosmic dust *See* MICROMETEORITE

cosmic microwave background (CMB) Remnant radiation from the creation of the Universe. Early cosmologists predicted theoretically that the Universe originally began as a singularity that expanded into a small, hot 'soup' of radiation and elementary particles. As it expanded, the temperature dropped and the nuclei of atoms became stable. Finally, after the first 300,000 years, the Universe was cold enough for the radiation to decouple from the matter. As the Universe continued to expand and cool, the radiation and matter evolved independently. The radiation cooled (redshifted) as the Universe expanded but remained homogeneous and isotropic on a large scale; the matter clumped owing to gravitational interactions on smaller scales. Predictions were made that at the present stage of the Universe, the temperature of this background radiation field should have cooled to around 5 K.

In 1965 Arno PENZIAS and Robert Wilson attempted to rid the large Bell Telephone Laboratory microwave antenna of background noise. This noise appeared to be constant and, if it originated outside the antenna, was extraterrestrial and isotropic and represented a background temperature of 3 K. They realized that they were indeed observing the cosmic microwave background radiation that is a remnant of the early Universe.

Astronomers realized that this cosmic microwave background held clues to the physical conditions early in the evolution of the Universe. NASA launched a satellite, the COSMIC BACKGROUND EXPLORER (COBE), to investigate this radiation in detail. COBE mapped the radiation in frequency as well as spatially with unprecedented accuracy and confirmed the background temperature at 2.73 K. It also mapped the temperature variations in all directions. This mapping provided evidence for the inflationary scenario of the early Universe.

cosmic rays Charged particles (protons and the nuclei of heavier atoms) that arrive at the top of the atmosphere and come from cosmic space. Their energies extend up to 10^{20} electron volts (eV) per particle and thus represent the highest individual particle energies known. The particles that come from space and hit the top of the atmosphere are called primary cosmic rays, and the particles that hit the ground are called secondary cosmic rays. Initially it was thought that cosmic rays comprised some form of ultra-penetrating gamma radiation, hence the term cosmic rays, but later work showed that particles, mainly protons, were responsible. Telescopes have been operated in many different environments, from satellites to the bottom of deep mines; directional telescopes in mines have shown upward-moving cosmic rays caused by NEUTRINOS that have penetrated the whole Earth before interacting to cause detectable secondary rays. Special ground-based telescopes can detect the extensive air showers that occur when a primary particle hits an oxygen or nitrogen nucleus in the upper levels of the atmosphere. The initial secondary particles are pions, but these decay to the more stable muons. At ground level the showers contain mainly electrons with perhaps 5% muons.

The discovery of cosmic rays is credited to the Austrian physicist Victor HESS, who, in 1912, carried electrometers aloft in a balloon in an attempt to discover why it had proved impossible to eliminate completely a small residual background reading in the electrometers at ground level. Hess found that, after first falling, the reading started to increase as the balloon ascended and he made the remarkable claim that there was a need 'to have recourse to a new hypothesis; either invoking the assumption of the presence at great altitudes of unknown matter or the assumption of an extraterrestrial source of penetrating radiation'. After arguments lasting some years the extraterrestrial origin idea finally won; Hess was awarded the Nobel Prize in 1936.

Since cosmic rays are charged particles, they are deflected by magnetic fields in the Galaxy and beyond, so it is sometimes difficult to be certain about their origins.

However, cosmic rays and gamma rays (*see* GAMMA-RAY ASTRONOMY) are both high-energy signatures, so gamma rays can help locate the origins of cosmic ray sources. It has been suggested that CYGNUS X-3 provides a significant fraction of the cosmic rays in the Galaxy, but the galactic centre also produces cosmic rays, as may other objects such as supernovae. Some of the highest-energy cosmic rays come from the VIRGO CLUSTER; it contains the galaxy M87, which has a JET showing that there may be a supermassive BLACK HOLE at its centre. Low-energy cosmic rays are emitted by the Sun. One of the unusual features of these is the presence of comparatively large fluxes of the elements lithium, beryllium and boron, which are identified as fragments of heavier cosmic ray nuclei that have struck nuclei of the interstellar medium. The presence of a very small fraction of radioactive beryllium-10 (^{10}Be), taken with other data, leads to an estimated lifetime of low-energy cosmic rays of at least 20×10^6 years. Low-energy cosmic rays cause the production of radioactive carbon (^{14}C) and beryllium (^{10}Be) in the atmosphere, which leads to the possibility of studying intensity variations over 10^3–10^7 year periods from studies of ^{14}C in tree rings and ^{10}Be in deep-sea sediments.

cosmic year Time taken for the Sun to complete one revolution about the centre of the Milky Way galaxy – about 220 million Earth years. The entire Galaxy is rotating, gradually changing shape as it does so, and the Sun is located some 28,000 l.y. from the galactic centre.

cosmochemistry Study of the chemical reactions occurring in space and of the products of those reactions. The atoms forming most elements are synthesized inside stars (*see* ASTROCHEMISTRY) and then a small proportion is returned to interstellar space during SUPERNOVA explosions. The exceptions to this are principally hydrogen and helium-4, and to a much lesser extent deuterium (hydrogen-2), helium-3 and lithium-7, which were created in the first few seconds of the Big Bang. The roughly 25% (by mass) of the matter in the Universe that is helium-4, for example, resulted from reactions such as

$$n + e^+ \rightarrow p + \nu_e$$

where n is neutron, e^+ is a positron (positive electron), p is a proton and ν_e is an electron neutrino. Such reactions were in equilibrium during the early stages of the Big Bang and left a balance of 1 neutron to 6 protons when they ceased. The surviving neutrons (for the neutron is an unstable particle) eventually combined with protons to produce the helium nuclei. Hydrogen and helium form about 98% by mass of all the material in the interstellar medium, with carbon, nitrogen, oxygen, neon, magnesium, silicon, sulphur, argon and iron making up most of the balance. Many other elements have been detected, and we may expect all those found on Earth, plus some of those that do not occur naturally, such as technetium and plutonium, to occur in space. However, the molecules that can currently be observed are mostly combinations of hydrogen, carbon, nitrogen and/or oxygen (*see* INTERSTELLAR MOLECULES). One significant reason for this is that it is only molecules in the form of gases that can be identified with certainty (from their radio spectra). The composition of solid materials, such as interstellar dust particles (see INTERSTELLAR MATTER), can be only roughly characterized.

Once formed, the atoms in interstellar space can take part in chemical reactions, but the conditions differ so much from that of a terrestrial chemistry laboratory, that the reactions and their products are often very unusual by 'normal' chemistry's standards.

Some molecules, such as TiO, CN and C_2, are robust enough to exist in the outer atmospheres of cool stars. However, most molecules will be dissociated by the ultraviolet radiation in starlight if they float freely in space. The interiors of GIANT MOLECULAR CLOUDS (GMCs) are thus the sites where the vast majority of interstellar molecules are to be found. There they are sheltered by dust from the stellar ultraviolet radiation.

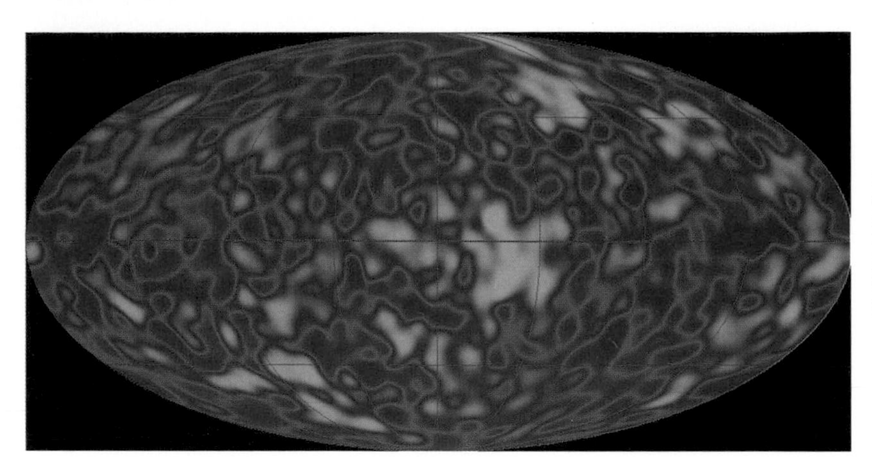

▲ **cosmic microwave background** This all-sky temperature map from COBE, from which the effects due to the motion of the Earth have been subtracted, shows 'ripples' in the cosmic microwave background. Although the extent of these variations is tiny – 30 millionths of a degree – they are believed to have been sufficient to trigger formation of the first galaxies after the Big Bang.

Even though GMCs are much denser than the average for the interstellar medium, they are still very hard vacuums by terrestrial laboratory standards. Thus interactions usually take place between two atoms in isolation, and this can make it difficult for stable chemical reactions to occur. For example, if two neutral hydrogen atoms encounter each other and join to form a molecule, the molecule will be highly excited because of the kinetic energy possessed by the atoms. It will thus be unstable, and under terrestrial conditions would lose its excess energy to a third particle to form the stable hydrogen molecule. That third particle is unlikely to be nearby within a GMC, and so the molecule will rapidly dissociate back to the two original hydrogen atoms. A possible way out of this would be for the excited hydrogen molecule to radiate away its excess energy, but most of the lower energy transitions of the hydrogen molecule are forbidden, and so this is unlikely to occur.

Stable reactions can occur when one of the particles is ionized, perhaps as a result of a collision with a COSMIC RAY particle. Thus water can be formed from an ionized hydrogen molecule in combination with oxygen through the following set of reactions:

$$H_2^+ + H_2 \rightarrow H_3^+ + H$$
$$H_3^+ + O \rightarrow OH^+ + H_2$$
$$OH^+ + H_2 \rightarrow H_2O^+ + H$$
$$H_2O^+ + H_2 \rightarrow H_3O^+ + H$$
$$H_3O^+ + e^- \rightarrow H_2O + H$$

It is thought that many molecules are formed only on the surfaces of the dust particles within the GMC. Individual atoms are adsorbed on to the dust particle and are then in such close proximity that relatively normal reactions can occur. The daughter molecule may be ejected back into space by the energy released during the reaction, or perhaps a cosmic ray particle passing through the dust particle may heat it sufficiently to evaporate all the molecules that have accumulated on its surface.

cosmogony Scientific discipline dealing with the formation of the Solar System or planetary systems in general. Cosmogony as a science began with the NEBULAR HYPOTHESIS of Immanuel KANT and Pierre Simon LAPLACE in the late 18th century; the PLANETESIMAL hypothesis was developed a century later by T.C. CHAMBERLIN and F.R. MOULTON. Although some features of their hypotheses have been incorporated into modern theories, most current research is based on the work of V.S. SAFRONOV. Until the latter half of the 20th century, cosmogony was primarily a field for speculation, with little possibility for testing theories. That situation has changed significantly for several reasons: computers have been developed that allow detailed numerical simulations of proposed processes; a wealth of data about bodies in our own Solar System has been provided by spacecraft; and EXTRASOLAR PLANETS have been discovered orbiting nearby stars. Cosmogony is now an active field of research conducted at a high level of detail and sophistication.

There is currently a 'standard model' that gives a general explanation of the formation of the Solar System. Many details remain to be worked out, and some parts of the model are likely to be proved wrong, but it provides a reference frame for scientific investigation.

It is generally accepted that planetary systems are not rare, but are a common product of star formation. Stars form from clouds of interstellar gas, mostly hydrogen and helium, containing a small fraction of heavier elements, mostly in the form of dust grains. If the gas reaches a critical density it begins to collapse under its self-gravity. This collapse may be initiated by a number of phenomena: cooling of the gas; loss of a supporting magnetic field; or an external perturbation that compresses the gas. Most star formation is observed to occur in clusters, as portions of a large cloud, with a total mass perhaps thousands of times that of the Sun, collapse over an interval of millions of years. The most massive stars have short lifetimes and explode as SUPERNOVAE within the cloud, mixing their material back into the medium and triggering further collapses by their shock waves. As a stellar-mass portion of the cloud collapses, it loses energy by infrared radiation from dust grains and molecules. ANGULAR MOMENTUM is conserved, and any initial rotation increases in speed as the cloud contracts. Clouds with larger amounts of angular momentum may produce a BINARY STAR. For lesser amounts, turbulence and magnetic coupling redistribute angular momentum, allowing most of the mass to fall towards the centre, while a portion remains in a rotating PROTOPLANETARY DISK orbiting the PROTOSTAR.

The protoplanetary disk is heated by the kinetic energy of its infall. Its inner part becomes hot enough to vaporize the dust, and the gas nearest the protostar is ionized. In the outer part, grains of interstellar dust may survive; as the disk cools, heavier elements recondense in the inner region. Dust grains collide and stick together to produce centimetre-sized aggregate particles, which settle towards the central plane of the disk, where they form a thin, dense particle layer. Particles in this layer are subject to drag exerted by the gas, which causes bodies of different sizes to move at different rates; they may also collide due to turbulence. They continue to grow, forming PLANETESIMALS of kilometre size. At this stage they are less affected by the drag of the nebular gas, and gravitational interactions become important. Their perturbations stir up eccentricities and inclinations, causing the orbits of the planetesimals to cross, which results in frequent collisions. These collisions may produce both growth and fragmentation, but the energy loss in inelastic collisions ensures that growth dominates. By this process of ACCRETION, a small number of large bodies, or PROTOPLANETS, forms. Collisions among them continue for millions of years until the remaining bodies are few and widely separated, yielding planets on stable orbits. The Moon is believed to have formed from debris after a Mars-sized protoplanet collided with the still-growing Earth. The final sweeping up of smaller bodies takes tens of millions of years.

Farther from the protosun, cooler temperatures allow ice to condense, providing more material and allowing growth of larger protoplanets. If these protoplanets reach a critical size, several times Earth's mass, they can capture gas from the protoplanetary disk, growing into giant planets like Jupiter and Saturn. After a few million years the gaseous component of the disk is driven off by activity of the newly formed star as it goes through its T TAURI phase. Still farther out, the growth times of the protoplanets are too long, and the disk may dissipate before they can capture large amounts of gas; planets like Uranus and Neptune result, their makeup being dominated by the accretion of icy cometary planetesimals.

This model for planetary formation provides a plausible qualitative explanation for the major features of the Solar System, although some parts are uncertain and many details remain to be worked out. The degree to which our planetary system is typical or unique is not known. The existence of extrasolar planets in systems quite unlike our own shows that other outcomes are possible. The final form of a planetary system probably depends on both systematic factors, such as the mass and angular momentum of a protostellar cloud, its temperature and composition, and so on, as well as on random events, such as the proximity of other stars and the outcomes of collisions between protoplanets.

cosmological constant Constant in Einstein's FIELD EQUATIONS of GENERAL RELATIVITY, originally added to stabilize the universe. If one writes down the general relativistic equation for the universe, one finds that the universe cannot be static, but must either expand or contract. When Einstein derived his 'universe' equation, astronomers had no idea the Universe was actually in a state of expansion. Therefore, Einstein added another term in the field equations multiplied by a constant:

$$R_{\mu\nu} - \tfrac{1}{2}g_{\mu\nu} + \Lambda g_{\mu\nu} = 8\pi G/c^2 T_{\mu\nu}$$

where Λ is the cosmological constant, $R_{\mu\nu}$ is the Riemann tensor, $g_{\mu\nu}$ is the metric tensor, G is Newton's universal constant of gravitation, c is the speed of light and $T_{\mu\nu}$ is the stress-energy tensor.

This cosmological constant can provide an attractive or repulsive force that operates throughout the entire universe and causes expansion, contraction or stability depending on its value. Shortly after Einstein published the addition to his theory, Edwin HUBBLE announced his observations of galaxies that showed the Universe was indeed in a state of constant expansion. Einstein quickly retracted his cosmological constant as being unnecessary and announced it as his biggest blunder. Even after astrophysicists realized that no cosmological constant was needed, some kept it and gave it a very small value, or a time-dependent value (in the case of inflationary theory). Lately, new observations of universal acceleration combined with the low mass density seen in the Universe indicate that it might be necessary after all. *See also* BIG BANG THEORY

cosmological distance scale Methods and scales with which astronomers determine the distance between galaxies and clusters in the visible Universe. In order to understand our Universe, the distances to galaxies must be determined. In order to measure these tremendous distances, astronomers rely on a variety of methods.

Distances in the Solar System can be measured by radar, and nearby stellar distances can be measured by direct stellar PARALLAX. In order to determine the distances to more distant stars, model-dependent methods like main-sequence fitting and the period–luminosity relationship for CEPHEID VARIABLES must be used. Measuring the distances to nearby galaxies relies upon detection of Cepheid variables or other bright stellar phenomena such as supernovae in those galaxies. These methods are fairly accurate for nearby galaxies, but become more difficult and uncertain for more distant galaxies. Other methods such as the TULLY–FISHER RELATION can be used for distance estimates of spiral galaxies, while distances to elliptical galaxies require assumptions about the average size of giant ellipticals in a cluster. Farther and farther out into the Universe, distance measurement becomes more and more uncertain. Until recently this has prevented us from determining what type of universe we live in.

This series of methods is also called the cosmic distance 'ladder', where each rung represents a method to determine the distance to more distant objects. Each method also depends on the previous one for accurate calibration. The final rung of the cosmological distance scale is the HUBBLE LAW. This relationship is used for objects so far away that we cannot image individual stars, molecular clouds, or even individual supernovae, and it is essential to have a reliable value for the HUBBLE CONSTANT to use it.

cosmological model Any of several theories or ideas that attempt to describe and explain the form and evolution of our UNIVERSE. The original cosmological model held that the Earth was the centre of the Universe, with the Sun, Moon and planets orbiting around us and the stars affixed to the CELESTIAL SPHERE. This geocentric (Earth-centred) model was in favour for over 1000 years after PTOLEMY first wrote it down. This model failed to explain the elliptical shapes of planetary orbits, the westward drift or retrograde motion of the planets, and the phases of Venus. The geocentric model was subsequently replaced by the more powerful heliocentric or 'Sun-centred' model. The heliocentric model was capable of explaining retrograde motion much more easily than its predecessor. Although COPERNICUS was the first to show that the heliocentric model was far better in explaining the observations than the geocentric model, it was not until after GALILEO's time that it became politically acceptable to publish such an idea. The current accepted cosmological model, called the BIG BANG, is based on Einstein's theory of GENERAL RELATIVITY and maintains that at one moment SPACETIME was collapsed into an infinitely small point, and subsequently the entire Universe expanded and continues to expand. Other cosmological models include the anisotropic models, the STEADY-STATE THEORY and BRANS–DICKE THEORY.

cosmological principle Assumption that SPACETIME is homogeneous and isotropic on the largest of scales and that we do not live in a preferred place in the Universe. This sweeping principle allows us to solve Einstein's FIELD EQUATIONS and to build COSMOLOGICAL MODELS that can be tested with observations.

cosmological redshift Shift of spectral lines due to the expansion of the Universe. Galaxies and clusters are located in SPACETIME, which is uniformly expanding and carrying the galaxies along with it; thus the galaxies exhibit a velocity relative to a distant observer. The 'Doppler-type' effect is not a result of the sources' velocity in spacetime, but is caused by the expansion of spacetime itself. Edwin HUBBLE showed in 1929 that the Universe is expanding, and further that the cosmological velocity is linearly related to the distance of the source. *See also* HUBBLE LAW

cosmology Study of the structure of the Universe on the largest scale. Contained within it is COSMOGONY.

Our early view of the Universe was prejudiced by the belief that we occupied a special place within it – at the centre. Only in the 20th century have we realized that the Earth is but a small planet of a dim star, located in the outer suburbs of a typical galaxy. Perhaps the most important astronomical discovery of the early 20th century was HUBBLE'S realization that the dim NEBULAE he observed were in fact enormous systems of thousands of millions of stars lying far outside our Galaxy. Soon after this discovery, astronomers realized that these galaxies were all receding from the Earth. Hubble, along with other astronomers, obtained optical spectra of many galaxies and found that their spectral lines were always shifted towards the red (longer wavelengths). He interpreted these REDSHIFTS as being a universal DOPPLER EFFECT, caused by the expansion of the Universe. Furthermore, the speed he inferred was found to be proportional to the galaxy's distance, a relationship known as the HUBBLE LAW.

It is now known that both the shift and the speed–distance proportionality follow naturally from an overall expansion of the scale size of the Universe. Galaxies are redshifted because the Universe has a different scale size now compared with the size it was when the light was emitted from the galaxies. Nevertheless, time has shown that Hubble was correct in his interpretation that the speed of recession is proportional to distance. Today, the constant of proportionality bears his name (*see* HUBBLE CONSTANT).

Close to the Sun, the distance of galaxies can be determined from the properties of some VARIABLE STARS they contain – such as the Cepheids – or from the size of HII regions. As we move farther out into the Universe, however, these methods become increasingly inaccurate. Eventually, distances can only be estimated by measuring the redshift and relying on the accuracy of the Hubble relation.

Unfortunately, for some of the most distant objects, such as QUASARS, we do not have adequate confirmation that this procedure for determining distances is valid. Some astronomers believe that at least part of the quasar redshifts may originate from unknown 'non-cosmological' causes. Attempts to determine the nature of the reshifts and the expansion rate have occupied much of the available time on large telescopes. Today the question is still unresolved.

However, perhaps the most important cosmological problems that remain at the beginning of the 21st century are to determine the rate at which the universal expansion is taking place (determining the Hubble constant), how it has expanded in the past and how it will continue to behave in the future. To these must be added the question of whether the overall geometry of the Universe is 'closed' or 'open'. In an OPEN UNIVERSE, the total volume of space is infinite, the universe has no boundary and will expand for ever. CLOSED UNIVERSES contain a finite amount of space, may or may not have boundaries and will eventually collapse back on themselves.

Attempts to obtain a grand view of the Universe have led to the construction of cosmological models. A starting point for many cosmologists has been the finding that the Universe appears much the same in all directions (the so-called ISOTROPY) and at all distances (HOMOGENEITY). However, the expansion of the Universe would at first seem to suggest that the overall density of the distribution of galaxies must decrease so that they become more sparsely distributed as time goes on.

Cosmological models have included the STEADY-STATE THEORY of Hermann BONDI, Thomas GOLD and Fred HOYLE in which the Universe is the same not only in all places but also at all times. It therefore had no beginning, will have no end and never changes at all when viewed on the large scale. This theory required matter to be created as the Universe expanded in order that the overall density of galaxies should not decrease. For this reason it is also referred to as the CONTINUOUS CREATION model.

On the other hand, according to supporters of the BIG BANG models originally proposed by George GAMOW, Ralph Alpher (1921–) and Robert Herman (1914–97), the whole Universe was created in a single instant about 20 billion years ago and is presently expanding (the modern consensus value is about 15 billion years). In the future it may continue to expand or possibly collapse back on itself depending on the total amount of matter and energy in it, that is, whether or not the Universe is open or closed. An important cosmological question is the MISSING MASS PROBLEM: the amount of matter we see in the Universe is far smaller (by a factor of about 100) than the amount we infer from the motions of the galaxies.

Definitive observations to discriminate between cosmological models are hard to make. The most informative parts of the Universe are those farthest away. Unfortunately, the objects we observe in such regions are faint and their nature is unknown. It is extremely difficult to tell to what extent quasars, for example, are similar to the nearer – and more familiar – objects. And if we cannot make comparisons, we cannot use them as standards to test cosmological models. It is also not known if our Earth-derived physical laws are applicable in the Universe at large.

Cosmologists have made several important discoveries, including an attempt to determine the Hubble constant. The DECELERATION PARAMETER, which determines whether the Universe will expand for ever or eventually collapse back on itself, has also been estimated. Recent observations of supernovae have shed some light on the values of these important cosmological parameters. They indicate that the Universe is probably accelerating and will never collapse back again.

Another important discovery in cosmology was the COSMIC MICROWAVE BACKGROUND radiation, which provided strong evidence against the steady-state theory. Its discovery also brought with it problems of its own: we do not understand how this background radiation can be so uniform in all directions when it comes from different parts of the Universe that have never been in communication with each other. Attempts have been made to invoke a very rapid period of expansion in the Universe's history in order to remove this difficulty (the so-called inflationary universe), but these attempts appear to many cosmologists to be less than convincing (*see* INFLATION).

▼ cosmological constant A supernova (arrowed) in a high-redshift galaxy, imaged by the Hubble Space Telescope. Observations of objects such as this support the suggestion that a cosmological constant might, after all, be required to describe the expansion of the Universe.

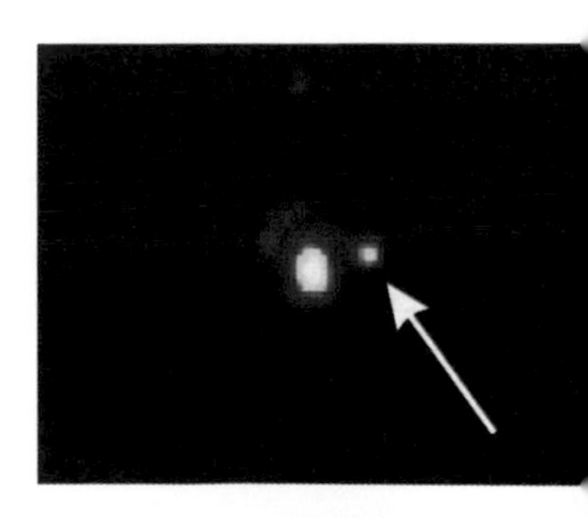

▶ **coudé focus** In this optical configuration, mirrors direct light from the telescope along the polar axis to a fixed observing position. This has several advantages for observations that require use of heavy or bulky detectors.

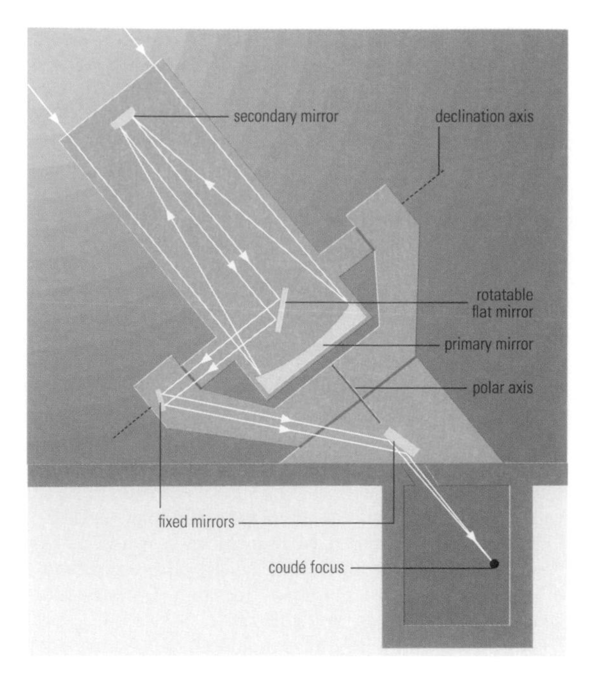

At the present time, the best observational evidence favours an inflationary Big Bang model for the Universe with a Hubble constant of around 68 km/s/Mpc.

Cosmology is presently based on a considerable amount of speculation fuelled by relatively little observational material. Future generations of cosmologists will be presented with plenty of problems – and opportunities. Even if we were to obtain a good understanding of the present and future behaviour of the Universe, we would still be far from comprehending what happened before the Big Bang.

Cosmos satellite Blanket name used by the former Soviet Union for most of its scientific and military satellites. By the end of 2001, 2386 had been launched.

coudé focus Focal point of an equatorially mounted telescope in which the light path is directed along the polar axis to a fixed position that remains stationary, regardless of the orientation of the telescope. Instruments such as high-dispersion SPECTROSCOPES, which are too large or heavy to be mounted on a moving telescope, may be placed at the coudé focus, which is often located in an adjacent room or even separate floor. A series of auxiliary mirrors is used to direct the converging beam of light from the SECONDARY MIRROR, down the hollow polar axis of the telescope mount, to the slit of the spectrograph. The word coudé is French for 'elbow', and describes the bending of the light path.

▼ **Crab Nebula** The remnant of a supernova that exploded in the constellation of Taurus in 1054. The expanding gas cloud has filamentary structure and a pulsar at its heart.

counterglow Alternative name for GEGENSCHEIN

Cowling, Thomas George (1906–90) English mathematician and theoretical astrophysicist who pioneered the study of magnetic fields in stars. Cowling worked out mathematical models that demonstrated the importance of RADIATION and CONVECTION in making and transporting nuclear energy through a star's atmospheric layers.

CPD Abbreviation of *CAPE PHOTOGRAPHIC DURCHMUSTERUNG*

Crab Nebula (M1, NGC 1952) SUPERNOVA REMNANT some 6500 l.y. away and located in Taurus (RA $05^h 34^m.5$ dec. $+22°00'$) about a degree from ζ Tau.

On 1054 July 4 Chinese astronomers recorded a 'guest star' in what is now the constellation Taurus. The star shone about as brightly as the planet Venus, being visible even in daylight. Surprisingly few records of so prominent an object have been found elsewhere in the world, though diligent searches are turning up some, including two rock engravings in the southwestern United States that may depict the supernova near the crescent moon.

Today the remnant of that explosion can be seen as an EMISSION NEBULA, a cloud of gas that originated in the star itself and now has expanded to a size of about 8 × 12 l.y. That nebula bears various names: Messier 1, NGC1952, Taurus A and Taurus X-1. The name by which it is best known, however, is the Crab Nebula. Lord ROSSE named it for its visual resemblance to a crab when observed through his 72-inch (1.8-m) reflector.

The Crab Nebula is a supernova remnant – the nebula, radio and/or X-ray source left over from a supernova. Supernova remnants usually take the form of an expanding, hollow shell. The Crab Nebula, however, emits from its centre outwards, and is a member of a very rare group of remnants known as PLERIONS or filled supernova remnants.

A supernova's material is ejected at speeds of typically 10,000 km/s (6000 mi/s), so that it rams violently into the surrounding interstellar gas. The effect of this continuous collision is to keep the gas hot long after the supernova explosion. The gas is warmed to several million degrees, emitting light and X-rays. In so doing, the gas is slowed down, so that eventually the remnant ceases to glow. Depending on its environment, a supernova remnant may shine for tens of thousands of years. The Crab Nebula is, therefore, a relatively young specimen.

The nature of the explosion is such that the gas forms filamentary structures. The filaments give old supernova remnants a wispy appearance. In the Crab, the filaments are indeed present: on colour photographs they glow with the characteristic red of hot hydrogen, like strands of red cotton wrapped around a soft yellow glow. It is the yellow glow that is distinctive in the case of the Crab Nebula. Within the filaments is an ionized gas in which electrons, freed from their parent atoms, are spiraling in intense magnetic fields and emitting SYNCHROTRON RADIATION.

The Crab Nebula is exceptional amongst plerions because the synchrotron radiation extends from the radio domain to the visible. It is the only synchrotron nebula that can be seen in a small telescope. In order to emit at such short wavelengths, the electrons must be very energetic indeed. The very act of producing synchrotron radiation removes energy from the electrons, so in the Crab Nebula there has to be a continuing supply of energetic electrons now, 900 years after the star was seen to explode. The source of these energetic electrons is the NEUTRON STAR or PULSAR found at the centre of the nebu-

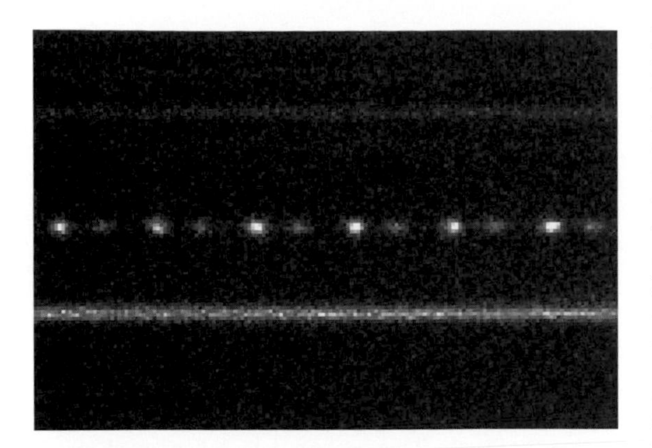

◀ **Crab Pulsar** Optical variations in the Crab Pulsar are seen in this series of High Time Resolution (HIT) mode images from the Very Large Telescope (VLT). As the Pulsar's image was allowed to drift across the CCD during the 2.5 second exposure, its light clearly varied in a regular pattern, whilst that of a neighbouring star (lower trace) remained constant.

la. The CRAB PULSAR, spinning at the rate of 30 times per second, continuously sprays out both radiation and electrons to replenish the synchrotron radiation.

The Crab Nebula holds other surprises yet. Only recently a very faint extension was noted from its northern edge – a broad, parallel-sided jet of gas. The origin of this jet is still obscure. Various theories have been proposed, but none seems quite convincing at present. The Crab Nebula certainly will retain a vital place in the study of supernovae and supernova remnants for a considerable time.

Crab Pulsar Pulsar with a period of $0^s.033$ at the heart of the CRAB NEBULA (Ml). It is the remains of the supernova of 1054, and its discovery confirmed that PULSARS are related to NEUTRON STARS. The Crab Pulsar is one of very few pulsars to be identified optically, and it has been seen to pulse in the optical, X-ray (with the EINSTEIN OBSERVATORY) and gamma-ray (with the COMPTON GAMMA RAY OBSERVATORY) regions, with the same period. In between the main pulse there is another weak pulse, called the inter-pulse. The HUBBLE SPACE TELESCOPE has taken a remarkable picture (at optical wavelengths) of the pulsar in which knots of gas can be seen in the pulsar jet, and wisps of gas are seen like ripples on a pond or a whirlpool. These are very variable.

Crabtree, William (1610–44) English cloth-merchant, apparently self-taught in astronomy. By 1636 he had become conversant with the researches of Galileo, Johannes Kepler, Pierre Gassendi and René Descartes (1596–1650), owned a telescope and several other astronomical instruments, and had become the centre of a group of astronomical correspondents in north-west England that included Jeremiah HORROCKS, William GASCOIGNE and Christopher Towneley (1604–74). Crabtree used his observations of eclipses, occultations and sunspots to advance the Copernican system. In 1639 he and Horrocks observed, from Salford, Lancashire, the first recorded transit of Venus across the Sun's disk.

Crater *See* feature article

crater Circular depression on the surface of a planet, satellite or other Solar System body. The term is typically applied to features of volcanic or impact origin. Volcanic craters are steep-walled depressions at the top or on the flanks of a volcano; they are often the major vents for eruption. They are the result of explosions or collapse at the top of a volcanic conduit. Impact craters are observed on all kinds of surface; they are formed by high velocity collisions of interplanetary bodies. A primary impact crater is typically surrounded by a concentric zone of blocky and hummocky terrain formed by EJECTA. Larger ejecta fragments may produce secondary craters.

On bodies lacking an atmosphere, even MICROMETEORITES may produce impact craters, known as microcraters, less than a micrometre in diameter. Where an atmosphere is present, its thickness determines the minimum possible crater size. On Venus, with its thick, dense atmosphere, the smallest primary impact craters are a few kilometres in diameter. Mars' rarefied atmosphere, however, leads to a minimum size of between 10 and 15 metres. High-velocity passage through an atmosphere often results in fragmentation of an impactor. If break-up happens close to the planet's solid surface, so that the impactor fragments' high velocity is retained, primary impact craters smaller than the above limits may form. Impact craters show changes in structure as a function of their size. Relatively small craters (usually less than 1 km in diameter) are typically bowl-shaped. Larger ones have an uplifted floor. Still larger craters have a central peak on their floor. The largest (tens, hundreds to a few thousand kilometres across) have concentric rings on their floor and are known as multi-ring BASINS. The diameter at which there is a transition from one type of crater structure to another one depends on the nature of the target material (for example rock or ice) and on the target body's gravitational acceleration. For example, the onset diameter of central-peaked craters is 8–15 km (5–9 mi) on Venus, 10–20 km (6–12 mi) on Mercury, 5–15 km (3–9 mi) on Mars, and 25–40 km (16–25 mi) on the Moon (*see* CRATER, LUNAR).

crater, lunar Roughly circular depression seen on all parts of the lunar surface. Craters range from micrometre-size features (microcraters), which are known from examination of lunar samples returned to Earth by APOLLO astronauts and LUNA robotic missions, to so-called multi-ring BASINS, the largest of which is about 2500 km (1600 mi) in diameter. On the whole Moon, numbers of CRATERS larger than 1000, 100, 10 and 1 km in diameter are, respectively, two, more than 200, *c.*10^4 and *c.*10^6. This close to power function extends to smaller sizes.

The shapes and structures of most craters show systematic variations with crater size. Craters with diameters greater than 10 km (6 mi) generally have prominent rims raised above the surrounding terrain. Some large craters are quite pristine and show morphological changes as their sizes increase. Craters with diameters smaller than 15–20 km (9–12 mi) are bowl-shaped. Above this size, pristine craters have flattened floors with hummocky surfaces. Craters larger than 25–40 km (16–25 mi) have a mountain at the centre of the flat floor and are known as central peak craters. At larger diameters, greater than 100–120 km (62–75 mi), the central peak is surrounded by a fragmentary ring of smaller peaks; such craters are known as central peak basins. Craters with diameters greater than 200–300 km (120–190 mi) have one or more concentric rings of peaks but lack a central peak; they are known as peak-ring and multi-ring basins. Bowl-shaped craters are often called simple craters, and those with a flattened floor and central peaks are called complex craters. Inner slopes of complex craters are steep, often with characteristic terraces formed by slumping and sliding. Many large craters are filled with EJECTA from other craters, which partly or completely bury their interior structures.

The outer flanks of the rims of large lunar craters are gently sloping; they consist of hummocky terrain, which grades back to the general (pre-crater) level. Beyond the hummocky terrain region are sometimes found large numbers of small, irregular SECONDARY CRATERS, which often occur in groups and lines elongated roughly radial or concentric to the main, or primary, crater. Even farther out from the primary crater rim, roughly radial patterns of brightening of the surface, known as RAYS, are visible when the Sun is high over the horizon. In plan view, the primary crater rims may range from nearly circular, through polygonal, to quite irregular.

Small craters, in the size range from 10 km (6 mi) down to a few centimetres, generally have simple bowl-like shapes. The inner walls are usually smooth, with occasional outcrops of bedrock; they show only minor modification

CRATER (GEN. CRATERIS, ABBR. CRT)

Faint constellation south of Leo representing a cup or chalice, linked with the legend of neighbouring Corvus. Its brightest star is δ Crt, mag. 3.56, spectral type G9 or K0 III, distance 195 l.y.

► C ring Saturn's C ring imaged by Voyager 2 during its flyby in 1981 August. Seen as single by Earth-based observers, the C ring was revealed by Voyager to be made up of hundreds of separate component ringlets.

by slumping. Flat floors and central mounds are rare. In plan view, the majority of small craters are near circular; some are elongated in a manner that is typical for secondary craters and for craters associated with lunar SINUOUS RILLES. Small craters show obvious progression from fresh-looking, morphologically prominent and relatively deep forms to subdued shallow features that are obviously a result of crater degradation due to impact 'GARDENING' and slumping. Some of the smallest craters (microcraters) seen on fragments of rocks and minerals have smooth interiors and rounded rims, showing evidence of plastic flow.

The morphological features of most craters – raised rims and hummocky exteriors – are consistent with the idea that they were formed by some explosive process that excavated material from the depression and deposited it to form the crater rim and ejecta blanket. The depletion in volatiles of lunar rocks argues against a volcanic-type explosion, while the systematic presence of meteoritic material in the ejecta of small (REGOLITH) and large (highland BRECCIA) craters suggests that the crater-forming explosions resulted from HYPERVELOCITY IMPACTS of METEOROIDS. This idea agrees well with the close to random distribution of primary craters on the homogeneously aged surface of the Moon. Large fragments of the target material, thrown out at a relatively high speed, form secondary craters, while smaller particles, thrown out with even higher speed, produce the bright rays.

Volcanic eruptions almost certainly occurred on the Moon, producing the extensive areas of dark deposits that can be seen thinly covering the surface in some places. These deposits are sometimes observed in association with kilometre-size shallow craters, which are known as dark-haloed craters. These craters are good candidates for volcanic landforms, especially those located along rilles or in straight lines, similar to those located on the floor of the crater ALPHONSUS, or those spatially associated with source areas of the youngest volcanism. Some kilometre-size and smaller rimless craters, especially those in association with fractures and cracks, may have formed by drainage of material into subsurface openings.

It is now considered certain that the great majority of lunar craters are of impact origin, although many large craters were subsequently modified by volcanic and TECTONIC processes. The most obvious example is the extensive flooding by lava flows of the large basins on the side facing the Earth. This process, which happened mainly between 4000 and 3000 million years ago, led to the formation of circular lunar maria, for example Mare IMBRIUM and Mare CRISIUM.

The relative numbers of craters of a given size seen on different parts of the surface has been used to obtain relative ages of the terrains: the older a surface is, the greater the degree of cratering. Also, when ejecta from a crater is seen to lie on top of other features it provides a time marker for those features. By combining these relative ages with absolute ages measured on rocks returned to Earth, the calibration function of transformation of relative numbers of craters of a given size into millions and billions of years has been worked out. This calibration curve can be applied to other bodies and serves as the only technique now available to estimate absolute ages of different terrains throughout the Solar System; major events on different planets and satellites can be correlated in time.

Crêpe ring Common name for Saturn's C RING.

crescent PHASE of the MOON between new and first quarter, or between last quarter and new, or of an inferior planet between inferior CONJUNCTION and greatest ELONGATION, when less than half its illuminated side is visible.

Cressida One of the small inner satellites of URANUS, discovered in 1986 by the VOYAGER 2 imaging team. Cressida is *c*.66 km (*c*.41 mi) in size. It takes 0.464 days to circuit the planet, at a distance of 61,800 km (38,400 mi) from its centre, in a near-circular, near-equatorial orbit.

Crimean Astrophysical Observatory (CrAO) Principal astronomical institution in the Ukraine, and one of the largest scientific centres in the republic. CrAO began before World War I as a station of the PULKOVO OBSERVATORY, but its modern history began in 1945 when the Soviet government decided to build a new facility at Nauchny, 12 km (8 mi) south-east of Bakhchisaraj; this continues as the observatory's main centre. Its main instruments are the 2.6-m (104-in.) Shajn Telescope (commissioned in 1961), two 1.25-m (50-in.) telescopes, and a large gamma-ray telescope having 48 mirrors and a total collecting area of 54 m^2 (580 sq ft). The observatory also operates a 22-m (72 ft) radio telescope near Simeiz.

C ring Innermost of the rings of SATURN visible from Earth. It lies inside the B RING at a distance of between 74,500 km (46,300 mi) and 92,000 km (57,200 mi) from the planet's centre.

Crisium, Mare (Sea of Crises) Lunar lava plain, roughly 500 km (310 mi) in diameter, located in the north-east quadrant of the Moon. The region's geology reveals that an enormous impact struck the area, producing a multi-ring impact BASIN. At a later time, lavas flooded the inner areas of the basin, but were generally contained by an inner ring, which is the shape of the present mare. This feature appears oblong due to foreshortening, though its actual shape is more hexagonal. A shelf lies just inside the margin, marking an inner ring of Crisium.

CRUX (GEN. CRUCIS, ABBR. CRU)

Smallest of all the 88 constellations, but one of the most famous and distinctive. Its four brightest members form a shape that has been likened more to a kite than a cross; its symmetry is disturbed by the off-centre of ε Cru. Crux was formed from stars in the hind-legs of Centaurus during the 16th century. The long axis of the cross points to the south celestial pole. α Cru (ACRUX) is a sparkling double for small telescopes with a wider binocular companion of mag. 4.9. γ Cru (Gacrux) is a wide binocular double; μ Cru is an easy double for small telescopes of mags. 4.0 and 5.1. Objects of particular interest in Crux include NGC 4755, known as the jewel box or KAPPA CRUCIS CLUSTER, and the dark COALSACK nebula. *See also* BECRUX

BRIGHTEST STARS

	Name	RA h m	dec. ° ′	Visual mag.	Absolute mag.	Spectral type	Distance (l.y.)
α	Acrux	12 27	−63 06	0.77	−4.19	B0.5 + B1	321
β	Becrux	12 48	−59 41	1.25	−3.92	B0.5	353
γ	Gacrux	12 31	−57 07	1.59	−0.56	M4	88
δ		12 15	−58 45	2.79	−2.45	B2	364

critical density Amount of mass needed to make the Universe adopt a flat geometry. The size of the critical mass depends on several cosmological parameters and is given by:

$$\rho_c = 3H_0^2/8\pi G$$

where H_0 is the HUBBLE CONSTANT and G is the universal constant of gravitation. If the Hubble constant is 70 km/s/Mpc, then ρ_c is 9.2×10^{-27} kg/m^3. The actual measured mass-energy density compared with this theoretical value determines whether the Universe is open, critical or closed. *See also* DENSITY PARAMETER; OPEN UNIVERSE; CLOSED UNIVERSE

Crommelin, Andrew Claude de la Cherois (1865–1939) English astronomer, born in Ireland, expert on the orbits of periodic comets and asteroids. With Philip Herbert Cowell (1870–1949), he used improved methods, based on a combination of mathematical theory and actual observations, to predict the perihelion date of Halley's Comet at its 1910 apparition. The prediction was accurate to within 3 days, far exceeding the accuracy of predictions for prior returns of the comet. Crommelin was a skilled observer, and took part in four major expeditions to observe total eclipses of the Sun, helping to confirm Einstein's general theory of relativity.

Crommelin, Comet 27P/ Periodic comet, first discovered by Jean Louis PONS in 1818 and again, independently, by Jérôme Eugène Coggia (1849–1919) and Friedrich Winnecke (1835–97) at the later apparition of 1873. The identity of this as a single object seen at different returns was not established until 1928, when a comet found by G. Forbes was shown by Andrew CROMMELIN to have the same orbital elements. Initially designated Pons–Coggia–Winnecke–Forbes, the comet was renamed Crommelin in 1948. The comet was recovered in 1956 only 10° from the expected position by Ludmilla Pajdusáková (1916–79) and Michael Hendrie (1931–). 27P/Crommelin has an orbital period of 27.4 years, and its most recent return in 1983–84 was used to test observing programmes ahead of 1P/Halley's 1986 perihelion.

crossed lens Convex lens designed to minimize SPHERICAL ABERRATION. A crossed lens has two convex spherical surfaces, the radii of which are in the ratio of six to one. This arrangement minimizes spherical aberration for a parallel beam of light.

crossing time Measure of the internal dynamics of a STAR CLUSTER, GALAXY or CLUSTER OF GALAXIES, given by the ratio of a relevant radius to the particle velocity. The times taken for a system to relax into a well-mixed configuration, for a galaxy merger to play itself out, or for the initial conditions of a stellar system to be erased, all scale proportionally to the crossing time. For stars in our part of the Milky Way, the crossing time is about 60 million years.

cross-staff Simple 14th-century instrument consisting of two attached pieces of wood in the shape of a cross and used as a sighting device for determining the angular distance between two objects.

Cruithne EARTH-CROSSING ASTEROID with an orbital period very close to one year; number 3753. Discovered in 1986, the peculiar orbital properties of Cruithne were not identified until 1997. There are no known TROJAN ASTEROIDS of the Earth occupying the LAGRANGIAN POINTS, but Cruithne has a regular dynamical relationship to our planet, delineating a horseshoe-shaped path relative to us as it orbits the Sun. The Saturnian moons JANUS and EPIMETHEUS also follow horseshoe orbits, but centred on that planet. Cruithne is about 5 km (3 mi) in size. *See* table at NEAR-EARTH ASTEROID

crust Outermost layer of a rocky planet; it is composed of the lightest silicate minerals that float to the surface during DIFFERENTIATION. Earth's crust is divided into continental (granitic) and oceanic (basaltic) provinces, with thicknesses of order 20–50 km (10–30 mi) and 5–10 km (3–6 mi), respectively. The crust is compositionally distinct from the denser material of the MANTLE. The Moon's crust is *c*.60 km (*c*.40 mi) thick.

Crux *See* feature article

cryovolcanism Form of low-temperature VOLCANISM on icy planetary bodies in which the eruptive fluid (magma) is water, water-ammonia mixture or brine rather than liquid silicate. The fluid originates in a warm layer at depth and escapes through fractures. It freezes on the surface, creating flow features and flood plains. Cryovolcanic features have been identified on larger icy satellites of the outer planets.

CSIRO Abbreviation of COMMONWEALTH SCIENTIFIC AND INDUSTRIAL RESEARCH ORGANISATION

C star Abbreviation of CARBON STAR

cubewano TRANS-NEPTUNIAN OBJECT that does not have a resonant orbit and is, therefore, distinct from a PLUTINO; cubewanos are members of the EDGEWORTH–KUIPER BELT. The term cubewano is derived from the pronunciation of the preliminary designation of the first trans-Neptunian object discovered, 1992 QB1.

culmination Passage of a celestial body across the meridian due south or due north of the observer. CIRCUMPOLAR STARS cross the meridian twice, the events being known as upper culmination (between pole and zenith) and lower culmination (between pole and horizon). Non-circumpolar objects obtain their maximum altitude above the horizon at culmination. *See also* TRANSIT

Cunitz, Maria (1610–64) German astronomer and mathematician who published a simplified version of Johannes Kepler's tables of planetary motion, making them much more widely available than before. For this work she earned the title 'Urania Propitia', meaning 'she who is closest to Urania', the muse of astronomy.

Curtis, Heber Doust (1872–1942) American astronomer who correctly described the 'spiral nebulae' as other galaxies lying far beyond the Milky Way. He joined the Lick Observatory in 1902, becoming expert at astrophotography and spectroscopy. With Lick's director Wallace CAMPBELL he undertook a project to measure the radial velocities of all northern-hemisphere stars brighter than visual magnitude 5.5, in the process discovering many spectroscopic binaries. Using Lick's Crossley Reflector, Curtis amassed the finest collection of photographs ever

◄ **Crux** One of the southern sky's most distinctive constellations. The long axis points to the south celestial pole, while the dark nebulosity of the Coalsack can be seen at lower left.

▼ **cryovolcanism** Water erupting from a possible ice volcano produced this flow-like deposit on Jupiter's satellite Ganymede. Imaged in 1997 by the Galileo orbiter, this feature is about 55 km (34 mi) long and up to 20 km (13 mi) wide.

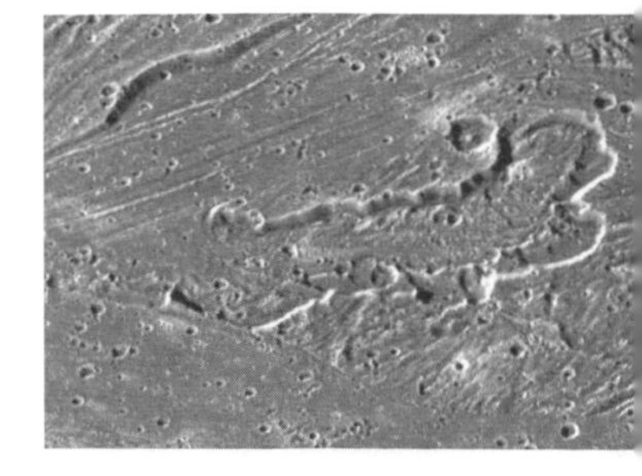

C

taken of nebulae. He found that 'spiral nebulae' were scarce along the galactic plane (the ZONE OF AVOIDANCE) and that many spirals showed dark dust lanes along their horizontal planes, similar to the Milky Way's. These observations and his spectroscopic work convinced him that the bright, diffuse nebulae resided inside our Galaxy, but that the 'spiral nebulae' were other galaxies lying at much greater distances. This was at variance with the 'metagalaxy' model proposed by Harlow SHAPLEY (*see* GREAT DEBATE).

Curtis became director of the Allegheny Observatory (1920–30), where he designed and built a wide variety of astronomical equipment, including 'measuring engines' to obtain precise positions of celestial objects from photographic plates, and specialized instruments for observing solar eclipses. During the 1925 total eclipse expedition to New Haven, Connecticut, he became the first to obtain an infrared spectrum of the solar corona and chromosphere. Curtis continued his solar astronomy at the University of Michigan's observatory, which he directed from 1930 to 1942, and at its McMath-Hulbert Observatory, which he helped to found.

curvature of spacetime Distortion of the spacetime CONTINUUM due primarily to the presence of mass. Einstein's theory of GENERAL RELATIVITY is a geometric theory in which the FIELD EQUATIONS describe how much effect mass-energy has on the shape of the four-dimensional SPACETIME continuum in which we live. Mass-energy curves or distorts spacetime, and particle paths are constrained to follow this curvature. If a particle is observed approaching the Sun, it appears to be attracted to the Sun by some force, or, equivalently, its path is seen to alter because of the curvature of spacetime around the Sun.

cusp Either of the two pointed extremities, or 'horns', of the Moon or an inferior planet when at its CRESCENT phase. The term literally means the point of meeting of two curves, in this case the curves being the limb of the Moon or planet and the TERMINATOR (the boundary between the illuminated and non-illuminated hemispheres).

cusp caps of Venus Bright hoods or caps at the north and south points of the planet, especially conspicuous at the latter. They were discovered 1813 December 29 by the German astronomer Franz Paula von Gruithuisen (1744–1852), who imagined them to be the polar snows. They have been extensively observed (1877–82) by Étienne Léopold Trouvelot (1827–95) and many others since. At first thought to be illusory, the cusp caps feature conspicuously on photographs taken in ultraviolet by Frank E. Ross at Mount Wilson during the favourable evening elongation of 1927. They are now identified with the bright polar cloud swirls imaged by the MARINER 10 and PIONEER VENUS space probes. The caps vary in brightness, shape, size and position; sometimes they are seen at both poles, at others at one alone.

CYGNUS (GEN. CYGNI, ABBR. CYG)

Large and easily recognized constellation, also known as the Northern Cross, representing the swan into which the god Zeus turned himself before seducing Leda, the queen of Sparta. Its brightest star, DENEB, is an immensely luminous supergiant, a member of the so-called SUMMER TRIANGLE. β Cyg, better known as ALBIREO, is a beautiful coloured double for the smallest apertures, while 61 Cyg is an easy pair of orange dwarfs (*see* SIXTY-ONE CYGNI). CHI CYGNI is a red MIRA STAR, ranging between mags. 3.3 and 14 with a period of about 400 days; P CYGNI is a highly luminous variable. Cygnus lies in the Milky Way and contains rich starfields. Its largest nebula is the CYGNUS RIFT, a dark cloud that bifurcates the Milky Way as it heads south. Near Deneb is the NORTH AMERICA NEBULA, NGC 7000, while near ε Cyg is the VEIL NEBULA, a supernova remnant. M39 is a large open star cluster easily seen in binoculars and just on the naked-eye limit. NGC 6826 is an 8th-magnitude planetary nebula popularly known as the BLINKING PLANETARY. Two celebrated objects in Cygnus within reach only of professional instruments are CYGNUS X-1, a black hole X-ray source that is in orbit around a visible star, and CYGNUS A, a strong radio source thought to be two galaxies in collision.

BRIGHTEST STARS

	Name	RA h m	dec. °	Visual mag.	Absolute mag.	Spectral type	Distance (l.y.)
α	Deneb	20 41	+45 17	1.25	−8.73	A2	3230
γ	Sadr	20 22	+40 16	2.23	−6.12	F8	1520
ε	Gienah	20 46	+33 58	2.48	0.76	K0	72
δ		19 45	+45 08	2.86	−0.74	B9.5	171
β	Albireo	19 31	+27 58	3.05	−2.31	K3	386
ζ		21 13	+30 14	3.21	−0.12	G8	151

CV Serpentis ECLIPSING BINARY star associated with the bright hydrogen nebula E41. It has a period of $29^{d}.64$, and its magnitude ranges from 9.86 to 10.81(mB). It has Wolf-Rayet and O8-O9 stellar components. CV Serpentis exhibits strange behaviour in that eclipses are sometimes observed when the Wolf-Rayet component eclipses the O-type star, and sometimes when the O-type star is the eclipsing body. Also, predicted eclipses do not always occur.

Cygnus *See* feature article

Cygnus A Strongest source of radio emission outside the Galaxy, and the first RADIO GALAXY to be detected. It emits a million times more energy in the radio part of the spectrum than the Milky Way. Cygnus A is thought to be two galaxies in collision; it has two radio LOBES centred on the parent galaxy. It is also an X-ray source.

Cygnus Loop *See* VEIL NEBULA

Cygnus Rift (Northern Coalsack) Northern part of a large DARK NEBULA that runs along the Milky Way from Cygnus through to Ophiuchus. The nebula as a whole is known as the Great Rift. The clouds forming the Great Rift are about 2000 l.y. away and have a mass of about one million solar masses. The Great Rift causes the Milky Way to appear to split just below Deneb (α Cyg), with the main branch running down through Altair (α Aql) and a spur forking towards β Oph.

Cygnus X-1 Binary star system around 6000 l.y. (2 kpc) away that is a strong source of X-rays. It is associated with a faint star (HDE 226868), a blue supergiant, with an orbital period of 5.6 days. The blue supergiant is around 30 times as massive as the Sun, and the invisible companion has a mass of around 10 solar masses, which is too big for a white dwarf or a neutron star, so it may be a black hole. The X-rays sometimes vary in intensity on a timescale of milliseconds. Cygnus X-1 is also a radio source.

Cygnus X-3 Source of X-rays, gamma rays and cosmic rays (and small radio flares), emitting more energy than all but a very few other objects in the Galaxy. Cygnus X-3 consists of a NEUTRON STAR orbiting a larger companion, with an orbital period of 4.8 hours. It is situated on the edge of the Galaxy, 36,000 l.y. from Earth.

Cyrillus Polygonal lunar crater, 97 km (60 mi) in diameter. With CATHARINA, it is a member of the THEOPHILUS chain. Its north-east wall is overlapped by the younger impact crater Theophilus. Cyrillus' east rim is straightened, and its interior walls gradually slope to the floor, which is broken at two places by lengthy landslips parallel to the outer ramparts. The west rim of Cyrillus is cut by a long north–south cleft; a long, curvy ridge abuts the north-east walls. Cyrillus adjoins Catharina through a wide, shallow valley that cuts through its south wall. The floor of Cyrillus shows much detail, including a group of three mountains lying east of centre. Although quite large, these mountains are lower and more heavily eroded than Theophilus' central mountains, thus proving that Cyrillus is the older crater. A sinuous rille originates at a volcanic vent beyond Cyrillus' south wall; it tops that wall, continues down to the south half of the crater's floor and extends towards a craterlet just south-west of centre.

Dactyl Small moon of asteroid (243) IDA; it is *c.*1.5 km (*c.*1 mi) in size. Dactyl was discovered in 1993 August in images returned by the GALILEO spacecraft. *See also* PETIT-PRINCE

D'Alembert Montes Range of mountains near RICCIOLI on the Moon's west limb. Some peaks attain altitudes of 6000 m (20,000 ft). Its most famous peak is Table Mountain.

Dall–Kirkham telescope Variation of the CASSEGRAIN TELESCOPE in which the primary is a concave prolate ELLIPSOID and the secondary is a convex sphere. This design is optically inferior to the Cassegrain, suffering from more severe COMA and FIELD CURVATURE. However, the spherical curve on the secondary is far easier to manufacture than the convex hyperboloid in the Cassegrain. The English master optician Horace Edward Stafford Dall (1901–86) invented this design in 1928. It was the subject of correspondence between Oregon amateur telescope-maker Allan Kirkham and Albert G. INGALLS, then editor of *Scientific American*, who named it the Dall–Kirkham.

Damocles Large ASTEROID, number 5335, discovered in 1991; it occupies a comet-like orbit. Damocles is about 10 km (6 mi) in size. It has perihelion at just below 1.6 AU, meaning that it reaches Mars' orbit, and aphelion at 22 AU, just beyond Uranus. This orbit is quite unlike that of any asteroid previously known, and its large inclination (62°) is also unusual. Damocles displays none of the activity characteristic of a comet, however, and so may be a devolatilized cometary nucleus. Dynamically, it may represent a later state of the orbital evolution of the CENTAURS. *See also* HIDALGO; LONG-PERIOD ASTEROID; RETROGRADE ASTEROID

Danjon, André-Louis (1890–1967) French astronomer who directed the observatories at Strasbourg and Paris, and improved instruments, especially for positional astronomy. From the 1920s he developed accurate photometers (and established the DANJON SCALE), and in 1951 he invented the **Danjon astrolabe**, an accurate transit instrument used for measuring star positions.

Danjon scale Fairly arbitrary five-point scale devised by the French astronomer Andre Danjon, used to describe the relative darkness of LUNAR ECLIPSES. The scale ranges from 0, a very dark eclipse, during which the Moon almost becomes invisible, to 4, which indicates a bright eclipse, where the Moon may show a strong red colour but remains easily visible. Variations in the eclipsed Moon's brightness from one event to another result mainly from the changing dust and cloud content of Earth's atmosphere. Eclipses that follow major volcanic eruptions are often notably dark.

dark matter General term used to describe matter that is not seen in astrophysical systems but that is thought to be present for other reasons. The rotation curves of galaxies indicate that more mass is present in the galactic haloes than we actually see, so this matter must not be emitting large amounts of light – hence the term 'dark' matter. Another use of dark matter is in closing the Universe. *See also* CLOSED UNIVERSE; CRITICAL DENSITY; DENSITY PARAMETER

dark adaptation Physiological alteration of the eye's response to light in low-light environments. On moving from a well-lit area to darkness, the eye's pupil dilates to a maximum diameter of about 7 mm in a few seconds, but one still cannot see very well. Almost immediately the eye begins to manufacture rhodopsin (also called visual purple), a chemical that increases the eye's sensitivity by a factor of many thousands. In uninterrupted darkness rhodopsin will continue to build up for as long as 2 hours, but dark adaptation is largely complete after 30 minutes.

dark nebula (absorption nebula) Cloud of dust and cool gas that is visible only because it blocks off the light of stars and nebulae beyond it. Dark nebulae range in size from minute, more or less spherical Bok GLOBULES, usually seen in photographs against the bright glow of EMISSION NEBULAE, through larger features like the HORSEHEAD NEBULA to the naked-eye clouds of the southern COALSACK and the gigantic RHO OPHIUCHI DARK CLOUD, which affects 1000 square degrees (2%) of the sky.

The clouds consist of a mixture of dusty particles and gas, the whole having a composition similar to the standard COSMIC ABUNDANCE of about 75% hydrogen and 23% helium with the rest heavier elements. The dust particles are probably less than a micrometre in size and comprise only about 0.1% of the mass of a cloud. The particles are, however, believed to play an important part in the formation of molecules in space. The outside of the dust particles provides a surface to which atoms within the cloud can adhere and perhaps combine with others to form simple molecules, such as H_2, or much more complicated compounds, such as formaldehyde or even amino acids (*see* COSMOCHEMISTRY and INTERSTELLAR MOLECULES). These relatively fragile compounds are protected from energetic ultraviolet radiation from stars by the cloud itself. The interiors of these molecular clouds are thus very cold, typically only 10 K, which allows them gradually to collapse under their own gravity, and eventually to begin the process of star formation.

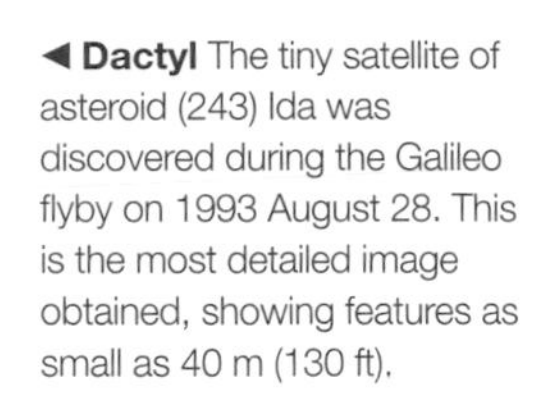

◄ **Dactyl** The tiny satellite of asteroid (243) Ida was discovered during the Galileo flyby on 1993 August 28. This is the most detailed image obtained, showing features as small as 40 m (130 ft).

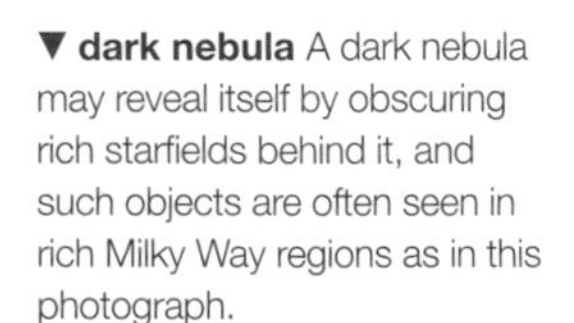

▼ **dark nebula** A dark nebula may reveal itself by obscuring rich starfields behind it, and such objects are often seen in rich Milky Way regions as in this photograph.

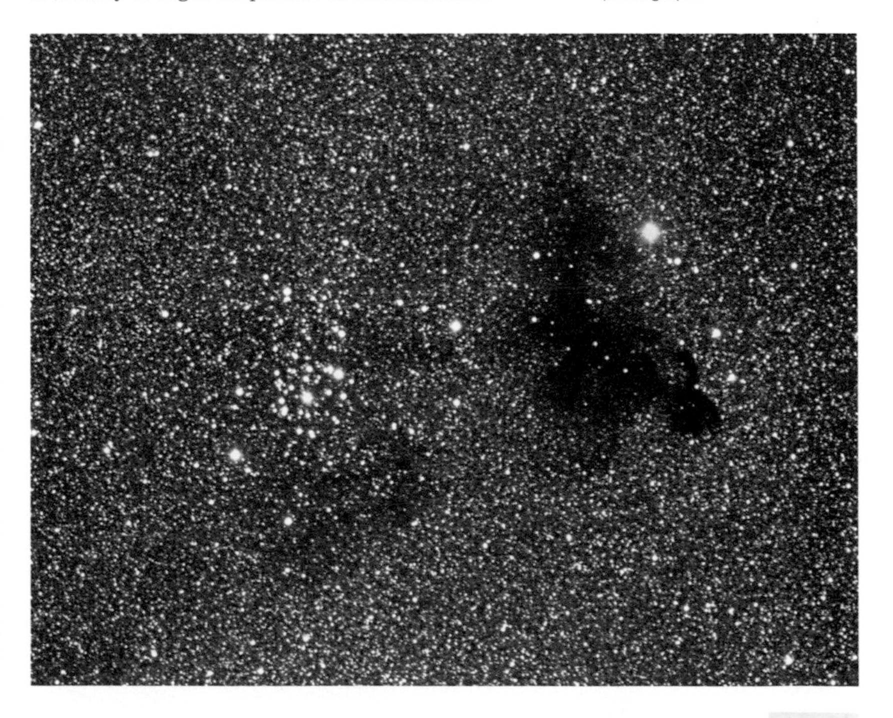

D

▲ **Deep Impact** An artist's impression of the Deep Impact probe releasing its projectile toward Comet 9P/Tempel. It is hoped that the mission will allow study of pristine cometary material exposed by the projectile's impact.

D'Arrest, Comet 6P/ Short-period comet discovered by Heinrich D'Arrest, Leipzig, Germany, in 1851. The comet has an orbital period of 6.51 years and is usually faint. The 1976 August return, however, was unusually favourable, with perihelion and closest approach to Earth coinciding, and the comet reaching magnitude +4.9. Comet 6P/D'Arrest has been seen at 17 returns up to that of 1995.

D'Arrest, Heinrich Louis (1822–75) German astronomer who in 1846, working with Johann GALLE, found Neptune, based upon theoretical orbital calculations by Urbain LE VERRIER. D'Arrest discovered three comets and one asteroid. In 1858 he became professor of astronomy at Copenhagen, and director of the observatory there. He published measurements of nebulae in 1858 and 1867.

data centre, astronomical Organization that collects astronomical data, from either ground- or space-based observatories or both, and organizes it into searchable databases, thus acting as an information hub, or provides on-line bibliographic, abstracting and other services. The archiving function of data centres is also important. A prime purpose of astronomical data centres is to maintain and correlate electronic versions of the thousands of CATALOGUES of celestial objects. The first such centre was established at Strasbourg, France, with the aim of collating all the various identifications of the same objects in different catalogues. Now called the CENTRE DE DONNÉES ASTRONOMIQUES DE STRASBOURG, it maintains the world's main database for astronomical objects.

NASA has a number of data centres. They include the astronomical data center; the Infrared Processing and Analysis Center (IPAC), at Caltech, Los Angeles, which maintains the National Extragalactic Database (NED); and the National Space Science Data Center (NSSDC), which provides access to a wide variety of astrophysics, space physics, solar physics, lunar and planetary data from NASA missions. It also maintains the Astrophysics Data System (ADS), a major abstracting service. Other centres include the Canadian Astronomy Data Centre (CADC), and the Astronomical Data Analysis Center (ADAC) run by the National Astronomical Observatory of Japan.

Davida Fifth-largest MAIN-BELT ASTEROID; number 511. Davida was discovered in 1903. It has a diameter of about 325 km (202 mi).

David Dunlap Observatory (DDO) Research centre of the University of Toronto, Canada, situated at Richmond Hill, Ontario some 20 km (12 mi) north of Toronto, opened in 1935. The DDO was funded through a bequest from the estate of David Dunlap (1863–1924), a wealthy Ontario amateur astronomer. Its principal instrument is a 1.88-m (74-in.) reflector, which was the world's second-largest telescope when it was built. Today's University of Toronto astronomers carry out their research with major international facilities such as the CANADA–FRANCE–HAWAII TELESCOPE and the GEMINI OBSERVATORY.

Dawes, William Rutter (1799–1868) English physician, clergyman and amateur double-star observer who devised the empirical formula used to measure telescope resolution (*see* DAWES' LIMIT). He established a private observatory in Lancashire (1829) where he measured the distance and separation of more than 200 double stars, acquiring the nickname the Eagle-eyed because of his acute vision. Later, Dawes made systematic observations of the planets, especially Mars and Jupiter, and in 1850 independently discovered Saturn's semi-transparent Crepe Ring (the C RING).

Dawes' limit Theoretical limit of RESOLVING POWER for a telescope, given by the formula

$$R = 115.8/D \text{ or } R = 4.56/D'$$

where R is the angular resolution in arcseconds, D the diameter in millimetres and D' the diameter in inches. Dawes' limit applies to the angular separation of two stars of equal magnitude under conditions of perfect seeing. In practice, Dawes' limit is never reached, mainly because of atmospheric turbulence.

day Time taken for the Earth to complete one rotation on its axis. It can be measured in a number of different ways.

A SIDEREAL DAY is the interval between two successive passages across the MERIDIAN of the FIRST POINT OF ARIES (the zero of RIGHT ASCENSION) and is equivalent to $23^h\ 56^m\ 4^s.091$. It is considered to be a true measure of the rotation period because the stars, which are used as reference points, are so far away that in this context they may be regarded as infinitely remote.

An **apparent solar day** is the interval between two successive passages across the meridian of the true Sun. Its length is not uniform, however. Due to the Earth's elliptical orbit, and the fact that the Sun appears to move along the ecliptic rather than the celestial equator, an apparent solar day varies by as much as 16 minutes during the course of a year (*see* APPARENT SOLAR TIME).

A **mean solar day** is the interval between two successive passages across the meridian of a fictitious, or MEAN SUN (*see* MEAN SOLAR TIME). Because of the Sun's movement relative to the background stars, at a rate of about one degree per day in an easterly direction, the mean solar day is slightly longer than a sidereal day, at $24^h\ 3^m\ 56^s.555$.

A **civil day** begins and ends at midnight and comprises two 12-hour periods, am and pm. These are never used in an astronomical context, however, where the 24-hour clock is always quoted.

daylight meteor stream METEOR STREAM that produces activity from a RADIANT too close to the Sun in the sky to allow direct (visual or photographic) observations. Instead, the METEOR SHOWER associated with the stream has to be observed by radio methods. Notable examples include the BETA TAURIDS and the Daytime Arietids, both of which are active during June.

decametric radiation RADIO WAVES having wavelengths of a few tens of metres or a few decametres, that is, from 10 up to 100 m (30 to 3 MHz). It includes most HF radio frequencies and the upper part of the MF band.

deceleration parameter Measure of the change in the expansion velocity of the Universe as a function of time. The deceleration parameter can be calculated by

$$q_0 = 4\pi\rho G/2H_0^2 = -(d^2R_0/dt^2)R_0H_0^2$$

where H_0 is the HUBBLE CONSTANT and ρ is the density of the Universe. If $q_0 = 0$ the Universe is flat and expands at a constant speed for ever. *See also* BIG BANG THEORY

de Chéseaux, Comet (C/1743 X1) Brilliant multitailed comet of 1743–44, discovered independently by Dirk Klinkenberg (1709–99) on December 9 and Philippe de Chéseaux (1718–51) on December 13. It reached apparent magnitude −7 around its 1744 March 1 perihelion.

decimetric radiation RADIO WAVES having wavelengths of a few tenths of a metre or a few decimetres, that is, from 0.1 up to 1 m (3 to 0.3 GHz). It includes most UHF radio frequencies and the upper part of the VHF band.

declination (dec., δ) Measure of angular distance, north or south of the CELESTIAL EQUATOR; it is one of the two coordinates of the EQUATORIAL COORDINATE system, the other being RIGHT ASCENSION. Designated by the Greek letter δ, declination is measured from 0° to +90° between the celestial equator and the north celestial pole, and from 0° to −90° between the celestial equator and the south celestial pole.

deconvolution Removal of imperfections in the signal received by an instrument. When an astronomical object is observed with any instrument (such as a telescope plus detector, for example), the instrument smears the radiation in its own characteristic way. The most obvious examples are the spiking seen on the optical images of bright stars, caused by the spider support for the secondary mirror, and the broad spots on the image rather than single points for the stellar images. The process of removing the effect of the instrument is called deconvolution, and it involves understanding and modelling the effects very thoroughly with a computer. The instrumental effect (the instrumental profile) is often approximated by a Gaussian distribution function because this is easy to correct computationally.

decoupling era Period when matter began to evolve separately from the COSMIC MICROWAVE BACKGROUND. During the early stages of the BIG BANG, the universe was much smaller, denser and hotter than it is now, and it was essentially radiation dominated. The photon density was so high that the formation of normal atoms and molecules was not possible. As the universe expanded and cooled, the radiation density decreased more rapidly than the matter density, and after several thousand years the universe became optically thin to radiation, and matter evolved separately from the background radiation. This transition is called decoupling and this era can be studied through the cosmic microwave background radiation.

Deep Impact NASA Discovery-class mission, to be launched in 2004 January, to impact a 498-kg instrumented cylindrical copper projectile into the nucleus of Comet 9P/Tempel 1 in 2005 July. The impact, at a speed of 32,200 km/h (20,000 mph), will create a 100-m (330-ft) diameter, 8-m (26-ft) deep crater, exposing pristine interior material and debris from the comet, which will be studied by a high-resolution camera and an infrared spectrometer on the mother craft and by ground- and space-based observatories.

deep sky Expression used to define a celestial body that lies outside the Solar System. It can be applied to both galactic and extra-galactic objects such as star clusters, nebulae and galaxies, which are referred to as **deep sky objects**; it is rarely used to describe individual stars.

Deep Space 1 (DS 1) First of NASA's NEW MILLENNIUM programme missions, launched in 1998 October, to demonstrate new technologies, including an ion propulsion system, an autonomous navigation system and an autonomous spacecraft computer, nicknamed the Controller. The NATIONAL AERONAUTICS AND SPACE ADMINISTRATION (NASA) hopes such technologies will enable future science craft to be smaller, less expensive, more autonomous and capable of independent decision making, relying less on ground control. DS 1's rendezvous with the asteroid 9969 BRAILLE at a distance of 15 km (9.3 mi) in 1999 July accomplished the prime objective of the mission, although an imaging system malfunction prevented any good images being taken. DS 1 was then directed towards a bonus target – a rendezvous with Comet 19P/Borrelly in 2001 September. On the way to the comet, however, DS 1 experienced a serious problem when its star tracker was lost. A remarkable recovery by engineers, who reconfigured computer software on the spacecraft, enabled control to be restored.

Deep Space Network (DSN) Three Deep Space Communications Complexes used by NASA to track and communicate with space probes; they are operated by the JET PROPULSION LABORATORY. The first to be completed, in 1966, is the Goldstone Tracking Station in California's Mojave Desert, near Goldstone Dry Lake, about 70 km north of Barstow. The other two are at Robledo de Chavella, 60 km west of Madrid, Spain, and at Tidbinbilla Nature Reserve, 40 km south-west of Canberra, Australia. Between them, they can maintain contact with space probes as the Earth rotates. Each complex has two 34-m (110-ft) antennas, one 26-m (85-ft) antenna and one 70-m (200-ft) antenna, which can also be used as radio telescopes and for radar observations of nearby planets.

defect of illumination Proportion of the disk of a planetary body that does not appear illuminated to an observer on Earth; measured as an angle in arcseconds. Also known as the TERMINATOR phase, the defect of illumination defines how much of a geometrical planetary disk is in darkness. It can also be expressed in terms of the angle of the terminator or the ratio of illuminated to non-illuminated areas. At OPPOSITION, there is no perceptible defect of illumination.

deferent In the PTOLEMAIC SYSTEM and earlier systems, the large circle centred on the Earth, around which was supposed to move the centre of a small circle, the EPICYCLE, around which a planet moved.

degenerate matter Matter that is so dense, atomic particles are packed together until quantum effects support the material. Material within WHITE DWARFS and NEUTRON STARS is degenerate.

At the enormous temperatures and pressures that exist at the centres of stars, atoms are stripped of almost all of their electrons, leaving a gas of nuclei and electrons. Because these particles are much smaller than atoms, such a gas can be compressed to much higher densities than an

▼ **Deep Space 1** Test firing of the xenon ion engine of Deep Space 1. Launched in 1998 October, the spacecraft had a highly successful encounter with Comet 19P/Borrelly in 2001 September.

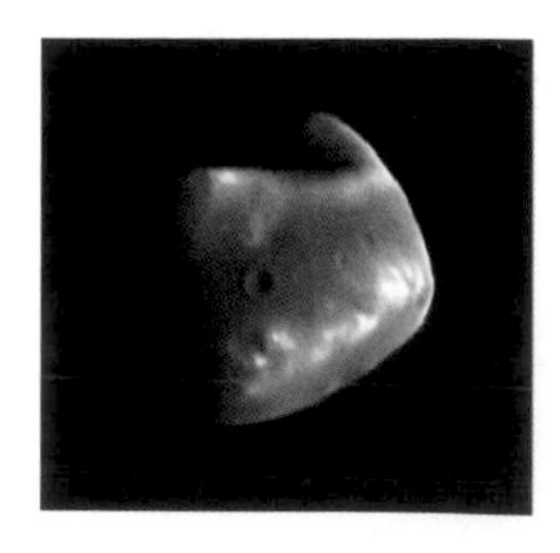

▲ **Deimos** A Viking Orbiter image of Deimos, obtained in 1977. Deimos is the smaller of Mars' two natural satellites and is believed to be a captured asteroid.

ordinary gas. However, the PAULI EXCLUSION PRINCIPLE, which states that no FERMION can exist in the same quantum state, prevents particles getting too close together.

In white dwarfs, or the centres of GIANT STARS and low-mass DWARF STARS, the electrons are degenerate; in neutron stars it is the neutrons that exert the degeneracy pressure. In degenerate matter the motions of the constituent particles are not determined by temperature, as they are in a normal gas, so a degenerate gas does not obey the usual gas laws. Degeneracy pressure is a function only of density.

If energy is put into a degenerate gas the temperature rises, this being achieved by a fraction of the particles with previously restricted velocities moving to higher and unrestricted states. At a sufficiently high temperature all of the particles are removed from their degenerate condition. Degeneracy, therefore, can exist only at temperatures below what is called the Fermi temperature.

Degree Angular-Scale Interferometer (DASI) Thirteen-element interferometer array designed to measure ripples in the cosmic microwave background radiation. It is located at the US National Science Foundation's Amundsen–Scott South Pole station.

degree of arc Unit of angular measure equal to 1/360 part of a circle. The degree (symbol °) is subdivided into ARCMINUTES (′) and ARCSECONDS (″), being 1/60 part and 1/360 part of a degree respectively. Angular measure is used widely in astronomy to determine the diameter or separation of celestial objects.

Deimos Smaller of the two moons of MARS, the other being PHOBOS, discovered in 1877 by Asaph Hall. Deimos is irregular in shape, measuring about 15 × 12 × 10 km (9 × 7 × 6 mi), with a mass of 1.8×10^{18} kg, giving it an average density of 1.7 g/cm^3. It has a near-circular orbit tilted at an angle of 1°.8 to Mars' equator, at a distance of 23,460 km (14,580 mi) from the centre of the planet. This gives it an orbital period of 1.262 days, compared to the 1.026 days it takes Mars to rotate on its axis; in consequence, an observer on the surface of the planet would see Deimos moving slowly westwards across the sky, whereas Phobos would appear to move more rapidly eastwards.

Deimos is a dark body (albedo 0.05), slightly red in coloration, and so is thought to be a captured ASTEROID with a composition similar to a CARBONACEOUS CHONDRITE. It is locked in SYNCHRONOUS ROTATION with Mars, such that its long axis is radially aligned with the planet. It shares these characteristics with its companion, Phobos.

De La Rue, Warren (1815–89) British printer and amateur astronomer, born in Guernsey, a pioneer of solar astronomy and ASTROPHOTOGRAPHY. Realizing that photography could bring out detail invisible to the eye, and provided permanent, objective records of features that could be drawn only with questionable reliability, De La Rue sought to improve photography of the Sun and Moon. He adapted his telescope to hold newly developed wet collodion photographic plates, and in the 1850s took remarkably detailed images of the Moon. In 1858, to photograph the Sun, he built an instrument he called a photoheliograph, and used it to prove that the prominences observed during total solar eclipses were produced by the Sun itself.

DELPHINUS (GEN. DELPHINI, ABBR. DEL)

Small but conspicuous northern constellation, representing a dolphin, between Pegasus and Aquila. Its brightest stars are α Del (Sualocin), mag. 3.8, and β Del (Rotanev), mag. 3.6; backwards their names read 'Nicolaus Venator', the Latinized form of the Italian name of Niccolò Cacciatore (1780–1841), who was assistant and successor to Giuseppe PIAZZI, in whose catalogue of stars (1814) the names first appeared. γ Del is a fine visual binary with orange and bluish-white components, mags. 4.3 and 5.2, separation 9″.8. Deep-sky objects in Delphinus include NGC 7006, an 11th-magnitude globular cluster, which, at about 115,000 l.y. away, is one of the most distant known.

Delaunay, Charles Eugène (1816–72) French engineer and mathematician, an expert in celestial mechanics. Twenty years' work on the lunar theory allowed Delaunay to reckon the latitude, longitude and parallax of the Moon to within 1″. In 1870 he was appointed director of the PARIS OBSERVATORY.

Delphinus *See* feature article

Delta Workhorse of the US satellite launcher programme since 1960, originally based on a Thor intermediate-range ballistic missile first stage. The original Thor Delta vehicle has been upgraded many times during a 40-year period of almost 300 satellite launches. Today, the Delta II and III models are operational. A new Delta IV is under production as a US Air Force Evolved Expendable Launch Vehicle, which will eventually be available commercially. The Delta II carries mainly US Air Force Navstar navigation satellites and NASA payloads, including Mars probes. The Delta III and IV are being marketed for commercial launches, mainly to geostationary transfer orbit (GTO). By 2002 the carrying capacity of these Deltas to GTO ranged from 3.8 to 13 tonnes.

Delta Aquarids METEOR SHOWER active between July 15 and August 20 each year. The RADIANT has northern and southern components, the latter (near δ Aquarii) being the more active at its July 29 peak, with zenithal hourly rate (ZHR) around 20. The northern branch, from close to the 'Water Jar' asterism, peaks around August 6.

Delta Scuti star (DSCT) VARIABLE STAR that pulsates with period of 0^d.01 to 0^d.2 and for which Delta Scuti is the prototype. The range of light amplitudes is from mag. 0.003 to 0.9 V, but is usually only several hundredths of a magnitude. Delta Scuti stars are of spectral types A0–F6 and luminosity classes III–V. Light-curve shape, amplitude and period vary, with some stars having only sporadic variations, which on occasions cease entirely as a result of strong amplitude modulations. Delta Scuti stars have radial and non-radial pulsations. The maximum expansion of surface layers does not lag behind maximum light for more than 0^P.1. These stars are found in the lower part of the Cepheid INSTABILITY STRIP. They are sometimes referred to as ultra-short-period Cepheids or as dwarf Cepheids, although the latter term is also applied to the AI VELORUM STARS. *See also* SX PHOENECIS

Deneb The star α Cygni, visual mag. 1.25, distance about 3200 l.y. Deneb is a blue-white supergiant of spectral type A2 Ia, with a luminosity nearly 500,000 times that of the Sun. It is the most distant and most luminous of all first-magnitude stars. Its name comes from the Arabic word *dhanab*, meaning 'tail', since it marks the tail of Cygnus, the swan.

Deneb Kaitos (Diphda) The star β Ceti, visual mag.2.04, distance 96 l.y. It is a yellow giant of spectral type G9 III. The name comes from an Arabic phrase referring to the tail of the sea-monster (Cetus), which is where it is positioned. An alternative name is *Diphda*, from the Arabic for 'frog', which is what it once represented to Arab astronomers.

Denebola The star β Leonis, visual mag. 2.14, distance 36 l.y., spectral type A3 V. Its name comes from the Arabic meaning 'lion's tail', which is where it lies.

Denning, William Frederick (1848–1931) English amateur observer of meteors and comets who identified and mapped the radiants for many meteor showers. In 1899 he published a comprehensive catalogue of radiants and was able to prove that the radiant for the Perseid meteor shower slowly drifted eastwards as the Earth moved through the meteor stream. Denning discovered four comets, and Nova Cygni 1920. His studies of Jupiter's GREAT RED SPOT suggested that this, or a very similar, feature had been seen by Robert Hooke and Giovanni Cassini in the 1660s, and by others as late as 1713, after which it was not again widely observed until 1878.

density Ratio of mass to volume for a given material or object. The average density of a body is its total mass divided by its total volume. Substances that are light for their size have a low density and vice versa. Water has a density of 1000 kg/m^3 under standard conditions. A wide variation in density is found throughout the Universe, ranging from about 10^{-20} kg/m^3 for the interstellar gas to over 10^{17} kg/m^3 for neutron stars. The average density of matter in the entire Universe is thought to be about 10^{-28} kg/m^3. A typical white dwarf star has a density of between 10^7 and 10^{11} kg/m^3. A normal matchbox containing white dwarf material would have a mass of some 250 tonnes, while if it were filled with material from a neutron star its mass would be 5,000,000,000 tonnes.

density parameter Ratio of the critical density ρ_c to the observed density ρ_o of the universe. The CRITICAL DENSITY is the mass-energy density necessary to close the universe. The density parameter is usually designated with a capital Lambda (Λ):

$$\Lambda = \rho_o/\rho_c$$

If Λ is greater than one, the universe would be closed and eventually collapse back upon itself. If Λ is equal to one then the universe is flat but will continue to expand forever. If Λ is less than one, the universe is open and will expand forever faster and faster. *See also* CLOSED UNIVERSE; OPEN UNIVERSE

density wave theory Explanation for the common occurrence of spiral arms in GALAXIES, when any arms that are simply lines of material should wind up in a relatively short cosmic time and lose their identity. The density wave theory, worked out by Chia Chiao Lin (1916–) and Frank Shu (1943–), identifies kinds of wave patterns that can propagate as coherent spirals through disks. In this theory, spiral arms are patterns through which stars and gas move, with the pattern moving at a different speed. The arms would appear enhanced because the extra mass in the arms keeps stars and gas there longer than they would be in the absence of the extra density, and star formation is triggered in molecular clouds by the denser environment. This theory probably applies to 'grand-design' spiral galaxies, many of which have close companions whose gravitational perturbations would excite the requisite density wave.

De revolutionibus orbium coelestium ('On the Revolution of the Heavenly Spheres') Nicholas COPERNICUS' masterpiece in which he set forth what is now called the COPERNICAN SYSTEM, his heliocentric theory of the Solar System. Although it was published in 1543, the year of Copernicus' death, it took more than a century for its cosmology to become widely accepted. Publication was overseen by the German theologian Andreas Osiander (1498–1552), who without Copernicus' knowledge inserted a preface stating that the heliocentric Solar System was not real, but merely an aid to calculating planetary positions. But the clear intention of the work itself is to promote a Sun-centred system. Much of *De revolutionibus* is concerned with geometrical derivations and tables for predicting the positions of the Sun, Moon and planets. Because the elliptical shape of planetary orbits had not yet been discovered, complicated epicycles were still required, as in the PTOLEMAIC SYSTEM, and in that respect the new theory was not much of an improvement on the previous one.

Descartes Concentric lunar crater (11°.5S 15°.5E), 50 km (31 mi) in diameter. The outline of Descartes is difficult to discern; scattered hills are the sole remnants of the impact site's south rim. A sizeable bowl crater sits atop the south-west rim. The floor of Descartes is accented by several curved ridges that are roughly concentric with its south-east and north-west walls. Apollo 16 landed near Descartes.

▲ **density wave theory** Blue young stars and red HII emission nebulae clearly mark out regions of star formation in the face-on spiral galaxy M83 in Hydra. The spiral structure is thought to result from density waves in such galaxies.

descending node (☋) Point at which an ORBIT crosses from north to south of the reference plane used for the orbit. For the planets the reference plane is the ECLIPTIC. For satellites it is usually the equator of the planet. *See also* INCLINATION; ORBITAL ELEMENTS

Desdemona One of the small inner satellites of URANUS, discovered in 1986 by the VOYAGER 2 imaging team. Desdemona is about 58 km (36 mi) in size. It takes 0.474 days to circuit the planet, at a distance of 62,700 km (39,000 mi) from its centre, in a near-circular, near-equatorial orbit.

de Sitter, Willem (1872–1934) Dutch astronomer and theoretical physicist who publicized relativity and helped to found 20th-century theoretical cosmology. His work was in CELESTIAL MECHANICS, including a re-determination of the fundamental constants of astronomy and refinement of the orbits of Jupiter's satellites. His double-star observations showed that the speed of light did not depend on source velocity (1913). The **de Sitter model**, which arose from his solution (1917) of Einstein's FIELD EQUATIONS of general relativity, was based on a hypothetical universe containing no matter. It suggested that the Universe was expanding, a conclusion that Einstein did not initially accept, choosing instead to 'prevent' the expansion mathematically by introducing the COSMOLOGICAL CONSTANT. The **Einstein–de Sitter model**, proposed by the two scientists in 1932, is a BIG BANG universe that expands for ever.

Deslandres, Henri Alexandre (1853–1948) French astrophysicist, director of the Meudon and Paris Observatories (1908–29). From studies of radial velocities and their effects on spectral lines, Deslandres made significant contributions to the understanding of molecular spectra as they appear in stars and planetary atmospheres that were not fully appreciated until the QUANTUM THEORY became

▲ **diamond-ring effect** Sunlight shining through a valley on the lunar limb at the last moments before totality begins, or just as it ends, at a total solar eclipse produces the spectacular diamond ring. The dazzling bead of exposed photosphere appears as if mounted on the ring of the corona surrounding the dark, eclipsing body of the Moon.

more fully developed. He determined accurate rotation periods for Jupiter, Saturn and Uranus. In 1894 Deslandres invented the SPECTROHELIOGRAPH, conceived independently at the same time by George Ellery HALE.

Despina One of the small inner satellites of NEPTUNE, discovered in 1989 by the VOYAGER 2 imaging team. Despina is about 150 km (90 mi) in size. It takes 0.335 days to circuit the planet, at a distance of 52,500 km (32,600 mi) from its centre, in a near-circular, near-equatorial orbit. It appears to act as an interior SHEPHERD MOON to the planet's Le Verrier and Lassell rings.

Destiny US laboratory module for the INTERNATIONAL SPACE STATION.

detached binary BINARY STAR system in which the two stars are gravitationally bound but physically separate. *See also* SEMIDETACHED BINARY

deuterium Heavy isotope of HYDROGEN. A deuterium nucleus contains one proton and one neutron and is denoted either by $^{2}_{1}$H or $^{2}_{1}$D.

De Vaucouleurs, Gerard (1918–95) French-American astronomer (born Gérard Henri de Vaucouleurs), who spent much of his career (1965–95) at the MCDONALD OBSERVATORY, and specialized in galaxies and other aspects of observational cosmology. He measured the brightness and distance for hundreds of galaxies, working largely at MOUNT STROMLO OBSERVATORY (1952–57). From these photometric measurements, he was able to map clusters of galaxies that themselves aggregated to form a 'local supercluster'. De Vaucouleurs' observations of galaxy distances suggested a high value for the Hubble constant, around 100 km/s/Mpc, implying a universe that was 50% 'younger' than widely believed.

de Vico, Comet 122P/ Short-period comet discovered on 1846 February 20 by Francesco de Vico (1805–48), Rome, Italy. During its discovery apparition, the comet reached a peak magnitude +5.0, but no definitive orbit was calculated. A comet discovered independently on 1995 September 17 by three Japanese observers (Yuji Nakamura, Masaaki Tanaka and Shonju Utsunomiya) was found to be identical to that seen in 1846; the previous return in 1922 April had been missed. Comet 122P/de Vico reached perihelion on 1995 October 6, again attaining peak magnitude +5.0. The period is now known to be 74.41 years.

de Vico–Swift, Comet 54P/ Short-period comet discovered on 1844 August 23 by Francesco de Vico (1805–48), Rome, Italy. Its short-period orbit was soon recognized, but poor weather and bad viewing geometry prevented further observations until it was discovered again by Edward Swift on 1894 November 21. Comet 54P/de Vico–Swift was next recovered in 1965 following calculations by Brian Marsden (1937–) and Joachim Schubart (1928–). A close passage (0.16 AU) to Jupiter in 1968 increased the comet's orbital period to 6.3 years; its perihelion distance of 1.62 AU makes it unfavourably faint.

dew cap Full-aperture extension placed around the end of a telescope (usually a REFRACTING TELESCOPE or SCHMIDT–CASSEGRAIN) that shields the front element to limit radiative cooling and subsequent condensation of moisture from the air upon the cold glass. Dew may be prevented from forming if the front element is surrounded by an electrically resistive wire loop giving out about 1–3 W of heat. Such 'dew zappers' are popular with amateur users of Schmidt–Cassegrain telescopes, and are also used to protect the lenses of cameras used for long-duration exposures in, for example, meteor photography.

Dialogues Shortened English name of GALILEO's *Dialogo sopra i due massimi sistemi del mondo* ('Dialogue on Two Chief World Systems'). One of Galileo's most famous works, it was published in 1632 and contained his chief defence of the Copernican viewpoint. It was written in Italian and took the form of a conversation between three friends – Salviati, who represented Galileo's own point of view, Sagredo, an intelligent gentleman of Venice, and Simplicio, a rather dense Aristotelian. Its publication led to his trial before the Inquisition, for Copernicanism had been proscribed by a papal edict of 1616.

diamond-ring effect Spectacle seen at SECOND CONTACT or THIRD CONTACT – at the beginning or end of TOTALITY – in a total SOLAR ECLIPSE, when a single dazzling point of the solar disk remains or becomes visible shining through a depression on the Moon's limb, appearing as if mounted on a ring formed by the Sun's corona.

DIB *See* DIFFUSE INTERSTELLAR BAND

dichotomy Exact half-phase of an inferior planet when 50% of the disk is illuminated and the TERMINATOR appears as a straight line. Although generally applied only to Mercury and Venus, the term is occasionally used to describe the half Moon.

Dicke, Robert Henry (1916–97) American radio physicist who, in 1964, postulated that the Universe is pervaded by a 'background glow' of radiation at the microwave frequency that is a remnant of the BIG BANG. This was confirmed in 1964 by Arno PENZIAS and Robert Wilson, who detected radiation at the predicted wavelength of 3.2 cm. In the 1940s Dicke had pioneered the development of microwave radar, and in 1944 he invented the microwave radiometer, a key component of contemporary radio telescopes. He was also an accomplished theorist, making many refinements to general relativity, including the BRANS–DICKE THEORY of gravitation.

differential rotation Rotation of a gaseous or fluid body at a rate that differs with latitude: the equatorial regions rotate more quickly, and revolve faster, than higher-latitude polar ones. The differential rotation of the Sun persists from the PHOTOSPHERE down to the base of its convective zone (*see* SOLAR INTERIOR). Differential rotation is also shown by the GIANT PLANETS. A solid body like the Earth cannot undergo differential rotation: it must rotate so that the angular velocity and period of rotation is the same everywhere.

differentiation Process by which a planetary body develops a layered structure. In a body that grows to a sufficient size by ACCRETION, energy supplied by RADIOACTIVITY, supplemented by kinetic energy of impacting bodies and gravitational compression, will cause melting in the interior. Under the action of gravity, denser material will sink towards the centre, and less dense material will rise towards the surface. These displacements release gravitational potential energy, heating the body further and causing additional melting; once begun, differentiation proceeds to completion on a short timescale. Differentiation of a terrestrial planet yields a nickel–iron CORE and rocky MANTLE and CRUST. The larger ice-rich satellites of the outer planets have differentiated into rocky cores and icy mantles. The distinct compositions of some METEORITES, consisting of iron and/or igneous minerals, indicate that their parent asteroids differentiated and solidified before being disrupted by impacts.

diffraction Slight spreading of a beam of light as it passes a sharp edge. If a beam of light strikes a hole, the beam that passes through the hole spreads slightly at the edges. A similar effect can be observed when waves in the sea hit the narrow opening of a harbour: as they pass into the harbour they begin to spread out. Because their wavelength is large compared to the harbour entrance the spread is large. The very short wavelength of light means that the effect is rarely noticed unless the opening is extremely small.

However, the diffraction of light at the edges of a telescope aperture gives rise to the AIRY DISK observed when a star image is examined at high magnification. The larger the aperture, and the shorter the wavelength, the smaller the Airy disk will be.

Diffraction is used to good effect in a DIFFRACTION GRATING, where many very narrow adjacent holes (usually in the form of long slits) cause the spreading light beams to interfere with each other thus producing a SPECTRUM.

diffraction grating Optical device used to disperse light into a SPECTRUM by means of a series of closely spaced, equidistant, parallel grooves ruled on to its surface. A typical grating will have several thousand such grooves per centimetre; light striking each individual slit is dispersed by DIFFRACTION into a spectrum. The grooves may be ruled on glass, producing a TRANSMISSION GRATING, or on metal to give a REFLECTION GRATING. Diffraction gratings produce quality high-dispersion spectra and are used in astronomical SPECTROSCOPES and SPECTROGRAPHS.

diffuse interstellar band (DIB) Broad ABSORPTION LINES found on ultraviolet, visible and infrared spectra of distant stars (some definitions restrict the term to the lines in the visible range of the spectrum, where the strongest lines are at 443 and 618 nm). Diffuse interstellar bands (DIBs) are found especially when the light from the star has been strongly reddened by interstellar dust absorption along the line of sight. Over 200 such lines have now been observed. Some of the DIBs are very broad. For example, the line at 220 nm is some 40 to 50 nm wide. Others are narrower but still broad compared with the stellar lines: for example, the DIB at 578 nm is about 2 nm wide, while the stellar lines are 0.1 nm or less. DIBs are clearly closely associated with INTERSTELLAR MATTER but their origin is still a matter of debate. Current possible explanations include graphite particles, OH^- radicals on the surfaces of silicate grains, silicate grains themselves, polycyclic aromatic hydrocarbons (molecules with carbon and hydrogen in hexagonal arrays) and complex carbon molecules such as Buckminsterfullerene (C_{60}). An alternative hypothesis suggests that the bands originate when hydrogen molecules absorb two photons simultaneously, the photons can then have a range of energies (that is, the absorption lines will be broad) since it is only the total energy of the two photons that has to equal the exact energy required for the transition. It is likely that there is more than one cause of the DIBs, so several of these suggestions could be correct. *See also* INTERSTELLAR MOLECULES

diffuse nebula Term used in the mid-19th century for any nebulous object in the sky that did not resolve into stars when observed at high resolution. It thus included SUPERNOVA REMNANTS, PLANETARY NEBULAE, REFLECTION NEBULAE, distant galaxies and hot gas clouds.

Nowadays it is taken to refer only to the hot gas clouds, and it is thus a somewhat archaic synonym for HII REGIONS. Diffuse nebulae absorb ultraviolet radiation emitted by hot stars embedded within them. This IONIZES the atoms within the nebula; the subsequent RECOMBINATION of an ion and electron leads to the emission of the optical radiation. The nebulae vary considerably in general appearance depending upon the distribution of gas and stars within them, but the ORION NEBULA and ROSETTE NEBULA are good examples. Nebulae are often the sites of current or recent star formation.

Digges, Leonard (*c.*1520–59) and **Digges, Thomas** (*c.*1545–95) English father and son who applied mathematical methods to surveying, navigation and ballistics. In 1553 Leonard Digges published *A General Prognostication*, one of the earliest astronomical almanacs. As well as astronomical data, it contained practical information on instruments and methods, such as finding local time from star positions.

Thomas Digges published revised editions of his father's works. *Pantometria* (1571) covers surveying and cartography and contains what some have interpreted as a description of a reflecting telescope. *Prognostication Everlasting* (1576) reiterates Leonard's support of the Sun-centred COPERNICAN SYSTEM, and also makes a case for an INFINITE UNIVERSE. Thomas' careful observations of Tycho's Star, the supernova of 1572, coupled with Tycho's own observations of it, have allowed modern astronomers to identify it with a known supernova remnant.

Digitized Sky Survey (DSS) *See PALOMAR OBSERVATORY SKY SURVEY*

diogenite *See* HOWARDITE–EUCRITE–DIOGENITE ASSOCIATION

Dione Fourth-largest SATELLITE of SATURN. It was discovered in 1684 by G.D. CASSINI. It is known to share its orbit with a small satellite, HELENE, which occupies a stable LAGRANGIAN POINT 60° ahead of it. Helene thus bears the same relationship to Dione as do some of the TROJAN ASTEROIDS to Jupiter. Dione is similar in appearance and size to TETHYS, which has the adjacent orbit inside that of Dione. However, Dione has a wider range of crater densities; this is partly attributable to at least two separate localized episodes of cryovolcanic flooding (*see* CRYOVOLCANISM), widely spaced in time, which affected different regions. These episodes may be a result of TIDAL HEATING caused by orbital interaction with ENCELADUS. However, the resurfacing events on Dione must have taken place far longer ago than the most recent resurfacing on Enceladus, because even Dione's least cratered areas have more craters than the most cratered areas on Enceladus. *See* data at SATURN

Diphda Alternative name for the star β Ceti. *See* DENEB KAITOS

dipole Pair of equal and opposite electric charges or magnetic poles separated by a finite distance. A dipole is also a design of ANTENNA for a radio telescope. An ordinary bar magnet is a dipole.

▼ **differential rotation** In a fluid body, such as a star or gas giant planet, the equatorial regions rotate more rapidly than the poles. As shown, a consequence of this is that a set of points lined up on the central meridian will become spread out in longitude over the course of a rotation. Points close to the equator will return to the central meridian earlier than those near the poles.

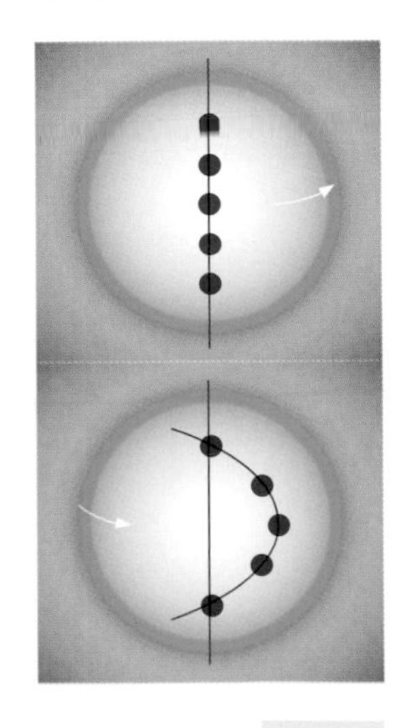

D

direct motion (prograde motion) Orbital or rotational motion in the same direction as that of the Earth around the Sun, that is, anticlockwise when viewed from above the Sun's north pole. For motion to be classed as direct, the inclination of the axis of rotation or orbital motion must be less than 90°. The motion of all planets in the Solar System, and most of the satellites, is direct. Direct motion is also the term given to the regular west-to-east motion of bodies in the Solar System when viewed from Earth, relative to the background stars.

disconnection event Shearing of a COMET's ion tail, resulting from a change in the polarity of the interplanetary magnetic field. Disconnection events often occur when a comet encounters a sector boundary in the SOLAR WIND; they are usually followed by the development of a new, differently oriented ion tail. Comets that appear around the time of sunspot maximum, when the solar wind is particularly turbulent, are especially prone to disconnection events.

Discovery Name of the third SPACE SHUTTLE orbiter; it first flew in 1984.

Discovery Name of a NATIONAL AERONAUTICS AND SPACE ADMINISTRATION (NASA) science programme with the aim of flying diverse missions with a 'faster, better, cheaper' approach.

disk galaxy Name given to any kind of GALAXY in the form of a substantial thin disk of stars in roughly circular orbits about the centre; the brightness of the galaxy decreases with increasing distance from the centre. Disk galaxies include SPIRAL GALAXIES (ordinary and barred) and LENTICULAR (S0) systems. Some IRREGULAR GALAXIES also appear to have thin disks in ordered rotation, though their brightness profiles are less regular. Disks are thought to have been formed while material was still in the form of interstellar clouds, rather than dense stars, so that cloud collisions could remove energy and flatten the distribution into the existing disk.

▼ **disk galaxy** NGC 4945 is a classic disk galaxy at a distance of 13 million l.y. Disk galaxies have their stars mainly confined to a flat, thin circular region surrounding the nucleus.

dispersion Separation of light into its constituent wavelengths by refraction or DIFFRACTION; it is produced by a lens, a prism or a grating. Chromatic dispersion also occurs in fibre-optics, where the speed of propagation is dependent on the colour.

dispersion measure Column density of free electrons along the line of sight to an object. In practice it is of concern mainly for PULSARS. The radio waves from the pulsar are delayed by interactions with the free electrons in space, with the lower frequencies being delayed more than the higher ones. Thus an individual pulse from the pulsar will have its high-frequency component received on Earth before the lower-frequency component. The delay between the receptions of the components of a pulse is called the differential dispersion and it increases as the dispersion measure increases. The unit for dispersion measure is m^{-2} but pc cm^{-3} is often to be found in the literature on the subject.

distance modulus Difference between APPARENT MAGNITUDE (corrected for interstellar absorption) and ABSOLUTE MAGNITUDE of a star or galaxy. It is a measure of the distance to an object, though absorption by the interstellar medium must be considered (*see* REDDENING).

diurnal motion Apparent daily motion of a heavenly body across the sky from east to west, caused by the axial rotation of the Earth. At the poles, a celestial body appears to describe a circle in the sky parallel to the horizon. At other latitudes it describes an arc from horizon to horizon, the length of which varies with latitude and declination; its angle to the horizon varies with latitude alone.

diurnal parallax (geocentric parallax) Difference in the apparent position of a celestial body as measured by an observer at the Earth's surface and by a hypothetical observer at the centre of the Earth. The Earth's daily rotation on its axis causes the positions of nearby objects to appear to shift in relation to the background stars. The effect is maximized when an object is on the horizon, but at a minimum if it is on the meridian. It is so small that it is only noticeable for objects within the Solar System. *See also* ANNUAL PARALLAX; SOLAR PARALLAX

divided telescope (heliometer) Historical instrument for measuring very small angles, consisting of a REFRACTING TELESCOPE with an objective lens cut in half. Each half was mounted on a graduated rack-and-pinion mechanism, so that when the two semi-lenses were slid against each other out of a common line of collimation, the observer saw a split image. The half-lenses were mounted with high precision, and the superimposition of their respective images could then be used to accurately measure very tiny angles. John DOLLOND in London began to make divided achromatic lenses for angle-measuring purposes in the 1760s, although it was in the superb HELIOMETERS by Adolf Repsold (1793–1867) of Munich after 1835 that the instrument reached its peak of excellence.

Dobsonian telescope NEWTONIAN TELESCOPE equipped with a low, stable ALTAZIMUTH MOUNTING; its designer, by John Lowry Dobson (1915–), refers to it as a **sidewalk telescope**. For amateur instruments, the 'Dob' has advantages of economy and is stable enough for useful observation. The design can be applied to large telescopes, and has led to modern amateurs using instruments with far greater apertures – up to a metre (3 ft) or even more – than were common when it first emerged in the 1950s. A Dobsonian requires no great skill to build, and can be made from very simple parts.

The basic features of the mounting are a ground board (or baseplate), topped by Teflon pads on which rests a cradle (or rocker-box). A single loose bolt in the centre of the ground board keeps the cradle in place but allows it to turn in azimuth. The sides of the cradle have Teflon-lined semicircular cut-outs, to accommodate rings bolted to the side of a box or tube containing the primary

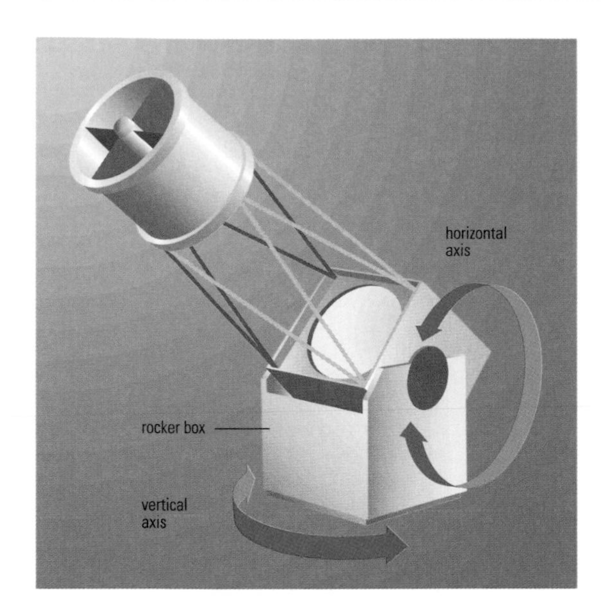

◄ **Dobsonian telescope** This simple altazimuth mounting has enabled many amateur astronomers to construct large-aperture instruments comparatively cheaply. The Dobsonian offers considerable advantages of portability.

mirror. The secondary may be in the same tube, or can be attached, with other components, before observing. Typically, the components are all made from lightweight materials, and the instrument can be disassembled into separate modules for added portability.

Dollond, John (1706–61) English optician who invented the ACHROMATIC LENS. He combined lenses of two different types of glass – crown and flint – to greatly reduce CHROMATIC ABERRATION. Dollond, originally a silk-weaver, joined with his son, the optician **Peter Dollond** (1730–1820), in 1752 to form Dollond & Son, the first company to make achromatic telescopes in any great quantity. The firm lasted through four generations of the family. John Dollond also invented the heliometer, or DIVIDED TELESCOPE, with which the angular separation of two stars could be reckoned.

dome, lunar Small (2–25 km/1–16 mi), gently sloping, generally circular feature found in the lunar maria. Domes are probably small shield volcanoes; they often have a small central pit crater, representing collapse around the vent. Lunar domes are basaltic in composition.

Dominion Astrophysical Observatory (DAO) Canadian national centre for astronomical research at optical wavelengths, situated 10 km (6 mi) north of Victoria, British Columbia. A facility of the Herzberg Institute of Astrophysics, it was founded by John PLASKETT in 1916. Its 1.85-m (72-in.) telescope saw first light in 1918 May and was briefly the largest telescope in the world; a 1.2-m (48-in.) reflector was added later. Today's DAO astronomers also use such major international facilities as the CANADA–FRANCE–HAWAII TELESCOPE and the GEMINI OBSERVATORY.

Dominion Radio Astrophysical Observatory (DRAO) Canadian national centre for astronomical research at radio wavelengths, and a facility of the Herzberg Institute of Astrophysics. It is located at Penticton, British Columbia, 250 km (150 mi) east of Vancouver. The DRAO operates a seven-antenna synthesis telescope and a 26-m (85-ft) fully steerable antenna.

Donati, Comet (C/1858 L1) Bright long-period comet discovered by G.B. DONATI on 1858 June 2. As it approached perihelion (0.58 AU), on 1858 September 30, Donati's Comet became spectacular. In early October, it reached magnitude −1, with a striking, curved dust tail stretching for 60° at the time of closest approach to Earth, on October 9. Two narrow ion tails were also evident. The comet was the first to be photographed (by William Usherwood, Surrey, England). It has an orbital period of about 2000 years.

Donati, Giovanni Battista (1826–73) Italian astronomer who discovered six comets between 1854 and 1864. The most famous, Donati's Comet of 1858, was the most widely observed object of its kind, and beautifully detailed drawings of it encouraged astronomers to make the first serious studies of cometary morphology. Donati made the first spectroscopic observations of comets, from which, in 1864, he concluded that comet tails were made of gases.

Doppler effect Phenomenon whereby the pitch of sound or the wavelength of light is altered by the relative velocity between the observer and the emitting object. The effect was first described for sound waves in 1842 by the Austrian physicist Christiaan Doppler (1803–53), who was director of the Vienna Physical Institute and professor at the University of Vienna. The French physicist Hippolyte Fizeau (1819–96) suggested the extension of the principle to light waves in 1848.

The Doppler effect can be visualized by imagining a transmitter emitting pulses regularly. Each pulse travels out in all directions, producing a pattern of circles. If the transmitter is moving, each circle has a different centre. In the diagram, circles A, B, C and D were emitted when the source was at a, b, c and d. In the forwards direction the circles lie closer together, whilst behind the transmitter they are stretched apart. In the case of light, if an object approaches, the light it emits has shorter wave separations and so becomes bluer; if the object recedes, the light is reddened. The latter is the much referred-to REDSHIFT.

The convention is used that radial velocity is positive when directed away, and negative when directed towards the Earth, so the Doppler formula is:

$$\frac{\lambda_0 - \lambda_L}{\lambda_L} = \frac{v}{c}$$

where λ_0 is the observed wavelength, λ_L is the laboratory (that is, unshifted) wavelength, v is the line-of-sight velocity and c is the velocity of light. Hence the line-of-sight velocity of an object can be determined from measurements of the changed wavelengths of spectrum lines. Most objects have some motion across the line of sight, but the Doppler effect cannot measure these, except at relativistic velocities when the transverse Doppler effect appears as a result of relativistic TIME DILATION.

Distant galaxies and quasars recede at speeds approaching that of light itself. The relativistic Doppler formula must then be used, as follows:

$$\frac{\lambda_0 - \lambda_L}{\lambda_L} = \frac{1 + \frac{v}{c}}{\sqrt{1 - \frac{v^2}{c^2}}} - 1$$

Dorado *See* feature article, page 118

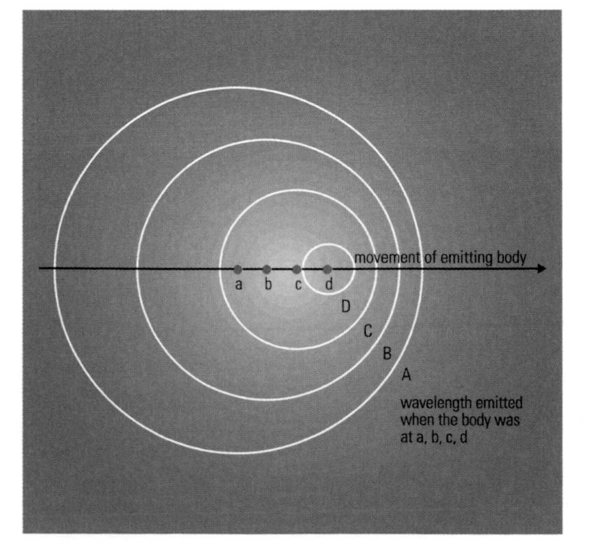

◄ **Doppler effect** Electromagnetic waves emitted from a moving point source appear compressed ahead of, and more spread out away from, the direction of motion. The result is that, for light, the wavelength is shortened (blueshift) ahead of a source approaching the observer and lengthened (redshifted) behind a receding source such as a distant galaxy.

DORADO (GEN. DORADUS, ABBR. DOR.)

Small, inconspicuous southern constellation, representing a goldfish or swordfish, between Pictor and Reticulum. Dorado was introduced by Keyser and de Houtman at the end of the 16th century. Its brightest star, α Dor, is mag. 3.3; β Dor is one of the brightest of all CEPHEID VARIABLES (range 3.5–4.1, period 9.84 days). The constellation contains most of the LARGE MAGELLANIC CLOUD, in which lie the extremely luminous eruptive variable S Dor (range 8.6–11.5; *see* S DORADUS STAR) and the TARANTULA NEBULA (NGC 2070); it was also the site of SUPERNOVA SN 1987A.

BRIGHTEST STARS

	RA h m	dec. ° ′	Visual mag.	Absolute mag.	Spectral type	Distance (l.y.)
α	04 34	−55 03	3.30	−3.6	A0	176
β	05 34	−69 29	3.5–4.1	−4.0–−3.4	F6	1039

Double Cluster (NGC 869 and 884) Pair of open star clusters in the northern constellation PERSEUS. The pair is visible to the naked eye as a hazy patch sometimes referred to as the 'Sword Handle'; it is also catalogued as h and Chi Persei. Each cluster contains several hundred stars. NGC 869 (RA 02^h $19^m.0$ dec. + 57°09′) is the younger, with an estimated age of 6 million years. NGC 884 (RA 02^h $22^m.4$ dec. + 57°07′) is thought to be 14 million years old. Both clusters are products of a burst of recent star formation, which has given rise to the Perseus OB1 association in the next spiral arm of our Galaxy outwards from that occupied by the Sun. The distances of NGC 869 and NGC 884 are 7200 l.y. and 7500 l.y. respectively. Each has an apparent diameter of 29′. NGC 869 has overall magnitude +5.3 and NGC 884 magnitude +6.1.

Double Double (Epsilon Lyrae) Quadruple star system near Vega (RA 18^h $44^m.3$ dec. +39°40′), with two well-matched pairs separated by 208″ (3′.5), making the star a naked-eye double. Telescopic examination shows each of the naked eye components to be double. $Epsilon^1$ (pair A–B) has a separation of 2″.6, components of mags. +5.4 and +6.5 and an orbital period of 600 years. $Epsilon^2$, the more southerly (C–D) pair, has components of mags. +5.1 and +5.3, separated by 2″.3, and an orbital period of 1200 years. The system lies 180 l.y. away.

double star Two stars that appear close together on the sky. An OPTICAL DOUBLE is a chance alignment: in reality the two stars lie at great distances from each other, but they appear to be close together as seen from the Earth. In physical doubles, or BINARY STARS, the two stars lie close enough together to be gravitationally bound, orbiting about their common centre of mass (*see* BARYCENTRE).

doublet Composite LENS made up of two simple lenses. The lenses can be cemented together or separated by a small air gap; doublets are used as ACHROMATS in refracting telescopes and as elements in EYEPIECES.

DQ Herculis star CATACLYSMIC VARIABLE in which the white dwarf component has an extremely strong magnetic field and rotates more rapidly than the orbital revolution (that is, it has non-synchronous rotation). Mass loss occurs from the main-sequence component, and although an ACCRETION DISK may be present, accretion on to the white dwarf occurs primarily through a column at the magnetic poles. There is persistent flickering, together with eclipses in those systems that are aligned with the line of sight. DQ Herculis systems are sometimes known as 'intermediate polars'.

The type star, DQ Herculis, is a 14th-magnitude variable star. It is the remnant of Nova Herculis, which rose to first magnitude in 1934 and was classed as a slow nova. It was particularly notable for a pronounced dip and recovery on the decline. Modern photographs show an expanding shell of gas, which was ejected with a velocity of about 1000 km/s (620 miles/s). DQ Herculis is an ECLIPSING BINARY with an orbital period of 4^h 39^m. The primary component is a white dwarf star and the secondary is a normal K or M dwarf (main-sequence star). In addition to eclipses and flickering, DQ Herculis shows regular variations in brightness with a range of a few per cent and a period of $71^s.0745$; the variations are thought to be caused by rotation of the white dwarf.

Draco *See* feature article

draconic month Time taken for the Moon to complete a single revolution around the Earth, measured relative to its ASCENDING NODE; it is equivalent to 27.21222 days of MEAN SOLAR TIME. *See also* ANOMALISTIC MONTH; MONTH; SIDEREAL MONTH; SYNODIC MONTH; TROPICAL MONTH

Draconids Alternative name for the GIACOBINIDS

Drake, Frank Donald (1930–) American radio astronomer who pioneered SETI, the Search for Extra-Terrestrial Intelligence. Working at the NATIONAL RADIO ASTRONOMY OBSERVATORY, he played a key role in the first SETI survey, called Project Ozma, which in 1960 looked for artificial radio signals from the vicinity of the nearby stars τ Ceti and ϵ Eridani. He introduced the **Drake equation** as an attempt to estimate the number of planets in our Galaxy capable of supporting intelligent life.

Draper, Henry (1837–82) American astronomer and equipment-maker. He took some of the best early photographs of the Moon, planets, comets, nebulae and the solar spectrum with a 28-inch (0.71-m) telescope he completed in 1872. That year, he took the first photograph of the spectrum of a bright star, Vega. To honour his work with photographic spectroscopy, his name was given to the *HENRY DRAPER CATALOGUE*, funded by a donation from his widow and astronomical assistant, **Mary Palmer Draper** (1839–1914). Henry's father was **John William Draper** (1811–82), an English scientist who emigrated to America in 1832, and some time in the winter of 1839/40 obtained the first photographic image (a Daguerrotype) of the Moon.

drawtube Moving portion of a telescope's focuser. The drawtube holds the eyepiece, and the remainder of the focuser assembly permits the drawtube to move in and out along the optical axis.

Dreyer, John Ludvig Emil (1852–1926) Danish astronomer, noted for his catalogues of deep-sky objects, who spent most of his career (1882–1916) as director of ARMAGH OBSERVATORY. During the 1870s he was an assistant at Birr Castle, where, with the 72-inch (1.8-m) reflec-

▶ **Double Cluster** Visible to the naked eye as a hazy patch at the northern end of Perseus, the Double Cluster NGC 869 and 884 is resolved into a pair of rich masses of stars in binoculars or a small telescope. A low-power view like this gives the best impression of the sheer numbers of stars in the two clusters.

tor (*see* BIRR CASTLE ASTRONOMY), he began a systematic study of star clusters and 'nebulous objects' (in the 19th century, galaxies could not be distinguished from true nebulae). In 1888 he published the *NEW GENERAL CATALOGUE* (NGC), following this up in 1895 and 1908 with two *INDEX CATALOGUES* (IC). To this day, astronomers use Dreyer's NGC and IC numbers.

D ring Innermost part of SATURN's ring system. Not a well-defined ring, it is not observable from Earth.

drive Mechanism by which a TELESCOPE is moved quickly (slewed) from one part of the sky to another, or moved slowly (tracked) to follow celestial objects in their sidereal motion.

A tracking motor can be use to turn the telescope about the polar axis of an EQUATORIAL MOUNTING at the sidereal rate of one revolution per 23^h 56^m. Early versions used clockwork or falling weights to power the drive. Modern drives use electric motors, precisely regulated by quartz crystal oscillators. Typically, the motor drives a worm gear – a spiral cut on the surface of a cylindrical rotating shaft – which engages the teeth of a large gearwheel attached to the polar axis. The gear must be carefully engineered to minimize 'play', which is the back-and-forth motion (giving rise to the so-called periodic error) that occurs as first one tooth and then the next touches the worm.

The residual errors of the drive mechanism have to be eliminated by the astronomer, who views a GUIDE STAR in an eyepiece and continually makes small positional adjustments by pressing buttons on a handset. These adjustments are necessary to compensate for flexure of the telescope and its mounting as its position changes, for ATMOSPHERIC REFRACTION, and for errors in the gear construction.

The most advanced telescope drives use computer control. In such systems, the computer rapidly and repeatedly calculates the required position of the telescope, determining for the time given by a quartz clock the hour angle and altitude of the object that it has been instructed to follow. The computer reads the telescope position from encoders attached to the two mounting axes (*see also* GO TO TELESCOPE).

It is possible in principle to attain a pointing accuracy of 1″ by these methods. The final adjustment of the telescope's position is made by visual inspection of the field of view, or by measuring the position of a guide star in the field, using a CCD camera. Both professional and large amateur telescopes are now commonly equipped with CCD-based AUTOGUIDERS.

DSS Abbreviation of Digitized Sky Survey. *See PALOMAR OBSERVATORY SKY SURVEY*

Dubhe The star α Ursae Majoris, visual mag. 1.81, distance 124 l.y., spectral type K0 III. It is a binary star with a close companion of magnitude 4.8 that orbits it every 44 years. Dubhe is the brighter of the two Pointers to the Pole Star. Its name comes from the Arabic *dubb*, meaning 'bear'.

Dumbbell Nebula (M27) Brightest PLANETARY NEBULA in the sky. M27 (NGC 6853) takes its popular name from a description by Lord Rosse, who observed its hourglass shape in his 72-inch (1.8-m) reflector. The Dumbbell has an apparent diameter of 350″ and overall magnitude +7.3. Lying in Vulpecula (RA 19^h $56^m.6$ dec. +22°43′), the Dumbbell Nebula is easy to find by sweeping 3° directly north from γ Sagittae, which marks the tip of Sagitta's Arrow. The Dumbbell lies at a distance of 1250 l.y.

Dunér, Nils Christofer (1839–1914) Swedish astronomer who worked at the Lund and Uppsala observatories. A skilled observer, he designed and built a high-resolution spectroscope, incorporating a diffraction grating made by Henry ROWLAND, and used it to analyse the solar spectrum, confirming the Sun's differential rotation from Doppler shifts in the spectrum. He also made numerous measurements of the distances and position angles of double stars, culminating in a comprehensive catalogue (1876) of those objects.

DRACO (GEN. DRACONIS, ABBR. DRA)

Eighth-largest constellation, representing the dragon Ladon in Greek mythology. It lies between Ursa Major and Cepheus, near the north celestial pole, and surrounds Ursa Minor on three sides. An inconspicuous constellation, its brightest star is ETAMIN (or Eltanin), mag. 2.2. ν Dra is a fine visual binary with bluish-white/white components, both mag. 4.9, separation 62″. In about 2800 BC the pole star was α Dra (THUBAN), mag. 3.7. Draco's deep-sky objects include the CAT'S EYE NEBULA (NGC 6543), a 9th-magnitude planetary nebula in which, in 1864, the so-called nebular lines typical of a mass of hot gas were observed for the first time. The GIACOBINID (Draconid) meteor shower radiates from the constellation.

BRIGHTEST STARS

	Name	RA h m	dec. ° ′	Visual mag.	Absolute mag.	Spectral type	Distance (l.y.)
γ	Etamin	17 57	+51 29	2.24	−1.0	K5	148
η		16 24	+61 31	2.73	0.6	G8	88
β	Rastaban	17 30	+52 18	2.79	−2.4	G2	362
δ	Altais	19 13	+67 40	3.07	0.6	G9	100
ζ		17 09	+65 43	3.17	−1.9	B6	340
ι	Edasich	15 25	+58 58	3.29	0.8	K2	102

Dunsink Observatory Oldest scientific institution in Ireland, founded in 1783, 8 km (5 mi) north-west of the centre of Dublin. Traditionally, its director was the Andrews' Professor of Astronomy in Trinity College; the best known of its past directors was William Rowan Hamilton (1805–65), eminent mathematician and second Royal Astronomer for Ireland. The observatory now houses the astronomy section of the School of Cosmic Physics in the Dublin Institute of Advanced Studies. Both observational and theoretical studies are carried out, with an emphasis on X-ray and infrared satellite data.

dust *See* INTERPLANETARY DUST; INTERSTELLAR MATTER

dwarf Cepheid *See* AI VELORUM STAR; DELTA SCUTI STAR

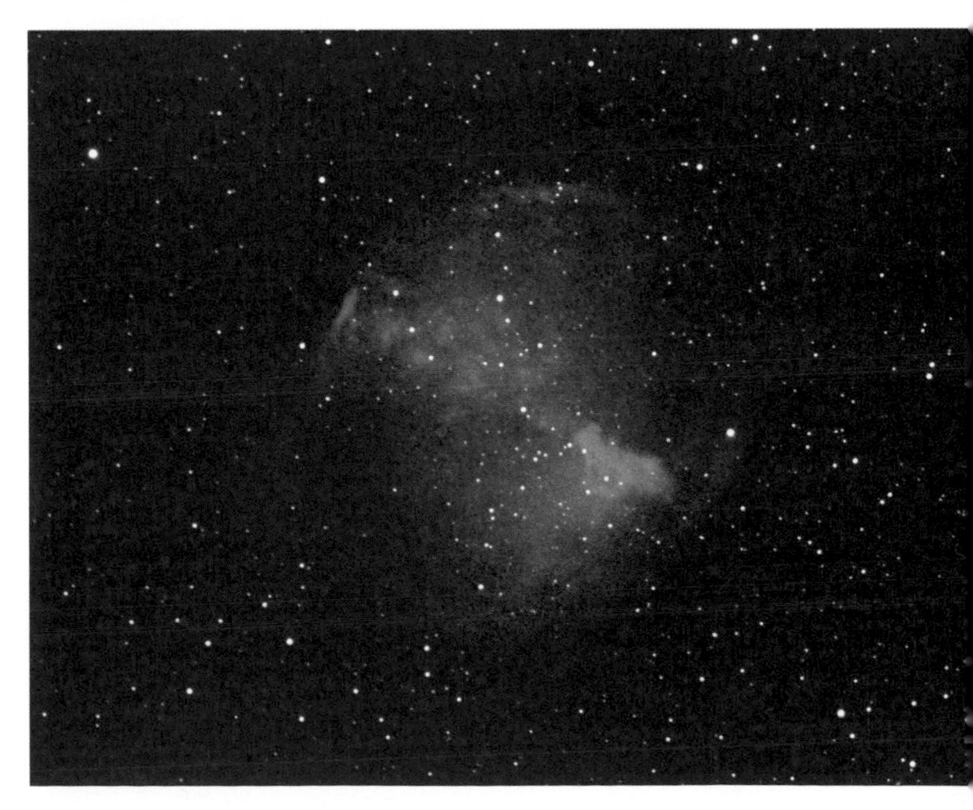

▼ **Dumbbell Nebula** The brightest of the planetary nebulae, M27 the Dumbbell takes its name from the two prominent lobes of material to either side of the central, illuminating star.

D

dwarf galaxy GALAXY that is much smaller and intrinsically fainter than the familiar spiral and giant elliptical systems. A common definition has absolute visual magnitude fainter than −18; by this criterion the MAGELLANIC CLOUDS are not dwarfs but all the other non-spiral members of the LOCAL GROUP are. Dwarf galaxies are usually spheroidal or irregular in structure, and are often found clustered around brighter galaxies.

dwarf nova (DN) CATACLYSMIC VARIABLE that undergoes repeated outbursts. Outbursts are generally similar to those found in a NOVA, but with smaller amplitudes, typically two to six magnitudes. The rise from minimum to maximum takes place in one or two days, and the return to minimum generally takes several days. The average interval, in days, between consecutive maxima is called the mean cycle. Although varying from star to star, each star has its own characteristic mean cycle, which often alters within fairly wide limits. Approximately 370 dwarf novae are known.

Physically, dwarf novae consist of a DWARF STAR or SUBGIANT K or M-type star that fills its ROCHE LOBE and transfers material towards a WHITE DWARF companion, which is surrounded by an ACCRETION DISK. In many systems a hot spot may be observed where the stream of material impacts on the outer edge of the accretion disk.

In the variable-star classification scheme, dwarf novae are designated UG. They are divided into three main subtypes in accordance with their optical behaviour. Each subtype is designated by the name of a type star. The first subtype (UGSS) comprises the SS Cygni or U GEMINORUM stars, which are named after the first two dwarf novae discovered. The cycles for stars of this subtype range from ten to several hundred days. Outbursts are termed either normal (short) or wide (long). The first have durations of a few days whereas the second last for up to about 20 days. Occasionally a star will take several days to rise to maximum; such maxima are called anomalous outbursts.

Stars of the SU Ursae Majoris subtype (UGSU) have two kinds of outburst: normal outbursts last from one to four days; supermaxima last from ten to 20 days, during which time the star becomes from half to one magnitude brighter than at normal outbursts. For almost all maxima, stars of this type rise to near maximum brightness in 24 hours or less. Some have a short pause at an intermediate magnitude on the rise for an hour or two. They also have periodic oscillations (superhumps), with amplitudes of mag. 0.2 to 0.5, superimposed on the light-curve. Their period is about 3% longer than the orbital period.

Z CAMELOPARDALIS STARS (UGZ) make up the third subtype of dwarf nova. They differ from other types in occasionally remaining at an intermediate magnitude between maximum and minimum for several cycles. These 'standstills' occur at unpredictable intervals. Z Camelopardalis stars may sometimes take several days to rise to maximum. Their mean cycles are shorter than those of the other two subtypes, ranging from ten to 40 days.

dwarf star Star lying on the MAIN SEQUENCE. The term 'dwarf' is derived from the early history of the HERTZSPRUNG–RUSSELL DIAGRAM, when stars were divided into the main-sequence 'dwarfs' and the more luminous 'giants'. The term 'main-sequence star' is more usually used today, except for stars of mass lower than about 0.8 solar mass, which are called RED DWARFS. Stellar-like bodies below 0.08 solar mass are called BROWN DWARFS. Main-sequence stars are distinct from WHITE DWARFS and BLACK DWARFS, which are evolved stars in the final stages of their lives. *See also* STELLAR EVOLUTION; SUBDWARF

Dwingeloo Radio Observatory Dutch national centre for radio astronomy, situated some 45 km (30 mi) south of Groningen in northern Holland. It is home to the 25-m (82-ft) Dwingeloo radio dish, one of the oldest purpose-built radio telescopes in the world. The facility is operated by the Netherlands Foundation for Research in Astronomy, and its research includes surveys for radio emission from galaxies behind the Milky Way.

dynamical parallax Method of measuring the distance to a visual binary star based on the estimated masses of the two components, the size of the orbit and the period of their revolution around one another. The angular diameter of the orbit of the stars around each other is observed, together with their apparent brightness, and by applying KEPLER'S LAWS and the MASS–LUMINOSITY RELATION, the distance to the binary star can be determined.

dynamical time System of time used since 1984 as the variable in gravitational equations of motion used for computing the predicted positions, or EPHEMERIDES, of the Sun, Moon and planets. Terrestrial dynamical time (TDT), an atomic timescale based on the SI second, is a continuation of EPHEMERIS TIME, the system previously used in calculations of geocentric ephemerides. Barycentric Dynamical Time (TDB) is used in equations of motion of planetary bodies relative to the BARYCENTRE of the Solar System.

dynamics Branch of mechanics devoted to the study of forces, accelerations and the resulting motions of bodies. NEWTON'S LAWS OF MOTION are the basis of dynamics, and the application of these together with NEWTON'S LAW OF GRAVITY results in KEPLER'S LAWS of planetary motion. A body undergoing circular motion with a radius, r, and at a velocity, v, will be accelerated towards the centre of the circle by an amount given by v^2/r.

It therefore experiences a force, F_c, called the CENTRIPETAL FORCE, which is perpendicular to its velocity and is given by $F_c = \frac{mv^2}{r}$, where m is the mass of the body. For an elliptical orbit, the centripetal force will vary as the distance from the BARYCENTRE changes, and the force will no longer always be perpendicular to the velocity, but the orbital motion is still governed by Newton's laws.

dynamo effect Generation of a planet's magnetic field by electromagnetic induction arising from convective motions in the electrically conducting core of the planet. The power sources must be sufficient to overcome the natural decay of electric currents in a conductor of finite conductivity (that is, the magnetic Reynolds number must exceed a critical value). The motions must have an asymmetry (just as the field coil in a self-exciting dynamo is wound in one sense), and this is attributed to the CORIOLIS FORCE arising from the planet's rotation. Cores in the terrestrial planets are iron-rich and thus electrically conducting; in Jupiter and Saturn they may be formed of metallic hydrogen, whereas for Uranus there may be a conducting electrolyte at its centre. The fields produced are strongly dipolar and roughly aligned along the rotational axes of the planets, except for Uranus. Reversals of the polarity of the magnetic field appear to be an inherent property of the core dynamo.

Dyson, Frank Watson (1868–1939) English astronomer, the ninth ASTRONOMER ROYAL (1910–33). He was co-leader, with Arthur EDDINGTON, of the 1919 total solar eclipse expedition that confirmed Einstein's theory of GENERAL RELATIVITY by detecting the Sun's gravitational deflection of starlight. Dyson supervised the ROYAL GREENWICH OBSERVATORY'S contribution to the massive sky-mapping project known as the *CARTE DU CIEL* and studied stellar proper motion.

◀ **Eagle Nebula** A European Southern Observatory image of the central part of the Eagle Nebula, showing the 'Pillars of Creation'. These dark fingers of nebulosity are associated with continuing star formation.

Eagle Nebula (M16) Bright cluster of recently formed stars and associated EMISSION NEBULA (HII region), catalogued by astronomers as NGC 6611 and IC 4703 respectively. The object lies in Serpens Cauda (RA 18^h $18^m.8$ dec. $-13°47'$), near the border with Scutum. The Eagle Nebula contains the famous 'Pillars of Creation', fingers of dark nebulosity associated with ongoing star formation, imaged from the HUBBLE SPACE TELESCOPE in 1995. The nebula and cluster are about 6500 l.y. away. The Eagle Nebula has an apparent diameter of $35' \times 28'$.

early-type star High-temperature star of spectral class O, B or A. The term was devised when astronomers erroneously believed that the sequence of spectral types was also an evolutionary sequence. The usage persists, 'early' being a common astronomical synonym for 'hot' (for example, a G2 star is 'earlier' than a G5 star). *See also* SPECTRAL CLASSIFICATION

Earth Third planet from the Sun; the largest and most massive of the four TERRESTRIAL PLANETS and the only planet known to support life. Its mean distance from the Sun is defined as one ASTRONOMICAL UNIT, a convenient standard for the description of distances within the Solar System. Its orbit around the Sun defines the plane of the ECLIPTIC, to which the orbits of the other planets, minor planets and comets in the Solar System are referred. The intersection of the ecliptic with the celestial equator at which the Sun appears to move from south to north defines the FIRST POINT OF ARIES, the zero point for the determination of RIGHT ASCENSION.

The Earth has one natural satellite, the MOON, which is believed to have formed by ACCRETION in orbit of material ejected following the impact of a Mars-sized body with the proto-Earth. As a planet, Earth is distinguished by its abundant water, nitrogen–oxygen ATMOSPHERE, and its persistent and general geological activity, including continuing and widespread plate TECTONICS.

SEISMOLOGY has revealed the Earth's internal structure, with a solid inner CORE, radius 1390 km (860 mi), which is believed to consist primarily of nickel and iron but probably containing additional elements such as oxygen or sulphur. This solid inner core is surrounded by a liquid outer core of similar material, with an external radius of 3480 km (2160 mi). The bulk of the planet lies in the MANTLE, which consists of iron- and magnesium-rich silicate materials in a semi-plastic state. Above the mantle is found the rigid LITHOSPHERE, which has two components, the outermost of which is a low-density CRUST. The crust consists of granitic continental rocks, extending to a depth of approximately 50 km (30 mi), and basaltic oceanic rocks, which are only 5–10 km (3–6 mi) in thickness. Denser lithospheric material underlies both continents and oceans to depths of approximately 100–200 km (60–120 mi) and 60 km (40 mi), respectively. The uppermost layer of the mantle, immediately below the lithosphere, has a much lower viscosity than the rest of the mantle, and is known as the ASTHENOSPHERE.

At the Earth's centre, the pressure is estimated to be approximately 4000 kbar and the temperature to lie within the range 5500–7500 K. (The uncertainty arises from difficulties in determining what elements may be present apart from iron and nickel.) The temperature of the liquid outer core is 4000–5000 K. This internal heat comprises residual heat from the original planetary accretion process together with heat produced by the radioactive decay of isotopes of elements such as uranium, rubidium and thorium. Radioisotope dating techniques suggest that the age of the Earth is approximately 4.5×10^9 years, slightly greater than the oldest dates established for lunar rocks (4.44×10^9 years). Because of Earth's level of geological activity, the greatest ages yet discovered are 4.276×10^9 years for zircon crystals from Australia, and 3.96×10^9 years for the Acasta gneiss from Canada. All these ages are exceeded by those found for some components of certain METEORITES, which predate the formation of the Solar System.

The solid inner core rotates slightly faster (by *c.*0.66 second) than the rest of the planet, and it is this, together with motions in the liquid core, that generates the Earth's moderately strong magnetic field. The field is in the form of a magnetic dipole, the axis of which is inclined by *c.*11°.5 to the planet's rotational axis. It has a field strength at the surface of *c.*0.3 gauss at the equator and approximately double this value at the magnetic poles. The magnetic field defines the Earth's MAGNETOSPHERE, and its interaction with the SOLAR WIND forms RADIATION BELTS (the VAN ALLEN BELTS) around the planet. The study of the palaeomagnetism fossilized in the crustal rocks indicates that the positions of the magnetic poles have migrated over geological time relative to the positions of the continents, and that there have been relatively regular reversals in the polarity of the field. Both of these effects, taken in conjunction with radioisotope dating, were instrumental in gaining acceptance of the theory of plate tectonics. The concept of continental drift was originally mooted by the Austrian meteorologist Alfred Wegener (1880–1930), who first lectured on the subject in 1912, but it did not gain widespread approval. The imprint of magnetic reversals in the basaltic rocks of the ocean floor allowed geophysicists to assess the rate and pattern of seafloor spreading, establishing that no part of the ocean floor was more than 200,000 years old, thus confirming the reality of plate motion.

Plate tectonics explains how the Earth's crust consists of about twelve major lithospheric plates and numerous

EARTH: DATA	
Globe	
Diameter (equatorial)	12,756 km
Diameter (polar)	12,714 km
Density	5.52 g/cm^3
Mass	5.974×10^{24} kg
Volume	1.083×10^{12} km^3
Sidereal period of axial rotation	$23^h\ 56^m\ 04^s$
Escape velocity	11.2 km/s
Albedo	0.37
Inclination of equator to orbit	23° 27′
Surface temperature (average)	290 K
Orbit	
Semimajor axis	1 AU = 149.6×10^6 km
Eccentricity	0.0167
Inclination to ecliptic	0° (by definition)
Sidereal period of revolution	$365^d.256$
Mean orbital velocity	29.8 km/s
Satellites	1

► **Earth** En route to the Moon in 1972 December, the Apollo 17 crew obtained this photograph of an almost full Earth. Africa and the Indian Ocean are well presented in this view.

smaller ones, the majority of which are in constant motion relative to one another. These rigid plates move over the more fluid asthenosphere. Plate boundaries are of three kinds: divergent, where adjacent plates are moving away from one another; convergent, where they are colliding; and conservative, where adjacent slabs are sliding past one another. The boundaries between plates are marked by dynamic activities, such as seismic disturbances, volcanism, major faulting and, over long periods of time, by the creation of new mountain ranges.

New oceanic crust is created at divergent boundaries or spreading centres, which are located at oceanic ridges and broader rises, such as the Mid-Atlantic Ridge and the East Pacific Rise. Hot basaltic magma rising from the Earth's mantle accretes on the edges of the diverging plates, which move laterally, creating new seafloor at rates that vary between 0.5 and 10 cm per year. The creation of new oceanic crust is largely balanced by destruction at a type of convergent boundary known as a subduction zone. A subduction zone is a steeply inclined belt where oceanic crust plunges beneath the adjacent plate, which may be of either continental or oceanic type.

Complex processes occur at subduction zones; the subducting slab carries with it some of the overlying sediments and both are eventually melted at depth. Some of the sedimentary veneer is scraped off the plunging slab, and it accretes as highly deformed rocks on the edge of the overlying plate. Magma, generated by frictional heat and the increasing temperatures with depth, is less dense than the overlying rocks and penetrates the upper plate. At the edge of a continental plate this volcanic activity gives rise to major mountain ranges such as the cordilleran chains of North, Central and South America. When the overriding slab is of generally oceanic type, volcanically active island-arcs are produced, as in the Aleutian islands, Japan, Indonesia and New Zealand. Subduction zones are normally marked by deep oceanic trenches, and they are the sites of major earthquake activity, including deep-focus earthquakes occurring at depths as great at 700 km (430 mi). The majority of volcanic and earthquake activity on Earth occurs around the destructive margins of the 'Ring of Fire', which girdles the Pacific Ocean. Major mountain-building also occurs where two continental plates converge. The Himalayas and the high plateau of Tibet were raised by the collision of the Indian and Eurasian plates. They form part of the other major belt of volcanic and earthquake activity, which runs west through Central Asia and the Mediterranean.

Improved seismographic techniques and analysis in recent years have provided considerable information about structures within the mantle. Cold subducting slabs may descend as far as the core–mantle boundary, where rising superplumes of hot mantle material originate. A continental plate that moves over a former subduction zone may be dragged vertically downwards by as much as 200–300 m (650–1000 ft). A continental plate above a superplume may be raised by a similar amount, as is occurring at present under southern Africa. Superplumes are also thought to be responsible for generating major and prolonged volcanic activity, such as that exhibited by the Deccan traps, the Siberian volcanic province and the Columbia River flood basalts.

The Earth's climate is determined by a number of factors. The overall pattern is set by the differing intensities of insolation at different latitudes. The resulting temperature differences set up pressure gradients, which in turn generate the complex atmospheric circulation. The latter is a principal factor in the oceanic circulation, although temperature and salinity gradients are also involved. Because the oceans act as a heat reservoir, they exert a great influence over decade-long changes in weather and climate. Over the course of geological time there have been significant changes in the Earth's climate, most notably shown by recurrent periods of major glaciation, or ice ages, as well as periods of widespread warmth. Plate tectonics has played a major part in initiating some of these climatic changes, because of its effects upon the oceanic circulation and the distribution of land masses at high or low latitudes. The immediate cause, however, appears to have been fluctuations (as well as secular change) in the CO_2 concentration in the atmosphere, with astronomical factors (such as variations in the eccentricity of the Earth's orbit, changes in inclination to the ecliptic,

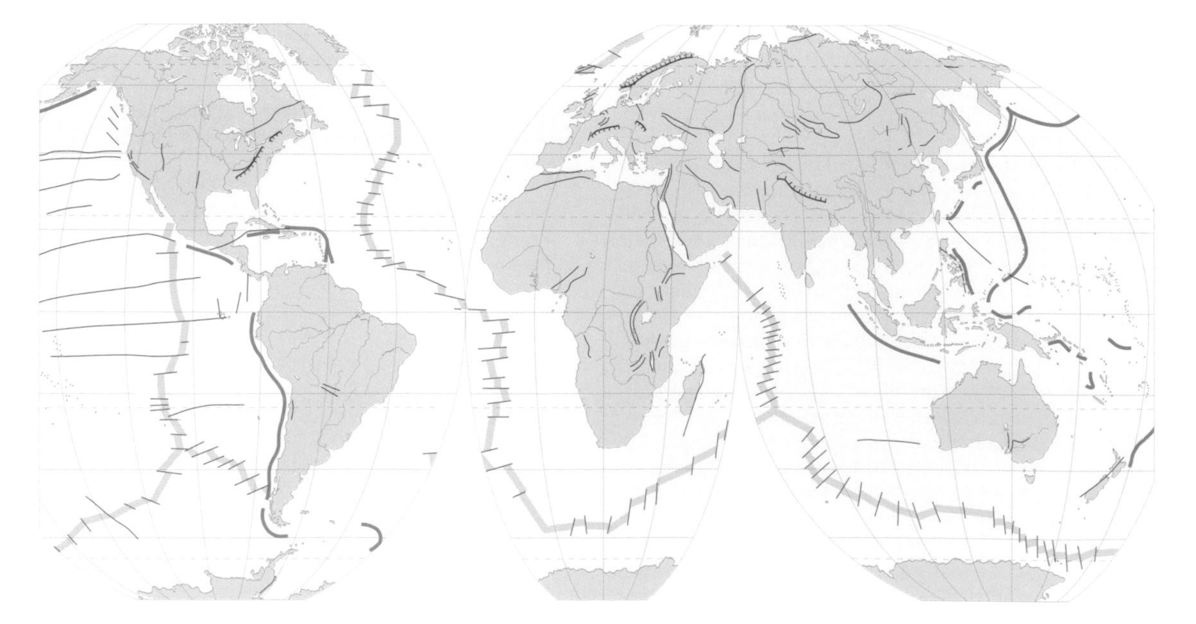

► **Earth** A geological map of Earth, showing the globe-encircling mid-oceanic ridges.

alterations in the timing of perihelion through precession, and possible changes in the solar constant) acting as the principal forcing mechanisms for cyclic change.

Earth-crossing asteroid ASTEROID with an orbit that crosses that of the Earth. *See also* APOLLO ASTEROID; ATEN ASTEROID; NEAR-EARTH ASTEROID; POTENTIALLY HAZARDOUS ASTEROID

Earthgrazer ASTEROID or COMET with a path that brings it close to the Earth's orbit. *See also* POTENTIALLY HAZARDOUS ASTEROID

earthshine Illumination by sunlight reflected off the surface and atmosphere of the Earth of that part of the Moon's disk that is in shadow. The effect is best observed either just before or soon after new moon, when the faintly illuminated unlit area can be seen nestling in the thin, brightly lit crescent, a phenomenon popularly described as 'the old moon in the new moon's arms'. As the illuminated portion of the lunar disk increases in size the earthshine becomes too faint in comparison to be detected. At the same time, the phase of the Earth, as seen from the Moon, is shifting from full to gibbous, meaning that less light is reflected from it.

eccentric Point introduced by Ptolemy to improve the description of planetary orbits. It is the centre of the deferent circle, but is near the Earth, not coincident with the Earth as in earlier geocentric systems (*see* PTOLEMAIC SYSTEM).

eccentricity (symbol e) One of the ORBITAL ELEMENTS; it describes the degree of ELONGATION of an elliptical orbit. The eccentricity is obtained by dividing the distance between the foci of the ELLIPSE by the length of the major axis. A circular orbit has $e = 0$; a parabolic orbit is the extreme case of an ellipse, with $e = 1$.

echelle grating High-dispersion DIFFRACTION GRATING that produces spectra with a high degree of resolution over a narrow band of wavelengths. Like a standard diffraction grating, an echelle grating uses a number of parallel grooves ruled on its surface to disperse light into a SPECTRUM. The difference with an echelle grating is that these grooves are stepped and their spacing relatively wide. Light is directed on to the grating at a large angle, producing a number of high-resolution overlapping spectra. These are then separated by a second, lower dispersion grating or a prism. Echelle gratings are used in echelle SPECTROGRAPHS.

eclipse Passage of one body through the shadow cast by another. A LUNAR ECLIPSE, where the Moon passes through Earth's shadow, is a good example. The Galilean satellites undergo eclipses in Jupiter's shadow: at certain times, twice in a jovian year, the orbits of the satellites align with the Sun such that one may eclipse another. Strictly speaking, a SOLAR ECLIPSE is an OCCULTATION, and the Moon's shadow covers only a relatively small part of the Earth's surface.

eclipse year Interval of 346.62 days, which is the time between successive passages of the Sun through a node of the Moon's orbit. Eclipses can only occur when the Sun and Moon are aligned at the node. The eclipse year is shorter than the sidereal year of 365.25 days because the nodes move westwards (regress) by about 19° per year. A SAROS contains 19 eclipse years.

eclipsing binary Subset of the CLOSE BINARY star systems – BINARY STARS whose separation is usually of the order of ten times, or less, the average size of the components. The likelihood of observing eclipses in binary systems is greatest when the mean separation of components is just a few times greater than the mean radius of the stars themselves.

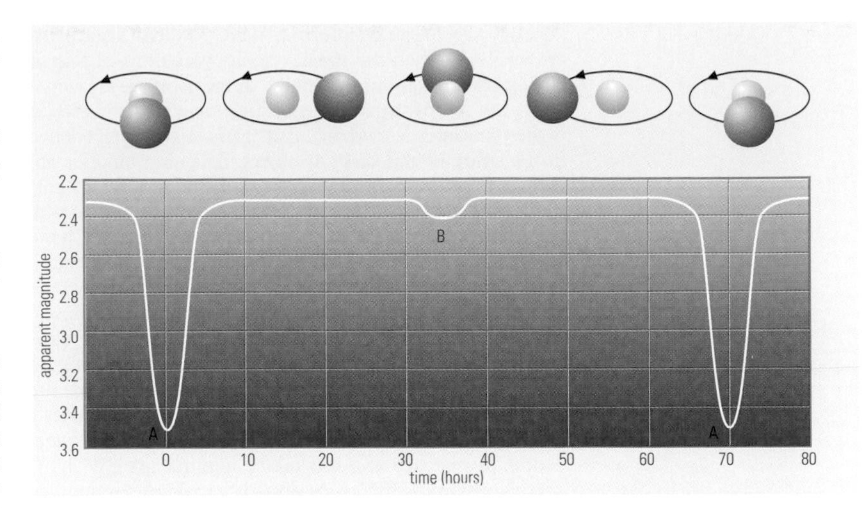

▲ **eclipsing binary** The light-curve (bottom) of an Algol-type binary can be accounted for by partial eclipses of each star by the other (top). At the deep primary minimum, the brighter, smaller star is largely covered by the fainter. When both stars are fully visible, light is at a maximum, while a less prominent secondary minimum occurs midway through the cycle as the brighter star passes in front of the fainter.

The component stars of most classical eclipsing pairs have similar radii. The light from a system in which one star is very much larger than the other would tend to come predominantly from that large star, so that the eclipse of the small star (OCCULTATION), or its passage in front of the large star (TRANSIT), would tend to have an effect too small to be readily observed, unless that small star were relatively very bright (per unit surface area) compared to the large star.

Eclipsing binaries undergo a regular cycle of variation in apparent brightness because their orbital motions take place in a plane that is oriented at a relatively low angle to the line of sight. For eclipsing binaries, the INCLINATION is usually not far from 90°, and, in any case, greater than about 58°. There will be a succession of two alternating photometric minima – one a transit, the other an occultation (unless the stars are exactly equal in size). Eclipses may be total, partial or annular.

Because there are two stars in the system, two eclipses may be expected during the course of an orbital cycle, though it sometimes happens that one of these events causes so slight a loss of overall light from the system as not to be noticeable. The deeper eclipse minimum, associated with the eclipse of the brighter star, is usually called the primary minimum. The star so eclipsed would tend to be called the primary star, and, in practice, this star is very often the more massive. The shallower secondary minimum is associated with the eclipse of the secondary component.

The components of most eclipsing binary systems revolve in orbits that are essentially circular. Elementary geometry will then show that the same effective area of stellar surface is eclipsed at both minima, and the minima are also of equal overall duration (for circular orbits). Since the eclipsed areas are equal at corresponding phases of either minimum, the ratio of depths in the minima at these phases is proportional to the brightness per unit area. The primary star is thus the star of greater surface temperature; though it is possible for the secondary, though cooler, to be bigger in size (the primary minimum would then be of occultation type), and actually to emit more light overall than the primary. This situation may be confusing when SPECTROSCOPY is being combined with PHOTOMETRY, in which case the spectroscopic primary (source of greater fraction of overall light) would then differ from the photometric primary (the hotter star).

Eclipsing binary systems attract attention because their light-curves can be analysed to determine key parameters of great astrophysical significance. From such systems, information on the sizes, luminosities, surface temperatures, masses, and, to some extent, composition and structure of the component stars may be derived. Such absolute parameters are not derived from a single monochromatic light-curve (though surface temperatures may be inferred from two such curves at different wavelengths), but if spectroscopic evidence is available on the system (such as the spectral type and spectral luminosity class of the brighter, or perhaps both components), then absolute values can be inferred. More powerful arguments are available when the radial velocity variations of one or both stars throughout the course of the orbit are known.

As the use of the word 'eclipsing' suggests, however, it is the photometric evidence that tends to take precedence in the study of this class of object. Eclipsing binaries are classified into three basic types on the basis of their light-curves. Historically, these were described as the ALGOL type (EA), the BETA LYRAE type (EB), and the W URSAE MAJORIS type (EW). EA types have an essentially constant (to within 0.1 mag.) light level outside of eclipses, which are thus clearly defined (at least for the primary eclipse). EB-type light-curves show a continuous pattern of variability. The abrupt variations that mark out the eclipse regions are still distinct, but proximity effects render the regions outside the minima (sometimes called 'shoulders') markedly curved. EW systems show more or less equal depths of primary and secondary minima. In EW light-curves proximity effects show such a scale of variation as to merge smoothly into the eclipse minima, which can no longer be clearly discerned. The proximity effects are explained in terms of two basic types of interaction – the reflection effect and gravitational interactions. Both effects increase with reduced separation of the components.

As well as classification based on light-curve morphology, schemes have been proposed that are influenced by the physics or evolution of the stellar components. A common scheme classifies eclipsing binaries into detached, semidetached and CONTACT BINARY systems, based on the relationship of the member stars to the ROCHE LOBE. If a binary component were to expand beyond this limiting surface, a dynamical instability of some kind would ensue. The detached systems are generally seen as unevolved, each star lying fully within its own Roche lobe. Probably the most frequently observed type is the semidetached kind, of which Algol may be regarded as typical. In these systems the secondary is very close to or in 'contact' with its surrounding Roche lobe. Physical Algol-type close binary systems do not always correspond with the EA-type light-curves: the physical Algols show light-curves that are actually a subset of the EA kind. The contact binaries are characterized by both stars filling their Roche lobes; they would thus be in contact with each other. It is now thought that these binaries, which, as a class, coincide closely with systems showing the EW-type light-curve, overflow beyond their common (inner) Roche limiting volume into a surrounding common envelope.

In addition to these major groupings, there are various smaller subgroups. Some DWARF NOVA systems exhibit eclipses, and the special geometrical circumstance may be particularly revealing on physical properties. In a more extreme physical condition are the X-RAY BINARIES; the eclipse effect may sometimes affect the X-ray radiation itself, thus providing insight into the geometrical arrangement of the components.

Though there is no definitely established case of an eclipsing binary member of a globular cluster, eclipsing systems have been observed in the MAGELLANIC CLOUDS. In principle, they offer the means of comparing stellar properties, in a direct way, over very large reaches of space.

About 6000 eclipsing binary systems are known, including many that also exhibit other forms of variability.

ecliptic Apparent yearly path of the Sun against the background stars. Caused by the Earth's orbital motion, the ecliptic is really a projection of the Earth's orbital plane on to the CELESTIAL SPHERE and, because of the tilt of the Earth's axis, is inclined to the CELESTIAL EQUATOR at an angle of approximately 23°.5, which is known as the OBLIQUITY OF ECLIPTIC. The celestial equator intersects the ecliptic at two points on the celestial sphere, the FIRST POINT OF ARIES and the FIRST POINT OF LIBRA, also sometimes referred to as the EQUINOXES. The ecliptic poles lie 90° from the ecliptic, in the constellations of DRACO and DORADO.

The annual path of the Sun takes it through the familiar 12 constellations of the zodiac, but the effects of PRECESSION, together with re-definitions of constellation boundaries, mean that the ecliptic now also passes through the constellation of OPHIUCHUS. The apparent paths of the planets also closely follow the ecliptic since the planes of their orbits are only slightly inclined to that of the Earth. The name 'ecliptic' derives from the fact that eclipses of the Sun or Moon can only take place when the Moon passes through this plane.

ecliptic coordinates System of locating points or celestial objects on the CELESTIAL SPHERE using the ECLIPTIC as the plane of reference and the measures of CELESTIAL LATITUDE and CELESTIAL LONGITUDE (also known as ecliptic latitude and ecliptic longitude). Celestial latitude is measured in degrees from the ecliptic (0°) in a positive direction to the north ecliptic pole (+90°) and a negative direction to the south ecliptic pole (−90°). Celestial longitude is measured around the ecliptic from 0° to 360°, eastwards of the FIRST POINT OF ARIES, the intersection of the ecliptic with the celestial equator.

ecliptic limits Greatest angular distance around the ECLIPTIC that the new moon or the full moon can be from the Moon's nodes for an ECLIPSE to take place. The limits are 18°.5 for the new moon at a SOLAR ECLIPSE, and 12°.2 for the full moon at a LUNAR ECLIPSE.

E corona Extended form of the solar CORONA. It is produced by interactions of ions with electrons and protons, and radiation from the PHOTOSPHERE, which result in production of EMISSION LINES. The E (emission line) corona is present in the visible part of the spectrum, where about 20 emission lines are present, and also at short, ultraviolet and X-ray wavelengths.

ecosphere Spherical region around a star, also known as the circumstellar habitable zone (CHZ), within which life can exist. Its inner and outer limits may be set at those distances from the star at which the effective energy flux is respectively ~1.2 and ~0.4 times the present value of the solar flux at Earth's orbit (1360 W/m^2). This range is determined by the tolerance of known forms of terrestrial life and the extremes of the Earth's surface BIOSPHERE; it applies only to a planetary body of suitable mass and chemical composition (similar to the Earth's) that orbits entirely within the ecosphere. For some subsurface micro-organisms (*see* LIFE IN THE UNIVERSE), the limiting radii of the Solar System ecosphere extends from ~0.4 to ~40 AU, a zone almost 75 times wider than the ecosphere.

In general, the limiting radii of the ecosphere for a planetary system depend on the luminosity and surface temperature of the primary star. For the Sun, a main-sequence G-type dwarf, the present ecosphere extends from just inside the Earth's orbit to just within the orbit of Mars. The width and distance of the ecosphere changes over time with the evolution of the star. The luminosity of the Sun is steadily increasing, and the limits of the ecosphere are migrating outwards. A star more massive and hotter than the Sun will have a larger and wider ecosphere, while a less massive and cooler star will have a smaller and narrower one.

Eddington, Arthur Stanley (1881–1944) English astrophysicist known for his advocacy of Einstein's general theory of relativity and his discovery of the MASS–LUMINOSITY RELATION. In 1905 he joined the ROYAL GREENWICH OBSERVATORY, where he successfully reduced photographic observations of the asteroid Eros to obtain a more precise calculation of the SOLAR PARALLAX. Eddington used statistical methods to discover the distribution of stars in the Milky Way and their motions. In 1913 Cambridge University named him Plumian Professor of Astronomy, and the next year he was appointed director of Cambridge Observatory.

Eddington soon became interested in Einstein's theory of GENERAL RELATIVITY and its prediction that light rays would be deflected in a gravitational field. The best test of this theory was to observe a total eclipse of the Sun, the nearest body possessing a massive gravitational field. If the positions of stars that appeared very close to the Sun's limb during an eclipse were altered slightly from their 'normal' positions as observed when the Sun was nowhere in their vicinity, this would prove Einstein's theory. For the 1919 total solar eclipse, Eddington organized an expedition to the West African island of Principe and obtained several photographic plates of stars near the solar limb. Back in England, he announced that careful measurement and reduction of these stars' positions confirmed the predicted bending of sunlight.

Eddington's greatest contributions were to the theory of energy production in stellar interiors and stellar evolution, summarized in his classic book *The Internal Constitution of the Stars* (1926). He discovered (1924) that for MAIN-SEQUENCE stars there exists a linear relationship between their absolute bolometric magnitudes (a function of the logarithm of the star's luminosity) and the logarithms of their masses. Eddington correctly modelled the abundance of hydrogen in solar-type stars, and evolved a theory to explain how CEPHEID VARIABLE stars pulsate.

Eddington limit Upper limit to the ratio of luminosity to mass that a star can have before RADIATION PRESSURE overcomes the GRAVITATIONAL FORCE. It was proposed by Sir Arthur Eddington. Early estimates gave a theoretical limit to the mass of stars of around 60 solar masses, while modern estimates give a limit as high as 440 solar masses. *See also* STELLAR MASS

Edgeworth–Kuiper belt (outer asteroid belt) Broadly planar zone, stretching from the orbit of Neptune at 30 AU out to about 1000 AU, populated by a substantial complex of planetesimals currently classed as ASTEROIDS. The existence of this band was proposed on theoretical grounds by Kenneth Edgeworth (1880–1972) in 1949 and independently by Gerard KUIPER in 1951. The basis of their idea was that a vast number of COMETS or planetesimals must have formed in the outer Solar System from the SOLAR NEBULA and that they should persist beyond Neptune, never having accreted to produce a fully fledged planet. The concept was resurrected in the 1980s as an explanation of the origin of SHORT-PERIOD COMETS: such a belt could explain the generally low orbital inclinations of such comets, whereas the spherical OORT CLOUD model could not. Recently it has been realized that in the 1930s Frederick Leonard (1896–1960) had put forward similar ideas to those of Edgeworth and Kuiper.

The first member of the belt was discovered in 1992. Through to late 2001 more than 500 objects had been found in this region. Their sizes are mostly in the range of 100 to 500 km (60 to 300 mi), although larger bodies exist, for example VARUNA. There is no reason to believe that there are not also many smaller members of this belt.

Members of the Edgeworth–Kuiper belt are often called TRANS-NEPTUNIAN OBJECTS (TNOs). Many of them have been found to occupy the 2:3 RESONANCE against the orbital period of Neptune, as does Pluto, and these are collectively termed PLUTINOS; other resonant orbits also exist. Other TNOs, following semi-random orbits, are often termed CUBEWANOS.

Although categorized as asteroids in terms of their naming and cataloguing, members of the Edgeworth–Kuiper belt are thought to be similar to comets in terms of their physical nature. Their spectral reflectivities show them to be reddish, continuing the trend seen from the inner border of the MAIN BELT of asteroids through to its outer edge, and then through the region of the giant planets. Edgeworth–Kuiper belt objects do not show comet-like activity because they are at such huge distances from the Sun. If they leak inwards into planet-crossing orbits they may become CENTAURS, in which case cometary outgassing can begin to occur, as observed in the case of CHIRON. It is thought likely that Neptune's moon TRITON and many of the small outer satellites of the giant planets may have begun their lives in this belt.

effective focal length Working focal length of an optical system that uses several elements to bring light to a focus; the quantity comes into play when, for example, eyepieces are used to increase the IMAGE SCALE of an instrument.

effective temperature Temperature of an object found by assuming that its total emission over all wavelengths is that of a BLACK BODY. The temperature is obtained using the STEFAN–BOLTZMANN LAW. It is similar to the RADIATION TEMPERATURE, except that the latter uses the emission from the object at a particular wavelength, or over a small wavelength region, and hence the Planck equation (*see* PLANCK DISTRIBUTION).

Effelsberg Radio Telescope *See* MAX-PLANCK-INSTITUT FÜR RADIOASTRONOMIE

Egg Nebula (CRL 2688) Young PLANETARY NEBULA in the constellation CYGNUS, near ϵ (RA 21^h $02^m.3$ dec. $+36°42'$). The Egg Nebula may be less than 500 years old. Images from the HUBBLE SPACE TELESCOPE show multiple arcs of material in shells surrounding the central star, indicating that mass ejection during planetary nebula formation is episodic. The Egg Nebula lies at a distance of 3000 l.y.

Egyptian astronomy Astronomy as practised by the ancient Egyptians, until the time of the Persian conquests in the first millennium BC. Although there are historical records dating back to around 3000 BC, there is no evidence that the Egyptians attained the technical level of BABYLONIAN ASTRONOMY. With a less developed geometry and number system, their knowledge of celestial motions was inferior. They did, however, possess a complex pantheon of deities that was closely linked to their constellations and creation myths, and solar rituals.

As in other ancient cultures, the main reason for making astronomical observations was to keep track of the passing of the year and the hours of day and night. Their agriculture depended on the annual flooding of the Nile, which deposited a layer of silt that fertilized the land. This annual event of vital importance had to be predicted, but the original Egyptian calendar was based on a fixed number of lunar months, and soon became out of step with the solar year. The calendar was reconciled with the true solar year by observing the heliacal rising of Sirius (then called Sothis), which heralded the flooding of the Nile. If a heliacal rising occurred in the 12th month, an extra, intercalary month was added, the first time in history that such a step was taken.

By perhaps as early as 3000 BC, the year had been standardized to 365 days, and 12 months of 30 days were each divided into three weeks of 10 days. With each week was associated one of 36 small constellations whose heliacal risings were separated by about 10 days. During the course of an average night (excluding twilight periods), 12 of these constellations were observed to rise, leading to the division of the night into 12 hours. Daytime was likewise divided into 12 hours, and thus arose the 24-hour clock.

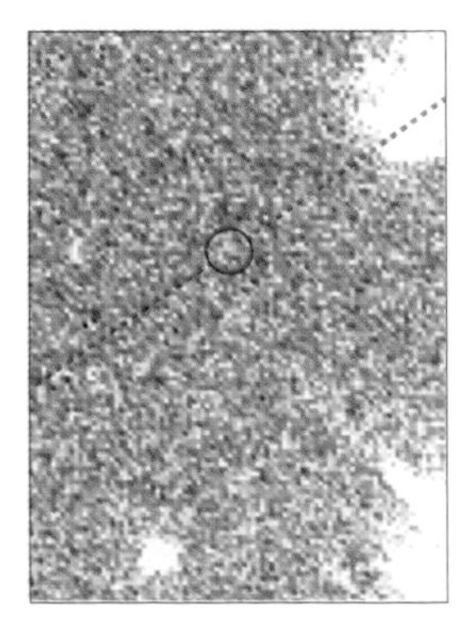

◀ **Edgeworth–Kuiper belt** A pair of Hubble Space Telescope images showing a faint, distant comet nucleus (circled) a few kilometres in diameter in the Edgeworth–Kuiper belt beyond the orbit of Neptune.

E

The passage of time during the day was marked by a T-shaped GNOMON, or shadow stick; later, in the 14th century BC, the clepsydra, or water-clock, was invented.

That the Egyptians were accomplished observers and surveyors is evident from the construction of pyramids and temples, some of which are aligned north–south to an accuracy of a few arcminutes. Several of the pyramids, including the Great Pyramid at Gizeh, have long, narrow shafts angled to the horizontal, which have been conjectured to align on certain stars at their culmination.

Einstein, Albert (1879–1955) German-American theoretical physicist who gained worldwide fame for his theories of SPECIAL RELATIVITY and GENERAL RELATIVITY linking matter, space and time, which had a profound effect on COSMOLOGY. At school and college, he showed no special aptitude; Hermann MINKOWSKI, one of his professors at the Technische Hochschule at Zürich, later expressed his astonishment that such an ordinary student had become a genius. Einstein became a Swiss citizen in 1901 and, while working at the Patent Office in Bern, explained the photoelectric effect by using Max PLANCK's radiation law to describe light in terms of 'quanta', discrete bundles of energy, later called PHOTONS. For this work, Einstein was awarded the 1921 Nobel Prize for Physics.

In 1905 Einstein introduced his theory of special relativity, which held that the velocity of light in a vacuum is constant and that the laws of physics must remain the same in all inertial reference frames. The special theory described space and time as inseparable entities (which he called SPACETIME), predicting the phenomena of time dilation for bodies moving at velocities near that of light and the Lorentz contraction, whereby such an object will appear foreshortened in the direction of its relativistic motion. From the special theory, Einstein derived his famous formula $E = mc^2$ to describe the relation between energy and relativistic mass. By 1909 he was gaining recognition as a leading theoretical physicist, and he became a professor of physics at the University of Zürich. His fame in the international scientific community spread, and in 1914 he was appointed director of the Kaiser Wilhelm Institute, Berlin.

Einstein's theory of GENERAL RELATIVITY, first published in 1915, introduced the gravitational EQUIVALENCE PRINCIPLE – that gravitational fields and uniform acceleration were equivalent. This concept, which defined gravitational fields as the warping of SPACETIME by massive objects, predicted that a massive body like the Sun would produce a gravity field capable of slightly deflecting the paths of light from stars near the solar limb. The best way of testing this prediction was to observe such stars during a total solar eclipse, and in 1919, eclipse expeditions led by Arthur EDDINGTON and others confirmed this effect. The effects of the curvature of space are most evident for the LARGE-SCALE STRUCTURE of the Universe, which led Einstein to apply general relativity to cosmology; he developed a model of a finite but unbounded universe, introducing the concept of a COSMOLOGICAL CONSTANT. From this time onwards, his main scientific interest was an unsuccessful quest for a unified field theory linking gravitation and electromagnetism.

Relativity made Einstein world-famous, a media star like no other scientist before or since, his face a 20th-century icon. As a Jew, Einstein found conditions in Germany intolerable after Hitler's accession in 1933, and he remained in the USA, where he had been visiting Caltech. He became the first professor at Princeton University's Institute for Advanced Study, where he remained for the rest of his life, taking US citizenship in 1940.

Einstein Observatory Name given after its launch in 1978 November to NASA's second High Energy Astronomical Observatory satellite. Its payload comprised four nested grazing incidence X-RAY TELESCOPES with a maximum diameter of 0.6 m (2 ft) and focal length 3.4 m (11 ft)

With an angular resolution of 4″, this satellite revolutionized X-RAY ASTRONOMY. Operational for two and a half years, it detected thousands of new X-ray sources. Its observations confirmed that almost all types of stars emit X-rays. Knowledge of the X-ray properties of QUASARS was greatly enhanced by the Einstein Observatory, while its observations of numerous GALACTIC CLUSTERS transformed our view of these extended X-ray sources.

Einstein ring Distorted image of a distant light source, seen around a GRAVITATIONAL LENS. Einstein's theory of GENERAL RELATIVITY predicted that rays of light should bend in a gravitational field, thus allowing nearby objects to focus the light of more distant objects. In the special case where the source is directly behind the lensing object, its light is spread out into a ring. In practice, perfect rings are rare, though incomplete rings – arcs – are commonly seen through GALAXY CLUSTERS.

▼ **Einstein ring** A Hubble Space Telescope image of B1938 +666, in which a foreground galaxy behaves as a gravitational lens. Light from a more distant galaxy is seen as an almost perfect ring surrounding the lensing object.

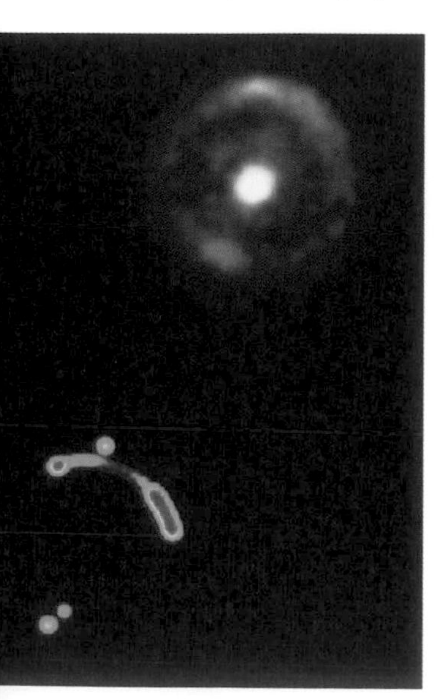

Einstein shift REDSHIFT (increase in wavelength) of spectral lines in radiation emitted from within a strong gravitational field. *See* GRAVITATIONAL REDSHIFT

ejecta Material, such as ash, lapilli and bombs, that is violently thrown out during the eruption of a volcano. Also the impact debris and sometimes impact melt that is ejected in the process of impact CRATER formation. Volcanic ash may be distributed by atmospheric currents up to hundreds or thousands of kilometres from the source. Lapilli and bombs fall in the neighbourhood of the eruption vent. Excavated material forms a concentric ejecta blanket around an impact crater, typically composed of the crushed target rocks with a minor admixture of impactor material. Impact-melted target material is significant in the ejecta from large impact craters. On Mars, ejecta from many impact craters show morphologic evidence of flow after they landed on the surface; they are known as fluidized ejecta. This flow is caused by the presence in the target of ice and water, which lubricate the ejecta and enable it to flow like a mud. On Venus many craters show the presence of ejecta outflows. These radar-bright flows originate within blocky and hummocky ejecta blankets. Ejecta outflows are believed to be flows of superheated impact melt and/or the fine-grained fraction of ejecta debris suspended in Venus' dense atmosphere.

Elara One of the small satellites in JUPITER's intermediate group, discovered by Charles Perrine (1867–1951) in 1905. Elara is about 76 km (47 mi) in size. It takes 259.6 days to circuit the planet, at an average distance of 11.74 million km (7.30 million mi) from its centre. It has a substantial inclination (near 27°) and moderate eccentricity (0.217). Although Elara has the same orbital period as LYSITHEA, their differing orbital characteristics mean they are not usually regarded as being CO-ORBITING SATELLITES in the same way as are other closer, dynamically well behaved, satellites of the giant planets. It seems likely that this pair originated through the break-up of a larger captured body that also spawned LEDA and HIMALIA.

E-layer One of the distinct layers within the Earth's IONOSPHERE where the density of free electrons is higher than average. It is more correctly described as a region, because it lies between 90 and 150 km (about 60–90 mi) above the Earth's surface. It is present during the daytime, but RECOMBINATION occurs at night and the layer weakens or disappears. Radio waves of sufficiently low frequency are reflected by the E-layer. Sometimes, a poorly understood phenomenon known as Sporadic-E occurs, principally in the months of May to August for the northern hemisphere. This effect causes VHF radio waves with frequencies up to 200 MHz to be reflected.

electromagnetic radiation (e-m radiation) RADIATION consisting of periodically varying electric and magnetic fields that vibrate perpendicularly to each other and travel through space at the speed of light. Light, radio waves and X-rays are all examples of electromagnetic radiation (*see* ELECTROMAGNETIC SPECTRUM).

◄ **ejecta** Material thrown out during the formation of the lunar crater Copernicus blankets the surrounding Mare Imbrium. The ejecta is well shown in this oblique-angle Lunar Orbiter 5 image.

The theory of electromagnetic waves was developed by James Clerk MAXWELL, who deduced that an oscillating charged particle would emit a disturbance that would travel through the electromagnetic field at the speed of light. He identified light as a form of electromagnetic radiation.

Electromagnetic radiation is a transverse wave motion, vibrating perpendicularly to the direction of propagation, like the wave in a guitar string. The distance between successive wave crests is the wavelength, and wavelengths range from less than 10^{-14} m (for the shortest gamma rays) to kilometres or more (for the longest radio waves). The number of wave crests per second passing a fixed point is the frequency. The relationship between wavelength (λ), frequency (f or v) and the speed of light (c) is $\lambda f = c$.

High frequency corresponds to short wavelength and low frequency to long wavelength. The unit of frequency is the hertz (Hz), where 1 Hz = 1 wave crest per second. Thus a wavelength of 1 m is equivalent to a frequency of 3×10^8 Hz; a wavelength of 1 nm is equivalent to 3×10^{17} Hz, and so on.

Electromagnetic radiation may be polarized. The state of POLARIZATION describes the orientations of the vibrations. An unpolarized beam consists of waves vibrating in all possible directions perpendicular to the direction of propagation, while a plane-polarized beam consists of waves vibrating in one plane only. In circularly polarized radiation the direction of the vibration rotates at the frequency of the radiation as the wave moves along, while in elliptically polarized radiation, the direction and the amplitude of the vibration change.

Electromagnetic radiation may also be described as a stream of particles, called PHOTONS, each of which is a little packet, or 'quantum', of energy. The energy of a photon, E, is related to wavelength and frequency by:

$$E = hf = hc/\lambda$$

where h is PLANCK'S CONSTANT. Thus the shorter the wavelength (or the higher the frequency), the greater is the energy of the photon. Gamma rays and X-rays have the highest energies while radio waves have the lowest energies.

Light and other forms of electromagnetic radiation exhibit wave properties in some circumstances (*see* REFRACTION, INTERFERENCE and DIFFRACTION) and particle properties in others, notably the photoelectric effect.

Electromagnetic radiation is emitted or absorbed by many processes, and the ones that predominate vary according to the wavelength regions. Thermal radiation is produced at all wavelengths, but the peak of the emission depends upon the temperature (*see* WIEN'S LAW). As a reasonable generalization, for the radio region there is direct interaction with electric currents, FREE–FREE RADIATION and SYNCHROTRON RADIATION, in the microwave and infrared, the interaction is with rotating and vibrating molecules, in the optical and ultraviolet it is with electrons in atoms and molecules, while at the shortest wavelengths the interaction is directly with nuclei.

Almost all our knowledge of the Universe is obtained from electromagnetic radiation emitted from objects in space. The only exceptions are direct sampling of Solar System bodies, COSMIC RAYS, NEUTRINOS, and eventually perhaps GRAVITATIONAL WAVES. *See also* ABSORPTION LINE; ČERENKOV RADIATION; EMISSION LINE; INFRARED ASTRONOMY; RADIO ASTRONOMY; SPECTRUM; ULTRAVIOLET ASTRONOMY; X-RAY ASTRONOMY

electromagnetic spectrum Complete range of ELECTROMAGNETIC RADIATION from the shortest to the longest wavelength, or from the highest to the lowest energy. Although there is no fundamental difference in nature between, say, X-rays and radio waves, the complete range of wavelengths has been divided into a number of sections. The principal divisions (and their approximate wavelength regions) are gamma rays (less than 0.01 nm), X-rays (0.01–10 nm), ultraviolet (10–400 nm), visible (400–700 nm), infrared (700 nm–1 mm), microwaves (1 mm–0.3 m) and radio waves (greater than 0.3 m). Further subdivisions that may be encountered include hard X-rays, soft X-rays, EXTREME-ULTRAVIOLET (EUV), NEAR-INFRARED, middle-infrared, FAR-INFRARED, submillimetre and millimetre wave.

The Earth's atmosphere prevents most wavelengths from penetrating to ground level. Gamma rays, X-rays and the great bulk of ultraviolet radiation are absorbed at altitudes of tens or even hundreds of kilometres, where they excite atoms and molecules, dissociate (break them apart) or ionize atoms or molecules (expel electrons from them).

Radiation of between about 300 nm and 750 nm can penetrate readily to ground level and this range of wavelengths is termed the 'optical window'. Water vapour and

▼ **electromagnetic spectrum** Spanning from short, gamma-ray wavelengths to long-wavelength radio emissions, the electromagnetic spectrum covers a wide range. Selection of appropriate wavelengths allows different processes and classes of objects to be observed. The visible region comprises only a tiny part of the electromagnetic spectrum.

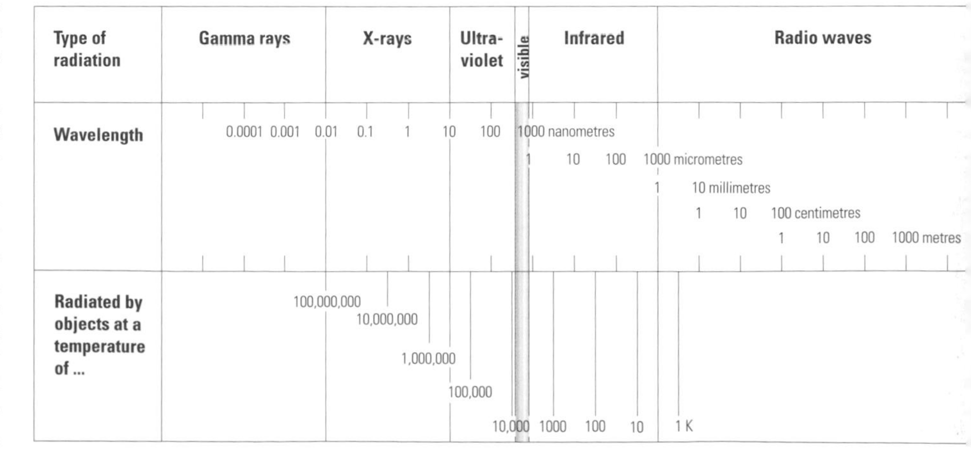

ELEMENTARY PARTICLES

Quarks		Leptons		3 Generations	Force Carriers	
u	up	v_e	e neutrino	III	g	gluon
d	down	e	electron			
c	charm	v_μ	μ neutrino	II	y	photon
s	strange	μ	muon		W	boson
t	top	v_τ	τ neutrino	I		
b	bottom	τ	tau		Z	boson

carbon dioxide are the major absorbers of infrared radiation, but there are a few narrow wavebands at which infrared can penetrate to ground level or, at least, to mountain-top observatories, notably at around 2.2, 3.5 and 10 μm (*see* INFRARED ASTRONOMY).

In the far-infrared there are no atmospheric windows, but a small proportion of submillimetre and millimetre wave radiation can penetrate to high-altitude sites. Radiation in the wavelength range from about 20 mm to nearly 30 m can penetrate to sea level, and this spectral band is referred to as the radio window. Waves longer than 30 m are reflected back into space by the IONOSPHERE.

Following early balloon-borne and rocket-borne experiments, instrumentation carried on orbiting satellites, high above the obscuring effects of the atmosphere, has opened up the entire electromagnetic spectrum and revolutionized our knowledge of the gamma-ray, X-ray, ultraviolet and infrared Universe. *See also* MILLIMETRE-WAVE ASTRONOMY; RADIO ASTRONOMY; SUBMILLIMETRE-WAVE ASTRONOMY; ULTRAVIOLET ASTRONOMY; X-RAY ASTRONOMY

electron Stable ELEMENTARY PARTICLE, one of the family of LEPTONS. Electrons are the constituents of all atoms, moving around the central, far more massive, nucleus in a series of layers or electron shells. Electrons can also exist independently as free electrons. The electron has a mass of 9.1×10^{-31} kg and a negative charge of 1.6×10^{-19} C. The antiparticle of the electron is called the positron. *See also* ATOMIC STRUCTURE

electron volt (symbol eV) Non-SI unit of ENERGY. It is defined as the energy gained by an electron when accelerated by a voltage of 1 volt. 1 eV = 1.602×10^{-19} J. It is often a convenient unit for the energies of PHOTONS and subatomic particles. Thus the visible spectrum ranges from photons with an energy of 1.8 eV (red) to 3.6 eV (violet).

electroweak force Part of the GRAND UNIFIED THEORY (GUT) of the FUNDAMENTAL FORCES. Elementary particle theory suggests that in the early universe when the temperature was near 10^{15} K the electromagnetic force separated from the weak force and they became distinct independent forces.

element Substance that cannot be broken down into simpler substances by chemical means. An element is composed solely of ATOMS with identical atomic numbers (that is, they all have the same number of protons within their nuclei), but it may contain a number of different ISOTOPES. So far 107 elements have been identified, and 90 of the 92 so-called naturally occurring elements have been found in nature, either free, for example copper (Cu), iron (Fe), gold (Au) and sulphur (S), or combined with other elements, for example silicon dioxide (SiO_2) and calcium carbonate ($CaCO_3$). The missing two are technetium, identified in the cyclotron transmutation products of molybdenum in 1937, and promethium, isolated in 1945 from nuclear fission products. The remainder have been synthesized by nuclear bombardment or fission.

The elements fall into three classes according to their physical properties: metals, non-metals and metalloids. Metals are the largest class and are physically distinguished as malleable, ductile, good conductors of heat and electricity, have a high lustre, a close-packed arrangement of atoms, and are usually strong. Non-metals are abundant in Earth's crust and are important biologically. Generally they are poor conductors of heat and electricity, brittle, hard, lack lustre, and have very high melting points. This class though does also include the noble gases (helium, neon and so on). Elements included in the metalloids are boron, silicon, germanium, arsenic, antimony, and tellurium; they have both metallic and non-metallic properties.

All elements have some isotopes that are radioactive. A radioactive element is one that has no stable isotopes. Such elements may exist in nature because they have very long HALF-LIVES, for example uranium-238, which has a half life 4.5×10^9 years, or as decay products of longer-lived radioactive elements, for example radon, whose longest half-life is 3.8 days for radon-222. *See also* ASTROCHEMISTRY; ATOMIC STRUCTURE

element, orbital *See* ORBITAL ELEMENTS

elementary particle Subatomic particles that are thought to be indivisible. Atoms that make up ordinary matter are not the smallest indivisible units of mass. Each atom can be divided up into parts called electrons, protons and neutrons. These smaller, more elementary particles can be arranged to explain all known elements. It was eventually noticed that protons and neutrons were not truly elementary either, but were made up of still smaller particles called quarks. We now know that there are dozens of elementary particles, so many that particle physicists call it the 'particle zoo'. The study of elementary particle physics endeavours to classify and understand all of the known particles, including massless particles like photons. The table lists all of the families of elementary particles known. Some underlying theory must be able to explain the existence and interactions of all of the known elementary particles.

The QUANTUM THEORY was devised to try to explain all of the known elementary particles and their interactions and fields. As work on the quantum theory progressed, more detailed treatments of the various particles and the discovery of new particles necessitated a revision of the standard quantum interpretation. The standard model linked the electromagnetic force with the strong force and explained many of the particles and their various interactions. Later theories like supersymmetry attempted to combine the standard model with the weak force to form a GRAND UNIFIED THEORY (GUT). A more sophisticated theory that would link the final force, gravity, to the three GUT forces is sometimes called the THEORY OF EVERYTHING (TOE).

ellipse One of a family of curves known as CONIC SECTIONS, with the significance for astronomy that the unperturbed orbits of the planets and satellites are ellipses around a primary body located at one of the foci of the ellipse. The other focus is empty. The longest line that can be drawn through the centre of the ellipse is the major axis; the shortest line is the minor axis. These two axes are perpendicular to each other. The two foci lie inside the ellipse on the major axis. The sum of the distances from any point on the ellipse to the two foci is

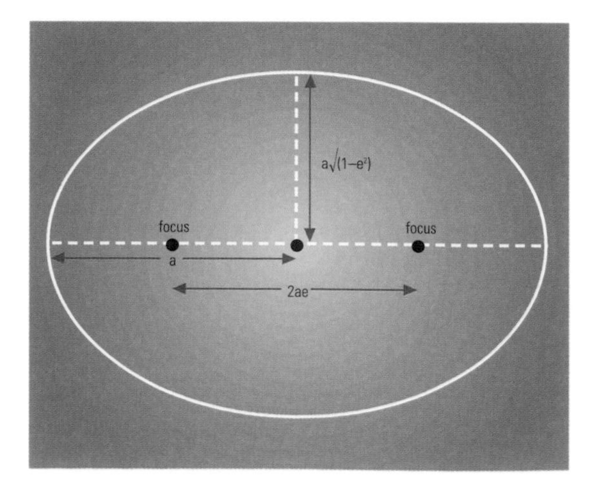

► **ellipse** The principal features of an ellipse are its two foci, its semimajor axis (a), its short (minor) axis and its eccentricity (e). In the case of an ellipse used to define a planetary orbit, the Sun occupies one of the foci while the other is empty. The planets' orbits are reasonably close to circular, but for most comets the orbital ellipse is extremely eccentric, with a very long major axis.

constant, and equals the length of the major axis. The two parameters that are used to describe the size and shape of an ellipse are the semimajor axis a, which is equal to half of the major axis, and the ECCENTRICITY e. The distance between the two foci is equal to $2ae$. The eccentricity can range from 0 for the special case of a circle, where the two foci coincide at the centre of the circle, to almost 1, where the ellipse becomes very elongated, and the part close to either of the foci is virtually a PARABOLA. *See also* ANOMALY; ORBITAL ELEMENTS

ellipsoid Three-dimensional 'egg-shaped' geometrical figure. Mathematically, it is defined as a closed surface whose cross-sections are ellipses or circles. An ellipsoid has three semi-axes: a semimajor axis (a), a semimean axis (b) and a semiminor axis (c), all at right angles to one another. Generally they will be of different lengths ($a > b > c$). If all are equal, the figure is a sphere. If $b = c$ the figure is called a prolate spheroid. If $a = b$ it is an oblate spheroid. An ELLIPTICAL GALAXY is an ellipsoid. A close satellite of a planet is distorted by rotational and tidal forces into a Roche ellipsoid, with the longest axis aligned towards the planet and the shortest in the direction of its orbital pole. A rapidly rotating self-gravitating body may be distorted into an equilibrium figure called a Jacobi ellipsoid. The properties of ellipsoids have been important in the study of the shape and evolution of elliptical galaxies. Ellipsoidal components are found in various optical systems.

ellipsoidal variable (ELL) CLOSE BINARY star system, the components of which are almost in contact, but non-interacting, with the result that they are distorted to non-spherical shapes. Eclipses do not occur, because of the orbital inclination to the line of sight, but variations in the projected area of the stars produce variations in the overall magnitude. Periods are less than one day, with ranges in magnitude of less than 0.8. Such stars are of spectral type F to G.

elliptical galaxy GALAXY that is elliptical in observed shape. Elliptical galaxies are generally composed of only old stars, with little dense gas available for additional star formation. Dynamical measurements show that many are triaxial, geometric figures with different radii along each axis, and thus their shapes reflect the distribution of stellar orbits rather than being produced by a net rotation. In the HUBBLE CLASSIFICATION, ellipticals are described according to apparent shape, from E0 for circular images to E7 for the most flattened ellipticals. Elliptical galaxies span a vast range in size and luminosity, from 100 million stars in a dwarf elliptical to 10 million million in the largest giant ellipticals.

ellipticity (oblateness) Measure of the amount by which an ellipse deviates from a circle or by which a spheroid deviates from a perfect sphere. It is defined as the ratio of the difference between the major (longer) and minor (shorter) axes of an ellipse to the length of the major axis, or in the case of a sphere, the ratio of the difference between the equatorial and polar radii of a body to its equatorial radius. The ellipticity of a celestial body is an indicator of its speed of rotation because the faster it spins, the more it 'bulges' at the equator.

Elnath Alternative form of ALNATH, the name of the star β Tauri.

elongation Angular distance between the Sun and a planet, or between the Sun and the Moon, as viewed from Earth. More accurately, the difference in the CELESTIAL LONGITUDE of the two bodies, measured in degrees. An elongation of 0° is called CONJUNCTION, one of 90° QUADRATURE, and one of 180° OPPOSITION. For an INFERIOR PLANET, the maximum angular distance, east or west, reached during each orbit is known as the GREATEST ELONGATION. When Mercury and Venus are at eastern elongation they may be visible in the evening sky; when at western elongation they may be seen in the morning. *See* illustration at CONJUNCTION

◄ **elliptical galaxy** M87 (NGC 4486) is a giant elliptical galaxy in Virgo. Such objects are usually the largest in galaxy clusters and have probably consumed other, smaller galaxies by cannibalism.

Elst–Pizarro Small Solar System body that has been given dual COMET and ASTEROID designations. It occupies an orbit characteristic of a main-belt asteroid but was observed in 1996 to display a dust tail like that of a comet. It is listed both as asteroid (7968) Elst–Pizarro and periodic comet 133P/Elst–Pizarro. It is approximately 5 km (3 mi) in size. Its orbital period is 5.61 years, and its perihelion distance 2.63 AU. *See also* CHIRON; WILSON–HARRINGTON

ELT Abbreviation of EXTREMELY LARGE TELESCOPE

Eltanin Alternative spelling of ETAMIN, a name given to the star γ Draconis.

Elysium Fossae Aligned valleys crossing the flanks of Elysium Mons on MARS (24°.8N 213°.7W).

embedded cluster Group of stars in the process of formation, still embedded within the interstellar medium from which they formed. These embedded clusters can be detected best in the infrared because the material surrounding them is opaque at other wavelengths.

emersion Re-emergence of a star or planet from behind the Moon's trailing (westerly) limb at an OCCULTATION. The term may also be used to describe the Moon's exit from Earth's UMBRA at a LUNAR ECLIPSE.

emission line Bright point (usually narrow) in a SPECTRUM, either by itself or superimposed on a continuum. An emission line is caused by the radiation of photons of specific wavelength by some species of atom, ion or molecule. Emission lines are produced when electrons associated with the various species jump to lower energy levels and thereby radiate their energy. *See also* EMISSION SPECTRUM

emission nebula Generic term for any self-luminous interstellar gas cloud. The term embraces SUPERNOVA REMNANTS, PLANETARY NEBULAE and HII REGIONS. The emitted radiation is normally taken to be in and around the visible region, thus DARK NEBULAE and REFLECTION NEBULAE are not included in the term even though they do emit microwave and radio radiation.

The energy source for the nebulae varies from one type to another. In HII regions it is the ultraviolet radiation from hot young stars within the nebula. This radiation ionizes the atoms within the nebula and the subsequent RECOMBINATION of an ion and electron leads to emission of optical radiation. A similar mechanism operates within planetary nebulae, although the star involved is old and collapsing down towards becoming a WHITE DWARF. In supernova and NOVA remnants the energy is initially

E

derived from the radioactive decay of nickel and cobalt to iron. Later the energy may be drawn from the kinetic energy of the material, which would have been hurled outwards at velocities of up to 20,000 km/s (12,000 mi/s) by the initial explosion. The collision of this material with the surrounding interstellar medium, or turbulent motions within the expanding material, leads to the heating of the gas and so to radiation being emitted. Collisionally induced emission also causes the radiation from HERBIG–HARO OBJECTS, as the jets of material from the protostar encounter the surrounding nebula. In some supernova remnants, notably the CRAB NEBULA, the central PULSAR provides a continuing source of high-energy electrons, and the nebula emits at least partially through SYNCHROTRON RADIATION.

emission spectrum SPECTRUM that contains EMISSION LINES. A hot solid or high-pressure gas emits a CONTINUOUS SPECTRUM at wavelengths that depend on temperature. A hot, low-density gas produces emission lines that may or may not be superimposed on a continuum. Emission spectra are produced by the solar corona, certain stars, planetary and diffuse nebulae, quasars and other objects. Depending on temperatures and physical processes, they are seen in the range from gamma rays to radio waves. Emission spectra can be used to determine densities, temperatures, velocities and chemical compositions.

Enceladus Small spherical SATELLITE of SATURN. It was discovered by William HERSCHEL in 1789. It is similar in size to its inner neighbour, MIMAS, but in other ways it could hardly be more different. Mimas has an old, heavily cratered surface, whereas Enceladus shows abundant signs of a prolonged active geological history. On images sent back by VOYAGER 2 in 1981, some apparently craterless (and therefore very young) regions are smooth, while other craterless tracts are traversed by sets of curved ridges, which may be cryovolcanic extrusions (*see* CRYOVOLCANISM). Elsewhere, the crater density varies greatly from one region to another, but even the most densely cratered areas have only about the same density of craters as the least cratered areas on Saturn's other satellites. Parts of some old craters at terrain boundaries have been obliterated, probably by cryovolcanic resurfacing processes. Enceladus has clearly experienced several periods of activity, presumably powered by TIDAL HEATING.

Enceladus is currently in a 2:1 orbital RESONANCE with the next-but-one satellite out, DIONE, and in a 1:2 orbital resonance with the small co-orbital satellites JANUS and EPIMETHEUS, the orbits of which lie within that of Mimas. Despite these orbital resonances, Enceladus probably experiences very little tidal heating today because the ECCENTRICITY of its orbit is currently too slight. However, its orbital eccentricity is likely to vary cyclically over tens or hundreds of millions of years, which could account for the multiple episodes of resurfacing that have left their traces on the surface of this remarkable satellite.

Enceladus has a brighter surface than any other icy satellite, giving it an albedo of virtually 1. This brightness could be caused by cryovolcanic frost distributed across the globe by explosive eruptions. Alternatively, it could be a result of continual impacts on to the surface by particles from Saturn's outermost ring (the E ring), within which the orbit of Enceladus lies. *See* data at SATURN

▼ emission nebula The Lagoon Nebula, M8, in Sagittarius is a classic example of an emission nebula. The Lagoon shows strong HII emission, stimulated by ultraviolet radiation from young stars forming in its midst.

Encke, Johann Franz (1791–1865) German astronomer, famous for calculating the orbit of the comet that bears his name. This comet had been observed in the 18th century by Pierre MÉCHAIN and Caroline HERSCHEL, and in 1805 and 1818 by Jean-Louis PONS. Pons suspected that his 'two' comets were actually the same object, and Encke's improved orbital calculations confirmed this suspicion. He also discovered the ENCKE DIVISION in Saturn's rings. Encke supervised the building of the new Berlin Observatory, which he directed from 1825 to 1865.

Encke, Comet 2P/ Short-period comet, first discovered as a naked-eye object by Pierre MÉCHAIN in 1786. Comets discovered in 1795 by Caroline HERSCHEL, in 1805 by Jean Louis PONS, J.S. Huth and Alexis Bouvard (1767–1843), and in 1818 by Pons again were shown by Johann ENCKE in 1819 to be further returns of the same object. Encke's Comet returned as predicted in 1822. To 1999 the comet has been observed at 58 returns, which is far more than any other comet. Its orbital period of 3.3 years is the shortest for any periodic comet. Small changes in the orbital period result from NON-GRAVITATIONAL FORCE produced by gas ejected from the comet's nucleus. Modern detectors allow observation of 2P/Encke around its entire orbit.

Encke Division Main division in Saturn's A RING; it was discovered by J.F. ENCKE. It is not a difficult feature to observe telescopically when the rings are suitably placed. The Encke Division's mean distance from the centre of Saturn is 133,600 km (83,000 mi). As with the CASSINI DIVISION, it is not empty but represents a zone where ring particles are few. This and the other divisions are controlled by gravitational perturbations by the satellites.

Endeavour Name of the fifth SPACE SHUTTLE orbiter; it replaced the *Challenger* lost in 1986. *Endeavour* first flew in 1992.

Endymion Lunar crater (55°N 55°E), 117 km (73 mi) in diameter, with rim components reaching 4500 m (15,000 ft) above its floor. Because it is near the limb, foreshortening makes Endymion appear oblong, though it is actually round. The floor is smooth and dark, having been flooded by lava. This resurfacing has erased most of the inside details, though more recent impacts and ejecta are visible. Because of its age, Endymion has walls that are deeply incised by impact scars.

energy Ability of a body to do work (that is, to induce changes to itself or to other bodies). The unit for energy is the joule (J), defined as the work done by a force of one newton moving through a distance of one metre.

Energy takes many forms: the energy stored by a body

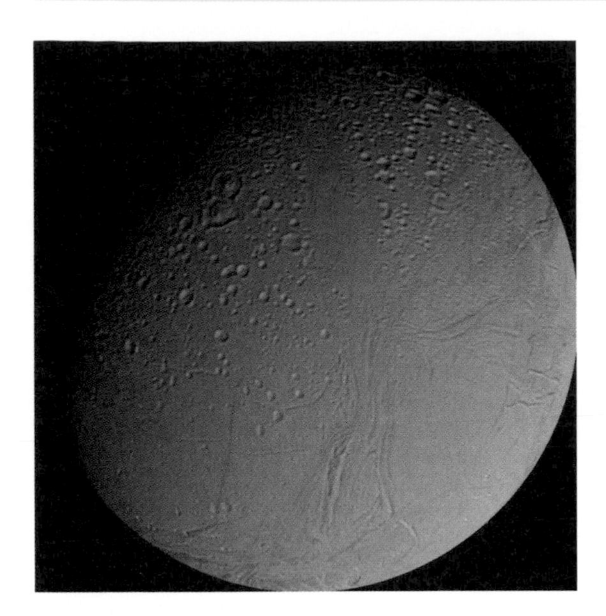

◄ **Enceladus** A composite Voyager 2 view of Saturn's satellite Enceladus. While heavily cratered in some regions, the surface of Enceladus has also been smoothed over by episodes of cryovolcanism.

by virtue of its position, shape or state, is called POTENTIAL ENERGY; the energy that is a consequence of a body's motion through space is called KINETIC ENERGY. Potential energy includes: energy stored in a body that is within a gravitational field and is higher than its equilibrium position; the energy of an electron in an excited state; the energy in a radioactive nucleus before it decays; and mass (via $E = mc^2$). Kinetic energy, as well as the obvious macroscopic motion of bodies (where $KE = \frac{1}{2}mv^2$), includes thermal energy produced by the motion of atoms and molecules. In a closed system energy (or more correctly mass and energy) is conserved, that is to say, the total amount of energy is constant although it may be converted from one form to another within the system.

The energy radiated by stars arises from different processes at different points in their lives. During the formation of a star, gravitational potential energy is being released; throughout most of its active life the energy then comes from nucleosynthesis reactions, while at the end, it is radiating stored thermal energy. Gravitational potential energy is negative since it is clearly zero when two bodies are separated by an infinite distance, and energy is released as they come closer together. For the Universe as a whole it is possible that the total amount of gravitational potential energy balances all other forms of energy so that the net energy of the Universe is zero.

English mounting Form of EQUATORIAL MOUNTING for telescopes.

Ensisheim Meteorite that fell as a shower of stones over Alsace, France, in 1492 November; approximately 127 kg of material was recovered. Ensisheim is an LL-group ORDINARY CHONDRITE, historically significant because it is the first well-documented meteorite fall.

enstatite chondrite (E) Meteorite group that comprises the most reduced of all CHONDRITES. Enstatite chondrites contain abundant iron–nickel metal (up to *c*.15% by volume), very small CHONDRULES and abundant sulphide minerals. E chondrites are further subdivided into two groups on the basis of their iron content – EH (high iron content) and EL (low iron content).

entropy (symbol *S*) Measure of the ENERGY within a system that is not available to do work. It is also often interpreted as the degree of disorder of a system, for example entropy is lower in a crystal of salt and a drop of pure water, and higher in the less-structured system obtained when the salt crystal is dissolved in the water.

When work is done by a system, the entropy increases by an amount equal to the energy released divided by the operating temperature. When work is done on a system the entropy similarly decreases. In an ideal reversible system undergoing equal and opposite changes, the increase in entropy during one change would be balanced by the decrease during the other. The entropy would then be the same at the end of the series of changes as it was at the beginning. All real processes, however, are to some extent irreversible, and therefore such a cycle results in the entropy being higher after the changes than before. Thus in any closed system entropy increases (or at best remains constant) with time. The level of disorder of the system, or its ability to do work, likewise increases with time.

If the Universe is a closed system, then the increase of entropy with time leads to the idea of the HEAT DEATH OF THE UNIVERSE. In other words, the Universe will eventually become a uniform mix of basic particles at a uniform temperature and with any one part of it being quite indistinguishable from any other.

ephemeris Table giving the predicted positions of a celestial object, such as a planet or comet, at some suitable time step, so that the position at intermediate times can be obtained by interpolating between tabular values. *See also* ALMANAC

Ephemeris Time (ET) Timescale used by astronomers from 1960 until 1984 in computing the predicted positions, or ephemerides, of the Sun, Moon and planets. Because the rotation of the Earth on its axis is not regular, neither MEAN SOLAR TIME nor SIDEREAL TIME are accurate enough for computing the gravitational equations of motion required for ephemerides. Ephemeris Time was based in principle on the Earth's revolution around the Sun, and in practice on the orbital motion of the Moon, since these are less affected by unpredictable forces than is the rotation of the Earth. It therefore provided a timescale that was more uniform in the long term than others but was replaced in 1984 by DYNAMICAL TIME.

epicycle In the PTOLEMAIC SYSTEM and earlier systems, a small circle around which a planet was supposed to move; the centre of the epicycle moved, in turn, on a larger circle, the DEFERENT, centred on the Earth.

Epimetheus One of the inner satellites of SATURN; it was discovered in 1978 by John Fountain (1944–) and Stephen Larson (1945–) and confirmed in 1980 in VOYAGER 1 images. It is spheroidal in shape, measuring about 140 × 120 × 105 km (87 × 75 × 65 mi). Epimetheus has a near-circular orbit, with a very slight tilt to Saturn's equator (inclination only 0°.34), at an average distance of 151,400 km (94,100 mi) from the planet's centre. Its orbital period is 0.695 days. *See also* JANUS

epoch Date at which a set of ORBITAL ELEMENTS had the particular values quoted. An orbit is specified by six orbital elements. Five of them define the size, shape and orientation of the orbit, and vary only slowly as a result of perturbations by other bodies. It does not matter much, therefore, if these five elements are used unchanged to compute the position at a time away from the epoch. The sixth element defines the position of the body in the orbit at some particular time. For planets and satellites it is usual to

◄ **Encke Division** Voyager 2 imaged the Encke Division, an apparent gap in Saturn's A ring, during its 1981 flyby. Voyager observations revealed the presence of a small clumpy ring within the Encke Division.

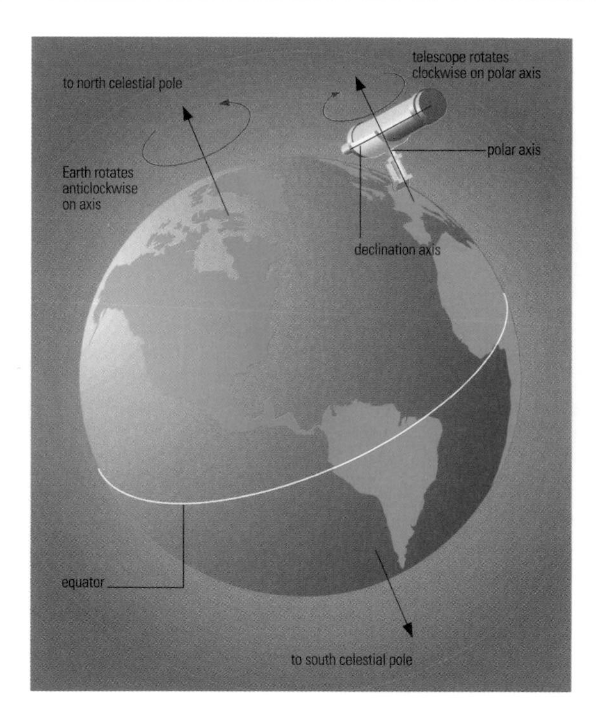

► **equatorial mounting** A useful means of mounting a telescope so that it can be driven to follow the apparent motion of the stars due to Earth's rotation, the equatorial is based around two main axes. The polar axis, around which the driving motor turns the telescope, is aligned parallel to Earth's axis of rotation. North and south movement is made along the declination axis.

take this element as the value of the mean LONGITUDE (or sometimes the mean ANOMALY) at the epoch of the elements, so clearly this quantity will be very much different at times away from the epoch, and this time difference from epoch has to be taken into account. For comets and newly discovered asteroids the sixth element is taken as the time of PERIHELION passage. This property of the orbit should be independent of the date (epoch) at which a set of elements is determined. So for this form of elements used for comets and asteroids the epoch of the elements does not have the significance that it does when the sixth element is the mean longitude or mean anomaly.

Epsilon Aurigae ECLIPSING BINARY star that after more than 150 years of observation and scientific research is still not fully understood. The variability of Epsilon Aurigae was first observed in 1821 by a German pastor, J.M. Fritsch, but it was spectroscopic observations by Hans Ludendorff (1873–1941) at the turn of the 20th century that produced the first significant data. The period is approximately 9892 days, which is the longest known for an eclipsing binary.

The primary component of Epsilon Aurigae is a luminous SUPERGIANT star that exhibits an F0-type spectrum. It is a naked-eye object with apparent magnitude +3.0 at maximum brightness and +4.1 during primary eclipse. Recent reports associate semiregular pulsations with the primary, but it is on the whole not a particularly remarkable star. All the interest in Epsilon Aurigae is associated with the mysterious companion star. The primary F0 spectrum is observed at all phases of the system's revolution (even at primary minimum) and, more surprisingly, no single feature has yet been observed that may be assigned to the secondary.

The light-curve of Epsilon Aurigae is also unusual. Primary eclipse has been observed in 1928–30, 1955–57 and 1982–84. This minimum is 'flat-bottomed' such that the total phase lasts approximately 400 days within an entire eclipse phase of approximately 700 days. The flat-bottomed nature of the primary minimum normally implies that the primary component is eclipsed by a larger companion. At primary eclipse the apparent magnitude of Epsilon Aurigae decreases by 0.8 mag. Such a considerable reduction in brightness would normally imply that the two components have similar surface brightness. A similar loss in brightness, therefore, would be expected when the secondary component is eclipsed by the primary, in which case a secondary minimum of comparable depth to the primary minimum should be observed. In the case of Epsilon Aurigae, however, no secondary minimum is observed at all.

Thus, evidence concerning the nature of the secondary component of Epsilon Aurigae is inconclusive and apparently contradictory. More sophisticated observations have been made and various models proposed in order to explain them.

Early models concentrated on the possibility of a secondary component larger than the primary. One idea proposes an object some 3000 times larger than the Sun. It is suggested that the object is cool (perhaps less than 1000 K), emitting strongly at infrared wavelengths but remaining sufficiently dim at optical wavelengths to explain observations. Furthermore, this object would be tenuous and therefore semitransparent, explaining why the primary component spectrum is observed unchanged even during primary eclipse.

The major problem facing models of this kind is that such large tenuous bodies are not capable of blocking enough radiation from the primary component during eclipse to reduce the apparent brightness by the observed amount. Infrared photometry of Epsilon Aurigae in 1964 by Frank Low (1933–) and R. Mitchell showed no evidence for a cool object of the type envisaged in these models.

Earlier, the Italian astronomer Margherita Hack (1922–) proposed a model for Epsilon Aurigae in which a a small secondary component might explain the complete absence of a secondary minimum in the light-curve. If this object were hot and luminous, however, it would be difficult to understand the absence of a secondary contribution to the observed spectrum. Moreover, with a small secondary object there is difficulty once again in explaining the extent to which brightness is reduced during primary eclipse. It is not clear how the secondary component could block so much radiation from the primary and yet not be noticeable as it, in turn, is eclipsed.

From detailed analysis of the spectrum of Epsilon Aurigae, Otto Struve pointed out that the primary component seemed to possess an extended atmosphere. This led to interpretations based on the idea of a small dark secondary star and a region of dusty matter situated between the two components. A disk of dust viewed edge-on, for example, could conceivably mask radiation from the primary star to the observed degree.

Attention was once again lavished upon Epsilon Aurigae during the eclipse of 1982–84. Observations suggest that the plane of the occulting disk does not coincide with the binary's orbital plane and that the disk itself is warped. Such a configuration could be sustained if the invisible secondary were actually a pair of white dwarf stars. The problem is unlikely to be resolved before the next eclipse takes place in 2009.

Epsilon Boötis *See* IZAR

equant Point that was introduced by Ptolemy to improve the description of planetary orbits. Viewed from the equant, the centre of the planetary EPICYCLE appears to move around the sky at a uniform rate (*see* PTOLEMAIC SYSTEM).

equation of the centre Variation of angular position in an orbit from uniform motion due to the ECCENTRICITY of the orbit. It is the difference between the true ANOMALY f and the mean anomaly M. For small eccentricity e it is given by

$$f = M + 2e \sin M + 5/4\, e^2 \sin 2M + \dots$$

EQUULEUS (GEN. EQUULEI, ABBR. EQU)

Inconspicuous constellation, the second-smallest of them all, representing a little horse, lying near the celestial equator between Delphinus and Pegasus. Its brightest star, Kitalpha, is mag. 3.9. ε (or 1, 'one') Equ is a quadruple star system whose brightest components are both pale yellow, mags. 6.1 and 6.4, separation 0″.9, period 101 years.

equation of time Difference at any point during the year between MEAN SOLAR TIME, as measured by a clock, and APPARENT SOLAR TIME, as shown on a sundial. Because the Sun does not appear to move across the sky at a uniform rate, mean solar time was introduced to provide a uniform civil timescale. On four occasions, April 15, June 14, September 1 and December 25, the difference between it and apparent solar time (and hence the value of the equation of time) is zero; it rises to a maximum of 16 minutes in November.

equatorial coordinates Most commonly used system for locating points or celestial objects on the CELESTIAL SPHERE, using the CELESTIAL EQUATOR as the plane of reference and the measures of RIGHT ASCENSION and DECLINATION. Right ascension is measured along the celestial equator in hours, minutes and seconds eastwards from the FIRST POINT OF ARIES, the intersection of the celestial equator with the ECLIPTIC. Declination is measured from 0° to +90° between the celestial equator and the north CELESTIAL POLE, and from 0° to −90° between the celestial equator and the south celestial pole. HOUR ANGLE and POLAR DISTANCE can be used as alternative measures.

equatorial mounting Telescope mounting in which one axis, the **polar axis**, is parallel to Earth's axis of rotation, and the telescope can be moved about the other axis, the **declination axis**, which is perpendicular to the polar axis. With an equatorial mount, it is possible to track objects across the sky by using a DRIVE mechanism to turn the polar axis at the sidereal rate of one rotation per $23^h\ 56^m$. Smooth, efficient driving depends on having the telescope carefully balanced by appropriate counterweights to minimize stress on the motor and its gear assembly.

There are a number of variations on the equatorial mounting. In the **German mounting** (invented by Johann von FRAUNHOFER), the centre of mass of the telescope lies near one end of the declination axis. Medium-sized amateur REFLECTING TELESCOPES are commonly placed on German mountings, which can be quite portable. The main limitation is in access to the sky in the direction of the celestial pole, where the mounting itself may get in the way of the telescope. Another limitation is the transmission of vibrations from the mounting's pier, particularly to smaller, lighter telescopes.

An alternative is the **English mounting**, where a cradle (yoke) aligned with the polar axis holds the telescope, allowing movement in declination on two pivots. A further variation replaces the cradle with a polar-axis bar, to which the telescope is attached at a single pivot. The English equatorial mounting allows greater access to the sky around the pole than does the German mounting.

To gain access to the areas of sky below the pole, however, a further variation is necessary. In such **horseshoe mountings**, the upper part of the polar axis is designed with a gap sufficiently large to accommodate the telescope. **Fork mountings** are similar to the cradle form of English equatorial, with the telescope held between two arms of a fork aligned with the polar axis. Such mountings are often used for short-tubed SCHMIDT–CASSEGRAIN instruments.

Although equatorial mounts were for a long time the standard for large telescopes, developments in computers and automated drive systems have led to modern professional instruments being mounted on ALTAZIMUTH MOUNTINGS. Similarly, many amateur GO TO TELESCOPES are provided with altazimuth mounts.

equinoctial colure Great circle on the CELESTIAL SPHERE that passes through the two CELESTIAL POLES and intersects the CELESTIAL EQUATOR at the FIRST POINT OF ARIES and the FIRST POINT OF LIBRA (also known as the equinoctial points). *See also* SOLSTITIAL COLURE

equinox Two instances each year at which the Sun appears to cross the CELESTIAL EQUATOR. The spring or VERNAL EQUINOX occurs around March 21, when the Sun is overhead at the terrestrial equator, crossing from south to north. The AUTUMNAL EQUINOX occurs when it crosses from north to south, around September 23. At the time of the equinoxes, day and night are of equal length all over the world (the word equinox means 'equal night'), the Sun rising due east and setting due west on those days. The term is also used to describe the two points on the CELESTIAL SPHERE where the celestial equator intersects with the ECLIPTIC – the FIRST POINT OF ARIES and the FIRST POINT OF LIBRA. The effects of PRECESSION cause these points to move westwards along the ecliptic at a rate of about one-seventh of an arcsecond per day. The equinox defined for a particular day is known as the true or apparent equinox and the inclusion of leap days into the CALENDAR every four years ensures that the true equinoxes always occur around the same date.

equivalence principle Principle stating that the gravitational mass of an object is exactly equal to the inertial mass of the object. Another formulation states that in any enclosed volume of spacetime, one cannot differentiate between an acceleration caused by gravity and one caused by any other force. This principle is one of the pillars upon which GENERAL RELATIVITY is built.

Equuleus *See* feature article

Eratosthenes Lunar crater (15°N 11°W), 61 km (38 mi) in diameter, with rim components reaching 4880 m (16,000 ft) above its floor. Eratosthenes is located at the end of the Montes APENNINES. It is a complex crater, with central peaks and side-wall terracing. Its ejecta is overlain by that of COPERNICUS, and its ray pattern has been eroded by micrometeorite bombardment; deeper impressions of its ejecta are still visible at low Sun angles.

Eratosthenes of Cyrene (276–194 BC) Greek geographer, mathematician and astronomer who was the first accurately to determine the Earth's circumference. He was the third librarian at Alexandria, where the library contained hundreds of thousands of literary and scientific works of classical antiquity. He learned that on the date of the summer solstice, the noonday Sun cast no shadow at Syene (now Aswan, Egypt), and illuminated the bottom of a vertical well there. He observed that, on the same day, the Sun cast a measurable shadow at Alexandria, located due north on approximately the same meridian as Syene. Using a GNOMON (a vertical stake) Eratosthenes found the Sun's zenith distance at Alexandria to be about 1/50th of a complete circle. The Earth's circumference

ERIDANUS (GEN. ERIDANI, ABBR. ERI)

Sixth-largest constellation, representing a river – possibly that into which Phaethon fell after losing control of the chariot owned by his father, the Sun-god Helios, in Greek mythology; alternatively the Nile or Po. Eridanus lies between Cetus and Orion and extends from the celestial equator to within 32° of the south celestial pole. ACHERNAR, its brightest star, marks its southern end. ACAMAR is a fine visual binary with bluish-white components, mags. 3.3 and 4.4, separation 8″.3. o^2 Eri is a triple system with an orange primary and white dwarf secondary, mags. 4.4 and 9.5, separation 93″, period about 248 years – the faint companion is the most easily visible white dwarf in the sky. ε Eri, mag. 3.7, is a nearby Sun-like star (distance 10.5 l.y.), orbited by a Jupiter-sized planet with a period of about 6.8 years (*see* EXTRASOLAR PLANET). Deep-sky objects in Eridanus include NGC 1291, a 9th-magnitude spiral galaxy; and NGC 1535, a 10th-magnitude planetary nebula.

BRIGHTEST STARS

	Name	RA h m	dec. ° ′	Visual mag.	Absolute mag.	Spectral type	Distance (l.y.)
α	Achernar	01 38	−57 14	0.45	−2.8	B3	144
β	Cursa	05 08	−05 05	2.78	0.6	A3	89
$\theta^1\theta^2$	Acamar	02 58	−40 18	2.88	−0.6	A4/A1	161
γ	Zaurak	03 58	−13 30	2.97	−1.2	M1	221

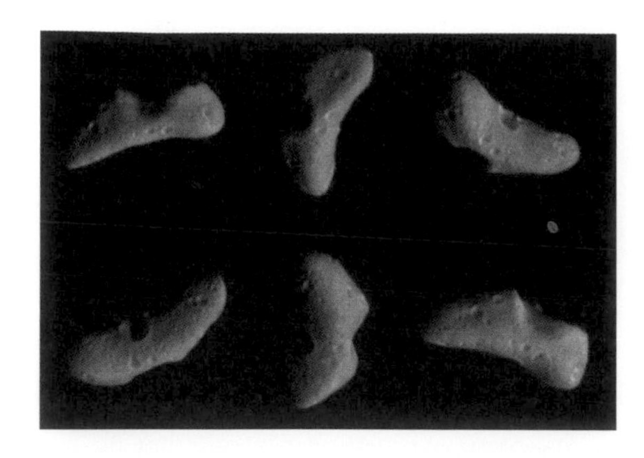

► **Eros** A series of images of (433) Eros, obtained from the NEAR–Shoemaker spacecraft as it made its close approach before going into orbit around the asteroid in 2000 February. The images show Eros' rotation.

was therefore about 50 times the distance between Syene and Alexandria. It is arguable how well he knew this distance, but his estimate in modern terms is about 47,000 km (29,000 mi), quite close to the correct value of 40,075 km (24,900 mi).

Eratosthenes also measured the tilt of the Earth's axis of rotation as 23°51′15″, and compiled a catalogue of 675 naked-eye stars. His mathematical work led him to develop many essential basic concepts of geometry. In geography he divided the Earth's globe into five zones – a torrid zone bordered by two temperate zones, which in turn were bordered by two frigid regions. He was also a chronologist, being the first Greek scientifically to date historical events; he constructed a calendar that contained leap years.

Erfle eyepiece Telescope EYEPIECE designed to have low power and a wide field of view. The design consists of three lenses, at least one of which is a DOUBLET, and it provides an apparent field of view of over 60°. It is named after its inventor, the German optician Heinrich Erfle (1884–1923).

Eridanus *See* feature article, page 133

Eros First NEAR-EARTH ASTEROID to be discovered, in 1898; number 433. With perihelion at 1.13 AU and aphelion at 1.78 AU, the orbit of Eros encompasses that of Mars, and it is classed as an AMOR ASTEROID. It is likely the second-largest object in that class (after GANYMED). Eros' ability to come fairly close to us makes it a particularly bright asteroid. In 1931 Eros passed within 23 million km (14 million mi) of the Earth, and a global programme of measurements of its position relative to background stars was organized in order to determine its parallax and thus its distance. From this effort an improved determination of the ASTRONOMICAL UNIT was expected, but the value derived was slightly too high. In 1975 another concerted campaign of observations was organized for a subsequent close approach by Eros.

Eros was the main target of the NEAR SHOEMAKER space probe, which orbited the asteroid for a full year beginning in 2000 February, before being guided to make a soft landing early in 2001. Many close-up images of this asteroid were returned, plus other science data. The size of Eros was determined to be about 33 × 13 × 13 km (21 × 8 × 8 mi), and its mass 6.7×10^{20} kg. Its rotation period is 5.27 hours. Eros is an S-type asteroid, with an ALBEDO of 0.16. *See also* MATHILDE

eruptive variable Member of a large heterogeneous class of VARIABLE STARS. The class contains many little-understood objects, ranging from SUPERNOVAE, through interacting binary systems, to flare activity on single stars and the ejection of shells of matter by a variety of objects.

The complex phenomena in CATACLYSMIC VARIABLE, SYMBIOTIC STAR and NOVA systems (such as rapid variations, large outbursts and X-ray emission) are primarily caused by the effects of orbital motion, varying rates of mass exchange, and thermonuclear burning of hydrogen that the WHITE DWARF (or possibly NEUTRON STAR) in interacting binary systems has acquired from its companion.

Related objects include X-ray transients such as A0538-66 in which a neutron star is in a highly eccentric orbit about a hot (B-type) giant. When the outer envelope of the hot star is sufficiently distended, the neutron star passes through it, giving rise to a burst of optical and X-ray radiation. Supernovae owe their outbursts to the rapid release of thermonuclear energy, which leads to the ejection of large amounts of mass and the formation of a neutron star. In the case of Type I supernovae, MASS TRANSFER from a companion to a white dwarf in a binary system may initiate its explosion.

UV CETI STARS, for example PROXIMA CENTAURI, are RED DWARF stars showing flare activity (*see* FLARE STAR). Some red dwarfs, such as BY DRACONIS STARS, show light variations that can be attributed to large starspots (roughly analogous to SUNSPOTS). The rotation of the star carries the spots periodically in and out of view, causing the observed light variations. Starspots are a magnetic phenomenon. Magnetic fields constrain the flow of matter and energy near the stellar surface, and lower temperature regions (the spots) form where the magnetic tubes of force cluster. Some red dwarfs are simultaneously BY Draconis and UV Ceti stars. Starspots are also found on the giant and subgiant components in some CLOSE BINARY systems (the RS CVB stars).

T TAURI STARS are very young (pre-MAIN SEQUENCE) objects generally found in regions of recent star formation. They show flaring activity and a strong, variable, chromospheric-type emission spectrum. As in UV Ceti stars, such activity is probably caused by the very strong magnetic field produced in a rapidly rotating star.

S DORADUS STARS are hot, very luminous, supergiants with dense expanding envelopes. Similar, but somewhat fainter, are the GAMMA CASSIOPEIAE STARS, which often show irregular variability because outflow of matter in their equatorial planes forms disks and rings.

R CORONAE BOREALIS variables have evolved past the red-giant stage and have ejected their hydrogen-rich envelopes to expose a carbon-rich, hydrogen-poor core. Their irregular declines in brightness (by up to about 9 magnitudes) are caused by obscuration by clouds of carbon particles, each cloud covering only a limited area of the stellar surface (perhaps about 3%). Declines occur when a cloud is ejected in our line of sight; clouds ejected in other directions can be detected spectroscopically and in the infrared. The process of mass ejection may be related to the pulsations which at least some eruptive variables undergo.

ESA Abbreviation of EUROPEAN SPACE AGENCY

escape velocity Minimum velocity that an object, such as a rocket or spacecraft, must attain in order to overcome the gravitational attraction of a larger body and depart on a trajectory that does not bring it back again. It is, in effect, the velocity required to achieve a parabolic orbit, and is given by $\sqrt{(2GM/R)}$, where M is the mass of the larger body, R is the distance from the centre of the larger body, and G the gravitational constant. It is a factor of $\sqrt{2}$ larger than the CIRCULAR VELOCITY at the same location.

ESCAPE VELOCITIES OF BODIES IN THE SOLAR SYSTEM

Body	Escape velocity (km/s)
Sun	617.0
Mercury	4.4
Venus	10.4
Earth	11.2
Moon	2.4
Mars	5.0
Jupiter	59.5
Saturn	35.5
Uranus	23.5
Neptune	21.3
Pluto	1.1

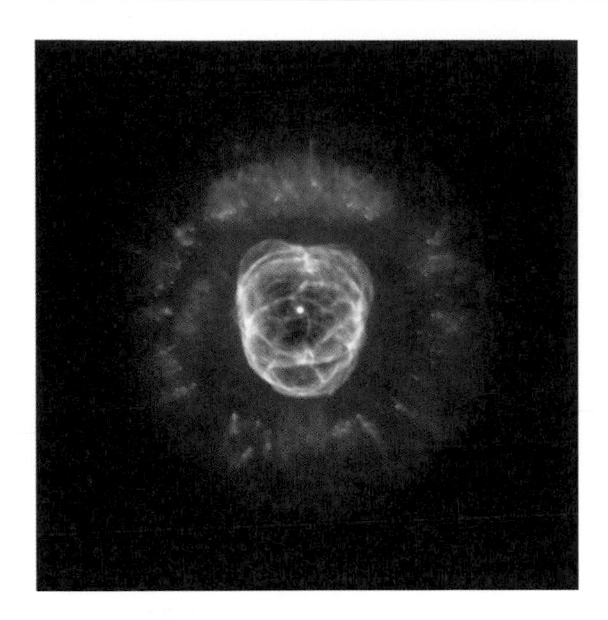

◀ **Eskimo Nebula** Following its 1999 December refurbishment, the Hubble Space Telescope obtained this extremely detailed image of the planetary nebula NGC 2392. The object is popularly known as the Eskimo Nebula, with the outer parts resembling the fur-lined hood of a parka in early photographs.

It is the escape velocity at the surface of a body rather than the mass that determines which gases the body is able to retain in its atmosphere. At the surface of the Earth the escape velocity is 11.2 km/s (6.9 mi/s).

Eskimo Nebula Bright, bluish PLANETARY NEBULA in the constellation GEMINI, a couple of degrees south-east of Delta (RA $07^h 29^m.2$ dec. +20°55′). The nebula has overall magnitude +9.2 and a diameter of 15″. Long-exposure photographs give the impression of a face surrounded by a fur hood, hence the nebula's popular name.

ESO Abbreviation of EUROPEAN SOUTHERN OBSERVATORY

Eta Aquarids METEOR SHOWER with a RADIANT near the 'Water Jar' asterism in Aquarius. It occurs between April 24 and May 20, with a broad maximum over a couple of days around May 4–5. The shower, with peak zenithal hourly rate (ZHR) as high as 50, is, like October's ORIONIDS, produced by debris from Comet 1P/HALLEY. At this encounter, Earth meets the METEOR STREAM at its descending node. Eta Aquarids are swift, at 67 km/s (42 mi/s), and yellowish; they often leave persistent TRAINS. The shower is best seen from southerly latitudes.

Eta Carinae One of the most luminous stars in our Galaxy. When the southern stars were catalogued, Eta Carinae was a third-magnitude star, but from 1833 it began to vary irregularly, becoming at its brightest second only to SIRIUS. What makes this performance remarkable is the great distance of the star, which was then unknown but is now recognized to be about 8000 l.y. Since it is obscured by a bright, diffuse nebula (the ETA CARINAE NEBULA), the star's distance cannot be found with any certainty. At its peak brightness, however, Eta Carinae shone some four million times as brightly as the Sun. The star then faded to just below naked-eye visibility. At the same time, a nebulous patch, called the HOMUNCULUS, formed and expanded around it. The nebula has a double-lobed structure, with an expansion velocity of about 30 km/s (20 mi/s) near the centre but up to 1000 km/s (600 mi/s) at the outer edge. The total amount of mass in the nebula is about ten times the mass of the Sun.

Although optically hidden by the nebula, Eta Carinae is the brightest infrared source in the sky (apart from Solar System objects), which shows that it continues to shine about as brightly to this day as it did last century. The drop in its magnitude was caused by the surrounding nebula, and since the mid-1900s the star has gradually brightened in visual light as the extinction by dust decreases.

Studies of the light reflected from the dust cloud show that the central object varies in emission lines, especially those of hydrogen and neutral helium, with a period of 5.52 years. This suggests that the central object is in fact a BINARY STAR system, consisting of two stars of about 80 solar masses orbiting each other at a distance of about 20 AU. The shape of the bipolar nebula may have been caused by the fact that it is a binary system.

Eta Carinae Nebula (NGC 3372) EMISSION NEBULA surrounding the variable star ETA CARINAE (RA $10^h 45^m.0$ dec. −59°50′). It lies at a distance of about 8000 l.y. and subtends an area of 2° × 2° in the southern sky. Lanes of dark dust split the nebula. It shows much structure, including the bright HOMUNCULUS NEBULA and dark KEYHOLE NEBULA. Changes in structure have been observed over timescales of decades. The Eta Carinae Nebula is the brightest in the sky, with an apparent integrated magnitude of +1.0.

Eta Geminorum Bright semiregular VARIABLE STAR and ECLIPSING BINARY with a range of mag. 3.15 to 3.9 V and a period of 232.9 days. The eclipses recur at intervals of 2938 days. Both components are RED GIANTS with M3 spectra.

Etamin The star γ Draconis, visual mag. 2.24, distance 148 l.y., spectral type K5 III. From observations of this star, the English astronomer James BRADLEY discovered the phenomenon of the aberration of starlight. Its name, which is also spelled *Eltanin,* comes from *al-tinnīn,* meaning 'the dragon', the Arabic name of the constellation Draco.

eucrite *See* HOWARDITE–EUCRITE–DIOGENITE ASSOCIATION

Eudoxus of Cnidus (408–355 BC) Greek astronomer and mathematician who developed a cosmology based on Plato's model of the Solar System. He built an observatory

▼ **Eta Carinae Nebula** The brightest emission nebula in the sky, NGC 3372 surrounds the eruptive variable star Eta Carinae. A spectacular object for observers in the southern hemisphere, the nebula lies 8000 l.y. away.

E

at Cnidus, Asia Minor (in what is now modern Turkey) and another at Heliopolis, Egypt. He described his observations in two books, *The Mirror* and *The Phaenomena*, both unfortunately now lost but referred to by HIPPARCHUS. Eudoxus was the first astronomer to make a specific note of the bright southern star Canopus.

Like PLATO before him, Eudoxus' cosmology consisted of a system of planets fixed to the surfaces of 'homocentric' crystalline spheres rotating on axes passing through the Earth's centre. He adjusted Plato's model to fit better the observed planetary motions by postulating that each of the spheres had its poles set to those of the next sphere, and that the axes of rotation were not fixed in space. Additional spheres were constructed to portray the movements of the planets against the fixed stars and the apparent daily rotation of the stars. In this way he was able to explain the retrograde motions of Jupiter and Saturn, but not of Mars. Although Eudoxus probably considered the homocentric crystalline spheres merely as a mathematical model, ARISTOTLE imagined that they had a physical existence.

Eugenia MAIN-BELT ASTEROID; number 45. In 1998 it was found to possess a small moon, PETIT-PRINCE. Eugenia is about 215 km (134 mi) in size; it orbits at about 2.72 AU. *See also* CAMILLA; PULCOVA

Euler, Leonhard (1707–83) Swiss mathematician who worked at the Berlin Academy and the St Petersburg Academy of Science, making significant contributions to almost every branch of mathematics. In astronomy, he greatly advanced the partial solution of the THREE-BODY PROBLEM, and developed a theory of the Moon's orbit (1772) that allowed Tobias MAYER to construct his tables of lunar positions. Euler developed formulae for finding the orbits of planets and comets from observations of their positions relative to background stars. His investigations in optics influenced the later development of telescopes.

Eureka First MARS-TROJAN ASTEROID to be discovered, in 1990; number 5261.

Europa Smallest of the GALILEAN SATELLITES of JUPITER; it is only slightly smaller than Earth's Moon and is thus a substantial world in its own right. It is in orbital RESONANCE with both IO and GANYMEDE, such that for every four complete orbits by Io, Europa makes two orbits and Ganymede one. The resulting TIDAL HEATING makes Europa an active world, and it is a prime target for exploration for extraterrestrial life in a salty ocean inferred to exist below its thin icy shell. In contrast to the more strongly tidally heated (and ice-free) Io, however, no surface changes have been seen to occur on Europa during the era of close-up imaging by VOYAGERS 1 and 2 (1979) and the GALILEO orbiter (1995–2001).

Europa's density, of 2.97 g/cm^3, almost puts it in the terrestrial planet league, but its exterior is icy down to a depth of about 100 km (60 mi). It is not known whether the ice is solid throughout, or whether its lower part is liquid, which raises the fascinating possibility of a global ocean sandwiched between the solid ice and the underlying rock. Gravity data from Galileo show that, like Io, Europa has a dense, presumably iron-rich CORE (about 620 km/390 mi in radius) below its rocky MANTLE. Europa has its own magnetic field, but it is unclear whether this is generated by convection within a liquid core or within a salty ocean beneath the ice.

Europa has a highly reflective surface with an albedo of about 0.7. Since the 1950s it has been known from spectroscopic studies that the surface is composed essentially of clean water-ice. More detailed recent observations by Galileo and the HUBBLE SPACE TELESCOPE have revealed regions where the ice appears to be salty; they have also revealed the presence of molecular oxygen (O_2) and ozone (O_3). The oxygen and ozone are thought to result from the breakdown of water molecules in the ice by exposure to solar ultraviolet radiation and charged particles from the SOLAR WIND. Hydrogen thus liberated would escape rapidly into space, as has been observed on Ganymede though not on Europa. It is not known whether the oxygen and ozone detected on Europa constitute an extremely tenuous atmosphere or whether they are mainly trapped within the ice.

Europa's surface is relatively smooth and much younger than that of other icy satellites, to judge from the paucity of impact craters. This paucity demonstrates that Europa experiences a rapid rate of resurfacing, driven by tidal heating, though it is not so rapid as on Io.

Neither of the Voyager probes passed close to Europa, so even the best Voyager images revealed few features smaller than a few hundred metres across. They were adequate, however, to show a bright surface with low topographic relief, criss-crossed by a complex pattern of cracks filled by darker ice. There are several places where the pattern of these bright plains becomes blotchy, and these were dubbed 'mottled terrain' by the Voyager investigators. The high-resolution images sent back by Galileo show that the bright plains are amazingly complex in detail, being composed of multiple families of straight or slightly curved ridges, each usually bearing a central groove. The appearance of these parts of Europa has been described as resembling the surface of a ball of string, a graphic description that casts no light on how the surface was created. Each grooved ridge probably represents a fissure, where some kind of icy lava was erupted during an episode of CRYOVOLCANISM.

Although by far the most abundant component in Europa's ice is water, it is likely to be contaminated by various salts (such as sulphates, carbonates and chlorides of magnesium, sodium and potassium) and possibly by sulphuric acid, resulting from chemical reactions between water and the underlying rock. Contaminants such as these could make any melt liberated from the ice behave in

► **Europa** A mosaic of images obtained by the Galileo orbiter, showing 'ice rafts' on the surface of Jupiter's satellite Europa. The reddish regions may be associated with internal geological activity, while relatively old ice appears blue.

a much more viscous manner than pure water. If erupted as a liquid, this type of lava would not necessarily spread far before congealing, especially if confined by a chilled skin of the sort likely to form upon exposure to the vacuum of space. Contaminants also allow the ice to begin to melt at a much lower temperature than pure water-ice: salts can depress the melting temperature by a few degrees, and sulphuric acid by as much as 55 K.

Europa's ridges are perhaps highly viscous cryovolcanic flows fed from lengthwise fissures. Alternatively, rather than flowing across the ground, the cryovolcanic lava could have been flung up from the fissure in semi-molten clods by mild explosive activity and fallen back to coalesce as a rampart on either side of the fissure.

Each fissuring event must represent the opening of an extensional fracture in the crust. This cannot happen across an entire planetary body unless the globe is expanding, which seems highly unlikely. There must, therefore, be some regions on Europa where the surface has been destroyed at a rate sufficient to match the crustal extension elsewhere. Galileo images revealed likely candidates for such regions in some of the mottled terrain imaged by Voyager. Here, in regions such as Conamara Chaos, the crust has been broken into 'rafts' that have drifted apart, maybe because an underlying liquid ocean broke through to the surface. The areas between rafts are a random jumble reminiscent of refrozen sea-ice on Earth. Some rafts can be fitted back together, but many pieces of the 'jigsaw puzzle' seem to be missing, having perhaps been sunk or dragged down beneath the surface.

Other regions of mottled terrain are not totally disrupted into chaos, but instead contain a scattering of domes up to 15 km (9 mi) across; these were perhaps caused by 'pods' of molten or semi-fluid low-density material rising towards the surface. In some cases the upwelling pod has ruptured the surface to form a miniature chaos region bearing raftlets of surviving crust.

Although little is understood about the processes that have shaped Europa's surface, it is clear that the satellite has had a complicated history. We cannot tell how old each region of the surface is, but there are abundant signs that there is, or at least has been, a liquid zone below the surface. A salty ocean below several kilometres of ice is not necessarily a hostile environment for life. Indeed, life could be richer and more complex than anything that is likely to have survived on Mars. In the depths of Earth's oceans are living communities that are independent of photosynthetic plants and depend instead on bacteria that feed on the chemical energy supplied by springs of hot water (hydrothermal vents). Since Europa is tidally heated, perhaps water is drawn down into the rocky mantle, where it becomes heated, dissolves chemicals out of the rock and emerges at hydrothermal vents that are surrounded by life. The possibility of a life-bearing ocean below the ice has set the agenda for the future exploration of Europa, beginning with NASA's proposed EUROPA ORBITER mission.

Europa Orbiter NASA spacecraft that may be launched in about 2005 to EUROPA, the fourth-largest satellite of JUPITER and one that has gained the rank of the highest priority target since the discovery of a possible subsurface ocean beneath a sheet of ice from data and images collected by the GALILEO orbiter. Internal heating may be melting the relatively young and thin 1-km (3300-ft) ice pack, forming an ocean of water underneath.

NASA's proposed Europa Orbiter mission (scheduled to orbit Europa from about 2007 to 2009) has the primary goal of verifying the existence of an ocean below the ice. It will identify places where the ice is thin enough for future landing missions to send robotic submarines – 'hydrobots' – through the ice to hunt for hot springs and their attendant life. It will achieve this by using an altimeter to determine the height of the tide raised on Europa by Jupiter. The tide should be only 1 m (3 ft) if the ice is solid throughout, but about 30 m (100 ft) if there is 10 km (6 mi) of ice overlying a global ocean. It will also use ice-penetrating radar to map ice thickness. High-resolution conventional images will identify sites of recent eruptions.

▲ **European Southern Observatory** The European Southern Observatory at Cerro Paranal in Chile's Atacama Desert. At the time this distant photograph was taken, the Very Large Telescope (VLT) was under construction.

European Southern Observatory (ESO) Major astronomical institution supported by ten European nations, founded in 1962 to operate world-class astronomical facilities in the southern hemisphere and to further international collaboration in astronomy. The partner countries are Belgium, Denmark, France, Germany, Italy, the Netherlands, Portugal, Sweden, Switzerland and, since 2002, the UK. ESO operates two high-altitude sites in the Atacama desert in northern Chile. They are the LA SILLA OBSERVATORY – location of several optical telescopes with diameters up to 3.6 m (141 in.), together with a 15-m (49-ft) submillimetre radio telescope – and the PARANAL OBSERVATORY, home of the VERY LARGE TELESCOPE. At the organization's headquarters in Garching, near Munich, technical development programmes are carried out to provide the Chilean facilities with advanced instrumentation, and there is also an astronomical DATA CENTRE.

European Space Agency (ESA) Principal European space organization whose programme includes launchers, space science, telecommunications, Earth observation and manned spaceflight. It began work in 1975, two years after ten European countries had agreed to combine the aims of the European Launcher Development Organisation (ELDO) and the European Space Research Organisation (ESRO) in a single agency. Today, ESA has 15 member states in Europe and includes Canada as a Cooperating State. It has laboratories throughout Europe and overseas. Its headquarters are in Paris, while its Space Operations Centre (ESOC) is at Darmstadt in Germany. The European Space Research and Technology Centre (ESTEC) is ESA's biggest establishment, situated at Noordwijk in the Netherlands.

ESA's early years were spent regaining ground lost by the failure of its predecessors to develop a European launch vehicle. A turning point came in 1986 with the outstanding success of the agency's GIOTTO spacecraft, which was launched on a European rocket. Since then, ESA has become a major player in the space industry. Its ARIANE rockets now command the global market in space launches, especially for communications satellites. It has developed some of the world's most advanced telecommunications and Earth-monitoring satellites, and is a full partner in the INTERNATIONAL SPACE STATION, building on the achievements of ESA astronauts on MIR and the SPACE SHUTTLE.

The agency is a partner with the NATIONAL AERONAUTICS AND SPACE ADMINISTRATION (NASA) in high-profile space projects such as the HUBBLE SPACE TELESCOPE and the SOHO solar observatory, and the Huygens lander for

► **Explorer** Seen here with a model of Explorer 1 are Dr William H. Pickering, Dr James A. van Allen and Dr Wernher von Braun. Explorer 1 was America's first successful artificial satellite.

E

the CASSINI mission. In the future, the ESA MARS EXPRESS (2003) will carry out the most detailed scrutiny of Mars yet attempted, while the HERSCHEL SPACE OBSERVATORY and PLANCK SURVEYOR will map the skies from 2007.

European VLBI Network (EVN) Powerful interferometric array of radio telescopes across Europe and beyond, formed in 1980. The network consists of 18 individual instruments, among them some of the largest and most sensitive radio telescopes in the world, such as the LOVELL TELESCOPE and the Effelsberg Radio Telescope. The EVN is administered by the European Consortium for VLBI (*see* VERY LONG BASELINE INTERFEROMETRY), and incorporates 14 major institutes.

EUVE *See* EXTREME ULTRAVIOLET EXPLORER

eV *See* ELECTRON VOLT

evection Largest periodic PERTURBATION of the Moon's longitude, caused by the Sun, displacing it from its mean position by $\pm 1^{\circ}16'26''.4$ with a period of 31.812 days. This perturbation was discovered by PTOLEMY around AD 140. Prior to this the knowledge of the Moon's orbit was limited to the discovery of the ECCENTRICITY of the orbit and the rotation of the line of apsides by HIPPARCHUS in around 140 BC. In the years following Ptolemy, various refinements to this model of the evection were proposed, but no further major improvement in the knowledge of the Moon's orbit was made until the discovery of the variation by Tycho BRAHE in around AD 1580.

evening star Another name for the planets VENUS or MERCURY when either appears, following superior conjunction, as a bright object in the western sky after sunset. Because of its appearance in both the evening and morning skies, at different times, the ancient Greeks thought Venus was two separate objects: an evening star, which they named Hesperus, and a MORNING STAR, which they called Phosphorus. Mercury under the same circumstances was known as Hermes (evening star) and Apollo (morning star).

event horizon Boundary of a BLACK HOLE beyond which no information is available because no form of radiation can escape. It is the point at which the ESCAPE VELOCITY of the object becomes equal to the speed of light. In a non-rotating black hole with no charge, the event horizon is equal to the SCHWARZSCHILD RADIUS.

Evershed, John (1864–1956) English astronomer who in 1908 discovered radial motions of gases in sunspots, now known as the EVERSHED EFFECT. Originally an amateur astronomer who distinguished himself by discoveries made with home-made equipment on expeditions to observe total solar eclipses, Evershed was a founding member of the British Astronomical Association (1890) and directed its solar and spectroscopic sections. He designed a high-dispersion solar spectrograph using large liquid prisms as the dispersive element, and was able to discover new spectral lines in the Sun's corona. He also made major contributions to the study of cometary structure, using spectroscopy to show that their nuclei contain high concentrations of cyanogens.

Evershed effect Radial, outwards flow of gases in the penumbra of SUNSPOTS at the level of the PHOTOSPHERE. It was discovered in 1908 by John Evershed from the DOPPLER EFFECT of lines in the spectra of sunspots. The mean flow velocity is 2 km/s (1.2 mi/s). At higher levels in the solar atmosphere, the flow is inwards and downwards, and of higher velocity (around 20 km/s or 12 mi/s).

EVN Abbreviation of EUROPEAN VLBI NETWORK

evolved star Star that has finished burning hydrogen in its core and has thus evolved off the MAIN SEQUENCE. Evolved stars include GIANT STARS, SUPERGIANTS and WHITE DWARFS. *See also* STELLAR EVOLUTION

excitation State of a body when its internal energy is higher than the minimum that may be required. The term is usually used in the context of ELECTRONS in ATOMS, IONS and molecules being raised to a higher energy level than the GROUND STATE. Electrons may be excited by gaining energy from photons or by collisions with other atoms; they normally return to the ground state by emitting radiation. *See also* BOLTZMANN'S EQUATION

exit pupil For a telescope, the image of the primary mirror or lens as seen through the eyepiece. The point at which this image is formed is the optimum position for the observer's eye, so for comfort it should not be too close to the eyepiece itself (*see* EYE RELIEF). For any eyepiece–telescope combination, the size of the exit pupil can be found by dividing the eyepiece's FOCAL LENGTH by the FOCAL RATIO of the telescope. On an *f*/8 instrument, for example, a 24-mm eyepiece will have an exit pupil of 3 mm. Ideally, to deliver most light, the exit pupil should be around 7 mm, equivalent to the maximum pupillary dilation of the dark-adapted eye (*see* DARK ADAPTATION).

exobiology *See* LIFE IN THE UNIVERSE

Exosat Name given to a EUROPEAN SPACE AGENCY X-ray astronomy satellite launched in 1983 May. Smaller than the EINSTEIN OBSERVATORY, it contained two X-RAY TELESCOPES and non-imaging X-ray detectors. Placed in a highly elliptical Earth orbit, observations lasting up to 90 hours were possible. The long observing times permitted major advances in studies of X-RAY BINARIES, stellar CORONAE, quasi-periodic oscillations in X-ray sources and the properties of QUASARS and SEYFERT GALAXIES. The mission ended in 1986. *See also* X-RAY ASTRONOMY

exosphere Outermost region of a planetary ATMOSPHERE, from which atoms and molecules may escape into interplanetary space. The pressure of an atmospheric layer decreases with increasing altitude so that, eventually, the atmosphere is so thin that the collisions between the gaseous species become very infrequent. (That is, the mean free path becomes extremely long.) Ultimately, there is a level at which there are no collisions; the fast-moving atmospheric constituents may then escape the gravitational influence of the planet altogether or perhaps go into orbit around it. For the Earth's atmosphere, this region begins at an altitude of 200–700 km (120–430 mi), depending on the degree of solar activity. The exosphere, therefore, overlaps with the outermost part of the THERMOSPHERE, but includes the RADIATION BELTS and extends to the MAGNETOPAUSE, where it meets the interplanetary medium.

expanding universe Outward movement (REDSHIFT) of galaxies. That the Universe is expanding was first demonstrated observationally in 1929 by Edwin HUBBLE, although the possibility had been suggested earlier on theoretical grounds by Willem de Sitter (1917), Aleksandr Friedmann (1922) and Georges Lemaître (1927). The evidence Hubble found was that if he took a spectrum of a galaxy, the spectral lines would always be redshifted by an amount proportional to the distance he estimated for the galaxy. This relationship is commonly called the HUBBLE LAW.

For distant galaxies, this expansion velocity is an appreciable portion of the speed of light. Recessional velocities of this magnitude would be difficult to understand if the galaxies were moving through space, but make perfect sense if the Universe itself were expanding. GENERAL RELATIVITY interprets this expansion as the expansion of SPACETIME itself, and even predicts that it should obey a linear law such as the Hubble law. If this expansion is extrapolated backwards until spacetime was a singular point, then the age of the Universe can be estimated. These ideas set the stage for the BIG BANG THEORY. Predictions made from these ideas include the temperature of the COSMIC MICROWAVE BACKGROUND and the COSMIC ABUNDANCE.

Explorer Series of small US scientific satellites, mostly launched into Earth orbit. Explorer 1, instrumented by James VAN ALLEN, discovered in 1958 the radiation belts girdling the Earth. Of the 72 successful Explorer missions launched by the end of 2000, five had operated for 10 years or more.

extinction Loss of light from the line of sight as it passes through a medium. The loss may be by SCATTERING, in which case the light energy is taken up by the medium and promptly re-emitted at the same energy but redirected out of the line of sight. Or the loss may be by ABSORPTION, in which case the light energy is eventually re-emitted in a different form altogether, for example as heat.

Typical examples of extinction in astronomy are the loss of starlight as it passes through the interstellar medium and is scattered and absorbed by INTERSTELLAR GRAINS, and the loss of starlight as it passes through the Earth's atmosphere and is scattered by air molecules.

extrasolar planet (exoplanet) Planet not associated with our Solar System. The existence of other worlds, especially inhabited ones, has been the subject of debate since ancient times. In the 20th century, stars with irregular proper motions were suspected of having planetary companions; BARNARD'S STAR was a famous candidate. In 1992, after centuries of speculation and decades of false announcements, Dale Andrew Frail (1961–) and D. Wolszcan detected the first extrasolar planets, orbiting the pulsar PSR 1257 + 12. In 1995 October, Michel Mayor (1942–) and Didier Queloz (1966–) found the first undisputed instance of a planet orbiting another main-sequence star, 51 Pegasi. By late 2001, about 85 planetary systems had been discovered around a wide diversity of stars up to 180 l.y. away; the table gives orbital parameters for ones of particular interest. Many of these **substellar objects** are in very close orbits and are many times the mass of Jupiter. The lower theoretical mass limit for a BROWN DWARF is about 0.013 solar mass, which is approximately 13 Jupiter masses (13 M_J); by convention, this is taken as the upper mass limit for a giant planet.

There are now ten times as many planets known to orbit other stars than there are major planets in our Solar System. Taking into account stars with no detectable planets, about 5% of main-sequence stars appear to have a giant planet orbiting at less than 2 AU. The first extrasolar giant planets were found 100 times closer to their parent star than expected. In standard models of planetary formation, Jupiter-sized planets form at a distance of 4–5 AU from solar-type stars. The very close orbits can be explained by a process known as orbital migration. Half the planets have orbits with unexpectedly large eccentricities. These could result from migration: if two planets are migrating inwards at different rates, one could expel the other from the system and have its own orbit perturbed.

There are three methods of detecting extrasolar planets. Direct imaging is possible in principle, either by a planet's reflected light from the parent star or by its infrared emission. But such reflected light or infrared radiation is extremely faint, and because the planet is extremely close to the star, it is very difficult to separate them; also, the image of the planet lies within the AIRY DISK of the star as seen through a telescope. There are two ways round this difficulty: reduce the size of the Airy disk so that the planet is outside it, or block the starlight. Shrinking the Airy disk would require a telescope with an aperture of 10–100 m (33–330 ft) or a multi-aperture interferometer, preferably in Earth orbit. Starlight can be blocked by a coronagraph, a physical obstruction at the telescope's focus, or by a technique in which the star's light is made to interfere destructively with itself without affecting the region around it.

The second method is to search for the planet's gravitational perturbation of its parent star. Both planet and star orbit their common centre of mass, and the small periodic change in the star's proper motion is apparent in its radial velocity (a DOPPLER EFFECT) and position on the sky. If the planet's orbital plane coincides with our line or sight, the variation of radial velocity is greatest. Radial velocity and position vary cyclically with a period equal to the planet's orbital period and an amplitude proportional to the planet's mass. The precision in radial velocity measurements is presently 3 m/s, which would enable the detection of planets similar in mass to Saturn. But there is no hope yet of detecting an Earth-like planet since it would give rise to a variation in the star's radial velocity of just 0.1 m/s.

The third method of detection is by transits: if a planet crosses the face of its sun, it can produce, under favourable circumstances, a small drop for a few hours in the star's light. The depth of the drop is given by the star-to-planet surface ratio: 1% for a jupiter, 0.01% for an earth. The planet's orbital plane must be favourably oriented: the probability that an orbit is sufficiently close to edge-on is 0.5% for a planet 1 AU from its star and

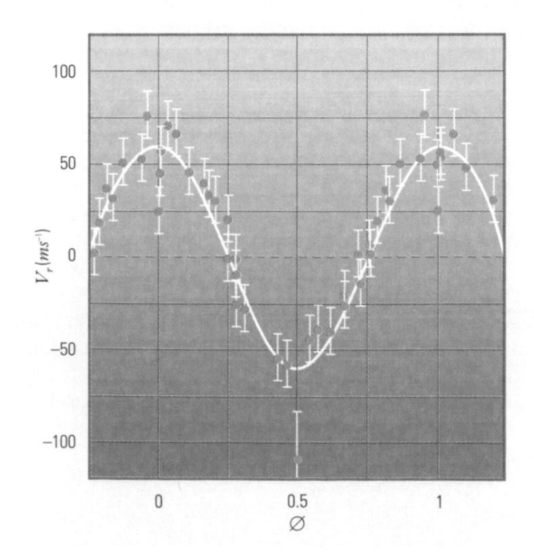

◀ **extrasolar planet** The 50 m/s periodic variation of the radial velocity of 51 Pegasi. One cycle takes 4.23 days – the orbital period of the planet, which has a mass of 0.47 M_J.

NOTEWORTHY EXTRASOLAR PLANETS				
Star	$M \sin i$ (M_J)	Period (d)	Semimajor axis (AU)	Eccentricity
HD 83443 b	0.35	2.986	0.04	0.00
HD 83443 c	0.17	29.830	0.17	0.42
HD 209458	0.63	3.524	0.05	0.02
51 Peg	0.46	4.231	0.05	0.01
υ And b	0.68	4.617	0.06	0.02
υ And c	2.05	241.300	0.83	0.24
υ And d	4.29	1308.500	2.56	0.31
GJ 876 b	1.89	61.020	0.21	0.10
GJ 876 c	0.56	30.120	0.13	0.27
HD 168443 b	7.64	58.100	0.29	0.53
HD 168443 c	16.96	1770.000	2.87	0.20
HD 114762	10.96	84.030	0.35	0.33
HD 82943 b	1.63	444.600	1.16	0.41
HD 82943 c	0.88	221.600	0.73	0.54
16 Cyg B	1.68	796.700	1.69	0.68
HD10697	6.08	1074.000	2.12	0.11
47 UMa b	2.56	1090.500	2.09	0.06
47 UMa c	0.76	2640.000	3.78	0.00
HD 74156 b	1.55	51.600	0.28	0.65
HD 74156 c	7.46	2300.000	3.47	0.40
ϵ Eri	0.88	2518.000	3.36	0.60

Some noteworthy extrasolar planets. It includes the first two discovered (HD 114762 and 51 Pegasi), the planet discovered by the transit method (HD 209458) and the closest one (ϵ Eridani). The others are all the multi-planet systems presently known. (Note that the radial velocity method can only give the product $M \sin i$, where i is the inclination of the planetary orbit to the plane of the sky.)

decreases (increases) with increasing (decreasing) distance. Jupiter-size planets can be detected in this way from the ground, and Earth-sized planets could be detected from space. The transit method was used to measure the radius of the planet orbiting the star HD 209458. The planet has a mass of 0.6 M_J and its radius is 30% larger than Jupiter. In 2000 the Hubble Space Telescope (HST) detected absorption lines from the planet's atmosphere during a transit that indicated the presence of sodium – the first indirect detection of an atmospheric component of an extrasolar planet.

The search for extrasolar planets is being stepped up, with many projects in the pipeline or under study. Astrometric space missions include NASA's Space Interferometric Mission (SIM), due for launch in 2009, and ESA's GAIA project. The planned 2009 launch of the NEXT GENERATION SPACE TELESCOPE will increase the potential for discovery. Interferometric missions aimed at detecting Earth-like planets have been motivated by the search for LIFE IN THE UNIVERSE. They include ESA's Darwin, consisting in its initial configuration of five orbiting 1.5-m (60-in.) telescopes separated by 20–100 m (65–330 ft), and NASA's Terrestrial Planet Finder (TPF).

COROT is an ESA/CNES mission to be launched in 2004 to detect planetary transits. Its 250-mm (12-in.) aperture telescope should be able to detect tens of planets down to twice the Earth's radius, some of which should orbit within the ECOSPHERE of their parent star. NASA's Kepler project, due for launch in 2007, has a 1-m (40-in.) telescope that will view 100 square degrees of sky at a time and should be able to detect a few hundred Earth-like planets in stellar ecospheres.

Finally, there is some evidence that not all extrasolar planets are orbiting stars: high-resolution infrared images of star-forming regions suggest the existence of what are called **unbound** or **free-floating planets**.

Extremely Large Telescope (ELT) Proposal for a 29-m (95-ft) optical telescope, announced in 1996 by a consortium of several US institutions. The design is a scaled-up version of the HOBBY–EBERLY TELESCOPE, using 169 mirror segments and copying the HET's simplified mounting to minimize costs. Since the ELT was proposed, the term 'extremely large telescope' has been applied to any optical/infrared telescope in the 25–30 metre class. *See also* CALIFORNIA EXTREMELY LARGE TELESCOPE; GIANT SEGMENTED-MIRROR TELESCOPE; OVERWHELMINGLY LARGE TELESCOPE; SWEDISH EXTREMELY LARGE TELESCOPE

extreme-ultraviolet Vaguely defined wavelength band that extends roughly from the end of the soft (low-energy) X-ray band near 10 nm to the edge of the FAR-ULTRAVIOLET near 75 nm.

Extreme Ultraviolet Explorer (EUVE) NASA spacecraft launched in 1992 to carry out a full-sky survey at EXTREME-ULTRAVIOLET wavelengths – a region of the spectrum not covered by previous satellites. It carried three EUV telescopes, each sensitive to a different wavelength band. A fourth telescope performed a high-sensitivity search over a limited part of the sky in a single EUV band. Its mission ended in 2000 December.

extrinsic variable VARIABLE STAR in which the changes in light occur as a result of some external (that is, mechanical) cause, rather than from internal changes, such as pulsation or eruptive processes. Typical sources of such variation are rotation (*see* ALPHA^2 CANUM VENATICORUM STARS and BY DRACONIS STARS), orbital motion (*see* ECLIPSING VARIABLES, ELLIPSOIDAL VARIABLES, REFLECTION VARIABLES) and obscuration (certain NEBULAR VARIABLES and R CORONAE BOREALIS STARS). *See also* INTRINSIC VARIABLE

eyepiece (ocular) System of lenses in an optical instrument through which the observer views the image. The eyepiece is usually a composite LENS that magnifies the final real image produced by the optics. There are many different designs of eyepiece, all attempting to achieve the best result for a particular application. Most incorporate an **eye lens** that is closest to the eye and a **field lens** that is closest to the OBJECTIVE.

The magnification provided by a telescope is calculated by dividing its focal length by the focal length of the eyepiece used with it. Thus the magnification can be changed simply by changing the eyepiece; astronomical eyepieces come in standard sizes with a simple push-fit to facilitate this. Standard sizes are 0.965 inch (24 mm), 1.25 inch (31 mm) and 2 inch (51 mm), with 1.25 inch by far the commonest.

When an eyepiece with a shorter focal length is used and the overall magnification increases, the **field of view** – the area of sky the observer can see – will decrease. However, some eyepiece designs provide a larger field of view than others, and manufacturers often quote an apparent field of view determined by the **acceptance angle** of the eyepiece, which is typically around 40°. The telescopic field of view is the acceptance angle of the eyepiece divided by the magnification. Wide-angle eyepieces can be designed to have acceptance angles up to 80°, but this additional width comes at the cost of increased distortion, particularly towards the edges of the field.

Also important to eyepiece performance are the EXIT PUPIL and EYE RELIEF. A good eyepiece will be designed to limit various forms of optical ABERRATION. *See also* BARLOW LENS; ERFLE EYEPIECE; HUYGENIAN EYEPIECE; KELLNER EYEPIECE; NAGLER EYEPIECE; ORTHOSCOPIC EYEPIECE; PLÖSSL EYEPIECE; RAMSDEN EYEPIECE

eye relief In an optical instrument such as a telescope or binoculars, the distance between the eye lens of the EYEPIECE and the EXIT PUPIL. When the observer's eye is placed at the exit pupil, the whole of the eyepiece's field of view is visible. The eye relief must be sufficiently large for comfortable viewing, around 7–10 mm (0.25–0.375 in.).

Faber–Jackson relation Correlation, first noted by US astronomers Robert Jackson (1949–) and Sandra Faber (1944–), between the velocity dispersion of stars in an elliptical galaxy and the luminosity of the galaxy. Refined with a third parameter into the Fundamental Plane, this has become a valuable distance indicator, analogous to the TULLY–FISHER RELATION for SPIRAL GALAXIES.

Fabricius, David (1564–1617) and **Fabricius, Johannes** (1587–1616) German father and son astronomers who made the first regular telescopic observations of sunspots, published in 1611. They noticed that these spots rotated with the Sun, so must be a part of the visible solar surface rather than small orbiting objects, and that the Sun itself must rotate, a theory proposed by Johannes KEPLER, with whom David Fabricius corresponded regularly. David Fabricius discovered the first known variable star, Mira, in 1596.

Fabry–Pérot interferometer OPTICAL INTERFEROMETER in which the two parts of the incoming beam are recombined after multiple partial reflections between parallel, lightly silvered, glass plates; the construction is called a Fabry–Pérot etalon. INTERFERENCE of the light waves creates a very narrow-bandpass filter, which can be tuned by changing the spacing between the plates. It is capable of making extremely accurate measurements of the wavelengths of emissions in atomic spectra.

faculae Exceptionally bright, irregular regions on the Sun visible in the white light of the PHOTOSPHERE, best seen in the regions near the limb where LIMB DARKENING increases the contrast. Faculae are usually associated with SUNSPOTS, and are brighter than the surrounding medium by virtue of their higher (by about 300 K) temperatures; they are closely associated with the chromospheric network and bright PLAGE regions in the CHROMOSPHERE. Faculae can precede the appearance of a sunspot, in the same place, by many hours or even a few days, and can persist in the same location for many months after the sunspots have gone. The number of faculae varies in step with the approximately 11-year SOLAR CYCLE, with a greater number when there are more sunspots. The increased solar brightness as a result of faculae is greater than the decrease caused by sunspots, so there is an overall increase in the SOLAR CONSTANT at the maximum of the sunspot cycle. Polar faculae, occurring at much higher latitudes than the sunspot-forming regions, are common in the rising phase of the solar cycle.

falling star Popular (chiefly US) non-scientific description of a METEOR.

False Cross Cross-shaped asterism (star pattern) in the southern sky made up of ι and ε Carinae and κ and δ Velorum, all of second magnitude. It is often mistaken for the SOUTHERN CROSS, but is larger and dimmer.

far-infrared Region of the ELECTROMAGNETIC SPECTRUM between around 25 and 350 μm. This region is only observable from space, because far-infrared radiation is absorbed by Earth's atmosphere. Very cool objects (most notably active galaxies), and the dust in interstellar space, emit strongly in this range. *See also* INFRARED ASTRONOMY

far-ultraviolet Region of the ELECTROMAGNETIC SPECTRUM that extends roughly from the end of the EXTREME-ULTRAVIOLET near 75 nm to the edge of the NEAR-ULTRAVIOLET at around 300 nm (the near-ultraviolet extending to the optical).

Far Ultraviolet Spectroscopic Explorer (FUSE) Another in a series of NATIONAL AERONAUTICS AND SPACE ADMINISTRATION (NASA) astronomical satellites, developed and operated for NASA by Johns Hopkins University. FUSE was launched in 1999 to extend the observations of the Universe by previous ultraviolet explorers, with greater sensitivity and resolving power. FUSE will help to answers some fundamental questions about the first few minutes of the BIG BANG, studying the cosmic abundance of deuterium, the heavy hydrogen isotope. It will also study the dispersal of chemical elements in galaxies and the properties of hot interstellar gas clouds.

Fast Auroral Snapshot Explorer (FAST) NASA satellite launched in 1996 August, carrying four instruments to study the creation and content of aurorae. Plasma and electron temperature data are collected by electric field detectors, magnetic field data by a magnetometer and data on electrons and ions by electrostatic analysers.

Faulkes Telescope Project to build two fully robotic 2-m (79-in.) telescopes for use principally by British educational institutions over the Internet. The telescopes will be located on the island of Maui in Hawaii and at SIDING SPRING OBSERVATORY in Australia, and will provide coverage of the night sky during UK daytime.

fault Fracture or fracture zone along which there has been displacement of the fracture sides relative to one another. A fault is typically formed by TECTONIC deformation of the outer solid parts of planets and satellites. Depending on the character of the displacement, several varieties of faults are distinguished. Normal faults are produced when displacement is essentially vertical and the hanging wall is depressed relative to the footwall. Strike-slip faults result when displacement of one side relative to the other is essentially horizontal.

Faye, Comet 4P/ Short-period comet found by Hervé Faye (1814–1902), a French amateur astronomer, in 1843, and recovered in 1850 following a prediction by Urbain LE VERRIER. The period is 7.34 years, and the comet was last seen in 1999–2000. The orbit has undergone some modification following close passages to Jupiter, and the comet is apparently becoming progressively fainter.

F corona Outer part of the white-light CORONA of the Sun. It is caused by sunlight scattered from solid dust particles in interplanetary space. The F corona is responsible for the greatest part of the brightness of the corona beyond about two solar radii, and it extends to far beyond the Earth. The spectrum of the F corona is that of the Sun, including FRAUNHOFER LINES; the 'F' stands for Fraunhofer. At large distances from the Sun, the dust particles in the F corona merge with those causing the ZODIACAL LIGHT. *See also* E CORONA; K CORONA; T CORONA

fermion Elementary particle that has a spin of $\frac{1}{2}$, is anti-symmetric and represent the material side of the quantum world. Fermions obey a particular type of statistics characterized by a Fermi distribution.

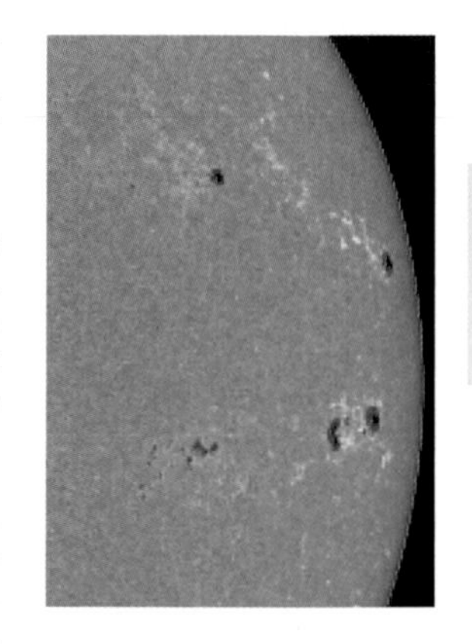

▲ **faculae** Bright regions marking the location of clouds of hot hydrogen above active regions on the Sun, faculae are best seen, as here, close to the limb. Faculae are commonest around sunspot maximum.

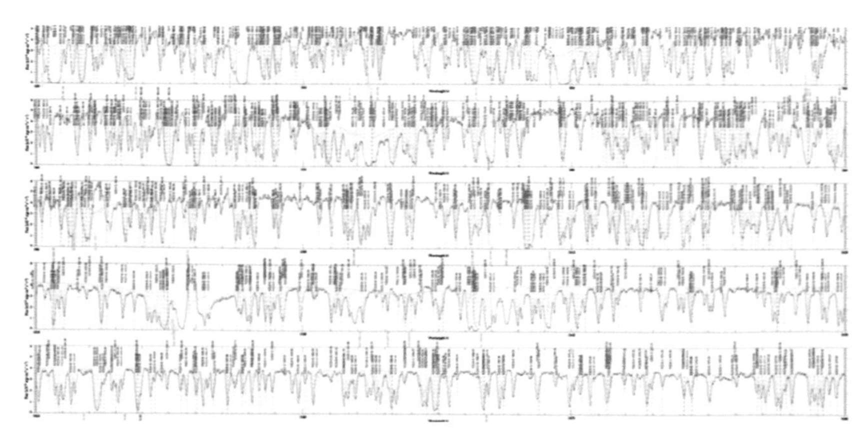

▼ **Far Ultraviolet Spectroscopic Explorer** (FUSE) Part of a spectrum taken by FUSE of the hot white dwarf star at the centre of the Dumbbell Nebula.

F

FG Sagittae Very unusual VARIABLE STAR, having changed from a blue SUPERGIANT to a red supergiant during the 20th century. Before 1900 FG Sagittae was a normal blue star evolving to become a white dwarf. In 1901, however, it suddenly started to brighten in the photographic (visual) and blue magnitude, and it brightened during most of the 20th century. The temperature dropped from over 10,000 K to about 4500 K, and at one point (around 1965) it was dropping by 250 K per year. The star has been variable for several decades, passing through the Cepheid INSTABILITY STRIP. In 1992 FG Sagittae's brightness faded by several magnitudes in a manner similar to the fadings of the R CORONAE BOREALIS STARS (although the star is not as deficient in hydrogen as are the R Coronae Borealis variables). There have been several more fading episodes since 1992. The dust clouds, which dim the light during these fading episodes, have a temperature between 700 K and 1000 K. FG Sagittae is a solar-mass star, and although similar to the unusual star SAKURAI'S OBJECT it evolves along a different track in the HERTZSPRUNG–RUSSELL DIAGRAM because it is less massive. Convection currents in the star's atmosphere have brought up material such as yttrium, zirconium, cerium and lanthanum (S PROCESS elements) from deep inside the star, so its composition appears abnormal. The star is surrounded by the PLANETARY NEBULA it created around 6000 to 10,000 years ago, and the composition of the nebula is completely normal.

fibril Linear dark structure in the solar CHROMOSPHERE seen in association with ACTIVE REGIONS. Fibrils are thought to mark out the chromospheric magnetic field. They are visualized in SPECTROHELIOGRAMS taken in the light of the HYDROGEN-ALPHA LINE.

field curvature Distortion produced by an astronomical telescope, leading to differences in scale across an image's width. Field curvature results from the fact that the image formed is not really flat, but is in fact a curved 'surface' on a concave sphere. An EYEPIECE will focus best at differing positions across the field. Photographic plates on SCHMIDT CAMERAS used for astrometry are curved to match the field curvature of the instrument.

field equations Term generally used to refer to Einstein's general relativistic field equations. These equations form the basis for Einstein's laws of gravity and essentially relate the curvature of SPACETIME to the presence of mass and energy. In his theory of SPECIAL RELATIVITY, Einstein showed that matter and energy are equivalent, and related by $E = mc^2$, which implies that energy also has to be included in any gravity calculation. So in the field equations, both mass and energy are represented by a quantity called the stress-energy tensor $T_{\mu\nu}$. This tensor is the source of the gravitational field. The field equations also describe how spacetime is warped by the presence of the gravitational sources. The curvature information is contained in two quantities, the Riemann tensor ($R_{\mu\nu}$) and the Riemann scalar R. The geometry of the coordinate system is also included in the field equations and is represented by the quantity $g_{\mu\nu}$, called the metric tensor. Together, the field equations are

$$R_{\mu\nu} - \tfrac{1}{2}Rg_{\mu\nu} = 8\pi G/c^4 T_{\mu\nu}$$

where G is Newton's universal constant of gravitation and c is the speed of light. The indices μ and ν represent each of the dimensions: length, width, height and time. Thus the equation above is actually 16 independent equations (4^2) that must be solved simultaneously. The major solutions of the field equations are the Schwarzschild solution for spherical geometry, the Kerr–Newmann solution for rotating black holes, and the Robertson–Walker metric for the entire universe.

field of view Angular diameter of the area of sky visible through an optical instrument. *See* EYEPIECE

field rotation Circular movement of a telescope's field of view as a result of DIURNAL MOTION when using an instrument that is either on an altazimuth mounting or an equatorial mounting whose polar axis is not properly aligned on the celestial pole. An image taken using such a system will show short star trails concentric with the field centre. A device called a **field de-rotator** is used to counter-rotate the image plane to allow long exposures through instruments with altazimuth mounts, as used with the majority of large modern professional telescopes.

filament Dark, elongated, ribbon-like structure visible in SPECTROHELIOGRAMS of the Sun taken in the light of the HYDROGEN-ALPHA LINE or the H and K lines of calcium. A filament is a PROMINENCE seen in projection against the bright solar disk.

▼ **filament** Seen in projection against the bright solar disk in spectroheliograms, prominences appear as dark filaments by contrast, as at top right in this image.

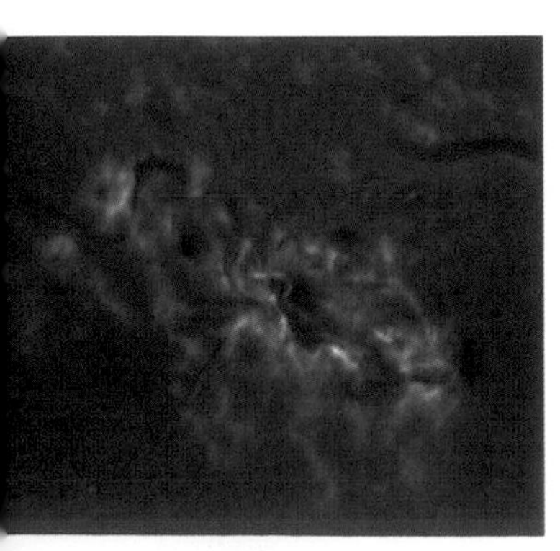

filamentary nebula Interstellar nebula in which the material is in the form of highly elongated and often highly twisted intertwined clouds. Many nebulae of many types have filamentary structures within them, so the term filamentary nebulae is reserved for those nebulae in which the filaments make up the whole or the largest part of the nebula. The VEIL NEBULA, a SUPERNOVA REMNANT, is the best-known filamentary nebula.

filter Optical element that absorbs some wavelengths of light whilst transmitting others. Filters are used to great effect in astronomy to isolate certain colours or wavelengths, making them easier to observe or analyse. The most common are filters made of coloured glass or gelatin. In visual astronomy these are used to improve the contrast in planetary features. For example, if a light blue filter is used to observe Jupiter, the Great Red Spot will appear darker than normal so will stand out against the surrounding body of the planet. Similarly, a red filter will show up dark features on the surface of Mars.

Colour filters are also used with monochromatic detectors such as photographic plates and CCD cameras. Taking three separate images through red, green and blue filters, then recombining them in the dark room or electronically will produce true-colour images.

Interference filters are made of many layers of dielectric deposited on a glass substrate. Thirty or forty layers may be used to produce the exact characteristics needed. In this way, sky glow caused by mercury and sodium street-lights can be rejected whilst hydrogen and oxygen light from faint nebulae is allowed through. Interference filters designed to pass only one sharply defined wavelength are called monochromatic filters.

Special solar filters are available for observing the Sun directly. High-density filters that cut out 99.999% of the incoming sunlight are used at the entrance aperture of a telescope to enable sunspots, faculae and surface granulation to be observed directly. Interference filters are used to reveal flares and prominences. Observing the Sun directly carries risk so advice should be sought from a specialist retailer or observatory before using a solar filter.

finder Small, low-powered TELESCOPE with a wide field of view, attached to the tube of a larger instrument for the purpose of more easily locating celestial bodies. The narrow field of view of a large telescope can make locating faint objects in particular quite difficult. The finder allows the observer to study a wide area of sky, and the two telescopes should be aligned so that when the object in question is centred on the cross wires of the smaller telescope, it appears in the centre of the field of view of the larger one.

fine structure Splitting of orbital energy levels. An atomic energy level for an orbiting electron is finely divided into substates as a result of the electron's ability to have different quantized angular momenta; to these substates is added splitting caused by the magnetic coupling between the electron's spin and its orbit. Each orbital level n of hydrogen has n substates, each of which (except for the first) is split in half. Multiple electrons' effects are combined. Transitions between fine structure states create the complex spectra of atoms heavier than hydrogen.

fireball METEOR of magnitude −5 or brighter, exceeding the brilliance of the planet Venus. On average, about one meteor in a thousand attains fireball status. Such events are quite commonly seen during the major METEOR SHOWERS such as the PERSEIDS or GEMINIDS. Fireballs are produced by larger METEOROIDS; they are associated with the arrival of substantial (centimetre-sized and upwards) fragments of asteroidal debris, which can on rare occasions survive atmospheric passage to be recovered on the ground as a METEORITE. An exploding fireball producing a sonic boom may be described as a BOLIDE.

FIRST Abbreviation of the Far Infrared Submillimetre Telescope, renamed the HERSCHEL SPACE OBSERVATORY.

first contact In a SOLAR ECLIPSE, moment when the Moon's leading (easterly) limb first appears to align tangentially with the westerly limb of the Sun. In a LUNAR ECLIPSE, first contact refers to the moment when the Moon's easterly limb first encounters the western edge of the UMBRA of Earth's shadow. In either instance, first contact is the beginning of the eclipse.

first light Term used to describe the first occasion on which a newly constructed telescope is used to observe a celestial object. This will occur only after all the optical components have been fully aligned but it will usually be some considerable time before the instrument is brought into full use, since a large amount of testing and calibrating will still be required. Whilst normally used to refer to an optical telescope, the term is also applied to instruments operating at other wavelengths; for example the first time that a radio telescope detects signals from space.

First Point of Aries Position on the CELESTIAL SPHERE where the Sun's centre crosses the celestial equator from south to north at the time of the VERNAL EQUINOX, on or around March 21 each year. This marks the intersection of the ECLIPTIC with the CELESTIAL EQUATOR at the Sun's ASCENDING NODE; it is used as the zero point for measuring CELESTIAL LONGITUDE, RIGHT ASCENSION and SIDEREAL TIME. Thus, when the First Point of Aries is exactly on an observer's MERIDIAN, the sidereal time is $0^h\ 0^m\ 0^s$.

The small annual change in position of the celestial pole, known as PRECESSION, has caused this point on the celestial sphere to move slowly westwards along the ecliptic by about one-seventh of an arcsecond each day, taking a period of around 25,800 years to complete a single revolution. Because of this, although originally located in the constellation of Aries, from whence it derived its name, it is has now drifted into neighbouring Pisces, although the name has remained unchanged.

First Point of Libra Position on the CELESTIAL SPHERE where the Sun's centre crosses the celestial equator from north to south at the time of the AUTUMNAL EQUINOX, on or around September 23 each year. The effects of PRECESSION have caused it to move slowly westwards along the ecliptic by about one-seventh of an arcsecond each day, taking a period of around 25,800 years to complete a single revolution. This means that it is no longer located in Libra, but instead has drifted into the constellation of Virgo.

first quarter *See* PHASES OF THE MOON

Fish's Mouth Region of dark nebulosity within the ORION NEBULA, close to the TRAPEZIUM stars (RA $05^h\ 35^m.3$ dec. $-05°23'$).

fission, nuclear NUCLEAR REACTION that involves a more massive nucleus splitting into two or more lighter nuclei or subatomic particles. For nuclei heavier than iron, the BINDING ENERGY per nucleon will increase during fission and so the reaction may result from spontaneous radioactivity. Thus, for example, uranium-235 decays to thorium-231 and helium-4 (an α-particle). The binding energy per nucleon, however, will decrease during the fission of iron and lighter nuclei so that more energy is required to cause the reaction than is released during it. Hence light element fission can only occur if energy is supplied from an external source such as a high energy collision or a gamma-ray photon. In this way nuclei between sulphur and iron may be fragmented by photo-dissociation during SUPERNOVA explosions; in the interstellar medium, impacts with COSMIC RAY particles cause the break-up of elements like carbon, nitrogen and oxygen to produce lithium, beryllium and boron. *See also* FUSION, NUCLEAR; SPALLATION

Fitzgerald contraction (Lorentz–Fitzgerald contraction) Shrinkage in the lengths of moving objects that was suggested in 1889 by the Irish physicist George Francis Fitzgerald (1851–1901) and, independently, by Hendrick LORENTZ, to explain the null result of the MICHELSON–MORLEY EXPERIMENT.

fixed star Term used in antiquity to describe those stars that appeared to remain static, relative to one another, on the celestial sphere, as opposed to the 'wandering stars' or planets. It was believed that the Earth lay at the centre of the Universe, surrounded by a solid crystal sphere to which the 'fixed stars' were attached. Nowadays, the term is used to describe those stars that have no detectable proper motion.

Fizeau, (Armand) Hippolyte (Louis) (1819–96) French physicist who in 1849 made the first successful measurement of the velocity of light in an experiment carried out wholly on the Earth. He used a rapidly rotating toothed wheel which interrupted a fine beam of light before and after reflection from a distant mirror more than 8 km (5 mi) away. Fizeau collaborated with Léon FOUCAULT in his optics experiments.

Flame Nebula EMISSION NEBULA in the constellation Orion, immediately east of ζ (RA $05^h\ 41^m.9$ dec. $-01°51'$). The Flame Nebula has a diameter of 30′ and is split into two sections by a dark lane running north–south. Difficult to see visually thanks to its proximity to ζ, the nebula is conspicuous on long-exposure photographs of the Orion's Belt region.

Flammarion, (Nicolas) Camille (1842–1925) French astronomer and prolific author of popular astronomy books, including *L'Astronomie populaire* (1879). He worked first (from 1858) at Paris Observatory, then from 1883 at his private observatory at Juvisy-sur-Orge, where he made serious studies of double stars and the planets using a 9.5-inch (240-mm) Bardou refractor. Flammarion's epic two-volume work *La Planète Mars et ses conditions d'habitabilité* (1892, 1909) is an authoritative history of Mars observing which speculates about the possibility of life there.

Flamsteed, John (1646–1719) English astronomer and clergyman, the first ASTRONOMER ROYAL and director of the Royal Observatory, Greenwich. He was a youthful convert to Copernicanism, admired Galileo and Kepler, and was deeply influenced by Tycho Brahe's insistence upon precise observation as the basis for theory. In 1675 he was requested by the Royal Society to examine a method for finding longitude at sea proposed by the sieur de St Pierre. Flamsteed's report, in which he stated that the longitude could not be reckoned until superior astronomical tables became available, caught the attention of King Charles II, who ordered the creation of the post of Astronomer Royal and the founding of the Greenwich Observatory. In spite of the royal appointment, Flamsteed had to pay for the observatory's instruments out of his own pocket.

At Greenwich, Flamsteed's task was to make accurate positional measurements of the Moon and stars for use in navigation. He pioneered three new precision instruments that transformed observation: the telescopic sight and the micrometer, both invented by William GASCOIGNE, and Christiaan Huygens' pendulum clock for timing star transits. He was a meticulous observer, and with instruments made by the clockmaker Thomas Tompion (1639–1713) he was able to measure the positions of stars and planets to within 10″, over 40 times more accurately than Tycho Brahe had managed with his equipment.

Observing every clear night for over 40 years at Greenwich, Flamsteed produced a body of data that set new standards in astronomical research. However, as something of a perfectionist he was slow to release his results, to the irritation of Isaac NEWTON, and objected when the Royal Society published his uncorrected observations. The catalogue of his corrected observations, the *Historia coelestis brittanica*, was not published until 1725, after his death; the numbers assigned to stars in this catalogue (FLAMSTEED NUMBERS) are still in use. While his tables were still not sufficiently accurate to find longitude, they proved fundamental to Newton's work on the lunar theory.

Flamsteed numbers Sequence of numbering of the stars by constellation in order of right ascension, for purposes of identification. The numbers were assigned by later astronomers to stars in John FLAMSTEED's posthumously published star catalogue. The most westerly star listed in Leo, for example, was designated 1 Leonis. Many stars are still best known by their Flamsteed designations.

flare Sudden, violent explosion on the Sun, lasting from a few minutes to a few hours; the local effects may persist for several days. Flares accelerate charged particles out into interplanetary space and down into the Sun. They emit ELECTROMAGNETIC RADIATION across the full spectrum. The frequency and intensity of solar flares increases near the maximum of the SOLAR CYCLE, and they are associated with ACTIVE REGIONS.

Flares originate in the low CORONA, where magnetic energy is built up and stored in the magnetic fields of active regions. The flares are frequently triggered in compact structures just above the tops of coronal loops, where magnetic fields of opposite polarity are drawn together, releasing large amounts of stored energy (up to 10^{25} Joule) by MAGNETIC RECONNECTION.

Every second, the PHOTOSPHERE emits light with an energy that is at least one hundred times greater than the total energy emitted by any flare. Therefore, only exceptionally powerful flares can be detected in the glare of visible sunlight. The first such white-light flare to be observed and recorded was seen by two Englishmen, Richard C. CARRINGTON and Richard Hodgson (1804–72), who noticed an intense brightening lasting for just a few minutes near a complex group of sunspots on 1859 September 1.

Routine visual observations can be made by tuning into the red emission of the HYDROGEN-ALPHA LINE. Light at this 656.3 nm wavelength originates in the CHROMOSPHERE. When viewed in this way, solar flares appear as two extended parallel ribbons on either side of the magnetic neutral line. Flares have been classified according to their area in hydrogen-alpha light, with an importance ranging from 1 to 4 that corresponds to an area from 2 to more than 25 square degrees.

CLASSIFICATION OF FLARES – SOFT X-RAYS

Importance[a]	Peak flux at 0.1 to 0.8 nm (watts per square metre)
A	10^{-8} to 10^{-7}
B	10^{-7} to 10^{-6}
C	10^{-6} to 10^{-5}
M	10^{-5} to 10^{-4}
X	10^{-4} and above

[a] A number is added as a multiplier, giving the peak flux in the first unit. For example, X5.2 stands for a peak X-ray flux of 5.2×10^{-4} or 0.00052 watts per square metre.

Since flares reach temperatures of tens of millions of degrees, they emit the bulk of their energy at X-ray wavelengths of about 0.1 nm. A large flare can briefly outshine the Sun in X-rays. The X-ray radiation of flares is emitted as Bremsstrahlung (*see* FREE–FREE TRANSITION). Hard X-rays (in the region 10–100 keV) are emitted during the impulsive onset of a solar flare, while soft X-rays (1–10 keV) gradually build up in strength and peak a few minutes after the impulsive emission. During the impulsive phase, high-speed electrons are accelerated to energies of 10 to 100 keV and hurled down the magnetic conduit of coronal loops into the dense chromosphere, where they create double, hard X-ray sources. Plasma heated to temperatures as high as 40 million K then expands upwards, by a process known as chromospheric evaporation, into the low corona along magnetic field loops that shine brightly in soft X-rays.

Because solar X-rays are totally absorbed in the Earth's atmosphere, X-ray flares must be observed from spacecraft. Flares can be characterized by their brightness in soft X-rays, as observed by monitoring satellites such as the Geostationary Operational Environmental Satellites (GOES). The classification shown here provides a good indication of the energy output and likely terrestrial effects of a flare.

Solar flares beam energetic charged particles along a narrow trajectory that follows the spiral shape of the interplanetary magnetic field in the SOLAR WIND. If a flare occurs in just the right place, west of the central meridian and near the solar equator, the accelerated particles can follow the field to Earth, threatening astronauts and satellites. The time taken for particles to reach Earth depends on their energy, taking roughly an hour for a particle energy of 10 MeV.

RADIO WAVES from flares are valuable for studying their physical properties. Flares can outshine the entire Sun at radio wavelengths. There are two main types of metre-wavelength radio flares, designated as type II and type III radio bursts. Type II bursts are thought to be caused by shockwaves moving outwards at velocities of about 1000 km/s (600 mi/s). Type III bursts are interpreted in terms of electron beams moving outwards through the corona at 0.05 to 0.2 times the speed of light. Type II bursts occur much more rarely than type III. The energetic electrons that produce the impulsive, flaring, hard X-ray emissions also emit synchrotron radiation at microwave (centimetre) wavelengths.

flare star (UV) Red dwarf star that exhibits sudden, short-duration flares with amplitudes ranging from a few tenths of a magnitude to 6–7 magnitudes at visible wavelengths. Flare stars are of late spectral types, dKe–dMe. During flares, the stellar emission lines become enhanced and a strong blue-ultraviolet continuum appears, giving a greater amplitude in the ultraviolet. The rise time is a few seconds to a few tens of seconds, and the decline lasts a few minutes to a few tens of minutes. The initial decline is usually rapid and may be followed by a short standstill and then a quasi-exponential decay. Small precursors or pre-flares are sometimes observed and occasionally slow rising flares are also seen. The stellar flare light is seen against the relatively faint light from the normal quiescent star, which has a cool photosphere at 2800–3600 K. At maximum optical brightness the larger flares may contribute more than 10 to 100 times the total output of the whole visible stellar hemisphere.

Since the photographic discovery by Ejnar HERTZSPRUNG in 1924 of DH Carinae, the first flare star, over 1500 have been found. Good examples of this class of object (and their apparent visual magnitudes) are UV CETI (13.0), AD Leonis (9.4), EV Lacertae (10.1), PROXIMA CENTAURI (11.0) and BY Draconis (8.2). Membership of binary star systems appears to be moderately common among flare stars, notably among those stars that also show activity similar to that of the BY DRACONIS STARS. Classical photographic spectroscopy has shown that during flares the Balmer emission lines of hydrogen become greatly enhanced, as do many other lines, for example of helium, calcium and other metallic elements. Modern techniques, using fast pulse-counting photometry, spectroscopy with sensitive CCD detectors, as well as satellite observations from above the Earth's atmosphere, have recently revealed that flare stars are amongst the most active stellar objects in our Galaxy, and that flare energies are often 1000 times their solar counterparts. In the variable-star classification, flare stars are designated 'UV' after UV Ceti, the prototype.

Collaborative international observations are an essential part of present-day flare star research, where large numbers of hours are devoted to a few selected stars, and it is necessary to collect simultaneous spectral, optical, radio and X-ray data. In the far-ultraviolet, the International Ultraviolet Explorer (IUE) satellite registered strong emission enhancements of doubly ionized magnesium, the carbon and silicon ionic species up to CIV and SiIV and also the high-temperature nitrogen line, NV, reminiscent of the solar outer atmosphere. Furthermore, micro-flaring in a few dMe stars has been observed simultaneously in both optical spectra and in coronal X-rays, which suggests that some stars are frequently active at a timescale of a few seconds to great heights above the stellar surface. Flare stars are thought by some to be spotted stars with overlying active atmospheric layers that correspond to the extremely wide range of observed temperatures, from 3000 K in the spots, through 10,000 K to 200,000 K in the chromosphere and transition region, to more than 1 million degrees in the corona.

The flare mechanism is poorly understood, although magnetic fields associated with differential stellar rotation are probably the cause. The low mass of the nearby dMe stars (many are less than 0.1 solar mass), and their relatively young age, means that steady thermonuclear burning in the interior, which maintains normal MAIN-SEQUENCE stars with radiative cores, may not yet have established itself. Wholly convective interiors may occur in the youngest stars of the lowest mass, and flaring appears to be a consequence of these instabilities, linked with complex magnetic fields. Difficult theoretical problems need to be solved as to the source and storage of energy prior to flares and their role in heating the intense stellar coronae.

A related subtype of flare stars is designated UVN. These stars are closely associated with nebulosity and are somewhat similar to the Orion variables, RW AURIGAE and T TAURI-type stars found in the region of the ORION NEBULA, the PLEIADES and most young galactic clusters (age a few million years and less). UVN stars abound in regions rich in dust and gas where star formation is still in progress. Because they are generally very faint (apparent visual magnitude 12 to 20), these flare stars have almost invariably been discovered with large Schmidt telescopes, and those known today required several thousand hours of patient observations and searching of photographic plates. Flares of amplitude 4–6 mags. in the ultraviolet are fairly common among the fainter of these stars. There are more than 400 such flare stars in and around the Orion Nebula and approximately the same number in the Pleiades. Flares have also been observed in Praesepe, the Hyades, and the Cygnus and Taurus dark clouds. These abundant objects occur at earlier spectral types (dK0 to dM6) than other flare stars and were consequently differentiated by the name 'flash stars'. Although little is known about these stars, there is good evidence that flares occur in quite a variety of objects, including, for example, the more massive RS CANUM VENATICORUM STARS, which are binaries with G- or K-type subgiant components.

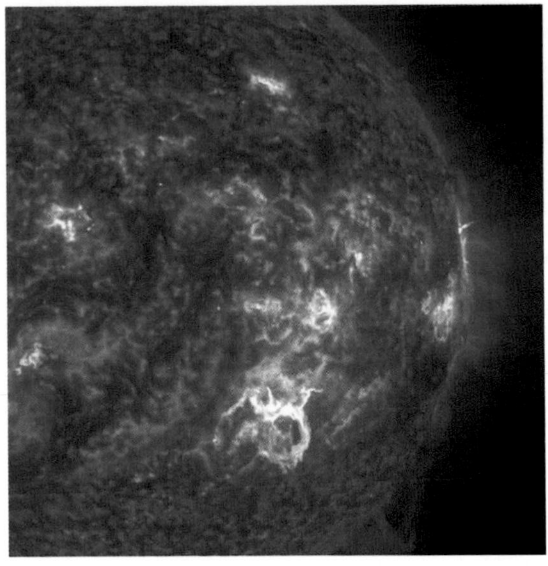

◀ **flare** Violent releases of magnetic energy in the inner solar atmosphere above active regions are seen in spectroheliograms as bright patches or ribbons, described as flares. Flares are most common at times of high sunspot activity.

flash spectrum Emission spectrum of the solar CHROMOSPHERE briefly seen during a total solar eclipse at the beginning and end of totality. It flashes into view for only a few seconds when the Moon completely covers the PHOTOSPHERE. The lines of the flash spectrum correspond to those seen as dark FRAUNHOFER LINES in the spectrum of the bright photosphere.

Fleming, Williamina Paton (1857–1911) Scottish-born astronomer who worked at Harvard College Observatory where, despite her lack of a formal higher education in astronomy, she developed a comprehensive system for classifying stellar spectra. Fleming and her husband emigrated to the USA in 1878, but he soon abandoned her, leaving her pregnant and destitute. Edward C. PICKERING, director of Harvard College Observatory, originally hired Fleming in 1881 as his housekeeper. Pickering, dissatisfied with the quality of work of his male assistants, complained that his maid could do a better job, and paid her as a temporary 'computer' at the observatory. He had just embarked on a major project to obtain large numbers of stellar spectra using an OBJECTIVE PRISM, and required a new system of classifying the many spectra that were being recorded.

Fleming's astronomical computing work was of such high quality that Pickering quickly gave her a permanent position. When she developed her own scheme for classifying stars according to their different spectra, he placed her in charge of the project, the outcome of which was the *HENRY DRAPER CATALOGUE*. She supervised many other female computers, using the 'Pickering–Fleming' alphabetical classification scheme (later refined by Annie Jump CANNON) to catalogue the spectra of 10,351 stars, published in 1890 as the *Draper Catalogue of Stellar Spectra*. During this time, she developed the first rigorous standards for measuring the photographic magnitudes of variable stars. In 1898 Fleming took charge of Harvard's rapidly growing collection of photographic plates, which she examined to discover 59 new nebulae, 222 variable stars and 10 novae – a major accomplishment, given that only 28 novae were known at the time. Fleming also discovered 94 WOLF–RAYET STARS and the first examples (1910) of WHITE DWARFS.

flickering Rapid variations in brightness over times of minutes, or even seconds, observed in many close BINARY STARS, especially NOVAE and DWARF NOVAE.

flocculi Bright, coarse, mottled pattern marking the chromospheric network. It is seen in SPECTROHELIOGRAMS of the Sun and is most noticeable in the H and K lines of ionized calcium, CaII.

Flora MAIN-BELT ASTEROID discovered in 1847; number 8. Flora is *c*.162 km (*c*.101 mi) in size. It is the archetype of a well-known HIRAYAMA FAMILY.

flux density Measure of radiation arriving from a source at a particular frequency – the energy received from an

FORNAX (GEN. FORNACIS, ABBR. FOR)

Small, inconspicuous southern constellation, representing a furnace, between Sculptor and Eridanus, introduced by Lacaille in the 18th century. Its brightest star, α For, is a binary with pale yellow components, mags. 3.8 and 7.0, separation 4″.5, period 314 years. Deep-sky objects include NGC 1365, a 9th-magnitude barred spiral galaxy; NGC 1316, another 9th-magnitude spiral, the optical counterpart of radio source Fornax A; and the Fornax Dwarf Galaxy, a dwarf spheroidal galaxy in the LOCAL GROUP, about 450,000 l.y. away.

astronomical source per unit area per unit time. Normally used in RADIO ASTRONOMY, with watts per square metre per hertz as the unit of flux density (*see* JANSKY).

focal length Distance between the centre of a lens or mirror and its focal point or FOCUS. The focal length of a converging lens can be measured by forming an image on a sheet of paper of a window on the opposite side of a room. The distance from the lens to the paper is the approximate focal length.

focal plane Plane through the focal point at right angles to the optical axis in which the image of a distant object will be formed. Some optical systems, such as the SCHMIDT CAMERA, produce an image that is in focus on a curved surface rather than on a flat plane.

focal ratio Ratio between the FOCAL LENGTH of a lens or mirror and its effective APERTURE. The focal ratio is often called the f-number. For example, a mirror of diameter 100 mm and focal length 800 mm would have a focal ratio of eight. This is often written as *f*8 or *f*/8. It is a measure of how quickly light rays at the edge of the aperture converge as they approach the focus; astronomers sometimes refer to optical systems with low f-numbers as 'fast' and those with high f-numbers as 'slow'.

focus Point at which parallel light rays meet after passing through a converging optical system, or, the point from which parallel light rays leaving a diverging optical system appear to originate. *See also* LENS; MIRROR

following (*f*) Term used to describe a celestial body, or feature on a body, that trails behind another in their apparent motion. Following features come into view later than the reference feature. For example, in a group of sunspots, the **following spot** is the last to appear as the Sun rotates, while the **following limb** of the Moon is the one facing away from its direction of travel. Any feature that appears earlier than a reference feature is said to be PRECEDING.

Fomalhaut The star α Piscis Austrini, visual mag. 1.16, distance 25 l.y., spectral type A3 V. Infrared observations have shown that it is surrounded by a PROTOPLANETARY DISK of dust from which a planetary system may be forming. The name is an abbreviation of the Arabic term meaning 'mouth of the southern fish', which is what it represents.

forbidden lines EMISSION LINES found particularly in the spectra of gaseous nebulae and in the corona of the Sun that are absent in terrestrial spectra because the required physical conditions cannot be satisfied in the laboratory.

Emission lines result from transitions of an atom from a high energy level to a lower level (as a result of collisions or radiative decay from a yet-higher level), resulting in the emission of photons with characteristic energies or wavelengths. Transition probabilities depend on how the individual quantum states interact with each other. Under simplified mathematical treatment, only transitions between certain kinds of states are allowed, while others are strictly forbidden. Under more detailed calculations, the 'forbidden lines' are in fact allowed, but are millions if not billions of times less probable.

Within gaseous nebulae and the solar corona, the states responsible for forbidden lines are populated by electron–ion collisions, which establish an equilibrium among the relevant states. The production rate of forbidden line photons depends on the instantaneous level populations multiplied by the decay probabilities. Given low density and enough mass (far more than can exist in the laboratory), forbidden line intensities can be very high, higher even than RECOMBINATION line intensities.

Forbidden lines are indicated by square brackets. The first found in gaseous nebulae were the 495.9 and 500.7 nm [O III] lines, which were eventually ascribed to the element NEBULIUM, while the first found in the solar corona, the 530.3 nm [Fe XIV] line, was assigned to CORONIUM; neither element exists. Analysis of forbidden lines in gaseous nebulae is crucial to the determination of electron temperatures, densities and chemical compositions.

Forbush decrease Reduction of the flux of high energy, extrasolar COSMIC RAYS observed at the surface of the Earth during high activity on the Sun. It occurs because solar activity produces enhancements in the magnetic field and plasma density around Earth, serving to scatter incoming cosmic rays. Such events thus occur more frequently at sunspot maximum. These events are named after Scott Forbush (1904–84), who first noted them in 1954.

fork mounting Type of EQUATORIAL MOUNTING in which the telescope is held between two arms parallel to the polar axis. The telescope's declination axis pivots between the arms.

Fornax *See* feature article

Forty-seven Tucanae (47 Tucanae; NGC 104) Second-brightest GLOBULAR CLUSTER, after OMEGA CENTAURI, at mag. +4.0. 47 Tucanae is relatively close, at a distance of 13,000 l.y., and has an apparent diameter of 31′. Its actual diameter is 120 l.y. 47 Tucanae is found 2°.5 west of the Small Magellanic Cloud (RA 07ʰ 24ᵐ.1 dec. −72°05′) but is some 15 times closer. Spectral analysis of its stars suggests 47 Tucanae to be younger than most other globular clusters.

forward scattering Reflection of light back towards the direction of the light source by an angle less than or equal to 90°. Light may be scattered from its direction of travel by fine particles of matter, either gas or dust. For particles significantly larger than the wavelength of the incident light, reflection or BACKSCATTERING of the light occurs.

Foucault, (Jean Bernard) Léon (1819–68) French physicist who in 1851 used a long pendulum of his own design to demonstrate the Earth's rotation. A FOUCAULT PENDULUM precesses in a clockwise direction with an angular speed equal to $\omega \sin \theta$, where ω is the angular speed of the Earth's rotation and θ is the observer's latitude. Working at Paris Observatory, Foucault invented the **Foucault knife-edge test** for mirrors and lenses, and silver coatings for mirrors. He was the first to photograph the Sun, and he invented the HELIOSTAT. He also devised an accurate method for measuring the speed of light.

Foucault pendulum Device to show that the Earth is truly rotating in space. Léon FOUCAULT noticed that a pendulum in the form of a weight suspended on a wire continued to swing in the same plane while its support was rotated. In 1851 he set up a 67-m (220-ft) pendulum in the Paris Pantheon, the swing of which moved clockwise by about 2 mm between each complete back and forth movement. This apparent rotation of the pendulum is actually due to the Earth rotating beneath the pendulum while the latter, through inertia, maintains its orientation in space.

fourth contact (last contact) In a SOLAR ECLIPSE, moment when the trailing limb of the Moon exits the solar disk. In a LUNAR ECLIPSE, fourth contact occurs when the Moon re-emerges completely into sunlight from Earth's UMBRA. In either case, fourth contact is the end of the eclipse.

Fowler, William Alfred (1911–95) American astrophysicist who explained the thermonuclear reactions that take place inside stars, and calculated their rates. He spent his entire career (1933–95) at the California Institute of Technology and its Kellogg Radiation Laboratory, where the staff made many fundamental discoveries in nuclear physics by accelerating positive ions to high velocities to produce neutrons, positrons and gamma rays by bombardment. These experiments enabled Fowler to make the first measurement (in 1934 while he was still a graduate student) of the reactions involved in the CARBON–NITROGEN–OXYGEN CYCLE in stars – he detected gamma rays emitted when carbon nuclei bombarded with protons captured one of the protons to make ^{13}N, which itself emitted a positron. Within two years, Fowler's group had discovered many more examples of these 'radiative captures', which proved to be of fundamental importance in understanding the energy processes of stars. In 1939 Hans BETHE and Carl von WEIZSÄCKER independently demonstrated how the carbon and nitrogen isotopes identified by Fowler and his colleagues were involved in converting hydrogen into helium inside stars.

After World War II, Fowler resumed his investigations of the thermonuclear power sources of the Sun and similar stars, developing experimental proof of the chain reactions that produce the 'heavy elements' (beyond helium). In 1957 Fowler co-authored, with Margaret and Geoffrey BURBIDGE and Fred HOYLE, the seminal paper 'Synthesis of the elements in stars', published in *Reviews of Modern Physics* and dubbed 'B^{2}FH' after its authors, which set forth the basic mechanisms by which stars synthesize chemical elements heavier than hydrogen and helium. In the 1960s and 1970s, Fowler's research interests included the synthesis of hydrogen, deuterium and helium in the 'hot' Big Bang, massive stars and supernovae, solar neutrinos, quasars and cosmology. He was awarded the 1983 Nobel Prize for Physics for his measurements of the thermonuclear rates in stars.

Fracastorius Lunar crater (21°S 33°E), 97 km (60 mi) in diameter. Fracastorius is flooded with the lavas that flooded the Nectaris Basin and only the highest parts of the northern wall are still visible. The floor is smooth from the lava resurfacing, but the highest part of the central peak is still visible. A long shallow rille nearly bisects Fracastorius. The walls are deeply incised by later impacts, revealing this crater's great age.

fractional method Way of visually estimating the magnitude of a VARIABLE STAR. It consists of estimating the brightness of the variable as a fraction of the interval between two comparison stars (A and B, say). If it were judged to be exactly midway between them, the estimate is written as A(1)V(1)B; if it is one quarter of the way from A (the brighter) to B (the fainter), and hence three-quarters of the way from B to A, the estimate is recorded as A(1)V(3)B. The magnitude may subsequently be calculated from the known magnitudes of the comparisons. The method has the advantage that the magnitudes of the comparisons need not be known in advance, but it is best suited to sequences that have comparisons at approximately 0.5 mag. intervals, because for physiological and psychological reasons, magnitude intervals (however great) cannot be subdivided beyond a ratio of approximately 1:4 or 1:5.

frame of reference Position against which measurements can be made. It is usually advisable to use a non-accelerated or inertial frame of reference when making measurements. The laws of physics can be simpler in certain frames of reference as compared with other frames, so the choice of the reference frame contributes dramatically to the ease with which equations can be solved.

Franklin-Adams charts First wide-field photographic SURVEY of the complete sky, accomplished by John Franklin-Adams (1843–1912) using a specially constructed wide-field 250-mm (10-in.) refractor known as the **Franklin-Adams camera**. The survey, published in 1913–14 as *Photographic Chart of the Sky*, consists of 206 plates of about 15° × 15° each, taken between 1903 and 1912 from Godalming, England, and Johannesburg, South Africa. With a limiting magnitude of around 16, these charts represent an historical record that is still used for studies of long-period variables.

Fraunhofer, Joseph von (1787–1826) Bavarian optician and pioneer of solar spectroscopy. In 1806 he joined the Munich optical firm of Reichenbach, Utzschneider u. Leibherr and began to make achromatic objectives for surveying instruments, improving the methods for polishing and grinding them. His work was of such high quality that by 1809 Joseph von Utzschneider (d.1839) had placed him in charge of the firm's glass-making facility at Benediktbeuern. There, Fraunhofer began detailed investigations of the optical properties of different kinds of glass, commencing a lifelong search for the perfectly achromatic lens. His methods of glass- and lens-making allowed his firm to produce superb refracting telescopes of greater apertures than had previously been feasible. Among them were the Dorpat (now Tartu, Estonia) 240-mm ($9\frac{1}{2}$-in.) refractor used by Wilhelm von STRUVE to discover many double stars, and the Königsberg 158-mm ($6\frac{1}{4}$-in.) heliometer with which Friedrich Wilhelm BESSEL measured the first stellar parallax. Fraunhofer also constructed one of the earliest examples of the 'German' EQUATORIAL MOUNTING.

In 1814–15, while studying the refractive indices of various types of glass, Fraunhofer examined the solar spectrum produced by a narrow slot placed in front of a low-power telescope, searching for the pair of bright orange-yellow lines of sodium he had already observed in

▼ **Forty-seven Tucanae** The splendidly rich, bright southern globular cluster 47 Tucanae is surpassed only by Omega Centauri. 47 Tucanae is readily visible to the naked eye.

F

the light from a candle. Instead of bright emission lines, he saw dark absorption lines at the same place in the solar spectrum. William Hyde WOLLASTON had first observed these dark features in 1802, but because Fraunhofer studied them in much greater detail they came to be called FRAUNHOFER LINES. This discovery led to the development of astronomical SPECTROSCOPY. To further these researches, Fraunhofer designed a dividing engine to manufacture diffraction gratings with 4000 parallel lines per centimetre, which he used to measure the wavelengths of over 500 different solar spectral features. He found that the most prominent lines are produced by atoms and ions of calcium, hydrogen, sodium, magnesium and iron.

Fraunhofer lines ABSORPTION LINES in the SPECTRUM, first noted and catalogued in sunlight by Joseph von FRAUNHOFER in the early 19th century. Lettered from A to K, the Fraunhofer lines are the strongest in the solar spectrum, and are prominent in the spectra of other stars.

In spite of possible confusion with chemical symbols, the various Fraunhofer designations are still in common use, particularly 'D' for neutral sodium at 589.6 and 589.0 nm (D1 and D2, respectively), 'G' for a band of CH (carbon and hydrogen, chemically bound, plus other absorptions) at 430.0 nm, and 'H' and 'K' for ionized calcium at 396.8 and 393.4 nm. Small 'g' and 'D3' were added later to indicate neutral calcium at 422.7 nm and neutral helium at 587.6 nm.

Fred L. Whipple Observatory (FLWO) Largest facility of the SMITHSONIAN ASTROPHYSICAL OBSERVATORY, situated on Mount Hopkins near Amado, 60 km (35 mi) south of Tucson, Arizona; it has been in operation since 1968. The FLWO is home to the MMT OBSERVATORY, whose 6.5-m (256-in.) telescope occupies the highest point on the site at an elevation of 2600 m (8550 ft). Other instruments are 1.5-m (60-in.) and 1.2-m (48-in.) reflectors and a 10-m (33-ft) gamma-ray telescope. The 1.3-m (51-in.) infrared telescope of the TWO MICRON ALL SKY SURVEY (2MASS) is also located on Mount Hopkins.

free–free transition Radiation emitted when a free electron is decelerated in the vicinity of an ion without being captured. It is also known as Bremsstrahlung ('braking radiation').

Friedman, Herbert (1916–2000) American astrophysicist who helped to found X-ray astronomy, using rocket-borne detectors to demonstrate the existence of extraterrestrial X-ray sources. After carrying out important studies of the Sun's X-ray emissions in the 1940s and 1950s, in 1964 Friedman discovered the first X-ray source outside the Solar System, finding that the Crab Nebula, a supernova remnant, is a powerful X-ray emitter.

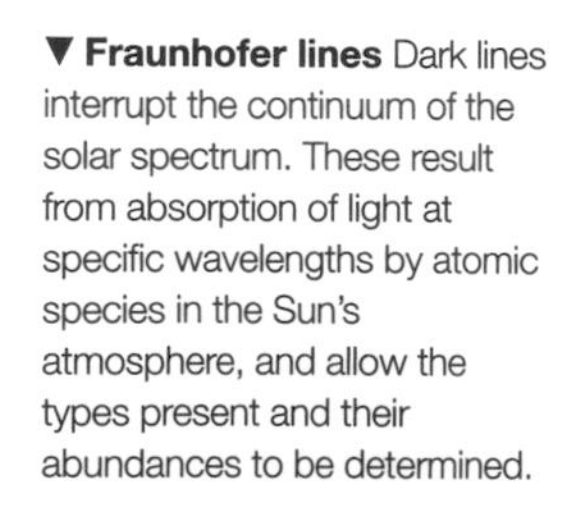

▼ **Fraunhofer lines** Dark lines interrupt the continuum of the solar spectrum. These result from absorption of light at specific wavelengths by atomic species in the Sun's atmosphere, and allow the types present and their abundances to be determined.

Friedmann, Aleksandr Aleksandrovich (1888–1925) Russian mathematician who in 1922 worked out the basic details of the expanding universe. He solved Einstein's field equations for the general theory of relativity, obtaining several solutions all suggesting a universe that was isotropic, or uniform in all directions, but was not static. This set of models is known collectively as the FRIEDMANN UNIVERSE. Friedmann's work was not influential at the time, and it remained unknown to Edwin HUBBLE.

Friedmann universe Cosmological model developed by Aleksandr FRIEDMANN and others, based on Einstein's theory of GENERAL RELATIVITY. It described a universe that is homogeneous and isotropic. Specifically, mass and energy are evenly distributed at the largest of scales. Following these assumptions, the FIELD EQUATIONS can be solved for the entire universe. This type of universe provides the basis for modern-day BIG BANG models.

frequency Number of oscillations of a vibrating system per unit of time. The normal symbol is f or ν, and the units are hertz (Hz). For an electromagnetic wave, the frequency is the number of wave crests passing a given point per unit time. *See also* ELECTROMAGNETIC RADIATION; ELECTROMAGNETIC SPECTRUM; WAVELENGTH

F ring Narrow outer ring of SATURN; it was first observed from PIONEER 11 in 1979 and in more detail during the VOYAGER encounters. The F ring has a narrow, 'braided' structure, governed by the gravitational influence of small satellites orbiting close to it. It lies 140,200 km (87,100 mi) from the planet's centre.

fringes Patterns of light and dark produced by the INTERFERENCE of light waves that are alternately in and out of phase with each other. A simple example can be produced by placing a convex lens on a sheet of plane glass. Interference fringes in the form of rings will be seen around the point of contact. Fringes provide the optical engineer with a very powerful tool for measuring the quality of optical surfaces because minor variations can cause large changes in the shape or position of fringes.

F star Any member of a class of yellow-white stars, the spectra of which are characterized by weakened, though still strong, hydrogen lines, strong ionized calcium (Fraunhofer H and K), and other ionized metal lines. MAIN-SEQUENCE dwarfs range from 6000 K at F9 to 7000 K at F0, with zero-age masses from 1.05 to 1.6 solar masses, zero-age luminosities from just above solar to 5 solar, and dwarf lifetimes from just under 10 billion to 2.5 billion years. Evolved F dwarfs can have masses up to twice that of the Sun. Class F GIANTS and SUPERGIANTS (which can have much higher masses) are distinguished by the ratios of the strengths of ionized to neutral metal absorptions. F SUBDWARFS have weakened metal lines, lower metal abundances, and appear hotter than they really are. Chemically peculiar AM STARS and AP STARS run into class F, where they are known as 'Fm' and 'Fp'.

Class F stars are distinguished by fundamental changes in stellar behaviour. Envelope convection diminishes with increasing mass and disappears near F0, where the carbon cycle begins to dominate the proton–proton chain. Rotation speeds increase as the convection layer and the associated magnetic braking die away. The Sun spins at only 2 km/s (1.2 mi/s), but by F5 the typical speed is more like 30 km/s (19 mi/s), and at F0 it has climbed to 100 km/s (62 mi/s), although some stars of course still rotate slowly).

Class F is the great bastion of the CEPHEID VARIABLES, which occupy an INSTABILITY STRIP in the Hertzsprung–Russell diagram extending from supergiants in class G down through F supergiants and giants. Where the strip cuts across the Population II horizontal branch, the RR LYRAE STARS of classes F and A are found. Population II Cepheids (W VIRGINIS STARS) are also F bright giants.

Bright examples of F stars include Porrima (γ Virginis) F0 V, Procyon F5 IV, Canopus F0 II and Polaris F5 Ia.

full moon *See* PHASES OF THE MOON

fundamental catalogue Catalogue of FUNDAMENTAL STARS whose positions and proper motions are accurately known and against which the relative position of other celestial objects may be measured.

The first such catalogue was prepared by F.W. BESSEL in 1830. Particularly important has been the FK (*Fundamental Katalog*) series compiled in Germany, the latest of which is the *FK5 Part II*. Also known as the *FK5 Extension*, this catalogue provides mean positions and proper motions at equinox and epoch J2000.0 for 3117 new fundamental stars. It extends the fundamental system defined by *FK 5 Part I*, which was published in 1988, to about magnitude 9.5. More than 200 existing catalogues were used in the compilation of the *FK5 Extension*.

Work is currently underway on the *Sixth Catalogue of Fundamental Stars* (*FK6*), which combines observations from the HIPPARCOS astrometric satellite with the ground-

◀ **F ring** Imaged by the Voyager 1 spacecraft during its 1980 flyby, the 'braided' outer F ring of Saturn is shepherded by a pair of small, near co-orbiting satellites.

based data of the basic *FK5*. The *FK6* will be the catalogue with the most accurate proper motions yet published.

fundamental forces In all of physics, every particle discovered and every interaction observed can be explained by the existence of just four fundamental forces of nature. These forces are the **electromagnetic** force, the **gravitational** force, the **strong** force and the **weak** force. The electromagnetic force and the gravitational force are most evident in our everyday lives since they describe chemical reactions, light and the Earth's gravitational field.

It first seemed that electrical forces and magnetic forces were separate and independent, but investigations into each phenomenon led to underlying similarities. James Clerk MAXWELL finally combined the forces and wrote down his famous equations, which described all electrical and magnetic phenomena and explained how magnetic fields could be generated by electrical currents and vice versa. This system of differential equations also yielded a wave solution, and the wave propagated at the speed of light. Thus light was finally understood in terms of electric and magnetic fluctuations that needed no medium through which to propagate.

Both the electromagnetic force and the gravitational force are inverse square forces in the sense that the strength of the field declines with the inverse square of the distance, and both forces have infinite range. A major difference between the electromagnetic force and the gravitational force is that the former is easily quantized, with the exchange particle being the photon. Gravity on the other hand has defied quantification, and the most successful gravitational theory to date is Einstein's theory of GENERAL RELATIVITY, which is a CONTINUUM theory. Attempts have been made to quantize gravity but with little success; a proposed particle of gravitational quanta is called the graviton.

The strong force and weak force are both very strong forces, but have extremely limited range. The strong force operates inside the atomic nucleus and essentially holds the protons and neutrons together by the exchange of particles called gluons. The actual exchange of gluons is between the quarks that make up protons and neutrons. The strong force is the strongest of the four fundamental forces, but its range is limited to distances of the order of 10^{-15} m. The weak force also operates inside the nucleus and is important in radioactive decays. The intermediate vector boson is the exchange particle and its range is confined to a distance of the order of 10^{-16} m.

FUNDAMENTAL FORCES

Force	Relative strength	Exchange particle	Range
strong	1	gluon	10^{-15} m
electromagnetic	1/137	photons	infinite
weak	10^{-4}	intermediate vector bosons	10^{-16} m
gravitational	6×10^{-39}	gravitons	infinite

Although the strong and weak force are relatively strong compared with the gravitational force, their limited range makes them unimportant in cosmological concerns, except close to the beginning of the Big Bang. The electromagnetic force is much stronger than the gravitational force, but since it comes with roughly equal numbers of opposite charges which neutralize each other, on cosmological scales it is less important than gravity. Thus the weakest of the four fundamental forces dominates the evolution of the universe.

All of the forces except for gravity are easily understood within the framework of QUANTUM THEORY. This fundamental difference between gravity and the other forces comes into play in the early universe. It is thought that under the hot and dense conditions of the early universe the electromagnetic force and the weak force combined into one force. Some time later this super-force then combined with the strong force. At very early times, shortly after the Big Bang, gravity must have joined and become one with the strong electroweak force. However, owing to the nature of general relativity, no QUANTUM THEORY OF GRAVITY has been found, and the mathematics describing the fundamental forces disallows this unification. The unification of the four fundamental forces of physics is the ultimate goal of physics (*see* THEORY OF EVERYTHING). Even Einstein spent some time working on the problem to no avail. Recent ideas such as SUPERSTRING theory claim to have accomplished unification, but at a price of postulating many unobserved dimensions of our Universe.

fundamental stars Reference stars whose positions and proper motions have been observed and established over a long period of time to a high degree of accuracy. A measurement of the position of a star (that is, its RIGHT ASCENSION and DECLINATION) is defined as being 'absolute' if it does not make use of the already known positions of other stars. Absolute positions are obtained from specialist instruments such as the CARLSBERG MERIDIAN TELESCOPE and the HIPPARCOS astrometric satellite. The coordinates and proper motions of fundamental reference stars, selected by the International Astronomical Union, are contained in a FUNDAMENTAL CATALOGUE. The system of fundamental stars provides the best available approximation to an inertial frame of reference, relative to which positions of all other celestial objects can be measured.

FU Orionis star (FU, fuor) Young star that increases in brightness by some 5–6 mag. over a period of months, following which it may remain in a permanent bright state or suffer a slight decline. Objects in this small group are believed to represent one of the evolutionary stages exhibited by young T TAURI STARS. All known examples (about 10 in number) are associated with cometary nebulae.

FUSE *See* FAR ULTRAVIOLET SPECTROSCOPIC EXPLORER

fusion, nuclear Process in which two light atomic nuclei join together to form the nucleus of a heavier atom. In the very high temperatures found in the core of the Sun (about 15.6 million K) two hydrogen nuclei are fused together to give a deuterium nucleus, plus a positron and a neutrino. Later stages in this fusion process eventually produce the nucleus of helium. Such processes are called NUCLEAR REACTIONS. With light elements, the BINDING ENERGY per nucleon increases during nuclear fusion and so vast amounts of energy are released. The fusion of elements up to the iron-peak results in the production of energy. The fusion of nuclei heavier than iron requires there to be an input of energy to keep the reaction going, and thus mainly occurs during SUPERNOVA explosions. Fusion is the energy-producing process in most stars. The formation of elements up to iron by fusion is called NUCLEOSYNTHESIS. *See also* FISSION, NUCLEAR

G

Gacrux The star γ Crucis (of which 'Gacrux' is a contraction); visual mag. 1.59 (but slightly variable), distance 88 l.y., spectral type M4 III.

Gagarin, Yuri Alexeyevich (1934–68) Russian cosmonaut who became the first human in space, completing a single orbit of the Earth in the spacecraft Vostok 1 on 1961 April 12 in a flight that lasted 1 hour, 48 minutes. He was the first to experience weightlessness for more than just a few minutes, and the first to see sunrise from outer space. Gagarin, honoured as a Hero of the Soviet Union, was killed during a test flight of a jet aircraft.

galactic centre Centre of the Galaxy. It is located in Sagittarius, about 4° west of γ Sgr (RA 17^h $45^m.6$ dec. −29°00′). Its distance is about 26,000 l.y. Dust clouds along the line of sight to it prohibit direct optical observation, although there are partial clearings of the dust clouds, such as that at BAADE'S WINDOW. X-ray, infrared and radio observations show a complex structure. A radio source, SAGITTARIUS A*, right at the centre, is thought to be a BLACK HOLE with a mass of about 2.5 million solar masses. Surrounding that black hole, and within a few light-years of it, are a dense star cluster, an irregular ring of molecular gas, a gas-free cavity and mini spiral arms. The interactions between these objects are complex and not yet fully understood.

galactic cluster Old term for an OPEN CLUSTER of stars.

galactic coordinates System used in the study of the distribution of objects in the Milky Way galaxy, using the plane of the Galaxy and the galactic centre as its reference points and the measures of galactic latitude and galactic longitude. Galactic latitude is measured from 0° at the galactic equator to +90° at the north galactic pole, while galactic longitude is measured from 0° at the galactic centre (RA 17^h 46^m, dec. −28°56′), eastwards to 360° along the galactic equator. *See also* CELESTIAL COORDINATES

galactic pole Either of two points that are diametrically opposite each other in the sky and for which the connecting line is perpendicular to the plane of our Galaxy. The north galactic pole is in Coma Berenices about 5° west of β Com (RA 12^h $51^m.4$ dec. +27°7′); the south galactic pole is in Sculptor, 2° northwest of α Scl (RA 00^h $51^m.4$ dec. −27°7′). The line joining the poles approximately marks the rotational axis of the Galaxy.

▼ **galactic centre** A wide-field view of the centre of our Galaxy in the direction of Scorpius, Centaurus and Sagittarius. The dense star clouds of the Milky Way are in places heavily obscured by dark nebulosity.

Galatea One of the small inner satellites of NEPTUNE, discovered in 1989 by the VOYAGER 2 imaging team. Galatea is about 160 km (100 mi) in size. It takes 0.429 days to circuit the planet, at a distance of 62,000 km (39,000 mi) from its centre, in a near-circular, near-equatorial orbit. It appears to act as an interior SHEPHERD MOON to the planet's Adams Ring, and it orbits just beyond the exterior edges of the Lassell and Arago rings.

galaxies, classification System of grouping similar GALAXIES together based on the properties of the galaxies' structure. One widely used system is the HUBBLE CLASSIFICATION.

galaxy Huge, gravitationally bound, assemblage of stars, gas, dusta and DARK MATTER, an example of which is our own GALAXY. Such objects span a wide range of size, luminosity and mass, with the largest CD GALAXIES being a million times brighter than the faintest known DWARF GALAXIES. The forms of most galaxies fall into the three main types recognized in the HUBBLE CLASSIFICATION scheme. ELLIPTICAL GALAXIES have smooth, elliptically symmetric brightness distributions, and generally have little or no ongoing star formation or cold interstellar gas. SPIRAL GALAXIES may be ordinary or barred; they include central bulges much like embedded elliptical galaxies, surrounded by thin disks that contain cold, dense gas and have hosted star formation over most of the galaxy's history. IRREGULAR GALAXIES have no distinct structure, are generally quite gas-rich, and form stars at a brisk rate. There are also transition S0 (or LENTICULAR) galaxies, which share the bulge and disk structure of spiral galaxies with the old stellar populations of ellipticals.

A defining property of galaxies seems to be that their internal motions suggest the presence of much more mass, in a more extended distribution, than can be accounted for by all the material that can be directly detected. This material is known as dark matter. Unless there are major modifications to the properties of gravity as we understand them, this enigmatic material makes up 90% of the mass in typical galaxies. The visible (and otherwise detectable) components of galaxies are such minor contributors by mass that the behaviour of dark matter must set the terms for formation of galaxies and the relations among size and mass.

Most galaxies collapsed to approximately their present forms and began star formation during a fairly brief epoch in the early Universe. Detailed study of nearby galaxies has suggested a picture in which an enormous gas cloud collapsed from a roughly spherical shape while forming stars, with remaining gas forming the thin disk of spirals in which subsequent star formation has taken place. This monolithic collapse ('top down') picture explains some aspects of galaxy structure well, but other aspects require a hierarchical buildup ('bottom up') of galaxies from smaller units (which can be seen today as CANNIBALISM). Galaxy evolution can be observed in several guises. Interactions and mergers can trigger bursts of star formation, resulting in winds that can sweep a galaxy free of gas, and sometimes leave one galaxy in place of two. Tidal encounters may also trigger ACTIVE GALACTIC NUCLEI, particularly QUASARS.

Many galaxies belong to groups or GALAXY CLUSTERS. Rich clusters today contain almost exclusively elliptical and lenticular galaxies, while such clusters at significant redshifts (seen as they were several thousand million years ago) can be rich in spirals. This shift, the Butcher–Oemler effect, shows that the cluster environment can drive galaxy evolution. Tidal encounters or interactions with the hot intracluster medium may be responsible for this change.

Galaxy, the Star system or GALAXY that contains our Solar System; the capital G distinguishes it from the many others that can be observed. It is also known as the

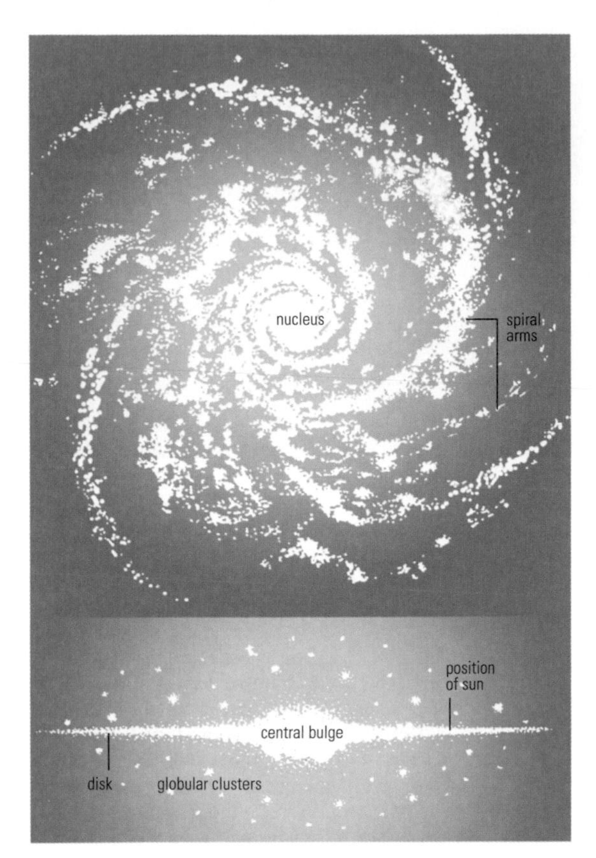

◀ **galaxy** A general outline for the structure of a spiral galaxy, of which our Milky Way is an example. Seen face-on, the galaxy has a fairly compact nucleus surrounded by an extensive disk containing star-rich spiral arms. Seen edge-on, the nucleus is revealed as a bulge, with the spiral arms confined to the flat plane of the disk. The galaxy as a whole is surrounded by a halo of globular clusters. As indicated, the Sun is located midway out in the disk of our Galaxy.

MILKY WAY after its visual guise: from the northern hemisphere it can be seen as a faint, luminous band that meanders from the deep south in Sagittarius, glides through Aquila and Cygnus, embraces the W of Cassiopeia, then fades gradually past Perseus and Orion to dip once more below the southern horizon in Puppis.

Seen from the southern hemisphere the Milky Way is far brighter and more spectacular. From Puppis it flows into Carina, where it is richly textured and bright enough to be seen from within large cities. Continuing bright, it sweeps past the stars of Crux, where an intervening cloud of dusty gas has removed a black scoop from it, the COALSACK. From there a narrow dark streak bisects the Milky Way as it passes through Centaurus, swells and brightens through the tail of Scorpius, and enters Sagittarius. When seen against a dark sky, and when it rides nearly overhead, the Milky Way in Sagittarius reaches its widest, at more than 30° in breadth, still split by a dark central band. That band continues as the Milky Way narrows northwards through Aquila to Cygnus. Northern hemisphere dwellers have termed this dark band the CYGNUS RIFT, but its full extent, one-third of the entire sky, can be appreciated only from the south. The fact that the Milky Way completely circles the sky clearly indicates that we live within it. It is our Galaxy.

External galaxies basically have three forms – spiral, elliptical and irregular. The Milky Way is quite symmetrical about the constellation Sagittarius, so it is not irregular. Elliptical galaxies are almost spherical blobs of stars, so if we lived inside one we would see its constituent stars in every direction. Instead, the narrowness of the band of the Milky Way shows that our Galaxy is basically flat, and as such it must be a spiral galaxy.

When viewed from the southern hemisphere, a spiral classification is readily appreciated. Our Solar System lies among the spiral arms, so that at least one wraps round outside us like a cloak, while others curl between the centre and us. And, like other spiral galaxies, our Galaxy contains bands of opaque dust clouds along the spiral arms and a central bulge like a miniature elliptical galaxy within the orbit of the arms. That bulge is in Sagittarius, which is why the Milky Way widens there. In front of the bulge passes a prominent spiral arm, which produces the great brightness of the Centaurus–Crux–Carina region. This arm curls away from us in Carina, where, seen effectively end-on, a great accumulation of stars and nebulae causes the extreme brightness and complexity. Immediately outside this arm, and therefore superimposed in front as we view it, is the band of gas and dust, a region where stars have yet to form, that causes the dark rift from Cygnus to Crux. Beyond Carina we see the fainter arm that wraps itself round outside the Sun. It crosses the northern sky, gaining brightness as we follow it gradually inwards towards Cygnus.

Although the trained eye easily sees these features, it took a long time for the Galaxy's structure to be deduced. Even today, the details remain contentious, for instance of how many spiral arms it has. RADIO ASTRONOMY has been of paramount importance. In 1931 Karl JANSKY discovered radio emission from Sagittarius, and before long the first maps of the radio sky were made. Apart from a few individual spots of radio emission, which were later to prove of immense importance, these maps showed a band exactly following the Milky Way. The mechanisms producing visible light and radio emission are grossly different, and very few objects liberate both in comparable amounts. Nonetheless, the Galaxy contains sources of both.

In the northern sky the radio emission was found to be weak, but it rose to a crescendo in Sagittarius. The difference between the radio and the visible Milky Way was, quite simply, that the radio waves passed unimpeded through the dark clouds. They showed what the Galaxy would be like if we could see all the stars. In particular, they focused attention on the very centre of the Galaxy (*see* SAGITTARIUS A and GALACTIC CENTRE).

As radio astronomy matured it began to make more sophisticated observations. One of the most important of these was the study of radiation by hydrogen atoms at a wavelength of 21 cm (*see* TWENTY-ONE CENTIMETRE LINE). Observations at this wavelength could pick out the gas clouds associated with spiral arms, revealing them irrespective of whether optically opaque clouds intervened. The radio astronomers put together a map of the Galaxy revealing its spiral arms. In practice the Galaxy appeared highly complex, and spiral arms could be sketched through the map only with the eye of faith. More recently, however, similar observations made from Australia, of the less ubiquitous gas carbon monoxide, have painted a much clearer canvas; this map suggests that there are four arms to our Galaxy. The best evidence now suggests that the Galaxy is a BARRED SPIRAL with a HUBBLE CLASSIFICATION of SBc.

Our view from within the disk of the spiral arms can never be adequate to disentangle the full structure of the Galaxy. The following description, therefore, is based in part on studies of our own Galaxy and in part on what is known of other spiral galaxies.

About 100,000 million stars populate the Galaxy, making it slightly above average in size. It is nearly 100,000 l.y. across, though its outer boundary is not well defined, and we live 26,500 l.y. from the centre. The bulge of stars in the central part comprises those that formed relatively early in the Galaxy's history. These stars have evolved

SUMMARY OF THE MAIN PARAMETERS OF THE MILKY WAY GALAXY

Hubble class	barred spiral type SBc
nucleus diameter	15,000 l.y. in the plane of the Galaxy; 6000 l.y. perpendicular to the plane of the Galaxy
disk diameter	80,000 to 100,000 l.y.
disk thickness	5000 l.y. near the nucleus to a few hundred l.y. at the outer edge and about 1000 l.y. at the position of the Sun
mass of nucleus and disk	2×10^{11} solar masses
halo diameter	150,000 l.y.
mass of halo	2×10^{11} to 4×10^{11} solar masses
corona diameter	600,000 l.y.
mass of corona	2×10^{12} to 6×10^{12} solar masses
position of the Sun	26,500 l.y from the centre and about 14 l.y. above the central plane of the galactic disk

all figures quoted are approximate

G

G

▲ **Galilean satellites** Images from the Galileo orbiter, showing the four large satellites of Jupiter, arranged in order of distance from the planet: (left to right) Io, Europa, Ganymede and Callisto.

away from the MAIN SEQUENCE, many becoming cool giants; those that remain main-sequence stars are small and cool. As a result the average colour of these stars is yellow, and they are collectively referred to as POPULATION II stars. By contrast the disk that spreads even beyond the Sun's orbit, and contains the spiral arms, is a site of star formation, as seen in the ORION NEBULA and similar locations. The integrated light of young stars is dominated by the hottest, brightest specimens, and is therefore blue. These stars are referred to as POPULATION I stars. Interspersed among the stars, and especially between the curving spiral arms, are dense clouds that will continue to form new stars for thousands of millions of years.

Just like planets round the Sun, the stars of our Galaxy orbit the central bulge. Each star does so at a speed dictated by the mass of material (stars, gas and any unseen objects such as BLACK HOLES) within its orbit. The stellar orbits are not quite circular, so stars shuffle relative to one another. In the bulge, some orbits are quite elliptical whereas others swing up and down in complex gyrations that may be visualized by drawing a sine wave around the surface of a slightly squashed cylinder. Our Sun circles the galactic centre at about 250 km/s (160 mi/s), taking some 200 million years per orbit.

If the motions of the outermost stars of the Galaxy are measured, the total mass can be calculated. This measurement is only partially feasible, because velocities can only be determined along the line of sight (using the DOPPLER EFFECT) and not across it. A further complication is our own motion, itself not very accurately known. Various statistical analyses of the measured motions suggest that the Galaxy's mass is nearly one million million times as much as the Sun, which is nearly ten times the mass of the visible constituents. This extra mass forms a spheroidal halo around the Galaxy some 150,000 l.y. across, which is in turn embedded within a spherical corona, which may be up to 600,000 l.y. in diameter and which includes the MAGELLANIC CLOUDS, the GLOBULAR CLUSTERS and some other dwarf galaxies. The Milky Way may have formed the halo and corona through the accretion and tidal break-up of other small galaxies.

galaxy cluster Group of GALAXIES. Galaxies are usually found as members of clusters. Rich clusters can have thousands of members and poor clusters may have only dozens of galaxies. The clustering of galaxies is an important constraint on cosmological models and the degree of clustering in the Universe today is related to the anisotropies in the matter distribution of the early universe. Margaret GELLER and colleagues at Princeton University have mapped a portion of the Universe and attempts are currently being made to characterize the degree of galaxy clustering using CORRELATION FUNCTIONS.

Galilean satellites Collective term for the four largest SATELLITES of JUPITER, discovered in 1610 by GALILEO GALILEI. They were the first proof that not all motion in the Solar System is centred on the Earth, and it paved the way for the acceptance of the HELIOCENTRIC THEORY.

Little was known about the Galilean satellites until they were seen from close range during flybys by the two VOYAGER probes in 1979. More detailed images were obtained during the 1995–2003 orbital tour by the GALILEO spacecraft, and repeated encounters cast light on their internal density distributions and detected their magnetic fields. The Galilean satellites show a progressive decrease in density, corresponding to an increase in the ratio of ice to rock, with distance from Jupiter. This observation reflects the higher temperatures that prevailed in the inner part of Jupiter's PROTOPLANETARY DISK while the satellites were accreting. The innermost satellite, IO, has the same structure as a TERRESTRIAL PLANET, with an iron-rich core surrounded by a rocky mantle. The next satellite, EUROPA, has a similar structure overlain by an outer layer of ice. GANYMEDE, the largest of the four, is composed of a roughly 40:60 ice-rock mixture, and its interior appears to be fully differentiated (*see* DIFFERENTIATION), with an iron-rich inner core, a rocky outer core and an icy mantle. CALLISTO, the outermost Galilean satellite, is slightly less dense and appears to be only weakly differentiated. TIDAL HEATING is responsible for present day volcanic activity on Io and the recent break-up of Europa's surface, where there may well be an ocean below the ice. It may also account for the substantially different evolution of Ganymede and Callisto.

Galilean telescope REFRACTING TELESCOPE having a plano-convex objective and a plano-concave eyepiece. It forms an erect image and gives a small field of view. This type of instrument, first employed by GALILEO to make astronomical observations, is no longer used in astronomy, but the same optical system is still used in opera glasses.

Galileo First spacecraft to orbit the giant planet JUPITER and to deploy an atmospheric probe into the Jovian atmosphere, in 1995 December. It was launched by the NATIONAL AERONAUTICS AND SPACE ADMINISTRAION (NASA) in 1989 October aboard the Space Shuttle and deployed initially in Earth orbit. The launch had originally been planned for 1986 May but was delayed by safety concerns about flying Galileo with a cryogenic upper stage on the Shuttle after the *Challenger* accident. Soon after deployment, it was discovered that the spacecraft's large, high-gain antenna had not deployed fully, jeopardizing the success of the mission's ability to transmit data. Corrections to software and systems meant that the spacecraft's loss could be compensated in part, although transmission of data took longer and contained less information. Galileo flew to Jupiter via a gravity assist flyby of Venus in 1990 February and two flybys of the Earth in 1990 December and 1992 December. The spacecraft's path allowed it also to visit two asteroids, marking a space first in 1991 October with a flyby of GASPRA. The asteroid IDA was explored in 1993 August. Close-up images of both asteroids at a minimum distance of 1600 km (1000 mi) and 2410 km (1500 mi) respectively were taken.

In 1995 July, the atmospheric probe was deployed, some 80 million km (50 million mi) from Jupiter, and both spacecraft headed towards a rendezvous with the planet. The probe entered the Jovian atmosphere on 1995 December 7 and transmitted data for 57 minutes during a 152-km (95-mi) descent under a single parachute, transmitting data to Galileo as it was preparing to enter orbit the following day. Transmission of the probe data to Earth was completed by 1996 April. The data revealed an intense radiation belt 50,000 km (31,000 mi) above the clouds, a Sun-like hydrogen–helium ratio, some organic compounds, one cloud layer and 640 m/s (2000 ft/s) winds below the cloud deck. Thunderstorms many times larger than Earth's result from the vertical circulation of water in the top layers, leaving large areas where air descends and becomes dry and other areas where water rises to form thunderstorms.

Galileo began its tour of the Jovian satellites, which continued until 2002, with a flyby of GANYMEDE, followed by several further encounters with this and the other main moons. Galileo established that the ring system is made of small grains blasted off the satellites' surfaces by meteoroid impacts. IO was found to be the Solar System's most volcanically active body, while EUROPA has a possible salty ocean containing more water than does Earth. Ganymede has a magnetic field and CALLISTO may also have a water ocean. In 2000 December, Jupiter was for the first time in space history explored by two spacecraft simultaneously, with Galileo's observations being complimented by those from CASSINI en route to Saturn during the 'Millennium Flyby'.

Galileo Galilei (1564–1642) Italian astronomer, physicist and mathematician. He was one of the first to use the telescope for astronomical observations, discovering mountains and craters on the Moon, and four satellites of Jupiter now known as the Galilean satellites. He observed the phases of Venus, and studied sunspots, from whose motion he deduced that the Sun rotates. Galileo concluded that ARISTOTLE's picture of the world, still widely believed in his time, was wrong and he championed COPERNICUS' heliocentric theory. This brought him into conflict with the Catholic Church and led to his trial and house arrest for the last eight years of his life.

Born in Pisa, the son of a lawyer and musician Vincenzio Galilei, Galileo was educated at the universities of Pisa and Padua. At first he studied medicine, but abandoned it for mathematics. Early in his career, he developed the fondness of controversy that would play a major part in precipitating his condemnation in 1633. One body of opinion in the universities of his day, which Galileo came first to criticize and then to openly ridicule, was the interpretation of all natural phenomena in terms of the philosophy of Aristotle, with its axiomatic assumption that the Earth stood fixed at the centre of the Universe (*see* GEOCENTRIC THEORY). It is all too easy for us to sympathize with Galileo's attacks upon Aristotelianism, and take his polemical language at face value when he called his Aristotelian contemporaries 'simpletons'. Yet even by the time of his death in 1642, let alone in 1600, there was no known physical proof of the Earth's motion in space, and Copernicus' theory hinged more upon mathematical elegance than upon observed fact.

Galileo's youthful anti-Aristotelianism was confirmed by a series of experiments in physics. The first of these, dating from his time in Padua in 1583, made him realize that when a pendulum swings, the determining physical factor is not the heaviness of the bob, but the length of the pendulum itself. Likewise, the results of his researches in Padua, where he became professor of mathematics in 1592, further contradicted Aristotle, for it was clear that the rate of acceleration of a falling body is governed by a fixed mathematical law that has nothing to do with the weight of the body. In Aristotle's philosophy there was no place for such laws. Terrestrial events were caused by an intermixing of earth, water, air and fire; only the heavens displayed mathematical regularities. Yet Galileo had shown that terrestrial as well as celestial motions conformed to exact mathematical laws, and he can be said to have established physics as a discipline founded on mathematical theory and experiment.

But it was his astronomical work that made Galileo famous, to become by the end of 1610 Europe's first scientific celebrity. Even before 1600, he was convinced of the truth of the heliocentric COPERNICAN SYSTEM. His conviction derived from Copernicanism's mathematical elegance in explaining the retrograde motions of planets, combined with his own anti-Aristotelian conclusions in physics. But when he first used the newly invented telescope to observe the night sky over the winter of 1609/10, he saw things that he felt secured the Copernican case. He found that the Moon's terrain was mountainous, not the smooth surface Aristotle had predicted. He also found that even under the ×30 magnification of his most powerful telescope, the Milky Way fragmented into millions of individual stars, and that instead of a few thousand stars being attached to the inside of a black dome, as the ancients had taught, there seemed to be countless millions extending throughout space. And then, in 1610 January, he discovered that the classical 'wandering star', Jupiter, was a spherical world, with four tiny satellites orbiting it. (Simon MARIUS later claimed to have observed them first.) Very obviously, the Earth was not the only centre of rotation in the Universe.

These and other telescopic discoveries by Galileo were published in 1610 March as *Siderius nuncius*, 'The Starry Messenger', and won him renown across Europe. All these discoveries, along with that of the phases of Venus (which showed that Venus revolved around the Sun) and, in 1611, of sunspots (which showed that the Sun was not an unblemished golden sphere), added to the weight of evidence against Aristotle's ideas, yet failed to provide physical proof that the Earth was in motion. For that, a stellar parallax or some similar phenomenon was needed, and instruments would not become sufficiently accurate to measure such a parallax until 1838.

Galileo's advocacy of Copernicanism thus rested not upon being able to prove the heliocentric theory, but upon his skill in undermining the objections of conservative astronomers who still adhered to Aristotle and Ptolemy. And as he could be very mocking to his opponents, Galileo made enemies, especially in the Jesuit order, whose Christoph SCHEINER had co-discovered sunspots, and yet had interpreted them in accordance with Aristotelian criteria. In the wake of his 1610 discoveries, however, Galileo had won the patronage of the Medici family, who had invited him to their Grand Ducal Court in Florence. The now famous Galileo was elected to the prestigious Accademia dei Lincei ('Academy of Lynxes') and left Padua to become a scientific courtier and holder of a research chair at Pisa. In Florence, he was

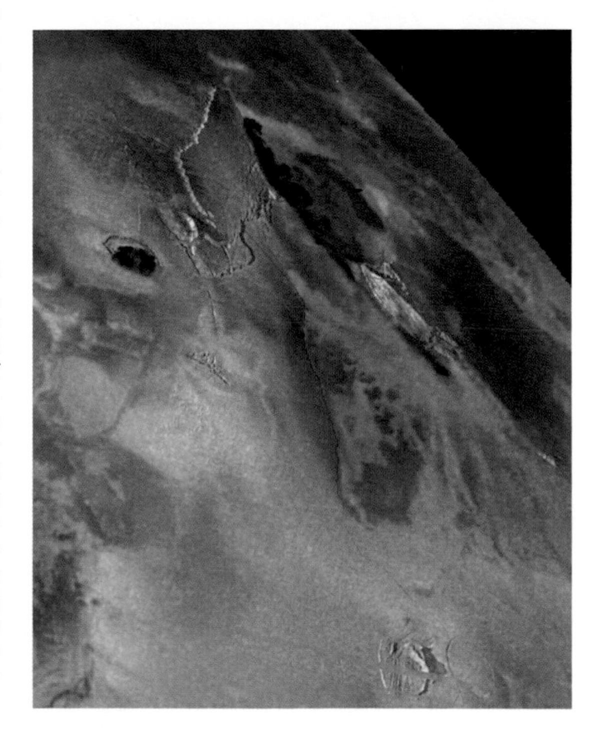

◀ **Galileo** Repeated passes close to the four large satellites of Jupiter by the Galileo orbiter since 1996 have revealed these in more detail than ever before. This image shows the volcano Zal Patera on Io.

lionized by the Medici and moved on easy terms amongst ambassadors, cardinals and bishops.

In 1616, however, Galileo was officially prohibited by Cardinal Bellarmine from teaching Copernicanism as the truth, as opposed to a hypothesis. Wisely, Galileo – who was a sincerely devout Catholic – complied, but when his friend Cardinal Barbarini became Pope Urban VIII in 1623, he attempted unsuccessfully to have the 1616 prohibition lifted. Nonetheless, in 1632, when the Church granted permission for his *DIALOGUES* to be published, he believed that he had the right publicly to discuss Copernicanism. However, Galileo's ensuing condemnation in 1633 was not for a serious heresy (the Church then had no official line for or against Copernicanism) but was disciplinary, for breaching the prohibition of 1616. Even so, the elderly Galileo's condemnation to house arrest at his villa at Arcetri, Florence (where he also discovered the lunar libration) shocked scientists across Europe. But no further Copernicans were punished, and one wonders how far Galileo's provocative personal style, the enemies he made in the Jesuit order, and the undoubted pride of Pope Urban VIII (who felt personally slighted by Galileo), rather than the science itself, were responsible for his fate. The Catholic Church re-opened the Galileo affair in 1822, and in 1992 Pope John Paul II proclaimed Galileo's complete exoneration, and offered the Church's apologies.

Galileo National Telescope (TNG) In Spanish, the Telescopio Nazionale Galileo, a 3.5-m (138-in.) optical/infrared telescope completed in 1997 as an Italian national astronomy facility. It is located at the ROQUE DE LOS MUCHACHOS OBSERVATORY on the island of La Palma at an elevation of 2358 m (7740 ft). It is operated for the Consorzio Nazionale per l'Astronomia e l'Astrofisica (CNAA) by the Centro Galileo Galilei. The TNG is closely modelled on the ESO's NEW TECHNOLOGY TELESCOPE.

Galle, Johann Gottfried (1812–1910) German astronomer who, while at the Berlin Observatory, was the first to identify Neptune, on 1846 September 23, though others saw the planet before him but failed to recognize it as such. Galle, who was aided in his search for the new planet by Heinrich D'Arrest, used orbital calculations by Urbain LE VERRIER. He also discovered three comets and Saturn's semi-transparent Crêpe Ring (the C Ring) (1838), though George Phillips BOND and William Rutter DAWES often receive credit for re-discovering this ring years later. Galle made the first reliable distance estimates for asteroids by measuring their parallaxes, first with minor planet (8) Flora in 1873

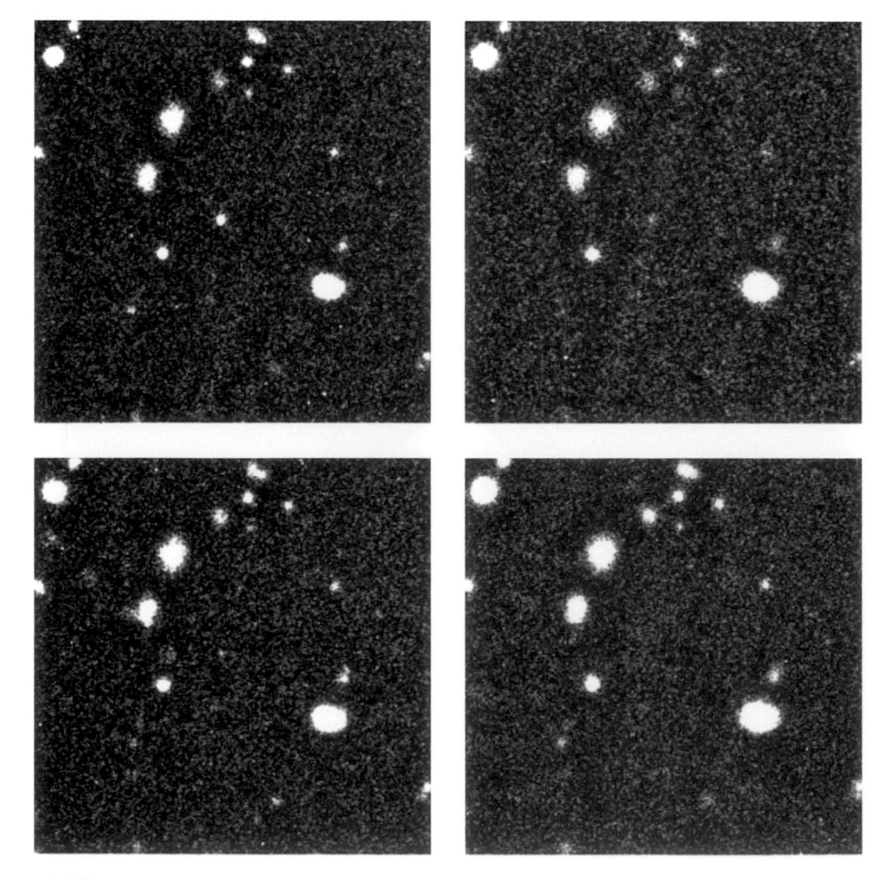

▼ **gamma-ray burst** GRB 00131 was detected by several spacecraft on 2000 January 31. This series of images obtained with the Very Large Telescope (VLT) shows the fading visible-wavelength afterglow, just left of centre.

Gamma Cassiopeiae star (GCAS) Rapidly rotating VARIABLE STAR of class Be III–V; it occasionally sheds material into surrounding space, generally from the equator, at apparently random intervals, in what are termed 'shell episodes'. Gamma Cassiopeiae stars may be subdivided into BE STARS, where the ejected material forms a ring around the star, and SHELL STARS, where it forms a circumstellar shell. They are related to, but less extreme than, the S DORADUS STARS.

gamma-ray astronomy Research covering the energy range of the ELECTROMAGNETIC SPECTRUM higher than 30 keV (wavelength less than 0.04 nm), essentially the last part of the spectrum to be explored. The extreme penetration of gamma rays makes them valuable for probing regions of the Galaxy and beyond where other radiation is absorbed. COSMIC RAYS were first thought to be a form of gamma radiation, and many gamma rays are produced by interactions of cosmic rays with the gas in the interstellar medium. However, generally gamma rays are associated with the most energetic mechanisms in the cosmos. There are several places where gamma rays are produced: cosmic rays interacting with gas in molecular clouds; SUPERNOVA REMNANTS, where cosmic rays accelerated in the supernova explosion interact with nearby gas; massive stars producing gamma rays in their winds; and PULSARS.

There were some early balloon experiments, but gamma-ray astronomy did not really start until 1967, with the launch of the American satellite OSO III. GAMMA-RAY BURSTS were first discovered around this time, by satellite-borne detectors meant to monitor violations of the Nuclear Test-Ban Treaty. In the 1970s SAS II and COS B were launched and they found a few tens of sources. The subject was revolutionized with the launch of the NASA COMPTON GAMMA RAY OBSERVATORY (CGRO) in 1991, which operated until 2000. New satellites are planned – INTEGRAL is due for launch by ESA in 2002 and GLAST is due for launch by NASA in 2006. At the highest gamma ray energies (around 0.2 TeV and above), ground-based telescopes can detect CERENKOV light from the passage of gamma rays in Earth's atmosphere. There are now several Atmospheric Cerenkov Imaging Telescopes (ACITs) which record the flashes (some ACITs have several dishes). Most of their results have been received in the 300 GeV to 30 TeV range.

The gamma-ray view of the sky in this energy range looks rather similar to the visible picture, in that there is a general concentration of emission in the galactic plane and an increase in the general direction of the galactic centre. Galactic gamma-ray radiation will be produced mostly by cosmic rays interacting with the gas in the interstellar medium. The lower-energy gamma rays (below about 100 MeV) are produced by electrons, and the electron intensity is higher in the inner Galaxy than locally and in the outer Galaxy. However, the poor angular resolution of the gamma-ray telescopes has led to many problems, particularly with the nature of the 'hot spots' of emission. Some of these are genuine discrete sources in that they can be identified by their time profile with pulsars, for example the CRAB PULSAR and the VELA PULSAR. The X-ray binary SS433 was detected in gamma rays, including some spectral lines. The Energetic Gamma Ray Experiment Telescope (EGRET) instrument on CGRO found around 170 high-energy (above 100 MeV) sources in the Galaxy; the brighter sources were in the galactic plane, the fainter ones in the solar neighbourhood, and a large number of sources were found to lie in GOULD'S BELT, a series of local star-forming regions. Several of the sources may be older pulsars, or other types of compact objects (such as BLACK HOLES) from past massive star formation, or massive hot stars. Gamma rays have been detected from extragalactic sources such as ACTIVE GALACTIC NUCLEI (AGNs), quasars and BL LACERTAE OBJECTS. EGRET also discovered a class of objects called BLAZARS, which are quasars emitting most of their radiation in the 30 GeV to 30 MeV range, and which are variable on a timescale of days. 3C 279, at a distance of 4 billion l.y., is a blazar (but no longer regarded as typical) and in 1991 it was one of the brightest gamma-ray objects in the sky.

The Imaging Compton Telescope (COMPTEL) on CGRO produced a map of the sky in the aluminium-26 line (1.809 GeV), in which the Cygnus, Vela, Carina and the inner Galaxy regions appear brightest. This radioactive element is produced during NUCLEOSYNTHESIS in supernovae. Another map of the sky was created in the 511 keV line, from positron–electron annihilation, which shows that this is occurring in a region around 10° in diameter centred on the galactic centre. Apart from the considerable strength, the emission is variable on the scale of a year – a possible explanation is there is a black hole at the galactic centre that somewhat irregularly consumes interstellar gas.

gamma-ray burst (GRB) Short-duration (typically 1 to 10 seconds) outpouring of gamma rays, usually in the energy range from 100 keV to 1 MeV, from a discrete source. Some gamma-ray bursts may be shorter or longer (up to a few hundred seconds), while energies as high as the TeV range may occasionally be recorded. The optical counterparts are found to be young galaxies at high redshifts, which makes GRBs extremely energetic events. GRBs were discovered by satellite-borne detectors meant to monitor violations of the Nuclear Test-Ban Treaty in 1967. In 1997 astronomers found that the GRBs are followed by an X-ray 'afterglow', which enabled their positions to be accurately pin-pointed. The BATSE instrument on the COMPTON GAMMA RAY OBSERVATORY (CGRO) detected 2704 GRBs using NaI detectors, sensitive to energy between 20 keV and 600 keV. In 1999 a GRB detected by CGRO was observed 22 seconds later by a ground-based robotic telescope (due to the automatic alert system), thus providing an accurate position, which meant that observations could be made in X-rays and optically with the HUBBLE SPACE TELESCOPE, PALOMAR and KECK. The 24th-magnitude galaxy had a peculiar shape and was very blue, suggesting that many new stars were being formed. It had a redshift of 1.6, which showed that it was around 10 billion l.y. away. It has been suggested that gamma-ray bursts occur when neutron stars collide or when there is a hypernova.

Gamow, George (1904–68) Ukrainian-American nuclear physicist and cosmologist, born Georgy Anthonovich Gamov, known for his theory of the atomic nucleus, the application of nuclear physics to problems of stellar evolution, and his model of a 'hot' BIG BANG; and for his later work in molecular biology. From his early work as a nuclear physicist in Russia, Göttingen and Copenhagen came his theory of α-decay (1928) and the 'liquid-drop' model of the nucleus (1931), which later proved essential for explaining nuclear fission and fusion. He emigrated to the USA in 1933, becoming professor at George Washington University (1934–56) and the University of Colorado (1956–68). Gamow solved the problem of how the relatively low collision velocities of nuclei in the interior of stars could trigger NUCLEAR REACTIONS. He showed that stars heat up as they exhaust their nuclear fuel. In World War II, he worked on the US Manhattan Project to develop the atomic bomb.

After the war, Gamow turned to the Big Bang theory. He visualized the early universe as composed of hot dense matter he called 'ylem', a soup of neutrons, protons, electrons and radiation, from which the first elements were

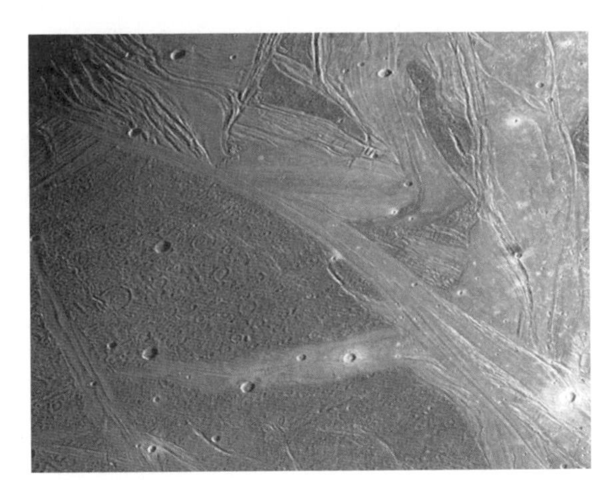

◀ **Ganymede** A Galileo orbiter view of Marius Regio and Nippur Sulcus on Ganymede. These alternating areas of light grooved terrain and darker cratered terrain are typical of the satellite's surface.

created. With Ralph Asher Alpher (1921–) and Hans BETHE, Gamow further developed this model, subsequently dubbed 'α-β-γ', to show that hydrogen nuclei could fuse to form helium nuclei (α-particles) in quantities that reflected the Universe's observed helium abundance (1 part in 12), but that heavier elements were not produced in the Big Bang. Gamow's greatest insight was that subsequent expansion of the universe would cool the 'cosmic fireball' of radiation left over from the ylem stage to about 10 K in the present. The 3 K COSMIC MICROWAVE BACKGROUND radiation discovered in 1965 by Arno PENZIAS and Robert Wilson confirmed Gamow's prediction.

Ganymed Possibly the largest near-Earth asteroid. It is estimated to be about 50 km (30 mi) in size based on an assumed albedo of 0.15. *See also* EROS; table at NEAR-EARTH ASTEROID

Ganymede Largest of the GALILEAN SATELLITES of JUPITER; it is the largest satellite in the Solar System. Ganymede's diameter of 5262 km (3270 mi) makes it 8% larger than Mercury, but it is a mainly icy body, with a density of only 1.94 g/cm^3, and has less than half Mercury's mass. It has a complex, but fairly heavily cratered, icy surface, attesting to major resurfacing events in the distant past, thanks to Ganymede's 1:2 orbital RESONANCE with EUROPA, which presumably caused TIDAL HEATING.

Ganymede's density suggests that, like CALLISTO, it consists of a 40:60 mixture of rock and ice. Its icy surface is much dirtier than Europa's, with an albedo of only 0.45. Gravity measurements made by the GALILEO spacecraft indicate that Ganymede has undergone full DIFFERENTIATION, with an iron inner core (filling 22% of its radius), a rocky outer core (filling 55% of its radius) and an icy mantle. Callisto, in contrast, is only weakly differentiated. Ganymede has a magnetic field with about 1% the strength of the Earth's; it could be generated either in the core or in a salty ocean within the ice layer.

Spectroscopic studies show that Ganymede's surface is dominantly water-ice, with scattered patches of carbon dioxide ice, and that the darkening is caused by silicate minerals (probably in the form of clay particles) and tarry molecules called tholins. There are also traces of oxygen and ozone, apparently trapped within the ice, and an extremely tenuous (and continually leaking) atmosphere of hydrogen, which is presumably the counterpart to the oxygen produced by the breakdown of water molecules.

There are two distinct terrain types on Ganymede, one darker than the other. It is apparent that the pale terrain (which commonly consists of multiple sets of grooves) must be younger than the dark terrain, because belts of pale terrain can be seen to cut across pre-existing tracts of dark terrain. However, the density of impact craters on both terrain types shows that each must be very old, perhaps as much as 3 billion years.

Intricate details of the grooved parts of the pale terrain hint that each belt of pale terrain might require just as complex a series of events to explain it as parts of the bright plains on Europa. On Europa there is abundant evidence for the two halves of a split tract of terrain having been moved apart to accommodate the new surface that has formed in between. On Ganymede, however, signs of lateral movement are scarce. Ganymede's pale terrain appears to occupy sites where the older surface has been dropped down by fault movements, allowing cryovolcanic fluids to spill out (*see* CRYOVOLCANISM). It is unclear whether the grooves within each belt of pale terrain represent constructional cryovolcanic features or reflect subsequent deformation, perhaps related to underlying faults. *See also* data at JUPITER

gardening Process whereby impacts of small METEOROIDS cause turnover of the outer few metres of lunar REGOLITH in geologically short timescales, of the order of tens of millions of years.

Garnet Star Name given by William HERSCHEL to the star μ Cephei on account of its strong red colour. It is a red supergiant, spectral type M2 Ia, and a semiregular variable (subtype SRC), ranging between about mags. 3.5 and 5.1 every 2 years or so. It lies about 2800 l.y. away.

Gascoigne, William (1612–44) English inventor of the micrometer, apparently self-taught in astronomy. Like his associates Jeremiah HORROCKS and William CRABTREE he seems to have been a Copernican. Gascoigne was the first to solve the problem of how to insert a cross-hair into the focus of a Keplerian refracting telescope so that it could be used as a sighting instrument. Around the same time (*c.*1640), he invented the eyepiece micrometer, in which a precision screw was used to move two pointers in the telescope's field of view. By enclosing the body being observed between these pointers, its angular diameter could be measured to within a few arcseconds from a knowledge of the focal length of the object glass and the pitch of the micrometer screw. Such precision measurements were crucial in quantifying changes in the lunar and solar diameters, and correlating them with Kepler's elliptical orbits, to match theory with physical evidence.

gas giant Collective term used to describe JUPITER, SATURN, URANUS and NEPTUNE, also described as the 'jovian planets'. These planets have thick opaque atmospheres and no solid surface visible. They are thought to contain a small solid core, with the majority of their mass made up of gas. Because of this the planets' densities are low, indeed that of Saturn is less than that of water. Uranus apart, the gas giants emit more radiation than they receive from the Sun, implying an internal heat source due in part to the effects of rotation and gravity. Saturn has long been known to have a ring system. Very much fainter ring systems have been found around the other planets since 1977. The term gas giant is also used descriptively for larger planets with similar characteristics thought to orbit around other star systems.

▼ **Gaspra** En route to Jupiter, the Galileo spacecraft made a flyby of asteroid (951) Gaspra in 1991, revealing it to be an irregular-shaped, heavily cratered body. Gaspra lies in the main belt.

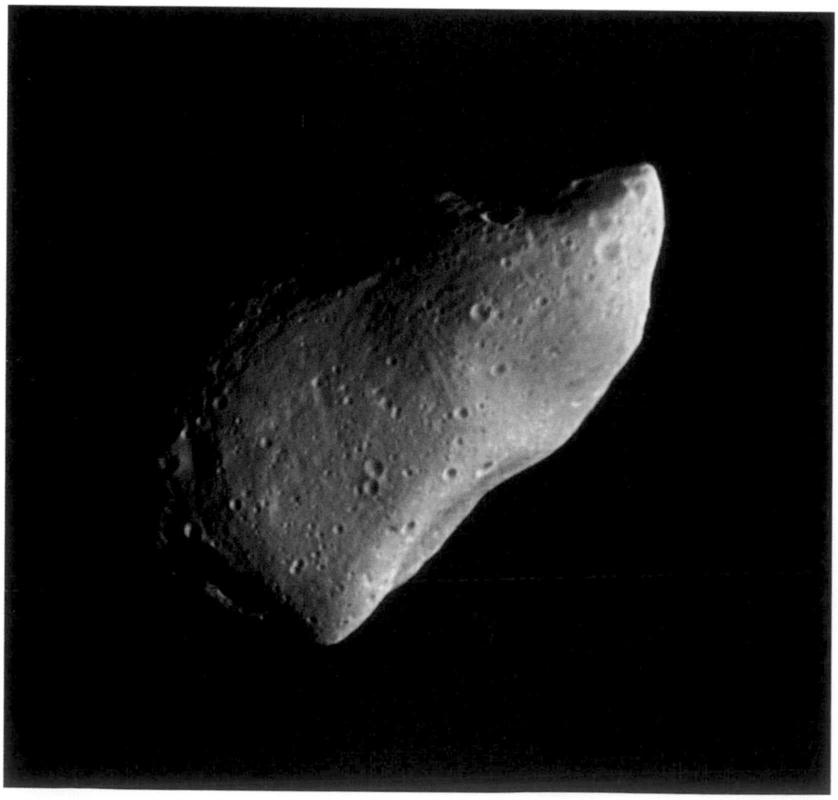

Gaspra Inner MAIN-BELT ASTEROID and member of the FLORA family; number 951. Gaspra was encountered by the GALILEO spacecraft in 1991 October while en route to Jupiter; it was the first asteroid to be examined in this way.

Gaspra is angular and thus non-spherical in shape, measuring about 18 × 11 × 9 km (11 × 7 × 6 mi). It is peppered with small craters but lacks medium to large craters, suggesting that its surface is relatively young (much younger than the Solar System). This view is supported by its planar surfaces, which may indicate where portions of the original body were sheared off by one or more major collisions some billions of years ago. Crustal cracks similar to those seen on PHOBOS are also apparent. It is believed that Gaspra is one of the daughter products of the inter-asteroid collision that produced Flora and the other members of its HIRAYAMA FAMILY.

Gaspra's spectral reflectance indicates it to be an S-class asteroid, similar in gross composition to IDA, the other asteroid examined during flyby from Galileo. Images show significant colour variations over Gaspra's surface, with some craters and ridge tops – where the REGOLITH might be expected to be thinner or absent – appearing bluer than the surrounding areas. This variation may indicate that long term space weathering, through exposure to micrometeoroid impacts, is the cause of the redness of the surface, with the solid body of Gaspra beneath being less red.

Gassendi, Pierre (1592–1655) French mathematician and astronomer. In 1631 he used the eyepiece projection method with a Galilean telescope to safely make the first observation of a transit of MERCURY across the Sun. Johannes KEPLER, with whom Gassendi corresponded, had predicted this transit would take place, using his then-newly conceived theories of planetary motion.

Gassendi Lunar crater (18°S 40°W), 89 km (55 mi) in diameter, with rim components up to 2700 m (9000 ft) above its floor. Gassendi has numerous rilles. Its central peaks formed from rebound of the floor after impact. Gassendi's rim is uneven, dipping in its southern region to the point where lavas have flowed through gaps to enter the crater's interior.

Gauss, Carl Friedrich (1777–1855) German mathematician and physicist who made major contributions to celestial mechanics. He made a comprehensive study of methods of determining the orbits of planets and comets, taking into account the effects of gravitational perturbations. He then studied the theory of errors of observation, which led him to develop a technique known as the method of least squares for estimating true values from observations subject to error.

In 1801 Gauss used his least-squares method successfully to reconstruct the orbit of the first asteroid, CERES, discovered by Giuseppe PIAZZI but subsequently lost. Piazzi obtained only three positions for Ceres, covering only a tiny 9° arc of its orbit. Gauss sent his predictions for the future positions of Ceres to Franz von ZACH, who was able to recover the asteroid on 1801 December 7. Gauss' method of computing minor perturbations to planetary orbits was later employed by Urbain LE VERRIER and John Couch ADAMS in their computations of the orbit of Neptune.

In 1807 Gauss was appointed director of the new observatory at Göttingen, completed in 1816. During this period he published a massive treatise on celestial mechanics, *Theoria motus corporum coelestium* (1809). He demonstrated that Earth has two magnetic poles and was the first accurately to locate the south magnetic pole. The gauss, a unit for measuring magnetic field strength, is named after him.

GCVS Abbreviation of *GENERAL CATALOGUE OF VARIABLE STARS*

Geller, Margaret Joan (1947–) American astrophysicist who, working with John HUCHRA at the Harvard-Smithsonian Center for Astrophysics (CfA), has made groundbreaking surveys of galaxy redshifts to map the LARGE-SCALE STRUCTURE of the Universe. The CfA Redshift Survey mapped the distribution of over 18,000 galaxies in three dimensions. The most famous map constructed by Geller's team, which looked something like a 'stick-man' figure, disclosed the structure known as the GREAT WALL (1989). Geller was only the second woman to be awarded a doctorate by Princeton University.

Geminga One of the most powerful gamma-ray objects in the sky, around 500 l.y. (160 pc) from Earth. It has a period of 0.237 second, so it is most likely a short-wavelength equivalent of a PULSAR, and thus the closest pulsar to Earth. No radio waves have been detected from Geminga, so it is assumed that the JETS are not pointed at Earth. The name Geminga comes from **Gemini** gamma-ray source, since the object is in the constellation Gemini, and its Italian discoverers thought it a good name since 'geminga' in the Milanese dialect means 'it does not exist'.

Gemini *See* feature article

Geminids Most consistently active of the annual METEOR SHOWERS, occurring between December 7 and 16. At peak, the RADIANT lies just north of Castor, and the shower produces zenithal hourly rates (ZHR) as high as 120. The Geminids have been observed since 1838 and have become gradually more active over the years. In the 1970s, the peak was narrow with ZHR around 70; returns in the late 1990s, however, showed a much higher and broader maximum. The changing activity pattern is consistent with computer models of the METEOR STREAM, the orbit of which is being pulled closer to that of the Earth by gravitational perturbations. The models suggest that peak activity will become progressively later during the 21st century and will eventually begin to diminish in intensity. At present, the rise to maximum is gradual, with a more rapid decline.

Geminid meteors are relatively slow, with a velocity of 35 km/s (22 mi/s). A detailed analysis by G.H. Spalding (Didcot, UK) showed evidence for particle sorting in the stream, with bright events – produced by larger meteoroids – being more numerous after Geminid maximum. Geminid meteoroids have long been known to have a higher density (2 g/cm^3) than those from other (cometary) sources, and they survive ABLATION down to lower altitudes in Earth's atmosphere. This can be accounted for by their origin from asteroid (3200) PHAETHON, discovered in 1983. Asteroid and meteor stream have an unusual orbit, with a very small perihelion distance of 0.14 AU.

Gemini Observatory Major international facility consisting of two identical 8.1-m (27-ft) optical/infrared instruments, the **Gemini Telescopes**, one at MAUNA KEA OBSERVATORY (Gemini North), dedicated in 1999 June, and the other on the 2715-m (8910-ft) high ridge of Cerro Pachón in the Chilean Andes (Gemini South), which became operational in 2002 January. The seven partner countries supporting the Gemini Observatory are the USA, UK, Canada, Australia, Argentina, Brazil and Chile.

GEMINI (GEN. GEMINORUM, ABBR. GEM)

Prominent northern zodiacal constellation, representing Castor and Pollux, the mythological twin sons of Queen Leda of Sparta; their brotherly love was rewarded by a place among the stars, between Auriga and Canis Minor. CASTOR is a remarkable sextuple system, consisting of a pair of bluish-white stars, mags. 1.9 and 3.0, separation 3″.1, period about 467 years, both of which are SPECTROSCOPIC BINARIES, and an eclipsing binary, YY Gem (range 8.9–9.6, period 0.81 day). η Gem is one of the brightest CEPHEID VARIABLES (range 3.6–4.2, period 10.15 days). Gemini also contains the long-period variable R Gem (range 6.0–14.0, period about 370 days) and the DWARF NOVA U Gem (range 8.2–14.9, mean period about 105 days). Deep-sky objects include M35 (NGC 2168), an open cluster of about 200 stars fainter than 8th magnitude, and the CLOWN FACE NEBULA (or Eskimo Nebula, NGC 2392), a 9th-magnitude planetary nebula. The GEMINID meteor shower radiates from the constellation.

BRIGHTEST STARS

Name	RA h m	dec. ° ′	Visual mag.	Absolute mag.	Spectral type	Distance (l.y.)
β Pollux	07 45	+28 02	1.16	1.1	K0	34
α Castor	07 35	+31 53	1.58	0.6	A1/A2	52
γ Alhena	06 38	+16 24	1.93	−0.6	A0	105
μ Tejat	06 23	+22 31	2.7–3.0	−1.6–1.3	M3	232
ε Mebsuta	06 44	+25 08	3.06	−4.2	A3	903
ξ	06 45	+12 54	3.35	2.1	F5	57
η Propus	06 15	+22 30	3.1–3.9	−2.0–−1.2	M3	349
δ Wasat	07 20	+21 59	3.50	2.2	F0	59

The two telescopes employ new-technology thin mirrors with ACTIVE OPTICS to provide stable images under all conditions of telescope attitude and wind loading. They also use state-of-the-art ADAPTIVE OPTICS to compensate for atmospheric turbulence, yielding images of very high resolution. These are analysed with a comprehensive suite of optical and infrared instruments. The telescopes are particularly suited to research on star formation within dust clouds and studies of the very distant Universe. As well as the telescope facilities, the observatory has a project office in Tucson, Arizona, and sea-level operations centres at Hilo, Hawaii and La Serena, 400 km (250 mi) north of Santiago in Chile. The Gemini Observatory is managed jointly by the ASSOCIATION OF UNIVERSITIES FOR RESEARCH IN ASTRONOMY and the National Science Foundation.

Gemini Project Series of 12 US spacecraft launched by Titan II rockets between 1964 and 1966. Gemini 3–12 carried two-man crews and played a vital role in preparations for the Apollo missions to the Moon. Experiments included orbital rendezvous, docking and EVA (extra-vehicular activity). The longest mission (Gemini 7) lasted for $330^h 35^m 17^s$.

Gemini Telescopes *See* GEMINI OBSERVATORY

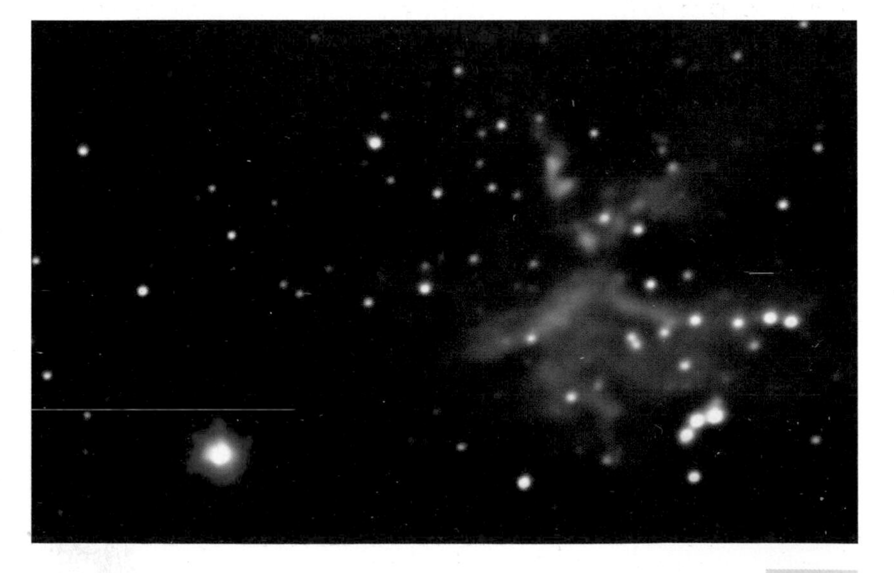

▼ **Gemini Observatory** This infrared view of a star-forming region in the Milky Way was taken using the Gemini North telescope on Mauna Kea. Adaptive optics helped sharpen the image.

Geminus Lunar crater (35°N 57°E), 90 km (56 mi) in diameter, with a western rim rising to 4880 m (16,000 ft) above its floor. Geminus is a complex crater, with central peaks and wall terracing. The crater appears oblong due to foreshortening but is actually more nearly round. Micrometeorite erosion has removed its bright rays, though its EJECTA are still visible.

Gemma Alternative name for the star α Coronae Borealis. *See* ALPHEKKA

General Catalogue of Variable Stars (GCVS) Principal listing of stars discovered to be variable, compiled by Pavel Nikolaevich Kholopov (1922–88) and others. The first edition appeared in 1948; the 5th edition of 1998 contains 28,484 variables. The catalogue is produced at the Sternberg Astronomical Institute, Moscow, where an updated electronic version is maintained.

G

general relativity New theory of gravity derived by Albert EINSTEIN in 1916 that generalized his already successful special theory of relativity to include accelerations and gravity. The fundamental postulates of this new theory are: (i) SPACETIME is a four-dimensional CONTINUUM; and (ii) the PRINCIPLE OF EQUIVALENCE of gravitational and inertial mass.

The basis of his theory suggests that mass-energy distorts the fabric of spacetime in a very predictable way, and the altered paths of particles in this warped spacetime are deflected as if a force were exerted on them. Einstein needed a type of mathematics called differential geometry in order to express the relationship between mass-energy and curvature. His FIELD EQUATIONS describe the curvature produced in spacetime by a gravitational source.

Some of the consequences of general relativity are that in strong gravity time slows down, light paths are altered by gravity, and gravity in strong gravitational fields no longer obeys Newton's inverse square law. These effects are minor for most planetary orbits, but were measurable even in Einstein's time for the orbit of Mercury. General relativity has opened up new fields of astronomy such as the study of black holes, gravitational waves and cosmology. Unlike SPECIAL RELATIVITY, which is universally accepted by physicists, many were dubious of the application of differential geometry and awaited observational confirmation of this new gravitational theory. Sir Arthur EDDINGTON was the first to measure the effects of gravity on light from distant stars during a total solar eclipse. Since then, many tests have been done and general relativity has passed every one of them. In spite of general relativity's success, there are several competing theories of gravity that make similar but not identical predictions as Einstein's.

The major problem with general relativity is it is a continuum theory and to date it has resisted quantization. A merger of the quantum theory and general relativity predicts on length scales of 10^{-43} m (the quantum realm), relativity fails. Although some assign a classical particle called a graviton to general relativity, it is not compatible with current elementary particle physics. The search for a QUANTUM THEORY OF GRAVITY is the search for a GRAND UNIFIED THEORY (GUT) or the THEORY OF EVERYTHING (TOE).

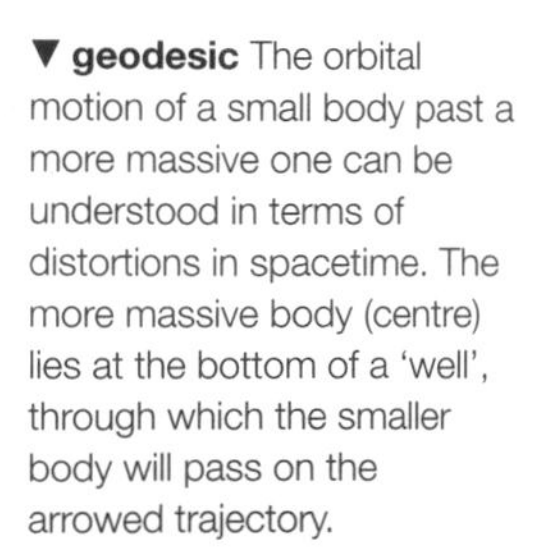

▼ **geodesic** The orbital motion of a small body past a more massive one can be understood in terms of distortions in spacetime. The more massive body (centre) lies at the bottom of a 'well', through which the smaller body will pass on the arrowed trajectory.

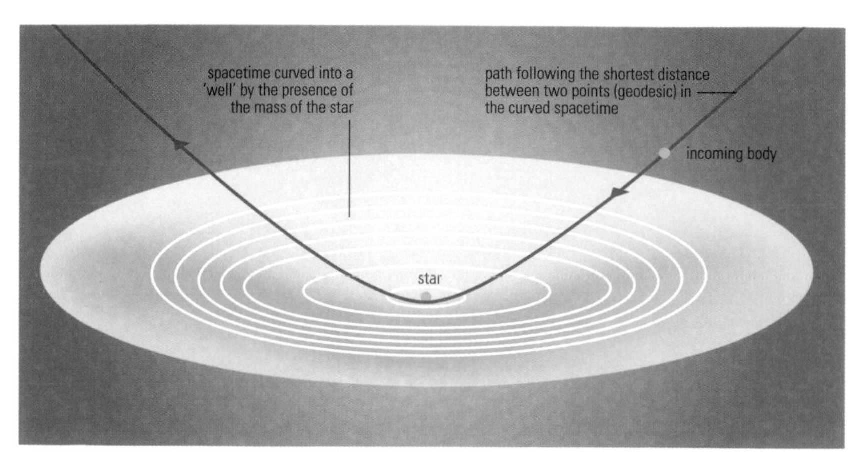

Genesis NASA Discovery-class mission to collect samples of the SOLAR WIND and return them to Earth for analysis. Genesis was launched in 2001 June and is stationed at the L_1 LAGRANGIAN POINT between the Earth and Sun. It will collect samples for two years and return a capsule with the solar material to Earth in 2004 June.

Genesis rock Lunar rock that is notable for its high content of the mineral plagioclase; it was collected on the Apollo 15 mission. Rocks with this mineral composition are called ferroan anorthosites; they represent the mineral composition of the Moon's early crust.

gegenschein (Ger. counterglow) Faint and diffuse patch of light seen opposite the Sun in the night sky under the most transparent, moonless conditions. The gegenschein is slightly oval in shape, with its long axis extending along the ecliptic, and has a diameter of about 10°. In brightness, the gegenschein is comparable to the fainter parts of the Milky Way, and it is best seen when high in the sky. The best viewing times in the northern hemisphere are in December and January, while in the southern hemisphere the gegenschein is best seen in August (being masked by the brightness of the Milky Way, against which it then lies, in June and July).

The gegenschein results from the scattering of sunlight by tiny dust particles in interplanetary space. Photopolarimetric measurements from the Pioneer and Voyager spacecraft indicate that the gegenschein is essentially absent beyond the orbit of Jupiter. The reflective material belongs to the zodiacal dust complex. Under very clear, dark conditions, the gegenschein can be seen to be connected to the rather brighter ZODIACAL LIGHT by narrow faint extensions, known as ZODIACAL BANDS, along the ecliptic.

geocentric parallax Alternative name for DIURNAL PARALLAX. *See also* ANNUAL PARALLAX; SOLAR PARALLAX

geocentric theory Model of the Universe in which a stationary Earth takes central place, the world-view of many early civilizations. The Sun, Moon and planets were regarded as revolving around the Earth. By the 5th century BC, the Greeks were aware that the apparent motion of the Sun could be represented as the net result of a daily rotation around the Earth and a yearly journey along the ecliptic, the two motions being inclined to each other by about 24°. All motions were believed to proceed at uniform rates along perfectly uniform circles.

The first theory of planetary motion was put forward by EUDOXUS, who proposed a system of 26 revolving 'homocentric' (Earth-centred) spheres which carried the Sun, Moon and planets through space. By invoking contorted arrangements for the rotational axes of these spheres, many of the observed phenomena, including retrograde motions, could be reproduced. Further complexity was introduced when ARISTOTLE added even more spheres. Because the spheres are not visible, yet had to be capable of supporting the planets, they were regarded as 'crystalline' – invisible, or glass-like. It was Aristotle's model, as modified by Ptolemy, that survived until the end of the Renaissance and that was defended so vigorously by the Catholic Church, largely thanks to his theory of motion, which was the nearest and most coherent approach to physics before the age of Galileo Galilei and Isaac Newton.

In an alternative geocentric model proposed by APOLLONIUS around 200 BC, the Earth remains at rest in the centre of the Universe, but the Moon, Sun and planets move in circular motion about centres displaced from the Earth. This allows to some extent for the non-uniformity of planetary motion. Apollonius showed that these 'displaced' circular orbits can just as well be described in terms of motion on a circle (the epicycle) whose centre itself moves on a larger circle (the deferent), centred on the Earth.

Geocentric theory reached its highest development in the work of Ptolemy (*see* PTOLEMAIC SYSTEM), which held sway for about 1500 years. Nicholas Copernicus' heliocentric theory (*see* COPERNICAN SYSTEM) did not immediately dislodge it; that came only with the improved observations of Tycho Brahe, as systematized by Johannes Kepler and formed into a consistent heliocentric theory by Newton, and the telescopic discoveries of Galileo.

geochemistry Division of chemistry concerned with the abundances of elements and their distribution within planetary bodies. The most primitive, least altered METEORITES are composed of minerals that contain the non-volatile elements (excluding hydrogen, helium and the noble gases) in relative proportions very similar to their abundances in the Sun, as deduced from SPECTROSCOPY. Larger bodies that formed from such material were altered to various degrees by heating and DIFFERENTIATION. The most abundant elements separated into metallic CORE and silicate MANTLE and CRUST. The less abundant elements were partitioned among them according to their chemical and thermodynamic properties. The elements classified as siderophile, lithophile or chalcophile have affinities for metallic iron, silicates and sulphides, respectively. Chemical analysis of one component of a planetary body can provide information about the rest; for example, a sample of the body's crust can reveal much about the composition of its core. Moreover, some elements have affinities that vary with temperature, pressure and the oxidizing/reducing conditions of their surroundings; the distribution of these elements among various mineral phases in a rock reveals the conditions of its formation. A closely related discipline is COSMOCHEMISTRY, which includes the behaviour of elements in interstellar space and the SOLAR NEBULA.

geodesic Shortest distance between two points, given by a straight line on a flat surface and by a great circle on a sphere. In general relativistic curved spacetime, a geodesic is the path giving smallest separation between two events (*see* TIME). The orbital motion of planets and other astronomical objects is then understood as the objects following geodesic tracks through a spacetime that has been distorted from Euclidean geometry by the presence of masses.

Geographos APOLLO ASTEROID discovered in 1951; number 1620. Radar imaging has shown Geographos to have a highly elongated shape. *See also* table at NEAR-EARTH ASTEROID

geomagnetic storm Large-scale disturbance of Earth's MAGNETOSPHERE resulting from the arrival via the SOLAR WIND of energetic, highly magnetized plasma thrown from the Sun during CORONAL MASS EJECTION or FLARE events. The most visible effects of a geomagnetic storm take the form of enhanced auroral activity, including the spread of the AURORA to lower latitudes than normal. Magnetometer equipment allows detection of other effects, which can, during severe storms, be global in extent.

The onset of a geomagnetic storm, as ejected solar material impacts on the magnetosphere, is detected by magnetometers as a sudden storm commencement (SSC). During the SSC, compression of the magnetosphere produces an abrupt intensification of the ground-level magnetic field, the horizontal east–west component of which shows a rapid change in orientation. During the main phase of the storm, which can last for 24–36 hours, the magnetic field continues to fluctuate rapidly before gradually returning to normal. The fluctuations can be correlated with episodes of increased auroral activity. They result from the build-up and release of stress produced as plasma from the solar wind penetrates the magnetosphere, disturbing the resident population.

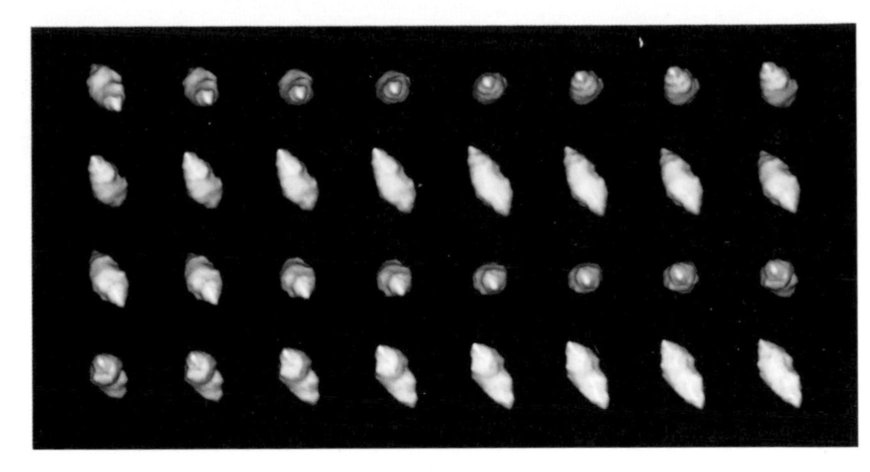

▲ **Geographos** A series of radar images showing the irregular shape of the Apollo asteroid (1620) Geographos. The observations were made from Goldstone in 1994. [S. Ostro (JPL/NASA)]

Ground-level electrical currents induced by geomagnetic storms have in the past disrupted telegraph communications, and they can cause corrosion in high-latitude oil and gas pipelines. During the huge storm of 1989 March 13–14, these currents caused a power blackout affecting Quebec Province in Canada. Ionospheric currents pose a hazard to artificial satellites, leading to a build-up of charge which can damage electrical components. Satellites that have been lost following geomagnetic storms include Marecs-B (used in navigation), Anik-1 and Anik-2 (telecommunications) and ASCA (astronomical research).

Georgian planet Obsolete name for URANUS, originally suggested by William HERSCHEL, who discovered the planet in 1781, in honour of his patron King George III. The name was in use in the Nautical Almanac until 1850.

geostationary orbit *See* GEOSYNCHRONOUS ORBIT

geosynchronous orbit Orbit in which a satellite completes one revolution around the Earth in the same time as the Earth's SIDEREAL rotation period of $23^h\ 56^m\ 4^s$. If this orbit is circular and in the plane of the Earth's equator (that is, with an inclination of 0°), the satellite appears to be virtually stationary in the sky, so it is described as **geostationary**. A geostationary orbit is at an altitude of 35,900 km (22,380 mi) above the Earth. A satellite in a geosynchronous orbit inclined to the Earth's equator appears to trace out a figure-of-eight shape over the course of a day.

Geotail Japanese satellite launched in 1992 as part of the INTERNATIONAL SOLAR TERRESTRIAL PHYSICS programme, involving the US WIND and POLAR spacecraft and the European SOHO and CLUSTER satellites. Geotail measures global energy flow and transformation in the Earth's magnetotail to increase understanding of the processes in the MAGNETOSPHERE. The craft used two gravity assist flybys of the Moon to enable it to explore the distant part of the magnetotail. Two years later, the orbit was reduced so that the spacecraft would study the near-Earth magnetotail processes.

German mounting Type of EQUATORIAL MOUNTING used for telescopes.

ghost crater CRATER that was flooded with lava such that its ejecta, floor and rim were completely covered. Because lava above the crater's rim remained relatively thin, the later subsidence of the lava placed stress over the rim. This stress is expressed as a mare ridge over the rim; the ridge is circular, hence the appearance of a 'ghost crater'.

ghosting Production of one or more faint, displaced secondary images – **ghost images** – in an optical system caused by unwanted internal reflections. Ghosting can also be caused by an imperfect diffraction grating, causing spurious spectral lines.

Ghost of Jupiter (NGC 3242) PLANETARY NEBULA in the constellation Hydra, a couple of degrees south of μ (RA 10^h $24^m.8$ dec. −18°38′). The object has an apparent diameter of 16″, and a pronounced blue-green colour. It takes its name from dark bands seen within its disk, giving it a resemblance to the planet Jupiter. NGC 3242 has overall mag. +7.8, with a central star of mag. +11.4. It lies 1400 l.y. away.

Giacconi, Riccardo (1931–) Italian-American physicist who pioneered X-RAY ASTRONOMY. In the late 1950s, Giacconi designed and built the first grazing-incidence X-ray telescopes, which were flown on rockets above the Earth's atmosphere. In 1962 his group used these detectors to discover Scorpius X-1, the first powerful X-ray source found outside the Solar System. Giacconi and colleagues designed and built the UHURU X-ray satellite, which made the first comprehensive survey of X-ray sources, and the EINSTEIN OBSERVATORY, which made the first highly detailed contour maps of individual X-ray sources.

Giacobinids (Draconids) Periodic METEOR SHOWER associated with Comet 21P/GIACOBINI–ZINNER. Showers of Giacobinids have been seen only in those years when the parent comet has returned, and they have durations of a few hours. Possible low activity was seen in 1915 and 1926, with substantial displays on 1933 October 9 and 1946 October 10. During a 4½-hour period, the 1933 return produced Giacobinid rates of between 50 and 450 per hour. In 1946, despite strong moonlight, visual rates of about 5000 Giacobinids per hour were reported; this display was also observed using radio echo methods at Jodrell Bank. Japanese observers recorded another significant outburst on 1985 October 8. Until 1998 reported Giacobinid displays had occurred when Earth followed the comet to the NODE where its orbit intersects that of the Earth; the strong display of 1998 October 8, however, with Giacobinid rates reaching 50–100 per hour, was somewhat surprising because, on this occasion, Earth crossed the node 44 days ahead of the comet, which indicated a greater spread of meteoroids in the space around the nucleus than had previously been suspected. Giacobinid meteors come from a radiant near the 'head' of Draco; they are notably slow, with an atmospheric velocity of only 20 km/s (12 mi/s).

▼ Giacobini–Zinner, Comet This comet was imaged on 1998 November 1 by the Kitt Peak 0.9 m (36 in.) telescope. The teardrop-shaped coma shows strong bluish CO emission.

Giacobini–Zinner, Comet 21P/ Short-period comet discovered by Michel Giacobini (1873–1938), Nice Observatory, France, on 1900 December 20, and re-discovered two returns later on 1913 October 13 by Ernst Zinner (1886–1970) at Remeis, Germany. The comet's period is 6.61 years, with favourable apparitions at every second return. Comet 21P/Giacobini–Zinner has been observed at 15 returns up to 1998, with only the 1953 apparition having been missed. The comet is the parent of the GIACOBINID meteors. Comet 21P/Giacobini–Zinner became the first comet to be investigated at close range by a spacecraft: the International Comet Explorer (ICE) flew through its tail on 1985 September 11.

giant impactor theory Theory that addresses the origin of the MOON. Prior to the Apollo missions, there were three 'classical theories' of lunar formation. First, the fission hypothesis, in which the Moon was 'spun out' from a rapidly rotating, but still molten, Earth. Secondly, the binary accretion hypothesis, in which the Earth and Moon formed as a 'double planet', and so remained rotating around each other. The problem with this notion is that it should produce two bodies with the same composition, which is not the case. Thirdly, the capture hypothesis whereby the Moon was formed elsewhere in the Solar System and then captured by Earth's gravity. The Moon is, however, so large that, with any significant difference in approach velocity, the Earth would be unable to capture it unless a highly improbable set of orbital coincidences were satisfied.

After the Apollo missions returned samples of the lunar surface, the above theories were discounted. In the 1970s, a new mechanism of lunar formation was proposed – the giant impactor theory. In this scenario, the Earth was struck on its side by a Mars-sized planetesimal. This glancing blow propelled large amounts of Earth's mantle and of the impactor into orbit. The impactor's core later decelerated and returned to Earth, but the rest of the material remained in orbit, forming a transient ring. This material rapidly accumulated by impact processes into the Moon.

The giant impactor theory offers an explanation for the Moon's unique mineral composition. The Moon is depleted in iron because the Earth had already partially differentiated (the iron was in the core), and so little iron was in the mantle's ejecta. The high temperatures of the impact process explain the Moon's depletion in volatiles (such as potassium and water) and its enrichment in refractory elements (such as europium). The Moon has the same isotopic compositions as Earth because much of it came from the Earth's mantle. The giant impactor theory also explains the angular momentum of the Earth–Moon system by proposing a glancing blow. Finally, the accumulation of orbital materials into the Moon by high temperature impact processes offers an explanation for the early Moon's near global melting (producing a global magma ocean, which accounts for the mineralogical stratification of the Moon).

Giant Metre-wave Radio Telescope (GMRT) Major facility operated by the National Centre for Radio Astrophysics at Pune, India. The GMRT consists of 30 fully steerable 45-m (150-ft) antennae spread over distances of up to 25 km (16 mi) and located about 80 km (50 mi) north of Pune.

giant molecular cloud (GMC) MOLECULAR CLOUD with a mass that exceeds 10,000 solar masses. About half of all the interstellar matter in the Galaxy is in the form of GMCs. Numerous complex molecules are found inside the clouds (*see* INTERSTELLAR MOLECULES) and some of these form naturally occurring MASERS. The larger GMCs are gravitationally bound and in most of them parts of the cloud are collapsing to form new stars. The ORION NEBULA is such a star-forming region within the nearest GMC to Earth. The formation of stars, especially the luminous massive ones, eventually heats up the GMC and the stars' ultraviolet radiation dissociates the molecules. GMC lifetimes are thus only around a few tens of millions of years.

giant planet Term used to describe the planets JUPITER, SATURN, URANUS and NEPTUNE to distinguish them from the smaller, less massive rocky terrestrial planets of the Solar System. They are also known as the Jovian planets or gas giants.

Giant Segmented-Mirror Telescope (GSMT) Project of the New Initiatives Office of the ASSOCIATION OF UNIVERSITIES FOR RESEARCH IN ASTRONOMY for a 30-m (100-ft) ground-based optical telescope (formerly known as MAXAT) to be commenced by 2010. The GSMT is intended to complement the ATACAMA LARGE MILLIMETER ARRAY and the NEXT-GENERATION SPACE TELESCOPE.

giant star Star with radius between 10 and 100 solar radii; it is placed well above the main sequence in the HERTZSPRUNG–RUSSELL DIAGRAM. Giant stars include ARCTURUS, CAPELLA and ALDEBARAN. A giant star will have a larger radius than a main-sequence star (DWARF STAR) of the same temperature; however, a giant star of intermediate temperature is smaller in size than the hottest dwarf. Thus the term 'giant' refers to the relative luminosity of a star rather than its size.

Giant stars represent late stages in STELLAR EVOLUTION, when the stars have exhausted hydrogen in their cores and SHELL BURNING is occurring. They have dense central cores and tenuous atmospheres. RED GIANTS are normal stars that are undergoing hydrogen shell burning. Higher-mass stars can continue to burn other elements in their cores and in concentric shells in their atmosphere.

As the mechanism that creates energy within a star switches from one set of nuclear reactions to another, its radius, temperature and luminosity change. Thus giant stars move around the giant region and HORIZONTAL BRANCH of the Hertzsprung–Russell diagram. Most stars pass through the INSTABILITY STRIP at some point, becoming CEPHEID VARIABLES or RR LYRAE VARIABLES. In its final phase, a giant star becomes brighter and moves to the asymptotic giant branch.

Because of their high intrinsic luminosity, giants are quite common among naked-eye stars but are in fact relatively rare in space.

gibbous phase PHASE of a planet or the Moon when between first or last quarter and full. The word gibbous means 'bounded by convex curves' and describes the appearance of a body when more than half but less than all of its illuminated side is visible.

Gilbert, William (1540/44–1603) English doctor, the first scientist carefully to investigate magnetic phenomena, including the Earth's magnetic field, as described in his *De magnete* ('On Magnetism', 1600). He believed the planets were held in their orbits by magnetic attraction. Gilbert was one of the first English supporters of Copernicus' ideas, and he prepared a pre-telescopic Moon map.

Gill, David (1843–1914) Scottish-born astronomer who served as the fifth director of the Cape Observatory (1879–1907). Gill's early work involved determining the solar parallax from the 1874 and 1882 transits of Venus, and from observations of Mars and asteroids. He made the most accurate pre-photographic measurements of stellar parallaxes, computing accurate distances to 22 stars in the southern sky. His photographs of the Great Comet of 1882 showed so many stars that he proposed photography as the best method of acquiring data for star catalogues. In collaboration with the Dutch astronomer Jacobus KAPTEYN, who compiled data from Gill's photographic plates, he produced the massive *CAPE PHOTOGRAPHIC DURCHMUSTERUNG*, which lists the positions and magnitudes of 455,000 stars in the southern hemisphere brighter than 10th magnitude. Gill was one of the major forces behind the all-sky photographic survey known as the *CARTE DU CIEL*.

Giotto EUROPEAN SPACE AGENCY's first deep-space probe and the first spacecraft to make close-range studies of two COMETS. It was named after the Italian painter Giotto di Bondone (*c*.1267–1337), who depicted Halley's Comet as the 'Star of Bethlehem'.

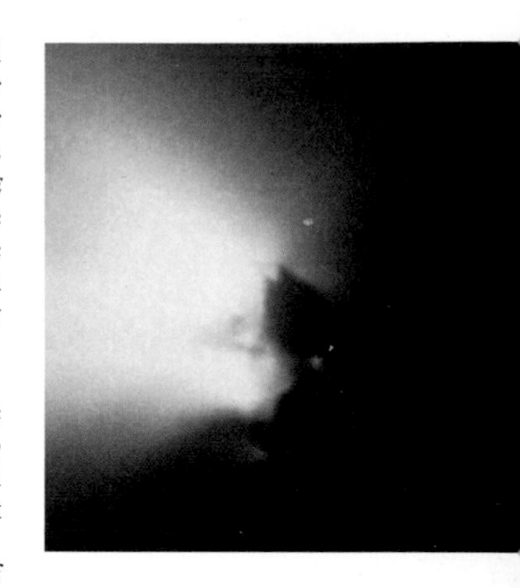

▲ **Giotto** The ESA Giotto probe recorded this close-up view of the nucleus of Comet 1P/Halley on 1986 March 13. Shortly after this image was obtained, collision with a dust particle from the comet disabled the camera.

Giotto carried ten experiments to study all aspects of the comet and its environment. It was launched from French Guiana on 1985 July 2 by an ARIANE 1 rocket and encountered Comet HALLEY on 1986 March 14. Giotto was peppered by 12,000 dust impacts on its approach to the nucleus. Only 15 seconds before closest approach, the spacecraft was sent spinning by an impact with a large particle. Contact was eventually restored 32 minutes later, by which time Giotto had passed the nucleus on the sunward side at a distance of 596 km (370 mi).

Giotto's images revealed a black nucleus about 15 km (9 mi) long and 7–10 km (4–6 mi) wide. It was pockmarked by craters, and bright jets of material could be seen on the sunlit side. About 80% of this material was water, with substantial amounts of carbon monoxide (10%), carbon dioxide (2.5%), methane and ammonia. The density of the nucleus was 0.3 g/cm^3 (one-third the density of water).

Although the spacecraft and half of its experiments were damaged, Giotto completed the first Earth gravity assist by a spacecraft coming from deep space on 1990 July 2. It was then retargeted to encounter Comet GRIGG–SKJELLERUP on 1992 July 10.

During the closest ever cometary flyby, Giotto passed by on the night side at a distance of about 200 km (125 mi). Dust caused very little damage to the spacecraft, owing to its slow relative approach speed of 14 km/s (9 mi/s), compared with 68 km/s (109 mph) for Halley, and a much lower dust production rate. The experiments were shut down on July 11.

Glenn, John Herschel (1921–) First American to orbit the Earth, in the Mercury spacecraft named *Friendship 7*, on 1962 February 20. After distinguished careers as a World War II combat pilot and a test pilot, Glenn was selected as one of the seven original Mercury astronauts in 1959, and less than three years later he completed his three orbits of the Earth in 5 hours. On 1998 October 29, Glenn returned to space for nine days aboard the Space Shuttle *Discovery*, becoming the oldest astronaut in history.

Glenn Research Center In full, John H. Glenn Research Center at Lewis Field, a NATIONAL AERONAUTICS AND SPACE ADMINISTRATION (NASA) laboratory specializing in propulsion, power and communications technologies, situated next to Hopkins International Airport, Cleveland, Ohio. It was founded in 1941 as the Aircraft Engine Research Laboratory, and was later known as the Lewis Research Center. It is NASA's lead centre for aeropropulsion and its centre of excellence in turbomachinery. It also conducts research in microgravity, fluid physics and materials science.

glitch Sudden increase in the rotation rate observed in a number of younger PULSARS, such as the Crab and Vela pulsars. The largest glitches have been seen in the Vela Pulsar; they occur every few years and typically decrease the period by a few millionths of the period. The glitch relaxes on a timescale of a few days. It is believed that glitches are caused by sudden changes in the internal structure of the NEUTRON STAR, for example by quakes in the solid components, such as the crust, or by vortices (whirlpools) in the superfluid layers. These vortices move outwards, but because they are pinned at the top and bottom they are not able to move outwards freely; they move jerkily when the pinning breaks.

Global Oscillation Network Group (GONG) Group that studies the structure and dynamics of the SOLAR INTERIOR using HELIOSEISMOLOGY. Six solar telescopes around the Earth obtain nearly continuous observations of the solar oscillations. The sites in the GONG network are the Big Bear Solar Observatory, the Learmonth Solar Observatory, the Udaipur Solar Observatory, the Observatorio del Teide, the Cerro Tololo Interamerican Observatory and the Mauna Loa Observatory.

Global Positioning System (GPS) Navigation satellite system, operated by the US Air Force but also widely used by civilians, that enables users equipped with GPS receivers accurately to locate their positions on the Earth's surface. The system comprises 24 satellites, the first of which was launched in 1978.

globular cluster Almost spherical, compact, gravitationally bound cluster of stars, containing around ten thousand to a million members. The core of a globular cluster is very densely packed and cannot generally be resolved into individual member stars. Globulars range in diameter from a few tens to over 300 l.y.

The Milky Way has fewer than 150 known globular clusters, but it is expected that there are about 200 in total, the others being obscured by the dust and gas in the galactic plane and the galactic centre. The farthest known globular clusters, for example NGC 2419, lie beyond the galactic disk at distances of around 300,000 l.y.

Most globular clusters in the Milky Way lie in the galactic HALO, surrounding the galactic centre; they have highly elliptical orbits of periods about a hundred million years or longer. The fact that they are centred on the galactic centre allowed the American astronomer Harlow SHAPLEY to estimate the distance to the galactic centre. Around 20% of globulars lie in the galactic disk and move in more circular orbits; these are known as disk globulars.

▼ **Global Oscillation Network Group** (GONG) A computer representation of one of the almost 10 million modes of sound wave oscillation in the Sun, as monitored by helioseismologists working under the umbrella of GONG. Solar regions that are receding are shown in red, those that are approaching in blue.

Globular clusters contain a mix of stars of various sizes, all of which evolved from the same cloud of gas at the same time. As all the stars within a cluster lie at the same distance from us, the HERTZSPRUNG–RUSSELL DIAGRAM (HR diagram) can be plotted for them simply by measuring their colour and their APPARENT MAGNITUDE (brightness), knowing that there is a constant difference between the apparent and the ABSOLUTE MAGNITUDE according to the cluster's distance. It is possible to deduce the cluster's distance by comparison with a standard HR diagram because the MAIN SEQUENCE will appear in the correct place only if the distance is right.

The HR diagram of a globular cluster reveals more than its distance. As time passes, the brightest stars in a cluster exhaust their hydrogen and become RED GIANTS. This happens progressively to fainter and fainter stars. The HR diagram clearly shows which stars have begun their transformation to red giants, and thus the age of the cluster. Many globular clusters have been dated about 10 billion years old; they must have formed before the Universe attained half its present age (*see* TURNOFF POINT).

In those remote times the chemical mix of our Galaxy was quite distinct from that of the present. Stars, especially those that turn into SUPERNOVAE, create elements such as carbon, oxygen, silicon and iron out of their hydrogen. Over the years, the proportion of these elements, which astronomers call metals, has increased, so we expect to see more metals in young stars than in the old globular clusters. In general, the stars within globular clusters are old, metal-poor, POPULATION II stars.

Further detailed examination of galactic globular clusters suggests distinct subgroups: the halo globulars, further divided into 'old halo' and 'young halo' clusters; the disk globulars, which are younger and more metal-rich than halo globulars; and the bulge globulars, which lie in the more central parts of the Galaxy and are fairly metal-rich, with almost solar-type chemical abundances.

The oldest globular clusters probably formed 13 to 14 billion years ago in the initial collapse of the forming Galaxy. The younger, more metal-rich disk globulars and the bulge globulars formed significantly later. The young halo globulars, such as M54, may have formed outside the Galaxy, perhaps in nearby DWARF GALAXIES, which were subsequently pulled into the Milky Way. M54 is believed to have belonged to the Sagittarius Dwarf Galaxy, discovered in 1994, which is currently merging with the Milky Way.

Another important role of the globular clusters has been to give us a means of estimating distances. Globular clusters contain a particular type of variable, the RR LYRAE STARS, the intrinsic brightnesses of which can be determined. Thus, when these stars are recognized in other locations, the distances can be derived for different objects.

Globular clusters can be seen around other galaxies. Some galaxies, such as M87, contain several thousand globular clusters, whereas other galaxies, like the two Magellanic Clouds and the Pinwheel Galaxy, appear to have globulars that are only a few billion years old. Recent observations of some globulars suggest star formation may be occurring within them.

globule Dense compact dust and gas cloud that strongly absorbs the light from objects behind it. A globule is the smallest of the DARK NEBULAE. There are two main types of globule – the COMETARY GLOBULE and the Bok globules. The latter are named after Bart Jan BOK, who first suggested that the globules might be the forerunners of PROTOSTARS.

Bok globules are roundish dark clouds that may have diameters as small as 0.1 l.y. At such small sizes, they can only be seen when they lie in front of a bright emission nebula. Larger globules are seen as dark patches against the stellar background of the Milky Way; they may have sizes of up to 3 l.y. The mass of dust in a globule can be estimated from the EXTINCTION of background light, but it is certain that the dust makes up only a small fraction of the total mass of a globule. Being well shielded from ultraviolet stellar radiation, and at a tem-

◀ **globular cluster** Containing at least 300,000 ancient stars, the globular cluster G1 orbits the Andromeda Galaxy. This Hubble Space Telescope image shows G1 to have a rich, compact core and less densely packed periphery.

perature of 10 K, most of the gas in a globule will be in molecular form. The expected main constituent, molecular hydrogen, is very difficult to observe, but the much less abundant molecules of carbon monoxide, formaldehyde and the hydroxyl radical have been observed at radio wavelengths. Total estimated globule masses thus range from about 0.1 solar mass for the smallest, to about 2000 solar masses for the RHO OPHIUCHI dark cloud. There is no sharp cut-off in the mass range, so that at the larger sizes the globules simply become the smaller dark nebulae such as the COALSACK.

Such a large mass of gas and dust within the small radius of a globule at a low temperature will undergo gravitational contraction to form first protostars and then normal main-sequence stars. The close positional association of globules with T TAURI STARS and HERBIG–HARO OBJECTS, both phenomena associated with star formation, tends to confirm this view.

gluon Particle exchanged when two quarks interact by the strong force. The strong force is the strongest of the four FUNDAMENTAL FORCES, and the exchange of gluons between quarks holds nuclear particles such as protons and neutrons together, as well as holding them in atomic nuclei. The gluon is a spin-1 boson.

GMC *See* GIANT MOLECULAR CLOUD

GMT *See* GREENWICH MEAN TIME

gnomon Ancient instrument for measuring the altitude of the Sun, consisting of a vertical stick or rod fixed in the ground. The length of the shadow indicates the Sun's altitude, and thus the time of day. The term is also used for the pointer on a SUNDIAL.

Goddard, Robert Hutchings (1882–1945) American astronautics pioneer, who flew the first liquid-propellant rocket, on 1926 March 16. In 1920 he set forth most of the basic principles of modern rocket propulsion. He later set up a rocket research and development test facility in the New Mexico desert, where he developed stabilizing and guidance systems, including the first rocket gyroscopes, and parachute systems that could be used to recover heavy payloads. Although Goddard's achievements went largely unrecognized during his lifetime, German rocket scientists led by Werner von BRAUN used his designs to build the V-2 rockets that, after their capture at the end of World War II, kick-started the American space programme.

Goddard Space Flight Center NATIONAL AERONAUTICS AND SPACE ADMINISTRATION (NASA) major space science laboratory, located in Greenbelt, Maryland, near Washington D.C. Established in 1959 May, it is named in honour of the American rocket pioneer Robert GODDARD. The centre specializes in astronomy, solar–terrestrial relations, and environmental studies of, for example, the Earth's ozone layer and global warming. Goddard operates its own spacecraft, and conducts research in spaceflight technology. It also houses the National Space Science Data Center and the ASTRONOMICAL DATA CENTER.

Gold, Thomas (1920–) Austrian-American cosmologist known for his advocacy of non-mainstream theories, including the steady-state cosmological model. He left Austria for England before Word War II, receiving his education at Cambridge, then moving to Greenwich (1952–56) under Martin RYLE. He emigrated to the United States in 1956, working first at Harvard, then at Cornell. With Fred HOYLE and Hermann BONDI, Gold conceived a theory of cosmology where, though the universe expands, the density of matter is kept constant by the continuous creation of new matter. Gold was the first correctly to explain that pulsars are rotating neutron stars with powerful magnetic fields.

Goldberg, Leo (1913–87) American astronomer, an expert on stellar evolution, who specialized in solar astronomy and the study of stellar atmospheres and mass loss. At the University of Michigan's McMath–Hulbert Observatory, he used a variety of imaging techniques to capture the structure and motion of solar flares and prominences. Spending much of his career at Harvard, he designed and helped to build an innovative spectroheliograph that was flown aboard the fourth Orbiting Solar Observatory, launched in 1967.

Golevka APOLLO ASTEROID; number 6489. Radar images show it to be irregular in shape. Golevka makes regular close approaches to the Earth, every four years. *See also* illustration, page 164; table at NEAR-EARTH ASTEROID

GONG Acronym for GLOBAL OSCILLATION NETWORK GROUP

▼ **globule** Seen in contrast against bright emission nebulosity, dark globules such as these are probably forerunners of protostars.

▲ **Golevka** A series of radar images showing the irregular profile of near-Earth asteroid (6489) Golevka. The asteroid has a long dimension of about 0.8 km (0.5 mi). [S. Ostro (JPL/NASA)]

G

Goodricke, John (1764–86) English variable-star observer, born in Groningen, Holland. In 1782, working with his friend Edward Pigott (1753–1825), and using a $3\frac{3}{4}$-inch (95-mm) refracting telescope and a good clock, Goodricke established the $2^{d}\ 20^{h}\ 49^{m}\ 8^{s}$ periodicity of the star Algol. Goodricke and Pigott went on to propose – correctly, as later astronomers would discover – that Algol's light was being eclipsed by a dimmer body rotating around it. Then in 1784, Goodricke measured the light variation of δ Cephei and realized that this variation was not the result of an eclipsing body.

GO TO telescope Telescope, usually on an ALTAZIMUTH MOUNTING, whose axes are fitted with position encoders that can be directed to find objects by entering simple commands from a handset. On-board computers on GO TO telescopes can store databases of many thousands of objects. Following initial set-up, which usually requires the observer to centre a couple of reference stars in the field of view, the mount will automatically slew the telescope to objects of choice. This greatly eases object location for novice amateur observers, and has also been used to increase the throughput in more advanced PATROLS for supernovae in external galaxies.

Göttingen Observatory Scientific facility within the physics faculty of the Georg-August-Universität at Göttingen, 90 km (55 mi) south of Hanover. Its origins go back to 1750; construction of the present observatory (just outside the old city walls) began in 1803. Its first director, Carl Friedrich GAUSS, was appointed in 1807, and it became a major centre for astrometry. Its reputation continued into the 20th century under the directorship of Karl SCHWARZSCHILD. Today, the observatory specializes in solar physics, X-ray astrophysics, extragalactic astrophysics and instrumentation for 8-m (25-ft) class optical/infrared telescopes.

Gould, Benjamin Apthorp (1824–96) American astronomer who catalogued the southern stars from the CÓRDOBA OBSERVATORY, Argentina, which he founded and directed (1870–85). He studied astronomy under Carl Friedrich Gauss at Göttingen, and on his return to America he founded the *Astronomical Journal* in 1849. After directing Dudley Observatory (Albany, New York) from 1856 and his own observatory at Cambridge (Massachusetts), he moved to Córdoba, where he catalogued tens of thousands of stars visible only from the southern hemisphere, publishing several catalogues (1879–96). Gould applied photography to the study of star clusters. GOULD'S BELT is named after him.

Gould's Belt Belt of stars and gas about 3000 l.y. across tilted to the galactic plane by about 16°. It contains many O- and B-type stars. Gould's Belt probably represents the local spiral arm of which the Sun is a member. It is named after Benjamin Gould, who identified it in 1879.

GPS Abbreviation of GLOBAL POSITIONING SYSTEM

graben Structure formed by a pair of inward dipping normal FAULTS. A graben is typically a linear depression in the surface relief. It forms as the result of stretching in the rocks. In planetary studies, the presence of grabens, which are easily distinguished on the remote sensing images, indicates an extensional TECTONIC environment at a given location. Grabens are observed on practically all planets and satellites as large as or larger than the Earth's Moon.

Graham-Smith, Francis (1923–) British radio astronomer, the thirteenth ASTRONOMER ROYAL (1982–90), who did pioneering work using Michelson interferometers, helping to develop the phase-switching technique. He helped to confirm the nature of the strong radio sources Cygnus A and Cassiopeia A, and was an integral part of the team that compiled the *THIRD CAMBRIDGE CATALOGUE* (3C) of radio sources in the 1950s. Much of Graham-Smith's research is on pulsars, for which he conceived the relativistic beaming theory.

Granat Soviet gamma-ray and X-ray spacecraft launched in 1989 December, carrying a French Sigma gamma-ray instrument, with contributions from Denmark and Bulgaria. With a spacecraft bus based on the Soviet Astron astrophysical platform, Granat had completed, within the first three years, over 700 specific observations lasting typically 24 hours at a time, some as long as 46 hours. Sigma observations in 1991 were coordinated with the newly launched NASA COMPTON GAMMA RAY OBSERVATORY.

Grand Tour In the late 1960s, US planners and scientists realized that the four major outer planets – Jupiter, Saturn, Uranus and Neptune – would be in an approximate alignment in the late 1970s/1980s that is repeated only once every two centuries. Ambitious plans were made for two gravity-assist launches to Jupiter/Saturn/Pluto in 1976–77 and similar launches to Jupiter/Uranus/Neptune in 1979, the so-called Grand Tour Missions. Political and economic constraints led to a reduced two-spacecraft mission targeted to only Jupiter and Saturn. Begun in 1972 as Mariner-Jupiter-Saturn (MJS), the project's name was changed in 1977 to VOYAGER. Voyager 2 eventually accomplished a Grand Tour when its mission was extended to Uranus (in 1986) and then Neptune (in 1989).

grand unified theory (GUT) Theory that attempts to 'unify' three of the four forces of nature, that is, to demonstrate that they are different manifestations of a single force. In descending order of strength, the four FUNDAMENTAL FORCES are: the strong nuclear interaction (which binds atomic nuclei); the electromagnetic force (which holds atoms together); the weak nuclear interaction (which controls the radioactive decay of atomic nuclei); and gravitation.

The weak and strong interactions operate over tiny distances within atomic nuclei while the other two are infinite in range. With the exception, so far, of gravitation, all the forces can be described by quantum field theories, which imply that forces are communicated by means of particles: the more massive the force-carrying particle, the shorter its life and the shorter the range of the force. Grand unified theories (GUTs) attempt to unify the weak, strong and electromagnetic forces. Many physicists believe that a more elaborate theory will eventually be found which will unify all four forces.

The first step towards unification was the Weinberg–Salam–Glashow theory of the 'electroweak' force, published in 1967, which implied that at particle energies greater than about 10^{12} eV, the electromagnetic and weak forces should be equal in strength and behave in the same way. The theory was confirmed in 1983 when the predicted force-carrying particles – the massive W and Z bosons – were detected.

GUTs imply that at energies greater than about 10^{24} eV the electroweak and strong nuclear interactions merge into one common force. Such energies far exceed the capabilities of our particle accelerators, but would have existed everywhere in the Big Bang universe during the first 10^{-35} seconds after the initial event. Thereafter, as the universe expanded and cooled, the unified force would have split into the strong force and the electroweak force which, in turn, split into the electromagnetic and weak forces.

The simplest GUT, proposed in 1973 by Sheldon Glashow (1932–) and Howard M. Georgi (1947–), requires the existence of a set of supermassive 'X-particles' that can transform quarks into leptons and vice versa. Since quarks are the basic constituents of protons, the theory predicts that protons themselves must eventually decay into lighter particles. The half-life of a proton is believed to be at least 10^{32} years, but given enough protons it should be possible at any time to detect the sporadic decay of a few. Experiments are under way to test this prediction; if confirmed, it implies the eventual disintegration of all the familiar chemical elements.

Gran Telescopio Canarias (GTC) Spanish 10.4-m (34-ft) telescope under construction at the ROQUE DE LOS MUCHACHOS OBSERVATORY on the island of La Palma, due for first light in 2003. With its segmented mirror, the instrument closely resembles the 10-m telescopes of the W.M. KECK OBSERVATORY. The project is led by the INSTITUTO DE ASTROFÍSICA DE CANARIAS with participation from Mexico and the University of Florida.

granulation Mottled, cellular pattern that is visible in high-resolution images of the Sun in the white light of the PHOTOSPHERE; it is a result of hot plasma rising from the SOLAR INTERIOR by CONVECTION. The granulation consists of about one million bright convection cells, called granules. The bright centre of each granule indicates hot plasma rising to the photosphere, and its dark edges indicate cooled material descending towards the interior. Individual granules are often polygonal, and they appear and disappear on time scales of about ten minutes. Each granule has an average size of approximately 1000 km (600 mi). Larger convection cells, each about 30,000 km (20,000 mi) across, make up the supergranulation.

graticule System of reference wires or fine lines placed in the focal plane of a TELESCOPE and used for measuring the positions and separations of celestial objects within the field of view. The graticule may take the form of parallel vertical wires or cross-wires, or a grid of squares (a reticule). The wires have to be so fine that spider threads have been utilized for this purpose in the past. *See also* FILAR MICROMETER

grating *See* DIFFRACTION GRATING

gravitational constant (G) Fundamental constant in Newton's law of gravitation (*see* GRAVITATIONAL FORCE). Its value has been determined as $G = 6.67259 \times 10^{-11}$ m³/kg s², with an uncertainty of about 0.00085×10^{-11}.

gravitational force Mutual attraction between two masses; it is the fourth of the physical forces. In his *Principia*, NEWTON announced that the force of gravity between any two masses, m_1 and m_2, is proportional to the product of their masses and inversely proportional to the square of the distance, r, between them:

$$F = \frac{G\, m_1\, m_2}{r^2}$$

where G is the GRAVITATIONAL CONSTANT. Thus, he was able to link the acceleration with which objects, such as the apocryphal apple, fall to the Earth's surface with the orbital motion of the Moon.

Gravity is very weak at close range, compared with electromagnetic and nuclear forces, but it becomes important on larger scales because it has a longer range than the nuclear forces and because, unlike electric charges, all mass is positive and attracts every other mass. Thus gravitation is the dominant force on large scales. For this reason it is astronomical observations that have provided the tests of theories of gravitation, such as Newton's original theory, Einstein's GENERAL RELATIVITY and the BRANS–DICKE THEORY.

Astronomers often determine the masses of astronomical objects such as the Sun, planets, stars and galaxies from their orbital dynamics. But in fact what they actually find is the product of mass and G. The gravitational constant can only be determined by measuring the attraction between known masses in the laboratory. There is no way to screen out forces from masses outside the laboratory so the experimenter must measure a disturbance on the object being tested which is a tiny fraction of the Earth's attraction. Thus the accuracy with which G is known is often the limiting factor in measuring the masses of stars and planets and other astronomical objects.

gravitational lens Configuration of an observer, an intervening mass and a source of light in which the mass acts as a lens or imaging device for light from the distant source. GENERAL RELATIVITY predicted that a mass (star, galaxy or black hole) situated directly in front of a distant light source (galaxy or quasar) will bend the light and possibly form images of the distant source. The deflection of starlight around the Sun during a solar eclipse was the very first observational confirmation of general relativity, made in 1917 by Sir Arthur Eddington. This prediction and confirmation paved the way for more complex situations, in which light from distant galaxies and quasars might be bent enough by foreground galaxies or stars to form images. This consequence was championed by Fritz ZWICKY in 1937 and was verified in 1979 with the discovery by Dennis Walsh, Robert Carswell and Ray Weymann of the first of the known 'double quasars' (0957+561).

Many more gravitational lens systems have now been observed, including distant galaxies and quasars lensed by nearer galaxies; these distant galaxies would not be visible

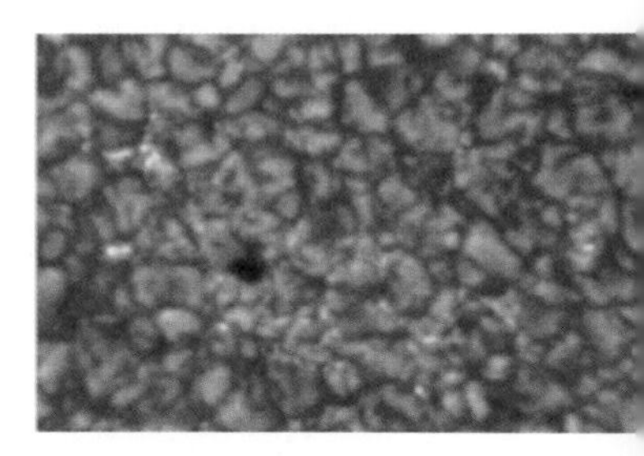

▲ **granulation** Under good seeing conditions, fine mottling of the Sun's photosphere can be observed. This granulation is due to convection, with bright hot material rising in the centre, while the darker edges of the granules indicate regions where cool material descends.

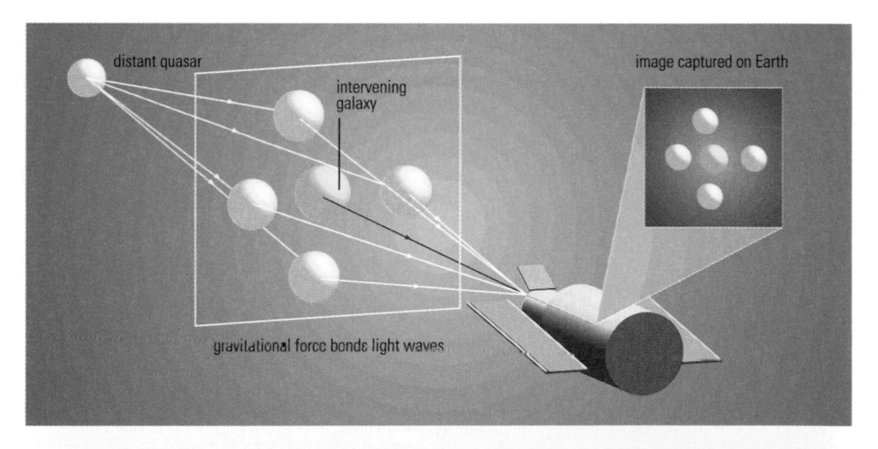

▼ **gravitational lens** (top) Light from a distant quasar may be bent in the gravitational field of a foreground galaxy to produce a distorted or multiple image. In this example, the quasar's light is split into four separate images, each arriving at the telescope along a slightly different path. (bottom) The massive galaxy cluster Abell 2218 in Draco acts as a gravitational lens, producing distorted, arc-shaped images of more distant objects in the same line of sight.

G

were it not for the amplification provided by the gravitational lens. Under certain conditions, a compact lens that lies directly along the line of sight forms multiple images that overlap. The resulting magnified image appears much brighter than the original source. This situation is known as **microlensing**. Microlensing might be important in quasars when intervening objects (stars or black holes) move across the line of sight to distant quasars and cause an increase of brightness in the quasar. These 'microlensed' outbursts are symmetric in time and frequency independent. Dark matter in our galactic halo might also 'microlens' stars in nearby galaxies, thus revealing their presence. *See also* EINSTEIN RING

gravitational redshift Increase in the wavelength of a photon emitted from an object within a gravitational field. The name arises because in the visible region the wavelength shift is towards the red end of the spectrum. This is not true at longer wavelengths, so the alternative name, Einstein shift, may be preferred. The change in wavelength is:

$$\frac{G m \lambda}{c^2 R}$$

where G is the gravitational constant, m the mass of the body, λ the original wavelength, c the velocity of light, and R the distance of the emitting region from the mass.

gravitational wave Periodic disturbance in a gravitational field, which, according to EINSTEIN'S theory of GENERAL RELATIVITY, propagates through space at the speed of light.

Any changes in a gravitational field are expected to travel at the speed of light so that, for example, if the Sun were to be annihilated, $8^m\ 18^s$ would elapse before the Earth ceased to 'feel' the gravitational influence of the Sun. Just as an accelerating or oscillating charged particle emits electromagnetic waves, so an accelerating, oscillating, or violently-disturbed mass is expected to radiate wave-like gravitational disturbances, which are known as gravitational waves.

Because gravitation is by far the weakest of the forces of nature, gravitational waves are expected to be weak and difficult to detect. The most likely sources of these waves include: rapidly spinning bodies distorted from perfect sphericity; close binary stars, particularly those involving collapsed objects like NEUTRON STARS or BLACK HOLES; SUPERNOVAE; and events involving massive black holes. *See also* GRAVITATIONAL WAVE DETECTOR

gravitational wave detector Device intended for the detection of GRAVITATIONAL WAVES. A passing gravitational wave is expected to stretch and squeeze a solid body or to cause small displacements in the separation between two independent masses. In principle, these effects could be measured. In practice, few, if any, events are likely to produce gravitational waves causing displacements (strains) larger than 1 part in 10^{19}. Two masses separated by a metre would thus be moved by an amount far smaller than an atomic nucleus.

Current detectors include solid bars (cooled, to minimize noise) and Michelson interferometer systems. The Michelson systems attempt to measure the relative displacement of two or more mirrors attached to test masses, 'freely' suspended in a vacuum, with multiple reflections being used to increase the effective separation between the test masses. Several interferometer-based gravitational wave detectors are currently nearing completion and are expected to be able to detect strains of 10^{-22}. So if predictions of the magnitudes of gravitational waves to be expected from astronomical events are correct, then these waves should soon be detected.

gravity gradient Rate of change of the strength of a gravitational field with distance. Tidal forces arise through gravity gradients (*see* TIDES). The gravity gradient at a distance, R, from a mass, M, is proportional to M/R^3. Thus the greatest gravity gradients are found close to condensed bodies such as NEUTRON STARS and BLACK HOLES. In the latter case they give rise to HAWKING RADIATION. An artificial satellite with an elongated shape will align itself along the gravity gradient; this can be used to orient the spacecraft.

Gravity Probe NASA spacecraft that will be launched in about 2003 to test Einstein's theory of GENERAL RELATIVITY by measuring the precession of gyroscopes in Earth orbit. The development spacecraft has been delayed several years and is still under budgetary scrutiny. The spacecraft will slowly roll about the line of sight to a guide star viewed by a reference telescope, while four gyroscopes, cooled by liquid helium, will compare data from the telescope to an accuracy of 0.1 milliarcsecond. The polar-orbiting satellite is to look for the expected 6.6 arcseconds/year precession to the Earth's rotational axis and the 0.042 arcsecond/year parallel to it.

Great Attractor Object that is apparently pulling our LOCAL GROUP of galaxies as well as some nearby clusters towards it. Study of slight temperature differences in the COSMIC MICROWAVE BACKGROUND has revealed that the Local Group is actually moving towards a tremendous concentration of mass nearly 150 million l.y. away, in the direction of the constellation Leo. This concentration of galaxies and dark matter is gravitationally attracting our Local Group of galaxies and thousands of other galaxies in this part of the Universe. The velocity of the Local Group is currently measured as 620 km/s towards the Great Attractor. It is not clear what makes up the dark matter, or exactly how much of the attractor is in dark matter. The only thing that is clear is the tremendous concentration of mass in that part of the Universe.

great circle Circle on the surface of a sphere whose plane passes through the centre of the sphere and which divides the sphere into two equal hemispheres. Examples of great circles on the celestial sphere are the CELESTIAL EQUATOR and any MERIDIAN (a great circle passing through the poles).

Great Debate Culmination, in 1920, of the rivalry between opposing views on the size of the Milky Way and

▼ **Great Attractor** A European Southern Observatory (ESO) image of the central region of ACO 3627, a huge galaxy cluster thought to form part of the Great Attractor. The Great Attractor's gravitational pull is dragging all the galaxies in our local cluster towards it.

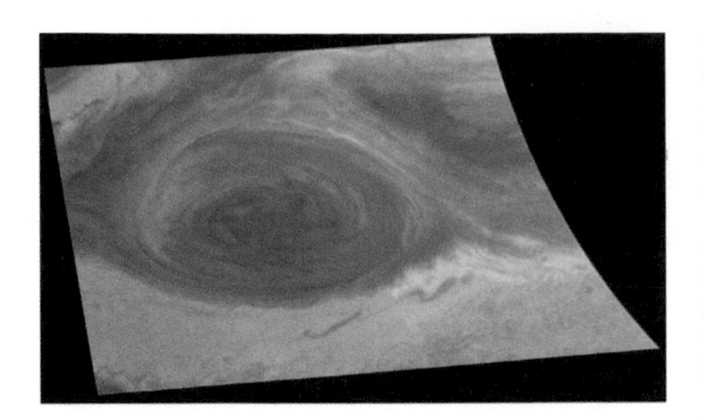

◄ **Great Red Spot** A 1996 Galileo orbiter image shows this huge storm, which has raged in Jupiter's atmosphere for hundreds of years.

the nature and distance of the 'spiral nebulae'. The Great Debate itself, organized by George Ellery HALE, took place on 1920 April 26 at the National Academy of Sciences in Washington, D.C. The proponents of the rival theories were Heber CURTIS and Harlow SHAPLEY.

From his Lick photographic surveys, Curtis had discovered that 'spiral nebulae' were scarce along the galactic plane – the zone of avoidance; also, many spirals showed dark dust lanes along their horizontal planes, as were apparent in the Milky Way. These observations and his spectroscopic work convinced Curtis that bright, diffuse nebulae were relatively close, lying inside the Milky Way galaxy, but that the spirals were other galaxies, 'island universes' or 'external galaxies', at much greater distances.

This theory was at variance with the 'metagalaxy' model proposed by Shapley, which, though it correctly placed the Solar System far from the galactic centre, held that the Galaxy was an immense 300,000 l.y. across (Curtis' model, following Jacobus KAPTEYN's, called for a smaller Milky Way.) Spiral nebulae belonged to our own Galaxy, Shapley maintained, partly because measurements of their rotational velocities by Adriaan VAN MAANEN – later shown to be in error – indicated that if these systems were as distant as Curtis believed, their outer regions must be rotating well in excess of the speed of light.

The matter was resolved in 1924–25, when Edwin HUBBLE used CEPHEID VARIABLE stars in the Andromeda Galaxy (M31) to show that this galaxy was much farther than even the radius of Shapley's 'metagalaxy', vindicating Curtis' views.

greatest elongation Maximum angular distance, east or west, measured between an INFERIOR PLANET and the Sun, as viewed from Earth. This varies between 18° and 28° for Mercury, and 45° to 47° for Venus. When at greatest ELONGATION east, the two planets set at their latest time after the Sun, and when at greatest elongation west, they rise at their earliest time before it.

Great Observatories NASA's description of a major astronomical programme involving four space telescopes – the HUBBLE SPACE TELESCOPE (1990), the COMPTON GAMMA RAY OBSERVATORY (1991), the CHANDRA X-RAY OBSERVATORY (1999) and the SPACE INFRARED TELESCOPE FACILITY (2002).

Great Red Spot (GRS) Largest and longest-lasting atmospheric feature on JUPITER. It is a huge anticyclonically rotating meteorological eddy, which currently is 24,000 km (15,000 mi) long and 11,000 km (7000 mi) wide and centred at a latitude of 22°S. The Great Red Spot has been observed by ground-based observers for more than 300 years. First reported by Robert Hooke in 1664, it was followed for several years by G.D. Cassini. Drawings by Heinrich Schwabe in 1831, William Dawes in 1851, and Alfred Mayer and the fourth Earl of Rosse in the 1870s all show the Red Spot. The feature became quite prominent in the 1890s, and British Astronomical Association records contain extensive data from that time to the present.

Green Bank Observatory Location at Green Bank, West Virginia, of the 100-m (330-ft) Robert C. Byrd Green Bank Telescope (GBT), the world's largest fully steerable radio dish, a facility of the NATIONAL RADIO ASTRONOMY OBSERVATORY. Dedicated in 2000 August, the GBT replaced a 90-m (300-ft) dish built in 1962 that collapsed without warning in 1988. It has an unusual design, with a main dish measuring 100 × 110 m (330 × 360 ft) that reflects the incoming radiation to an off-axis feed arm. There is no central obstruction as in a conventional radio telescope, and the response pattern is greatly improved. The two thousand panels making up the reflective surface are individually controlled by computer-driven actuators.

green flash Purely atmospheric phenomenon in which the last visible remnant of the setting Sun may suddenly appear to change from red to vivid green for a few seconds as it sinks below the horizon. At higher latitudes, where the Sun sets at a shallow angle, the green flash may last for several seconds. Visibility of the green flash is strongly dependent on atmospheric conditions; stable layering of air over a clear sea horizon during the summer appears to particularly favour its occurrence. A green flash can also occur at sunrise. Like the Sun, the bright planet Venus can also be seen to produce a green flash under favourable conditions as it sets.

greenhouse effect Phenomenon whereby the surface of a planet is heated by the trapping of infrared radiation by the ATMOSPHERE. This mechanism is important in the atmospheres of the Earth, Venus and Titan. On Earth, the most important greenhouse gases are carbon dioxide and water vapour, with significant contributions from methane, nitrous oxide, ozone, and anthropogenic hydrofluorocarbons, perfluorocarbons and sulphur hexafluoride. These gases are transparent to the short-wave radiation from the Sun but absorb longer wavelength infrared radiation from the surface, re-emitting this radiation to the atmosphere and surface, thus raising the overall temperature. In the absence of an atmosphere, the Earth would have an average temperature of *c.*256 K instead of 290 K. Carbon dioxide (CO_2) is thought to have played a significant role in the early atmospheres of all the terrestrial planets, and it is currently responsible for the extremely high surface temperature of Venus (737 K). The burning of fossil fuels is rapidly increasing the amount of CO_2 in the Earth's atmosphere.

Greenstein, Jesse Leonard (1909–) American astronomer who spent his career at Yerkes Observatory and CalTech's Mount Wilson and Palomar Observatories. He used spectroscopy to study stars with peculiar spectra, the abundances of chemical elements in stellar atmospheres, the interstellar medium and the

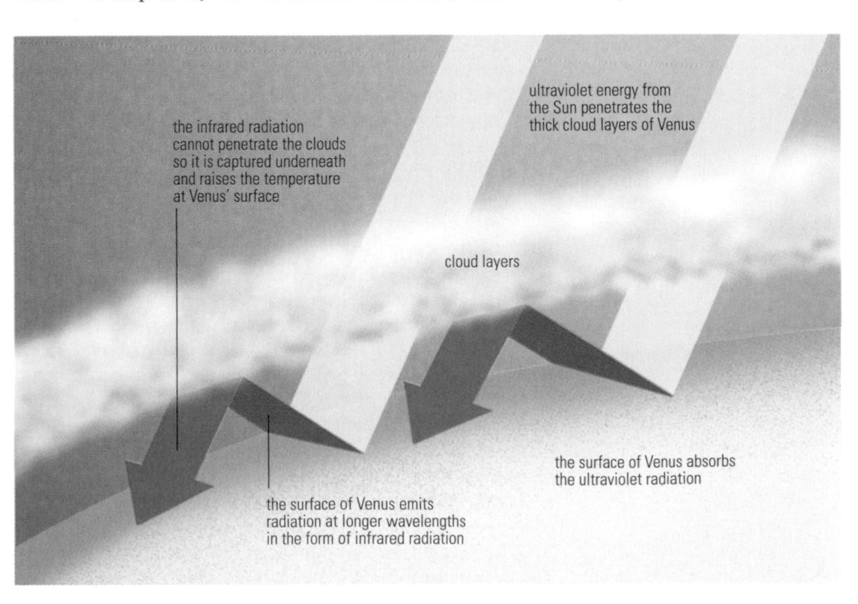

▼ **greenhouse effect** Short-wave radiation from the Sun readily penetrates the atmosphere to reach the ground. Longer wavelength radiation re-emitted from the ground is trapped by atmospheric constituents such as carbon dioxide, raising the planetary temperature. This greenhouse effect is significant in keeping Earth habitable, but has led to inhospitably high temperatures on Venus.

composition of nebulae. From these studies he developed a theory of FREE–FREE TRANSITIONS in hot, highly ionized gases. Greenstein became an authority on stars made from degenerate matter, including white dwarfs, studying their masses and luminosities. He helped Maarten SCHMIDT to make the first definitive identifications of quasars' optical counterparts.

Greenwich Mean Time (GMT) Name given to MEAN SOLAR TIME at the Greenwich meridian; it is the popular name for civil time kept in the UK. The need for a standard time across the country came with the development of the railways, from 1825 onwards. For example, passengers travelling from Bristol to London would find a ten-minute difference between the time kept in the two cities because local time, measured according to the position of the Sun in the sky, varies with longitude. Eventually the railways adopted Greenwich time, which became known as 'railway time', as their standard and in 1880 it became the legal time for the whole country. In 1928 the INTERNATIONAL ASTRONOMICAL UNION (IAU) recommended that Greenwich Mean Time should become the standard timescale for scientific purposes and be known as UNIVERSAL TIME (UT). Today, Greenwich Mean Time is still the standard against which the system of TIME ZONES around the globe is measured, each one being an integral number of hours ahead of or behind it.

Greenwich Observatory Britain's former Royal Observatory, founded in 1675 by Charles II to make accurate positional measurements of stars for use in maritime NAVIGATION. The purpose-built observatory at Greenwich, London by the architect-scientist Christopher Wren (1632–1723) also provided accommodation for the first Astronomer Royal, John FLAMSTEED, who was paid £100 per year yet had to fund the cost of instruments and assistants from his own pocket. In 1766 the first *Nautical Almanac* was published under the guidance of Nevil MASKELYNE, fulfilling King Charles' ambition.

Throughout most of the 18th and 19th centuries, the Royal Observatory remained dedicated to positional astronomy. Investigations of the physical nature of celestial objects began in the late 19th century under the directorship of William CHRISTIE. He installed a 0.71-m (28-in.) refractor in the famous Onion Dome, and founded the Physical Observatory, equipping it with a 0.66-m (26-in.) photographic refractor and a 0.76-m (30-in.) reflector. To these were added the 0.91-m (36-in.) Yapp Reflector in 1934, but by then the observing conditions in London skies were deteriorating.

Plans to move away from the capital were not acted upon until 1948, when Harold Spencer JONES took up residence at Herstmonceux Castle, Sussex, at the start of a ten-year transfer from Greenwich. The institution became the ROYAL GREENWICH OBSERVATORY (RGO), and continued most of the activities previously carried out at Greenwich, including the time service. The old observatory buildings, known today as the Royal Observatory Greenwich, were developed into a museum as part of the National Maritime Museum. Its main attraction for many visitors is the Prime Meridian of the world.

Gregorian calendar System devised by Pope Gregory XIII in 1582, and still in use today, which corrected errors in the previously used JULIAN CALENDAR. By 1582 the Julian calendar, which approximated the actual solar year of 365.24219 days to a CALENDAR year of 365.25 days with a leap year every fourth year, had accumulated an error of 10 days. Pope Gregory XIII therefore removed 10 days from the calendar, so that in 1582 October 15 followed October 4. He instigated a new system in which only century years divisible by 400 (that is, 1600, 2000 and so on) would be leap years. The start of the year was also changed from March 25 to January 1.

The Gregorian calendar was not adopted in the UK, however, until 1752, by which time the error had increased to 11 days, so that September 14 followed September 2. Today the Gregorian calendar is used throughout the Western world and parts of Asia. It is also known as the Christian calendar since it uses the birth of Christ as a starting point, subsequent dates being designated *anno domini* (in the year of our Lord) and preceding dates being BC (before Christ). The accumulated error between the Gregorian year and the true solar year now amounts to just three days in 10,000 years.

Gregorian telescope Type of REFLECTING TELESCOPE designed in 1663 by Scottish astronomer James GREGORY. A parabolic primary mirror with a central hole reflects light to a concave elliptical secondary placed outside the focus, and thence back through the hole to an eyepiece behind the primary where it gives an erect image. It was popular in the 17th and 18th centuries with telescope-makers such as James SHORT, as the secondary was easier to make than the convex secondary required for the CASSEGRAIN TELESCOPE, and the tube length was shorter than that of a NEWTONIAN TELESCOPE. It is not much used today.

Gregory, David (1659–1708) Scottish mathematician and nephew of James GREGORY. He proposed a variety of telescope designs, including ways of making a telescope free from chromatic aberration. Gregory was a strong supporter of Isaac Newton's theories of calculus and mechanics, publishing a comprehensive summary of those ideas, entitled *Astronomiae physicae et geometricae elementa* (1702). His main astronomical interests lay in the newly founded study of optics, also greatly advanced by Newton.

Gregory, James (1638–75) Scottish mathematician who developed many of the fundamental tools of calculus. In his *Optica promota* (1663) he covered such diverse astronomical topics as elliptical orbits, transits of planets and the determination of parallax, and described a design for a new kind of telescope (*see* GREGORIAN TELESCOPE). Gregory also experimented with the very first example of a DIFFRACTION GRATING, using a fine sea bird's feather to project the image of the spectrum of sunlight on to a wall in a darkened room. David GREGORY was his nephew.

Grigg–Skjellerup, Comet 26P/ Short-period comet, initially discovered on 1902 July 23 by John Grigg (1838–1920), New Zealand. Poor weather restricted observations at the discovery apparition and prevented derivation of a true orbit. A comet discovered on 1922 May 17 by John Francis Skjellerup (1875–1952), South Africa, was found to be identical, and an elliptical orbit determined. The comet is usually faint. It has had its perihelion distance gradually increased to 0.99 AU following encounters with Jupiter. The current period is 5.11 years. Close approaches to Earth have given rise to periodic displays of Pi Puppid meteors around April 23, as in 1972 and 1977. The GIOTTO spacecraft passed through the tail of 26P/Grigg–Skjellerup on 1992 July 10, finding a lower-than-expected dust density in the comet's vicinity.

Grimaldi Lunar impact basin (6°S 68°W), 193 km (120 mi) in diameter, with rim components reaching 2750 m (9000 ft) above its floor. Grimaldi has two rings, both so deeply eroded by later impacts that in some places they are difficult to identify. The floor of Grimaldi is smooth, dark and almost featureless as a result of lava resurfacing. Indeed, it is one of the darkest regions on the lunar surface. A number of rilles, called Rimae Grimaldi, are found on Grimaldi's floor.

GRO Abbreviation of COMPTON GAMMA RAY OBSERVATORY

ground state Lowest energy state for a system. The term is usually used in the context of the electron structure of atoms, ions and molecules. It is when the

G

electrons occupy those energy levels that give the atom, ion or molecule the lowest total energy. *See also* ATOMIC STRUCTURE; EXCITATION

GRS Abbreviation of GREAT RED SPOT

Grubb, Parsons & Co. Irish telescope-making firm formed by the merger (1925) of firms founded by Thomas and Howard Grubb and Charles Parsons. **Thomas Grubb** (1800–1878) began making small astronomical telescopes in the 1840s, experimenting with reflector designs at a time when refractors were more popular. He became proficient at constructing sturdy equatorial mountings for large instruments. In 1834 Thomas Romney Robinson (1792–1882) ordered from Grubb a 15-inch (380-mm) Cassegrain reflector for the ARMAGH OBSERVATORY. This was the first large Cassegrain instrument, and the first large reflector to be carried on a polar axis equipped with a clock drive.

In 1865 **Howard Grubb** (1844–1931) joined his father as a telescope-maker, and took over the running of the business in 1868. The next year, the Grubbs completed the 48-inch (1.2-m) Melbourne Reflector. Another of their famous instruments was the 27-inch (0.69-m) refractor for the Vienna Observatory (1880). In 1901 they finished the Radcliffe Telescope for Oxford University's Radcliffe Observatory, consisting of two refractor tube assemblies, a 24-inch (0.6-m) instrument for visual observations and a 17½-inch (0.45-m) photographic telescope, carried on a German equatorial mounting. Several of the 'astrograph' photographic refractors used for the *CARTE DU CIEL* sky-mapping project were made by Grubb.

In 1925 Howard retired and his telescope making business was taken over by **Charles Algernon Parsons** (1854–1931), son of the Third Earl of ROSSE and famous as the inventor of the steam turbine engine. The company was renamed Grubb, Parsons & Co. and continued to make many large telescopes, the last of which (1987) was the Royal Greenwich Observatory's 4.2-m (165-in.) WILLIAM HERSCHEL TELESCOPE erected on La Palma, in the Canary Islands.

Grus *See* feature article

GSC Abbreviation of *GUIDE STAR CATALOGUE*

G star Any member of a class of yellow-white stars, the spectra of which are characterized by relatively weak hydrogen lines, strong ionized calcium lines (FRAUNHOFER LINES H and K), and, as temperature declines, the beginning of strong neutral calcium and sodium lines (Fraunhofer g and D). MAIN-SEQUENCE dwarf stars range from 5300 K at G9 to 5900 K at G0, with zero-age masses from 0.9 solar mass to just over solar, zero-age luminosities from 0.5 to just above solar, and dwarf lifetimes from 10 billion years to greater than the age of the Galaxy. Giants and supergiants, which, because of lower densities, are about 800 K cooler than the dwarfs, are more difficult to distinguish spectroscopically than in other classes. However, the Wilson–Bappu effect (*see* K STAR) can yield luminosities. G SUBDWARFS have weakened metal lines and lower metal abundances. The Sun is a G2 dwarf, allowing a good understanding of G stars.

G stars have deep convection zones, the Sun's occupying its outer third. Although they tend to be slow rotators (unless in binary systems where they are tidally spun up), G stars have active magnetic fields, which produce CHROMOSPHERES and hot CORONAE. The solar magnetic field is amplified within the Sun. Ropes of magnetic flux float upwards and produce SUNSPOTS where they exit and re-enter the Sun. Interacting and collapsing magnetic fields create bright solar FLARES, releasing the magnetically confined corona in CORONAL MASS EJECTIONS. Chromospheres produce EMISSION LINES. Stellar chromospheres can be recognized by emissions set within the calcium H and K absorptions; coronae can be confirmed through X-ray emission. About two-thirds of the G dwarfs exhibit Sun-like activity.

The highest luminosity (and longest period) Cepheids are class G supergiants. Numerous planets have been found orbiting G stars. Bright examples of G stars include the Sun (G2 V), Alpha Centauri-A (G2 V) and Capella (G0 III + G8 III).

GTC Abbreviation of GRAN TELESCOPIO CANARIAS

Guardians of the Pole Name sometimes given to the stars β and γ Ursae Minoris (KOCHAB and Pherkad), which form part of the bowl of the Little Dipper, Ursa Minor.

guide star Star on which a GUIDE TELESCOPE is tracked to ensure that the main telescope is accurately aligned on its target for the duration of an astronomical exposure.

Guide Star Catalogue (GSC) Catalogue created for the control and target acquisition of the Hubble Space Telescope in 1990, containing positions and magnitudes for about 20 million stars; it is mainly used for astrometric calibration. A second-generation catalogue, the *GSC-II*, will include the positions, proper motions, magnitudes in three colours and star/galaxy classifications for about 1 billion objects.

guide telescope Smaller telescope mounted on the side of a main instrument and used for viewing a field star, the guide star, near the target object. By watching the position of the guide star relative to cross-wires in the guide telescope's eyepiece, the observer can make adjustments to the DRIVE motion, ensuring that the target remains centred. An OFF-AXIS GUIDER may also be used, while modern instruments (both professional and amateur) are increasingly being equipped with CCD-based AUTOGUIDERS.

Gum Nebula Extensive faint nebula in the southern constellations of Puppis and Vela. It spans an angular diameter of 36°, making it the largest nebula in the sky. Named after its discoverer, the Australian astronomer Colin Stanley Gum (1924–60), the nebula lies at a distance of 1300 l.y. and has a true extent of 840 l.y. The ancient Vela supernova remnant is part of the Gum Nebula.

GUT Abbreviation of GRAND UNIFIED THEORY

Guth, Alan Harvey (1947–) American cosmologist and particle physicist at Massachusetts Institute of Technology, with Andrei LINDE one of the principal conceivers of INFLATION, a process hypothesized to have operated in the very early universe (up to just 10^{-37} seconds after the BIG BANG). By providing an initial period of rapid expansion, inflation accounts for observed properties of the Big Bang Universe. Guth has also developed a theory of cosmological phase transitions to explain how the Universe's symmetry altered in its earliest moments.

GRUS (GEN. GRUIS, ABBR. GRU)

Small but not inconspicuous southern constellation, representing a crane, between Piscis Austrinus and Tucana. Grus was introduced by Keyser and de Houtman at the end of the 16th century. Its brightest star, Alnair, is mag. 1.7. The constellation contains two naked-eye optical double stars: δ Gru, which has yellow and red components, mags. 4.0 and 4.1, separation about 13′; and μ Gru, two yellow stars, mags. 4.8 and 5.1, about 17′ apart.

BRIGHTEST STARS

	Name	RA h m	dec. ° ′	Visual mag.	Absolute mag.	Spectral type	Distance (l.y.)
α	Alnair	22 08	−46 58	1.73	−0.7	B7	101
β		22 43	−46 53	2.07	−1.5	M5	170
γ		21 54	−37 22	3.00	−1.0	B8	203
ε		22 49	−51 19	3.49	0.5	A3	130

G

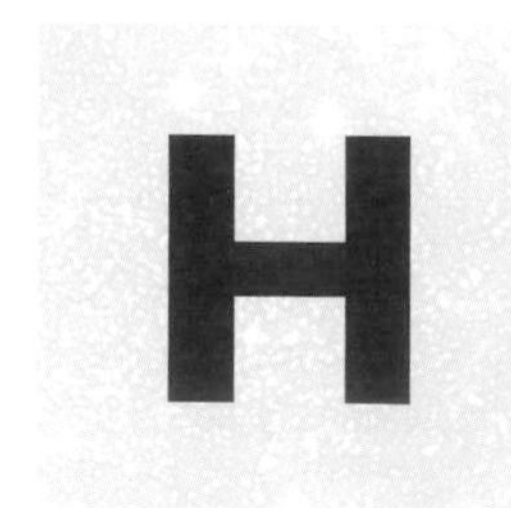

HA Abbreviation of HOUR ANGLE

habitable zone *See* ECOSPHERE

Hadar (Agena) The star β Centauri, at visual mag. 0.61 the 11th-brightest star, although it is actually a pulsating variable of β Cephei type with a range of a few hundredths of a magnitude and a period of 3.8 hours. It lies 525 l.y. away and its spectral type is B1 III. The name comes from an Arabic expression referring to one member of a pair of stars, although the origin of the designation is unknown. Its alternative title, *Agena*, is also of obscure origin, although it may come from the Latin word *genu*, meaning 'knee', since it marks one of the knees of the centaur, Centaurus.

Hadley cell One of the primary circulation cells in the Earth's ATMOSPHERE, in which air rises over the tropics and descends at the subtropical high-pressure regions located approximately at latitudes 30°N and 30°S. At the surface, the Hadley cells are represented by the trade winds on either side of the equator. The circulation was originally proposed by George Hadley (1685–1768) in 1735, as a single cell in each hemisphere, with warm air ascending at the equator and cold air descending at the poles. For dynamical reasons such a circulation cannot exist at tropospheric levels on Earth, although a similar circulation does occur in the STRATOSPHERE in winter. It has been proposed that the atmospheric circulation on Venus may be of this form.

hadron Elementary particle that interacts through the strong force. All hadrons have nucleon numbers of 1, 0 or −1, and can be divided into the subclasses baryons, anti-baryons and mesons. Examples of hadrons are protons, anti-protons and pions.

Haedi Nickname given to the stars ζ and η Aurigae, near Capella. The name is Latin for 'kids' – in mythology, they were the offspring of Capella, the She-Goat. The goat and kids are visualized as being carried on the arm of Auriga, the charioteer.

halation Spreading of a point image on a photograph caused by light being reflected, scattered and diffused within the emulsion during exposure. The image of a star can spread into a small disk, and sometimes a **halation ring** is produced around it.

Hale, George Ellery (1868–1938) American pioneer of solar astrophysics who discovered the magnetic fields of sunspots and facilitated the construction of very large telescopes: the 40-inch (1-m) refractor at Yerkes Observatory, the 60-inch (1.5-m) and 100-inch (2.5-m) reflectors at Mount Wilson Observatory, and the 200-inch (5-m) reflector on Mount Palomar, later named the HALE TELESCOPE in his honour.

After graduating from the Massachusetts Institute of Technology in 1890, where he constructed the first spectroheliograph (1890–96), Hale carried out solar research from his private Kenwood Observatory, equipped with a fine 12-inch (300-mm) refractor. He brought these and other instruments to the University of Chicago, where a new, major observatory was being built – Hale had convinced Chicago streetcar magnate Charles Tyson Yerkes (1837–1905) to fund the world's largest refracting telescope and the observatory to house it. The 40-inch refractor was first displayed at the 1893 Columbian Exposition, then permanently installed at YERKES OBSERVATORY when that facility opened four years later.

In 1905 Hale determined that sunspots are cooler than the surrounding chromosphere, and by observing the ZEEMAN EFFECT, a splitting of the Sun's spectral lines, he discovered (1908) that sunspots are associated with strong magnetic fields. He went on to show (1919) that solar magnetic fields reverse their polarity twice every 22–23 years. As early as 1904 Hale had persuaded the Carnegie Institution to build an observatory devoted to solar research atop Mount Wilson, near Pasadena, California. The SNOW TELESCOPE was the first major instrument installed at Mount Wilson Observatory, which Hale directed from 1904 to 1923. Mount Wilson became the world's leading observatory after it acquired, through Hale's fundraising efforts, the 60-inch reflector in 1908 and, in 1919, the 100-inch HOOKER TELESCOPE.

Hale–Bopp, Comet (C/1995 01) Long-period comet, the brightest of the 1990s, discovered independently by American amateur astronomers Alan Hale (1958–) and Thomas Bopp (1949–) on 1995 July 22–23. As with Comet KOHOUTEK in 1973, the comet was remarkably bright (mag. +10) at discovery, given its distance at that time of 7 AU from the Sun. Comet Hale-Bopp, however, lived up to the optimistic forecasts for its performance around perihelion, reached on 1997 April 1 at a distance of 0.91 AU from the Sun. The comet brightened steadily during 1996 and, following conjunction with the Sun, re-emerged into the morning sky in February 1997 as an already-prominent naked-eye object with a distinctive, fanned dust tail and a fainter ion tail. From mid-March, Comet Hale–Bopp became a magnificent object in early evening skies, peaking in brightness at mag. −0.5 around perihelion, at which time the dust and ion tails reached lengths of 20° and 25° respectively. Close examination of the COMA revealed concentric 'shells' of material being ejected from the NUCLEUS, indicating a rotation period of 11.5 hours. The nucleus is large, with an estimated diameter of 40 km (25 mi). Comet Hale–Bopp showed a strong ANTITAIL around 1998 January 5. In early 2001, the comet had faded to mag. +14.5, but it continued to show strong

▼ **Hale–Bopp, Comet** Photographed on 1997 March 10, Comet Hale–Bopp showed a clearly distinct bluish ion tail and yellowish dust tail. Reaching magnitude −0.5, this magnificent object was the most spectacular comet of the 1990s.

dust emission activity at a distance of 13 AU (midway between the orbits of Saturn and Uranus) from the Sun. The comet has a very highly inclined (89°.4) orbit, with a period of about 2400 years.

Hale Telescope PALOMAR OBSERVATORY's famous 200-inch (5-m) reflector, for more than a quarter of a century the world's largest optical telescope. It is named after George Ellery HALE, whose organizational genius lay behind the project's commencement in the 1920s. Although he witnessed the successful casting of the mirror blank in 1934 December (at the second attempt), Hale died a decade before the telescope was finally declared finished in 1949 November. The telescope has had a distinguished career in all areas of astronomy and continues to provide front-line service, despite increasing LIGHT POLLUTION from nearby San Diego.

half-life (symbol $t_{\frac{1}{2}}$) Time taken during an exponential decay process, such as RADIOACTIVITY, for half the available reactions to have occurred. Half-lives can range from tiny fractions of a second to billions of years. For example, the oxygen-15 isotope that is involved in the CARBON–NITROGEN–OXYGEN CYCLE has a half-life of 124 seconds. The term can be used for related processes, such as a neutron splitting into a proton and electron (half-life of 12 minutes). The average, or mean, lifetime of a particle is 1.44 times its half-life.

Hall, Asaph (1829–1907) American astronomer who, in 1877 August, discovered Phobos and Deimos, the two tiny satellites of Mars. He used the United States Naval Observatory's 26-inch (0.66-m) refracting telescope, at the time the largest in the world, to detect the satellites visually. He had previously used the same instrument to discover a white spot in Saturn's atmosphere, allowing him accurately to determine that planet's rotational period. Hall also found that the orbit of Saturn's satellite Hyperion was precessing by about 20° every year.

Halley, Comet 1P/ Brightest and probably most famous SHORT-PERIOD COMET; it is the only one to have been seen regularly with the naked eye. Comet 1P/Halley has been observed at each return since 240 BC, and it was first reliably noted in Chinese annals from the winter of 1059–1058 BC. The comet has a highly elliptical orbit, with a period of about 76 years. Its most recent perihelion, 0.587 AU from the Sun, between the orbits of Mercury and Venus, was on 1986 February 9. Aphelion carries 1P/Halley out beyond the orbit of Neptune, to a distance of 35.295 AU from the Sun. The orbit is retrograde, with an inclination of 162°.2.

The comet is named after the second British Astronomer Royal, Edmond Halley, who was first to calculate its orbit, making the connection between a comet he had seen in 1682 and previous recorded apparitions in 1531 and 1607. Halley successfully predicted its next return, in 1759.

Historically, Halley's Comet has made a great impression on observers at a number of past returns. Its appearance in 1066, when it passed 0.1 AU from Earth and reached mag. −4, shortly preceded the Norman Conquest of England, and the comet is depicted in the Bayeux Tapestry with the Saxon courtiers looking on in horror as King Harold totters on his throne. In 1301 the comet was seen by the Florentine painter Giotto di Bondone, who used it as a model for the Star of Bethlehem in his 1304 fresco *Adoration of the Magi* in the Arena Chapel in Padua.

The comet's apparitions vary considerably in terms of how favourably it may be viewed. In AD 837, 1P/Halley passed 0.034 AU (510,000 km/320,000 mi) from Earth, with the COMA shining as brightly as Venus and its tail spanning 90° of sky. At the 1910 return, closest approach was 0.15 AU, with the comet reaching mag. 0; Earth actually passed through the comet's tail on May 20 of that year. The most recent return, in 1985–86, however, was relatively unfavourable, with 1P/Halley never closer than 0.42 AU (April 11) and being fairly close to the Sun in the sky when at perihelion. In the early months of 1986, the comet reached mag. +3.0 and showed a 10° tail.

Although unfavourable in many respects for Earth-bound observers, the 1985–86 return was extremely rewarding scientifically. Observations across a range of astronomical disciplines were co-ordinated by an International Halley Watch, with the results being archived centrally at the Jet Propulsion Laboratory, Pasadena, USA. Early in 1986 March a flotilla of spacecraft passed close to the comet, sending back an immense quantity of data on its structure and composition. The nucleus was imaged by the European Giotto spacecraft and found to be an irregular body with an extremely dark (albedo 0.04) outer crust and dimensions of roughly 15 × 8 km (9 × 5 mi). An estimated 3.1 tonnes of dust was being ejected from the nucleus each second at around the time of the Giotto close approach on 1986 March 13; even at this rate of loss, 1P/Halley may have an active lifetime of more than 100,000 years.

Large ground-based telescopes were able to follow the comet on its way out from perihelion until it was at the distance of Uranus' orbit in 1994. Surprisingly, 1P/Halley showed an outburst of activity in February 1991, when 14.3 AU from the Sun (beyond Saturn's orbit). The cause of this brightening is uncertain, but it may have been connected with shockwaves from a CORONAL MASS EJECTION propagated through the SOLAR WIND.

The comet's next return, in 2061, will be even less favourable than that of 1985–86, with 1P/Halley at perihelion while close to conjunction on the far side of the Sun from Earth.

Comet 1P/Halley is the parent of the METEOR STREAM that produces the ETA AQUARIDS and ORIONIDS.

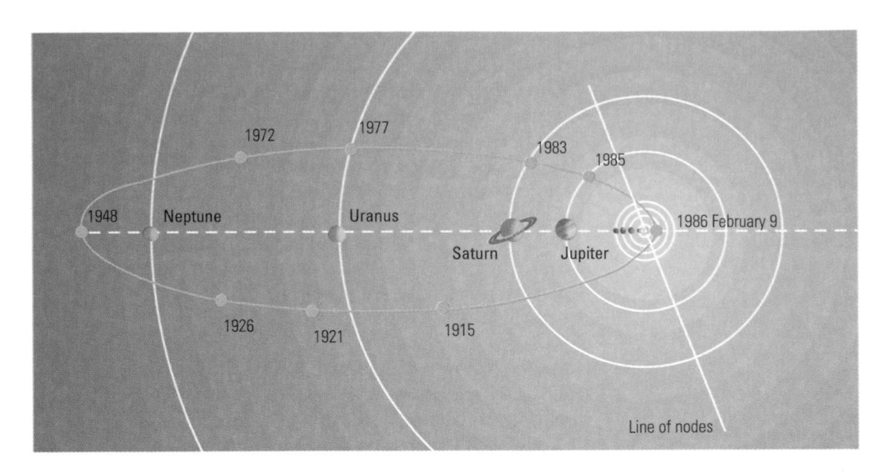

▲ **Halley, Comet 1P/** The elongated elliptical orbit of Halley's Comet takes it from aphelion beyond the orbit of Neptune to perihelion between the orbits of Venus and Mercury. The comet's orbit is retrograde, and 1P/Halley was last at perihelion on 1986 February 9.

Halley, Edmond (1656–1742) English scientist, the second ASTRONOMER ROYAL, famous for the long-period comet named after him, his observation and cataloguing of southern stars, his part in the publication of Newton's *Principia*, and his discovery of stellar proper motions. Outside astronomy, his accomplishments were numerous. He founded geophysics, charting variations in the Earth's magnetic field and establishing the magnetic character of the aurora borealis. He showed that atmospheric pressure decreases with altitude, and studied monsoons and trade winds. Halley was one of the first to use mortality statistics to cost life assurance policies.

He matriculated at Oxford University in 1673, where he made his earliest published observations, including one of a lunar occultation of Mars in 1676. At that time he also began assisting the first Astronomer Royal, John FLAMSTEED, with observations of star positions at the Greenwich Observatory. Halley never completed his formal studies at Oxford, opting in 1676 November to voyage to the South Atlantic island of St Helena to map the stars of the southern hemisphere. Despite poor weather, he made observations from which he compiled a catalogue of 341 southern stars, observed a transit of Mercury, and carried out comprehensive meteorological and hydrographic studies. Also, he confirmed that the period of a pendulum located closer to the equator was longer than for a similar device back in England, indicating that the Earth is oblate. When Halley returned, King Charles II bestowed upon him a degree from Oxford. Halley then published his southern star catalogue, *Catalogus stellarum australium*, and devised a method for determining the solar parallax from his observations of the transit of Mercury.

In the 1680s, as an outgrowth of his interest in the INVERSE-SQUARE LAW of gravitation – which he managed to derive from Kepler's third law – and its implication that the planets must move in elliptical orbits, Halley befriended Isaac NEWTON. He encouraged and paid for the publication of Newton's *PRINCIPIA*, and even corrected the book's proofs. From 1686 to 1693 Halley edited the famous *Philosophical Transactions* of the Royal Society. In the 1690s, he began a mathematical investigation of the orbits of comets, paying close attention to the bright comets of 1531, 1607 and 1682. He concluded that these three comets, as well as similar objects appearing in 1305, 1380 and 1456, were just different apparitions of the same comet, which had a period of 76 years. Taking into account the perturbations of Jupiter, he calculated an orbit for the comet indicating it would return in 1758. Although he did not live to see his prediction come true, he achieved immortality when it appeared on Christmas Day of that year. Comet 1P/HALLEY has since become the best-studied of all comets.

After further scientific voyages (1698–1701), Halley was appointed Savilian Professor of Geometry at Oxford in 1704, where he continued his astronomical work. In 1710, after noting that the observed positions of some stars differed significantly from their coordinates given in Ptolemy's catalogue, he confirmed PROPER MOTION in three bright stars – Sirius, Arcturus and Procyon. In 1720 he published a paper in which he discussed what is now called OLBERS' PARADOX, arguing for the infinity of space. Also in that year, Halley succeeded Flamsteed as Astronomer Royal, a post he held for 21 years. During his tenure he modernized the Greenwich Observatory, installing the first transit instrument and devising a method for determining longitude at sea by using lunar observations.

halo Ring of light seen round a celestial body. A halo is an atmospheric phenomenon produced as a result of refraction by ice crystals in thin cirrus cloud at altitudes of 10–15 km (6–9 mi), near the top of the TROPOSPHERE. The commonest form of halo is a 46° diameter ring of light with the Sun or Moon at its centre. Lunar haloes are usually only seen when the Moon is within five days of full; at other times its light is not sufficiently bright. Solar haloes often show some colour, with a blue outer and red inner edge; the sky inside the halo is noticeably darker than that outside.

halo (galactic halo) Extensive distribution of stars in the outer regions of a GALAXY. The term also refers to the invisible, extended distribution of DARK MATTER extending beyond the visible part of a galaxy.

halo population stars Oldest, most metal-poor stars, found in the halo of the Galaxy and in GLOBULAR CLUSTERS. *See also* POPULATIONS, STELLAR

Hα *See* HYDROGEN-ALPHA LINE

Hamal The star α Arietis, visual mag. 2.01, distance 66 l.y., spectral type K2 III. Its name comes from an Arabic word meaning 'the lamb'.

Hamburg Observatory Former national observatory, dating from the beginning of the 19th century. It moved to its present location at Bergedorf, 20 km (12 mi) east of Hamburg, in 1909. The observatory operates several small telescopes, including a 1-m (40-in.) reflector, and has been part of the University of Hamburg since 1968. Today, the observatory's research centres around theoretical astrophysics, together with studies of metal-poor stars, the interstellar medium, quasars and gravitational lenses.

Harriot, Thomas (1560–1621) English mathematician, astronomer and surveyor. He came to move in the circle of John Dee (1527–1608), and under the patronage of Sir Walter Raleigh, whom he had tutored, spent 1585 as a member of a survey expedition to Virginia. He and his Welsh friend William Lower (*c.*1570–1650) began to use the telescope for astronomy some time before Galileo. By 1609 he had studied Jupiter's satellites, observed sunspots (deriving the Sun's rotation period) and mapped the Moon, though he never published his results.

Harrison, John (1693–1776) English instrument-maker who invented the MARINE CHRONOMETER to solve the problem of finding longitude at sea. When in 1714 the British Government offered a £20,000 prize for the solution of the problem, Harrison devoted his life to the perfection of a sea-going 'chronometer'. Between 1735 and 1761, he produced four chronometers, each of which was an improvement on its predecessor, and in due course received the prize money. Contrary to popular mythology, Harrison was a highly respected scientist, not an outsider at odds with the British scientific establishment.

Hartebeesthoek Radio Astronomy Observatory (HartRAO) South African national research facility in the Magaliesburg Hills 65 km (40 mi) north-west of Johannesburg. Built as a NASA tracking station in 1961 and converted for radio astronomy in 1975, the observatory became a national facility in 1988. It has a 26-m (85-ft) fully steerable dish – the only major radio telescope in Africa.

Hartmann, Johann Franz (1865–1936) German astronomer who made many important breakthroughs in astrophysics while working at the observatories at Potsdam (1896–1909), Göttingen (1909–21) and La Plata (1921–34), often with instruments he designed and built himself. One such device was a microphotometer, which used photoelectric cells to measure the density of light passing through photographic plates of starfields; this intensity was then compared to images of stars to be photometrically measured. He also devised a method for testing lenses that is named after him. Hartmann discovered the presence of interstellar matter by studying the spectrum of the spectroscopic binary δ Orionis in 1904.

Harvard classification Classification system for stellar spectra developed at Harvard College Observatory by Edward PICKERING, Williamina FLEMING, Antonia MAURY and Annie CANNON between 1890 and 1901. Stars were lettered from A to Q primarily on the basis of the strength of the HYDROGEN SPECTRUM. Several letters were then dropped as unnecessary, while B was placed in front of A, and O in front of B, such that other absorption lines (principally helium) formed a continuous sequence. The result was the spectral sequence OBAFGKM, which correlates with both colour and temperature. *See also* SPECTRAL CLASSIFICATION

Harvard College Observatory (HCO) One of the first observatories established in the United States, founded in 1839; W.C. BOND was its first director. Located at Cambridge, Massachusetts, the observatory is the home of a 15-inch (38-cm) refractor dating from 1847 that was for twenty years the largest telescope in the United States. It established itself as a major astronomy research centre under its fourth director, Edward C. PICKERING. Today, the HCO carries out a broad programme of research in collaboration with the Smithsonian Astrophysical Observatory (*see* HARVARD–SMITHSONIAN CENTER FOR ASTROPHYSICS).

Harvard–Smithsonian Center for Astrophysics (CfA) Major US institution that combines the resources and research facilities of the HARVARD COLLEGE

OBSERVATORY and the SMITHSONIAN ASTROPHYSICAL OBSERVATORY (SAO) to study the basic physical processes that determine the nature and evolution of the Universe. The symbiosis between these institutions began when the SAO moved its headquarters to Cambridge, Massachusetts, in 1955, and was formalized by the establishment of the Center in 1973. The CfA has pioneered many new areas of astronomical instrumentation, and participated in the landmark CfA Redshift Survey of the distances to some 20,000 galaxies.

harvest moon Full moon nearest to the time of the autumnal equinox in the northern hemisphere (around September 23). At this time of year, the inclination of the Moon's path to the horizon is very low, causing it to rise no more than 15 minutes later each evening, as opposed to the more usual half-hour or more, providing a succession of moonlit evenings. The name derives from the fact that this was of great benefit to farm workers bringing in the harvest. In the southern hemisphere, the harvest moon is the full moon nearest to the vernal equinox, around March 21. *See also* HUNTER'S MOON

Hat Creek Observatory Radio observatory near Redding, California, some 400 km (250 mi) north of Berkeley. Its principal instrument is the BIMA (BERKELEY ILLINOIS MARYLAND ASSOCIATION) Millimeter Array, an aperture synthesis telescope that operates at wavelengths of 3 mm and 1 mm. It has ten antennae 6.1 m (20 ft) in diameter that can be located at various stations along a T-shaped track.

Haute Province, Observatoire de (OHP) French national facility, established in 1937 on a 645 m (2100-ft) high plateau at Saint-Michel, some 100 km (60 mi) north of Marseilles. The OHP's telescopes include a 1.93-m (76-in.) reflector, operating since 1958, and a 1.52-m (60-in.) reflector built in 1967. Both these instruments are used for spectroscopy, while two smaller reflectors and a Schmidt camera are used for imaging. A search for extrasolar planets is being undertaken with the 1.93-m telescope.

Hawking, Stephen William (1942–) English theoretical physicist who has applied general relativity and quantum theory to cosmology, greatly advancing the understanding of BLACK HOLES and SPACETIME. While still a graduate student at Cambridge University, where he has spent his entire career, Hawking was diagnosed in 1963 with a degenerative neuromuscular disease and told he had only 18 months to live; against these odds, he has survived to become the foremost researcher of theoretical relativistic cosmology.

In the late 1960s, Hawking demonstrated that a singularity must have characterized the beginning of the Universe in the BIG BANG. In the early 1970s, he developed mathematical proofs of the 'no hair theorem' originated by physicist John Archibald Wheeler (1911–), which states that a black hole's gravitational field is defined by three quantities – its mass, angular momentum and electric charge.

In 1971 Hawking showed that the Big Bang could have given rise to numerous 'mini black holes', exotic objects the size of a proton but with a mass of about a billion tonnes. In 1974 he discovered that the physical temperature of a black hole is not necessarily absolute zero. Hawking's model of black holes showed that they can emit a kind of thermal radiation by the simultaneous creation of pairs of matter and anti-matter particles. This process, called HAWKING RADIATION, implies that the black hole shrinks and may eventually evaporate, since it is gradually converting its own mass into thermal energy.

Cambridge University appointed Hawking Lucasian Professor of Mathematics in 1979. In 1983 he put forth his 'no-boundary proposal': that although both space and time are infinite, these quantities have no physical boundaries. Hawking has popularized many of his theories through a series of magazine articles, books and television programmes; his book *A Brief History of Time* (1988) spent 4 years on the bestseller lists.

Hawking radiation Radiation that is emitted from a BLACK HOLE; the process was proposed by Stephen HAWKING. From quantum field theory (the theory developed by Max PLANK in 1900, which proposes that various properties of a system, including energy, can only change by discrete amounts), it can be deduced that vacuum fluctuations occur throughout space whereby pairs of virtual particles are created. If such a pair is created near the EVENT HORIZON of a black hole, one particle may lie just on the inside, one on the outside, and the outside particle would be free to radiate away. A black hole of one solar mass will evaporate over 10^{67} years. A black hole at the centre of a galaxy will take around 10^{97} years to evaporate. Primordial black holes created in the Big Bang will have already disappeared.

Hayashi track Path on a HERTZSPRUNG–RUSSELL DIAGRAM that a pre-MAIN-SEQUENCE star follows as it evolves on to the main sequence. It is named after the Japanese astrophysicist Chushiro Hayashi (1920–), who studied stellar evolution during the 1950s and 1960s.

PROTOSTARS lie to the right of the main sequence. An initial rapid collapse of the protostar, which is poorly understood, moves the star leftwards out of the so-called Hayashi forbidden zone. This collapse is followed by a large decrease in LUMINOSITY, caused by the contracting radius. This line on the diagram is called the Hayashi track. During this phase the star is almost entirely convective, with only a small radiative core. Finally, the star becomes radiative and approaches the main sequence along the HENYEY TRACK.

Haystack Observatory Multidisciplinary research facility of the Massachusetts Institute of Technology (MIT) at Westford, Massachusetts, engaged in radio astronomy, geodesy, atmospheric science, radio interferometry and radar imaging, established in 1964 originally as a military facility. The principal instrument is a 37-m (120-ft) fully steerable antenna, enclosed in the world's largest space-frame radome. The telescope can be used either for single-dish observing or as part of a VERY LONG BASELINE INTERFEROMETRY network.

▼ halo The most prominent components of spiral galaxies like the Milky Way are the stars, gas and dust populating their nuclear bulge and disk. Many are also surrounded by tenuous halos of hot plasma. This X-ray image of a Milky Way-type galaxy reveals how vast such a halo can be.

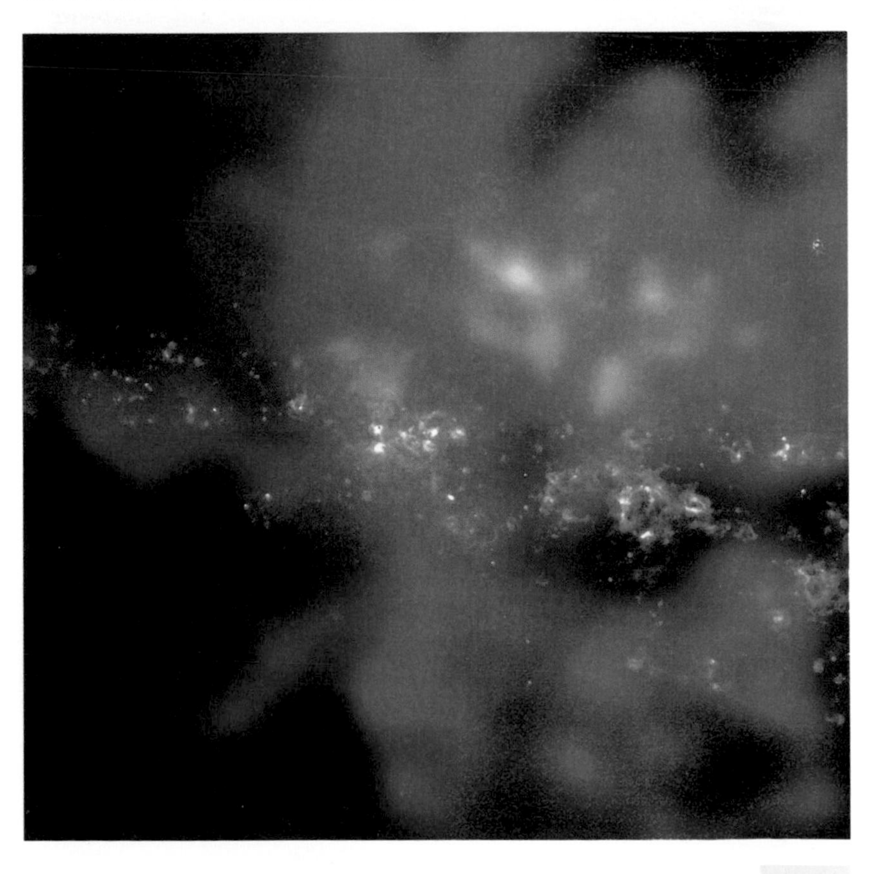

HD Abbreviation of *HENRY DRAPER CATALOGUE*

HDF Abbreviation of HUBBLE DEEP FIELD

heat death of the Universe Eventual fate of the Universe, perceived in the 19th century to be the logical outcome of applying the second law of thermodynamics to the whole Universe. This law states that in a closed system, ENTROPY can only increase or remain constant – or that heat is transferred only from warmer masses to cooler ones. The Universe 'dies' once when all matter has reached the same temperature, for then the Universe's entropy will have reached a maximum, and there will be no energy available. This 'running down' was pointed out by Lord KELVIN, who preferred to avoid such a fate by invoking an omnipotent force that created living beings. In terms of modern cosmology, the heat death would not occur in a closed finite universe in which the DENSITY PARAMETER is greater than 1, but it would occur in a universe that expanded for ever.

Heaviside layer *See* KENNELLY–HEAVISIDE LAYER

heavy elements All the chemical elements apart from hydrogen and helium. Within astronomy, the term 'metals' is sometimes used as a synonym for heavy elements. The concept of the heavy elements is valuable because in terms of relative abundance (*see* ASTROCHEMISTRY) hydrogen and helium account for around 98% of the mass of the observable Universe, while the remaining heavier elements amount to just 2%. Hence it is often useful to consider them all under one heading.

Hebe Large MAIN-BELT ASTEROID discovered in 1847; number 6. Hebe is about 186 km (116 mi) in diameter.

HED Abbreviation for HOWARDITE–EUCRITE–DIOGENITE ASSOCIATION

Heinrich Hertz Telescope Joint facility of STEWARD OBSERVATORY and the MAX-PLANCK-INSTITUT FÜR RADIOASTRONOMIE run by the Submillimeter Telescope Observatory. The enclosed 10-m (33-ft) telescope is located on Mount Graham, Arizona, and operates in the 0.3–2-mm wavelength range. It is named after the German physicist Heinrich Rudolf Hertz (1857–94), pioneer investigator of electromagnetic waves.

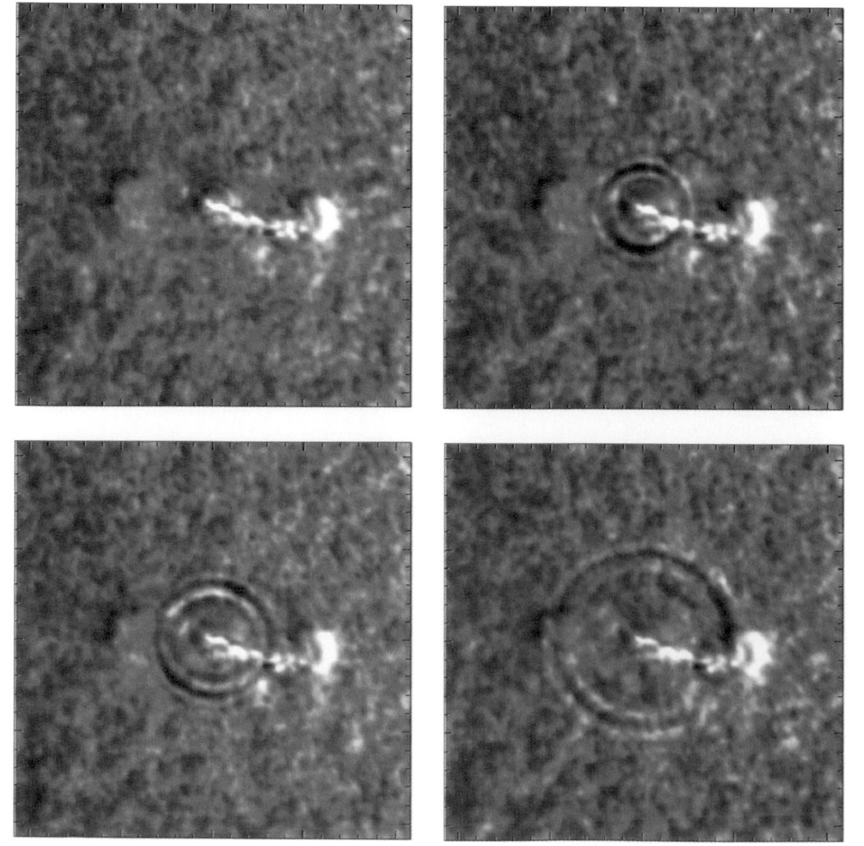

▼ **helioseismology** This series of SOHO images shows shockwaves propagating away from a flare associated with an active region on the Sun.

Hektor Third TROJAN ASTEROID to be discovered, in 1907; number 624. It is the largest of all Trojans, at approximately 225 km (140 mi) in size. Variations in Hektor's brightness indicate that it has an elongated shape, perhaps like a dumbbell.

Helene Small satellite of SATURN; it is co-orbital with DIONE. It was discovered by Pierre Laques (1934–) and Jean Lecacheux (1944–) in 1980. It is spheroidal in shape, measuring about 32 km (20 mi). Helene has a near-circular, near-equatorial orbit, remaining close to the L_4 LAGRANGIAN POINT associated with Dione's orbit around Saturn. It takes 2.74 days to make a circuit of Saturn at a distance of 377,400 km (234,500 mi).

heliacal rising Strictly, the rising of a celestial body simultaneously with the Sun. More commonly, it refers to the date when the body first becomes visible in the dawn sky, rising just before the Sun. The heliacal rising of SIRIUS, the brightest star in the sky, was used by the ancient Egyptians to predict the annual flooding of the River Nile, and to denote the beginning of the new year. The Nile flood was an important event in the Egyptian calendar, heralding the first of three seasons; after the river subsided, planting of crops could commence, followed by the harvest.

heliocentric parallax Alternative name for ANNUAL PARALLAX. *See also* DIURNAL PARALLAX; TRIGONOMETRIC PARALLAX

heliocentric theory Model of the Solar System in which the Sun is at the centre and the planets revolve around it. A heliocentric theory was proposed by the Greek astronomer ARISTARCHUS in the 3rd century BC, but it seemed counter-intuitive at the time and was not widely adopted. GEOCENTRIC THEORY, championed by Aristotle and Ptolemy and by later scholars and theologians, was to remain supreme for nearly 1500 years.

It was Nicholas Copernicus who re-invigorated heliocentric theory with his COPERNICAN SYSTEM, published the year of his death. Opposed by both Roman Catholic and Protestant ecclesiastics, its influence spread only very slowly. However, the great improvements in observational accuracy introduced by Tycho BRAHE showed up the inadequacy of geocentric theory. Brahe himself favoured a hybrid, now referred to as the TYCHONIC SYSTEM, in which the Earth remained at rest and was orbited by the Sun, but the other planets orbited the Sun. His observations showed that the stars were much more distant than the planets and he also realized that comets moved in orbits that would have taken them through the 'crystalline spheres' of Aristotle's geocentric theory. Thus the spheres could not be real.

The telescopic discoveries by GALILEO – that a body other than the Earth, namely Jupiter, had moons in orbit around it, that the planet Venus exhibited phases like the Earth's Moon and that the Sun was covered in spots which made it less than 'perfect' – was evidence enough to put an end to the geocentric theory. Galileo's open advocacy of the Copernican position placed him in conflict with elements within the Catholic Church. His main publication in support of heliocentrism, the *DIALOGUES* of 1632, led to his trial before the Inquisition. When Johannes Kepler, using data accumulated by Brahe, discovered his three laws (*see* KEPLER'S LAWS), the need for the complicated epicyclic motions was swept aside by the simplicity of elliptical orbits.

In the decades following Kepler's and Galileo's discoveries, heliocentric theory gained wide acceptance. Isaac Newton, with the publication of his *PRINCIPIA* in 1686–87, showed that the assumption of an INVERSE-SQUARE LAW of gravitational attraction would account for Kepler's laws. Following his work, the heliocentric theory of the Solar System could no longer be questioned.

heliometer *See* DIVIDED TELESCOPE

Helios Either of two probes launched in 1974 and 1976 to study the Sun and interplanetary space. They both approached the Sun to within 43 million km (26.7 million miles).

helioseismology Study of the structure and dynamics of the SOLAR INTERIOR by the analysis of sound waves that propagate through the Sun and manifest themselves as oscillations of the PHOTOSPHERE. Helioseismology is a hybrid name combining the Greek words *helios* (for Sun), *seismos* (meaning quake or tremor) and *logos* (for reasoning or discourse). The periods of the vertical oscillations are usually around five minutes, but they can range to a few hours long. They are most often detected by the DOPPLER EFFECT of a solar ABSORPTION LINE formed in the photosphere, but they can also be detected by light variations of the photosphere. There are millions of distinct, resonating, sound waves whose periods depend on their propagation speeds and the depths of their resonant cavities. Examination of the photosphere oscillations that the sound waves produce enables helioseismologists to infer the temperature, chemical composition and motions at different depths within the Sun.

heliosphere Vast region of space surrounding the Sun where the interplanetary magnetic field and SOLAR WIND have a dominant influence on the movements of electrons and ions. The heliosphere, immersed in the local interstellar medium, defines the extent of the Sun's influence. It extends well beyond the planets to the heliopause, which is located at a distance from the Sun of about 100 AU. This outer boundary marks the place where the solar wind pressure balances that in interstellar space.

heliostat Flat, equatorially mounted mirror that is driven to follow the Sun's apparent motion across the sky and to direct its light into a fixed solar telescope. Because of the long focal length of a SOLAR TELESCOPE, and the fact that the heat of the Sun creates a layer of warm, turbulent air close to the ground, the heliostat is usually mounted on the top of a tall tower. As it follows the Sun across the sky, the image it forms slowly rotates during the course of the day and, because of this, the more sophisticated COELOSTAT is sometimes preferred. *See also* SIDEROSTAT

helium (symbol He) Second lightest chemical element; it is the second most abundant element in the Universe (*see* ASTROCHEMISTRY). Helium-4 forms almost 100% of the naturally occurring element and has an atomic number of 2 and an atomic mass of 4.0026 amu. Helium is the lightest of the 'noble' gases – those elements that are virtually inert chemically. Its boiling point is 4.2 K. There are five known isotopes, three of which (helium-3, helium-4 and helium-5) are stable. The nucleus of helium-4 is also called the α-particle (alpha particle).

Helium is thought to be generated by NUCLEOSYNTHESIS in stars. The basic reaction, which powers MAIN-SEQUENCE stars, involves the FUSION of hydrogen to form helium via the PROTON–PROTON REACTION and the CARBON–NITROGEN–OXYGEN CYCLE. However, the present abundance of helium in the Universe cannot be explained solely by nucleosynthesis in stars. It is now generally accepted that most of the helium observed today was synthesized during the first few minutes of the BIG BANG. The sequence of nucleosynthetic reactions continues after the formation of helium with the TRIPLE-α PROCESS, in which three helium-4 nuclei collide almost simultaneously to form carbon-12. The reaction requires a temperature of around 100 million K for its initiation.

Helium was first identified in the solar spectrum by Jules JANSSEN and Joseph LOCKYER in 1868 as a set of lines that did not correspond with those for any element then known. It was named after *Helios*, the Greek for 'Sun'. It was found on Earth in 1895 as a gas released by radioactivity from the mineral clevite.

Helium is used to pressurize the propellant tanks of some liquid-fuelled rockets and, in liquid form, to cool many infrared and radio detectors.

▲ **Helix Nebula** One of the closest planetary nebulae, NGC 7293 is visually faint because its light is spread over a fairly large area of sky.

helium flash Theoretically predicted event in the evolution of a lower-mass star (around 1 to 2 solar masses) whereby helium fusion occurs explosively, with changes in the central regions of the star occurring over timescales of minutes.

After the hydrogen in the core of a MAIN-SEQUENCE star has been exhausted, the star becomes a RED GIANT, with hydrogen SHELL BURNING occurring. The core collapses until temperatures of 10^8 K are reached, when helium burning can start. In stars of low mass, this event occurs explosively and is known as the helium flash. In stars of higher mass, helium fusion will commence gradually because the temperature in the core can reach this point before the core becomes degenerate. *See also* DEGENERATE MATTER

helium star Non-white dwarf star, the outer layers of which contain more helium than hydrogen. A helium star is an EVOLVED STAR that has lost its hydrogen-rich envelope, possibly by the effect of a companion in a CLOSE BINARY system, thereby exposing its helium core. The R CORONAE BOREALIS stars are examples of helium stars. The term 'helium star' was also originally used for B-type stars.

Helix Nebula (NGC 7293) PLANETARY NEBULA in southern Aquarius (RA 22^h $29^m.6$ dec. $-20°48'$). Lying less than 300 l.y. away, the Helix is one of the closest planetary nebulae and, as a result, subtends a large angular diameter of 13′. Ostensibly bright, at mag. +7.3, the nebula is spread over such a wide area that it has low surface brightness and is consequently rather difficult to observe visually. The central star has mag. +13.6.

Hellas Planitia Largest impact basin on MARS (43°.0S 290°.0W); it has an average diameter of 1800 km (1100 mi). Hellas Planitia appears on early maps of Mars as a roughly circular region of ochre hue. It is situated south of the Martian equator, in a hemisphere predominantly well above datum (an arbitrary height at which the pressure is 6.2 millibars). Hellas Planitia itself lies well below this level, and at its deepest point, near to the western rim, it descends to 5 km (3 mi).

Hellas' rugged hilly rim is between 50 and 400 km (30–250 mi) in width and is associated with a concentric pattern of fracturing that extends as far as 1600 km (1000 mi) from the centre of the floor. The floor of the basin is occupied by plains materials, which are believed to be volcanic in origin, and is mantled by windblown debris. Large sections of the rim are either missing or difficult to trace due either to erosion or to later burial by younger deposits. The most continuous section lies on the western side, where it passes into the blocky mountains of Hellespontus Montes. Spacecraft imagery reveals there to be several annular, inward-facing scarps outside of the main rim, together with a small number of large ancient volcanic structures, such as Amphitrites Patera and Hadriaca Patera. The floor of Hellas Planitia is occupied by rather complex plains units, on the surface of which are small impact craters and some short channel systems.

H

Helwân Observatory Egyptian observatory at Helwân, about 30 km (20 mi) south of Cairo. Dating from the turn of the 20th century, the observatory was equipped with a 0.76-m (30-in.) reflector in 1905. A new observing station was set up at Kottamia, 70 km (45 mi) north-east of Helwân in the 1950s, and a 1.9-m (74-in.) telescope began work there in 1964; it was upgraded in the late 1990s.

Henderson, Thomas (1798–1844) Scottish astronomer, second Astronomer at the Cape (1831–33) and first Astronomer Royal for Scotland (1834–44). His observations from South Africa in 1832 provided the first parallax for α Centauri, though he did not announce his result (0″.93, corresponding to a distance of 3.5 l.y.) until 1839, the year after Friedrich Wilhelm BESSEL's measurement of the parallax of Sixty-one Cygni. Henderson made over 60,000 observations of star positions, leading to the compilation of several important star catalogues.

Henry, Paul Pierre (1848–1905) and **Henry, Prosper Mathieu** (1849–1903) French brothers whose achievements in ASTROPHOTOGRAPHY helped to stimulate the *CARTE DU CIEL* sky survey. The Henrys discovered 14 asteroids, beginning with (175) Liberatrix in 1872. Their searches required the preparation by hand of detailed star charts for the ecliptic, so that asteroids would not be confused with stars – a labour-intensive and error-prone procedure. Their successful photography of star clusters, including the Double Cluster in Perseus, a ¾-hour exposure that recorded stars as faint as 12th magnitude, convinced them that good charts could be prepared photographically.

The Henrys designed and built a 13.5-inch (340-mm) refractor for Paris Observatory specifically for astrophotography, completed in 1885. At the Astrographic Congress on 1887 April 18, observatories round the world agreed to undertake an all-sky photographic survey, the *Carte du Ciel*, using 'astrographs' – telescopes identical to the 13.5-inch at Paris – about half of which were made by the Henrys.

► **Hellas Planitia** The largest impact structure on Mars, Hellas Planitia is one of the most obvious features in the planet's southern hemisphere. The roughly circular basin has a diameter of 1800 km (1100 mi).

Henry Draper Catalogue (HD) Catalogue of stellar spectra compiled by Annie CANNON at Harvard College Observatory. It was named after American pioneer of astrophotography Henry DRAPER, whose widow supported the work financially. The catalogue, completed in 1924, classified about 225,000 stars to 10th magnitude according to the Harvard system: O, B, A, F, G, K, M, in order of decreasing surface temperature. The *Henry Draper Extension* (HDE), with spectral classifications of 47,000 11-th magnitude stars, appeared in 1936. Stars are still widely known by their HD or HDE numbers. *See also* SPECTRAL CLASSIFICATION

Henyey track Evolutionary path on a HERTZSPRUNG–RUSSELL DIAGRAM that a pre-MAIN-SEQUENCE star follows from the base of the HAYASHI TRACK to the main sequence. It is named after the American astrophysicist Louis Henyey (1910–70), who studied stellar evolution during the 1950s.

Along the Henyey track both EFFECTIVE TEMPERATURE and LUMINOSITY increase such that luminosity, L, is proportional to effective temperature, T_e (L is proportional to $T_e^{4/5}$). During this phase, energy transfer within the star is largely radiative rather than convective, in contrast to the HAYASHI TRACK. The length of the Henyey track (the increase in T_e) is greater for high-mass stars than low-mass stars, although the time spent on this track is shorter for the high-mass stars. Stars of mass less than 0.5 solar mass do not have a Henyey track, instead they evolve directly on to the main sequence from the Hayashi track. Before reaching the main sequence, the star's luminosity dips slightly due to the expansion of the star caused by the onset of nuclear burning.

Hephaistos APOLLO ASTEROID discovered in 1978; number 2212. It is one of the largest Apollos, with a size of about 8 km (5 mi). *See* table at NEAR-EARTH ASTEROID

Heraclides, Promontorium Southern termination of the Moon's Sinus IRIDUM; it rises to 1220 m (9000 ft) and appears as a cape-like structure jutting out into Mare IMBRIUM. With the northern termination, Promontorium Laplace, it forms the outer boundaries of a much larger impact structure. This structure was inundated with lava, covering the eastern rim. Multiple mare (wrinkle) ridges lie between the two promontories.

Herbig–Haro object (HH object) Any of a class of small, faintly luminous nebulae discovered, independently, by George Herbig (1920–) and Guillermo Haro (1900–1990) in the 1950s. HH objects are irregular in outline and contain bright knots; they are found in regions rich in interstellar material. Their spectra reveal a weak continuum dominated by emission lines from hydrogen, oxygen, nitrogen and iron. HH objects are produced when high-velocity material, ejected from young stars, interacts with surrounding material.

The stars that produce HH objects are thought to be young, pre-main-sequence stars that are accreting material via an ACCRETION DISK. The whole system is surrounded by the material from which the star is forming. High-velocity ejections are then produced, although the mechanism by which this occurs is not yet fully understood, and they interact with the surrounding material producing shock waves. The shock waves heat and ionize the surrounding gas, producing the emission lines observed in HH objects.

HH objects are roughly split into low and high-excitation objects, according to their spectra. The difference is thought to be caused by the velocities of the shock waves. High-excitation HH objects, with strong emission lines of highly ionized material, suggest shock wave velocities of 200 km/s (120 mi/s), while low-excitation HH objects

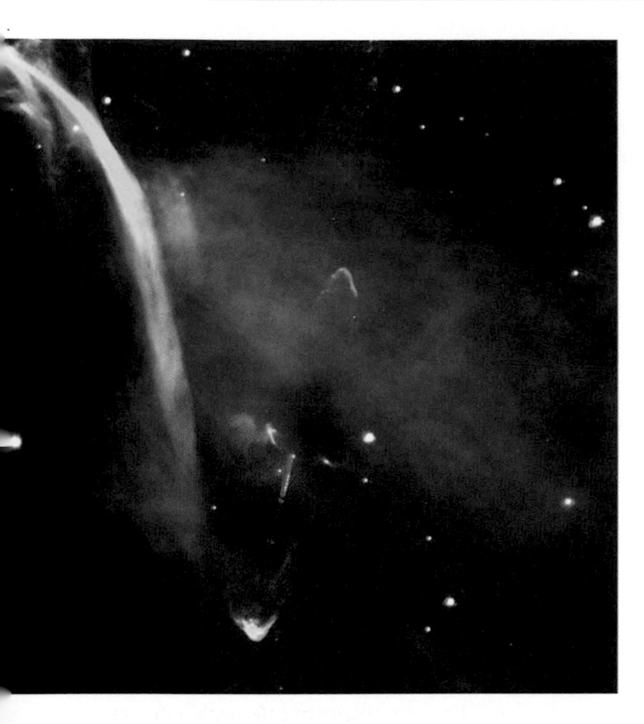

◀ **Herbig–Haro object** High-speed jets from the young object HH 34, near right centre, impact on the surrounding interstellar material to produce V-shaped shockwaves. HH 34 lies in Orion, in a region of ongoing star formation.

have shock wave velocities of only around 20 km/s (12 mi/s). Many HH objects show bright emission knots located in two diametrically opposing lobes, moving in opposite directions away from the source. This is consistent with a high-velocity jet of material being ejected as a bipolar outflow. *See also* T TAURI STAR

Hercules *See* feature article

Hercules X-1 Strong X-ray source and an ECLIPSING BINARY system; it comprises a visible star, HZ Herculis, and an accreting neutron star that is an X-ray pulsar. The pulsar has a period of 1.24 seconds, emitting in X-rays and visible light. The orbital period of the binary is 1.7 days, and there is also a 35-day modulation of the X-ray emission possibly caused by precession of the accretion disk. Since Hercules X-1 is an eclipsing binary, the stellar masses can be determined – 0.98 solar mass for the neutron star and 1.99 solar mass for the companion.

Hermes Third APOLLO ASTEROID to be discovered, in 1937, when it was passing within about 800,000 km (500,000 mi) of the Earth. This close approach led to the popular assumption of a name despite the orbit not being sufficiently well determined to allow Hermes to be added to the list of numbered asteroids. Hermes was observed for only a few days and then lost. It was catalogued as 1937 UB.

Herodotus Lunar crater (23°N 50°W), 37 km (23 mi) in diameter, with rim components reaching 1220 m (4000 ft) above its floor. Herodotus' floor has been resurfaced with lava, rendering it nearly featureless. Its northern wall appears to be degraded, with a gap pointing towards Vallis SCHROTERI. The Aristarchus Plateau, surrounding Herodotus, has many dark patchy areas, representing volcanic fire fountaining.

Herschel family Dynasty of English astronomers, originally from Hanover, Germany. Its founder was Friedrich Wilhelm Herschel (1738–1822), who settled in England in 1757 and assumed the name William HERSCHEL when he became a naturalized Englishman. In 1772 he was joined by his youngest sister, Karoline Lucretia Herschel (1750–1848), who became his astronomical assistant (*see* Caroline HERSCHEL). William was also aided by his son, John Frederick William HERSCHEL (1792–1871), who went on to become one of the foremost English scientists of his day. John's son **Alexander Stewart Herschel** (1836–1907), born during his father's sojourn at the Cape of Good Hope, determined a number of meteor radiants and their association with cometary orbits. The male line of the British Herschels is now extinct.

Herschel, Caroline Lucretia (1750–1848) German astronomer, sister of and assistant to William HERSCHEL, and aunt of John HERSCHEL. Karoline Herschel (the German form of her name) was born in Hanover and worked as a family drudge until 1772, when William brought her to England, where she became first his domestic companion, then his musical protégée, and finally his astronomical assistant. When they moved to Datchet and Slough she started sweeping the sky herself, discovering several nebulae (including the Sculptor Galaxy, NGC 253, in 1783) and eight comets. When William died in 1822, she returned to Hanover, where she received many honours.

Herschel, John Frederick William (1792–1871) English scientist and astronomer, the only child of William HERSCHEL's marriage to Mary Pitt in 1788. After studying mathematics, and an abandoned attempt at training for the legal profession, he became his father's protégé and successor in astronomy, discovering several hundred new nebulae and clusters from England. John Herschel's main goal was to extend his father's sky surveys, undertaken for the purpose of fathoming the 'construction of the heavens', what we would call cosmology. In 1834 he took one of William's telescopes to the Cape of Good Hope (modern South Africa) to survey the southern skies, and found 1200 new double stars and 1700 new nebulae and clusters.

By 1840 Herschel's observing career was effectively over. The rest of his life was devoted to interpreting his, and his father's, cosmological work, and acting as an international consultant for all things astronomical and physical. He compiled the *General Catalogue of Nebulae and Clusters* from his and his father's observations; this would form the basis for J.L.E. Dreyer's *NEW GENERAL CATALOGUE*. Herschel's *Outlines of Astronomy* (1849) remained a standard textbook for many decades. He was also a pioneer of early photographic chemistry and astrophotography.

Herschel, (Frederick) William (1738–1822) German-born English astronomer and musician. He took up astronomy in the 1770s, making his own telescopes and mirrors, and won fame in 1781 for his discovery of the planet Uranus. He discovered two satellites of Uranus

HERCULES (GEN. HERCULIS, ABBR. HER)

Fifth-largest constellation, lying in the northern sky between Lyra and Corona Borealis, but with no particularly bright stars. In Greek mythology, Hercules was the strongman who undertook twelve labours. In the sky he is depicted on one knee, with his left foot on the head of Draco, the dragon. The constellation's most recognizable feature is a quadrilateral known as the KEYSTONE, formed by η, ζ, ε and π Her. On one side of the Keystone lies the finest globular cluster in northern skies, M13, appearing as a hazy star to the naked eye under clear skies and easily found in binoculars. It is estimated to contain 300,000 stars and lies 25,000 l.y. away. Another globular, smaller and fainter, is M92. α Her is a variable (2.7–4.0) red giant known as Rasalgethi, and is also a double star. η Her is a close binary, mags. 2.8 and 5.5; easier doubles for small telescopes are δ, κ, ρ, 95 and 100 Her.

BRIGHTEST STARS

	Name	RA h m	dec. ° ′	Visual mag.	Absolute mag.	Spectral type	Distance (l.y.)
α	Rasalgethi	17 15	+14 23	2.7–4.0	−2.6–−1.3	M5	382
β	Kornephoros	16 30	+21 29	2.78	−0.50	G8	148
ζ		16 41	+31 36	2.81	2.64	F9	35
δ		17 15	+24 50	3.12	1.21	A3	78
π		17 15	+36 49	3.16	−2.10	K3	367
μ		17 46	+27 43	3.42	3.80	G5	27
η		16 43	+38 55	3.48	0.80	G8	112

▶ **Herschel, (Frederick) William** Originally trained as a musician, William Herschel made many important astronomical discoveries, including that of the planet Uranus in 1781. His sky surveys yielded extensive catalogues of stars, clusters and nebulae.

(1787) and two of Saturn (1789). He observed and catalogued many double stars, nebulae and clusters. Herschel realized that the Milky Way is the plane of a disk-shaped stellar universe, the form of which he calculated by counting the numbers of stars visible in different directions, thus establishing for the first time what he called the 'construction of the heavens'.

Wilhelm Friedrich Herschel, as he was christened, was born in Hanover, the son of an army band musician. As a boy he received a surprisingly good education at the garrison school, especially in music, and visited England in 1756 with his regiment. Finding that England offered abundant well-paid opportunities for skilled musicians, he returned in 1757, resolved to seek his fortune. After several years of hard work, his reputation won him the prestigious post of organist at Bath's fashionable Octagon Chapel in 1766. In 1772 he brought his sister Caroline Lucretia HERSCHEL from Germany to become his domestic companion and to allow her to pursue her own musical career.

By 1771 William Frederick (as he now called himself) was earning £400 a year from music – a very handsome income for the time – and, as Caroline recorded, William's interest in the mathematical harmonics of music aroused in him a curiosity about the mathematical basis of light and astronomy. In 1773 he started to make reflecting telescopes with mirrors made from speculum (Latin for 'mirror') metal, and used them to observe the heavens from the Herschel's house at 19 New King Street, Bath (now the Herschel Museum). He soon showed himself to be a brilliantly gifted practical optician, for it was later declared by Nevil MASKELYNE, the Astronomer Royal, that his reflecting telescopes were superior to any professionally made instruments then on sale in London. Herschel was thrust into international celebrity in 1781 when he discovered a 'comet' that soon turned out to be the planet Uranus – the first new planet in the Solar System to be discovered in recorded history.

This find, however, was the unexpected by-product of one of Herschel's surveys of the sky with a 6-inch (150-mm) reflecting telescope. He had begun systematically to 'sweep' the skies on a zonal basis, primarily as part of a strategy aimed at discovering star clusters, binary stars and nebulae, and trying to understand their distribution on the sky. By the mid-1780s, indeed, this technique had led him to conclude that the Milky Way consisted of a flat plane of stars running through space, so that when one looked through the long axis of the plane one saw dense starfields all around, whereas when one looked at right angles out of the plane, one saw relatively few stars. This observation still holds good for the general structure of our Galaxy, though not for the Universe in general.

Herschel's discovery of Uranus fundamentally changed his career. The international fame that it brought won him (and his sister Caroline) royal patronage, William's fellowship of the Royal Society, and an annual pension of £200, which enabled him to give up music and devote himself full-time to astronomy. This drop in income was more than made up, however, by commercial demand for his telescopes. Herschel produced the optics, and employed joiners to make the elegant mahogany tubes and stands. A Herschel 6-inch refractor of 7 ft (2.1 m) focal length sold for £105, while bigger instruments, built for European royalty, could cost as much as £3150.

Herschel's truly creative decade was the 1780s, for in the wake of the MESSIER CATALOGUE of 1784, his sky surveys increased the number of known deep-sky objects to over 2000. He was the first scientist to realize that the cosmos is not unchanging, but dynamic, as stars were now understood to form clusters under gravity, disintegrate into nebulous matter, and then re-condense into incandescent stars. Herschel also realized that the light from dim objects gathered by his 18¾-inch (0.48-m, 1783) and 48-inch (1.2-m, 1788) aperture telescopes of 20 and 40 ft (6 and 12 m) focal length had been travelling through space for so long that when observing them, one was looking into the past. By 1795 Herschel had pushed contemporary optical technology as far as it could go, and he spent the remaining 27 years of his life trying to obtain new data to make further breakthroughs. But not until the invention of ASTROPHOTOGRAPHY and astronomical spectroscopy in the next century would this be possible. In 1788 Herschel married Mary Pitt, and in 1792 John Frederick William HERSCHEL was born to them.

Herschelian telescope Type of telescope developed by William HERSCHEL in the late 18th century. It uses a PARABOLOIDAL primary, as in the NEWTONIAN TELESCOPE, but the primary is tilted so that the image is formed to one side of the incoming light, obviating the need for a secondary mirror: there is thus no central obstruction. The EYEPIECE is aimed down the tube, with the observer facing the main mirror. This optical arrangement, also known as the front-view system, introduced image distortions, and was soon superseded by the achromatic refractor and the silver-on-glass Newtonian reflector.

Herschel–Rigollet, Comet 38P/ Comet found by Caroline Herschel on 1788 December 21 and unexpectedly recovered in 1939 July by Roger Rigollet (1909–81). At each return, the comet attained peak mag. +7.5. Initially thought to be parabolic, the comet's elliptical orbit has a current period of 155 years.

Herschel Space Observatory (Far Infrared and Submillimetre Telescope, FIRST) EUROPEAN SPACE AGENCY (ESA) observatory to be launched in 2007. It will orbit at the L_2 Lagrangian point between the Earth and the Sun, 1.5 million km (0.9 million mi) from the Earth. Named after William HERSCHEL, who discovered infrared light, the craft is equipped with a 3.5-m (11-ft) diameter primary mirror. It will cover the full FAR-INFRARED and SUBMILLIMETRE waveband with the prime objective of studying the formation and evolution of stars. Herschel will be working in tandem with another ESA astronomical observatory, PLANCK.

Hertzsprung, Ejnar (1873–1967) Danish astronomer after whom the HERTZSPRUNG–RUSSELL DIAGRAM is named. Hertzsprung spent most of his career (1919–44) at the Leiden Observatory, which he directed for nine years. He was a prolific gatherer of stellar data, making more than a million photographic observations of binary stars and finding the brightnesses, spectral types and proper motions of thousands of stars in the Pleiades. He began to establish the mathematical form of Henrietta Leavitt's period–luminosity relation for Cepheid variable stars, and used it to determine accurately the distance to the Small Magellanic Cloud. Hertzsprung's work spanned the extremes of stellar types: the Cepheids are highly luminous yellow supergiants; and, in contrast, he was the first to classify, in 1911, the giant and dwarf subdivisions of late-type stars.

The culmination of this work, based in part on the work of pioneering astronomer Antonia MAURY, was his invention of the Hertzsprung–Russell diagram, so called because the American astronomer Henry Norris RUSSELL later rediscovered the relation. Hertzsprung found, by plotting luminosity against temperature for many individual stars, that most 'normal' stars, including the Sun, fall along a well-defined curve called the main sequence. His results were published in 1905 and 1907 in an obscure journal, and were unknown to Russell.

Hertzsprung gap Region to the right of the MAIN SEQUENCE on the HERTZSPRUNG–RUSSELL DIAGRAM where very few stars are seen. The gap exists because stars move very rapidly through this region. In the gap, the stars are evolving off the main sequence, burning hydrogen in shells before helium burning starts in the core. They travel on near-horizontal tracks with near-constant luminosity as the radius expands with a corresponding drop in effective temperature.

Hertzsprung–Russell diagram (HR diagram) Plot of the ABSOLUTE MAGNITUDE of stars against their spectral class. It is equivalent to a plot of the luminosity of stars against their surface temperature. The Hertzsprung–Russell diagram is an extremely valuable tool wherein observational and theoretical information are blended to produce a deep understanding of STELLAR EVOLUTION. It is a very powerful diagram, able to represent almost the entire evolution of any star. There are a finite number of equations defining a star's structure at every point in its life. These equations are interpreted on an HR diagram, so that determining the position of any star on an HR diagram will give information on its structure and evolutionary phase.

In the early 1900s Ejnar HERTZSPRUNG in Denmark and Henry Norris RUSSELL in the United States independently began to consider how the brightnesses of stars might be related to their spectra. In 1911 Hertzsprung plotted the APPARENT MAGNITUDES of stars in several clusters against their spectral class. In 1913 Russell plotted the absolute magnitudes of stars in the solar neighbourhood against their spectral class. Both astronomers emphasized the non-random patterns they saw in the arrangement of the data on these graphs. Such graphs have become extremely important in modern astronomy and are today called Hertzsprung–Russell diagrams.

A Hertzsprung–Russell diagram, or HR diagram, is any graph on which a parameter measuring stellar brightness is plotted against a parameter related to a star's surface temperature. For example, apparent magnitude, absolute magnitude, absolute BOLOMETRIC MAGNITUDE or LUMINOSITY may be plotted along the horizontal axis, while spectral class, COLOUR INDEX or EFFECTIVE TEMPERATURE may be plotted along the vertical axis.

Three HR diagrams are shown with this article, with diagram A being a schematic showing all the different regions. Diagram B resembles Russell's original graph. Each dot represents a star whose absolute magnitude and spectral class have been determined from observations. Most of the data lie along a broad line called the MAIN SEQUENCE, which extends diagonally across the diagram. The Sun (absolute magnitude around +5, spectral type G2) is a typical main-sequence star. A second prominent grouping of data points represents very large, cool stars called RED GIANTS. In the lower left corner, a third grouping identifies compact, hot stars called WHITE DWARFS.

Diagram C shows an HR diagram with luminosity of stars plotted against their surface temperature. Notice that temperature increases towards the left. The spectral classes OBAFGKM constitute a temperature sequence, and the first HR diagrams were drawn with the hot O stars on the left and the cool M stars on the right because the astronomers did not realise exactly what they were plotting. Although this means the temperature scale increases to the left, HR diagrams are still plotted this way today.

These HR diagrams show that the total range in stellar brightness is around 27 magnitudes (corresponding to a factor of 10^{11} in luminosity), and the range in the surface temperature of stars is from 2200 K to 50,000 K. The size of a star is related to both its luminosity and its surface temperature, as indicated by the dashed lines. Most main-sequence stars are roughly the same size as the Sun. White dwarfs are about the same size as the Earth, while red giants can be as big as the Earth's orbit.

The statistical distribution of data on an HR diagram demonstrates that stars fainter than the Sun are far more numerous than those brighter than the Sun. About 90% of stars are main-sequence stars, about 10% are white dwarfs and about 1% are red giants or SUPERGIANTS. The three main groupings on the HR diagram correspond to the three very different stages through which a typical star passes during its life. Computer models reveal how the luminosity and surface temperature of a star change as it evolves. With this information, the path followed by an evolving star can be plotted on an HR diagram.

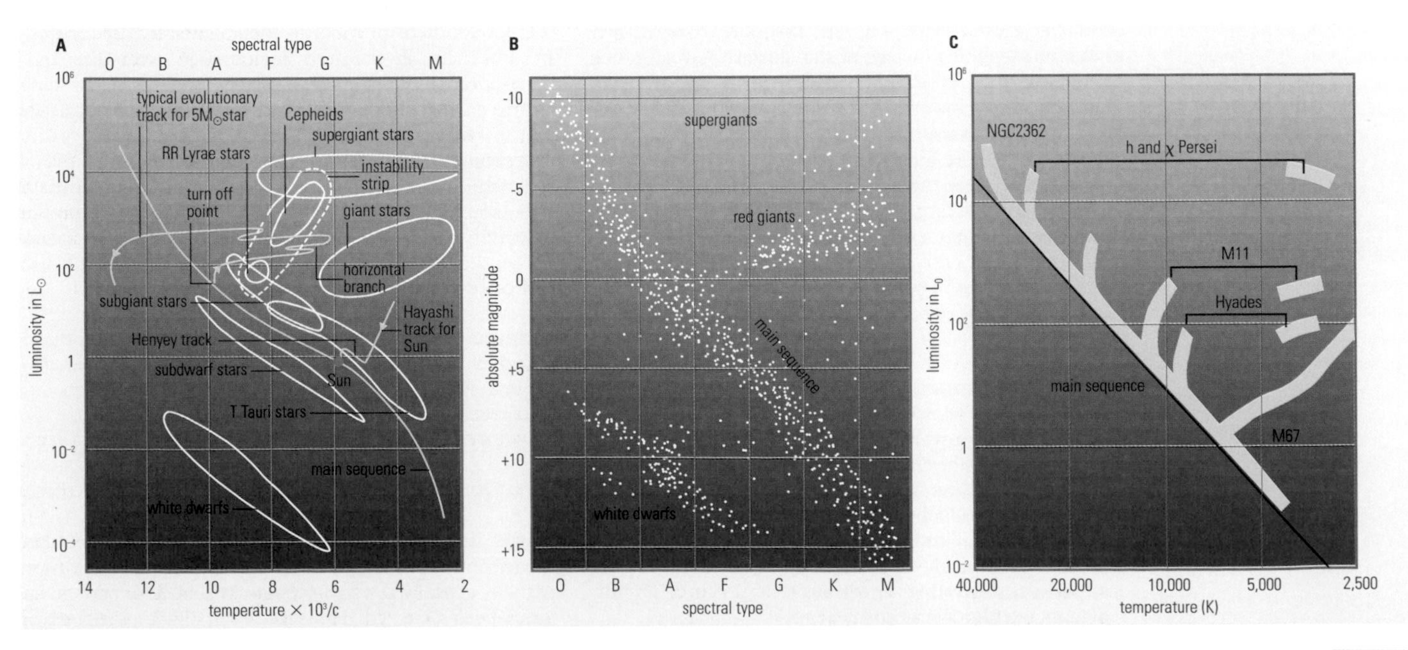

▼ **Hertzsprung–Russell diagram** A plot of stars' luminosity against temperature. Most stars lie on the main sequence running from top left to bottom right. Part B shows the principal regions of the Hertzsprung–Russell diagram, with giant stars above the main sequence to the top right, and dwarfs to the lower left. Part C illustrates some of the evolutionary tracks that stars may take as they branch away from the main sequence. For example, hot, massive O-class stars (top left) evolve to become red supergiants (top right).

PROTOSTARS are formed in fragmenting molecular clouds. As they evolve, their pre-main-sequence phase is represented on the Hertzsprung–Russell diagram as HAYASHI and HENYEY TRACKS. Contracting, pre-main-sequence stars change rapidly as gravitational contraction compresses and heats their cores. When the central temperature reaches several million degrees, nuclear FUSION begins, in the form of hydrogen burning, and the evolutionary track stops at the main sequence. Main-sequence stars are stable stars, burning hydrogen at their cores. A star will spend most of its life on the main sequence.

Pre-main-sequence evolutionary tracks deposit newborn stars on the main sequence in a position consistent with the MASS–LUMINOSITY RELATION, so that the more massive a star is, the brighter it is. Thus, the main sequence is a sequence in mass, as well as temperature and luminosity. These tracks also show that the more massive a star is, the more rapidly it evolves.

As a star consumes hydrogen, it gradually becomes slightly brighter and cooler. When the hydrogen supply is exhausted, the star's core contracts, its atmosphere expands, and it becomes a red giant or supergiant. It evolves off the main sequence at a point called the TURNOFF POINT. As it burns the hydrogen in its shell, it is on the red giant branch. If it is massive enough, the temperature of the collapsing core can become hot enough for helium burning to start. When a star is burning helium in its core, it is a HORIZONTAL BRANCH STAR.

Between the horizontal branch stars and the red giant branch is a gap. When the evolutionary track of a low-mass star takes it across this region, the star pulsates as an RR LYRAE VARIABLE. This gap is at the lower end of the so-called INSTABILITY STRIP, a vertical region in the middle of the HR diagram where stars are unstable against radial oscillations. When a massive star passes through this region, it pulsates and is called a CEPHEID VARIABLE.

The asymptotic giant branch has stars with both hydrogen and helium shell burning. PLANETARY NEBULAE are ejected at this stage. A star less massive than about 3 solar masses can eject as much as half its mass at the end of its life, producing such a planetary nebula. The exposed, burned-out core of the star then contracts to become a white dwarf. Detailed calculations of this process do indeed produce evolutionary tracks that rapidly take low-mass stars from the red giant region in the upper right of the HR diagram to the white dwarf region in the lower left of the diagram. Massive stars can end their lives in SUPERNOVA explosions, which leave behind NEUTRON STARS or BLACK HOLES. As these are detected at non-visible wavelengths, they are not represented on an HR diagram.

Stellar evolution calculations relate the HR diagram to time, and thus these graphs can provide information about the ages of stars. The HR diagram is particularly useful for estimating the age of star clusters. A very young cluster, such as NGC2362, consists primarily of main-sequence stars. However, the more massive a star is, the more rapidly it consumes its core supply of hydrogen to become a red giant. Since the main sequence is a progression in stellar mass, the stars at the upper end of the main sequence are the first to leave. As the cluster ages, the turnoff point for the evolving stars moves back along the main sequence. M67, for example, is an old cluster because only those stars less massive and less luminous than the Sun still remain on the main sequence. *See also* SPECTRAL CLASSIFICATION

Hess, Victor Francis (1883–1964) Austrian-American physicist (born Viktor Franz Hess), one of the discoverers of cosmic rays, for which he shared the 1936 Nobel Prize for Physics. Hess began investigating the phenomenon of atmospheric ionization by launching electroscopes in high-altitude balloons. Instead of this ionization decreasing with altitude, he found (1912) that it actually increased by a factor of eight in the upper levels of the atmosphere, suggesting an extraterrestrial source for this radiation, later known as cosmic rays.

HET Abbreviation of HOBBY–EBERLY TELESCOPE

Hevelius, Johannes (1611–87) German astronomer and instrument-maker, also a wealthy brewer and city magistrate, born in Danzig (modern Gdansk, Poland). In 1641 he built at Danzig what for a time was the world's leading observatory; called Sternenburg, it was destroyed by fire in 1679. His second wife, **Elizabeth Margarethe Hevelius** (1646/7–93), assisted him in much of his observational work, and she edited and published his magnum opus, *Prodromus astronomiae*, published posthumously in 1690. Hevelius was the first astronomer to describe the bright features of the solar photosphere known as faculae, which he found were connected with the formation of sunspots. He also derived an accurate value for the Sun's rotation period. In 1644 he observed the phases of Mercury that Nicholas Copernicus had predicted the planet would show.

Hevelius discovered four new comets, and suggested that these objects, which he called 'pseudo-planetae', moved about the Sun in parabolic orbits. He was also a skilled observer of the Moon and planets. In *Selenographia* (1647), the first truly detailed lunar atlas depicting how the appearance of lunar features changed with the Moon's phases, Hevelius coined the term 'mare' to describe the Moon's dark areas. He measured the heights of lunar mountains, obtaining results more accurate than Galileo's. His catalogue of 1564 stars and the accompanying *Uranographia* star atlas was based on many years of naked-eye observations with sighting instruments similar to those used by Tycho BRAHE; Hevelius was the last major astronomer not to use a telescope.

Hewish, Antony (1924–) English radio astronomer and recipient, with Martin RYLE, of the 1974 Nobel Prize for Physics for their role in discovering pulsars, and also for their development of the radio-astronomy technique of APERTURE SYNTHESIS. Hewish was part of the team at Cavendish Laboratory, Cambridge, that mapped the radio sky during the 1950s and 1960s, compiling the *Cambridge Catalogues* of radio sources. In the 1960s he made extensive studies of radio scintillation, the variation of radio waves' intensities, quantifying how Earth's atmosphere, solar radiation and even the interstellar medium all contribute to this effect. Although he received credit for discovering pulsars, it was actually his graduate student Jocelyn BELL BURNELL who first noticed the pulsed emissions.

hexahedrite *See* IRON METEORITE

Hey, (James) Stanley (1909–2000) British physicist and radar pioneer whose work during World War II led to the development of modern radio astronomy techniques. In 1942 he investigated interference with England's defence radar, which many experts thought was 'jamming' by the enemy; instead, he discovered that solar radiation was to blame. In later years Hey studied radio emission from sunspots, as well as variations in the Sun's overall radio emission. After the war, Hey turned his attention to radio sources outside the Solar System, and was the first to identify the exact location of the Cygnus A radio galaxy.

HH object Abbreviation of HERBIG–HARO OBJECT

Hidalgo Asteroid number 944; it is unusual in that it occupies a comet-like orbit stretching from perihelion near 2 AU out to aphelion at 9.6 AU. Hidalgo was discovered in 1920. It is about 30 km (20 mi) in size. *See also* CENTAUR; DAMOCLES; JUPITER-CROSSING ASTEROID

Hida Observatory *See* KWASAN AND HIDA OBSERVATORIES

Higgs field Field due to the Higgs boson. The standard model of elementary particles leaves room for one more particle, possibly the spin-0 Higgs boson. This particle has never been detected. However, Alan Guth postulated in the early universe, the high temperature left spacetime in a false vacuum state. As it cooled, the symmetry of the field associated with the Higgs boson spontaneously broke, dumping energy into spacetime and causing the INFLATION of the universe. *See also* BIG BANG THEORY

highlands, lunar Ancient elevated crust of the Moon. Because of their elevation, the highlands have not been covered by later lava flows. The highlands have a different composition from the maria (*see* MARE) and appear lighter due to a higher reflectance of sunlight. As the highlands are much older than the maria, they are more deeply cratered.

high-redshift quasar Most distant QUASARS observed, as determined from their REDSHIFTS and the HUBBLE LAW. Quasars are intrinsically so bright that some can be seen out to extremely large distances. Although the distribution of quasars peaks around z = 2.5, many quasars have been detected as far as z ~4.1. These extremely luminous, distant sources are seen as they were several billions of years ago.

high-velocity cloud (HVC) Cloud of mostly neutral hydrogen that is moving 100 to 200 km/s (60–120 mi/s) faster than the expected circular orbital velocity for its distance from the centre of the Galaxy. The clouds are normally only detectable from their radio emissions. The best-known example is an enormous bridge of material between our Galaxy and the MAGELLANIC CLOUDS. Known as the Magellanic Stream, it probably resulted from a close passage of the Magellanic clouds with the Galaxy at some time in the past. Other HVCs may be the result of SUPERNOVA explosions or may be infalling intergalactic material still being accreted by the Galaxy. The origin of most HVCs is not known.

high-velocity star Star whose space motion exceeds about 100 km/s (60 mi/s). There are several quite distinct classes. Many nearby Population II stars have high velocities by virtue of their galactic orbits, which may be neither circular nor close to the galactic plane; the nearby red dwarf binary 61 Cygni is an example. RUNAWAY STARS are young massive O or B stars moving at very high velocities, thought to have been ejected from binary systems or clusters. Somewhat older A stars of high velocity include many binary systems, making them less likely to have been ejected from clusters, and have orbits that inhabit a thick disk. They may trace the remnants of smaller galaxies that have merged with the Milky Way and not yet settled into its thin gas-rich disk. High-velocity neutron stars originate in the disruption of a binary during a supernova explosion; asymmetric explosions can impart a velocity 'kick'.

Hill, George William (1838–1914) American mathematician who applied his skills to celestial mechanics. He is most famous for his work on lunar theory, a working out of the Moon's complex motion and the perturbations exerted upon it. Hill solved many complicated problems involving the orbits of Jupiter and Saturn and made innovative strides towards solving the THREE-BODY PROBLEM.

Hilda asteroid Any of a group of ASTEROIDS with orbital periods close to two-thirds that of JUPITER. The archetype is (153) Hilda. Unlike the resonances represented by the KIRKWOOD GAPS, this 2:3 resonant position attracts objects and provides for long-term dynamical stability. *See also* HIRAYAMA FAMILY

Himalia One of the small satellites in JUPITER's intermediate group, discovered by Charles Perrine (1867–1951) in 1904. Himalia is *c*.186 km (*c*.116 mi) in size, making it the largest of Jupiter's family of satellites apart from the GALILEAN SATELLITES and AMALTHEA. It takes 250.6 days to circuit the planet at an average distance of 11.46 million km (7.12 million mi) from its centre. It has a substantial inclination (near 27°) and moderate eccentricity (0.162). *See also* ELARA

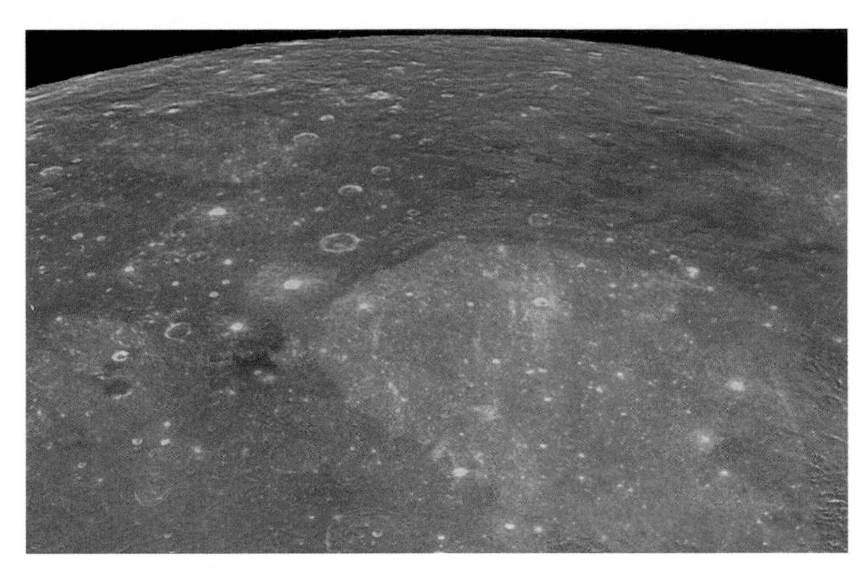

▲ **highlands, lunar** As it passed Earth, picking up speed for its long trip to Jupiter, the Galileo spacecraft obtained this false-colour image of the lunar highlands. The light highlands contrast with the dark maria, and are heavily cratered.

Hind's Crimson Star Name given to the deep-red variable star R Leporis, the colour of which was likened by the English astronomer John Russell Hind (1823–95) to 'a drop of blood'. R Leporis is a MIRA STAR, ranging from 12th magnitude at minimum to a peak of mag. 5.5 with a period of 14 or 15 months. It is a carbon star of spectral type C6 II.

Hipparchus Lunar crater (6°S 5°E), 135 km (84 mi) in diameter, with rim components reaching 1220 m (4000 ft) above its floor. An ancient crater, Hipparchus has been severely modified by meteorite erosion. Indeed, this erosion has removed all remnants of its ejecta blanket and has brought its walls to near level with the surrounding surface. The floor of Hipparchus is relatively smooth because of lava flooding.

Hipparchus of Nicea (*c*.190–120 BC) Greek astronomer, geographer and mathematician who made observations from Nicea (in what is now modern Turkey), Alexandria, Egypt and the Greek island of Rhodes. Little is known of his life, and the only book of his to survive is a commentary on the work of ARATUS. Most information about him comes from Ptolemy's *ALMAGEST*, from which it is clear that Hipparchus was the greatest observational astronomer of antiquity. He is best known for his discovery of the PRECESSION OF THE EQUINOXES. Hipparchus measured the length of the TROPICAL YEAR to within 6½ minutes; this allowed him to compare the dates of the equinoxes with the dates given by ARISTARCHUS of Samos and other earlier astronomers, and he found that the equinoxes had moved westwards around the ecliptic. His catalogue of some 850 stars – believed by some scholars to form the basis of the catalogue in the *Almagest* – was the first to classify stars by their apparent visual magnitudes.

Hipparchus developed mathematical models describing the apparent motion of the Sun and Moon, introducing epicycles, basing his work on his own and older Babylonian observations. He determined the sizes and distances of the Sun and Moon from observations of solar and lunar eclipses. Finding the Sun's parallax too small to measure, he assumed a value based on the accuracy of his observing equipment. This was almost five times too large, giving a value for the Sun's distance almost five times too small, and a figure for the Sun's diameter less than one-sixteenth of the true value, but these results were a vast improvement on previous estimates. Hipparchus' calculations of the Moon's average distance as 59–67 Earth radii (modern value 60.4) and its diameter as 0.33 times the Earth's (modern value 0.27) were more accurate.

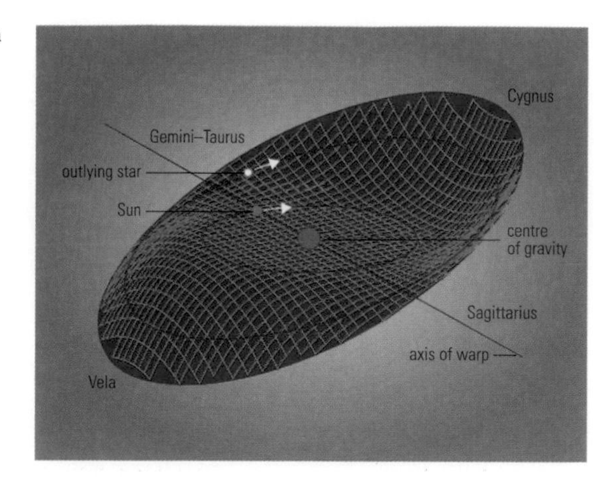

► **Hipparcos** Astrometric data from the Hipparcos satellite revealed warping of the disk of the Galaxy. The scale of the warping is exaggerated in this artist's rendition.

H

Hipparcos (High Precision Parallax Collecting Satellite) EUROPEAN SPACE AGENCY satellite, the first satellite to be designed for ASTROMETRY. The launch in 1989 was only partially successful, placing Hipparcos in a highly elliptical orbit instead of the intended geostationary orbit. However, the satellite was able to meet its scientific objectives before operations ended in 1993 August.

During systematic scanning of the sky, Hipparcos used direct triangulation to measure the relative positions, annual PROPER MOTIONS and TRIGONOMETRIC PARALLAXES of some 120,000 selected stars down to 11th magnitude – nearly every star within 250 l.y. of Earth.

The satellite's payload was a telescope that could view simultaneously in two directions 58° apart. The satellite rotated at approximately 12 revolutions per day about an axis perpendicular to the two fields of view. The direction of this axis of rotation was changed slowly so that the whole sky was scanned many times during the mission.

The results were eventually published in 1997 as the *HIPPARCOS CATALOGUE*. The Hipparcos survey more than doubled the number of known variable stars and discovered over 10,000 new binary or multiple star systems. It also led to the creation of the first accurate three-dimensional picture of the bright stars in our stellar neighbourhood. Distance measurements of CEPHEID VARIABLE stars indicated that the Universe must be 10% older than previously thought. Other measurements showed that the Milky Way is slightly warped and in the process of changing shape.

Hipparcos Catalogue (HIP) Catalogue, published in both paper and electronic forms, compiled from data gathered by the HIPPARCOS astrometry satellite. It lists 118,218 selected stars down to mag. 12.4 with highly accurate positions (to around 0″.001), parallaxes, proper motions and photometry. The *Tycho Catalogue* (TYC) gives less accurate positions (to around 0″.05) but lists 1,058,332 stars to mag. 11.5 with two-colour photometric measurements. The catalogue was published in 1997; it was followed in 2000 by the *Tycho-2 Catalogue*, which made use of data from the *Astrographic Catalogue* (*see* CARTE DU CIEL) to derive very accurate proper motions.

Hirayama family Group of ASTEROIDS characterized by having closely similar orbital elements. The existence of such families was pointed out by Japanese astronomer Kiyotsugu Hirayama (1874–1943) in 1928. About a hundred such families are now recognized, each being named usually after its largest member. Some of the more populous families may have hundreds of constituents. About half of all MAIN-BELT ASTEROIDS are members of one Hirayama family or another.

Four of the main families are the FLORA, Eos, Themis and Koronis groups. The fact that in many such families the members have similar spectral reflectivities suggests that they originated through the collisional break up of a larger progenitor. GASPRA, for example, is a member of the Flora family, and images of it are indicative of its liberation as an independent body in the astronomically recent past (during the last hundreds of millions of years as opposed to billions). Such disintegrations are known to have occurred from other lines of evidence; for example iron meteorites must once have been part of much larger bodies, which were sufficiently massive for internal heating to have occurred and caused the liquefaction and chemical fractionation that is obvious from the meteorites' appearance. Similarly, the family associated with VESTA appears to be the source of several distinct meteorite types, in that the reflectivities of these meteorites are matched by the surface regions of that asteroid and its cohorts. Bands of INTERPLANETARY DUST believed to be associated with certain families may be debris produced by collisions in relatively recent times (within the past few million years).

Other types of asteroid family may be recognized on the basis of orbital similarity, but with that similarity having been produced by dynamical segregation and accumulation around certain sets of orbital parameters, rather than by a common genesis. The TROJAN ASTEROIDS represent an obvious example, another being the HILDA ASTEROIDS; other groupings include the HUNGARIA ASTEROIDS.

HI region Interstellar gas cloud in which the hydrogen is in the form of neutral atoms. The name derives from the convention whereby ionized atoms are labelled with a Roman numeral one larger than the number of electrons lost: ionized hydrogen is HII (*see* HII REGION) and neutral hydrogen is therefore HI. The sizes of HI regions can vary considerably, but typically they are 10 to 20 l.y. across and have a mass of 10 to 100 solar masses. Their temperature is around 100 K, and the particle density around 10^7 m^{-3} (ten times denser than the average for INTERSTELLAR MATTER). The density is too low for molecular hydrogen to form in significant quantities, and any that did form would soon be dissociated by stellar radiation.

HI regions do not emit visible radiation, but they can be detected from their TWENTY-ONE CENTIMETRE radio emissions. Large radio telescopes can thus detect HI regions even in distant galaxies. Different kinds of galaxy contain different proportions of HI. Our own Galaxy and other spiral or irregular galaxies contain an appreciable fraction of their total mass in HI. Elliptical galaxies contain no (or extremely little) detectable HI.

In our Galaxy, HI clouds are heavily concentrated to a flattened disk in which a spiral pattern occurs. Spiral arms are made conspicuous, even in very distant galaxies, by the groups of hot, high mass, blue stars lying along the arms, like pearls on a string. These groups contain stars that have formed recently from the HI clouds. Not all HI clouds are capable of forming stars; only the densest ones that also contain molecular cores can do so (*see* GIANT MOLECULAR CLOUDS). A number of HI clouds lie out of the galactic plane and are very diffuse. Even dense hydrogen clouds are much more tenuous than the best laboratory vacuum achievable on Earth.

HII region Interstellar gas cloud in which the hydrogen is in the form of ions (*see* STRÖMGREN SPHERE). The name derives from the notation, HII, used to denote ionized hydrogen (*see* HI REGION). The hydrogen is ionized by ultraviolet radiation from hot young stars embedded within the region, and the gas is heated to betweeen 8000 and 10,000 K. HII regions are often embedded within HI regions, but have densities 10 to 100,000 times greater than those of the latter. The high temperatures and densities of HII regions mean that the gas pressure inside these regions is far higher than that of the surrounding HI region or interstellar medium. All HII regions are thus increasing in size at expansion velocities of around 10 km/s (6 mi/s). Visible radiation from an HII region results from the RECOMBINATION of ions and electrons, and hot dust causes infrared emission; FREE–FREE TRANSITIONS lead to emission by the region at radio wavelengths. On images, HII regions often appear red because of the HYDROGEN-ALPHA LINE at 656 nm;

however, they can look greenish when seen visually because of a prominent pair of FORBIDDEN LINES at 496 and 501 nm emitted by oxygen.

Three main classes of HII regions are recognized: ultra-compact HII Regions (UCHIIRs), which are less than about 0.1 l.y. in size; compact HII regions, which are about 1 l.y. across; and classical HII regions, which may be several tens of light-years in size. Giant and supergiant HII regions, with sizes up to several hundred light-years, are found in some external galaxies. UCHIIRs and compact HII regions occur around single stars, but the larger classes usually have from a few to 100,000 stars within them. The amount of ionized gas in the cloud ranges from a small fraction of a solar mass to millions of solar masses. The gas density similarly ranges from 10^7 to 10^{11} particles m^{-3}, with the largest clouds being the least dense. The smaller HII regions are often, but not always, roughly spherical in shape. The larger regions can have very complex morphologies arising from the distribution of gas and stars within them. The ORION NEBULA is a good example of a classical HII region.

Regions of ionized hydrogen can occur for many reasons, such as ultraviolet emission from old hot stars or from nova or supernova explosions, or they may form as a result of high velocity jets impacting the material of the interstellar medium. Conventionally, however, these are not included in the term HII region, but are instead given their own specific names (*see* PLANETARY NEBULA, SUPERNOVA REMNANT and HERBIG–HARO OBJECT).

HII regions are stellar nurseries. The hot stars that ionize the gas have recently been formed through gravitational collapse within a GIANT MOLECULAR CLOUD. The life of an HII region is only a few million years. The presence of such regions, along with the hot stars, marks out the arms in spiral galaxies. HII complexes also characterize STARBURST galaxies, in which high-mass star-forming activity is occurring strongly.

HM Nautical Almanac Office Principal organization in the UK for providing astronomical almanacs and other numerical data. With the US NAVAL OBSERVATORY it is responsible for producing *The ASTRONOMICAL ALMANAC*, *The Nautical Almanac* and *The UK Air Almanac*. The Office dates back to the foundation of the ROYAL GREENWICH OBSERVATORY (RGO) in 1675; with the closing of the RGO in 1998, it moved to its present location at the Rutherford Appleton Laboratory.

Hoba Largest single METEORITE known. Hoba is approximately 3 × 3 m (10 × 10 ft) across, and at least 1 m (3.3 ft) deep; it is estimated to weigh *c*.60 tonnes. Hoba is an ataxite IRON METEORITE. It was found in 1920, near Grootfontein, Namibia, where it still lies in the ground, preserved as a national monument.

Hobby–Eberly Telescope (HET) Large optical telescope of unusual design at an elevation of 2025 m (6640 ft) on Mount Fowlkes at MCDONALD OBSERVATORY. It is run by a consortium consisting of Penn State University, Stanford University, Ludwig-Maximilians Universität München, and Georg-August Universität Göttingen. With its innovative simple mounting, it was only a fraction of the cost of a conventional 8-metre class telescope. It entered operation in 1997. It is named after William P. Hobby and Robert E. Eberly, American supporters of public education.

The HET utilizes a 'tilted arecibo' concept: the telescope tube is fixed at 55° from the horizontal; the telescope itself remains stationary during observations, though the whole structure can rotate through 360° of azimuth on air bearings. This single rotational axis, combined with the apparent rotation of the heavens, gives the HET access to 70% of the sky visible from McDonald Observatory over the course of the year. To follow the stars, there is a tracker mounted in the focal plane of the spherical primary mirror; this carries a Spherical Aberration Corrector (SAC).

The mirror consists of 91 identical segments forming a hexagon 10 × 11 m (33 × 36 ft) in size. The working aperture of 9.2 m (30 ft), determined by the optics of the SAC, is smaller than the primary mirror to allow the beam to track across the mirror during observation. An imager and a spectrograph are mounted at the prime focus, while larger spectrographs are mounted below the telescope and connected by fibre-optic feeds.

Hoffleit, (Ellen) Dorrit (1907–) Pioneering American woman variable-star astronomer and compiler of the Yale *BRIGHT STAR CATALOGUE*. Working at Harvard College Observatory, she discovered over a thousand variable stars, mostly in the southern sky. Between 1948 and 1956 Hoffleit determined very accurate spectroscopic absolute magnitudes for many southern stars, which allowed other astronomers to map the Milky Way's large-scale structure. In 1956 she joined Yale University and undertook a thorough revision of the *Bright Star Catalogue*, first compiled by Frank SCHLESINGER. She has produced three editions of this catalogue, which contains positional, spectral and other basic data for over 9000 stars.

Hoffmeister, Cuno (1892–1968) German astronomer, founder, and for 42 years the director, of the Sonneberg Observatory, who specialized in variable stars. Using a battery of fourteen small cameras, he secured over 100,000 photographic plates of the night sky. His careful examination of these plates resulted in the discovery of over 10,000 new variables. Hoffmeister also conducted long-term visual meteor counts to refine the positions of shower radiants and the parameters of meteor streams. He made detailed studies of the zodiacal light and gegenschein.

Hogg, Helen (Battles) Sawyer (1905–93) American astronomer who spent her entire professional career in Canada, investigating variable stars in globular clusters and popularizing astronomy. In 1930 she married Canadian **Frank Scott Hogg** (1904–51), and they both joined the Dominion Astrophysical Observatory. From 1931 to 1935, she used DAO's 72-inch (1.82-m) reflector to photograph variable stars, work that would lead to the compilation of an important catalogue of those objects. In 1935 the Hoggs moved to David Dunlap Observatory, where Helen used the 74-inch (1.88-m) telescope to discover hundreds of new variable stars.

Holden, Edward Singleton (1846–1914) American astronomer, director of Washburn (University of Wisconsin) Observatory (1881–85) and first director of Lick Observatory (1888–97). At Washburn, Holden studied Saturn's rings, prepared catalogues of red stars and southern stars, and encouraged Sherburne Wesley BURNHAM to search for new double stars using

▼ **Hoba** The world's largest single meteorite, Hoba lies where it fell in Namibia. This iron meteorite is now preserved as a national monument.

H

HOROLOGIUM (GEN. HOROLOGII, ABBR. HOR)

Obscure southern constellation, next to the southern end of Eridanus, introduced by Lacaille in the 18th century. It represents a pendulum clock. α Hor, mag. 3.85, spectral type K1 III, is its only star above 4th magnitude. R Hor is a MIRA STAR, ranging from 5th to 14th magnitude every 13 months or so.

Washburn's 15-inch (380-mm) Clark refractor. He also made some of the earliest statistical studies of stellar distribution that would later show the Milky Way to be a disk-shaped spiral galaxy with a central bulge. Holden was the major force behind the founding, in 1889, of the ASTRONOMICAL SOCIETY OF THE PACIFIC.

Holmes, Comet 17P/ Short-period comet discovered in 1892 during what seems to have been an unusually bright and active apparition, when it became a naked-eye object and may have undergone nuclear fragmentation. It was seen again in 1899 and 1906 but was then lost until 1964, when it was below mag. +18. Comet 17P/Holmes has since been recovered at every apparition, most recently in 2000, but remains very faint. It has an orbital period of 7.07 years.

Homestake Mine Site of an underground astronomical observatory, at Lead in the Black Hills of South Dakota, opened in 1965. At its depth, 1500 m (4850 ft), cosmic rays are blocked and experiments can be conducted to observe SOLAR NEUTRINOS, which were discovered here in 1968. In 2001 it ceased to be an operating mine (Homestake Gold Mine had been the largest US producer of gold), putting the future of the facility in doubt, given the existence of deeper sites such as Canada's Sudbury Neutrino Observatory, some 2070 m (6800 ft) below the surface.

▼ Horsehead Nebula Silhouetted against the emission nebula IC 434, the Horsehead Nebula (B 33) in Orion is a very difficult object visually. Photographs such as this show the striking profile of the intruding dark nebula in the foreground.

homogeneity Principle that in any particular volume of spacetime the properties of the Universe on average are identical to the same volume anywhere else in the Universe. On small scales, this is obviously not correct, but on scales larger than galactic clusters this idea holds.

Homunculus Nebula Bright dust-cloud (REFLECTION NEBULA) produced by material ejected during outbursts from Eta Carinae beginning in 1843. Located within the ETA CARINAE NEBULA, the Homunculus (RA $10^h 45^m.1$ dec. $-59°41'$) is named after its resemblance to a human outline. It spans an area of 17″ × 12″.

Hooker Telescope Name of the 100-inch (2.5-m) telescope at MOUNT WILSON OBSERVATORY, immortalized in the 1920s as the instrument with which Edwin HUBBLE discovered the distance scale of the Universe. It is named after the financier John D. Hooker (1837–1910), who was persuaded by George Ellery HALE to fund its construction. It was completed in 1917, and was the largest reflector in the world until the HALE TELESCOPE became operational in 1948. The telescope can be used in various optical configurations.

The Hooker Telescope is the work of George W. RITCHEY, the American optician who pioneered the development of large telescope optics in the early 20th century. The 4.5-tonne mirror blank was cast in 1908, and two years later Ritchey began the five-year task of polishing and figuring it. The completed instrument was first turned on the sky in 1917. It soon made history when, in 1920, Francis G. PEASE and John A. ANDERSON used the STELLAR INTERFEROMETER built and fitted to the 100-inch by Albert Abraham Michelson (1852–1931) to make the first measurements of star diameters.

After a long and illustrious career the telescope was declared inactive in 1986, but in 1994 it was reinstated as a research instrument, and underwent major upgrades to its control systems, including a state-of-the-art ADAPTIVE OPTICS system.

Hooke, Robert (1635–1703) England's first great experimental physicist, best known for his law of elasticity. In 1662 he became curator of instruments at the Royal Society, and the same year tried to measure the gravitational constant using a pair of weights, and in 1665 was observing Jupiter's belts and the craters of the Moon with a telescope of focal length 36 ft (11 m). He realized that the nucleus was the active part of a comet, and in 1669 made the first serious attempts to measure the parallax of a star, γ Draconis. He pioneered the use of William GASCOIGNE's micrometer, and designed instruments for the Royal Observatory, Greenwich.

horizon GREAT CIRCLE on the CELESTIAL SPHERE 90° away from the ZENITH, or overhead point, of an observer located on the Earth's surface. Any point on the celestial sphere greater than 90° from the observer's zenith is below their horizon and therefore invisible to them. The term is also used to describe a boundary in SPACETIME beyond which events cannot be observed (an EVENT HORIZON) or particles cannot yet have travelled (a particle horizon). The term is also used to describe the farthest distance from which light could possibly reach us given the age of the Universe and the finite speed of light.

horizontal branch star Star that is burning helium in its core. Such stars lie on a horizontal strip of the HERTZSPRUNG–RUSSELL DIAGRAM to the right of the MAIN SEQUENCE and to the left of the RED GIANT branch.

When a star has finished burning hydrogen in its core, it starts hydrogen SHELL BURNING. The star expands off the main sequence to become a red giant. If the star has sufficient mass, helium burning can occur in the core. When a star is burning helium in its core, it is on the horizontal branch. Low-mass stars start helium core burning with a HELIUM FLASH, and the star moves rapidly to the horizontal branch. If the mass of the star is small, it lands on the left extension of the horizontal branch; if the mass of the star is larger, it will fall farther to the right on the horizontal branch.

Hertzsprung–Russell diagrams of GLOBULAR CLUSTERS show differences in the form of their horizontal branches from cluster to cluster. As yet there is no consensus of opinion as to what causes the differences, although the metal abundance of the stars and the age of the cluster appear to be two contributing factors. RR LYRAE VARIABLES are formed where the INSTABILITY STRIP crosses the horizontal branch. On some Hertzsprung–Russell diagrams, variable stars are not shown and thus there is a gap in the horizontal branch.

horizontal coordinates System of locating points or objects on the CELESTIAL SPHERE using the horizon as the plane of reference and the measures of ALTITUDE and AZIMUTH. Altitude is the angular distance of a celestial body above an observer's horizon. It is measured vertically from 0° at the horizon, along the GREAT CIRCLE passing through the object, to a maximum of 90° at the ZENITH. Azimuth is the angular distance of an object measured westwards along the horizon, with north as the zero point, to the vertical circle (meridian) running through the object.

Horologium *See* feature article

Horrocks, Jeremiah (1619–42) English astronomer, the first to build significantly on the work of Galileo and Johannes Kepler. Born at Toxteth, Liverpool, into a farming family, he studied at Emmanuel College, Cambridge, between 1632 and 1635, where he discovered the errors in existing planetary tables, and set about correcting them from instruments of his own devising, including an 'astronomical radius', a solar projector and a small telescope. In 1636 he became acquainted with William CRABTREE of Salford, and soon afterwards William GASCOIGNE of Leeds and Christopher Towneley (1604–74). Horrocks' surviving writings abound with references to the importance of primary observation. In 1637 he and Crabtree observed the sudden occultation of the stars of the Pleiades by the Moon, which led him to believe that the Moon must be airless. By 1638 he had demonstrated that the lunar orbit is a Keplerian ellipse, and had measured the changing angular diameters of the Sun and Moon, matching them with elliptical orbits. His measurement of the solar parallax gave a Sun–Earth distance of around 100 million km (about 60 million mi), which would remain the best estimate for many years. His successful prediction and observation of the transit of Venus on 1639 November 24 won him posthumous immortality. From it, Horrocks calculated the angular diameter and parallax of Venus and suggested that they were consistent with a Keplerian orbit.

Horsehead Nebula (B33) Dark nebula seen in silhouette against the emission nebula IC 434 near ζ Orionis (RA $05^h\ 41^m.0$ dec. −02°24′). It is part of the extensive gas and dust complex in the Orion region, lying at a distance of 1600 l.y. Covering an area of 6′ × 4′, the Horsehead is named after its resemblance in long-exposure images to a chess knight. Visually, the Horsehead shows little contrast with IC 434, and it is very difficult to see even in large telescopes.

horseshoe mounting Variation on the EQUATORIAL MOUNTING, used for large telescopes.

Horseshoe Nebula *See* OMEGA NEBULA

hot dark matter Type of matter that has been proposed as the missing mass component of galaxies (*see* MISSING MASS PROBLEM). Dark matter was postulated to be responsible for closing the universe (*see* CLOSED UNIVERSE) and for explaining galaxy formation. Hot dark matter could consist of massive neutrinos moving near the speed of light (hence the term hot). However, predictions for galaxy formation assuming the presence of hot dark matter do not seem to agree with current observations.

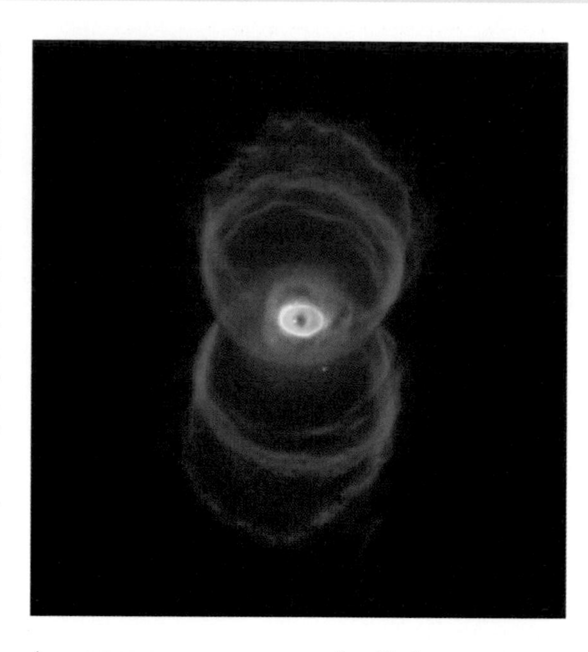

◄ **Hourglass Nebula** A Hubble Space Telescope image from 1996, showing delicate loops in the twin lobes from which the Hourglass Nebula takes its name.

hour angle (HA) Measure of the elapsed SIDEREAL TIME since a given celestial object last crossed the observer's meridian. Hour angle is also a measure of the angle between the HOUR CIRCLE passing through a celestial body and the observer's local CELESTIAL MERIDIAN, measured westwards from the meridian.

hour circle Any circle on the CELESTIAL SPHERE that passes through both CELESTIAL POLES, perpendicular to the CELESTIAL EQUATOR. Each half of an hour circle, taken from pole to pole, is a line of constant RIGHT ASCENSION, and the DECLINATION of a celestial object is measured along its hour circle. The zero hour circle coincides with the observer's meridian. The term is also used to describe the scale on the mounting of an equatorial telescope that indicates the hour angle at which the telescope is pointing.

Hourglass Nebula (MyCn 18) Young PLANETARY NEBULA in the southern constellation of Musca (RA $13^h\ 40^m$ dec. −67°23′). The Hourglass lies 8000 l.y. away and has a apparent diameter of 4″. Images from the Hubble Space Telescope in 1996 showed loops of ejected material surrounding the central star.

howardite *See* HOWARDITE–EUCRITE–DIOGENITE ASSOCIATION

howardite–eucrite–diogenite association (HED) One of the ACHONDRITE meteorite subgroups. Howardite–eucrite–diogenite association is a suite of generally brecciated igneous rocks formed by volcanism on their parent body. Diogenites are coarse-grained orthopyroxenites. Eucrites are cumulates and fine-grained basalts, composed of pyroxene and plagioclase. Howardites are regolith breccias (asteroidal soils), rich in both solar wind gases and clasts of carbonaceous material. The HEDs all have similar oxygen isotopic compositions. Candidates for the HED parent body are the asteroid VESTA or its HIRAYAMA FAMILY.

Hoyle, Fred (1915–2001) English astrophysicist and cosmologist known for his origination and advocacy of the STEADY-STATE THEORY and for his fundamental explanation of how stars synthesize elements heavier than helium. In 1957 he co-authored, with Margaret and Geoffrey BURBIDGE and William FOWLER, the seminal paper 'Synthesis of the Elements in Stars'. Dubbed 'B^2FH' for the initials of its authors' surnames, this paper describes how stars synthesize chemical elements heavier than hydrogen and helium. Hoyle's major contribution to this theory of stellar nucleosynthesis was his prediction of an excited state of the ^{12}C nucleus. In 1948, with Hermann BONDI and Thomas GOLD, Hoyle conceived a cosmological theory describing the Universe as flat space expanding at a constant rate, the expansion being balanced by the continuous creation of matter so that its mean density remains unchanged. This steady-state theory was proposed as an alternative to the BIG BANG, a term Hoyle coined to poke fun at the model accepted by most cosmologists.

Hoyle later helped Geoffrey Burbidge to develop a 'quasi-steady-state theory', which, unlike the original version, accounts for the COSMIC MICROWAVE BACKGROUND radiation. This theory suggests that the isotope ^4He is produced, not in the Big Bang, but by stellar hydrogen burning, as are the isotopes ^2H (deuterium), ^3He, ^6Li, ^7Li, ^9Be, ^{10}B and ^{11}B.

Hoyle modelled the collapse of interstellar dust clouds to form stars, and in the 1940s he was the first to describe mathematically the process of accretion, by which stars accumulate INTERSTELLAR MATTER. He suggested that the dust clouds are composed of inorganic carbon graphite grains or organic forms of carbon, the latter forms possibly imported to Earth with other organic molecules where they formed the building blocks of life, thus reviving the PANSPERMIA theory. With this and other non-mainstream theories, Hoyle often courted controversy.

For many years Hoyle was Plumian Professor of Astronomy at Cambridge University, founding its Institute of Theoretical Astronomy (1966).

HR diagram Abbreviation of HERTZSPRUNG–RUSSELL DIAGRAM

HST Abbreviation of HUBBLE SPACE TELESCOPE

H2 Booster by which Japan achieved its own independent launch into geostationary orbit. The H2 booster was first launched, two years late, in 1994. It flew five successful missions but there were two failures, and high costs forced a cancellation of the programme. The new H2A was intended to launch in 1999, but the programme was delayed by technical problems until 2001. The standard H2A will replicate the capability of the original booster, carrying 4 tonnes to geostationary transfer orbit; larger versions will increase this capability to 7.5 tonnes in 2005.

Hubble, Edwin Powell (1889–1953) American astronomer who proved that the spiral 'nebulae' were galaxies lying far beyond our own Milky Way, established a widely used scheme for the classification of galaxies, and who derived the velocity–distance relation for these objects, thus allowing the scale of the Universe to be reckoned for the first time. It is difficult to overstate Hubble's impact on observational cosmology; it has been said that what Nicholas Copernicus did for the Solar System, and William Herschel did for the Galaxy, so Hubble did for the Universe.

A native of Missouri, Hubble worked briefly (1914–17) as a research assistant at Yerkes Observatory, then, following a stint in the United States Infantry from 1917 to 1919, he joined the staff of MOUNT WILSON OBSERVATORY, where he remained for the rest of his career. He was also involved with the construction of the 200-inch (5-m) HALE TELESCOPE at Palomar Observatory, which he used in his latter researches (1948–53). For most of his career, Hubble used Mount Wilson's 60-inch (1.5-m) and 100-inch (2.5-m) reflecting telescopes to study CEPHEID VARIABLES in the Andromeda Galaxy (M31), M33 and NGC 6822, the distance and nature of which were the subjects of the so-called GREAT DEBATE in 1920. From 1925 to 1929 he published three important papers showing that these nebulae were galaxies lying at distances much greater than any object in the Milky Way, confirming the 'island universe' model of Heber CURTIS. In 1932 he discovered the first globular clusters outside the Milky Way, in M31.

Extending the earlier work of Vesto M. SLIPHER, Hubble measured the spectral REDSHIFTS of 46 galaxies, attributing the redshifts to a recession of the galaxies and using the results to find their recessional velocities. Comparing these velocities to the distances of the galaxies, in 1929 he announced what is now called HUBBLE'S LAW: the farther away a galaxy is, the faster it is receding. When these two quantities, velocity and distance, were plotted against each other, the result was an almost perfectly linear fit – the slope of the line is the HUBBLE CONSTANT (H_0), and its reciprocal, the **Hubble time,** is the age of the Universe since the Big Bang. These discoveries provided powerful evidence that the real Universe resembled the kind of expanding universe derived theoretically by Albert EINSTEIN and Willem DE SITTER.

From 1929 to 1936, Hubble and his assistant, Milton HUMASON, published a series of studies showing conclusively that for velocities of recession up to 40,000 km/s (25,000 mi/s), the velocity–distance relation holds true. The two astronomers also invented the first (1925) systematic scheme for the classification of galaxies (now called the HUBBLE CLASSIFICATION) based on their morphologies, labelling these objects as elliptical, 'normal' spiral or barred spiral galaxies, which they further divided into subtypes. Hubble believed that his scheme indicated an evolutionary sequence for galaxies, from ellipticals to spirals, though this is now known not to be the case.

Hubble's surveys of bright galaxies, carried out in the 1930s and 1940s, showed them to be distributed isotropically on the largest scales of the Universe. On more local scales, galaxies were often found in clusters composed mostly of elliptical galaxies – Hubble was the first to measure accurately the surface brightness profiles of ellipticals. His counts of galaxies confirmed that they are comparatively scarce in the ZONE OF AVOIDANCE along the plane of the Milky Way, where they are obscured by our Galaxy's disk of dust and gas. Hubble's studies of the Milky Way's spiral arms (1935–43) provided the first strong evidence that they 'trail' as the Galaxy rotates, instead of opening outwards 'ahead' of the galactic rotation, as some

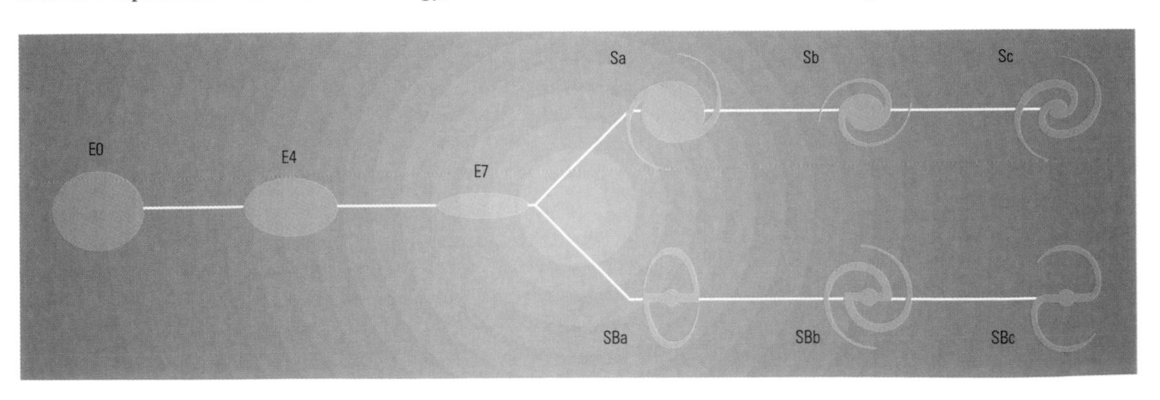

▶ **Hubble classification** A useful way of describing galaxy shapes, the Hubble classification splits these into three main classes. Ellipticals (E) are subdivided on the basis of how far they are from spherical. The classification branches (and is hence sometimes known as the 'tuning-fork diagram') to afford description of spiral (S) and barred spiral (SB) galaxies.

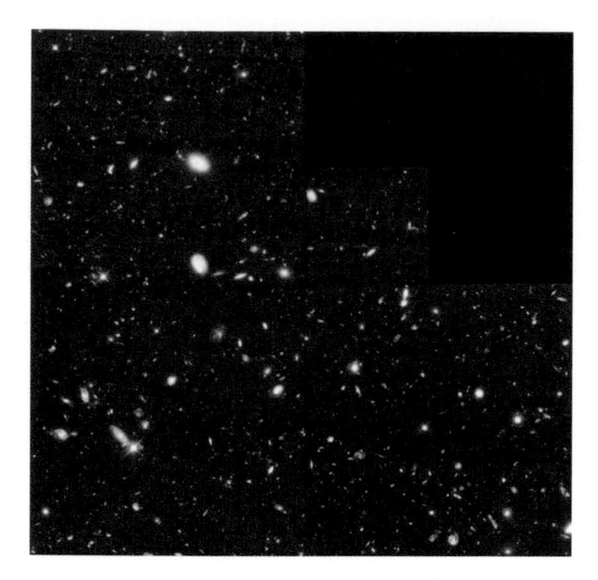

◄ **Hubble Deep Field** A composite of many long exposures, the Hubble Deep Field shows extremely faint and remote galaxies. Some of these formed very early in the history of the Universe.

astronomers had hypothesized. He was the first (1922) to divide the bright diffuse nebulae of the Milky Way into reflection and emission nebulae, giving the proper astrophysical explanations for their different spectra. Three decades after his death, the HUBBLE SPACE TELESCOPE (HST) was named to honour him.

Hubble classification System for classifying GALAXIES according to their shape on photographs; it was introduced by Edwin HUBBLE in 1925, with extensions and revisions by several later workers, including Allan SANDAGE, Gerard DE VAUCOULEURS, and Sidney Van den Bergh (1929–).

ELLIPTICAL GALAXIES are denoted E with a numeral describing the apparent shape, from E0 for a circular image, through E5 for an image with an axial ratio of 2, to the most flattened ellipticals at E7. Ordinary (nonbarred) SPIRAL GALAXIES are denoted S with a letter a,b,c,d to indicate a stage along the TUNING-FORK DIAGRAM. Sa galaxies have large and bright central bulges, with tightly wound arms. Sc galaxies have small and faint bulges, with prominent and loosely wound arms. Intermediate types Sab and Sbc can also be used. A parallel sequence of BARRED SPIRALS includes SBa, SBb and SBc. IRREGULAR GALAXIES are denoted Irr or simply I. Originally two types, Irr I and Irr II, were used, but Irr II galaxies appear to be temporary results of gravitational collisions and not a distinct kind of galaxy. The extension to the system by de Vaucouleurs recognized inner ring-like or spiral-shaped structures, as well as external rings. The van den Bergh classification added the use of the spiral pattern as a rough indicator of galaxy luminosity.

While the Hubble classification was originally designed to be descriptive, and based on blue-light photographs of a particular exposure range, it has retained enormous utility because these designations correlate well with physically interesting galaxy properties. Stellar and gas content and star formation rate, for example, change systematically between Hubble types.

Hubble constant Parameter in the HUBBLE LAW that relates the velocity of a galaxy to its redshift. In GENERAL RELATIVITY, it is the rate at which the scale size (R) of the Universe is changing with time. It is also the rate at which distances between all 'co-moving' objects – that is, those objects with no individual motion relative to the fabric of space – are increasing. The units for the Hubble constant are such that its inverse is a time, which can very loosely be thought of as an 'age' for the Universe. More usually, however, the units employed are kilometres per second per megaparsec since, for small redshifts, galaxy recession velocities increase by this amount for each megaparsec of distance.

The Hubble constant is normally designated H_0, the zero subscript specifying that it is the expansion rate at the present epoch that is meant. The relative expansion rate changes with time and the Hubble parameter and its rate of change are related to the deceleration parameter, q_0, by

$$H_0 = \dot{R}/R$$

$$q_0 = \ddot{R}/(RH_0^2)$$

both values being calculated for the present epoch.

The search for accurate values for H_0 and q_0 has been the driving force behind much of observational cosmology in the last 80 years. Many different methods of obtaining values for H_0 have been devised. But they all have in common redshift and distance measurements of a set of similar objects carefully chosen to eliminate selection and measurement bias to the greatest extent possible. A key project for the Hubble Space Telescope was the task of deriving an accurate value for the Hubble constant. After many years of work, the best value of H_0 available at present from the Space Telescope Key Project is 68 ± 6.0 km per second per megaparsec.

Hubble Deep Field (HDF) Long exposure of a small area of sky in the north of Ursa Major obtained by the HUBBLE SPACE TELESCOPE in 1995. Just 2′.5 across, the HDF is a composite of 342 exposures accumulated over 100 hours in four wavelength ranges in the visible, near-ultraviolet and near-infrared. In 1998 the *Hubble Deep Field South* (HDF-S) was imaged: a 2′ sample of a region in the constellation Tucana. Both fields show over 2000 galaxies, including some of the most distant – and thus the youngest – ever imaged.

Hubble diagram Graph in which the apparent magnitude of galaxies is plotted against the redshift of their spectral lines. It is a straight line, demonstrating the linear relation between redshift and distance, as embodied in the HUBBLE LAW.

Hubble law Law proposed by Edwin HUBBLE in his landmark paper of 1929 claiming a linear relation between the distance of galaxies from us and their velocity of recession, deduced from the redshift in their spectra. The law can be stated as:

$$v = H_0 d$$

where v is the radial velocity of a galaxy, d is the distance to the galaxy and H_0 is a constant now known as the HUBBLE CONSTANT. This linear velocity–distance relation could be readily explained if the Universe as described by general relativity were expanding.

Hubble–Sandage variable *See* S DORADUS STAR

Hubble Space Telescope (HST) Telescope put into orbit by the Space Shuttle *Discovery* STS 31 in 1990 April. The Hubble Space Telescope (HST) is in an orbit inclined to the equator by 28°.5, which is almost circular at an altitude of about 607 km (380 mi). HST was

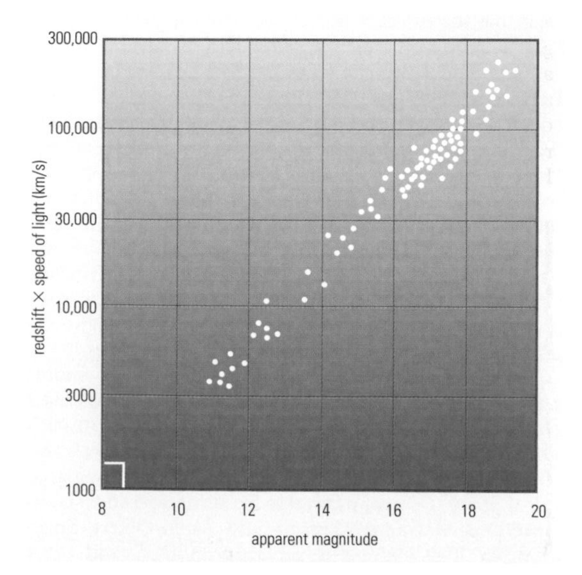

◄ **Hubble diagram** Plotting the apparent magnitude of galaxies relative to redshift produces a straight line, indicating that objects with higher redshifts are more distant.

▲ **Hubble Space Telescope** Launched in 1990, the Hubble Space Telescope has provided astronomers with many results of fundamental importance. Here it is seen soon after release following the second servicing mission, carried out by Space Shuttle astronauts in 1997.

designed to be serviced in orbit by later Space Shuttle crews and to have instruments removed and replaced, including its twin 12.19-m-long (40-ft-long) solar arrays. The 11,600 kg telescope is 13 m (43 ft) long and, at its widest, 4.2 m (14 ft) in diameter. It is equipped with two high-gain antennae, which enable it to transmit direct to the ground using a tracking and data relay satellite system, and two low-gain antennae. HST has a data management system and a fine-pointing system, which allows it to be aimed and remain locked on to any specific target to within 0″.01. The optical telescope assembly is configured in such a way that the telescope is the equivalent of 57.6 m (189 ft) long, but compacted to 6.4 m (21 ft). Light entering the aperture door travels down the tube on to a 2.4-m (94-in) primary mirror and is reflected on to a secondary mirror 0.3 m (12 in) in diameter, where it is reflected through a hole in the centre of the primary mirror on to the focal plane. HST's various science instruments then receive the light.

HST was originally equipped with a Faint Object Camera, Wide-Field/Planetary Camera (WFPC), Goddard High-Resolution Spectrograph (GHRS), Faint Object Spectrograph (FOS), High Speed Photometer (HSP) and Fine Guidance Sensors (FGS). Once the telescope had reached orbit, and despite the fact that some good images of the Universe were taken, it soon became clear that the primary mirror was suffering from SPHERICAL ABERRATION, which apparently arose during manufacture. The Corrective Optics Space Telescope Axial Replacement (COSTAR) unit was flown to the HST aboard Space Shuttle STS 61 in 1993 December and fixed inside the telescope during one of a series of five spacewalks. The dual teams of astronauts also replaced a solar panel, repaired electronics, installed a new computer processor, replaced the WFPC with a new unit, installed magnetometers, removed the HSP and installed a redundancy kit for the GHRS. The results of installing COSTAR were startling: HST images were now truly spectacular and the telescope immediately caught the imagination of the public worldwide.

Another planned HST servicing mission by Shuttle astronauts was launched in 1997 February. The STS 82 *Discovery* crew made comprehensive changes to Hubble's suite of instruments and conducted routine spacewalk servicing. The GHRS and FOS were removed and replaced with a Space Telescope Imaging Spectrograph and a combined Near-Infrared Camera and Multi-Object Spectrometer. The astronauts replaced an FGS and other equipment, installed an optical electronics enhancement kit, and changed the HST stabilization and fine-pointing reaction wheel assemblies. Other work included the laying of new thermal insulation blankets.

Hubble worked on, entering its tenth year of service when it was time to launch a new Shuttle servicing mission. This was planned originally for 2000 but when gyroscopes on the telescope failed to a critical level, the mission was brought forward to 1999 December and split into two, with the second half of the mission being shifted to 2001. STS 103 *Discovery* was launched in 1999 December, and the mission installed six new gyros and voltage/temperature kits, a new computer 20 times faster and with six times the memory, a new digital tape recorder and a replacement FGS and a radio transmitter. The spacewalking astronauts also placed some new insulation over part of the HST. STS 109 *Columbia* in 2002 March installed new solar arrays and an Advanced Camera for Surveys, a Cosmic Origins Spectrograph and a third WFPC.

Hubble's Variable Nebula REFLECTION NEBULA in northern Monoceros (RA 06^{h} 39^{m}.2 dec. +08°44′); it covers a V-shaped area of 3′.5 × 1′.5. The nebula's brightness and appearance change in response to variations in the light output of the illuminating star, R Monocerotis. Hubble's Variable Nebula has the distinction of being the first object photographed with the 200-inch (5-m) Hale Telescope at Mount Palomar, California, in 1949. It lies at a distance of 2600 l.y.

Huchra, John Peter (1948–) American astrophysicist and observational astronomer at the Harvard–Smithsonian Center for Astrophysics (CfA) who has collaborated with Margaret GELLER and others to map the three-dimensional large-scale distribution of galaxies in the Universe. He has taken part in the CfA Redshift Survey, which showed that galaxies are arranged in sheets and clumps, separated by large soap bubble-like voids tens of millions of light-years wide. Huchra is also involved with the TWO MICRON ALL SKY SURVEY (2MASS). His other major research interest is the extragalactic distance scale.

Huggins, William (1824–1910) English amateur astrophysicist and pioneer of astronomical spectroscopy. He used an inheritance to build an observatory at Tulse Hill, south London with a 12-ft (3.7-m) dome, equipped with a 5-inch (130-mm) equatorial by John DOLLOND, and a separate transit telescope room. In 1858 he acquired an 8-inch (200-mm) refractor with an objective by ALVAN CLARK & SONS. Huggins became interested in astronomical spectroscopy after learning of Gustav KIRCHHOFF's spectroscopic investigations of the Sun. With his neighbour, William Allen Miller (1817–70), professor of chemistry at King's College, Huggins used two dense flint glass prisms to construct a new instrument, the stellar SPECTROSCOPE, which they attached to the Clark refractor. To compare stellar spectra with the spectral lines produced by known chemical elements, Huggins equipped his observatory with batteries, Bunsen burners and other chemical apparatus, making it unlike any other astronomical observatory then in existence.

By 1863 Huggins and Miller had published their first observations of stellar spectra. The next year, Huggins found that some nebulae displayed monochromatic spectra typical of gases, whereas the 'spiral nebulae' showed continuum, stellar-type spectra. He concluded that, contrary to what was thought at the time, not all nebulae were aggregations of stars too faint to be resolved. In 1866 he was the first to examine the spectrum of a nova, which displayed a series of bright hydrogen lines produced by the very hot gaseous emissions typical of these objects. In 1868 he made the first detailed spectroscopic investigation of a comet, confirming the bright C_2 'Swan bands'. That same year, Huggins and Miller made the first stellar RADIAL VELOCITY measurement, obtaining a value of 47 km/s (29 mi/s) for Sirius.

In 1875 Huggins married Margaret Murray, herself an accomplished amateur astronomer and telescope-maker. Lady **Margaret Lindsay Huggins** (1848–1915) collaborated with her husband for 35 years. As early as 1863, Huggins had photographed the spectra of Sirius and Capella, but was frustrated by the impracticalities of the 'wet collodion' plates then available. In 1876 he photographed the spectrum of Vega on a gelatine dry plate, the first such application by an astronomer. The dry plates allowed longer exposures and thus fainter details could be captured, including features in the near-ultraviolet part of the spectrum. Over the next 30 years, Huggins obtained numerous photographic spectra proving that many of the chemical elements known on Earth (for example, hydrogen, calcium, sodium and iron) are also present in stars and nebulae.

Hulse, Russell Alan (1950–) and **Taylor, Joseph Hooton, Jr** (1941–) American physicists whose 1974 discovery of the first binary pulsar provided strong evidence for the gravitational waves predicted by Einstein's general theory of relativity. Using Arecibo Observatory's 305-m (1000-ft) radio telescope, they found a pulsar designated PSR 1913+16, a neutron star of 1 solar mass and diameter 10 km (6 mi), with a close, equally massive companion. Observations showed that this exotic stellar system has a decaying orbit, and Hulse and Taylor deduced that the binary pulsar is losing energy by emitting gravity waves. For their discovery they received the 1993 Nobel Prize for Physics.

Hulst, Hendrik Christoffel van de (1918–2000) Dutch physicist and pioneer radio astronomer who in 1944 predicted that interstellar HI regions (gas clouds composed of neutral hydrogen) emit radio waves at a wavelength of 21 cm. Radio studies of the Milky Way (and other galaxies) by observing the 21-cm emission, first done by Edward Mills Purcell (1912–97) and Harold Irving Ewen (1922–) in 1951, have allowed galactic structure to be mapped more accurately than by using optical methods alone. Van de Hulst also researched the nature of the solar corona and the interstellar medium.

Humason, Milton Lasalle (1891–1972) American astronomer who made many of the earliest redshift measurements of galaxies. He had no formal training, and was at first a mule driver and janitor at MOUNT WILSON OBSERVATORY, but in 1917 became a night assistant on the 60-inch (1.5-m) telescope there, working under George Ellery HALE. In early work, Humason became an expert observer and was Edwin HUBBLE's chief assistant, measuring the redshifts for many 'nebulae' and helping to show that they were actually galaxies similar to but far beyond the Milky Way. This work, begun in 1928, led to the discovery that the Universe is expanding. Over the next three decades, Humason collaborated with astronomers such as Allan SANDAGE, using the largest telescopes in the world – the 100-inch (2.5-m) at Mount Wilson and the 200-inch (5-m) at Mount Palomar – to measure the redshifts of hundreds of fainter, more distant galaxies. The results of this work, published in 1956, showed that Hubble's law was valid out to a recessional velocity of 60,000 km/s (37,000 mi/s).

Humorum, Mare (Sea of Humours) Lunar lava plain located in the south-west quadrant of the Moon. The geology of this region reveals that it was struck by an enormous object, producing a multi-ring impact BASIN. Lava later flooded the inner parts, while impact erosion deeply degraded the outer rings, though these are still visible in the western region. An inner mare ridge forms a circle, revealing the presence of an inner ring. Prominent arcuate rilles lie in both the eastern and western regions of Humorum.

Hungaria asteroid Any member of a group of ASTEROIDS with orbits similar to (434) Hungaria. Found close to the inner edge of the MAIN BELT, these asteroids are objects with low eccentricity orbits and relatively high inclinations. When viewed from the Earth, Hungaria asteroids display large declination motions but lower right ascension movements than other main-belt asteroids.

hunter's moon First full moon following the HARVEST MOON. This usually occurs in October in the northern hemisphere, providing a succession of moonlit evenings. In the southern hemisphere, it occurs in April or May.

Huygenian eyepiece Basic EYEPIECE consisting of two simple plano-convex elements, commonly used on small refractors. The eye relief is good, the angular field of view is large and the eyepiece is relatively free from CHROMATIC ABERRATION. It is named after Christiaan and Constantijn HUYGENS, who invented it in 1703.

Huygens, Christiaan (1629–95) Dutch physicist and astronomer, born into a prominent family in The Hague. After studying mathematics and law at the universities of Leiden and Breda, Huygens worked at the Bibliothèque Royale in Paris (1666-81), before returning to The Hague. He and his brother **Constantijn Huygens** (1628–97) became skilled opticians and telescope-makers, designing the achromatic HUYGENIAN EYEPIECE. Christiaan Huygens' first telescope, which he made himself, had a focal length of 3.5 m (11.5 ft). He used it in 1655 to discover TITAN, determining that this satellite revolved about Saturn in 16 days.

With a better instrument of 7-m (23-ft) focal length, Huygens made observations of Saturn's rings that were detailed enough for him to deduce their true nature. GALILEO had seen these changing features as lobes or 'arms', but Huygens correctly described them as a single, thin ring system, detached from Saturn's globe, and published his results in *Systema saturnium* (1659).

Huygens also made many fundamental contributions to physics that are relevant to astronomy. He explained the optical phenomena of refraction and reflection, and was the first to describe the wave nature of light in his magnum opus, *Traité de la lumière* (1678). He derived the formula $F = mv^2/r$ to describe centrifugal force. In 1656 he patented the first pendulum clock, which would prove useful in astronomical timekeeping and for finding longitude at sea.

Huygens European spacecraft scheduled to land on TITAN, Saturn's largest moon, in 2004. Huygens will ride to the Saturn system aboard CASSINI, landing on the surface of Titan in 2004 November. During its parachute descent, Huygens will study Titan's atmosphere, currently understood as an orange-red smog that rains liquid methane. Huygens may still be operating when it

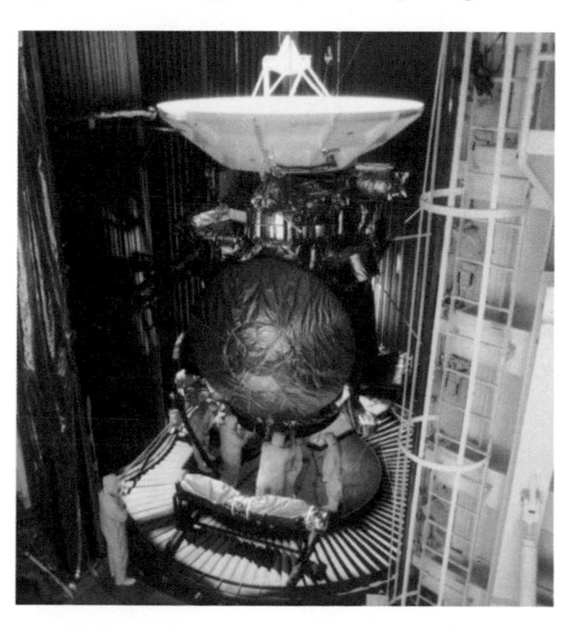

◄ **Huygens** The conical Huygens probe (lower centre) is seen here attached to the Cassini spacecraft during pre-launch testing. Huygens is scheduled to make a descent to the surface of Saturn's moon Titan in 2004.

H

lands on the surface – becoming the first probe to do so on a moon other than our own – the nature of which remains uncertain.

Hyades Second nearest OPEN CLUSTER to the Sun. The Hyades has more than 200 member stars, most of which lie within a distance of 20 l.y. It covers an area of about 6° in the constellation of TAURUS. The brightest stars, together with Aldebaran, which is a foreground star and not a member of the cluster, form a distinctive V shape. Recognized since antiquity, and even mentioned in Homer's *Iliad*, the Hyades derived its name from a group of daughters of Atlas and Aethra, half-sisters of the PLEIADES. The cluster is about 600 million years old.

In the 19th century, from studies of stellar PROPER MOTIONS, it was realized that the Hyades cluster moves in a general south-easterly direction, together with numerous other stars in the surrounding parts of the sky, in what is called the Taurus stream. Refined proper motion measurements subsequently revealed that the Hyades shares a common motion, to within very narrow limits, with an appreciable number of stars in its vicinity. This cluster was designated the Taurus moving cluster; the name Hyades was often used to denote the central denser clustering, which includes the original stars. Today the name Hyades is generally used for the whole cluster, and individual members are sometimes termed Hyads.

The distance to the Hyades has been determined in several ways, most recently by the Hipparcos satellite. It lies at a distance of around 151 l.y. This distance can be used to derive absolute magnitudes for cluster members of different kinds, and these values are in turn used as calibrators for determinations of distances to other stellar aggregates, also containing intrinsically brighter objects. In this way the distance to the Hyades forms one crucial step on the distance ladder of the Universe.

The majority of the Hyades cluster member stars belong to the MAIN SEQUENCE, the brightest of these being of spectral type A2. It is from this fact that the age estimate is derived, since more massive stars have had time to evolve away from the main sequence. A few such stars are known in the Hyades; they are now yellow GIANT STARS. Also belonging to the cluster are a score of WHITE DWARF stars. A considerable share of the cluster members is believed to consist of BINARY STAR systems.

The central concentration, primarily of the more massive stars, indicates that the cluster is nearly relaxed, meaning that the stars have had time to distribute the available kinetic energy equally amongst themselves. The relaxation is, however, not complete, as is seen from the somewhat ragged distribution of stars in the outskirts of the cluster.

Due to the close distance of the Hyades, there is no apparent EXTINCTION of light by cosmic dust in front of the stars, and also no dust has been found within the cluster. *See also* URSA MAJOR MOVING CLUSTER

Hyakutake, Comet (C/1996 B2) Long-period comet discovered on 1996 January 31 by Japanese amateur astronomer Yuji Hyakutake (1950–2002). The comet made a remarkably close passage (0.10 AU) to Earth on March 25, becoming a prominent naked-eye object. Around the time of closest approach, C/1996 B2 Hyakutake was particularly well placed for observers in the northern hemisphere, lying close to Polaris, with a bluish ion tail stretching for up to 70° through the Plough and beyond. Several DISCONNECTION EVENTS were observed in the ion tail. The COMA reached a peak magnitude of −1. Comet Hyakutake faded somewhat after its near-Earth approach, but brightened again in late April as it neared its May 1 perihelion, 0.23 AU from the Sun. The comet, possibly making its first visit from the OORT CLOUD to the inner Solar System, has an orbital period of about 14,000 years.

Around 1996 May 1, the ULYSSES spacecraft recorded anomalies in the SOLAR WIND, later determined to be due to an encounter with the ion tail of Comet Hyakutake. Stretching some 3.8 AU downstream from the nucleus, the comet tail is by far the longest yet found.

Hydra *See* feature article

hydrocarbons *See* INTERSTELLAR MOLECULES

hydrogen (symbol H) First and lightest of the chemical elements; it is the most abundant element in the Universe (*see* ASTROCHEMISTRY). Hydrogen-1 (that is, an atom with a single proton as its nucleus with a single electron attached to it) forms almost 100% of the naturally occurring element and has an atomic number of 1 and an atomic mass of 1.007825 amu. Its boiling point is 20.29 K and its melting point is 14.02 K. Hydrogen-2, which is also known as DEUTERIUM, is present on Earth in the proportion of 0.015%. A third isotope, hydrogen-3, also known as tritium, is radioactive, with a half-life of 12.26 years; it is produced in nuclear reactors. The name hydrogen derives from the Greek *hydro* (water) and *genes* (forming), and was first used by the French chemist Antoine Lavoisier (1743–94). Hydrogen was recognised as a distinct substance, which he called 'inflammable air', in 1766 by the English chemist Henry Cavendish (1731–1810). In the visible region, hydrogen produces the familiar BALMER LINE spectrum, which arises from transitions to and from hydrogen's first excited level. Transitions to and from the ground state produce the ultraviolet LYMAN SERIES; from higher excited levels, a number of sets of lines in the infrared are produced, starting with the PASCHEN SERIES.

Our most familiar contact with hydrogen is in the form of one of its two oxides, water (H_2O, the other is hydrogen peroxide, H_2O_2), or more rarely as the gas that appears when an acid and a metal react. It is widely used in industry, to make metals and margarine, to fix nitrogen via the Haber process and, in liquid form, as a convenient, if hazardous, rocket fuel. Its presence in the molecules of carbohydrates also makes it fundamental to the existence of life.

Beyond our immediate environment, the everyday chemistry of hydrogen is of little consequence. Within the

▼ **Hyakutake, Comet** Reaching a peak brightness of magnitude −1 in 1996 March, Comet Hyakutake made a close approach to Earth. The comet was particularly notable for its strong bluish ion tail, which showed much fine structure and underwent several disconnection events.

Sun, for example, the immense pressures and temperatures of the interior ionize the atom into free protons and electrons and so turn hydrogen into a nuclear fuel (*see* NUCLEOSYNTHESIS). At pressures in excess of 10^{11} Pa, solid molecular hydrogen starts to behave like a metal, and in this form it is thought to make up the central regions of the Jovian planets. The Milky Way contains huge amounts of non-luminous material that is mostly atomic and molecular hydrogen. About half of the mass of our Galaxy is hydrogen gas and most of the rest is in the form of stars that are again largely hydrogen. Hydrogen between the stars is difficult to detect because it is usually at a low temperature, and the lowest energy spectrum lines from both atomic and molecular hydrogen are FORBIDDEN LINES. Nonetheless, atomic hydrogen is so abundant that the forbidden emission line at 21 cm (the TWENTY-ONE CENTIMETRE LINE), which results from the electron switching its direction of spin, is easily picked up by radio telescopes. The presence of hydrogen can also be inferred through its association with INTERSTELLAR DUST particles, which absorb light very strongly and are therefore easily detectable. If the hydrogen gas is in the vicinity of hot stars, ultraviolet radiation from those stars will ionize the hydrogen, producing an HII REGION embedded within the surrounding cold HI REGION. *See also* HYDROGEN SPECTRUM; STRÖMGREN SPHERE

hydrogen-alpha line (Hα line) Spectrum line at 656.3 nm caused by the transition between orbits (energy levels) 2 and 3 of the hydrogen atom. In emission, the electron jumps from orbit 3 to orbit 2, in absorption it jumps from 2 to 3. *See also* BALMER LINES; HYDROGEN SPECTRUM

hydrogen emission region Cloud of hot gas in interstellar space that is principally visible because of the emission of the Balmer series of spectrum lines from hydrogen. The strongest BALMER LINE is the HYDROGEN-ALPHA LINE (H-α), which produces the red glow observed from many EMISSION NEBULAE. *See also* HII REGION

hydrogen spectrum Set of EMISSION LINES or ABSORPTION LINES produced by the neutral hydrogen atom. Hydrogen is surrounded by an infinite number of potential quantized orbits, each with different energy. While the orbital radii increase in size from the ground state orbit ($n = 1$) according to n^2, their energies approach a limit at 13.6 eV above the ground state.

Upwards or downwards transitions between the energy states form different series of absorption or emission lines, which are named and lettered according to the lowest state's number. The lines that arise from, or land on, the ground state ($n = 1$) belong to the LYMAN SERIES. The line connecting level 1 to 2 is Lyman α (at 121.6 nm), 1 to 3 Lyman β, and so on, with the series converging to the Lyman limit at 91.2 nm. Photons short of the Lyman limit ionize planetary and diffuse nebulae.

Transitions underlain by level 2 are called the BALMER LINES; they begin at Hα (656.3 nm) and go through Hβ, Hγ and Hδ at 486.1, 434.0 and 410.1 nm, ending at the Balmer limit at 364.6 nm. The infrared PASCHEN SERIES begins and ends on level 3 (Paschen α at 1875.1 nm, with the limit at 820.3 nm), the Brackett series on level 4 (from 4.0512 μm to 1.4584 μm), the Pfund series on level 5, and the Humphries series on level 6. Radio transitions between high states arise from diffuse and planetary nebulae; the first found was the H109α line (level 110–109) at 5007 MHz. Radio lines have been observed nearly to level 300.

Any single-electron ion will have a hydrogen-like spectrum, except that the wavelengths will be shifted downward by Z^2, where Z is the charge on the nucleus. The same nomenclature is used for such ions. The α line of the Paschen series of ionized helium falls at 468.6 nm, while its Brackett series (the PICKERING SERIES) overlays the hydrogen Balmer series.

Hydrus *See* feature article

HYDRA (GEN. HYDRAE, ABBR. HYA)

Largest of the 88 constellations, extending over a quarter of the way around the sky from its head, which adjoins Canis Minor, to its tail, next to Libra. Hydra represents the many-headed water-snake killed by Hercules. For all its size, its only star above mag. 2 is α Hya, known as Alphard. ε Hya is a close binary for apertures over 75 mm (3 in.), mags. 3.4 and 6.7. R Hya is one of the brightest MIRA STARS, reaching mag. 3.5 at maximum, with a minimum of 10th magnitude and a period of 390 days, while U Hya is a semiregular variable ranging from mag. 4.2 to 6.6 in just under 4 months. M48 is a large open cluster well seen through binoculars, and NGC 3242 is a 9th-magnitude planetary nebula known as the GHOST OF JUPITER. Perhaps the best-known object in Hydra is M83, a face-on spiral galaxy some 15 million l.y. away.

BRIGHTEST STARS

	Name	RA h m	dec. ° ′	Visual mag.	Absolute mag.	Spectral type	Distance (l.y.)
α	Alphard	09 28	−08 40	1.99	−1.69	K3	177
γ		13 19	−23 10	2.99	−0.05	G8	132
ζ		08 55	+05 57	3.11	−0.21	G8	151
ν		10 50	−16 12	3.11	−0.03	K1	139
π		14 06	−26 41	3.25	0.79	K2	101
ε		08 47	+06 25	3.38	0.29	G5	135

Hygeia Fourth-largest MAIN-BELT ASTEROID, discovered in 1849; number 10. Hygeia has a diameter of about 408 km (254 mi). It is a C-type asteroid and is particularly dark, with an albedo of 0.04.

Hyginus, Rima Lunar feature composed of a series of volcanic collapse pits set in a generally V-shaped rille that intersects the crater Hyginus. The rille is beset with many craterlets, which formed where the magma withdrew.

hyperbola CONIC SECTION obtained when a right circular cone is cut by a plane that makes an angle with its base greater than that made by the side of the cone. It is an open curve (not closed like a circle or ellipse), with an ECCENTRICITY greater than 1. If an object is in a hyperbolic orbit then it will escape from the primary body. Long-period comets enter the inner Solar System in highly elliptical or parabolic orbits, but never hyperbolic. For a few of them the perturbations by the planets change the departing orbit into a hyperbola, and thus these comets are ejected from the Solar System. When a spacecraft has a close flyby to a planet it enters a temporary hyperbolic orbit around the planet.

hyperboloid Shape that is obtained when a HYPERBOLA is rotated about its axis. Convex hyperboloid mirrors are conventionally used as the SECONDARY MIRROR in a reflecting CASSEGRAIN telescope in conjunction with a parabolic PRIMARY MIRROR.

hyperfine structure Splitting of atomic energy levels by the magnetic interaction between the spins of the electron and nucleus. The best-known example is the splitting of the ground state of hydrogen into two closely spaced levels. In the upper level, the electron and proton

HYDRUS (GEN. HYDRI, ABBR. HYI)

Small, far-southern constellation, between the two Magellanic Clouds, introduced by Keyser and de Houtman at the end of the 16th century. It represents a small water-snake. There is little for the casual observer apart from π Hyi, a binocular double of mags. 5.6 and 5.7.

BRIGHTEST STARS

Name	RA h m	dec. ° ′	Visual mag.	Absolute mag.	Spectral type	Distance (l.y.)
β	00 26	−77 15	2.82	3.45	G2	24
α	01 59	−61 34	2.86	1.16	F0	71
γ	03 47	−74 14	3.26	−0.83	M2	214

spin in the same direction; in the lower, they spin oppositely. The transition between the two levels creates the TWENTY-ONE CENTIMETRE LINE of interstellar neutral hydrogen. Hyperfine splitting in hydroxyl (OH) underlies the interstellar OH lines at 1667, 1712, 1612 and 1665 MHz, the last two being powerful MASERS associated with star-forming regions.

Hyperion Largest, irregular-shaped SATELLITE of SATURN; it measures 370 × 280 × 225 km (230 × 174 × 140 mi). Hyperion was discovered in 1848 jointly by W.C. and G.P. BOND in the United States and William LASSELL in England. It is dark and reddish and may be a fragment of the interior of a larger satellite destroyed by a collision. Hyperion's reddish colour and irregular shape could also be consistent with it being perhaps a captured cometary planetesimal or asteroid. Hyperion is one of the few satellites to have an orbital period that is not synchronous with its orbit (*see* SYNCHRONOUS ROTATION). In fact, as Hyperion progresses round its orbit it tumbles in a seemingly random way, with both its rotation period and the axis of rotation itself changing in a chaotic fashion. This random movement is probably caused by Hyperion's eccentric orbit and elongated shape, which lead to a strongly variable tidal pull from Saturn. *See* data at SATURN

hypersensitization Technique in which film for long exposures in ASTROPHOTOGRAPHY is heated or treated chemically to overcome RECIPROCITY FAILURE. One way to increase the performance of film is to bake it for several hours, but this has now been superseded by chemical hypersensitization (known as hypering), in which the film is soaked for several hours in hydrogen gas, or a gas that contains a proportion of hydrogen.

hypervelocity impact Impact with velocity high enough to generate a stress of an order of magnitude larger than the compressive strength of the target material. Velocities at which impacts attain hypervelocity are different for different materials but are typically in the range of 1 to 10 km/s (0.6–6 mi/s). Hypervelocity impact into solid surfaces of planets and satellites results in the formation of an impact CRATER. Accompanying phenomena include impact melting and vaporization of the impactor and target material, and formation of high-density mineral phases, such as diamond at the expense of graphite.

◀ **Iapetus** A distant Voyager 2 view of Saturn's moon Iapetus, showing features as small as 19 km (12 mi) across. Iapetus is remarkable in having a very dark leading hemisphere which contrasts with the brighter trailing hemisphere.

Iapetus Outermost of the main SATELLITES of SATURN. It is the third-largest but was the second to be discovered, by G.D. CASSINI in 1671, thanks to its great distance from the glare of the planet. It has the remarkable property of appearing much fainter when lying to the east of Saturn in the sky, when we see its leading hemisphere, than when lying to the west, when we see its trailing hemisphere. This observation can be explained by the fact that most of the leading hemisphere has a sooty coating with an albedo of less than 0.1, whereas the trailing hemisphere is relatively clean ice with an albedo of about 0.5. The best VOYAGER images revealed that the trailing hemisphere is heavily cratered, but they were not capable of showing any features in the opposite hemisphere. It has been suggested that the dark material is dust originating from PHOEBE, a dark red irregular satellite in a retrograde orbit outside that of Iapetus, but the spectroscopic match is rather poor. *See* data at SATURN

IAPPP Abbreviation of INTERNATIONAL AMATEUR–PROFESSIONAL PHOTOELECTRIC PHOTOMETRY

IAU Abbreviation of INTERNATIONAL ASTRONOMICAL UNION

IC Abbreviation of *INDEX CATALOGUE*

Icarus APOLLO ASTEROID discovered in 1947. Icarus possesses a particularly small PERIHELION distance (0.187 AU), which led to its being named after the myth of the son of Daedalus, who flew too close to the Sun. The early determination of its orbit led to the recognition that Icarus makes repeated close approaches (within 0.04 AU) to the Earth. It gained wide fame as a result of the publication in 1967 of Project Icarus, a student exercise at the Massachusetts Institute of Technology in which the feasibility of deflecting this object, assumed to be heading for our planet, was studied.

ICE Abbreviation of INTERNATIONAL COMETARY EXPLORER

Ida MAIN-BELT ASTEROID, number 243, examined by the Jupiter-bound GALILEO spacecraft in 1993 August. Ida was the second asteroid (after GASPRA) for which such imagery was obtained.

Ida is elongated in shape, with a narrow waist, and measures about 56 × 24 × 21 km (35 × 15 × 13 mi). It is heavily cratered from impacts by smaller bodies. Its spectral reflectance indicates it to be an S-class asteroid, composed largely of metal-rich silicates of the type that make up the meteorites known as ordinary chondrites. Galileo images revealed Ida to be accompanied by a small moon, DACTYL, the orbital distance of which is consistent with Ida having an average density of between 2 and 3 g/cm^3. Ida's surface is heavily cratered and has linear grooves and a scattering of large blocks or boulders. It is thought that Ida is coated with a thick regolith. Its interior comprises a number of discrete components, making the entire object a type of rubble pile.

IDA Abbreviation of INTERNATIONAL DARK-SKY ASSOCIATION

igneous rock Rock formed as a result of cooling of molten material. Formerly, rocks formed from the cooling of impact melts were considered to be igneous, but geologists now only consider rocks to be igneous if they are derived from melts formed in the interiors of planets or satellites, usually as a result of decompression of high temperature solid material from the interior, as, for example, in a hot mantle PLUME. Igneous rocks formed on the surface of a planet or satellite, where the melt cools rather fast, are known as volcanic rocks; they typically contain varying proportions of silicate glass together with mineral crystals. Igneous rocks formed at depth in the interior, where the cooling is much slower, are known as plutonic rocks; they are fully crystalline. The principal minerals in igneous rocks are olivine $(Mg,Fe)_2SiO_4$, pyroxene $(Mg,Fe,Ca)_2Si_2O_6$, feldspars $CaAl_2Si_2O_8$-$(Na,K)AlSi_3O_8$ and quartz SiO_2. The most widespread volcanic igneous rock on the surface of TERRESTRIAL PLANETS and non-icy satellites is BASALT. Ices on the surface of satellites of giant planets, if formed from non-impact melts, should be considered as igneous rocks.

◀ **Ida** Asteroid (243) Ida imaged by the Galileo spacecraft in 1993 August. Ida's tiny satellite Dactyl is visible at right.

Ikeya–Seki, Comet (C/1965 S1) Spectacular, bright comet, discovered independently by the Japanese observers Kaoru Ikeya (1943–) and Tsutomu Seki (1930–) on 1965 September 18. The comet, a KREUTZ SUNGRAZER, reached perihelion on October 21, 0.008 AU (1,200,000 km/750,000 mi) from the Sun. At perihelion, Comet Ikeya–Seki had an estimated peak mag. −10, and could be seen in daylight. During perihelion passage, the nucleus broke into three fragments. The tail reached a maximum length of 60° soon after perihelion and was a spectac-

◀ **Ikeya–Seki, Comet** Photographed in the pre-dawn skies of 1965 October, this spectacular sungrazing comet had a long, bright tail. As it passed perihelion, the nucleus split into three fragments.

ular sight in the pre-dawn sky for observers at southerly latitudes. The orbit is elliptical, with a period of 880 years.

IMAGE Acronym for IMAGER FOR MAGNETOPAUSE TO AURORA GLOBAL EXPLORATION

image intensifier Opto-electronic device for amplifying the brightness of an image. The optical image from a telescope is focused on to a photocathode, causing electrons to be liberated, the number of which depends on the intensity of the incoming optical signal and its wavelength. The electrons are accelerated through a potential of around 40 kV and focused by an electric or magnetic field on to a phosphor screen. In addition to being accelerated, the number of electrons may be increased by cascading successive photocathode stages (*see also* CASCADE IMAGE TUBE). A drawback of the system is the slight distortion of the image caused by the electronic focusing.

image photon-counting system (IPCS) Electronic imaging device sensitive enough to detect individual photons of light. The system comprises a high-gain IMAGE INTENSIFIER on the front of which is mounted a photocathode. A photon falling on the photocathode releases an electron, which produces a cascade of further electrons through the image instensifier, which are detected by a CCD. An important feature of the system is the centre detection logic, which identifies the position on the photocathode of each photon event, thus providing increased spatial resolution. As each individual photon is counted, a two-dimensional image is built up in the detector memory. The system is used mainly for high-dispersion SPECTROSCOPY of faint objects.

image processing Science and practice of extracting information from digital images, or from images on traditional photographic emulsion which have been scanned and converted to digital form. Most professional and increasing numbers of amateur astronomers carry out image processing to glean the maximum amount of information contained in their astronomical images. With the ready availability of low-cost computers and CCDS, amateur astronomers are able to apply complex processing routines and produce results that match or exceed those achieved by their professional counterparts only a decade ago.

In general terms, image processing is the manipulation of information recorded by the pixels (light-sensitive areas) that make up a modern astronomical detector. The most basic form of image processing is **linear scaling**, where individual pixels on an image are manipulated to enhance brightness and contrast. This type of processing is useful to remove the effects of LIGHT POLLUTION from the sky background or to enhance faint features such as the arms of spiral galaxies. Another method of boosting the level of information is by the co-addition or stacking of multiple images to improve the SIGNAL-TO-NOISE RATIO of faint astronomical objects.

More complex is the use of **logarithmic scaling** to boost selectively or withhold brightness values on images that contain a wide range of intensity levels. When applied to images of globular clusters, for example, a logarithmic scale extracts information contained in the bright core region, which would normally be overexposed, yet maintains a high level of detail in the outer reaches of the cluster. In a new technique called **digital development processing**, a dual routine simultaneously compresses the brightness levels contained in an image and applies UNSHARP MASKING to produce unprecedented levels of detail.

Spatial filters are used to sharpen images or to blur as a means of reducing instrument noise or processing artefacts. Other useful routines include positional translations, notably rotation and scaling of image size. Many astronomers use these functions when combining images for the analysis of transient events, for example in supernova and asteroid searches. One of the most powerful of all image-processing procedures is the use of **fast Fourier transforms**. In principle, an image is broken down into a discrete spread of spatial frequencies, to which mathematical operations are applied before the image is reconstructed. By varying the type of operations applied, many different forms of image manipulation are possible. Perhaps the best known of these is **deconvolution**. Indeed, this method was used to correct images that were taken by the original faulty mirror of the Hubble Space Telescope. For ground-based astronomy, deconvolution is used to improve resolution that has been degraded by viewing through an unsteady atmosphere.

Most CCD detectors can only take images in shades of grey, so to build a colour picture it is necessary to take multiple exposures of astronomical objects through different coloured FILTERS. A common method is to employ red, green and blue filters and to take an exposure through each of these in turn, although other filter combinations are possible. Once the images are stored on a computer, they can be combined and image processing applied to synthesize a colour image and to extract useful scientific information. If care is taken during the combination and processing of these image sets, it is possible to produce

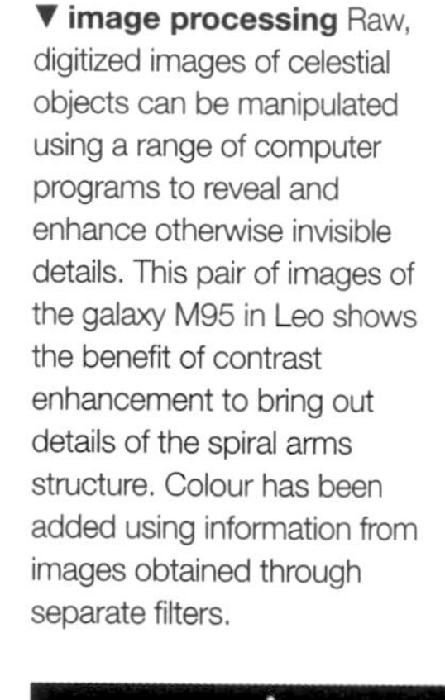

▼ **image processing** Raw, digitized images of celestial objects can be manipulated using a range of computer programs to reveal and enhance otherwise invisible details. This pair of images of the galaxy M95 in Leo shows the benefit of contrast enhancement to bring out details of the spiral arms structure. Colour has been added using information from images obtained through separate filters.

true-colour images, which is very difficult to do using modern photographic films.

Imager for Magnetopause to Aurora Global Exploration (IMAGE) NASA science satellite launched into a polar Earth orbit in 2000 March. IMAGE uses neutral atom, ultraviolet and radio-imaging techniques to study the magnetic phenomena involved in the interaction of the SOLAR WIND with the Earth. Mission objectives are to identify the dominant mechanisms for injecting plasma into the Earth's magnetosphere, to determine the response of the magnetosphere to changes in the solar wind and to discover how and where magnetospheric plasmas are energized and transported.

image scale (plate scale) Relationship between angular measure in the sky and linear measure at an image plane, usually as recorded on film or CCD, typically measured in arcseconds per millimetre. Its value in degrees is given by $57.3/f_{eff}$, where f_{eff} is the EFFECTIVE FOCAL LENGTH of the optical system.

Imbrium, Mare (Sea of Showers) Lunar lava plain, roughly 1300 km (800 mi) in diameter, located in the north-west quadrant of the Moon. It represents flooding of a multi-ring impact basin, produced by the impacts of an asteroid approximately 100 km (60 mi) in diameter. The lava was generally contained by the outer wall, but it flooded over the inner rings. The highest peaks of these inner rings are still visible above the lava (for example Mons Piton), and the rings are marked by circular mare ridges. Apollo 15 visited Hadley Rille in Mare Imbrium.

immersion Disappearance of a star or planet behind the Moon's leading limb at an occultation. The term may also be used to describe the Moon's entry into Earth's umbra at a lunar eclipse.

IMO Abbreviation of INTERNATIONAL METEOR ORGANIZATION

impact feature Any feature formed by METEOROID impacts on the surface of planets and satellites. Impact features include impact CRATERS, both primary and secondary, of any size from microcraters to multi-ring BASINS, all varieties of EJECTA from impact craters, as well as impact-generated fractures and faults. The dark spots resulting from the impact of fragments of Comet SHOEMAKER–LEVY 9 on Jupiter in 1994 are examples of transient impact features.

inclination (of an orbit) Angle between the plane of an ORBIT of a body and a suitable fixed reference plane. In the Solar System the reference plane is chosen on dynamical considerations. PERTURBATIONS on an orbit cause the LINE OF NODES to regress around the plane of action of the dominant perturbing forces while maintaining constant inclination to this plane, and thus this is the most suitable plane to use as the reference plane. For the planets, the main perturbations come from the other planets; as these all lie close to the ECLIPTIC plane this is the chosen reference plane. For artificial satellites of the Earth and most satellites of other planets, the equatorial bulge (oblateness) of the planet is the dominant perturbation, and so the equatorial plane is chosen as the reference plane. The Moon is relatively very distant from the Earth (in units of Earth radii), and solar perturbations are much larger than the oblateness perturbation, so the ecliptic is chosen as the reference plane. For a few satellites the oblateness and solar perturbations are of similar size, and then the appropriate reference plane, named the Laplacian plane, lies somewhere between the equator and the planet's orbit plane. This effect was first noted by Pierre LAPLACE in 1805 for the Saturnian satellite Iapetus. The effect also occurs for the Earth's geostationary satellites, for which the Laplacian plane is inclined at 7° to the equator.

▲ **impact feature** A satellite image of the circular Manicouagan impact structure in Quebec, Canada. The 70 km (44 mi) diameter feature was formed by an asteroid impact about 212 million years ago.

For binary stars the inclination of the orbit is measured relative to the plane at right angles to the line of sight. The angle is zero if the orbit is seen in plan, and 90° if seen in profile.

inclination (of a planet's equator) Angle between the equatorial plane of a planet or satellite and its orbital plane; this is equivalent to the angle between the axis of rotation and the perpendicular to the orbital plane. For the Earth this angle is named the OBLIQUITY OF ECLIPTIC.

Index Catalogues (IC) Two supplements to the *NEW GENERAL CATALOGUE* (NGC), the *Index Catalogue* (1895) and *Second Index Catalogue* (1908), between them adding another 5386 galaxies, nebulae and clusters (in one continuous numbered sequence) to the NGC's 7840. Like the NGC, the *Index Catalogues* were compiled by J.L.E. DREYER; some objects in them are still referred to by their IC numbers.

Indian astronomy Astronomy as practised in India from ancient times until the 18th century, when Western European astronomy became prevalent. The precise origins of Indian astronomy are unknown, but probably date from the Indus Valley civilization (*c.*2000 BC). Some practical astronomy existed during the Vedic period, which preceded the foundation of the Buddhist and Jain religions around 500 BC.

The early Indian calendar was luni-solar, based on 12 lunar months with extra months inserted as necessary to keep it in step with the solar year.

The beginning of scientific astronomy in India, based partly on Greek knowledge, dates from about AD 500, during the lifetime of ĀRYABHATA, who introduced the use of sines and other mathematical techniques to the prediction of solar, lunar and planetary positions. These he extrapolated from observations made during his own lifetime. His and other early astronomical writings were written in verse and are exceedingly cryptic in form, which led to later scholars doubting they contained anything of scientific value (modern opinion has removed this doubt). Āryabhata believed that the Earth rotated on its axis once a day, but this view was not accepted by later writers.

BRAHMAGUPTA was the other great Indian astronomer of this period, and his works and those of Āryabhata remained influential. At this time Indian astronomy began to spread to other parts of Asia. From the start of the Moghul period in the 16th century, Islamic astronomy gained influence. The Moghul ruler Jai Singh II (1688–1743) had a special interest in Islamic astronomical knowledge. The mammoth instruments he constructed at

INDUS (GEN. INDI, ABBR. IND)

Southern constellation, between Grus and Pavo, introduced by Keyser and de Houtman at the end of the 16th century, representing a Native American. Its brightest star is α Ind, mag. 3.11, distance 101 l.y., spectral type K0 III. ϵ Ind, an orange dwarf of visual mag. 4.69 and spectral type K5 V, is among the closest naked-eye stars, distance 11.8 l.y. θ Ind is a double for small telescopes, mags. 4.5 and 7.0.

Jaipur for determining the positions of celestial objects are still standing.

Although Western astronomical techniques accompanied the European domination of India, traditional calendars are still in use for astrological and religious purposes, and large numbers of traditional almanacs are published each year.

Indus *See* feature article

inequality Departure from uniform orbital motion of a body due to PERTURBATIONS by other bodies. The most notable is the great inequality that affects Jupiter and Saturn, as a result of the close 5:2 COMMENSURABILITY between their mean motions. This inequality causes perturbations of period about 938 years in their longitudes, of amplitude $\pm0^{\circ}.36$ for Jupiter and $\pm0^{\circ}.87$ for Saturn.

inertia Tendency of a body to resist acceleration; it is the tendency to remain at rest or to continue moving at a constant velocity in a straight line unless acted on by a force. Inertia is quantified by the MASS of the body. *See also* NEWTON'S LAWS OF MOTION

inferior conjunction Alignment of an inferior planet between the Earth and the Sun. If the alignment takes place at a favourable node, then the inferior planet may be seen to transit the Sun's disk. CONJUNCTION occurs when two bodies have the same celestial longitude as viewed from the Earth. For an inferior planet, this can also occur when it is on the far side of the Sun to the Earth, at which point it is said to be at SUPERIOR CONJUNCTION. Because of the inclination of the orbits of Mercury and Venus to the ecliptic, they rarely transit the Sun's disk at inferior conjunction, instead passing either north or south. *See* diagram at CONJUNCTION

inferior planet Planet with an orbit that lies closer to the Sun than does the orbit of the Earth; the inferior planets are MERCURY and VENUS. Since both planets' orbits are within that of the Earth, they do not appear far from the Sun in the sky and are, therefore, best observed as early morning or early evening objects. Again, because of their orbital positions, both planets display distinctive phases like those of the Moon.

infinite Universe, idea of Concept of a cosmos with no bounds. Since ancient Greek times, astronomers had generally considered the planetary Universe to be enclosed within a sphere of fixed stars. By the 15th century AD, however, some scientific thinkers within the Catholic Church, such as Cardinal Nicholas of Cusa (1401–64), were coming to argue that there was no reason why an infinite, all-powerful God should not have created an infinite Universe, containing a multitude of stars and planets. But it was the implications of the COPERNICAN SYSTEM after 1543 that re-opened the debate about infinity, for if the Earth and planets were revolving around the Sun, why could not other stars also have planets rotating around them? In 1576, indeed, the English Copernican Thomas DIGGES spoke of the starry 'sphere' as extending 'infinitely' into space. Yet all of this speculation about an infinite Universe suddenly acquired a physical ground in 1610, when GALILEO resolved the Milky Way into individual stars with his telescope. Thereafter, every increase in telescopic power revealed yet more dim stars, creating the impression that our Sun was but one star in an infinite three-dimensional Universe. In 1695, moreover, Christiaan and Constantijn HUYGENS even wondered whether intelligent creatures in space were looking at *us* with powerful telescopes! By the 1780s William HERSCHEL realized, from the then known velocity of light, that when we look into deep space we look into 'times past'. By Herschel's time, however, extensive telescopic observation had confirmed that the Universe appeared to be infinite, and after about 1950, that it was also expanding. *See also* COSMOLOGY; OLBERS' PARADOX; PERFECT COSMOLOGICAL PRINCIPLE

▼ **Infrared Astronomical Satellite** (IRAS) An all-sky map of infrared sources recorded by IRAS in 1983. The bright band running across the middle is the plane of the Milky Way.

inflation Faster-than-light expansion of the universe very early in its evolution. The BIG BANG theory had several problems, specifically the horizon problem and the flatness problem. The horizon problem occurred because measurements of the background temperature from every direction are nearly identical, but these regions of the Universe had to be out of causal contact, and therefore would not be expected to have identical temperatures. The flatness problem arose because the background radiation implied a critical density very close to one, and thus at early times in the universe the density had to be very close to one. The inflationary model of Alan Guth proposed that at approximately 10^{-35} s after the Big Bang the universe expanded exponentially for about 10^{-24} s and increased in size by a factor of 10^{50}. The energy to drive this expansion came from the condensation HIGGS FIELD as the temperature of the universe fell below a critical value. This idea solved the horizon problem by assuming that the opposite sides of the universe were in causal contact before inflation, but were propelled out of causal contact by the inflationary expansion. It also solved the flatness problem by decreasing our view of the universe to a small portion so that it would appear to be flat, even if the entire universe were generally not flat. Inflation has been refined and survived a number of observational tests, and is now on a solid theoretical and observational footing.

Infrared Astronomical Satellite (IRAS) International project involving the USA, the Netherlands and the UK. Its main objective was to make the first all-sky infrared survey from space, searching for objects with temperatures between 10 and a few hundred kelvin. Its 60-cm (23-in.) reflecting telescope focused incoming radiation on to an array of 64 semiconductor detectors, which were cooled to 1.8 K by liquid helium.

IRAS was launched by a DELTA rocket from Vandenberg Air Force Base, California, on 1983 January 26. It was placed in a 900-km (560-mi) Sun-synchronous, circular orbit inclined at 99° to the equator.

During the main survey in 1983, between early February and the end of August, most of the sky was observed twice. A second sky survey began in September and continued until the helium coolant ran out on 1983 November 22. By then IRAS had scanned 98% of the sky and carried out many thousands of targeted observations. The IRAS catalogue published in 1984 contained 245,000 infrared point sources, more than 100 times the number known before launch.

Many discoveries were made both during and after the mission. Solar System discoveries included: six new comets, notably Comet IRAS–ARAKI–ALCOCK; huge invisible tails on Comet Tempel 2 and other comets; several asteroids, including the unusual object 3200 PHAETHON; and bands in the ZODIACAL DUST.

Dust shells, possibly related to planetary formation, were discovered around VEGA and several other stars. Star formation regions in DARK NEBULAE were studied in great detail and many PROTOSTARS were discovered. The star BETELGEUSE was found to have ejected three huge dust shells, and clouds of dust named infrared cirrus were discovered all over the sky. IRAS also studied the galactic centre in great detail. Beyond the Milky Way, IRAS observed that many galaxies are powerful emitters of infrared radiation and some of these, the STARBURST GALAXIES, emit much more infrared than visible light. *See also* INFRARED ASTRONOMY

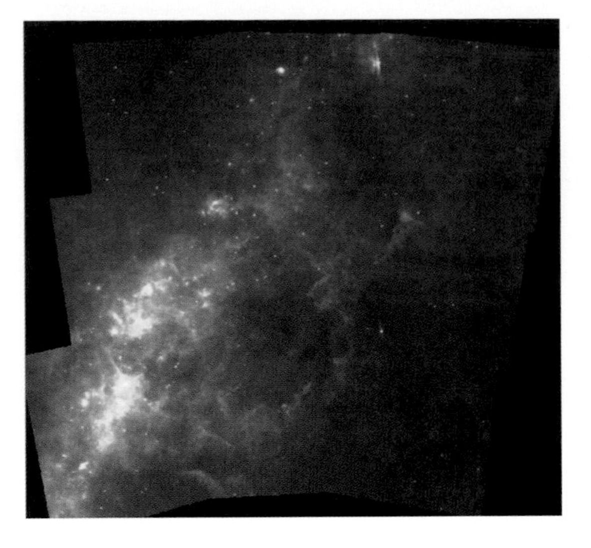

◄ **infrared cirrus** Infrared emission from dust grains heated by stellar radiation produces this wispy structure in the Puppis–Vela region of the Milky Way.

infrared astronomy Study of astronomical objects at infrared wavelengths. INFRARED RADIATION penetrates dust clouds more easily than that at optical wavelengths, so astronomers using the infrared can study stars forming deep inside dense dust clouds, the centre of the Galaxy, and other galaxies (both normal and peculiar). Astronomical sources of infrared radiation also include the planets and other Solar System bodies, stars and the dusty regions themselves. Wien's law shows that relatively low-temperature objects (less than 3000 K) emit most strongly, and so are easiest to observe, in the infrared.

Astronomers on the ground, using the infrared, must confine their observations to the atmospheric 'windows' where there is very little absorption by water vapour or carbon dioxide, and infrared telescopes are usually sited on mountain tops (above about 3000 m/10,000 ft) to take advantage of the windows at 0.75–2.5 μm, 3–5 μm, 7.5–14 μm and 16–21 μm. Mauna Kea in Hawaii and Atacama in Chile are examples of good sites. To avoid the Earth's atmosphere, infrared observations have been made from high-flying aircraft, notably the KUIPER AIRBORNE OBSERVATORY (KAO) and the new STRATOSPHERIC OBSERVATORY FOR INFRARED ASTRONOMY (SOFIA), from unmanned balloons, rockets, and from Earth-orbiting satellites. The INFRARED ASTRONOMICAL SATELLITE (IRAS) surveyed the entire sky at several infrared wavelengths cataloguing almost a quarter of a million objects. More recently the INFRARED SPACE OBSERVATORY (ISO) targeted over 30,000 astronomical objects for detailed study. There were several other early surveys, two of which were particularly noteworthy: one from the ground, the 2μm Sky Survey by Gerry Neugebauer (1932–) and Robert Leighton (1919–97), producing the IRC catalogue published in 1969; and one using rockets, the Air Force Geophysics Laboratory survey by Steve Price and Russ Walker, producing the AFGL catalogue published in 1976.

Astronomical sources can emit infrared radiation either throughout the ELECTROMAGNETIC SPECTRUM (according to the PLANCK DISTRIBUTION of BLACK BODY RADIATION) or in one or more spectral lines or bands, depending upon the temperature and composition of the emitting gas (*see* EMISSION SPECTRUM). The emission in the infrared lines is sometimes the main mechanism for cooling gas/dust clouds (such as the dense clouds where new stars are forming). Among the most important infrared atomic lines are the FORBIDDEN LINES of carbon ([C II] at 157 μm) and oxygen ([O III] at 88.35 μm). There are very important molecular bands found in emission and ABSORPTION, in particular hydrogen (H_2), water (H_2O), carbon monoxide (CO) and carbon dioxide (CO_2). Infrared observations from space, with ISO, have shown that water is very common in the Solar System, in the Galaxy and in other galaxies. The infrared has revealed the existence of another type of molecule, polycyclic aromatic hydrocarbons (PAHs), the emission bands of which were originally known as the Unidentified Infrared Bands (UIBs). The molecular bands are not only indicative of chemical composition but also of the physical processes in progress, such as the passage of shock waves through regions in which star formation is occurring.

The infrared region of the electromagnetic spectrum has many broad features due to the solid material (dust) in interstellar space or around stars. The dust grains range greatly in size, but a typical value is a few tens of micrometres. Very small dust grains (with sizes of a few micrometres) can be transiently heated to about 1000 K and emit MID-INFRARED radiation. Silicate dust has often been found in crystalline form as well as the amorphous (astronomical) silicate form, giving sharp spikes in the normally very smooth broad emission features. Water-ice and carbon dioxide ice have also been detected in the infrared, in star-forming regions. Carbon-rich dust, such as silicon carbide, can also be observed via its infrared emission features. Both old and young astronomical objects show these dust features.

Probably the most studied and most famous molecular cloud is in Orion's sword, where star formation is observed in progress in the densest parts of the ORION MOLECULAR CLOUDS. Temperatures in this cloud reach about 2000 K in regions where shock waves excite molecular hydrogen, but other regions range from a few tens to a few hundred degrees. The infrared region of the spectrum is very sensitive to the slightest heating of the dust cloud as it starts to collapse. IRAS showed that there is a great deal of star formation occurring in Orion, not just in the sword region.

The huge infrared emission output when stars are forming means that these regions can be detected at large distances; the STARBURST GALAXIES are one example. Another type that is very bright in the infrared, and hence can be detected at large distances, is a galaxy with an ACTIVE GALACTIC NUCLEUS (AGN), which is thought to be powered by a BLACK HOLE. The infrared can be used to detect the difference between these two types of galaxy, since the starburst galaxy lacks some of the most energetic forbidden lines found in the AGN spectrum.

Closer to home, Solar System scientists have used the techniques of infrared astronomy to work out the basic chemistry of the planets, their satellites and the asteroids. In the NEAR-INFRARED, molecules in planetary atmospheres exhibit a rich absorption spectrum, the analysis of which permits the composition and temperature of the atmospheres to be established. By careful selection of a precise infrared wavelength, different layers in the atmosphere can be probed in detail because of the variations in temperature. Observations of the outer planets from ISO have shown the presence of water in their upper atmospheres due to icy material falling on to them from interplanetary space.

infrared cirrus Tenuous cold dust in interstellar space which emits faintly in the infrared. When FAR-INFRARED data are displayed as maps, the emission resembles the

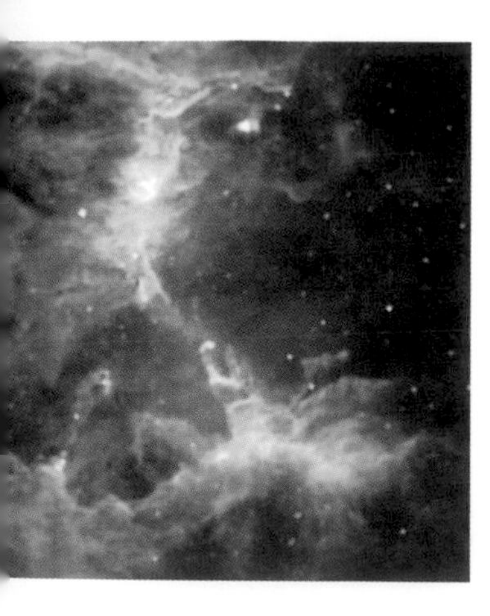

▲ **Infrared Space Observatory** (ISO) A composite ISO image of the Eagle Nebula. Shorter infrared wavelengths are rendered in blue, longer wavelengths in red.

wispy structure of cirrus clouds in the sky. The infrared cirrus clouds were first mapped with the INFRARED ASTRONOMICAL SATELLITE (IRAS) and they have been detected everywhere in the Galaxy. The temperature of the dust is typically around 20–30 K.

Infrared Imaging Surveyor (IRIS) Japan's Astro F series spacecraft, to be launched in 2003–2004 into a 750-km (460-mi) Sun-synchronous Earth orbit. The satellite will survey the infrared sky with greater sensitivity than any previous Astro mission, using a 70-cm (28-in) telescope cooled with liquid helium. It will investigate the formation and evolution of galaxies, stars and planets.

infrared radiation Portion of the ELECTROMAGNETIC SPECTRUM in the wavelength range 0.75–350 μm, lying between optical and radio wavelengths. Historically, wavelengths between 1 μm and 1 millimetre (mm) were considered to be infrared. Recent developments in detectors mean that the submillimetre region is now considered to start at 350 μm. Most objects emit some infrared radiation, but according to WIEN'S LAW those with temperatures less than 3000 K emit most intensely in the infrared. Many molecules (for example molecular hydrogen, H_2) have important spectral features in the infrared. *See also* INFRARED ASTRONOMY

Infrared Space Observatory (ISO) EUROPEAN SPACE AGENCY (ESA) spacecraft launched in 1995 November to observe the sky with enhanced sensitivity and resolution using a 60-cm (24-in.) diameter primary mirror. It had four science instruments – an infrared camera, a photopolarimeter, and two spectrometers, provided by France, Germany, the Netherlands and the UK, cooled by a cryostat of liquid helium. The craft was able to operate fully for an additional eight months until 1998 May when the coolant was depleted. A short-wavelength spectrometer, however, was used until 2001.

initial mass function Distribution of stellar masses at birth, which is taken to be when nuclear fusion begins in the stars' cores. Determined by Edwin SALPETER in 1955, and sometimes called the Salpeter mass function, the initial mass function $\phi(M)$ represents the number of stars with mass M at birth per unit volume of space. In the solar vicinity, $\phi(M)$ is approximately equal to $M^{-2.35}$, but there are deviations from this law for massive stars. It is also difficult to estimate how many low-mass stars (less than 0.1 solar mass) exist, even in the solar neighbourhood.

inner planet Term used to describe any planet, the orbit of which lies inside that of the ASTEROID BELT. The inner planets are Mercury, Venus, Earth or Mars. *See also* TERRESTRIAL PLANETS

Innes, Robert Thorburn Ayton (1861–1933) Scottish-born double star observer and discoverer of PROXIMA CENTAURI, director of the Union (later Republic) Observatory in Johannesburg (1903–27). Innes emigrated to Australia in 1884 and again in 1896, this time to the Cape of Good Hope (South Africa), where he became a first-rate observer at the observatory there. Innes observed and catalogued many new southern double stars, culminating in the *Southern Double Star Catalogue* (1927). He also revised the *Cape Photographic Durchmusterung*, a massive catalogue of stars visible from the southern hemisphere. Innes was an ardent advocate of southern hemisphere astronomy, urging many older observatories to establish southern stations. He was one of the first to recognize the astronomical usefulness of the blink comparator, which he used in 1915 to discover Proxima Centauri (known for a while as *Innes' Star*).

insolation Total amount of radiant energy from the Sun falling on to a body per unit area perpendicular to the direction of the Sun, in unit time. For the Earth, the insolation is also called the SOLAR CONSTANT. At the top of the Earth's atmosphere it has a value of 1366.2 Wm^{-2}.

instability strip Part of the HERTZSPRUNG–RUSSELL DIAGRAM where pulsating stars are located. It is a narrow strip, extending from the CEPHEIDS through the RR LYRAE VARIABLES, DELTA SCUTI VARIABLES and dwarf Cepheids, down to the pulsating white dwarf (ZZ CETI) stars.

Most stars will pass through this region at some time in their lives: they become pulsating variables of some type when they have a small imbalance in the gravitational force and the outward internal pressure so that they are not in hydrostatic equilibrium. The part of the instability strip that a star passes through depends on its mass.

Institut de Radio Astronomie Millimétrique (IRAM) Multi-national scientific institute that operates two major facilities: a 30-m (98-ft) telescope on Pico Veleta in the Sierra Nevada, southern Spain, and an array of five 15-m (49-ft) telescopes on the Plateau de Bure in the French Alps. Tragedy struck the Plateau de Bure site in 1999 July when a cable car fell to the ground, killing all 20 scientists and engineers on board.

Institute for Astronomy, University of Edinburgh Research institute within the Department of Physics and Astronomy of Edinburgh University, located in the grounds of the ROYAL OBSERVATORY EDINBURGH. Areas of specialization include cosmology, the Universe at high redshifts, X-ray surveys, galaxy formation and studies of the intergalactic medium. The Institute's Wide-Field Astronomy Unit supports the UNITED KINGDOM SCHMIDT TELESCOPE and other wide-field telescopes. It is responsible for the operation of the SuperCOSMOS plate measuring machine and the overall management of the 6dF Galaxy Survey.

Institute for Astronomy, University of Hawaii Research institute founded at the University of Hawaii in 1967 to manage the Haleakala Observatory and MAUNA KEA OBSERVATORY and to pursue its own programme of fundamental astronomical research. Its main base is at Manoa on the island of Oahu, close to the main campus of the university. Its sea-level telescope operations and instrument-development facility for Mauna Kea is at Hilo on the Big Island of Hawaii.

Institute of Astronomy, University of Cambridge (IoA) Department of the University of Cambridge engaged in teaching and research in theoretical and observational astronomy. It came into being in 1972 with the amalgamation of the Cambridge University Observatory (founded 1823), the Solar Physics Observatory (1912) and the Institute of Theoretical Astronomy (1967). Some of the best-known names in modern astronomy have been associated with the IoA, including Fred Hoyle and Martin Rees.

Institute of Space and Astronautical Science (ISAS) Japanese institute for space science research, operating its own launch vehicles, scientific satellites, planetary probes and balloons. ISAS had its origins in the University of Tokyo in the 1950s, but took its present form in 1981. Its work complements that of the NATIONAL SPACE DEVELOPMENT AGENCY OF JAPAN, which operates applications satellites and their launch vehicles.

Instituto Argentino de Radioastronomia (IAR) Principal institution for radio astronomy in Argentina, created in 1962, located near Buenos Aires. It operates two 30-m (98-ft) radio telescopes.

Instituto de Astrofísica de Canarias (IAC) International research centre in the Canary Islands, comprising the Instituto de Astrofisica, La Laguna, and the Observatorio del Teide (both situated on the island of Tenerife) and the ROQUE DE LOS MUCHACHOS OBSERVATORY on La

Palma. Together they constitute the European Northern Observatory. Both observing sites are noted for their exceptional sky-quality. The IAC is the host organization for the GRAN TELESCOPIO CANARIAS.

INT Abbreviation of ISAAC NEWTON TELESCOPE

Integral (acronym for International Gamma Ray Astrophysics Laboratory) EUROPEAN SPACE AGENCY (ESA) Horizon 2000 mission to be launched on a Russian PROTON booster in 2002 into a high (40,000 km/25,000 mi) Earth orbit. Developed in collaboration with NASA and Russia, Integral is dedicated to fine spectroscopy observation and high-resolution imaging of celestial gamma-ray sources, with concurrent source monitoring in X-ray and visible wavelengths.

integrated magnitude Total brightness of an extended body. Stellar and planetary bodies typically have brighter and darker regions and the integrated MAGNITUDE is the sum of all of these.

intensity interferometer Instrument used to study an astronomical object by means of INTERFEROMETRY (*see also* INTERFERENCE) in order to obtain more detail in the map of the object. The first interferometer was developed by Albert Michelson (1852–1931) in the 1920s, and it worked at optical wavelengths. It consisted of mirrors at either end of a steel beam placed across the aperture of the 100-inch (2.5-m) telescope at Mount Wilson, and with it Michelson measured diameters of a few large stars by examining the interference pattern formed in the eyepiece. Modern systems link two telescopes, either electronically or by laser beams, and use electronic devices, such as PHOTOMETERS, to record the signals. The technique has been used with great success at radio wavelengths (*see also* RADIO INTERFEROMETER). The two telescopes receive the signal at different times because the waves of the electromagnetic radiation have to travel farther to one of the telescopes than to the other. This delay is slightly different for separate, but adjacent, points on the sky, so that for an extended object, such as a large galaxy or nebula, the interference pattern is washed out.

interacting galaxies Pairs or groups of GALAXIES whose forms are distorted by the gravitational influence between them, sometimes leading to a merger. These interactions can set long tidal tails of stars and gas into motion well away from the original galaxy (some pieces of which can eventually clump together to form DWARF GALAXIES). Galaxy interactions are though to cause STARBURST activity, including what appear to be newly formed GLOBULAR CLUSTERS, and perhaps some kinds of ACTIVE GALACTIC NUCLEI.

Interamnia Sixth-largest MAIN-BELT ASTEROID; number 704. It has a diameter of 316 km (196 mi). Interamnia was not discovered until 1910 due to its low albedo (0.07).

interference Effect observed when two trains of waves of the same wavelength meet. If maxima (crests) of the waves arrive simultaneously at the same place, their maxima add together to produce a wave of larger amplitude. This is constructive interference. If the maxima of one train coincide with the minima (troughs) of the other, this is destructive interference. They cancel totally if the amplitudes of the two wave trains are identical, otherwise they cancel partially. Thus two wave trains crossing each other produce an **interference pattern**, with alternate lines of constructive interaction and of destructive interaction. This applies to any wave motion – electromagnetic radiation or waves on the surface of liquids.

interferometry Study of point-like (unresolved) astronomical objects using INTERFERENCE to reveal more detail in the object via the interference pattern produced. There are two types of interferometry, using SPECKLE INTERFEROMETRY or an INTENSITY INTERFEROMETER, to determine spatial detail, and there are other types to study spectral lines in detail (*see* FABRY–PERÓT INTERFEROMETER). An interferometer works on the principle that ELECTROMAGNETIC RADIATION (usually optical or radio) will follow two paths to produce the interference, either because of Earth's atmosphere in the case of speckle interferometry, or through two paths in the same instrument (as in the case of Fabry–Perót), or via two telescopes (intensity interferometer). If the two signals are combined correctly they will either reinforce or cancel, depending on the delay: when the signals are in phase (that is, they either have no delay or a delay corresponding to a whole number of wavelengths) then the maximum combined signal will be obtained. *See also* STELLAR INTERFEROMETER; VERY LONG BASELINE INTERFEROMETRY

intergalactic matter Matter in the space between GALAXIES. There is no significant amount of dust in intergalactic space, but there is ample evidence for several kinds of gas. Within galaxy clusters, the hot intracluster medium is at temperatures of typically 20 million K and has been chemically enriched by supernovae. Outside these clusters, observations of absorption lines, above all from material in front of distant and luminous QUASARS, shows an intricate medium tracing the large-scale distribution of ordinary matter (which must be close to that of the DARK MATTER as well, since its gravity will be the dominant force). Hydrogen and helium absorption shows that this intergalactic matter is highly ionized everywhere, and more so in the least dense regions, where particle collisions that could lead to RECOMBINATION are less frequent. Because it is so highly ionized, and only the tiny neutral fraction is observed, the amount and chemistry of this matter are still very uncertain. Even intergalactic matter in the lowest-density regions, generally most remote from luminous galaxies, has been enriched to some degree with atoms synthesized in stars, as shown by the presence of highly ionized oxygen traced in observations from FUSE and the Hubble Space Telescope. Thus the intergalactic medium is not simply leftover material that never formed galaxies: it has, at least in the early Universe, participated in the stellar recycling inside galaxies.

International Amateur–Professional Photoelectric Photometry (IAPPP) Organization that fosters partnerships among amateur, professional and student astronomers who wish to make precise brightness observations of celestial objects. Formed in 1980, the IAPPP was far ahead of its time in promoting such collaborations, which are becoming more important as the pace of astronomical discovery quickens and the need for continuing follow-up observations grows.

International Astronomical Union (IAU) Principal coordinating body of world astronomy. Its mission is to

▼ **interacting galaxies** A Hubble Space Telescope image of the edge-on galaxy ESO 510-G13 in Hydra. The dark, dusty disk shows warping indicative of a recent collision with another galaxy.

promote and safeguard the science of astronomy through international cooperation. Founded in 1919, the IAU has 11 scientific divisions and 40 commissions covering all aspects of astronomy. Its membership includes most of the professional community active in astronomical research and education at PhD level and beyond, and amounts to some 8300 individuals in 67 countries.

The IAU is perhaps best known as the sole authority responsible for naming celestial bodies and their surface features. However, its remit extends far beyond that, and ranges from the definition of fundamental astronomical constants to strategic planning of future large-scale facilities. It holds a General Assembly every three years at which its long-term policy is defined, and sponsors about a dozen high-profile symposia and colloquia each year. The IAU also promotes education research in developing countries through its International Schools for Young Astronomers. The organization's headquarters are at the Institute d'Astrophysique in Paris, where a permanent secretariat is based.

International Atomic Time (TAI) Continuous and uniform time scale derived from atomic clocks and used for scientific purposes. TAI is based on the SI second (*see* ATOMIC TIME) and is formed retrospectively by intercomparing data from around 200 atomic clocks, or frequency standards as they are known, at around 40 laboratories across the globe. Each of these atomic clocks should be accurate to within one second in three million years, but a large number are used to form the time scale in order to reduce the likelihood of the results from any rogue timepieces affecting the overall combined mean. The resultant TAI time scale is then used as a standard against which other clocks can be measured.

TAI has run continuously, without adjustment, since $0^h\ 0^m\ 0^s$ GMT on 1958 January 1 and is co-ordinated by the International Bureau of Weights and Measures (BIPM) in Paris. Because it is both uniform and continuous, it is ideal as a time scale for scientific purposes but not practical for everyday use, since it is not linked to the rotation of the Earth. For the purposes of forming an accurate civil time scale COORDINATED UNIVERSAL TIME (UTC) was introduced. This is still derived from atomic clocks but is kept in step with the Earth's rotation through the periodic introduction of LEAP SECONDS. Because of this, UTC differs from TAI by an integral number of seconds.

Every major industrial nation contributes to International Atomic Time. In the UK, responsibility for maintaining the national time service is held by the National Physical Laboratory (NPL) at Teddington. Prior to 1984 the Time Service was the responsibility of the Royal Greenwich Observatory. *See also* TIMEKEEPING

▼ **International Space Station** Seen from the Space Shuttle *Endeavour* is the International Space Station under construction in 2001 December.

International Cometary Explorer (ICE) NASA spacecraft originally launched in 1978 as ISEE-3 (International Sun–Earth Explorer-3); it was renamed when it was diverted by means of a lunar gravitational assist to fly through the tail of Comet GIACOBINI–ZINNER. The comet flyby – the first ever made – took place on 1985 September 11. ICE later passed the sunward side of Halley's Comet in 1986 March at a distance of 28 million km (17 million mi).

International Dark-Sky Association (IDA) International non-profit organization, based in the USA, founded in 1988 to campaign against the adverse impact of LIGHT POLLUTION on optical astronomy. The IDA seeks to raise public awareness about good and bad outdoor-lighting practices, including aesthetic, security and economic issues. The IDA is also building awareness of other threats to the astronomical environment, such as from radio-frequency interference and space debris and from other pollutants such as aircraft contrails.

International Date Line Imaginary, irregular line, close to and sometimes coincident with that of 180° longitude, marking the point on the Earth where the date changes; points east of the line being one day earlier than those west of it.

Located mainly in the Pacific Ocean, therefore avoiding places of habitation, the line avoids crossing land by skirting around Siberia, the Aleutian Islands, the Fiji Islands and New Zealand. It was adopted by international agreement in 1885.

International Geophysical Year (IGY) Period of intensive, multi-nation collaborative research, including studies of solar and auroral process, meteorology and oceanography, that was was carried out between 1957 July 1 and 1958 December 31.

International Meteor Organization (IMO) Organization dedicated to the study of meteors and their parent dust particles, established in 1988, with a worldwide membership. The IMO collects data taken visually, photographically, by video and by radio. It maintains a database of visual observations extending back to 1984 and publishes *WGN*, a bimonthly journal.

International Occultation Timing Association (IOTA) International organization founded in 1975 to encourage and facilitate the observation of occultations and eclipses. It provides predictions for grazing occultations of stars by the Moon and of stars by asteroids and planets, and acts as a coordinating body for reports of such events. IOTA is based at Topeka, Kansas, and has a European Section.

International Solar Terrestrial Physics (ISTP) Missions Major multi-agency, multi-spacecraft programme of the 1990s aimed at exploring the terrestrial MAGNETOSPHERE and near-Earth space. The overall goal of the programme is to improve our understanding of the solar–terrestrial interaction, particularly the coupling of matter and energy between the SOLAR WIND and the magnetosphere. Central elements include WIND, a NASA satellite instrumented to study the solar wind upstream of the Earth and thus provide information on the likely external conditions affecting the magnetosphere. A sister NASA spacecraft, POLAR, in an eccentric polar orbit with apogee over the northern pole, observes activity of the AURORA and makes plasma measurements within the cusps of the magnetosphere. The Japanese space agency ISAS provided the spacecraft GEOTAIL, which is designed to make measurements of the nightside magnetotail region of the Earth. Data from this spacecraft are particularly relevant to studies of the PLASMA SHEET and

MAGNETOSPHERIC SUBSTORMS. The European element, CLUSTER, was finally launched in the summer of 2000, after the original four-spacecraft mission was lost to launch failure in 1996. This mission is designed to study boundaries, such as the BOW SHOCK and MAGNETOPAUSE. A number of other missions and facilities are associated with ISTP and provide supporting and context information. These missions include SOHO (the SOLAR AND HELIOSPHERIC OBSERVATORY), the Russian Interball missions, Equator-S, and many ground-based radars, all-sky cameras and magnetometer networks located around the northern auroral zones.

International Space Station (ISS) SPACE STATION, which when completed in 2006 – budgets, schedule and technology permitting – will be a space superlative, measuring 111.32 m (365 ft) from end to end. The space station was originally given the go-ahead in 1984, in response to the Soviet Union permanent presence in orbit. This US programme, first called Freedom, which was to have been operational in 1994, became embroiled in politics and financial problems. Following the collapse of the Soviet Union, Russia had no money to build a new space station. The US government was on the point of cancelling Freedom, but the station was saved when Russia joined the programme. The reconfigured ISS was to have been declared operational in 2001 but that deadline has already slipped to 2006. Still known as the International Space Station, although having an unofficial name, Alpha, the project is subject to further delays due to financial and technical difficulties.

It is planned that the ISS will eventually be crewed by up to seven people, but that is unlikely before 2008. It will have 1624 m^3 (46,000 cubic feet) of pressurized space in several modules – the equivalent of a Boeing 747 Jumbo jet – and will be equipped with four photovoltaic modules, each with two arrays 34.16 m (112 ft) long and 11.89 m (39 ft) wide, generating 23 kW, and with a surface area of about half an acre. The electrical power system will be connected by 12.8 km (8 mi) of wire. A major external part of the ISS will be the Canadian remote manipulator system comprising two robot arms and a mobile transporter travelling along the length of the station.

The ISS will concentrate on five main areas of science research – life sciences, space science, Earth science, engineering research and technology, and space product development. It will consist of six major scientific modules serviced by connecting passageways (called nodes), service and control modules, living quarters, an airlock, the manipulator system, logistics vehicles and crew transfer vehicles and propulsion modules. A major benefit of the ISS will be the cooperative work of so many nations of the world. The ISS is going to be the largest international civil, cooperative programme ever attempted, involving 16 nations – the USA, Russia, Canada, Japan, Brazil, Belgium, Denmark, France, Germany, Italy, the Netherlands, Norway, Spain, Sweden, Switzerland and the UK. The likelihood is that the ISS will not quite resemble what has been designed and will develop more on a step-by-step basis, according to the state of delays and finances, with redesigns and compromises being continually made. It may not be declared operational until 2010.

International Sun–Earth Explorer (ISEE) Series of three NASA–ESA scientific satellites launched in 1977–78 in an international project to study the near-Earth environment (particularly the magnetosphere) and its interaction with the SOLAR WIND.

International Ultraviolet Explorer (IUE) Joint project of the NATIONAL AERONAUTICS AND SPACE ADMINISTRATION (NASA), the EUROPEAN SPACE AGENCY (ESA) and the UK. It was the longest-lived astronomical spacecraft ever flown, and arguably the most important ultraviolet space observatory so far launched.

IUE was launched by a DELTA rocket in 1978 January. It carried a 45-cm (18-in.) telescope and spectrographs equipped with ultraviolet-sensitive cameras to study Solar System objects, stars and galaxies. The observing time on IUE was initially shared out between the three agencies in approximate proportion to their contributions, with NASA getting two-thirds of the time and the UK and ESA one-sixth each. After 1981 the combined European share was assigned on scientific merit alone.

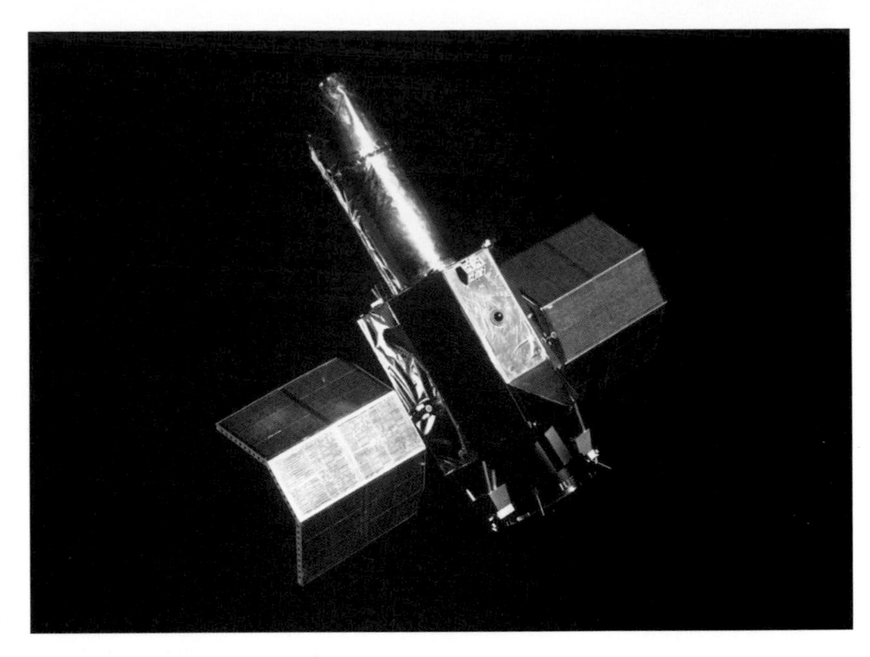

▲ **International Ultraviolet Explorer** (IUE) An artist's impression of the IUE. Launched in 1978, this highly successful satellite operated until 1996.

IUE was inserted into a geosynchronous elliptical orbit between 26,000 and 45,000 km (16,000–28,000 mi) high and inclined at 28°.4 to the equator. As a result, it was visible from the ESA ground station in Spain for up to 12 hours a day and permanently visible from the NASA ground station in Maryland. Control of IUE was transferred permanently to Europe in 1995 October.

During its lifetime, IUE made 104,000 ultraviolet observations. The data are kept in three archives, one operated by each of the collaborating agencies, easing access for thousands of astronomers around the world.

In the Solar System, IUE observed dozens of comets, including Comet HALLEY, Comet HYAKUTAKE and Comet ENCKE, which was studied during several returns. It was also used to monitor asteroids, the atmospheres of the giant planets and the cloud of ions associated with Jupiter's moon Io.

Beyond the Solar System, IUE provided new insights into stellar winds and energy transport in the atmospheres of hot, massive stars. Its flexibility opened the door to studies of unpredictable novae and Supernova 1987A. Ultraviolet emissions from 'normal' galaxies, Seyfert galaxies and quasars helped to unravel the processes taking place in accretion disks around massive black holes. *See also* ULTRAVIOLET ASTRONOMY

International Years of the Quiet Sun (IQSY) Period of collaborative research involving scientists from many nations; it was organized as a follow-up to the INTERNATIONAL GEOPHYSICAL YEAR in order to study the same phenomena under conditions of minimal SOLAR ACTIVITY during 1964–65.

interplanetary dust (zodiacal dust) Lens-shaped cloud of dust particles centred on the Sun and with its major axis lying in the ecliptic plane. The cloud consists of dust particles a few micrometres in size, and it extends for at least 600 million km (370 million mi). It is extremely tenuous: at the Earth's orbit its average density is equivalent to a single particle, two micrometres across, in a region of space one hundred cubic metres in volume.

Sunlight scattered by the cloud is responsible for the ZODIACAL LIGHT. In the scattering process the dust particles absorb energy from the Sun on their illuminated side but re-radiate it isotropically. This process, known as the

I

POYNTING–ROBERTSON EFFECT, results in a loss of kinetic energy and produces a deceleration, causing the particles to spiral slowly in towards the Sun over a long period of time. As they approach the Sun, the particles are eroded, chiefly by evaporation and the effect of the SOLAR WIND.

The rate of depletion of the dust particles indicates that the cloud must be continually replenished by some mechanism, otherwise it would have completely dispersed by now. Collisions between ASTEROIDS had long been suspected as a possible source of replenishment, and in 1983 the INFRARED ASTRONOMICAL SATELLITE (IRAS) located three bands of dust in the region of the asteroid belt between the orbits of Mars and Jupiter. These bands are believed to be the result of collisions between asteroids. Over a long period of time, the particles within the bands slowly decelerate, moving inwards from the asteroid belt towards the Sun and spreading out into the background cloud of interplanetary dust.

Asteroid collision alone cannot produce sufficient material to replenish the cloud at the required rate, however, and analysis of data from both the IRAS and the COSMIC BACKGROUND EXPLORER (COBE) satellites indicates that main-belt asteroids are the source of approximately only 33% of the interplanetary dust.

Comets are another source of interplanetary dust, producing dust each time they enter the inner Solar System. The largest particles produce METEOR STREAMS, whilst the smallest are incorporated into the dust cloud. Even this does not account for all the new material required, though, and the remainder is thought to be cosmic in origin. The issue as to the ultimate source of the dust is still open to debate, but satellite observations of the sky brightness and in-situ measurements of the dust particles are advancing our understanding of the relative contributions from comets, asteroids and the interstellar medium.

interplanetary dust particle *See* MICROMETEORITE

interplanetary magnetic field *See* SOLAR WIND

interplanetary medium Tenuous mixture of INTERPLANETARY DUST, charged atomic particles and neutral gas that occupies the space between the planets. The dust is believed to have originated from collisions between asteroids and from comets entering the inner Solar System. The charged particles – electrons, protons and helium nuclei (alpha particles) – stream out from the Sun as the SOLAR WIND, while the neutral gas exists in the form of hydrogen and helium atoms. As the Sun moves through space it passes through interstellar matter, resulting in a continual stream of neutral atoms through the Solar System. Interstellar matter is quickly ionized by the solar wind. *See also* ZODIACAL LIGHT

interstellar absorption ABSORPTION or EXTINCTION of light by interstellar dust (*see* INTERSTELLAR MATTER). On average, the extinction amounts to one stellar magnitude per kiloparsec of distance.

interstellar dust *See* INTERSTELLAR MATTER

interstellar grain Volumetrically insignificant, but scientifically critical, component within CHONDRITE meteorites. Interstellar and circumstellar grains comprise several populations of nanometre-sized diamond particles and micrometre-sized silicon carbide, graphite and aluminium oxide particles. The presence of the grains was first inferred in the late 1970s to early 1980s on the basis of the isotopic composition of noble gases found in mineralogical analyses of chondrites. The unusual isotopic signatures of the noble gases implied the existence of several different hosts; analyses of acid-resistant residues suggested that the hosts might be carbon-rich. Together with the noble gas results, carbon and nitrogen isotope data imply a variety of extra-solar sources for the grains, including supernovae and red giant stars. Currently, grains from at least 15 different extra-solar sources have been isolated from chondritic meteorites. These grains were presumably introduced into the pre-solar nebula prior to its collapse and the onset of protoplanet formation.

interstellar matter Molecules that exist in interstellar MOLECULAR CLOUDS. The main constituents of interstellar matter are everywhere the same, mostly hydrogen with some helium, but the proportions of the minor constituents differ widely between different locations. The total density and other properties of this matter also differ widely from place to place. Mostly, the matter is gaseous but a small proportion is embodied in the form of minute solid dust particles. On average there are about 10^6 atoms m^{-3} and one dust grain per 100,000 m^3 of space (for comparison, there are about 3×10^{25} molecules m^{-3} in Earth's atmosphere at sea level). Although the density of interstellar matter is extremely low, the volume of the space in a galaxy is so great that the total quantity of interstellar material is very considerable. Our own Galaxy contains about 10^{10} solar masses of material between the stars, making up about 10% of its total mass. Most of this matter is distributed in the spiral arms and the disk of our Galaxy and is confined to a layer only a few hundred light-years thick.

The most obvious manifestations of interstellar material are nebulae. Several types have been classified – REFLECTION NEBULAE, HII REGIONS, PLANETARY NEBULAE, SUPERNOVA REMNANTS, GLOBULES and DARK NEBULAE. The differences between these stem mainly from the way the material is illuminated or the way light from other sources is obscured; they also vary in density or recent history. Other types of nebulae not apparent on optical photographs are molecular clouds, which can be detected by their emission or absorption of radio, microwave or infrared radiation rather than of visible radiation.

The existence of gas between the stars was discovered in 1904 by Johann HARTMANN through the observation that a few of the absorption lines recorded in the spectrum of the binary star δ Orionis did not change in wavelength (by the Doppler effect) as the star moved around its orbit. This followed the observation in 1874 by William HUGGINS that certain nebulae had a spectrum characteristic of rarefied gas. Since that time interstellar absorption lines have been recorded in the spectra of many stars. In the optical region these lines are few in number and are usually much narrower than the stellar features themselves. Frequently they have multiple Doppler-shifted components, arising from clouds with different line-of-sight velocities. The strongest optical lines are due to neutral sodium and singly ionized calcium atoms. A large number of atoms in various states of ionization have their absorption lines at ultraviolet wavelengths and have been studied by means of telescopes borne on balloons, rockets or spacecraft. The Lyman-α transition (*see* LYMAN SERIES) of neutral hydrogen, falling at 121.6 nm wavelength, is by far the strongest of all absorption lines observed. While most of the hydrogen is neutral, some elements exist in the interstellar medium primarily in an ionized state. The ionization of such elements arises largely from the presence of energetic photons from stars. The low density of the general interstellar medium ensures that the time an atom in the ionized state has to wait before it can recombine with a free electron is quite long. Hydrogen molecules are detected at far-ultraviolet wavelengths from absorption lines in the spectra of hot stars. Some 120 other INTERSTELLAR MOLECULES have now been identified, many quite complex and mostly to be found in the depths of GIANT MOLECULAR CLOUDS.

The total number of atoms or molecules of each kind between a background star and us can be calculated from the shape and strength of the characteristic absorption features in the star's spectrum. Significant differences in the abundances of the heavier elements relative to hydrogen are found between the interstellar medium and the stars. For example, in some cases the interstellar gas seems to contain only a hundredth of the iron and a thousandth of the calcium that is commonly present in the atmospheres of stars. The missing proportions are in the interstellar dust grains (small grains of matter, typically about 100 nm in diameter, in interstellar space). These dust grains are very effective at absorbing and scattering visible and ultraviolet light, thus making distant stars appear fainter and redder. Because the grains produce polarization of starlight, they are believed to be elongated particles aligned by the galactic magnetic field. Most grains appear to be composed of silicates or graphite; some have icy mantles. The grains may form by condensing from gas flowing out of the atmospheres of cool stars.

Most of our knowledge about the large-scale distribution of the interstellar gas in our Galaxy has come from the study of the emission and absorption of radiation detected at the TWENTY-ONE CENTIMETRE LINE of neutral atomic hydrogen. From the Doppler shifts of the lines, clouds with different velocities along the same line of sight can be distinguished, allowing the distribution and motion of the neutral gas in distant parts of the Galaxy to be studied. Such observations show that neutral hydrogen clouds (HI REGIONS) are clustered mostly along the spiral arms.

Even when taken together, molecular clouds and neutral hydrogen clouds very far from fill the volume of interstellar space. Although it is still a subject of speculation, much of the gas between the clouds is thought to be very hot and very tenuous, at a temperature of around a million K and a density of a few thousand particles per cubic metre. It is thought to be the outcome of the expansion of numerous supernova remnants.

interstellar molecules Molecules that occupy the space between the stars. In this harsh environment few molecules are able to survive. The ultraviolet radiation from stars will quickly cause many of them to dissociate. Some molecules, however, such as titanium oxide (TiO), cyanogen (CN) and diatomic carbon (C_2), are sufficiently stable to survive in the immediate vicinity of stars, or even in the outer layers of cool stars. These molecules can be observed through the absorption bands that they produce in the optical and infrared spectra of the stars.

Most interstellar molecules, however, require shelter from ultraviolet radiation in order to survive. The molecules themselves provide this shelter, since they can exist in either gaseous or solid form (no liquids are known or to be expected in interstellar space). The solid takes the form of interstellar dust particles (*see* INTERSTELLAR MATTER), which appear to be composed of silicates or graphite and sometimes possess outer layers of frozen gases. The composition of the particles is difficult to establish with certainty, but they are known to be a few hundred nanometres in size and possibly needle-shaped; they absorb starlight very efficiently (*see* DARK NEBULAE). Concentrations of particles can shelter gaseous interstellar molecules from the stellar ultraviolet radiation. Thus the main sites where molecules can be found are inside cold dense gas and dust clouds (*see* MOLECULAR CLOUDS).

Prior to the advent of radio astronomy, a few molecules were discovered from the ABSORPTION LINES that they produced in the optical spectra of stars. These molecules were CH, CH+ and CN, and they could only be detected within clouds that were thin enough for stars beyond the cloud to be observed. Such clouds would not be dense enough to provide shelter for more complex interstellar molecules.

Most interstellar molecules have been detected from their emissions or absorptions at radio wavelengths. The reason for this is that molecules can emit or absorb radiation by three separate processes. In the optical region, most spectrum lines from molecules are due to changes in the energy state of electrons within the molecule. Energy can also be stored by molecules in the form of the vibrations of its constituent atoms and by the rotation of the molecule itself. Just as the energies of electrons are quantized, so also can these vibrational and rotational energies only take specific values. Thus when the molecule changes its vibrational or rotational energy it does so in discrete steps and produces emission or absorption lines. The energy stored in the form of molecular vibrations or rotations is much less than that stored by electrons, and so the resulting spectrum lines are at long wavelengths. Vibrational lines mostly appear in the infrared, and rotational lines in the microwave and radio regions. Inside molecular clouds the temperature is low, and so the molecules typically emit or absorb only via the lowest energy rotational transitions.

The most abundant molecule in molecular clouds is expected to be that of hydrogen (H_2). However, the hydrogen molecule is symmetrical, like a miniature barbell, which means that the rotational transitions are FORBIDDEN and so molecular hydrogen does not produce spectrum lines within the molecular clouds. It can, however, sometimes be observed weakly when temperatures reach 500 K or more, for example in shock fronts between colliding gas clouds. In 1963 the first interstellar molecule detected by radio astronomy was hydroxyl (OH). Detection of the second most abundant molecule, carbon monoxide (CO), soon followed, from its lines at 115, 230 and 340 GHz (wavelengths of 2.6, 1.3 and 0.9 mm). The carbon monoxide lines are now widely used as an easily observed tracer for molecular clouds.

Detection of polyatomic molecules such as water (H_2O) and ammonia (NH_3) came in 1968. It was soon followed by the discovery of some of the hydrocarbons (compounds of hydrogen and carbon such as methane, CH_4). More recently, much more complex atoms including, possibly, the amino acid glycine (NH_2CH_2COOH), which forms one of the building blocks for our form of life, have been found. Currently a thirteen-atom molecule, $HC_{11}N$, is the most complex found. About 120 interstellar molecules are now known and more are being added at the rate of a few every year.

interstellar planet PLANET that wanders in interstellar space, not gravitationally bound to any star. In recent years candidate objects have been discovered in deep space, for example in the Orion Nebula. There is debate as to whether these objects should properly be classed as planets or whether they should be categorized as stars, even though many of them have insufficient mass ever to initiate hydrogen or even deuterium fusion in their cores, making them sub-brown dwarfs. The terms 'planetar' or 'grey dwarf' have been coined to label them. Considering bodies of a quite different size and nature much closer to home, there are several TRANS-NEPTUNIAN OBJECTS with aphelion distances so large (some hundreds of astronomical units) that they may be on the verge of escaping into interstellar space. An example is 2000 CR105, which is about 250 km (155 mi) in size and has perihelion at 44 AU, aphelion at 413 AU.

interstellar reddening REDDENING of starlight passing through interstellar dust (*see* INTERSTELLAR MATTER). It

KNOWN INTERSTELLAR AND CIRCUMSTELLAR MOLECULES (AS OF 2001)

Two atoms		Three atoms		Four atoms	Five atoms	Six atoms	Seven atoms	Eight atoms	Ten atoms
H_2	AlCl	H_3^+	C_2S	NH_3	CH_4	CH_3OH	CH_3NH_2	$HCOOCH_3$	$(CH_3)_2CO$
CH	AlF	N_2O	c-SiC_2	H_3O^+	SiH_4	CH_3SH	CH_3C_2H	CH_3C_3N	CH_3C_5N
CH^+	PN	CH_2	SO_2	H_2CO	CH_2NH	C_2H_4	CH_3CHO	C_7H	NH_2CH_2COOH?
NH	SiN	NH_2	CO_2	H_2CS	C_3H_2	CH_3CN	CH_2CHCN	C_6H_2	
OH	SiO	H_2O	OCS	C_2H_2	c-C_3H_2	CH_3NC	C_6H	CH_3COOH	
C_2	SiS	H_2S	MgNC	$HCNH^+$	H_2C_2N	HC_2CHO	HC_5N		
CN	CO^+	C_2H	MgCN	H_2CN	NH_2CN	NH_2CHO	c-CH_2O CH_2	**Nine atoms**	**Eleven atoms**
CO	SO^+	HCN	NaCN	C_3H	CH_2CO	HC_3NH^+	C_7		
SiC	HF	HNC		c-C_3H	HCOOH	H_2C_4		$(CH_3)_2O$	HC_9N
CP		HCO		HCCN	C_4H	C_5H		C_2H_5OH	
CS		HCO^+		HNCO	HC_3N	C_5S		C_2H_5CN	
NO		HOC^+		$HOCO^+$	HC_2NC	C_5N		CH_3C_4H	**Thirteen atoms**
NS		N_2H^+		HNCS	HNC_3			HC_7N	
SO		HNO		C_3N	C_4Si			C_8H	$HC_{11}N$
HCl		HCS^+		C_3O	C_5				
NaCl		C_3		C_3S	H_2COH^+				
KCl		C_2O		c-SiC_3					

I

arises because the dust is less effective at attenuating long-wave (red) light than short-wave (blue) light.

interstellar scintillation Analogue of the twinkling of stars seen with the naked eye that occurs for unresolved radio sources. The source of the scintillation is irregularities in the electron and ion density of the INTERSTELLAR MATTER. Movements within the medium, or more usually of the Earth and the radio source, cause the radio source to fluctuate in brightness on a time scale of a few minutes to a few hours as the scattering and delay within the interstellar medium alters.

intraterrestrial asteroid ASTEROID that has an orbit entirely interior to that of the Earth; it thus has an aphelion distance of less than 0.9833 AU. By analogy, intravenusian and intramercurian asteroids might also be defined. No such bodies are known, although they have been suggested in the past (*see* VULCAN). The discovery of such bodies would be difficult using ground-based telescopes because they are always on the sunward side of our planet.

intrinsic variable VARIABLE STAR in which the variations in brightness arise from processes that cause an actual change in the amount of radiation emitted, rather than modulate a fixed output. Typical processes are pulsation (both radial and non-radial) and eruptions (both from the accretion of material, as in a CATACLYSMIC VARIABLE, and from internal mechanisms as in a FLARE STAR). *See also* EXTRINSIC VARIABLE

invariable plane Plane of reference that is at right angles to the total ANGULAR MOMENTUM vector of the Solar System; it is unaffected in orientation by any of the PERTURBATIONS between the bodies in the Solar System. It is inclined at 1°.577 to the ECLIPTIC of EPOCH J2000. As most of the angular momentum of the Solar System is contained in the orbit of Jupiter, the invariable plane has an INCLINATION of just 0°.324 to Jupiter's orbital plane. The angular momentum of a body depends on its position, velocity and mass. The positions and velocities of the planets at any instant relative to the ecliptic are well known, but the masses of some planets are not so well known, and thus the calculation of the location of the invariable plane is a little uncertain and it is not used much as a reference plane. It is, however, used in the definition of the north and south poles of planets and satellites, which are defined as north if they are above the invariable plane.

inverse-square law Frequently encountered relationship whereby the magnitude of a physical quantity diminishes in proportion to the square of distance. Common examples of the inverse square law include Newton's law of gravity:

$$F = \frac{Gm_1 m_2}{r_2}$$

Coulomb's law, giving the force between two charged particles:

$$F = \frac{Q_1 Q_2}{4\pi\varepsilon d^2}$$

where Q_1 and Q_2 are the charges on the particles, d is their separation and ϵ, the permitivity of the intervening medium.
Also the brightness, I_d, of a point light source a distance, d, away from the observer is given by an inverse-square law:

$$I_d \propto \frac{1}{}$$

This last expression results in Pogson's equation relating the magnitudes, m_1 and m_2, of two objects to their brightnesses:

$$m_1 - m_2 = -2.5\log_{10}\left(\frac{I_1}{I_2}\right)$$

invisible astronomy Astronomical research at wavelengths of ELECTROMAGNETIC RADIATION invisible to the human eye. This covers GAMMA-RAY ASTRONOMY, X-RAY ASTRONOMY, ULTRAVIOLET ASTRONOMY, INFRARED ASTRONOMY, RADAR ASTRONOMY, RADIO ASTRONOMY, and the study of particles such as COSMIC RAYS and NEUTRINOS. By studying celestial bodies across the whole of the ELECTROMAGNETIC SPECTRUM, and not just at visible wavelengths, astronomers are able to build up a more complete picture of the Universe.

Io Innermost of the GALILEAN SATELLITES of JUPITER. Its size, mass and density are all only a little greater than those of the Moon, yet strong TIDAL HEATING by Jupiter makes it by far the most volcanically active body in the Solar System. Unlike the other Galilean satellites, Io is ice-free, and it has a sulphur-stained rocky surface.

Of the many revelations from the VOYAGER tours of the outer Solar System, the discovery of active volcanoes on Io probably ranks among the most important. Before Voyager, most people had assumed that bodies of Io's size, whether rocky or icy, would be geologically dead like the Moon. It is now realised that the orbital RESONANCE between the three innermost Galilean satellites results in tidal heating. The effect is greatest for Io because it is closest to Jupiter and hence experiences the strongest tidal forces.

There are often more than a dozen volcanoes erupting on Io at any one time. These are identified either by a visible 'eruption plume' powered by the explosive escape of sulphur dioxide, and rising 100–400 km (60–250 mi) above the surface, or by infrared detection of a hot spot. The record for the highest local temperature is at least 1700 K, whereas the normal daytime surface temperature on Io is only 120 K.

Io's density indicates that it is predominantly a silicate body, like the TERRESTRIAL PLANETS. Gravity and magnetic observations by the GALILEO orbiter confirm that it has a dense, presumably iron-rich core below its rocky mantle. Spectroscopic data show that Io's surface is covered by sulphur, sulphur dioxide frost and other sulphur compounds. However, these are no more than thin, volatile veneers resulting from volcanic activity, and the crust as a whole is some kind of silicate rock.

Io's tenuous atmosphere contains sulphur dioxide, as well as atoms of oxygen, sodium and potassium. The sur-

▼ **Io** A composite of images from the Galileo orbiter, showing the volcanically active surface of Jupiter's moon Io. Intense volcanism driven by tidal stresses leads to rapid resurfacing of Io.

face pressure is less than a millionth of the Earth's but nearly a billion times greater than the atmospheric pressure at the surface of the Moon or Mercury. Io's atmosphere continually leaks away into space, contributing to a 'cloud' of sodium and potassium atoms falling inwards towards Jupiter and into a magnetically confined belt of ionized sulphur that stretches right round Jupiter, concentrated around Io's orbit (the Io torus). The atmosphere is replenished by a combination of volcanic activity and collisions on to Io's surface by high-speed ions channelled by Jupiter's magnetic field. When Io passes into the shadow of Jupiter its atmosphere can be seen faintly glowing in an auroral display caused by these same magnetospheric ions impinging on the atmosphere.

Io's surface is totally dominated by the results of volcanic activity. There are lava flows up to several hundred kilometres in length and vast swathes of mostly flat terrain covered by fallout from eruption plumes. Most of the lava flows are now believed to have formed from molten silicate rock, which is often discoloured by a sulphurous surface coating, but there could also be some flows that formed from molten sulphur. The theory that prevailed for several years after the initial Voyager observations – that all or most of Io's lava flows are formed from sulphur – has been disproved by temperature measurements of eruption sites by the Galileo spacecraft and by infrared telescopes operating from Earth. Indeed, some of the temperatures detected are not only too high to represent molten sulphur, which would boil away at less than 700 K, but also too high to be characteristic of most types of molten rock, including basalt, which is the most common lava on the Earth and Moon. The exceptionally hot sites on Io might possibly be where a silica-poor, magnesium-rich relative of basalt known as komatiite is able to reach the surface.

In some places volcanoes rise above the general level of Io's plains. Their summits are occupied by volcanic calderas, up to 200 km (120 mi) across, which are formed by subsidence of the roof of the volcano after magma has been erupted from within. No impact craters are visible, even on the most detailed images, because the volcanic eruptions deposit fresh materials across the globe at an average rate of approximately a centimetre thickness per year.

More than 500 volcanoes have been identified on Io, and about 100 of these have been seen to erupt. The long duration of the Galileo mission enabled many changes on Io's surface to be documented, including deposits left by fallout from eruption plumes and new lava flows. Io's volcanoes appear to be randomly distributed, and Io certainly lacks the kind of well-defined global pattern displayed by the Earth. Unlike Earth, which releases internal heat by plate TECTONICS, and Venus, where heat escapes by conduction, probably punctuated by bouts of resurfacing every half billion years or so, Io's heat escapes through a multitude of volcanoes. One factor that probably influences the difference between the Earth and Io is that, to maintain a steady temperature, Io has to lose heat at a rate of about 2.5 W/m^2 compared with only 0.08 W/m^2 for the Earth. Possibly, the tidal heating experienced by Io is sufficient to keep a large fraction of its MANTLE partially molten. *See* date at JUPITER

ion An ATOM that has lost (or gained) one or more electrons compared with the normal, or 'neutral', atom. A positive ion has fewer electrons and a negative ion has more electrons than a neutral atom. *See also* IONIZATION

ionization Name given to any process by which normally electrically neutral atoms or molecules are converted into IONS, through the removal or addition of one or more electrons. This gives them a positive or negative electrical charge. An ion can itself be ionized, by losing or gaining a second electron. The minimum energy required to remove an electron from an atom, ion or molecule is called its ionization potential. For negative ions, the degree of ionization is denoted by a Roman numeral that is one larger than the number of lost electrons. For example, neutral iron is Fe I, singly ionized iron (one electron missing) is Fe II, doubly ionized iron is Fe III, and so on. An alternative notation is to indicate the net charge of the ion by superscript '+' and '−' signs. Using this system gives Fe^+ for singly ionized iron, Fe^{++} for doubly ionized iron, and so on. This latter system has the advantage of incorporating the negative ions, as in H^-, but it becomes cumbersome for high levels of ionization. Ionization normally occurs through the absorption of radiation or through collisions with other atoms and ions.

The material that forms stars, planetary nebulae and HII REGIONS is almost completely ionized. In other regions some atoms may be ionized while others remain as neutral atoms; for example, in the Earth's IONOSPHERE only about one atom in one million is actually ionized.

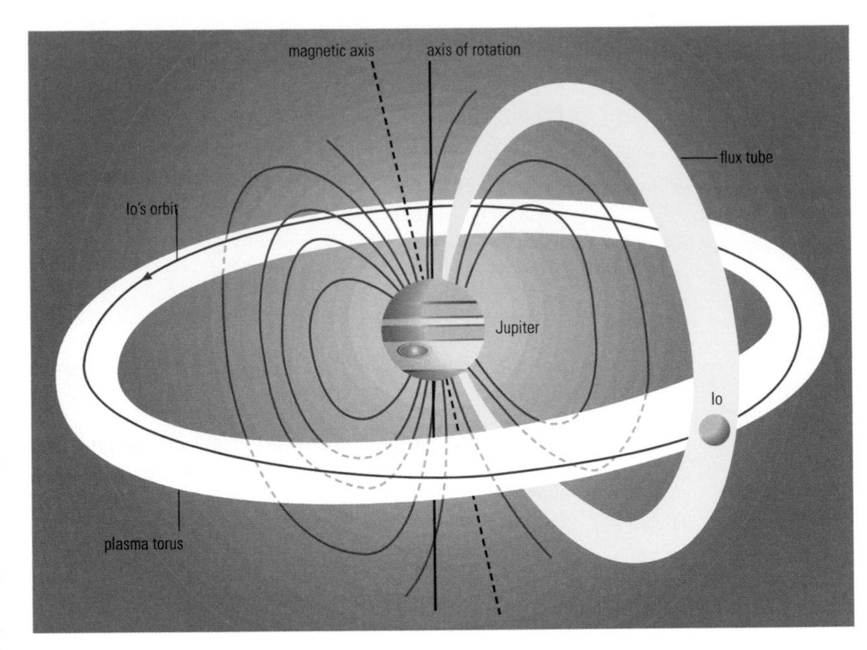

▲ **Io** Material ejected from Io's volcanoes forms a torus around Jupiter. A magnetic flux tube links Io and Jupiter: particles ejected during the satellite's volcanic eruptions can lead to enhancements of the Jovian aurorae.

ionosphere Region in the Earth's ATMOSPHERE that extends from a height of about 60 to 500 km (40–310 mi) above the surface. Within this layer most of the atoms and molecules exist as electrically charged ions. This high degree of ionization is maintained by the continual absorption of ultraviolet and X-ray radiation from the Sun. These free electrons and ions can disturb the transmission of radio waves through the ionosphere. There are several distinct ionized layers, which are known as the D, E, F_1, F_2 and G layers. The layers are rather variable: at night the D layer disappears, the E layer weakens or disappears, and the F_1 and F_2 layers merge. The D layer is situated at a height of between 60 and 90 km (40–60 mi), the E layer at 90–150 km (about 60–90 mi), the F_1 layer at 200 km (120 mi) and the F_2 layer at 300–400 km (190–250 mi). The free electrons in the E and F layers strongly reflect some radio waves: they enable long-distance radio communications by successively reflecting the waves between the layer and the ground. In RADIO ASTRONOMY, the presence of the E and F layers makes ground-based observations almost impossible below 10 MHz. The D layer, where collisions between the molecules and ions are more frequent, tends to absorb radio waves rather than reflect them.

IOTA Abbreviation of INTERNATIONAL OCCULTATION TIMING ASSOCIATION

Iota Aquarids Minor METEOR SHOWER active during July and August, with peak around August 6. It has a zenithal hourly rate (ZHR) no greater than 10. The meteors are generally swift and faint.

IPCS Abbreviation of IMAGE PHOTON COUNTING SYSTEM

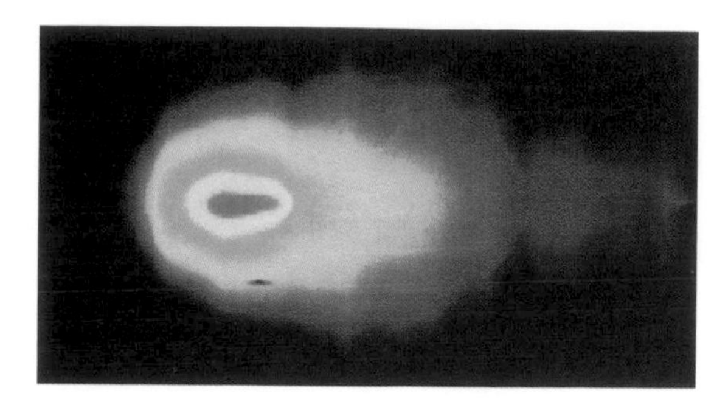

► **IRAS–Araki–Alcock, Comet** An infrared image of Comet IRAS–Araki–Alcock, which made a close passage to Earth in 1983 May.

IRAS Abbreviation of INFRARED ASTRONOMICAL SATELLITE

IRAS–Araki–Alcock, Comet (C/1983 H1) Bright long-period comet that passed remarkably close (0.031 AU) to Earth on 1983 May 11. The comet was discovered by Japanese amateur astronomer Genichi Araki (1954–) and, independently, English amateur astronomer George Alcock (1912–2000) on May 3, having earlier been detected by the Infrared Astronomical Satellite (IRAS). At closest approach, IRAS–Araki–Alcock reached mag. +2.0, showing a diffuse 2° diameter coma. Due to its proximity, the comet moved rapidly across the northern sky. Perihelion, 0.99 AU from the Sun, was reached on 1983 May 21. The orbital period is roughly 1000 years.

Iridum, Sinus (Bay of Rainbows) Lunar lava plain (45°N 31°W), 255 km (160 mi) in diameter. Sinus Iridum was formed by a large impact in a ring of the IMBRIUM multi-ring impact basin. Later, when lunar basalts poured into Mare Imbrium, they flooded this crater as well, covering the eastern side of the rim so that it is no longer visible. Multiple mare (wrinkle) ridges are visible in Sinus Iridum, including the circular mare ridge of an inner Imbrium basin ring.

IRIS Acronym for INFRARED IMAGING SURVEYOR

Iris Large MAIN-BELT ASTEROID discovered in 1847; number 7. Iris is *c.*200 km (*c.*124 mi) in diameter. Because of its orbit, near the inner edge of the main belt, this asteroid appears particularly bright, with only VESTA, CERES and PALLAS exceeding it.

iron (symbol Fe) Seventh most abundant element by numbers of atoms, and the fifth most abundant in terms of the mass content of the Universe. It is a metal. Iron's properties include: atomic number 26; atomic mass of the natural element 55.847 amu; melting point 1808 K; boiling point 3023 K. It has 10 isotopes, with iron-54 (5.8%), iron-56 (91.7%), iron-57 (2.2%) and iron-58 (0.3%) being stable and occurring naturally on Earth.

Iron is the major constituent of the cores of the planets Mercury, Venus and Earth, and of IRON METEORITES. Iron grains are believed to make up a proportion of the interstellar dust population.

Elements up to and including iron can be built up by energy-releasing FUSION reactions, which act as energy sources in stars. Iron, however, has the highest BINDING ENERGY per nucleon, and so these NUCLEOSYNTHESIS reactions halt when it has been produced. *See also* METALS

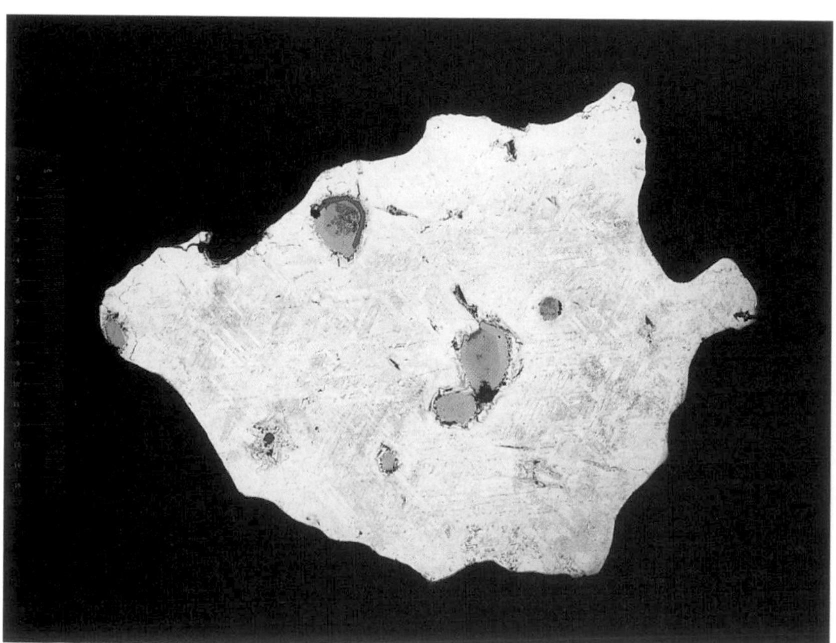

▼ **iron meteorite** An etched cross-section through an iron meteorite, showing the Widmanstätten pattern. This pattern indicates that iron meteorites crystallized slowly in the pre-solar nebula.

iron meteorite METEORITE composed of iron metal, generally with between 5 and 20% by weight nickel. Iron meteorites account for approximately 5% of all observed meteorite falls. The mineralogy of iron meteorites is dominantly an intergrowth of the two iron–nickel alloys kamacite and taenite. Kamacite (αFe,Ni) has a body-centred cubic structure and a nickel content less than 7% by weight. Taenite (γFe,Ni) is face-centred cubic and *c.*20–50% by weight nickel.

Iron meteorites are highly differentiated materials, the products of extensive melting processes on their parent bodies. They can be divided into magmatic irons and non-magmatic irons. Magmatic irons are those that have solidified by fractional crystallization from a melt. Non-magmatic irons are those that seem not to have completely melted; they may have formed during impact processes.

The iron meteorites are subdivided into 13 different groups on the basis of nickel and trace element chemistries (gallium, germanium and iridum contents). Each separate group of magmatic irons has a fairly restricted range of nickel contents but a wide range of trace element abundances; these trends are consistent with fractional crystallization from a melt. In contrast, the non-magmatic irons show a wide range in nickel contents but less variation in trace element composition; these trends can be better explained by formation by partial melting. Many irons defy chemical classification and simply remain 'ungrouped'. It is thought that each chemical group derives from its own parent asteroid.

Prior to classification on the basis of trace-element chemistry, iron meteorites were classified in terms of their metallographic structure. Laths of kamacite intergrown with nickel-rich phases form the 'Widmanstätten pattern' revealed in polished and etched iron meteorites. The width of the kamacite lamellae allows classification of iron meteorites into five structural groups: the coarsest, coarse, medium, fine and finest octahedrites. Plessitic octahedrites are transitional between octahedrites and ataxites. Ataxites are nickel-rich, with more than 20% by weight nickel, and are mainly taenite. Hexahedrites have nickel less than 6% by weight and comprise kamacite only. Neither hexahedrites nor ataxites display a classic Widmanstätten pattern. Meteorites from an individual chemical group can display a range of structural types.

irradiation Process by which a region of space or material is subjected to radiation, whether by light, radio, infrared or other forms of ELECTROMAGNETIC RADIATION, or by energetic particles such as protons and electrons.

irregular galaxy GALAXY that shows no symmetry. Some irregular galaxies are smaller than spiral and elliptical galaxies, contain much gas, and are undergoing star formation. Some are classified as irregular for the only reason that they do not fit into other categories of galaxy. Many irregular galaxies have overall rotation and a relatively thin, gas-rich disk, sometimes including a bar (as in the LARGE MAGELLANIC CLOUD), forming a continuation of the HUBBLE CLASSIFICATION beyond the spirals of types Sc and SBc.

irregular variable VARIABLE STAR that displays no periodicity in its light changes; there are two broad types.

One type (I) includes many poorly understood variables as well as the various forms of NEBULAR VARIABLE. The second type (L) consists of slowly varying pulsating stars that are otherwise very similar to various types of LONG-PERIOD VARIABLE and SEMIREGULAR VARIABLE. Although many types of variable star (*see* CATACLYSMIC VARIABLE and FLARE STAR) exhibit fluctuations at random, unpredictable intervals, these types are not defined as 'irregular' under the classification scheme.

Isaac Newton Group of Telescopes (ING) Group of instruments consisting of the 4.2-m (165-in.) WILLIAM HERSCHEL TELESCOPE, the 2.54-m (100-in.) ISAAC NEWTON TELESCOPE and the 1.0-m (39-in.) JACOBUS KAPTEYN TELESCOPE at the ROQUE DE LOS MUCHACHOS OBSERVATORY on the island of La Palma. There is a sea-level base at Santa Cruz de La Palma. The ING is funded by the UK's PARTICLE PHYSICS AND ASTRONOMY RESEARCH COUNCIL, the NWO (Nederlanse Organisatie voor Wetenschappelijk Onderzoek) in the Netherlands, and Spain's INSTITUTO DE ASTROFÍSICA DE CANARIAS.

Isaac Newton Telescope (INT) Optical 2.54-m (100-in.) telescope, part of the ISAAC NEWTON GROUP on La Palma, which can be used either for wide-field imaging or spectroscopy. It was installed at the ROYAL GREENWICH OBSERVATORY, Herstmonceux, in 1967 and was originally fitted with a 2.50-m (98-in.) mirror made from a Pyrex blank cast in 1936 for Michigan University Observatory. However, atmospheric conditions at Herstmonceux were poor, and when the UK's participation in ROQUE DE LOS MUCHACHOS OBSERVATORY was proposed in the 1970s, it was decided to move the INT there. Equipped with a new mirror, the telescope began operation on La Palma in 1984.

ISAS Abbreviation of INSTITUTE OF SPACE AND ASTRONAUTICAL SCIENCE

ISEE Abbreviation of INTERNATIONAL SUN–EARTH EXPLORER

Ishtar Terra Continent-sized highland block of VENUS. It covers an area comparable to Australia and drops steeply to the surrounding plains, especially on its south-western flank. It is unique among Venus' upland regions inasmuch as its perimeter rises several kilometres above the interior. Western Ishtar comprises Lakshmi Planum (a vast plateau encircled by a series of mountain belts), Freya Montes to the north, Akna Montes in the west, and the Danu Montes, which extend some 1200 km (750 mi) to the south and south-east. Immediately east are the Maxwell Montes, rising to 17 km (11 mi) above the mean planetary radius, the highest point on Venus. Like the mountains around Lakshmi Planum, the Maxwell Montes have a complex banded structure clearly seen on radar images. Eastern Ishtar takes the form of a rather hummocky plateau extending outwards between about 100 and 1000 km (60–600 mi). These highly deformed zones are below the level of the mountain chains; they slope down towards the exterior plains or terminate in steep scarps.

Isidis Planitia Impact basin on MARS (13°.0N 273°.0W). It is situated between SYRTIS MAJOR PLANITIA and the Elysium Rise and is approximately 1100 km (680 mi) across. The basin is poorly defined on its eastern side, where it merges on to the plains associated with the Elysium volcanoes.

Islamic astronomy Astronomy as practised in the Middle East, North Africa and Moorish Spain during the flowering of the Islamic Empire, from around the 8th to the 14th century. Although it is sometimes referred to as **Arab** or **Arabic astronomy**, some of its practitioners were from other ethnic or linguistic groups, and the unifying cultural force in this region and during this period was Islam. Two circumstances fostered the growth of astronomy under Islamic rule. The first was that the seats of ancient learning lay within or just outside the bounds of the empire, and Islam was tolerant of scholars from other creeds. A second impetus came from Islamic religious observances, which gave rise to many problems in mathematical astronomy, mostly related to timekeeping. In solving these problems, Islamic astronomers went far beyond the Greek mathematical methods and provided essential tools for the creation of Western RENAISSANCE ASTRONOMY.

Following the foundation of Baghdad, the new capital of the Abbasid dynasty, in AD 762, there began a massive effort to translate into Arabic all the major scientific texts of antiquity. The most vigorous patron was the Caliph al-

▼ **Isaac Newton Group of Telescopes** High above the clouds on the Roque de Los Muchachos peak on La Palma, the ING telescopes enjoy some of the best observing conditions in Europe. Largest of these instruments is the Isaac Newton Telescope itself (large dome at left), a 2.5 m (98 in.) reflector.

Ma'mūn. Shortly after he came to power in 813 he founded the *Bayt al-Hikma* (House of Wisdom) in Baghdad. There, scholars of all creeds worked to translate manuscripts acquired from the ancient libraries that now lay within the empire, which stretched from Spain to India, to stock what was to become one of the world's great academies.

The chief scholar of this great enterprise was Abu'l-Hasan Thābit ibn Qurra (*c*.835–901), who wrote over a hundred scientific treatises, including a commentary on Ptolemy's *ALMAGEST*. Another astronomer (and geographer) in 9th-century Baghdad was Abu'l-Abbas al-Farghānī (*c*.825–61), whose *Elements* helped to spread the more elementary and non-mathematical parts of Ptolemy's geocentric astronomy to the West. By 900 the stage was set for the spread of scientific knowledge throughout the Empire, with a single language, Arabic, as its vehicle. This knowledge later diffused into Christian Europe via Spain, where there was considerable scholarly interaction with visiting European translators until the defeat of the Moors in the 12th century (*see* MEDIEVAL EUROPEAN ASTRONOMY).

The times and dates of Islamic religious activities are regulated according to a lunar calendar, and the first appearance of the new moon is of great importance. Predictions of this and the preparation of almanacs regulating the hours of prayer led to a considerable interest in spherical trigonometry, and the development of the modern trigonometric functions (although some originated in India) and the identities between them.

Islamic astronomers did not make exhaustive observations of the sky. They restricted their sightings, or at least those they chose to record, primarily to measurements that could be used for re-deriving key parameters of solar or planetary orbits. An impressive example of an Islamic astronomer working strictly within a Ptolemaic framework but establishing new values for Ptolemy's parameters was Muhammad AL-BATTĀNĪ, a younger contemporary of Thābit ibn Qurra. Al-Battānī's *Zīj* ('[Astronomical] Tables') was one of the most important works of astronomy between the time of Ptolemy and the Renaissance – Nicholas COPERNICUS cites his 9th-century predecessor no fewer than 23 times.

By contrast, one of the greatest astronomers of medieval Islam, 'Ali ibn 'Abd al-Rahmān ibn Yūnus (950–1009), remained virtually unknown to European astronomers until around 1800. Working in Cairo a century after al-Battānī, Ibn Yūnus wrote a major astronomical handbook called the *Hakimi zīj*. Unlike other Islamic astronomers, he prefaced his *zīj* with a series of more than a hundred observations, mostly of eclipses and planetary conjunctions.

Ptolemy's *Almagest* had contained a catalogue of over a thousand stars. The first critical revision of this catalogue was carried out in the 10th century by Abu'l-Husain AL-SŪFĪ. His *Kitāb suwar al-kawākib al-thābita* ('Book on the Constellations of the Fixed Stars') followed Ptolemy's often faulty list, but it did give improved magnitudes. The book's splendid pictorial representations became known in the Latin West; it also contains the first known representation of the Andromeda Galaxy.

Although most Islamic astronomers remained securely within the geocentric framework of Ptolemy and Aristotle, some criticized particular technical details of the PTOLEMAIC SYSTEM which seemed to violate the ancient belief that only uniform circular motion can explain the movements of celestial bodies. One of the first critics was the physicist Ibn al-Haytham (965?–*c*.1041), known in the West as Alhazen, who held the planetary models of Ptolemy's *Almagest* to be false. Later, Muhammad ibn Rushd (1126–98), known in the West as Averroës, declared that Ptolemy's eccentrics and epicycles were 'contrary to nature'.

A fresh attack on the Ptolemaic system was undertaken in the 13th century by Nasīr al-Dīn AL-TŪSĪ. A prolific writer with 150 known titles to his credit, al-Tūsī constructed a major observatory at Marāgha (in present-day Iran). Other astronomers at the Marāgha observatory also offered new arrangements of circles, but a fully acceptable alternative (from a philosophical point of view) did not come until the work of Ibn al-Shātir (1304–75/6) at Damascus around 1350. Although al-Shātir's solution, as well as the work of the Marāgha observatory, remained generally unknown in the West, this and other Islamic criticisms of Ptolemy may have had an influence on Copernicus.

Conspicuous examples of modern astronomy's Islamic heritage are found in its vocabulary in terms such as 'nadir' and 'zenith', but in particular in the names of stars: Betelgeuse, Rigel, Aldebaran and Altair are just a few of the star names that are Arabic in origin or are Arabic translations of Ptolemy's Greek descriptions. (Some of the names as we have them today bear little resemblance to their original forms, having been corrupted in transliteration and by centuries of transcription.) Many of these star names begin with 'Al-' because *al* is Arabic for 'the'. The Arabic star nomenclature entered the West by another route – the making of ASTROLABES, on which star names were inscribed. The earliest dated astrolabe in Arabic dates from 927–928. It is primarily from Spain that astrolabe-making, together with Arabic names for the stars, reached the West via England in the late 13th and 14th centuries.

isostasy Principle that recognizes there to be a state of balance between topographic masses and the underlying materials that support them. Higher elevations tend to be supported by material of lower density; for example, continental crust is less dense than oceanic crust and is also thicker, and mountain ranges have 'roots' of crustal material extending downwards towards the mantle. The depth of such features can be deduced from precise measurements of local variations in the gravity field. The concept owes much to the work of Sir George AIRY.

ISO Abbreviation of INFRARED SPACE OBSERVATORY

isotherm Line drawn on a weather map, joining all places that are, at a given moment of time, experiencing the same temperature. Isotherms can also be drawn on plots of radio or infrared emission from astronomical bodies.

isothermal process Process in thermodynamics in which a change in a system occurs with transfer of heat to or from the environment so that the system remains at a constant temperature. The collapse of a PROTOSTAR (*see* HAYASHI TRACK) and the collapse of a star to form a WHITE DWARF are both isothermal processes.

isotopes Atomic nuclei that have the same atomic number but different atomic masses; that is, they contain the same number of protons but different numbers of neutrons. *See also* ATOM

isotropy Idea that the UNIVERSE looks the same in all directions.

ISS Abbreviation of INTERNATIONAL SPACE STATION

Istituto di Radioastronomia, Bologna Institute of the Italian National Research Council (CNR) which operates two stations, respectively in Medicina (Bologna) and Noto (Siracusa). Its three radio telescopes, the 600 × 600-m (2000 × 2000-ft) Northern Cross array and two single-dish 32-m (105-ft) antennae, are used mainly for VERY LONG BASELINE INITERFEROMETRY.

IUE Abbreviation of INTERNATIONAL ULTRAVIOLET EXPLORER

Izar (Pulcherrima) The star ε Boötis, visual magnitude 2.35, distance 210 l.y. It is a striking double star of mags. 2.5 and 4.6, difficult to resolve in the smallest telescopes because of the closeness of the components, less than 3″ apart. Their spectral types are K0 II or III and A0 V, producing a beautiful colour contrast of orange and blue. The name Izar comes from the Arabic meaning 'girdle' or 'loincloth'. Its alternative title, *Pulcherrima*, is Latin for 'most beautiful', from its telescopic appearance.

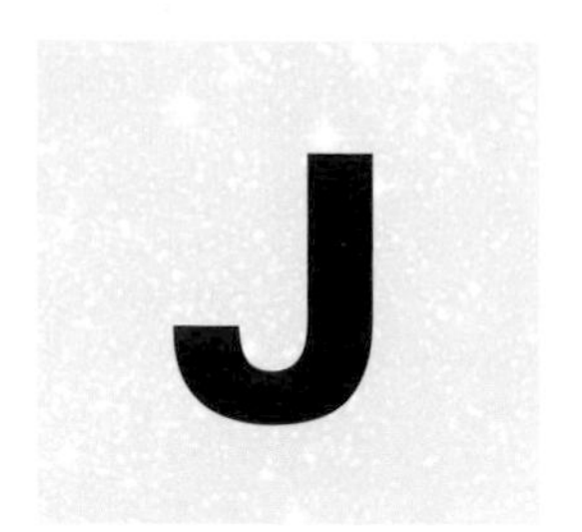

Jacobus Kapteyn Telescope (JKT) Smallest of the three Anglo-Dutch telescopes of the ISAAC NEWTON GROUP on La Palma. The JKT has a 1.00-m (39-in.) diameter mirror and is used exclusively for astronomical imaging. It was installed at the ROQUE DE LOS MUCHACHOS OBSERVATORY in 1985.

James Clerk Maxwell Telescope (JCMT) World's largest radio telescope capable of working at submillimetre wavelengths (0.3–2 mm), operated by the JOINT ASTRONOMY CENTRE. The JCMT is situated at MAUNA KEA OBSERVATORY in Hawaii, where it is above 97% per-cent of the water vapour in the atmosphere – an essential requirement for this waveband. The primary mirror, 15 m (49 ft) across, consists of 276 individually adjustable aluminium panels. An opening enclosure protects the instrument. Since its inauguration in 1987, the JCMT has undertaken pioneering work at submillimetre wavelengths, including the first observations made with an imaging bolometer array.

Jansky, Karl Guthe (1905–50) American radio engineer, whose detection of radio waves from the Milky Way marked the beginning of RADIO ASTRONOMY. In 1928 Jansky joined the Bell Telephone Laboratories in Holmdel, New Jersey, after graduating from the University of Wisconsin. He made his historic discovery with a rotating antenna that received both very short radio waves (20.5 MHz) and much longer waves (45 kHz), which he was using to search for the cause of static that interfered with ship-to-shore and other kinds of telecommunications. He identified three kinds of static. The first two were associated with thunderstorms: localized storms, and more distant storms whose radio noise was reflected by the ionosphere. The third kind, a steady hiss, was initially a mystery, but Jansky noticed that it reached a maximum intensity every $23^h 56^m$, the SIDEREAL PERIOD.

Jansky noted that during the solar eclipse on 1932 August 31, the intensity of the waves did not decrease, as it would have done if the radio emission came from the Sun. Also, the hiss was strongest along the galactic plane, especially at a right ascension of around 18^h, in Sagittarius – the likely centre of the MILKY WAY already identified by Harlow SHAPLEY and Jan OORT. Janksy concluded that giant ionized clouds of INTERSTELLAR MATTER in the Milky Way were producing the static, marking the birth of the new science of radio astronomy. Although this made the front page of the *New York Times*, there was little interest from the scientific community. Jansky himself failed to see the potential of his discovery. It was taken up, initially single-handedly, by Grote REBER, and only after World War II did radio astronomy really take off. To honour Jansky, the IAU in 1973 adopted his name for the unit of flux density of extraterrestrial radio emissions.

jansky (symbol Jy) Unit of FLUX DENSITY adopted by the International Astronomical Union in 1973 in honour of Karl JANSKY. One jansky is equal to 10^{-26} watts per square metre per hertz.

Janssen, (Pierre) Jules (César) (1824–1907) French astronomer who invented the spectroheliscope and, independently of Norman LOCKYER, discovered the spectral line of helium in the solar chromosphere. Originally a businessman, he turned to solar astronomy after hearing of the spectroscopic discoveries of Gustav KIRCHHOFF. Janssen built an observatory in 1862 atop his house at Montmartre to investigate the dark bands in the Sun's spectrum. Using a five-prism spectroscope, Janssen was able to resolve the bands into fine lines that were most intense during sunrise and sunset, when the Sun was observed through the thickest layers of the Earth's atmosphere, and identified water vapour in the atmosphere as the source of these TELLURIC LINES.

On an expedition to India, Janssen used a spectroscope to study solar prominences visible during the 1868 August total eclipse, demonstrating that the bright lines in their spectra indicated a gaseous nature. He re-examined these lines by constructing a spectrohelioscope, with which the prominences could be observed spectroscopically without waiting for a total eclipse. Janssen observed three bright yellow lines in their spectra, two of which were already identified with the element sodium; the third, previously undiscovered, was attributed by Lockyer to a new chemical element, called helium, not discovered on Earth until 1895. Two years later, during the Franco–Prussian War, Janssen escaped the besieged city of Paris in a balloon, intending to observe the1870 December total solar eclipse in Algeria. During the 1874 and 1882 transits of Venus, he obtained multiple images of the planet in rapid succession using a technique that was a forerunner of cinematography. His collection of more than six thousand superb solar photographs taken between 1876 and 1903 was published as *Atlas du photographies solaires* (1904). Janssen helped to establish the MEUDON OBSERVATORY (1875) and another at Mont Blanc (1893–95), which was later known as the PIC DU MIDI OBSERVATORY.

Janus One of the inner satellites of SATURN, identified in 1980 in VOYAGER 1 images. It may have been observed earlier, in 1966, by Audouin Dollfus. It is spheroidal in shape, measuring about 200 × 180 × 150 km (120 × 110 × 90 mi). Janus has a near-circular orbit, with a very slight tilt to Saturn's equator (inclination only 0°.14), at an average distance of 151,500 km (94,100 mi) from the planet's centre, giving it an orbital period of 0.695 days. It is co-orbital with EPIMETHEUS; the two may be derived from a single split body. *See also* CRUITHNE

Jeans, James Hopwood (1877–1946) English theoretical physicist who developed advanced mathematical models to attack many fundamental astrophysical problems. Jeans' first major theoretical work, published while he was at Trinity College, Cambridge, used statistical mechanics to explain the behaviour of gas molecules. Since many astrophysical objects, including stars and nebulae, are largely composed of gas, this kinetic theory of gases became widely used by astronomers investigating the structure of stellar interiors and the interstellar medium. A report written by Jeans in 1914 helped the QUANTUM THEORY to become accepted.

He then began a long study of rotating fluid masses, building on previous work by, among others, George Howard Darwin (1845–1912) and Édouard ROCHE. This led him to reject the NEBULAR HYPOTHESIS of Pierre Simon de LAPLACE by showing that the rotating pear-shaped mass required by this theory was dynamically unstable. Jeans proposed instead that the Solar System formed after another star passed close by our Sun and violently removed some of its matter, which subsequently coalesced into the planets. This tidal theory was later extended by Harold JEFFREYS. In 1928 Jeans proposed CONTINUOUS CREATION, a concept that laid the foundations for the STEADY-STATE THEORY of cosmology, though it was left to others, including Fred HOYLE, fully to develop that model. Jeans concluded his life in astronomy by writing popular books, including *The Universe Around Us* (1929), *The Mysterious Universe* (1930) and *Through Space and Time* (1934), and radio broadcasts aimed at a general audience.

Jeffreys, Harold (1891–1989) British physicist who developed models for the structure and evolution of the Earth and the Solar System. His work in hydrodynamics and geophysics led him to extend James JEANS' tidal hypothesis for the origin of the Solar System. This theory suggests that the passage of stars close to the Solar System triggers the formation of planets. Jeffreys also proposed models of the circulation of Earth's atmosphere and the atmospheres of the outer planets.

Jet Propulsion Laboratory (JPL) Federally funded research and development facility managed by the Cali-

▼ **Janus** Saturn's small moon Janus imaged from Voyager 1 during its 1980 flyby. Janus is seen here against the background of the planet.

J

fornia Institute of Technology (Caltech) for the NATIONAL AERONAUTICS AND SPACE ADMINISTRATION (NASA). It owes its origin (and its name) to rocket experiments carried out by Theodor VON KÁRMÁN in the 1930s in a canyon near Pasadena, California. Spurred by the onset of World War II, the laboratory conducted missile experiments on the site and it grew rapidly, eventually flying the first successful US spacecraft, EXPLORER 1, in January 1958. When NASA was created the following December, JPL was transferred to the new space agency from US Army control, bringing expertise in building and flying spacecraft, propulsion systems, guidance, systems integration and space telecommunications.

JPL is NASA's centre of excellence for deep-space systems, and its broad remit covers most aspects of unmanned US space missions. It has been responsible for the high-profile planetary probes of the past few decades, and continues this role with such future projects as deep impact and EUROPA ORBITER. It also plays a major role in space observatories such as the HUBBLE SPACE TELESCOPE and the SPACE INFRARED TELESCOPE FACILITY, as well as Earth science satellites such as Terra, CloudSat and GRACE (the Gravity Recovery And Climate Experiment).

jets Long, thin gaseous structures characterized by radio emission and/or optical nebulosity. There are always two jets, but one may be hidden by the central object or disrupted by the medium through which it travels. Jets are usually collimated (kept straight) and confined by strong magnetic fields or winds, and the rotation of the central object plays a significant role. Jets arise from two vastly different situations.

In highly luminous RADIO GALAXIES (double-lobed radio sources) and QUASARS, two jets of highly supersonic, energetic electrons bore into the ambient intergalactic medium. The jets remain collimated unless the surrounding gas causes wiggles and bends by refraction, or until they impact upon sufficiently dense material so that they dissipate much of their energy in shocks, creating radio 'hot spots' of SYNCHROTRON RADIATION within the large radio LOBES. Sometimes galaxies even generate knots along their jets that can be seen as optical nebulae. The jet of M87 is 8000 l.y. long, and the Hubble Space Telescope is able to image knots and features as small as 10 l.y. across in the jet. The jets connect the radio lobes to the centre of the galaxy, and they are thought to be powered by a BLACK HOLE at its centre. The central engine is able to pump the jets in the same two directions for periods of order 100 million years. At times the velocity of the material in the jets of M87, and from several quasars, appears to be greater than the speed of light. This SUPERLUMINAL motion is actually a geometric effect, and it requires the jet to be pointing almost directly at Earth (within 10°). This may mean that more objects have jets than expected, but they are not see because they are not pointing towards Earth.

A second generator of jets is a low-luminosity PROTOSTAR. The material around the fledgling protostar collapses on to the star, helping to initiate the nuclear fusion reaction that turns hydrogen into helium. This reaction reverses the infall, and eventually an outflow occurs. The wind is millions of times stronger than the SOLAR WIND, and is usually supersonic, so that shock waves are produced in the interstellar medium where the jet impacts. The shocked clumps of material are called HERBIG–HARO OBJECTS; they can be as small as the size of the Solar System, and the mass of the Earth. The clumps have been found to be variable: they can be seen to move and change over several years. The protostar itself is invisible, hidden behind a thick cocoon of dust and probably with a large dust disk, but it is assumed that the jets leave the system along the rotational axis of the star. The material in the jets can be detected in the optical, infrared and radio regions, with the radio observations highlighting the presence of magnetic fields.

▶ **Jodrell Bank** The 76-m (250-ft) dish of the Lovell Radio Telescope at Jodrell Bank. Built in 1957, this is the world's second-largest fully steerable radio telescope.

jet stream Narrow belt of high-speed wind in a planetary ATMOSPHERE. On Earth, the jet streams that have the greatest influence on the surface weather, including the development, motion and decay of depressions, are two westerly jets, the subtropical and polar jet streams, in each hemisphere. These typically occur as discontinuous segments around the Earth, thousands of kilometres long, hundreds of kilometres wide, and a few kilometres deep. They lie in the upper TROPOSPHERE and lower STRATOSPHERE, often being located where there is a break in the TROPOPAUSE. Seasonal jets, such as the easterly equatorial and the westerly polar-night jet streams, also occur within the stratosphere. On Earth, jet stream wind speeds generally lie in the range 160–320 km/h (100–200 mph), with a recorded extreme of 656 km/h (408 mph). Far higher speeds, sometimes in excess of 1400 km/h (870 mph), have been observed on Jupiter, Saturn and Neptune.

Jewel Box Name given to the KAPPA CRUCIS CLUSTER by John HERSCHEL.

Jodrell Bank Observatory Major radio astronomy facility of the University of Manchester and the location of Britain's most famous scientific instrument, the **Lovell Telescope**. The Jodrell Bank site, 32 km (20 mi) south of Manchester, was selected by Bernard LOVELL in 1945 as a suitable location from which to observe cosmic rays. He used ex-army radar equipment to try to detect the rays, but instead found reflections from meteors. Building on this experience, he constructed a 66-m (218-ft) fixed parabolic aerial in 1947 and, in 1957, the fully steerable 76-m (250-ft) dish known originally as the Mark I telescope that now bears his name. Both telescopes were the largest of their kind in the world when built, and the 250-ft telescope immediately became famous through its tracking of SPUTNIK 1. Outstanding debts on the telescope were eventually cleared by the Nuffield Foundation, and what had been called the Jodrell Bank Experimental Station became the **Nuffield Radio Astronomy Laboratories**.

A second fully steerable radio telescope, the 26-m (85-ft) Mark II, was completed in 1964 and, together with a third telescope at Nantwich, began pioneering work on long-baseline interferometry. The Multi-Telescope Radio-Linked Interferometer project of 1976 extended this work to several other sites with telescopes operated by Jodrell Bank. In the early 1990s, this array (now renamed MERLIN) was extended to include a new 32-m (105-ft) telescope at Cambridge. In the meantime, the Mark I had been upgraded and renamed the Lovell Telescope on its 30th anniversary. Further upgrades to the Lovell in 2000–2002 ensure that the telescope will continue to play a leading role in astronomy both as a single-dish instrument and as part of MERLIN. Its state-of-the-art receivers make it 30 times more sensitive than when it was first built.

Johnson, Harold Lester (1921–80) American astronomer who co-invented the UBV PHOTOMETRY system used to analyse the physical properties of stars and galaxies. Johnson was an electronics expert who designed and built many kinds of photometer. He improved the sensitivity of the detectors used in these instruments and found innovative ways of reducing the 'noise' that interfered with photometric observations. Johnson also designed and built a prototype of the first multiple mirror telescope, installed in the 1970s at the Mexican National Observatory.

Working with William W. MORGAN in the early 1950s, he developed the UBV SYSTEM for measuring stellar magnitudes in three basic colours or wavebands: U (ultraviolet), B (blue) and V (visual, or yellow). Later, Johnson extended the system to include the R (red) and I (near-infrared) bands, then to additional bands out to 4 μm, known as J, K and L. Johnson and his colleagues measured thousands of 'reference stars' with the UBV and UBVRI photometry systems, and they were able to calibrate the spectral energy distributions ('absolute energy curves'), resulting in more accurate stellar temperature scale. Using his new system of photometry, Johnson evolved powerful tools for analysing stars and clusters of stars by plotting their apparent visual magnitudes against B–V differences or 'indices' (colour–magnitude diagram) or U–B index versus B–V index (colour–colour diagram). He was able to estimate the ages of clusters from their colour–magnitude diagrams. The same principles were applied to studies of individual stars, and using these tools Johnson made many investigations involving carbon stars, subdwarfs, Cepheids and M stars, as well as structures like circumstellar shells.

Johnson Space Center (JSC) In full, the Lyndon B. Johnson Space Center, the NATIONAL AERONAUTICS AND SPACE ADMINISTRATION (NASA) operational headquarters for all US manned space missions, and home to NASA's Astronaut Corps. The JSC lies about 40 km (25 mi) south-east of Houston, Texas. It was founded in 1961 as the Manned Spaceflight Center, but its name was changed in 1973 February to honour the former US President. The JSC's remit also includes spacecraft design, development and testing; the selection and training of astronauts; mission planning; and participation in the various experiments conducted in space. Its facilities include the Neutral Buoyancy Laboratory, a giant pool that can accommodate full-size mock-ups of the INTERNATIONAL SPACE STATION and allow astronauts to gain experience working in a weightless environment.

Joint Astronomy Centre (JAC) Organization based in Hilo, Hawaii, that operates the JAMES CLERK MAXWELL TELESCOPE and the UNITED KINGDOM INFRARED TELESCOPE on behalf of the three partner countries (the UK, Canada and the Netherlands).

Joint Institute for Laboratory Astrophysics (JILA) Interdisciplinary institute for research and graduate education in the physical sciences, operated jointly by the University of Colorado (CU) and the National Institute of Standards and Technology (NIST). Founded in 1962, it is located on the main CU campus in Boulder, Colorado. JILA's research areas include atomic and molecular interactions, new states of matter, cooling and trapping of matter and nanotechnology. Its laboratories host 'the coldest place in the Universe'.

Jones, Harold Spencer (1890–1960) English astronomer, the tenth ASTRONOMER ROYAL, who arranged the move of GREENWICH OBSERVATORY to Herstmonceux, and improved the value of the SOLAR PARALLAX. From 1923 to 1933 Jones served as His Majesty's Astronomer at the Cape, South Africa. There he supervised studies of lunar occultations, using data accumulated between 1880 and 1922 to verify and correct the lunar constants calculated by E.W. BROWN. He also measured the parallaxes, and hence distances, to hundreds of stars in the Milky Way. In 1933 Jones returned to England to become Astronomer Royal, retiring in 1955.

Jones' major lifetime project was a highly detailed analysis of tiny variations in the Earth's rotation rate. He is most famous for his refinement of the solar parallax, and hence of the ASTRONOMICAL UNIT. In 1928 the IAU appointed him president of a commission to determine the solar parallax from observations to be made of the asteroid (433) Eros, which would be at opposition in 1931. Using thousands of visual and photographic observations from some fifty observatories in both hemispheres, Jones began a ten-year process of reduction and calculation that ranks as one of the most impressive calculational feats of the pre-computer era. The result, announced in 1941, was 8″.790, corresponding to a value for the AU of 149,670,000 km.

Jovian Pertaining to the planet Jupiter. The term Jovian planet is also used to describe the Solar System's giant planets or gas giants: Jupiter, Saturn, Uranus and Neptune. The atmospheres of the Jovian planets represent a significant proportion of their total mass, in contrast to those of the inner, terrestrial planets.

Joy, Alfred Harrison (1882–1973) American astronomer who used spectroscopy to study variable stars and novae, and to find stellar distances and radial motions. He spent almost 60 years at the MOUNT WILSON OBSERVATORY, which he joined in 1915, working under George Ellery HALE. He used the technique of spectroscopic parallax developed by Walter S. ADAMS to calculate the distances to stars. Joy measured the radial velocities for thousands of stars, including many Cepheid variables, which allowed him to determine the distance of the Solar System to the centre of the Milky Way. He also was the first to classify the T Tauri stars.

JPL Abbreviation of JET PROPULSION LABORATORY

Julian calendar CALENDAR devised by Julius Caesar and the Greek astronomer Sosigenes of Alexandria in 45 BC. The system was based on a year of 365.25 days, fixed at 365 with a leap year of 366 days every four years to compensate. The year commenced on March 25. The calendar, which was completed by Caesar Augustus in 44 BC and adopted throughout the Roman Empire, established the present-day names, lengths and order of the months.

The Julian calendar was in general use in the West until 1582, by which time the approximation to the true solar year of 365.24219 days had accumulated an error of ten days. It was superseded by the GREGORIAN CALENDAR but was still used in the UK until 1752.

Julian date Number of days that have passed since noon GMT on 4713 BC January 1 (the JULIAN DAY number), plus the decimal fraction of a mean solar day that has elapsed since the preceding noon. For example, the Julian date at 18.00 UTC on 2000 January 1 was 2,451,545.25. The modified Julian date (MJD) is a shortened form, using only the last five digits of the full Julian date and commencing at 0^h UTC on 1858 November 17. The modified Julian date at 18.00 UTC on January 1 therefore was 51544.75, the 0.5 difference being because the Julian date changes at noon but the MJD at midnight UTC.

Julian Day Method of easily accounting for the number of days that have elapsed over a long period of time, independent of calendarial systems. Each day is allotted a Julian Day number, which runs consecutively from

J

4713 BC January 1 and commences at noon, 12 hours later than the corresponding civil day. The starting point was arbitrarily chosen by the French mathematician Joseph Scaliger (1540–1609); the name 'Julian' is in honour of his father Julius and has no connection with Julius Caesar. This numbering system helps to reconcile disparate historical CALENDARS and makes it possible to calculate easily the time in days between the recurrence of celestial events. *See also* JULIAN DATE

Juliet One of the small inner satellites of URANUS, discovered in 1986 by the VOYAGER 2 imaging team. Juliet is about 84 km (52 mi) in size. It takes 0.493 days to circuit the planet, at a distance of 64,400 km (40,000 mi) from its centre, in a near-circular, near-equatorial orbit.

June Bootids Alternative name for the PONS–WINNECKIDS

Juno Large MAIN-BELT ASTEROID discovered in 1804; number 3. Juno is about 268 km (167 mi) in diameter. Despite its early discovery, there are seven main-belt asteroids larger than Juno.

Jupiter Largest planet in the Solar System; it is 318 times as massive as the Earth and revolves about the Sun in an orbit with an average radius of 5.203 AU. The orbital period is 11.86 years. At opposition, when Jupiter appears on the observer's meridian at midnight, it subtends a diameter of about 47″. Earth-based observers' ability to resolve the disk allowed the polar and equatorial radii to be established at an early date and, using the mass derived from Newton's law of gravity, the average density was determined. A volume equal to 1335 times that of the Earth and an average density of 1.31 g/cm^3, less than one-fourth that of the Earth, indicated that internally Jupiter was not earthlike. Jupiter is markedly oblate, with equatorial diameter 142,800 km (88,800 mi) and polar diameter 133,500 km (83,000 mi).

▼ **Jupiter** A view of Jupiter, assembled from four images obtained by the Cassini spacecraft on 2000 December 7. The smallest features visible are about 144 km (89 mi) across. The Great Red Spot is seen at lower right, while the shadow of the satellite Europa is a dark spot at lower left.

JUPITER: DATA	
Globe	
Diameter (equatorial)	142,800 km
Diameter (polar)	133,500 km
Density	1.33 g/cm^3
Mass (Earth = 1)	317.8
Volume (Earth = 1)	1320
Sidereal period of axial	
rotation (equatorial)	9^h 50^m 30^s
Escape velocity	59.5 km/s
Albedo	0.52
Inclination of equator to orbit	3° 07′
Temperature at cloud-tops	125 K
Surface gravity (Earth = 1)	2.69
Orbit	
Semimajor axis	5.203 AU =
	778.3 × 10^6 km
Eccentricity	0.048
Inclination to ecliptic	1° 18′
Sidereal period of revolution	11.86y
Mean orbital velocity	13.06 km/s
Satellites	27

Early spectroscopic studies revealed that Jupiter's spectrum was highly similar to that of the Sun. In 1932, however, Rupert Wildt (1905–76) identified absorption features of methane and ammonia in its spectrum. Multi-atom molecules dissociate at high temperatures, so this discovery indicated that the observed spectrum was reflected sunlight from a cool planet with a molecular atmosphere. Hydrogen molecules and helium atoms do not possess readily observable absorption features in visible light; with access to infrared data, however, it has been ascertained that the composition of Jupiter, with respect to the relative abundance of hydrogen, helium, carbon and nitrogen, is solarlike.

Because Jupiter revolves at an average distance from the Sun of 5.2 AU, the effective solar heating per unit area is reduced by a factor of 0.037 relative to the Earth. Utilizing the data from the PIONEER and VOYAGER missions, the total light scattered in all directions by the planet can be accurately determined; thus, the total absorbed energy is known. When this is compared to the non-solar infrared component of Jupiter's radiation, the ratio of emitted to absorbed solar radiation is 1.668 ± 0.085; that is, Jupiter emits 1.668 times as much heat as it receives. Hence, heat generated internally from decay of unstable isotopes or slow gravitational contraction enters the atmosphere from the interior of the planet.

Even with its internal heat source, the outer regions of Jupiter's atmosphere are extremely cold, with the cloud-deck temperature near 150 K. The visible cloud deck occurs at a depth of about one bar of pressure and is composed mainly of ammonia ice. Beneath this deck temperatures and pressures increase inwards. At a depth of 5–10 bars the water cloud cycle should occur, moving energy from the deeper regions by convection and condensing and releasing the energy at higher levels. At still deeper levels the gaseous atmosphere becomes a liquid ocean, which eventually changes into a region composed chiefly of hydrogen and helium. Under the large pressures due to gravitational force on the overburden, this region exhibits a crystalline structure and is a good conductor of heat. It is referred to as the 'metallic hydrogen' core. At the centre of the planet is thought to be a dense central core of about 15 Earth masses, comprising the heavy elements, which have migrated inwards. Results from the GALILEO orbital mission have improved our knowledge of the Jovian system dramatically.

The first detection of radio emissions from Jupiter was made in 1955 by B.F. Burke and K.L. Franklin. These emissions are concentrated in wavelengths of tens of metres (called decametric) and tenths of metres (decimetric). The decametric emission is caused by electrical discharges along the Jovian field lines when IO crosses them. The rotation rate of the observed radio signal, presumed to be associated with the rotation of the conductive core, is approximately $9^h\ 55^m.5$. Based on this data, a standard rotating co-ordinate system, SYSTEM III, has been selected by the International Astronomical Union (IAU). The zero longitude has been chosen, corresponding to the central meridian of the planet at 0^h UT on 1965 January 1. Longitude has been defined to increase with time at a rate of 870°.536 per 24 hours. Within this system, the latitude and longitude of small eddies within the cloud deck can be measured as a function of time; thus the prevailing winds can be derived. These measurements are interpreted as atmospheric motions relative to the interior of the planet, and, if fully understood, they could reveal a great deal about energy transport within the planet.

Average zonal winds are characterized by strong eastwards winds within 10° of the equator and a series of alternating westwards and eastwards jets extending towards the poles. At latitudes equatorwards of 40°, bright zones are bounded on the equatorwards side by a westwards jet and on the polewards side by an eastwards jet. These regions have anticyclonic flow and are equivalent to upwelling regions in the Earth's atmosphere. The brown intervening belts have cyclonic shear and, by analogy, should be regions of downwards flow where ammonia ice melts. Detailed photometric studies indicate that there is an aerosol haze extending from high altitudes to below the cloud deck. This haze contributes a yellowish-brown coloration to the planet. The size of the particles and the depth to which the line-of-sight penetrates play a large role in determining the hue and shade of a specific cloud feature. The number of times the reflected photons have been scattered off particles and the amount of ammonia ice located at high elevations both play a role in the observed ALBEDO.

Individual cloud features within the Jovian atmosphere are large and long-lived. A low effective temperature causes Jupiter to lose the excess energy that is brought up by convection cells more slowly than does the Earth's atmosphere; however, this does not explain the extreme longevity of some cloud systems. The most well-known feature is the GREAT RED SPOT, centred at about 22°S. Nested between a westwards wind along the northern edge and an eastwards wind on the south, this giant elliptical cloud system rotates in a counter-clockwise, or anticyclonic, sense. To a large extent, the visibility of the Great Red Spot is determined by the amount of turbulence present in the westwards jet deflected around its northern perimeter. White eddies enter the feature during periods of active convection, rendering the contrast so low that detection is difficult. Actually, the Great Red Spot reflects no more red light than do the surrounding white clouds. Its uniqueness lies in the fact that the circulation of the feature brings to the surface some unknown constituent that is a strong ultraviolet and violet absorber; hence observing through a broad-band violet haze filter enhances the visibility of the feature. Small red spots, displaying the same ultraviolet absorber,

JUPITER: NOMENCLATURE OF BELTS AND ZONES

Name	Abbreviation
North polar region	NPR
North north temperate zone	NNTZ
North north temperate belt	NNTB
North temperate zone	NTZ
North temperate belt	NTB
North tropical zone	NTropZ
North equatorial belt	NEB
Equatorial zone	EZ
South equatorial belt	SEB
South tropical zone	STropZ
South temperate belt	STB
South temperate zone	STZ
South south temperate belt	SSTB
South south temperate zone	SSTZ
South polar region	SPR

◀ **Jupiter** Three white oval storms (lower centre) can be seen approaching the Great Red Spot in this 1995 image. Jupiter's turbulent atmosphere shows constant change and interactions between features such as these.

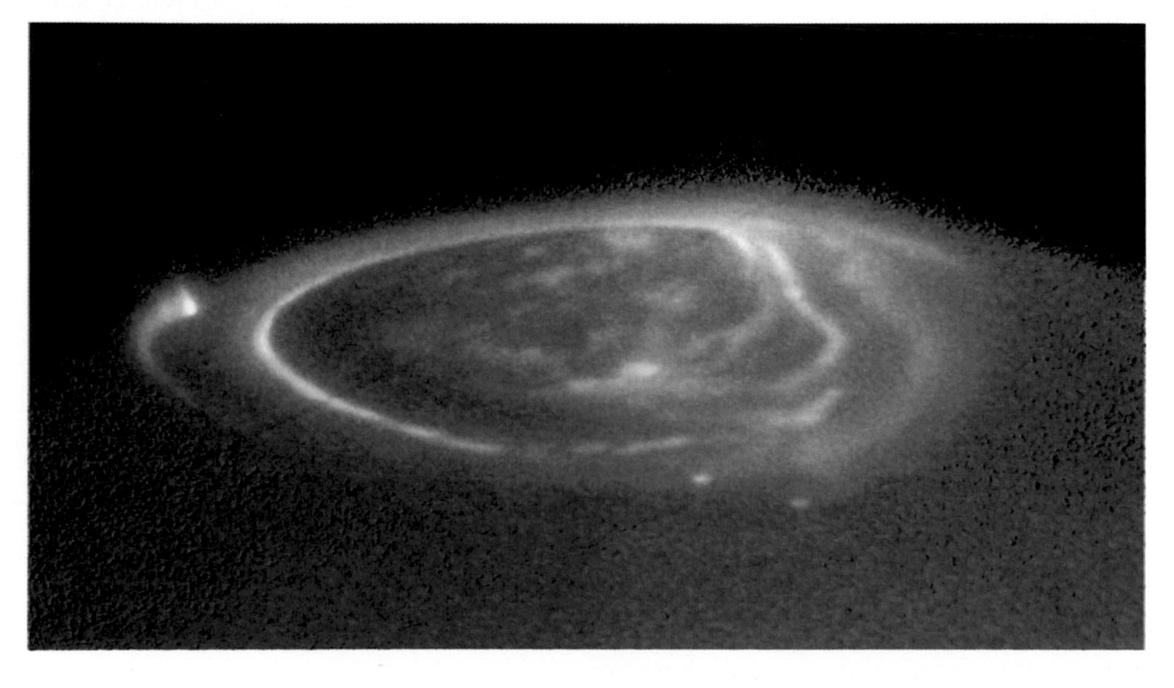

◀ **Jupiter** An ultraviolet image of Jupiter's north polar region, showing auroral activity in 1998 November. Bright spots in the auroral oval mark out magnetic 'footpoints' connected to the satellites Io, Ganymede and Europa.

JUPITER: SATELLITES AND RINGS				
Rings				
ring or gap	**distance from centre (thousand km)**		**width (km)**	
Halo	100–122.8		22800	
Main	122.8–129.2		6400	
Gossamer	129.2–214.2		85000	
Satellites	**diameter (km)**	**distance from centre of planet (thousand km)**	**orbital period (days)**	**mean opposition magnitude**
Metis	60 × 34	128	0.29	17.5
Adrastea	25 × 20 × 15	129	0.30	19.1
Amalthea	270 × 165 × 150	181	0.50	14.1
Thebe	110 × 90	222	0.67	15.7
Io	3630	422	1.77	5.0
Europa	3138	671	3.55	5.3
Ganymede	5262	1070	7.15	4.6
Callisto	4800	1883	16.69	5.6
Leda	16	11,170	238.7	20.2
Himalia	186	11,460	250.6	14.8
Lysithea	36	11,720	259.2	18.4
Elara	76	11,740	259.6	16.8
Ananke	30	21,280	631 R	18.9
Carme	40	22,600	692 R	18.0
Pasiphae	50	23,620	735 R	17.0
Sinope	36	23,940	758 R	18.3

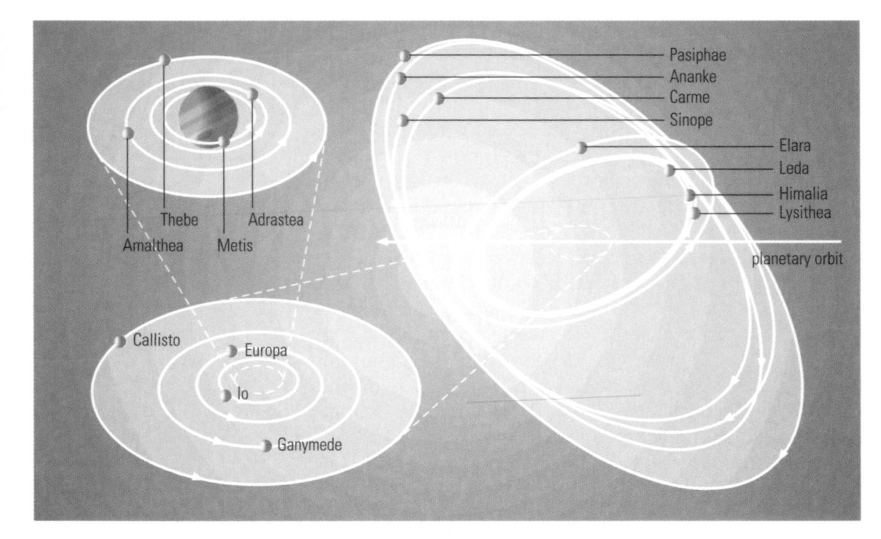

appear at a similar latitude in the northern hemisphere; however, they do not grow to over-fill their windspace and become long-lived features. Many models have been proposed to explain the long-lived well-defined cloud features. These models deal with stable wave solutions and energy and momentum transport.

Direct sampling of the Jovian atmosphere by the entry probe released from Galileo in 1995 December allowed measurements of its composition to be made to an accuracy better than one percent. Hydrogen and helium are present in essentially solar proportions, along with molecular species including CH_4, NH_3 and H_2S. These molecular species, together with S_2 and various metals, were dredged up from depth by the SHOEMAKER–LEVY 9 impacts in 1994 July.

A ring system was discovered during the 1979 VOYAGER 1 encounter, with the brightest component, comprising micrometre-sized particles lying in the equatorial plane, being 7000 km (4000 mi) wide, and located 129,000 km (80,000 mi) from the planet's centre.

Jupiter's magnetic field is similar to that of the Earth. To a first order of approximation, it can be represented as a tilted dipole that is 10 times stronger than the Earth's field. Deviations from a dipole shape and failure of the magnetic axis to be aligned with the rotation axis or to pass through the centre of mass indicate that complex structure may occur deep in the interior. Interaction of the magnetic field with the SOLAR WIND causes particles to decelerate and become entrapped in the magnetic field, creating a complex region around the planet, the MAGNETOSPHERE. The high velocities of the entrapped particles create a hazard for exploring spacecraft, because high-speed electrons, protons and alpha particles can penetrate the craft and cause electronic or radiation damage. The shape of the magnetic field has been sampled only near the equatorial plane by the Pioneer, Voyager and Galileo spacecraft. Advance planning for a Jovian polar orbiter has begun.

Ongoing research to obtain a model that is self-consistent involves several major regions of the planet. Within the deep core, problems concerning the nature of the contaminated hydrogen–helium alloy and the manner in which it conducts heat outwards are challenging. How this region interfaces with a convective envelope, and the degree to which the rapid rotation introduces an organized cylindrical circulation about the axis of rotation within the envelope, is under investigation. The manner in which the envelope couples with the atmosphere, and the degree to which the dynamics of the atmosphere is driven from below, are an area of active research and controversy. As we continue to explore the Solar System, comparison of Jupiter with the other gas giants – Saturn, Uranus and Neptune – will enhance our understanding of this giant planet.

Jupiter-crossing asteroid Any of a group of asteroids with orbital paths that cross some part of the orbital path of JUPITER. Because of Jupiter's huge mass, lifetimes on such orbits are very limited (of the order of tens of thousands of years only), with most of these asteroids quickly being diverted on to radically altered trajectories or ejected from the Solar System entirely. The result is that very few Jupiter-crossing asteroids exist in any epoch. *See also* DAMOCLES; HIDALGO; LONG-PERIOD ASTEROID

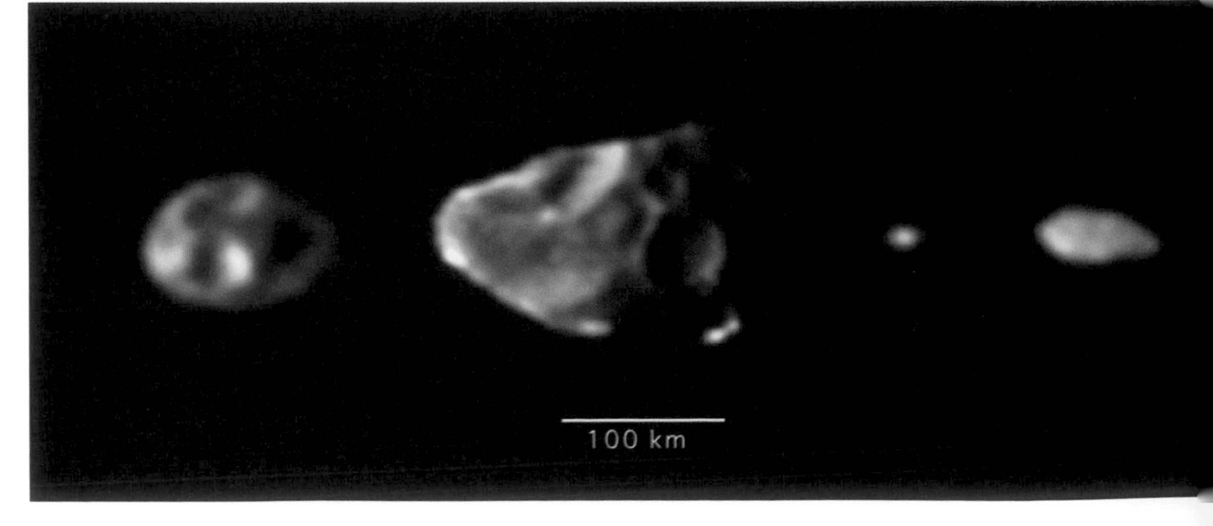

▶ **Jupiter** A set of images showing Jupiter's small inner moons as seen from the Galileo orbiter in 1996/7. These are shown at their correct relative sizes: (left to right) Thebe, Amalthea, Adrastea and Metis. The accompanying diagram shows the scale of Jupiter's extensive satellite system. Many of the more distant satellites are probably captured asteroids or comet nuclei.

Kant, Immanuel (1724–1804) German philosopher whose cosmological speculations anticipated later discoveries about our Galaxy and others. Forced to work for room and board as a tutor after his father died, Kant studied the philosophy of science at the University of Königsberg, where he later became professor of logic and metaphysics. In 1754 he suggested that the Moon's tide-raising forces cause its 'near' hemisphere always to face the Earth and that these same frictional forces have the long-term effect of slowing down the Earth's rotation rate. In 1755 Kant published *Universal Natural History and Theory of the Heavens*, the first comprehensive model of the Milky Way's structure. Commenting on the cosmological theories of Thomas WRIGHT of Durham, who argued that the bright band of light crossing the night sky consisted of innumerable suns in a disk-shaped distribution, Kant concluded that the Solar System lies in that plane of stars or very near it. He likened our Galaxy to the structure of the Solar System, where the planets are also approximately in the same plane. The Milky Way, he said, was lens-shaped, its appearance determined by our perspective in space. Kant is usually credited with coining the term 'island universe' to describe our Galaxy as it related to the other patches of nebulous light that we now know are other galaxies. In his magnum opus, *Critique of Pure Reason* (1781), he maintained that cosmologists cannot ascertain with absolute certainty the finiteness or infinity of space and time.

Kappa Crucis Cluster (NGC 4755) Open star cluster in the southern constellation Crux (RA 12^h $53^m.6$ dec. $-60°20'$); it was first recorded by Nicolas LACAILLE in 1751–52. The cluster, also known as the Jewel Box because of its colourful stars, is young (estimated age 7.1 million years) and contains several luminous O- and B-class giants. Its apparent diameter is $10'$ and mag. +4.2. The Kappa Crucis Cluster lies at a distance of 7600 l.y.

Kappa Cygnids Minor METEOR SHOWER, with a zenithal hourly rate (ZHR) less than 10, active between August 17 and 26, with a peak around August 20. The shower is noted for producing occasional bright, slow-moving FIREBALLS.

Kapteyn, Jacobus Cornelius (1851–1922) Dutch astronomer renowned for his work in photographic astrometry and for introducing statistical methods in stellar astronomy. As the founder of the University of Groningen's Observatory in 1878, Kapteyn found himself with neither equipment nor research facilities. He therefore set about participating in international programmes of observation so as to accumulate large amounts of observational data, and developing mathematical techniques for reducing the data and statistical methods of handling them. A major collaboration (1885–97) was with David GILL to reduce the photographic positions of almost half a million southern-hemisphere stars. For this project, which yielded the *CAPE PHOTOGRAPHIC DURCHMUSTERUNG*, Kapteyn invented a PLATE-MEASURING MACHINE that automatically converted photographic plate coordinates into real star positions and allowed accurate comparisons of stellar magnitudes from one plate to another. He made unprecedentedly accurate distance measurements for large numbers of stars in the Hyades and Perseus Double Cluster, allowing him to determine exactly which stars belonged to these clusters. In the course of 10,000 measurements of proper motions he discovered KAPTEYN'S STAR, at a distance of only 12.7 l.y. the nearest subdwarf to the Solar System.

Kapteyn's statistical investigations led to his discovery in 1904 of two 'star streams' in the Milky Way moving in opposite directions. Later astronomers such as Jan OORT and Bertil LINDBLAD concluded that this streaming meant that the Galaxy had DIFFERENTIAL ROTATION about its central nucleus. With Pieter Johannes van Rhijn (1886–1960), Kapteyn studied the distribution of stars in the Galaxy by counting numbers of stars visible in different directions, by essentially the same method as used by William HERSCHEL. Their model of the Milky Way was too small by about one-third because they were unable to account correctly for the effects of interstellar reddening of starlight, which makes stars look closer than they really are. In 1906 Kapteyn announced his plan to make systematic statistical studies of stars in 206 so-called Selected Areas (later extended to 252) to gain a better picture of the structure of our Galaxy.

Kapteyn's Star Star with the second-largest proper motion known, 8″.7 per year, covering a distance on the celestial sphere equivalent to the apparent diameter of the Moon in just over 200 years. It is a red dwarf of spectral type M0 V, with only 0.4% of the Sun's luminosity. Its visual magnitude is 8.86 and it lies 12.8 l.y. away in the southern constellation Pictor. The star's rapid motion was discovered in 1897 by Jacobus KAPTEYN.

Karl Schwarzschild Observatory Optical observatory 350 m (1150 ft) above sea level at Tautenburg near Jena, Germany. The principal telescope, inaugurated along with the observatory itself in 1960, is a 2-m (78-in.) reflector that can be used in various configurations, including as a SCHMIDT TELESCOPE. In this format, it is the world's largest Schmidt.

Kasei Valles Major outflow channel area that enters CHRYSE PLANITIA on MARS.

K corona Inner part of the Sun's white-light CORONA; it is caused by light from the PHOTOSPHERE that has been scattered by free electrons in the corona. The K corona is responsible for the greatest part of the brightness of the corona out to about two solar radii. It is polarized and

▼ **Kappa Crucis Cluster** Also known as the Jewel Box, the Kappa Crucis Cluster NGC 4755 is one of the brightest in southern hemisphere skies. It is comparatively young, and contains a number of coloured stars.

decreases rapidly in intensity with distance from the Sun, extending to about 700,000 km (400,000 mi) from the photosphere. The corona's electron density can be inferred from white-light observations of the K corona. The K corona emits a CONTINUOUS SPECTRUM without absorption lines, and the K stands for the German word *Kontinuum* or *Kontinuierlich*. *See also* E CORONA; F CORONA; T CORONA

Keck Telescopes *See* W.M. KECK OBSERVATORY

Keeler, James Edward (1857–1900) American astronomer who was director of both the ALLEGHENY OBSERVATORY (1891–98) and the LICK OBSERVATORY (1898–1900). He specialized in spectroscopy, the study of nebulae and planetary research. His spectroscopic study of Saturn (1895) revealed different Doppler shifts in the inner and outer parts of the ring system, showing that it did not rotate at a single rate, supporting the theory of James Clerk MAXWELL that the rings were composed of many individual particles.

Kellner eyepiece Eyepiece with large EYE RELIEF, which makes it widely used for binoculars as well as telescopes. It is essentially a RAMSDEN EYEPIECE with one element replaced by an achromatic doublet. Like the Ramsden, the Kellner is prone to GHOSTING, caused by internal reflections within the eyepiece. It is, however, popular with many observers. It was invented in 1849 by the German optician Carl Kellner (1826–55).

Kelvin, Lord (William Thomson) (1824–1907) Scottish mathematical physicist who gave his name to the Kelvin TEMPERATURE SCALE, based on absolute zero. When just fifteen years old, the University of Glasgow awarded Kelvin its Gold Medal for his 'Essay on the Figure of the Earth', an amazingly sophisticated work on geodetics, and in 1846 he was elected professor there. He proposed his temperature scale in 1848 during the course of thermodynamical studies with the physicist George Gabriel Stokes (1819–1903). Apart from his major influence on physics, Kelvin made contributions to astronomy. He analysed the second law of thermodynamics, concluding (1849–52) that if ENTROPY always increases, this would result in the HEAT DEATH OF THE UNIVERSE, from which it would not be possible to observe or measure any information about its physical state. He made one of the first properly scientific attempts at estimating the age of the Earth, based on known cooling rates of materials, although his result (20–400 million years) was far too low.

▼ **Kennedy Space Center** The 1994 November 3 launch of Atlantis on Space Shuttle Mission STS-66. NASA's fleet of four Space Shuttles is based at the Kennedy Space Center.

Kennedy Space Center (KSC) NATIONAL AERONAUTICS AND SPACE ADMINISTRATION (NASA) facility for the preparation and launch of US manned space missions and its spaceport technology centre. Located at the Cape Canaveral Spaceport, Florida (on Merritt Island, north of Cape Canaveral itself), the site was selected for guided missile tests in 1947, but the main complex was commissioned in 1961 as the launch facility for the APOLLO PROGRAMME. It was named after former president John F. Kennedy. The KSC is NASA's centre of excellence for launch and payload processing systems and the lead centre for the management of expendable launch vehicle services. It is best known for handling the check-out, launch and landing of the SPACE SHUTTLE and its payloads. The HUBBLE SPACE TELESCOPE, the CHANDRA X-RAY OBSERVATORY and the modules of the INTERNATIONAL SPACE STATION all passed through the facility.

Kennelly–Heaviside layer (Heaviside layer) Term formerly used to denote either the whole of Earth's IONOSPHERE or a single region (the E-LAYER) within it. After Guglielmo Marconi's successful transmission of radio signals across the Atlantic Ocean, Arthur Edwin Kennelly (1861–1939) and Oliver Heaviside (1850–1925) suggested that an electrically conducting region existed in the ATMOSPHERE at high altitudes capable of reflecting radio waves back to Earth.

Kepler, Johannes (1571–1630) German astronomer and mathematician, remembered for his three laws of planetary motion. A convert to Copernicanism, he became in 1600 assistant to Tycho BRAHE, who died the next year. Kepler then began the task of completing tables of predicted planetary positions begun by Tycho. From Tycho's accurate observations Kepler concluded that Mars moves in an elliptical orbit, and went on to establish in 1609 the first of his laws of planetary motion (KEPLER'S LAWS). Speculations on magnetism led him to the second of these laws, and his desire to match celestial and musical harmony led to the third, published in 1619. The *Rudolphine Tables*, based on Tycho's observations and Kepler's laws, appeared in 1627 and remained the most accurate planetary tables until the 18th century. Kepler also studied the supernova of 1604 (known as KEPLER'S STAR), and made important contributions to optics and the theory of the telescope.

Johann von Kappel (though he is now always known by the Latinized version of his name) was born at Weil der Stadt, near Stuttgart. His father was a mercenary who was apparently killed while fighting in Holland when Johannes was only five, and the boy was raised by his mother at her father's inn. A precocious child, Kepler obtained a scholarship in 1587 to Tübingen University. There he showed himself to be an able mathematician, and though he began studying Protestant theology after gaining his master's degree, he was nominated provincial mathematician and teacher of mathematics at the Lutheran school in Graz. Despite his mixed views on astrology (he called it 'the foolish little daughter of astronomy'), Kepler followed fashion and provided astrological predictions about the weather, peasant uprisings and Turkish invasions – all of which happened to be fulfilled. Casting horoscopes would later bring Kepler the income he needed to pursue his astronomical work.

At Tübingen, he was influenced by the professor of astronomy, Michael Mästlin (1550–1631), who was well versed in Copernicus' HELIOCENTRIC THEORY. Kepler was fired with enthusiasm for the Copernican idea that the Earth and the other planets orbit the Sun. Despite being twice excommunicated from the Lutheran Church, Kepler remained a lifelong believer in a divine plan for the cosmos, and came upon what seemed to him a proof of the idea. He found that the sizes of the spheres upon which the planets were then supposed to be carried were such that the five regular solids of classical geometry could be inserted between them, in most cases almost exactly. Between the spheres of Saturn and Jupiter, a cube could be fitted; between Jupiter and Mars, a tetrahedron (solid with four faces); and so on. This grand design was published in 1596 as *Mysterium cosmographicum* ('The Mystery of the Universe'), establishing Kepler's reputation as a brilliant mathematician.

Religious upheavals in central Europe forced Kepler to leave Graz, and he visited Tycho in Benatky, outside Prague. Kepler recognized the superb precision of Tycho's observations, and that their mathematical analysis would improve planetary theory. Tycho did not accept the Copernican theory (*see* TYCHONIAN SYSTEM), but was so impressed with Kepler's abilities that in 1600 he invited him to become his assistant. Kepler accepted and, when Tycho died the next year, was appointed his successor as Imperial Mathematician. In 1604 a supernova appeared in the constellation Ophiuchus, and in *De stella nova* Kepler described its astronomical and astrological aspects – his duties as Imperial Mathematician required him to provide the Holy Roman Emperor Rudolf II with astrological interpretations of astronomical phenomena. During this period Kepler continued his analysis of Tycho's observations of Mars, which had the greatest eccentricity, and hence most difficult orbit to explain, of all the known planets except Mercury.

Kepler found that Tycho's accurate data fitted neither the original version of the Copernican theory or Tycho's own planetary scheme; after years of painstaking calculations, he concluded in 1609 that Mars orbits the Sun, not in a perfect circle but in an ellipse. This was a fundamental break with tradition, for previously everyone had followed the Greeks in believing that the planets move in circular orbits and, moreover, at an unvarying rate. Kepler also showed this to be untrue. The orbital velocity of Mars was greatest when closest to the Sun in its elliptical path, at perihelion, and slowest when at its most distant, at aphelion. Kepler went on to examine Tycho's body of observations to determine whether the other planets moved in elliptical orbits. He found that they did – a result he formulated in what we now know as Kepler's first law of planetary motion.

Having taken the important step of letting observations determine his planetary theories, Kepler next sought a physical explanation for the planets' elliptical orbits. He speculated about a magnetic force emanating from the Sun, sweeping the planets round as the Sun rotated. This was incorrect, but it led Kepler to formulate an equation for orbital velocity which, though accurate only at aphelion and perihelion, brought him to his second law of planetary motion: that the line from Sun to planet (the radius vector), sweeps out equal areas of the ellipse in equal times. He published these first two laws in *Astronomia nova* (1609). With the idea of a divinely harmonious Universe in mind, he managed to fit the different orbital velocities at aphelion and perihelion to the musical scale in an attempt to achieve a 'celestial harmony', and derived what seemed to him an example of divine law – the relationship between a planet's distance from the Sun and the time it takes to complete an orbit. First appearing in *Harmonices mundi* ('Harmonies of the World', 1619), Kepler's third law states that for any two planets the ratio of the squares of their orbital periods equals the ratio of the cubes of their mean orbital radii.

The seven-volume *Epitome astronomiae Copernicanae* (1617–21) was Kepler's longest and most influential book. Reprinted in 1635, it was the one of the few accessible systematic accounts of the Copernican theory. It also contained detailed expositions of Kepler's laws and his theory of the Moon's motion. His *Rudolphine Tables* (1627) are generally regarded as the first modern truly scientific ephemeris of planetary positions: they synthesized the first two laws of planetary motion with Kepler's reductions of Tycho's data, and remained in use until the 18th century.

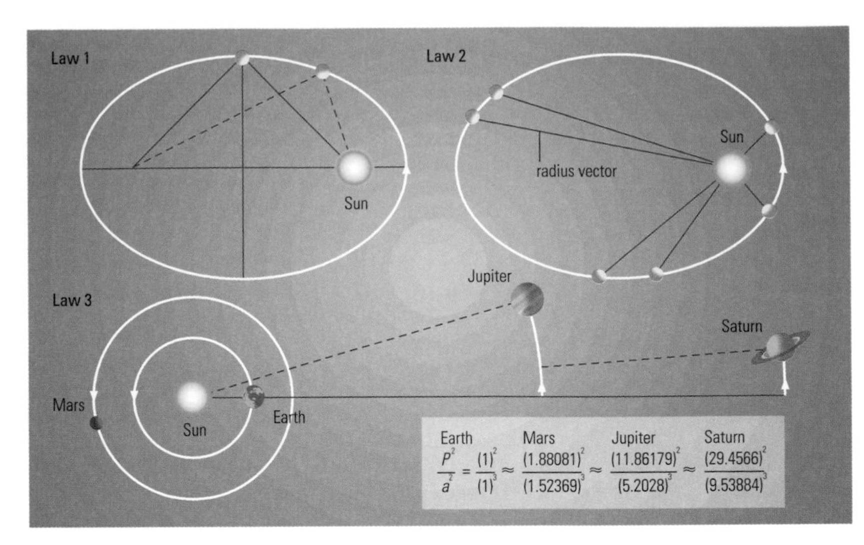

▲ **Kepler's laws** Planetary motions around the Sun are described by Kepler's three laws. Law 1 states that each planet's orbit is an ellipse with the Sun at one focus. Law 2 states that the radius vector sweeps out equal areas in equal times, regardless of position in the orbit. Law 3 relates the square of the planets' sidereal periods to the cube of their distances from the Sun.

Kepler wrote two important works on optics, including *Astronomia pars optica* (1604), which contains the first application of the technique known as ray-tracing to analyse optical systems. The second, *Dioptrice* (1611), was a comment on GALILEO's *Siderius nuncius* in which he proposed his own design of telescope (now called the *Keplerian telescope*, still used in opera-glasses) which employed two convex lenses. His reviews of Galileo's observations of Jupiter's moons caused Kepler to coin the term 'satellite' (1610) to describe these minute bodies. Kepler wrote one of the first science fiction novels, the *Somnium*, which describes an imaginary trip to the Moon.

All his life Kepler suffered from poor health. When his mother was tried for witchcraft, he went to help her, defending her with great success to his own advantage as well as hers, for her conviction would have damaged his career. Kepler ended his days in Regensburg, chronically short of money because of delays in the payment of his Imperial stipend.

Kepler Lunar crater (8°N 38°W), 35 km (22 mi) in diameter, with rim components reaching 3050 m (10,000 ft) from its floor. It is a recent complex crater, with central peaks and wall terracing. Kepler still has a brilliant RAY system, which is most well developed towards the west; it is best observed at high Sun angles. The inner part of the EJECTA blanket is also easily identifiable, and best observed at low Sun angles.

Kepler's laws Three fundamental laws of planetary motion, announced by Johannes KEPLER in 1609 and 1618, deduced from observations made by Tycho BRAHE. (1) The planets move in elliptical orbits, the Sun being situated at one focus of the ellipse. (2) The radius vector, an imaginary line joining the centre of the planet to the centre of the Sun, sweeps out equal areas in equal times (thus, a planet moves fastest when closest to the Sun). (3) The squares of the sidereal periods of the planets are proportional to the cubes of their mean distances from the Sun.

These laws, which Kepler saw as of divine origin, disproved the PTOLEMAIC SYSTEM and vindicated the COPERNICAN SYSTEM. The laws apply to all bodies in closed orbits around the Sun, and to satellites orbiting planets. Strictly, the laws only apply in the absence of perturbations by other planets and other sources.

Kepler's Star Supernova that appeared in 1604 in OPHIUCHUS, only 32 years after TYCHO'S STAR. No galactic supernova has definitely been seen since that time, although the supernova remnant known as Cassiopeia A has been identified with a star observed by John Flamsteed in 1680.

Reaching an apparent magnitude of −3, the 1604 star was extensively observed in Europe and the Far East. Both Johannes KEPLER and the official astronomers of

Korea made fairly systematic brightness estimates for up to a year. The smooth light-curve closely resembles that of a Type I supernova. Accurate measurements by both Kepler and David FABRICIUS fixed the position of the star to better than one arcminute. The remnant is a powerful source of electromagnetic radiation.

Keyhole Nebula (NGC 3324) Region of dark nebulosity seen in silhouette against the ETA CARINAE NEBULA (RA 10^h $44^m.3$ dec. $-59°53'$). It was named after its shape by John HERSCHEL in the 19th century. It covers an area of about $3' \times 11'$.

Keystone Popular name for the distinctive quadrilateral marked out by the four stars ϵ, ζ, η and π Herculis. The Keystone represents the pelvis of Hercules.

kiloparsec Unit of distance measurement equivalent to one thousand parsecs; a PARSEC being the distance at which a star would subtend a PARALLAX of one arcsecond.

kinematics Branch of mechanics devoted to the study of the motions of bodies. Forces are not considered within kinematics but form a separate branch of mechanics called DYNAMICS. Thus, for example, the kinematic PARALLAX (also known as the MOVING CLUSTER METHOD) of stars is obtained by knowing only the PROPER MOTIONS of stars in a STAR CLUSTER.

kinetic energy (KE) ENERGY that a body possesses because of its motion. It is given by

$$KE = \frac{mv^2}{2}$$

where m is the mass of the body and v its linear velocity or by

$$KE = \frac{I\omega^2}{2}$$

where I is the moment of inertia of the body and ω its rotational velocity. It is equal to the energy that would be released (or the work done) if the body were brought to a halt. *See also* POTENTIAL ENERGY

Kirch family German astronomers, noted for compiling calendars, who played a leading role in founding astronomy in what was then Prussia. **Gottfried Kirch** (1639–1710), who was taught by Johannes HEVELIUS, was the first astronomer at the Berlin Observatory (1700). He made the calculations for, and published, many astronomical ephemerides and calendars, and made detailed observations of the 1699 comet.

Gottfried's wife, **Maria Margarethe Kirch** (1670–1720), assisted her husband in much of his ephemeris and calendar work, and also discovered a bright comet in 1702. Maria had taken up astronomy before meeting her husband, learning the science from the self-taught Christoph Arnold (1650–95). After her husband's death in 1710, Maria worked with her son, **Christfried Kirch** (1694–1740), at the Berlin Academy. Christfried carried on the calendrical work of his parents and restored the observatory of Hevelius, from where he discovered the bright comet of 1723. He was assisted in his astronomical computations by his mother and his sister **Christine Kirch** (*c.*1696–1782).

Kirchhoff, Gustav Robert (1824–87) German physicist, born in Königsberg (in modern Russia), a leading physicist of his time who, with Robert Wilhelm Eberhardt Bunsen (1811–99), developed the techniques and principles of astronomical spectroscopy. After teaching at the universities of Berlin and Breslau, Kirchhoff followed Bunsen to Heidelberg to become professor of physics. The two men established the principles of analysing absorption and emission spectra, but it was Kirchhoff who saw its potential for astronomy. Even more relevant to astrophysics are his three gas laws, all announced in 1860. The first provides that a dense object produces a continuous 'blackbody' spectrum when heated. His second gas law says that a low-density excited gas will produce a bright emission-line spectrum as electrons lose energy. Kirchhoff's third gas law states that a continuous spectrum source will produce dark absorption lines as it passes through a cooler low-density gas. Using these laws Kirchhoff discovered the elements caesium and rubidium, named for the blue and red colours of their emission lines. Applying his gas laws to the solar atmosphere, Kirchhoff reached four important conclusions: (1) solar sodium atoms produce the orange-yellow Fraunhofer emission lines; (2) the Sun's light is produced by a high-temperature gas; (3) dark absorption lines must come from a hot gas, but one which is cooler than the interior layers of our star; and (4) the dark absorption lines in the solar spectrum originate as the cooler chromosphere absorbs photons from the photosphere below.

Kirchhoff's laws Laws of SPECTROSCOPY discovered by the German chemists Gustav KIRCHHOFF and Robert Bunsen (1811–99) and published in 1856. In essence they state that an incandescent solid, liquid or high-density gas emits a continuous spectrum, whereas a low-density gas emits or absorbs light at particular wavelengths only. Each element has its own characteristic pattern of lines. Thus stellar spectra have absorption lines arising from their low density outer layers; gaseous nebulae have emission line spectra.

Kirkwood, Daniel (1814–95) American expert on celestial mechanics who discovered (1866) the gaps in the asteroid belt, located at 2.5, 2.95 and 3.3 AU, now known

▼ **Keyhole Nebula** Taking its name from its shape, the Keyhole Nebula is a dark region seen in silhouette against the bright emission of the Eta Carinae Nebula.

K

as KIRKWOOD GAPS. While classifying the asteroids into families defined by their orbital parameters, Kirkwood noticed, as early as 1857, that very few asteroids had orbits in certain regions. He was able to show that perturbations induced by the powerful gravity of Jupiter preclude long-term stable orbits at the distances corresponding to these gaps. Kirkwood applied this same analysis to explain the gaps in Saturn's ring system, calculating the tendency of Saturn's satellites to constrain ring particles in well-defined zones, while excluding them from the gaps.

Kirkwood gaps Largely vacant regions within the main belt of asteroids, in terms of their heliocentric distances, kept clear by gravitational perturbations caused by Jupiter. These distances are equivalent to orbital periods that are simple sub-dividers of Jupiter's period, resulting in resonance effects that repel asteroids. These gaps were noticed by Daniel KIRKWOOD in 1857, their explanation in terms of celestial mechanics coming a decade later.

Kitt Peak National Observatory (KPNO) US national research facility for ground-based optical astronomy. Located in the Quinlan Mountains 90 km (55 mi) south-west of Tucson, Arizona, Kitt Peak was selected by the National Science Foundation (NSF) as an observatory site in 1958 after a nationwide site survey. It is operated jointly by the NSF and the ASSOCIATION OF UNIVERSITIES FOR RESEARCH IN ASTRONOMY. KPNO is one of the four divisions of the NATIONAL OPTICAL ASTRONOMY OBSERVATORY, and has headquarters in nearby Tucson. It now hosts the most diverse collection of telescopes on Earth for optical and infrared astronomy and daytime studies of the Sun. Of these, KPNO itself operates the 4-m (158-in.) Mayall Telescope (dedicated in 1973), the 3.5-m (138-in.) WIYN TELESCOPE (on behalf of the WIYN Consortium) and a 2.1-m (84-in.) reflector optimized for infrared as well as optical observations.

As well as these large telescopes, there is a 0.9-m (36-in.) instrument (known as the Coudé Feed Telescope) which feeds light to the main spectrograph at the 2.1-m telescope, another 0.9-m (36-in.) telescope and the Burrell Schmidt telescope. A further 16 optical telescopes and two radio telescopes share the site; these are operated by the national solar observatory, the NATIONAL RADIO ASTRONOMY OBSERVATORY and various other university consortia. Largest among them are the STEWARD OBSERVATORY's 2.3-m (90-in.) Bok Telescope (1969) and the 2.34-m (92-in) Hiltner Telescope, built for the Michigan-Dartmouth–MIT Observatory in 1986.

Kleinmann–Low Nebula Nebula containing several strong infrared sources; it is situated behind the ORION NEBULA. It was first observed in 1967 by the American astronomers Douglas Kleinmann (1942–) and Frank Low (1933–). The Kleinmann–Low Nebula is a large dusty star-forming region. It has several distinct infrared sources within it, including the BECKLIN–NEUGEBAUER OBJECT. These sources are thought to be individual PROTOSTARS. Strong outflows within the nebula may be powered by the strongest of the infrared sources, IRc2.

Kleopatra MAIN-BELT ASTEROID; number 216. Radar imaging has shown that Kleopatra has a bizarre shape somewhat like a dog's bone, 217 km (135 mi) long and 93 km (58 mi) wide.

Kochab (Kocab) The star β Ursae Minoris, visual magnitude 2.08, distance 126 l.y., spectral type K4 III. Its name is of uncertain derivation.

Kohoutek, Comet (C/1973 E1) Long-period comet discovered in 1973 by Luboš Kohoutek (1935–), Hamburg, Germany, on photographic plates. Its relative brightness at the time of discovery, while still as distant from the Sun as Jupiter (5 AU), engendered forecasts that Comet Kohoutek would be a spectacular object in late December and early January. Perihelion (0.14 AU) was reached on 28 December, at which time the comet reached mag. −3, and images were obtained using a coronagraph by Skylab astronauts. As it emerged into the evening sky after perihelion, Comet Kohoutek was a rapidly fading mag. +3.0 object. While it failed to live up to expectations, the comet was still a respectable binocular object, with a tail up to 25° long. The orbit of it is hyperbolic.

▲ **Kitt Peak National Observatory** An aerial view of Kitt Peak, showing the many telescopes sited there. Largest, at the top right, is the 4-m (158-in.) Mayall Telescope. The triangular building in the foreground is the McMath–Pierce Solar Facility, the world's largest solar telescope.

Konkoly Observatory National observatory of Hungary, operated by the Hungarian Academy of Sciences. Its benefactor, Miklós Konkoly-Thege (1842–1916), donated his astronomical instruments to the state in 1899, and the observatory was established in Budapest after World War I. A 0.6-m (24-in.) reflector, still in use, began operations in 1928. In 1958 a dark-sky site, Piszkéstetôi Observatory, was set up in the Mátra Mountains about 120 km (75 mi) north-east of Budapest; three telescopes are currently operated there, including a 1.0-m (39-in.) built in 1974.

Kopff, Comet 22P/ Short-period comet discovered photographically by August Kopff (1882–1960), Heidelberg, Germany, on 1906 August 23. The orbit has been modified by passages close to Jupiter in 1942 and 1954, and it has a current period of 6.45 years. It was last seen, as a diffuse magnitude +7.0 object, in 1996.

Kottamia Observatory *See* HELWÂN OBSERVATORY

▼ **Kleopatra** A series of radar images of asteroid (216) Kleopatra, showing its rotation. The twin-lobed structure is commonly compared to a dog's bone! [S. Ostro (JPL/NASA)]

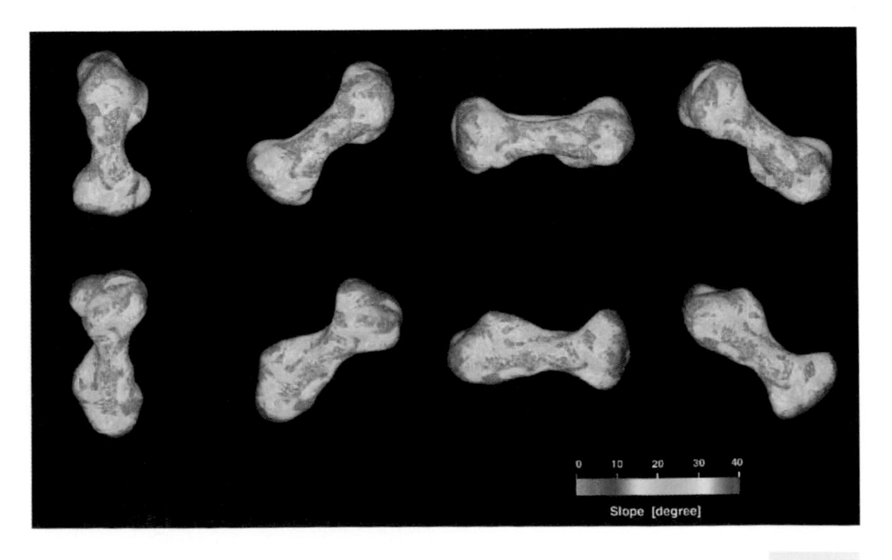

KPNO Abbreviation of KITT PEAK NATIONAL OBSERVATORY

Kraus, John D. (1910–) American radio astronomer noted for his design of radio antennae and receivers and his work with Ohio State University's Big Ear radio telescope. Big Ear, built near Delaware, Ohio (1956–63), consisted of a large, flat surface tilted about its horizontal axis and reflecting signals to a fixed paraboloidal reflector. It was demolished in 1998. From 1965 Kraus used Big Ear to make the most detailed map of the radio sky in history, discovering over 20,000 new sources of radio waves. In 1970, after completing the 'Ohio Survey', Kraus and his collaborators used Big Ear to search for extraterrestrial life.

KREEP Specific type of lunar material enriched in potassium (K), rare-earth elements (REE) and phosphorus (P). It was first found among the samples brought by APOLLO 12 astronauts. In its chemistry and bulk mineralogy KREEP is a variety of lunar BASALT. It is believed to be formed from melt residue from the crystallization of large masses of crust and upper mantle materials of the Moon.

Kreutz sungrazer Any of a group of COMETS with similar orbits that at perihelion lie within 0.01 AU (1.5 million km/1 million mi) of the Sun. The group was first recognized by the German astronomer Heinrich Carl Friedrich Kreutz (1854–1907) in 1888. Kreutz sungrazers have RETROGRADE orbits with periods of 500–1000 years; there are two principal subgroups. Kreutz sungrazers are believed to result from fragmentation in the remote past of a single large progenitor comet. The numerous smaller members of the group have been produced from further break up of the parent comet. SOHO spacecraft coronagraph observations have revealed large numbers of very small fragments on Kreutz sungrazer orbits. Larger fragments have produced some of the brightest comets in history, with C/1965 S1 IKEYA–SEKI being a noteworthy example. Their orbital inclination of 141°.9 means that Kreutz sungrazers are usually best observed after perihelion from southerly latitudes. Many sungrazers fail to survive the intense solar heating experienced close to perihelion.

KSC Abbreviation of KENNEDY SPACE CENTER

K star Any member of a class of yellow-orange stars, the spectra of which are characterized by weak hydrogen lines, with strong ionized calcium lines (Fraunhofer H and K) in the warmer K stars, and powerful neutral calcium (Fraunhofer g) and neutral sodium (Fraunhofer D) in the cooler ones. Molecular bands of CH (Fraunhofer G) and CN are also strong. Main-sequence dwarfs range from 4000 K at K9 to 5200 K at K0, with zero-age masses from 0.5 to nearly 0.9 solar mass, and luminosities from 0.1 to 0.5 solar. Dwarf lifetimes are longer than the age of the Galaxy. K giants and supergiants (because of lower densities typically 600 K cooler than the dwarfs) are distinguished from dwarfs by the prominent weakening of the CN absorption bands. Stellar luminosities are also found from the 'Wilson–Bappu effect', in which luminosities are directly tied to the width of a chromospheric emission line set within the broad ionized calcium H and K absorptions.

K supergiants, which are rare because of rapid evolution through the class, can range to high masses. Prominent among K giants are visually numerous helium-burning 'clump stars' of modest mass, which on the HERTZSPRUNG–RUSSELL DIAGRAM clump together near absolute magnitude zero. K subdwarfs have weakened metal lines and lower metal abundances. Except where they are members of short-period binary systems, most K stars are slow rotators. Some K giants are carbon rich: class R CARBON STARS, which begin at G5 temperatures, track fairly well the range of class K. Moderate carbon abundances combined with heavy-element excesses are found in the BARIUM STARS, which are enriched as a result of mass transfer from companions that have evolved to white dwarfs.

Bright examples of K stars include Dubhe and Pollux KO III, Arcturus K1 III, Epsilon Pegasi K2 Ib, Alpha Centauri-B K1 V and Epsilon Eridani K2 V.

Kuiper, Gerard Peter (1905–73) Dutch-American astronomer (born Gerrit Pieter Kuiper) responsible for revitalizing Solar System astronomy in an era dominated by stellar and galactic research. After studying astronomy at the University of Leiden with Bart BOK, Kuiper emigrated to the United States in 1933, joining the staff of Lick Observatory, where he concluded from studying stars of large parallax that at least half of all nearby stars are double or multiple systems. He invented the term 'contact binary' to describe the very close double star BETA LYRAE. Moving to the Yerkes Observatory in 1938, he tightened the definition of the mass–luminosity relation, and concluded that white dwarfs are highly dense objects that behave differently from main-sequence stars. He was later director of Yerkes and of the McDonald Observatory, established by Yerkes in the darker skies of west Texas in 1939.

During World War II Kuiper used the large telescopes at Yerkes and McDonald for photographic studies of the planets and their satellites at wavelengths in the visual and near-infrared; his 1944 discovery of methane on Titan proved for the first time that planetary satellites could have atmospheres of their own. A string of Solar System discoveries followed: a tenuous CO_2 atmosphere on Mars, a fifth satellite of Uranus (Miranda), a second satellite of Neptune (Nereid) and methane in the atmospheres of Uranus and Neptune. In 1960 he fulfilled an early ambition to set up an institute devoted to Solar System studies – the lunar and planetary laboratory (LPL) at the University of Arizona. Intimately involved in most of NASA's space programmes, he was Chief Experimenter for the pioneering Ranger missions to the Moon. First at Yerkes and then at LPL, Kuiper directed the preparation of several important photographic lunar atlases that ensured the choice of safe sites for the Apollo landings. His search for superior observatory sites led to the CERRO TOLOLO INTER-AMERICAN OBSERVATORY in Chile and MAUNA KEA OBSERVATORY in Hawaii. He also initiated the use of telescopes in high-flying aircraft for infrared observations (*see* KUIPER AIRBORNE OBSERVATORY).

Kuiper Airborne Observatory (KAO) Lockheed C-141 transport aircraft containing a 91-cm (36-in.) short-focus Cassegrain reflecting telescope designed for INFRARED ASTRONOMY. The KAO was named in honour of Dutch-American astronomer Gerard KUIPER. It was operated from NASA's AMES RESEARCH CENTER in California from 1975 to 1995. The KAO was used to discover the rings of Uranus, water vapour in Jupiter's atmosphere and Halley's Comet, condensed cores in BOK GLOBULES, and more than 60 previously unobserved spectral features in the interstellar medium. Its successor is the STRATOSPHERIC OBSERVATORY FOR INFRARED ASTRONOMY (SOFIA).

Kuiper belt *See* EDGEWORTH–KUIPER BELT

Kwasan and Hida Observatories Observatories of Kyoto University, Japan. Kwasan Observatory dates from 1929, when it was established on Mount Kwasan in Kyoto city for solar and stellar research. Increasing light pollution led to Hida Observatory being set up in 1968 at Kamitakara, 350 km (220 mi) north-east of Kyoto, at an altitude of 1275 m (4180 ft). Hida has a 0.65-m (26-in.) refractor (the largest refractor in Asia, built in 1972), a 0.6-m (24-in.) reflector and the 0.6-m (24-in.) Domeless Solar Telescope (DST), built in 1979.

Lacaille, Nicolas-Louis de (1713–62) French astronomer, known as the 'father of southern astronomy'. He led an expedition to the Cape of Good Hope (1750–54) where he formed 14 new southern constellations which he named after items of scientific equipment. The major achievement of this expedition was the cataloguing of over 10,000 stars, published posthumously as *Coelum australe stelliferum* (1763). Using only small refractors, Lacaille also discovered 42 previously unknown nebulae, which he correctly deduced were objects within the Milky Way. He also postulated that the Magellanic Clouds are detached parts of our Galaxy. With Jérôme LALANDE, he measured the lunar parallax.

Lacerta *See* feature article

Lagoon Nebula (M8, NGC 6523) Bright EMISSION NEBULA surrounding the young open star cluster NGC 6530 in the constellation Sagittarius (RA 18^h $03^m.3$ dec. $-24°23'$). The eastern half of the nebula is illuminated by hot O-class stars in NGC 6530, while the western half is illuminated by 9 Sagittarii. The Lagoon Nebula is divided by a broad lane of obscuring dark dust. First recorded by John FLAMSTEED in 1680, the Lagoon has angular dimensions of $45' \times 30'$ and mag. +6.0. It lies 5200 l.y. away in the direction of the galactic centre, and covers an area of 60×38 l.y.

Lagrange, Joseph Louis (1736–1813) French mathematician, regarded as the 18th century's greatest, born in Turin, Italy, where he did his early work. After serving as director of the Berlin Academy (1767–87), he emigrated to Paris, where he spent the rest of his life. His contributions to astronomy were in celestial mechanics, including the Moon's libration, the gravitational interactions of Jupiter and its Galilean satellites, and the three-body problem – for which he found special solutions that suggested the existence of five equilibrium positions, now known as the LAGRANGIAN POINTS.

Lagrangian points Points in the orbital plane of two massive bodies at which a third body can remain in equilibrium, so that all three bodies remain in a fixed geometrical configuration. There are five such points, as shown in the diagram. The Lagrangian points L_4 and L_5 form equilateral triangles with M_1 and M_2, and L_1, L_2 and L_3 are collinear with M_1 and M_2. Orbits close to L_4 and L_5 are stable and perform slow oscillations around the equilibrium points. The points L_1, L_2 and L_3 are unstable, and any objects in these orbits would very slowly drift away (although some particular orbits in these vicinities are stable). The TROJAN ASTEROIDS are close to the L_4 and L_5 points of Jupiter's orbit around the Sun, with oscillation amplitudes around the equilibrium points of up to 30°. Similar orbits occur in Saturn's satellite system. The tiny satellites Telesto and Calypso are close to the L_4 and L_5 points of Tethys' orbit around Saturn, and Helene is close to the L_4 point of Dione's orbit. The spacecraft SOHO has been placed into a stable orbit close to the L_1 point of the Earth. This point is located between the Earth and the Sun at about 1,500,000 km (930,000 mi) from the Earth, which is about four times the distance of the Moon and one hundredth of the distance to the Sun.

Lalande, (Joseph) Jérôme Le François (1732–1807) French astronomer, skilled mathematician and observer and popularizer of astronomy, who became director of PARIS OBSERVATORY (1768). He compiled many accurate tables of planetary positions and a catalogue of over 47,000 stars. Lalande organized expeditions to observe the transits of Venus in 1761 and 1769, preparing detailed reports. He collaborated with Nicolas-Louis de LACAILLE in determining a highly improved value for the lunar parallax, Lalande observing from Berlin, Lacaille from South Africa.

Lambda Boötis star Population I galactic disk star that is severely depleted in certain metals. The prevailing theory is that when young, these stars accreted interstellar gas that was depleted in refractory (high-condensation temperature) metals.

LAMOST Abbreviation of LARGE-AREA MULTI-OBJECT SURVEY TELESCOPE

lander Spacecraft that reaches the surface of a planet or other solid body. The first spacecraft to achieve a landing on another Solar System object was the Soviet LUNA 2 (Lunik 2) in 1959. The first controlled landing was by Luna 9 in 1966. The first manned lander was the Lunar Module Eagle during the APOLLO 11 mission

◄ **Lagoon nebula** A high-resolution Hubble Space Telescope image of the Lagoon Nebula's centre, showing 'twister' structures. Globules, precursors of star formation, are also found in abundance here.

LACERTA (GEN. LACERTAE, ABBR. LAC)

Small, inconspicuous northern constellation, representing a lizard, between Cygnus and Andromeda. Lacerta was introduced by Hevelius in 1687. Its brightest star, α Lac, is mag. 3.8. Deep-sky objects include NGC 7243, an open cluster of several dozen stars of mag. 8.5 or fainter; and BL Lac, the variable nucleus (range 12.4–17.2) of a remote elliptical galaxy and the first object of its type (*see* BL LACERTAE OBJECT) to be recognized.

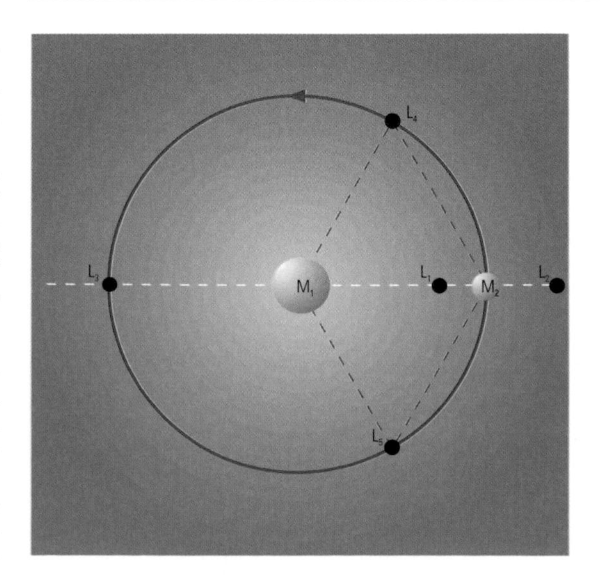

◄ **Lagrangian point** In a system where a body orbits a more massive primary, there are five key positions where the gravity of the two balances out such that a third body can remain there stably. The L_1 position, close to the less massive body on the line towards the primary, is one such Lagrangian point; many spacecraft are stationed around the L_1 point between Earth and the Sun. The L_4 and L_5 (60° ahead of and behind the planet) Lagrangian points of Jupiter's orbit are occupied by Trojan asteroids.

▲ **lander** A 360° panoramic view of the Martian surface taken from the Mars Pathfinder lander in 1997. The Sojourner rover vehicle can be seen next to the large boulder at centre.

in 1969. *See also* LUNAR PROSPECTOR; MARS PATHFINDER; VENERA; VIKING MISSIONS

Langley Research Center NATIONAL AERONAUTICS AND SPACE ADMINISTRATION (NASA) facility at Hampton, Virginia, for research in advanced aerospace technologies such as airframe systems. More than half of Langley's research is in aeronautics, with applications not only to military and civilian aircraft but also to spacecraft. Studies of the behaviour of materials in non-terrestrial atmospheres will ensure that next-generation spacecraft will be able to survive extreme environments.

Langrenus Lunar crater (9°S 61°E), 132 km (82 mi) in diameter, with rim components rising 2750 m (9000 ft) above its floor. A nearly circular crater, it is near the limb of the Moon, appearing oblong as a result of foreshortening. Langrenus has bright ray EJECTA, which is fragmental because of its age. The floor contains several central peaks, due to rebound after impact. The side walls are heavily terraced.

L

Langrenus, Michael Florentius (or Michiel Florenz van Langren) (1600–1675) Flemish engineer and cartographer who, in 1645, compiled the first published map of the Moon. Langrenus began mapping the Moon as part of a project to find longitude at sea by noting the precise times at which specific lunar mountains were illuminated or darkened by the Sun as observed from different locations. He introduced the practice of naming lunar features after famous persons.

La Palma Observatory *See* ROQUE DE LOS MUCHACHOS OBSERVATORY

Laplace, Pierre-Simon de (1749–1827) French mathematician who did much to establish celestial mechanics as a major scientific discipline; his *Méchanique céleste* is, after Isaac Newton's *PRINCIPIA*, the most important work on the subject. His instructors at Caen University, where he matriculated at the age of sixteen, soon recognized his talents for complex mathematical methods, and three years later Laplace secured a position as professor of mathematics at the École Militaire. By 1771 he had published a fundamental study on the orbits and perturbations of the planets that was the cornerstone of his later studies of Solar System dynamics.

This work was followed by a three-part study (1784–86) of the orbital mechanics of Jupiter and Saturn that utilized methods first developed by Joseph Louis de LAGRANGE to demonstrate that not only those two giant planets, but all the Solar System's bodies, will tend to assume orbits of small eccentricity and small inclination over long time periods. This was the first major post-Newtonian work to show that the Solar System is stable under the precepts of Newtonian gravity. Many of these ideas were more fully explicated in his five-volume masterwork *Traité de méchanique céleste* (1799–1825). In *Exposition du système du monde* (1796) Laplace proposed his NEBULAR HYPOTHESIS. His studies of comets led him to issue a warning that comets could collide with the Earth, which would have catastrophic consequences for life on our planet.

▼ **Langley Research Center** This model of the Space Shuttle is undergoing wind tunnel tests at NASA's Langley Research Center to simulate the stress of re-entry.

Large-Area Multi-Object Survey Telescope (LAMOST) Chinese optical telescope of unusual design, located at the Xinglong station of the BEIJING ASTRONOMICAL OBSERVATORY and expected to become operational in 2004. A 5.7 × 4.4-m (220 × 170-in.) reflective Schmidt corrector feeds light to a fixed spherical mirror 6.7 × 6.0 m (265 × 236 in.) in size. Both components use segmented-mirror technology, and the corrector moves to allow the telescope to track objects across the sky. Fibre-optic tubes transmit light from some 4000 objects in LAMOST's 5° field of view to stationary spectrographs.

Large Binocular Telescope (LBT) Unique facility for optical astronomy at MOUNT GRAHAM INTERNATIONAL OBSERVATORY consisting of two 8.4-m (331-in.) telescopes mounted in parallel on the same structure with an axial separation of 14.4 m (47 ft). Due for completion in 2004, it will have the light-gathering power of a single 11.8-m (465-in.) telescope. The binocular configuration makes interferometric imaging easier than with separately mounted telescope arrays like the VERY LARGE TELESCOPE. The LBT project is a partnership between the University of Arizona, the ARCETRI ASTROPHYSICAL OBSERVATORY, a consortium of German institutions and other US universities.

Large Magellanic Cloud (LMC, Nubecula Major) Small GALAXY that is the nearest external galaxy to our own. It is about 170,000 l.y. away and has just 5% to 10% of the mass of our Galaxy. It extends over about 8° of the sky within the constellations of Dorado and Mensa (RA $05^{h}\ 30^{m}$ dec. −68°), where it is easily visible to the naked eye, appearing like a detached portion of the MILKY WAY. The Large Magellanic Cloud (LMC), along with the SMALL MAGELLANIC CLOUD, is named after Ferdinand Magellan, who observed them during his voyage around the world in 1519, but they were known before that date. The LMC's physical diameter is 25,000 to 30,000 l.y. At first sight it appears to be an irregular galaxy, but it has a central bar and possibly a spiral arm, so it is generally classed with the spiral galaxies. The TARANTULA NEBULA complex is found within the LMC and is one of the largest HII REGIONS known. The complex is a vigorous star-forming region, and this, together with other star-forming regions, may be the result of a close passage between the LMC and our Galaxy some 200 million years ago. There is also an enormous bridge of material between the Galaxy and the MAGELLANIC CLOUDS. Known as the Magellanic Stream, this bridge is one of the largest HIGH-VELOCITY CLOUDS known and probably resulted from the same close passage. The LMC orbits our Galaxy roughly at right angles to the plane of the Milky Way, and it may at some time in the future be disrupted and captured by our Galaxy.

Young objects, such as hot stars and Cepheids, in the LMC lie in a thin disk, which is seen nearly face-on (an angle to the plane of the sky of about 27°). Older objects, such as planetary nebulae, form a somewhat thicker disk structure, but there is no evidence at present for a spherical halo such as is formed by very old objects in our own Galaxy. In our Galaxy, GLOBULAR CLUSTERS are all old objects. However, the cluster NGC1866 in the LMC, though globular in form, is a young system. Several such 'blue globulars' exist in the clouds. No similar objects are known in our own Galaxy.

large-scale structure Overall distribution of galaxy clusters and galaxies in the visible Universe. This distribution has been mapped in two wedges by astronomers up to a redshift of $z = 0.04$. A relevant cosmological model must explain the large-scale structure of matter in the Universe, essentially the distribution of galaxies and galaxy clusters. This is usually done by comparing the CORRELATION FUNCTIONS derived from theoretical models utilizing COLD DARK MATTER or HOT DARK MATTER with the correlation function of the observed large-scale distribution of galaxies. There are several features of note in the observed distribution: VOIDS or regions generally devoid of galaxies; walls or linear distributions of galaxies and clusters; and SUPERCLUSTERS such as the Coma Cluster of galaxies.

Larissa Second-largest (after PROTEUS) of the inner satellites of NEPTUNE. Larissa was first detected in 1981 by Harold Reitsema and colleagues when it occulted a star; it was confirmed in 1989 in images returned by VOYAGER 2. Larissa is near-spherical in shape, measuring about 208 × 178 km (129 × 111 mi). It takes 0.555 days to circuit the planet, at a distance of 73,500 km (45,700 mi) from its centre, in a near-circular, near-equatorial orbit.

Las Campanas Observatory Major optical observatory in the southern Atacama Desert 100 km (60 mi) north of La Serena, Chile, at an altitude of 2400 m (7900 ft). It was established by the CARNEGIE OBSERVATORIES in 1969, and remains their primary observing site. The main instruments are the two 6.5-m (256-in) MAGELLAN TELESCOPES. The 2.5-m (98-in.) Irénée du Pont Telescope and the 1-m (39-in) Henrietta Swope Telescope, completed in 1977 and 1972 respectively, both have unusually wide fields of view.

laser (acronym for 'light amplification by stimulated emission of radiation') Device that produces a beam of high-intensity, coherent (all waves in phase), monochromatic (all of one wavelength) ELECTROMAGNETIC RADIATION at infrared, optical or shorter wavelengths. A similar device at microwave frequencies is known as the MASER.

The stimulated emission of a photon of particular wavelength occurs when an electron in a high-energy level of an atom is induced to drop to a lower level by an encounter with a photon of energy exactly equal to the difference in energy between the upper and lower levels. The emitted photon has exactly the same wavelength and direction of propagation as the stimulating photon, and the emitted radiation is said to be coherent. In a laser, large numbers of electrons are pumped up into a higher energy level by means of a suitable energy source so that there are more electrons in the higher level than the lower level; this distribution of electrons is known as an 'inverted population'. The electrons are then stimulated to emit a beam of coherent radiation.

Laser action can be produced in solids, liquids or gases, and lasers may be pulsed or continuous in operation.

Lasers are used in various astronomical contexts. For example, very precise measurements of the lunar distance, and the Moon's slow rate of recession, have been made by bouncing laser beams off reflectors placed on the lunar surface by the Apollo astronauts, and laser beams reflected from orbiting satellites allow the rate of continental drift to be measured. Lasers are also used to check and maintain the alignment of optical components, particularly in very large telescopes with segmented primary mirrors.

Laser Interferometer Gravitational-wave Observatory (LIGO) Giant laser interferometer at Livingston, Louisiana, about 80 km (50 mi) north-west of New Orleans, consisting of mirrors suspended at the corners of an L-shaped vacuum vessel 4 km (2.5 mi) on a side. It is designed to sense the minute motions of the mirrors that would be caused by GRAVITATIONAL WAVES from supernova collapses and collisions of neutron stars, and the gravitational remnants of the Big Bang. LIGO is a joint project of the CALIFORNIA INSTITUTE OF TECHNOLOGY and the Massachusetts Institute of Technology, and is sponsored by the National Science Foundation.

La Silla Observatory First of the two observing sites of the EUROPEAN SOUTHERN OBSERVATORY, inaugurated in 1969. It is near the southern extremity of the Atacama Desert, 160 km (100 mi) north of La Serena, Chile, at an altitude of 2400 m (7900 ft). ESO's smaller telescopes were the first on the site: the 1-m (39-in.) and the 1.5-m (59-in.) were both installed in the late 1960s. The 1.0-m (39-in.) Schmidt telescope began work in 1971 and was joined in 1976 by the 3.6-m (142-in.), the largest instrument at La Silla. The 1.4-m (55-in.) Coudé Auxiliary Telescope (CAT) was built to provide an alternative feed to the 3.6-metre's high-resolution spectrograph, while in 1984 the 2.2-m (87-in.) began operation. (The CAT and the Schmidt have since been decommissioned.) The NEW TECHNOLOGY TELESCOPE and SWEDISH ESO SUBMILLIMETRE TELESCOPE followed in the late 1980s. In addition to the ESO telescopes, La Silla hosts a number of national telescopes operated by ESO member states, the largest being the Geneva Observatory's 1.2-m (48-in.) telescope.

Lassell, William (1799–1880) English 'Grand Amateur' astronomer. By the age of 25, he had founded the brewery that would finance his lifelong career in astronomical research. Lassell studied primarily the bodies of the outer Solar System. Following the announcement of the Berlin discovery of Neptune on 1846 September 30, he quickly located the new planet and on October 10 identified its large satellite, Triton; five years later he found Uranus' satellites Ariel and Umbriel (1851). He also co-discovered (with G.P. and W.C. BOND at Harvard) Saturn's Crepe (C) Ring and its satellite Hyperion (1848).

Lassell's enduring reputation comes from his transformation of the big reflecting telescope from the timber and rope altazimuth constructions developed by William Herschel to beautifully engineered iron equatorials with superlative optics. His 9-inch (230-mm) and 24-inch (0.6-m) Newtonians were the first sizeable instruments of their type to be mounted equatorially. With his engineer-astronomer friend James Nasmyth (1808–90), Lassell pioneered the prototype steam-powered polishing machine for making 48-inch (1.2-m) mirrors, and con-

▼ **Larissa** During its 1989 August flyby, Voyager 2 imaged Neptune's inner satellite Larissa.

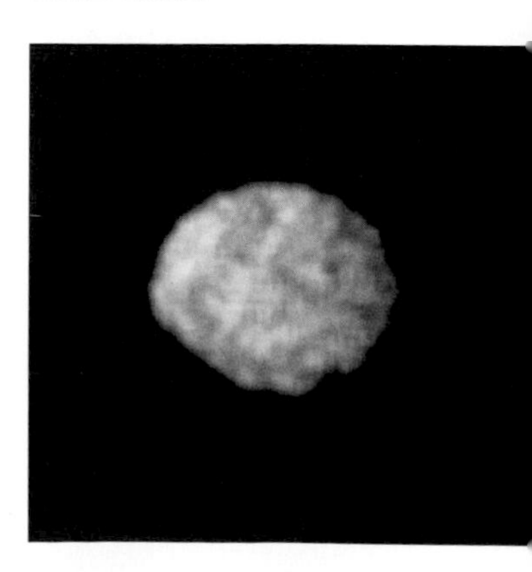

LEO (GEN. LEONIS, ABBR. LEO)

Large, distinctive northern constellation and one of the signs of the zodiac. It represents the Nemean lion killed by Hercules as the first of his twelve labours in Greek mythology, and lies between Cancer and Virgo. Leo is easily recognized by the asterism known as the SICKLE, a reversed question mark formed by six of the constellation's brightest stars, with the brightest of all, REGULUS, mag. 1.4, at its base. β Leo, DENEBOLA, marks the lion's tail. γ Leo, ALGEIBA, is a fine visual binary with orange and yellow components, mags. 2.5 and 3.6, separation 4″.6, period about 600 years. The constellation also contains Wolf 359 (also known as CN Leo, a variable star of the UV CETI type, range 11.5–17.1), the third-closest star to the Sun, just 7.8 l.y. away. Deep-sky objects in Leo include five 9th-magnitude spiral galaxies: M65 (NGC 3623), M66 (NGC 3627), M95 (NGC 3351), M96 (NGC 3368) and M105 (NGC 3379), the first two and last three of which are grouped close together in the sky. Leo I (10th magnitude) and Leo II (12th magnitude) are a pair of dwarf spheroidal galaxies in the LOCAL GROUP, about 815,000 and 685,000 l.y. away, respectively. The Leonid meteor shower radiates from the constellation.

BRIGHTEST STARS

	Name	RA h m	dec. ° ′	Visual mag.	Absolute mag.	Spectral type	Distance (l.y.)
α	Regulus	10 08	+11 58	1.36	−0.5	B7	77
γ	Algeiba	10 20	+19 51	2.01	−0.9	K0	126
β	Denebola	11 49	+14 34	2.14	1.9	A3	36
δ	Zosma	11 14	+20 31	2.56	1.3	A4	58
ϵ		09 46	+23 46	2.97	−1.5	G0	251
θ	Chertan	11 14	+15 26	3.33	−0.4	A2	178
ζ	Adhafera	10 17	+23 25	3.43	−1.1	F0	260
η		10 07	+16 46	3.48	−5.6	A0	2138

structed a 48-inch Newtonian on a fork mounting. Then, having developed the most technically exact and versatile reflecting telescopes of the age (and published the details of their construction), he spent over five years in Malta, using them in a 'prime sky' location.

last contact *See* FOURTH CONTACT

last quarter *See* PHASES OF THE MOON

late heavy bombardment Period of intense cratering in the inner Solar System near the end of the formation of the TERRESTRIAL PLANETS. Evidence for the bombardment is seen in the distribution of craters on Mercury, Mars and the Moon. Ages of lunar rocks indicate that the large mare BASINS were formed between about 4.0 and 3.8 billion years ago. Undoubtedly, the early Earth experienced impacts of comparable or larger magnitude, but much of the evidence has been erased by erosion and TECTONICS. It is not clear whether the bombardment was simply the final stage of a continuously declining flux of impacts at the end of ACCRETION of the planets, or whether it was a distinct 'cataclysm' with a temporary large increase in the impact rate. If the latter, the projectiles may have been asteroids perturbed from the MAIN BELT by Jupiter, or comets scattered from the outer Solar System due to the formation of Uranus and Neptune.

late-type star Lower temperature star of spectral class K, M, R, N or S. The term was devised when astronomers erroneously believed that the sequence of SPECTRAL TYPES was also an evolutionary sequence. The usage persists, 'late' being a common astronomical synonym for 'cool' (for example, a G5 star is 'later' than a G0 star).

latitude Angular measure from an equatorial great circle or plane in a spherical coordinate system. On the surface of the Earth, latitude is measured in degrees north or south of the equator, reaching a maximum at the poles. **Celestial latitude** (or ecliptic latitude) is the angular distance north or south of the ecliptic to the ecliptic poles. **Galactic latitude** is a measure of angular distance from the galactic equator to the galactic poles. **Heliocentric latitude** is the same as celestial latitude, but corrected for a hypothetical observer located at the centre of the Sun, while **heliographic latitude** is measured on the Sun's disk, north or south of its equator.

Lauchen, Georg Joachim von *See* RHAETICUS

launch window Period of time during which it is possible to launch a particular space mission.

leap second Periodic adjustment by one second of COORDINATED UNIVERSAL TIME (UTC) to ensure that it remains linked to the rotation of the Earth.

The braking action of the tides causes the Earth's period of rotation to slow by between 1.5 and 2.0 milliseconds per day per century. In order that time as kept by atomic clocks (UTC) and that derived from observations of the stars (UT1, *see* UNIVERSAL TIME) do not differ by more than 0.9 seconds, an extra second is introduced to UTC at midnight on either December 31 or June 30 as required.

This has the effect of producing an accumulated difference between UTC and INTERNATIONAL ATOMIC TIME (TAI) of an integral number of seconds. The most recent leap seconds were inserted as follows:

Date	TAI–UTC
1989 December 31	+25.0 sec
1990 December 31	+26.0 sec
1992 June 30	+27.0 sec
1993 June 30	+28.0 sec
1994 June 30	+29.0 sec
1995 December 31	+30.0 sec
1997 June 30	+31.0 sec
1998 December 31	+32.0 sec

leap year Year containing 366 instead of the usual 365 days. The CALENDAR is based on a solar year of 365.25 days but fixed at 365. An extra day is therefore added at the end of February every fourth year to compensate. However, this still only approximates to the true solar year of 365.24219 days and so century years (1800, 1900, and so on) are not leap years unless exactly divisible by 400 (1600, 2000, and so on).

Leavitt, Henrietta Swan (1868–1921) American astronomer who discovered 2400 variable stars, and whose studies of Cepheid variables in the Small Magellanic Cloud led to the discovery of the period–luminosity relationship used to measure cosmic distances. Edward C. PICKERING hired Leavitt, who was deaf, to work at Harvard College Observatory, originally as an unpaid volunteer, on the tedious task of measuring the brightness of star images on Harvard's massive collection of photographic plates. In 1902 Leavitt's work was recognized when she was appointed head of Harvard's Department of Photographic Photometry.

She began to specialize in variable stars (in all she would discover 2400 – half the total known in her lifetime), studying images of CEPHEID VARIABLES on plates taken at Harvard's southern station at Arequipa, Peru. While analysing images of stars in the Magellanic Clouds, Leavitt noticed that the periods of stars that appeared to be Cepheids depended on their apparent magnitudes – the more luminous they were, the longer their periods. In her 1912 paper summarizing this work, she gave the periods for 25 Cepheids in the Small Magellanic Cloud. By 1918 Ejnar HERTZSPRUNG and Harlow SHAPLEY had recast the PERIOD–LUMINOSITY LAW in terms of absolute, rather than apparent, magnitudes, which allowed Leavitt's Cepheid variables to be used as reliable yardsticks for measuring distances within the Milky Way galaxy and, once they had been discovered in the Andromeda Galaxy, beyond.

Leavitt's other studies included preparatory work on the NORTH POLAR SEQUENCE, a standard reference for photographic magnitudes she helped to establish by comparing nearly three hundred plates of the region around the north celestial pole taken through different telescopes.

Leda One of the small satellites in JUPITER's intermediate group, discovered by Charles Kowal (1940–) in 1974. Leda is only *c.*16 km (*c.*10 mi) in size. It takes 238.7 days to circuit the planet, at an average distance of 11.17 million km (6.98 million mi) from its centre. Leda has a substantial inclination (near 27°) and moderate eccentricity (0.164). *See also* ELARA

Lemaître, Georges Édouard (1894–1966) Belgian Catholic priest and cosmologist whose theory of the 'primeval atom' foreshadowed the Big Bang theory. He began his work on the expansion of the Universe during a postdoctoral fellowship at Harvard (1924–25). Lemaître heard that Edwin HUBBLE had used CEPHEID VARIABLES in the Andromeda Galaxy to determine that it lay far beyond the Milky Way, and was aware of Vesto M. SLIPHER's earlier finding that the radial velocities of galaxies showed a redshift in their spectra. He realized that galaxies must be speeding away from one another. Independent of similar ideas proposed by Aleksandr FRIEDMANN (which were not well known outside Russia), he considered how this expansion would vary with time according to GENERAL RELATIVITY, taking into account the effect of the pressure of matter, and assuming a non-zero value of the COSMOLOGICAL CONSTANT. This LEMAÎTRE UNIVERSE, as it became known, was the first to attempt to reconcile a cosmological model with the observed expansion of the Universe.

Lemaître became professor of astronomy at Louvain in 1927. That year he published a paper stating that the more distant a receding galaxy, the greater its velocity of recession – a concept not explained by Friedmann. Arthur EDDINGTON publicized Lemaître's results at the 1927 meeting of the Royal Astronomical Society. In 1931 Lemaître proposed that the Universe was born from a 'primeval atom' that disintegrated over time to produce gaseous clouds, stars and cosmic rays, an idea we now know to be incorrect. The Belgian priest is often considered the father of the BIG BANG theory (which Lemaître himself colourfully called the 'fireworks theory') of cosmology.

Lemaître universe Same as the FRIEDMANN UNIVERSE, except that Lemaître assumed the value for the COSMOLOGICAL CONSTANT was non-zero.

Le Monnier, Pierre Charles (1715–99) French astronomer and physicist whose scientific career spanned many different disciplines. In celestial mechanics, he analysed the perturbing effect of Saturn on Jupiter's orbit and devised a theory to explain the Moon's complex motions. His contributions to celestial cartography included the measurement of many star positions and compilation of the *Atlas céleste*. Le Monnier observed Uranus before it was seen by its discoverer, William Herschel, but failed to recognize it as a planet.

lens Basic optical component made of transparent material through which light passes in order to produce or modify an image. Astronomical lenses range in size and complexity from tiny spherical sapphire lenses attached to the ends of fibre optics, through to large OBJECTIVE lenses used to collect and focus light in refracting telescopes. Lenses are also used to shape and direct light as it passes through astronomical instruments for analysis and measurement, but the most common example as far as the amateur astronomer is concerned is the telescope EYEPIECE.

There are two basic types of lens, converging and diverging. Converging lenses are thicker in the centre than at the edges and cause parallel light passing through them to bend towards a common point, that is, to converge. The simplest example of a converging lens is the magnifying lens. Diverging lenses cause parallel light to spread out and to appear to come from a common point behind the lens, that is, to diverge. Diverging lenses are thicker at the edges than at the centre and the most common example is the spectacle lens used to correct short-sightedness.

Converging lenses often have two convex surfaces and so are sometimes called convex lenses. Diverging lenses often have two concave surfaces and so are called concave lenses. However, both types can have one convex surface and one concave so this terminology can be slightly misleading.

A lens can produce a focused image because light is bent (refracted) as it passes from air to glass and then from glass back to air. It is the shape of the air–glass surface that determines how the light is bent and what effect this has on the image produced. Surfaces are usually spherical, although more complex shapes are used to reduce optical aberrations. Several manufacturers now advertise binoculars with aspheric optics which offer superior performance.

Single lenses produce ABERRATIONS in the images they produce. This is inevitable even with lenses that are perfectly shaped. It is caused by fundamental geometric effects and the fact that the different wavelengths of light (the different colours) are bent by different amounts as they pass through a lens. Optical designers use multi-element lenses and materials with different properties to minimize aberrations. Even a simple telescope eyepiece will contain two, three or four separate pieces of glass; complex ones can contain seven, eight or more. *See also* ACHROMAT; ANASTIGMAT; APOCHROMAT

lenticular galaxy GALAXY that appears lens-shaped when seen edge-on. Lenticular galaxies are of a type intermediate in form between the much more common elliptical and spiral types. They are classified S0, which indicates that they have the flattened form of spirals but no spiral arms. Lenticular galaxies contain both central bulges and surrounding disks of stars, and the disks include significant amounts of dust in some cases.

Leo *See* feature article

Leo Minor *See* feature article

Leonids Interesting METEOR SHOWER, active between November 15–20 and usually peaking around November 17. The RADIANT lies within the 'sickle' of Leo. The shower produces reasonable activity each year, with peak ZENITHAL HOURLY RATE (ZHR) about 15. It attracts most attention, however, in the ten-year span centred on the perihelion return of the parent comet, 55P/TEMPEL–TUTTLE, during which time rates are greatly elevated, perhaps reaching meteor storm proportions on some occasions. Past Leonid storms have been separated by roughly 33-year intervals: those on 1799 November 12 and 1833 November 12 did much to stimulate scientific interest in the study of meteors. The later time of maximum at the modern epoch results from regression of the meteor stream orbit's descending node.

Following the 1833 return, Hubert Newton (1830–96) of Princeton, United States, investigated the origin of the Leonids. He found historical records of the shower going back to AD 902, including exceptional displays in 934, 1002, 1101, 1202, 1366, 1533, 1602 and 1698. Newton surmised that these were caused by a dense cloud of meteoroids encountered by Earth at intervals of about 33 years; he made the bold prediction, confirmed by Heinrich OLBERS, that another storm might occur in 1866. This prediction was duly borne out on

LEO MINOR (GEN. LEO MINORIS, ABBR. LMI)

Small, inconspicuous northern constellation, representing a small lion (compared to Leo), between Ursa Major and Leo. Leo Minor was introduced by Hevelius in 1687. Its brightest star, 46 LMi (there is no α Leo Minoris), is mag. 3.8. β LMi is a very close binary with yellow components, mags. 4.4 and 6.1, separation 0″.2. Deep-sky objects include NGC 3245, an 11th-magnitude lenticular galaxy.

LEPUS (GEN. LEPORIS, ABBR. LEP)

Small southern constellation representing a hare (sometimes said to be the quarry of Canis Major, one of Orion's two dogs), between Orion and Columba. Its brightest star, ARNEB, is mag. 2.6. γ Lep is a double star, with yellow and orange components, mags. 3.6 and 6.2, separation 97″. The constellation also contains HIND'S CRIMSON STAR (designated R Lep, a long-period variable, range 5.5–11.7, period about 427 days). Lepus' deep-sky objects include M79 (NGC 1904), an 8th-magnitude globular cluster; and IC 418, a 9th-magnitude planetary nebula.

BRIGHTEST STARS

	Name	RA h m	dec. ° ′	Visual mag.	Absolute mag.	Spectral type	Distance (l.y.)
α	Arneb	05 33	−17 49	2.58	−5.4	F0	1284
β	Nihal	05 28	−20 46	2.81	−0.6	G5	159
ε		05 05	−22 22	3.19	−1.0	K4	227
μ		05 13	−16 12	3.29	−0.5	B9	184

1866 November 13–14, when Leonid rates are estimated to have reached 10,000/hr over western Europe.

Subsequent gravitational perturbations by Jupiter and Uranus pulled the densest parts of the Leonid stream away from close encounter with Earth, and the much-anticipated 1899 return proved a relative disappointment: while rates were again enhanced (200–300/hr), the shower did not produce a storm. Comparatively modest returns also followed in 1932 and 1933. The strongest historical displays had usually been seen when the comet/stream orbit's node lay inside Earth's orbit, and Earth arrived there within 2500 days of the comet. Further gravitational perturbations restored such conditions in time for the 1966 return, when estimated Leonid rates of 140,000/hr were seen in a 40-minute interval by observers in the western United States on the morning of November 17.

The Leonids were again the subject of much study around the 1998 perihelion of 55P/Tempel–Tuttle. Activity began to increase markedly in 1994, and many attempts were made to forecast when any storm might occur at this return. No storm was seen in 1998, when Earth followed the comet to the node by 258 days. Instead, to the great surprise of all, an outburst comprising mainly bright Leonids occurred some 16 hours ahead of node-passage, with ZHR around 250 for some 10–12 hours. The 1999 shower, however, did produce storm activity, albeit at lower levels than many of the great events of the past: around node-passage on November 18^d 02^h UT, ZHR reached about 3000, with activity on this occasion consisting mostly of faint Leonids.

Perhaps the most successful forecasts of Leonid activity close to the comet's 1998 perihelion have been those of Robert McNaught (1956–) of the Anglo-Australian Observatory and D.J. Asher of Armagh Observatory, Northern Ireland. They based their stream model on discrete filaments or arcs of debris that remained close to the nucleus following their release at separate perihelion returns of 55P/Tempel–Tuttle. Using this model McNaught and Asher have been able to predict, with reasonable accuracy, the times of several peaks during the 1999 and 2000 displays. More importantly, they were successful in forecasting substantial peaks around 10^h and 19^h UT on 2001 November 18, each of which produced activity similar to that in 1999. The strong activity in 2001 persisted for several hours centred on these maxima, unlike the short-lived, sharp peak two years earlier.

In the longer term, forecasts for the Leonids' next round of elevated returns around 2032 are not particularly favourable, as their orbit is once again pulled away from Earth. Expectations are that the most active displays at that time may be about double those seen in most years from the PERSEIDS.

LIBRA (GEN. LIBRAE, ABBR. LIB)

Inconspicuous southern zodiacal constellation, representing a pair of scales, between Virgo and Ophiuchus. The ancient Greeks grouped its brightest stars into a constellation named Chelae (the Claws – of the Scorpion); the name Libra was introduced by the Romans in the 1st century BC, when it contained the autumnal equinox, and day and night were of the same length (in balance). It is the only zodiacal constellation not representing an inanimate object. Its brightest star, Zubeneschamali (Arabic for 'the northern claw'), is mag. 2.6; Zubenelgenubi ('the southern claw') is a wide double, with bluish-white and pale yellow components, mags. 2.8 and 5.2. ι Lib is a multiple system consisting of two close binaries. The brightest deep-sky object in Libra is NGC 5897, a 9th-magnitude globular cluster.

BRIGHTEST STARS

	Name	RA h m	dec. ° ′	Visual mag.	Absolute mag.	Spectral type	Distance (l.y.)
β	Zubeneschamali	15 17	−09 23	2.61	−0.8	B8	160
α^2	Zubenelgenubi	14 51	−16 02	2.75	0.9	A3	77
σ		15 04	−25 17	3.25	−1.5	M3/M4	292

Leonov, Alexei Arkhipovich (1934–) Soviet cosmonaut, born in Siberia, who was the first man to 'walk' in space, during the second orbit of the Voskhod 2 spacecraft on 1965 March 18. This first 'extra-vehicular activity' lasted just ten minutes. He went on to train for the abortive Soviet manned lunar programme. In 1975 Leonov commanded the Soyuz spacecraft that successfully docked with a US Apollo vehicle in the Apollo–Soyuz Test Project.

leptons Spin $\frac{1}{2}$ fermions that do not interact through the strong force. Some examples of leptons are electrons, positrons and neutrinos. Each of these particles carries a quantum number called the lepton number of +1, while their antiparticles have a lepton number of −1. They are restricted to weak interactions and, if electrically charged, to electromagnetic interactions. The neutrinos take part in weak interactions only. Leptons do not join together to form other particles, and do not appear to be made up of quarks.

Lepus *See* feature article

Le Verrier, Urbain Jean Joseph (1811–77) French mathematician who specialized in celestial mechanics, predicting the existence of Neptune from the perturbations of Uranus. Trained as an experimental chemist, he shifted his research interests to astronomy after accepting a post at the École Polytechnique in 1837. As director of Paris Observatory (1854) he introduced major reforms; he was dismissed from office in 1870 but reinstated after the death of his successor, Charles DELAUNAY, in 1872.

Le Verrier analysed the stability of orbits of planets and minor bodies of the Solar System, including periodic comets. He investigated the advance in the perihelion of Mercury, which he attributed to a hypothetical planet prematurely named VULCAN. But it was his careful analysis – independently of John Couch ADAMS – of discrepancies between predicted and observed positions of Uranus that brought Le Verrier his greatest fame. He was able to compute the position of a hitherto undiscovered planet orbiting beyond Uranus that explained the irregularities in terms of gravitational perturbations by the new body. Using Le Verrier's calculations, Berlin Observatory astronomers Johann GALLE and Heinrich D'ARREST located the new planet on 1846 September 23, within 1° of Le Verrier's predicted position.

Lexell, Anders Johan (1740–84) Swedish astronomer, born in Åbo (modern Turku, Finland), who spent most of his career (1768–80, 1782–84) at the St Petersburg Academy of Science and specialized in computing cometary orbits. He computed the orbit of a comet discovered by Charles Messier, and found it to have a period of just 5.5 years – the first known example of a 'short-period' comet. Lexell correctly deduced that comets are much less massive than the major planets, whose orbits they did not perturb, even during close encounters. He was the first to prove the planetary nature of Uranus (it was originally suspected to be a comet).

Lexell, Comet (D/1770 L1) First short-period comet on record. Discovered by Charles MESSIER, it is one of the best-known cases of a cometary orbit having been altered considerably after being perturbed by the gravitational pull of the massive planet Jupiter. In 1770 this comet, in its approach towards perihelion, passed between the satellites of Jupiter. Later, on 1770 July 1, it came within 2.5 million km (1.6 million mi) of the Earth. As it passed by, the transient gravitational grasp of the Earth caused a decrease in the comet's period by almost three days, but the change in the period of the far more massive Earth was so small as to be immeasurable. The orbit of the comet was investigated by Anders LEXELL, who found it to have a period of 5.6 years. He showed that the comet had been highly perturbed by Jupiter in 1767 May, when its orbit had been changed from a much larger ellipse to its present shape, which explained why it had never been seen previously. The comet was never seen again after 1770. The reason for this was provided by Pierre LAPLACE, who showed that a second close approach to Jupiter had occurred in 1779. On this occasion it had passed so close to Jupiter that its orbit was altered dramatically, with large changes in its orbital period and perihelion distance.

Libra *See* feature article

libration Small oscillation of a celestial body about its mean position. The term is used most frequently to mean the Moon's libration. As a result of libration it is possible to see, at different times, 59% of the Moon's surface. However, the areas that pass into and out of view are close to the LIMB and therefore extremely foreshortened, so in practice libration has less of an effect on the features that can be clearly made out on the Moon's disk than this figure might suggest.

Physical libration results from slight irregularities in the Moon's motion produced by irregularities in its shape. Much more obvious is **geometrical libration**, which results from the Earth-based observer seeing the Moon from different directions at different times. There are three types of geometrical libration. Libration in longitude arises from a combination of the Moon's SYNCHRONOUS ROTATION and its elliptical orbit. As a result, at times a little more of the lunar surface is visible at the eastern or western limb than when the Moon is at its mean position. Libration in latitude arises because the Moon's equator is tilted slightly from its orbital plane, so that the two poles tilt alternately towards and away from the Earth. A smaller effect is diurnal libration, by which the Earth's rotation allows us to see more of the Moon's surface at its western limb when it is rising, and more at the eastern limb when it is setting.

Lick Observatory Major optical observatory, and the first in the United States to be built on a mountain-top site. Its telescopes are on Mount Hamilton, about 32 km (20 mi) east of San Jose, California, at an altitude of 1280 m (4200 ft). Its first director was Edward S. HOLDEN. The observatory was the result of a bequest to the University of California by an eccentric millionaire, James Lick (1796–1876), whose body lies beneath the pier of the 0.9-m (36-in.) Lick Refractor. Inaugurated in 1888, the telescope was for nine years the largest in the world, and it remains the world's second-largest refractor. With it, Edward E. BARNARD discovered Amalthea in 1892, the first Jovian satellite to be found since 1610. The front element of its objective lens was refigured in 1987, having suffered long-term damage from atmospheric corrosion.

Several other telescopes are on Mount Hamilton, including the Crossley 0.9-m (36-in.) reflector, once owned by Andrew COMMON, and the 1-m (39-in.) Nickel Reflector, built in 1983. The largest telescope at Lick is the C. Donald Shane Telescope 3.05-m (120-in.), completed in 1959 and named after a former director. Lick Observatory is still operated by the University of California Observatories, with headquarters in Santa Cruz.

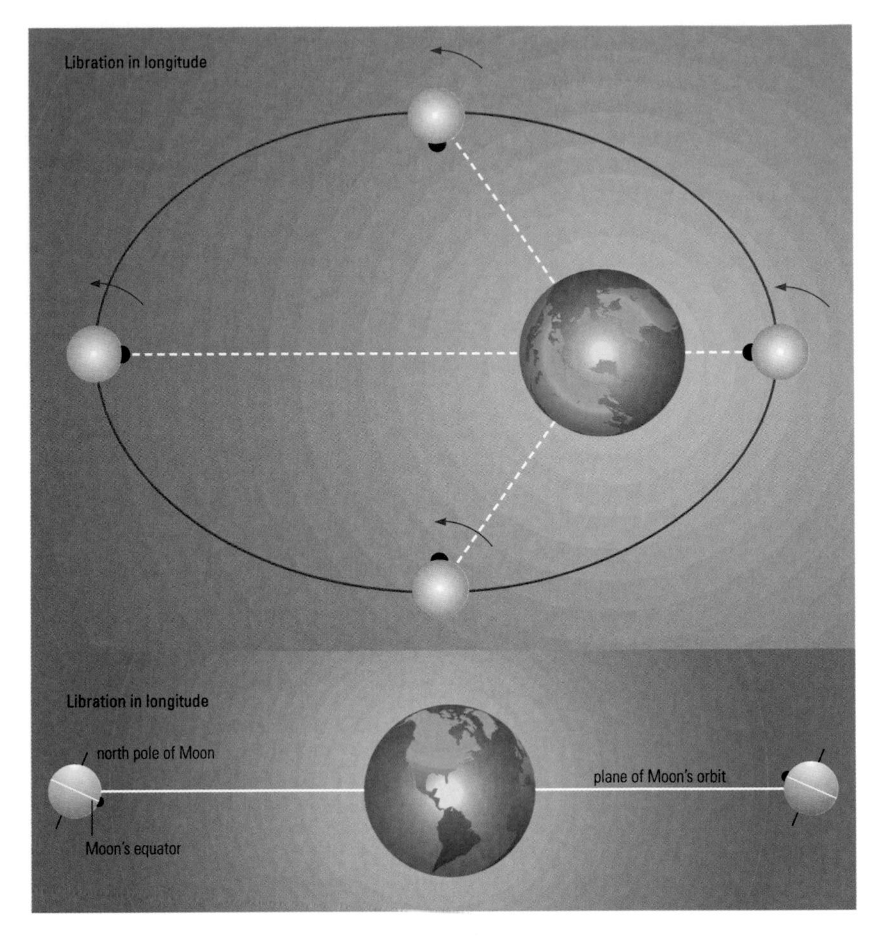

▲ **libration** Although the Moon is in synchronous rotation with respect to the Earth, ground-based observers can still, over the course of several lunations, see significantly more than just half of its surface. Libration in longitude (top) allows the observer to see slightly around the mean east or west limb, while the tilt of the lunar axis relative to the Moon's orbital plane allows regions beyond either pole to be seen on occasion.

life in the Universe The study of the possible existence of life in the Universe beyond the Earth is variously termed **bioastronomy, astrobiology** and **exobiology**. It covers questions on the existence of life on other planetary bodies in the Solar System, such as Mars and Europa; the study of organic molecules in giant molecular clouds and circumstellar material (*see* INTERSTELLAR MOLECULES) and in comets and meteorites; the study of planetary surfaces and atmospheres; questions of the origins of life, whether on the Earth or elsewhere (*see also* PANSPERMIA), and the range of conditions under which it can survive; and the search for life in space, by 'listening' for intelligent signals (*see* SETI) and investigating extrasolar planets.

The first truly scientific ideas (as opposed to philosophical speculation on the 'plurality of worlds') on the possibility of extraterrestrial life followed the invention of the telescope. Early ideas were entirely anthropocentric and shaped largely by contemporary religious beliefs. In the late 18th century Immanuel Kant (1724–1804) formalized these ideas, proposing that all planets would have their own appropriate inhabitants matched to the conditions that prevailed upon them. There were suggestions that the Moon, with its 'terrae' and 'maria', and even the Sun, were inhabited. With improvements in telescope design during the 19th century, Mars became a focus of attention. The planet was soon being described as a world of reddish 'continents', dark 'seas' and white icy polar caps that grew and shrank with the passing of the Martian seasons. Some concluded that Mars might support life or even be inhabited. Schiaparelli's identification of dark linear streaks – *canali* – on the Martian surface sparked a major debate on whether these features were natural or artificial (*see* CANALS, MARTIAN). By the mid-20th century new instruments and techniques had revealed Mars as a world incapable of supporting any form of reasonably advanced terrestrial

L

► **life in the Universe** An artist's impression of the Darwin spacecraft, which will be used for observations of planets around other stars. To be stationed at the L_1 Lagrangian point, these craft are designed to operate as a free-flying optical interferometer, with six 1.5 m (49 in.) telescopes sending their collected light to a central hub.

life. However, a small number of scientists continued to hope that relative simple forms of primitive life such as lichens might be found there. That hope died in 1964, when the first successful flyby of Mars, by the Mariner 4 spacecraft, revealed a cold, arid Moon-like surface showing no visible evidence of life.

To search for extraterrestrial life, astronomers must first decide exactly what to look for. Only carbon seems able to build the backbones of the complicated molecules required for life. (Although silicon can form long-chain molecules, its ability to do so is limited and there is no evidence that silicon-based life is possible.) In addition to carbon there are just five other elements that are found in all metabolizing living organisms, hydrogen, nitrogen, oxygen, phosphorus and sulphur – CHNOPS for short. Other elements – sodium, magnesium, chlorine, potassium, calcium, iron, manganese, copper and iodine – are regularly but not always found in living cells. Altogether, two-thirds of the 92 naturally occurring elements are found in various living organisms, but never all together in the same one. Clearly, living material is distinguished from the inanimate not by its constituent atoms, but by its complex organized structure and active metabolism.

As the CHNOPS elements are among the commonest and most abundant in the Universe, it is probably safe to assume that life, wherever it exists, is based on these same elements. Life can be broadly defined as a chemical process, seemingly unique to carbon, that draws energy from its local environment to generate and maintain structural complexity. In so doing it reduces local entropy, thus increasing the total entropy of the surrounding Universe. The search for life is therefore a search for evidence of departures from thermodynamic and chemical equilibrium that are associated with biological activity.

Until recently, the list of Solar System objects considered worth searching for evidence of life automatically excluded worlds devoid of an atmosphere, and those outside the Sun's ECOSPHERE (or habitable zone), whose limits correspond quite closely to the orbits of Venus and Mars. Although Venus has a dense atmosphere, space probes have revealed it to be hostile to life in every way. Mars, at the outer limit of the Sun's ecosphere, appears to be a possible if uncertain abode of life. The surface chemistry of Mars at two different sites was investigated in 1976 by the sophisticated landers of the two VIKING missions. They revealed no unequivocal evidence for the existence of Martian life. It was concluded that, at the present epoch at least, Mars is devoid of life, implying that any extraterrestrial life would have to be sought beyond the Solar System (*see* EXTRASOLAR PLANET).

Interest in searching for life within the Solar System has been revitalized by a succession of major discoveries about the extreme limits of life on Earth. It was formerly assumed that the region extending several kilometres below the Earth's surface was essentially sterile. In the late 1970s, the first hot hydrothermal plumes and 'black smokers' – hydrothermal outflows rich in heavy metal sulphides – were discovered rising from the slopes of seafloor ridges 2–3 km (1.25–2 mi) down. Several hundred smokers are now known, they have temperatures of as much as 620 K (at this depth, water boils at ~670 K). The regions surrounding these vents harbour one of the most extraordinary ecosystems on Earth. Under enormous pressure, in complete darkness without energy from the Sun, are microorganisms that feed on the chemicals that rise from deep inside the Earth, many living at temperatures of ~380 K.

In 1993 new species of microorganisms were collected from rocks 2.7 km (1.7 mi) below ground – anaerobic organisms that die when exposed to oxygen, having little resemblance to the typical microbes found on the surface, thriving at an ambient temperature of ~350 K. (Similar thermophilic microbes colonize volcanically heated geysers and springs, such as those in Yellowstone National Park, and hydrothermal plumes on the ocean floor.) Present evidence indicates that subsurface life may extend to ~4 km (2.5 mi) below the continental crust and ~7 km (4.5 mi) below the oceanic crust.

An entirely new library of names has entered biology for microorganisms that occupy a new branch of the tree of life – the Archaea. This branch includes microbes that thrive at very high temperatures – **thermophiles** and **hyperthermophiles**, at very high pressures – **baryophiles**, and in hyper-saline environments – **halophiles**; organisms that metabolize methane or sulphur compounds – **methanogens** and **thiophiles**: microbes that thrive at very low temperatures – **cryophiles**; organisms that live within rocks – **lithophiles**; and even organisms that live in the most acid and alkaline environments that would rapidly dissolve and destroy most other life forms. Other microorganisms show extremely high levels of tolerance to intense UV and nuclear radiation, some can resist almost total dehydration, and some can survive being in a vacuum for biologically long periods of time. (Interestingly, hyper-thermophiles may be the oldest surviving form of terrestrial life, and clearly would have been well adapted to the conditions that prevailed more than 4 billion years ago, shortly after the Earth formed.) Microorganisms living in extreme environments are collectively referred to as **extremophiles**.

The discovery that terrestrial life seems to have filled every possible environmental niche, above and below the surface, suggests that subsurface life may exist on other planetary bodies in the Solar System, even if their surfaces are totally inhospitable. The essential requirement is that water is present and that the internal heat flow and pressure allow liquid water to exist at some depth, forming a subsurface **ecozone**. Subsurface conditions suitable for some form of extremophile life range from small circumpolar regions on MERCURY – (if water/ice is present there) to the distance of Pluto. Pluto and its satellite Charon are locked by spin-orbit coupling, and solar tides squeezing Charon could be maintaining a subsurface layer of liquid water, providing a suitable environment for cryophiles. This range extends far beyond the limits of the conventionally defined ecosphere.

For the foreseeable future, only targets in the Solar System will be accessible for close observation from orbit or direct investigation of their surfaces, though the search for extrasolar planets may reveal some clues. The four planetary size satellites of Jupiter and Saturn's large moon Titan are major candidates in the search for life. Titan's thick atmosphere is mostly nitrogen, with measurable quantities of methane and ethane, and there is potential for some form of cryophilic microbial life there. Results from the Galileo mission to Jupiter show that Europa and Ganymede, and possibly Callisto, possess subsurface oceans. Both Europa and Ganymede possess internal heat sources, in part derived from tidal heating, and are strong candidates for the existence of some form of widespread or localized submarine hydrothermal vent activity – perhaps capable of supporting chemosynthetic ecosystems (*see also* EUROPA ORBITER). The search for traces of past or extant microbial life on Mars is already under way, and the question whether surface or subsurface life once existed or still exists there may be answered in the first half of the 21st century.

Subsurface environments conducive to chemosynthetic microbial life cannot be limited to the Solar System, and can be expected to be common throughout the Universe. The most widespread and abundant forms of life in the Universe will most likely have more in common with microorganisms than with the complex multicellular organisms that characterize the life on the Earth's surface.

light, velocity of The value for the velocity of light (and of all other forms of ELECTROMAGNETIC RADIATION) in a vacuum, usually symbolized as '*c*', was officially recognised as 299,792,458 m/s by the International Astronomical Union (IAU) in 1976. For many purposes, however, the approximate value of 300,000 km/s (3×10^8 m/s) is sufficient.

One of the first attempts to measure the velocity of light was made by GALILEO in 1600, prior to that, most scientists had expected its velocity to be infinite. Galileo and an assistant stood about a kilometre apart with lanterns. When the assistant saw a flash from Galileo's lantern, he briefly opened the cover on his own. Galileo then attempted to measure the time lag between uncovering his own lantern and seeing the flash from the other lantern. Needless to say, the experiment failed since the reaction times involved were ten thousand times longer than the light flight times. The first reasonably successful determination of *c* was made be Ole RÖMER in 1676. He had been observing and timing transits and occultations of Jupiter's Galilean satellites, when he noticed that the events were occurring later than he had predicted on the basis of earlier observations. Römer realised that this was because the Earth and Jupiter had moved around their orbits and were further apart than when he had made the previous observations. The delays in the occurrence of the satellite phenomena therefore arose because of the time it took light to cross the increased distance between the planets. From his measurements, Römer determined a velocity of 240,000 km/s for light.

Hippolyte FIZEAU made the first 'laboratory' determination of *c* in 1849. He shone light on to a mirror placed nearly 9 km (6 mi) away and looked at the reflection. Both the outgoing and returning beams passed between the teeth of a rotating gearwheel. At slow rotations, the returning beam would pass through the same gap between the teeth of the gearwheel as the outgoing beam, and so could be seen. When the speed of rotation was increased, there would come a point where the returned beam was intercepted by the next tooth, and so could not be seen. Increasing the speed still further enabled the returning beam to be seen again as it passed through the gap next to the one used by the outgoing beam. The number of teeth on the gearwheel (720) and its rotational velocity (1500 rpm) when the returning beam was seen for the second time, provided Fizeau with the time taken by light for the return journey, and from this he found a velocity for light of 315,000 km/s.

As his apparatus was refined, Fizeau was able to show that light passing through moving water had a higher velocity when it went with the flow rather than against it, and in 1853 Leon FOUCAULT showed that the velocity was lower in water than in air. Experiments like these led to the expectation that the velocity of light would vary depending on whether a light beam was travelling in the same direction through space as the Earth, or in the opposite direction. So in 1887 the German-American physicist Albert Michelson (1852–1931) and the American chemist Edward Morley (1838–1923) conducted their famous experiment (*see* MICHELSON–MORLEY EXPERIMENT). Their failure to detect the Earth's motion by finding a changing velocity for light led directly to the idea that the velocity of light was constant for all observers, whatever their motion, and so to EINSTEIN'S theory of SPECIAL RELATIVITY.

Michelson later (1926) determined a value of 299,796 km/s for the velocity of light, using a variation of Fizeau's method, with the light beam reflected out to and back from a mirror 35 km (22 mi) away via a rotating octagonal prism.

For astronomers, a consequence of the finite velocity of light is that objects are seen as they were when they emitted the light that is now received, not as they may be today. An object one light-year away is seen as it was a year ago and so on. For more distant objects, this 'look back' time, can be very large, as much as 10^{10} years or so for the most distant quasars. Some objects may thus have changed radically or even no longer exist, though they can still be seen in the sky. Thus the 1987 SUPERNOVA in the Large Magellanic Cloud could still be seen as a star 170,000 years after it had actually exploded.

light, visible ELECTROMAGNETIC RADIATION, with wavelengths from about 380 to 700 nm, that may be seen by the human eye.

The nature of light is now thought to be fairly well understood, yet it was the subject of controversy for more than 200 years, as experimenters uncovered conflicting evidence of its nature. Some evidence suggested that light was composed of individual particles, other evidence suggested that it was composed of waves. It was even found necessary to invent a substance to carry these waves, the so-called luminiferous ether, which is now known not to exist. Not until the pioneering experiments of the Scottish physicist Thomas Young (1773–1829) was the wave-like nature of light firmly established for the scientific community. EINSTEIN, however, in his 1905 explanation of the photoelectric effect then showed that light also behaved as though it were a stream of particles.

It is now known that light is both particle and wave. That is, that light comes in small packets, called PHOTONS. This is true whether light, radio waves or X-rays are considered (all of which are types of electromagnetic radiation). This packet is composed of an electric field and a magnetic field, each of which 'oscillates' at right angles to its direction of motion. The interaction of these fields was first described in mathematical form by James Clark MAXWELL. His laws are now known to all students of physics as 'Maxwell's laws' and are the basis of the study of electromagnetism, the study of all phenomena related to the interactions of electric and magnetic fields.

Despite the opening-up of other regions of the ELECTROMAGNETIC SPECTRUM in recent years, our picture of the Universe is still largely based on what we actually see, that is, the information brought to us by light.

light cone In a spacetime diagram, line or curve that represents the path of a light ray. Spacetime diagrams are graphical three-dimensional representations of four-dimensional SPACETIME. Any journey through spacetime is represented by a line or curve in such diagrams. The path of a light ray emanating from a particular point forms a cone with an angle of 45° from the vertical; these

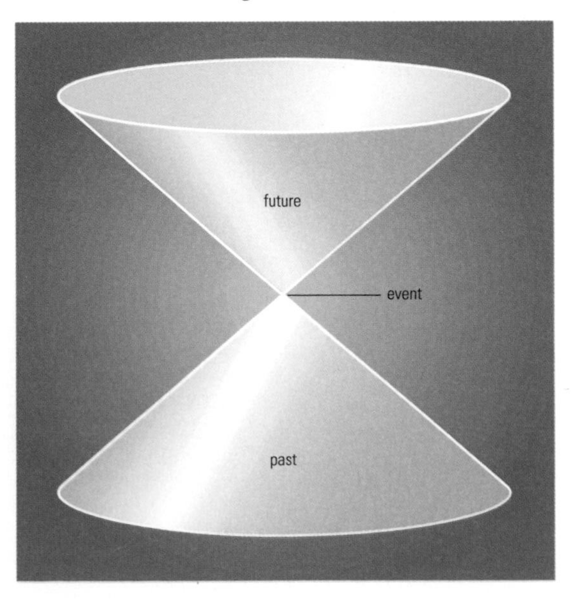

◀ **light cone** Events in spacetime can be mathematically modelled in terms of a double cone. In this model, the event lies at the tip of the cones. Time runs vertically, so that the upward-directed cone coming away from the event represents the future, while the downward-directed cone is the past. The past converges upon, while the future diverges from, the event.

L

▲ **light-curve** A plot of the magnitude (brightness) of Nova Delphini 1967 against time. The nova was slow to arrive at peak brightness, showing several maxima over a twelve-month period, before its light-curve entered a long, gradual decline.

cones are called light cones. Any journey by matter through spacetime must be slower than the speed of light, thus forming a line or curve making an angle less than 45° with the vertical time axis.

light-curve Graph of the variation with time of the brightness of an astronomical body (usually a VARIABLE STAR).

light pollution Effect of street and other lighting concentrated mainly in urban areas on astronomical observing and imaging. For amateur astronomers, it limits the celestial objects (particularly extended objects) that can be seen. For professional astronomers its effects can so impede or distort scientific observations as to result in the eventual closure of observatories enveloped by urban sprawl in the years since their foundation.

light-time (look back time) Time taken for light to travel between the object and the observer. When making accurately timed measurements of, for example, variable stars, the varying time taken for light to reach the Earth at different points in its orbit must be taken into account. Such measurements are often given in heliocentric time: the time determined by an hypothetical observer on the Sun. *See also* LIGHT, VELOCITY OF

light-year (l.y.) Unit of distance measurement equal to that travelled by a ray of light, or any electromagnetic radiation, in a vacuum in one TROPICAL YEAR. Light travels at a speed of 300,000 km/s and so a light-year is equivalent to 9.4607×10^{12} km, 0.3066 PARSECS, or 63,240 ASTRONOMICAL UNITS.

LIGO Abbreviation of LASER INTERFEROMETER GRAVITATIONAL-WAVE OBSERVATORY

limb Extreme edge of the visible disk of the Sun, Moon or a planet. The term could refer to any celestial body that shows a detectable disk. The limb on the same side as the direction of travel of a body is called the PRECEDING limb, and that farthest away, the FOLLOWING limb.

limb brightening Increase in brightness from the centre to the LIMB of an astronomical body. Observing the effect suggests looking at the surface of an object rather than its atmosphere, since atmospheres typically display LIMB DARKENING.

limb darkening Decrease in brightness of an astronomical body, particularly the Sun, from the centre to the edge, otherwise known as the LIMB. In the case of the Sun, the observer's line of sight to the centre of the disk is through a smaller volume of the Sun's light-absorbing, light-scattering atmosphere than that to the limb. *See also* LIMB BRIGHTENING

limiting magnitude (1) Faintest MAGNITUDE visible or recordable by photographic or electronic means. It depends on APERTURE, sky transparency, SEEING, exposure and sensitivity of the eye or recording apparatus. (2) Faintest magnitude of objects shown in a star atlas or listed in a catalogue.

Lindblad, Bertil (1895–1965) Swedish astronomer who directed Stockholm Observatory (1927–65). He identified the Galaxy's spiral structure and its DIFFERENTIAL ROTATION. He correctly determined that the distance to the galactic centre is about 50,000 l.y. and confirmed Harlow SHAPLEY's model that located the galactic centre in Sagittarius. Lindblad's other major research involved studies of star clusters and their structure, and the use of spectroscopy to establish luminosity criteria for stars that allowed main sequence and giant stars to be differentiated.

Lindblad resonance Resonance that occurs when the rotational period of the spiral pattern of a spiral galaxy is such that the stars are at the same point in their epicyclic motions each time the star passes through one of the arms. The inner Lindblad resonance occurs where the stars are moving faster than the spiral pattern, and the outer Lindblad resonance occurs where they are moving slower. For our GALAXY, these resonances occur at about 12,000 l.y. and 50,000 l.y. from the centre of the Galaxy respectively. The Lindblad resonances mark the limits of the region where spiral density waves can form spiral arms within a galaxy.

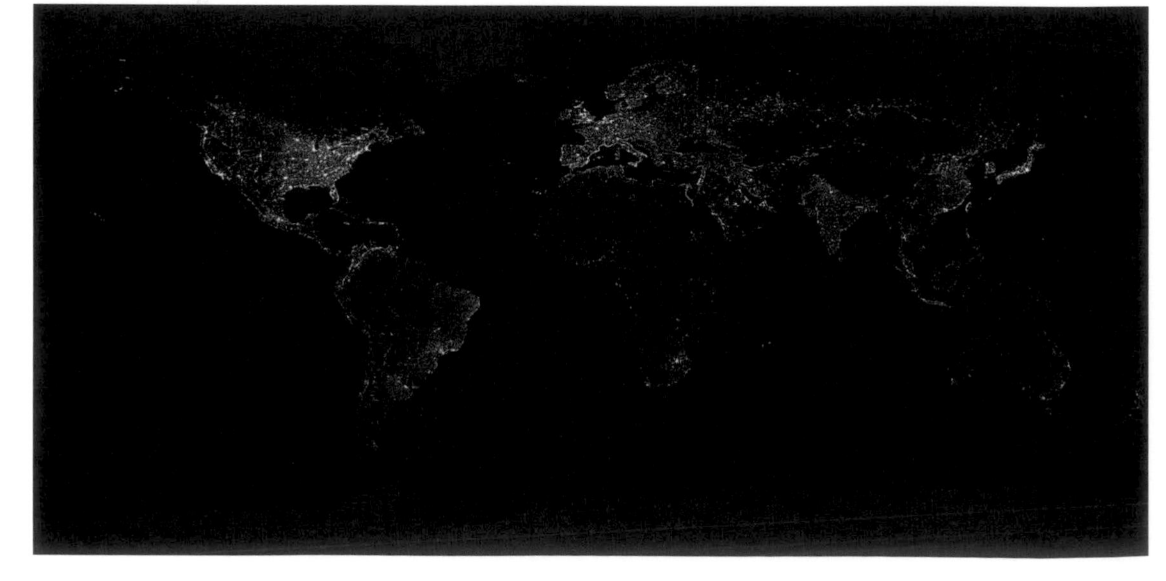

► **light pollution** Data from the US Defense Meteorological Satellite Program show the extent of light spillage into the night sky over cities worldwide. Eastern North America, Europe and Japan, with their high population densities, are particularly affected by this light pollution, which degrades astronomical observing conditions.

Linde, Andrei Dmitrievitch (1948–) Russian cosmologist, professor of physics at Stanford University, with Alan GUTH one of the originators of the INFLATION, a model of the very early Universe (up to just 10^{-37} s after the BIG BANG). By causing an initial rapid expansion, inflation accounts for certain observed properties of the Big Bang Universe. Linde advocates 'chaotic inflation', which allows many 'self-reproducing' or 'parallel' universes to branch off from a parent universe; these separate universes may differ completely from one another in their physical properties.

LINEAR, Comets Comets discovered by the MIT Lincoln Laboratory Near Earth Asteroid Research Project (LINEAR). Established in 1998 March, LINEAR employs two 1-m (0.39-in.) telescopes sited at White Sands, New Mexico, USA, to search for potentially hazardous minor planets. The patrol, using wide-field CCD cameras, has been very successful and has also led to the discovery of several comets: as of late 2001 the comet discovery total stood at over 60.

Among the most notable of these comets was C/1999 S4 (LINEAR), discovered on 1999 September 27. Orbital calculations suggested that this comet would become a reasonably bright object around perihelion, 0.77 AU from the Sun, on 2000 July 26. The comet showed unusual activity in early July, shedding at least one fragment from its nucleus. Following a slight brightening around July 20–21, the comet broke into a cloud of smaller fragments, and, as the released gas and dust dispersed, it faded rapidly from view. Among other LINEAR discoveries, C/2001 A2 became a bright binocular object prone to outbursts in June and July of 2001, and 2000 WM1 became bright in late December 2001.

line of nodes Line of intersection of the orbit plane of a body with the reference plane used for the orbit. It is the line joining the ASCENDING NODE and the DESCENDING NODE. For the planets the reference plane is the ECLIPTIC, and for satellites it is usually the equator of the planet. *See also* INCLINATION; REGRESSION OF THE NODES

LINER (acronym for low-ionization nuclear emission-line region) Common type of galactic NUCLEUS defined by the ratios of (usually weak) emission lines. Some LINERs are clearly low-power ACTIVE GALACTIC NUCLEI, judged by the presence of weak broad lines similar to those in type 1 SEYFERT GALAXIES or associated radiation in the ultraviolet, radio or X-ray regions. Other LINERs are connected to STARBURST GALAXIES, either by having gas excited by shocks in the winds blown from starbursts or photo-ionized by unusually hot stars as the starburst progresses. At low levels, most luminous galaxies have a LINER nucleus.

line spectrum SPECTRUM that exhibits ABSORPTION LINES superimposed on a continuum; EMISSION LINES superimposed on a continuum; or emission lines with no continuous background. Line spectra are found within the entire ELECTROMAGNETIC SPECTRUM, from the gamma-ray domain to radio. Stars have mostly absorption spectra, though some have emission spectra as well. Nebulae have emission spectra.

Linné Lunar crater (28°N 12°E), 1.5 km (1 mi) in diameter. Located in Mare SERENITATIS, Linné is one of the youngest visible craters on the Moon (a few tens of millions of years old). It is bowl-shaped and circular, with a bright collar of RAY material surrounding it, which in Earth-based telescopes is lacking in discrete rays. Linné has been the focus of much debate over supposed changes across time, though none of these have been substantiated.

Lippershey, Hans (or Lipperhey, Hans) (c.1570–c.1619) German-born lens-maker who worked in Middelburg, in the Netherlands. In the early 1600s he began to experiment with combinations of spectacle lenses, and eventually crafted a small refracting telescope of the kind subsequently used by GALILEO. He was one of three Dutch opticians who claimed to have invented the telescope – the others were Jacob Adriaanzoon (1571–1635) and Zacharias Janszoon (1580–*c.*1638) – but Lippershey was the first (in 1608 October) to apply for a patent for the invention.

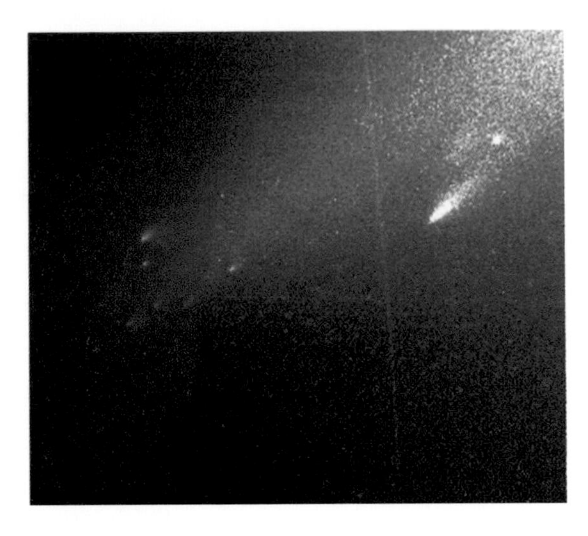

◄ **LINEAR, Comets** Comet 1999 S4 LINEAR broke up into a shoal of small fragments around the time of its 2000 July 26 perihelion. This Hubble Space Telescope image shows the rapidly evaporating remains of the breakup.

lithosphere Outer semi-rigid shell of a planetary body; it is capable of supporting significant stresses. Its definition is based on mechanical properties, unlike the CRUST, which is defined by composition. On Earth, the lithosphere comprises the whole of the crust, together with the uppermost layer of the underlying MANTLE.

lithium (symbol Li) Third element and the lightest of the metals; it is about 80th in order of cosmic abundance. Its properties include: atomic number 3; atomic mass of the naturally occurring element 6.941 amu; melting point 453.7 K; boiling point 1620 K. Lithium has five ISOTOPES, two of which are stable and form the naturally occurring element (lithium-6, 7.4%, and lithium-7, 92.6%).

The abundance of lithium is far higher in the interstellar medium than it is in stars. This is because lithium is produced in the interstellar medium by cosmic ray impacts with carbon, oxygen and other nuclei. It is quickly destroyed in stars because it undergoes nuclear reactions at temperatures as low as 500,000 K. Lithium thus acts as an important initiator in PROTOSTARS to start the hydrogen-burning reactions.

Little Dumbbell (M76, NGC 650 and NGC 651) PLANETARY NEBULA in northern Perseus (RA $01^h\ 42^m.4$ dec. +51°34′). At mag. +10.1, it is reckoned to be the most difficult to observe of the Messier objects. Like the DUMBBELL, M76 shows two principal conspicuous lobes of nebulosity, each of which has been given its own NGC designation. M76 has an apparent diameter of 65″, and lies at a distance of 3400 l.y. It is sometimes described as the Butterfly Nebula.

Liu Xin (or Liu Hsin) (c.AD 50) Chinese school minister and astronomer who was one of the earliest cataloguers of stars in recorded history. Predating Ptolemy's *Almagest* by almost 100 years, Liu Xin's catalogue of over a thousand bright stars assigned stars to six different magnitude categories. He very accurately calculated the true length of a year as 365.25 days.

Liverpool Telescope Fully robotic, 2.0-m (79-in.) telescope at the ROQUE DE LOS MUCHACHOS OBSERVATORY operated by Liverpool John Moores University. Unlike the identical FAULKES TELESCOPES, it was built primarily for university research and teaching, and uses both imaging and spectroscopy. It became operational in 2002.

LMC Abbreviation of LARGE MAGELLANIC CLOUD

lobe Region around an ANTENNA where reception (or transmission) varies in intensity. An antenna neither radiates nor receives radio waves equally in all directions; similarly a RADIO TELESCOPE receives radiation preferentially in one direction but also from other directions too. A number of distinct lobes exist, resembling the petals on a daisy. The main lobe is in the direction of best reception (or transmission), where the telescope/antenna is pointing. All the others are called side lobes and are usually unwanted. When a radio telescope points at a faint astronomical object in the main lobe, the side lobes may pick up electrical interference from ground-based appliances such as microwave ovens. A good design keeps the magnitude of these side lobes to a minimum in most cases.

lobe Region, usually one on each side, of radio emission from high-energy particles. The radio lobes are mostly associated with RADIO GALAXIES, and there is one on each side of the galaxy, connected to it by a JET.

local arm Segment of one of our GALAXY'S spiral arms within which the Sun is located. It is also known as the ORION ARM.

local bubble Volume of interstellar space that encloses the Solar System and other nearby stars. The gas density within the bubble is only about 10% of the average for the interstellar matter. The bubble may be the result of a supernova explosion some time in the past. The bubble is about 300 l.y. across and the Solar System is about 40 l.y. from its edge.

Local Group Small group of about 40 galaxies that includes our GALAXY, the two MAGELLANIC CLOUDS, the ANDROMEDA GALAXY and the TRIANGULUM GALAXY; most are DWARF GALAXIES. They are distributed over a roughly ellipsoidal space, about 5 million l.y. across. All the members are gravitationally bound so that, unlike more distant galaxies, they are not receding from us or from one another.

local standard of rest (LSR) Frame of reference, centred on the Sun, in which the average motion of the stars in the immediate vicinity is zero. Each star, including the Sun, in a volume of space about 330 l.y. in diameter, is moving relative to the local standard of rest, with the Sun moving in the direction of the solar APEX. The zero point for the local standard of rest is an imaginary point in the galactic plane moving clockwise in a circular orbit around the galactic centre at a velocity of about 220 km/s (140 mi/s) with a period equal to that of the Sun. Astronomers measure a star's velocity components with respect to the LSR rather than to the Sun, because the Sun's orbit around the galaxy is slightly non-circular.

local supercluster Group of galactic clusters that contains our LOCAL GROUP of galaxies and several other clusters. The Virgo, Centaurus and Ursa Major Clusters are among the clusters contained in our local supercluster. The local supercluster contains more than 4500 galaxies.

THE LOCAL GROUP: THE MAIN MEMBERS

Galaxy	Apparent mag.	Absolute mag.	Distance (thousand l.y.)
Andromeda Galaxy (M31)	3.4	−21.1	2365
Milky Way	—	−20.6	—
Triangulum Galaxy (M33)	5.7	−18.9	2590
Large Magellanic Cloud	0.1	−18.1	160
IC 10	10.3	−17.6	2675
M32	8.2	−16.4	2365
Barnard's Galaxy (NGC 6822)	9	−16.4	1760
M110 (NGC 205)	8.0	−16.3	2365
Small Magellanic Cloud	2.3	−16.2	190
NGC 185	9.2	−15.3	2020

local thermodynamic equilibrium (LTE) Simplifying approximation used when constructing computer models of stellar atmospheres. The equations describing the properties of gases and plasmas, such as pressure and excitation, are characterized by one temperature; the equations describing the radiation are characterized by another, different, temperature.

local time Time at any given location on the Earth's surface, measured either as MEAN SOLAR TIME or APPARENT SOLAR TIME. Local time varies around the globe with longitude, with a difference of 15° longitude equating to a difference of one hour in local time. By convention, the Earth is divided into 24 TIME ZONES, each one hour apart from its neighbours, but for convenience the civil time kept in these zones may vary from actual zone time, depending on centres of habitation. In the summer, many countries also adopt daylight saving time, advancing their clocks by one hour.

Lockyer, (Joseph) Norman (1836–1920) English amateur astronomer who discovered the previously unknown chemical element helium in the Sun. From 1857 he worked at Britain's War Office, taking up astronomy and other scientific researches in his spare time and constructing a private observatory at his home in Hampstead. From 1864 Lockyer began to study the Sun almost exclusively after hearing of Gustav KIRCHHOFF's discovery in 1859 that the solar spectrum contained lines indicating the presence of sodium atoms. After discussions with William HUGGINS, Lockyer attached a small dispersion spectroscope to his own telescope in the hope of making similar discoveries.

In 1868 he found, independently of Jules JANSSEN, that solar prominences produced bright, yellow emission lines. One of these lines did not appear in any laboratory spectrum produced by a known terrestrial chemical element, so Lockyer suspected that it was produced by a 'new' element, which he named 'helium', after the Greek word for 'Sun'. It was not until 1895 that William Ramsay (1852–1916) managed to isolate helium from a terrestrial source. This discovery proved the worth of astronomical spectroscopy.

In 1869 Lockyer founded the scientific journal *Nature*, which he edited for 50 years. Combining astronomy and archaeology, he was able accurately to date the megalithic Stonehenge monuments by reckoning the changing position of the summer solstice sunrise since the monuments were built. In 1890 he established the Solar Physics Laboratory at South Kensington, serving as its director until 1911. He spent the rest of his scientific life at his Hill Observatory at Sidmouth, Devon, now known as the Norman Lockyer Observatory.

Loki Active volcano on Jupiter's satellite IO. It was discovered on VOYAGER 1 images in 1979. Since then, continuing eruptions have been documented by infrared telescopic observations from Earth and by the GALILEO spacecraft orbiting JUPITER. Loki is the most persistently strong thermal source on Io. It is, on average, the site of about 25% of Io's global heat output. Most of the thermal radiance comes from a probable crusted lake of molten lava, to the north of which is a fissure where plumes of sulphur dioxide are propelled up to hundreds of kilometres skywards.

Lomonosov, Mikhail Vasilievich (1711–65) First Russian scientist to make notable contributions to astronomy. By 1745 he was a full member of St Petersburg's Imperial Academy of Sciences, and was later appointed Professor of Chemistry at the university there. He conceived early versions of the wave theory of light and of the law of conservation of matter.

Lomonosov was a strong supporter of Copernicus' theory (then unpopular in Russia). In 1761 he observed a transit of Venus, correctly deducing that the planet had a dense atmosphere.

long baseline interferometry Several radio telescopes, separated by large distances, operating together as a radio interferometer, to observe astronomical objects with better resolution and to determine the positions more accurately. *See also* RADIO INTERFEROMETER; VERY LONG BASELINE INTERFEROMETRY

longitude Angular measure from a chosen reference point around an equatorial circle in a spherical coordinate system. On the Earth, longitude is measured around the equator from the Greenwich meridian, while CELESTIAL LONGITUDE (or ecliptic longitude) is measured around the CELESTIAL EQUATOR, eastwards from the FIRST POINT OF ARIES. **Galactic longitude** is measured from 0° at the galactic centre, eastwards to 360° along the galactic equator. **Heliocentric longitude** is the same as celestial longitude but corrected for a hypothetical observer located at the centre of the Sun. **Heliographic longitude** is measured from the solar meridian that passed through the ascending node of the Sun's equator on the ecliptic on 1854 January 1 at 12.00 UT. *See also* CARRINGTON ROTATION

Long March Family of satellite launchers developed and operated by China and based on an intercontinental ballistic missile first stage. In 1970 April, a Long March 1 (LM 1) launched China's first satellite; today, the country operates various LM 2, 3 and 4 boosters, some of which provide international commercial satellite launches. The LM2F has flown two tests flights of the Shen Zhou spacecraft, which will carry two Chinese astronauts into orbit in about 2002–2003. The LM3, 3A and 3B fleet offers flights to geostationary transfer orbit for payloads weighing between 1.4 and 5 tonnes. The LM4A and 4B are used to launch payloads into polar orbit.

long-period asteroid ASTEROID with an orbital period greater than about 20 years but a high eccentricity and perihelion interior to Jupiter, making it similar in terms of dynamics to a long-period comet. The first such object to be discovered was DAMOCLES in 1991. Other examples include 1996 PW (period 5900 years) and 1997 MD10 (period 140 years). By late 2001 15 such asteroids were known. *See also* CENTAUR; JUPITER-CROSSING ASTEROID; RETROGRADE ASTEROID

long-period comet COMET for which the interval between successive perihelion returns is greater than 200 years. About 750 long-period comets are known. Many of the most spectacular comets fall within this category, with examples including WEST, HALE–BOPP, HYAKUTAKE and the KREUTZ SUNGRAZERS.

long-period variable Alternative name for a type of pulsating giant or supergiant VARIABLE STAR known as a MIRA STAR, and designated 'M' in the classification scheme. In the past the term was sometimes used for any variable that had a period in excess of 100 days, but this led to confusion and such usage is now discouraged. The abbreviation 'LPV' is commonly used in discussion of stars of this type.

look back time *See* LIGHT-TIME

Loop Nebula *See* TARANTULA NEBULA

Lorentz, Hendrik Antoon (1853–1928) Dutch mathematical physicist who won the 1902 Nobel Prize for Physics for his mathematical model of the electron. He suggested, independently of George Francis Fitzgerald (*see* FITZGERALD CONTRACTION), that bodies moving at relativistic speeds (that is, near the speed of light) contract. He developed LORENTZ TRANSFORMATIONS (1904) to describe this contraction, the tendency of these bodies to increase in mass, and the phenomenon known as time dilation – all important aspects of the theory of SPECIAL RELATIVITY.

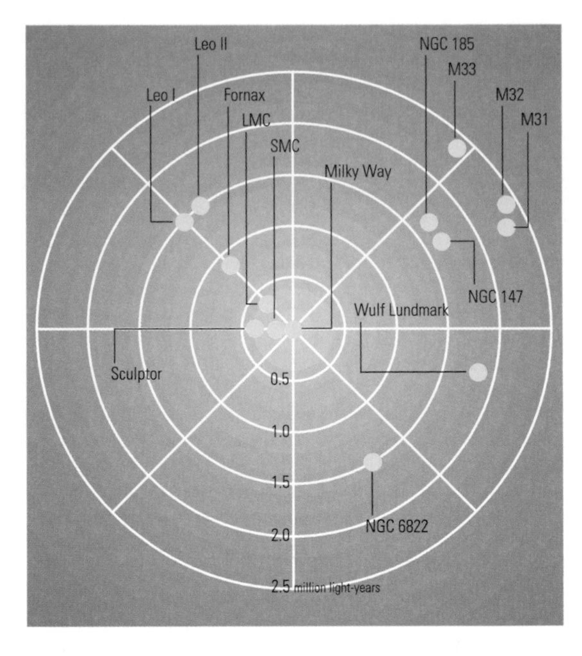

◀ **Local Group** A map of galaxies in the neighbourhood of the Milky Way. Members of this Local Group are gravitationally bound to one another. As shown in this plot, centred on our Galaxy, the closest neighbours are the Magellanic Clouds and Sculptor Dwarf.

Lorentz transformations Set of equations, worked out by Hendrik LORENTZ, which relate space and time coordinates in frames of reference that are in uniform relative motion (*see* RELATIVITY). The space and time coordinates (x_1, y_1, z_1, t_1) in the first frame of reference are related to those in the second, (x_2, y_2, z_2, t_2) by:

$$x_2 = \frac{x_1 - vt}{\sqrt{1 - \frac{v^2}{c^2}}}$$

$$y_2 = y_1$$

$$z_2 = z_1$$

$$t_2 = \frac{t - \frac{vx}{c^2}}{\sqrt{1 - \frac{v^2}{c^2}}}$$

where v is the relative velocity of the two frames of reference (assumed to be along the x axis), and c is the velocity of light.

Lovell, (Alfred Charles) Bernard (1913–) British radio astronomer responsible for building the radio telescope at Jodrell Bank, near Manchester. Early in his career, Lovell studied cosmic rays; after World War II he built on his wartime radar research by using a 4.2-m (14-ft) radar telescope to study the ionized radio echoes produced by meteors. In 1950 he offered the first definitive explanation for the phenomenon of SCINTILLATION produced by Earth's ionosphere, which can propagate and amplify incoming or distant radio waves. This phenomenon, also known as SCATTERING, is useful for studying the solar wind and for measuring the angular sizes of extraterrestrial radio sources. Lovell built a 66.5-m (218-ft) wire bowl-shaped transit radio telescope that was used by Robert Hanbury BROWN and others to detect the first radio waves from the Andromeda Galaxy.

In 1951 Lovell was appointed professor of radio astronomy at Manchester University, and six years later his dream of a radio-astronomy observatory was realized when Jodrell Bank, now the NUFFIELD RADIO ASTRONOMY LABORATORY, was opened. Its primary instrument is a 76-m (250-ft) fully steerable parabolic dish receiver on an altazimuth mount, completed in 1957. That was the same year that the Russian SPUTNIK 1 satellite was launched, and Lovell was able to justify the expense of the great radio telescope – originally known as the Mark I, but renamed the LOVELL TELESCOPE in 1987 – by using it successfully to track the then-feared spaceship. In the 1960s he collaborated with Fred WHIPPLE to make important studies of FLARE STARS simultaneously at radio and optical wavelengths, an observational technique now routinely used on all types of astrophysical objects.

Lovell Telescope *See* JODRELL BANK OBSERVATORY

Lowell, Percival (1855–1916) Wealthy American diplomat and amateur astronomer from a distinguished family; he built the LOWELL OBSERVATORY and became famous for his observations of Mars and his theory that the planet was inhabited by intelligent beings. Lowell graduated from Harvard in 1876 with a distinction in mathematics. During the 1880s and 1890s, he spent time on diplomatic missions to Japan and Korea. Around 1890 he became enthused by the Mars observations of Giovanni SCHIAPARELLI and the theories that the planet might be inhabited by sentient life forms made popular by Camille FLAMMARION and others. Four years later, largely to determine whether these theories were true, Lowell built a private observatory at Flagstaff, Arizona, just in time for the 1894 opposition of Mars.

For the next two decades, Lowell used his observatory's refractors (the largest a 24-inch (600-mm) completed by ALVAN CLARK & SONS in 1896) to study Mars closely. He concluded that the network of straight, dark lines that Schiaparelli had called *canali* was a global irrigation system, constructed by intelligent beings determined to save their planet from drought by transporting water from the planet's polar icecaps to its arid temperate and equatorial regions. He set out his theories in three popular books – *Mars* (1895), *Mars and its Canals* (1906) and *Mars as the Abode of Life* (1908) – and produced intricate maps of the canal network. Today, the Martian CANALS are regarded as a mere optical illusion and historical curiosity.

Although he was criticized for his bold, even unscientific approach to the question of life on Mars, Lowell left a rich astronomical heritage through his observatory. During the last eight years of his life he initiated a search for a trans-Neptunian planet that led to Clyde TOMBAUGH's discovery of Pluto in 1930. It was Lowell who in 1914 urged Vesto M. SLIPHER to make the observations that led to the discovery of galaxy REDSHIFTS two years later.

Lowell Observatory Observatory founded in 1894 by Percival LOWELL at Flagstaff, Arizona, at an altitude of 2200 m (7300 ft). It is managed by a sole trustee who is a Lowell descendant, while an internal advisory board and director supervise day-to-day operations. It is famous for Lowell's investigations of Mars, Vesto M. SLIPHER's discovery of galaxy redshifts, and Clyde TOMBAUGH's discovery of Pluto. The observatory's first major telescope was a 0.6-m (24-in.) refractor built in 1896. A second site at Anderson Mesa, south-east of Flagstaff, hosts the Ohio Wesleyan University's 1.83-m (72-in.) reflector, operated jointly with Ohio State University.

lower culmination Passage of a CIRCUMPOLAR STAR across the meridian between the pole and the horizon. At this point the star's HOUR ANGLE is exactly 12^h. Circumpolar stars do not set below the observer's horizon and therefore cross the meridian twice during each SIDEREAL DAY, once at lower CULMINATION and once at UPPER CULMINATION, between the pole and the ZENITH.

low-surface-brightness galaxy GALAXY whose surface brightness falls well below those of the familiar luminous spirals and elliptical galaxies, either because it contains very few stars or because its stars are spread over an unusually large area. Such galaxies range from dwarfs to very luminous systems, and may be quite gas-rich. The type example, Malin 1, is a spiral behind the VIRGO CLUSTER; its disk is invisible on typically exposed images. Although difficult to detect, low-surface brightness galaxies may be quite common and account for an important fraction of the total mass in GALAXY CLUSTERS.

L star New spectral class formally added in 1999 to the cool end of the SPECTRAL CLASSIFICATION system (as were T STARS), making the sequence OBAFGKMLT. The class was needed as a result of the ability of new infrared technologies to find cool, red – really infrared – stars. L stars are characterized by deep red colours and no hydrogen lines; metallic hydrides (CrH, FeH) replace the metallic oxides (TiO and VO) of M STARS. TiO is still present in the warmer L types, but it precipitates into solid mineral form and disappears completely in the cooler types. L stars exhibit powerful, remarkably broad resonance lines of the alkali metals sodium, potassium, rubidium and caesium. Temperatures range from 2000 K at L0 down to around 1500 K at L9, and are so low that visual radiation is effectively non-existent (that is, L stars in general have no visual magnitudes, only red and infrared magnitudes).

The masses of L stars range downwards from about 0.08 solar mass, which is near the critical mass required for full hydrogen fusion to operate. The warmer L stars are a mixture of real stars just off the end of class M and BROWN DWARFS. The cooler L stars are all brown dwarfs. Unlike the traditional classes, there are no giant or supergiant L stars because giants and supergiants cannot become so cool.

Like cool M stars, L stars are convective and should possess no magnetic fields (which are thought to be anchored on radiatively stable cores). Also like cool M stars, L stars have been seen to flare, which is the only form of variation known among them.

LTE Abbreviation of LOCAL THERMODYNAMIC EQUILIBRIUM

luminosity Total amount of ENERGY emitted by a star per second in all wavelengths. It is dependent on the radius of the star and its temperature. Luminosity is usually expressed in units of watts.

luminosity class Parameter describing the size of a STAR. It is usually combined with the spectral class to add a second dimension in the Yerkes System. The classes range from I, for a SUPERGIANT, through III, for a GIANT, to V, for a DWARF STAR. There are a number of subdivisions to these to provide a finer description. *See also* SPECTRAL CLASSIFICATION

luminosity function Numerical distribution of stars (or galaxies) among different values of LUMINOSITY or ABSOLUTE MAGNITUDE. Values of the luminosity function are usually expressed as the number of stars per cubic parsec within a given range of luminosity or absolute magnitude.

The value of the luminosity function in the solar neighbourhood increases to a maximum at absolute magnitudes of 14–15, and decreases again at higher (fainter) magnitude values. This implies that faint M-type dwarfs of about one ten-thousandth of the solar luminosity are the most abundant type of star locally.

Luminosity functions can also be plotted for special categories of object, such as stars of particular spectral classes, stellar clusters, galaxies or clusters of galaxies.

luminous blue variable *See* S DORADŪS STAR

Luna Series of 24 spacecraft used (1959–76) by the Soviet Union for robotic exploration of the Moon. The first three were named Lunik. The programme was marked by a number of notable successes. Lunik 3 returned the first views of the lunar farside in 1959 October. Luna 9 became the first spacecraft to achieve a soft landing on the Moon in 1966. Later that year, Luna

10 became the first spacecraft to enter lunar orbit. In 1970 Luna 16 completed the first automated sample return from the Moon and Luna 17 delivered the first roving vehicle, Lunokhod. Approximately 320 g of lunar material were brought back by Luna 16, 20 and 24.

Lunar A Japanese dual-lunar penetrator mission to be launched in about 2002–2003 to study the lunar interior using seismometers and heat-flow probes. The penetrators will be impacted on the nearside and farside of the Moon. The Lunar A spacecraft will first be inserted into a lunar orbit with a low point of 40 km (25 mi) and the penetrators released. Small retro rockets on each penetrator will be fired and the craft will free-fall towards targets located on the nearside close to either the Apollo 12 or 14 landing sites – for comparisons between Lunar A and Apollo data – and a position near the antipodal point of some deep moonquake foci on the farside. The penetrators will impact at an estimated 8000 g and are predicted to penetrate to depths of up to 3 m (10 ft). The mother ship will fly into a 200-km (125-mi) orbit and act as the penetrators' data relay satellite.

Lunar and Planetary Laboratory (LPL) Research and teaching establishment based at the University of Arizona, Tucson, and operated in conjunction with the university's Department of Planetary Sciences (PtyS). The LPL, the oldest institute devoted to planetary science, was founded in 1960 by Gerard KUIPER, and the PtyS in 1973. PtyS/LPL is concerned with the formation and evolution of the Solar System; it conducts spacecraft missions, laboratory-based and theoretical research, and astronomical observations.

lunar eclipse Passage of the Moon through Earth's shadow. Lunar eclipses, which may be total or partial, can occur only at full moon when Sun, Earth and Moon are in line. Since the Moon's orbit is inclined at 5°.15 to that of the Earth, lunar eclipses do not occur every month; they can only happen when full moon occurs at or close to the ascending or descending node of its orbit. The westwards movement – regression – of the nodes means that lunar eclipses occur in 'seasons', separated by an ECLIPSE YEAR of 346.62 days. Lunar eclipses usually precede or follow SOLAR ECLIPSES by a fortnight.

As the Sun is an extended light source, Earth's shadow has two components – a dark, central UMBRA, where the Sun is completely obscured, and a lighter outer PENUMBRA, within which the obscuration is partial. During a lunar eclipse, the Moon passes first through the penumbra, taking about an hour, moving eastwards by its own diameter in this time, to reach the western edge of the umbra. During the penumbral phase, the light of the full moon is only slightly reduced. In some cases, where the Moon is only just too far north or south of the node to pass through the umbra, a PENUMBRAL ECLIPSE is seen.

Following FIRST CONTACT, the Moon's more easterly limb will show a growing dark 'nick' as it begins to enter the umbra more deeply. At a total lunar eclipse, the Moon takes about an hour to become completely immersed in the umbra. To begin with, there is little to see beyond a darkening of part of the Moon, but as the eclipsed portion increases, some colour may become evident. The curved edge of the umbra may show a slight blue fringe. Refraction of sunlight – particularly at the red end of the spectrum – through Earth's atmosphere means that the umbra is not completely dark. Consequently, the Moon does not usually disappear from view; instead, it is dimmed and often takes on a reddish colour, which becomes increasingly apparent as the umbral phase advances. At totality, the Moon – completely immersed in shadow – normally shows red, orange and yellow colours, often with some gradation. Those parts of the Moon closer to the edge of the umbra often appear somewhat brighter and more yellowish than those towards the centre, which are redder.

At the distance of the Moon's orbit, the dark shadow cone of Earth's umbra subtends a diameter of about 9200 km (5700 mi). The duration of totality in a lunar eclipse depends on how nearly centrally the Moon passes through the umbra: a central passage gives a maximum duration of 1 hour 47 minutes. During totality, any lighter border on the Moon will appear to move gradually round from the westerly to the easterly limb. Totality ends when the Moon's easterly limb re-emerges into sunlight, and over the next hour or so the full moon exits the umbra to regain its usual brilliance, thereafter clearing the penumbra.

No two lunar eclipses are ever quite the same, and although some dismiss them as of less interest than solar eclipses – particularly total solar eclipses, where the events centred on totality occur rapidly, in contrast with the more stately progress of a lunar eclipse – they are fascinating to observe. Lunar eclipses also have the advantage of being visible from the entire hemisphere of the Earth where the Moon is above the horizon, rather than being most impressive along a narrow track. The main difference from one event to another is the degree to which the totally eclipsed Moon is darkened, and how colourful it might be.

The level of darkening is often described in terms of the DANJON SCALE. Total lunar eclipses following major volcanic eruptions, which greatly increase the dust load and opacity of the upper atmosphere, can often be notably dark, as was the case with the 1992 December 9–10 eclipse, a year after the Mount Pinatubo eruption in the Philippines. Not only are these events relatively dark, they also lack much of the colour seen at other times. By contrast, eclipses that occur when the upper atmosphere is fairly dust-free are often bright and strongly coloured, an example being the 1989 August 17 event. Lunar eclipse brightness can also be influenced by the amount of cloud in Earth's atmosphere.

The diameter of Earth's umbra is thought by some to change in response to solar activity, perhaps decreasing at solar maximum as the upper atmosphere becomes slightly more extended in response to increased ultraviolet and X-ray emissions. Many observers try to measure the precise extent of the umbra during eclipses by timing the passage of its edge across identifiable lunar features such as craters.

Unless a lunar eclipse is very dark, it is usually quite easy to make out the dark maria, contrasting with the brighter, cratered highlands, with the naked eye. Telescopic observations during totality sometimes suggest that some craters, notably Aristarchus, appear markedly brighter than their surroundings.

lunar meteorite METEORITE that originated on the Moon. There are currently 18 known lunar meteorites, 15 of which have been collected in Antarctica. Several are gabbroic or basaltic in nature, but the majority are anorthositic regolith breccias. The rocks were removed from the Moon by impact events.

Lunar Module Section of an Apollo spacecraft that carried two astronauts on to the surface of the Moon. It comprised two stages, each with its own propulsion

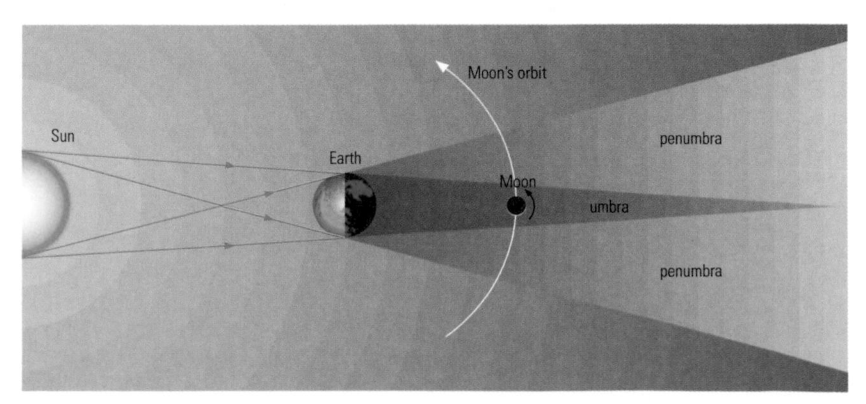

▼ **lunar eclipse** Earth casts a shadow to the right in this diagram. The shadow has two components – a dark central umbra, and a lighter area of partial shadow, the penumbra. If the full moon – opposite the Sun in Earth's sky – passes through the central cone of the umbra, a lunar eclipse will result.

▲ **Lunar Module** Alan Bean begins his descent from the Lunar Module 'Intrepid' to join Charles Conrad on the Moon's surface on the second, Apollo 12, landing in 1969 November.

system. The lower, descent stage, which was used as a launching platform for the ascent stage, was left on the lunar surface. Nine Lunar Modules were flown on manned missions and six of them completed successful landings on the Moon. In the case of the crippled Apollo 13, the power and consumables provided by the Lunar Module enabled the astronauts to return safely to Earth. *See also* APOLLO PROGRAMME

Lunar Orbiter Series of five NATIONAL AERONAUTICS AND SPACE ADMINISTRATION (NASA) spacecraft that were designed for lunar photography in order to aid selection of suitable landing sites for the Apollo missions. They were launched at three-monthly intervals between 1966 August and 1967 August. Images were recorded on fine-grain photographic film which was processed on board and scanned for transmission back to Earth. The highest spatial resolution achieved was about 2 m (80 in.). Almost all of the Moon's farside was imaged, and many scientifically interesting formations on the nearside were photographed. *See also* APOLLO PROGRAMME

Lunar Prospector Third of NASA's low-cost, Discovery-class missions; it was intended to provide global data on key characteristics of the MOON. The 296-kg spacecraft entered lunar orbit four days after its launch on 1998 January 6. From a 100-km (60-mi) circular, polar-mapping orbit, the spacecraft's suite of five science instruments was able to map the entire Moon and update existing data, largely from the Apollo era, which concentrated on the equatorial regions.

One of the most dramatic 'discoveries' was the announcement that the neutron spectrometer had found a definitive signal for water ice at both of the lunar poles. Analysis of the data indicated that a significant quantity of water ice, possibly as much as 3000 megatonnes, was mixed into the REGOLITH at each pole, with a greater quantity existing at the north pole. Further analysis showed that the suspected water ice appeared to be in discrete deposits buried under the lunar soil.

Another intriguing result was the discovery of localized magnetic fields on the Moon, which were located diametrically opposite young, large-impact basins. Gravity measurements detected a number of new mass concentrations, while other data indicated that the Moon's iron core is very small.

Lunar Prospector's six-month-long extended mission began in 1999 January. Swooping to within 10 km (6 mi) of the surface, it provided scientists with much higher resolution data. By the end of this phase, the spacecraft had provided a high-quality gravity map of the Moon, global magnetic field maps, and global absolute abundances of 11 key elements, including hydrogen.

The dramatic finale came on 1999 July 31, when the spacecraft was deliberately deorbited and crashed into a permanently shadowed crater near the Moon's south pole. Scientists hoped that the impact might create a cloud of water vapour, proving once and for all that water exists on the Moon. However, no positive results were obtained.

Lunar Rover (Lunar Roving Vehicle) Battery-powered vehicle that was provided on the last three Apollo missions. It greatly improved astronaut mobility and dramatically increased the area that could be explored. The 220-kg rover was carried to the Moon in a storage bay on the LUNAR MODULE descent stage. Distances covered by each rover were: Apollo 15, 27 km (17 mi); Apollo 16, 27 km (17 mi); and Apollo 17, 36 km (22 mi). *See also* APOLLO PROGRAMME

lunar transient phenomenon (LTP) Short-duration (usually a few seconds to a few minutes) change observed to occur in the appearance of lunar surface features. Most LTPs are flashes or glows of light or dark obscurations of lunar features, or portions of them, documented by visual observers, but some, notably bright flashes of light that may be meteoric impacts, have been imaged by photography or CCDs.

At least three different mechanisms may account for such events – impact flashes, release of gas from a vent, and the electrostatic elevation of dust. Impact events are the most easily distinguished. These are visualized as bright flashes of extremely short duration, and are caused by the vaporization of impactor and target materials. The

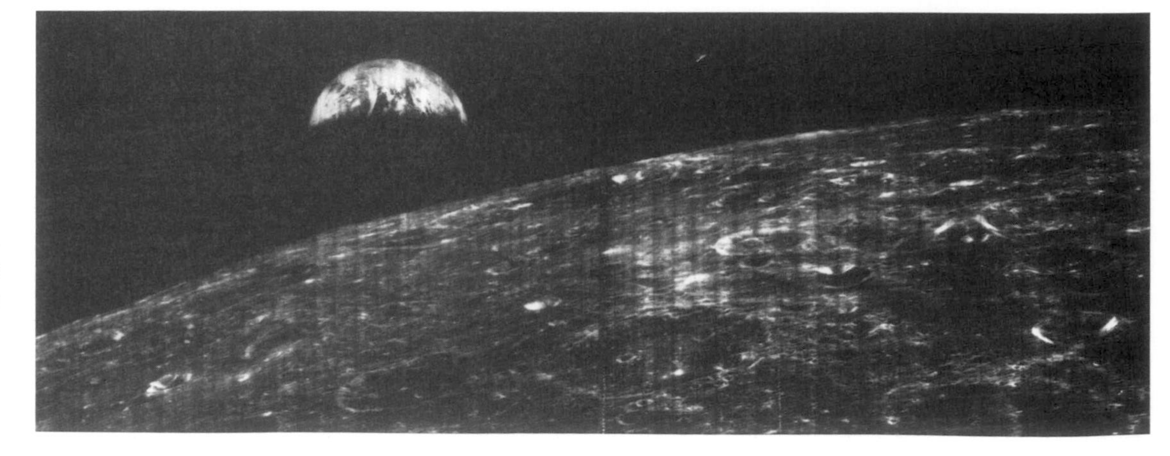

► **Lunar Orbiter** Lunar Orbiter 1 obtained this first-ever view of Earth from a spacecraft orbiting the Moon on 1966 August 23. By obtaining detailed mapping data, the Lunar Orbiter missions paved the way for the Apollo Moon landings.

duration of these events varies with the size of the impactor, but for the small impacts occurring now, it is in the millisecond range. While Apollo seismometers revealed numerous impact events, most were far below the visible detection limit. The larger impacts, however, may be visualized, and have recently been recorded on videotape images of the Moon during strong displays of the LEONIDS.

The release of gas, accompanied by surface dust, may also produce a temporary brightening. This would appear more as a 'glow', and would last longer (until the gas dissipated and the dust settled). One such LTP was spectrographically photographed in 1958 (*see* ALPHONSUS), and later Apollo missions detected radon gas from orbit. A suggested mechanism is that radon gas accumulates in crustal pockets, increasing in pressure until release occurs via a conduit (fault) to the surface. This correlates with the data that such emissions are associated with maria, are locally repetitive, and occur where faults are expected to occur (the three most common areas being ARISTARCHUS, PLATO and Alphonsus).

The electrostatic elevation of dust may produce a transient glow with obscuration of surface features. Under laboratory conditions, irradiation of lunar surface materials with ultraviolet and visible light produced an increase in electrical conductivity of several orders of magnitude. As the TERMINATOR crosses the surface, conditions may occur for sufficient electrostatic charge to levitate dust from the surface. Such events must be uncommon.

lunation Time interval between one new moon and the next. It corresponds to the SYNODIC MONTH of 29.53059 days.

Lundmark, Knut Emil (1889–1958) Swedish astronomer, from 1929 director of Lund Observatory, who studied the nature and distances of galaxies. In 1919 he found evidence suggesting that the Solar System moves in the plane of the Milky Way, concluding from further studies (1924) that this movement was rotational. Lundmark's measurements of their radial velocities (1925) suggested that galaxies are receding at velocities that increase with increasing distance – a conclusion later codified in HUBBLE'S LAW (1929). These insights were made possible by Lundmark's use of novae and supernovae (which he showed to be intrinsically very different) in other galaxies as distance indicators, based on the assumption that they had luminosities similar to their counterparts in the Milky Way.

Lund Observatory Swedish observatory that can trace its roots back to 1672. Located in central Lund, it is operated by Lund University's Astronomy Department. In 1966 it opened a branch station with a 0.6-m (24-in.) reflector at Jävan, 18 km (11 mi) south-east of Lund. Today, Lund Observatory astronomers use facilities such as the NORDIC OPTICAL TELESCOPE and the instruments of the EUROPEAN SOUTHERN OBSERVATORY.

Lunik *See* LUNA

lunisolar precession Combined effect of the gravitational pull of the Sun and the Moon on the Earth, which causes the gradual PRECESSION of the rotational axis. Precession causes the CELESTIAL POLES slowly to drift, describing a circle in the sky of radius 23°.5 (the inclination of the Earth's axis) over a period of some 25,800 years, the effect only being noticeable over a few decades. Lunisolar precession amounts to around 50″.4 per year and the Moon's contribution is about twice that of the Sun because it is that much closer. *See also* PLANETARY PRECESSION

Lunokhod Name given to two Soviet automatic roving vehicles sent to the Moon. All operations were controlled from the Earth. The first was soft-landed by LUNA 17 in north-west Mare Imbrium in 1970 November. The second was carried by Luna 21 and touched down on the eastern Mare Serenitatis in 1973 January.

▲ **Lunar Rover** Astronauts on the later Apollo lunar landings were able to explore greater areas of the surface aboard Lunar Roving Vehicles. Here, Eugene A. Cernan carries out checks on the Lunar Rover on the Apollo 17 mission at the Taurus–Littrow landing site in 1972 December.

Lupus *See* feature article

Luyten, William (1899–1994) Dutch-American astronomer (born Willem Jacob Luyten) who spent most of his career at the University of Minnesota (1931–94), becoming an authority on stellar kinematics and discovering the majority of known white dwarfs. Luyten was famous for his many surveys of PROPER MOTIONS, which he used to derive the stellar LUMINOSITY FUNCTION – this, in turn, allowed him accurately to map the stellar population of the solar neighbourhood. Luyten found that although only 0.5% of the Milky Way's naked-eye stars are less luminous than the Sun, 96% within 2250 l.y. are less luminous. We see these stars only because they are close, but basing a star census on this population would give a misleading notion of the Galaxy's overall stellar make-up.

Luyten used a quantity known as reduced proper motion to relate the apparent and absolute magnitudes of stars to their proper motions, calibrating the mean proper motions using parallaxes. Between 1927 and 1963, he calculated proper motions for 120,000 stars in the southern hemisphere – the Bruce Proper Motion Survey. During this survey he found several hundred white dwarf stars, many SPECTROSCOPIC BINARIES, and UV CETI – the first flare star to be identified. In the 1950s he initiated the National Geographic Society's PALOMAR OBSERVATORY

LUPUS (GEN. LUPI, ABBR. LUP)

Small but not inconspicuous southern constellation, representing a wolf, between Centaurus and Scorpius. Its brightest star, α Lup, is mag. 2.3. γ Lup is a very close binary with bluish-white components, mags. 3.5 and 3.6, separation 0″.8; ξ Lup is another binary with bluish-white components, mags. 5.1 and 5.6, separation 10″.4; and μ Lup is a triple system consisting of bluish-white components, mags. 5.0 and 7.2, separation 23″, the former with a bluish-white close companion, mag. 5.1, separation 1″.1. Lupus contains NGC 5822, an open cluster of more than 100 stars of mags. 9–12; and NGC 5986, a 9th-magnitude globular cluster.

BRIGHTEST STARS

	RA h m	dec. ° ′	Visual mag.	Absolute mag.	Spectral type	Distance (l.y.)
α	14 42	−47 23	2.30	−3.8	B1.5	548
β	14 59	−43 08	2.68	−3.3	B2	523
γ	15 35	−41 10	2.80	−3.4	B2	567
δ	15 21	−40 39	3.22	−2.8	B1.5	510
ϵ	15 23	−44 41	3.37	−2.6	B2	504
ζ	15 12	−52 06	3.41	0.7	G8	116
η	16 00	−38 24	3.42	−2.5	B2.5	493

LYNX (GEN. LYNCIS, ABBR. LYN)

Faint northern constellation, representing a lynx, between Ursa Major and Auriga. It was introduced by Hevelius in 1687, pointing out that one would have to be lynx-eyed to see it! Its brightest star, α Lyn, is visual mag. 3.14 (position RA 09^h 21^m, dec. +34° 24′; absolute mag. –1.0; spectral type M0; distance 222 l.y.); 38 Lyn is a binary with bluish-white components, mags. 3.9 and 6.6, separation 2″.7. Deep-sky objects include NGC 2419, a relatively bright (10th-magnitude) yet unusually remote (310,000 l.y.) globular cluster; and NGC 2683, a 10th-magnitude spiral galaxy.

LYRA (GEN. LYRAE, ABBR. LYR)

Small northern constellation, representing the lyre invented by Hermes and given by Apollo to Orpheus in Greek mythology, lying between Hercules and Cygnus. It is easily recognized by virtue of VEGA, at mag. 0.0, which is the fifth-brightest star in the sky and marks one corner of the SUMMER TRIANGLE. BETA LYRAE (Sheliak) is a multiple system consisting of an eclipsing binary (range 3.25–4.36, period 12.91 days – *see also* BETA LYRAE STAR) and a fainter, mag. 7.2 companion, separation 46″. Another multiple star is ε Lyr (also known as the DOUBLE DOUBLE), which consists of two close binaries whose components are easily seen through a small telescope. RR Lyr (range 7.06–8.12, period 13.7 hours) is the brightest member of a class of pulsating variables (*see* RR LYRAE STAR). Deep-sky objects in Lyra include the RING NEBULA (M57, NGC 6720), a 9th-magnitude planetary nebula; and M56 (NGC 6779), an 8th-magnitude globular cluster. The Lyrid meteor shower radiates from the constellation.

BRIGHTEST STARS

	Name	RA h m	dec. ° ′	Visual mag.	Absolute mag.	Spectral type	Distance (l.y.)
α	Vega	18 37	+38 47	0.03	0.6	A0	25
γ	Sulafat	18 59	+32 41	3.25	−3.2	B9	635

SKY SURVEY (POSS I). Despite the loss of an eye in a tennis accident, Luyten made his discoveries by eye, using a BLINK COMPARATOR device.

Lyman-α forest Set of hundreds of redshifted ABSORPTION LINES that appear to the short-wave side of the Lyman-α EMISSION LINE in the spectra of high-redshift quasars. The quasar back-lights neutral hydrogen gas in the intergalactic medium. Each absorption line of the 'forest' is produced by a discrete filament of some sort that lies between Earth and the distant quasar, and thus has a different REDSHIFT caused by the expansion of the Universe. Absorptions of higher-order Lyman lines and similar systems for other elements are also seen.

Lyman series Series of spectrum lines in the ultraviolet HYDROGEN SPECTRUM resulting from electron transitions down to or up from the lowest energy level (the ground state). The Lyman lines are named with Greek letters: Lyman α connects levels 1 and 2 at 121.6 nm; Lyman β connects levels 1 and 3 at 102.5 nm; Lyman γ connects levels 1 and 4 at 97.2 nm; and so on. The series ends at the Lyman limit at 91.2 nm.

Lynx *See* feature article

Lyot, Bernard Ferdinand (1897–1952) French astrophysicist and instrument designer who worked at Meudon and Pic du Midi Observatories. He invented a coronagraph and a series of monochromatic filters (*see* LYOT FILTER) with transmissions as fine as 0.1 nm (1933–39). He also developed polarimeters to measure the polarization of light reflected by the surfaces of the Moon and planets, and used them to discover that Mars has dust storms and that the lunar surface is largely covered by volcanic basalts. Lyot was the first to use cinematography to capture the motions of solar prominences and to probe the structure of the chromosphere.

Lyot filter Narrow waveband filter used for making observations of the Sun. The filter consists of alternating layers of polaroid sheets and quartz plates. The polarized light undergoes double refraction by the quartz plate and, when realigned by a further polaroid sheet, interference occurs, some wavelengths being cancelled and some reinforced. This results in a number of very narrow (0.1 nm) but widely spaced wavebands, which allows the desired wavelength to be easily isolated. It was devised by Bernard LYOT in 1933.

Lyra *See* feature article

Lyrids METEOR SHOWER active between April 19 and 25, with peak around April 21 or 22. The shower, the radiant of which lies 10° south-west of Vega, usually produces comparatively modest rates, with maximum zenithal hourly rate (ZHR) 10–15. Stronger displays have been seen on occasion, notably on 15 BC March 27 when 'stars fell like rain', and in 1803 and 1922. An outburst on 1982 April 21–22 produced rates of 75–80 Lyrids per hour. Lyrid meteors are quite fast (49 km/s or 30 mi/s), with a reasonable proportion of bright events. The shower is associated with the long-period Comet C/1861 G1 Thatcher.

Lysithea One of the small satellites in JUPITER's intermediate group, discovered by Seth Nicholson (1891–1963) in 1938. Lysithea is about 36 km (22 mi) in size. It takes 259.2 days to circuit the planet at an average distance of 11.72 million km (7.28 million mi) from its centre. Lysithea's orbit has a substantial inclination (near 28°) and moderate eccentricity (0.112). *See also* ELARA

M

M *See* MESSIER NUMBERS

McDonald Observatory Major US optical observatory 720 km (450 mi) west of Austin, Texas, in the Davis Mountains, on the twin peaks of Mount Locke (2070 m/ 6790ft) and Mount Fowlkes (1980 m/6500 ft). It is a facility of the University of Texas at Austin, and operates the 9.2-m (360-in.) HOBBY–EBERLY TELESCOPE and several smaller instruments, all equipped with state-of-the-art instrumentation. These include the 2.7-m (107-in) Harlan J. Smith Telescope, opened in 1968, and the 2.1-m (82-in.) Otto Struve Telescope, dating from 1939. The observatory has one of the first and most productive lunar ranging stations.

Mach, Ernst (1838–1916) Austrian physicist and philosopher. He was the first (1877) to discover the shock waves produced by projectiles moving faster than the speed of sound, and his name is best known in connection with the **Mach number**, the ratio of an object's velocity to the velocity of sound. His philosophy of science freed EINSTEIN from the theoretical restrictions imposed by Newtonian spacetime and helped him develop his general theory of relativity. **Mach's principle** holds that all the inertial properties of a piece of matter are in some way attributable to the influence of all the other matter in the Universe.

MACHO *See* MASSIVE COMPACT HALO OBJECT

McMath–Pierce Solar Telescope World's largest solar telescope, located at KITT PEAK NATIONAL OBSERVATORY, incorporating three separate optical instruments with apertures of 1.6 m (63 in.) for the main telescope and 0.9 m (36 in.) for the two auxiliaries. Unusually, these instruments have all-reflective optics with no windows, lenses or central obscuration, and produce a very high image quality. The McMath–Pierce is operated by the NATIONAL SOLAR OBSERVATORY. Completed in 1962, it is used mainly for solar spectroscopy, polarimetry and imaging, but also for planetary work and observations of comets. It also provides a unique facility for monitoring the Earth's atmosphere.

Mädler, Johann Heinrich (1794–1874) German astronomer who collaborated with amateur astronomer Wilhelm BEER to produce *Mappa selenographica* (1837), the first truly accurate and comprehensive map of the Moon. They constructed their map by making numerous micrometer measurements of feature positions and sizes, using only a 3.75-inch (95-mm) refractor. They also produced the first map of Mars to show its albedo features (1830). From 1840 Mädler directed Dorpat Observatory, Estonia, observing and cataloguing hundreds of double stars with the fine 9.6-inch (245-mm) refractor there.

Maffei Galaxies Two relatively nearby galaxies in Cassiopeia. Lying close on the sky to the plane of the Milky Way, they are almost completely obscured by galactic dust and can only be detected at red or infrared wavelengths. They were discovered in 1968 by the Italian astronomer Paolo Maffei (1926–) as two infrared sources in line of sight close to the bright galaxy IC 1805. Their nature as faint, extended objects with the attributes of nearby galaxies raised the exciting possibility that they were newly discovered members of the LOCAL GROUP.

Maffei 1 (RA 02^h 36^m.3 dec. +59°39′) is now known to be a giant elliptical galaxy 4 million l.y. away on the edge of the Local Group. It has an absolute magnitude of −20.

Maffei 2 (RA 02^h 41^m.9 dec. + 59°36′) lies far beyond the Local Group, 20 million l.y. away, and is an Sb spiral galaxy.

Magellan NATIONAL AERONAUTICS AND SPACE ADMINISTRATION (NASA) mission to map the planet VENUS at high resolution, using synthetic aperture radar (SAR). The spacecraft was launched from the Space Shuttle *Atlantis* on 1989 May 4 and inserted into a near-polar elliptical orbit around Venus, with a periapsis altitude of 294 km (183 mi), on 1990 August 10. Each mapping orbit typically imaged an area 20 km (12 mi) wide by 17,000 km (10,600 mi) long. The raw SAR data were then processed into image strips that were assembled into mosaics.

The mission was divided up into 'cycles'. Each cycle lasted 243 days, the time necessary for Venus to rotate once under the Magellan orbit. The first three cycles (1990 September to 1992 September) were dedicated to radar mapping of the surface. At the completion of this phase, 98% of the surface had been imaged at resolutions better than 100 m (330 ft), and many areas were imaged several times. The fourth and fifth cycles were mainly devoted to obtaining gravity data. During the final stages of the mission, scientists carried out a 'windmill' experiment to study the effects of atmospheric drag on the spacecraft.

Although its primary objectives were to map the surface of Venus and determine its topographic relief, Magellan also collected radar emissivity, radar reflectivity, slope and radio occultation data.

◀ **McMath–Pierce Solar Telescope** The main shaft of the McMath–Pierce Solar Telescope is 152 m (500 ft) long. Light is reflected from a heliostat down the shaft to a 1.5-m (5-ft) mirror, back up to a second mirror and then down to the underground observation room, to form an image of the Sun 85 cm (34 in) across.

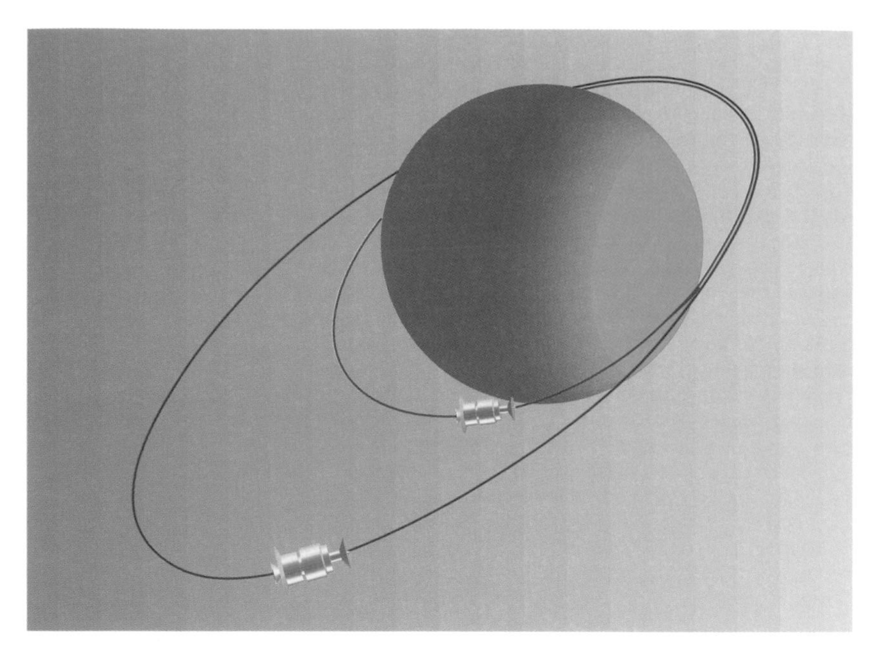

▲ **Magellan** During the first phases of NASA's Magellan mission, its orbit around Venus was highly elliptical (blue), and it used radar to map almost the whole planet. After three years, the orbiter used aerobraking techniques to reach a lower, circular orbit (red); it then mapped Venus' gravity field.

Magellan revealed an Earth-sized planet with no evidence of Earth-like plate tectonics. At least 85% of the surface was found to be covered with volcanic flows, with the remainder marked by highly deformed mountain belts. More than 1000 impact craters and 1100 volcanic features were discovered, as well as sinuous valleys and unique geological structures known as coronae and arachnoids. A strong correlation was found between the gravity field of Venus and the surface topography, suggesting that processes deep in the interior play a major role in influencing the distribution of highlands and lowlands. Radio contact with Magellan was lost on 1994 October 12. It burned away in Venus' atmosphere.

Magellanic Clouds Two small GALAXIES that are the nearest external galaxies to our own. Both clouds are easily visible to the naked eye, appearing like isolated offshoots of the MILKY WAY with apparent diameters of about 6° (the LARGE MAGELLANIC CLOUD) and 3° (the SMALL MAGELLANIC CLOUD). Their distances are estimated to be about 170,000 and 190,000 l.y. respectively. The clouds are sufficiently close to the Earth for detailed observations to be made of stars and nebulae; such observations are not possible for more distant galaxies. For example, the PERIOD–LUMINOSITY RELATIONSHIP for CEPHEID variables was first established in 1912 by Henrietta LEAVITT'S studies of the Small Magellanic Cloud.

Magellan Telescopes Major southern-hemisphere optical astronomy facility consisting of two identical 6.5-m (256-in.) telescopes at LAS CAMPANAS OBSERVATORY in Chile. Its construction was a collaboration between the CARNEGIE OBSERVATORIES (which operate it) and other US universities, including the University of Arizona, whose STEWARD OBSERVATOR Mirror Laboratory fabricated the unusual optics. The spun-cast primary mirrors are steeply curved, with a FOCAL RATIO of only *f*/1.25. The two telescopes can be used independently or in combination, and are named after the astronomer Walter BAADE and the benefactor Landon T. Clay (1926–). They were completed in 2000 and 2002.

Maginus Lunar crater (50°S 60°W), 177 km (110 mi) in diameter, with rim components reaching 4270 m (14,000 ft) above its floor. An ancient crater, it has been severely modified by impact erosion, having walls nearly level with the surrounding terrain. The rim is difficult to identify in places, while the EJECTA blanket has been completely eroded away. The walls are deeply incised by impact scars. The floor contains several small central peak remnants and later impact craters.

magma Molten rock, predominantly silicate in composition; upon solidification it yields IGNEOUS ROCKS.

magnetic field One of the fundamental forces of nature. Magnetic flux density, symbol B, is measured in units of tesla (T). Magnetic field strength, H, is related to magnetic flux density by the magnetic permeability, μ, of the medium containing the magnetic field, with $B = \mu H$. It is measured in A/m. On Earth a weak magnetic field exists capable of swinging a compass. In astronomical objects the field can be more than 10^{12} times stronger, and it then can control the motions of gases and the shape of objects. Planetary magnetic fields are often discussed in units of Gauss (G); 1 G = 0.0001 T.

magnetic monopole Elementary particle that contains only one pole of a magnetic field. The existence of magnetic monopoles was predicted by symmetry considerations in elementary particle models and they were thought to be a single pole of a magnetic field. The fact that magnetic monopoles have never been detected in the Universe presented a minor problem for the BIG BANG THEORY. The theory of INFLATION, if correct, eases the constraints on the detection of magnetic monopoles and explains why they have yet to be seen.

magnetic reconnection Mechanism for the exchange of energy from a magnetic field to a plasma, quoted as a possible acceleration mechanism in a wide variety of astrophysical phenomena, from planetary MAGNETOSPHERES, SOLAR FLARES and CORONAL MASS EJECTIONS, to ACCRETION DISKS and ACTIVE GALACTIC NUCLEI. Two regions of space or astrophysical plasma containing magnetic fields with different orientations are generally separated by a thin current layer; for example, the magnetopause separates a planetary magnetosphere from the SOLAR WIND. The two regions are thus not generally able to mix. Magnetic reconnection is thought to occur on such a current layer if the fields on either side are close to antiparallel. Under these circumstances, the fields on either side break and reconnect with their counterparts on the other side of the boundary. A magnetic link is created through the boundary, which allows the mixing of plasmas from each side along the reconnected field lines. In addition, magnetic energy is liberated from the system, and appears as thermal and kinetic energy of the plasmas on the reconnected field lines. The actual physical process involved in magnetic reconnection is not well understood. However, observations of plasmas at energies compatible with the reconnection process are widespread within the terrestrial magnetosphere, providing strong circumstantial evidence of its occurrence within astrophysical plasmas.

magnetic star Variety of star that has a strong magnetic field, well beyond the typical level of the Sun. The most common examples are the magnetic AP STARS, which have odd chemical compositions (enhanced silicon, strontium and rare earths) and starspots; their related fields run from a few hundred into the tens of thousands of gauss (the Sun's field has a strength of 2000–4000 gauss over active regions).

Two classes of degenerate stars are strongly magnetic. A small percentage of DA (hydrogen-rich) WHITE DWARFS have fields in the megagauss (or even hundreds of megagauss) range. These stars tend to be relatively massive and are suspected to be the evolved descendants of the Ap stars. NEUTRON STARS are by their nature even more magnetic. Typical field strengths for newly born pulsars are a million million (10^{12}) gauss. A small percentage of neutron stars, suspected to derive from higher mass supernovae, the 'magnetars', have fields a hundred times greater, up to 10^{14} gauss.

magnetohydrodynamics (MHD) Study of the interactions between a PLASMA and a MAGNETIC FIELD. The science of magnetohydrodynamics was founded largely by the pioneering work of Hannes Olof Gösta ALFVÉN. A plasma is an ionized gas. It contains free electrons and positively charged nuclei. In most astronomical plasmas, the nuclei will be predominantly free protons. A plasma may also contain a proportion of neutral atoms, but their behaviour, in particular their motions, will be governed by that of the plasma as a whole. The plasma is highly electrically conducting so that if it moves across a magnetic field, an electric current is induced within it in exactly the same way that a current is induced in the copper wire of a dynamo moving through its magnetic field. The electric current in the plasma sets up its own magnetic field, which interacts with the original magnetic field in such a way that the relative motion between the field and plasma is opposed. The interaction may be sufficient to halt almost completely the relative motion of the plasma across the magnetic field lines. The magnetic field is then said to be 'frozen-in' to the plasma. The movement of the plasma is then constrained to follow the magnetic field lines or to move with the magnetic field if it should change; or the magnetic field is dragged with the plasma as it moves.

Most of the material in the Universe is in the form of plasmas; therefore, magnetohydrodynamical behaviour is involved wherever there is also a magnetic field. On the Sun, for example, the shapes of both quiescent and active PROMINENCES clearly show linkage to magnetic fields; SUNSPOTS are cool and appear dark because the magnetic field within them inhibits convective transfer of heat; solar FLARES arise through magnetic field–plasma interactions, although there is still some uncertainty in understanding the exact process. There is also some uncertainty in exactly how the solar sunspot cycle arises, but it is again clearly linked to magnetic field interactions. Other magnetohydrodynamic behaviour occurs in the formation of planetary MAGNETOSPHERES, within ACCRETION DISKS and STELLAR WINDS. Oscillations in the plasma and/or magnetic field produce ALFVÉN WAVES. These move at the Alfvén speed

$$\sqrt{\frac{B^2}{\mu\,\rho}}$$

where B is the magnetic field strength, ρ is the plasma density and μ is the magnetic permeability of the plasma. Such waves may be involved in transferring energy out to the solar CORONA and are thus involved in maintaining its enormously high temperature.

magnetopause Boundary separating a planetary MAGNETOSPHERE from the external SOLAR WIND. The magnetopause is a thin sheet of electrical current, only a few thousand kilometres thick. It is a crucial boundary at which the exchange of energy and matter between the solar wind and the magnetosphere occurs. This coupling is particularly strong at times when the solar wind magnetic field is directed antiparallel to the planetary magnetic field just inside the boundary, a situation probably caused by MAGNETIC RECONNECTION at the magnetopause. *See also* SPACE WEATHER

magnetosphere Region surrounding a magnetized planet that is occupied by the planetary magnetic field; it acts to control the structure and dynamics of plasma populations and ionized particles within it. A magnetosphere can be considered as the magnetic sphere of influence of the planet. The size of the magnetosphere is controlled by the interaction of the planetary field with the SOLAR WIND, which compresses the upstream side while dragging the downwind side out into an extended magnetotail, resembling a windsock. The magnetosphere represents an obstacle to the supersonic solar wind flow from the Sun, with the solar wind having to flow around the magnetosphere. To do this, the wind is slowed, deflected and heated at a BOW SHOCK, standing in the flow upstream of the magnetosphere. The shocked solar wind plasma and associated magnetic field downstream of the bow shock is known as a magnetosheath. This region is separated from the magnetosphere proper by a layer of electrical current known as the MAGNETOPAUSE. In the solar direction, the magnetopause occurs at the point where the pressure associated with the planetary magnetic field balances the pressure of the impinging solar wind flow. As the solar wind pressure is highly variable, these boundaries expand outwards and contract inwards in response to these variations. In the terrestrial system, however, the subsolar points on the bow shock and magnetopause stand at average distances of 15 Earth radii (R_E) and 11 R_E towards the Sun respectively. On the nightside, the deflected solar wind flow effectively creates a vacuum, which the planetary magnetic field expands to fill. MAGNETIC RECONNECTION occurring on the dayside magnetopause results in magnetic flux being peeled from the dayside and added to the nightside region. These two effects create the magnetospheric tail, or magnetotail. In the terrestrial case, this feature is approximately cylindrical, with a diameter of order 50–60 R_E and a length that extends many thousands of Earth radii.

A planetary magnetosphere contains a number of regions with distinct plasma or energetic charged particle populations. Closest to the planet, these regions include the RADIATION BELTS (the VAN ALLEN BELTS at the Earth) and the PLASMASPHERE. The radiation belts contain trapped energetic particles captured from the solar wind or originating from collisions between upper atmosphere atoms and high-energy COSMIC RAYS. The plasmasphere is populated by the polar wind, an upflow of plasma from the upper IONOSPHERE. Farther out, the PLASMA SHEET occupies the central portion of the magnetotail and contains a hot plasma of solar wind origin. Each of these regions, indeed the magnetosphere as a whole, can be extremely dynamic, especially during times of enhanced solar wind–magnetosphere interaction. Magnetic storms and MAGNETOSPHERIC SUBSTORMS are disturbances that cause global restructuring of these magnetospheric regions and significant energization of the particle populations within them. In addition, there is significant coupling between the magnetosphere and the ionosphere. At the Earth this coupling is achieved by the flow of large-scale electrical currents from the magnetosphere through the polar ionosphere, which results in heating of the ionosphere, particularly at disturbed times. It is also associated with spectacular displays of the AURORA.

▼ Magellanic Clouds The Large Magellanic Cloud in Dorado is the second closest external galaxy. The LMC and SMC are joined to the Milky Way by the Magellanic Stream, an arc of neutral hydrogen gas thought to be dragged from the smaller galaxies through tidal interactions.

Each of the other magnetized planets, Mercury, Jupiter, Saturn, Uranus and Neptune, is known to possess a magnetosphere. The magnetic field of Mercury is relatively weak (only one-hundredth of the terrestrial field), and thus it has a very small magnetosphere. However, it still provides a definite interaction with the solar wind, with a well-defined bow shock and magnetopause. MARINER 10 observations suggest that magnetospheric substorms occur on timescales of minutes, rather than over one to two hours as occurs at Earth. Mercury has no significant atmosphere or ionosphere, although, curiously, these regions are known to be important for substorms in the terrestrial magnetosphere. The planet itself also occupies a large fraction of the magnetosphere, such that the regions that would constitute the radiation belts lie below the surface.

The Jovian magnetosphere is the largest in the Solar System: if it could be visualized from the Earth, it would be seen to subtend a cross-section equivalent to the Moon's diameter in the sky. It is extremely variable in size, extending to distances of between 50 and 100 R_J in the direction of the Sun as a result of the changes in the solar wind. The magnetotail of Jupiter is so large that it extends over more than 5 AU and can interact with Saturn. The high-energy particles in Jupiter's magnetic field also form radiation belts, which are 1000 times more intense than those of the Earth. Seven of Jupiter's innermost satellites and the rings are in this hostile region. These satellites are constantly bombarded by the high-energy particles, which erode the surfaces and alter their chemistry. In addition, the structure of the Jovian magnetosphere is influenced by the moon IO, which adds plasma to the inner regions by volcanic processes, creating a cloud of sodium, potassium and magnesium, known as the Io torus, stretching around Io's orbit. The rapid rotation of Jupiter accelerates plasma from the torus to high velocities, and the resulting centrifugal force distorts the magnetosphere outwards, in its equatorial regions, to form the Jovian magnetodisk. Io and Jupiter are also connected by a magnetic flux tube, which carries a current of order 5 million amps across a potential difference of 400,000 volts. This electrical energy plays an important role in the local heating of the surface of Io and the generation of the volcanic activity. Jupiter emits waves at radio frequencies, corresponding to decametric and decimetric regions of the electromagnetic spectrum, which are greater in intensity than any extraterrestrial source other than the Sun. The interactions between the charged particles and the Jovian atmosphere produces polar aurorae, which are about 60 times brighter than their terrestrial counterparts.

The Saturnian magnetosphere is intermediate in size between those of Jupiter and the Earth, and, similar to those, it is sensitive to variations of the solar wind. The VOYAGER 1 spacecraft crossed the bow shock five times, at distances ranging from 26.1 to 22.7 R_S. A major difference between the Saturnian magnetosphere and that of Jupiter is the concentration of highly electrically charged dust in the ring plane. It is possible that this material is the source of the high-energy discharges in the rings. At Saturn the large system of rings and many of the satellites have the effect of sweeping a path through the charged particle environment. The rings and the satellites are very efficient in absorbing the protons in the magnetosphere. The Saturnian magnetosphere is a major target of the CASSINI mission, due to arrive at Saturn in 2004.

Uranus has an unusual magnetosphere in which the dipole magnetic field axis is tilted by an angle of 60° with respect to the planetary rotation axis, with a centre that is offset by 0.3 Uranus radii. At the time of the Voyager 2 encounter the Uranian magnetosphere extended 18 R_U, and the magnetotail had a radius of 42 R_U at a distance of 67 R_U. The extreme tilt of the magnetic axis, combined with the tilt of the rotational axis, causes the field lines in the Uranian magnetotail to be wound up into a corkscrew configuration. Voyager 2 found radiation belts at Uranus of intensity similar to those at Saturn, although they differ in composition. The moons Miranda and Ariel modulate the flux of the trapped radiation. It is possible that this radiation environment is responsible for the creation of the unusually dark material found on the satellites and rings of Uranus. There are also radio emissions from the magnetosphere of Uranus, although weaker than those from Saturn. A further magnetospheric interaction with the atmosphere, which is not known on Earth, is called the electroglow. This is only seen on the dayside of Uranus, where the aurorae are not observed.

Neptune's magnetic field is tilted 47° from the planet's rotation axis, and is offset at least 0.55 R_N from the planet's centre. As a result of this unusual arrangement, Neptune's magnetosphere goes through dramatic changes as the planet rotates in the solar wind. During each rotation, the magnetosphere moves from an orientation similar to that at Earth to one in which the south pole points head on into the solar wind, and then back again. The pole-on orientation of the magnetic field was observed when Voyager 2 flew into the southern cusp of Neptune's magnetosphere in 1989. The spacecraft subsequently remained in the magnetosphere long enough to observe two of these rotation cycles. Voyager 2 detected periodic radio emissions generated in the magnetosphere. These waves provided the first accurate estimation of the rotation rate of the planet's interior. Voyager 2 also detected weak aurorae in Neptune's atmosphere. Because of Neptune's complex magnetic field, the aurorae occur over wide regions of the planet, not just near the planet's magnetic poles.

Finally, it is possible for unmagnetized bodies to exhibit magnetospheric-like behaviour. For example, the ionosphere of Venus and cometary comae also provide obstacles to the solar wind flow. The slowing of the flow in the vicinity of the body creates a draping of the solar wind magnetic field around the obstacle, thus creating an induced magnetotail. Regions of remnant surface magnetization on the Moon and Mars also interact with the solar wind on a small scale, thus creating the 'mini-magnetospheres' detected, for example, by the Mars Global Surveyor and Lunar Prospector missions.

▼ **magnetosphere** The Earth's moving iron core gives it an active magnetic field that extends well beyond the planet. The arrows indicate the direction of the magnetic field. The magnetosphere protects us from solar radiation.

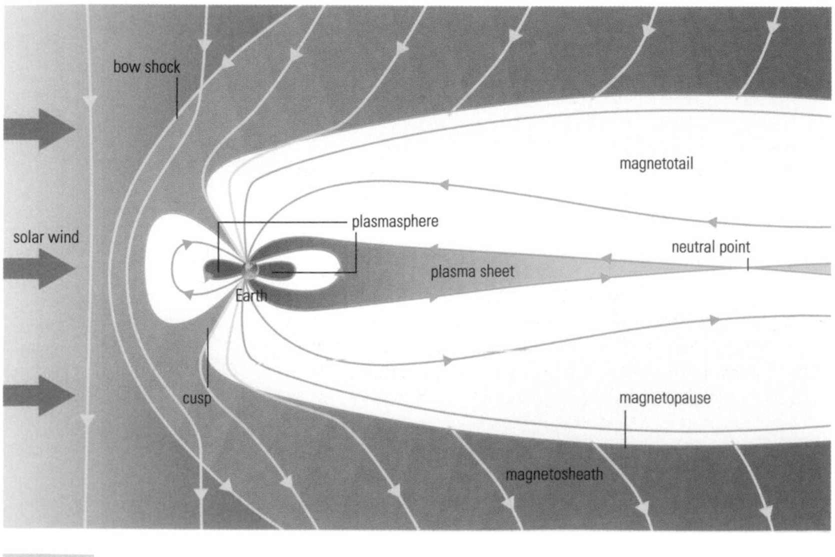

magnetospheric substorm Major reconfiguration of the terrestrial MAGNETOSPHERE resulting in an explosive release of energy stored in the magnetotail and subsequent deposition of this energy into the polar IONOSPHERE, RADIATION BELTS and downstream SOLAR WIND.

The 'growth phase' of a magnetospheric substorm begins with a period of enhanced coupling between the solar wind and the magnetosphere, which may last several hours. MAGNETIC RECONNECTION at the dayside MAGNETOPAUSE results in a build-up of magnetic energy within the magnetotail, reaching levels of 10^{16}–10^{17} Joules. A poorly understood instability occurring within the magnetotail PLASMA SHEET results in this energy being rapidly released at the onset of the substorm 'expansion phase'. This energy is converted into fast plasma jets within the magnetotail plasma sheet and is also dissipated by a shearing off of a large portion of the tail plasma sheet to form a plasmoid, which is ejected away from the Earth and out into the solar wind. In the near-Earth region, energy is deposited as an injection of energetic particles into the radiation belts. A large, substorm-associated current connects the magnetosphere to the ionosphere, and flows across the ionosphere at auroral latitudes. This current dissipates energy by heating the ionosphere, and it is also associated with a major expansion of the nightside AURORA, both in the poleward and east–west directions. Spectacular high-latitude auroral displays are thus observed from the ground during substorms. Typically, after about 30–60 minutes activity dies away, and the magnetosphere and ionospheric systems relax to their pre-substorm state during the 'recovery phase'. Magnetospheric substorms are thought to occur at a much faster rate at Mercury, and probably also occur in the other magnetized planets.

magnification Factor by which an optical system increases the apparent size of an object. It is a measure of the increase in the angle subtended by the object at the observer's eye. It is usually designated by '×' used before or after the magnification power; for example, eight times magnification would be written as ×8 or 8×. In the specification of a pair of binoculars two numbers are usually given, separated by ×; the first is the magnification. For example, 10×50 binoculars have a magnification of ten times and an entrance aperture 50 mm in diameter.

The magnification of a telescope can be calculated by dividing the FOCAL LENGTH of the OBJECTIVE by the focal length of the EYEPIECE. A telescope with a focal length of 1000 mm would give a magnification of 50× when used with an eyepiece of focal length 20 mm.

magnitude Brightness of an astronomical body. APPARENT MAGNITUDE 1 is exactly 100 times brighter than magnitude 6, each magnitude being 2.512 times brighter than the next. Magnitudes brighter than 0 are minus figures, thus Sirius is −1.4, and the Sun is −26.8. The faintest objects yet photographed are about magnitude 26. *See also* ABSOLUTE MAGNITUDE; BOLOMETRIC MAGNITUDE; PHOTOGRAPHIC MAGNITUDE; PHOTOVISUAL MAGNITUDE; VISUAL MAGNITUDE

main-belt asteroid One of the many ASTEROIDS that occupy the region of space between Mars and Jupiter. Most such bodies are dynamically stable, and they are believed to have remained in that part of the Solar System since their formation more than 4.5 billion years ago. The main-belt asteroids represent chunks of solid material, mostly rock and metal, which failed to form a major planet because of the effect of the gravitational perturbations of Jupiter. Their total remnant mass is small – only about 5% the mass of the Moon.

main sequence Region of the HERTZSPRUNG–RUSSELL DIAGRAM where most stars are found. It runs diagonally from top left (high temperature, high luminosity stars) to bottom right (low temperature, low luminosity stars). Stars spend most of their lifetime on the main sequence, and a main-sequence star is one in which energy is primarily produced from the fusion of hydrogen into helium in its core.

The original line formed by stars shortly after their ignition is called the ZERO-AGE MAIN SEQUENCE. As compositional changes alter the internal structures of these stars, the points that represent them on the HR diagram will move away from the main sequence and trace out an evolutionary track on the diagram *(see also* STELLAR EVOLUTION).

Main-sequence stars are the most common type of star in the GALAXY: they constitute some 90% of the approximately 10^{11} stars in the Galaxy and contribute about 60% of its total mass. From a theoretical point of view main-sequence stars are common because stars spend most of their evolutionary lives on the main sequence.

The time for which a star remains on the main sequence is related to the ratio of its mass to its luminosity. Using the MASS–LUMINOSITY RELATION, this gives rough main-sequence lifetimes proportional to $M^{-3.7}$ for solar-type stars and $M^{-0.6}$ for very high-mass stars (where M is the mass of the star). Very low-mass stars are fully convective and can continue core hydrogen fusion for much longer than its mass–luminosity ratio would suggest, giving a star of mass one tenth of a solar mass a main-sequence lifetime of around 10^{13} years.

The internal structure of a main-sequence star consists of a core depleted in hydrogen (to a degree determined by the time it has spent on the main sequence; that is, the time since NUCLEAR REACTIONS first started in its interior), surrounded by a hydrogen-rich envelope. In stars heavier than the Sun the principal nuclear reaction network is the CARBON–NITROGEN–OXYGEN CYCLE and the core is mixed by convection. In lower-mass stars the nuclear energy generation comes from the PROTON–PROTON REACTION and there is no convection in the core. An outer convection zone, extending in from the surface of the star, increases in depth for lower-mass stars. For stars of SPECTRAL TYPE A0 (about 3 solar masses) there is little or no convective envelope; at the mass of the Sun the convection zone reaches in about a quarter of the way to the centre; at about 0.3 solar mass the zone reaches the centre.

The relatively short lifetimes of the hottest main-sequence stars result in their being very rare in the Galaxy, whereas the lifetimes of the lowest mass (spectral type K and M or RED DWARFS) are so long that none has evolved from the main sequence since the Galaxy was first formed. This strong lifetime dependence, together with the fact that the processes of star formation produce far more low-mass than high-mass stars, results in the lower-mass dwarfs being much more populous.

Most main-sequence stars have the same abundances of hydrogen and helium as the Sun, and have abundances of the heavier elements which are within a factor of two of those in the Sun. For stars with masses less than 0.08 times that of the Sun, the core temperatures do not become high enough to start hydrogen-burning nuclear reactions; as a result the main sequence terminates at stars with surface temperatures of about 2500 K. However, lower-mass stars are formed, probably in even greater numbers than the M red dwarfs, and become BROWN DWARFS.

An upper limit to the masses of main-sequence stars is determined by the increase of radiation pressure in their interiors: above about 150 solar masses the outward force exerted by radiation exceeds the attractive force of gravity, preventing such a star from being formed, or making it unstable with a short lifetime if it does form (*see also* STELLAR MASS). At various places along the main sequence variability of luminosity is found, such as in Beta Cephei, Delta Scuti, oscillating Ap and flare stars *(see also* VARIABLE STARS).

As well as the intrinsic properties of main-sequence stars, their space velocities show some interesting correlations with other stellar properties. The dispersion of their velocities increases towards lower masses. This is a consequence of the increase in average age of stars at lower masses: stars of all masses are probably formed with approximately the same dispersion in space velocities, but the mechanisms that act to increase the dispersion have longer to work on those stars that remain as main-sequence stars for the longest time. Thus O and B stars evolve off the main sequence before their velocity dispersions have changed much from their initial values, but among the M main-sequence stars there is a spread of ages from that of the Galaxy itself down to the most recent ones formed. Over long periods of time, encounters with massive interstellar clouds (and, to a lesser extent, other stars) change the orbits of stars in the Galaxy, increasing their velocity dispersion. Thus the oldest M main-sequence stars have greatly increased dispersions. *See also* HIGH-VELOCITY STARS; RUNAWAY STARS; STELLAR POPULATIONS

► **mare** Mare Humorum is a lava-filled basin on the Moon, some 370 km (230 mi) in diameter. It has fewer impact craters than the areas around it, indicating its relative youth.

major planet Name given to the nine main planetary members of the Solar System: Mercury, Venus, Earth, Mars, Jupiter, Saturn, Uranus, Neptune and Pluto. The term distinguishes them from the ASTEROIDS or minor planets. There is some debate as to whether Pluto, because of its size, warrants classification as a major planet.

Maksutov, Dmitri Dmitrievich (1896–1964) Soviet optician and telescope-maker, born in the Ukraine, who designed the type of reflecting telescope now named after him. He organized a laboratory of astronomical optics at the State Optical Institute in Leningrad (now St Petersburg), where he designed and then built (1941) the first of several MAKSUTOV TELESCOPES. The meniscus-shaped optical surface employed by Maksutov was later used to build ultra-fast, ultra-wide-angle systems. Maksutov also was a principal designer of the SPECIAL ASTROPHYSICAL OBSERVATORY's 6-m (236-in.) reflector, at one time the world's largest telescope.

Maksutov telescope Modification of the SCHMIDT TELESCOPE design, using a spherical corrector plate whose convex side faces the PRIMARY mirror. It was devised independently by Albert Bouwers (1893–1972) in 1940 and Dmitri MAKSUTOV in 1944. The design has the advantage of being cheap and easy to produce, making it popular for short-tube amateur telescopes. Silvering of a spot in the centre of the corrector plate allows visual and photographic use in a CASSEGRAIN TELESCOPE configuration. Maksutov telescopes are free from COMA and ASTIGMATISM, and almost free from CHROMATIC ABERRATION, while their enclosed tubes obviate problems with internal air currents.

mantle Intermediate region of a planetary body that has experienced DIFFERENTIATION; the mantle is the volume between the CORE and CRUST. Earth's mantle extends from the base of the crust, at depths of a few tens of kilometres, to the the core–mantle boundary at about 2700 km (1700 mi); it comprises about two thirds of the mass of the planet. The mantle is divided into layers, which increase in density with depth, probably due to changes in composition and pressure-induced changes in crystal structure. Although it is mostly solid, material in Earth's mantle moves slowly (a few centimetres per year) because of convection driven by internal heat; it is this movement that drives continental drift. The other terrestrial planets and the Moon also have silicate mantles.

The term mantle is also used for the thick layer of ice that overlies the core of a differentiated icy body.

many-body problem (n-body problem) Problem in CELESTIAL MECHANICS of finding how a number (n) of bodies move under the influence of their mutual gravitational attraction. The nine planets plus the Sun form a ten-body problem, which is far too complicated to tackle using traditional analytical methods. Instead, the problem of determining Saturn's orbit can be treated as a series of THREE-BODY PROBLEMS: Saturn orbiting the Sun perturbed by Jupiter, Saturn orbiting the Sun perturbed by Uranus, and so on. One instance in which nine bodies (Pluto excluded) are considered together is in the determination of the secular PERTURBATIONS of their orbits. This problem is simplified by ignoring the rapid longitude motions of the planets, and regarding all ECCENTRICITIES and INCLINATIONS as small. With the advent of powerful computers it is now perfectly feasible to tackle the ten-body problem by numerical integration of their orbits, and this is the way that the current planetary ephemerides are generated.

Maraldi, Giacomo Filippo (1665–1729) Italian cartographer and astronomer, also known as Jacques Philippe, who worked at Paris Observatory (1687–1718). The nephew of Giovanni Domenico CASSINI, he collaborated (1700–1718) with Jacques CASSINI to determine the Paris longitudinal meridian. Maraldi was also an astute planetary observer. In 1704 he noticed the Martian polar caps and later mapped their variations, concluding that they were made of some kind of ice. He is known to historians as Maraldi I to distinguish him from Maraldi II – his nephew, the astronomer Giovanni Domenico Maraldi (1709–88), also known as Jean Dominique.

mare Dark plain on the MOON; the name comes from the Latin meaning 'sea', and refers to the belief that the dark patches on the Moon were ancient oceans. The maria are now known to represent lava flows. They stand as one of the major geologic units of the Moon, the other one being the HIGHLANDS. Because the maria and highlands have a different mineral composition, they stand out from one another: the highlands are light, while the maria are dark because of their lower reflectance of sunlight.

Lunar lava formed in the Moon's mantle, where radioactive elements decayed producing heat. When that heat reached a certain level, it melted the surrounding rocks (olivine, pyroxene), producing magmas that were low in silica but high in magnesium and iron (called 'mafic'). These magmas, lower in density than the surrounding rocks, began to rise towards the surface. Upon nearing the crust, the magmas tracked up the deep faults that had been created by large impacts.

The largest impacts on the Moon produced BASINS, which are immense structures, generally 300–1200 km (190–750 mi) in size, characterized by having two or more ring-like walls surrounding a flat central region. More importantly, these impacts deeply fractured the bedrock, thus providing conduits for ascending magmas and topographically low regions for lavas to pond. Thus, the maria are generally found in basins, and are usually restrained by one of the basin's rings. The extent of flooding was variable, however, and so produced a variety of shapes for the final basin-mare form: (1) the lava may have been restrained by a basin ring (for example, the ring around Mare CRISIUM); (2) the lavas may have flooded over an inner ring(s) producing a 'circular' mare ridge system; and (3) the lavas may have broken through the outer ring to flood adjacent low-lying areas (for example, Mare IMBRIUM, Mare SERENITATIS).

The set of features that is especially associated with basin rings is the circular mare ridge system, within and roughly concentric to a basin wall. When lava floods a basin, it subsides by compaction and mass loading. As the lava sags, it generates compressive forces that are expressed where the lava thins, such as over basin rims, and there the mare ridges form. Thus, these circular mare ridges mark the underlying basin rings, and they

M

are often the only method available for identifying their location. Examples are visible in Mare Serenitatis and Mare HUMORUM. Significantly, the lavas failed to cover the highest peaks in two of the Imbrium Basin's inner rings (for example, Mons Piton and Mons PICO). Thus, here the inner rings are marked by both mare ridges and individual peaks.

One other volcanic feature often associated with the maria are dark mantling deposits. These are considerably darker areas located around the edges of certain maria. Apollo 17 visited one such site in Mare Serenitatis, where astronauts found glass beads in the regolith. This kind of glass is produced by a process called fire fountaining, which occurs when magma containing dissolved gasses ascends. As the confining pressure is released, the gasses come out of solution. If there is enough gas in the fluid magma, it will both disrupt the magma and act as a propellant, lobbing magma droplets high above the surface. Here the droplets rapidly cool, forming a volcanic ash composed of small glass beads. As the lunar variety of glass is high in certain metals (such as titanium), it assumes a dark colour.

Maria Mitchell Observatory Small observatory on Nantucket Island, off the coast of Massachusetts, founded in 1902 to commemorate Maria MITCHELL, the first woman professor of astronomy in the United States. It is now operated by the Maria Mitchell Association, which aims to increase knowledge and public awareness of the Universe and the natural world. The Association supports both public astronomy and astronomical research, and since 1997 has presented the annual Maria Mitchell Women in Science Award.

marine chronometer Mechanical clock capable of keeping time accurately for long periods on board a small ship at sea. While latitude is easily found by observing the Sun or stars, finding longitude at sea requires an accurate knowledge of the time. Although land-based pendulum clocks had achieved the necessary precision by the 17th century, they could not be used at sea because the motions of a ship disturbed the pendulum too much, not to mention variations in temperature and humidity.

In 1714 the British Parliament passed the Longitude Act, which offered a prize of £20,000 for a satisfactory solution to this problem. A useful marine chronometer which depended on a balance-wheel escapement was ultimately devised by John HARRISON. Thereafter, the development of chronometers proceeded rapidly. Chronometers remained essential until the general availability of reliable radio time signals in the 1920s and 1930s. Nowadays, most practical navigation makes use of the GLOBAL POSITIONING SYSTEM (GPS) satellites.

Mariner spacecraft Series of ten US planetary probes, of which seven were successful. Mariner 2 was the first probe to reach VENUS (1962). Mariner 4 achieved the first successful flyby of MARS, sending back pictures of a cratered surface (1964). Mariner 5 made a close flyby of Venus in 1967. Mariners 6 and 7 extended the mapping of Mars, including the south polar region (1969). Mariner 9 went into orbit around Mars, becoming the first spacecraft to be placed in orbit around another planet (1971). Mariner 10 flew past Venus before becoming the first spacecraft to study MERCURY.

Marius, Simon (1570–1624) German mathematician and astronomer whose career was characterized by controversy. In 1608 he learned of Hans LIPPERSHEY's design for a telescope and quickly made one from spectacle lenses. He was the first to use a telescope to observe the ANDROMEDA GALAXY (1612), but could not resolve its spiral structure. In 1614 he claimed to have discovered the four brightest satellites of Jupiter in 1609 November. Because of the five-year delay, he was accused of falsifying his observations by GALILEO, whom the scientific community had credited for this discovery. However, reliable records show that Marius observed the Galilean moons at least as early as December 1610.

Markab The star α Pegasi, visual mag. 2.49, distance 140 l.y., spectral type A0 III. Its name is the Arabic word for 'saddle'.

Markarian galaxy GALAXY found to show excess ultraviolet radiation in a survey carried out at the Byurakan Observatory under the direction of the Armenian astronomer Benjamin Eghishe Markarian (1913–85). Markarian galaxies included the first large examples of SEYFERT GALAXIES, as well as many STARBURST GALAXIES and some QUASISTELLAR OBJECTS and BL LACERTAE OBJECTS.

Mars Fourth major planet from the Sun, popularly known as the Red Planet. Mars is intermediate in size between the Earth and Mercury. Mars' coloration is a function of widespread regions of reddish dust, which on occasions may be raised high into the tenuous Martian atmosphere by winds. A great dust storm was observed at close quarters in 1971 by the American spacecraft MARINER 9 as it approached the planet and eventually went into orbit around it. This storm reached global proportions, taking months to abate. On the surface the dust collects into vast sand-sheets and dune fields.

Telescopically, Mars shows bright regions at the poles, sometimes bright areas at the LIMB or TERMINATOR, and various dark markings. Most early observers believed the dark features to be seas, but when, in the 19th century, it became apparent that the atmosphere was too tenuous for this to be the case, the consensus view became that they were dried-up sea beds supporting lowly vegetation. Together with the Martian CANALS theory, this supposition has been long rejected by scientists.

The atmosphere of Mars is extremely tenuous, being only 1/150th that of Earth. Ninety-five per cent of it is carbon dioxide, the rest being nitrogen (2.7%), argon (1.6%) and very small amounts of oxygen, carbon monoxide and water vapour (0.03%). The thin atmosphere, coupled with Mars' eccentric orbit and axial tilt, leads to a wide temperature range, from 136 K (over the winter south polar cap) to 299 K (after noon in summer), with a mean of 250 K. Most of the VOLATILES are frozen out of the atmosphere in the polar caps, but there is sufficient water vapour for water ice clouds to form at altitudes from 10 to 15 km (6–9 mi) in

▼ **Mars** The atmosphere on Mars is extremely tenuous. This is clearly shown in this true-colour Mars Pathfinder image of the Martian sunset where there is very little refraction of light.

M

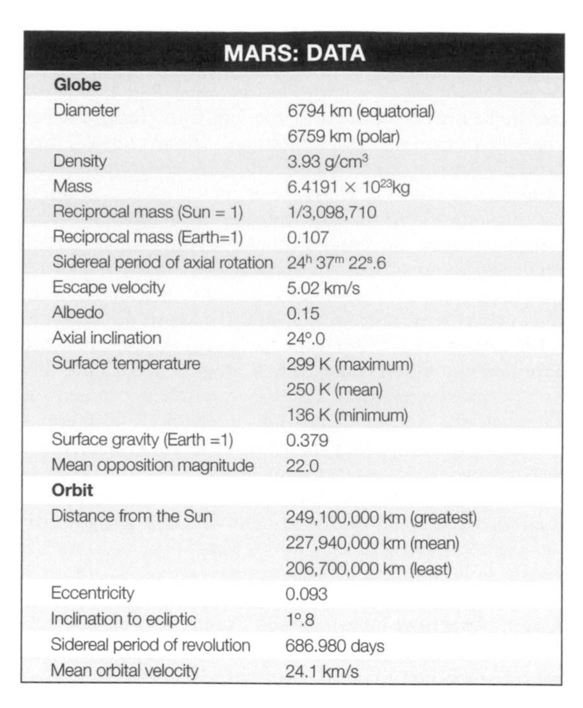

MARS: DATA	
Globe	
Diameter	6794 km (equatorial)
	6759 km (polar)
Density	3.93 g/cm^3
Mass	6.4191×10^{23}kg
Reciprocal mass (Sun = 1)	1/3,098,710
Reciprocal mass (Earth=1)	0.107
Sidereal period of axial rotation	$24^h\ 37^m\ 22^s.6$
Escape velocity	5.02 km/s
Albedo	0.15
Axial inclination	24°.0
Surface temperature	299 K (maximum)
	250 K (mean)
	136 K (minimum)
Surface gravity (Earth =1)	0.379
Mean opposition magnitude	22.0
Orbit	
Distance from the Sun	249,100,000 km (greatest)
	227,940,000 km (mean)
	206,700,000 km (least)
Eccentricity	0.093
Inclination to ecliptic	1°.8
Sidereal period of revolution	686.980 days
Mean orbital velocity	24.1 km/s

▲ **Mars** The Viking spacecraft gave unprecedentedly sharp images of Mars. The prominent dark area left of centre is Syrtis Major, and the bright area below it is Hellas Planitia.

the early morning and evening, and over specific topographic features such as the giant volcanoes. Dust storms are also common, though only the largest can be seen from Earth. The most extensive storms occur near perihelion, when the temperature is relatively high and the southern polar cap is rapidly evaporating and releasing volatiles to increase the atmospheric pressure. Wind speeds during these phenomena may reach as much as 400 km/hr (250mi/hr) but average a few tens of km/hr. Many major Martian dust storms have begun in the Hellas, Argyre, Chryse and Isidis basins, and in and around the VALLES MARINERIS complex. Dust deposition causes the contours of the dark regions of the planet to change slightly with time, and colours the Martian sky pink.

Since the Mariners of the 1960s, the VIKINGS of the 70s, MARS PATHFINDER in 1997 and the ongoing MARS GLOBAL SURVEYOR, the topography and geology of Mars has become known in some detail. The planet is believed to have an iron-rich core, some 2900 km (1800 mi) across, and it exhibits a magnetic field hundreds of times weaker than Earth's. The MANTLE may be 3500 km (2200 mi) thick, and the crust some 100 km (60 mi) deep.

There is a hemispheric asymmetry to the planet: much of the southern hemisphere is up to 3 km (2 mi) above datum (datum being an arbitrary height at which the pressure is 6.2 millibars) and is heavily cratered. The high density of craters suggests that this terrain is as old as the highland regions of Earth's Moon. The two large impact basins HELLAS PLANITIA and ARGYRE PLANITIA are located in this hemisphere. In several places this ancient cratered terrain is cut by gullies and complex channel systems, which appear to have been incised by water. Several tongues of the cratered surface extend into the northern hemisphere.

Most of the northern hemisphere is less heavily cratered and much is below datum. The surface looks altogether 'smoother' on the large scale. The most impressive features of this hemisphere are the vast shield volcanoes of THARSIS MONTES and the extensive valley network known as Valles Marineris, both first revealed in Mariner 9 images. The Tharsis region is in the nature of a huge bulge in the Mars lithosphere; it has a size roughly the same as that of Africa south of the Congo river, 4000 × 3000 km (2500 × 1900 mi). The entire Tharsis region rises to some 9 km (6 mi) above datum, but it rises to heights of up to 18 km (11 mi) in the large shield volcanoes ASCRAEUS MONS and PAVONIS MONS. The great volcano OLYMPUS MONS, 1500 km (930 mi) to the west, rises to 24 km (15 mi) above datum. A similar but smaller bulge resides in the region known as Elysium, containing the volcanoes Elysium Mons, Hecates Tholus and Albor Tholus.

These shield volcanoes, in terms of profile, are analogous to Earth's Hawaiian volcanoes, such as Mauna Loa. They are, however, substantially larger and have vast summit depressions (calderas) crowning them. Very extensive lava flows can be traced radiating out from these constructs. Radiating out from the Tharsis Montes

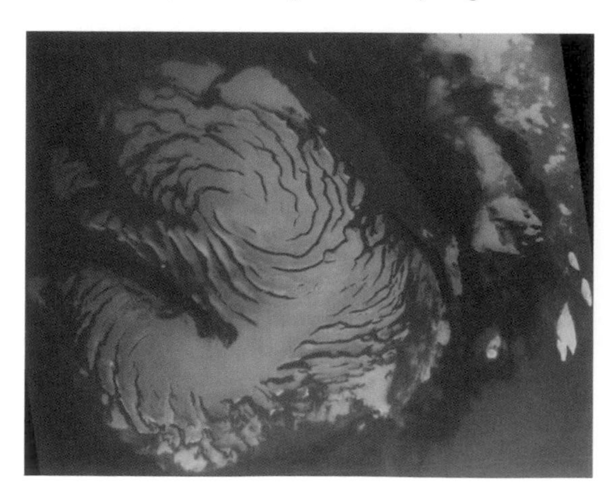

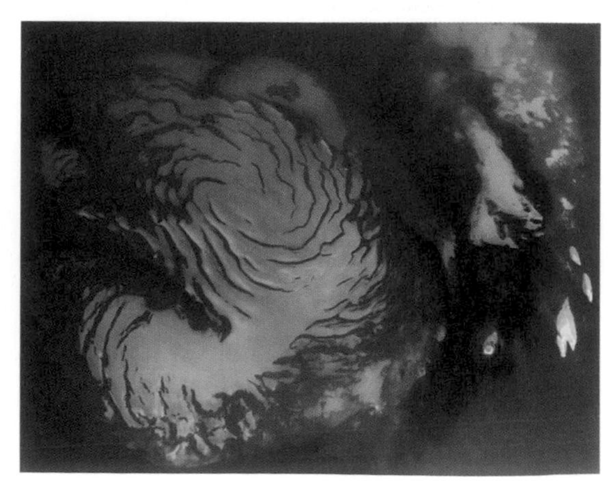

► **Mars** The polar caps on Mars are not permanent. These two Mars Global Surveyor images of the northern polar cap were made exactly one Martian year apart, in early northern summer. Close examination reveals that there was less frost in the later image.

and Elysium regions is a vast array of fractures, which must have developed in response to the formation of the two crustal bulges. Mars has no plate TECTONICS, and any volcanic hot-spot will remain in one place. Continued activity over a long period, aided by the low Martian gravity, has enabled high volcanic mountains to form. The planet is probably geologically inactive today. Olympus Mons seems relatively young, perhaps 30 million years old, but the oldest volcanoes are estimated to be up to 3.4 billion years old.

To the east of Tharsis is an impressive canyon system named Valles Marineris. It extends from near the summit of the bulge at 5°S 100°W for 4500 km (2800 mi) eastwards, eventually merging into an immense area of 'chaotic terrain' between Aurorae Planum and Margaritifer Terra. Valles Marineris marks a vast faulted area up to 600 km (370 mi) wide at the widest point; in places it falls 7 km (4 mi) below the rim.

The oldest recognisable features are the roughly circular impact basins, of which the most prominent is Hellas Planitia. Also prominent are Argyre Planitia and ISIDIS PLANITIA. A number of other, generally less well-defined basins have been recognized on spacecraft images. For instance, CHRYSE PLANITIA is believed to occupy an ancient multi-ringed basin, although the surrounding ramparts are not so clearly defined as those of Argyre and Hellas Planitia.

The most exciting of all Martian features are the 'outflow channels', in no way related to the illusory canali. These channels can be several hundred kilometres long and tens of kilometres wide. They generally start abruptly, without tributaries. Most channels are located north of the great canyon system and converge on the plain known as Chryse Planitia, although others appear to be associated with Elysium's north-west edge. They seem to be related to a period of flooding in the distant past, when the climate was quite different from that of today and liquid water (probably ice-covered) could have existed in quantity on the Martian surface. Standing water could have extended beyond Chryse, but the matter is controversial.

Mars Global Surveyor imaging greatly surpassed the resolution of the Viking Orbiters, enabling several intriguing discoveries to be made. High resolution images revealed many dozens of small channels within craters, each apparently carved by running water. These features are generally more than 30° from the equator. Each feature resembles a terrestrial gully, exhibiting boulders, a winding channel and debris at the bottom of the flow. The lack of associated small impact features implies a geologically recent history. The channels might have been eroded by water suddenly released behind a plug of ice. There are also many examples of what look like small, dried-up Martian lakes, which display apparent sedimentary layering: these have mostly been imaged at low latitudes in canyons and deep craters. In this instance, however, dust deposits could also have masqueraded as sedimentary strata.

The Viking Lander probes took panoramic photographs of the two landing sites, Chryse Planitia and UTOPIA PLANITIA, which are both plains areas. They recorded dune-like features and a reddish dusty surface crowded with dark rocky blocks. Many of the blocks were vesicular and appeared similar in aspect to terrestrial basaltic rocks. Chemical analyses carried out on the surface revealed that the soil was probably the result of weathering and oxidation of basaltic rocks. Pulverization by meteorite impacts has created fine micrometre-sized dusty particles. The deep reddish coloration is due to the presence of iron (III) oxide. Mars Pathfinder's landing site in ARES VALLIS looked broadly similar. Its Sojourner robot vehicle showed by X-ray spectrometry that the boulders were of various types, some similar to terrestrial basalt and andesite, and others apparently of a sedimentary nature. These findings implied that the rocks had been transported from geologically different places, underscoring the long-range impression from orbital imagery that the region is an ancient flood-plain. The Thermal Emission Spectrometer aboard Mars Global Surveyor indicated the presence of the iron-containing minerals haematite and olivine in the basaltic bedrock areas observable in Mars' southern hemisphere.

The Vikings found the Martian soil to be sterile, bereft of organic matter, though long exposure to solar ultraviolet radiation would have destroyed any evidence of such matter in time. The so-called 'microfossils' described in a Martian meteorite (ALLAN HILLS 84001) in 1996 are now thought by a consensus of scientists to be simply crystals of inorganic origin, rather than past Martian life-forms. This meteorite was blasted from the surface of Mars by a nearby impact over 16 million years ago and later picked up in Antarctica.

In addition to the volcanic shields and the surrounding lightly cratered plains, the heavily cratered terrain of the southern hemisphere and the polar caps, there are extensive regions adjacent to the poles that have been dissected to reveal strongly laminated deposits. Such lamination points to a geological history of alternating warmer and colder climates. This can be explained by the fact that the inclination of the Martian axis is known to oscillate from 14°.9 to 35°.5, and the planet's orbital eccentricity is known to vary from 0.004 to 0.141 over tens of thousands of years.

The polar caps are believed to represent the visible summits of more extensive subsurface permafrost. Each consists of a permanent cap perhaps several kilometres thick (water ice in the north, a mixture of water and carbon dioxide in the south) and an overlying seasonal cap of carbon dioxide. The seasonal caps undergo cyclic changes, occupying a diameter of over 60° in latitude in

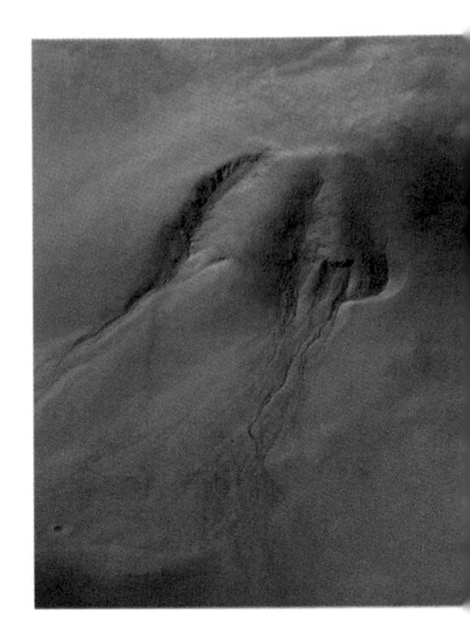

▲ **Mars** The existence of water on Mars, whether as subsurface liquid or as ice, is vital for the question of whether there was, or is, life on Mars. A number of outflow channels in crater walls, including this example in Noachis Terra, were imaged by the Mars Global Surveyor, and appear to resemble closely gullies on Earth.

SELECTED MARTIAN FEATURES

Name	Latitude	Longitude	Comments
ACIDALIA PLANITIA	47°.0N	22°.0W	large low-albedo area
ALBA PATERA	40°.5N	109°.9W	low-profile shield volcano
Antoniadi	21°.5N	299°.2W	large crater, diameter 381 km (237 mi)
ARES VALLIS	9°.7N	23°.4W	flood-plain valley, landing site of Mars Pathfinder
ARGYRE PLANITIA	50°.0S	44°.0W	impact basin
ARSIA MONS	8°.4S	121°.1W	shield volcano
ASCRAEUS MONS	11°.9N	104°.5W	shield volcano
Cassini	23°.8N	328°.2W	large crater, diameter 415 km (258 mi)
Chasma Boreale	83°.0N	45°.0W	rift in north polar cap
CHRYSE PLANITIA	27°.0N	40°.0W	plains region, Viking 1 landing site
Cimmeria Terra	35°.0S	215°.0W	low-albedo feature
Deuteronilus Mensae	45°.7N	337°.9W	Martian mesae area
ELYSIUM FOSSAE	24°.8N	213°.7W	aligned valleys flanking Elysium Mons
Elysium Mons	25°.0N	213°.1W	volcano
Ganges Catena	2°.8S	68°.7W	crater chain
Hadriaca Patera	32°.4S	268°.2W	one of the oldest recognisable Martian volcanic structures
Hecates Tholus	32°.7N	209°.8W	shield volcano
HELLAS PLANITIA	43°.0S	290°.0W	largest Martian basin
Huygens	14°.3S	304°.6W	large crater, diameter 456 km (283 mi)
Isidis Planitia	13°.0N	273°.0W	low elevation plain
Ismenia Fossa	38°.9N	326°.1W	long shallow depression
Ius Chasma	7°.0S	85°.8W	canyon in western part of Valles Marineris
KASEI VALLES	24°.6N	65°.0W	major outflow channel area
Lunae Planum	10°.0N	67°.0W	plain
Margaritifer Terra	5°.0S	25°.0W	region of chaotic terrain
Noctus Labyrinthus	7°.0S	102°.2W	maze of intersecting canyons
Ogygis Rupes	34°.1S	54°.9W	ridge
OLYMPUS MONS	18°.6N	134°.0W	giant shield volcano
Ophir Chasma	4°.0S	84°.7W	part of central Valles Marineris canyon system
PAVONIS MONS	0°.8N	113°.4W	large shield volcano
PROTONILUS MENSAE	44°.2N	309°.4W	Martian mesae area
Sabaea Terra	2°.0N	318°.0W	low-albedo feature
Sirenum Terra	40°.0S	150°.0W	low-albedo feature
SOLIS PLANUM	25°.5S	86°.5W	low-albedo feature
SYRTIS MAJOR PLANITIA	9°.5N	290°.5W	plateau and low-albedo feature
Tempe Terra	40°.0N	71°.0W	desert area
THARSIS MONTES	1°.2N	112°.5W	Mars' main volcanic region
Tithoniae Catena	5°.4S	82°.4W	crater chain
Tyrrhena Terra	15°.0S	270°.0W	low-albedo feature
UTOPIA PLANITIA	50°.0N	242°.0W	plain, Viking 2 landing site
VALLES MARINERIS	13°.9S	59°.2W	great canyon system
VASTITAS BOREALIS	68°N		Northern circumpolar plain

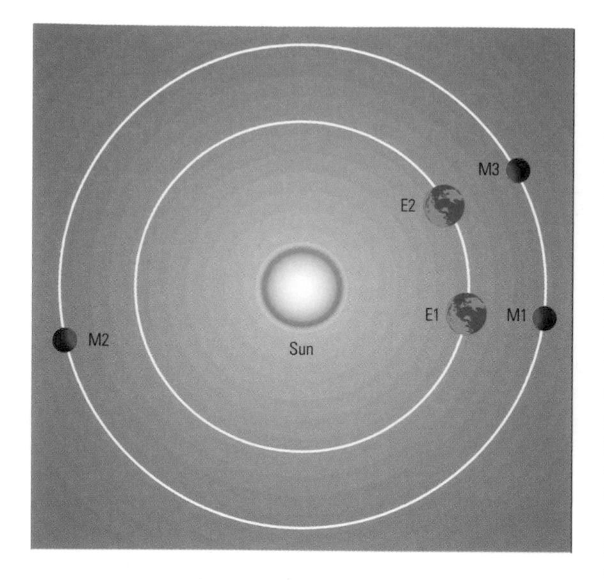

► **Mars** Because Mars takes longer to orbit the Sun than Earth, the mean period between oppositions (the closest approach) is more than two years. With the Earth at E1 and Mars at M1, Mars is in opposition. A year later the Earth has come back to E1, but Mars has only reached M2. The next opposition occurs only after 780 days when Earth has caught up with Mars, with the Earth at E2 and Mars at M3.

winter, and shrinking in summer to just 10° or less. It is estimated that the complete melting of the polar caps, which would only be possible given an adequate atmospheric pressure, would cover the planet with 10 metres of water – sufficient for the supposed rivers and larger bodies of water of Mars' geological past.

Mars-crossing asteroid Any of a group of ASTEROIDS with orbits that cross some part of the orbit of Mars. In general this implies a perihelion distance of less than about 1.67 AU, although some NEAR-EARTH ASTEROIDS may be disqualified from this group because their aphelia lie within Mars' perihelion distance of 1.38 AU. Asteroids with perihelia less than 1.30 AU are classed as AMOR ASTEROIDS, unless they actually cross the Earth's orbit, in which case they are either APOLLO ASTEROIDS or ATEN ASTEROIDS.

Mars Exploration Rover NATIONAL AERONAUTICS AND SPACE ADMINISTRATION (NASA) mission, originally known as Mars 2003, to land two roving vehicles on Mars. Although it is possible that NASA will be forced to reduce the mission to one rover, the plan in 2001 involved two launches in 2003 June, with landings in 2004 January and February, using an inflatable airbag method used successfully for MARS PATHFINDER in 1997. The craft could bounce a dozen times and could roll as far as 1 km (0.6 mi). The bags will deflate and retract and petals will open up, bringing the lander to an upright position and revealing the rover. The landings will be made in two separate locations to be identified using MARS GLOBAL SURVEYOR images. The criteria will be that the sites reveal clear evidence of surface water, such as lakebeds or hydrothermal deposits. The identical rovers will carry a suite of instruments, including three spectrometers and a microscopic imager, to search for evidence of liquid water that may have been present in the planet's past. The rocks and soil will be collected by a sampler, while a rock abrasion tool will expose fresh rock surfaces for study. The 1500-kg rovers will be equipped with a 360° panoramic camera, giving three times the resolution of that carried on the Mars Pathfinder. This will be used to enable scientists to select samples. The rovers will be able to travel about 100 m (330 ft) a day with an operational lifetime expected to last 90 Martian days.

Mars Express EUROPEAN SPACE AGENCY (ESA) mission to be launched in 2003 June and to enter orbit around Mars in 2003 December. Mars Express was originally envisaged as part of a wider international effort to return samples of Mars to Earth in 2004 and 2006. The craft was to help track the US ascent vehicles. Such missions are unlikely to take place until the 2010 timeframe. Nonetheless, Mars Express will carry the UK's BEAGLE 2 Mars lander – a late addition to the craft – and will be equipped with a suite of seven instruments to study the surface and atmosphere in detail, including a 10-m (33-ft) resolution photo-geology imager and a spectrometer with a resolution of 100 m (330 ft) to map the mineralogy of the planet. The spacecraft will operate in a polar orbit and will complement the new US MARS EXPLORATION ROVER and MARS ODYSSEY missions.

Mars Global Surveyor (MGS) One of the USA's most successful Mars probes. It was launched in 1996 November and entered orbit around Mars in 1997 September. The spacecraft was still operating in 2001 for special observations after its primary mission had ended. The MGS took over 300,000 high-resolution images of the Martian surface, including those that revealed what seemed to be clear evidence that at one time liquid water flowed across the surface. MGS was equipped with five of the seven instruments that flew on the Mars Observer, which failed to orbit the planet in 1993, probably because of an engine explosion. The images returned by the MGS – and the MARS PATHFINDER, which landed in 1997 – freely and widely available on the Internet, represented a new era in wide public access to and personal involvement with the exploration of space.

Mars Odyssey NASA's 2001 Mars Odyssey is the former Mars Surveyor 2001 craft renamed in honour of the English writer Arthur C. Clarke. Launched in 2001 April, Mars Odyssey entered an initial orbit around Mars in 2001 October and for the following 76 days used aerobraking in the upper atmosphere to manoeuvre into its operational orbit. Mars Odyssey is equipped with a thermal emission imager, a gamma-ray spectrometer and a radiation environmental experiment. The spectrometer is identical to the instrument lost when the Mars Observer mission failed in 1993.

Mars Pathfinder First of NASA's 'faster-cheaper-better' Discovery-class missions, mainly designed as a technology demonstration. It was launched on 1996 December 4 and arrived at Mars on 1997 July 4. Airbags were used to cushion the landing. The spacecraft impacted the surface at a velocity of about 18 m/s (60 ft/s) and bounced about 15 m (50 ft) into the air. After bouncing another 15 times it eventually came to rest about 1 km (0.6 mi) from the initial impact site. The landing took place on an ancient flood plain in the Ares Vallis region.

Two days later, a small rover known as Sojourner rolled down a ramp on to the rocky surface. The 10-kg

THE MARTIAN SATELLITES

	Phobos	Deimos
mean distance from Mars (km)	9378	23,459
mean sidereal period (days)	0.3189	1.2624
orbital inclination(°)	1.068	0.8965
orbital eccentricity	0.01515	0.0003
diameter (km)	27 × 22 × 18	15 × 12 × 10
density (g/cm³)	2.0	1.7
magnitude at mean opposition	11.6	12.8

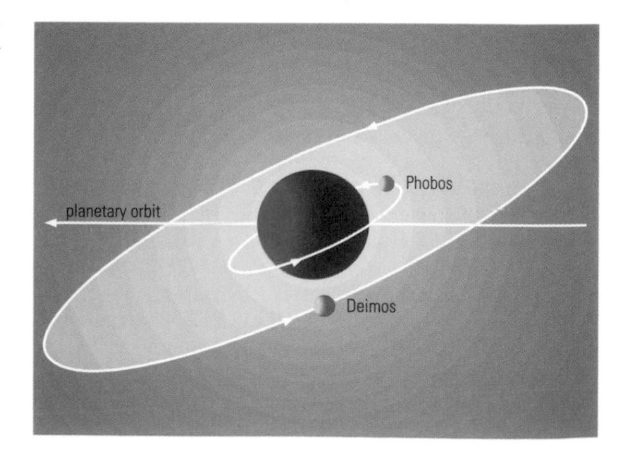

► **Mars** The Martian satellites Phobos and Deimos are thought to be captured asteroids. They orbit close to the planet's surface – Phobos at 9378 km (5827 mi) and Deimos at 23,459 km (14,577 mi) – in almost circular orbits.

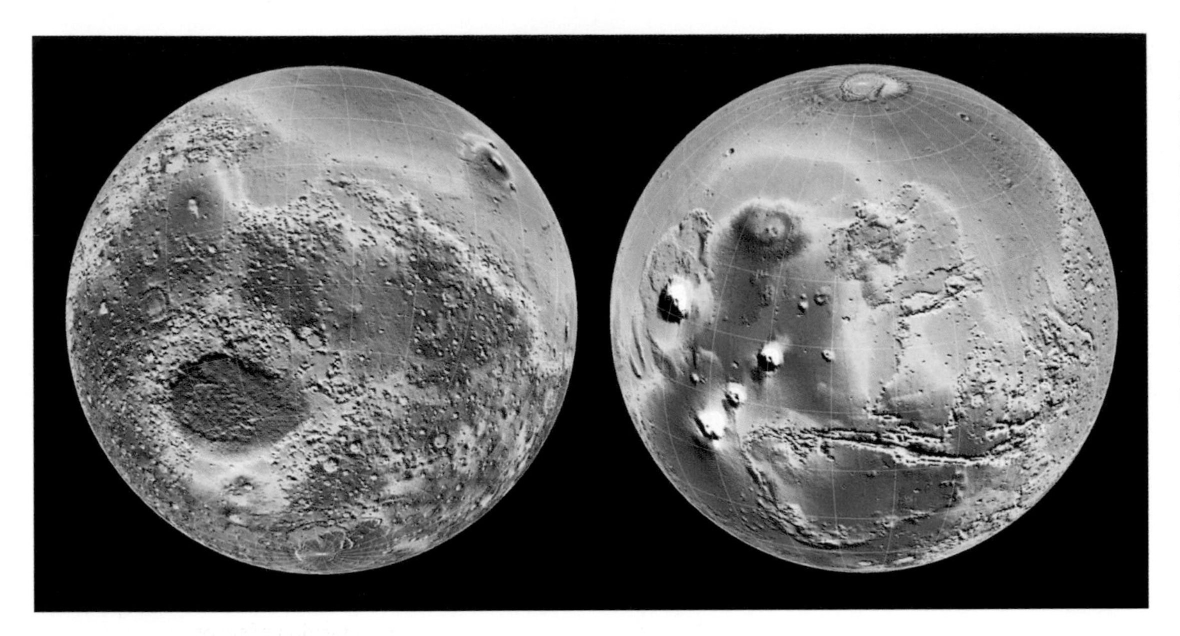

◀ **Mars Global Surveyor** (MGS) The Mars Orbiter Laser Altimeter (MOLA) on the MGS confirmed there is a marked difference between the northern and southern halves of the planet.This false-colour image shows that the north consists mainly of low-lying ground (blue) and the south of highlands (green, yellow, red and white). The volcanoes of the Tharsis region are clearly visible, as is the deep Hellas Planitia impact basin.

rover carried a stereo camera system and an alpha-proton X-ray spectrometer to study the composition of soil and rocks. Panoramic views showed a landscape of broad, gentle ridges overlain with numerous dark-grey rocks of various shapes and sizes. Close-up images of the rocks revealed pitted, layered and smooth surfaces. It was suggested that some of the rocks were conglomerates – sedimentary rocks that had formed when liquid water existed on the planet's surface.

By the time rover operations ceased on 1997 September 27, Sojourner had travelled about 100 m (330 ft), carried out 16 chemical analyses of rocks and soil and taken 550 images. The lander, which was renamed the Sagan Memorial Station in honour of planetary scientist Carl SAGAN, relayed an unprecedented 2.3 gigabits of data, including 16,500 lander images and 8.5 million measurements of atmospheric pressure, temperature and wind.

Mars sample return missions Attempts to return samples of Mars to the Earth to help answer finally the questions about possible life on Mars. The first sample return mission was to have been made by a Mars Surveyor mission in 2005. The Mars Surveyor class lander would collect a modest 2 kg of samples and place them into a small ascent capsule, which would be fired off the surface and on a course to Earth. The failures of the Mars Climate Orbiter and Mars Polar Landing missions in 1999 were a milestone in Martian exploration, resulting in major re-evaluation of future plans and the realization that such a mission is unlikely to take place until the 2010 timeframe. The high cost and level of technology required for a mission are major stumbling blocks in times of low budgets, cut-price spacecraft and the technological disasters of losing two craft in a matter of months.

Mars-Trojan asteroid ASTEROID that has the same orbital period as Mars, positioned around one of its LAGRANGIAN POINTS, L_4 or L_5. By late 2001 there were five Mars Trojans known, all of which are associated with L_5. *See also* EUREKA; TROJAN ASTEROID

Martian canals *See* CANALS, MARTIAN

Martian meteorite METEORITE that originated on Mars. Martian meteorites are also known as the SNCs, after the type specimens of the three original subgroups (SHERGOTTY, NAKHLA and CHASSIGNY). The collection of additional Martian meteorites from Antarctica and the Sahara Desert has extended the number of subgroups to five. The different subgroups are igneous rocks that formed in different locations at, or below, the Martian surface. The groups have different mineralogies and chemistries, and cannot all have come from a single impact event. At least three craters, with minimum diameters of *c.*12 km (*c.*7 mi), are required to produce the variety of Martian meteorite types. As of summer 2001, there are 18 separate meteorites that almost certainly originated on Mars. The Martian origin rests on the age, composition and noble gas inventory of the meteorites.

mascon (mass concentration) Concentration of mass below the Moon's surface. Gravity mapping of the Moon has revealed mascons over many of the lunar basins, including IMBRIUM, SERENITATIS and NECTARIS.

The gravity field over a planet is never completely even. Places of large mass concentrations exist, which are usually due to accumulation of either heavy materials (for example high-density minerals) or the piling up of materials (such as mountains). Planetary bodies usually even out these mass concentrations through redistribution of mass, in order to achieve isostatic equilibrium. On the Moon, this occurs through movement (plastic flow) of the asthenosphere, which underlies the rigid lithosphere.

Lunar mascons are always associated with basins, the largest impact structures on the Moon. As enormous amounts of surface material were removed by the cratering process, the asthenosphere moved towards the surface to replace the mass loss and to maintain isostatic equilibrium. At a later time, lavas, which are much denser than crustal rock, poured into and filled the basins to depths of 2–4 km (1.2–2.5 mi). These two processes together formed the large mass concentrations.

Early in the Moon's history, the lithosphere was relatively thin, and mass concentrations were mostly smoothed out. Later, however, the lithosphere thickened because of continued cooling, so that new mass concentrations were maintained.

▼ **Mars Pathfinder** The Sojourner rover was deployed two days after Pathfinder landed on Mars. It spent 12 highly successful weeks examining a variety of rocks strewn across the flood plain of Ares Vallis.

maser (acronym for 'microwave amplification by the stimulated emission of radiation') Celestial object in which radio emission from molecules stimulates further radio emission at the same energy from other molecules; a maser is analogous to a LASER but in the radio region of the ELECTROMAGNETIC SPECTRUM. A maser can be created in an astronomical source when one energy level of a molecule is preferentially populated by the radiation environment, and the molecule cools by stimulated emission. Water (H_2O), the hydroxyl radical (OH), silicon monoxide (SiO) and several other molecules can produce maser emission in galactic and extragalactic objects. The name 'maser' is also used for the astronomical object itself. Since the molecular emission is amplified, the objects are easier to detect, and over 10,000 are known in the Galaxy. Masers occur in places where new stars are forming, and in the atmospheres of evolved stars, or variable stars with high mass loss which are in the process of becoming planetary nebulae or supernovae. Some of the first masers to be discovered were the OH–IR stars, which radiate in the 1612 MHz line (*see* OH). A star will have many masers in its envelope, and they can be tracked using MULTI-ELEMENT RADIO-LINKED INTERFEROMETER NETWORK or VERY LONG BASELINE INTERFEROMETRY, showing that the conditions in the envelope change over months and years. OH, CH and H_2O megamasers have been found in galaxies such as the Large Magellanic Cloud, infrared luminous galaxies, active galaxies, starburst and Seyfert galaxies.

Maskelyne, Nevil (1732–1811) English astronomer and clergyman, the fifth and longest-serving ASTRONOMER ROYAL (1765–1811). After studying mathematics at Cambridge University, he became James Bradley's assistant at GREENWICH OBSERVATORY. In 1761 he took part in a Royal Society expedition to the island of St Helena to observe a transit of Venus. The voyage inspired him to apply astronomy to navigational problems, and in his *British Mariner's Guide* (1763) he explained how to determine longitude at sea by comparing the Moon's position as observed aboard ship to its position as observed from a known terrestrial longitude. He travelled to Barbados in 1764 to test John Harrison's newly invented MARINE CHRONOMETER, which proved to be more practical and accurate than the lunar method of finding longitude. Two years later Maskelyne founded the *NAUTICAL ALMANAC*. In 1774 he measured the deflection of a large pendulum erected atop the Scottish mountain of Schiehallion; from the deflection he calculated the mountain's mass and thence the gravitational constant and the Earth's density. His value of 4.87 times the density of water is close to the accepted modern value of 5.52.

▼ **massive compact halo object** Astronomers hope to detect MACHOs through their gravitational effects on the light from objects beyond them. In this microlensing event, the MACHO is a nearby red dwarf star (arrowed). Most of these objects, however, must be invisible.

mass Measure of an object's INERTIA. The mass defined in this way is called the inertial mass. Mass may also be defined from the GRAVITATIONAL FORCE that it produces, leading to the gravitational mass. Experiments have shown that the two masses are identical to better than one part in 10^{12}. This has led to the principle of equivalence 'no experiment can distinguish between the effects of a gravitational force and those of an inertial force in an accelerated frame', which underlies EINSTEIN's theory of GENERAL RELATIVITY.

massive compact halo object (MACHO) Dark object, perhaps a BROWN DWARF or BLACK HOLE, postulated to explain the missing mass seen in galactic rotation curves.

mass–luminosity relation For stars on the MAIN SEQUENCE there exists a one-to-one relationship between luminosity and mass, expressed approximately as $L \propto M^x$. This relation enables mass to be determined from absolute magnitude. The exponent x varies with mass, being 1.6 for stars of around 100 solar masses, 3.1 for stars of around 10 solar masses, 4.7 for stars of a solar mass and 2.7 for low-mass stars of around a tenth of a solar mass. These differences are due primarily to the opacity of the stars caused by their different interior temperatures.

mass transfer Process that occurs in CLOSE BINARY stars when one star fills its ROCHE LOBE and material transfers to the other star through the inner LAGRANGIAN POINT. Material can be transferred directly to the other star but is more usually transferred via an ACCRETION DISK.

The material flowing through the inner Lagrangian point moves very like a free particle, but when the material impacts the accretion disk around the other star, it stops moving like a free particle, since gas cannot flow freely through other gas, and the energy is dissipated via shock waves.

The observed period of the outbursts from the shock waves gives information about the time it takes material to spiral from the secondary on to the accreting primary. For dwarf novae, the time is typically a few days. These observations suggest that material accretes a lot faster than should be allowed by the natural viscosity of gas. If it is to accrete on to the primary, the material must first lose its original ANGULAR MOMENTUM. It may be that the gas in an accretion disk behaves like a fluid with a very high viscosity, with the viscosity slowing the motion of the gas.

Masursky MAIN-BELT ASTEROID; number 2685. It was imaged in 2000 January with the camera on board the Saturn-bound CASSINI space probe. Masursky is *c.*15 km (*c.*9 mi) in size.

Masursky, Harold (1922–90) American geologist and planetary scientist who played a major role in most of NASA's 1960s–1980s missions to explore the Solar System. He designed the scientific experiments and helped select the landing sites for the RANGER 8 and 9 lunar probes, helped to coordinate the LUNAR ORBITER programme, was responsible for the APOLLO 8 and 10 geochemical investigations, and coordinated the geological experiments carried out by Apollos 14–16. Masursky led the team that built MARINER 9, was largely responsible for selecting the two VIKING landing sites, and led the Imaging Radar Group for the two PIONEER VENUS ORBITERS. His career climaxed with his involvement with the VOYAGER missions.

Mathilde MAIN-BELT ASTEROID; number 253. Mathilde was visited by NASA's NEAR SHOEMAKER probe in 1997, as the spacecraft headed for rendezvous with EROS. The information returned shows that Mathilde has a mean density somewhat less than that of stony meteorites, indicating that it may contain voids and may thus be an agglomeration of smaller rocky components held together by self-gravity.

Mauna Kea Observatory World's largest astronomical observatory. It is situated at the 4205-m (13,796-ft) summit of Mauna Kea (which means 'White Mountain'), a dormant volcano on the island of Hawaii and the highest point in the Pacific Basin. It is above 40% of the Earth's atmosphere and 97% of the water vapour in the atmosphere, resulting in very high atmospheric transparency and freedom from cloud. Laminar airflow over the mountain produces superb image quality.

The observatory, operated by the University of Hawaii, hosts nine telescopes for optical and infrared astronomy and two for submillimetre astronomy. They include the largest single-mirror optical/infrared telescopes in the world, the twin reflectors of the W.M. KECK OBSERVATORY; and the largest submillimetre telescope in the world, the JAMES CLERK MAXWELL TELESCOPE. Also on Mauna Kea is the CALTECH SUBMILLIMETER OBSERVATORY, while a third submillimetre instrument (the SUBMILLIMETER ARRAY) is under construction. The westernmost antenna of the VERY LONG BASELINE ARRAY is nearby.

Astronomy on Mauna Kea began in the 1960s, when the University of Hawaii placed a 0.6-m (24-in.) telescope there, followed in 1970 by their 2.2-m (88-in.) reflector. Three more telescopes were built on the site in 1979: the 3.2-m (126-in.) NASA Infrared Telescope Facility (IRTF), the 3.6-m (142-in.) CANADA–FRANCE–HAWAII TELESCOPE and the 3.8-m (150-in.) UNITED KINGDOM INFRARED TELESCOPE. Mauna Kea's exceptional observing conditions attracted four 8–10 metre class telescopes in the 1990s: the two Keck Telescopes, the Japanese SUBARU TELESCOPE, and the northern telescope of the GEMINI OBSERVATORY. Further development of the site is likely.

Astronomers and technicians working on Mauna Kea must first acclimatize to the altitude. A mid-level facility at Hale Pohaku, at an altitude of 2800 m (9300 ft), provides accommodation and other facilities; it is named the Ellison Onizuka Center for International Astronomy to honour Hawaiian astronaut Ellison Shoji Onizuka (1946–86), who died in the *Challenger* disaster.

Maunder, (Edward) Walter (1851–1928) British solar astronomer and a founder of the BRITISH ASTRONOMICAL ASSOCIATION (BAA). He was appointed Photographic and Spectroscopic Assistant at the Royal Observatory, Greenwich in 1873, where he used various instruments to observe sunspots, faculae and prominences. Over the next thirty years, he compiled the most complete record of sunspot activity, supplementing the Greenwich data with observations from overseas. By plotting the mean heliographic latitude of sunspot groups against time, Maunder created the first BUTTERFLY DIAGRAM and determined that the Sun showed DIFFERENTIAL ROTATION. He also researched historical sunspot records, discovering the paucity of sunspot activity during the years 1615–1745 now known as the MAUNDER MINIMUM.

Maunder organized and participated in many expeditions to observe total solar eclipses, on which he was often accompanied by his second wife, **Annie Scott Dill Maunder** (1858–1947), who also worked as an astronomer at Greenwich. Annie Maunder took fine photographs during these eclipses – on one image the corona could be traced to a distance from the limb of six solar radii. The Maunders were instrumental in the formation of the BAA (1890), which soon became the world's foremost group of amateur astronomers.

Maunder minimum Period between 1645 and 1715 when very few SUNSPOTS were observed. It has been named after Walter MAUNDER who in 1922 provided a full account of the 70-year dearth in sunspots, first noticed by Gustav Friedrich Wilhelm SPÖRER in 1887–89. The scarcity of sunspots during this period has been substantiated by other indicators of low solar activity, such as the amount of radioactive carbon-14 in old tree rings and the occurrence of low-latitude aurorae. The Maunder minimum coincided with years of sustained low temperatures in Europe, from about AD 1400 to 1800, known as the Little Ice Age.

Mauritius Radio Observatory Location of the Mauritius Radio Telescope, constructed to make a southern-hemisphere survey to complement the 6C (sixth Cambridge) radio survey of the MULLARD RADIO ASTRONOMY OBSERVATORY. It also observes selected southern pulsars.

Maurolycus Ancient lunar crater with deeply incised walls (42°S 14°E), 105 km (65 mi) in diameter. Continued meteoritic erosion by meteorites has worn away most elements of its EJECTA. Its central peaks, produced by rebound of the floor, are still visible. Maurolycus lies over several older craters, the rims of which are visible to the south-west and north-west. At high Sun angles, the bright ray system of TYCHO can be seen on Maurolycus' floor.

Maury, Antonia Caetana (1866–1952) Pioneer American woman astronomer at the HARVARD COLLEGE OBSERVATORY. Maury, a niece of Henry Draper, helped to compile the *HENRY DRAPER CATALOGUE*, named as a memorial to him. This work, carried out at Harvard from 1887 under the direction of Edward C. PICKERING, required Maury to examine and classify thousands of stellar spectra, in the course of which she refined the classification system to account for the sharpness of spectral lines. In 1905 Ejnar HERTZSPRUNG showed that Maury's modification could be used to distinguish between giant and dwarf stars. Maury also became expert at identifying spectroscopic binaries, helping Pickering to prove the true nature of the first star to be identified as such, Mizar (ζ Ursae Majoris), and in 1889 calculated its 104-day period.

Maximum Aperture Telescope (MAXAT) Former name of the GIANT SEGMENTED-MIRROR TELESCOPE

maximum entropy method (MEM) Mathematical technique applied to the inverse problem of how to make

▲ **Mathilde** Asteroid 253 is a C-type, carbonaceous asteroid, with a composition typical of the outer parts of the main asteroid belt. It is about 55 km (34 mi) long, and has several craters *c*.20 km (*c*.12 mi) across.

M

▼ **Mauna Kea Observatory** The most recent generation of telescopes on Mauna Kea, including Gemini North (foreground), the twin Keck telescopes (right) and the Japanese Subaru (beyond the Keck domes), give results that are second to none.

reliable deductions from noisy and uncertain data. It derives its name from the concept of an increase in ENTROPY as equivalent to a decrease in the level of structure in a system. The MEM deduction is therefore the one that has the least amount of structure within it (that is, the maximum entropy) and yet is still consistent with the data. The technique is mostly applied to the interpretation of noisy images.

Max-Planck-Institut für Astronomie (MPIA) Institute located in Heidelberg, Germany, which conducts research in astronomical instrumentation, stellar and galactic astronomy, cosmology and theoretical astrophysics. The institute is responsible for operating the CALAR ALTO OBSERVATORY.

Max-Planck-Institut für Astrophysik (MPA) Institute located at Garching, near Munich, Germany, which undertakes research in theoretical astrophysics. Its scientists work closely with astronomers from the EUROPEAN SOUTHERN OBSERVATORY, whose headquarters are close by. Also nearby is the Max-Planck-Institut für extraterrestrische Physik (MPE), which conducts observations in spectral regions accessible only from space (for example, far-infrared, X-ray and gamma-ray).

Max-Planck-Institut für Radioastronomie (MPIfR) Institute located in Bonn, Germany, with the primary purpose of operating the fully steerable 100-m (330-ft) telescope at Bad Münstereifel-Effelsberg, 40 km (25 mi) south-west of the city. The Effelsberg Radio Telescope is a lightweight paraboloidal dish commissioned in 1972, and it is used at centimetre wavelengths both as a stand-alone instrument and as an element of global VERY LONG BASELINE INTERFEROMETRY experiments. The MPIfR collaborated with the STEWARD OBSERVATORY to build the 10-m (33-ft) HEINRICH HERTZ TELESCOPE for submillimetre-wavelength observations.

Maxwell, James Clerk (1831–79) Scottish physicist who unified the forces of electricity and magnetism and showed that light is a form of ELECTROMAGNETIC RADIATION. In 1859 he demonstrated mathematically that Saturn's rings could not be stable if they were completely solid or liquid, and most likely consisted of small solid, particles – as was proved by James KEELER in 1895.

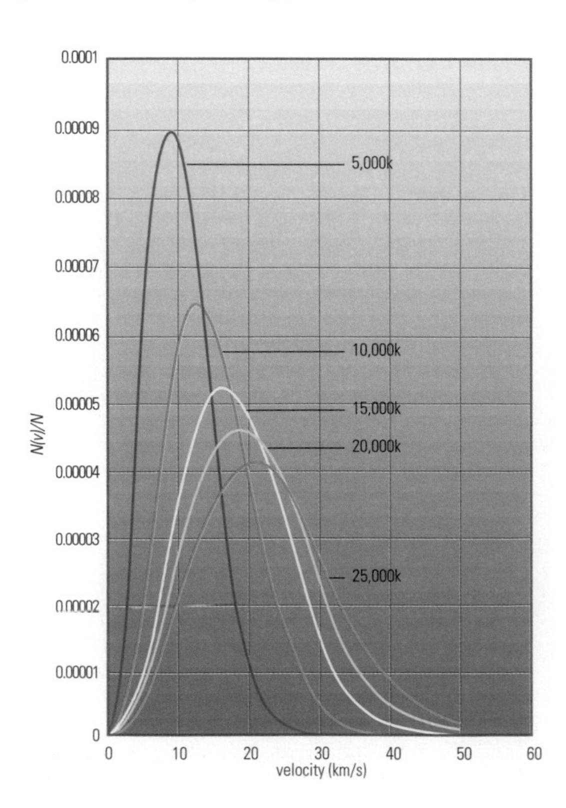

► **Maxwell–Boltzmann distribution** The velocity of atoms within a gas is determined by the temperature and density. Maxwell–Boltzmann's distribution shows that for any given density, atoms move more rapidly as the temperature increases.

Maxwell–Boltzmann distribution Law determining the speed of particles in a gas. It takes the form:

$$\frac{N(v)\,dv}{N} = \left(\frac{2m^3}{\pi k^3}\right)^{1/2} T^{-3/2}\, v^2\, e^{-mv^2/2kT} dv$$

where $N(v)$ is the number density of particles with velocities in the range v to $v + dv$, N is the total number density of the particles, m is the particle mass, T the temperature and k is BOLTZMANN'S CONSTANT. It gives a bell-shaped distribution, the peak of which moves to higher velocities as the temperature increases. The onset of degeneracy (*see* DEGENERATE MATTER) in, for example, WHITE DWARFS is indicated by deviations of particle speeds from the Maxwell–Boltzmann distribution.

Mayer, (Johann) Tobias (1723–62) German mathematician, astronomer and cartographer who produced precise tables of lunar positions, taking into account lunar libration, that were used by later astronomers to reckon longitude at sea. He produced the first map of the Moon based upon micrometer measurements (1750). From his observations he concluded that the Moon has no appreciable atmosphere. In 1753 Mayer began publishing his lunar tables, and he developed a mathematical method of estimating longitude to within half a degree. He also derived an improved method for calculating solar eclipses.

Mayall Telescope *See* KITT PEAK NATIONAL OBSERVATORY

Mayan astronomy *See* NATIVE AMERICAN ASTRONOMY

mean anomaly *See* ANOMALY

mean solar time Local time based on a fictitious, or mean, Sun which is defined as moving around the celestial equator at a constant speed equal to the average rate of motion of the true Sun along the ecliptic. Because the true Sun does not appear to move across the sky at a uniform rate, the length of the solar day as measured by APPARENT SOLAR TIME varies throughout the year by up to 16 minutes. For the purposes of establishing a uniform civil time, it was therefore necessary to invent a MEAN SUN. The difference between apparent solar time and mean solar time is the EQUATION OF TIME.

mean Sun Fictitious Sun conceived to provide a uniform measure of time equal to the average APPARENT SOLAR TIME. The mean Sun is deemed to take the same time as the real Sun to complete one annual revolution of the celestial sphere, relative to the FIRST POINT OF ARIES, but it moves along the celestial equator at a constant speed equal to the average rate of motion of the true Sun along the ecliptic. *See also* MEAN SOLAR TIME.

Méchain, Pierre François André (1744–1805) French astronomer at France's Marine Observatory, best known as a colleague of Charles MESSIER who contributed to the famous MESSIER CATALOGUE of 'nebulae', and for his own observations of comets. He discovered ten comets between 1781 and 1799, often computing their orbits himself. One of these discoveries (1786) was the famous Comet ENCKE. Like Messier, Méchain began cataloguing celestial objects that appeared 'nebulous' in order to avoid misidentifying them as new comets. He discovered 21 of the nebulous objects that Messier included in his final list published in 1781, and with Messier discovered the six objects added in the 20th century as M104–109.

medieval European astronomy Astronomy as practised in Western Europe between about AD 400 and 1500. Modern scholarship is fundamentally changing the belief that no serious astronomy took place in Europe in this period. But medieval astronomy differs from that of later ages in its attitude to new knowledge. Medieval scientists looked back to the writers of antiquity for their definitive standards in observation and explanation, generally believing that the Greeks had already uncovered the great truths of nature. Preservation rather than progress was the aim, yet much useful observation and invention came from them.

Between the end of the Roman world in the 5th century AD and the 12th century, only fragments of PTOLEMY, ARISTOTLE and other Greek scientific writers were available in the West. Astronomy was learned from encyclopedic digests, such as those by Pliny the Elder (AD 23–79) and Boethius (*c.* 480–524). Nonetheless, the basic structures of the classical cosmos were familiar to all educated people. The Earth was not flat, but a sphere, set motionless at the centre of a series of 'crystalline' spheres that carried the Moon, Sun, planets and stars, and rotated around us at different speeds; the Moon's sphere in 28 days, Saturn's in 29 years. The stars were all the same distance away, and were gathered into Ptolemy's 48 constellations, including the 12 zodiacal signs.

One of the main reasons for the cultivation of astronomy in medieval Europe was the refinement of the calendar, and in particular, the accurate determination of the date of Easter, the most sacred of Christian festivals, which was calculated from a formula governed by the 'Paschal' or full moon following the vernal equinox. 'The Venerable' BEDE became Britain's first astronomer when he developed superior techniques for calculating Easter. Throughout the medieval period, the requirements of the calendar, and, to a lesser extent, of observing the daily motions of the stars to determine the times for monastic prayers, kept astronomy and the Church closely wedded. Not until 1582, by which time astronomers had sufficient data on calendrical errors, and established the Earth's rotation period to within seconds of the modern value, could they refine calendrical calculations to produce the GREGORIAN CALENDAR we still use today.

It was only after the mid-12th century that astronomy came to be extensively cultivated in Europe, partly as a result of contacts with ISLAMIC ASTRONOMY in Spain and Palestine. The Arabs had already translated Ptolemy, Aristotle and other writers into Arabic. These works, in turn, came to be translated into Latin, so that for the first time European scholars had access to complete versions of the leading classical texts. They also acquired Latin translations of the original researches of various Arab astronomers. Europeans, such as the Frenchman Gerbert of Aurillac (*c.* 940–1003), visited Muslim Spain, and it is Gerbert who is credited with introducing the ASTROLABE into Europe.

The large number of astronomical books being translated into Latin gave astronomy an assured place in the curricula of Bologna, Paris, Oxford, and the other emerging universities of the 12th century. In the quadrivium, students were instructed in astronomy, geometry, arithmetic and music: the four 'sciences' of mathematical proportion. Johannes de Sacrobosco (d.*c.*1256) wrote the best-selling *De sphaera mundi* ('On the Sphere of the World') around 1240, which would be a reference for students of astronomy for the next 400 years. Through the universities especially, astronomical knowledge became widespread in educated society. The poet Geoffrey Chaucer (*c.*1340–1400) wrote *Treatise on the Astrolabe* (*c.*1381), the first technological book in the English language, being a practical manual describing the use of the astrolabe.

One of the most adventurous branches of medieval astronomical thought was cosmology. While celestial mechanics was explained in terms of Aristotle's spheres and Ptolemy's epicycles, several theologian-astronomers had some remarkably modern-sounding ideas about time and space. Archbishop of Canterbury Thomas Bradwardine (*c.*1290–1349), Bishops Jean Buridan (*c.*1295–*c.*1358), Nicole de Oresme (*c.*1323–82) and Nicholas of Cusa (1401–64), and others asked such questions as, could time have existed before God created the Universe? Could there be such a thing as an INFINITE UNIVERSE? And was motion relative? Space, time and infinity fascinated medieval scholars. No one was burnt at the stake for asking such questions, for the academic clergy saw them as lying within the legitimate bounds of university discussion. Apart from an ultimately unsuccessful attempt to ban aspects of Aristotelian science in Paris in 1277, the medieval Church had no specific policies on astronomy, and would not have until the 17th century.

From the sheer number of manuscripts and, after 1460, printed astronomical books, astrolabes, dials and artefacts in libraries and museum collections, it is clear that astronomy had a high profile in medieval European culture. It was essential to Church administration, it was a major component of the university curriculum and it even penetrated vernacular literature. It was also suspicious of astrology. Where it differed essentially from the astronomy of the scientific revolution, however, was in its conservative, as opposed to the latter's progressive, approach. Without an already established astronomical culture, the developments of RENAISSANCE ASTRONOMY could not have taken place.

MENSA (GEN. MENSAE, ABBR. MEN)

Small and very faint southern constellation between Hydrus and Volans. Mensa was named Mons Mensae (Table Mountain) by Lacaille in the 18th century, because the southern part of the Large Magellanic Cloud in the northern part of the constellation reminded him of cloud overlying Table Mountain, South Africa. The brightest star is a dim mag. 5.1.

Megrez The star δ Ursae Majoris, visual mag. 3.32, distance 81 l.y., spectral type A2 V. Its name comes from the Arabic *maghriz*, meaning 'root' (of the tail), referring to its position in Ursa Major.

MEM Abbreviation of MAXIMUM ENTROPY METHOD

meniscus lens Thin LENS usually having one convex and one concave surface and resembling the shape of the meniscus at the surface of a liquid such as water. Common examples are contact lenses. In astronomy, meniscus lenses are used to improve image quality in reflecting telescopes. Examples are the corrector plate in SCHMIDT–CASSEGRAIN, Maksutov–Cassegrain and Maksutov–Newtonian telescopes. In all of these, a large meniscus lens with little optical power is mounted at the entrance to the optical tube; it is often referred to as the corrector plate. As the light enters the telescope its path is altered slightly by the meniscus lens so that it hits the main mirror at the optimum angle for forming sharp images right across the whole field of view. In the Schmidt–Cassegrain telescope the meniscus lens appears to be a flat plate, although it has a mild aspheric shape to one surface. In the MAKSUTOV TELESCOPE the meniscus lens is steeply curved.

Menkalinan The star β Aurigae, visual mag. 1.90, distance 82 l.y., spectral type A1 IV. It is an eclipsing binary of period 3.96 days, undergoing two minima of 0.1 mag. in each orbital cycle. Its name comes from the Arabic *mankib dhi al-'inān*, meaning 'shoulder of the charioteer'.

Menkar The star α Ceti, visual mag. 2.54, distance 220 l.y., spectral type M2 III. Binoculars show an apparent companion of mag. 5.6, but this is an unrelated background star. Its name comes from the Arabic *mankhar*, meaning 'nostril' or 'nose'.

Mensa *See* feature article

Menzel, Donald Howard (1901–76) American solar astronomer, astrophysicist and astronomy administrator who directed Harvard College Observatory (1952–66)

M

▲ **Mercury** Mariner 10 made three sets of observations of Mercury. This mosaic of images shows that the small planet is heavily cratered, indicating it has undergone no resurfacing for many millennia.

and helped to found several modern observatories. Menzel and his predecessor at Harvard, Henry Norris RUSSELL, strongly influenced the course of American astrophysics.

At Lick Observatory (1924–32), Menzel analysed flash spectra of the Sun obtained by William CAMPBELL, developing quantitative spectroscopy in the process. He used the new quantum physics to interpret stellar absorption and emission lines, and derived the abundances of the chemical elements in the solar chromosphere, paying close attention to the lines of neutral and ionized helium observable in the upper chromosphere and in prominences. Menzel's findings that these lines originated at temperatures greater than 20,000 K, together with his discovery of more ionized hydrogen than had been expected, proved that the chromosphere was a distinct layer of the Sun and not an extension of the photosphere, as had been thought.

Menzel applied his knowledge of astrophysics to the nature of diffuse nebulae. In the mid-1930s he developed the quantitative analysis of the spectra of nebulae, culminating in the publication of *Physical Processes in Gaseous Nebulae*. He refined the theory of RADIATIVE TRANSFER and made important studies of the late stages of stellar evolution and associated phenomena, including planetary nebulae around white dwarf stars. He derived a formula for calculating the temperature of a planetary nebula's central star.

MERCURY: DATA	
Globe	
Diameter	4879.4 km
Density	5.43 g/cm^3
Mass (Earth = 1)	0.055
Volume (Earth = 1)	0.0562
Escape velocity	4.4 km/s
Inclination of equator to orbit	0°.1
Sidereal period of axial rotation	58.646 days
Surface gravity (Earth = 1)	0.378
Surface temperature	700 K (day) 100 K (night)
Albedo	0.11
Mean visual magnitude (greatest elongation)	0.0
Orbit	
Semimajor axis	0.387 AU
Eccentricity	0.206
Inclination to ecliptic	7°.0
Sidereal period of revolution	87.969 days
Mean synodic period	115.88 days
Mean orbital velocity	47.87 km/s
Mean longitude of perihelion	77°.455
Satellites	none

During his directorship of Harvard College Observatory, Menzel engineered the transfer of the Smithsonian Astrophysical Observatory to Cambridge, Massachusetts, where it became the HARVARD–SMITHSONIAN CENTER FOR ASTROPHYSICS. He was involved with the founding of many observatories, including the High Altitude Observatory at Climax, Colorado, the solar observatory at Sacramento Peak, New Mexico, the NATIONAL RADIO ASTRONOMY OBSERVATORY and KITT PEAK NATIONAL OBSERVATORY.

Merak The star β Ursae Majoris, visual mag. 2.34, distance 79 l.y., spectral type A0m. It is the southern and fainter of the two POINTERS to Polaris, the North Pole Star. The name comes from the Arabic *marāqq*, meaning 'loins' or 'groin'.

Mercator Telescope *See* ROQUE DE LOS MUCHACHOS OBSERVATORY

Mercury Series of one-man spacecraft in which the USA first gained experience of space flight. The first flights, by modified Redstone boosters, were suborbital. On Mercury-Redstone 3, Alan SHEPARD became the first US astronaut to fly in space (1961 May 5). The capsule reached an altitude of 186 km (116 mi) and landed 478 km (297 mi) downrange. After a repeat suborbital flight by Virgil Grissom (1926–67), four orbital missions were flown using Atlas D boosters. On 1962 February 20, John GLENN became the first American to orbit the planet, completing three Earth revolutions in $4^h\ 55^m\ 23^s$. The last and longest Mercury mission was flown by Gordon Cooper (1927–) on 1963 May 15–16.

Mercury Innermost planet of the Solar System. Mercury is often described as the elusive planet, which is a misconception. Mercury is, in fact, easily picked up with the naked eye around the time of its greatest elongation from the Sun, when it can be followed without optical help for several days. Admittedly the planet is difficult to observe telescopically, mainly because of its small angular size and its proximity to the Sun; indeed it can never be seen more than 28° from the Sun since its orbit is inside that of the Earth. Nevertheless bright and dusky markings are visible in moderate-sized telescopes and observers have synthesized their findings into a reasonably reliable albedo chart of named features.

Mercury is the innermost and smallest of the Solar System's major planets. It is also the densest. The unusual density to radius relationship is inferred to indicate an exceptionally large metallic core, which is thought to be about 1800 km (1100 mi) in radius, only some 600 km (370 mi) beneath the surface. Mercury's core is far larg-

◀ **Mercury** Because Mercury is nearer to the Sun than the Earth, it displays phases. At (1) it is new, at (2) it is half-phase, at (3) full, and at (4) half-phase again.

er, by proportion, than those of the other terrestrial planets. The evolution of the planet must, therefore, be strongly influenced by core formation, probably from what is believed to be highly refractory material. The magnetic field is approximately a dipole aligned along the axis of rotation and can probably be attributed to dynamo action in a presumed fluid core. Alternatively, if the core is actually solid, Mercury's magnetic field may – uniquely among the terrestrial planets – be a remnant of magnetization of the crust that was acquired during the planet's formation.

Mercury orbits the Sun in 87.969 days and revolves about its axis in two-thirds of the orbital period, 58.646 days. The slow rotation (*see also* VENUS) may have come about through the retarding tidal action of the Sun. The two-thirds resonance requires the 'trapping' of Mercury in this state, and a non-hydrostatic bulge is inferred, probably caused by convection in its MANTLE occurring by solid state creep.

The lack of atmosphere, the intensity of solar radiation and the length of the planet's day lead to immense temperature contrasts. On the equator at perihelion the noonday temperature soars to 700 K. At night it plunges to 100 K.

Although very little was known about Mercury prior to the three MARINER 10 encounters in 1974 and 1975, Earth-based photometric data did hint at a rough and uneven surface analogous to that of the Moon. Mariner 10 confirmed the supposition. It revealed an airless world, the surface of which, in the form of craters, bright ray systems, ridges (dorsa), valleys (valles) and smooth lava plains (planitia), showed the imprint of a violent past. Craters are named after artists, authors and musicians; valleys after prominent radio observatories; and the planitia after Mercury in various languages and after ancient gods with a role similar to the Roman god Mercury. The rupes or scarps commemorate ships associated with exploration and scientific research.

Mariner 10 imaged only 35% of Mercury's surface. Of this, 70% was found to be ancient, heavily cratered terrain. The most significant formation is the CALORIS PLANITIA, which is most likely the result of an impact by a body similar to those that formed the multi-ring basins (circular maria) on the Moon. Characteristic surface features found on the planet are long, sinuous cliffs or lobate scarps. These scarps are steep, with an average height of 1 km (0.6 mi), extending for hundreds of kilometres. They may have formed as the crust cooled and shrank. In places the scarps cut across craters, intercrater plains and smooth plains. No strike-slip faults are seen on Mercury and there is no evidence of plate tectonics. Compressional features are seen, however, which could be explained by a contraction caused by cooling since the planet's formation. Estimates of the contraction required are about one part in 1000 or 10,000 and would be consistent with temperature decreases of a few hundred degrees. Solidification of a once-fluid core would be very effective.

Astonishingly, radar data indicate the possible existence of small polar ice deposits. This is not as improbable as it seems. It is just one of the mysteries that await resolution when the next planned space mission starts to probe the battered surface of our thin-shelled neighbour world.

mercury–manganese star Chemically peculiar late B STAR that has greatly enhanced abundances of mercury, manganese and other elements, including rare earths. In these class 3 CHEMICALLY PECULIAR (CP3) stars, mercury can be enhanced 100,000 times or more. The odd element composition is caused by diffusion in the quiet

▼ **Mercury** Discovery Rupes is one of the most prominent lobate scarps imaged by Mariner 10. It is obviously younger than many of the other features in the image because it cuts through them.

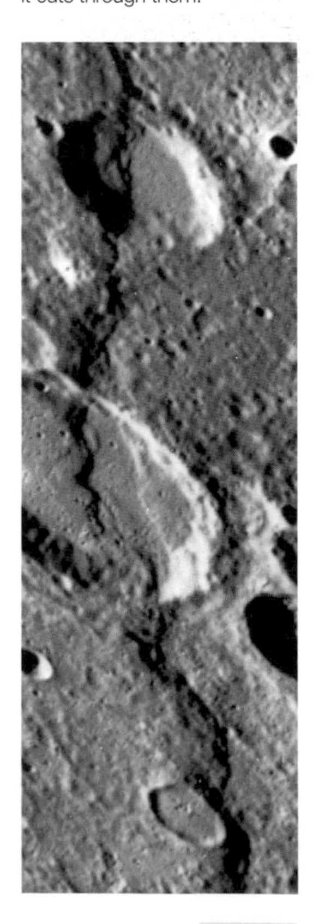

SELECTED SURFACE FEATURES OF MERCURY

Name	Latitude	Longitude	Description
Antoniadi Dorsa	25°.1N	30°.5W	scarp
Arecibo Vallis	27°.5S	28°.4W	valley
Beethoven	20°.8S	123°.6W	crater, 625 km (388 mi) in diameter
Borealis Planitia	73°.4N	79°.5W	the 'Northern Plain'
Budh Planitia	22°.0N	150°.9W	plain
CALORIS PLANITIA	30°.5N	189°.8W	largest multi-ring structure on Mercury,1300 km (800 mi) in diameter as imaged by Mariner 10; highest temperatures near this position
Caloris Montes	39°.4N	187°.2W	mountainous rim of Caloris Planitia
Chao Meng-Fu	87°.3S	134°.2W	south polar crater, diameter 167 km (104 mi)
Chekhov	36°.2S	61°.5W	crater, 199 km (124 mi) in diameter; it has an inner mountain ring
Discovery Rupes	56°.3S	38°.3W	ridge
Goethe	78°.5N	44°.5W	crater, diameter 383 km (238 mi), near border of the Borealis Planitia
Haystack Vallis	4°.7N	46°.2W	valley; named after radio telescope facility in Massachusetts
Heemskerck Rupes	25°.9N	125°.3W	ridge adjoining Budh Planitia
Homer	1°.2S	37°.0W	crater, 314 km (195 mi) in diameter
Hun Kal	1°.6S	21°.4W	crater, 13 km (8 mi) in diameter
Kuiper	11°.3S	31°.1W	crater, 62 km (39 mi) in diameter; it intrudes into the 130 km (81 mi) crater Murasaki. Kuiper is a ray-centre and was the first crater to be recognized on Mercury during the approach of Mariner 10 in 1974
Mozart	8°.0N	190°.5W	crater, 270 km (168 mi) in diameter, at the very edge of the region surveyed from Mariner 10
Murasaki	12°.6S	30°.2W	crater, 130 km (81 mi) in diameter; its wall is broken by the ray-crater Kuiper
Petrarch	30°.6S	26°.2W	prominent crater, 171 km (106 mi) in diameter
Pourquoi Pas Rupes	58°.1S	156°.0W	major ridge
Schubert	43°.4S	54°.3W	crater, 185 km (115 mi) in diameter
Sobkou Planitia	39°.9N	129°.9W	plain to east of the Caloris Planitia
Suisei Planitia	59°.2N	150°.8W	plain north-east of the Caloris Planitia
Tir Planitia	0°.8N	176°.1W	plain south of the Caloris Planitia
Tolstoy	16°.3S	163°.5W	apart from Beethoven, the largest named crater on the planet, 390 km (242 mi) in diameter
Zeehaen Rupes	51°.0N	157°.0W	major ridge east of Caloris Planitia

atmospheres of slowly rotating stars, some chemical elements sinking under the force of gravity, with others being lofted upwards by radiation.

meridian North–south reference line, particularly a GREAT CIRCLE on the Earth's surface that runs through both poles and that connects points of the same longitude. A **local meridian** is the line that passes through an observer, connecting both poles. The **prime meridian** is the line that passes from pole to pole through the Greenwich Observatory and is the zero point for measuring LONGITUDE. The CELESTIAL MERIDIAN is the great circle on the celestial sphere that passes through the north and south CELESTIAL POLES, together with the ZENITH and the NADIR.

MERLIN Abbreviation of MULTI-ELEMENT RADIO-LINKED INTERFEROMETER NETWORK

Merrill, Paul Willard (1887–1961) American astronomer who specialized in spectroscopy, pioneering infrared spectroscopy at Mount Wilson Observatory (1919–52). His most famous discovery was made in the early 1950s, during his studies of S stars, including the variable star R Andromedae. Merrill identified the chemical element technetium in these stars (*see* TECHNETIUM STAR), providing observational support for the *s*-process of nucleosynthesis of elements in stars.

mesosiderite One of the two subdivisions of STONY-IRON METEORITES. They are a much more heterogeneous group of meteorites than the PALLASITES, which comprise the second stony-iron subdivision. Mesosiderites are a mixture of varying amounts of iron–nickel metal with differentiated silicates, the whole assemblage of which seems to have been brecciated. Mesosiderites are subclassified on the basis of textural and compositional differences within the silicate fraction of the meteorites. Like main group pallasites, mesosiderites have oxygen isotope compositions similar to HOWARDITE-EUCRITE–DIOGENITE ASSOCIATION achondrites.

mesosphere Layer in the Earth's ATMOSPHERE directly above the STRATOSPHERE, in which the temperature falls with height to reach the atmospheric minimum of 110–173 K at its upper boundary, the mesopause. The mesopause's altitude shows two distinct values, 86±3 km (53±1.9 mi) and 100±3 km (62±1.9 mi), the higher value being encountered near the poles in summer, when there is up-welling at high latitudes. Above the mesopause, in the THERMOSPHERE, the temperature rises again. Within the mesosphere, heating by the absorption of solar ultraviolet radiation by ozone declines from the high value found in the upper stratosphere.

Messenger NASA Discovery programme spacecraft that will resume exploration of the planet MERCURY in 2008 January, after an interval of 33 years since MARINER 10. Messenger, which will make an initial flyby of the planet, will later orbit around Mercury in 2009 September, marking a first in space exploration. The orbit will be 200 km (125 mi) by 15,190 km (9430 mi), with an inclination of 80°. The spacecraft, designed to monitor Mercury's surface, space environment and geochemistry, as well as ranging, will study the surface composition, geological history, core, mantle, magnetic field and very tenuous atmosphere of the planet. It will search for possible water ice and other frozen volatiles at the poles over a nominal orbital mission of one year. The path of Messenger to Mercury will involve a gravity assist flyby of the Earth in 2005 August, and two Venus flybys in 2006 October and 2007 June. It will also make two flybys of Mercury, in 2008 January and October, prior to its orbital insertion in 2009.

Messier, Charles Joseph (1730–1817) French astronomer known for his discoveries of comets and his compilation of the famous MESSIER CATALOGUE of deep-sky objects. Little is known of his life before 1751, when he was hired by Joseph-Nicolas Delisle (1688–1768) of Paris Observatory to record observations in his 'neat, legible hand'. Delisle, who had found Messier a post at the observatory at the Hôtel de Cluny, calculated positions for the 1758/9 return of Halley's Comet, and instructed his assistant to search for it. Messier recovered the comet on 1759 January 21.

Messier, inspired by his success with Halley's Comet, searched for more. Between 1758 and 1801 he found about twenty comets (14 of which were sole discoveries), increasing his fame and status – in 1770 he was elected to the prestigious Paris Academy of Sciences, and was dubbed the 'comet ferret' by King Louis XV. But Messier's lasting fame rests on the numbered list he compiled of fuzzy, permanent objects that might be mistaken for comets. He had independently discovered the first two objects on this list, the Crab Nebula (M1) and a globular cluster in Aquarius (M2), in 1758 and 1760. His first sole discovery, the globular M3 in 1764, prompted him to make a systematic search, and by the end of that year he had observed and recorded objects up to M40. The list (up to M45) was first published in 1774, and longer versions appeared in 1780 (to M68) and 1781 (to M103). Today's Messier Catalogue of 110 objects includes seven others known to have been observed by Messier.

Messier Catalogue Listing of nebulae, star clusters and galaxies, numbering over 100, begun by Charles MESSIER. Messier drew up the list so that he and other comet-hunters would not confuse these permanent, fuzzy-looking objects with comets. The first edition of the catalogue was published in 1774, with supplements in 1780 and 1781. Not all the objects in the catalogue, known as **Messier objects**, were discovered by Messier himself; several were found by Pierre MÉCHAIN, and others added by later observers.

Objects in the catalogue are given the prefix M, and are still widely known by their Messier numbers. The Messier objects (*see* the accompanying table, page 258–59), including as they do many of the showpiece deep-sky objects visible from northerly latitudes, make popular targets for amateur observers. An attempt to observe as many as possible during the course of one night is known as a 'Messier marathon'. *See also* CALDWELL CATALOGUE

Messier numbers Numbers allocated to clusters, nebulae and galaxies listed by Charles Messier in his MESSIER CATALOGUE of comet-like objects.

Me star M STAR that has hydrogen EMISSION LINES in its spectrum; the 'e' is appended to the luminosity class. Me stars are found at all luminosities. Among the dwarfs (dMe stars), the emissions become more prevalent towards later (cooler) subtypes. PROXIMA CENTAURI (M5 Ve) is a good example. The emissions signify notable chromospheres, which in turn indicate strong magnetic dynamos. As a result, dMe stars can produce powerful solar-like flares involving much or even all of the star. FLARE STARS extend to the K dwarfs as well, and even into class L.

Among the class M giants, hydrogen emission – caused by pulsation-generated shock waves – is seen in both oxygen-rich and carbon-star Mira variables; Mira itself (M7 IIIe), Chi Cygni (S6 IIIe) and R Leporis (Hind's Crimson Star, C7 IIIe) are fine examples. In the most luminous class M supergiants (Mu Cephei M2 Iae, VV Cephei M2 Iaep), the emission is related to powerful enveloping winds; such stars are also IRREGULAR VARIABLES.

metals Term used by astronomers to describe all elements heavier than helium. Thus the metal content of a star, denoted by Z, is the combined mass fraction of all elements heavier than helium. Mass fractions of hydrogen and helium are denoted by X and Y respectively; thus a

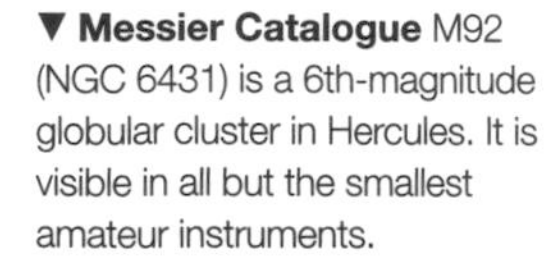

▼ **Messier Catalogue** M92 (NGC 6431) is a 6th-magnitude globular cluster in Hercules. It is visible in all but the smallest amateur instruments.

typical star may have $X = 0.75$, $Y = 0.24$ and $Z = 0.01$. While the values of X and Y are relatively constant from one unevolved star to another, Z changes dramatically over a range approximately 10^{-4} to 0.02. Indeed, stars are postulated (Population III stars) with $Z = 0$ (*see* POPULATIONS, STELLAR).

The relative abundances of the elements are usually quoted as a ratio to the amount of hydrogen, whose value is arbitrarily set at 12.0. The following are the figures for the solar abundances for the more common elements: H (12.00), He (10.93), O (8.82), C (8.52), N (7.96), Ne (7.92), Fe (7.60), Si (7.52), Mg (7.42) and S (7.20). These figures probably reflect the relative abundance of the heavy elements on a cosmic scale. The abundance of elements in a star is measured from the absorption lines in its SPECTRUM.

For stars other than the Sun the abundances of the elements are not generally so well known or easily measured. Thus Fe/H, the iron to hydrogen ratio, is often used to measure the 'metallicity' (total abundance of elements heavier than helium) of stars, it being assumed that the other heavy elements will be in the same proportion to iron as found in the Sun. Formally, Fe/H is log (NFe/NH)(star) − log (NFe/NH)(Sun).

The reason for the large variation in metal content, Z, of stars is that the heavy elements are not primordial. Only hydrogen and helium (and very small quantities of lithium) are formed in the BIG BANG. All other elements are synthesized in the NUCLEAR REACTIONS that take place in the centres of stars. This processed material is eventually returned to the interstellar medium by stellar mass loss, of which the most dramatic example is the SUPERNOVA explosion. Thus the interstellar medium is continually being enriched with 'metals'. It also follows that young stars that have recently formed from the interstellar medium will have large Z. Thus Population I stars, which are young, have large Z whereas Population II stars, which formed first and are now old, have low Z values. The cosmic lithium abundance is an important indicator of the nature of the early Universe. Unfortunately, there are many reasons why stellar lithium abundances will not accurately reflect primordial lithium abundances.

The metallicity Z has important consequences for a star. The most important effect is that a larger value of Z gives rise to a greater opacity of the stellar gas. The increase in opacity is due to the many possible lines, free-bound transitions and free electrons produced by ionized and partially ionized heavier elements. This will radically affect the structure of a star at any stage and also its evolution. Probably the most striking example of this is that Population II MAIN-SEQUENCE stars (low Z) are both hotter and more luminous than Population I stars of the same mass. Thus the Population II main sequence lies leftwards of the Population I main sequence in the HERTZSPRUNG–RUSSELL DIAGRAM (HR diagram). Furthermore, Population II stars will spend a shorter time on the main sequence than their Population I counterparts. Thus a good knowledge of Z is essential for judging the age of star clusters from the main sequence. Likewise, Z must be well known if the main sequences of two clusters are to be compared for the purpose of measuring cluster distances. A similar effect occurs on the HORIZONTAL BRANCH in the HR diagram. Thus low metallicity RR LYRAE STARS are thought to be intrinsically brighter than high-metallicity RR Lyrae stars.

The original metallicity of a star can also affect its structure and evolution by modifying the nuclear burning in its centre. The most obvious example of this is that the abundance of carbon and nitrogen will affect the CARBON–NITROGEN–OXYGEN CYCLE for hydrogen burning in massive main-sequence stars.

Some stars have particular elements or groups of elements that are relatively either over- or under-abundant. For example, AP STARS show over-abundances of manganese or europium, chromium and strontium. This is thought to be caused by some diffusion process which brings these elements up from the interior. Similarly AM STARS show a slight over-abundance of elements near iron and an under-abundance of calcium and scandium. Again, a diffusion mechanism is thought to operate.

◄ **meteor** Meteors sometimes leave a trail of ionized gas in the atmosphere as they burn up. This is known as a 'train', and may last for some minutes.

Amongst the cool giant stars, CARBON STARS (C stars) show carbon to oxygen ratio four times that of normal stars. S STARS show an over-abundance of zirconium, yttrium, barium and even technetium, which has a half-life of only 2×10^6 years (*see* TECNETIUM STAR). Some G and K giants also show over-abundance of barium, strontium and other elements produced by the S PROCESS as well as over-abundant carbon. These stars are called BARIUM STARS.

metastable state Relatively long-lived excited state (*see* EXCITATION) of an atom that has only forbidden transitions (*see* FORBIDDEN LINES) to lower levels. In the rarefied interstellar medium an electron can remain in such a state for its natural lifetime, but under terrestrial conditions atoms are very soon knocked out of metastable states by collisions. *See also* LASER; TWENTY-ONE CENTIMETRE LINE

meteor Brief streak of light seen in a clear night sky when a small particle of interplanetary dust, a METEOROID, burns itself out in Earth's upper atmosphere. As the meteoroid collides with atoms and molecules of air, a large quantity of heat energy is produced, which usually vaporizes the particle completely by a process of ABLATION. Vaporized atoms from the ablating meteoroid make further collisions, causing first excitation, then ionization as electrons are stripped from air atoms and molecules.

An ablating meteoroid thus leaves behind it a trail of highly excited atoms, which then de-excite to produce the streak of light seen as a meteor. Ionization produces a trail of ions and electrons which can scatter or reflect radio waves transmitted from ground-based equipment, causing a radio meteor. The trail of ionization is only a few metres wide, but may typically be 20–30 km (12–19 mi) long.

Most meteors appear at altitudes between 80–110 km (50–70 mi), where the air density becomes sufficiently high for ablation to occur. The altitude of this meteor layer varies slightly over the sunspot cycle, being greater at times of high solar activity. A typical meteor reaches its maximum brightness at an altitude of 95 km (59 mi). Usually, a visual meteor will persist for between 0.1 and 0.8 seconds. Brighter meteors sometimes leave a faintly glowing TRAIN or wake after extinction, and may show bursts of brightening (flares) along their paths.

Meteoroids enter the atmosphere at velocities between 11 and 72 km/s (7–45 mi/s). At the lower end of this range, the velocity is simply that of a particle in free fall hitting the Earth. The greatest value is obtained by summing the maximum heliocentric velocity of the meteoroid at a distance from the Sun of 1 AU (42 km/s or 26 mi/s) with Earth's mean orbital velocity of 30 km/s (19 mi/s).

A typical naked-eye meteor around magnitude +2 is produced by ablation of a meteoroid 8 mm in diameter, and with a mass around 0.1 g. Over the whole Earth, 100 million meteors in the visual range down to magnitude +5 occur each day.

M

Meteors can occur at any time, with the bulk of the annual influx of meteoroidal material (estimated at 16,000 tonnes) comprised of random, background SPORADIC METEORS. At certain times of year, numbers are enhanced by the activity of METEOR SHOWERS, which are produced as Earth passes through streams of debris laid down by short-period COMETS.

Meteor Crater (Arizona Crater, Barringer Crater) First crater to be recognized as being caused by METEORITE impact. It is situated on a flat plateau between Flagstaff and Winslow, Arizona, USA. The crater is a basin-shaped depression some 1200 m (3940 ft) in diameter. The walls surrounding the crater rise 37–50 m (121–164 ft) above the surrounding plain. The outer slopes are quite gentle, but the inner slopes are steep, being as much as 80° in the southern sector. A whole 600-m (2000-ft) section of the pre-existing sedimentary rocks has been lifted about 30 m (100 ft) to form the south wall of the crater.

The feature first created interest in 1891, when large quantities of meteoritic iron were discovered on the surrounding plain. In 1905 boreholes and shafts were sunk in the centre of the crater in an attempt to find the main mass of the meteorite. After passing through crushed sandstone and rock flour, undisturbed rocks were found at a depth of 185 m (607 ft). In 1920 attention was concentrated on the southern rim without success. It is now known that at times of such impacts, the meteorite is either vaporized or shattered to extents that depend on

MESSIER CATALOGUE

M	NGC	RA (2000.00) h	m	Dec. °	′	Mag (vis.)	Size (′)	Constellation	Description
1	1952	05	34.5	+22	00	8.4	6 × 4	Taurus	CRAB NEBULA
2	7089	21	33.5	−00	49	6.3	12	Aquarius	globular cluster
3	5272	13	42.2	+28	23	6.4	19	Canes Venatici	globular cluster
4	6121	16	23.6	−26	32	6.4	23	Scorpius	globular cluster
5	5904	15	18.6	+02	05	6.2	20	Serpens	globular cluster
6	6405	17	40.1	−32	13	4.2	33	Scorpius	BUTTERFLY CLUSTER
7	6475	17	53.9	−34	49	4:	50	Scorpius	open cluster
8	6523	18	03.3	−24	23	6:	45 × 30	Sagittarius	LAGOON NEBULA
9	6333	17	19.2	−18	31	7.3	6	Ophiuchus	globular cluster
10	6254	16	57.1	−04	06	6.7	12	Ophiuchus	globular cluster
11	6705	18	51.1	−06	16	6.3	12	Scutum	WILD DUCK CLUSTER
12	6218	16	47.2	−01	57	6.6	12	Ophiuchus	globular cluster
13	6205	16	41.7	+36	28	5.7	23	Hercules	globular cluster
14	6402	17	37.6	−03	15	7.7	7	Ophiuchus	globular cluster
15	7078	21	30.0	+12	10	6.0	12	Pegasus	globular cluster
16	6611	18	18.8	−13	47	6.4	8	Serpens	gaseous nebula
17	6618	18	20.8	−16	11	6.0	20 × 15	Sagittarius	OMEGA NEBULA
18	6613	18	19.9	−17	08	7.5	7	Sagittarius	open cluster
19	6273	17	02.6	−26	16	6.6	5	Ophiuchus	globular cluster
20	6514	18	02.3	−23	02	6.3	20	Sagittarius	TRIFID NEBULA
21	6531	18	04.6	−22	30	6.5	12	Sagittarius	open cluster
22	6656	18	36.4	−23	54	5.9	17	Sagittarius	globular cluster
23	6494	17	56.8	−19	01	6.9	27	Sagittarius	open cluster
24	6603	18	16.9	−18	29	4.6	4	Sagittarius	open cluster
25	IC4725	18	31.6	−19	15	6.5	35	Sagittarius	open cluster
26	6694	18	45.2	−09	24	9.3	9	Scutum	open cluster
27	6853	19	59.6	+22	43	7.3	8 × 4	Vulpecula	DUMBBELL NEBULA
28	6626	18	24.5	−24	52	7.3	15	Sagittarius	globular cluster
29	6913	20	23.9	+38	32	7.1	7	Cygnus	open cluster
30	7099	21	40.4	−23	11	8.4	9	Capricornus	globular cluster
31	224	00	42.7	+41	16	3.4	180 × 80	Andromeda	ANDROMEDA GALAXY
32	221	00	42.7	+40	52	8.7	3 × 2	Andromeda	galaxy
33	598	01	33.9	+30	39	5.7	67 × 42	Triangulum	PINWHEEL GALAXY
34	1039	02	42.0	+42	47	5.5	30	Perseus	open cluster
35	2168	06	08.9	+24	20	5.3	29	Gemini	open cluster
36	1960	05	36.1	+34	08	6.3	16	Auriga	open cluster
37	2099	05	52.4	+32	33	6.2	24	Auriga	open cluster
38	1912	05	28.7	+35	50	7.4	18	Auriga	open cluster
39	7092	21	32.2	+48	26	5.2	32	Cygnus	open cluster
40*		12	22.4	+58	05	9.0, 9.3	–	Ursa Major	double star
41	2287	06	47.0	−20	44	4.6	32	Canis Major	open cluster
42	1976	05	35.4	−05	27	4:	66 × 60	Orion	ORION NEBULA
43	1982	05	35.6	−05	16	9:		Orion	ORION NEBULA
44	2632	08	40.1	+19	59	3.1	95	Cancer	PRAESEPE
45	–	03	47.0	+24	67	1.6	120	Taurus	PLEIADES
46	2437	07	41.8	−14	49	6.0	27	Puppis	open cluster
47*	2422	07	36.6	−14	39	5.2	25	Puppis	open cluster
48*	2548	08	13.8	−05	48	5.5	35	Hydra	open cluster
49	4472	12	29.8	+08	00	8.6	4 × 4	Virgo	galaxy
50	2323	07	03.2	−08	20	6.3	16	Monoceros	open cluster
51	6194-5	13	29.9	+47	12	8.4	11 × 7	Canes Venatici	WHIRLPOOL GALAXY
52	7654	23	24.2	+61	35	7.3	13	Cassiopeia	open cluster
53	5024	13	12.9	+18	10	7.6	14	Coma Berenices	globular cluster
54	6715	18	55.1	−30	29	7.3	6	Sagittarius	globular cluster
55	6809	19	40.0	−30	58	7.6	15	Sagittarius	globular cluster

* identification uncertain

the characteristics of the particular event. It is, therefore, concluded that no large mass exists.

Over 30 tonnes of iron meteorite, known as Canyon Diablo, have been found around the crater. The meteorites consist mainly of iron with just over 7% nickel and 0.5% cobalt. In addition to the irons, oxidized iron shale balls were found intermingled with the local rock debris. Silica glass and very finely divided white sand (known as rock flour), together with forms of quartz known as coesite and stishovite, all point to the structure having been formed by meteoritic impact.

Studies of the distribution of the meteoritic material around the crater have led to the conclusion that the meteoroid responsible for the crater was travelling from north-north-west to south-south-east. This is consistent with the evidence gained from studies of the tilt of the rock layers forming the rim.

Many attempts have been made to ascertain the age of the crater. Early attempts suggested 2000 to 3000 years; current estimates give an age of about 50,000 years.

meteorite Natural object that survives its fall to Earth from space. It is named after the place where it was seen to fall or where it was found. About 30,000 meteorites are known, of which *c*.24,000 were found in Antarctica, *c*.4000 in the Sahara Desert and *c*.2000 elsewhere.

When an object enters the atmosphere, its velocity is greater than Earth's escape velocity (11.2 km/s or 7 mi/s), and unless it is very small (*see* MICROMETEORITE) frictional heating produces a FIREBALL. This fireball may rival the

MESSIER CATALOGUE (CONTINUED)

M	NGC	RA (2000.00)		Dec.		Mag (vis.)	Size (′)	Constellation	Description
		h	m	°	′				
56	6779	19	16.6	+30	11	8.2	5	Lyra	globular cluster
57	6720	18	53.6	+33	02	8.8	1 × 1	Lyra	RING NEBULA
58	4579	12	37.7	+11	49	8.2	4 × 3	Virgo	galaxy
59	4621	12	42.0	+11	39	9.3	3 × 2	Virgo	galaxy
60	4649	12	43.7	+11	33	9.2	4 × 3	Virgo	galaxy
61	4303	12	21.9	+04	28	9.6	6	Virgo	galaxy
62	6266	17	01.2	−30	07	8.9	6	Ophiuchus	globular cluster
63	5055	13	15.8	+42	02	8.5	9 × 5	Canes Venatici	BLACK EYE GALAXY
64	4826	12	56.7	+21	41	6.6	8 × 4	Coma Berenices	galaxy
65	3623	11	18.9	+13	05	9.5	8 × 2	Leo	galaxy
66	3627	11	20.2	+12	59	8.8	8 × 2	Leo	galaxy
67	2682	08	50.4	+11	49	6.1	18	Cancer	open cluster
68	4590	12	39.5	−26	45	9:	9	Hydra	globular cluster
69	6637	18	31.4	−32	21	8.9	4	Sagittarius	globular cluster
70	6681	18	43.2	−32	18	9.6	4	Sagittarius	globular cluster
71	6838	19	53.8	+18	47	9:	6	Sagittarius	globular cluster
72	6981	20	53.5	−12	32	9.8	5	Aquarius	globular cluster
73	6994	20	58.9	−12	32	9.0	3	Aquarius	asterism
74	628	01	36.7	+15	47	10.2	8	Pisces	galaxy
75	6864	20	06.1	−21	55	8.0	5	Sagittarius	globular cluster
76	650 + 651	01	42.4	+51	34	10.1	1 × 1	Perseus	LITTLE DUMBBELL
77	1068	02	42.7	−00	01	8.9	2	Cetus	galaxy
78	2068	05	46.7	+00	03	8.3	8 × 6	Orion	gaseous nebula
79	1904	05	24.5	−24	33	7.9	8	Lepus	globular cluster
80	6093	16	17.0	−22	59	7.7	5	Scorpius	globular cluster
81	3031	09	55.6	+69	04	7.9	16 × 10	Ursa Major	galaxy
82	3034	09	55.8	+69	41	8.8	7 × 2	Ursa Major	galaxy
83	5236	13	37.0	−29	52	10.1	10 × 8	Hydra	galaxy
84	4374	12	25.1	+12	53	9.3	3	Virgo	galaxy
85	4382	12	25.4	+18	11	9.3	4 × 2	Coma Berenices	galaxy
86	4406	12	26.2	+12	57	9.7	4 × 3	Virgo	galaxy
87	4486	12	30.8	+12	24	9.2	3	Virgo	galaxy
88	4501	12	32.0	+14	25	10.2	6 × 3	Coma Berenices	galaxy
89	4552	12	35.7	+12	33	9.5	2	Virgo	galaxy
90	4569	12	36.8	+13	10	10.0	6 × 3	Virgo	galaxy
91*	4548	12	35.4	+14	30	10.2	5 × 4	Coma Berenices	galaxy
92	6341	17	17.1	+43	08	6.1	12	Hercules	globular cluster
93	2447	07	44.6	−23	52	6.0	18	Puppis	open cluster
94	4736	12	50.9	+41	07	7.9	5 × 4	Canes Venatici	galaxy
95	3351	10	44.0	+11	42	10.4	3	Leo	galaxy
96	3368	10	46.8	+11	49	9.1	7 × 4	Leo	galaxy
97	3508	11	14.8	+55	01	9.9	3	Ursa Major	OWL NEBULA
98	4192	12	13.8	+14	54	10.7	8 × 2	Coma Berenices	galaxy
99	4254	12	18.8	+14	25	10.1	4	Coma Berenices	galaxy
100	4321	12	22.9	+15	49	10.6	5	Coma Berenices	galaxy
101	5457	14	03.2	+54	21	9.6	22	Ursa Major	galaxy
102*	–	–	–	–	–	–	–	–	–
103	581	01	33.2	+60	42	7.4	6	Cassiopeia	open cluster
104	4594	12	40.0	−11	37	8.0	7 × 4	Virgo	SOMBRERO GALAXY
105	3379	10	47.8	+12	35	9.2	2 × 2	Leo	galaxy
106	4258	12	19.0	+47	18	8.6	20 × 6	Ursa Major	galaxy
107	6171	16	32.5	+13	03	9.2	8	Ophiuchus	globular cluster
108	3556	11	11.5	+55	40	10.7	8 × 2	Ursa Major	galaxy
109	3992	11	57.6	+53	23	10.8	7	Ursa Major	galaxy
110	205	00	40.4	+41	41	8.0	17 × 10	Andromeda	galaxy

* identification uncertain

Sun in brightness. For example, a brilliant fireball on 1890 June 25, at 1pm, was visible over a large area of the mid-west of the United States; the CHONDRITE fall at Farmington, Kansas, was the result. If an object (a meteoroid) enters the atmosphere at a low angle, deceleration in the thin upper atmosphere may take tens of seconds. The fireball of 1969 April 25 travelled from south-east to north-west and was visible along its 500 km (310 mi) trajectory from much of England, Wales and Ireland. As commonly occurs, towards the end of its path the fireball fragmented. Sonic booms were heard after its passage, and two meteoritic stones were recovered, some 60 km (37 mi) apart, the larger at Bovedy, Northern Ireland, which gave its name to the fall. A meteorite that fell at Pultusk, Poland, in 1868, after fragmenting in the atmosphere, is estimated to have had a total weight of 2 tonnes among some 180,000 individual stones. Large meteoroids of more than *c.*100 tonnes that do not break up in the atmosphere are not completely decelerated before impact. On striking the surface at hypersonic velocity, their kinetic energy is released, causing them to vaporize and produce explosion craters, such as METEOR CRATER.

Photographic observations of fireballs indicate that more than 19,000 meteorites heavier than 100 g land annually, but, of these, most fall in the oceans or deserts and fewer than 10 become known to science. Photographic observations and visual sightings of meteorite-producing fireballs show that they have orbits similar to those of EARTH-CROSSING ASTEROIDS. It is apparent that most meteorites come from the ASTEROID BELT, but a few come from the Moon (LUNAR METEORITES) or from Mars (MARTIAN METEORITES). Meteorites can be divided into three main types, according to their composition: STONY METEORITES (CHONDRITES, ACHONDRITES); STONY-IRON METEORITES (MESOSIDERITES, PALLASITES); and IRON METEORITES. There is not always a clear-cut distinction between types: for example, many iron meteorites contain silicate inclusions related to chondritic and achondritic meteorites.

Some 95% of the meteorites seen to fall are stony meteorites, being composed dominantly of stony minerals. Iron meteorites constitute the bulk of the remainder, while meteorites composed of equal-part mixtures of iron–nickel metal and stony material, known as stony-iron meteorites, are very rarely seen to fall. However, many more iron and stony-iron meteorites have been found than were observed to fall, which reflects their resistance to erosion and their distinctive appearance relative to terrestrial rocks, rather than a change in the composition of the meteorite flux with time. Since 1969 meteorites have been found in large numbers on the surface of the ice in parts of Antarctica. The small number of Antarctic iron meteorites relative to stony types is similar to the ratio in observed falls.

Although frictional heating during atmospheric flight causes the outside of a meteoroid to melt, the molten material is swept into the atmosphere as droplets. The bulk of the heat is removed with the melt, and the inside of the object stays cold. Only the melt during the last second of hypersonic flight solidifies on the object's surface as it falls to Earth under gravity. The solidified melt is known as fusion crust. On most stony meteorites it is dull black, but on many achondrites it is a glossy black. Iron–nickel metal conducts heat more efficiently than stone, so some of the heat generated in atmospheric flight may penetrate to the interior of an iron meteorite. Stony meteorites, however, preserve a record of their history before their encounter with our planet. Meteorites often record shock or thermal events when they were part of their parent bodies. For example, many L-group ordinary chondrites were shock-reheated 500–1000 million years ago. Many chondrites preserve evidence of conditions in the Solar System of 4560 million years ago, and none has an age or isotopic signature consistent with an origin outside it. They provide important clues to the origin and history of the Solar System, as well as records of conditions in inter-planetary space.

Various ages of meteorites can be measured. The **formation interval** is the period between stellar processing and the incorporation of an element into a meteorite. Chemical elements heavier than hydrogen and helium are synthesized in stars, and many meteorites preserve a record of these processes. Chondrites often contain the decay products of short-lived radionuclides, such as plutonium, indicating that this element was present in the matter from which the Solar System formed. From the quantity of plutonium that must have been present relative to other elemental abundances, the plutonium must have been formed within about 200 million years of the formation of the Solar System.

Most meteorites or their components, such as CHONDRULES, went through a high-temperature event early in their history. The **age of formation** is the time, to the present, since a meteorite first cooled to become a closed chemical system. Uranium, for example, decays to lead at a fixed rate, and the uranium–lead age of a meteorite is the time that has elapsed since the uranium and lead were able to exchange freely, when the body was hot. The lead formed from the decay of uranium can be measured; the quantity is proportional to the uranium content, also measured, and to time, which usually is close to 4560 million years.

When an object is broken from its parent asteroid, it continues to orbit the Sun. Cosmic rays from the Sun, and beyond, bombard its surface. The **exposure age** is the time during which this bombardment takes place. Radiation damage can be measured in various ways, including the content of a substance produced in nuclear reactions, such as a radioactive isotope of aluminium. (The levels of radioactivity in meteorites are so low that specially prepared, ultra-sensitive counting equipment is required for their measurement.) Exposure ages range from a few hundred thousand years for some stony meteorites to 1000 million years for a few iron meteorites. These ages reflect the susceptibility of stony types to erosion by impact in space, compared with the durability of iron–nickel metal.

The **terrestrial age** is the time since a meteorite landed on Earth. The meteorite with the longest terrestrial age known is a chondrite within a 460-million-year-old Swedish Ordovician limestone. The second-longest terrestrial age may be that of an iron meteorite found in 300-million-year-old coal in Russia. Apart from these examples, most meteorites have much shorter terrestrial ages. Some meteorites from Antarctica have lain in the ice for up to a million years, but these are exceptional.

SOME IMPORTANT ANNUAL METEOR SHOWERS AND COMETARY ASSOCIATIONS

Stream	Maximum	Normal limits	ZHR	Radiant RA h	m	Dec. °	Parent comet
Quadrantids	Jan 3–4	Jan 1–6	100	15	28	+50	96P/Macholz 1 ?
Lyrids	Apr 21–22	Apr 19–25	15	18	08	+32	C/1861 G1Thatcher
*Pi-Puppids	Apr 23–24	Apr 21–26	40?	07	18	−44	26P/Grigg–Skjellerup
Eta Aquarids	May 5–6	Apr 24–May 20	50	22	20	−01	1P/Halley
Daytime Beta Taurids	Jun 29–30	Jun 23–July 05	20	03	40	+15	2P/Encke
Alpha Capricornids	Aug 2–3	July 15–Aug 25	8	20	36	−10	45P/Honda–Mrkós–Pajdusaková
Delta Aquarids	July 28–29	July 15–Aug 20	20	22	36	−17	96P/Macholz 1 ?
Perseids	Aug 11–12	July 25–Aug 21	90	93	04	+58	109P/Swift–Tuttle
*Giacobinids	Oct 8–9	Oct 7–10	variable	17	23	+57	21P/Giacobini–Zinner
Orionids	Oct 21–22	Oct 16–30	25	06	24	+15	1P/Halley
Taurids	Nov 4–5	Oct 20–Nov 30	10	03	44	+14	
				03	44	+22	2P/Encke
**Leonids	Nov 17–18	Nov 15–20	15	10	08	+22	55P/Tempel–Tuttle
Geminids	Dec 13–14	Dec 7–15	120	07	28	+32	Asteroid 3200 Phaethon
Ursids	Dec 21–22	Dec 17–25	10	14	28	+78	8P/Tuttle

*The Pi-Puppids and October Draconids are periodic streams, only giving good displays when the parent comets return to perihelion.
**The Leonids vary in peak activity from a ZHR of 15 in years between perihelion returns of the parent comet, to storms, with estimated ZHR as high as 140,000 to 200,000 under favourable circumstances at roughly 33-year intervals

M

meteoroid Small natural body in orbit around the Sun. Meteoroids may be cometary or asteroidal in origin. The distinction between a large meteoroid and a small ASTEROID is rather vague; rocky bodies of, say, 10 m (30 ft) in diameter could fit into either category. Millimetre-sized dusty cometary meteoroids entering Earth's atmosphere are completely destroyed by ABLATION, producing short-lived streaks of light seen in the night sky as METEORS. Cometary meteoroids have typical densities of 0.2 g/cm^3.

meteor shower Enhancement of METEOR activity produced when Earth runs through a METEOR STREAM. About 25 readily recognisable meteor showers occur each year, the most prominent being the QUADRANTIDS, PERSEIDS and GEMINIDS. Meteor showers occur at the same time each year, reflecting Earth's orbital position, and its intersection with the orbit of the particular meteor stream whose meteoroids are being swept up. Shower meteors appear to emerge from a single part of the sky, known as the RADIANT. Some showers are active only periodically as a result of uneven distribution of stream meteoroids, the GIACOBINIDS perhaps being the best example. The accompanying table lists some of the more important meteor showers; the zenithal hourly rate (ZHR) can vary, and values given here are intended only as a rough guide.

meteor stream Trail of debris, usually from a COMET, comprised of METEOROIDS that share a common orbit around the Sun. Passage of Earth through such a stream gives rise to a METEOR SHOWER. Meteor streams may also be associated with some asteroids – most notably 3200 PHAETHON, the debris from which produces the GEMINIDS.

Meteor streams undergo considerable evolution over their lifetimes, which are probably measured in tens of thousands of years. Initially, when the parent comet has only recently been captured into a short-period orbit, debris (released only at perihelion) is found relatively close to the comet NUCLEUS. Since the nucleus rotates, meteoroids are ejected both behind and ahead of the comet. Over successive returns, more material is added to the near-comet meteoroid cloud, which begins to spread around the orbit.

Eventually, over an interval of only a few tens of years for a very short-period comet, up to thousands of years for those with longer periods, meteoroids will spread right around the orbit, completing loop formation. Many meteor streams are believed to have a filamentary structure, with interwoven strands of meteoroids, released at separate perihelion returns of the parent comet, running through them.

Gravitational perturbations by the planets, and the POYNTING–ROBERTSON EFFECT, serve to increase the spread of material in a stream. Stream meteoroids come to perihelion at very similar orbital positions (since this is where they are released from the parent nucleus) but show considerable spread in aphelion distances: consequently, meteor streams are much narrower and more concentrated around the perihelion point.

As a stream ages further, and its parent comet ceases to add new material, it begins to disperse, eventually merging into the general background of the zodiacal dust cloud permeating the inner Solar System.

methane (CH_4) First of the alkanes or saturated acyclic hydrocarbons. Its melting point is 90.7 K and its boiling point 109 K. It was discovered on Earth in 1778 by the Italian chemist Count Alessandro Volta (1745–1827). In 1935 it was detected in the atmosphere of Jupiter by Rupert WILDT from its spectrum lines. In 1977 it was identified in interstellar space through its emissions at a wavelength of 3.9 mm.

Methane is principally significant as a greenhouse gas. On the Earth, it presently makes a smaller contribution to global warming than does carbon dioxide. Potentially though, methane could become much more important, even closing the 8 to 12 μm 'window' if the vast quantities of the gas currently locked in permafrost regions are released as the Earth's temperature rises. Methane is also found in significant quantities on TITAN, TRITON and PLUTO. On Titan, methane is the second most abundant atmospheric gas after nitrogen, though it still forms only a few per cent of the atmosphere. It may, however, also be present in liquid and/or solid form on the surface. On Triton and Pluto methane is present in small amounts, along with other frozen gases, as a thin solid surface coating.

Metis One of the inner moons of JUPITER, discovered in 1979–80 by Stephen Synnott in images obtained by the VOYAGER project. It is irregular in shape, measuring about 60 × 34 km (37 × 21 mi). With ADRASTEA, Metis orbits within Jupiter's main ring, about 128,000 km (80,000 mi) from the centre of the planet, taking 0.295 days to complete one of its near-circular equatorial orbits.

Metis is the also name of a large MAIN-BELT ASTEROID, number 9, discovered in 1849. It has a diameter of about 190 km (120 mi).

Metonic cycle Interval of 19 years after which the phases of the Moon recur on the same days of the year. This occurs because 19 years (6939.60 days) is almost exactly equal to 235 lunar months (6939.69 days). Its discovery is often attributed to the Greek astronomer Meton around 433 BC and the cycle forms the basis of the Greek and Jewish calendars.

Meudon Observatory French observatory in a southern suburb of Paris, dating from 1876. It originated when Meudon's ancient royal estate was placed at the disposal of astronomer Jules JANSSEN to pursue his research away from urban pollution. A large refractor of 0.83 m (32 in.) aperture and a 1-m (39-in.) reflector were commissioned in 1893, and other instruments followed, including solar telescopes. In 1925 Meudon became part of PARIS OBSERVATORY, and saw further expansion. Solar physics flourished at the observatory in the 1960s, and a 36-m (118-ft) high solar tower was erected in 1969. Astronomers from Meudon have traditionally used French facilities such as the PIC DU MIDI OBSERVATORY, and today use EUROPEAN SOUTHERN OBSERVATORY and other international telescopes.

MGS Abbreviation of MARS GLOBAL SURVEYOR

MHD Abbreviation of MAGNETOHYDRODYNAMICS

Miaplacidus The star β Carinae, visual mag. 1.67, distance 111 l.y., spectral type A1 III. The origin of its name is unknown.

Mice (NGC 4676 A and B) Pair of interacting galaxies – a spiral and a lenticular – in the constellation Coma Berenices (RA 12^h 46^m.2, dec. +30°44′); they are also catalogued as IC 819 and IC 820. Tails of stars, gas and dust extend away from the pair, which have overall magnitude +14 and total angular size 4′. The Mice lie at a distance of 350 million l.y.

Michell, John (1724–93) English physicist and one of the first to suggest the possibility of black holes. In 1784 he made complex mathematical calculations describing the physical characteristics of a star with the same density as the Sun but having a much larger diameter, showing that the escape velocity for such a hypothetical object was the velocity of light. Michell concluded that light itself could therefore not escape from such a supermassive object.

Michelson–Morley experiment Experiment carried out by German-American physicist Albert Michelson (1852–1931) in 1881 and, with greater precision, by Michelson and the American chemist Edward Morley

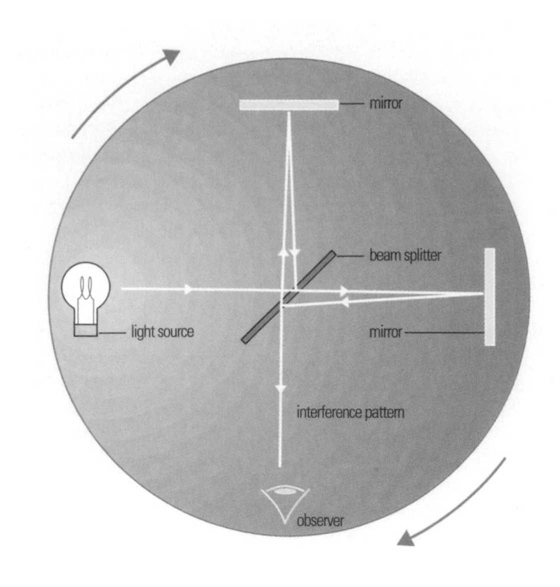

▶ **Michelson–Morley experiment** Light from a source is split into two beams, which travel at right angles to each other before being recombined in an interference pattern. The table is rotating, and if the Earth were moving through an 'ether', as was believed by some astronomers in the 19th century, the light in one path would take longer to reach the mirror and the interference patterns would be seen to change as the table rotated.

(1838–1923) in 1887. The experiment attempted to detect the motion of the Earth through the luminiferous ether (*see* LIGHT). Their apparatus split a beam from a common source into two parts, one travelling at right angles to the other. Both beams were reflected and recombined to produce an INTERFERENCE pattern. If one beam were to travel out and back along the direction of the Earth's supposed motion through the ether then, if light moved through the ether at a constant speed, it should take a marginally longer time to cover the same distance compared with a beam travelling at right angles to the Earth's motion. Rotation of the two beams should have produced a readily detectable change in the observed interference pattern. However, no change was detected. The null result of this experiment was explained by the theory of SPECIAL RELATIVITY. *See also* FITZGERALD CONTRACTION

Michelson stellar interferometer *See* STELLAR INTERFEROMETER

microdensitometer Computer-controlled measuring machine primarily used for retrieving data from astronomical photographic plates. The microdensitometer provides a measure of both plate position and photographic density, lending itself to applications in stellar ASTROMETRY and PHOTOMETRY. It operates by shining a beam of light through the plate and measuring the transmitted light. The plate is scanned in small strips, a massive x–y table being used to move it relative to the beam. A complete scan is built up by moving the table in the y direction with the x coordinate fixed; once a strip is complete, the table is stepped by a few millimetres in the x direction and the next strip measured. *See also* PLATE-MEASURING MACHINE

▼ **Milky Way** The plane of our Galaxy is visible in the night sky as the Milky Way. The red patches are star-forming regions within the spiral arms, including, towards the left, M8 (NGC 6523), the Lagoon Nebula.

micrometeorite (interplanetary dust particle (IDP), cosmic dust) Natural object from space that is sufficiently small (typically less than *c*.200 μm in diameter) to escape destruction during atmospheric flight. Micrometeorites arise from comets or from collisions between asteroids. The particles are dominantly silicates, with additional carbon, sulphides and metal. They are collected routinely from the atmosphere by research aircraft flying at altitudes between 18 and 22 km (11–14 mi). Micrometeorites may also be collected from localities on the Earth's surface where the terrestrial dust component is in low abundance. One of the most successful recovery programmes involves the melting of large volumes of Antarctic ice and the subsequent filtering of the water; the Antarctic micrometeorites so recovered are little altered by terrestrial processes.

micrometer Instrument attached to a telescope for measuring small angular separations of celestial objects. Several different types exist.

Both the **binocular micrometer** and the **comparison image micrometer** produce an artificial image of a pair of stars that can be viewed simultaneously with the real star images. The separation of the artificial pair can be adjusted until they match the real binary pair and the angle between them read on a scale.

The **cross wire micrometer** contains two wires, crossed at 90° in the focal plane of the telescope. By measuring the TRANSIT times of stars, including those of known position, the RIGHT ASCENSION and DECLINATION of an object can be determined.

A **double image micrometer** uses a split lens or prism to produce a doubling of the image. The orientation of the field may be adjusted to bring the line of the separated lenses into line with a pair of double stars, so determining their position angle relative to the north point or zero.

The **objective-grating micrometer** uses a coarse grating or bars to produce a diffraction pattern from which the separation of double stars can be measured, while the **reticulated micrometer** consists of a grid of parallel lines which cross each other at 90°. Any two stars in the field of view can be so lined up on the grid so that their relative positions can be estimated.

The **ring micrometer** consists of a ring of opaque material on glass placed in the focal plane of the telescope's objective or primary mirror. The ring, whose internal dimensions against the sky have been determined, is used to time any two stars close enough in the field of view to cross the ring so that their relevant angular separations can be calculated. *See also* FILAR MICROMETER

Microscopium *See* feature article

microwaves Part of the ELECTROMAGNETIC SPECTRUM between 1 mm and 30 cm (300 GHz and 1 GHz) wavelength, contained in the millimetre and radio region of the spectrum. Originally the equipment used to detect astronomical microwaves had been built to measure directly the microwave's varying electric field, but the term is not often used to describe instruments or telescopes, for example the ARECIBO radio telescope works at microwave frequencies. The COSMIC MICROWAVE BACKGROUND peaks at around 1 mm, although it was originally detected at 7 cm using a microwave receiver system. Molecules such as neutral hydrogen (21 cm) and the hydroxyl radical (18 cm) were first detected in the microwave region of the spectrum.

mid-infrared Region of the ELECTROMAGNETIC SPECTRUM between around 5 and 25 μm, part of which is observable from ground-based telescopes at very dry sites. Important spectral features from molecules and dust are found in this range. Warm objects (with temperatures of a few hundred kelvins), in particular many ASTEROIDS, emit strongly in the mid-infrared. *See also* INFRARED ASTRONOMY

midnight Sun Visibility of the Sun above the horizon at midnight, from inside the Arctic Circle (66°33′N latitude) or Antarctic Circle (66°33′S latitude) for a period centred on the summer solstice in the northern and southern hemispheres respectively. Visibility of the midnight Sun is restricted to the day of the solstice at the Arctic or Antarctic Circle. At either pole, however, the Sun is always above the horizon between the spring and autumn equinoxes.

Milky Way Faint luminous band that encircles the sky at an angle of about 63° to the celestial equator. It is irregular and patchy and varies from about 3° to 30° in width. It is easily visible to the naked eye from a good site on a clear moonless night, especially the southern section around Sagittarius. Through even a small telescope it can be seen to be composed of millions of faint stars, and it is in fact the part of our own GALAXY immediately surrounding the Sun (*see* ORION ARM). The distribution of stars is essentially uniform, and the patchiness results from absorption of starlight in some regions by DARK NEBULAE such as the COALSACK, PIPE NEBULA and CYGNUS RIFT. The term 'galaxy' derives from the Greek word for milk, γαλα (gala).

Miller, Stanley *See* UREY, HAROLD

millimetre-wave astronomy Study of extraterrestrial objects that emit in the region of the ELECTROMAGNETIC SPECTRUM between 1 mm and 10 mm (300 GHz and 30 GHz) wavelength, originally part of the radio and microwave region. It was defined as a separate region of the spectrum because of the special equipment and techniques needed to observe astronomical objects at these wavelengths. The science undertaken using the millimetre wavelength range focuses on studying the places where new stars form, molecules, and very cold material, but many types of astronomical object can be detected at millimetre wavelengths.

Millimetre-wave telescopes are built at high, dry sites, since millimetre waves are absorbed by the water vapour in Earth's atmosphere. The detectors are cooled to lower the BACKGROUND NOISE and to increase sensitivity in a manner similar to the techniques used for INFRARED ASTRONOMY. Many radio techniques can be applied at millimetre wavelengths, such as those used for RADIO INTERFEROMETERS and APERTURE SYNTHESIS. Since the wavelengths are smaller, the telescopes can be smaller and the distances between them less, to achieve the same resolution. Several large millimetre arrays have been built, such as INSTITUT DE RADIO ASTRONOMIE MILLIMÉTRIQUE (IRAM) in Europe, Owens Valley Radio Observatory (OVRO) and BERKELEY ILLINOIS MARYLAND ASSOCIATION (BIMA) in the United States, and NOBEYAMA in Japan; there is an international project to build an array of 64 telescopes in Chile called ATACAMA LARGE MILLIMETRE ARRAY (ALMA), due for completion around 2010 (first science observations in 2006). The arrays are used to map small structures such as cold dust disks around protostars. The energy distribution of the COSMIC MICROWAVE BACKGROUND peaks at 1 mm.

Several important molecules are observed in the millimetre region, the most important being CO (carbon monoxide) at 2.6 mm and 1.3 mm, the spectral emission lines from the lowest energy levels in the molecule. CO can be used to trace molecular hydrogen and thus CO surveys have mapped the distribution of the coldest gas in the Galaxy. The CO line is easily saturated because of the large number of molecules in the lines of sight in the Galaxy, but several isotopes (^{13}CO and $C^{18}O$) can also be detected and these have lower densities and so are not saturated. Whilst CO traces the low-density gas, CS and HCN can trace higher densities, so that regions of star formation can be probed using different molecules to map different temperature and density regimes. *See also* SUBMILLIMETRE ASTRONOMY

MICROSCOPIUM (GEN. MICROSCOPII, ABBR. MIC)

Small, inconspicuous southern constellation, representing a microscope, between Sagittarius and Piscis Austrinus/Grus. Microscopium was introduced by Lacaille in the 18th century. Its brightest stars, γ and ε Mic, are both mag. 4.7.

millisecond pulsar PULSAR with a period of a few milliseconds (ms). The first millisecond pulsar to be discovered had a period of 1.6 ms or 0.0016 s (as compared to the CRAB PULSAR, which has a period of 0.033 s). It is believed to be over a hundred million years old, and it has a weak magnetic field, which is why it has not slowed its rotation as expected. There are now more than forty millisecond pulsars known, with periods up to 300 ms; roughly half are in binary systems.

Mills, Bernard Yarnton (1920–) Australian engineer and radio astronomer, and inventor of the Mills Cross radio telescope. He designed and built many different types of radio telescope. Mills Cross, named after its designer, was built in 1954 and featured two 457-m (1500-ft) arrays. The culmination of Mills' career came with the design and construction of the Molonglo Cross, a huge 408-MHz radio telescope with an aperture of 1600 m (1 mi), completed in 1967.

Mills Cross One of the earliest radio arrays, designed by Bernard Mills of the Commonwealth Scientific and Industrial Research Organisation (CSIRO), Sydney, Australia. The antennae were arranged along two lines 1500 ft long (457 m) at right angles. It was used to map the Magellanic clouds. *See also* RADIO ASTRONOMY

Milne, (Edward) Arthur (1896–1950) English theoretical astrophysicist and cosmologist. At Cambridge's Solar Physics Observatory (1920–32) he solved fundamental problems of stellar atmospheres and structure. His most valuable work, involving the mathematical modelling of radiative equilibrium and thermal ionization in stellar atmospheres, enabled astronomers to predict a star's temperature and atmospheric pressures from measurements of the widths and intensities of its spectral lines, and by relating the optical depths, or opacities, of stellar atmospheres to spectral features. Milne was the first astrophysicist to correctly explain that instabilities in the balance between radiation pressure and gravity in the chromosphere of the Sun and similar stars could eject atoms at velocities of 1000 km/s (600 mi/s).

Mimas Innermost and smallest of Saturn's spherical SATELLITES. It was discovered by William HERSCHEL in 1789. Unlike its neighbour, the slightly larger ENCELADUS, Mimas appears to have been a totally passive world, with no sign of the effects of TIDAL HEATING. It has an ancient, heavily cratered, icy surface. There is one giant crater, named Herschel, which is 130 km (80 mi) across. Relative to the satellite's size, Herschel is almost as large as the crater Odysseus on TETHYS, but unlike Odysseus it has retained its original topography, including an impressive central peak. The impact event that created

▼ Mimas This Voyager 1 image of Saturn's tiny satellite Mimas from 425,000 km (264,000 mi), shows a heavily cratered world, including the giant crater Herschel.

M

▲ **Mir** The space shuttle *Atlantis* is seen here leaving Mir after mission STS-71. Despite many setbacks, Mir was a notable success for the Soviet/Russian space agency.

Herschel was almost large enough to break Mimas into fragments; indeed, it may be that Mimas re-accreted from fragments of a previous object that was shattered by a catastrophic collision. *See* data at SATURN

Mimosa Alternative name for the star β Crucis. *See* BECRUX

Minkowski, Hermann (1864–1909) German mathematician and professor of physics at Zürich and Göttingen who, in 1908, suggested that non-Euclidean space and time were intimately related and that events should be considered in the context of a four-dimensional spacetime continuum. Before Minkowski's theoretical work, space and time were held to be separate entities. Einstein incorporated the concept of the spacetime continuum into his theory of general relativity.

Minkowski, Rudolph Leo (1895–1976) German-American astrophysicist (born in Strasbourg, now in France) who greatly advanced the understanding of supernovae and sources of radio waves. He moved to the USA, joining the staff of Mount Wilson Observatory (1935–60), and later worked at the Radio Astronomical Laboratory in Berkeley, California (1960–65). Minkowski was an expert on the spectra of non-stellar objects and made an important study of the Orion Nebula (M42). In 1939 he classified supernovae into Types I and II based upon their spectral lines. He also studied supernova remnants such as the Crab Nebula (M1), which was found to contain a pulsar. He made extensive studies of planetary nebulae and set up his own survey which more than doubled the number known. With Walter BAADE, Minkowski was able to identify the optical counterparts of many of the strongest radio sources in the sky, including Cygnus A (1951). In 1960 he used the 200-inch (5-m) Hale Telescope to discover the galaxy that corresponded to the radio source 3C 295; for 15 years, the redshift he found for this object (z = 0.48) remained the highest known. Minkowski supervised the *PALOMAR OBSERVATORY SKY SURVEY* (POSS).

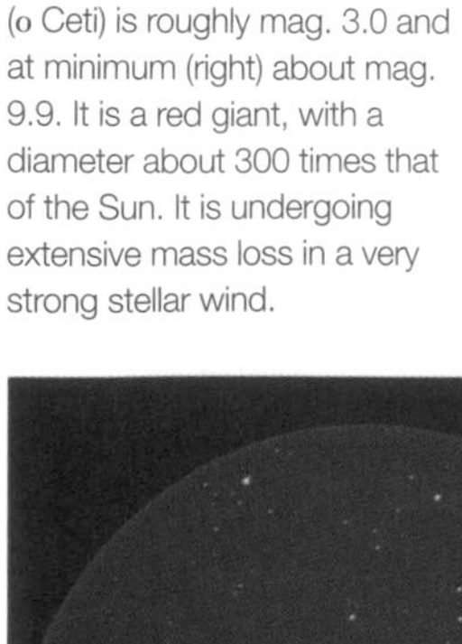

▼ **Mira** At maximum (left), Mira (o Ceti) is roughly mag. 3.0 and at minimum (right) about mag. 9.9. It is a red giant, with a diameter about 300 times that of the Sun. It is undergoing extensive mass loss in a very strong stellar wind.

Minnaert, Marcel Gilles Josef (1893–1970) Flemish astrophysicist who was an expert on the behaviour of light. Minnaert joined the University of Utrecht's (Belgium) solar physics group after World War I and later became director of Utrecht Observatory (1937–63). He compiled the Utrecht Atlas of the solar spectrum (1940) and devised a new way of diagnosing stellar atmospheres by constructing a 'curve of growth'.

minor planet Synonym for ASTEROID. 'Minor planet' is the term preferred by the INTERNATIONAL ASTRONOMICAL UNION.

Minor Planet Center (MPC) Organization based at the Smithsonian Astrophysical Observatory, responsible for collecting, computing, checking and disseminating astrometric observations and orbits for asteroids and comets. It operates under the auspices of the International Astronomical Union and publishes the *Minor Planet Circulars*, monthly in printed form and 'on demand' electronically.

Mintaka The star δ Orionis, visual mag. 2.25, distance about 2000 l.y., spectral type O9.5 II. The northernmost of the three stars that form the Belt of Orion, Mintaka is an eclipsing binary, an ALGOL STAR with a period of 5.37 days and a total range of just over 0.1 mag. It has a wide companion of mag. 6.9, visible through small telescopes or even good binoculars. The name comes from the Arabic *mintaqa*, meaning 'belt' or 'girdle'.

minute of arc (symbol ′) Unit of angular measure also known as an arcminute; it is equivalent to 1/60 of a degree. A minute of arc is subdivided into SECONDS OF ARC, or arcseconds, being 1/60 of a minute of arc and 1/360 of a degree. ANGULAR MEASURE is used widely in astronomy to determine the diameter or separation of celestial objects.

Mir ('Peace') Successor to the Soviet Union's SALYUT SPACE STATIONS. The core module (also known as the base block), which was launched on 1986 February 20, was the backbone of the Mir SPACE STATION. Its most notable innovation was the provision of six docking ports, which enabled additional modules to be attached to the station in 1987–96 as well as the temporary docking of manned and unmanned resupply ships.

The Mir core module was the principal control centre for the station, and it was equipped with its own orbital manoeuvring engines. Although these could not be used after the arrival of Kvant (the first station module), the base block still provided the principal propellant storage tanks and primary attitude control for the entire space station. The core section also provided 90 cubic metres (3200 cubic feet) of habitable space, and contained the main computers, communications equipment, kitchen and hygiene facilities, and primary living quarters.

The base block was divided into four compartments, designated as the working, transfer, intermediate and assembly compartments. All except the assembly compartment were pressurized. A small airlock was also available for experiments or for the release of small satellites or refuse. Exercise equipment was located in the conical portion of the working compartment. Spatial orientation was aided by dark-green carpet on the 'floor', light-green walls and a white 'ceiling' with fluorescent lamps. The Vozdukh electrolytic system was used to recycle station atmosphere, with a back-up chemical scrubbing system.

The first addition to the core section came in 1987: the 11-tonne Kvant (Quantum) astrophysical module contained a number of X-ray, gamma-ray and ultraviolet astronomy experiments. Next to arrive, in 1989, was the 19.6-tonne Kvant-2, which was designed to improve living conditions on the station as well as carry out scientific research. The 19.6-tonne Kristall followed in 1990, with a number of furnaces intended for materials processing. In 1995 the Spektr (Spectrum) module delivered many US scientific experiments to the station. This module was abandoned after a collision with a Progress spacecraft on 1997 June 25. The final Russian-made module to be attached to Mir was Priroda (Nature), which was devoted largely to Earth observation. A US interface module was added to Kristall in 1995, opening the way for the Space Shuttle to dock with Mir on nine occasions.

Cosmonauts on Mir carried out hundreds of experiments involving such diverse research as production of new alloys, protein crystal growth, production of semiconductors, Earth observation and astronomy. However, the space station was plagued by problems, including equipment failures, lack of power, decompression after a collision and at least one serious fire.

Mir was almost permanently occupied from 1986 February until 1999 August. During its lifetime it hosted 28 long-term crews and was visited by 104 cosmonauts/astronauts. Mir re-entered Earth's atmosphere in 2001 March.

Mira (Omicron Ceti) First star to be discovered to vary in a periodic manner. David FABRICIUS first noticed Mira as of third magnitude on 1596 August 13. He could not find it in star catalogues, atlases or globes. A few months later it was invisible, but he saw it again on 1609 February 15 at third magnitude. Johann BAYER, in 1603, lettered it Omicron and noted it as of fourth magnitude. It was observed from 1659 to 1682 and thought to be a new star, catalogued 68 of sixth magnitude. Johann Holwarda (1618–51), in 1638, noted that Mira became visible to the naked eye from time to time and invisible in between these times.

In 1660 the star was shown to vary in an approximate period of 11 months. Its period is 331.96 days, but subject to irregularities. Mean range in brightness is 3.0 to 9.9 mags. Maxima have been observed as bright as mag. 1.7 and as faint as mag. 4.9. Minima show a similar variation of from mag. 8.6 to 11.1. The magnitude of a future maximum or minimum is not predictable. Usually Mira has a protracted minimum phase followed by a steep rise and a slow decline.

The spectrum varies M5e to M9e, which is that of a cool star. There are strong bands of titanium oxide and also lines of neutral metals. An EMISSION SPECTRUM becomes superimposed on the ABSORPTION spectrum when Mira has reached seventh magnitude on the rise, with the intensities of the lines increasing to well after maximum light; they then disappear. The emission spectrum is mainly hydrogen.

Mira lies at a distance of 196 l.y. Robert AITKEN discovered a faint companion to Mira in 1923. The separation is 0″.61; the companion is now known as VZ Ceti, a variable with a range of mag. 9.5 to 12. It shows small variations in times of several hours on which are superimposed 10- to 15-minute variations. It also has very occasional flares, which last for about two minutes. It is possible that VZ Ceti is a white dwarf with an accretion disk. It orbits Mira in 1800 years. *See also* MIRA STAR

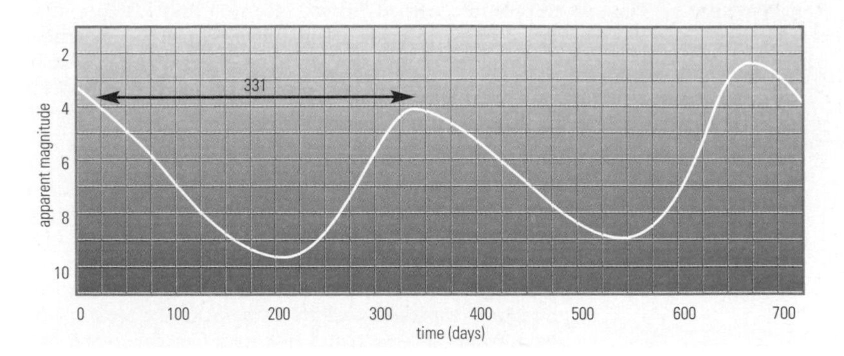

▲ **Mira** The light-curve of Mira (o Ceti) shows a slow fall in brightness followed by a long minimum and then a steep climb up to a short, sharp maximum.

Mirach The star β Andromedae, visual mag. 2.06, distance 199 l.y., spectral type M0 III. Its name is a corruption of the Arabic word *mi'zar*, meaning 'loin cloth'.

Miranda Smallest and innermost of the five SATELLITES of URANUS that were known before VOYAGER 2's flyby in 1986. It was discovered in 1948 by Gerard KUIPER. Miranda's dimensions of 480 × 468 × 466 km (298 × 291 × 290 mi) are just large enough for it to be pulled into a near-spherical shape by its own gravity. Miranda was expected to be a passive ice-ball, like Saturn's satellite MIMAS, and therefore not especially interesting. However, at the time of the Voyager 2 encounter, the orientation of Uranus and its satellite orbits was such that it was convenient to direct the probe closer to Miranda than to any of the other satellites, and Miranda turned out to be one of the most fascinating bodies in the Solar System.

Like all Uranus' satellites seen by Voyager 2, only the southern hemisphere of Miranda was sunlit at the time, but that was enough to show a world unlike any other known planetary body. Just over half the area seen is heavily cratered, but most of the craters are blurred, as if blanketed with dust or 'snow', perhaps dispersed from explosive cryovolcanic eruptions (*see* CRYOVOLCANISM). The rest of the visible hemisphere is occupied by three remarkable tracts of terrain, referred to as 'coronae'. They do not closely resemble one another and, apart from their vaguely concentric pattern, they have little in common

▼ **Miranda** Uranus' satellite Miranda was discovered by Gerard Kuiper in 1948. Its surface is unlike anything else in the Solar System.

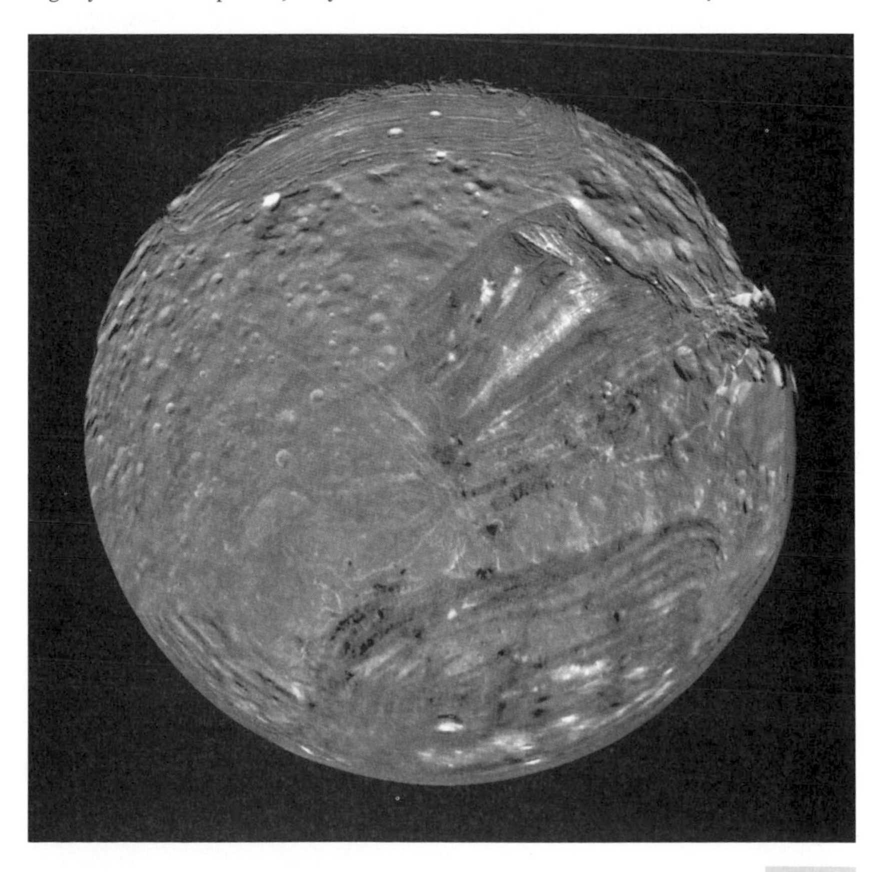

with the features called coronae on Venus. Miranda's coronae are named Arden, Inverness and Elsinore after the settings of some of Shakespeare's plays. Inverness Corona was informally referred to as 'the Chevron', for its V shape. Craters within these coronae are fresh and sharp in appearance, as are the younger craters elsewhere, so evidently the global mantling of the older terrain took place before the coronae were created.

It is unclear how Miranda's coronae originated. Arden and Elsinore Coronae were only partly visible, so their complete shape is unknown. When the images were first received it was suggested that Miranda is a body that re-accreted following collisional break-up. Each corona was thought to represent a discrete fragment, the tonal banding apparent in Arden and Inverness Coronae representing cross-sections through the interior of the original body. This explanation now seems unlikely, if only because of the much greater age of the terrain between the coronae, as judged by comparing the relative numbers of impact craters. The edges of Inverness Corona appear to be related to faults. An alternative explanation, however, is that this and other coronae overlie internal density anomalies inherited from a re-accretion event. If so, each corona could have been created by a past episode of TIDAL HEATING through orbital interaction with ARIEL or UMBRIEL. The bright patches on Arden and Inverness Coronae could perhaps be fresh powder erupted from explosive vents. On the other hand, Elsinore Corona lacks tonal variations and is more likely to be a product of multiple eruptions of cryovolcanic lava. In contrast with Saturn's satellites, where the presence of ammonia is possible but not required by models of Solar System formation, it is extremely likely that Uranus' icy satellites contain a significant proportion of ammonia mixed with the water ice. The lavas this mixture is likely to produce would be highly viscous, which could explain Elsinore Corona's bulbous ridges. *See* data at URANUS

M

Mira star (M) Class of VARIABLE STAR named after MIRA, the first star found to vary with an approximate period of several months. They are also called long-period variables and red variables, although the latter term is ambiguous, given that there are several other types of variable with late spectral classes. Mira stars are GIANTS or SUPERGIANTS belonging to the disk population and with periods ranging from 100 to 1000 days. The difference in their periods stems from whether they are Population I or II. The former generally have periods longer than 200 days, while the periods of Population II stars tend to be shorter and less than 200 days.

The visual light variations of Mira variables range from mag. 2.5 to 11, but in the infrared their amplitudes are much smaller, below mag. 2.5 and less than one magnitude for most of them. Each star has a mean cycle – that is, an average period during which it goes through a complete cycle from maximum and back to maximum again or from minimum to minimum. These periods for any one star may vary by about 15%. Thus their periods are to some extent irregular.

Amplitudes may also vary widely from one cycle to the next. Their light-curves show a wide diversity, but fall into three main divisions. There are those stars that have rises steeper than the fall; these tend to have wide minima and sharp, short maxima. As the asymmetry becomes greater, the period lengthens. Stars with symmetrical curves have the shortest periods. A third group shows humps on their curves or have double maxima and have long and short periods.

Mira variables have late-type emission spectra. The longer the period, the later the spectral type. They show a spectral (or temperature) relationship that is a continuation of the same relationship found in CEPHEID VARIABLES. Both Mira variables and Cepheids are hotter at maximum, Mira variables having the smaller range in temperature. Pulsation is thus the underlying cause of their variability. Pulsations send running waves through the surface layers.

Mirphak (Algenib) The star α Persei, visual mag. 1.79, distance 592 l.y., spectral type F5 Ib. Its name, which is also spelled *Mirfak*, comes from the Arabic *mirfaq*, meaning 'elbow'. Its alternative name, *Algenib*, comes from the Arabic *al-jānib*, meaning 'the side'.

mirror Reflecting surface used to shape or steer light beams. Mirrors have been used to collect and focus starlight since Sir Isaac NEWTON constructed the first successful reflecting telescope in 1668. His design is still in use today and the NEWTONIAN TELESCOPE remains one of the most popular types amongst amateur astronomers. It consists of a concave mirror that reflects the incoming light from a star towards a focal point immediately in front of the mirror. A second small flat mirror is used to intercept the light before it reaches the FOCUS and to steer it out to one side where the observer can use an EYEPIECE to magnify and examine the image.

Newton's first telescope used mirrors made of polished metal that had low reflectivity and tarnished easily. Nevertheless, it overcame the inherent problem of CHROMATIC ABERRATION that had affected early refracting telescopes. Improvements in materials made metal reflectors more useful, in particular the development of speculum metal, an alloy of 71% copper and 29% tin. This metal could be polished to provide fairly high reflectivity, but it still tarnished and needed periodic re-polishing.

The largest successful reflector using speculum metal was the 72-inch (1.8 m) Newtonian telescope built by the Earl of Rosse at BIRR CASTLE in Ireland in 1842.

Optical mirror technology took another big step forward with the development of coating techniques to place a highly reflective thin metal coating on to glass. Initially silver was used but aluminium is more common today as it produces a more durable surface. Almost all mirrors used in astronomy are coated in aluminium, although gold, silver and other materials are used where their special properties are needed. Amateur telescope mirrors often have a protective overcoat of silicon dioxide to increase the life of the aluminium coating. Professional telescopes usually dispense with the overcoat but are re-coated periodically, often every one or two years.

The largest single (monolithic) mirrors used in astronomy at present are just over 8 m (310 in.) in diameter. They are quite thin relative to their diameter (to keep weight down) and require elaborate support mechanisms to maintain their very accurate shapes as they point to different parts of the sky. Segmented mirrors of 10-m (393-in.) diameter are in use and even larger telescopes are planned using this technique.

The largest mirror in space at present is the 2.4-m (94-in.) diameter mirror of the HUBBLE SPACE TELESCOPE; the NEXT GENERATION SPACE TELESCOPE is planned to have a segmented mirror around 4 m (160 in.) in diameter.

Mirzam The star β Canis Majoris, visual mag. 1.98, distance 499 l.y., spectral type B1 II or III. It is a BETA CEPHEI STAR, a pulsating variable with a period of 6 hours and a total range of less than 0.1 mag. The name comes from the Arabic *murzim*, meaning 'announcer', thought to refer to the fact that it rises before the much brighter Sirius.

missing mass problem Quandary whereby most of the matter in the Universe seems to be either subluminous or else invisible. This dark matter is thought to surround individual galaxies as haloes and to pervade clusters of galaxies.

The first evidence of dark matter came from the work of Jan OORT, who, in 1932, determined the thickness of our Galaxy in the solar neighbourhood from the locations of red giants in the sky. These bright stars are sufficiently numerous that a map of their distribution in space reveals that the galactic disk has a thickness of about 2000 l.y. In addition, however, Oort used the distribution and speeds of the red giants in his study to calculate the vertical gravitational field of the Galaxy as well as the mass in our vicinity needed to produce it. A dilemma arose when Oort compared this calculated mass with the observed masses of all the stars and gas clouds in the solar neighbourhood. The observed mass is only about half that needed to confine the red giants to a disk 2000 l.y. thick. Thus half of the matter in our part of the Galaxy seems to be invisible.

Observations of the rotation of the Galaxy confirm the existence of this invisible matter. The rotation rate of the inner regions of our Galaxy is determined from bright stars and emission nebulae. In the dim outer regions, radio observations of hydrogen and carbon monoxide in giant gas clouds provide the required data. In all cases, Doppler shift measurements (*see* DOPPLER EFFECT) are combined with information about the position of a source to deduce its circular velocity around the galactic centre. The results are best displayed on a plot of circular velocity against distance from the galactic centre.

Beyond the confines of most of the Galaxy's matter, the orbital velocity of outlying stars should decrease according to Kepler's third law, just as the velocities of the planets decrease with increasing distance from the Sun. The dashed line in the accompanying diagram indicates such a Keplerian decline. However, the rotation curve of the Galaxy is nearly flat, even to a distance of 60,000 l.y. from the galactic centre. This means that astronomers have still not detected the edge of the Galaxy and thus a substantial quantity of invisible matter must exist beyond the observable stars, nebulae and gas clouds.

The conventional picture of our Galaxy is a flattened disk 100,000 l.y. in diameter surrounded by a spherical halo of old stars and globular clusters roughly 130,000 l.y. across. Recent analyses of rotation curve data strongly suggest that the galactic halo is embedded in an enormous 'corona' that is roughly 600,000 l.y. in diameter and contains at least 10^{12} solar masses of subluminous matter. Thus the corona is at least five times more massive than the disk and halo combined. Most other spiral galaxies also exhibit flat rotation curves and thus they too must be surrounded by large, massive coronae.

The distribution of matter in such a galaxy can be compared with its luminosity by constructing the mass-to-light ratio, M/L. In essence, M/L tells us how much matter there is in a given region compared with the radiation that it generates. By definition, M/L for the Sun is 1.0 (that is, 1 solar mass produces 1 solar luminosity), and 0.8 for a typical globular cluster.

From a rotation curve, astronomers calculate the rate at which the matter density declines with distance from a galaxy's centre. However, a galaxy's surface brightness decreases much more rapidly with distance from its centre. Since the brightness falls much more rapidly than the matter density, M/L climbs to 50 or more in the outer regions of such a galaxy. This dramatic increase in the mass-to-light ratio is the crux of the missing mass problem: an increasing proportion of non-luminous matter is found as one moves outwards from a galaxy's centre.

A partial solution of this aspect of the missing mass problem came with the detection of several faint BROWN DWARF stars in the halo of our Galaxy. If the brown dwarfs are distributed as evenly as they appear to be, then they might account for some of the missing mass in our Galaxy.

Apparently, the space between galaxies in a cluster is also dominated by non-luminous matter. In the 1930s, Fritz Zwicky and Sinclair Smith pointed out that the Virgo Cluster must contain a substantial amount of non-luminous matter, otherwise there would not be enough gravity to hold the cluster together.

The mass of a cluster of galaxies can be determined from the VIRIAL THEOREM, which relates the average speed of the galaxies to the size of the cluster. For example, the Coma Cluster contains roughly 1000 bright galaxies spread over a volume 10 million l.y. in diameter. Doppler shift measurements are available for 800 Coma galaxies and the average velocity relative to the cluster's centre is about 860 km/s (530 mi/s). These data along with the virial theorem give a total mass of 5×10^{15} solar masses for the Coma Cluster. If each of the 1000 brightest galaxies

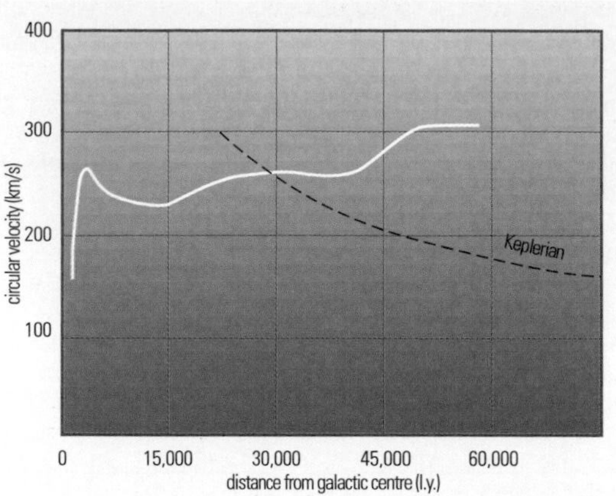

◄ missing mass problem The dotted curve shows the rate at which bodies in our Galaxy should be circling the centre, according to Kepler's third law, if it consisted wholly of the visible matter we see. The white line shows that objects farther from the centre of the Galaxy are, in fact, moving faster, and this can only be explained if there is a substantial amount of 'missing' dark matter surrounding the Milky Way.

has a mass of 10^{12} solar masses then we are able to account for only one-fifth of the cluster's mass.

Similar results are obtained for other rich clusters. Such clusters typically have mass-to-light ratios in the range of 200 to 350, clearly indicating a substantial presence of dark matter. This missing mass problem was also recently solved when X-rays from the centres of large galaxy clusters were seen. These X-rays indicated the presence of large amounts of hot gas that radiated primarily in the X-ray region of the spectrum, and thus were invisible in the optical region. The amount of this gas is generally estimated to be enough to bind the clusters together as required by the virial theorem.

The issue of missing mass also arises on the largest cosmological scales. According to the general theory of relativity, the expansion rate of the Universe must be slowing down because of the mutual gravitational attraction of all the matter in the Universe. Astronomers detect this cosmic deceleration by measuring the recessional velocities and distances of extremely remote galaxies. The purpose of such observations is to determine the so-called DECELERATION PARAMETER (q_0), which is directly related to the average density throughout space. Thus, by measuring the rate at which the cosmic expansion is slowing down, astronomers can deduce the average density of matter in the Universe.

Observations indicate that the average density of matter in the Universe is near the CRITICAL DENSITY of 5×10^{-30} g/cm^3, which is equivalent to about three hydrogen atoms per cubic metre of space. This density is called 'critical' because it ensures that the Universe will just barely continue expanding for ever, without ever collapsing back upon itself. However, the average density of matter that astronomers actually observe in space is about 3×10^{-31} g/cm^3. Thus the observed matter is less than a tenth that needed to account for the Universe's behaviour.

With the inflationary model (*see* INFLATION), and observations indicating that the Universe is actually accelerating, the need for non-luminous matter is eliminated.

Mitchell, Maria (1818–89) First American woman astronomer of note. Her father, William Mitchell, was a respected amateur astronomer, and from him she got her love of astronomy and her observing skills. She assisted him in observations he made, for example timing the contacts of an annular eclipse in 1831. In 1836 she became a librarian, and in the same year the US Coast Survey equipped the Mitchell family home as an outstation, installing an observatory. She began making regular observations, and achieved world fame with her discovery of a comet, C/1847 T1.

From 1849 to 1868 she was employed by the US Government Almanac Office to compute ephemerides for Venus. From 1865 until her death she worked at Vassar Female College at Poughkeepsie, NY (of which she had been a founding member in 1861), as both professor of astronomy and director of the observatory there.

Mitchell, known as 'the female astronomer', was widely honoured, and was elected the first woman member of the

MONOCEROS (GEN. MONOCEROTIS, ABBR. MON)

Inconspicuous equatorial constellation, representing a unicorn, between Orion and Hydra. Monoceros was probably introduced by the Dutch theologian and geographer Petrus Plancius in 1613. Its brightest star, β Mon, is a fine triple system with bluish-white components, mags. 4.5, 5.2 and 5.6 (combined mag. 3.8), separations 7″.2 and 10″.1. The constellation also contains PLASKETT'S STAR (HD 47129), a 6th-magnitude spectroscopic binary whose components are among the most massive stars known; and the variables U Mon (an RV TAURI STAR, range 6.1–8.8) and R Mon (a T TAURI STAR, range 11.0–13.8), which illuminates HUBBLE'S VARIABLE NEBULA. Deep-sky objects include the star clusters NGC 2264, with over 100 stars, including 15 Mon (also known as S Mon), mag. 4.7 (variable); and NGC 2244, which is embedded in the ROSETTE NEBULA (NGC 2237–9).

American Academy of Arts and Sciences. She was active in many campaigns for the advancement of women. The MARIA MITCHELL OBSERVATORY is named after her.

Mizar The star ζ Ursae Majoris, visual mag. 2.23, distance 78 l.y., spectral type A2 V. As well as being a spectroscopic binary with a period of 20.53 days, it has a binary companion of mag. 4.0 visible through a small telescope. Mizar also forms a much wider naked-eye double with ALCOR.

MK system Abbreviation of MORGAN–KEENAN CLASSIFICATION

MMT Observatory (MMTO) Major US optical astronomy facility operating the MMT Telescope at FRED L. WHIPPLE OBSERVATORY. It takes its name from the Multiple Mirror Telescope, which occupied the same housing between 1979 and 1998. The original MMT had six 1.8-m (72-in.) primary mirrors mounted together and feeding a common focus, with a combined light-gathering power equivalent to that of a single 4.5-m (177-in.) mirror. The new MMT has a 6.5-m (256-in.) mirror made at the STEWARD OBSERVATORY Mirror Laboratory. The refurbished instrument was dedicated in 2000. The MMTO is a joint facility of the Smithsonian Institution and the University of Arizona.

▼ **Moon** When it is at its brightest, at full moon, the differences between the bright lunar highlands and the dark, lava-filled basins become very apparent.

mock Sun Popular name for a PARHELION

modified Julian date (MJD) *See* JULIAN DATE

molecular cloud Cloud in which the gas is primarily in the form of molecules. Such clouds abound within the spiral arms of galaxies, including our own. The clouds' temperatures may be as low as 10 K, and their number densities as much as 10^{12} m^{-3}, which is a million times the average density of INTERSTELLAR MATTER. The clouds' sizes range from less than a light-year to several hundred light-years, and their masses range from a few solar masses to several million times the mass of the Sun. Clouds with masses that exceed 10,000 solar masses are called GIANT MOLECULAR CLOUDS.

The clouds are difficult to observe because of their low temperatures. In the visible region, they may sometimes be seen silhouetted against bright backgrounds, since the dust particles within them absorb light. In the radio region the most abundant molecule, hydrogen, does not emit, so the clouds have to be found from the emissions of their less abundant molecules such as carbon monoxide (*see* INTERSTELLAR MOLECULES).

Molonglo Observatory Synthesis Telescope (MOST) Radio astronomy facility near Canberra, Australia, developed in the late 1970s from the earlier One-Mile Mills Cross Telescope. It consists of two cylindrical PARABOLOIDS measuring 780 × 12 m (2560 × 39 ft), separated by 15 m (49 ft) and aligned east–west. The instrument is steered by rotating the cylinders about their long axis and 'phasing' the feed elements along the two arms. MOST is operated by the University of Sydney and is used for analysing complex structure in radio objects, and for major surveys of the radio sky.

monocentric eyepiece Solid EYEPIECE design, now largely obsolete, comprising three elements cemented together, and therefore free from GHOSTING. The limited field of view restricts the usefulness of the monocentric eyepiece to planetary or deep-sky observing.

Monoceros *See* feature article

Montanari, Geminiano (1633–87) Italian astronomer and mathematician who discovered the variability of Algol (1669). Using a specially made reticule, he drew one of the earliest maps of the Moon (1662). He made some of the earliest systematic magnitude estimates of variable stars. Montanari studied the effects of atmospheric refraction, taking these into account in his own observations, which formed the basis for ephemerides he published in 1665.

month Unit of TIME based on the period of revolution of the Moon around the Earth. This can be measured with respect to a number of different reference points, but the most commonly used is the SYNODIC MONTH, the period between successive new or full moons (29.53059 days of MEAN SOLAR TIME). The CALENDAR month is a man-made method of dividing the year into twelve, roughly equal, parts.

Because the total of complete lunar cycles in a year is not a whole number, synodic months cannot be simply reconciled with the common calendar. However, 235 lunar months is almost exactly equal to 19 years (*see* METONIC CYCLE) and while the modern civil calendar is based on the year, the dates of religious festivals (such as Easter) are still set by reference to the lunar month. *See also* ANOMALISTIC MONTH; DRACONIC MONTH; SIDEREAL MONTH; TROPICAL MONTH

Moon Earth's only natural satellite. Because of its proximity, it is the brightest object in the sky, apart from the Sun, being at a mean distance of only 384,000 km (239,000 mi). Like the Sun, its apparent diameter is about half a degree. With an actual diameter of 3476 km

(2160 mi) and a density of 3.34 g/cm^3, the Moon has 0.0123 of the Earth's mass and 0.0203 of its volume. The Earth and Moon revolve around their common centre of gravity, the BARYCENTRE. Although it is so bright (the full Moon has apparent visual magnitude of −12.7) its surface rocks are dark, and the Moon's albedo is only 0.07. It is a cratered world, in many ways a typical Solar System satellite. It is the only other world whose surface features can easily be seen from the Earth, and the only other world on which humans have walked.

As the Moon orbits the Earth, it is seen to go through a sequence of PHASES as the proportion of the illuminated hemisphere visible to us changes. A complete sequence, from one new moon to the next, is called a lunation. At new moon or full moon, eclipses can occur. An observer on the Earth always sees the same side of the Moon because the Moon has SYNCHRONOUS ROTATION: its orbital period (the SIDEREAL MONTH) around the Earth is the same as its axial rotation period. The visible side is called the nearside, and the side invisible from the Earth is the farside. In fact, the face the Moon presents to us does vary slightly because of a number of effects known collectively as LIBRATION.

The Moon has been studied by many space probes, including the LUNA and ZOND series launched by the former Soviet Union, and the US series RANGER, SURVEYOR and LUNAR ORBITER. The manned APOLLO missions and some of the later Luna probes returned samples of lunar material to the Earth for study. In the 1990s, the CLEMENTINE and LUNAR PROSPECTOR probes gathered much valuable new information.

The surface features may be broadly divided into the darker maria (*see* MARE), which are low-lying volcanic plains, and the brighter HIGHLAND regions, which occur predominantly in the southern part of the Moon's nearside and over the entire farside. There are impact features of all sizes. The largest are called BASINS, produced during the early history of the Moon when bombardment by impacting objects was at its heaviest. On the nearside, where the crust is thinner, some basins were subsequently filled with upwelling lava to produce the maria, which in turn became cratered. In the regions of CRATERS at the north and south poles, the interiors of which are in permanent shadow, water ice exists just below the surface. The smaller basins are similar to the largest craters, formerly called walled plains, which have flat floors and are surrounded by a ring of mountains. Other features are mountain peaks and ranges, valleys, elongated depressions known as RILLES, wrinkle ridges, low hills called domes, and EJECTA and bright RAYS radiating from the sites of the more recent cratering impacts.

Early lunar cartographers assumed that the darker lunar features were expanses of water, and named them after fanciful oceans (oceanus), seas (mare), bays (sinus), lakes (lacus) and swamps (palus). Other features are named after famous people, principally astronomers and other scientists. The old names are still in use on lunar maps.

At present, the most generally accepted theory of the Moon's origin is the GIANT IMPACTOR THEORY, according to which a Mars-sized body collided with the newly formed Earth, and debris from the impact accreted to form the Moon. Impacts during the accretion process melted the lunar surface to a depth of 300 km (190 mi), forming a magma ocean. As accretion slowed, the magma ocean began to cool, allowing minerals to crystallize. DIFFERENTIATION led to formation of a crust.

Many of the most prominent lunar features are scars from a final stage of accretion (LATE HEAVY BOMBARDMENT). Massive impacts produced basins such as Mare IMBRIUM and Mare ORIENTALE. Impact ejecta covered the lunar surface in a deep blanket of fragmented material, the mega-REGOLITH. Flooding of the basins by lava from the lunar mantle (kept molten by radioactive decay) led to formation of the darker maria.

Smaller impacts gave rise to the craters. A volcanic origin for the craters has long been dismissed by most astronomers, but Clementine photographed a volcanic crater near the large farside crater Schrödinger. Mild moonquakes occur at depths of roughly 700 km (450 mi), and LUNAR TRANSIENT PHENOMENA appear to be associated with them; otherwise, the Moon is now geologically inactive.

▲ **Moon** Much of the Moon's surface is covered in a fine dust, called the lunar regolith. Because there is no wind or weathering, this material remains just as it settles, and the marks of small impacts, as well as astronauts with lunar carts, remain more or less permanently.

Seismic measurements at Apollo landing sites provided some information on the Moon's internal structure. The lunar crust is about 100 km (60 mi) thick in the highland regions, but only a few tens of kilometres thick under the mare basins. Under the largest basins the underlying mantle has bulged upwards to form MASCONS. An iron-rich core a few hundred kilometres across might exist, with a temperature at the centre of about 1500 K. There is a small iron core at the Moon's centre, accounting for about 4% of its mass. Although there is now no significant overall magnetic field, there are distinct 'magnetic areas' extending for hundreds of kilometres.

Knowledge of lunar rocks and their compositions is based largely on laboratory analyses of the 382 kg of material returned to Earth by the Apollo missions. Planetary geologists have identified several types of lunar rocks. Lunar BASALTS are fine-grained igneous rocks that formed from cooled lavas; they contain two characteristic minerals – pyroxene and plagioclase. Ending about 3 billion years ago, the volcanic activity that produced these rocks consisted of very gradual, thin flows of lava with much lower viscosity than terrestrial lavas; these spread out in thin sheets covering millions of square kilometres instead of building up shield volcanoes of the kind seen on Earth and Mars.

Another important class of lunar rocks are the BRECCIAS found in the highlands. Breccias consist of a matrix of stony fragments (mostly granite), with finer mineral components gluing the fragments together. The breccias are rich in whitish-coloured calcium, magnesium and aluminium, which explains why the lunar highlands appear much brighter than the maria.

The Apollo missions also discovered two minerals unique to the Moon. ANORTHOSITE is a coarse calcium feldspar that is a major constituent of the lunar crust. The other, KREEP, consists of potassium (chemical symbol K), rare earth elements and phosphorous.

The Moon has only the most tenuous of atmospheres. Apollo instruments detected traces of gases such as helium, neon and argon. The atmosphere is probably made up of solar wind particles retained temporarily, and atoms sputtered (knocked off) from the surface by the solar wind. Consequently the surface temperature variation is extreme, ranging from 100 to 400 K.

moon Natural SATELLITE of one of the larger bodies in the Solar System. Mars, for example, possesses two moons, PHOBOS and DEIMOS. Each of the giant planets is accompanied by a large retinue of moons. Smaller moons may be termed moonlets, and the distinction between a moon/moonlet and a body too small to be so classified (for example, the billions of individual items comprising planetary ring systems) is arbitrary.

Moore, Patrick Alfred Caldwell (1923–) English author, television presenter and popularizer of astronomy, an accomplished lunar and planetary observer who has directed the Lunar Section of the British Astronomical Association. After serving in the Royal Air Force during World War II, he wrote the first of over 60 books on astronomy aimed at a popular audience. In the 1950s, with Hugh Percy Wilkins (1896–1960), he used Meudon Observatory's 33-inch (0.84-m) refractor to make a thorough survey of the Moon's topography, discovering many new features. Since 1957 Moore has presented 'The Sky at Night', the longest-running BBC television programme in history. The CALDWELL CATALOGUE is his extension of the Messier Catalogue.

Mopra Telescope *See* AUSTRALIA TELESCOPE NATIONAL FACILITY

Moreton waves Wave-like disturbances in the CHROMOSPHERE and lower CORONA by FLARES. They are named after Gail E. Moreton, who first observed them in 1960 as a circular hydrogen-alpha brightening that expanded away from a flare and across the solar disk. Moreton waves travel to distances of about a million kilometres at velocities of about 1000 km/s (600 mi/s).

MOON

Name	Position	Diameter	Comments
Abenezra	21°S 12°E	43 km	central peak and ridges
Abulfeda	14°S 14°E	64 km	craterlet string on floor
Aestuum, Sinus	12°N 9°W		'Bay of Heats'
ALBATEGNIUS	12°S 4°E	129 km	ancient walled plain
Aliacensis	31°S 5°E	82 km	paired with Werner
Almanon	17°S 16°E	48 km	makes pair with Abulfeda
ALPES, MONTES	49°N 0°		Alps Mountains
ALPES, VALLIS	49°N 2°E		Alpine Valley
ALPHONSUS	14°S 3°W	117 km	walled plain in Ptolemaeus chain
ALTAI, RUPES	24°S 22°E		scarp
ANAXAGORAS	75°N 10°W	51 km	young crater with rays
APENNINUS, MONTES	20°N 2°W		Apennine Mountains
ARCHIMEDES	30°N 4°W	82 km	largest crater in Mare Imbrium
Ariadaeus, Rima	7°N 13°E		fine rille
ARISTARCHUS	24°N 47°W	40 km	bright crater, often shows lunar transient phenomena (LTPs)
ARZACHEL	18°S 2°W	97 km	walled plain in Ptolemaeus chain
ATLAS	47°N 44°E	88 km	paired with Hercules
BAILLY	67°S 63°W	298 km	huge walled plain near south lunar pole
Blanconus	64°S 21°W	92 km	walled plain with central peak
Boscovich	10°N 11°E	43 km	dark, eroded crater
BULLIALDUS	21°S 22°W	63 km	crater with magnificent walls
Byrgius	25°S 22°W	64 km	near western limb
Casatus	75°S 35°W	105 km	high-rimmed crater
CATHARINA	18°S 24°E	88 km	part of Theophilus chain
CAUCASUS, MONTES	36°N 8°E		mountain range
Censorinus	0° 32°E	5 km	overlaps 100-km wide Klaproth
CLAVIUS	58°S 14°W	225 km	large crater with many secondary impacts
CLEOMEDES	27°N 55°E	126 km	walled plain with peaks and rilles
COPERNICUS	10°N 20°W	92 km	magnificent bright ray crater
CRISIUM, MARE	18°N 58°E	500 km	'Sea of Crises'
CYRILLUS	13°S 24°E	97 km	part of Theophilus chain
D'ALEMBERT, MONTES	7°S 87°W		mountain range near west limb
DESCARTES	12°S 16°E	50 km	Apollo 16 landing site
Dionysius	3°N 17°E	19 km	bright crater in south-west Mare Tranquillitatis
Doppelmayer	29°S 41°W	68 km	'bay' in Mare Humorum
ENDYMION	55°N 55°E	117 km	lava-floded crater near limb
ERATOSTHENES	15°N 11°W	61 km	crater with prominent central peak
FRACASTORIUS	21°S 30°E	97 km	lava-flooded crater
Fra Mauro	6°S 17°W	81 km	Apollo 14 landing site
Furnerius	36°S 60°E	97 km	ancient walled plain
GASSENDI	18°S 40°W	89 km	crater with central peak and rilles
GEMINUS	35°N 57°E	90 km	crater with cental peak and terraced walls
GRIMALDI	6°S 68°W	193 km	lava-filled impact basin
Hadley, Rima	26°N 3°E		rille; Apollo 15 landing site
Harpalus	53°N 43°W	52 km	one of the Moon's deepest craters
HERACLIDES, PROMONTORIUM	41°N 33°W		mountain range
Hercules	47°N 39°E	67 km	lava-flooded crater
HERODOTUS	23°N 50°W	37 km	lava-flooded crater
Hesiodus	29°S 16°W	45 km	lava-flooded crater
Hevelius	2°N 67°W	122 km	crater with rilles on floor
HIPPARCHUS	6°S 5°E	135 km	ancient, eroded crater
Humboldtianum, Mare	57°N 80°E	165 km	'Humboldt's Sea'

M

Morgan, William Wilson (1906–94) American astronomer whose work led to the Morgan–Keenan system of spectral classification and the UBV system of photometry. He spent his entire career at Yerkes Observatory (1926–94), briefly serving as its director (1960–63).

Assisted by Philip Childs Keenan (1908–2000) and Edith Marie Kellman (1911–), Morgan classified huge numbers of stars into six different LUMINOSITY CLASSES. This luminosity classification was combined with a new two-dimensional spectral classification in *An Atlas of Stellar Spectra With an Outline of Spectral Classification* (1943), still the standard work in its field. The MORGAN–KEENAN CLASSIFICATION (abbreviated to MK or MKK system) allowed 'normal' stars to be distinguished from peculiar A-type and metallic-line A stars. It also made possible vastly more accurate calibrations of stellar distances, especially for the nearer of the O and B stars.

In the late 1940s and early 1950s, Morgan and Harold JOHNSON developed a new system for measuring stellar magnitudes in three basic colours or wavebands: U (ultraviolet), B (blue) and V (visual, or yellow). They combined their UBV SYSTEM (which Johnson later extended) with the MKK system to improve distance estimates for stars and star clusters. In 1957 Morgan used these tools to devise the first system for classifying the integrated spectra of globular clusters. He went on to classify galaxies by comparing the galaxy's integrated spectrum with its bulge-to-disk ratio, in what was called the Yerkes system.

Morgan–Keenan classification (MK system, MKK system) Categorization of a stellar SPECTRUM from the visual appearance of its ABSORPTION LINES. In the early 1940s, William MORGAN, Philip Keenan (1908–2000) and Edith Kellman (1911–) extended the one-dimensional temperature-dependent HARVARD CLASSIFICATION (OBAFGKM) to two dimensions by adding Roman numerals to indicate luminosity – I for supergiant, II for bright giant, III for giant, IV for subgiant and V for dwarf. Brighter supergiants are Ia, lesser ones Ib. (The Roman letters are also commonly used to separate giants of different luminosities, with IIIa being brighter than IIIb.) R, N, C and S stars also use

MOON (CONTINUED)

Name	Position	Diameter	Comments
HUMORUM, MARE	23°S 38°W		'Sea of Humours'
HYGINUS, RIMA	8°N 6°E		rille
IMBRIUM, MARE	36°N 16°W	1300 km	'Sea of Showers'
IRIDUM, SINUS	45°N 31°W	255 km	'Bay of Rainbows'
KEPLER	8°N 38°W	35 km	bright ray crater
LANGRENUS	9°S 61°E	132 km	crater with central peak and terraced walls
LINNE	28°N 12°E	1.5 km	small crater in Mare Serenitatis
Littrow	22°N 32°E	35 km	Apollo 17 landing site
Longomontanus	50°S 21°W	145 km	ancient, eroded crater
MAGINUS	50°S 60°W	177 km	ancient, eroded crater
Marius	12°N 51°W	42 km	lava-flooded crater
MAUROLYCUS	42°S 14°E	105 km	ancient crater
Moretus	70°S 8°W	105 km	crater with central peak and terraced walls
Nebularum, Palus	38°N 1°E		'Marsh of Clouds'; light region in Mare Imbrium
NECTARIS, MARE	14°S 34°E	290 km	'Sea of Nectar'
PETAVIUS	25°S 61°E	170 km	crater with central peak and terraced walls
PICO	46°N 8°W		mountain peak
PLATO	51°N 9°W	97 km	dark-floored lava-flooded crater
Plinius	15°N 24°E	48 km	bright ray crater
POSIDONIUS	32°N 30°E	96 km	crater with central peak and terraced walls
PROCELLARUM, OCEANUS	10°N 47°W		'Ocean of Storms'
PROCLUS	16°N 47°E	29 km	bright ray crater
PTOLEMAEUS	14°S 3°W	148 km	huge walled plain, part of chain
PURBACH	25°S 2°W	120 km	ancient, eroded crater
PUTREDINIS, PALUS	27°N 1°W		'Marsh of Decay'
RECTA, RUPES	22°S 8°W		straight wall
RECTI, MONTES	48°N 20°W		straight range mountains
REGIOMONTANUS	30°S 48°E	117 km	irregular-shaped crater
Reichenbach	31°S 49°E	48 km	old, featureless crater
Rheita, Vallis	39°S 47°E		valley, 2.5 km deep
RICCIOLI	3°S 75°W	160 km	lava-flooded, eroded crater
SCHICKARD	44°S 54°W	202 km	ancient, eroded crater
SCHRÖTERI, VALLIS	26°N 52°W		Schröter's Valley
SERENITATIS, MARE	30°N 17°E		'Sea of Serenity'
Sirsalis	3°S 60°W	32 km	crater with central peak and terraced walls
SMYTHII, MARE	10°S 88°E		'Smyth's Sea'
SOUTH POLE–AITKEN BASIN	55°S 180°W	2500 km	largest basin on Moon
STADIUS	11°N 14°W	70 km	ghost crater
STOFLER	41°S 6°E	145 km	ancient, eroded crater
Theatetus	37°N 6°E	26 km	small crater with central peak
THEOPHILUS	12°S 26°E	101 km	crater with central peak and terraced walls
Timocharis	27°N 13°W	35 km	crater with central peak and terraced walls
TRANQUILLITATIS, MARE	9°N 30°E		Apollo 11 landing site
TRIESNECKER	4°N 4°E	23 km	crater with rille system
TSIOLKOVSKY	29°S 129°E	180 km	dark-floored farside crater
TYCHO	43°S 11°W	87 km	crater with extensive rays
VENDELINUS	16°S 62°E	165 km	ancient crater
WALTER	33°S 1°E	129 km	ancient, eroded crater
WARGENTIN	50°S 60°W	89 km	lava-flooded crater
Wilhelm Humboldt	27°S 81°E	193 km	extensive rille system on floor
Zupus	17°S 52°W	26 km	depression

M

MK luminosity classes. Keenan later added Arabic '0' for extra-luminous 'hypergiants'.

The MK system defines each spectral class by the spectrum of a representative star (B0 V by τ Scorpii, B1 V by η Orionis, and so on.). The essence of the system is a set of MK blue-violet photographic spectra of standard stars against which others stars are compared and which form the basis for the extension of the system to other wavelength domains.

Though commonly applied to stars of Population II, MK classification is formally appropriate only for solar-metallicity Population I. Outside the system, subdwarfs and white dwarfs are sometimes referred to as VI and VII.

morning star Popular name for the planet VENUS when it appears, prior to superior conjunction, as a bright object in the eastern sky before sunrise. The term is occasionally applied to MERCURY. *See also* EVENING STAR

Moulton, Forest Ray (1872–1952) American mathematician and astronomer who specialized in celestial mechanics. He was a long-time professor at the University of Chicago (1898–1926). With Thomas CHAMBERLIN, he conceived the planetesimal hypothesis – that the Solar System formed when a passing star caused the Sun to eject filaments of matter that condensed into protoplanetary bodies, which later accreted to form the planets.

Mount Graham International Observatory Major US optical observatory on Mount Graham in the Pinaleno Mountains, some 110 km (70 mi) north-east of Tucson, Arizona, at an altitude of 3260 m (10,700 ft), operated by the University of Arizona. The site is in the Coronado National Forest, and in 1988 controversy followed Congressional approval of the observatory's construction because the forest is home to the endangered Mount Graham red squirrel (the population of which has since increased). The observatory hosts the VATICAN ADVANCED TECHNOLOGY TELESCOPE, the HEINRICH HERTZ TELESCOPE of the Submillimeter Telescope Observatory and, when it is completed, the LARGE BINOCULAR TELESCOPE.

mounting Means of supporting the weight of a TELESCOPE and aiming it at different points in the sky. Of necessity, all large professional instruments are on permanent mounts; many amateur telescopes and their mounts are portable. Whether a mounting is permanent or portable, rigidity is a prime requirement. *See* ALTAZIMUTH MOUNTING; EQUATORIAL MOUNTING

Mount Palomar Observatory *See* PALOMAR OBSERVATORY

Mount Pleasant Radio Observatory Observatory run by the Physics Department of the University of Tasmania, Australia, located 20 km (12 mi) east of Hobart. It is equipped with two antennae: a 26-m (85-ft) donated to the University by NASA in 1985, and a 14-m (46-ft).

Mount Stromlo Observatory One of the oldest institutions in the Australian Capital Territory, at an altitude of 760 m (2500 ft) in the south-western suburbs of Canberra. It was founded in 1924 as the Commonwealth Solar Observatory, and in the 1920s and 1930s research was conducted there on solar and atmospheric physics. In 1957 the Observatory became part of the Australian National University. During the same decade, four new instruments came to the mountain, including the 1.88-m (74-in.) reflector (for twenty years the joint-largest telescope in the southern hemisphere) and a 1.27-m (50-in.) reflector that had started life in 1868 as the Great Melbourne Telescope. This instrument, now refurbished, is used for the MASSIVE COMPACT HALO OBJECT dark-matter project. During the 1960s, under the directorship of Bart BOK, Mount Stromlo founded SIDING SPRING OBSERVATORY to escape the light pollution of Canberra, and the institution became the Mount Stromlo and Siding Spring Observatories (MSSSO). In 1998 the name was changed again to the Research School of Astronomy and Astrophysics (RSAA).

Mount Wilson Observatory One of the most famous observatories in the world, located in the San Gabriel range near Pasadena, California, at an elevation of 1740 m (5700 ft). It was founded in 1904 as Mount Wilson Solar Observatory by George Ellery HALE, using funds from the Carnegie Institution of Washington, and the SNOW TELESCOPE for solar spectroscopy was installed the following year. Mount Wilson was developed as a site for stellar astronomy with the completion of a 1.5-m (60-in.) reflector by George W. RITCHEY in 1907. It was followed in 1917 by the 2.54-m (100-in.) HOOKER TELESCOPE, which dominated US astronomy until the completion of the HALE TELESCOPE at Palomar Observatory in 1948. This was the instrument Edwin HUBBLE used in 1919 to identify CEPHEID VARIABLE stars in the Andromeda Galaxy, and for subsequent work that led to the realization that the Universe is expanding. In 1917 the word 'Solar' was dropped from the observatory's name.

Today, Mount Wilson remains part of the CARNEGIE OBSERVATORIES. Although seriously affected by the lights of Los Angeles, 32 km (20 mi) to the south-west, the observatory has entered a new era with the installation of Georgia State University's CHARA Telescope Array. CHARA, the Center for High Angular Resolution Astronomy, consists of a Y-shaped arrangement of five 1-m (39-in.) telescopes within a 400-m (quarter-mile) radius linked interferometrically. It is possible that the Hooker and 60-inch telescopes will be incorporated into the array.

moving cluster MOVING GROUPS of stars close enough to have their distance determined by the moving cluster parallax method, as follows.

If the moving cluster is close to Earth, then projecting the spatial motions of the stars backwards will produce a set of paths that converge at one point. The direction of this convergent point is parallel to the space motion of the cluster. Observing individual stars within the cluster will give a value for the PARALLAX and hence the distance of the system. This is known as the moving cluster parallax method of distance determination. Using:

$$v_r = v\cos\theta$$

and

$$v_t = v\sin\theta = 4.74\mu/\pi$$

where v is the SPACE VELOCITY, v_r is RADIAL VELOCITY and v_t is the TANGENTIAL VELOCITY in km/s; μ is the PROPER MOTION in arcseconds per year; and π is the parallax in arcseconds. Measuring the radial velocity of individual stars and the angle θ to the convergent point gives a value for the space velocity and hence the tangential velocity. The proper motion of the individual stars then gives an idea of the parallax of the cluster.

The moving cluster parallax method has been used to determine the distances of several nearby OPEN CLUSTERS. The proximity of these clusters allows distances to be calculated with a high precision. The HYADES in particular is often used as the main calibrator of the distance scale of the Universe.

moving group Group of stars that share the same motion through space, and have the same age and similar chemical compositions. Moving groups can be in many forms, including OPEN CLUSTERS, GLOBULAR CLUSTERS, OB ASSOCIATIONS and T ASSOCIATIONS. If they lie close enough to Earth for the spatial motions of the individual stars to be determined, they are termed MOVING CLUSTERS and the cluster distance can be determined by the moving cluster parallax method. The nearest moving group to Earth is the URSA MAJOR MOVING CLUSTER.

MPC Abbreviation of MINOR PLANET CENTER

Mrkós, Comet (C/1957 P1) LONG-PERIOD COMET discovered as a naked-eye object at dawn on 1957 August 2; it became the second prominent comet of that year. When discovered by Antonín Mrkós (1918–96) from Czechoslovakia, the comet was one day past its perihelion (distance 0.35 AU). Closest approach to Earth came on August 13, and Comet Mrkós reached a peak magnitude +1.0 early in the month, with a dust tail extending for about 5°. The ion tail showed considerable activity, the sunspot cycle at this time being close to maximum. The orbit is elliptical, with a period of 13,200 years.

M star Any member of a class of orange-red stars, the spectra of which are defined by molecular absorption bands of metallic oxides, particularly titanium oxide. Vanadium oxide is present in addition to some hydrides and water. Hydrogen is very weak and disappears towards cooler subclasses. Neutral calcium is strong; its weakening with increasing luminosity helps to determine the MK class. Only classes L and T are cooler than M stars. Main-sequence subclasses range from 2000 K at M9.5 to 3900 K at M0; zero-age masses range from the hydrogen-fusing limit near 0.08 solar mass to 0.5; zero-age luminosities range from 2 3 1024 to half solar (most of the radiation emerging in the infrared). Lifetimes far exceed that of the Galaxy. Long lifetimes and high rates of low-mass star formation make M stars numerous. Ignoring classes L and T (whose populations are not yet assessed), class M contains 70% of all dwarf stars (none are visible to the naked eye).

The convective envelopes of cool dwarfs extend more deeply as mass decreases, and at 0.3 solar mass (class M5) they reach the stellar centres, forcing cool M stars below about 3000 K to be completely mixed. Despite slow rotation, the deep convective zones somehow produce stellar magnetic fields and Sun-like activity. A good fraction of M dwarfs (dMe stars) display chromospheric emission and powerful flares that can brighten the stars by one or more magnitudes and that radiate across the spectrum.

No giant or supergiant is cooler than class M. Having evolved from dwarfs of classes G to O, such stars are much more massive than M dwarfs. As a result of lower densities, giant surface temperatures are somewhat cooler than their dwarf counterparts. M giants divide into those with quiet helium cores that are climbing the giant branch for the first time, and asymptotic giant branch (AGB) stars with quiet carbon–oxygen cores that are climbing it for the second time. First-ascent M giants have early subtypes and can reach luminosities of 1000 times solar and radii of 100 times solar. Second-ascent giants can reach into later subtypes, and become brighter (over 10,000 solar) and larger, with the more massive rivalling the diameter of Mars' orbit.

The brighter AGB stars pulsate as Mira variables, changing by five or more visual magnitudes over periods that range from 100 to 1000 days. Much of the visual variation is caused by small temperature variations that send the radiation into the infrared and create more powerful obscuring TiO bands. Bolometric (true luminosity) variation is much less, at about a magnitude. The pulsations create running shock waves, which generate emission lines, rendering Miras a class of me star. When M giants dredge the by-products of nuclear fusion (including carbon) to their surfaces, they become s stars and then carbon stars. Pulsations, plus the large luminosities and radii, promote powerful winds and mass-loss rates of 1025 solar masses (or more) per year. Silicate dust condenses in the winds of M class Miras; carbon dust condenses in the winds of carbon stars. The winds of class M Miras create vast envelopes that can radiate powerful OH maser emission, whence they are known as OH/IR (infrared) stars.

Among lower luminosity giants are found 'semi-regular' (SR) variables. SRa stars (S Aquilae, R Ursae Minoris) are similar to Miras but exhibit smaller light variations and irregularities in their periods. SRb stars (R Lyrae, W Orionis) have periods that are less well defined. 'Lb' (irregular) M giants have no periods at all.

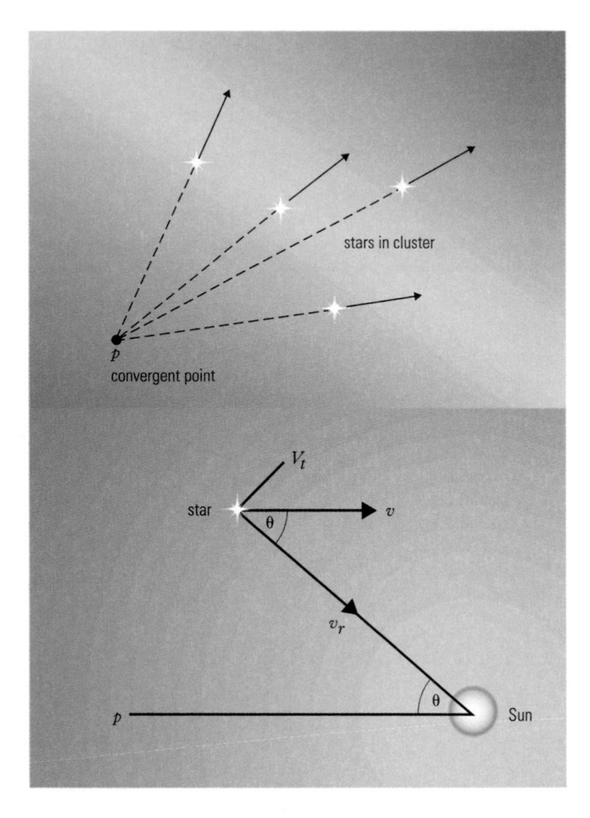

◄ **moving cluster** The velocity of stars in an open cluster can be determined if they are close enough for their parallax to be measurable over a short period. Their parallax is combined with their line-of-sight rate of motion towards or away from Earth to obtain their speed of movement at right angles to the line of sight. Combined with the measurement of the angle of movement, this allows astronomers to judge their distances.

The most luminous M supergiants develop from main-sequence O stars of up to 60 solar masses, most fusing helium in their cores. Bolometric luminosities can reach 750,000 times that of the Sun. Radii can approach 2000 times solar, rivalling the diameter of Saturn's orbit (nearly 10 AU). Many M supergiants are irregular (Lc) variables. betelgeuse (M2 Iab), for example, varies with a range of about half a magnitude on a timescale of years. Some of the lesser M supergiants (SRc stars) display some semi-regular periodicity.

Prominent examples of M stars include Proxima Centauri M5 V, Betelgeuse M2 Iab, Antares M1.5 Ib, Mu Cephei M2 Ia, VV Cephei and M2 Iaep 1 O8 Ve.

Mu Cephei *See* GARNET STAR

Mullard Radio Astronomy Observatory (MRAO) Radio astronomy observatory of the University of Cambridge, and part of the Cavendish Laboratory (the university's Department of Physics). Radio astronomy at Cambridge dates back to the earliest days of the science just after World War II, and flourished under the direction of Martin RYLE from 1945 to 1982. The first Cambridge radio telescopes were built on the western outskirts of the city and specialized in the study of 'radio stars', sources now known to range from relatively nearby supernova remnants to the most distant galaxies. Whereas early work at JODRELL BANK OBSERVATORY led to the development of large single dishes, the emphasis at Cambridge was on the design and construction of smaller, widely spaced aerials used as an interferometer to make better positional measurements.

In 1957, through the generosity of Mullard Ltd and with support from the Science Research Council, the MRAO was built on its present site at Lords Bridge, 8 km (5 mi) south-west of Cambridge. Today its instruments include the RYLE TELESCOPE (formerly the Five-Kilometre Telescope, dating from 1972), the CAMBRIDGE OPTICAL APERTURE SYNTHESIS TELESCOPE and the CAMBRIDGE LOW-FREQUENCY SYNTHESIS TELESCOPE. In the 1990s the observatory built the Cosmic Anisotropy Telescope (CAT) at Cambridge as a prototype instrument for ground-based mapping of the cosmic microwave background radiation.

MUSCA (GEN. MUSCAE, ABBR. MUS)

Small but distinctive southern constellation, representing a fly, between Carina and Circinus. It was introduced by Keyser and de Houtman at the end of the 16th century. Its brightest star, α Mus, is mag. 2.7. β Mus is a close binary with bluish-white components, mags. 3.5 and 4.0, separation 1″.2. Deep-sky objects include the 7th-magnitude globular clusters NGC 4372 and NGC 4833.

BRIGHTEST STARS

	RA h m	dec. ° ′	Visual mag.	Absolute mag.	Spectral type	Distance (l.y.)
α	12 37	−69 08	2.69	−2.2	B2	306
β	12 46	−68 06	3.04	−1.9	B2	311

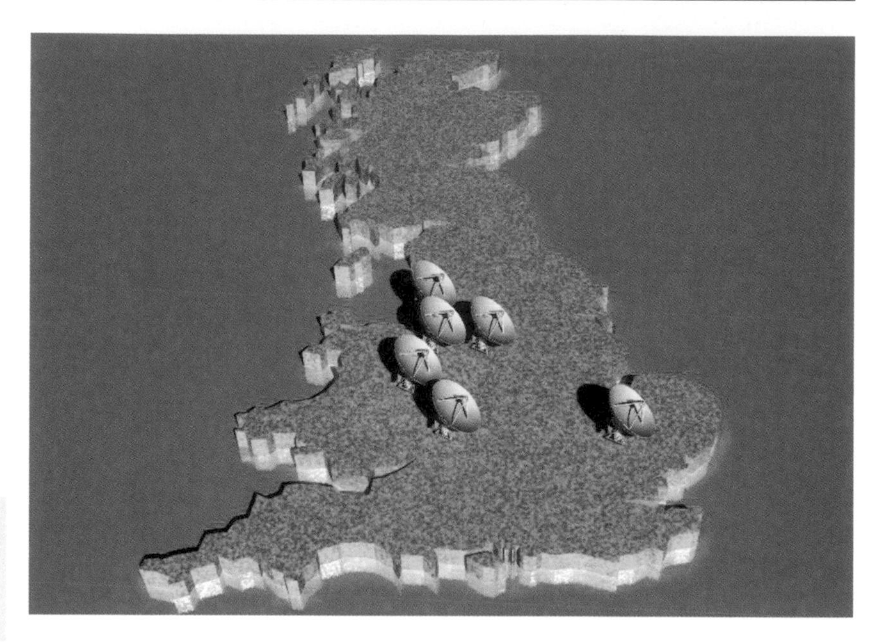

▲ **Multi-Element Radio-Linked Interferometer Network** (MERLIN) The long-baseline interferometry of the six radio telescopes that make up MERLIN give the network similar resolving power to that of the Hubble Space Telescope. The network can also be joined to other radio telescopes around the world for even greater accuracy.

In 2000 its successor, the VERY SMALL ARRAY, was installed on the island of Tenerife in the Canaries.

Müller, Johannes *See* REGIOMONTANUS

Multi-Element Radio-Linked Interferometer Network (MERLIN) Group of six RADIO TELESCOPES linked to JODRELL BANK Observatory, spread from Cambridge to Wales, UK. MERLIN acts as a RADIO INTERFEROMETER with a maximum baseline of around 220–230 km (137–143 mi). At a wavelength of 6 cm MERLIN has a resolution of 0″.05. One of the main tasks of MERLIN has been to study the structure of radio JETS from QUASARS.

multiple star Gravitationally connected group of stars with a minimum of three components. About one in five BINARY STAR systems is gravitationally bound to one or more other stars. The additional stars are often far enough from the binary not to affect its evolution significantly.

The similarity between multiple stars and star clusters suggests that they form in similar ways, that is, by condensation from interstellar clouds. A good example of this is the TRAPEZIUM in the ORION NEBULA. The Trapezium consists of four very young stars enmeshed in nebulosity, with a further five possible members nearby. The Trapezium is the prototype of a special type of quadruple star in which very little relative motion is observed. It may be that the stars will spend most of their lifetimes in this configuration. Many other examples of Trapezium-type systems are known. In hierarchical quadruple systems, the most common situation is two pairs revolving about each other, as, for example, in EPSILON LYRAE.

The point at which highly multiple stars and star clusters overlap is not clear, but systems such as CASTOR and Alpha^2 Capricorni, with six components each, are still regarded as multiple stars. In these systems there is a recognized hierarchy or order. In Castor, for instance, the two main pairs rotate about a common centre of gravity, whilst a pair of cool red dwarfs rotates about the same centre, but at a much larger distance.

mural quadrant Early instrument used for measuring declination. It consists of a graduated circle fixed to a wall (in Latin, *murus*) orientated north–south. A telescope (or, in early versions, a simple sighting tube) is mounted centrally on a pivot so that its position can be read off the circle. The eyepiece of the telescope contains a horizontal graticule along which the image of the star must move as the Earth turns. Setting of the telescope must be done quickly, as the star passes the meridian. It was superseded in the mid-19th century by the TRANSIT CIRCLE.

Murchison METEORITE that fell as a shower of stones in the state of Victoria, Australia, on 1969 September 28; more than 100 kg of material was collected. Murchison is classified as a CM2 CARBONACEOUS CHONDRITE. Rich in organic compounds, it was the first meteorite in which extraterrestrial amino acids were identified. It also contains abundant INTERSTELLAR GRAINS.

Musca *See* feature article

Muses Name of a series of Japanese science missions. When launched, they are given specific names.

Muses B Japanese spacecraft, named Haruka, which was launched in 1997 February and is the first astronomical satellite dedicated to VERY LONG BASELINE INTERFEROMETRY. The 8-m (26-ft) diameter antenna, made of Kevlar wire, with a pointing accuracy of 0°.01, is combined with ground-based radio telescopes to provide an extremely high resolution, particularly useful for observing active galactic nuclei and quasars.

Muses C Japanese spacecraft to be launched in 2002 June at the earliest to land on an ASTEROID in 2003 April, take a sample and return it to Earth in 2006 June. If successful, Muses C will be the first time a sample of an asteroid has been brought to Earth. The target for Muses C is the small asteroid 10302 1989ML, which is in an orbit between the Earth and Mars, so not situated in the large asteroid belt between Mars and Jupiter. The asteroid is thought to be 400 m (1300 ft) across.

MV Japanese solid propellant spacecraft and satellite launcher; it first flew in 1997. The MV launches Japan's science and planetary spacecraft. It can carry payloads weighing 2 tonnes to low Earth orbit and payloads weighing 0.5 tonne to planetary targets.

nadir Point on the CELESTIAL SPHERE directly below the observer and 180° from the ZENITH.

Nagler eyepiece Telescope eyepiece with a very wide apparent field of view, 80° or more, making it suitable for activities such as searching for comets or novae. The Nagler eyepiece offers excellent EYE RELIEF, and is relatively free from ABERRATION, COMA and FIELD CURVATURE. It is named after its designer, Al(bert) Nagler (1935–).

naked-eye Celestial object visible, or an observation made, without the use of any optical instruments such as a telescope. In theory, stars down to sixth magnitude should be visible to the naked eye under perfect conditions but this is rarely achieved.

naked singularity SINGULARITY that is not hidden by an EVENT HORIZON and so is visible to the outside world. *See also* BLACK HOLE

Nakhla METEORITE that fell as a shower of around 40 stones over the El Nakhla el Baharia village in Alexandria, Egypt, in 1911 June; approximately 10 kg of material was recovered. The legend that one of the stones of Nakhla struck and killed a dog is almost totally without foundation. Nakhla is the type specimen of the nakhlites, one of the main subgroups of the MARTIAN METEORITES (SNCs). Nakhlites are almost unshocked clinopyroxene-rich cumulate rocks. They formed at or near the Martian surface in a thick flow or sill. They have crystallization ages of *c.*1.3 billion years. Their secondary minerals (clays, carbonates, and sulphates) are tracers of low temperature aqueous processes and can be used to infer conditions on the Martian surface.

NaI detector Crystals that produce a flash of visible light (scintillation) when hit by gamma rays. NaI detectors were used in the BATSE instrument on the COMPTON GAMMA RAY OBSERVATORY to detect GAMMA-RAY BURSTS.

Nançay Radio Telescope Large transit-type radio telescope at the Nançay Radio Observatory, 200 km (125 mi) south of Paris and operated by PARIS OBSERVATORY. It has a fixed spherical reflector 300 × 35 m (985 × 115 ft) fed by a tiltable flat reflector 200 × 40 m (655 × 130 ft).

Nanjing Observatory *See* PURPLE MOUNTAIN OBSERVATORY

nanometre (symbol nm) Unit of length equivalent to one thousand millionth (10^{-9}) of a metre. The nanometre is commonly used as a measurement of the wavelength of electromagnetic radiation, having replaced the previously used ANGSTROM UNIT.

Naos The star ζ Puppis, visual mag. 2.21, distance about 1400 l.y. Its spectral type is O5 Ia, which makes it the hottest naked-eye star – its surface temperature is estimated to be around 40,000 K. The name is Greek for 'ship'.

Narlikar, Jayant Vishnu (1938–) Indian cosmologist whose research has often explored alternative cosmological models, including the steady-state theory originated by Fred HOYLE. Narlikar was educated at Cambridge University, where Hoyle was his advisor; later, they collaborated on a theory of 'self-gravitating systems' that explains the formation of astronomical objects ranging in size from a few kilometres to about 1 million l.y. Narlikar proposed a 'quasi-steady-state' model that takes into account recent findings on the anisotropic distribution of matter in space. As head of India's Inter-University Centre for Astronomy & Astrophysics, he has investigated black holes and the hypothetical particles known as TACHYONS.

NASA Acronym for NATIONAL AERONAUTICS AND SPACE ADMINISTRATION

NASDA Acronym for NATIONAL SPACE DEVELOPMENT AGENCY OF JAPAN

Nasireddin (or Nasiruddin) *See* AL-ṬŪSĪ

Nasmyth focus Focal point of an altazimuth-mounted reflecting telescope in which the converging light beam is reflected by means of a third mirror to a point outside the lower end of the telescope tube, one side of the altitude axis. This allows large or heavy instruments, such as spectrographs, to be mounted on a permanent platform which rotates in AZIMUTH with the telescope. An arrangement of prisms and mirrors compensates for the rotation of the image as the telescope tracks the object, holding it stationary relative to the instrument. This arrangement was first used by the Scottish engineer James Nasmyth (1808–90) and has been revived with the latest generation of altazimuth-mounted, large professional telescopes. *See also* COUDÉ FOCUS

National Aeronautics and Space Administration (NASA) US national aeronautics and space agency, with overarching responsibility for American civilian aerospace exploration and associated scientific and technological research. NASA was created in 1958, following the USSR's success with SPUTNIK, and inherited the work of the National Advisory Committee for Aeronautics (NACA) and other government organizations. It began working almost immediately on human spaceflight, and managed the successful MERCURY and GEMINI projects before embarking on the APOLLO lunar landing programme. After the SKYLAB missions and the joint Apollo–Soyuz Test Project in the 1970s came the SPACE SHUTTLE programme, starting in 1981, and today's INTERNATIONAL SPACE STATION.

NASA's unmanned space projects are no less well known. Scientific probes such as PIONEER, VOYAGER, VIKING, MARS PATHFINDER and MARS GLOBAL SURVEYOR have explored the planets, while orbiting observatories such as the HUBBLE SPACE TELESCOPE and the CHANDRA X-RAY OBSERVATORY have sharpened our view of the Universe. The agency has also managed landmark projects in communications and terrestrial remote sensing.

The headquarters of NASA are in Washington, D.C., and the agency operates a large number of specialist facilities. They include the AMES RESEARCH CENTER, the GLENN

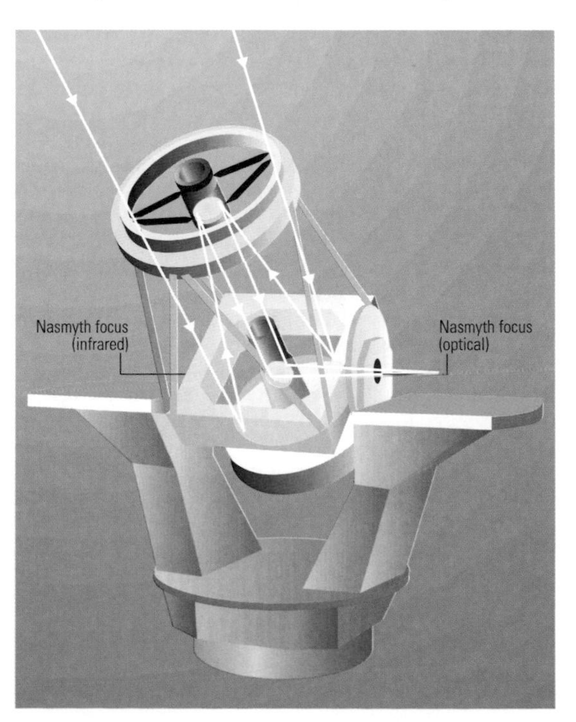

◄ **Nasmyth focus** A number of large, professional telescopes have instruments at their Nasmyth foci. In the William Herschel Telescope there are two foci, one used for infrared and the other for optical observations. They are sited at opposite ends of the declination axis, and the third mirror (the Nasmyth flat) can be moved to direct light to either of them.

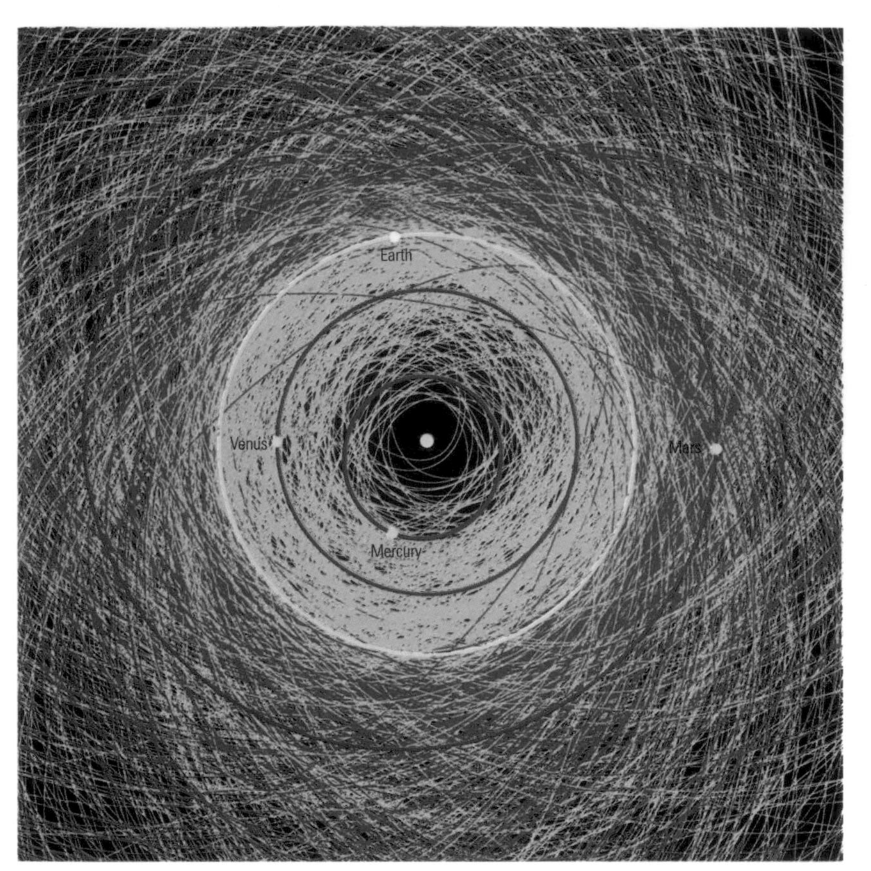

▲ **near-Earth asteroid** The inner Solar System is swarming with small, rocky bodies. Asteroids with Apollo- and Aten-type orbits (Earth-crossing) are shown in amber and Amor-type asteroids are shown as red lines.

N

RESEARCH CENTER, the GODDARD SPACE FLIGHT CENTER, the JET PROPULSION LABORATORY, the JOHNSON SPACE CENTER, the KENNEDY SPACE CENTER, the LANGLEY RESEARCH CENTER and facilities such as the Hugh L. Dryden Flight Research Center, the Marshall Space Flight Center and the White Sands Test Facility. NASA also maintains a number of DATA CENTRES and databases. Taken together, the remit of these facilities covers almost the full gamut of human scientific and technological endeavour.

NASA itself has experienced periodic crises of confidence, ranging from the CHALLENGER disaster in 1986 to the loss of Mars Climate Orbiter in 1999. The agency has always responded by revising its policies and systems accordingly, and it is fair to say that it retains the enthusiastic support of most American taxpayers.

National Astronomical Observatory of Japan (NAOJ) Principal research institute for astronomy in Japan, with headquarters in the Tokyo suburb of Mitaka. The NAOJ was created in 1988 by amalgamating the Tokyo Astronomical Observatory of the University of Tokyo with two other research institutions. Its biggest optical telescope then was the 1.88-m (74-in.) reflector of the Okayama Astrophysical Observatory on Mount Chikurin-Ji (elevation 372 m, 1220 ft) in south-western Okayama, opened in 1960. Today, the NOAJ operates the 8.2-m (320-in.) SUBARU TELESCOPE in Hawaii. Its Radio Astronomy Division operates the NOBEYAMA RADIO OBSERVATORY.

National Optical Astronomy Observatory (NOAO) Organization formed in 1982 to bring all US ground-based optical observatories managed by the ASSOCIATION OF UNIVERSITIES FOR RESEARCH IN ASTRONOMY (AURA) under a single directorship. Initially, the NOAO's facilities consisted of KITT PEAK NATIONAL OBSERVATORY (KPNO), CERRO TOLOLO INTER-AMERICAN OBSERVATORY (CTIO), and the NATIONAL SOLAR OBSERVATORY (NSO). Today, they also include the United States Gemini Program (USGP) – the gateway to the GEMINI OBSERVATORY for the US astronomical community. The NOAO is funded by the National Science Foundation and operated by AURA. Its headquarters are in Tucson, Arizona.

National Radio Astronomy Observatory (NRAO) Research facility of the US National Science Foundation providing radio telescope facilities for the astronomical community. The NRAO was founded in 1956, and has its headquarters in Charlottesville, Virginia. Its newest instrument is the ATACAMA LARGE MILLIMETER ARRAY; it also operates the VERY LARGE ARRAY, the VERY LONG BASELINE ARRAY, and the telescopes of GREEN BANK OBSERVATORY. Other facilities include one of the four NASA tracking stations (also at Green Bank) for the Very Large Baseline Interferometry Satellites.

National Solar Observatory (NSO) Principal US organization for solar astronomy. Its instruments include the MCMATH–PIERCE SOLAR TELESCOPE at KITT PEAK NATIONAL OBSERVATORY, and it also operates telescopes at Sacramento Peak, New Mexico. The NSO is a facility of the NATIONAL OPTICAL ASTRONOMY OBSERVATORY, and has headquarters in Tucson, Arizona.

National Space Development Agency of Japan (NASDA) Japan's space agency, established in 1969 as an umbrella organization for the nation's space activities and to promote the peaceful use of space. NASDA has developed and flown its own spacecraft and launch vehicles, and also participates in collaborative ventures with other countries. It is a partner in the INTERNATIONAL SPACE STATION, and Japanese astronauts have flown aboard the SPACE SHUTTLE.

native American astronomy Astronomy as practised in pre-Columbian America, until its cultures were eradicated by European invaders. The most advanced astronomy, of comparable sophistication to anything the Old World could offer despite their complete isolation from Europe and Asia, was possessed by the Maya of Central America, whose empire spread from the Yucatán Peninsula to cover other parts of modern Mexico, Guatemala and Belize and was at its height *c.*AD 300–900. Significantly, the Maya also had the best-developed mathematics in the New World. Some knowledge of their astronomy comes from a 'bark book' known as the Dresden Codex, written in hieroglyphics. Like most other indigenous American cultures, the Maya placed great importance on the motions of Venus and the Pleiades.

The Maya had an intricate system of interlocking calendars. One was based on a year of 365 days, with 18 months of 20 days each (20 was the base of their number-system) and a special period of five 'unlucky' days. Another 'year' was based on 360 days and a third sacred year had 260 days with 13 subdivisions of 20 named days each. An almanac in the Dresden Codex covers the cycles of its apparent revolutions with great precision. They also had some knowledge of the cycle of solar eclipses, though they were not able to predict their actual occurrence with certainty, only being able to give days on which eclipses might occur.

There is evidence of astronomy having been practised by other American cultures, though none reached the heights of the Mayan. Extensive systems of lines centred on the Inca Temple of the Sun display astronomical alignments, and huge pre-Inca figures marked in the desert at Nazca have some astronomical significance. Many other structures in South, Central and North America – for example the Medicine Wheel stone circle in Wyoming – also display astronomical alignments. The Aztecs' astronomy was largely bound up with ritual and sacrifice. In common with other city-cultures, including the Maya and the Toltec, they pictured a layered comos, the Earth occupying the lowest level and the creator the uppermost.

nautical twilight *See* TWILIGHT

navigational astronomy Use of astronomical observations to determine the geographical position of a vessel at sea or an aircraft. It is now largely of historical interest thanks to the GLOBAL POSITIONING SYSTEM (GPS) of navigational satellites.

The need for celestial navigation arose with the ever-increasing distances travelled by European seafarers from the 15th century onwards. Latitude was easy enough to determine by observing the altitude of Polaris (or that of the Sun) and consulting tables giving its declination for a given date. Simple instruments such as the CROSS-STAFF, QUADRANT and ASTROLABE were used for this purpose.

The determination of longitude depended on having accurate time available. Various methods, such as tables of the Moon's position among the stars or observations of Jupiter's satellites, were proposed for determining time at sea. However, they depended on the ability to predict positions with sufficient accuracy. An additional difficulty was that the observations required for comparison with positions in the tables were by no means easy to make on a rolling and pitching ship. The development of the MARINE CHRONOMETER in the 18th century and the annual publication of *The Nautical Almanac* in Britain from 1766 eventually made possible the determination of longitude at sea. Time signals derived from transit observations were provided at various ports so that chronometers could be accurately set and checked.

Radio time signals became common in the 1920s and 1930s, thus removing the need for independent timekeeping except at times of war; navigators on naval ships and military aircraft received training in 'astronavigation'. Further developments based on radio techniques were the Loran, Decca and Omega systems, which involved (in essence) timing the radio transmissions from fixed transmitters, multiplying by the speed of light to get the distances and plotting the result on a chart to get the position.

n-body problem *See* MANY-BODY PROBLEM

NEA *See* NEAR-EARTH ASTEROID

neap tide Low-level high TIDES that occur when the directions of the Sun and Moon are at a right angle.

NEAR Abbreviation for Near Earth Asteroid Rendezvous spacecraft, later known as NEAR SHOEMAKER.

near-Earth asteroid (NEA) ASTEROID with an orbit that brings it close to the Earth's orbit. NEAs are subdivided into three main classes – ATEN ASTEROIDS, APOLLO ASTEROIDS and AMOR ASTEROIDS. *See also* POTENTIALLY HAZARDOUS ASTEROID; NEAR-EARTH OBJECT

near-Earth object (NEO) ASTEROID or COMET with an orbit that brings it close to the orbit of the Earth. By analogy with the classes of orbit that comprise the near-earth asteroids, the criterion for an NEO is that the perihelion distance be less than 1.3 AU. It is usual, though, to regard NEOs as including only those objects with smaller perihelia, making very close approaches to the Earth (and possibly impacts) feasible in the near term.

near-infrared Region of the electromagnetic spectrum between around 0.75 and 5 μm. Near-infrared radiation can be observed from the ground at good dry sites, except for a few narrow regions that are absorbed by water vapour and carbon dioxide in Earth's atmosphere. Many objects, including stars, emit strongly in this range, and there are important atomic and molecular features present, notably due to hydrogen. *See also* INFRARED ASTRONOMY

NEAR Shoemaker First spacecraft to orbit and land on an asteroid, EROS, in 2000 February and 2001 February

SELECTED NEAR-EARTH ASTEROIDS

Catalogue number	Name	Discovery year	Perihelion distance (AU)	Orbital period (years)	Approximate size (km)	Notes
433	EROS	1898	1.13	1.76	25	first known near-Earth/Amor-type asteroid
719	ALBERT	1911	1.19	4.28	3	second Amor-type asteroid
887	Alinda	1918	1.08	3.91	8	third Amor-type asteroid
1036	GANYMED	1924	1.23	4.34	50	fourth Amor; largest near-Earth asteroid
1221	AMOR	1932	1.09	2.66	1.5	sixth Amor-type asteroid; class archetype
1566	ICARUS	1949	0.19	1.12	1.5	early Apollo discovery, small perihelion
1580	Betulia	1950	1.13	3.26	5	very high inclination Amor (52°)
1620	GEOGRAPHOS	1951	0.83	1.39	4	radar images show elongated shape
1627	Ivar	1929	1.13	2.54	9	fifth Amor-type asteroid
1685	TORO	1948	0.77	1.60	6	early, large, Apollo discovery
1862	APOLLO	1932	0.65	1.78	2.5	first known Earth-crosser; class archetype
1863	Antinous	1948	0.89	3.40	4	early Apollo discovery
1866	SISYPHUS	1972	0.87	2.61	10	largest known Apollo asteroid
1915	Quetzalcoatl	1953	1.09	4.05	0.8	orbital period one-third that of Jupiter
2062	ATEN	1976	0.79	0.95	1.5	archetype of Aten class
2063	BACCHUS	1977	0.70	1.12	1.5	radar images reveal non-spherical shape
2100	Ra-Shalom	1978	0.47	0.76	3	early Aten discovery, very short period
2101	ADONIS	1936	0.44	2.57	0.8	possible associated meteor shower
2102	Tantalus	1975	0.91	1.47	3	extremely high inclination (64°)
2201	OLJATO	1947	0.62	3.20	4	may be an extinct cometary nucleus
2212	HEPHAISTOS	1978	0.36	3.19	8	one of the largest Apollo asteroids
2340	Hathor	1976	0.46	0.78	0.6	early Aten discovery, very short period
3200	PHAETHON	1983	0.14	1.43	5	parent of Geminid meteor shower
3552	Don Quixote	1983	1.21	8.70	10	one of the largest near-Earth asteroids
3753	CRUITHNE	1986	0.48	1.00	5	same orbital period as Earth
4015	WILSON–HARRINGTON	1949/79	1.00	4.30	3	also known as Comet Wilson–Harrington
4179	TOUTATIS	1989	0.92	3.98	4	radar images reveal potato shape
4183	Cuno	1959	0.72	2.79	5	early, large, Apollo discovery
4581	ASCLEPIUS	1989	0.66	1.03	0.4	near-miss of Earth at time of discovery
4660	NEREUS	1982	0.95	1.82	1.0	candidate spacecraft target
4769	CASTALIA	1989	0.55	1.10	1.5	radar images reveal twin-lobed shape
4954	Eric	1990	1.10	2.83	13	one of the largest near-Earth asteroids
6489	GOLEVKA	1991	1.00	3.98	0.8	radar images show irregular shape
1937 UB	HERMES	1937	0.62	2.10	1.0	discovered during near-miss; lost
1954 XA	–	1954	0.38	0.62	0.8	first Aten-type discovery; lost

By May 2001 there were 109 Aten, 624 Apollo and 624 Amor asteroids known. The table lists some of the more notable individuals from this large discovered population.

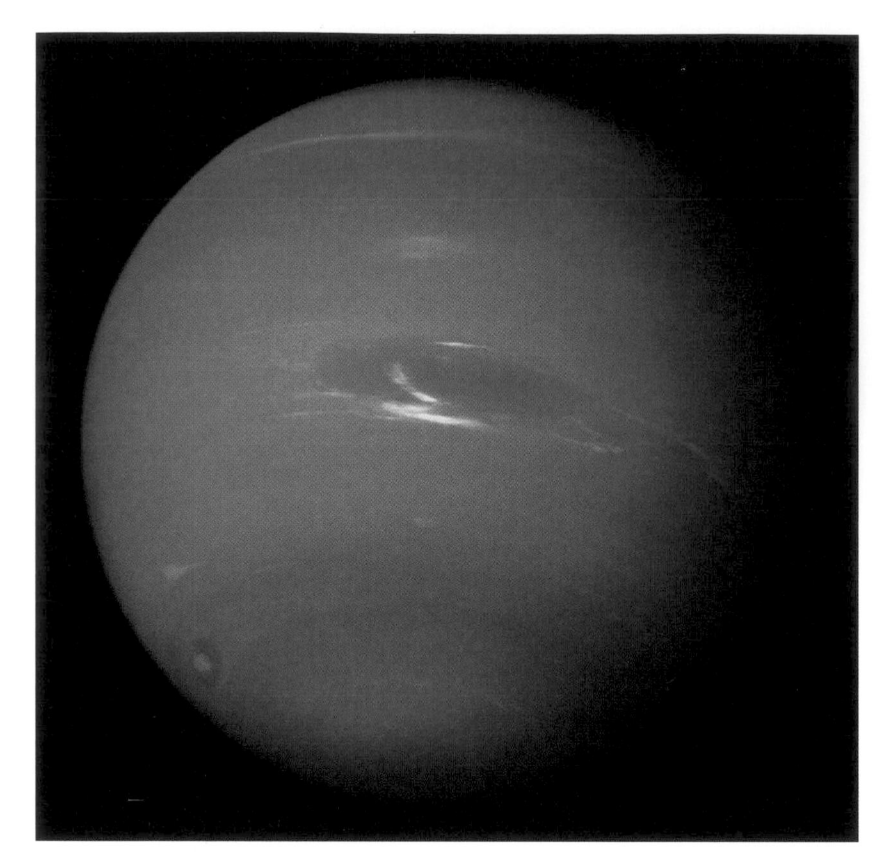

▲ **Neptune** Neptune appears bluish because methane in its atmosphere absorbs red light. Ice particles in its upper cloud decks also scatter blue light.

▼ **Neptune** The bands in Neptune's atmosphere are more difficult to detect than those of Jupiter or Saturn. In this colour-enhanced image, the highest clouds are red and yellow, with white ones lower in the atmosphere. The blue line just below the equator is the equatorial belt, where wind speeds reach almost 1500 km/h (900 mph).

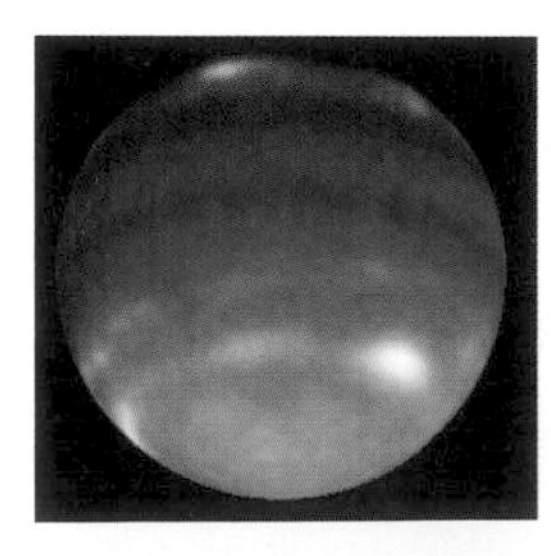

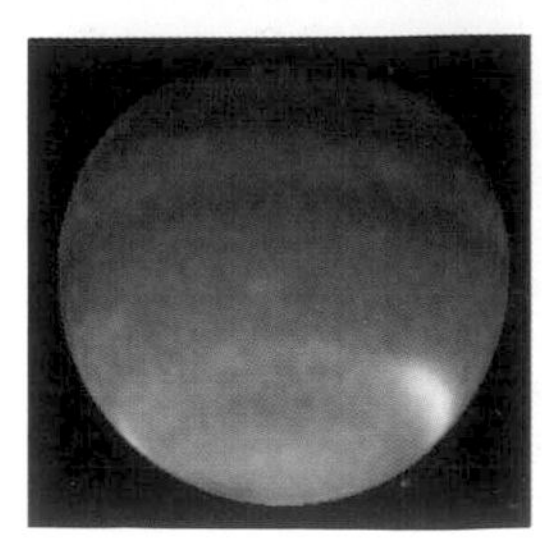

respectively. The Near Earth Asteroid Rendezvous (NEAR) spacecraft was also the first in NASA's DISCOVERY programme series to be launched, in 1996 February. The craft was renamed NEAR Shoemaker in honour of the astrogeologist Gene Shoemaker, who was killed in an accident while the spacecraft was en route. NEAR flew past the asteroid MATHILDE in 1997 June and made a gravity assist flyby of the Earth in 1998 January, setting itself up for the rendezvous with Eros. The first attempt to enter orbit in 1999 January was thwarted by an engine fault, but ingenious troubleshooting enabled a second attempt to be made in 2000 February. NEAR surveyed Eros and returned a wealth of data, including close-up images of the surface. Scientists and mission controllers arranged a spectacular finale to the mission, guiding the craft to a controlled touchdown on the surface in 2001 February. This originally unscheduled manoeuvre was a surprising success.

near-ultraviolet Region of the ELECTROMAGNETIC SPECTRUM between around 380 and 310 nm, which can be observed from the ground but is not visible to the naked eye.

NEAT Acronym for Near-Earth Asteroid Tracking, an asteroid research programme based at the JET PROPULSION LABORATORY in Pasadena, California, USA. Several comets discovered in this project have been given the name NEAT.

nebula Celestial object that appears larger and fuzzier than a star. Originally the term was used for any such object, but it is now restricted to those objects that are clouds of gas and dust and so cannot be resolved into stars.

Most EMISSION NEBULAE emit their light in a series of EMISSION LINES. In these nebulae the atoms are ionized by ultraviolet radiation from a hot star or stars, and the visible light is emitted as the electrons are recaptured. Hydrogen is the most abundant element in almost all nebulae, and because it is ionized such nebulae are called HII REGIONS. Hydrogen gives off its strongest light in the red at 656 nm, thus HII regions often appear red on images; however, they can appear greenish when seen visually because of a prominent pair of FORBIDDEN LINES at 496 and 501 nm

NEPTUNE: DATA	
Globe	
Diameter (equatorial)	49,528 km
Diameter (polar)	48,686 km
Density	1.64 g/cm^3
Mass (Earth = 1)	17.14
Volume (Earth = 1)	57.67
Sidereal period of axial rotation	16^h 07^m
Escape velocity	23.5 km/s
Albedo	0.41
Inclination of equator to orbit	29° 34′
Temperature at cloud-tops	55 K
Surface gravity (Earth = 1)	0.98
Orbit	
Semimajor axis	30.06 AU = 4497 × 10^6 km
Eccentricity	0.0097
Inclination to ecliptic	1° 46′
Sidereal period of revolution	164.79 years
Mean orbital velocity	5.43 km/s
Satellites	8

emitted by oxygen. If the star exciting an emission nebula is very hot, such that most of its radiation is in the ultraviolet, the nebula can appear much brighter than the star, since it processes the ultraviolet radiation to visible light. Although looking bright and almost solid, emission nebulae are very tenuous. Typically every kilogram of material is spread over a volume of a thousand million cubic kilometres. Nebulae that are SUPERNOVA REMNANTS often emit SYNCHROTRON RADIATION; this is usually in the radio region, but in a few cases, like the CRAB NEBULA, it can extend into the visible.

Some nebulae are seen because they scatter starlight. In this case we see not the gas itself but the myriad tiny motes of dust that normally are mixed with it. The dust is lit up like smoke in sunlight. These REFLECTION NEBULAE are always fainter than the star that illuminates them, unless a denser portion of the nebula hides the star from us. Reflection nebulae generally appear blue on images.

DARK NEBULAE, such as the COALSACK, are not lit by any star, thus they can be seen only when silhouetted against something brighter. *See also* COMETARY GLOBULE; GIANT MOLECULAR CLOUD; GLOBULE; HI REGION; INTERSTELLAR MATTER; MOLECULAR CLOUD; PLANETARY NEBULA; PLERION

nebular hypothesis Theory that the Solar System was formed from nebulous material in space. In 1755 the philosopher Immanuel KANT published his *Universal Natural History and Theory of the Heavens*, in which he suggested that the Sun and planets condensed out of diffuse primordial matter under the attraction of gravity. A similar theory was developed independently and in greater detail by Pierre Simon LAPLACE in his *Exposition du systeme du monde* (1796). He envisioned that the Sun originally had an extended hot atmosphere throughout the volume of the Solar System. As this cooled and contracted, it spun more rapidly and threw off a series of gaseous rings; the planets condensed from this material. A similar process on a smaller scale supposedly produced satellites from contracting planets.

This nebular hypothesis offered explanations for some properties of the Solar System, including the nearly circular and coplanar orbits of the planets, and their motion in the same direction as the Sun's rotation. Some elements of the nebular hypothesis are echoed in modern theories of planetary formation, but most have been superseded. *See also* COSMOGONY

nebular variable VARIABLE STAR associated with nebulosity. The most important are the FU ORIONIS STAR and the T TAURI STAR classes. These are very young stars that have started to shine but have not yet reached the MAIN SEQUENCE. They are believed to represent the last stage in the development of PROTOSTARS, before they

◀ **Neptune** In 1989 Voyager 2 imaged Neptune's rings. The material in them is unevenly distributed, leading to an observed 'clumpy' structure.

settle down on the main sequence. The T Tauri stars (classified as INT) vary irregularly in brightness, whereas the FU Orionis stars (FU) exhibit a single brightening, possibly as they assume a stable state. Nebular variables are F-, G- or K-type giant stars, many still surrounded by warm cocoons of dust and gas. There are a great number in the ORION NEBULA. While contracting, nebular variables probably lose large amounts of material in the form of a stellar or T-Tauri wind. The rate of mass loss is about one-ten millionth of a solar mass per year. Such stars are also rotating very rapidly, throwing off material at speeds of up to 300 km/s (180 mi/s). Subclasses of the nebular variables are named after their prototype stars – YY Orionis (extremely young), T ORIONIS and RW AURIGAE. The UVN subtype of FLARE STAR is also sometimes considered to be a nebular variable.

nebulium Hypothetical element once thought to exist in planetary and diffuse nebulae in order to explain unidentified green emission lines at 500.7 and 495.9 nm (as well as others later found). The nebulium lines were later discovered to be FORBIDDEN LINES of various ionization states of oxygen, nitrogen, neon and other atoms (the 500.7 and 495.9 nm lines coming from doubly ionized oxygen [OIII]).

Nectaris, Mare (Sea of Nectar) Lunar lava plain, roughly 290 km (180 mi) in diameter, located in the south-east quadrant of the Moon. This region was struck by an enormous object, producing a multi-ring impact BASIN. Basalts later flooded the inner regions. The outer basin ring is deeply eroded, but is still visible in the south-west as Rupes ALTAI. A prominent dark-haloed crater has excavated the dark lava in the western region. The basin's EJECTA scours the Moon to the south, producing the Janssen formation, including Vallis Rheita.

NEO Abbreviation of NEAR-EARTH OBJECT

Neptune Eighth planet from the Sun. It is the fourth largest (by diameter) and the third most massive planet. From 1979 to 1999 PLUTO, thanks to its eccentric orbit, was nearer to the Sun than Neptune: during this period Neptune was thus the ninth and farthest known planet.

At magnitude +7.8, Neptune is too faint to be seen with the unaided eye, and its discovery awaited the invention of the telescope. After URANUS was discovered by William HERSCHEL in 1781, its orbit was calculated in order to follow the planet's motion. Before about 1820, however, Uranus was found to get ahead of its predicted position and after this to lag behind. It was suggested that this anomaly was a result of the gravitational influence of a unknown planet orbiting farther from the Sun. From observed irregularities in the motion of Uranus, the mathematicians John Couch ADAMS, working in England, and Urbain LE VERRIER, working independently in France, both predicted the location of the unseen planet. A comedy of errors and ineptitude prevented English astronomers following up Adams' predictions, and it was Le Verrier who eventually persuaded the German astronomer Johann GALLE in Berlin to search the region of the sky in which he had calculated that the new planet would lie. On 1846 September 23, Galle and his student, Heinrich D'ARREST, located the new planet very close to the predicted position.

Scholars have largely given joint credit for the discovery to Adams and Le Verrier because their predictions were based on a sound analysis of data and a thorough knowledge of gravitational theory. At least, this was the situation for 134 years. In 1980, however, the astronomical world was astonished when an astronomer and Galileo scholar announced that Galileo himself had first seen Neptune with his telescope in 1613 January, some 233 years before it was identified by Galle. It was not recognized as a planet: Galileo's sketches showed its apparent motion, but his interest was concentrated on Jupiter and the satellites and so the discovery was delayed.

Although Neptune is sometimes described as a 'jovian' planet, its resemblance to JUPITER and SATURN is superficial. Neptune is, however, quite similar to Uranus in many ways, including diameter, composition and mass. Seen through the telescope, both planets are nearly featureless, in strong contrast to the abundant cloud structures on Jupiter and the banded appearance of Sat-

▼ **Neptune** The blue planet's atmosphere has many features, including this high-level cirrus. Most atmospheric structures on Neptune, such as the Great Dark Spot, appear to be short-lived.

NEPTUNE SATELLITES AND RINGS

Rings				
ring or gap	distance from centre (thousand km)		width (km)	
Galle	41.9–43.6		1700	
Le Verrier	53.2		15	
Lassell/Arago	53.2–59.1		5900	
Adams	62.9		50	
Satellites				
	diameter (km)	distance from centre of planet (thousand km)	orbital period (days)	mean opposition magnitude
Naiad	58	48	0.29	24.7
Thalassa	80	50	0.31	23.8
Despina	150	53	0.33	22.6
Galatea	160	62	0.43	22.3
Larissa	208 × 178	74	0.55	22.0
Proteus	436 × 416 × 402	118	1.12	20.3
Triton	2706	355	5.88 *R*	13.5
Nereid	340	5513	360.1	18.7

▲ **Neptune** Among Neptune's eight known satellites are some of the largest planetary satellites. Triton, which was discovered only a few days after the planet itself, orbits in a retrograde direction. Nereid is farther out has a highly elliptical orbit. The other six satellites are all closer to the planet than Triton.

▼ **Nereid** Neptune's outermost satellite was discovered in 1949 by Gerard Kuiper. The most detailed images of it were obtained by Voyager 2 from a distance of 4.7 million km (2.9 million mi).

urn's clouds. While Uranus is a faint greenish colour, Neptune is weakly tinted blue.

The atmosphere of Neptune shows faint bands rather like those of Jupiter and Saturn. The bands are not easily seen from Earth as the planet appears so small, subtending only 2″.5. The planet is composed mainly of molecular hydrogen and helium, with a small silicate and metallic core. The atmosphere is primarily methane and other hydrocarbons. During its 1989 encounter, VOYAGER 2 identified several dark spots, the largest of which was called the 'Great Dark Spot', though recent images from the Hubble Space Telescope show that it has now disappeared. There was also a bright, fast-moving spot, referred to as the 'Scooter'. Whether these spots have dissipated or faded is not known. Also identified were bands of light-coloured cirrus cloud, which from the shadows cast appeared to be some 50–100 km (30–60 mi) higher than the tops of the main cloud belts. The overall haziness of the atmosphere may be anti-correlated with the sunspot cycle, in that at times of maximum solar activity Neptune appears in our telescopes to be a little fainter than average, while at sunspot minimum it is brighter. The Sun's activity thus appears to influence the atmosphere of Neptune, some 4500 million km (2800 million mi) away, by the formation of a global haze affecting the apparent brightness of the planet as seen from Earth. However, a recent analysis of brightness observations of Neptune and also Uranus does not support this view in the long term.

By following the motions of the cloud top features, Voyager 2 and Earth-based astronomers (observing in the infrared) have measured the period of the planet's rotation. These measurements give periods in the region of about 17.7 hours for the rotation of the upper atmosphere. From radio emissions, Voyager 2 determined a rotation period of 16 hours and 7 minutes for the planet.

Being a GAS GIANT, Neptune has only a small solid core, which is thought to be of silicon, iron and other metals. The bulk of the planet is largely composed of hydrogen and helium, which under the temperature and pressure act as supercooled liquids. Neptune's mean density is 1.64 g/cm^3, which is somewhat higher than the other gas giants. Convection currents in the mantle carry heat up from the core region. As Neptune continues to contract, because of its own strong gravitational field, and as the constituents of the mantle gradually differentiate, with the heavier molecules settling to the bottom, energy is liberated from the planet. This energy is emitted as excess heat. The temperature in the stratosphere at the 100 mbar pressure level was measured to be 55 K, though the temperature predicted on the basis of Neptune's distance from the Sun and other parameters is a few degrees cooler. Voyager 2 also identified a weak magnetic field, inclined at 47° to the rotation axis, radio emissions, aurorae and radiation belts.

Neptune has eight known satellites, the brightest of which, TRITON, was discovered by William LASSELL shortly after Neptune's discovery was announced. The others are much fainter. Nereid was discovered in 1949, with a further six identified during Voyager's flyby. Analysis of observations during stellar occultations by Neptune suggested that several ring arcs surrounded the planet, and Voyager 2 imaged several rings of varying intensity. Four obvious rings were detected, with ring particles found in much of the space in between.

Nereid Second satellite of NEPTUNE to be discovered, by Gerard Kuiper in 1949. It is about 340 km (210 mi) in size. Despite being slightly smaller than PROTEUS, Nereid was discovered telescopically long before any of the inner Neptunian satellites were spotted (in VOYAGER 2 images), because of its much greater distance from the glare of the planet. Nereid takes 360 days to orbit Neptune, at an average distance of 5.51 million km (3.42 million mi) from the planet's centre. It has an unusual orbit, with a low inclination (near 7°) and the highest eccentricity of all known planetary satellites (0.751), indicating that Nereid is likely a captured body, previously a CENTAUR or a TRANS-NEPTUNIAN OBJECT.

Nereus APOLLO ASTEROID discovered by Eleanor Helin in 1982; number 4660. Nereus is a prime candidate for future visits by space probes because of its low speed relative to the Earth. Its name reflects this fact: it is 'near us' in terms of accessibility, if not distance. *See* table at NEAR-EARTH ASTEROID

neutral point Position between two bodies where their gravitational pulls are equal in strength; the neutral point is closer to the less massive body.

neutrino astronomy Study of bodies in the Universe via their emission of neutrinos.

The neutrino is an electrically neutral subatomic particle whose existence was first suggested in 1930 by the Austrian physicist Wolfgang Pauli (1900–1958) in order to account for the apparent non-conservation of energy in β-decays. The neutrino was required to take away a varying amount of excess energy during the reactions in an (almost) undetectable manner, so that the total energy released remained constant. The neutrino was discovered experimentally in 1955 by the American physicists Clyde Cowan Jnr (1919–) and Frederick Reines (1918–). Its interaction with other forms of matter is indeed weak: of all the neutrinos produced at the centre of the Sun, 99.99% escape without any contact with other particles during their 700,000 km (400,000 mi) journey to the Sun's surface. There are now known to be three varieties of neutrino, together with their antiparticles. The original neutrino is associated with normal electrons and is called the electron-neutrino. The other two neutrinos are associated with the heavy electrons (leptons) known as muons and tau particles and so are named muon-neutrinos and tau-neutrinos. It has recently been shown experimentally that one type of neutrino can metamorphose into another. This is of great significance for the SOLAR NEUTRINO problem and also implies that neutrinos have a mass, albeit perhaps only a millionth that of a normal electron.

Neutrinos from space are detected in two main ways, either through their taking part in nuclear reactions, or by their collisions with other particles. The first detector started operating in 1968 and continues to this day. Devised by the American chemist Raymond Davis, Jr (1916–), it operates by detecting the radioactive decay of argon-37. The argon-37 is produced by a neutrino interacting with chlorine-37, and so each decay corresponds to a neutrino capture. The detector is a tank containing 600 tonnes of tetrachloroethene; it is buried 1.5 km (0.9 mi) below ground at the HOMESTAKE MINE in Dakota in order to shield it from unwanted interactions. Two other detectors, SAGE (in the Caucasus) and GALLEX (in Italy), detect neutrinos in a similar fashion, but based upon the conversion by a neutrino of gallium-71 to radioactive germanium-71. The second class of detectors operates by detecting the ČERENKOV RADIATION from high energy electrons in water. The electrons result either from inverse beta decay, in which a neutrino and proton convert to a neutron and positron (positive electron), or gain their energy by direct collision between the neutrino and an electron. The water for this class of detectors is either contained in large tanks buried underground, as at IMB (the Irvine Michigan Brookhaven experiment) in the USA, Super-Kamiokande in Japan and SNO (Sudbury Neutrino Observatory) in Canada; or is a part of the ocean, for example DUMAND (Deep Undersea Muon and Neutrino Detection) in the Pacific Ocean and NESTOR in the Mediterranean Sea; or is part of the Antarctic icecap, for example the Antarctic Muon and Neutrino Detector Array (AMANDA).

Neutrinos are expected to be produced in many reactions, but to date have only been detected from three astronomical sources. The first is from the core of the Sun, where two neutrinos result from the formation of a helium nucleus (*see* NUCLEOSYNTHESIS); the second is from the 1987 SUPERNOVA in the Large Magellanic Cloud; and the third is as a result of COSMIC RAY interactions with nuclei at the top of the Earth's atmosphere. The potential importance of neutrinos, however, lies in their ability to

N

◀ neutrino astronomy When operating, neutrino detectors such as the Super-Kamiokande in Japan contain exceptionally pure water. Because neutrinos themselves are so difficult to detect most instruments search instead for radiation and decay particles resulting from collisions between the neutrinos and atoms in the water.

allow us to 'see' directly what is happening at the centres of stars, and in providing information on the Universe that is independent of ELECTROMAGNETIC RADIATION.

neutron Electrically neutral subatomic particle that along with the PROTON forms the nuclei of atoms. The British physicist Sir James Chadwick (1891–1974) discovered the neutron in 1932. Inside a nucleus it is stable, but as an independent particle it decays with a half-life of about 12 minutes to a proton, electron and anti-neutrino. Its mass is 1.674929×10^{-27} kg (1.008793 amu).

neutron star Densest and tiniest star known, composed of DEGENERATE MATTER and supported by neutron degeneracy pressure. Neutron stars are the end points of stars whose mass after nuclear burning is greater than the CHANDRASEKHAR LIMIT for white dwarfs, but whose mass is not great enough to overcome the neutron degeneracy pressure to become BLACK HOLES. The neutron, which along with the proton makes up the atomic nucleus, was discovered in 1932. At the end of 1933 Walter BAADE and Fritz ZWICKY tentatively suggested that in SUPERNOVA explosions, ordinary stars are turned into stars that consist of extremely closely packed neutrons. They called these stars neutron stars.

Neutron stars were thought to be too faint to be detectable and little work was done on them until 1967 November, when Franco Pacini (1939–) pointed out that if the neutron stars were spinning and had large magnetic fields, then electromagnetic waves would be emitted. Unbeknown to him, radio astronomer Antony HEWISH and his research assistant Jocelyn Bell (*see* BELL BURNELL) at Cambridge were shortly to detect radio pulses from stars that are now believed to be highly magnetized, rapidly spinning neutron stars, known as PULSARS.

Any star with an initial main-sequence mass of around 10 solar masses or above has the potential to become a neutron star. As the star evolves away from the main sequence, subsequent nuclear burning produces an iron-rich core. When all nuclear fuel in the core has been exhausted, the core must be supported by degeneracy pressure alone. Further deposits of material from shell burning cause the core to exceed the Chandrasekhar limit. Electron degeneracy pressure is overcome and the core collapses further, sending temperatures soaring to over 5 billion K. At these temperatures, photodisintegration (the breaking up of iron nuclei into alpha particles by high-energy gamma rays) occurs. As the temperature climbs even higher, electrons and protons combine to form neutrons, releasing a flood of neutrinos. When densities reach nuclear density of 4×10^{17} kg/m^3, neutron degeneracy pressure halts the contraction. The infalling outer atmosphere of the star is flung outwards and it becomes a Type II or Type Ib supernova. The remnant left is a neutron star. If it has a mass greater than about 2–3 solar masses, it collapses further to become a BLACK HOLE. Other neutron stars are formed within CLOSE BINARIES.

The surface of the neutron star is made of iron. In the presence of a strong magnetic field the atoms of iron polymerize. The polymers pack to form a lattice with density about ten thousand times that of terrestrial iron and strength a million times that of steel. It has excellent electrical conductivity along the direction of the magnetic field, but is a good insulator perpendicular to this direction.

Immediately beneath this surface the neutron star is still solid, but its composition is changing. Larger nuclei, particularly rich in neutrons, are formed, and materials that on Earth would be radioactive are stable in this environment, for example nickel-62. With increasing depth the density rises. When it reaches 400 thousand million times that of water the nuclei can get no larger and neutrons start 'dripping' out. As the density goes up further the nuclei dissolve in a sea of neutrons. The neutron fluid is a superfluid – it has no viscosity and no resistance to flow or movement.

Within a few kilometres of the surface the density has reached the density of the atomic nucleus. Up to this point the properties of matter are reasonably well understood, but beyond it understanding becomes increasingly sketchy. The composition of the core of the star is particularly uncertain: it may be liquid or solid; it may consist of other nuclear particles (pions, for example, or hyperons); and there may be another phase change, where QUARKS start 'dripping' out of the neutrons, forming another liquid.

A neutron star has a mass comparable to that of the Sun, but as it is only about 10 km (6 mi) in radius, it has an average density 1000 million million times that of water. Such a large mass in such a small volume produces an intense gravitational force: objects weigh 100,000 million times more on the surface of a neutron

N

▼ neutron star A hot X-ray source within the supernova remnant IC443 is thought to be the neutron star created in the explosion. Measurements of its movement through the nebula have led to the conclusion that light from the supernova reached Earth some 30,000 years ago.

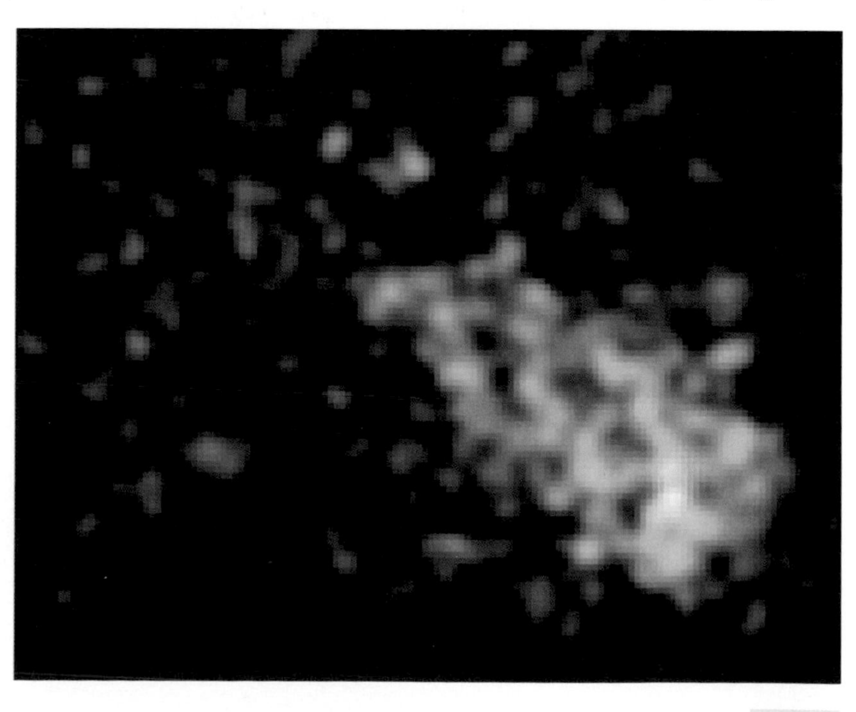

star than on the surface of the Earth. The intense gravitational field affects light and other electromagnetic radiation emitted by the star, producing significant redshift (z approximately equal to 0.2).

The strong gravitational attraction allows neutron stars to spin rapidly (hundreds of revolutions per second) without disintegrating. Such spin rates are expected if the core of the original star collapses without loss of angular momentum. If the original star has a magnetic field, then this too may be conserved and concentrated in the collapse to a neutron star. Pulsars, GAMMA-RAY BURST sources, and the neutron stars in some X-RAY BINARIES are believed to have magnetic fields of strength about 100 million Tesla (roughly a million million times the strength of the Earth's magnetic field).

There are thought to be of the order of 10^8 neutron stars in the Galaxy, but they can only be easily detected in certain instances, such as if they are a pulsar or part of a binary system. Non-rotating and non-accreting neutron stars are virtually undetectable, but the Hubble Space Telescope has observed one thermally radiating neutron star (RX J185635-3754). Its surface temperature is 6×10^5 K and its radius 12–13 km (7.5–8 mi). Neutron star mergers may produce gamma-ray bursts.

Newall Telescope Refractor of 0.63 m (25 in.) aperture built by Thomas Cooke & Sons for amateur observer Robert Stirling Newall (1812–89) of Gateshead, England. Completed in 1869, it was for a while the world's largest refractor; it is now in Greece, and is operated by the National Observatory of Athens.

Newcomb, Simon (1835–1909) American mathematical astronomer, born in Canada, who calculated highly accurate values for astronomical constants and tables of the motions of the Moon and planets and was the most influential astronomer of his day. In 1861 he became professor of mathematics at the US NAVAL OBSERVATORY (USNO), Washington. Studying the orbits of asteroids, he proved that they could not have originated from the break-up of a single body, a once-popular theory. Newcomb began a lifelong study of the Moon's motion. His results suggested that the discrepancy between the Moon's calculated and observed motion was caused by a gradual slowing of the Earth's rotation; this was finally proved by others, including E.W. BROWN.

Four years later, Newcomb became superintendent of the US Nautical Almanac Office, where he set about refining the orbits of the planets, using observations dating back to 1750. This was a major feat of celestial mechanics, carried out with the collaboration of E.W. Brown, G.W. HILL and others. The work required an accurate value of the SOLAR PARALLAX (π), which Newcomb had begun working towards some years before. To this end he observed the 1874 and 1882 transits of Venus, but decided that ultimately the accuracy of π depended on having an accurate value of c, the speed of light, to measure which he devised a method based on that of Léon FOUCAULT. He also established that the discrepancy between the observed and calculated values of the advance of Mercury's perihelion was 43″, exactly the difference later explained by the theory of GENERAL RELATIVITY.

Newcomb's planetary tables, published in 1899, remained in use until the 1950s. Equally enduring was his consistent system of refined astronomical constants, calculated jointly with Arthur Matthew Weld Downing (1850–1917) of the UK Nautical Almanac Office, and adopted worldwide in 1901.

New General Catalogue (NGC) In full, the *New General Catalogue of Nebulae and Clusters of Stars*, compiled by J.L.E. DREYER. At the suggestion of the Royal Astronomical Society, Dreyer combined what were then the three main catalogues of nebulae – John HERSCHEL's *General Catalogue of Nebulae* (which Dreyer had already revised, in 1879), and those of Heinrich D'ARREST and Lord ROSSE; and supplemented the amalgamated listing with more recent observations by a number of other astronomers. The NGC, published in 1888, contains nearly 7840 non-stellar objects (including galaxies, although their real nature was not then known) and over 5000 more are listed in two subsequent *INDEX CATALOGUES*. Objects are still referred to by their NGC numbers. A revision for the present epoch has been published as *NGC 2000.0* (1988).

New Millennium NATIONAL AERONAUTICS AND SPACE ADMINISTRATION (NASA) programme to pioneer new technologies for all types of spacecraft, from planetary explorers to Earth observation satellites, including new propulsion technologies and sensors. *See also* DEEP SPACE 1

New Technology Telescope (NTT) Innovative 3.6-m (142-in.) telescope at LA SILLA OBSERVATORY, operated by the EUROPEAN SOUTHERN OBSERVATORY. Completed in 1989, the NTT pioneered several aspects of modern telescope design, including the use of ACTIVE OPTICS and ADAPTIVE OPTICS, and is essentially a prototype of the VERY LARGE TELESCOPE. Frequently operated remotely, the instrument had its control system upgraded to VERY LARGE TELESCOPE standards in 1997. It is equipped with spectroscopic and imaging instruments designed to capitalize on its excellent image quality.

Newton, Isaac (1642–1727) English physicist and mathematician. Most of his theories – on gravitation, light and calculus, for example – he developed in basic form while in his early twenties. In 1668 he built the first reflecting telescope in order to avoid the chromatic aberration inherent in lenses. In 1684, urged by Edmond Halley, he began work on *Principia* (1687), in which he put forward his law of gravitation and laws of motion, and showed how gravitation explained Kepler's laws. In his *Opticks* (1704) he set out his theories of light, including how white light is made up of the colours of the spectrum. Newton's greatest mathematical achievement was his invention of calculus.

Born prematurely at Woolsthorpe, a hamlet near Grantham, Lincolnshire, three months after his father's death, Newton was left in the care of his grandmother, during which time his mother married again. Hating his stepfather, Newton suffered all his life from a sense of insecurity; it was this that later in life made him abnormally sensitive to criticism of his scientific work. When his stepfather died, Newton's mother determined that he should become a farmer and manage her estates, but he showed no interest in such a career, and she was persuaded to allow him to be prepared for university. In 1661 Newton entered Trinity College, Cambridge, as a subsizar – an impoverished undergraduate who performed menial tasks to help pay for his keep.

Newton graduated in 1665 April, having already invented his mathematical method of fluxions, though few people were aware of the fact. However, the university was then closed because of the plague and the scholars dispersed. Newton went back to Woolsthorpe where, except for a return for a couple of months to Cambridge, he consolidated his mathematical work and his nascent optical theory of the nature of colour. He also recognized that each planet was held in its orbit by a force emanating from the Sun which diminished according to the square of its distance – an inverse-square law. But he published nothing.

By 1671 Newton had built his first reflecting telescope (*see* NEWTONIAN TELESCOPE). At the time all telescopes were refractors, forming their image with a lens, not a mirror. To avoid CHROMATIC ABERRATION and to provide good sharp images, lenses had to be of very long focal length, making telescopes equally as long and thus very cumbersome. Newton's theory that white light is a mixture of the light of all colours persuaded him (wrongly) that the refractor could not be cured of this defect. He therefore designed a reflecting telescope – the first successful one ever to be made and it aroused much interest; not only was it free from chromatic aberration, but it was far shorter than any refractor. A duplicate instrument was sent to the Royal

Society in London at their request, and Newton was promptly elected a Fellow. At his own suggestion he then submitted a paper on his theory of light and colour. This historic paper was criticized, and the ensuing controversy severely upset him, though in 1675 he did submit two other optical papers. One was about the colours seen in very thin films of oil and other materials, the other on his theory that light was composed of tiny particles, or 'corpuscles'. These also caused controversy, and in the end, Newton shut himself away in Cambridge and shunned further publication.

Not until late in 1679 did his resume correspondence, with Robert Hooke, Honorary Secretary to the Royal Society and previously a severe critic of Newton's theories of light and colour. An exchange of letters led Newton to re-examine his work on planetary orbits, but again he published nothing; not until 1684 was he stimulated to consider making his ideas known. This was the result of a visit from Edmond HALLEY.

In London, Halley had been discussing the problem with Hooke and with the astronomer-architect Christopher Wren (1632–1723). No one had been able prove mathematically that an inverse-square law would result in elliptical planetary orbits, and Halley decided to visit Newton and ask him whether he knew the reason. Newton replied that he did, and had a mathematical proof. Unable to locate it among his papers, he promised to convey it to Halley later. This he did, sending a small tract with the title *De motu* ('On Motion'). When Halley received it, he was mathematician enough to realize that here was a document of the greatest significance. He therefore prevailed on the Royal Society to publish an entire book on the subject by Newton, and also persuaded Newton to continue expanding his ideas for publication.

There were difficulties, however. On the one hand the Royal Society was short of money and could not afford to publish; Halley himself had to edit the manuscript and pay the printer out of his own pocket. On the other hand, Hooke raised objections, claiming that Newton had stolen his results. Newton was furious – he went through his text deleting nearly every reference to Hooke, and refused to complete the book. It took all of Halley's undoubted diplomatic gifts to save the situation, and at last, in 1687 July, the volume appeared with the title *Philosophiae naturalis principia mathematica* ('The Mathematical Principles of Natural Philosophy'). The book, known by its shortened name, *PRINCIPIA*, was a *tour de force*, one of the most important scientific works ever written. It not only solved virtually all the questions and problems of traditional astronomy, but stimulated a vast amount of further research.

Newton was exhausted after this vast intellectual effort, and seems to have suffered a nervous breakdown. He withheld publication of his *Opticks* and his acceptance of the presidency of the Royal Society until after Hooke's death, and did no more major scientific work. Thereafter he spent his time researching biblical chronology and practising alchemy, as well as seeking public preferment. He was appointed Warden of the Mint in 1696 to take charge of a great recoinage scheme; later he became Master of the Mint. This work he carried out with outstanding ability, and was knighted in 1705 in recognition of his exceptional service.

Newton's later years were darkened by two major controversies. The first was over the publication of the observations of John FLAMSTEED, which he persuaded Halley to edit; the other was a priority dispute over the invention of calculus. Newton had expanded his fluxions into the technique now known to have been discovered independently by Gottfried Wilhelm Leibniz (1646–1716). Newton was encouraged by others to accuse Leibniz of stealing his ideas, and the controversy became very bitter, pursued even after Leibniz's death.

By the end of his life not only was Newton considered the doyen of British science, but his reputation had become international. He died on 1727 March 20 in London, and was buried in Westminster Abbey following a state funeral. His achievements were so important for astronomy that the phrase 'Newtonian revolution' is often used. Indeed, his influence on the physical sciences was profound and long-lasting: physics *was* Newtonian physics until the arrival of RELATIVITY and QUANTUM PHYSICS.

Newton European Space Agency (ESA) X-Ray Multi-Mirror (XMM) telescope launched in 1999 December as the second Cornerstone Horizon 2000 science mission. Newton carries high-throughput X-ray telescopes with an unprecedented effective area and an optical monitor, the first flown on an X-ray observatory. The large collecting area and ability to make long uninterrupted exposures provide highly sensitive observations.

Newtonian telescope REFLECTING TELESCOPE having a PARABOLOIDAL primary mirror, and a diagonal plane mirror positioned to divert the light path through 90° to a focus near the side and towards the upper end of the telescope tube. For apertures less than 150 mm (6 in.), simpler spheroidal primaries can be used without significant loss of optical performance. Before the rise of the Schmidt–Cassegrain telescope, the Newtonian was the reflector of choice for many amateur observers.

Newton's law of gravitation Law governing the mutual attraction between two masses that arises from the GRAVITATIONAL FORCE. In his *Principia*, published in 1687, Sir Isaac NEWTON announced that the force of gravity between any two masses, m_1 and m_2, is proportional to the product of their masses and inversely proportional to the square of the distance, r, between them:

$$F = \frac{Gm_1 m_2}{r_2}$$

where G is the gravitational constant ($= 6.6726 \times 10^{-11}$ N m^2 kg^{-2}). Although gravity is the weakest of the four forces (the others being the strong and weak nuclear forces and the electromagnetic force), it operates at long range and there are no negative masses to counter-balance the effect of normal mass (in the way that positive and negative electric charges cancel out each others' effects). Gravity is, therefore, the dominant force governing large scale phenomena within the Universe. It determines, amongst many other things, the shapes and sizes of satellites, planets and stars, the nature of their orbits, the shapes of galaxies and clusters of galaxies and their interactions.

Newton's laws of motion Laws governing the movement of bodies proposed by Sir Isaac NEWTON in his *Philosophiae naturalis principia mathematica* ('The mathematical principles of natural philosophy', usually known as the *PRINCIPIA*), which was published in 1687. The laws are: (1) A body continues in a state of rest or uniform motion in a straight line unless acted upon by an

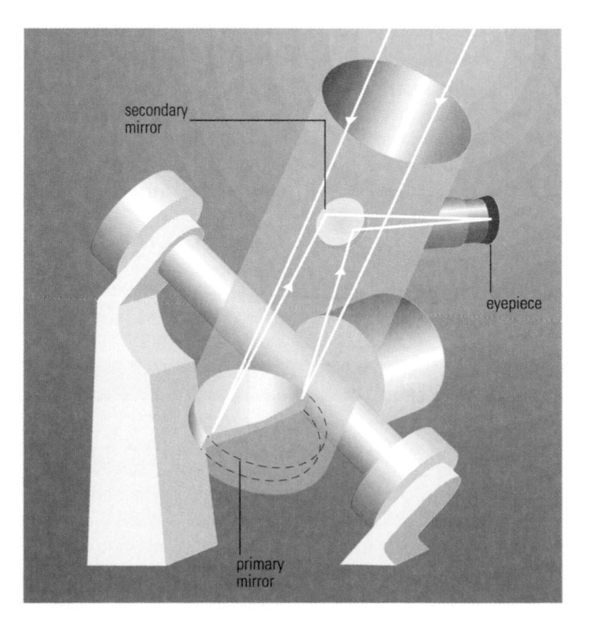

◄ **Newtonian telescope** In this form of reflecting telescope, light is reflected from a parabaloidal primary mirror to a flat secondary mirror (or sometimes a prism), placed diagonally within the tube, and directed to a side-mounted eyepiece. The resulting image is inverted.

▲ **noctilucent cloud** These beautiful high clouds can only be observed from between 50° and 65° north or south. This is because they do not occur closer to the equator and during summer, when they occur, twilight is too bright.

external force; (2) The acceleration of a body is proportional to the force acting on it, and the constant of proportionality is called the mass of the body; (3) For every force there is an equal and opposite force, called the reaction, acting on the body providing the force.

These laws, along with NEWTON'S LAW OF GRAVITATION, enable the motion of bodies in orbit around each other to be predicted. *See also* DYNAMICS

new moon *See* PHASES OF THE MOON

Next Generation Space Telescope (NGST) International successor to the HUBBLE SPACE TELESCOPE. The US$2 billion NGST will be an 8-m (26-ft) class instrument combining visible and infrared astronomy. It will orbit at the L_2 Lagrangian point between the Earth and the Sun, 1.5 million km (0.9 million mi) from the Earth. NGST will be launched after 2010 following a series of missions to help develop technology. The telescope will be operated by the SPACE TELESCOPE SCIENCE INSTITUTE and will be a key component of NASA's Origins programme, helping to answer key questions about the nature of the Universe, including its creation, physics, chemistry and potential biology, and to find out whether there are other Earth-like planets. NGST will be able to observe first-generation stars and galaxies and individual stars in nearby galaxies, and to penetrate dust clouds and discover thousands of objects in the Edgeworth–Kuiper Belt.

NGC Abbreviation of *NEW GENERAL CATALOGUE*

NGST Abbreviation of NEXT GENERATION SPACE TELESCOPE

N galaxy GALAXY designation introduced by William W. MORGAN to indicate a system whose light is dominated by a bright unresolved nucleus. N galaxies may be QUASISTELLAR OBJECTS, BL LACERTAE OBJECTS, RADIO GALAXIES or extreme SEYFERT GALAXIES. As the quality of images and other supporting data has improved, this classification is now mostly found in a historical context.

Nice Observatory *See OBSERVATOIRE DE LA CÔTE D'AZUR*

Nicholson, Seth Barnes (1891–1963) American astronomer who spent his entire career at Mount Wilson Observatory, specializing in the planets, asteroids and comets. Nicholson discovered four faint satellites of Jupiter, and used a thermocouple device to measure the temperatures of the Sun, Moon, planets and stars. He calculated an accurate temperature for sunspots, showing that they were cooler than the surrounding photosphere, and for the solar corona, showing that its 2 million K temperature was much hotter than the layers of the Sun's atmosphere immediately below it. Nicholson's precise measurements of the temperatures of giant stars in the solar neighbourhood allowed the first good estimates of those stars' true diameters.

NORMA (GEN. NORMAE, ABBR. NOR)

Small, inconspicuous southern constellation, representing a surveyor's level, between Lupus and Ara. It was named Norma et Regula (the Level and Square) by Lacaille in the 18th century. It brightest star, γ Nor, is a naked-eye double, with bluish-white components, mags. 4.0 and 5.0; ϵ Nor is a wide binary with components of mags. 4.8 and 7.5, separation 23″. Norma's brightest deep-sky object is NGC 6087, an open cluster of about 40 stars between mags. 6 and 10.

Nicol prism Optical device used to polarize plane light and to analyse polarized light. Invented in 1828 by William Nicol (1768–1851), the prism consists of two pieces of calcite or Iceland Spar cemented together using Canada Balsam. When light enters the prism it undergoes double REFRACTION, that is, the light is split into an ordinary ray and an extraordinary ray, each having the opposite POLARIZATION to the other. The ordinary ray is reflected at the Canada Balsam surface, while the extraordinary ray passes straight through. Thus any light beam can be split into its two polarized components. Astronomers use Nicol prisms to measure the degree of polarization present in light from the object under study.

nitrogen (symbol N) Element that in order of cosmic abundance is fifth by number of atoms and seventh in terms of the mass content of the Universe (*see* ASTROCHEMISTRY). Its properties include: atomic number 7; atomic mass of the naturally element 14.0067 amu; melting point 73.3 K; boiling point 77.4 K. It was discovered in 1772 by the Scottish chemist Daniel Rutherford (1749–1819), although several other chemists were close behind. Nitrogen has seven ISOTOPES, ranging from nitrogen-12 to nitrogen-18; two of these are stable and form the naturally occurring element (nitrogen-14, 99.6%, nitrogen-15, 0.4%).

Nitrogen comprises 78.08% by volume of dry terrestrial air (75.52% by mass). Under normal terrestrial conditions, it exists as a colourless, odourless gas formed of the diatomic molecule, N_2, and is chemically unreactive because the bond between each pair of atoms is hard to break. It is also contained in a wide range of compounds, including ammonia (NH_3), and is an important constituent of many organic molecules and foodstuffs for living organisms. Atomic and ionized nitrogen are produced in the upper atmosphere by the action of solar radiation; they react with oxygen and ozone to produce nitric oxide. Sudden enhancements of atomic and ionized nitrogen, resulting from the impact of energetic particles from SOLAR FLARES, can therefore result in significant temporary depletions of the terrestrial ozone layer.

Apart from the Earth, the only Solar System body to possess a nitrogen-rich atmosphere is TITAN, the giant satellite of SATURN. Nitrogen makes up 94% by volume of Titan's atmosphere.

Nitrogen plays an important catalytic role in the CARBON–NITROGEN–OXYGEN CYCLE, which is the major helium-producing nuclear reaction in stars more massive than the Sun. A variant of the carbon–nitrogen–oxygen cycle known as the CNO bi-cycle involves the capture of further PROTONS to produce nitrogen and helium as end products, and this process may have produced a large fraction of the cosmic abundance of nitrogen-14.

Liquid nitrogen is widely employed to cool sensitive astronomical detectors in order to reduce unwanted background noise.

NOAO *See* KITT PEAK, CERRO TOLOLO

Nobeyama Radio Observatory Facility of the NATIONAL ASTRONOMICAL OBSERVATORY OF JAPAN located at Nagano in central Japan. It operates the Nobeyama Radio Telescope, a 45-m (148-ft) millimetre-wave instrument equipped with a 25-beam focal plane camera. It

also operates a millimetre array with six 10-m (33-ft) antennae. Nobeyama contributes to both the VLBI Space Observatory Programme (VSOP) and the ATACAMA LARGE MILLIMETRE ARRAY.

noctilucent cloud Pearly white or silver-blue clouds formed during the summer months at high latitudes in either hemisphere. They are found in thin sheets at altitudes around 82 km (51 mi), close to the mesopause. The clouds are very tenuous, only becoming visible by contrast with the twilit sky when the Sun lies between 6° and 16° below the horizon. In addition to their colour and 'night-shining' nature, noctilucent clouds are distinguished by their highly banded wave structure. In the northern hemisphere, they are most often seen in late June and early July. Noctilucent clouds are believed to consist of water ice condensed on to small solid nuclei, probably of meteoric origin. The frequency and brightness of displays appears to have shown a secular increase since the 1950s.

node Either one of the two points at which an orbit of a body crosses the reference plane used for the orbit. It is the ASCENDING NODE if passing from south to north, and the DESCENDING NODE if passing from north to south. The line joining the two nodes is the LINE OF NODES. *See also* ORBITAL ELEMENTS; REGRESSION OF THE NODES

non-baryonic matter Matter that does not exhibit BARYON characteristics.

non-gravitational force Any force that is not gravitational acting on a celestial body, in particular a force that produces effects similar to gravity. The activity of gas jets emerging from a comet's NUCLEUS around the time of perihelion may cause acceleration or deceleration, leading to changes in a comet's orbital period. This non-gravitational force, sometimes described as a 'rocket effect', has undoubtedly contributed to difficulties in accurately predicting the movements of some comets, including 109P/SWIFT–TUTTLE and 23P/BRORSEN–METCALF.

Nordic Optical Telescope Pioneering 2.6-m (102-in.) optical/infrared telescope located at the ROQUE DE LOS MUCHACHOS OBSERVATORY. It is operated by the Nordic Optical Telescope Scientific Association (NOTSA), whose member states are Norway, Sweden, Denmark, Finland and Iceland. Completed in 1989, the telescope was the first large instrument to have a dome equipped with ventilating doors to improve local atmospheric turbulence. ACTIVE OPTICS provide high-resolution imaging.

Norma *See* feature article

North America Nebula (NGC 7000) Large, roughly triangular EMISSION NEBULA in Cygnus (RA $20^h\ 58^m.8$ dec. +44°20′). It covers an area of about 120′ × 100′. Under good conditions, the object is visible to the naked eye as an enhancement of the Milky Way east of Deneb. Long-exposure images show intruding dark nebulosity, which gives it a striking resemblance to the continent of North America, complete with the Gulf of Mexico. The dark nebulosity separates NGC 7000 from the PELICAN NEBULA to the west. The North America Nebula lies at a distance of 3000 l.y.

Northern Coalsack Region of dark nebulosity, which obscures the Milky Way in Cygnus, south-east of Deneb and γ. It marks the start of the Cygnus Rift, which splits the northern Milky Way into two branches. The Northern Coalsack has an apparent diameter of about 5° and is centred on RA $20^h\ 40^m$ dec. +42°.

Northern Cross Popular name for the shape formed by the main stars of the constellation CYGNUS. Deneb (α Cygni) marks the top of the cross, with Albireo (β Cygni) at its foot. The crossbar is formed by δ, γ and ϵ Cygni.

northern lights *See* AURORA

north polar distance Angular distance of a celestial object from the north celestial pole, measured along a great circle. Polar distance is the complement of DECLINATION, the sum of an object's polar distance and its declination always being 90°.

North Polar Sequence List of 96 stars near the north celestial pole, ranging in magnitude from 2 to 20. The PHOTOGRAPHIC MAGNITUDES have been accurately measured and are used for comparison purposes.

nova (N) Form of CATACLYSMIC VARIABLE that undergoes an unexpected, sudden outburst with a dramatic rise in luminosity in a matter of hours or days,

◄ **North America Nebula** NGC 7000 was discovered by William Herschel in 1786. It is a complex of absorption, emission and reflection nebulae.

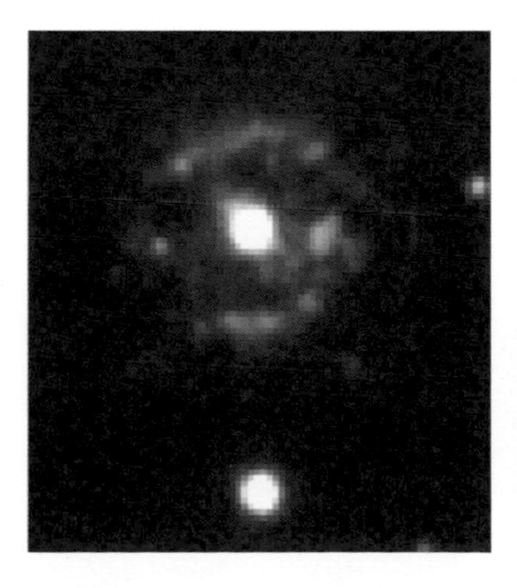

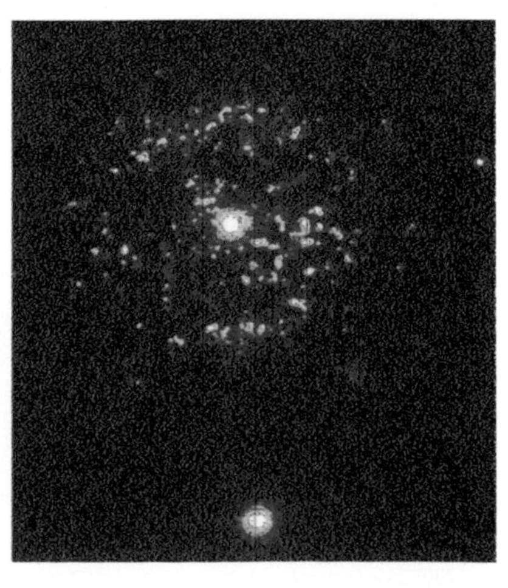

▼ **nova** The shells of gas around the recurrent nova T Pyxidis have been shown by the Hubble Space Telescope to consist of blobs rather than smooth rings. This is thought to be because material from successive explosions, travelling at different speeds, is colliding.

normally with an amplitude of 11–12 magnitudes. Nova Cygni 1975 (V1500 Cyg) had the exceptionally large amplitude of 19 magnitudes. The subsequent decline is much slower. Two subtypes are recognized on the basis of the time taken to fall 3 magnitudes from maximum: fast novae (NA) take less than 100 days; slow novae (NB) take 150 days or more. A further subtype (NC) shows both a slow rise (sometimes as long as a decade) and an extremely slow decline.

The appearance of a bright star-like point of light in the sky over only a few nights of observation, at a position where previously no star had been known, was a matter of great interest in ancient times. The phenomenon tended to be ignored by European chroniclers during the Middle Ages because the prevailing culture belief was that the heavens represented a perfect and therefore unchangeable creation by God. When change did in fact occur in the sky the matter was dealt with by ensuring that no written record of it was made. When astronomers today are concerned to know about what happened a thousand years ago they consult the records of cultures other than the European, notably of the Chinese (*see* CHINESE ASTRONOMY). The appearance of TYCHO'S STAR and KEPLER'S STAR (both actually SUPERNOVAE) in 1572 and 1604, respectively, were seminal events in bringing about a complete change in the European conception of the heavens.

Because astronomers nowadays have photographic plates of the whole sky on which even exceedingly faint stars may be seen, it is often possible to locate a nova's precursor. An even greater amount of information may be obtained by following the spectral features of a nova through its sudden outburst into its decline over a time interval of weeks and months. The outcome of such investigations is that novae are in the main, if not indeed wholly, WHITE DWARF stars that happen to be members of BINARY STAR systems, more specifically, that they are interacting binaries.

Novae may be observed in galaxies other than our own, and attempts have been made to use such observations as an indicator of the distances of other nearby galaxies. The great majority of observed cases are within our own Galaxy, where for the limited region of the solar neighbourhood they are found to occur at a rate of about two per year. The solar neighbourhood is determined by the fogging effect of INTERSTELLAR DUST, which cuts down the range of effective observation to about 5% of the whole Galaxy, for which the rate of occurrence of novae is estimated to be about 40 per year.

Our subjective perception of a nova is confined to light in the visual and photographic ranges of wavelength, omitting the ultraviolet. Because the ultraviolet is of variable importance, probably being of greater importance before outburst than during it, our subjective perception tends to exaggerate the contrast between the pre-nova state and the maximum emission of light during the outburst. The observed contrast for visible light is usually about 10,000 to 1, but if all wavelengths were included the contrast would probably be about 100 to 1.

The emission of visible light in the pre-nova stars is of a similar order to the emission from the Sun, whereas the emission at maximum outburst is of a similar order to that of a supergiant star of type F8, of which Wezen, the fourth-brightest star in the constellation of Canis Major is an example. A typical nova rises to its maximum in a few days, and declines thereafter in brightness by a factor of about 10 in 40 days, although cases of both slower and more rapid declines are known and studied.

Clouds of gas are ejected at high speeds during outbursts, speeds typically of about 1500 km/s (930 mi/s), which is more than sufficient for the expelled gases to become entirely lost into interstellar space, together with myriad fine dust particles that condense within the gases as they cool during their outwards motion. The total amount of material thus lost is estimated to be about one part in ten thousand of the total mass of the parent white dwarf star, although the amount in especially violent cases is almost certainly appreciably larger than this.

Much thought has been given in recent years to the cause of novae. The consensus of opinion is that the basic mechanism of the explosion of a nova is the same as for the explosion of a thermonuclear weapon. To begin with, energy is produced at a comparatively gentle rate by NUCLEAR REACTIONS in material that is covered (tamped) by a layer of relatively inactive other material. Because of the tamping material, the energy produced by nuclear reactions cannot escape and accumulates within the reacting material itself. Provided the tamping effect is strong enough, the rising temperature and pressure cannot be alleviated by expansion, causing the nuclear reactions to become progressively more violent. The same cycle of events is repeated with the energy released from the nuclear reactions, and it accelerates at an ever-increasing rate, until eventually the situation gets out of hand (in a 'thermonuclear runaway'), or until the tamping effect of the overlying material fails, and the material is expelled from the strong gravitational field of the white dwarf. Depending on the mass of the white dwarf, its chemical composition, and the amounts of the reactions and tamping materials, the details of explosion may vary in ways that are subject to mathematical calculation, and which have been found to agree with many of the observed features of novae.

Novae are now considered to be like other forms of cataclysmic variable in consisting of an interacting binary system. The secondary component is a class K or M giant, subgiant or dwarf star, which transfers matter on to the white-dwarf primary. Unlike certain supernovae, the outbursts do not disrupt the system, although material is expelled into space, and it is generally accepted that the system settles down into a POST-NOVA stage. Mass transfer is later resumed and after a long interval a second nova outburst may occur. The recurrence time is related to the mass of the white dwarf and (with reasonable assumptions about the mass-transfer rate) is of the order of 30,000 years for a white dwarf of 1.3 solar masses, and one million years for a white dwarf of 0.6 solar mass.

Certain stars that apparently resemble novae are known to undergo explosions repeatedly, for example T Coronae appeared as a low-amplitude nova in 1866 and 1946, with two smaller outbursts in 1963 and 1975, while T Pyxidis did so in 1890, 1902, 1922, 1944 and 1966. Although it was once thought that these systems (*see* RECURRENT NOVA) might represent a stage following a principal nova outburst, this now appears unlikely, because of their different physical characteristics.

The circumstances in which the nuclear reactions in a nova occur, especially if a supply of protons is available for mixing with carbon and oxygen, lead to the production of some nuclides that are not synthesized by nuclear processes occurring towards the centres of stars. Examples are ^{15}N and ^{26}A1. It is also possible that neutrons produced by reactions of alpha particles with ^{13}C participate in a form of R PROCESS, in which very heavy neutron-rich nuclei are synthesized. Condensing solid grains in the gases expelled by novae may be expected to contain such unusual nuclides. In particular, outbursts on CNeMg (carbon–neon–magnesium) white dwarfs pro-

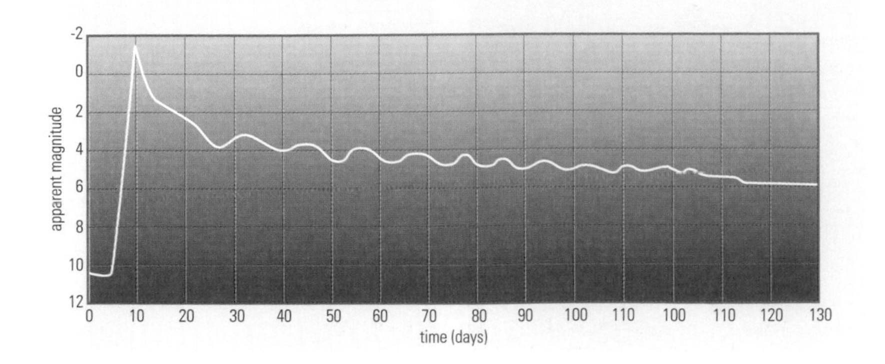

▼ **nova** The light-curve of a nova typically shows a steep rise in brightness by a factor of many thousands followed by a slow, irregular, decline. The rapid rise in brightness is caused by thermonuclear runaway – explosive hydrogen burning on the surface of the white dwarf component of the binary pair.

duce various nuclei, including the radioactive species ^{22}Na and ^{26}Al. Novae now appear to make a substantial contribution (together with supernovae and red giants) to the abundance of such isotopes in the interstellar medium. When the Solar System formed from interstellar material, some grains derived from novae would be present. The presence of ^{22}Ne and ^{26}Mg (the daughter products of ^{22}Na and ^{26}Al, respectively) in meteorites indicates that the protosolar nebula did indeed contain species from one or all of these sources.

The total energy of explosion of a typical nova has been estimated at 10^{45} ergs, which is about as much energy as the Sun emits in 10,000 years. It has been suggested that novae might be precursors for Type I supernovae. Spectroscopic evidence, however, indicates that some of the material from the white dwarf is lost at each outburst. This suggests that despite the accretion of material and the production of certain heavier elements in the outburst, the overall mass of the white dwarf is unlikely to increase sufficiently for it to reach the CHANDRASEKHAR LIMIT.

nova-like variable (NL) VARIABLE STAR subtype, a form of CATACLYSMIC VARIABLE. Nova-like variables are poorly known. Their general characteristics (particularly changes in light and spectra) resemble those of a NOVA or POST-NOVA rather than a DWARF NOVA or SYMBIOTIC STAR. Some show nova-like outbursts, but others have never displayed such activity. Once an object of this type is studied in detail it is normally possible to assign it to a more specific cataclysmic type.

Nozomi Japan's first MARS explorer. Nozomi was launched in 1998 July and was propelled towards a rendezvous with Mars in 2004 by a series of gravity-assisted lunar flybys. It will be placed into a low orbit around Mars, with the primary objective of studying the planet's upper atmosphere with emphasis on its interaction with the SOLAR WIND. Nozomi will measure the Martian magnetic field for the first time, investigate the ionosphere, take images of the weather and the moons, PHOBOS and DEIMOS, especially to confirm if there is a dust ring along the orbit of Phobos.

N star Member of a class of cool CARBON STARS. Class N, part of the original Harvard system of SPECTRAL CLASSIFICATION, was introduced to describe red stars that lack the class M absorption bands of TiO, and instead have strong absorptions produced by carbon compounds, particularly C_2, CN and CH. Class N second-ascent giants (asymptotic giant branch stars and some Miras) divide into subclasses N0 to N9, whose temperatures roughly track those of the M giants (3600 to 2000 K). The class, along with the warmer R STARS, has been subsumed into class C, with subtypes of roughly C6–C9.

Nubecula Major *See* LARGE MAGELLANIC CLOUD

Nubecula Minor *See* SMALL MAGELLANIC CLOUD

nuclear reactions Interactions between atomic nuclei or between atomic nuclei and subatomic particles or gamma rays. Nuclear reactions produce transformations from one ELEMENT or ISOTOPE to another. There are two basic types of reaction, FUSION and FISSION.

The main astronomical occurrences of fusion are within stellar interiors, novae and during the early stages of the BIG BANG, resulting in the build-up of elements up to iron from lighter nuclei. In stellar interiors and in SUPERNOVAE absorption of protons and neutrons by nuclei builds up elements heavier than iron. The fusion reactions build elements of heavier mass number from the lighter. The first of these, thermonuclear fusion, is the process that created most of the helium in the Universe in the first few minutes after the Big Bang; it provides the main energy output of the stars, and indeed it governs STELLAR EVOLUTION. It also provides the energy released in hydrogen bombs; and it may eventually be harnessed for power generation in fusion reactors.

The main astronomical occurrences of fission are within novae and supernovae, where very short wavelength gamma rays cause nuclei between about sulphur and iron to disintegrate to lighter nuclei (photo-disintegration). In addition, in the interstellar medium cosmic ray impacts fragment carbon, nitrogen and oxygen nuclei to lithium, beryllium and boron (*see* SPALLATION). Fission accounts for the production of some elements and isotopes that are not produced by other processes. It is also involved in many radioactive decay processes, and is the basis of current nuclear energy generation and atom bombs. *See also* ASTROCHEMISTRY; BINDING ENERGY; CARBON–NITROGEN–OXYGEN CYCLE; NUCLEOSYNTHESIS; P PROCESS; PROTON–PROTON REACTION; R PROCESS; S PROCESS

nucleosynthesis Process whereby HELIUM and the HEAVY ELEMENTS are created from HYDROGEN. There are four main areas involved in nucleosynthesis: the early stages of the BIG BANG; the energy-generating reactions inside stars; reactions during NOVA and SUPERNOVA explosions; and COSMIC RAY collisions within the interstellar medium.

The first nucleosynthesis reactions to be identified were the PROTON–PROTON REACTION and the CARBON–NITROGEN–OXYGEN CYCLE. Both of these were suggested in 1938 as sources for the energy of the Sun, the former by Hans BETHE, and the latter, independently of each other, by Bethe and Carl von WEIZSÄCKER. The processes both convert hydrogen into helium-4, the proton–proton reaction being most important for stars with masses up to that of the Sun and the carbon–nitrogen–oxygen cycle becoming dominant within more massive stars.

In 1956 and 1957 Geoffrey and Margaret BURBIDGE, William FOWLER and Sir Fred HOYLE identified most of the remaining reactions inside stars that synthesise the elements up to iron, though only the most massive stars are able to follow the whole sequence. The next stage after helium formation is not the production of beryllium-8, because that decays back to two helium-4 nuclei (α-particles) in 2×10^{-16} s, but it is the TRIPLE-α PROCESS producing carbon-12. The triple-α process involves two helium nuclei combining and then a third being added before the beryllium-8 can decay. It requires a temperature of about 10^8 K before it will start since the NUCLEAR REACTIONS involved are slightly endothermic (requiring the supply of energy) and therefore need the input of thermal energy to get them to occur. The triple-α process overall is exothermic: more energy is released by the process than is used in starting it, because the carbon-12 nucleus that is the result of the reaction is in an excited state and emits a high-energy gamma ray.

Once carbon-12 has been produced, helium-4 nuclei can be added to it (a process known as α-capture) to produce, successively, oxygen-16, neon-20, magnesium-24, silicon-28, sulphur-32 and argon-36. For the Sun, the sequence will probably come to a halt with the production of oxygen-16. α-capture can also start from other elements and this is particularly significant for the nitrogen-14 left by the carbon–nitrogen–oxygen cycle, leading to fluorine-18 and neon-22. As the temperature increases, carbon and oxygen start to combine directly and eventually, at about 3×10^9 K, silicon-28 combines with itself to produce nickel-56. The latter is radioactive and decays to cobalt-56 and then to the stable isotope iron-56, although those decay processes will occur outside the star, because the production of nickel at the star's centre occurs immediately prior to its explosion as a supernova. There are many more reactions than are listed here, and some result in the release of neutrons. Those neutrons combine with nuclei to produce intermediate elements and isotopes (*see* S PROCESS). The reactions up to the production of iron-56 are exothermic. The BIND-

N

▲ **nucleus, galactic** At X-ray wavelengths the nucleus of the Milky Way shows hundreds of white dwarf and neutron stars, together with black holes, surrounded by incandescent gas at a temperature of 10 million K. The bright white patch in the centre of the image is the site of the supermassive black hole at the heart of the galaxy.

▼ **nucleus** NASA's Deep Space 1 imaged the inactive nucleus of Comet Borrelly from 3417 km (2123 mi). The bright patches on the surface of the 8-km (5-mi) long body are thought to be the source of the material that forms a cometary coma when heated by the Sun.

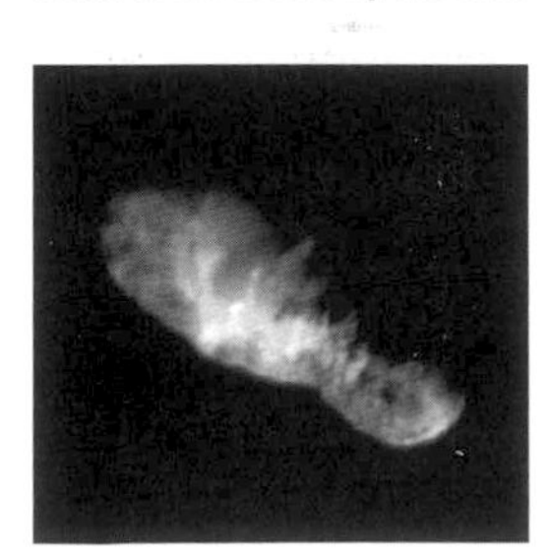

ING ENERGY per nucleon however is a maximum for iron, and so the formation of heavier elements requires more energy to be supplied than is produced. The elements above iron are therefore thought to be formed during supernova explosions by the successive addition of neutrons (the R PROCESS) and protons (the P PROCESS), with the energy of the explosion fuelling the reactions.

Although helium-4 is produced within stars, its abundance at about 25% by mass (*see* ASTROCHEMISTRY) throughout the Universe is far too high to be accounted for in this way. It is therefore thought that most of the helium-4, along with hydrogen-2 (deuterium), helium-3 and lithium-7, was generated during the early stages of the Big Bang. At a time a few seconds after the start of the Big Bang, the numbers of protons and neutrons would be in the ratio of about 87% to 13%. The abundance ratio of protons to neutrons is largely independent of the exact conditions in the Big Bang. This is because, in equilibrium, the ratio is determined by the balance between the number of protons and electrons combining to form neutrons and neutrinos and the number of neutrons and neutrinos combining to form protons and electrons, and this depends only upon the reaction rates. About 100 seconds after the Big Bang , the temperature would have fallen sufficiently for deuterium to form and to be stable. All the available neutrons would then quickly combine with protons to produce deuterium, and the deuterium would build up to helium-4. Since the 13% of the mass of the Universe in the form of neutrons combines with 13% in the form of protons during this process, the final amount of helium-4 produced this way is about 26%, which is convincingly close to the observed 24.9%.

Finally, some of the light elements exist in far greater abundance than might be expected from the amounts predicted by the above processes. They are also easily destroyed inside stars. Such nuclei as lithium-6, lithium-7, beryllium-9, boron-10 and boron-11 are therefore thought to be produced in the interstellar medium by SPALLATION reactions. In these reactions high-energy cosmic ray protons and α-particles collide with nuclei of carbon, nitrogen and oxygen, fragmenting them and producing the lighter nuclei. *See also* FISSION, NUCLEAR; FUSION, NUCLEAR

nucleus Central part of an ATOM, containing most of the mass of the atom and formed from PROTONS and NEUTRONS. The nucleus has a density of about 2.3×10^{17} kg/m^3 and is a few times 10^{-15} m across.

nucleus Small solid body containing the main mass of a COMET. Comet nuclei are comprised of icy, volatile material with an admixture of dust, surrounded by a dark outer crust. They are mechanically fragile, and many have been observed to undergo catastrophic disintegration when close to the Sun. Comet nuclei range in size from the relatively tiny bodies (only tens of metres in diameter) detected in large numbers by the SOHO coronagraphs, through to kilometre-sized objects, and up to much larger icy planetesimals. Among this last category, asteroid 2060 CHIRON (also given a periodic comet designation of 95P/Chiron) is thought by many astronomers to be a giant cometary nucleus, *c.*180 km (*c.*110 mi) in diameter, which will in some tens of thousands of years enter the inner Solar System and fragment to spawn a new group of short-period comets on similar orbits.

nucleus, galactic Central region of a GALAXY, meaning a region from a few thousand light-years in size down to the small volume of an ACTIVE GALACTIC NUCLEUS. This is the region of greatest density of stars and gas and least angular momentum in a galaxy, where massive objects may be formed. Bursts of star formation as well as ACCRETION on to massive BLACK HOLES can generate enormous amounts of energy.

Nuffield Radio Astronomy Laboratories *See* JODRELL BANK OBSERVATORY

Nunki The star σ Sagittarii, visual mag. 2.05, distance 224 l.y., spectral type B2.5 V. The name is of Babylonian origin, but its meaning is not known.

nutation Small, cyclical oscillation superimposed upon the 25,800-year PRECESSION of the Earth's rotational axis, caused by the combined gravitational effects of the Sun and the Moon. Discovered by James BRADLEY in 1747, nutation produces a slight 'nodding' of the Earth's axis; it changes with the continually varying relative positions of the Sun, Moon and Earth. Lunar nutation, which is caused by the 5° inclination of the Moon's orbit to the ecliptic, causes a variation of 9″ on either side of the mean position, in a period of 18 years 220 days. Solar nutation has a period of 0.5 of a TROPICAL YEAR. Fortnightly nutation has a period of 15 days.

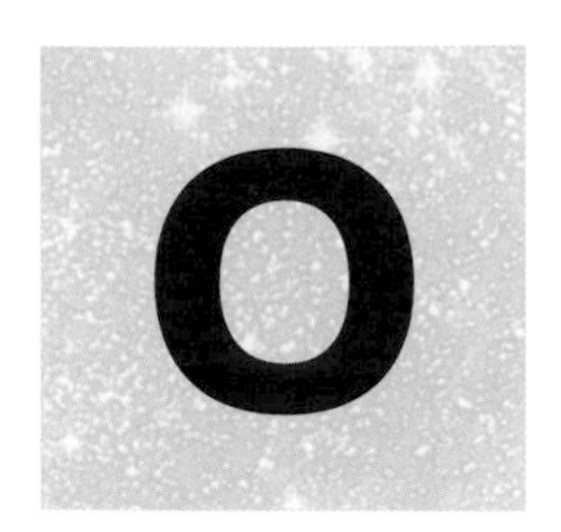

O and B subdwarf Hot star with luminosity less than that of O and B DWARFS but greater than that of a WHITE DWARF. They are generally stars in the final stage of collapse to white dwarfs.

OAO Abbreviation of ORBITING ASTRONOMICAL OBSERVATORY

OB association Group of stars of spectral type O and B. OB associations are regions of space where massive stars are currently being formed or have just recently been formed. They are one of the least dense type of star cluster.

The predominance of the very luminous high-mass O STARS and B STARS renders OB associations recognizable at great distances, even through the dusty haze of interstellar space. Tens or hundreds of these massive stars can occur in a single association. Low-mass (faint) stars may exist in abundance too, but they are far harder to see in distant OB complexes. The Great Nebula in Orion, with its core of hot stars, is part of an OB association that is relatively close to us, hence low-mass young stars are also detected.

Recent computer models suggest that OB associations could be self-perpetuating: that is, an appreciable population of massive stars can lead to the preferential formation of new massive stars. The mechanism involves the powerful stellar winds from the surrounding O stars. Vigorous stellar winds could disperse parental clouds, inhibiting the formation of lower-mass stars.

OB associations are a feature of the gas-rich spiral and irregular galaxies, but not of elliptical galaxies in which no young stellar population is apparent. The frequency with which stars of different mass occur in a galaxy (or part of one) is called the 'mass spectrum'. In our Galaxy, low-mass stars (like our Sun, and smaller) are prodigiously more abundant than high-mass stars and contribute most of the mass. However, hot blue stars are unrepresentatively more obvious because of their tremendous luminosities. The high-mass of these stars leads to the high incidence of supernovae observed within OB complexes. *See also* OPEN CLUSTER; T ASSOCIATION

Oberon Outermost of the five SATELLITES of URANUS that were known before the flyby by VOYAGER 2. It was discovered in 1787 by William HERSCHEL, who also found its neighbour TITANIA. It has a dark icy surface with an albedo of 0.25. Oberon was the most distant of the previously known satellites from Voyager's trajectory, with the result that the best images have a resolution no better than about 20 km (12 mi). This is sufficient to show that the whole of the visible area of its surface is fairly densely cratered, and that there are patches of bright ejecta surrounding what are presumably the youngest impact craters. The darkest parts of the surface are patches on the floors of some of the craters. Other notable features on Oberon are a mountain peak about 11 km (7 mi) high and a large trough with scalloped walls named Mommur Chasma. *See* data at URANUS

objective LENS nearer to the object being observed in a TELESCOPE or microscope. The main mirror in a reflecting telescope can also be called the objective but this use is now uncommon. In astronomy, objective almost always means the large lens that collects and focuses the light to produce an image. A single-lens objective suffers from several ABERRATIONS, the most noticeable being CHROMATIC ABERRATION. ACHROMATS, with two lenses, and APOCHROMATS, normally with three lenses, are used as objectives in astronomical and terrestrial telescopes.

objective prism Narrow-angled PRISM that is placed in front of the aperture of a telescope to produce a low-resolution SPECTRUM of every star in the field of view. Photographs taken using this instrument allow the easy and rapid spectral classification of a large number of stars to be made.

oblateness Measure of the amount by which a spheroid deviates from a perfect sphere; it is defined as the ratio of the difference between the equatorial and polar radii of a body to its equatorial radius. Any celestial body that rotates tends to bulge at its equator by an amount that depends both on its speed of rotation and on whether it is solid or fluid in composition. Oblateness is therefore a good indicator of the rotational speed of a star or planet. *See also* ELLIPTICITY

oblate spheroid *See* SPHEROID

obliquity of ecliptic Angle between the plane of the ECLIPTIC and the CELESTIAL EQUATOR, caused by the inclination of the Earth's rotational axis to the plane of its orbit. The present value of the obliquity of the ecliptic is 23°26′ but the effects of PRECESSION and NUTATION cause it to vary between 21°55′ and 24°18′ over a period of 40,000 years. It is currently decreasing by about 0′.47 per year and in about 1500 years time will begin to increase again. This value also represents the maximum angular distance the Sun can lie north or south of the equator. *See also* SEASONS

Observatoire de la Côte d'Azur French research institution formed in 1988 by amalgamating Nice Observatory with the Centre for Geodynamic and Astronomical Study and Research (CERGA, which dates from 1974). Nice Observatory was founded in 1881 and equipped with a 29-inch (0.74-m) telescope named after the observatory's sponsor, Raphael Louis von Bischhoffsheim (1823–1906), which was completed in 1886. CERGA operates a 0.9-m (35-in) Schmidt telescope. Today, the observatory specializes in theoretical and observational astrophysics using facilities such as the SOLAR AND HELIOSPHERIC OBSERVATORY and the VERY LARGE TELESCOPE.

observatory Any facility specifically for making celestial observations and measurements. Some structures such as Stonehenge (*see* ARCHAEOASTRONOMY) may well have been used for observations. The observatory proper evolved from Babylonian constructions set up for astrological observation from about 750 BC, and reached a high state of scientific development in the late 16th century with the observatories of Tycho BRAHE. The observatory as a telescope housing dates from the 17th century; notable examples are COPENHAGEN OBSERVATORY (1642), PARIS OBSERVATORY (1671) and GREENWICH OBSERVATORY (1675), founded principally for geodesy and navigation.

The era of observatories built for scientific research rather than the practical needs of the state began in the 18th century. It brought about the development of large facilities such as Lord Rosse's observatory in Ireland (1845; *see* BIRR CASTLE ASTRONOMY), and LICK OBSERVATORY in California (1888) and YERKES OBSERVATORY in Wisconsin (1897) with their great refractors. The first half of the 20th century saw a continuing dominance of the USA in large telescope building with MOUNT WILSON OBSERVATORY's 100-inch (2.5-m) and PALOMAR OBSERVATORY's 200-inch (5-m) reflectors, completed in 1917 and 1948, respectively.

With the arrival of frequent and rapid air travel in the 1960s, optical astronomers were no longer restricted in where to locate their observatories, and the next decade saw a proliferation of telescopes in the 4-metre class on excellent remote sites. The post-war years also saw a growth in the number of large radio telescopes, and the first proposals for observatories in space to overcome the limited range of wavelengths available to astronomers on the Earth's surface.

Today, the turn-of-the-century boom in 8–10-metre class telescopes is well under way, and a new generation of optical-astronomy institutions such as the GEMINI OBSERVATORY and the W.M. KECK OBSERVATORY is emerging. The modern observatory is often far from its telescopes, either because of the geographical remoteness of the observing site (as with the EUROPEAN SOUTHERN OBSERVATORY) or

▼ Oberon This is the closest image obtained of Uranus' satellite Oberon during the 1986 January Voyager 2 flyby. Oberon's surface shows several impact craters surrounded by bright rays.

because the observatory operates a space facility (as with the SPACE TELESCOPE SCIENCE INSTITUTE).

Increasingly important are observatories dedicated to popular astronomy. These range from the private observatories of amateur astronomers (some of which, like their 18th-century forebears, challenge the smaller professional observatories in the quality of their equipment) to major educational facilities such as the FAULKES TELESCOPE. Public education and outreach is giving a new lease of life to the older facilities of professional observatories, whose working astronomers now use newer telescopes elsewhere. A good example is the 0.91-m (36-in.) reflector of the ROYAL OBSERVATORY, EDINBURGH (1928), which is now the centrepiece of the observatory's Visitor Centre.

occultation Event in which one body is obscured by another – in other words, when a distant object is hidden by a nearer one. An obvious example is a SOLAR ECLIPSE, which is, strictly, an occultation of the Sun by the Moon.

The term occultation can be applied to a range of celestial objects, but the most commonly observed are lunar occultations, which occur as the Moon obscures stars during its passage along the band of the ECLIPTIC. The most readily observed lunar occultations are disappearance (immersion) events, occurring on the Moon's leading (easterly) limb before full. At this time, the leading limb is dark; occultations occurring when the Moon is a waxing crescent may be especially favourable for observation, with the limb dimly visible by EARTHSHINE as it approaches the target star. Inexperienced observers are often surprised by the suddenness with which the star vanishes: this is due to the Moon having no atmosphere and the star being at such a great distance that it is virtually a point source of light.

After full moon, reappearance (emersion) events are more readily observed, occurring at the dark trailing lunar limb. As with disappearances, these events occur very abruptly. Successful observation depends on good predictions, including an accurate forecast of the POSITION ANGLE on the Moon's limb at which reappearance will take place.

Visual study of lunar occultations has been undertaken for over a century, and still plays a role in astronomy. Precise timings of events, coupled with precise positional data for the observing location, provide useful information on the Moon's orbital motion. Timing accuracy can be further improved by the application of video recording techniques. High-speed photometry has also been used to obtain light curves for stars during immersion, allowing identification of previously unknown close double stars and measurement of the diameters of some of the nearer stars.

In the course of a year, the Moon may occult more than 4000 stars in the range of a fairly standard amateur telescope. Good star catalogues and modern computing capabilities make the calculation of predictions relatively straightforward. Observational reports are collected by the International Lunar Occultation Centre in Japan for reduction, and the resulting data are available for use by professional astronomers.

Stars are not the only bodies to be occulted by the Moon; on occasion, a planet or asteroid may also be occulted. In the case of Jupiter, the planet's own satellite system can be seen to be occulted. The major planets, being very much closer than the stars, show disks, so that occultations are gradual, lasting several seconds; the brightness of Venus, Mars, Jupiter and Saturn allows events occurring at the Moon's bright limb to be observed quite readily. Occultations of the planets, while rare and yielding less scientific data than those involving stars, can be spectacular to watch in a small telescope or binoculars.

A spectacular planetary occultation was observed from central England on 1980 October 4, when, in the pre-dawn sky, Venus was grazed by the Moon's limb, being only partially obscured. Lunar mountains and valleys were seen in profile against Venus' bright gibbous disk. Graze occultations for a star can be seen along a track no more than 1 km (0.6 mi) wide on Earth's surface and are consequently rare. Teams of observers spread perpendicular to a graze track can obtain multiple timings of events as the Moon's irregular limb alternately covers and reveals the star, gaining much useful information about the lunar profile.

Planets may also, on rare occasions, occult stars. Brightening and fading of stars during immersion or emersion can provide information about planets' atmospheric layering. Among noteworthy events was the occultation of 28 Sagittarii by Saturn and its rings on 1989 July 3. The following night, Saturn's satellite Titan occulted the star in an event seen extensively from Europe, with a brightening at mid-occultation resulting from refraction of starlight by Titan's dense atmosphere. The rings of Uranus were discovered during an occultation of the star SAO 158687 on 1977 March 10.

At certain times, when the plane of their orbits aligns with Earth, Jupiter's Galilean satellites may undergo mutual occultations. They are also, of course, frequently seen to be occulted by Jupiter itself as they pass behind it on their orbits.

Occultations of stars by asteroids are also rare, but provide extremely valuable opportunities to determine the profiles of these small bodies. Essentially, an asteroid casts a 'shadow' equivalent to its diameter (usually less than 200 km/120 mi) as it occults a star. The critical observation in these cases is whether an occultation occurs at all, and its duration (defining a cord across the asteroid's diameter). Secondary occultations have revealed the presence of satellites orbiting some asteroids, including (6) Hebe, (18) Melpomene and (532) Herculina.

occulting disk Small metal disk placed in the focal plane of a telescope to cover a bright object in order that a fainter one may be observed. An occulting disk is used in a CORONAGRAPH to produce an artificial eclipse of the Sun. It blocks the light from the Sun's disk and enables observations to be made of the fainter CORONA. The Infrared Astronomical Satellite (IRAS) also made use of an occulting disk to block out the direct light from stars, allowing it to image the infrared radiation from their surrounding regions.

ocean Great body of salt water on the surface of a planet. At the present geological epoch, oceans are known only on the Earth. There is strong evidence, however, that in the geological past an ocean covered the northern part of MARS. The Martian northern plains are very flat, and their borders with the highlands have features that resemble terrestrial oceanic shorelines. Furthermore, many presently dry Martian valleys of obvious fluvial morphology open into these plains. There is also evidence that in the recent geological past an ocean covered all the surface of the Jovian satellite EUROPA: its surface, which is flat on the large scale, is made of almost pure water ice

▼ **ocean** The Conamara Chaos region of Jupiter's satellite Europa, imaged from the Galileo orbiter in 1997. These irregularly shaped blocks of ice are thought to overlie and occasionally be moved by a deep liquid ocean beneath Europa's surface.

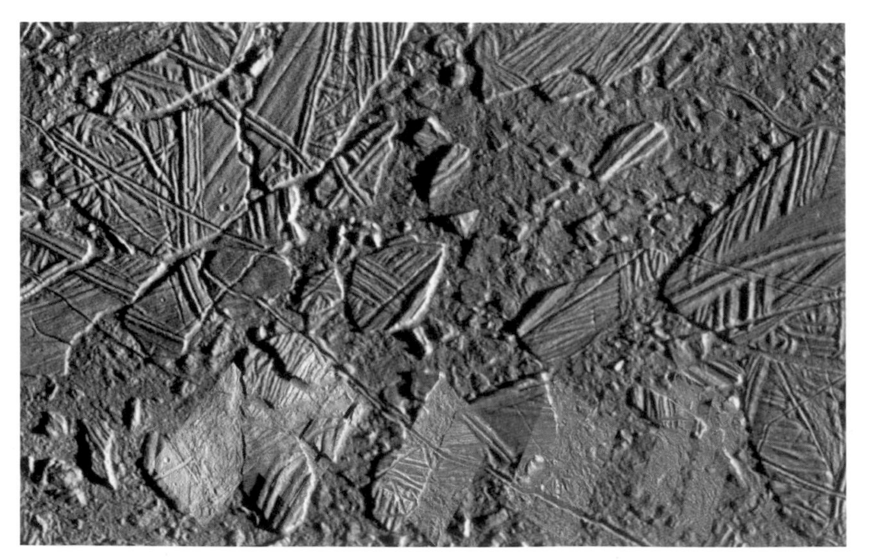

and is almost free of impact craters, indicating that it is geologically young. Some Europan landforms strongly resemble icebergs frozen into crushed ice. The youth of Europa's surface suggests that the currently observed icy crust may conceal a great body of liquid water. There is also evidence, from its electromagnetic response to Jovian magnetosphere fluctuations, that GANYMEDE may harbour great bodies of salt water.

octahedrite *See* IRON METEORITE

Octans *See* feature article

ocular Alternative name for an EYEPIECE, now seldom used.

Odysseus Impact CRATER on Saturn's satellite TETHYS; its diameter is 440 km (270 mi). Odysseus is the largest known crater in the Solar System, relative to the size of the body on which it is found.

off-axis guider Optical arrangement whereby a GUIDE STAR close to but outside a telescope's field of view can be used to ensure that the instrument remains accurately aligned with the target during imaging. A mirror or prism may be used to acquire the light of the guide star, which is then directed to a small telescope for manual guiding, or to an AUTOGUIDER. In off-axis guiding, the guide star cannot itself be imaged.

Of star O STAR whose spectrum contains EMISSION LINES of helium (at 468.6 nm) and doubly ionized nitrogen (463.4 and 464.0 nm) as a result of high luminosity and some chemical change caused by early evolution. O(f) stars have only a weak 468.6 nm line, while O((f)) stars have nitrogen in emission but ionized helium strongly in absorption. Of* stars display a strong [NIV] line.

OH Hydroxyl radical, the first INTERSTELLAR MOLECULE to be discovered at radio wavelengths (18 cm), in 1963. There are four components to the line (at frequencies 1612, 1665, 1667 and 1720 MHz). In some places one component of the line may be enhanced up to a million times the expected strength by MASER action. OH maser sources are associated with late-type stars and HII regions.

Olbers, Comet 13P/ Short-period comet discovered on 1815 March 6 by Wilhelm Olbers, Bremen, Germany. An elliptical orbit was calculated for the comet by Friedrich Bessel, this being revised following recovery, some months later than expected, by William Brooks (1844–1921) at the next return in 1887 August. At the most recent return, in 1956, 13P/Olbers reached peak magnitude +5.6. The orbital period is 69.56 years.

Olbers, (Heinrich) Wilhelm (Matthäus) (1758–1840) German physician and amateur astronomer, now best known for the paradox named after him. He was one of the CELESTIAL POLICE who searched for the supposed missing planet between the orbits of Mars and Jupiter, discovering two of the first four asteroids, Pallas (1802) and Vesta (1807). He also found several new comets, but more importantly invented a new method of calculating a comet's orbit from just three observations of its position. Although he was not the originator of OLBERS' PARADOX – Edmond Halley mentioned it, and the Swiss astronomer Philippe de Chéseaux (1718–51) gave essentially the same explanation as did Olbers – it was Olbers' name that became attached to it, following his discussion of it in a paper published in 1823. He was tireless in encouraging others, and set Friedrich Wilhelm BESSEL on his career.

Olbers' paradox Paradox discussed in 1826 by Heinrich OLBERS (although it had been raised earlier by, for example, Edmond Halley) when he posed the question 'Why is the sky dark at night?' If space were infinite and uniformly filled with stars, then in whatever direction the observer were to look he or she would eventually end up looking at the surface of a star and so the entire sky should be as bright as the surface of the Sun.

The paradox can be resolved in various ways. For example: if the Universe is not sufficiently old, light from the more remote objects cannot yet have reached us; or the expansion of the Universe ensures that radiation emitted by galaxies is weakened by the redshift and cannot be detected beyond a certain range.

Oljato One of the earliest APOLLO ASTEROIDS to be discovered, in 1947; it was numbered as 2201 after its recovery in 1979. Possible outgassing activity has been observed, and Oljato may perhaps have an associated METEOR SHOWER, suggesting that it may have originated as a cometary nucleus. *See also* ADONIS; PHAETHON; table at NEAR-EARTH ASTEROID

Olympus Mons Most spectacular shield volcano in the Solar System. Situated at 18°.6N 134°.0W on the northwestern flank of the THARSIS MONTES on MARS, it rises some 24 km (15 mi) above the surrounding plains and is more than 700 km (430 mi) in diameter. It comprises a central nested caldera 80 km (50 mi) across, from which the terraced flanks slope away at angles of only 4°. The flanks are traversed by numerous lava flows and channels with a roughly radial arrangement. In many respects Olympus Mons is similar to terrestrial basaltic shields like those of Hawaii, but with a volume that is between 50 and 100 times greater. Towards the north and south ends of the main shield, lava flows drape a prominent peripheral escarpment, which, in places, forms a cliff 6 km (4 mi) high. Numerous landslides degrade this cliff and spread out on to the adjacent plains. The flows themselves are very long compared with terrestrial flows, a phenomenon that may be explained either by very high

OCTANS (GEN. OCTANTIS, ABBR. OCT)

Constellation containing the south celestial pole, but otherwise unremarkable. Octans was created in the 18th century by Lacaille to commemorate the octant, an early form of navigational instrument that was the forerunner of the sextant. Its brightest star is ν Oct, visual mag. 3.73, distance 69 l.y., spectral type K0 III. SIGMA OCTANTIS, mag. 5.45, is the nearest naked-eye star to the south celestial pole.

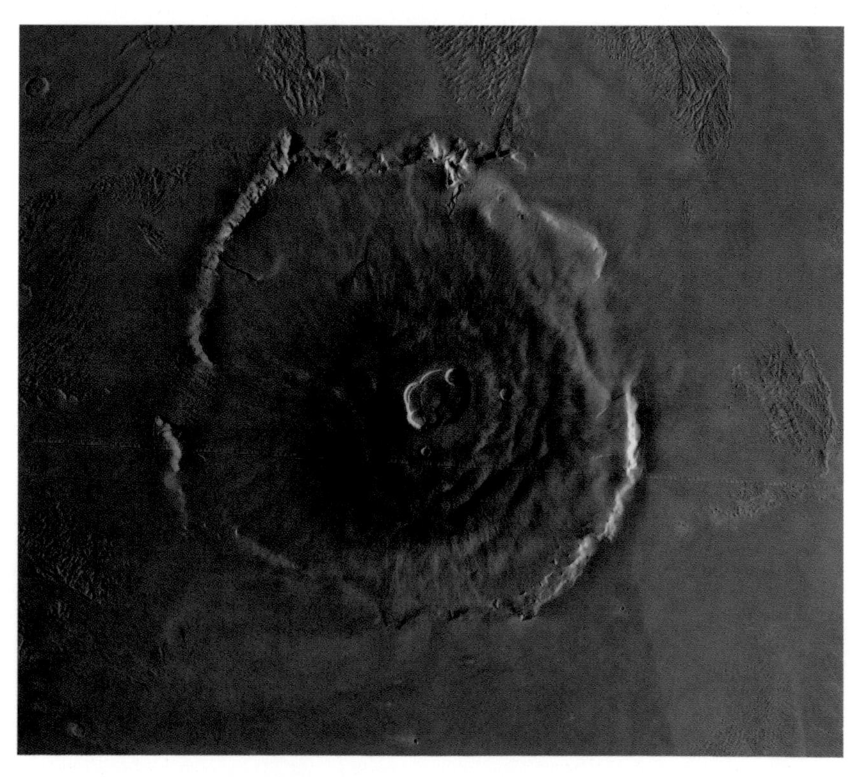

▼ **Olympus Mons** A Viking mosaic image of Olympus Mons, the Solar System's largest volcano. The summit caldera is about 80 km (50 mi) across, while the gently sloping flanks cover an area 700 km (430 mi) wide on the Tharsis Montes region of Mars.

► **Omega Centauri** The brightest globular cluster in the sky, Omega Centauri is a showpiece object for observers in the southern hemisphere. Several hundred thousand stars are packed into a volume only 200 l.y. across.

eruption rates, very large-volume eruptions, or a combination of the two. Surrounding the shield is a wide and very complex region of lobate ridged terrain which is termed the 'aureole'. In places the aureole extends 700 km (430 mi) from the basal scarp.

Omega Centauri (NGC 5139) Brightest GLOBULAR CLUSTER in the Galaxy, found in the southern constellation of Centaurus (RA 13^h $26^m.8$, dec. $-47°29'$). It is visible to the naked eye as a fuzzy star, and it received a stellar designation before its true nature was realized. Omega Centauri is of mag. 3.7, and it is about 40′ in diameter, with a broad, bright central region. Its brightest stars are 11th magnitude. At 17,000 l.y. away, it is one of the nearer globular clusters. It contains several hundred thousand stars in a volume 200 l.y. across. Omega Centauri is intrinsically the brightest globular cluster in the Galaxy. It is noted for wide variations in the heavy element ('metal') content of its stars, suggesting that they were formed at different times. Some 200 VARIABLE STARS have been found in the cluster.

Omega Nebula (M17, NGC 6618) EMISSION NEBULA and associated open star cluster in the constellation Sagittarius (RA 18^h $20^m.8$ dec. $-16°11'$). It is also known as the Swan Nebula or Horseshoe Nebula. The nebula has angular dimensions of $20' \times 15'$ and mag. 6.0. Its true diameter is 27 l.y., and it lies 6800 l.y. away.

▼ **Omega Nebula** Also known as the Swan Nebula, M17 in Sagittarius is part of a huge complex of emission nebulosity in the direction of the galactic centre. It contains about 800 solar masses of material.

Onsala Space Observatory Swedish national facility for radio astronomy located 48 km (30 mi) south of Gothenburg, and operated by Chalmers Institute of Technology. It has a 25-m (82-ft) telescope that can be used independently or in very long baseline interferometry (VLBI), and a 20-m (66-ft) dish for millimetre-wave observations. The observatory also operates the SWEDISH ESO SUBMILLIMETRE TELESCOPE.

Oort, Jan Hendrik (1900–1992) Dutch astronomer who investigated the structure and dynamics of our Galaxy and others, was one of the first to use radio astronomy to study the Galaxy, and put forward the 'Oort cloud' theory of the origin of comets. After studying at Groningen under Jacobus KAPTEYN he joined the University of Leiden, where he spent the rest of his working life, becoming professor of astronomy in 1935 and director of Leiden Observatory in 1945.

In the late 1920s, building on work by Kapteyn and Bertil LINDBLAD on the structure of the Galaxy and our position within it, Oort showed from statistical studies of stellar motions that the Sun is 30,000 l.y. (a better value than Lindblad's 50,000 l.y.) from the galactic centre and takes about 225,000 years to make one revolution. From this he calculated that the Galaxy contains around 100 billion solar masses. He also introduced the OORT CONSTANTS to describe stellar motion.

After World War II, Oort and Hendrik van de HULST recognized the potential of the newly emerging techniques of radio astronomy. In 1951 they discovered the TWENTY-ONE CENTIMETRE LINE in the spectrum of interstellar neutral hydrogen (HI REGIONS) that van de Hulst had predicted seven years before. Neutral hydrogen pervades the Galaxy, and the 21-cm emission passes through interstellar dust clouds that block visible light, so Oort and his colleagues were able to map the Galaxy, revealing its spiral arms.

In 1950 Oort proposed that comets reside in a huge shell-like region a light-year or so from the Sun, and that gravitational perturbations occasionally send a comet sunwards. This region is now known as the OORT CLOUD. In 1956 he and Theodore Walraven (1916–) found that radiation from the Crab Nebula is highly polarized, characteristic of SYNCHROTRON RADIATION.

Oort cloud Spherical halo of COMET nuclei gravitationally bound to and surrounding the Sun to a distance of 100,000 AU (a third of the way to the next closest star), proposed by the Dutch astronomer Jan Oort in 1950. While direct evidence for the existence of the Oort cloud is currently impossible to obtain, the idea is widely accepted as an explanation for the observed frequency and orbital characteristics of new LONG-PERIOD COMETS. The Oort cloud is believed to contain a population of up to 10^{12} comet nuclei. Gravitational perturbations by passing stars may dislodge Oort cloud nuclei from their distant orbits, causing them to fall sunwards and perhaps to pass through the inner Solar System, where they can become spectacular new comets.

A new comet of this nature may return to the far depths of space on a long-period orbit (as, for example, was the case with C/1996 B2 HYAKUTAKE, which is not expected to return for 14,000 years). If its path takes it close to one of the planets, particularly Jupiter, a comet may have its orbit modified to one of short period.

Refinements of the model from the 1950s onwards led to the suggestion that the Oort cloud may become more concentrated towards the ecliptic plane at distances of 10,000 to 20,000 AU from the Sun, extending inwards to join the EDGEWORTH–KUIPER BELT. Some Solar System dynamicists propose that residence in the Kuiper belt is an intermediate stage between the initial inwards fall from the Oort cloud and appearance as a new long-period comet.

Oort constants Two constants invented by Jan OORT to describe the motions of stars around the Galaxy. The orbital motions of stars mean that nearby stars are moving with respect to the Sun. For example a star that is closer to the centre of the Galaxy and behind the Sun will have a relative velocity that on average causes it to move towards the Sun and away from the centre of the Galaxy.

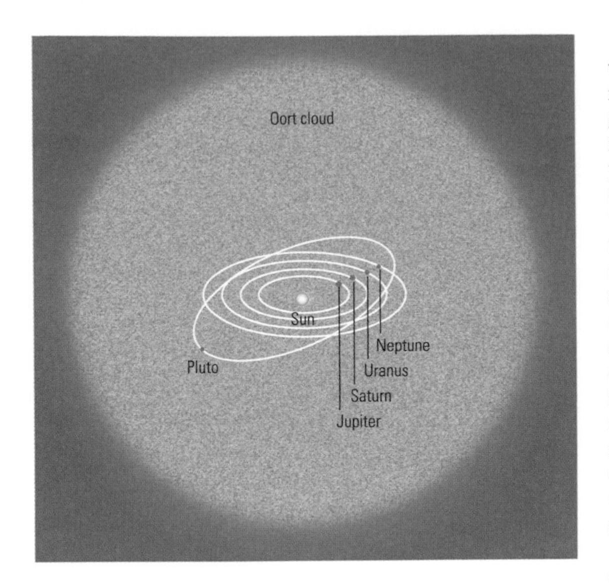

◀ **Oort cloud** An artist's impression of the spherical Oort cloud of cometary nuclei, surrounding the Solar System to a distance of 100,000 AU. Nuclei from this vast reservoir may be perturbed into orbits carrying them close to the Sun.

The radial velocity with respect to the Sun is given by $Ar\sin 2l$, while the velocity across the line of sight is $Br + Ar\cos 2l$. In these formulae, r is the distance of the star from the Sun in kilo parsecs (1 kpc = 3300 l.y.) and l is the star's galactic longitude (*see* GALACTIC COORDINATES). A (= 14 km s^{-1} kpc^{-1}) and B (= −12 km s^{-1} kpc^{-1}) are the Oort constants. *See also* STAR STREAMING

opacity Measure of the ability of a semi-transparent medium to absorb radiation. It is given by the ratio of the intensity of the emergent radiation to that of the incident radiation. The reciprocal of opacity is transmittance. A commonly used related quantity is the optical depth, τ, given by $\tau = -\log_e$. An optical depth of 1 corresponds to the apparent surface of a semi-transparent object or region such as the solar photosphere.

open cluster Loose irregularly shaped star cluster, containing from a few tens to a few thousand members. An open cluster may be regarded as a localized region of enhanced star density as compared to its immediate surroundings. A few open clusters have been known since antiquity, including the HYADES, the PLEIADES and PRAESEPE. Several star clusters were included in the MESSIER CATALOGUE, and still more are found in the *NEW GENERAL CATALOGUE* (NGC).

The stars belonging to a cluster not only lie close together in space but also share a common origin; they generally travel together through space, although they also to some extent move within the cluster. A sometimes severe problem in studies of stellar clusters, especially in star-rich regions, is the separation of true cluster members from field stars in the same region of the sky.

Open clusters are much less rich in stars than GLOBULAR CLUSTERS. They sometimes contain no more than a few tens of stars, and seldom more than a few thousand. A typical radius for an open cluster is about 10 l.y. The distribution of stars may vary considerably from cluster to cluster, but it is never as concentrated towards the centre as in globular clusters.

Open clusters are susceptible to disruption, not only through members becoming free of the gravitational influence of the group but also through tidal forces from the Galaxy or encounters with interstellar clouds. Only the richest clusters may survive more than 10^9 years, while the smallest and least tightly bound do not last more than a few million years. Their hazards are strengthened by their being found only near the galactic plane, which is the reason for the older designation 'galactic clusters'. The number of known open clusters in our Galaxy is nearly 1200, but this is certainly only a very small fraction of the total number. As all stars in a cluster take part in the general motion of the cluster, it is possible to derive the distances of nearby clusters by the MOVING CLUSTER method.

All stars are thought to be born in clusters, most of which, however, are gravitationally unbound and thus rapidly disrupt and spread their contents throughout the general field. In still recognizable clusters all members may thus be considered equally old (barring the youngest clusters and associations, where the ages are comparable to the spread in star formation time, of order 10^6 years). Studies of star clusters, therefore, yield good tests for theories of STELLAR EVOLUTION. Evolution proceeds more rapidly for the more massive stars, so these stars leave the MAIN SEQUENCE before the less massive ones. Accordingly, the position of the brightest main-sequence stars in a colour–magnitude diagram gives an estimate of the cluster age. For extremely young clusters, it is the faintest main-sequence stars that yield the age, as no stars have yet left the main sequence and the least massive ones have not yet got there. Such clusters are often still embedded in their parent clouds of gas and dust. *See also* EMBEDDED CLUSTER; OB ASSOCIATION; STELLAR ASSOCIATION; T ASSOCIATION

open universe Solution of Einstein's equations of GENERAL RELATIVITY in which the mass density of the universe is less than the CRITICAL DENSITY. This critical density is related to the HUBBLE CONSTANT and is estimated to be of the order of 9.2×10^{-27} g/cm^3. This implies that the universe will eventually accelerate and expand until the stars and radiation come into thermodynamic equilibrium, a state that is referred to as the HEAT DEATH OF THE UNIVERSE. *See also* BIG BANG THEORY; CLOSED UNIVERSE

Ophelia One of the small inner satellites of URANUS, discovered in 1986 by the VOYAGER 2 imaging team. Ophelia is about 32 km (20 mi) in size. It takes 0.376 days to circuit the planet, at a distance of 53,800 km (33,400 mi) from its centre, in a near-circular, near-equatorial orbit. With CORDELIA it acts as a SHEPHERD MOON to the planet's Epsilon Ring, Ophelia being just outside the orbit of that ring.

Ophiuchus *See* feature article

OPHIUCHUS (GEN. OPHIUCHI, ABBR. OPH)

Large but ill-defined constellation of the equatorial region of the sky. It represents Aesculapius, a son of Apollo and a mythical healer, holding a huge snake, a symbol of regeneration. The snake is represented by the constellation SERPENS, which is divided into two, one half each side of Ophiuchus. RASALHAGUE, the brightest star, is mag. 2.1. ρ Oph, mag. 4.57, is a fascinating multiple star; binoculars show two wide companions of 7th magnitude, one either side, while high magnification on a small telescope divides ρ Oph itself into two. An easy binary is 70 Oph, mags. 4.2 and 6.0, colours yellow and orange, which has an orbital period of 88 years. Another binary for small telescopes is 36 Oph, consisting of two orange dwarfs each of mag. 5.1 with an orbital period around 500 years. RS Oph is a RECURRENT NOVA with five recorded outbursts (a record it shares with T Pyxidis) in 1898, 1933, 1958, 1967 and 1985. Lying in the Milky Way towards the centre of our Galaxy, Ophiuchus is rich in star clusters, particularly globulars. The two best globular clusters are M10 and M12. NGC 6633 and IC 4665 are both large open clusters for binoculars. Ophiuchus also contains the nearby red dwarf BARNARD'S STAR, and KEPLER'S STAR appeared here in 1604.

BRIGHTEST STARS

	Name	RA h m	dec. ° ′	Visual mag.	Absolute mag.	Spectral type	Distance (l.y.)
α	Rasalhague	17 35	+12 34	2.08	1.30	A5	47
η	Sabik	17 10	−15 44	2.43	0.37	A2	84
ζ		16 37	−10 34	2.54	−3.20	O9.5	458
δ	Yed Prior	16 14	−03 42	2.73	−0.86	M1	170
β	Cebalrai	17 43	+04 34	2.76	0.76	K2	82
κ		16 58	+09 22	3.19	1.09	K2	86
ε	Yed Posterior	16 18	−04 42	3.23	0.64	G8	108
θ		17 22	−25 00	3.27	−2.92	B2	563
ν		17 59	−09 46	3.32	−0.03	K0	153

Ophiuchids Minor METEOR SHOWER (zenithal hourly rate no higher than 5) that peaks around June 20, with activity from late May until July. The radiant is just south of η Ophiuchi.

Öpik, Ernst Julius (1893–1985) Estonian astronomer, an expert on stellar evolution and meteors, who made contributions to almost every field of astronomy; many of his bold predictions were later proved to be true. He helped to found Tashkent (Uzbekistan) University during World War I, returning to Estonia to head the astronomy department at the University of Tartu. In 1948 he fled to Northern Ireland, and spent the rest of his career at Armagh Observatory. In 1910 Öpik discovered that o^2 Eridani was a white dwarf star; two decades later (1938), he showed how main-sequence stars like the Sun can evolve into red giants. From a study of the orbits and perturbations of comets, he predicted that they reside in a cloud extending to 60,000 AU from the Sun, an idea later revived by Jan Oort (*see* OORT CLOUD). Öpik studied the ablation of meteoroids passing through the Earth's atmosphere, knowledge later used to design heat shields that protect spacecraft during re-entry.

Oppolzer, Theodor (Egon Ritter) von (1841–86) Austrian mathematician and astronomer who calculated the time and track of every lunar and solar eclipse from 1207 BC to AD 2163. This work was published posthumously (1887) as the *Canon der Finsternisse* ('Canon of Eclipses'). Oppolzer was professor of celestial mechanics at the University of Vienna and published over 300 papers on the orbits of comets and asteroids. He devised a new theory to explain the lunar motion, which was taken into account by later researchers in this field, including E.W. BROWN.

opposition Position of a SUPERIOR PLANET when it lies directly opposite the Sun in the sky, as viewed from Earth. At this point, the three bodies are in exact alignment, with the Earth between the Sun and the planet. The term can also refer to the time at which this alignment occurs. When at opposition, a planet's ELONGATION is 180° and its phase is full. Opposition is the best time for observing a planet since it marks its closest approach to Earth although, because orbits are elliptical, this distance can vary between oppositions, particularly in the case of Mars. The term is equally applicable to asteroids. Inferior planets cannot come to opposition. See also CONJUNCTION

optical double DOUBLE STAR, the components of which appear to lie close together as seen from Earth but are actually totally unconnected. Optical doubles appear close together due to a line-of-sight effect, but in reality the stars often lie at great distances from each other.

optical interferometer Instrument that combines the light from two or more light paths to produce high-resolution images or spectra. When two or more light beams from the same object are combined, any small differences due to the separate paths along which they have travelled cause the beams to interfere with each other. This interference gives rise to patterns of dark and light which can be measured to analyse the differences producing them.

The earliest example of an optical interferometer used in astronomy is that which enabled Albert Michelson (1852–1931) and Francis PEASE to measure the angular diameter of Betelgeuse as 0″.047 in 1920. It consisted of a 6-m beam mounted on the front of the 100-inch (2.5-m) Mount Wilson reflector. Two movable mirrors on the beam directed two light beams from Betelgeuse on to the telescope, where they were combined in an interferometer. By moving the mirrors, Michelson and Pease determined the separation at which the fringes just disappeared and from this they calculated Betelgeuse's angular diameter.

Radio astronomers routinely use the INTERFEROMETRY principle to construct arrays of telescopes that can resolve detail equivalent to that attainable by a single reflector with a diameter equal to the separation of the telescopes in the array. The same method can be used by optical astronomers but is much more difficult to achieve. Recent developments have led to several observatories designing optical interferometers based on existing and new telescopes. For example, ESO's VERY LARGE TELESCOPE (VLT) will one day operate as an optical interferometer using its four 8-m (315-in.) telescopes in conjunction with an array of smaller telescopes around 1.8 m (71 in.) in diameter. Similarly, there are plans to combine the two 10-m (390-in.) Keck telescopes with several 2-m (79-in.) telescopes to form an interferometer.

orbit Path of a celestial body in a gravitational field generated by other bodies. There is usually a single dominant primary body, and the path is basically an ELLIPSE around the primary body, which is located at one of the foci of the ellipse. In rare cases the orbit may be a PARABOLA or a HYPERBOLA. The size, shape and orientation of the orbit are described by the ORBITAL ELEMENTS. The attractions of other bodies in addition to the primary body cause PERTURBATIONS to the orbit.

orbital elements Describing the size, shape and orientation of an ORBIT, and the position of the orbiting body in the orbit at some EPOCH. There are six orbital elements, and if the mass of the primary body is not know (such as in a binary star system) then the orbital period (or the mean motion) is also needed. In addition, the epoch of the elements is usually specified. The elements vary slowly with time as a result of perturbations by other bodies, and the epoch is the time at which they had the particular values specified.

The various angles defining an elliptical orbit are shown in the diagram. The point P is the position of a fictitious object that orbits with constant angular speed with the same period as the real object, and coincides with the real object at pericentre and apocentre. The orbital elements are:

a the semimajor axis (half the length of the major axis of the ellipse)
e the eccentricity
i the inclination to the reference plane
M the mean anomaly
ω the argument of pericentre
Ω the longitude of the ascending node

Note that M and ω are measured around the orbit plane, and Ω is measured around the reference plane. Two further composite angles are defined, which are measured

▼ **orbital elements** The orbit of a planet around the Sun can be defined by a number of characteristics, including the inclination (i) relative to the reference plane of the ecliptic, the longitude (Ω) of the ascending node measured in degrees from the First Point of Aries, and the argument of perihelion (ω) measured in degrees from the ascending node.

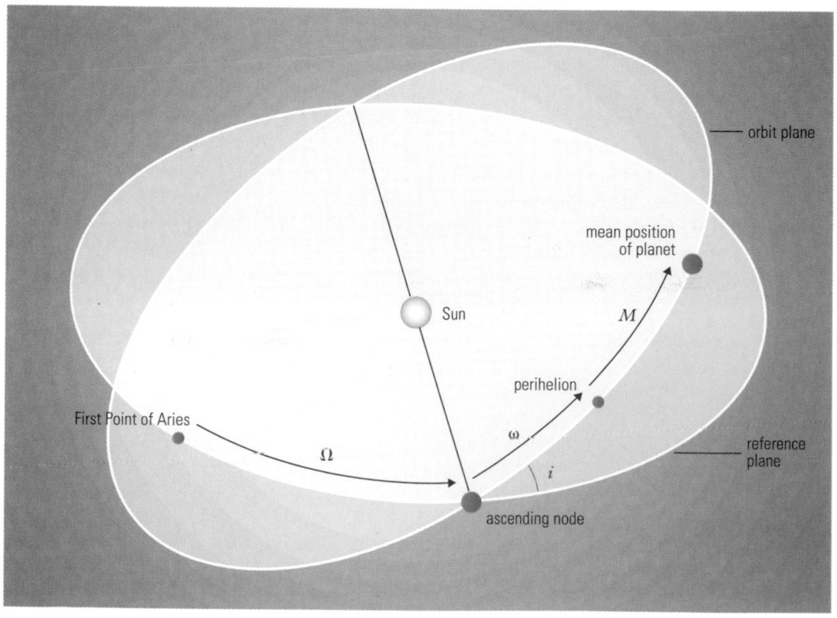

◀ **Orion** A wide-field view of the constellation Orion, showing the three distinctive Belt stars at centre. Orange-red Betelgeuse is the bright star at top left, while Rigel is the brilliant white star at bottom right. The diffuse red glow of the Orion Nebula (M42) is prominent below the middle star of the Belt.

partly around the reference plane and partly around the orbit plane:

$\varpi = \Omega + \omega$ the longitude of pericentre

$\lambda = \Omega + \omega + M$ the mean longitude

Further quantities that are sometimes used are:

T the time of pericentre passage, that is, the time at which $M = 0$

$q = a(1 - e)$ the distance from the primary at pericentre.

Only six of these quantities are needed to define an orbit. For the planets and natural satellites, it is usual to take as the orbital elements the quantities a, e, i, λ, ϖ, Ω. The reason for choosing the composite angles λ and ϖ rather than M and ω is that for many planets and satellites the inclinations are small, and so the NODE is not a well-defined point (for the Earth the inclination to the ecliptic of date is zero, and so the node is not defined at all). It is preferable, therefore, to measure all angles from the well-defined EQUINOX. For asteroids and artificial satellites of the Earth the inclinations are generally fairly large, and so the node is well defined. In this case the orbital elements are usually taken as the quantities a, e, i, M, ω, Ω. In either case, the quantities λ or M vary rapidly as the object moves around its orbit, and so the epoch of the elements is an important quantity.

For comets and newly discovered asteroids the perihelion distance and the time of perihelion passage are usually of particular interest, and so these are taken as two of the orbital elements. The full set used is T, q, e, I, ω, Ω. As the mean longitude or mean anomaly at epoch is not used, the epoch of the elements is of lesser importance than it is for planetary elements.

orbiter Any spacecraft that makes at least one revolution around a Solar System body. If the spacecraft is injected into a polar orbit, the whole surface of the body can, in general, be scanned by cameras and other instruments. The first spacecraft to orbit the Moon was the Soviet LUNA 10 in 1966. The first planetary orbiter was NASA's MARINER 9, which went into orbit around Mars in 1971.

Orbiting Astronomical Observatory (OAO) Series of US astronomical satellites launched 1966–72. The first was a failure, but OAO-2 and OAO-3 (known as COPERNICUS) made ultraviolet observations of stars and interstellar matter. OAO-3, which operated for nine years, also carried a UK X-ray instrument which made important observations.

Orientale, Mare (Eastern Sea) Farside lunar lava plain with rings extending on to the visible western limb of the Moon. It is the most extensively studied multi-ring impact BASIN on the Moon, because it is the youngest and so has the best-preserved ring system. The eastern aspects of the rings are visible during a favourable LIBRATION. Long after the impact, lunar basalts tracked up underlying faults and flooded the inner region of this basin, producing Mare Orientale. Extensive scarring from this basin's EJECTA is visible on western limb craters.

origin of planets *See* COSMOGONY

Orion *See* feature article

Orion Arm (Orion Spur) Segment of one of the spiral arms of the GALAXY within which our Solar System is situated. The arm curves away from the galactic centre, starting near Cygnus, passing Orion and ending near Puppis and Vela. But in fact all the brighter naked-eye stars, plus the ORION NEBULA, are within the arm. The Solar System is on the edge of the arm closest to the centre of the Galaxy. The Orion Arm is probably a branch off one of the main spiral arms of the Galaxy, and so is perhaps better called by its alternative name of the Orion Spur.

Orionids METEOR SHOWER, active between October 16 and 30, produced by debris from Comet 1P/HALLEY encountered at the ascending node of its orbit. The maximum is broad, over a couple of days around October 20–22, and shows some variation in timing from year to year, presumably as a result of the meteor stream's filamentary nature. Peak zenithal hourly rate (ZHR) is

ORION (GEN. ORIONIS, ABBR. ORI)

Generally regarded as the most splendid constellation of all, lying squarely on the celestial equator and hence equally well seen from both hemispheres. In Greek mythology, Orion was a great hunter, the son of Poseidon, but was stung to death by a scorpion, now represented by the constellation Scorpius. In the sky, Orion's dogs follow at his heels in the form of the constellations Canis Major and Canis Minor. Orion's leading star, RIGEL, is the seventh-brightest in the entire sky, while BETELGEUSE, tenth-brightest in the sky, is a red supergiant that varies unpredictably. The most distinctive feature is the Belt of Orion, a row of three 2nd-magnitude stars named ALNITAK, ALNILAM and MINTAKA. Other 2nd-magnitude stars are BELLATRIX and SAIPH.

South of the belt lies the **Sword of Orion**, marked by a complex of stars and nebulosity. The main part of the nebula was catalogued by Charles Messier as M42, but is better known as the ORION NEBULA. At its heart is the TRAPEZIUM, a multiple star. A subsection of the same cloud is catalogued as M43, while more nebulosity, known as NGC 1977, surrounds the 5th-magnitude star 42 Ori to the north. Farther north still is the large open cluster NGC 1981. The southern tip of the sword is marked by ι Ori, a wide double for small telescopes, mags. 2.8 and 6.9. σ Ori is a 4th-magnitude quadruple star divisible in small telescopes, with a fainter triple star, Struve 761, in the same field of view. Another of Orion's nebular treasures is the HORSEHEAD NEBULA, a dark cloud resembling a chess knight, silhouetted against a strip of bright nebulosity, IC 434, that extends south from ζ Ori. Unfortunately, the Horsehead is too faint to detect in amateur telescopes and is well seen only on long-exposure photographs.

BRIGHTEST STARS

	Name	RA h m	dec. ° ′	Visual mag.	Absolute mag.	Spectral type	Distance (l.y.)
β	Rigel	05 15	−08 12	0.18	−6.69	B8	773
α	Betelgeuse	05 55	+07 24	0.45(v)	−5.14	M2	427
γ	Bellatrix	05 25	+06 21	1.64	−2.72	B2	243
ε	Alnilam	05 36	−01 12	1.69	−6.38	B0	1300
ζ	Alnitak	05 41	−01 57	1.74	−5.26	O9.5	820
κ	Saiph	05 48	−09 40	2.07	−4.65	B0.5	720
δ	Mintaka	05 32	−00 17	2.25	−4.99	O9.5	2000
ι		05 35	−05 55	2.75	−5.30	O9	1300
π		04 50	+06 58	3.19	3.67	F6	26
η		05 25	−02 24	3.35	−3.86	B1 + B2	900
λ	Meissa	05 35	+09 56	3.39	−4.16	O8	1100

O

typically around 25 meteors/hr, but slightly higher activity has been seen on occasion, as in 1993. Outstanding displays may have occurred in AD 288 and 1651. Orionid meteors are swift (66 km/s or 41 mi/s) and a high proportion leave persistent trains. The shower RADIANT appears to have a complex structure, and it lies midway between Betelgeuse and γ Geminorum. A further encounter between Earth and the meteor stream in May produces the ETA AQUARIDS.

Orion molecular clouds Complex group of GIANT MOLECULAR CLOUDS in Orion. At a typical distance of 1500 l.y., they are the nearest such clouds to us. The individual clouds may be up to 100,000 times the mass of the Sun and 100 or more light-years in size. The clouds are the sites of active star formation, especially in their denser cores. These cores are only detectable from their infrared and molecular radio emissions. One such core, designated OMC-1, is associated with the ORION NEBULA and the young stars of the TRAPEZIUM. *See also* BECKLIN–NEUGEBAUER OBJECT; KLEINMANN–LOW NEBULA

Orion Nebula (M42, NGC1976) HII REGION that is easily visible to the naked eye as the central object in the sword of ORION (RA $05^h\ 35^m.4$ dec. $-05°27'$). The main part of the nebula is separated from a smaller part of the same cloud, known as M43 or NGC1982, by a dark absorbing region called the FISH'S MOUTH. The nebula surrounds, and is excited by, the four young stars of the TRAPEZIUM (also known as θ^1 Ori). The nebula is about 1500 l.y. away on the nearer side of the ORION MOLECULAR CLOUDS and is little more than the inside of an incomplete spherical hole in the side of the much larger dark nebula. The Orion Nebula is over a degree across on the sky and thus some 25 to 30 l.y. in physical size. Its mass is several hundred times that of the Sun.

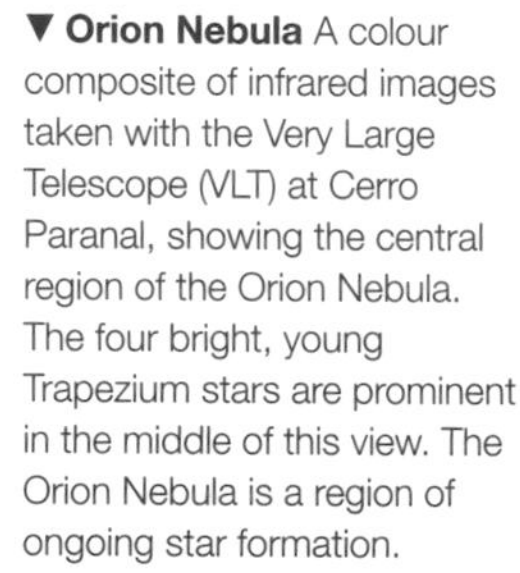

▼ **Orion Nebula** A colour composite of infrared images taken with the Very Large Telescope (VLT) at Cerro Paranal, showing the central region of the Orion Nebula. The four bright, young Trapezium stars are prominent in the middle of this view. The Orion Nebula is a region of ongoing star formation.

orrery Geared mechanical model of the Solar System. The first orrery, probably that preserved in the Adler Planetarium, Chicago, was made by the English instrument-maker George Graham (*c.*1673–1751) in about 1708. The name is usually attributed to John Rowley, who made a similar instrument for Charles Boyle (1676–1731), 4th Earl of Cork and Orrery, in *c.*1712. Many 'grand orreries' up to 1 m (3 ft) in diameter, showing the motions of the six known planets and their satellites, with clockwork movements, were made in England during the 18th century. From about 1775, hand-driven portable orreries came on the market, often with alternative fittings so that an orrery could be a 'planetarium', 'tellurium' or 'lunarium', as desired.

orthoscopic eyepiece Telescope EYEPIECE designed to give good definition, little geometric distortion and comfortable EYE RELIEF. It consists of two elements – a single eye lens that is normally plano-convex, and a cemented TRIPLET that is normally symmetrical. This arrangement produces an apparent field of view of 40° to 50°, which is smaller than other popular designs such as the PLÖSSL. Orthoscopic eyepieces are sometimes preferred by planetary observers looking for good performance at high magnification.

Oschin Schmidt Telescope World's first large SCHMIDT telescope, completed in 1948, with a 1.2-m (48-in.) corrector plate and a 1.8-m (72-in.) spherical mirror. Formerly known as the Palomar Schmidt, and renamed after the benefactor Samuel Oschin, the telescope is sited at PALOMAR OBSERVATORY and operated by the CALIFORNIA INSTITUTE OF TECHNOLOGY. It became famous in the 1950s for producing the PALOMAR OBSERVATORY SKY SURVEY. Photographic plates from the survey were used to identify targets for the nearby HALE TELESCOPE. In 1984 the Oschin Schmidt was fitted with a new achromatic corrector plate in preparation for a new survey of the northern sky. Today, it is used with a CCD camera to search for NEAR-EARTH OBJECTS.

oscillating universe Idea that if the universe is closed it will eventually collapse back to an infinitely small point, an event sometimes called the 'Big Crunch'. It is conceivable that a new BIG BANG could follow the Big Crunch and create a brand new universe. This cycle could in principle repeat itself, alternately creating, then destroying and finally creating a new universe each cycle.

osculating orbit ORBIT that a celestial body would continue in if all perturbing forces from other bodies were instantaneously cut off, leaving just the gravitational attraction of the primary body. The elements of this orbit are the osculating elements, and are mathematically equivalent to the instantaneous position and velocity. The current planetary ephemerides are generated by a numerical method that gives instantaneous position and velocity at each time step. These can be converted into osculating elements to give a more meaningful representation of the orbit. They can be used for calculating the position of the body at any required time, and are very accurate close to the EPOCH of the elements, and still reasonably accurate (0°.01) up to one year from the epoch. Osculating elements of the planets are tabulated in *The Astronomical Almanac.*

O star Any member of a class of blue-white stars, the spectra of which are defined by ABSORPTION LINES of ionized helium. Neutral helium lines are present in the cooler subclasses, while those of hydrogen remain strong. O stars lie at the extreme hot end of the main sequence. Dwarf subclasses range from 32,000 K at O9.5 to 50,000 K at O3 (the earliest defined); zero-age masses range from 16 to perhaps 120 solar masses (the upper limit is not well known); zero-age luminosities range from 30,000 solar to over 2 million; dwarf lifetimes are from 10 million years to only 2 million. Higher luminosity MK classes are recognized in part by the weakening of the ionized helium absorptions.

As O dwarfs evolve, the more massive ones become O giants and supergiants. The high luminosities result in line-driven winds that can eventually cut the stellar masses by large fractions. As a result, O stars commonly have

O

emission lines in their spectra. Those with hydrogen emission are called Oe stars, while those with ionized helium and doubly ionized nitrogen are OF STARS. Emission line and 'Of' behaviour increase with increasing luminosity. Ultraviolet (and even optical) P CYGNI LINES reveal wind speeds of up to 4800 km/s (3000 mi/s) and mass-loss rates of up to 10^{-5} solar masses per year.

As O stars evolve, the lower masses turn into class M helium-fusing supergiants. Above about 60 solar masses, however, high mass-loss rates make the stars' evolutionary paths stall as B supergiants and luminous blue variables, which then become Wolf–Rayet stars that have been stripped of their outer hydrogen envelopes. As evolution proceeds past helium fusion, O stars develop iron cores that collapse to produce Type II supernovae, whose stellar remains become neutron stars or black holes.

As a consequence of both low formation rates and short lifetimes, O stars are rare. Less than one of every 3 million main-sequence stars is an O star. The odds of finding one nearby are small, and most are very far away. Short lifetimes also mean that O stars do not move far from their birthplaces, and they are commonly associated with star-forming interstellar clouds of gas and dust. Since their high temperatures result in hard ultraviolet radiation, O stars can ionize their surroundings and create diffuse nebulae; the most famous is the Orion Nebula, which is ionized by the Trapezium quartet.

Because star formation takes place in the Galaxy's disk, O stars are found almost exclusively along the Milky Way, the relatively nearby ones helping to create Gould's Belt. Most O stars are members of unbound OB ASSOCIATIONS (from which they are moving away), the best known concentrating in Orion, Scorpius, Centaurus and Perseus. A good fraction of O and B stars form in binary systems, many with components of similar mass. Encounters between stars in multiple systems, or between pairs of close binaries, can eject individuals from the associations at high velocities. Supernovae may also create such RUNAWAY STARS. In a double star, the more massive star evolves first and explodes. Supernovae can explode off-centre, which can drive the resulting neutron star away at high speed, the companion fleeing in the other direction. It is estimated that as many as 20% of the O stars are runaways; the best known are μ Columbae and AE Aurigae, which were ejected from the Trapezium cluster 2.5 million years ago.

O stars, plus those of classes B0 to B1, produce much of the ionizing radiation that energizes the thinner interstellar clouds. Their high mass-loss rates also return a great deal of matter to the interstellar medium (red giant winds and planetary nebulae producing the bulk). Since the stars are being stripped by winds, much of this matter has been processed by energy-generating nuclear reactions to create new helium, nitrogen and carbon. The eventual supernovae then produce most of the heavy elements in the Universe, including all the iron. Moreover, the shock waves from wind-blown O star bubbles and from the eventual supernovae aid in the compression of the interstellar gases, and therefore help form new stars. Examples of O stars include ζ Puppis 05 Ia, ξ Persei O7.5 Iab, δ Orionis 09 V, ζ Orionis 09 Ib and ζ Ophiuchi 09 V.

outer asteroid belt *See* EDGEWORTH–KUIPER BELT

outer planets Any of the planets with orbits that lie beyond that of the ASTEROID BELT; the outer planets are Jupiter, Saturn, Uranus, Neptune and Pluto. With the exception of Pluto, the outer planets have low densities in comparison to the inner, TERRESTRIAL PLANETS and possess ring systems and comparatively large numbers of satellites. The differences between the two types of planet are thought to be caused by their respective distances from the Sun during the early formation of the Solar System.

outgassing Emission of VOLATILE material from planetary bodies. Terrestrial bodies, formed from primeval Solar System material, contained volatile material. When these bodies were heated by radioactivity, DIFFERENTIATION occurred, and some of the volatile material escaped to the surface. This process, termed 'outgassing', produced planetary atmospheres.

Overwhelmingly Large Telescope (OWL) Largest optical telescope yet proposed: a 100-m (330-ft) aperture instrument whose reflecting surface would be formed by 1600 hexagonal mirror-segments, each 2.3 m (91 in.) across. The technique of multi-conjugate ADAPTIVE OPTICS would enable OWL to provide diffraction-limited images with a resolution of 0″.001. This would reveal surface features on nearby stars and EMISSION NEBULAE in the most distant known galaxies. OWL is not yet funded, but was the subject of a concept study by the EUROPEAN SOUTHERN OBSERVATORY in 2000.

Owens Valley Radio Observatory (OVRO) World's largest university-operated radio observatory, near Bishop, California, some 400 km (250 mi) north of Los Angeles. Operated by the CALIFORNIA INSTITUTE OF TECHNOLOGY, OVRO has a 40-m (130-ft) telescope, a millimetre-wavelength array consisting of six 10.4-m (34-ft) dishes, and a solar array.

OWL Abbreviation of OVERWHELMINGLY LARGE TELESCOPE

Owl Nebula (M97, NGC 3587) PLANETARY NEBULA in Ursa Major, located a couple of degrees south-east of β, the lower of the 'Pointers' (RA 11^h $14^m.8$ dec. +55°01′). The Owl, at mag. +9.9, is one of the fainter objects in Messier's list. Its 3′.4 circular profile contains a couple of dark patches, or 'eyes', which give it a resemblance to an Owl's face, hence its popular name. The nebula has an estimated age of 6000 years, and it lies 2600 l.y. away. The central star is of mag. +16.

oxygen (symbol O) Element that is the third most abundant in the Universe (*see* ASTROCHEMISTRY). Its properties include: atomic number 8; atomic mass of the natural element 15.9994 amu; melting point 54.8 K; boiling point 90.2 K. It was discovered in 1774 by the English chemist Joseph Priestley (1733–1804). Oxygen has eight ISOTOPES ranging from oxygen-13 to oxygen-20; three of these are stable and form the naturally occurring element (oxygen-16 99.8%, oxygen-18 0.2% and oxygen-17 <0.1%).

Oxygen comprises 21% by volume of dry terrestrial air. Under normal terrestrial conditions, it exists as a colourless, odourless gas composed of the diatomic molecule O_2. In this form it is essential to aerobic life. It is also essential to life as water, when it is combined with hydrogen, and as the carbohydrates, when combined with hydrogen and carbon. Oxygen makes up about two-thirds of the human body. The allotrope of oxygen, O_3, known as ozone, is present in small quantities high in the atmosphere, where it absorbs solar ultraviolet light at wavelengths shorter than 320 nm. It thus protects life on the surface from that damaging radiation. The ozone layer though is currently being depleted to dangerous levels, particularly over the polar regions, through the action of man-made fluorocarbons and other pollutants.

Oxygen is a highly reactive element and is thus an important constituent of the Earth and other terrestrial planets in the form of oxides of silicon, calcium, iron, aluminium and others. It makes up 49% by mass of the Earth's crust. It is used in liquid form as part of the fuel for liquid-fuelled rockets.

Oxygen plays an important role in the CARBON–NITROGEN–OXYGEN CYCLE, which is the major helium-producing nuclear reaction in stars more massive than the Sun. Oxygen burning to produce silicon occurs at temperatures over 2×10^9 K (*see* NUCLEOSYNTHESIS) in high-mass stars.

Ozma, Project *See* SETI

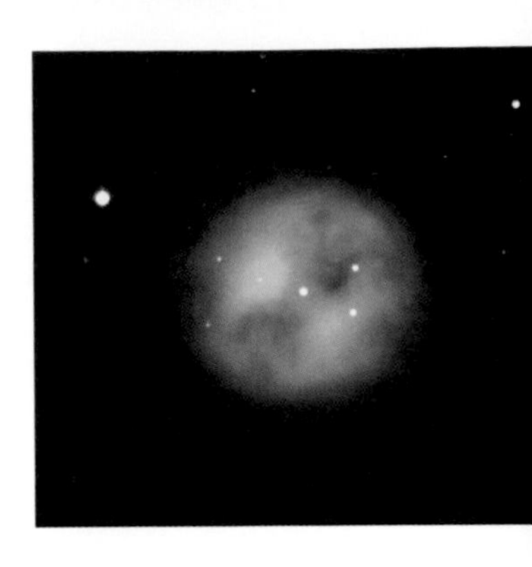

▲ **Owl Nebula** The dark 'eyes' of the Owl Nebula, M97 in Ursa Major, are well shown in this Kitt Peak 0.9 m (36 in.) telescope image. The Owl is a planetary nebula, produced as an aged star puffed off its outer layers.

P

PA *See* POSITION ANGLE

palimpsest Roughly circular ALBEDO spot found on the icy surfaces of Jupiter's satellites GANYMEDE and CALLISTO, presumably marking the site of a former impact CRATER and its rim deposit. Most, if not all, of the topographic structure has disappeared, but the visual distinction from adjacent areas remains. The topographic flattening of craters on their transformation into palimpsests might be brought about by their filling by water and/or the viscous flow of ice on the satellite surface layers.

Pallas Second MAIN-BELT ASTEROID to be discovered, and so numbered 2. It is the second-largest asteroid, after CERES. Pallas' mean density is about 4.2 g/cm^3, which is greater than that of stony meteorites or common terrestrial rocks, indicating that it contains a metallic component similar to nickel–iron meteorites. Despite its size, Pallas is by no means spherical in shape, measuring about 570 × 525 × 482 km (354 × 326 × 300 mi). Its mass is about 3.2×10^{20} kg (0.43% the mass of the Moon). Pallas is also noteworthy in that its orbital plane is tilted substantially relative to the ecliptic (inclination almost 35°).

pallasite One of the two subdivisions of STONY-IRON METEORITES, the other being the MESOSIDERITES. Pallasites are an approximately equal mixture of iron–nickel metal and silicates, predominantly olivine. They are presumed to represent material from the core–mantle boundary of their parent bodies. There are currently 50 known pallasites, almost all of which are Main Group (MG) pallasites. MG pallasites have oxygen isotope compositions similar to those of the HOWARDITE–EUCRITE–DIOGENITE ASSOCIATION achondrites.

Palomar Observatory Formerly known as Mount Palomar Observatory, this world-famous astronomical facility is at an elevation of 1706 m (5597 ft) in the San Jacinto Mountains 80 km (50 mi) north-east of San Diego, California. It was planned in 1928, when MOUNT WILSON OBSERVATORY's success with the 100-inch (2.5-m) HOOKER TELESCOPE was fresh in the minds of US astronomers. The Rockefeller Foundation was persuaded to fund a 200-inch (5-m) instrument, eventually named the HALE TELESCOPE, to be owned by the CALIFORNIA INSTITUTE OF TECHNOLOGY (Caltech).

The observatory was founded in 1938, but construction was delayed by World War II, and it was not until 1948 that the Hale Telescope began work, and another year before it was declared finished. Also in 1948, the 1.2-m (48-in.) Palomar Schmidt telescope (now known as the OSCHIN SCHMIDT TELESCOPE) became operational. During the 1950s and 1960s, these two instruments proved a formidable combination, giving the USA an unassailable position in world astronomy. Several smaller telescopes were also built at Palomar. The observatory is still owned by Caltech. While the LIGHT POLLUTION from San Diego limits the observational programmes, the Hale and Oschin telescopes remain effective instruments.

▼ **parhelion** This colourful 'Sun dog' was produced by refraction of sunlight through ice crystals in high cirrus cloud. A parhelion like this will most often be seen when the Sun is quite low in the sky.

Palomar Observatory Sky Survey (POSS) Systematic photographic SURVEY of the sky north of −30° made with the 1.2-m (48-in.) Oschin Schmidt telescope of the Palomar Observatory, jointly funded by the National Geographic Society. The *First Palomar Observatory Sky Survey* (POSS I) consists of about 2000 glass plates (1000 in each blue and red colours) observed between 1948 and 1957; it was completed in the 1970s by the ESO Schmidt Telescope in Chile (for the red part) and UK Schmidt Telescope in Australia (for the blue part). The *Second Palomar Observatory Sky Survey* (POSS II) was carried out in the 1990s in three colours (blue, red and infrared). Both surveys have been digitized and are known as the *Digitized Sky Surveys* (DSS I and DSS II).

Pan Innermost satellite of SATURN, discovered by Mark Showalter (1957–) in 1990 amongst images obtained a decade earlier by the VOYAGER 2 spacecraft. Pan is about 20 km (12 mi) in size. It has a circular equatorial orbit within the ENCKE DIVISION in Saturn's rings. It takes 0.575 days to complete a circuit of the planet, at a distance of 133,600 km (83,000 mi) from its centre.

The name **Pan** is also applied to a large crater, *c.*90 km (*c.*60 mi) long, on AMALTHEA.

Pandora One of the inner satellites of SATURN, discovered in 1980 by S.A. Collins and others in VOYAGER 1 images. It is irregular in shape, measuring about 110 × 85 × 60 km (68 × 53 × 37 mi). Pandora has a near-circular equatorial orbit at a distance of 141,700 km (88,100 mi) from the planet's centre, where its orbital period is 0.629 days. It appears to act as a SHEPHERD MOON to the outer rim of the F RING. *See also* PROMETHEUS

panspermia Theory that life on Earth did not originate here, but arrived from the depths of space. It originated with Svante ARRHENIUS, who in 1908 suggested that light from stars could blow microscopic germs from a world orbiting one star to another world orbiting another star. Panspermia never met with much support, but the idea never quite disappeared, and was rekindled in 1996 when a team of NASA scientists claimed to have found evidence for the presence of fossilized ancient microorganisms in a Martian meteorite (*see* ALLAN HILLS 84001).

Computer models have shown that rocks launched from Mars by a large meteorite impact can immediately enter Earth-crossing orbits if they are ejected just marginally faster than Mars' escape velocity, and that about 1 in 10 million of Martian meteorites that reach the Earth may have spent less than half a Martian orbital period, about one year, in space. Any microorganisms aboard such meteorites might just have survived the rigours of space travel and entry through Earth's atmosphere (microorganisms known as extremophiles are now known to have extraordinary survival attributes – *see* LIFE IN THE UNIVERSE). Over the lifetime of the Solar System, the inner planets are calculated to have exchanged tens of thousands of tonnes of material in this way.

Fred HOYLE and N. Chandra Wickramasinghe (1939–) have reformulated the idea of panspermia to suggest that life was brought to the Earth by a comet, and that interstellar dust 'grains' have a bacterial component. Their main argument is that the origin and evolution of life involves too many steps, each in itself inherently improbable, to have happened on Earth – instead, it needs all the 'resources of space'. Life will be carried around, mainly by comets, and take root wherever conditions are suitable.

Evidence is growing that the composition of giant molecular clouds, particularly in star-forming regions, is far richer that previously supposed (*see* INTERSTELLAR MOLECULES). Whether or not any prebiotic molecules that are formed survive long enough to be incorporated subsequently in planets orbiting newborn stars is not yet clear. However, only a tiny amount of such prebiotic material needs to survive to seed planets with the chemical basis of life. Life is more likely to have disseminated in this manner than by the transfer of living organisms by panspermia. Panspermia theories have, however, received support from the detection in 2001 of microorganisms in the stratosphere, at a height of 41 km (25 mi).

parabola Open curve, one of the CONIC SECTIONS, obtained by cutting a cone in a plane parallel to the side of the cone. It can be regarded as an ELLIPSE with only one focus, an infinite major axis and ECCENTRICITY of 1. A parabolic orbit is used for some of the LONG-PERIOD COMETS. These comets are observable over only a short arc of their orbits near perihelion, and it is often not possible to distinguish between exactly parabolic orbits and extremely elongated elliptical orbits.

paraboloid Three-dimensional surface generated by rotating a PARABOLA about its own axis. This shape is commonly used for the primary mirror in NEWTONIAN TELESCOPES in preference to a simple spherical surface. Images formed by a spherical mirror suffer from SPHERICAL ABERRATION because incident light at different distances from the axis come to a focus at different points. A paraboloid overcomes this problem and produces a diffraction-limited image for a point object, such as a star, that is in the centre of the field of view. Incoming light that is not parallel to the axis of the paraboloid (that is, from objects away from the centre of the field) produces images that suffer from COMA. Newtonian telescopes often have long focal lengths to reduce this effect.

parallax (symbol π) Change in apparent position of a celestial object, relative to its background, caused by a shift in the position of the observer. When observed six months apart from opposite sides of the Earth's orbit around the Sun, the position of a nearby star relative to the more distant background stars, will appear to have shifted by an angle $\Delta\theta$. Half of this angle, π, is the ANNUAL PARALLAX (heliocentric parallax), which is also a measure of the angular size of the radius of the Earth as seen from that star. If a star's parallax can be measured, then its distance can be determined. A unit of stellar distance is the PARSEC – the distance at which a star would have a parallax of one second of arc; it is equivalent to 3.2616 l.y.

This is an example of TRIGONOMETRIC PARALLAX, where a baseline of known length is used to make separate observations of a celestial object, the choice of baseline depending on the distance of the object in question. For objects in the Solar System, the radius of the Earth is used as a baseline, producing a measure of DIURNAL PARALLAX (geocentric parallax). SOLAR PARALLAX is the Sun's geocentric parallax, in other words the angular size of the Earth's equatorial radius from a distance of one astronomical unit (1 AU). Another method of determining stellar distances uses a star's spectral type and is known as SPECTROSCOPIC PARALLAX.

The astrometric satellite HIPPARCOS, which operated between 1989 and 1993, extended the range of accurate parallax distances by roughly 10 times, at the same time increasing the number of stars with good parallaxes by a much greater factor. Of the 118,218 stars in the Hipparcos catalogue, the distances of 22,396 are now known to better than 10% accuracy. Prior to Hipparcos, this number was less than 1000. The companion Tycho database, from another instrument on the satellite, provides lower accuracy for 1,058,332 stars. This includes nearly all stars to magnitude 10.0 and many to 11.0. *See also* DYNAMICAL PARALLAX; SECULAR PARALLAX

Paranal Observatory Second of the two observing sites of the EUROPEAN SOUTHERN OBSERVATORY (ESO), inaugurated in the early 1990s. Situated about 500 km (300 mi) north of its fellow ESO facility, LA SILLA OBSERVATORY, the site is at an elevation of 2635 m (8640 ft) on Cerro Paranal in northern Chile and offers superb atmospheric conditions for astronomy. It is home to the four instruments that comprise the VERY LARGE TELESCOPE (VLT). A nearby 2.5-m (100-in.) VLT Survey Telescope (VST) will provide imaging data at optical wavelengths, while the 4-m (157-in.) UK–ESO Visible–Infrared Survey Telescope for Astronomy being built at Paranal will provide infrared imaging data.

▲ **Parkes Radio Telescope** An evening view of the Parkes 64-m (210-ft) radio telescope in New South Wales, Australia. Constructed in 1961, the dish can record radio wavelengths between 5 mm and 2 cm.

parhelion (mock Sun, sundog) Atmospheric phenomenon produced by refraction of sunlight by ice crystals in cirrus clouds at altitudes of 10–15 km (6–9 mi), near the top of the TROPOSPHERE. A parhelion appears level in elevation above the horizon with the Sun, at an angular distance of 23° from it. There may be a single parhelion, or two parhelia lying to either side of the Sun in the sky. Sometimes parhelia are visible as brighter patches on a HALO around the Sun. As with the halo, parhelia can show strong colour, with red towards the Sun, blue away from it. Parhelia commonly appear as elongated bars of light.

Close to full moon, the equivalent lunar phenomenon, called a parselene, may be visible.

Paris Observatory Second-oldest observatory in the world, after COPENHAGEN OBSERVATORY. The 'Observatoire de Paris' was commissioned by King Louis XIV principally for geodesy, and built in 1667. Many famous 17th-century astronomers worked there, including Jean Picard, Ole RÖMER, Christiaan HUYGENS, Nicolas-Louis de LACAILLE and Urbain LE VERRIER. Four generations of the CASSINI FAMILY were observatory directors in the 17th and 18th centuries, and such was its importance to geodesy that the Paris meridian remained the origin of longitude until 1884. The observatory helped to create the metric system and instigated the *CARTE DU CIEL* project. Today, Paris Observatory houses a collection of astronomical instruments from the 16th century to the 19th; it is also a modern research institution, operating MEUDON OBSERVATORY and the NANÇAY RADIO TELESCOPE.

Parkes Observatory *See* PARKES RADIO TELESCOPE

Parkes Radio Telescope Major Australian radio astronomy facility 25 km (16 mi) north of Parkes in central New South Wales. The 64-m (210-ft) fully steerable dish was completed in 1961 and has undergone several upgrades, the most recent of which provides a unique multi-beam facility that allows the telescope to carry out direct radio imaging. Part of the AUSTRALIA TELESCOPE

P

PAVO (GEN. PAVONIS, ABBR. PAV)

Constellation of the southern sky representing a peacock, introduced at the end of the 16th century by Keyser and de Houtman. κ Pav is one of the brightest W VIRGINIS STARS, range 3.9–4.8, period 9.1 days. NGC 6744 is a large barred spiral often pictured in books while NGC 6752 is a 6th-magnitude globular cluster visible through binoculars, 15,000 l.y. away.

BRIGHTEST STARS

Name	RA h m	dec. ° ′	Visual mag.	Absolute mag.	Spectral type	Distance (l.y.)
α Peacock	20 26	−56 44	1.94	−1.81	B2	183
β	20 45	−66 12	3.42	0.29	A5	138

NATIONAL FACILITY, the telescope is notable for the discovery of hundreds of pulsars and for large-scale radio surveys. It also forms part of the Australia Telescope Long Baseline Array.

parsec (symbol pc) Unit of distance measurement defined as the distance at which a star would have an annual PARALLAX of one second of arc. When observed six months apart from opposite sides of the Earth's orbit, the position of a nearby star relative to more distant background stars will appear to have shifted by an angle $\Delta\theta$. Half of this angle, π, is its annual parallax. The parsec is equivalent to 3.2616 LIGHT-YEARS, 206,265 ASTRONOMICAL UNITS or 30.857×10^{12} km.

Parsons, Charles *See* GRUBB, PARSONS & CO.

Parsons, William *See* ROSSE, THIRD EARL OF

partial eclipse SOLAR ECLIPSE in which the Moon's passage across the Sun's disk does not result in complete obscuration. When a total solar eclipse occurs, a partial eclipse is visible in a broad band to either side of the narrow track of totality; before and after totality, the eclipse is, of course, partial. LUNAR ECLIPSES may also be partial: in such cases, only the northern or southern edge of the Moon enters the UMBRA of Earth's shadow.

particle physics Branch of physics that studies ELEMENTARY PARTICLES, fields and their interactions. Physicists use nuclear accelerators to speed up particles such as electrons or atomic nuclei and crash them into other atomic nuclei. The results of the collisions are examined to investigate the internal structure of the particles themselves. Linear accelerators use magnets to guide charged particles through tunnels, and the particles are accelerated using radio frequency energy. These particles can be accelerated to close to the speed of light. Other accelerators use large rings to store and accelerate particles, focusing and guiding the particles by a complex array of magnetic and electric fields until they are moving fast enough, and then guiding them to the targets.

Once the particles collide with the targets, the products of the collisions are detected and studied by a complex array of detectors. Some detectors are made up of cloud chambers in which a fast particle leaves a trail of condensation as it moves through the chamber. The shape and length of the condensation trail can be used to determine the speed, mass and charge of the particle. Other detectors are made of scintillators that emit photons as a fast elementary particle collides with the scintillating material. These photons are then detected by an array of phototubes trained on the scintillation chamber. The intensity and direction of the photons can be used to identify the particle.

Some of the most productive particle physics accelerators are the Stanford Linear Accelerator (SLAC), the CERN Accelerator, and the Jefferson Laboratory (formerly called CEBAF) Accelerator in Virginia. Facilities like these are being used to test the standard model of particles and fields. These accelerators are not powerful enough to test cosmologically interesting interaction energies. The superconducting super-collider (SSC) was a powerful accelerator facility to be built in the USA, but it was cancelled due to lack of funding. The SSC could have accelerated particles up to energies necessary to test elementary particle theories further. In order to test cosmologically interesting energies, an accelerator with a radius of 70 l.y. would have to be constructed!

The theoretical side of particle physics attempts to understand all of the particles resulting from collisions and all of the interactions by using the QUANTUM THEORY. Theorists work closely with experimentalists in trying to predict collision cross-sections, interaction strengths and resonances that may be observed in the particle accelerators. The standard model for elementary particles is being heavily tested in existing accelerators. Underlying all of this effort is the search for a correct GRAND UNIFIED THEORY (GUT) of elementary particles, and perhaps even a QUANTUM THEORY OF GRAVITY.

Particle Physics and Astronomy Research Council (PPARC) UK's funding agency for basic research in astronomy, planetary science and particle physics. It coordinates UK involvement in telescopes on La Palma, on Hawaii and in Chile, and it operates the ASTRONOMICAL TECHNOLOGY CENTRE and the MERLIN VLBI National Facility.

Paschen series Series of infrared EMISSION or ABSORPTION lines in the HYDROGEN SPECTRUM resulting from electron transitions down to or up from the third energy level of that atom. The Paschen lines are named with Greek letters. Paschen α, which connects levels 3 and 4, lies at 1875.1 nm; Paschen β, which connects levels 3 and 5, lies at 1281.8 nm; Paschen γ lies at 1093.8 nm, and so on, with the series ending at the Paschen limit at 820.3 nm.

Pasiphae One of JUPITER's outer moons, discovered in 1908 by Philibert Melotte (1880–1961). It is *c.*50 km (*c.*30 mi) in size. Pasiphae takes 735 days to orbit Jupiter, at an average distance of 23.62 million km (14.68 million mi), in an orbit of eccentricity 0.409. It has a retrograde path (inclination 151°), in common with other members of its group. *See also* ANANKE

Pathfinder *See* MARS PATHFINDER

pathlength Travel distance of RADIATION through an instrument from entry until it encounters the detector (instrument or eye).

The term pathlength is also used to describe the distance observed radiation travels when passing through the atmosphere. It is equal to one airmass in the zenith and two airmasses at an angle of 60° from the zenith.

Patroclus Second TROJAN ASTEROID to be discovered, in 1906; number 617. It is approximately 150 km (90 mi) in size.

patrol Systematic collection of repeated observations in the hope of making a discovery or recording a rare or unusual event. Among the most successful amateur patrols has been the UK Nova/Supernova Patrol operated by the BRITISH ASTRONOMICAL ASSOCIATION and *THE ASTRONOMER*. Many amateurs have now discovered supernovae in distant galaxies, taking advantage of computerized telescope guiding, CCD imaging, and the ability to rapidly check new on-screen images for stellar interlopers relative to reference images. Several amateur observers run photographic patrols in the hope of recording FIREBALLS associated with incoming meteorites.

Pauli exclusion principle Principle formulated in 1925 by the Swiss physicist Wolfgang Pauli (1900–1958) to explain the electron structure and the properties of the ELEMENTS. It requires that no two electrons within an atom have the same set of quantum numbers (*see* ATOMIC STRUCTURE).

Pavo *See* feature article

Pavonis Mons Massive shield volcano on MARS (0°.8N 113°.4W), situated along the crest of the THARSIS Rise. It has a large summit caldera surrounded by concentric graben and radiating volcanic flows.

Payne-Gaposchkin, Cecilia Helena (1900–1979) English astronomer who spent all of her life in America, mainly at Harvard, studying stellar evolution and composition. Cecilia Payne was at Cambridge University intending to study botany when, in 1919, her scientific interest was dramatically switched to astronomy when she heard Arthur EDDINGTON announce the results of the Greenwich eclipse expedition that confirmed a prediction of general relativity. In 1923 she left for America, Harlow SHAPLEY having offered her a post at Harvard College Observatory.

Payne's Harvard thesis on stellar atmospheres was hailed by Otto STRUVE as 'the most brilliant PhD thesis ever written in astronomy'. In it she showed how a star's spectral type was related to its surface temperature, thus adding another dimension to the HERTZSPRUNG–RUSSELL DIAGRAM. She also found that hydrogen and helium are the most abundant elements in stars (though there was initially doubt about this result), and that, these elements apart, the relative abundances of the elements in normal stars strongly match the composition of the Earth.

In 1934 she married **Sergei Illarionovich Gaposchkin** (1898–1984), an émigré astronomer from the Ukraine who had joined the staff of Harvard the previous year. Together they embarked on a huge programme of measuring the magnitudes of variable stars, the results appearing in a catalogue published in 1938. A similar study of variables in the Magellanic Clouds appeared in 1971.

P Cygni Highly luminous VARIABLE STAR of apparent visual magnitude 5. P Cygni is a galactic B1 supergiant 724,000 times more luminous than our Sun, with an effective surface temperature of 19,300 K. Its distance is 5900 l.y., its radius 76 solar radii and its mass 50 times that of the Sun. P Cygni appeared as a reddish star of visual magnitude 3 in 1600–1606, faded to magnitude 6 in 1620, rose again in 1655–59 and then fluctuated around minimum until 1715. In modern times P Cygni appears as a relatively constant, yellowish-white star. Its spectral signature, the 'P Cygni profile', is seen particularly in O, B and A supergiants showing episodes of catastrophic mass loss.

P Cygni's characteristic spectrum exhibits strong EMISSION LINES accompanied by equally strong variable blue-shifted ABSORPTION components. These P CYGNI LINE profiles are seen in the strongest BALMER hydrogen lines, in ionized helium, nitrogen, silicon, sodium and iron. The episodic nature of the mass loss, witnessed in the 17th century although not then understood, is supported by the detection of a radio arc interpreted as the remnant of an earlier gas shell.

Calculations suggest that P Cygni's evolutionary track across the HERTZSPRUNG–RUSSELL DIAGRAM may be exceptional. P Cygni lies on the border-line below which STELLAR EVOLUTION brings other massive stars into the red supergiant phase. It is classified as an S DORADUS STAR; it would have been the prototype for the class if its characteristics had been recognized earlier and if its name had not already been given to specific spectral features.

P Cygni line Stellar EMISSION LINE flanked to the shorter wavelength (blue) side by an ABSORPTION line; it is indicative of a powerful wind. The emission comes from a hot circumstellar gas. Since the gas flows in all directions, Doppler shifts average out, and the emission, though broadened, is centred at the line's laboratory wavelength (ignoring the star's radial velocity). The gas directly in front of the star, however, absorbs starlight and is Doppler shifted to the blue. The absorption's short-wave limit gives the wind's velocity and its strength the wind's density.

Peacock The star α Pavonis, visual mag. 1.94, distance 183 l.y., spectral type B2 IV. It is named after its parent constellation Pavo, which represents a peacock.

Pease, Francis Gledheim (1881–1938) American astronomer and instrument-maker who pioneered techniques of optical interferometry. He spent most of his career at MOUNT WILSON OBSERVATORY (1904–38), playing a major role in the design and construction of the observatory's most famous instruments, including its STELLAR INTERFEROMETER (1920). Pease used the interferometer to make the first measurement of the diameter of a star outside the Solar System, finding the diameter of Betelgeuse to be 0″.047.

peculiar A star *See* AM STAR; AP STAR

Peebles, (Phillip) James (Edwin) (1935–) Canadian astronomer and cosmologist who interpreted the COSMIC MICROWAVE BACKGROUND (CMB) and has studied the LARGE-SCALE STRUCTURE of the Universe. Peebles obtained his PhD at Princeton, studying under Robert DICKE, and has worked there ever since. In 1965, unaware of earlier work by George GAMOW and colleagues, Dicke, Peebles and others published a paper in which they interpreted the CMB, discovered by Arno PENZIAS and Robert Wilson, as radiation left over from the Big Bang. Peebles also calculated the levels of deuterium and helium that should now be present in a Big Bang Universe, results that agreed with the observed abundances.

In the early 1970s, Peebles and Jeremiah Ostriker (1937–) found evidence for dark matter. They calculated that the visible disk of a galaxy like the Milky Way could not remain stable for more than one galactic rotation unless it were constrained by the gravitational effect of an invisible halo of material. Peebles' studies of the present-day distribution of galaxies and how it relates to irregularities in the early expanding Universe resulted in the classic text *The Large-Scale Structure of the Universe* (1980).

In 1979 Peebles and Dicke identified the so-called flatness problem: how, in standard BIG-BANG THEORY, could the DENSITY PARAMETER Ω have been so close to 1 from so early in its history? If Ω had been just a little larger, the Universe would long ago have collapsed in a 'Big Crunch'; a little smaller, and by now the Universe would be a near-vacuum. This led Alan GUTH to develop the theory of INFLATION.

Pegasus *See* feature article

PEGASUS (GEN. PEGASI, ABBR. PEG)

Seventh-largest constellation, lying in the northern sky next to ANDROMEDA, representing the winged horse of Greek mythology born from the blood of the Medusa after she was decapitated by Perseus. Its most famous feature is the SQUARE OF PEGASUS marked out by four stars, one of which actually belongs to neighbouring Andromeda. β Peg, known as SCHEAT, is a variable red giant, while ε Peg is a wide double for the smallest telescopes, mags. 2.8 and 8.4. Even easier is the binocular double π Peg, mags. 4.3 and 5.6. M15 is one of the finest globular clusters, easily visible in binoculars looking like an out-of-focus star of 6th magnitude.

BRIGHTEST STARS

Name	RA h m	dec. ° ′	Visual mag.	Absolute mag.	Spectral type	Distance (l.y.)
ε Enif	21 44	+09 53	2.38	−4.19	K2	672
β Scheat	23 04	+28 05	2.44(v)	−1.49	M2	199
α Markab	23 05	+15 12	2.49	−0.67	B9.5	140
γ Algenib	00 13	+15 11	2.83	−2.22	B2	333
η Matar	22 43	+30 13	2.93	−1.16	G8 + F0	215
ζ Homam	22 41	+10 50	3.41	−0.62	B8.5	209

P

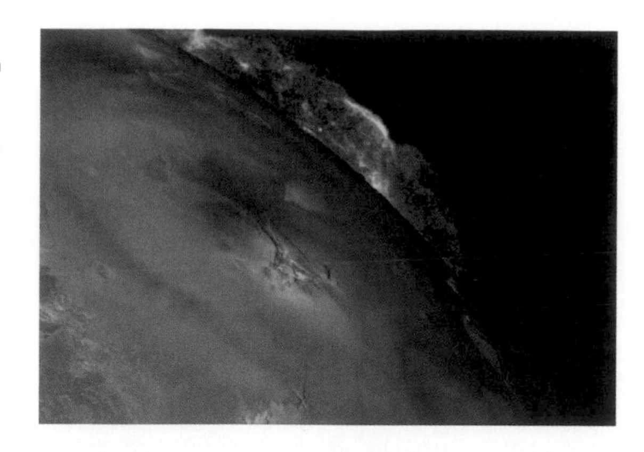

► **Pele** An eruption of the volcano Pele on Jupiter's moon Io can be seen in this image captured by Voyager 1 during its 1979 flyby. Material from the eruption was thrown 300 km (188 mi) above Io's surface.

Peiresc, Nicolas Claude Fabri (1580–1637) French astronomer and archaeologist who made the first recorded telescopic observation (1610) of the Orion Nebula. Soon after Galileo, he made some of the earliest observations of the four bright satellites of Jupiter (also 1610), becoming the first astronomer in France systematically to record their transits and eclipses. He recognized the value of these 'satellite events' for calculating terrestrial longitudes.

Peking Observatory *See* BEIJING OBSERVATORY

Pele Site of a large and persistent umbrella-shaped volcanic eruption plume on Jupiter's satellite IO. The plume was 300 km (190 mi) high when discovered on VOYAGER images in 1979; it was 400 km (250 mi) high in the late 1990s when seen by the HUBBLE SPACE TELESCOPE and the GALILEO orbiter.

Pelican Nebula (IC 5067 and IC 5070) Faint EMISSION NEBULA in the constellation Cygnus, just west of the NORTH AMERICA NEBULA (RA 20^h $50^m.8$ dec. $+44°16'$). It is best seen in long-exposure images of the region. The popular name comes from the resemblance of the nebula's profile to that of a pelican with the head and beak to the east. The Pelican Nebula covers an area of $80' \times 70'$.

P

penumbra (1) Area of partial shadow surrounding the main cone of shadow cast by the Earth or other nonluminous body in sunlight. In that area, a PARTIAL ECLIPSE is visible.

penumbra (2) Grey, less cool region surrounding a SUNSPOT. The central coolest, darkest region is termed the UMBRA.

penumbral eclipse LUNAR ECLIPSE at which the full moon passes just north or south of the dark UMBRA of Earth's shadow, producing little or no noticeable diminution in its brightness. During a penumbral eclipse, an observer on the Moon would see the Sun partially eclipsed by the Earth. Where the Moon's penumbral passage is very close to the umbra, it can sometimes show a slight darkening at its northern or southern edge; the effect is difficult to detect visually but can be recorded photographically.

Penzias, Arno Allan (1933–) and **Wilson, Robert Woodrow** (1936–) Respectively, German-American and American physicists turned radio astronomers who in 1964 discovered the COSMIC MICROWAVE BACKGROUND (CMB) radiation. Penzias' family emigrated to the United States in 1940; in 1961 he joined Bell Telephone Laboratories in Holmdel, New Jersey, to study satellite communications. Wilson helped set up CalTech's Owens Valley Radio Observatory in the early 1960s; he joined Bell in 1963.

That year, Penzias and Wilson began using a 6-m (20-ft) horn antenna to study terrestrial and extraterrestrial radio 'noise' that interfered with telecommunications. During a survey of the Milky Way's galactic halo, they detected a persistent signal with a wavelength of 7.35 cm that originated beyond the Milky Way and corresponded to an 'excess' antenna temperature of just 3 K. Theoretical astrophysicists, including Robert DICKE and Jim PEEBLES, and a group led by George GAMOW, had predicted and even searched for this CMB radiation as early as 1948. Penzias and Wilson were awarded the 1978 Nobel Prize for Physics for their discovery of one of the strongest pieces of observational evidence for the BIG BANG THEORY. After discovering the CMB, they went on to detect carbon monoxide and organic molecules in interstellar dust clouds.

perfect cosmological principle Hypothesis that on the large scale the UNIVERSE looks the same everywhere, in all directions and at all times. This principle was proposed in 1948 by Hermann BONDI and Thomas GOLD as the basis for their formulation of the STEADY-STATE THEORY (Fred HOYLE approached the theory from a different standpoint).

Most modern cosmological theories assume the validity of the ordinary 'cosmological principle' that on the large scale the Universe is homogeneous (looks the same everywhere) and isotropic (looks the same in every direction). The perfect cosmological principle went further in asserting that the large-scale appearance of the Universe remains the same at all times. Since the galaxies are moving apart, the perfect cosmological principle requires that new galaxies be formed at a rate just sufficient to maintain a constant average number of galaxies in each given volume of space. This implies the continuous creation of matter at a very slow but steady rate, in the region of 10^{-44} kg/m^3/s, that is, about one proton per cubic metre of space every ten years.

The perfect cosmological principle also implies that the Universe is infinite, with no beginning and no end, and that there never was a time when all matter was densely concentrated together (as the BIG BANG theory requires). It follows, too, that galaxies accelerate away from each other so that, despite being infinite, the Universe has a 'horizon' beyond which galaxies cannot be seen because their light is too severely redshifted. The principle also asserts, by definition, that the laws of nature do not change with time (some cosmologists contend that this need not be valid in an evolving universe).

The 'timeless' quality of the perfect cosmological principle made it attractive to many people, but it is now widely believed to be untenable in the face of the present observational data.

► **Pelican Nebula** This false-colour composite image of the Pelican Nebula in Cygnus was taken with the Kitt Peak 0.9 m (36 in.) telescope. Red signifies hydrogen-alpha emission, while blue-green comes from sulphur.

peri- Prefix referring to the nearest point of an orbit to the primary body, as in periastron, perihelion, perigee, pericentre, periapse. *See also* APSIDES

periastron Point in the orbit of a member of a double or multiple star system nearest to the primary star. *See also* APSIDES

perigee Point in the orbit of the Moon or an artificial satellite that is nearest to the Earth. *See also* APSIDES

perihelion Point in the orbit of a planet, comet or asteroid that is nearest to the Sun. *See also* APSIDES

periodic comet Any COMET, the elliptical orbit of which has been sufficiently well determined to allow accurate prediction of its future perihelion returns. Such comets are separated into LONG-PERIOD COMETS and SHORT-PERIOD COMETS, with an arbitrary division at 200 years between returns.

period–luminosity law Relationship, discovered by Henrietta LEAVITT in 1912, between the pulsation periods of CEPHEID VARIABLES and their median luminosities or absolute magnitudes. The brighter the Cepheid, the longer is its period. The importance of this relationship lies in the fact that observed periods are distance independent, whereas apparent magnitudes are dimmed by distance according to the inverse square law. Absolute magnitude is related to luminosity and is the apparent magnitude as measured from a standard distance. The measurement of period indicates absolute magnitude, which may then be compared with apparent magnitude to indicate the distance of the Cepheid. Similar relationships are found in other types of pulsating variable star, most notably in the MIRA STARS, certain SEMIREGULAR VARIABLES and W VIRGINIS STARS.

Perrine, Charles Dillon (1867–1951) American observational astronomer who worked at LICK OBSERVATORY (1893–1909) and later directed CÓRDOBA OBSERVATORY, Argentina (1909–36). At Lick, Perrine discovered Jupiter's sixth (Himalia, 1904) and seventh (Elara, 1905) satellites. In Argentina, he completed the final volume of the *CÓRDOBA DURCHMUSTERUNG*, a catalogue of about 600,000 stars, and made the first astrophysical observations of galaxies in the southern hemisphere. Perrine modernized Córdoba Observatory by installing 30-inch (0.75-m) and 60-inch (1.5-m) reflectors. He applied spectroscopic techniques to a wide variety of astronomical objects, including novae and Wolf–Rayet stars.

Perseids One of the most reliable annual METEOR SHOWERS, active between July 25 and August 20. Activity is substantial between about August 8 and 14, following a slow rise in the first week of the month. Maximum normally occurs around August 12, when observed rates of 60 meteors per hour may be found. At peak, the RADIANT lies close to the 'Sword Handle' at the northern end of Perseus. As first demonstrated in 1866 by Giovanni SCHIAPARELLI, the shower is associated with Comet 109P/SWIFT–TUTTLE.

Observations during the 1970s suggested increasing peak Perseid activity, further borne out by strong returns in 1980 and 1981 when zenithal hourly rate (ZHR) reached 120 (as opposed to the more normal 80–90). Still more significant was enhanced activity, manifesting as an additional peak 18 hours or so ahead of the established maximum, from 1988 onwards, associated with 109P/Swift-Tuttle's 1992 perihelion return. This early peak was a narrow feature, apparently resulting from encounter with a concentrated filament of recently released debris close to the comet, with enhanced activity occurring mainly in a 30-minute interval. In 1993 and 1994, the 'early' peak was at its most intense, producing numerous FIREBALLS and observed rates of up to 10 Perseids per minute (equivalent to a ZHR as high as 300). Unusual activity was still evident, albeit less pronouncedly, as late as 1997, some five years 'downstream' of the parent comet.

◀ **penumbra** (1) Mir cosmonauts took this photograph of the shadow cast by the Moon on to the Earth during the 1999 August 11 solar eclipse. The dark central umbra, under which the eclipse was total, is surrounded by a lighter outer penumbra, within which a partial eclipse was visible.

Thanks to the stream's high-inclination (113°.8) retrograde orbit, Perseid meteors are swift (66 km/s or 41 mi/s), and many are bright and leave persistent trains. The shower peak is usually a favourable time for meteor photography.

The Perseids have been known since antiquity. Chinese records as far back as AD 36 July 17 mention them, and there are frequent accounts in Chinese, Korean and Japanese annals from the 8th century onwards. European observers in the early 19th century were certainly familiar with the 'August meteors'. For many years, the Perseids were known traditionally as 'the Tears of St Lawrence' in memory of the Spanish martyr killed on 258 August 10, close to the date of the shower maximum.

Perseus *See* feature article, page 304

Perseus Arm Next spiral arm of the GALAXY after the ORION ARM, moving away from the centre of the Galaxy. At its closest it is about 5000 l.y. from the Sun, and it contains the CRAB NEBULA and ROSETTE NEBULA. It extends in the sky from Cassiopeia to Gemini. It may best be traced from its TWENTY-ONE CENTIMETRE radio emissions from hydrogen and by the presence of young blue stars.

personal equation Small, consistent error usually resulting from an observer's reaction time between seeing and recording an event. The personal equation is particularly applicable to visual OCCULTATION observations, where there can often be a delay between the observer seeing the disappearance or reappearance of the target star, and activating the stopwatch used to obtain the timing. Experienced observers who are aware of their personal equation

▼ **penumbra** (2) The lighter shading of the penumbra surrounding the dark central umbra is obvious in this SOHO image of a sunspot.

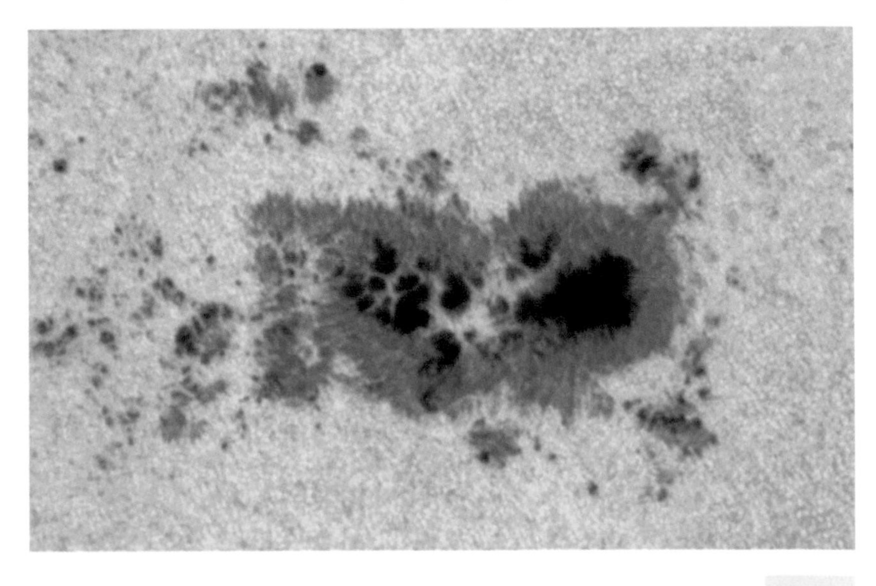

PERSEUS (GEN. PERSEI, ABBR. PER)

Major constellation of the northern sky, lying in the Milky Way between Auriga and Cassiopeia, representing the mythical Greek hero who saved princess ANDROMEDA from the jaws of a sea-monster. Its brightest star, MIRPHAK, is surrounded by a scattered cluster of bright stars covering six Moon diameters of sky. β Per is ALGOL, a famous eclipsing binary star. ε, ζ and η Per are all doubles with much fainter companions. Rho Per is a semiregular variable, ranging from 3.3 to 4.0 with a period of very roughly seven weeks. There are various open clusters, notably the DOUBLE CLUSTER, also known as the Sword Handle, designated NGC 869 and 884 or h and χ Persei. M34 is another open cluster easily visible in binoculars, but far less rich and condensed than the Double Cluster. M76 is a planetary nebula known as the LITTLE DUMBBELL, the faintest object in Messier's catalogue and hence a challenge for small telescopes. Perseus contains the CALIFORNIA NEBULA (NGC 1499), a large emission nebula that shows up well only on photographs. The year's best meteor shower, the Perseids, radiates from the northern part of the constellation, near γ Per, in August each year.

BRIGHTEST STARS

	Name	RA h m	dec. ° ′	Visual mag.	Absolute mag.	Spectral type	Distance (l.y.)
α	Mirphak	03 24	+49 52	1.79	−4.50	F5	592
β	Algol	03 08	+40 57	2.09(v)	−0.18	B8 + F	93
ζ		03 54	+31 53	2.84	−4.55	B1	982
ε		03 58	+40 01	2.90	−3.19	B0.5	538
γ		03 05	+53 31	2.91	−1.57	G5 + A4	256
δ		03 43	+47 47	3.01	−3.04	B5	528
ρ		03 05	+38 51	3.32(v)	−1.67	M3	325

(typically 0.1–0.2 second, determined by running mock events on a computer) allow for it in their reports.

perturbation Departure of an ORBIT from a fixed ELLIPSE around the Sun or a planet due to gravitational attractions from other bodies. According to KEPLER'S LAWS, the orbit of a planet around the Sun, or of a satellite around a planet, is an ellipse. In reality attractions by other planets and satellites, and by the non-uniform distribution of mass in the planets, cause a significant displacement of the position of a body from where it would be in pure elliptical motion. This displacement is the perturbation of the orbit, and it has periodic and secular components. The periodic components consist of fairly small short-period perturbations, with periods similar to the orbital periods, and much larger long-period perturbations, which are mostly caused by close COMMENSURABILITIES of the orbital periods. The most notable example is the so-called great inequality of Jupiter and Saturn, which is caused by their 5:2 commensurability. It causes variations of their longitudes with a period 938 years and amplitudes of 0°.36 and 0°.87 respectively. The secular perturbations are continual changes of an orbit in one direction, which generally cause the mean ellipse of the orbit to rotate in the direct sense, and cause the REGRESSION OF THE NODES around the reference plane of the orbit. These perturbations are the result of the long-term average attractions of the other bodies, and, for the satellites, by the flattened shape of their planets. For the planets there appear in addition to be steady changes of the ECCENTRICITY and INCLINATION of the orbit. In reality these changes are very long-period perturbations, with periods ranging from 25,000 to 2,000,000 years, but over the few hundred years of observations it is convenient to represent them as secular terms.

Perturbations can cause very large changes in the orbits of comets. All the short-period comets probably originated as long-period comets in the Edgeworth–Kuiper belt outside the orbit of Neptune; they were perturbed into their present orbits following many close approaches to the outer planets.

▼ **Phaethon** Phaethon's orbit (red) carries it from the inner part of the main asteroid belt at aphelion, to a perihelion closer to the Sun than Mercury. As an Apollo asteroid, Phaethon has an orbit that cuts across that of Earth.

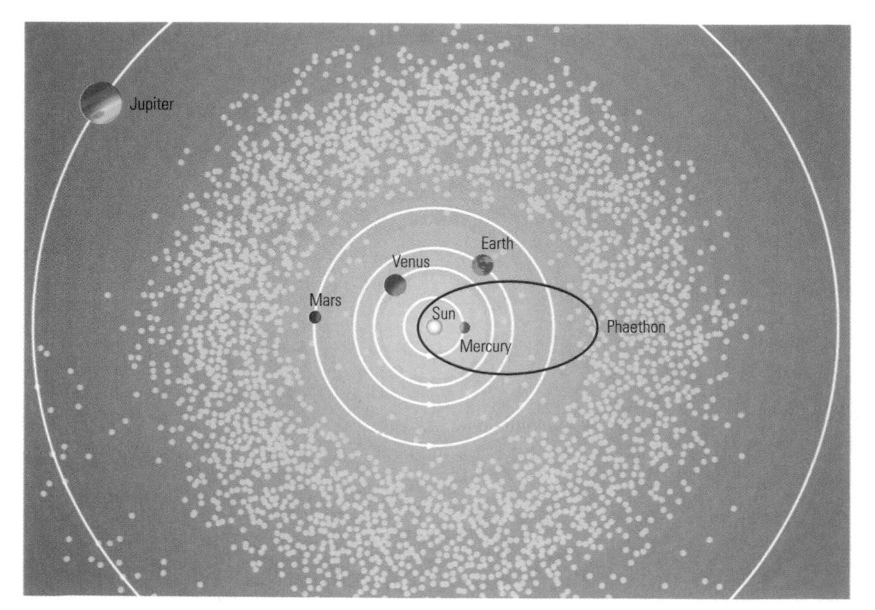

Petavius Lunar crater (25°S 61°E), 170 km (106 mi) in diameter, with rim components reaching 3350 m (11,000 ft) above its floor. It is a complex crater, with side-wall terracing and central peaks. Petavius is of sufficient age that meteorite erosion has removed its bright ray pattern. Its floor is fractured as a result of lava intrusion. Because of foreshortening, Petavius appears oblong, but its actual shape is more nearly circular.

Petit-Prince Moon of the MAIN-BELT ASTEROID (45) EUGENIA; it was the first such object to be discovered by direct telescopic observation from Earth, in 1998. (DACTYL, the small moon of (243) Ida, had earlier been discovered in images returned by the Galileo spacecraft. Occultation observations had previously revealed satellites around (6) Hebe, (18) Melpomene and (532) Herculina.) Petit-Prince is about 13 km (8 mi) across, and it orbits Eugenia once every 4.7 days at a distance of 1190 km (740 mi).

Peurbach *See* PURBACH

PHA *See* POTENTIALLY HAZARDOUS ASTEROID

Phaethon APOLLO ASTEROID with a particularly small perihelion distance (0.14 AU); number 3200. It was discovered in 1983 with data collected by the INFRARED ASTRONOMICAL SATELLITE. An origin as a cometary nucleus has been suggested, because it is securely associated with the GEMINID meteor shower on the grounds of orbital similarity. *See also* ADONIS; OLJATO; table at NEAR-EARTH ASTEROID

phase Extent to which the illuminated hemisphere of a Solar System body, particularly the Moon or an INFERIOR PLANET, is visible as viewed from the Earth. When close to QUADRATURE, Mars shows a distinct GIBBOUS phase. These bodies exhibit phases because they do not emit any light of their own and only shine by reflected sunlight. The observed phase, which is sometimes expressed as a percentage or decimal fraction, depends on the relative positions of the Sun, Earth and the body in question. When the whole of the illuminated side is visible, the phase is said to be full (100%). When the illuminated side is turned away from the Earth, it is new. Between the two extremes, a gradually increasing or decreasing proportion of the disk is seen. From new, the phases progress through crescent, first quarter, gibbous to full and then through gibbous, last quarter, crescent to new again. *See also* PHASES OF THE MOON

phase angle Angle between two lines joining the centre of a celestial object, usually a planet, asteroid or comet, to the Sun and the Earth. Phase angle corresponds to elongation of the Earth as seen from the object being observed. Thus, when the phase angle is 180°, the sunlit side of the object faces away from the Earth, as happens at INFERIOR CONJUNCTION. If the phase angle is 0°, the object is fully illuminated as seen from Earth, which happens at OPPOSITION. The term is also applicable when observing artificial satellites: the apparent magnitude of a satellite depends on its phase angle since, as this alters, so too does the amount of reflected light.

phase anomaly (Schröter effect) Difference of at least a day or so between theoretical and observed phase when VENUS is at DICHOTOMY (exact half-phase); during western (morning) elongations dichotomy is late, at eastern (evening) elongations it is early. The phenomenon was first remarked upon in the 1790s by Johann SCHRÖTER. Phase anomaly is unexplained, and it may be an atmospheric effect.

phases of the Moon Proportion of the Moon that appears sunlit from the viewpoint of an Earth-based observer. PHASE may be described as a percentage or decimal. Specific phases occurring at significant points during the LUNATION may be defined as new moon, first quarter, full moon and last quarter. New moon occurs when the Moon is at the same longitude in the sky as the Sun, such that its un-illuminated side is presented to the observer: the Moon at this phase is completely dark. After new, the Moon emerges into the evening sky (east of the Sun) as a gradually growing (waxing) crescent.

Approximately seven days after new, the Moon appears half lit and half dark; this phase is described as first quarter, and it occurs when the Moon is at QUADRATURE 90° east of the Sun.

As the Moon travels still further eastwards, its phase becomes gibbous (almond-shaped). Full moon occurs when the Moon is directly opposite (180° from) the Sun in Earth's sky, thus appearing completely illuminated. Telescopically, at full moon shadows are absent and crater outlines are difficult to see, although surrounding ray systems become prominent.

Following full, the Moon's phase begins to diminish (wane) through gibbous to last quarter (again at quadrature, this time 90° west of the Sun). Thereafter, the waning crescent draws ever closer to the Sun in the pre-dawn sky until new moon is reached and the next lunation commences.

Phecda The star γ Ursae Majoris, visual mag. 2.41, distance 84 l.y., spectral type A0 V. The name, also spelled *Phekda*, comes from the Arabic *fakhidh*, meaning 'thigh'.

Phobos Larger of the two moons of MARS, the other being DEIMOS, discovered in 1877 by Asaph Hall. Phobos is irregular in shape, measuring about 27 × 22 × 18 km (17 × 14 × 11 mi), with a mass of 1.1×10^{19} kg, giving it an average density of 1.9 g/cm^3. It has an orbit of low eccentricity (0.015) tilted at an angle of 1°.1 to Mars' equator, at an average distance of 9380 km (5830 mi) from the centre of the planet. This gives it an orbital period of 0.319 days, which is less than the 1.026 days it takes Mars to rotate on its axis. In consequence, an observer on the surface of the planet would see Phobos moving eastwards across the sky, rising in the west and setting in the east less than six hours later, whereas Deimos moves slowly in the opposite direction.

Physically, Phobos is quite similar to Deimos. It is a dark body (albedo 0.05), slightly red in coloration, and is thought to be a captured ASTEROID with a composition similar to a CARBONACEOUS CHONDRITE. Also like Deimos, it is locked in SYNCHRONOUS ROTATION with Mars, such that its long axis is radially aligned with the planet.

Spacecraft images show Phobos to have a heavily cratered surface. It shows cracks or lineations, which are thought to have originated from shocks produced by past impacts on this body by smaller asteroids, although a tidal cause is also feasible, given Phobos' proximity to Mars. Two large craters on Phobos are named Hall and Stickney, the surname of the discoverer and the maiden name of his wife, respectively; another crater is named after the French mathematician Édouard Roche.

Phoebe Outermost substantial satellite of SATURN, discovered in 1898 by William Pickering. It is near spherical in shape, measuring about 230 × 210 km (140 × 130 mi). With a RETROGRADE orbit (inclination 175°) and moderate eccentricity (0.164), it is likely to be a captured body related to the CENTAURS. It takes 550 days to orbit Saturn, at an average distance of 12.94 million km (8.04 million mi). Phoebe rotates on its axis once every 9.4 hours. Distant images obtained by the VOYAGER spacecraft show it to have a mottled surface. Material lost from Phoebe is thought to be the cause of variation in the brightness of IAPETUS: material spirals in

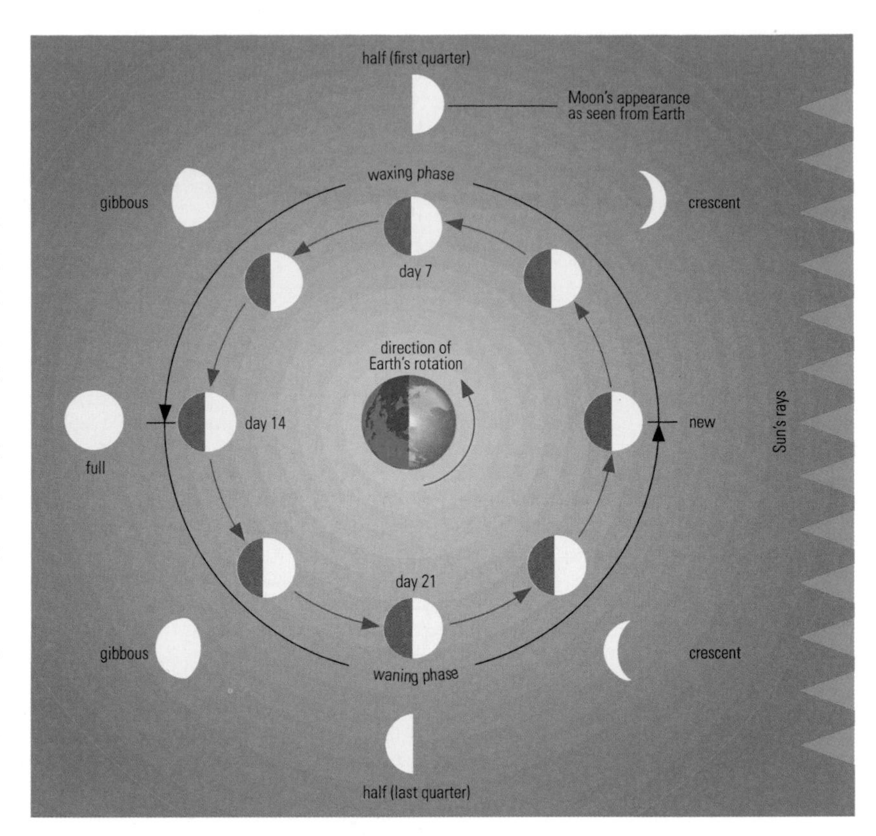

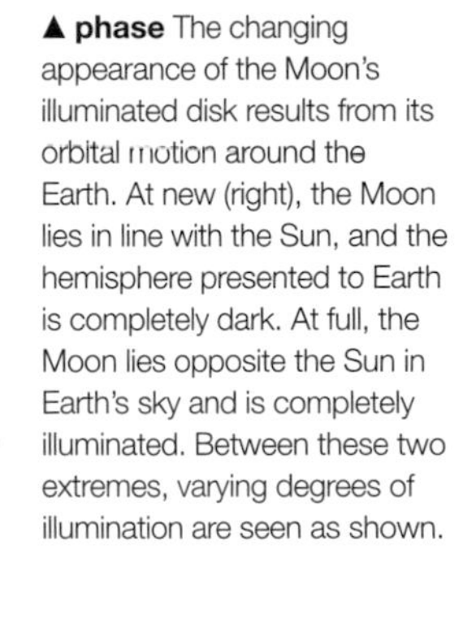

▲ **phase** The changing appearance of the Moon's illuminated disk results from its orbital motion around the Earth. At new (right), the Moon lies in line with the Sun, and the hemisphere presented to Earth is completely dark. At full, the Moon lies opposite the Sun in Earth's sky and is completely illuminated. Between these two extremes, varying degrees of illumination are seen as shown.

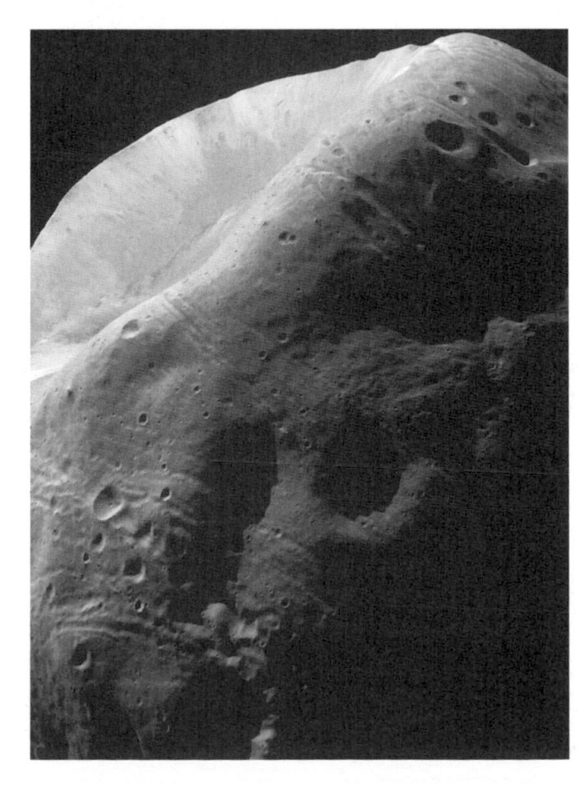

◄ **Phobos** Mars' larger moon, Phobos, imaged by the Mars Global Surveyor spacecraft in 1998. Believed to be a captured asteroid, Phobos is heavily cratered and shows linear structures resulting from impact shockwaves.

towards Saturn on a retrograde path and collects only on the leading face of Iapetus, the next moon inwards, because it is locked in a SYNCHRONOUS ROTATION state. As is also the case for Jupiter (*see* ANANKE), the population of known outer satellites of Saturn is increasing rapidly, with a dozen having been discovered since 1999.

Phoenix *See* feature article

Pholus CENTAUR object discovered in 1992. It is about 150 km (90 mi) in diameter. It has been catalogued as ASTEROID number 5145, although it may well turn out to have characteristics of a COMET. Pholus has perihelion near 8.7 AU and aphelion at 31.9 AU, meaning that it crosses the orbits of Saturn, Uranus and Neptune, rendering it dynamically unstable. Like the other Centaurs it is regarded as likely to have escaped from the EDGEWORTH–KUIPER BELT in the astronomically recent past (time scale of order a million years).

photoelectric cell Electronic device that converts a light signal, or a signal in the near-infrared or ultraviolet regions of the spectrum, into an electric current. The device utilizes the **photoelectric effect**, whereby electrons are liberated from a photocathode when struck by photons of electromagnetic radiation. If the energy of the photons is above a certain threshold level, the most weakly bound electrons can escape from their atoms and are attracted to a positively charged anode. This causes an electric current to flow which is proportional to the intensity of the radiation. Photoelectric detectors are therefore useful in measuring light intensity.

photoelectric magnitude MAGNITUDE of an object measured using an electronic detector and a combination of filters to match a standard passband. If the Johnson System is used, the filters are typically in the U (ultraviolet), B (blue), V (visual), R (red) or I (near-infrared) wavelengths.

photographic magnitude (mpg) MAGNITUDE of a star as recorded on a traditional photographic emulsion which has its peak response towards the blue end of the spectrum and records very poorly at wavelengths longer than 500 nm. The photographic magnitude differs significantly from the VISUAL MAGNITUDE (mv) recorded by the eye.

photographic zenith tube (PZT) Fixed, vertically mounted telescope used for the determination of time and latitude by the observation of the transit of FUNDAMENTAL STARS as they pass near to the ZENITH. The PZT contains a horizontal lens that focuses stellar images on to a pool of mercury from which they are reflected on to a photographic plate mounted on a moveable carriage. The carriage is driven during the exposure to compensate for the Earth's rotation; at each star transit, four exposures are recorded through two complete rotations of the instrument and the exact time recorded. The north and south displacements between image pairs provide latitude, while east and west displacements give the time of transit. By comparing the position of the images on the plate with the clock times at which they are observed it is possible to detect minor variations in the rate of rotation of the Earth. PZTs have now been superseded by SATELLITE LASER RANGERS.

photography, astronomical *See* ASTROPHOTOGRAPHY

photometer Instrument with which astronomers can measure the intensity (that is, the brightness) and various other properties of visible LIGHT, and also of infrared radiation and ultraviolet radiation. The earliest instruments were visual equalization photometers and extinction photometers. In **equalization photometers**, the intensity of a star image is steadily reduced until it is judged to be equal in brightness to that of another star (natural or artificial) in the same field of view. The magnitude difference of the two stars is then 2.5 times the logarithm of the star's reduced brightness expressed as a fraction of its original brightness. If the magnitude difference is great, a reduction in the intensity of the brighter source, by a known amount, to near equalization, may be required. This can be achieved by a reduction of aperture, use of polarized filters or use of an optical wedge. The eye can detect fairly slight differences in intensity between the astronomical source and the comparison source, so that reasonably accurate measurements can be made. Examples of this type of instrument are: the ZÖLLNER PHOTOMETER, where the brightness of the artificial source is changed using an optical wedge; the polarization photometer, where the reduction is carried out using polarized filters; and Danjon's cat's-eye photometer, where natural stars are used for comparison purposes, the brightness reduction being effected by using a combination of two prisms and a cat's-eye diaphragm. The flicker photometer is a true equalization photometer, in which the equality of the artificial and real light sources is assessed by the absence of flicker between the two images, when viewed in rapid succession from the same position.

In **extinction photometers**, an optical wedge or polarized filter, marked with a scale, is used steadily to reduce the intensity of the light source until it just becomes invisible. The position on the scale at which this occurs is noted, and in this way the brightness of the source can be determined.

Although visual photometers have been employed to obtain some useful results in the past, and the human eye has the ability to detect a wide range of light levels, it is comparatively insensitive to very small changes in brightness. In most cases, a difference in brightness of about one-tenth of a magnitude between the known brightness of the reference source and the unknown brightness of the source under study is the best that can be achieved with a single measurement. Greater precision may sometimes be achieved by taking the average of a series of measurements. However, for the far greater accuracies now required in astronomical research, or in cases where the intensity is constantly changing or varying very rapidly, physical photoreceptors such as the PHOTOELECTRIC CELL, PHOTOMULTIPLIER tube or BOLOMETER must be used. In all of these devices the incident electromagnetic radiation is converted into an electrical signal, the magnitude of which can be determined very precisely.

In optical, INFRARED ASTRONOMY and ULTRAVIOLET ASTRONOMY, magnitudes are determined using photometers over a selection of wavelength bands. Present systems such as the internationally accepted UBV SYSTEM use a suitable combination of a special filter and a photoelectric cell or photomultiplier tube (plus an electronic amplifier) to select light or other radiation of a desired wavelength band and measure its intensity. The UBV system of stellar magnitudes is based on photoelectric photometry. The photoelectric magnitudes denoted by U, B and V are measured at three broad bands, U (ultraviolet), B (blue) and V (visu-

PHOENIX (GEN. PHOENICIS, ABBR. PHE)

Southern constellation, introduced at the end of the 16th century by Keyser and de Houtman, representing the mythical bird that was reborn from its own ashes. The most interesting object for amateur observers is ζ Phe. Small telescopes show this to be a double with components of 4th and 8th magnitudes; the brighter of the two is also an ALGOL STAR, an eclipsing variable of range 3.9–4.4, period 1.67 days.

BRIGHTEST STARS

	Name	RA h m	dec. ° ′	Visual mag.	Absolute mag.	Spectral type	Distance (l.y.)
α	Ankaa	00 26	−42 18	2.40	0.52	K0	77
β		01 06	−46 43	3.32	−0.60	G8	198
γ		01 28	−43 19	3.41	−0.87	K5	234

al – that is, yellow-green), centred on wavelengths of 360 nm, 440 nm and 550 nm respectively. The light intensity can be measured for a particular wavelength band as a greatly increased (amplified) electrical signal. Magnitudes can also be measured for various infrared bands. Narrower-wavelength bands are used in the UVBY system. Photoelectric photometry can be used to study the light output from an isolated star or small region of the sky. Photomultipliers are available with photocathodes sensitive in the infrared, thus extending the wavelength bands investigated.

In a typical photoelectric photometer, the light from the telescope is fed via a pinhole and a Fabry lens to the photocathode of the photomultiplier tube (PMT). The pinhole is located in the focal plane of the objective mirror or lens of the telescope. It effectively excludes all light except that from the object under examination. The Fabry lens ensures that the image is correctly positioned on the photo-cathode. Provision must also be made between the pinhole and the Fabry lens for a viewing eyepiece, to ensure that the object is centred in the pinhole, and for a set of standard colour filters. These, together with suitable circuitry for the PMT, are enclosed in a light-proof box, termed the photometer head, which is placed at the eyepiece position of the telescope. The current from the PMT is fed to the measuring equipment via a suitable amplifier. In older photometer systems, a simple meter or a chart-recorder was used to measure the amplified output from the PMT. Nowadays, most photometer systems employ a microcomputer, which, with peripherals, is used to receive, store, process and print the data.

A wide variety of telescopes is used for carrying out photoelectric photometry. It is vital that the telescope and its mounting are sturdy enough to carry the photometer head, which can weigh a couple of kilograms. It is also essential that the object under study is kept in the centre of the pinhole during each observation, which will typically last for 30 seconds or more, so the telescope mounting must be accurately aligned and driven precisely at sidereal rate. Most astronomers use the technique of differential photometry, where the brightness of the object under study is compared with standard stars of constant, known brightness. Regular procedures are carried out whereby the object being studied, the comparison star(s), and the background sky alone are observed in a sequence. The contribution from the background sky is then deducted from the other readings, so that the true difference in brightness between the object and the comparison star(s) may be calculated. The time taken for each observation is typically a few tens of seconds, and usually a whole series of observations is made, and the average then taken. Of course, this may not be possible in the case of very rapidly varying sources. Where appropriate, the standard colour filters are used to obtain observations that can be reduced to one of the standard photometric systems (for example, UBV or UVBY), thus enabling data from one observatory to be accurately related to that from another.

In recent years the use of silicon solid state detectors (such as charge coupled devices – CCDs) has increased the passbands of data regularly determined.

photomultiplier Extremely sensitive electronic device used for measuring the intensity of light radiation. The photomultiplier, or photomultiplier tube (PMT), consists of a photocathode and a series of electrodes, known as dynodes, in a sealed evacuated glass enclosure. Photons striking the photocathode, which is kept at a high negative voltage, eject electrons as a result of the photoelectric effect. These electrons are accelerated towards the dynodes, which are maintained at successively less negative potentials. Additional electrons are released at each dynode, producing a cascading effect and thus amplifying the signal. Photomultipliers are used in many ground-based and space-borne instruments, including PHOTOMETERS, scanning SPECTROPHOTOMETERS and POLARIMETERS.

photon Discrete packet or 'quantum' of ELECTROMAGNETIC RADIATION. As well as having wave-like properties, radiation of frequency f can be regarded as a stream of photons each of energy E given by $E = hf$, where h is the PLANCK CONSTANT. A photon can thus be visualized as an elementary particle with zero rest-mass, zero charge and spin 1, which travels at the speed of light. The photon theory explains several aspects of the behaviour of electromagnetic radiation, including the spectrum of BLACK BODY RADIATION, the photoelectric effect and the COMPTON EFFECT.

photosphere Part of the Sun from which visible light is emitted; more generally, the region of any star that gives rise to its visible radiation. Most of the Sun's energy escapes into space from the photosphere. It is the lowest level of the solar atmosphere, below the CHROMOSPHERE, TRANSITION REGION and CORONA. About 500 km (300 mi) thick, the photosphere caps the convective zone within the SOLAR INTERIOR. The effective temperature of the Sun's photosphere is around 5780 K, varying through its depth from about 9000 K at the base to about 4500 K. The continuum radiation of the photosphere is absorbed at certain wavelengths by slightly cooler gas just above it, producing the dark ABSORPTION LINES (called FRAUNHOFER LINES), which are observed in the solar spectrum. SUNSPOTS, FACULAE and the solar GRANULATION are observed in the photosphere. The in-and-out heaving motions, or oscillations, of the photosphere are used in HELIOSEISMOLOGY to decipher the dynamics and structure of the SOLAR INTERIOR.

photovisual magnitude MAGNITUDE of an object measured photographically using a combination of filters and photographic emulsions that mimic the spectral response of the eye.

Piazzi, Giuseppe (1746–1826) Italian priest and astronomer who in 1801 made the first discovery of an asteroid. He directed the observatory at Palermo (1790–1826), and established the observatory at Naples (1817). On New Year's Eve 1800, he found a moving object and obtained three accurate positions for it. Carl Friedrich GAUSS used these observations and his own method for calculating orbits to determine that Piazzi's new object, later named CERES, orbited the Sun between Mars and Jupiter – where BODE'S LAW had predicted the existence of a planet. Piazzi also compiled a catalogue of 7646 stars (1803–14) and discovered the large proper motion of SIXTY-ONE CYGNI ('Piazzi's Flying Star'), which encouraged Friedrich Wilhelm BESSEL to use that star to make the first measurement of stellar parallax.

Pic du Midi Observatory French optical observatory built between 1878 and 1882 on the 2890-m (9470-ft) Pic du Midi de Bigorre in the southern Pyrenees on an isolated site that enjoyed exceptional atmospheric transparency and stability. Despite its inaccessibility, the observatory's 20th-century achievements include the mapping of planetary surfaces, the determination of Venus' rotation period, preparation for the Apollo Moon landings, and the exploration of the solar corona using the CORONAGRAPH made by Bernard LYOT. Access to the observatory was greatly improved in 1951 by the installation of a *téléphérique* (cable car). The largest telescope now on the site, dating from the 1970s, is the 2-m (79-in.) Bernard Lyot Telescope, enclosed in an unusual near-spherical dome. Now that most French astronomers use larger overseas telescopes, the site has only limited scientific usage and is being redeveloped for public outreach.

Pickering, Edward Charles (1846–1919) American astrophysicist, elder brother of William H. PICKERING, who as director of HARVARD COLLEGE OBSERVATORY (1876–1918) undertook a long-term programme of stellar photometry and spectral classification. These studies produced the *Harvard Photometry*, a catalogue of 4260 bright stars (1884), and the *Harvard Revised Photometry* (1908) for 45,000 stars. Under Pickering's

PICTOR (GEN. PICTORIS, ABBR. PIC)

Modest southern constellation near brilliant Canopus introduced by Lacaille in the 18th century, representing a painter's easel. Its brightest star is α Pic, visual mag. 3.24, distance 99 l.y., spectral type A7. ι Pic, appearing to the eye as mag. 5.2, is an easy double of 6th-magnitude stars for small telescopes. BETA PICTORIS is surrounded by a disk of dust thought to be a protoplanetary disk. Also in Pictor is KAPTEYN'S STAR, a red dwarf with high proper motion.

direction, Harvard astronomers amassed over 250,000 photographic plates of starfields, and he compiled the first *Photographic Map of the Entire Sky* (1903) using photographs taken at Harvard and the observatory's southern station at Arequipa, Peru.

Pickering invented the objective prism, which allowed spectra to be obtained quickly and for many more stars than previously possible. The Harvard collection of spectra was analysed by a large, mostly female staff of assistants, referred to at the time as 'Pickering's women' (or even more unkindly, 'Pickering's harem'), several of whom – including Annie Jump CANNON, Williamina FLEMING and Antonia MAURY – became accomplished astronomers in their own right. The outcome of this work was the *HENRY DRAPER CATALOGUE* (1918), containing the spectral classifications of 225,000 stars. An important extension of this work was Pickering's invention of the B – V COLOUR INDEX for stars, which he realized was related to its temperature. This system was later extended by William W. MORGAN and Harold JOHNSON to measure the magnitude differences for eight colours of starlight (the UBVRIJKL system). Pickering's studies of stellar spectra also resulted in his discovery (1889) of spectroscopic binary stars, the first being Mizar (ζ Ursae Majoris).

Pickering, William Henry (1858–1938) American astronomer who, like his elder brother Edward C. PICKERING, worked at HARVARD COLLEGE OBSERVATORY. In 1891 he organized the observatory's stations at Arequipa, Peru and Mandeville, Jamaica. He discovered Phoebe, the ninth satellite of Saturn (1898), and compiled the first photographic atlas of the Moon to use coordinate grids to identify features (1903). In 1894 he helped Percival LOWELL set up Flagstaff Observatory, and became involved in Lowell's Mars observation programme, and in particular in the quest for a trans-Neptunian planet, for which he published many predictions (1908–28).

▼ **Pinwheel Galaxy** This face-on spiral galaxy, M33 in Triangulum, has loosely wound spiral arms surrounding a relatively small nucleus. M33 belongs to the Local Group, which also includes the Milky Way and the Andromeda Galaxy.

Pickering series Series of EMISSION or ABSORPTION lines in the ionized helium spectrum resulting from electron transitions down to or up from the fourth energy level of that ion (corresponding to the BRACKETT SERIES of hydrogen). Pickering α, which connects levels 4 and 5, falls at 1012.4 nm. In the VISUAL SPECTRUM, Pickering β, which connects levels 4 and 6, is at 656.0 nm (nearly coincident with BALMER α). Pickering γ lies at 541.1 nm, Pickering δ at 485.9 nm (nearly coincident with Balmer β) and Pickering ϵ at 454.2 nm. The series ends at the Pickering limit (nearly coincident with the Balmer limit) at 364.4 nm.

Pico Lunar mountain located in the north-central region of the Moon. Mountains on the Moon are generally produced by impact processes. Mons Pico was formed when an enormous impact created the IMBRIUM multi-ring impact basin. Magma later poured into the basin, covering all but the highest peaks of the inner rings and being restrained by the outer ring. Mons Pico is one of the peaks that marks an inner ring. Other peaks include Montes Teneriffe and Montes Recti.

Pictor *See* feature article

Pinwheel Galaxy (M33, NGC 598) Face-on spiral galaxy in the constellation Triangulum (RA 01^h 33^m.9 dec. +30°39′); it is the third-largest member of the LOCAL GROUP. At magnitude +5.7, M33 is visible to the naked eye under exceptional conditions. Usually, however, its low surface brightness – the light is spread over an apparent area of 67′ × 41′.5 – makes detection more difficult than might be anticipated, even with optical aid. M33's popular name comes from its appearance, with loosely wound spiral arms surrounding a small nucleus. HII regions and star clouds are visible in the spiral arms, and some of these have been assigned their own NGC/IC numbers. M33 has a diameter of 50,000 l.y. and lies at a distance of 2.4 million l.y.

The 8th-magnitude M101 (NGC 5457) in Ursa Major is also sometimes referred to as the Pinwheel Galaxy; like M33, this is a face-on spiral galaxy with a compact nucleus.

Pioneer Series of US spacecraft used for a variety of purposes. They achieved several notable 'firsts'. Pioneer 1 was the first US probe to go anywhere near the Moon. Pioneer 6 was the first to make a direct investigation of a comet. Pioneers 6 and 7 set a 'longevity record' for probes in solar orbit. The PIONEER VENUS Orbiter (sometimes called Pioneer 12) was the first to obtain detailed radar maps of almost the entire surface of Venus.

Perhaps the best known are Pioneers 10 and 11, the first probes to pass through the asteroid belt. Pioneer 10 then became the first spacecraft to obtain close-range data from JUPITER, while Pioneer 11 made the first flyby of SATURN. Both of them are now well beyond the orbit of Pluto and heading out of the Solar System. They carry plaques in case they are ever picked up by some alien civilization.

Pioneer Venus NATIONAL AERONAUTICS AND SPACE ADMINISTRATION (NASA) mission to explore VENUS with an orbiter and four small atmospheric probes. The PIONEER Venus 1 orbiter (also known as Pioneer 12) was launched on 1978 May 20; it entered an elliptical orbit around Venus on December 4. From this 24-hour orbit, the radar mapper was able to observe almost the entire surface of the planet during one Venusian day. Eventually, the orbiter produced the first topographical map of Venus, covering 92% of the surface at a resolution of 50–140 km (30–90 mi). The maps revealed that most of Venus consisted of flat lowland plains, interspersed with continent-like highlands and volcanic mountains. The orbiter also returned ultraviolet images of the cloud tops and confirmed the absence of a magnetic field. The orbiter burned up in the atmosphere in 1992 October.

The Pioneer Venus 2 multiprobe mission (or Pioneer 13) was launched on 1978 August 8. The spacecraft and its four probes entered the planet's atmosphere on December 9. Although none of them was designed to reach the surface, one of them (called the Day Probe) survived for more than an hour after impact. The probes returned temperature and wind-speed profiles of the atmosphere and found that there were three main cloud layers. It was also confirmed that the clouds were made of concentrated sulphuric acid and that the atmosphere was 96% carbon dioxide.

Pipe Nebula DARK NEBULA, the shape of which resembles a smoker's pipe with smoke clouds emerging from the 'bowl' (RA 17^h 33^m dec. $-26°$). It is in Ophiuchus and may connect with the RHO OPHIUCHI DARK CLOUD.

Pisces *See* feature article

Piscis Austrinus *See* feature article

plage Bright, dense region in the CHROMOSPHERE that is at a higher temperature than the surrounding material. Plages are found above SUNSPOTS or other regions of enhanced magnetic field. Plages associated with sunspot groups can outlive the sunspots. Plages are best seen in a SPECTROGRAM or SPECTROHELIOGRAM that isolates the monochromatic light of the HYDROGEN-ALPHA LINE or the H and K lines of ionized calcium. Plages are associated with FACULAE, which occur just below them in the photosphere.

Planck, Max Karl Ernst Ludwig (1858–1947) German physicist who, in 1900, described the intensity of black-body radiation with the formula $E = h\nu$ by asserting that radiation is emitted in discrete quantities called quanta (*see* PHOTON). He thus originated the QUANTUM THEORY, for which he received the 1918 Nobel Prize for Physics.

Planck EUROPEAN SPACE AGENCY (ESA) Horizon 2000 mission to image the COSMIC MICROWAVE BACKGROUND radiation with unprecedented sensitivity and angular resolution. Named after Max Planck, the craft was first called Cobras/Samba and will be launched in 2007 rather than 2003 as originally planned. Budgetary problems resulted in Planck being designed to fly piggyback on another ESA spacecraft, FIRST (see HERSCHEL SPACE OBSERVATORY). Planck will operate at the L_2 Lagrangian point of the Earth–Sun system.

Planck constant (symbol h) Fundamental constant, named after Max PLANCK. It relates the quantum of energy of a PHOTON to its frequency. Its value is 6.626176×10^{-34} Js.

Planck distribution Bell-shaped distribution of energy radiated from a BLACK BODY. It is given by Planck's equation:

$$I_\lambda = \frac{2hc^2\mu^2}{\lambda^5(e^{hc/\lambda kT} - 1)}$$

where I_λ is the intensity at wavelength λ per unit wavelength interval, μ is the refractive index of the medium (normally unity), k is BOLTZMANN'S CONSTANT, and T is the temperature. The peak of the distribution shifts to shorter wavelengths as the temperature increases (*see* WIEN'S LAW). This leads to the common experience that at moderate temperatures objects glow a dull red; they then change colour successively through bright red, yellow, white to blue-white as the temperature is increased. The total emitted energy, F, increases rapidly with temperature, leading to the STEFAN–BOLTZMANN LAW.

At wavelengths significantly longer than that of the peak emission, the simpler Rayleigh–Jeans approximation (named after the British physicist Lord Rayleigh (1842–1919) and Sir James JEANS) may often be used in place of the Planck equation:

$$I_\lambda \approx \frac{2kc\mu^2T}{\lambda^4}$$

At wavelengths shorter than the peak, the Wien approximation may be used:

$$I_\lambda \approx \frac{2hc^2\mu^2}{\lambda^5 e^{hc/\lambda kT}}$$

Planck era *See* PLANCK TIME

Planck scale Distance light can travel during one PLANCK TIME. This works out to be about 3×10^{-35} m.

Planck time Timescale within which the known laws of physics break down. The time between 0 and 10^{-43} seconds roughly defines the **Planck era**. A QUANTUM THEORY OF GRAVITY is necessary to discuss the state of the BIG BANG universe at times earlier than 10^{-43} seconds after the initial event.

planet Large, non-stellar body orbiting the Sun or another star and shining only by reflected light. Planets

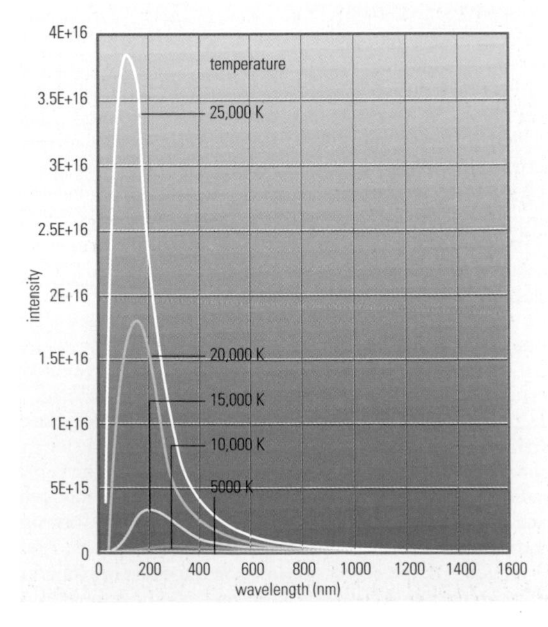

◀ **Planck distribution** The distribution of energy radiated by a black body shifts to shorter wavelengths at higher temperatures. Hence, for a hot star (25,000 K) most of the radiation is at short wavelengths, and it will appear white. A cooler star will emit mainly at yellow-red wavelengths and appear yellow or red.

PISCES (GEN. PISCIUM, ABBR. PSC)

Zodiacal constellation, not particularly prominent, extending north from the celestial equator between Aquarius and Aries. It represents two fish tied together by their tails. α Psc (ALRESCHA), mag. 3.82, marks the knot in the cords joining the fish. A Greek legend identifies these fish as Aphrodite and her son Eros. Alrescha is not the brightest star in Pisces – that honour goes to η Psc, mag. 3.62. Its most distinctive feature is a ring of seven stars of 4th and 5th magnitudes known as the **Circlet**. One of these stars, TX Psc, is an erratically variable red giant, range 4.8–5.2; another, κ Psc, is a wide binocular double, mags. 4.9 and 6.3. ζ and ψ^1 Psc are both easy doubles for small telescopes. TV Psc is a semiregular variable, range 4.7–5.4, with a period of about 50 days. M74 is a beautiful face-on spiral galaxy. Pisces contains the VERNAL EQUINOX, the point where the Sun crosses the celestial equator from south to north in March each year.

PISCIS AUSTRINUS (GEN. PISCIS AUSTRINI, ABBR. PSA)

Southern constellation, known since ancient times, representing a large fish at the southern end of the stream of water flowing from the urn of Aquarius. Its only star above mag. 4 is first-magnitude FOMALHAUT, which marks the mouth of the fish. β PsA is a wide double for small telescopes.

P

▲ **planetary nebula** A Hubble Space Telescope image of the Ant Nebula (Mz 3), a planetary nebula in the southern constellation Norma. The twin-lobed structure has been produced by ejection of gas from the aged star at the nebula's centre. The Ant Nebula lies about 3000 l.y. away.

may be either rocky in composition, such as Mercury, Venus, Earth and Mars, or mainly gaseous, such as Jupiter, Saturn, Uranus and Neptune. Nine major planets exist in our Solar System, together with thousands of minor planets or asteroids. The name is Greek in origin, meaning 'wandering star'. *See also* PLANET X

planetar *See* INTERSTELLAR PLANET

planetarium Hemispherical domed theatre housing a centrally mounted projector that can simulate the appearance of the constellations in the night sky at any time of the year and from any point on the surface of the Earth. There are now some 2500 planetariums around the world, visited each year by an estimated 92 million people. Planetariums provide a vital link to the general public about new developments in astronomy, and many of them act as regional resource centres for local news and informed interpretation of results from the research community.

Precursors of the planetarium were the moving celestial globe, and the mechanical orreries of the 18th century (some of which were known as 'planetaria'). From the mid-17th century were built hollow globes 4 to 5 m (13–16 ft) in diameter, from inside which a small audience could view stars as small holes in the globe. Walther Bauersfeld (1879–1959) built the first modern planetarium star projector while working for the firm of Carl Zeiss in Germany. The 'Model I' was completed in 1923 and demonstrated that year at the Deutsches Museum, Munich. The Model II was the first to use the familiar dumbbell configuration. The first public theatres were built in Vienna, Rome and Moscow. The 1930 World's Fair in Chicago saw the Adler Planetarium open its doors; it remains one of the world's leading planetariums. The London Planetarium, completed in 1958, was the first to have a concrete dome. Armand Spitz (1904–71) became the champion of the small planetarium in the United States when, during the 1960s race to the Moon, he installed hundreds of small planetariums. In Japan, the optical manufacturers Seizo Goto and Minolta installed hundreds of planetariums around the country. During the early 1980s Evans & Sutherland developed a digital star projector.

There are two general types of planetarium projector, opto-mechanical and digital. An opto-mechanical projector houses 'star-plates' in one of two spheres, one northern and one southern hemisphere. A lamp or a cluster of fibre optics feeds light through the holes in the star plates; the light is projected on to the reflective inner surface of the planetarium dome, thus recreating an accurate representation of the night sky. The two spheres holding the star plates are mounted at opposite ends of a cage housing individual planet projectors. Using a complex gear arrangement, the correct position and motion of planets at any time in the future or in the past can be recreated on the planetarium dome. The use of the fixed star plates in opto-mechanical projectors gives an inherently Earth-centred view of the sky, which is fine for most purposes.

A digital star projector holds a stellar database in computer memory and includes distances to the nearest few thousand stars. Star points are generated by an electron gun similar to those found in cathode ray tubes. A large phosphor plate displays the image, which is projected on to the dome via a large fisheye lens mounted above it. The projector can display the constellations as seen from Earth, or move to other stars to view the sky from different vantage points. In addition to constellations, the projector can create many geometrical shapes for projection on the dome. Digital projectors are more versatile than their opto-mechanical counterparts, but the star images they produce are not as sharp.

Planetariums come in many sizes, ranging from 3 to 23 m (10–75 ft) in diameter. Small inflatable domes with portable star projectors enable small communities to experience a simple planetarium demonstration. Some schools and colleges house small permanent planetariums. Larger facilities, often associated with public science centres, provide a rich variety of experiences and can be found in many major cities around the world.

The majority of modern planetariums incorporate video projection to insert moving computer graphics within a starfield, significantly enhancing the visual effect. Many traditional planetarium projectors are being replaced with digital projectors. Planetariums used to be places to view the night sky as seen from Earth; many can now immerse their audiences in a universe of three-dimensional realism. Faster computer power and real-time image generation enable elaborate computer models of galaxies, star clusters and nebulae to be combined with superb modern images from large telescopes.

planetary nebula EMISSION NEBULA associated with and deriving from a star, in principle having a disk-like appearance, similar to the telescopic view of a planet. About 1500 planetary nebulae are catalogued, perhaps one-tenth of the number in our Galaxy. Their radii range from that of our own Solar System to a few light-years. At the small end of the range the nebulae may appear point-like on direct images, and their true natures have to be inferred from their EMISSION LINE spectra.

William HERSCHEL coined the name in 1785 from the resemblance of some of the nebulae to planets (Herschel had discovered Uranus just four years earlier). Herschel thought at first that planetary nebulae were unresolved groups of stars like distant globular clusters, but in 1790 he observed NGC1514, a planetary nebula that looked like a 'star of about the eighth magnitude, with a faint luminous atmosphere, of a circular form .... The star is perfectly at the centre....' The particularly simple form of the nebula, and the strikingly central position of the very bright star, so much brighter than others nearby, convinced Herschel that true nebulosity existed, which, no matter how powerful his telescopes, could never be resolved into stars.

The central stars of planetary nebulae are extremely hot, their temperatures can range from 20,000 K to 100,000 K or more. Their brightnesses, however, are relatively low for such hot stars – just 0.1 to 100 solar luminosities – so they must, therefore, be physically small. In fact the stars are thought to be contracting down to become WHITE DWARFS. The material forming the nebula has been ejected from the star in the recent past, when it was a RED GIANT, and is expanding outwards at velocities of up to 60 km/s (40 mi/s). In some cases the central star subsequently develops a high velocity stellar wind that sweeps the central volume of the nebula clear of material and compresses it into a spherical shell. Projected against the sky, such a nebula can then appear similar to a smoke ring, like M57, the RING NEBULA.

Only some 10% of planetary nebulae have a circular shape. About 70% have a bipolar structure and two lobes. In other cases magnetic fields, rotation, binarity and/or the presence of other gaseous material can lead to far more complex shapes. There are several classification schemes for planetary nebulae. One of the more widely used ones employs six main classes labelled by Roman numerals, as follows:

I star-like
II smooth disk, subdivided into (a) brighter towards the centre, (b) uniform and (c) some ring structure.
III irregular disk, subdivided into (a) very irregular and (b) some ring structure.
IV ring
V irregular
VI anomalous

A planetary nebula shines because of ultraviolet light emitted by the hot central star. The atoms of the nebular gas can be ionized by ultraviolet photons of sufficient energy. Hydrogen atoms, for example, can be ionized by photons with a wavelength of 91.2 nm or shorter. When the ion recombines with an electron it emits photons of light from a structured ladder of energy steps. Each ultraviolet photon input into the nebula produces a spectrum of photons out.

Each central star has enough light output to ionize a certain volume of planetary nebula. There may be gas beyond the visible boundary of the nebula, but ultraviolet photons do not reach far enough to make it visible. In fact within each planetary nebula numerous sub-nebulae exist, each corresponding to the ionization of a given type of atom or ion. Thus pictures taken in different optical spectral lines from different ions show stratification – the images of the planetary nebula are nested like a set of Russian dolls made of coloured glass. The Ring Nebula (M57) is a good example of this: the blue, green and red images are successively larger because each is dominated by a particular spectral emission from a different atom or ion.

All the planetary nebulae are expanding. This can be determined from two techniques. The DOPPLER SHIFT of the optical spectral lines provides a measurement of the expansion velocity along the line of sight: values of 20 km/s (12 mi/s) are typical. The increase in size of the image of a planetary nebula on photographs taken over the years shows its expansion across the line of sight. The expansion in the radial and tangential directions (if assumed to be the same) can be combined to yield the distance of the planetary nebula. The fact that the nebulae are expanding as the central stars fade means that planetary nebulae are relatively transitory objects, with lifetimes measured in tens of thousands of years. As there are approximately 10,000 planetary nebulae in our Galaxy, several must form each year. *See also* BUTTERFLY NEBULA; DUMBBELL NEBULA; ESKIMO NEBULA; HELIX NEBULA; OWL NEBULA; SATURN NEBULA

planetary nomenclature Scheme used to assign names to surface features of planets and satellites in order to be able to identify them uniquely. Planetary nomenclature has been the responsibility of the International Astronomical Union (IAU) since 1919, when a committee was appointed to standardize the then chaotic lunar and Martian nomenclature. This nomenclature had developed since the 17th century, when the first selenographers had each devised their own systems for naming lunar features, a practice continued in the 19th century when the first maps of Mars were made. As a result of the IAU's work, the first systematic nomenclature for the Moon was published in 1935, and a standard list of light, dark and coloured Martian features followed in 1958.

The procedure assumed even greater importance with the dawning of the space age and the mapping of increasing numbers of planets and their satellites by space probes from the 1960s onwards. In 1973 the IAU established the present Working Group for Planetary System Nomenclature (WGPSN).

The system that is used assigns a binomial name to a feature, comprising a 'descriptor' and a specific name. The descriptor is a Latin or Greek generic term for the type of feature, such as Mons for mountain or volcano and Fossa for a long, narrow depression or ditch. The specific name may be that of a famous person, a mythical or literary character or a terrestrial place name. On each body, a particular theme is chosen for the specific names. For example, surface features on the moons of Uranus take their theme from the name of the parent satellite, which itself was chosen to be a character from the plays of William Shakespeare. Thus, on Oberon, features are named after tragic Shakespearean heroes, such as Macbeth and Hamlet, while those on Titania are named after female characters, for example the craters Elinor and Ursula.

When the first images of a body are obtained, an appropriate theme is chosen and the most prominent features assigned names, usually chosen by those working closely on the project. Once more detailed work is undertaken, anyone can propose a suitable name for consideration. The names of political or military leaders are prohibited while other guidelines ensure that names should be simple, clear and unambiguous, that the system should be international in its choice of names and that, where possible, themes established in the past should be expanded on.

planetary precession Small component of general PRECESSION caused by the GRAVITATIONAL FORCE exerted on the Earth by other planets. Precession causes the Earth's rotational axis slowly to describe a circle in the sky of radius 23°.5 over a period of some 25,800 years, the effect only being noticeable over a few decades. The

DESCRIPTORS FOR PLANETARY NOMENCLATURE

descriptors	plural	description
Astrum	Astra	radial-patterned feature on Venus
Catena	Catenae	chain of craters
Cavus	Cavi	hollow, irregular, steep-sided depression
Chaos	—	distinctive area of broken terrain
Chasma	Chasmata	deep, elongated, steep-sided depression
Colles	—	small hills or knobs
Corona	Coronae	ovoid feature
Crater	Craters	a circular depression
Dorsum	Dorsa	ridge
Facula	Faculae	bright spot
Farrum	Farra	pancake-like structure, or a row of such structures
Flexus	Flex\-us	very low curvilinear ridge with a scalloped pattern
Fluctus	Fluct\-us	flow terrain
Fossa	Fossae	long, narrow, shallow depression
Labes	Lab\-es	landslide
Labyrinthus	—	complex of intersecting valleys
Lacus	Lac\-us	small plain on the Moon
Lenticula	Lenticulae	small dark spot on Europa
Linea	Lineae	dark or bright elongate marking, may be curved or straight
Macula	Maculae	dark spot, may be irregular
Mare	Maria	large circular plain on the Moon
Mensa	Mensae	flat-topped prominence with cliff-like edges
Mons	Montes	mountain
Oceanus	—	very large dark area on the Moon
Palus	Pal\-us	small plain on the Moon
Patera	Paterae	irregular crater, or a complex one with scalloped edges
Planitia	Planitiae	low plain
Planum	Plana	plateau or high plain
Promontorium	Promontoria	headland on the Moon
Regio	Regiones	large area marked by reflectivity or colour distinctions from adjacent areas
Reticulum	Reticular	reticular (net-like) pattern on Venus
Rima	Rimae	fissure
Rupes	Rup\-es	scarp
Scopulus	Scopuli	lobate or irregular scarp
Sinus	Sin\-us	bay
Sulcus	Sulci	subparallel grooves and ridges
Terra	Terrae	extensive land mass
Tessera	Tesserae	tile-like, polygonal terrain
Tholus	Tholi	small domed mountain or hill
—	Undae	dunes
Vallis	Valles	valley
Vastitas	Vastitates	extensive plain

P

major cause of precession is the combined gravitational pull of the Sun and the Moon on the Earth (*see* LUNISOLAR PRECESSION), but the other planets in the Solar System also exert a small force, albeit in the opposite direction. The sum of planetary and lunisolar precession is called general precession.

Planetary Society International society promoting planetary exploration. Based in California, the Planetary Society was founded in 1980 by Carl SAGAN, Bruce Murray and Louis Friedman. It continues to fund novel research and to support public education in planetary science.

planetesimal Solid body that orbits a young star; it is one of the 'building blocks' in a PROTOPLANETARY DISK from which planets form by ACCRETION. The term is generally applied to bodies large enough for their motions to be controlled by gravitational forces rather than by the drag of gas in the disk; they are roughly of kilometre size or larger. Planetesimals form by coagulation of grains of metal and silicates (and ices in the colder outer part of the disk). ASTEROIDS and COMETS are remnant planetesimals that avoided collisions with planets and survived, albeit with some metamorphism and collisional processing, to the present time. *See also* COSMOGONY

Planet X Name given to a hypothetical tenth major planet, once believed to exist in the outer Solar System beyond the orbit of Pluto.

In the late 19th century, Percival Lowell, William H. Pickering and others calculated orbits for a large planet that they believed was responsible for gravitational perturbations to the orbit of Uranus unattributable to Neptune. Lowell called this unknown body 'Planet X'.

In 1930 Clyde Tombaugh discovered PLUTO, but although its position was only five degrees from that predicted by Lowell and its orbit was similar, it soon became obvious that the newly discovered planet was too small to be Planet X. Despite a continued search by Tombaugh, during which he surveyed the entire sky, no further candidate was found.

Over the following years a number of different orbital predictions were made for a tenth planet, the gravitational influence of which would explain the deflection of comets from the outer Solar System towards the Sun. However, apart from the discovery by Charles Kowal (1940–) in 1977 of CHIRON, which orbits between Saturn and Uranus and caused some brief excitement, nothing was discovered. Further searches for a solution to the discrepancies in the orbits of Uranus and Neptune also failed to reveal any new planets.

In 1989 the VOYAGER 2 spacecraft established a revised mass for Neptune and this, together with the discovery that the orbital anomalies of Uranus and Neptune could be the result of observational errors, removed the necessity for the existence of a tenth planet. This was reinforced by the discovery in 1992 of the EDGEWORTH–KUIPER BELT objects beyond the orbit of Neptune, none of which are planet-sized. The members of the Edgeworth–Kuiper belt are believed to be remnants of the original Solar System material, and some experts argue that Pluto and its satellite Charon really belong in this category.

On 2000 November 28, a large minor planet was discovered in the Edgeworth–Kuiper belt at a distance of 43 AU from the Sun. Its apparent magnitude of 20 made it the brightest-known trans-Neptunian object other than Pluto. Initially known as minor planet (20000) 2000 WR106, the object has since been named VARUNA by the International Astronomical Union. Subsequent observations have revealed that it is dark compared to Pluto, indicating little surface frost, and about half its diameter, making it also the largest as well as the brightest of the trans-Neptunian population discovered so far, after Pluto.

planisphere Easy-to-use, portable two-dimensional map of the CELESTIAL SPHERE, as seen from a particular latitude, showing which stars are visible at a given date and time. The planisphere comprises two disks, one a star map and the other an overlay containing an oval window. The two are connected at the centre and may be rotated relative to one another in order to select the desired date and time, which are displayed around their outer edges. The oval window in the overlay disk will then show those stars that are currently above the horizon.

Plaskett, John Stanley (1865–1941) Canadian astronomer who began his career as a mechanic at the University of Toronto, eventually becoming the first director (1917) of the DOMINION ASTROPHYSICAL OBSERVATORY (DAO). He was an expert instrument builder who won fame for a high-resolution spectrograph he made for the DOMINION OBSERVATORY's 15-inch (380-mm) refractor and for his design of DAO's 72-inch (1.8-m) reflecting telescope, the largest in Canada. He used the spectrographs he built to study stellar radial velocities, investigating spectroscopic binary stars and hot, young O and B stars; PLASKETT'S STAR is a massive spectroscopic binary system he discovered. His radial velocity studies helped to confirm the model of the Milky Way proposed by Jan OORT and Bertil LINDBLAD. His son, **Harry Hemley Plaskett** (1893–1980), was an accomplished solar spectroscopist who worked mostly in England.

Plaskett's Star (HD47129, V640 Mon) One of the most massive BINARY STARS known; it is located in the constellation MONOCEROS. In 1922 the Canadian astronomer John PLASKETT showed it to be a SPECTROSCOPIC BINARY with a period of around 14 days. The system comprises two blue SUPERGIANTS of SPECTRAL CLASS O. Each component has an estimated mass of around 50 solar masses.

plasma Almost completely ionized gas, containing equal numbers of free electrons and positive ions, moving independently of each other. Plasmas such as those forming the atmospheres of stars, or regions in gas discharge tubes, are highly electrically conducting but electrically neutral, since they contain equal numbers of positive and negative charges. The temperature of a plasma is usually very high. The electrical currents flowing within a plasma mean that it carries with it an intrinsic magnetic field (*see* MAGNETOHYDRODYNAMICS). A plasma has been described as 'the fourth state of matter', the others being solids, liquids and gases.

plasma sheet Region of hot plasma generally observed in the central regions of the extended magnetotail of a planetary MAGNETOSPHERE. The inner edge of the plasma sheet may extend to the dayside at high altitudes, just inside the MAGNETOPAUSE. The near-Earth plasma sheet plays a critical role in the dissipation of stored magnetotail energy during MAGNETOSPHERIC SUBSTORMS. The deposition of this energy into the polar IONOSPHERE drives spectacular auroral displays in these regions. During substorms, large sections of the distant plasma sheet are also sheared off. The resulting plasmoid is ejected at high speeds down the tail and into the SOLAR WIND.

plasmasphere Region of cold, dense plasma that co-exists in approximately the same region of space as the RADIATION BELTS within a planetary MAGNETOSPHERE. It is populated by the polar wind, an upflow of plasma from the upper ionosphere. In the terrestrial magnetosphere, the outer boundary, known as the plasmapause, is observed as a sharp boundary at altitudes of between 3 and 5 Earth radii, depending on the level of geomagnetic activity; the plasmasphere co-rotates with the Earth below.

plate-measuring machine Dedicated computer-controlled measuring machine used for accurately recording the positions and brightness of images on astronomical photographic plates. A typical plate-measuring machine consists of a solid but moveable base, known as an *x–y* table, on which the plate to be measured is mounted. The photographic plate is scanned in a series of strips by shining a beam from a light source through the plate and allowing the light transmitted to be collected by a detector such as a CCD. To record each strip, the plate is moved under the CCD in one direction (the *y* direction) by the *x–y* table. When each strip is complete, the table is stepped in the *x* direction by a single scan width ready for the next strip to be scanned. The entire operation is carried out in an environmentally controlled chamber providing clean-room conditions and strict thermal stability during the scanning process.

Plate-measuring machines, such as the Automatic Plate Measuring facility at Cambridge and the SuperCOSMOS machine at Edinburgh, are used for the systematic digitization of photographic plates from major sky surveys. These have been undertaken over many years using telescopes such as the United Kingdom Schmidt Telescope (southern sky) and the Palomar Schmidt (northern sky).

Photographic plates are also often scanned to try to correlate optical images with objects newly discovered at other wavelengths, such as X-rays. This can be important when attempting follow-up observations. Other work involves measuring the positions of radio sources in order to link the radio and optical reference frames. *See also* MICRODENSITOMETER

plate scale *See* IMAGE SCALE

Plato Lunar crater (51°N 9°W), 97 km (60 mi) in diameter. Plato lies in a ring of the Mare IMBRIUM. It is a complex crater that originally contained side-wall terracing and central peaks. Later, however, lava flooded the crater, covering the central peaks and rising high enough to cover most of the side-walls. On the western inner wall, a slumped block is visible, as are a few small craterlets in the lava floor.

Platonic year Period of around 25,800 years that it takes the celestial pole to describe a complete circle around the pole of the ecliptic, due to the effects of PRECESSION. It is named after the Greek philosopher Plato (*c*.427–*c*.347 BC).

Pleiades (Seven Sisters) Young OPEN CLUSTER in the constellation TAURUS. It has an apparent diameter of 1°50′, which corresponds to an actual diameter of 7 l.y. It gets its popular name from the fact that seven of its component stars, which are termed Pleiads, could once be seen easily with the unaided eye on a clear night. These seven stars are named after Atlas and his daughters: the brightest is called Alcyone (η Tauri), and the others are Maia, Atlas, Electra, Merope, Taygete and Pleione. Today it is easy to see six of the stars; Pleione is variable and may once have been brighter. It is thought to be unstable, because of its fast rate of rotation, and is throwing off shells of gas. The cluster probably contains as many as 3000 stars in total.

The brightest stars in the Pleiades are highly luminous blue-white stars (B or Be stars). The less brilliant ones are mainly A and F stars. In 1995 a brown dwarf was detected within the Pleiades. Designated PPl 15, it has a surface temperature of around 2000 K. Since its discovery, other brown dwarfs in the cluster, including Teide 1, have been observed. Gas and dust surround the brighter stars and reflect the starlight, thus producing faint REFLECTION NEBULAE around the stars. The Pleiades is thought to be about 410 l.y. away from us. It is around 76 million years old.

plerion (filled-in supernova remnant) Rare type of SUPERNOVA REMNANT (SNR), in which radiation is emitted from the whole area of the nebula, not just from the outer rim. The name derives from *pleres*, the Greek for 'full'. In addition to their filled-in appearance, plerions have no shell structure and emit strongly polarized SYNCHROTRON RADIATION, which gives them a characteristic green-yellow glow if they can be detected in the visible. In the radio region their spectra are flat. It is possible that plerions may divide into two subgroups on the basis of the ratio of their X-ray to radio fluxes. The best-known example of a plerion is the CRAB NEBULA; others include the radio source 3C 58 and the SNR in the nebula IC 443. The VELA SUPERNOVA REMNANT appears to be a hybrid between a plerion and more conventional SNRs.

The cause of the difference between plerions and other SNRs appears to be that the former have a continuing source of high-energy electrons at their centres in the form of a pulsar. The electrons spiral around the complex magnetic fields within the SNR and emit synchrotron radiation. The emission of the radiation reduces the electrons' energies, and so the pulsar must continually supply them in order for the nebula to maintain its emission.

The dearth of plerion-type SNRs indicates either that they result from unusual types of supernovae, or that they are only a brief phase during the life of a conventional SNR.

▲ **Pleiades** The hot, young blue stars of the Pleiades Cluster illuminate a shroud of interstellar dust. Marking the shoulder of Taurus, the Bull, the Pleiades is a distinctive naked-eye object, prominent on winter evenings in the northern hemisphere.

Plössl eyepiece Telescope eyepiece with good EYE RELIEF and wide field of view (acceptance angle about 40°). The commonest form has two identical achromatic doublets (*see* ACHROMAT), but some quite different designs have been called 'Plössls'. The eyepiece is named after the Austrian optician Simon Plössl (1794–1868).

Plough Popular (chiefly UK) name for the shape formed by the seven main stars of Ursa Major: α (Dubhe), β (Merak), γ (Phecda), δ (Megrez), ε (Alioth), ζ (Mizar), and η (Alkaid). The name comes from the group's resemblance to an old horse-drawn plough. In the USA these stars are usually known as the **Big Dipper**.

plume Rising, thermally buoyant material in the MANTLE of a TERRESTRIAL PLANET or large satellite. A plume pushes up LITHOSPHERE, causing uplifting of the planet surface. The upwelling plume material melts rock, and some of this magma may reach the surface and erupt as volcanic lava. Most plume material solidifies within the planet mantle and crust.

plutino TRANS-NEPTUNIAN OBJECT that happens to have an orbital period 50% greater than that of Neptune, so that it occupies the 3:2 mean-motion resonance, as does PLUTO itself.

P

PLUTO: DATA	
Globe	
diameter	2300 km
density	2.0 g/cm^3
mass (Earth = 1)	0.0022
sidereal period of axial rotation	6.3872 days
escape velocity	1.1 km/s
surface gravity (Earth = 1)	0.059
albedo	0.55
inclination of equator to orbit	119°.6
surface temperature (average)	45 K
Orbit	
semimajor axis	39.54 AU = 5914 × 10^6 km
eccentricity	0.249
inclination to ecliptic	17°.1
sidereal period of revolution	248.02 years
mean orbital velocity	4.74 km/s
satellites	1

▲ **Pluto** Pluto (left) and its satellite Charon (right) are sufficiently similar in size to be regarded as a 'double planet'. They are locked in synchronous rotation about the barycentre between them.

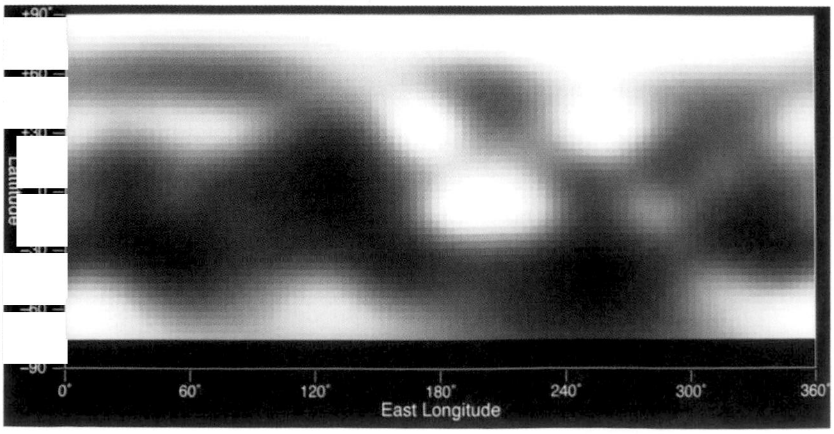

▼ **Pluto** This albedo map of Pluto was derived from occultation observations in the late 1980s, when Charon and Pluto alternately passed in front of one another in our line of sight. The precise nature of the light and dark areas remains uncertain.

Pluto Ninth and smallest planet in the Solar System. It is slightly smaller than Neptune's largest satellite, TRITON, with which it may have much in common. The realization that Pluto is much smaller than originally assumed, together with the fact that it shares many properties with EDGEWORTH–KUIPER BELT objects, led to speculation that it would be stripped of its planet status. However, the INTERNATIONAL ASTRONOMICAL UNION has decided that Pluto will continue to be classified as a planet.

Pluto, like Neptune, was discovered as the result of a deliberate search. Towards the close of the 19th century, it was suggested that minor deviations in the orbits of Uranus and Neptune were being caused by gravitational perturbations by a large outlying planet, which became known as PLANET X. In the early years of the 20th century, Percival LOWELL and others made repeated attempts to locate it. Lowell died in 1916, and in 1929 Clyde TOMBAUGH was recruited by the Lowell Observatory in Flagstaff, Arizona, to continue the search. Tombaugh used a BLINK COMPARATOR to reveal the motion of any planet relative to the background stars on pairs of photographic plates taken a few nights apart. On plates taken on 23 and 29 January 1930, Tombaugh found a faint (magnitude 14) object within a few degrees of the predicted position, which was moving by the amount expected of a trans-Neptunian planet. The newly discovered object was named Pluto after the classical god of the underworld.

Pluto was too distant and too faint for much to be learned about it during the next half-century. Even its size was a mystery. Before it had been found, Pluto was assumed to be about ten Earth-masses, but upon discovery its faintness caused its estimated mass to drop tenfold. The 1978 discovery of Pluto's satellite, CHARON, enabled Pluto's mass to be determined as only about 0.25% of the Earth's. Thus, Pluto is not massive enough to be responsible for the apparent deviations in the orbits of Uranus and Neptune, which are now known not to be real, but to have stemmed from tiny errors in 19th-century observations. Finding Pluto near the expected position for a massive trans-Neptunian planet was essentially a matter of luck. The post-1992 documentation of hordes of Edgeworth–Kuiper belt objects orbiting between 30 and 50 AU means that it is quite impossible for an undiscovered major planet to be in the same region.

Pluto's orbit is more eccentric than that of any of the major planets. When Pluto is near perihelion it is closer to the Sun than Neptune, as was the case between 1979 September 5 and 1999 February 11. Perihelion occurred on 1989 September 5, when Pluto was only 29.66 AU from the Sun. Its distance from the Sun will now increase until it reaches aphelion at 49.54 AU in 2114. Pluto's orbital period is exactly 50% longer than Neptune's, so the two are in 3:2 orbital RESONANCE. Although Pluto's orbit crosses Neptune's, the 17° inclination of Pluto's orbit means that their paths do not intersect. Pluto's inclined orbit takes it 8 AU north of the ecliptic at perihelion, and 13 AU south of it at aphelion. Its orbital resonance with Neptune means that the distance between the two bodies is never less than 17 AU.

Like Uranus, Pluto lies on its side, having an axial inclination of nearly 120° and, therefore, RETROGRADE rotation. This did not become fully apparent until after the discovery of Charon, which orbits in Pluto's equatorial plane. Imaging by modern telescopes has revealed that not only does Charon keep the same face to Pluto (being in SYNCHRONOUS ROTATION), but also Pluto itself always keeps the same face towards Charon. The two bodies are mutually tidally locked, so that Pluto's rotation keeps pace with Charon's orbital motion about it.

Pluto and Charon are probably a product of a giant impact between two Edgeworth–Kuiper belt objects. They are the nearest thing the Solar System can boast to a double planet. This honour was formerly thought to belong to the Earth and Moon. However, Charon's 19,636 km (12,202 mi) orbital radius is only about sixteen times the radius of its planet, whereas the Moon's orbital radius is about sixty times the Earth's radius. Moreover, Charon has 12% the mass of Pluto but the Moon has only 1.2% the mass of the Earth. Strictly speaking, no satellite orbits around the centre of its planet, but instead about the system's centre of mass (or barycentre). For all other planet-satellite systems this point lies inside the planet. In the case of Pluto–Charon the barycentre lies between the two bodies in open space, but closer to Pluto because it is the more massive of the two. Repeated searches for smaller satellites of Pluto have failed to find any, and if Pluto does have any other companions, they must be less than about 100 km (60 mi) across.

There is no direct information on the interiors of Pluto or Charon. However, the melting that would have been caused by a giant impact and the heating that must have occurred while tidal forces were bringing Charon into a near-circular synchronous orbit mean that both bodies are very likely to have become internally differentiated (*see* DIFFERENTIATION). Pluto's density suggests that below its icy mantle there is a rocky core containing about 70% of the planet's mass, with possibly an iron-rich inner core.

Thanks to modern telescopic and image-processing techniques there is some knowledge of Pluto's surface markings, even though it has never been visited by a spacecraft. Pluto's surface is bright but with a surprising degree of contrast between its brightest regions, where the albedo is as high as 0.7, and its darkest spots, where albedo drops to about 0.15. Spectroscopy has revealed that the surface consists of frozen nitrogen (presumably concentrated as frost in the brightest regions), methane and carbon dioxide. There are also traces of ethane apparently dissolved within the nitrogen ice. Because the dark areas absorb more solar heat than the bright areas, they have a temperature about 20 K higher, and they are warmer than anywhere on Triton (at least while Pluto is near perihelion). The bulk of Pluto's ice is likely to be water ice, but this has not been detected spectroscopically. Presumably it lies buried, even more completely than is the case on Triton, beneath the more volatile ices.

The presence of Pluto's atmosphere was confirmed when the planet passed in front of a star in 1988. Nitrogen, the most volatile of the surface ices, probably makes up the bulk of the atmosphere, with methane, carbon monoxide and ethane also present. The temperature contrast between the dark and bright regions is likely to drive ferocious near-surface winds. There is perhaps a tenuous low-altitude haze layer, which could be a photochemically induced smog like those on Titan and Triton, caused by such substances as hydrogen cyanide, acetylene and ethane. Although Pluto's atmosphere is insubstantial, the planet's feeble gravitational hold means that it is particularly extensive. For example, an imaginary shell enclosing 99% of Pluto's atmosphere would be about 300 km (200 mi) above the surface, whereas for the Earth the equivalent height is only 40 km (25 mi).

Near perihelion, Pluto's atmospheric pressure was comparable to that of Triton. Because of the eccentricity of Pluto's orbit, however, it receives nearly three times as much solar warmth at perihelion than it does at aphelion. As it moves farther from the Sun (and, coincidentally, one pole moves into season-long shadow), much of Pluto's present atmosphere is vulnerable to becoming frozen out on to the surface.

Pluto Kuiper Express Mission planned by NASA to be launched 2004 December on a 12-year flight to the planet Pluto, using the final opportunity of gravity-assisted flybys en route. The craft would fly past Pluto at about 15,000 km (9400 mi) and take high-resolution images of it and its moon, Charon. The spacecraft would also explore some of the Solar System's most distant bodies in the Edgeworth–Kuiper belt. If the craft reaches Pluto in 2016, the exploration of all the planets of the Solar System will have been completed within a timespan of 54 years, since Mariner 2 and Venus in 1962 December. Pluto Kuiper Express will study the geology, geomorphology and potential atmospheres of both Pluto and Charon.

Pogson, Norman Robert (1829–91) English astronomer who promoted a logarithmic scale of stellar magnitudes. Observing from the private observatories of John Lee (1783–1866) and George Bishop (1785–1861), he discovered twenty new variable stars; he later worked at Oxford University's Radcliffe Observatory (1852–60), and then as director of Madras Observatory. In the course of analysing historical light-curves for all known variable stars, he devised a way to make the magnitude scale mathematically rigorous (1856). Comparing the brightness ratios in the different magnitude scales then in use, he noticed that their average value was very nearly $100^{1/5}$, or 2.512, the ratio – known as **Pogson's ratio** – between successive whole numbers on the modern magnitude scale. Pogson also discovered eight asteroids.

Pogson step method Visual method of estimating the brightness of a VARIABLE STAR relative to fixed-magnitude comparisons. It is based upon training the eye to recognize steps of 0.1 magnitude and thus requires considerable experience. For example, if there are two comparison stars A (magnitude 7.0) and B (7.6) and the variable is midway between them, its magnitude will then be 7.3. If it is estimated as 0.1 magnitude below A and 0.5 above B, then its magnitude will be 7.1, and so on. The method is generally less satisfactory than either the ARGELANDER STEP METHOD or the FRACTIONAL METHOD, but may be of practical use when a comparison sequence has yet to be established (as in the case of a nova, for example), when a sequence has wide magnitude gaps between suitable comparison stars, or when a magnitude must be extrapolated outside an existing sequence.

Poincaré, (Jules) Henri (1854–1912) French mathematician who studied celestial mechanics, especially the three-body problem. He showed that the MANY-BODY PROBLEM can never be solved exactly, and in the process discovered chaotic orbits – ones that lack long-term stability. His major works included *Les Méthodes nouvelles de la mécanique céleste* (3 volumes, 1892–99) and *Leçons de mécanique céleste* (1905).

Pointers Popular name for the stars α and β Ursae Majoris (Dubhe and Merak), which point towards the North Pole Star, Polaris.

Polar One of two US contributions to the INTERNATIONAL SOLAR TERRESTRIAL PHYSICS programme. Polar was launched in 1996 February, with 11 instruments to provide complete coverage of the inner MAGNETOSPHERE and to obtain global images of the AURORA. Polar also measures the high-latitude entry of the solar wind, ionospheric plasma and the deposition of energy into the upper atmosphere. Polar was designed to have on-board instrument–communications interconnectivity, using on-board computers to share data among the instruments.

Polar Star that shows strongly variable linear and circular polarization and other features. Polars are binary systems. In the variable star classification scheme they are known as AM HERCULIS STARS. *See also* CATACLYSMIC VARIABLE; DQ HERCULIS STAR; DWARF NOVA

polar cap Accumulation of ice, snow or frost in the polar region of a planet or satellite. Polar caps are found on Earth, Mars, the Jovian satellite Ganymede and the Neptunian satellite Triton. Such caps may be seasonal, accumulating in winter and disappearing in summer, or perennial, lasting for many years or even geological epochs. Seasonal caps are observed on all four mentioned

▲ **Pluto** Shown to scale, Pluto is compared here with the Moon, Triton and Charon. Pluto is, indeed, slightly smaller than Triton and significantly smaller than the Moon. Some astronomers have argued that Pluto should be regarded as a large Edgeworth–Kuiper belt object, rather than as a planet in its own right.

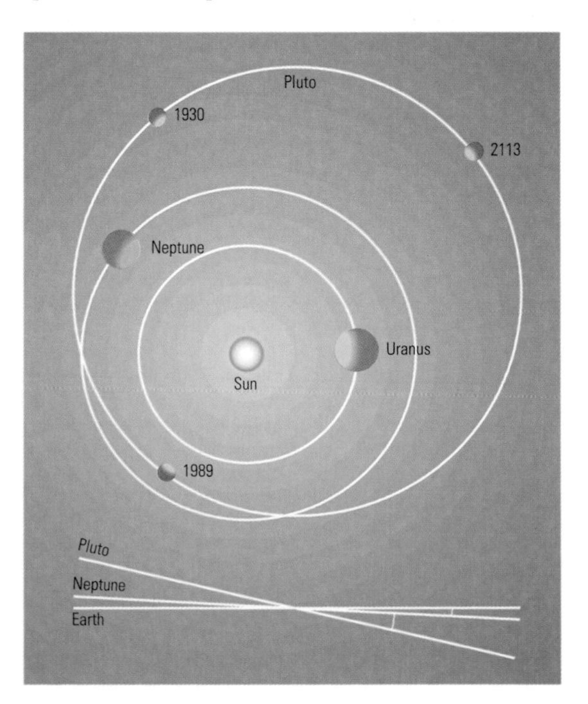

◄ **Pluto** Pluto's orbit is much more eccentric than that of the other planets. Between 1979 and 1999, Pluto was actually closer to the Sun than Neptune, reaching perihelion in 1989. The orbit of Pluto is markedly inclined relative to the ecliptic plane (bottom).

▲ **populations, stellar** Shown here is the face-on spiral galaxy NGC 1288 in Fornax. The differing stellar populations can be clearly seen: young, blue stars in the spiral arms (Population I) and older, yellow stars in the galaxy's central hub (Population II).

bodies. Permanent caps are known on Earth and Mars. Indirect evidence has recently been found for the existence of ice in permanently shadowed parts of polar areas of the Moon and Mercury.

polar distance Angular distance of a celestial object from the CELESTIAL POLE, measured along a great circle. Polar distance is the complement of DECLINATION, the sum of an object's polar distance and its declination always being 90°.

polarimeter Instrument used to measure the POLARIZATION of a beam of electromagnetic radiation. The polarimeter incorporates optically active components – optical elements that can change the polarization of the beam being tested. When unpolarized light is scattered by a rough surface, or by a cloud of gas or dust, it becomes polarized. By investigating the degree of polarization, it is possible to determine the characteristics of, for example, the surface or atmosphere of a planetary body.

Polaris The star α Ursae Minoris, visual mag. 1.97, distance 431 l.y., spectral type F5–8 Ib. It is a CEPHEID VARIABLE, but its fluctuations in output have slowly diminished over the past century and are now barely noticeable. The HIPPARCOS satellite detected a change in brightness of a few hundredths of a magnitude, with a period of 3.97 days. Polaris currently lies within 1° of the north celestial pole; precession will bring it to within half a degree of the pole around the year 2100, after which the separation will increase again. Small telescopes show that Polaris has an apparent companion of mag. 8.2.

polarization In the wave description of ELECTROMAGNETIC RADIATION, light consists of oscillating electric and magnetic fields at right angles to each other and to the direction of propagation. The direction of the electric vibrations can be at any angle around the direction of motion. Ordinary unpolarized light is composed of vibrations with the same amplitudes in all the directions. If the vibrations have greater amplitude in a preferred direction, then the light is said to be linearly polarized, ranging from partially to totally polarized. Some crystals transmit only one direction of vibration, thus converting unpolarized into linearly polarized light (Polaroid sunglasses use this property). Light reflected from a surface or scattered in a gas becomes partially polarized. Such polarization occurs in planetary atmospheres and through the action of interstellar dust grains.

If the plane of vibration rotates systematically in the direction of motion, then the light is circularly polarized. If the amplitude of the vibration also varies then it is elliptically polarized. Reflection from a metallic surface is a source of elliptical polarization.

In astronomy, the occurrence of circularly polarized light is an indication of the presence of a strong magnetic field in the region where the light is emitted. Examples are found in the radio and visible emissions from pulsars and SUPERNOVA remnants (for example the CRAB NEBULA) and also in MAGNETIC STARS.

polar motion Small and irregular movement of the Earth's geographic poles relative to the planet's crust. It has to be taken into account when correcting positional measurements of stars made with instruments such as a TRANSIT CIRCLE. Polar motion is caused by the fact that the Earth's axes of rotation and symmetry are not exactly coincident. Its magnitude is only of the order of around 0″.3. *See also* CHANDLER PERIOD

pole One of the two points on a sphere that lie 90° from its equator and that mark the extremities of its axis of rotation. The projection of the Earth's north and south poles on to the celestial sphere marks the CELESTIAL POLES, which lie 90° north and 90° south of the celestial equator. The ecliptic poles lie 90° from the plane of the ecliptic and the galactic poles are 90° from the galactic plane.

pole star Name given to the nearest naked-eye star to either of the CELESTIAL POLES, though most commonly used to describe the north pole star, POLARIS (α Ursae Minoris). Because of the effects of PRECESSION, the positions of the celestial poles are gradually drifting, and during the last 5000 years the north celestial pole has moved from Thuban (α Draconis) to now lie within 1° of Polaris, which will be at its closest to the pole in AD 2100. By AD 10,000 the north pole star will be Deneb and by AD 14,000 it will be Vega. The south celestial pole is currently marked by the fifth-magnitude star σ Octantis.

Pollux The star β Geminorum, visual mag. 1.16, distance 34 l.y., spectral type K0 III. It is named after one of the mythological twins commemorated by the constellation Gemini, the other being CASTOR. Despite its designation, Pollux is the brighter of the pair.

Pond, John (1767–1836) English astronomer, the sixth ASTRONOMER ROYAL (1811–35) and director of Greenwich Observatory, whose equipment he greatly improved. He succeeded Nevil MASKELYNE, whom he had convinced to replace Greenwich's old quadrant (which Pond showed was warped) with a more modern mural circle made by Edward Troughton (*see* COOKE, TROUGHTON & SIMMS). He also developed methods to reduce or eliminate instrumental errors in measuring star positions, much improving the accuracy of the observatory's star catalogues.

Pons, Jean-Louis (1761–1831) French comet-hunter who began his career (1789) as the door-keeper at Marseilles Observatory, where he learned astronomy. Using a telescope with a 3° field of view, and armed with excellent eyesight and a photographic memory for the starfields he constantly swept, he became one of the most successful comet-hunters of all time. Between 1803 and 1827 he discovered 37 new comets, including Comet Encke.

Pons–Brooks, Comet 12P/ Short-period comet found by Jean Louis PONS in 1812 and rediscovered by William Brooks (1844–1921) in 1883. It returned as predicted in 1954, just reaching naked-eye visibility. The orbital period is 70.9 years.

P

Pons–Winnecke, Comet 7P/ Short-period comet discovered by Jean Louis PONS on 1819 June 12. Initial attempts to derive an orbit proved difficult, and the comet was not recovered until 1858 March 9, when it was found by Friedrich Winnecke (1835–97). From observations at this return, an orbital period of 5.55 years was determined, and the comet was seen again at the next favourable return on 1869. In the early 20th century, the comet orbit came close to Earth, producing activity from the PONS–WINNECKID meteor shower in 1916, 1921 and 1927. Gravitational perturbation by Jupiter has since increased the perihelion distance and orbital period (currently 6.37 years), and 7P/Pons–Winnecke is now extremely faint.

Pons–Winneckids (June Bootids) METEOR SHOWER associated with Comet 7P/PONS–WINNECKE. The shower produced substantial displays in 1916 (100 meteors/hr), 1921 and 1927, each close to perihelion returns of the parent comet. Gravitational perturbations thereafter pulled the meteor stream orbit away from Earth. The shower was presumed to be essentially extinct until the occurrence of an unexpected outburst on 1998 June 27–28, when observed rates of a meteor per minute were seen for over 12 hours. The meteors are slow and come from a diffuse radiant in northern Boötes near the border with Draco. It remains to be seen whether further displays may occur.

population index Observationally determined factor that indicates the relative abundance of METEORS in adjacent whole-magnitude intervals; it is used in determining METEOR SHOWER zenithal hourly rates (ZHR). High population index (r) values indicate that a shower has many faint meteors produced by small particles. SPORADIC METEORS have r = 3.42; typically, PERSEIDS have r = 2.35, and GEMINIDS have r = 2.44.

populations, stellar Classification of types of stars, broadly based on age. There are two main categories: Population I and Population II, with Population I stars being the younger. A further Population III has been added to represent the oldest, very massive, stars which have already exploded as SUPERNOVAE.

The observational basis for the idea of stellar populations originated with Walter BAADE. Baade obtained photographs of galaxies, exposing blue- and red-sensitive emulsions in order to emphasize the contributions of different-coloured light. He discovered that the central regions of spiral galaxies were smooth distributions of red light, from innumerable faint stars, whereas the spiral arms were patchy distributions of bright blue stars and nebulae. Elliptical galaxies had the appearance of the central regions of spirals. These two contrasting kinds of stellar material were the original Populations I and II. The distinction between the two populations is now explained in terms of STELLAR EVOLUTION, with the spatial distribution of the different populations being explained by galaxy formation.

Population I stars are relatively young and metal-rich; they are the younger generation of stars. They are strongly concentrated in the plane of our Galaxy and have roughly circular orbits about the galactic centre. They are observed in the disks and spiral arms of other spiral galaxies. Population I stars are further categorized into extreme and intermediate Population I.

Extreme Population I stars are the youngest stars and have an uneven distribution within the spiral arms of spiral galaxies. They are stars with the greatest abundance of heavy elements. Extreme Population I includes stars that are in the process of forming, for example T TAURI STARS, stars in young OPEN CLUSTERS, OB ASSOCIATIONS, SUPERGIANTS and classical CEPHEID VARIABLES. The Sun is an intermediate Population I star, along with many GIANT STARS and stars in older open clusters.

Population II stars are older, with less abundance of heavy elements. They are further categorized into disk, intermediate and halo Population II stars. The disk Population II stars are the youngest and form a continuum with the oldest Population I stars. They are concentrated in the bulge of our Galaxy, with orbital eccentricities and inclinations to the galactic plane lying between those of the oldest Population I and the halo Population II stars. Short-period RR LYRAE stars are disk Population II stars. Intermediate Population II stars are metal poor and are more concentrated towards the galactic plane than the metal-poor halo stars. This group includes HIGH-VELOCITY STARS and MIRA STARS. The oldest, most metal-poor stars are the halo Population II stars, which include long-period RR Lyrae stars, stars in GLOBULAR CLUSTERS and W VIRGINIS STARS.

The heavy elements within Population II stars are thought to have been produced by a generation of massive and thus short-lived stars, which formed the hypothetical Population III. Being massive, Population III stars would have exploded as supernovae, producing the heavy elements observed in Population II stars. They would now exist as NEUTRON STARS and BLACK HOLES.

The distribution of populations in our Galaxy has been explained in terms of the collapse of the Galaxy during its formation. The oldest population of stars was made during the infall, which explains why Population II stars are found in the galactic halo and central regions of the Galaxy. These stars, including those in globular clusters, still 'remember' their infalling motion and continue to have elliptical orbits around the Galaxy. Stars that formed after the Galaxy had developed its flat rotation plane and spiral arms represent Population I, journeying in circular orbits. Between these two extremes lie intermediate populations, showing progressive flatter and flatter distributions, which represent successive stages in the collapse of the Galaxy.

pore Small, short-lived dark area in the photosphere out of which a SUNSPOT may develop. Pores are comparable in size to individual granules, at about 1000 km (600 mi).

Porrima The star γ Virginis, visual mag. 2.74, distance 39 l.y. It is a binary, with components each of mag. 3.5 and spectral type F, period 169 years. The two are closest around 2005, when they are difficult to split in amateur telescopes, but move apart thereafter. Porrima is named after a Roman goddess.

Porro prism Prism with a cross-section that is a right-angled isosceles triangle. Porro prisms are most commonly encountered in pairs inside each half of a binocular, where they are used to invert the light beam to produce an image the correct way up as seen by the user. The two prisms are placed at right-angles to one another with their long sides in contact, leaving half of each side exposed to allow light in and out. Light enters one prism and after four reflections leaves the other inverted. The distinctive shape of porro-prism binoculars is a result of the arrangement of the prisms.

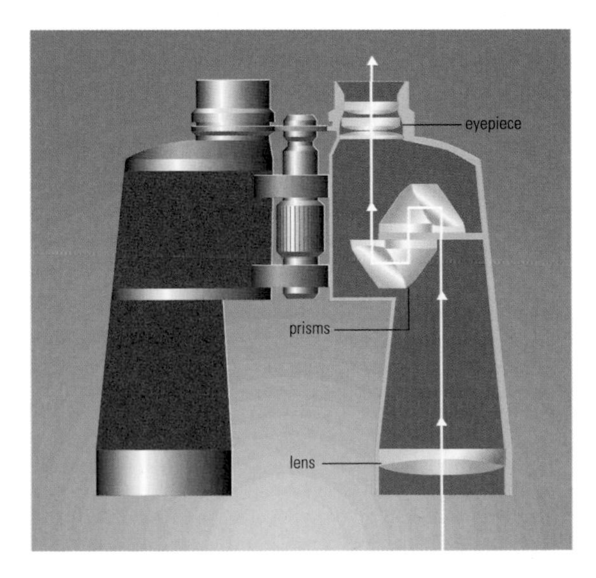

◀ **Porro prism** This schematic cutaway illustrates the optical layout in a pair of binoculars. Light collected by each objective passes through a pair of Porro prisms. The prisms allow portable, compact binocular designs.

P

Porter, Russell Williams (1871–1949) American telescope-maker who designed the horseshoe mounting for Mount Palomar Observatory's 200-inch (5-m) HALE TELESCOPE. Porter, together with Albert G. Ingalls (1888–1958), founded the amateur telescope-making movement in America during the 1920s, greatly expanding astronomy as an avocation. He started the Springfield Telescope Makers in 1920, which still holds its annual Stellafane exhibition of amateur-made astronomy instruments – the largest meeting of its kind. Porter conceived many ingenious designs, including the turret telescope, which sheltered the observer, and the Springfield mounting, which moves the telescope around a stationary eyepiece.

Portia One of the small inner satellites of URANUS, discovered in 1986 by the VOYAGER 2 imaging team. Portia is about 108 km (67 mi) in size. It takes 0.513 days to circuit the planet, at a distance of 66,100 km (41,100 mi) from its centre, in a near-circular, near-equatorial orbit.

Posidonius Lunar crater (32°N 30°E), 96 km (60 mi) in diameter. Posidonius is a complex crater, with central peaks. After its formation, lava intruded beneath the crater, updoming its floor and fracturing it. This action produced rilles in the floor and obscured most of the side-walls. In the western region, flowing lavas produced a SINUOUS RILLE. In the eastern region there is a large slumped wall region, which is nearly semicircular and concentric with the outer rim.

position angle (PA) Direction in the sky of one celestial body with respect to another, measured from 0° to 360° eastwards from the north point. For binary star systems, the position of the fainter component, relative to the brighter, is given. Position angle can also indicate the angle at which the axis or some other line of a celestial body is inclined to the HOUR CIRCLE passing through the centre of the body. That angle is also measured eastwards from the north.

POSS Abbreviation of *PALOMAR OBSERVATORY SKY SURVEY*

post-nova Quiescent stage that follows a NOVA eruption. There is a highly characteristic spectrum, visible even decades after the outburst, with hydrogen and helium emission, and various highly excited spectral lines of elements such as helium, carbon and nitrogen. These spectral features are seen in a few objects for which no outburst has been observed and which are assumed to be novae that escaped detection. It is assumed that MASS TRANSFER has yet to be re-established between the secondary and the white dwarf. It is possible that some NOVA-LIKE VARIABLES are post-novae at a slightly later stage of evolution when mass transfer has resumed.

potential energy ENERGY stored by a body by virtue of its position, shape or state. Potential energy is stored in a body that is within a gravitational field and is higher than its equilibrium position; the energy is given by mgh, where m is the mass, g the local gravitational acceleration, and h the height above the equilibrium position. Potential energy is also the energy of an electron in an excited state, the energy in a radioactive nucleus before it decays, and, via $E = mc^2$, mass. *See also* KINETIC ENERGY

potentially hazardous asteroid (PHA) ASTEROID with an orbit that brings it very close to the orbit of the Earth, making a collision feasible in the near term, the object being large enough to cause significant damage in such an event. The limit set for the minimum orbit intersection distance is generally taken as being 0.05 AU. The minimum size is regarded as being defined by an absolute magnitude H=22, corresponding to an equivalent diameter of between 110 m (360 ft), if the albedo were 0.25, and 240 m (790 ft), if the albedo were 0.05. Sometimes the term 'potentially hazardous object' (PHO) is used, mainly because the acronym provides a homophone of 'foe'. By analogy with the terms NEAR-EARTH ASTEROID and NEAR-EARTH OBJECT, PHOs include COMETS as well as asteroids.

Poynting–Robertson effect Non-gravitational force produced by the action of solar radiation on small particles in the Solar System; it causes the particles to spiral inwards towards the Sun. When particles in orbit around the Sun absorb energy from the Sun and re-radiate it, they lose kinetic energy and their orbital radius shrinks slightly. The effect is most marked for particles of a few micrometres in size. A similar effect makes particles in planetary rings and satellite systems slowly spiral inwards. The effect is named after John Poynting (1852–1914) who first described the effect, and Howard Robertson (1903–61), who established the relativistic theory of the effect.

PPARC Abbreviation of PARTICLE PHYSICS AND ASTRONOMY RESEARCH COUNCIL

p process Method of creating heavy, proton-rich, stable nuclei. In the p process, protons are added to nuclei formed by the R PROCESS and S PROCESS. The p process accounts for low-abundance heavy isotopes. *See also* NUCLEOSYNTHESIS

Praesepe (M44, NGC 2632) Bright open cluster in the constellation Cancer (RA $08^h\ 40^m.1$ dec. +19°59′). Known since ancient times, it was catalogued as a 'little cloud' by HIPPARCHUS in the 2nd century BC. With overall magnitude +3.1, Praesepe (also known as the Beehive) is readily visible to the naked eye in good conditions. Indeed the 1st century AD Roman writer Pliny described the use of the Praesepe Cluster as a weather-predictor by Mediterranean sailors – as high clouds arrive ahead of bad weather, Praesepe disappears from view. The cluster has an apparent diameter of 95′, and contains around 60 stars. It lies at a distance of 525 l.y. and is estimated to be about 700 million years old.

preceding (p) Term used to describe a celestial body, or feature on a body, that leads another in their apparent motion. Preceding features come into view earlier than the reference feature. For example, in a group of sunspots, the **preceding spot** is the first to appear as the Sun rotates, while the **preceding limb** of the Moon is the one facing its direction of travel. Any feature that appears later than a reference feature is said to be FOLLOWING. The term preceding can also be used to describe a star at a lower RIGHT ASCENSION than another.

precession Gradual circular motion of the Earth's axis of rotation which causes the position of the celestial poles to describe a circle over a period of 25,800 years.

Discovered by HIPPARCHUS OF NICEA about 150 BC, precession is largely caused by the gravitational pull of the Sun and Moon on the Earth's equatorial bulge. Were the Earth to be a perfect sphere precession would not occur, but its spin causes its equatorial radius to exceed its polar radius by about 0.3 per cent. The gravitational pull on the bulge produces an angular momentum in a perpendicular direction, according to the laws of dynamics. This makes the Earth 'wobble' or precess like a spinning top and causes the celestial poles to trace a circle of radius equal to the inclination of the Earth's axis (23°.5).

Precession comprises two components: the combined pull of the Sun and the Moon, known as LUNISOLAR PRECESSION, is the greater and amounts to 50″.40 per year. A smaller component, PLANETARY PRECESSION, caused by the gravitational effects of the other planets, amounts to 0″.12 per year, but in the opposite direction. The net effect of these, called general precession, is 50″.26 per year.

Precession causes the FIRST POINT OF ARIES and the FIRST POINT OF LIBRA to drift westwards around the ecliptic, relative to the background stars, and is therefore some-

P

times known as the precession of the equinoxes. They have moved about $1^h 45^m$ in RIGHT ASCENSION (RA) since Hipparchus discovered the effect, so that each point has now shifted into the adjoining constellation.

Precession is also responsible for making the TROPICAL YEAR, measured with respect to the equinoxes, about 20 minutes shorter than the SIDEREAL YEAR, which is measured with respect to the stars. Another effect is that 13,000 years from now ORION will be a summer constellation (in the northern hemisphere). Star positions published for a particular epoch, such as 2000.0, also have to be corrected for the effects of precession to the date of observation. *See also* NUTATION

prehistoric astronomy *See* ARCHAEOASTRONOMY

pre-main-sequence star Young star, still in the process of formation, that has finished accreting material from its parent molecular cloud, but in which nuclear fusion in the core has yet to begin.

PROTOSTARS form within the fragmenting molecular clouds. They are very large and cool, lying to the right of the MAIN SEQUENCE on the HERTZSPRUNG–RUSSELL DIAGRAM (HR diagram). As they evolve, they contract and their cores start to heat up. With their large size, the change in radius dominates any change in luminosity. They therefore follow almost vertical lines, known as HAYASHI TRACKS, on the HR diagram. At this point, the star's energy is almost entirely transported by means of convection, and it has only a small radiative core.

The star then starts to become radiative, and the luminosity starts to increase as the effective temperature increases. The star's evolution at this phase is represented by the HENYEY TRACK on the HR diagram. The core temperature rises until nuclear fusion can start in its core, at which point it becomes a main-sequence star. The pre-main-sequence phase of the Sun lasted several hundred million years. T TAURI STARS are examples of pre-main-sequence stars. *See also* STELLAR EVOLUTION

pre-nova Putative interacting BINARY STAR system that is in a quiescent state but undergoing active MASS TRANSFER prior to erupting as a NOVA. Technically such a system may be classified as a CATACLYSMIC VARIABLE, and presumably exhibits characteristics similar to those of such a system, most probably appearing as a NOVA-LIKE VARIABLE. Although images of nova precursors have been detected after the event, no previously known system has yet undergone a nova outburst. The reasons for this undoubtedly lie in the moderately low number of systems known, the relatively short period during which they have been studied, and the extremely long recurrence interval (thousands of years) between nova outbursts. A few pre-outburst light-curves have been obtained, and these appear to show a general rise in the magnitude of the system by some 1–5 magnitudes in the years immediately preceding the outburst.

pressure broadening Broadening of spectrum lines as a result of electromagnetic 'collisions'. Spectrum lines are broadened by mass motions that shift the lines by the Doppler effect, and by the smearing of energy levels by quantum uncertainty, magnetic fields (the ZEEMAN EFFECT) and electromagnetic effects induced by close-passing atoms. The incidence of electromagnetic collisions depends on the closeness of the atoms and, therefore, on the pressure of the gas. Pressure broadening causes dwarfs to have wider hydrogen lines than do giants and supergiants.

primary Largest of a system of celestial bodies. The Sun is the primary member of the Solar System. The term is also used to describe the more massive component in a binary or multiple star system, the primary member of such a system being the one about which the others rotate. It can also be used to describe a planet with respect to its moons, for example, the primary body of the Moon is the Earth.

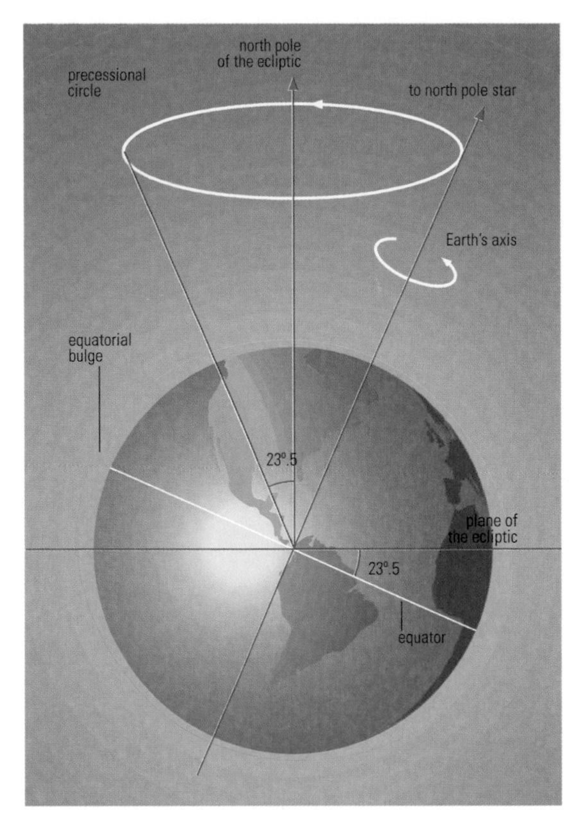

◀ **precession** Over the course of a 25,800-year cycle, the direction in which Earth's axis of rotation points describes a 47° diameter circle relative to the star background. One result of this is that different stars will, over time, become Pole Star.

primary Main, light-collecting mirror in a reflecting telescope.

prime focus Position at which an objective LENS or a primary MIRROR brings starlight directly to a focus (without the intervention of any additional lenses or mirrors). In large research telescopes astronomers use the prime focus of the main mirror to feed light to instruments that need a wide field of view. Some astrophotographers use the term to refer to images formed directly by a SCHMIDT–CASSEGRAIN TELESCOPE rather than from the arrangement where an eyepiece is also used to form the image. The term 'principle focus' would be more correct in this case.

Principia Shortened name of Isaac Newton's *Philosophiae naturalis principia mathematica* ('Mathematical Principles of Natural Philosophy'), summarizing his researches into physics and astronomy and published in three volumes in 1686–87. Most of the work it describes had been completed years before, and it was the persistence of Edmond HALLEY that persuaded the reclusive Newton to put it into print. Its appearance immediately established Newton's international fame, even though he wrote it (in Latin) in a deliberately abstruse manner 'to avoid being bated by little smatterers in mathematics', as he put it.

The *Principia* is a milestone in science, bringing together systematic observation and rigorous mathematical analysis. Its 'Newtonian mechanics' set the agenda for two centuries of enquiry into the physical world. It sets out NEWTON'S LAWS OF MOTION and NEWTON'S LAW OF GRAVITATION, and demonstrates their universality, from earthly motions to the celestial mechanics of the Solar System. It shows that an inverse-square law of gravitation leads inevitably to KEPLER'S LAWS governing planetary orbits. The motion of the Moon is analysed in detail, as is the phenomenon of the tides.

prism Transparent optical element with flat, polished sides used to bend or disperse a beam of light. Prisms can have a uniform cross-section, typically triangular as in the PORRO PRISM, or they can have a more complex shape, such as the penta prism used in a single-lens reflex camera. Porro prisms and penta prisms are both used to reflect light with little or no dispersion. Light enters one

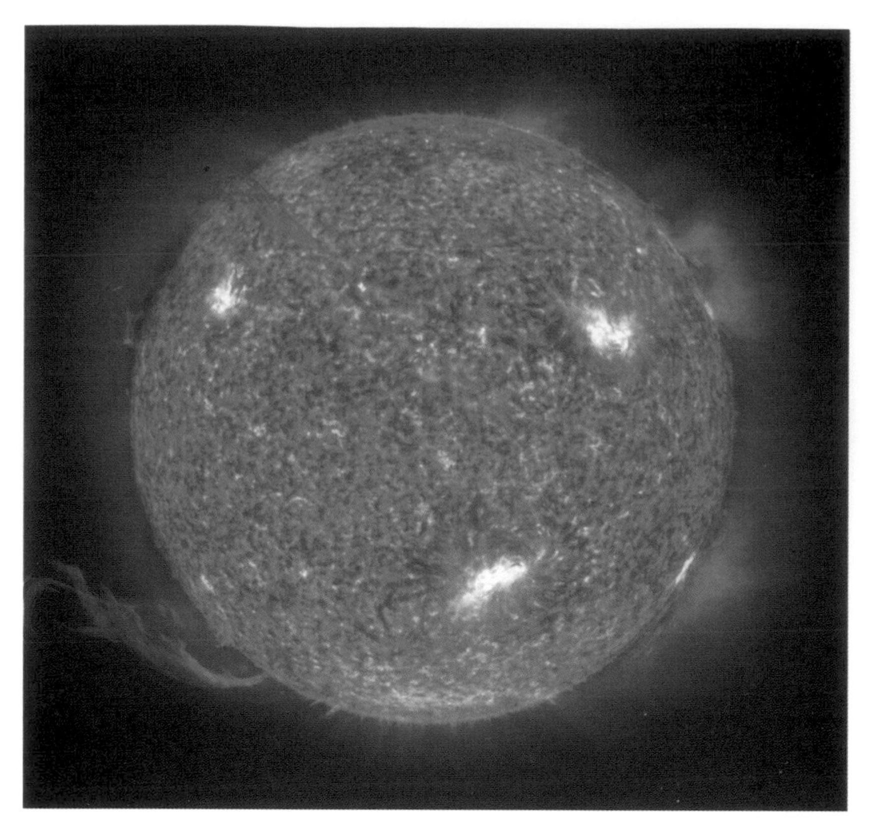

▲ **prominence** This SOHO image from 1997 September 14 shows at lower left a huge eruptive prominence high above the solar limb. The image is a spectroheliogram in the light of HeII at 30.4 nm in the extreme-ultraviolet.

face of the prism at right-angles and after two reflections leaves again, also at right-angles to the prism surface. The porro prism reflects the light through 180° whilst the penta prism reflects it through 90°. Prisms are often used in place of MIRRORS to produce a simple 90° reflection, for example in STAR DIAGONALS.

If light enters and/or leaves at any angle other than at right-angles to the surface then it will be dispersed into its component SPECTRUM. This dispersion allows the composition of the light to be measured and analysed. Early SPECTROGRAPHS used prisms as their main dispersing elements but modern ones are more likely to use DIFFRACTION GRATINGS because these can provide greater dispersion. Prisms are sometimes used in conjunction with ECHELLE GRATINGS to separate the different orders of spectra produced in high-dispersion spectrographs.

Large, thin prisms with a very small angle (typically one degree) between the two main sides are used as objective prisms to produce low-resolution spectra of all the objects in a telescope's field of view simultaneously. Provided the field is not so crowded that the spectra overlap, this technique can be used to survey a large number of objects in a short period of time. It is normally used to identify potential targets for further study at higher RESOLUTION.

Procellarum Oceanus (Ocean of Storms) Lunar lava plain in the western region of the Moon. While called an 'ocean', it is actually a vast MARE region. However, unlike the majority of maria, which formed within impact basins, Oceanus Procellarum formed in a topographically low region of the Moon. Many volcanic features are found here, including mare (wrinkle) ridges, domes and sinuous rilles. Unusual features include the magnetic Reiner Gamma and the volcanic ARISTARCHUS Plateau.

Proclus Lunar crater (16°N 47°E), 29 km (18 mi) in diameter, with rim components reaching 2440 m (8000 ft) above its floor. Proclus is intermediate in type between simple bowl-shaped craters and complex craters with central peaks. On its floor are a few small mounds, along with extensive non-terracing slumps. The bright ejecta are irregular because of an oblique impact, which produced a zone without bright ejecta to the south-west. The angle of impact was not low enough, however, to produce an elongated crater.

Proctor, Richard Anthony (1837–88) English astronomy writer and popularizer who collated and analysed the work of others, particularly in the areas of Solar System and stellar astronomy. By carefully reviewing drawings of the albedo features of Mars made between 1666 and 1878, he derived a rotation period of $24^h\ 37^m\ 22^s.7$, very close to the currently accepted value. Proctor studied the distribution of the Milky Way's stars, inventing graphical techniques to illustrate the arrangement of stars, clusters and nebulae. He charted the 324,000 stars in F.W.A. Argelander's *Bonner Durchmusterung* and constructed a map of the Milky Way and its structure.

Procyon The star α Canis Minoris, visual mag. 0.40 (the eighth-brightest in the sky), distance 11.4 l.y., spectral type F5 IV or V. Procyon has a white dwarf companion of mag. 10.7 which orbits it every 41 years, but it is far too close to be seen without a large telescope. The name Procyon comes from the Greek meaning 'preceding the dog', from the fact that it rises before Sirius, which is known as the Dog Star.

Prognoz satellites Series of ten Soviet spacecraft; they were launched from 1971–85. Most were designed to monitor solar activity and the interaction of the SOLAR WIND with the Earth.

prograde *See* DIRECT MOTION

Prometheus Long-lived 100-km-high (60-mi-high) volcanic eruption plume on Jupiter's satellite IO. Its location shifted by about 70 km (44 mi) between its detection on VOYAGER images in 1979 and its appearance on GALILEO images in the late 1990s. The source of the plume cannot therefore be a fixed vent and may instead be the end of a lava flow.

Prometheus One of the inner satellites of SATURN, discovered in 1980 by S.A. Collins and others in VOYAGER 1 images. It is irregular in shape, measuring about 150 × 90 × 70 km (90 × 60 × 40 mi). Prometheus has a near-circular equatorial orbit at a distance of 139,400 km (86,600 mi) from the planet's centre, where its orbital period is 0.613 days. It appears to act as a SHEPHERD MOON to the inner rim of the F RING. *See also* PANDORA

The name **Prometheus** is also applied to an active volcano on IO.

prominence Relatively cool and dense plasma suspended above the Sun's PHOTOSPHERE and contained by magnetic fields. The insulating effects of the magnetic fields are such that material in a quiescent prominence may have a temperature of 10,000 K while the surrounding corona is at 2,000,000 K. Prominences appear as bright features extending from the solar limb at a total eclipse or can be seen in a SPECTROGRAM or SPECTROHELIOGRAM of the CHROMOSPHERE taken in the HYDROGEN-ALPHA LINE or H and K lines of ionized calcium. When seen against the bright solar disk, prominences appear as dark filaments.

Prominences have been divided into two main classes – quiescent and active. **Quiescent prominences** occur mainly in two zones in either hemisphere, chiefly away from ACTIVE REGIONS. One zone lies just polewards of the sunspot-forming latitudes (migrating equatorwards behind the sunspots in accordance with SPÖRER'S LAW), while the other forms a high-latitude 'polar crown'. Quiescent prominences show no large motions, and can last for weeks or months. Typically, prominences are between 100,000 and 600,000 km (60,000–370,000 mi) in length, 5000 and 10,000 km (3000–6000 mi) wide, and can reach as high as 50,000 km (30,000 mi) above the photosphere.

Active prominences are short-lived features showing rapid motion; they are associated with active regions, sunspots and FLARES. They can appear as surges, sprays or loops, and can reach heights of about 700,000 km (400,000 mi) in just an hour. Quiescent prominences can develop from active prominences.

Quiescent prominences can become detached and disappear if disturbed by MORETON WAVES associated with flare activity. Such eruptive prominence events can arch a million kilometres outwards before bursting apart; their disappearance is often associated with a CORONAL MASS EJECTION.

proper motion (symbol μ) Apparent angular displacement of a star against the celestial sphere as a result of its motion over a year. Proper motion is a combination of a star's actual motion through space and its motion relative to the Solar System.

Stars in our Galaxy move relative to each other and relative to the Sun, but because of their great distances their apparent movements in the sky are very small. In fact most stars are so distant that their proper motions are negligible. The full space motion of a star is the combination of its proper motion, across an observer's line of sight, and its radial velocity, along the line of sight. The latter is usually measured from the Doppler shift in the star's spectrum.

Proper motion is usually denoted by the Greek letter Mu (μ) and is measured in seconds of arc (″) per year. About 300 known stars have proper motions larger than one arcsecond per year, but most proper motions are smaller than 0.1 arcsecond per year. Barnard's Star in the constellation Ophiuchus has the largest known proper motion, at 10″.27 per year.

Proper motions are measured by comparing the accurate positions of stars obtained at two or more epochs. This gives the components of the proper motion in right ascension, $\mu\alpha$, measured in seconds of time per year, and in declination, $\mu\delta$, measured in seconds of arc per year. The combined proper motion, μ, measured in seconds of arc per year, is given by:

$$\mu = (225\ \mu\alpha^2 \cos^2\delta + \mu\delta^2)^{1/2}$$

The transverse velocity in space of a star relative to the Sun, V_T (km/s), is related to its observed proper motion, μ (arcseconds/year) by:

$$V_T = 4.74 \frac{\mu}{\pi}$$

where π is the parallax of the star. For example, the parallax of Barnard's star is 0″.545, so its transverse velocity is 89.3 km/s (55.5 mi/s).

Proper motions have traditionally been determined using instruments such as transit circles or from photographic astrometry, but the astrometric satellite HIPPARCOS, which operated between 1989 and 1993, measured the proper motions for 118,218 stars with positional accuracies averaging hardly more than a thousandth of an arcsecond.

The detailed analysis of many observations are tabulated in a series of FUNDAMENTAL CATALOGUES, the most recent being called *FK5 Part II*, which contains the positions and proper motions of 3117 FUNDAMENTAL STARS brighter than magnitude 9.5. This catalogue defines the basic reference frame to which the kinematics of all stars in our Galaxy are ultimately referred.

Proper motions of stars have been measured with respect to galaxies. The proper motions of galaxies are negligible, so they provide a fixed frame of reference against which to measure stellar proper motions. This method is therefore independent of the positions and proper motions of the bright REFERENCE STARS measured by instruments such as meridian circles.

The local motion of the Sun relative to stars within a 20-parsec radius can be derived from proper motions. The average proper motion of the local stars depends on the angle between the direction of the Sun's motion and the direction to the stars. If the stars lie in the direction of the APEX (or ANTAPEX) of the Sun's motion, the average of their proper motions will be zero. The calculated position of the apex, and the solar velocity towards it, depends to some extent on the number and type of stars used in the analysis. The average result is that relative to the nearby stars the Sun is travelling about 20 km/s (12 mi/s) in the direction of the constellation Hercules. When the solar motion is subtracted from the observed proper motion of a star its peculiar proper motion is obtained. The apparent displacement of a star arising from the solar motion is called its parallactic motion. By measuring the parallactic motion of a homogeneous group of stars, the so-called SECULAR PARALLAX of the group is obtained. This method has been applied to estimate the average distances of groups of variable stars, such as RR LYRAE and CEPHEIDS, which are beyond the range of TRIGONOMETRIC PARALLAX.

Another important application of proper motions is in determining the distances of moving clusters. Since stars in a cluster have a common space motion, their proper motions appear to converge to a point on the celestial sphere. By combining the measurement of the convergent point for the cluster with measurements of its proper motion and radial speed, it is possible to calculate its distance.

Proper motion surveys also have an important application in investigating the distribution and luminosity of stars in the solar neighbourhood. Stars close to the Sun tend to have larger values of proper motion than more distant stars, so by selecting stars from proper motion surveys that are complete to some apparent magnitude, a statistical sample of the stars within a specified volume around the Sun is obtained. By combining this sample with the known parallaxes of some of the stars, and making allowances for the incompleteness of the sample, one can arrive at a statistical estimate of the distribution and luminosity of the stars in the vicinity of the Sun. This shows that most stars in the solar neighbourhood are intrinsically faint.

Proper motions have another important application in the subject of galactic rotation. They provide the only direct method of measuring the Oort constant B, which essentially describes how the rate of change of the angular velocity of rotation changes with distance from the centre of our Galaxy.

proplyd Acronym for PROTOPLANETARY DISK

Prospero One of the several small outer satellites of URANUS; it was discovered in 1999 by Matthew Holman (1967–) and others. Prospero is *c*.20 km (*c*.12 mi) in size. It takes 2037 days to circuit the planet at an average distance of 16.67 million km (10.36 million mi). It has a RETROGRADE orbit (inclination near 152°) with a substantial eccentricity (0.439). *See also* CALIBAN

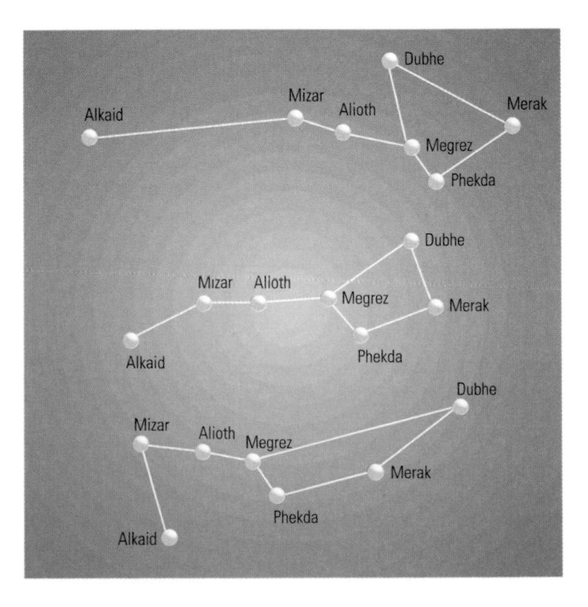

◀ **proper motion** Five of the stars making up the familiar pattern of the Plough are members of a cluster showing a common proper motion relative to the more distant stellar background. The two non-members, Alkaid and Dubhe, have different proper motions, and over the course of hundreds of thousands of years, the plough pattern will become distorted. The view in the centre shows the stars' relative positions today, with an earlier epoch represented at top, and the configuration in the distant future at bottom.

P

▶ **Proteus** Discovered by Voyager 2 during its 1989 flyby, Proteus is the largest of the inner satellites of Neptune. Proteus is notably dark, with an albedo of 0.07.

Proteus Largest of the inner satellites of NEPTUNE, and the second largest of Neptune's satellites overall, after TRITON. Proteus was discovered in 1989 by the VOYAGER 2 imaging team. It has a low albedo (0.07). Proteus is squarish in shape, measuring about 436 × 416 × 402 km (271 × 259 × 250 mi), and is heavily cratered. The most prominent feature on its surface is the large crater Pharos, which is 250 km (160 mi) across. Proteus takes 1.122 days to circuit the planet, at a distance of 117,600 km (73,100 mi) from its centre, in a near-circular, near-equatorial orbit.

protogalaxy Matter consisting of primarily hydrogen and helium gas which collapses by gravitational interaction to form a GALAXY. It is not clear if there are galactic seeds such as hot dark matter or cold dark matter, or even black holes around which this matter gravitates to form galaxies.

Proton Russian workhorse satellite and planetary space probe launcher. It started life as a two-stage booster, first flown in 1965, and has since made more than 300 flights in different models. With four stages, the Proton K was first launched in 1967, later proving its capability to place satellites directly into GEOSTATIONARY ORBIT (GEO). A three-stage version of the Proton K has been used to launch SALYUT, MIR and INTERNATIONAL SPACE STATION modules. The Proton K is now being marketed as a commercial launcher by a joint US–Russian company, International Launch Services, which also promotes the US Atlas booster. The launcher can place a satellite weighing 1.8 tonnes directly into GEO or a 4.9-tonne payload into geostationary transfer orbit (GTO). The new Proton M, with a powerful Breeze upper stage, was introduced in 2001, to place payloads weighing 2.9 tonnes into GEO or 5.5 tonnes into GTO.

proton Stable subatomic particle with a unit positive electric charge. Its mass is 1.672614×10^{-27} kg (1.007399 amu). It is also the nucleus of the H^1_1 atom and with the NEUTRON forms the nuclei of other atoms. Protons thus form about 87% of the mass of the Universe. The number of protons in a nucleus determines which ELEMENT it forms. Modern GRAND UNIFIED THEORIES suggest that a proton should eventually decay, but its lifetime may lie between 10^{35} and 10^{45} years, and the decay has yet to be confirmed experimentally.

Protonilus Mensa Mesa area on MARS (44°.2N 309°.4W). It is some 600 km (370 mi) long.

proton–proton reaction (p–p reaction) Set of NUCLEAR REACTIONS that results in the conversion of hydrogen into helium. 26.8 MeV (4×10^{-12} J) of energy is released during the formation of a single helium nucleus. This comes from the conversion of mass into energy, since the helium nucleus has a mass that is about 0.7% less than that of the four hydrogen nuclei (protons) that go to form it. The basic (PP I) set of reactions is:

1 $H^1_1 + H^1_1 \rightarrow H^1_2 + e^+ + v_e$
2 $H^1_2 + H^1_1 \rightarrow He^2_3 + \gamma$
3 $He^2_3 + He^2_3 \rightarrow He^2_4 + H^1_1 + H^1_1$

where e^+ is a positron (positive electron), v_e is an electron neutrino (*see* NEUTRINO ASTRONOMY), and γ is a gamma-ray photon.

The same end result is also reached via two lower probability alternative sets of reactions:

PP II
reactions 1 and 2 followed by
4 $He^2_3 + He^2_4 \rightarrow Be^4_7 + \gamma$
5 $Be^4_7 + e^- \rightarrow Li^3_7 + v_e$
6 $Li^3_7 + H^1_1 \rightarrow Be^4_8$
7 $Be^4_8 \rightarrow He^2_4 + He^2_4$

and PP III
reactions 1, 2 and 4 followed by
8 $Be^4_7 + H^1_1 \rightarrow B^5_8 + \gamma$
9 $B^5_8 \rightarrow Be^4_8 + e^+ + v_e$
10 $Be^4_8 \rightarrow He^2_4 + He^2_4$

The p–p chain is the dominant source of nuclear energy in MAIN-SEQUENCE stars with masses equal to or less than the Sun. *See also* CARBON–NITROGEN–OXYGEN CYCLE; NUCLEOSYNTHESIS

Proton satellites Four very heavy (12–17 tonnes) Soviet satellites launched 1965–68. They were used for monitoring cosmic and gamma rays.

protoplanet Planet in the making. The planet-forming process begins with small solid bodies, called PLANETESIMALS, in a PROTOPLANETARY DISK orbiting a star. Their mutual gravitational perturbations cause their orbits to intersect, resulting in collisions and growth by ACCRETION. The rate of collisions is enhanced by gravitational attraction, which is stronger for more massive bodies, thus the largest planetesimal in some local region of the disk becomes more effective at sweeping up mass than its smaller neighbours, which allows it to become still larger. This process results in rapid 'runaway growth' of a large body, called a planetary embryo, which dominates the region of the disk near its orbit and eventually exhausts the supply of planetesimals in that region. This process occurs at various distances from the star, producing hundreds of such bodies, of the order of the Moon's mass, on separated orbits. Because of their perturbations, these orbits are unstable on somewhat longer timescales, and collisions occur between the embryos. As they become larger in size and fewer in number, the largest may be called protoplanets; the final bodies at the end of this process are planets. In the inner Solar System, formation of embryos took less than a million years, while the late stage of giant impacts lasted for tens of millions of years.

The term protoplanet is also applied to self-gravitating condensations hypothesized to form by gravitational instability of the gaseous component of a protoplanetary disk. In some theories, this process is assumed to produce gas giant planets such as Jupiter and Saturn, after the gaseous protoplanet cools and contracts. *See also* COSMOGONY

protoplanetary disk (proplyd, protostellar disk) Material surrounding a newly formed star. A star forms by gravitational collapse of a cloud of interstellar gas. Almost invariably, such a cloud has some rotational motion, and conservation of angular momentum causes its spin rate to increase as it contracts. Material cannot fall directly into the centre, but instead forms a disk. Viscosity in the disk, produced by turbulence, magnetic fields, and/or gravitational interactions between regions of varying density, redistributes angular momentum towards the outer part, allowing most of the mass to flow inwards to form a PROTOSTAR. The remainder of the mass, perhaps a few per cent to a few tenths of the total, is left orbiting the star. The disk of gas and dust, which may persist for millions of years before dissipating, supplies the material for formation of planets. Protoplanetary disks, many of them much larger than our Solar System, have been observed by the Hubble Space Telescope.

protostar Young star, still accreting material, said to exist once fragmentation of its parent molecular cloud has finished. A protostar is the earliest phase of STELLAR EVOLUTION. This phase may take from 10^5 to 10^7 years, depending on the mass of the star. The Sun's protostellar stage took about 0.1–1 million years.

On a HERTZSPRUNG–RUSSELL DIAGRAM, protostars would lie above and to the right of the MAIN SEQUENCE. Protostars cannot be observed in the optical because they are embedded in the cloud of material from which they are forming, although their presence can be detected in the infrared. Once protostars become optically visible, they are known as PRE-MAIN-SEQUENCE STARS.

Proxima Centauri The closest star to the Sun, 4.22 l.y. away. It is a red dwarf of visual mag. 11.0, spectral type M5 V, with just 0.005% of the Sun's luminosity. Named because of its proximity to us, Proxima is a member of the ALPHA CENTAURI triple system but lies about 2° from the other two members. It is a FLARE STAR, which can brighten by up to a magnitude for several minutes at a time.

Ptolemaeus Ancient lunar crater which has deeply incised walls that show eroded rims (14°S 3°W); it is 148 km (92 mi) in diameter. Its features are the result of impact erosion, much of which came from the ejecta of the IMBRIUM multi-ring impact basin to the north-west. The scars from Imbrium's ejecta all trend from north-west to south-east. The floor of Ptolemaeus is smooth, also as a result of the ejecta from the Imbrium impact. The floor contains several ghost craters and a few recent impact scars.

Ptolemaic system GEOCENTRIC THEORY of the Universe as presented in Ptolemy's *ALMAGEST* of AD 140. The Earth is taken to be the centre of the Universe, with the Moon, Mercury, Venus, the Sun, Mars, Jupiter and Saturn revolving around it. Outside lies the sphere of the fixed stars.

The orbit of a planet as described by EUDOXUS was such that it moves around in a small circle, called the **epicycle**. The centre of the epicycle itself revolves about the Earth on a larger circle called the **deferent**. Ptolemy added two further points, the **eccentric** and the **equant**, to each orbit. The eccentric is at the centre of a line joining the Earth and the equant; the deferent is centred on the eccentric point rather than on the Earth itself. The centre of the epicycle thus moves around the deferent with a variable velocity that makes it appear to be moving with uniform angular velocity when viewed from the equant point – in direct conflict with the Aristotelian doctrine that the celestial motions had to be perfectly uniform.

The Ptolemaic system allowed the future positions of the planets to be predicted with reasonable accuracy and remained 'state of the art' until ousted by the COPERNICAN SYSTEM.

Ptolemy (Claudius Ptolemaeus) (2nd century AD) Egyptian astronomer and geographer. His chief astronomical work, the *Almagest*, was largely a compendium of contemporary astronomical knowledge, including a star catalogue, drawing on the work of HIPPARCHUS of Nicaea. It described the so-called PTOLEMAIC SYSTEM, a geocentric universe with the Earth fixed at the centre, and the Moon, Sun and planets revolving about it. Ptolemy's modification of the previous, simpler Greek theory based on epicycles and deferents reproduced the apparent motions of the planets, including retrograde loops, so well that it remained unchallenged until the revival of the HELIOCENTRIC THEORY by Nicholas Copernicus in the 16th century.

Little is known about Ptolemy's life. His dates and even his name are uncertain – 'Ptolemy' merely indicates that he had Greek or Greek-naturalized ancestors and lived in Egypt, which was then ruled by the Ptolemaic dynasty. He is sometimes referred to as Claudius, but this probably means simply that Ptolemy's Roman citizenship goes back to the time of the emperor Claudius I. It seems likely, though, that he lived and worked at the famous library and museum in Alexandria.

Ptolemy was a prolific author, but his earliest large work, the *Almagest*, was not only his greatest, but also the greatest astronomical work of antiquity. It is essentially a basic textbook, expecting its reader to know only the fundamentals of Greek geometry and some familiar astronomical terms; the rest Ptolemy explains.

The first two books form an introduction. They present an outline of the Ptolemaic system – a spherical universe with spheres carrying the planets – and give persuasive arguments that the Earth is stationary at its centre. The basic mathematics, making use of chords (the Greeks did not develop the sines and tangents of modern trigonometry), is described.

Book III discusses the apparent motion of the Sun, using observations mostly by Hipparchus, for whose work Ptolemy shows such reverence that of his own observations he seems to have selected only those that agreed with his predecessor's. A table of the Sun's motion is then constructed. The motion of the Moon according to Hipparchus is the subject of Book IV. The theory, based on eclipses, accounts for the Moon's motion at conjunction (new moon) and opposition (full moon), but is inadequate for intermediate positions in the orbit.

In Book V, Ptolemy develops his own lunar theory. Based on EPICYCLE and DEFERENT, it is extremely ingenious. His observations showed him that the Moon changed in apparent size, which he accounts for by a kind of crank mechanism operating on the centre of the Moon's epicycle. This theory accounts for the motion of the Moon at all positions of its orbit.

Book VI explains how to calculate every detail of eclipses. Books VII and VIII present tables of positions and brightnesses of the 'fixed' stars, and precession is mentioned.

The final books, IX to XIII, are concerned with the thorny problem of planetary motion. There were fewer observations for Ptolemy to use here, and very little past theoretical work to help him. The epicycle and deferent were not totally satisfactory in the planetary context, because on their own they could not account for variations in the apparent retrograde motion of a planet, nor could they show why retrograde motion occurred at seemingly irregular intervals.

▲ **protoplanetary disk** Seen here as a dark oval, about as wide as Pluto's orbit around the Sun, this protoplanetary disk contrasts with the background glow of the Orion Nebula. This is one of several such objects in the region, imaged by the Hubble Space Telescope.

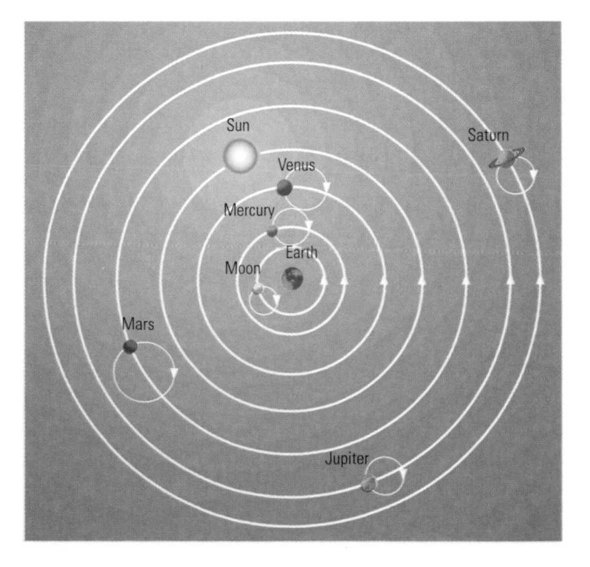

◄ **Ptolemaic system** Placing Earth at the centre, with the Sun orbiting between Venus and Mars, the Ptolemaic system was a prevailing world view for over a thousand years. To account for the planets' occasional retrograde (westwards) motion, an ever more complex pattern of epicycles – small circles superimposed on their orbits – had to be invoked. Eventually, the Ptolemaic system was superseded by the Copernican, which accounted more readily for observed planetary movements.

▲ **Puck** This distant view of Uranus' inner satellite Puck was taken from Voyager 2 during its 1986 January flyby. Puck has a diameter of about 154 km (96 mi).

Typically, Ptolemy tackled the problem in a truly scientific way. Using his own observations as well as those of Hipparchus and the Babylonians, he discovered that each planet had two irregularities or 'anomalies'. One depended on the planet's elongation, the other on its position along the ecliptic. Ptolemy had to explain these anomalies by a theory that fitted into the framework of the Greek geocentric universe.

Careful study showed him that the epicycle and deferent would account for the first anomaly. For the second he introduced the ECCENTRIC. Here the centre of the planet's epicycle moved around a point M whose distance from the Earth was determined by the planet's apparent eccentricity. It was another basic condition of Greek planetary motion that each planet should move at an unvarying rate about the centre of the Universe (even though observation showed that not one of them did). Ptolemy satisfied this condition by using another point, the EQUANT, on the opposite side of the Earth to M and an equal distance away. Motion with respect to the equant was uniform. This was an inspired solution, which, with an additional minor modification, explained the motions of all the planets then known.

The *Almagest* was not Ptolemy's only great text. He wrote the *Tetrabiblos* on astrology, much of it on 'natural astrology' – the physical effects of the Sun and Moon, for instance. Then he produced *Planetary Tables*, which were extracted from the *Almagest*, and a popular abridgement called *Planetary Hypotheses*, which, however, extends some of his theoretical ideas and in particular his measurements of the distances and sizes of the Sun and Moon. There was also his *Phases of the Fixed Stars*, which went into more detail about rising and setting of the stars just before dawn and just after sunset, and the *Analemma*, a book on constructing sundials. He also wrote on geography, music, mechanics, geometry and optics.

Puck Largest of the inner satellites of URANUS, discovered in images returned by VOYAGER 2 as it approached Uranus near the end of 1985. Puck is about 154 km (96 mi) across, roughly spherical in shape, and heavily cratered. It takes 0.762 days to circuit the planet, at a distance of 86,000 km (53,000 mi) from its centre, in a near-circular, near-equatorial orbit.

Pulcherrima Alternative name for the star ε Boötis. *See* IZAR

Pulcova MAIN-BELT ASTEROID; number 792. In 2000 it was found to be accompanied by a small moon. Pulcova is about 145 km (90 mi) in size, ten times as large as its satellite, which takes four days to orbit it at a distance of about 800 km (500 mi).

Pulkovo Observatory One of the oldest observatories in Russia, founded in 1839 with a 15-inch (400-mm) refractor that was then the largest refractor in the world. Members of the STRUVE FAMILY number among its past directors. It is located near St Petersburg (formerly Leningrad) and was completely destroyed in the siege of the city during World War II. In a symbolic gesture by the Soviet government, the observatory was rebuilt between 1945 and 1954. However, the observing conditions there are poor, and Pulkovo astronomers today use the optical and radio telescopes of the SPECIAL ASTROPHYSICAL OBSERVATORY and other better-sited facilities. The observatory carries out research in most branches of theoretical and observational astrophysics as well as solar physics and astronomical instrumentation.

P

▼ **pulsar** This schematic illustration of a pulsar shows the two beams of radiation directed from the collapsed star along its magnetic poles, which need not coincide with the axis of rotation. If the rotational axis is so aligned that the radiation beams sweep across our line of sight, the rapidly pulsing radio signal characteristic of a pulsar will be detected.

axis of radiation

beam of radiation

beam of radiation

pulsar Rapidly spinning NEUTRON STAR, emitting two beams of RADIO WAVES that are seen as pulses. The beam of radio waves emitted by the rotating pulsar sweeps past the Earth in a manner similar to the flash of light from a lighthouse beam. The first pulsar was discovered by Anthony HEWISH and Jocelyn BELL in 1967, and there are now over 1000 known. A few pulsars have been detected at wavelengths other than radio (for example visible light, X-ray and even gamma-ray). The pulsar periods range from milliseconds (*see* MILLISECOND PULSAR) to around 5 seconds, with the most common period being around 1 second. The pulses are not always exactly the same shape from cycle to cycle, or the same strength, and the pulse at each wavelength has a slightly different appearance, but the period is always the same. Some pulsars are found in binary systems (*see* BINARY PULSAR) and there is one case of a pulsar with planets (PSR B1257+12). The distribution of pulsars is concentrated towards the plane of the Galaxy, as is the concentration of supernovae, and there may be as many as 100,000 pulsars in the Galaxy.

A pulsar is formed following a supernova explosion, but only a few pulsars can be seen in their associated SUPERNOVA REMNANT. Most pulsars are over a million years old, so the supernova remnant will have long since dispersed and faded from view. Only something as small as a neutron star can spin fast enough to be a pulsar and not break up. The beams of radio waves come out from both magnetic poles, and the magnetic axis need not be the same as the rotation axis. The beam is a hollow cone, with the gas at the centre in different conditions as compared to the edge of the cone. In one pulsar the pulse shape is very complicated, suggesting that the rotating cone is made up of several beams. The first pulsar had a period of 1.337 s and was picked out during a survey at radio wavelengths for objects with signals that changed rapidly on a very short timescale (*see also* SCINTAR). Soon after the first discovery Bell and Hewish found three more pulsars, with periods of 0.253 s, 1.188 s and 1.274 s. The CRAB PULSAR has a very short period (0.033 s); it is at the heart of the Crab Nebula and the obvious result of a supernova. It is one of the few pulsars to be detected at visible wavelengths, and it has also been detected in X-rays and gamma rays. The beams of radio waves are pointing almost directly at Earth, so a weak pulse from the second beam can be seen between the main pulses (called the inter-pulse). The Hubble Space Telescope picture of the Crab Pulsar suggests there may be a JET (with knots of material) coming out from one pole, and waves of gas rippling outwards along the equatorial plane. The short period of the Crab Pulsar shows that it is a young pulsar, and it is very gradually slowing down. Almost all pulsars slow down because of the strong magnetic field, the except being the VELA PULSAR, which occasionally speeds up suddenly (an event called a glitch), producing a faster period.

There are around 50 pulsars, known as MILLISECOND PULSARS, with very short periods. Most millisecond pulsars are thought to have a weak magnetic field, which is why they have not slowed down. For some other millisecond pulsars, the short period is explained by proposing that the pulsar was once in a binary star system. The pulsar slowed down, but then speeded up again when material from the companion was accreted on to the pulsar (*see* BINARY PULSAR). One millisecond pulsar in a binary star system is called the 'black widow pulsar' because it is thought to have consumed most of its companion star. The pulses from the black widow pulsar disappear for 50 minutes every 9 hours, when the pulsar is eclipsed by its companion, and the eclipses show that the companion is very small (the mass is 2% of the solar mass although the size is 1.5 times that of the Sun).

Most pulsars have incredibly accurately determined periods, and the rate of change of the period is precisely known. This allows peculiarities to be identified easily. A pulsar was the first star to be identified as having planets orbiting around it. PSR B1257+12 has at least three planets in orbit, and there may be a fourth with a mass around that of Saturn. One pulsar was found to have a slightly irregular period, caused by the fact that it was in a binary system and the Doppler effect influenced the pulse period, by spreading the pulses apart slightly as the pulsar receded and jamming them together slightly as the pulsar approached.

pulsating star Star that expands and contracts, in particular, a class of VARIABLE STAR, including various types and subtypes. *See also* CEPHEID VARIABLE; IRREGULAR VARIABLE; MIRA STAR; RR LYRAE STAR; SEMIREGULAR VARIABLE

Puppis *See* feature article

Purbach (or Peurbach), Georg von (1423–61) Austrian mathematician-astronomer and student of Ptolemaic theory whose table of lunar eclipses, published in 1459, was still in use two centuries later. Purbach's real family name is unknown; he is identified by the name of the town in Austria where he was born. He revised and translated Ptolemy's *Almagest*, work completed after his death by his pupil REGIOMONTANUS. In his *Theoricae novae planetarum* ('New Theory of the Planets', 1460), Purbach attempted to explain the structure of the Solar System in terms of both Ptolemy's epicyclic orbits and the homocentric spheres of EUDOXUS and ARISTOTLE.

Purbach Lunar crater (25°S 2°W), 120 km (75 mi) in diameter, with rim components reaching 2440 m (8000 ft) above its floor. This ancient crater is from the earliest age of the Moon. Impact erosion has erased its ejecta blanket, rounded its rim, destroyed the north wall, and deeply incised its walls. Several ridges run through the central region of Purbach; these ridges represent the rims of dilapidated craters, perhaps in combination with central peak remnants.

Purcell, Edward Mills (1912–97) American physicist who won the 1952 Nobel Prize for Physics for his role in developing the technology of nuclear magnetic resonance. His major contribution to astronomy was his discovery (1951) of the TWENTY-ONE CENTIMETRE radio radiation from the spin-flip transition of neutral hydrogen atoms, predicted by Jan OORT. The 21-cm emission from H I regions enabled the structure of the Milky Way and other galaxies to be mapped.

Purkinje effect Change in the colour sensitivity of the eye that occurs when DARK ADAPTATION takes place. The normal level of sensitivity of the human eye lies in the 400–750 nm range, with the peak lying in the yellow to green region of the spectrum. As the level of illumination drops, this shifts more towards the green, sensitivity towards the red end of the spectrum decreasing correspondingly. The effect, which is named after the Czech physiologist Jan Purkinje (1787–1869), can also influence an observer's perception of star colours and magnitudes.

Purple Mountain Observatory Astronomical institution of the Chinese Academy of Sciences, also called Zijin Shan Observatory or Nanjing Observatory. It is located on Mount Zijin on the eastern outskirts of Nanjing, capital of Jiangsu Province, at an elevation of 267 m (880 ft). It was built in 1929 and began operation in 1934. Today, the observatory operates several optical telescopes up to 0.6 m (24 in.) in aperture and a 13.7-m (45-ft) millimetre-wavelength radio telescope at its Delinha station in Qinhai Province. The observatory runs research groups in celestial mechanics, astrophysics, radio and space astronomy and solar physics. It has four remote outstations within China, and possesses some of the Ming Dynasty instruments formerly at the ANCIENT BEIJING OBSERVATORY.

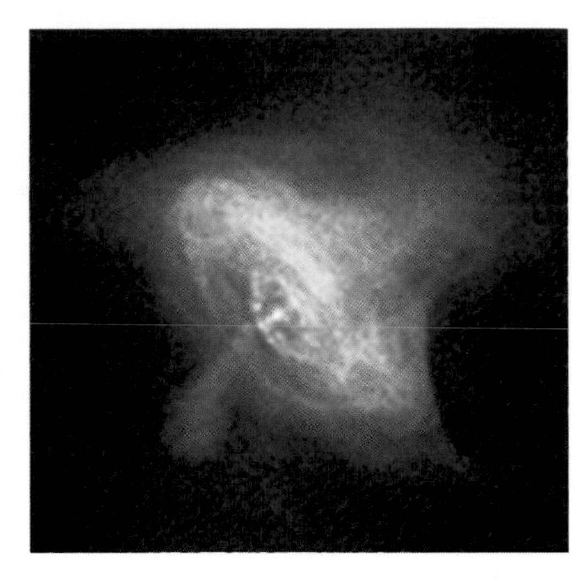

◄ **pulsar** This Chandra X-ray image shows the heart of the Crab Nebula, the site of a pulsar remnant from the supernova explosion of 1054. A thin jet almost 20 l.y. long emerges from the pulsar's south pole towards lower left.

Putredinis Palus (Marsh of Decay) Lunar lava plain located in the north-central region of the Moon. It is an irregularly shaped region within the Mare IMBRIUM. It contains numerous volcanic features, including a dark mantling deposit, produced by fire fountaining, and Hadley Rille, a SINUOUS RILLE that is an ancient lava channel. An arcuate rille cuts through part of this province. Putredinis Palus was selected for the Apollo 15 mission, which returned samples of pyroclastic 'green glass'.

Pyxis *See* feature article

PZT *See* PHOTOGRAPHIC ZENITH TUBE

PUPPIS (GEN. PUPPIS, ABBR. PUP)

Large southern constellation, part of the old Greek figure of ARGO NAVIS, the ship of the Argonauts; Puppis represents the ship's poop, or stern. The brightest stars of Argo are now in neighbouring Carina and Vela, so the lettering of the stars in Puppis begins with Zeta, its brightest, known as NAOS. ξ Pup is a wide binocular double, mags. 3.3 and 5.3, while k Pup consists of two nearly equal 5th-magnitude stars divisible in small telescopes. L^2 Pup is a semiregular variable ranging between 3rd and 6th magnitudes with a period of around 140 days; it forms a wide optical double with L^1 Pup, mag. 4.9. V Pup is an eclipsing binary, a BETA LYRAE STAR of range 4.4–4.9, period 1.45 days. Lying in the Milky Way, Puppis is rich in open star clusters. Among the best are M46 and M47, both just visible to the naked eye. M93 is a binocular cluster. NGC 2451 includes a mag. 3.6 star, c Pup. Next to it lies NGC 2477, which resembles a globular when seen through binoculars.

BRIGHTEST STARS

	Name	RA h m	dec. ° ′	Visual mag.	Absolute mag.	Spectral type	Distance (l.y.)
ζ	Naos	08 04	−40 00	2.21	−5.95	O5	1400
π		07 17	−37 06	2.71	−4.92	K3	1100
ρ	Tureis	08 08	−24 18	2.83	1.41	F5	63
τ		06 50	−50 37	2.94	−0.80	K0	183
ν		06 38	−43 12	3.17	−2.39	B8	420
σ		07 29	−43 18	3.25	−0.51	K5	184

PYXIS (GEN. PYXIDIS, ABBR. PYX)

Small southern constellation introduced by Lacaille in the 18th century. It lies on the edge of the Milky Way adjoining Vela and Puppis, and represents a magnetic compass as used aboard ships. Its brightest star is α Pyx, visual mag. 3.68, distance 845 l.y., spectral type B1.5 III. T Pyx is a RECURRENT NOVA that has undergone five recorded eruptions, in 1890, 1902, 1920, 1944 and 1966, although it reaches only 6th magnitude at maximum.

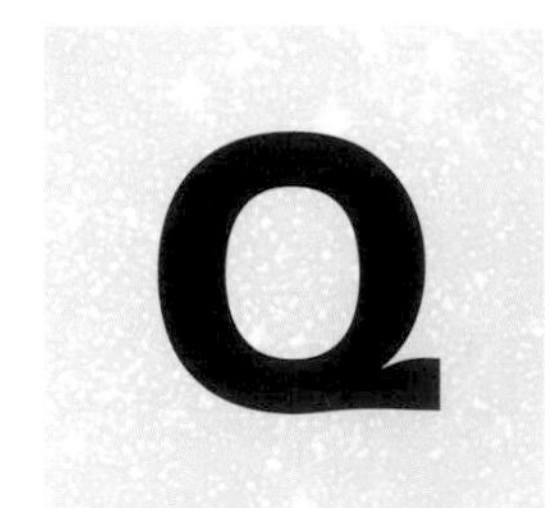

QSO Abbreviation of QUASISTELLAR OBJECT

quadrant Ancient hand-held navigational instrument consisting of a graduated 90° arc, resembling half a school protractor, and a plumb-line to mark the vertical. Observations were made by sighting a star along one side and reading the star's altitude where the plumb-line crossed the scale. It was succeeded by the SEXTANT. *See also* MURAL QUADRANT

Quadrantids Major annual METEOR SHOWER, active between January 1 and 6. The sharp maximum, on January 3 or 4, can produce zenithal hourly rates (ZHR) of up to 130 meteors/hr, but activity at other times is usually rather low. The shower is named after its RADIANT in the obsolete constellation of Quadrans Muralis, the stars of which are in what is now identified as northern Boötes.

Quadrantid meteors show a number of characteristics typical of a cometary origin, but identification of the stream parent has been rendered nigh-impossible by the stream's rapid orbital evolution. One possible candidate is the faint short-period (5.24 years) comet 96P/Machholz 1. Repeated close passage at aphelion of the stream orbit to Jupiter has led to a considerable spread of its component meteoroids. Computer modelling suggests that the stream was active from a radiant in Aquarius during July until the early centuries AD, before being perturbed away from encounter with Earth. Chinese and other annals provide evidence for this shower. Having been inactive for over a thousand years, the stream began to be encountered at the descending node of its orbit once more from about 1700 onwards, as the modern Quadrantid shower. Projecting the computer model into the future, the stream orbit is expected to be pulled away from Earth around 2200.

The January Quadrantids have been known since about 1835, with a more or less continuous observational record since the 1860s. There is apparent variation in the peak activity, with some remarkably high returns, notably in 1977 and 1992. Although widely spread at aphelion, the meteoroids are encountered by Earth close to their perihelion, where the stream is concentrated and has a narrow cross-section. Consequently, the peak activity is restricted to a comparatively brief period, less than 12 hours. At a given location, perhaps only one return in ten years may prove favourable, with the maximum occurring during the pre-dawn hours when the radiant is highest, and in the absence of clouds and moonlight.

The Quadrantids show evidence for particle mass-sorting, with faint meteors being proportionally more abundant in the early part of the peak, and bright events (produced by larger meteoroids) peaking later. This is consistent with the smaller meteoroids gradually spiralling towards shorter period orbits as a result of the POYNTING–ROBERTSON EFFECT. Dispersal of Quadrantid stream meteoroids is also reflected in the observation that the shower radiant is relatively diffuse, except close to maximum activity.

Quadrantid meteors are medium-paced at 41 km/s (25 mi/s). The fainter examples are often described as bluish, whereas the bright Quadrantids sometimes show a pronounced yellow-green colour.

quadrature Position of the Moon or a planet when at right angles to the Sun as seen from Earth. At this time, its ELONGATION, the angular distance between it and the Sun, is either 90° or 270°. The Moon is at quadrature when it is at first and last quarter. *See also* CONJUNCTION

quantum theory Theory introduced by Max PLANCK and Albert EINSTEIN at the beginning of the 20th century that showed that light really has a dual nature: it can interact with matter either as a particle (photon), exchanging energy and momentum, or as a wave, producing interference and diffraction phenomena. In 1924 Louis Victor de Broglie (1892–1987) introduced the concept that all forms of matter possess the dual behaviour of particles and waves. The de Broglie wavelength λ_d of a particle is given by $\lambda_d p = h$, where p is the momentum of the particle and h is the PLANCK CONSTANT. Clinton J. Davisson (1881–1958) and Lester Germer (1896–1971) experimentally confirmed the wave/particle duality of particles when they showed that electrons exhibit interference effects. Erwin Schrödinger (1887–1961) constructed a set of differential equations that describe the wave structure of atomic particles, and the laws that govern their interactions. Later developments in quantum mechanics, principally by Paul Dirac (1902–84) and Werner Heisenberg (1901–76), provided the backbone of the quantum theory of matter.

The quantum-mechanical approach pictures the electron as a wave structure that resides in energy levels where the amplitude of the wave is greatest in the vicinity of the classical Bohr electron orbit. In the quantum-mechanical picture, there is a finite probability that the electron can be anywhere in the Universe, but the probability density is highest at the classical orbit or energy level. The quantization of electron orbits emerges in a very straightforward manner in Schrödinger's wave formulation: the 'orbit' corresponds to the region where there are an integral number of waves around the atom. This three-dimensional pattern can also be thought of as the probability density of a cloud of negative charge (integrating to the charge on a single electron) around the atom.

The nucleus of the atom, being constructed of individual protons and neutrons, is also described by quantum mechanics. Its greater mass results in a more concentrated probability distribution than for an electron. Schrödinger's equation can be written down for any atomic system. It will include the effects of the electrostatic and magnetic interactions of all the constituent particles. The solution of this equation gives an exact description of the energy states of the system. Only in the simplest of atoms (hydrogen and helium) can accurate solutions – called wave functions or eigenfunctions – of the Schrödinger equation be obtained.

For complex structures (atoms containing many electrons, or the atomic lattice structure of crystals) only approximate solutions to the Schrödinger equation are, in general, possible. Modern high-speed computers are capable of furnishing solutions of wave equations for quite complex systems. In the area of astrophysics, such solutions can give, for example, the energy levels and hence emission-line wavelengths of highly ionized species or complicated molecules not observable in the laboratory.

The principal structures of atomic nuclei and the ways in which they interact are also governed by quantum mechanics. Some of the nuclear reactions of interest in the construction of theoretical models of stars cannot be measured in the laboratory because they occur at too low a rate. For such reactions, calculations of rates must be made, which involve solutions of the wave equations for the two nuclei as they approach, collide and recede from one another.

The probabilistic nature of quantum mechanics provides an explanation for the ability of particles to 'tunnel' from one state to another (of equal energy) without having to acquire the extra energy to pass over the potential barrier that separates the two states. Such tunnelling, for example, allows nuclear reactions to occur at lower energies (and hence lower temperatures) than they would otherwise require.

On a somewhat grander scale, attempts are currently being made to examine the properties of wave functions that represent the entire Universe, to see if the BIG BANG could be the result of tunnelling from a different state.

quantum theory of gravity Theory intended to possess the predictive power of GENERAL RELATIVITY and to exhibit quantum properties. Einstein's theory of general relativity is based on a form of mathematics called differential geometry and also on the idea that SPACETIME is a

CONTINUUM. Continuum theories differ from QUANTUM THEORY in several ways, the most basic being that there is some smallest size or unit in quantum theories, but not in continuum theories. Physically this means that a region of spacetime can be divided into infinitely small parts and there will never be a 'smallest' piece.

The mathematical apparatus of each of these types of theories is totally different. Theories of the strong, weak and electromagnetic forces are all quantum theories and they are described by wave functions and probabilities. Gravitational theory, on the other hand, is described by differential equations on a continuous manifold. There are no uncertainties in general relativity and no discrete quantum states. Many physicists, including Einstein himself, have endeavoured to find a quantum theory of gravity that would retain the predictive power of general relativity, but also exhibit quantum properties. If one were successfully to develop such a theory, one would need to postulate the existence of an exchange particle that mediates the GRAVITATIONAL FORCE. This BOSON, called a GRAVITON, must be mass-less so the force is proportional to $1/r^2$, and it must have a spin of 2. Individual gravitons have never been directly detected.

Every attempt to derive a complete quantum theory of gravity and to put it in a mathematical form similar to the other three FUNDAMENTAL FORCES has failed. Problems arise at very small distances, close to the PLANCK SCALE, where quantized theories of gravity have characteristics, described as 'quantum foam', that are not compatible with general relativity. This quantum foam represents fluctuations in the shape and form of the underlying spacetime and would lead to physical effects that are not observed.

SUPERSTRING THEORY and other TOEs (*see* THEORY OF EVERYTHING) are attempts to quantize gravity, and to make it compatible with the other three fundamental forces. Although these theories are progressing, they are plagued by a lack of testable predictions since the interaction energies involved are well beyond the energies achievable in nuclear accelerators. It is quite possible that cosmological observations might be the only way to test string theories of elementary particles.

quark Hypothetical subnuclear particles that are believed to be the fundamental building block of HADRONS (baryons and mesons). They have fractional electrical charges ($+\frac{2}{3}$ or $-\frac{1}{3}$, compared with the unit charge carried by a proton or an electron), and have spin values of $\frac{1}{2}$.

There are believed to be six types, or 'flavours', of quark, designated 'up', 'down', 'strange', 'charm', 'top' and 'bottom'. The up, charmed and top quarks have charges of $+\frac{2}{3}$, while the down, strange and bottom quarks have charges of $-\frac{1}{3}$. Up and down quarks have masses of 0.3 MeV, the others are heavier. Each flavour of quark has an antiquark with opposite properties.

Each baryon is thought to consist of a cluster of three quarks (or antiquarks), while mesons consist of quark–antiquark pairs. For example, a proton consists of two up quarks and one down quark ($u + u + d = +\frac{2}{3} + \frac{2}{3} - \frac{1}{3}$ = net charge of 1), a neutron consists of one up quark and two down quarks ($u + d + d = +\frac{2}{3} - \frac{1}{3} - \frac{1}{3}$ = net charge of zero), and a positive pi meson consists of an up and an anti-down quark ($+\frac{2}{3} + \frac{1}{3}$ = net charge of 1).

Quarks also have a property called 'colour', which is analogous to electrical charge and has nothing to do with colour in the conventional sense. There are three different colours (and three equivalent anti-colours for antiquarks). Hadrons have zero net colour, and this can be produced either by combining three quarks, each of different colour, or by combining two quarks, one with the anti-colour of the other. The requirement for zero net colour explains why baryons are composed of three quarks and mesons of two.

The interquark force is known as the colour force, and the interaction between quarks is carried by mass-less particles known as gluons. The strong nuclear interaction between hadrons themselves is believed to be a remnant of the interquark force. The colour force is weak when quarks are close together, but becomes extremely powerful as quarks move apart, preventing quarks from escaping from within hadrons. Almost certainly, therefore, quarks cannot exist outside hadrons as separate free particles.

quasar Most luminous category of ACTIVE GALACTIC NUCLEUS. The name is formed from the acronym QSRS (for quasistellar radio source) because the initial identification in 1963 was of several examples that had strong radio emission and appeared in the *Third Cambridge Catalogue* of radio sources.

Quasars are characterized by a very luminous core source, with a blue continuum and strong, broad emission lines, strong X-ray emission, and, for quasars proper, strong radio emission, either from the core itself or from a double radio source much like that around RADIO GALAXIES. Any surrounding galaxy must be very faint, so that it would not appear on conventional photographs. Quasars generally have redshifts (z) greater than 30,000 km/s (20,000 mi/s) out to the largest values yet observed. Most quasars are variable on a variety of timescales, which set limits to the size of the emitting region. These limits are a few light-days or even smaller, which is one of the reasons for the popular model in

▼ **quasar** This set of quasar images shows the bright nuclear activity that lies at the heart of these galaxies. Each of these quasars is probably fuelled by material falling into a massive central black hole.

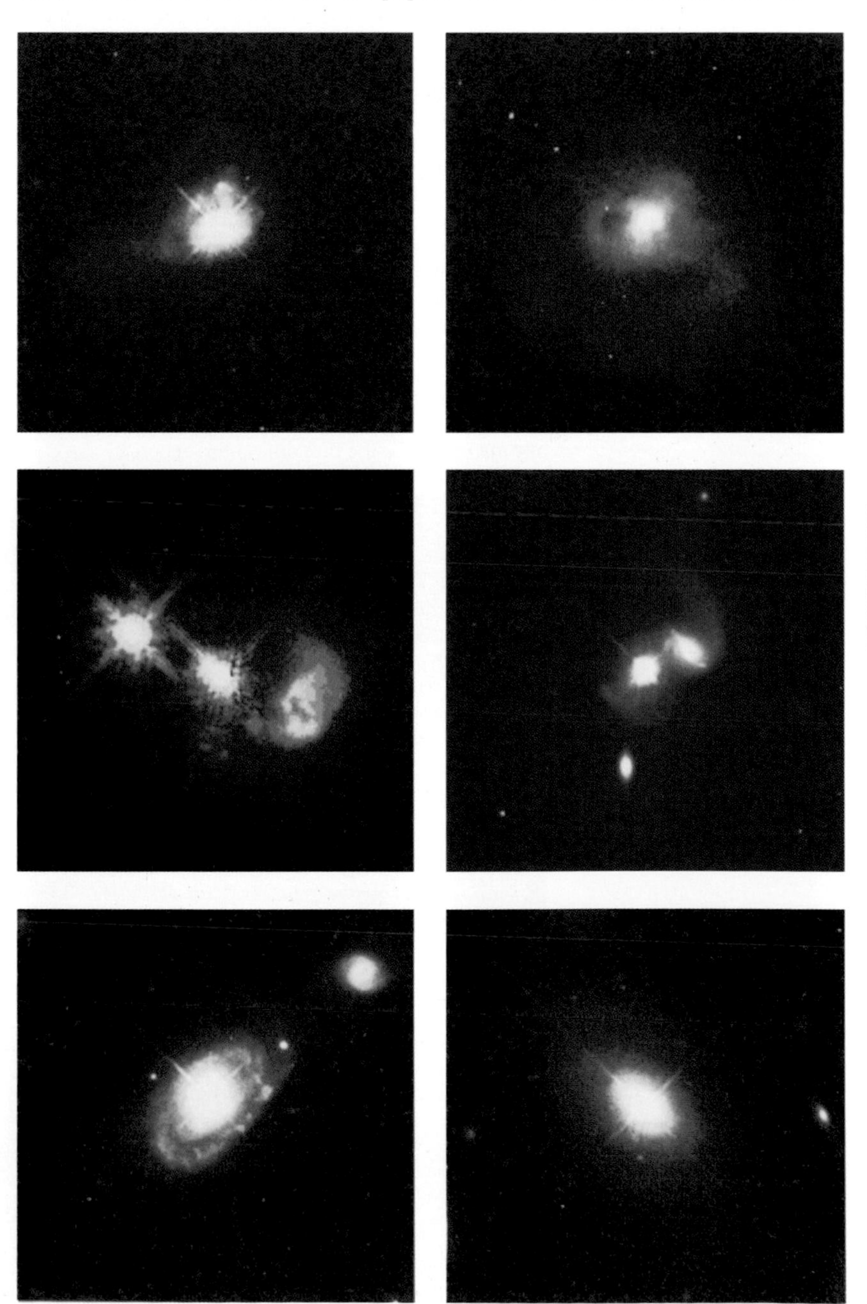

Q

► **quasistellar object** The Very Large Telescope (VLT) images of HE 1013-2136 show the quasar's intensely bright nucleus. The right-hand image, enhanced to show more of the quasar's surroundings, reveals probable tidal tails and knots in the host galaxy.

which the ultimate energy source is material accreting into a massive BLACK HOLE. The exact origin of the observed radiation remains unclear in this model, particularly where the continuum radiation arises with respect to the accretion disk.

Some information is now available on the host galaxies of quasars at moderate redshifts, thanks to data from the Hubble Space Telescope. Radio-quiet quasistellar objects (QSOs) may occur in spiral, elliptical or merging galaxies, while radio-loud objects occur in elliptical galaxies or strongly merging systems. There is a marked excess of galaxies with very close, compact companions, showing that not only can some kinds of interaction trigger the QSO phenomenon, but also that it can happen in episodes not much longer than the lifetime of such an interaction (or a statistical connection would not be seen). Quasars form a continuum in luminosity and spectroscopic properties with type 1 SEYFERT GALAXIES, so that there are some objects that have been treated as belonging to both categories because of their intermediate luminosity.

Various arguments have suggested that quasars and radio galaxies are related in a 'unified scheme'. This holds that most such objects have a core surrounded by a torus of obscuring material which absorbs radiation from infrared to soft X-rays very effectively, so that a quasar viewed near the plane of such a torus would appear as a radio galaxy.

The population of quasars has evolved strongly with cosmic time. Going out in redshift from our neighbourhood, the space density of quasars increases dramatically to redshift z=2.2. Beyond that (at earlier times), the density declines again, perhaps indicating that the objects actually turned on for the first time at this epoch.

Quasars are very important in the study of cosmology and galaxy evolution, not least because they furnish bright and distant background sources that allow detection of cool foreground gas (in and out of galaxies) via absorption lines. Quasars also allow the detection and measurement of foreground masses through GRAVITATIONAL LENSING.

quasistellar object (QSO) General designation for an object with the optical and X-ray properties of a QUASAR, but without the strong radio emission. QSOs by this definition outnumber radio-loud quasars by about a factor of ten, although in loose usage 'quasar' is often used to include both kinds.

Quetelet, (Lambert) Adolphe (Jacques) (1796–1874) Flemish statistician and astronomer, from 1828 director of the Royal Observatory, Brussels. Quetelet did important work on meteor showers and their radiants. In 1839 he published a *Catalogue des principales apparitions d'étoiles filantes*, which listed 315 meteor displays, calling attention to the recurrence of the Perseid meteors every August 12.

quiet Sun Term applied to the Sun when it is at the minimum level in the SOLAR CYCLE, as opposed to the active Sun. The quiet Sun shows reduced levels of FLARE, CORONAL MASS EJECTION, SUNSPOT and other activity.

quintessence Undetected particles proposed as a fifth fundamental force. Observations of distant supernovae suggest that the Universe is accelerating. This fact, coupled with the COSMIC MICROWAVE BACKGROUND results and the fact that there is not enough observed mass to close the Universe, suggests that the cosmological constant in Einstein's FIELD EQUATIONS is non-zero. The physical manifestation of this repulsive force could come in the form of previously undetected particles called quintessence. Quintessence (or fifth force) would have negative gravity and would cause the Universe to accelerate its expansion. There is no complete theory of quintessence particles and some particle physicists think that if they exist, they must be the size of galactic clusters.

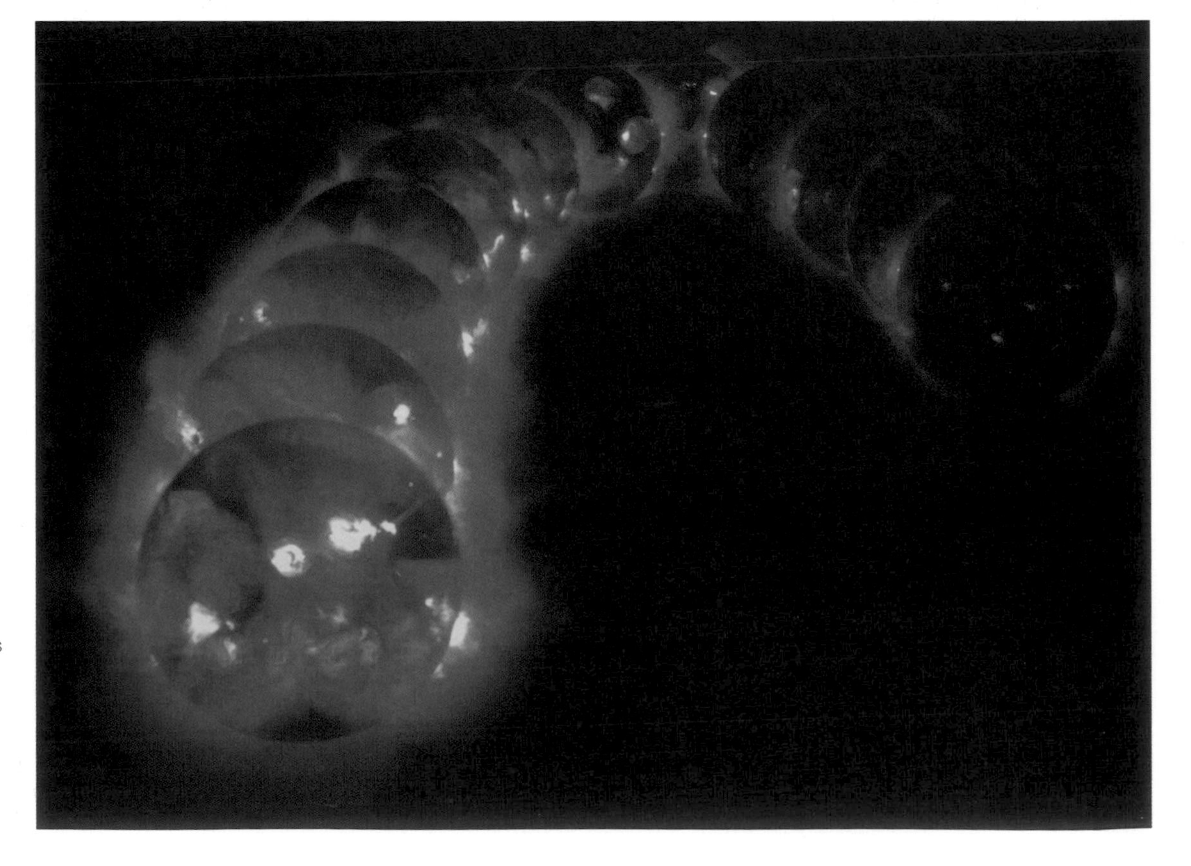

► **quiet Sun** This time series of YOHKOH X-ray images of the Sun shows its progression through the solar cycle. At maximum (left), the X-ray Sun is bright, with many strong magnetic loop structures evident in the inner corona above active regions. By solar minimum, the quiet Sun (right) shows only isolated X-ray bright spots.

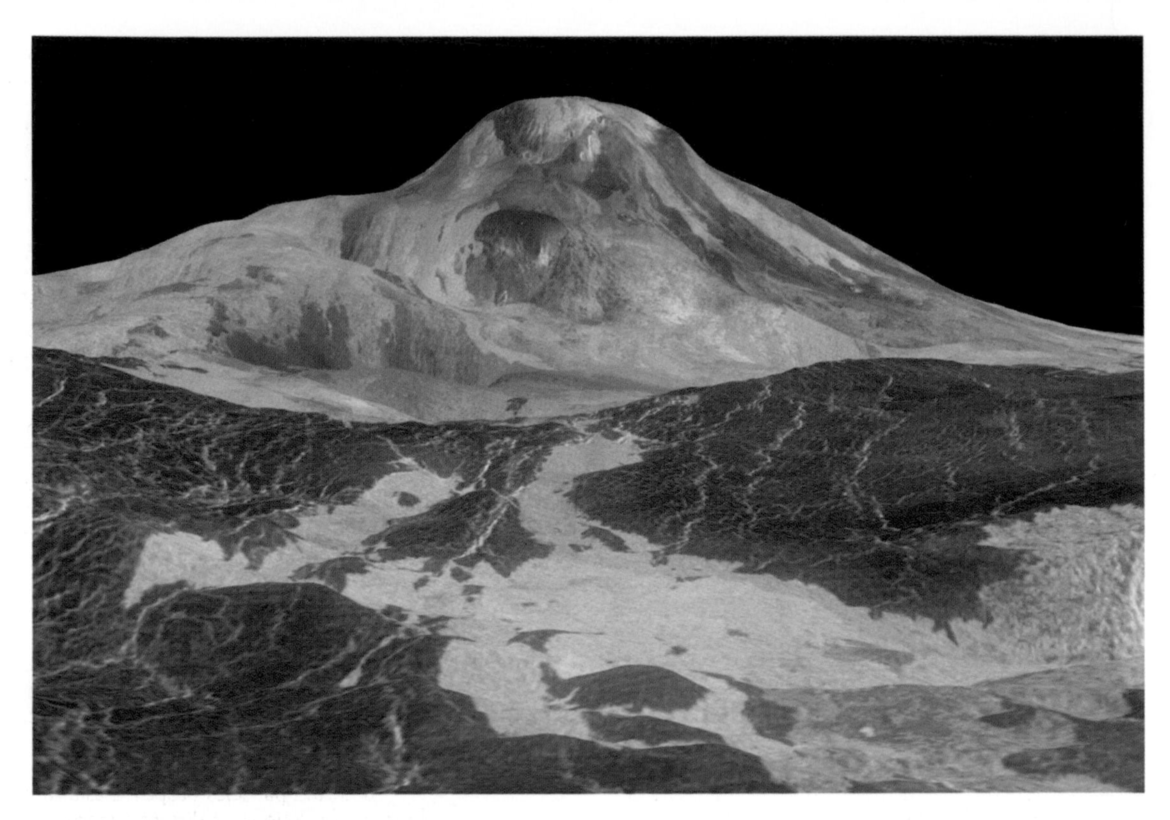

◀ **radar astronomy** This Magellan orbiter radar image shows Maat Mons on Venus. Radar observations have allowed planetary astronomers to penetrate Venus' dense atmosphere and obtain an idea of the planet's surface topography.

radar astronomy Category of research using continuous-wave or pulsed transmissions, from Earth or from space satellites, to study Solar System bodies. In 1944, when Germany began the bombardment of London by V2 rockets, Stanley HEY was asked to modify an anti-aircraft gun-laying radar in order to give early warning of the approach of a V2. By re-directing the aerial beam to an elevation of 60° radar echoes were obtained from the rockets, but many other short-lived radar echoes were observed that had no connection with the rockets. Hey concluded that these transient echoes were associated with the entry of METEORS into the upper atmosphere. The radar waves are reflected from the long, thin columns of ionization formed at altitudes of about 100 km (60 mi) when a meteoroid evaporates as it plunges into the Earth's atmosphere. The meteor trails allow the measurement of the wind systems in the 100 km region of the atmosphere. The radar technique allowed meteors to be observed during daylight and through clouds, revealing that major METEOR SHOWERS were active in summer daytime. Methods of measuring the velocity of the entry of meteors into the atmosphere were developed and a long-standing controversy about the origin of the SPORADIC METEORS was settled.

In 1947 November Frank Kerr (1918–2001) and Charles Shain (d.1960) of the Commonwealth Scientific and Industrial Research Organisation (CSIRO) in Australia investigated the erratic changes in the strength of the radar echoes scattered from the Moon. They made use of the Australian Government's short-wave transmitter, normally used for broadcasting to North America, and were able to distinguish two types of fading of the lunar echoes: a rapid fading with a period of seconds was superimposed on a longer-period fading of about 30 minutes. They suggested that the rapid fading was associated with the libration of the Moon and that the long-period fading might have an ionospheric origin. This was confirmed by JODRELL BANK, and it was discovered that the long-period fading was caused by the rotation of the plane of polarization of the radar waves as they passed through the Earth's IONOSPHERE (Faraday rotation). Further research showed that the radar waves were not reflected uniformly from the whole disk of the Moon, but from a region at the centre of the visible disk that had a radius of only one-third of the lunar radius. Another significant result of lunar radar work has been the systematic measurement of the total electron content between the Earth and the Moon derived from an analysis of the longer-period fading of the lunar echoes.

A radar measurement of the distance of Venus settled the ambiguities in the value of the SOLAR PARALLAX (hitherto known only to 0.1%) and the problem of the rate of rotation of the planet. Success was first achieved by NASA equipment at Goldstone on 1961 March 10 using a continuous-wave system. Pulsed radar contact was achieved at Jodrell Bank on April 8, soon followed by other successes in the United States and the Soviet Union. From these distance measurements a definitive value of the solar parallax was agreed. The spin of the planet affects the spectrum of the scattered radar waves, and in 1962 and 1963 measurements in the Soviet Union and the United States first established that the planet was in retrograde rotation with a period of 243 days. In 1962 the Soviet Union achieved radar contact with Mercury, followed by the Jet Propulsion Laboratory in the United States in 1963 May. In 1963 February radar contact was established with Mars by both the United States and the Soviet Union. Procedures were developed for radar mapping of the planetary surfaces. In the case of Mercury and Mars, optical imaging from space probes soon provided superior data, but the radar mapping of the perpetually cloud-covered surface of Venus has been of outstanding importance. Radar maps have been obtained of the planet's surface with a resolution of a few kilometres using APERTURE SYNTHESIS. In 1979 a computer analysis of several months of the data from the radar altimeter carried in the American PIONEER VENUS Orbiter produced a topographical map of more than 90% of the surface of Venus. The Soviet VENERA 15 and 16 spacecraft placed in orbit around Venus in 1983 carried an aperture synthesis radar system that mapped the whole planetary surface above 30° north latitude with a resolution of 1 to 2 km (*c*.1 mi). NASA's MAGELLAN satellite, launched in 1989, mapped 98% of Venus' surface with a resolution of around 100 m. Other objects in the Solar System detected by radar include a number of asteroids, the four Galilean satellites of Jupiter, and Saturn's rings. Radar echoes were obtained from Comet IRAS–ARAKI–ALCOCK by the Goldstone and Arecibo antennae when it made a close approach to Earth (0.031 AU) on 1983 May 11.

radial velocity Component of velocity directed along the line of sight. Radial velocities are found from the Doppler effect, in which the spectrum lines from an approaching body are shifted to shorter wavelengths ('to the blue') and those from a receding body to longer wavelengths ('to the red'). At low velocities compared with the speed of light (such that relativity theory is not required), the velocity relative to the speed of light is directly proportional to the relative shift, and is considered positive for a receding body, negative for an approaching one.

radiant Circular area of sky, usually taken as a matter of observational convenience to have a diameter of 8°, from which a METEOR SHOWER appears to emanate. Radiant positions are normally expressed in terms of right ascension and declination. Meteor showers usually take their name from the constellation in which the radiant lies – the PERSEIDS from Perseus, GEMINIDS from Gemini, and so on. The radiant effect, with meteors appearing anywhere in the sky with divergent paths whose backwards projection meets in a single area, is a result of perspective: in reality, shower members have essentially parallel trajectories in the upper atmosphere. Meteors close to the radiant appear foreshortened, whilst those 90° away have the greatest apparent path lengths. Meteor shower radiants move eastwards relative to the sky background by about one degree each day thanks to Earth's orbital motion.

A meteor shower radiant position may be determined from the backwards projection of plotted or photographed trails. The radiant can also be found from parallactic observations of a single meteor recorded at geographical locations separated by a few tens of kilometres.

radiation Energy transmitted in the form of electromagnetic waves or photons, or by subatomic particles. Radiation provides us with almost all the information we have about the Universe. In particular, the study of radiant energy is fundamental to our understanding of stellar structure, because it enables us to deduce the temperatures and luminosities of stars as well as their chemical constitutions.

Radiation behaves in two different ways depending upon the interaction with which it may be involved. During INTERFERENCE, DIFFRACTION and POLARIZATION it behaves like a wave, while during the photoelectric effect and the COMPTON EFFECT it behaves as though it were formed of a stream of particles. This wave–particle duality is not unique to radiation; entities such as ELECTRONS and PROTONS that are normally regarded as particles also exhibit a wave nature, as for example in the operation of the electron microscope.

As a wave, radiation has a wavelength, λ, and a frequency, f or v, which are related by the equation $\lambda f = c$, where c is the velocity of LIGHT (*see* ELECTROMAGNETIC SPECTRUM). As a particle, called a PHOTON, or quantum, each particle has an energy given by $E = hf$, where h is the PLANCK CONSTANT.

Of particular interest to astronomy is the relationship between the temperature of the surface of a hot body and the way that surface emits radiation, which is affected by the nature of the surface itself. It is well known that not all surfaces raised to the same temperature radiate energy in the same way. Some reflect well, but others may be highly absorbing. By KIRCHHOFF'S LAW the efficiency of the emission by a heated object at a particular wavelength is proportional to the efficiency of its absorption at the same wavelength. Thus an object that absorbs with 100% efficiency over the whole SPECTRUM, known as a black body, will also be the most efficient when it comes to emitting radiation. The emission from a black body, known as BLACK BODY RADIATION, is a good approximation to the emission from many astronomical objects, including stars (SPECTRUM lines are mostly minor deviations from the overall emission). Black body radiation follows a bell-shaped curve (*see* PLANCK DISTRIBUTION). The peak of the distribution shifts to shorter wavelengths as the temperature increases (*see* WIEN'S LAW), which leads to the common experience that at moderate temperatures objects glow a dull red, then change colour successively through bright red, yellow, white to blue-white as the temperature is increased. The total emitted energy increases rapidly with temperature, leading to the STEFAN–BOLTZMANN LAW. At wavelengths significantly longer than that of the peak emission, the simpler Rayleigh–Jeans approximation may often be used, while at wavelengths shorter than the peak, the Wien approximation can be used.

The operation of these laws leads to a simple relationship between the colour of a star and its surface temperature, T. For example:

$$T \approx \frac{8540}{(B-V)+0.865}$$

where (B – V) is the COLOUR INDEX. *See also* BREMSSTRAHLUNG; ČERENKOV RADIATION; ELECTROMAGNETIC RADIATION; FREE–FREE TRANSITION; LIGHT; LIGHT, VELOCITY OF; SYNCHROTRON RADIATION

radiation belts Regions of a planetary MAGNETOSPHERE that contain energetic ions and electrons trapped by the planet's magnetic field. These belts are usually doughnut-shaped regions. The particles trapped in the belts spiral along the magnetic field lines and bounce backwards and forwards between reflection points encountered as they approach the magnetic poles. The electron motion produces SYNCHROTRON RADIATION, a characteristic emission from the individual planetary systems. The particles are captured from the SOLAR WIND or are formed by collisions between COSMIC RAYS and ions in the planet's outer atmosphere. The Earth, Jupiter, Saturn, Uranus and Neptune are all known to possess radiation belts, while the magnetosphere of Mercury is too small compared to the planetary radius to support this phenomenon.

The radiation belts of Jupiter are vast in spatial scale and are additionally supplied by accelerated particles originating from the volcanic moon IO. The resulting particle flux is about 1000 times more intense than in the VAN ALLEN BELTS that surround the Earth. They present a formidable obstacle for spacecraft, such as GALILEO, since in the region of the most intense radiation the probe would receive an integrated dose of about 200,000 rads from the electrons and 50,000 rads from the protons. This may be compared with a dose of 500 rads, which is sufficient to kill a man. The intense bombardment of a spacecraft can saturate the sensitive instruments, upset the on-board computer systems and interfere with its communications with the Earth (*see also* SPACE WEATHER).

The radiation belts of Saturn were discovered by the PIONEER 11 spacecraft in 1979. The extensive system of rings and the satellites residing inside the magnetosphere of Saturn sweep away the charged particles, such that the fluxes maximize just outside the rings. The radiation belt particles around Uranus and Neptune also interact with

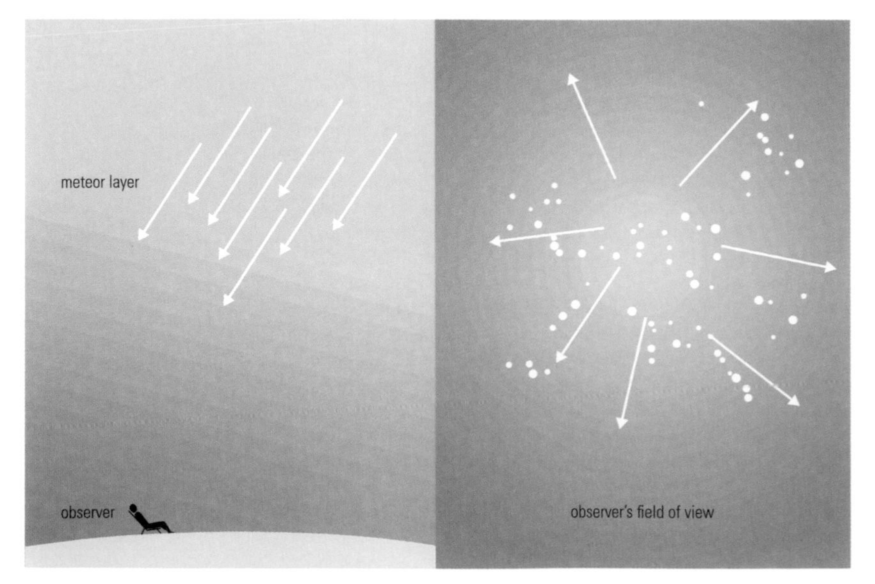

▼ **radiant** Meteors from a common source, occurring during a shower such as the Perseids, enter the atmosphere along parallel trajectories, becoming luminous at altitudes around 80–100 km (50–60 mi). If an observer on the ground plots the apparent positions of meteors on a chart of the background stars, they will appear to diverge from a single area of sky, the radiant. The radiant effect is a result of perspective.

the ring and satellite systems, but the fluxes are relatively low, probably due to the weak interaction between the solar wind and the planetary magnetospheres at these distances from the Sun.

radiation era Time before the universe became transparent to photons about 300,000 years after the BIG BANG. The period before this transition is called the radiation era since radiation dominated the thermal evolution of the universe up until that time. After the universe became transparent, matter began to dominate and atoms, stars and galaxies could form.

radiation pressure Pressure exerted on a surface by incident ELECTROMAGNETIC RADIATION. This pressure arises because each incident photon transfers a tiny quantity of momentum to the surface. For BLACK BODY RADIATION, the pressure is given by:

$$P = \frac{4\sigma T^4}{3c}$$

where σ is the STEFAN–BOLTZMANN constant, T the temperature and c the velocity of light. One well-known observed effect concerns the dust tails of COMETS, which are affected by the Sun's radiation pressure and therefore always point more or less away from the Sun, so that when moving outwards, after perihelion, the dust tail of a comet precedes the nucleus. Radiation pressure also limits the maximum mass of a star to about 100 solar masses. At that size, the temperature of the star will be sufficiently high that the radiation pressure pushing outwards balances the inward forces of gravity. Since both gravity and radiation follow INVERSE SQUARE LAWS, once equality is reached at one point within the star, it will be equal throughout the star. Any material moving outwards will therefore continue to do so, and the star will become unstable, shedding material until its temperature is lowered sufficiently for it to become stable again.

radiation temperature Temperature of an object found by assuming that its emission at a particular wavelength, or over a small wavelength region, is that of a BLACK BODY. The temperature is obtained using the Planck equation (*see* PLANCK DISTRIBUTION). It is similar to the EFFECTIVE TEMPERATURE, except that the effective temperature uses the total emission at all wavelengths from the object and hence the STEFAN–BOLTZMANN LAW.

radiative transfer Most important process by which heat energy is transported from a star's hot interior to its surface. In radiative transfer, high-energy photons lose energy as they travel outwards through the hot plasma. This loss of energy occurs as the photons are scattered, mainly by free electrons. The photons can also lose energy if they are absorbed by an ion, and photoionization occurs. These processes take place in the radiative zones of the stars.

radioactive age dating Method of determining age by measuring the RADIOACTIVITY of a nuclide with a known half-life. The principle behind radioactive age dating is the fixed rate with which an unstable radioactive isotope (the parent) decays to a stable isotope (its daughter). The time taken for a radionuclide to decay to half its initial abundance is the half-life ($T_{1/2}$) of the system. Several isotope systems with different half-lives are used to measure different events within Solar System history. Long-lived radionuclides, such as uranium or rubidium, are used to date the age of formation of meteorites. Short-lived radionuclides, such as plutonium, aluminium or manganese, date the formation interval between stellar processing of material and its incorporation into meteoritic components.

radioactivity Spontaneous emission of radiation from an atomic nucleus. An atom of a given chemical element is characterized by the number of protons in its nucleus, which is known as its atomic number. Nuclei of the same element may have different numbers of neutrons and thus different masses; these are called ISOTOPES of the element. Each nucleus with a particular number of protons and neutrons is called a nuclide. The sum of the numbers of protons and neutrons is the atomic weight. Some nuclides are unstable, generally because they contain too few or too many neutrons relative to the number of protons. These nuclides decay spontaneously into more stable nuclides. This process is called radioactivity or radioactive decay.

A heavy nucleus with too many protons may emit an alpha particle, consisting of two protons and two neutrons (a helium nucleus). A lighter nucleus may reduce the number of protons by capturing an electron or emitting a positron, thereby converting a proton to a neutron. If there are too many neutrons, one of them may be converted into a proton by emitting an electron (beta particle). Some of these transformations also produce gamma rays, which are highly energetic photons (X-rays). These processes transmute an unstable nucleus of one element into another that is stable; they are known as parent and daughter nuclides respectively. Each kind of unstable nuclide has a characteristic rate of decay. The time taken for half of the atoms of an isotope to decay is called its half-life; various nuclides have half-lives ranging from fractions of a second to billions of years.

The decay rate is unaffected by the chemical state of the atoms or by physical factors such as temperature and pressure; this property makes radioactivity an excellent chronometer for measuring ages of geological materials. For example, three different isotopes of uranium decay to three isotopes of lead with different half-lives, the longest of which is about 4.5 billion years. Precise measurements of abundances of these isotopes in minerals in the most primitive meteorites show that they formed (coincidentally) about 4.56 billion years ago, which defines the age of our Solar System. Other useful parent–daughter systems with long decay times include ^{40}K–^{40}Ar (the superscript is the atomic weight of the nuclide), ^{87}Rb–^{87}Sr and three nuclides of Sm–Nd. Shorter-lived nuclides produced by a nearby SUPERNOVA were present in the early Solar System. These nuclides are now extinct, but they have left evidence in the form of their stable daughter nuclides, which provide relative chronometers. The most important of these is ^{26}Al, which decays to ^{26}Mg with half-life 700,000 years. An excess ^{26}Mg in aluminium-bearing minerals indicates that they formed before all of the ^{26}Al had decayed; the amounts give relative formation times over a span of the first few million years of the Solar System's history. Other extinct parent–stable daughter pairs include ^{53}Mn–^{53}Cr and ^{129}I–^{129}Xe.

Energy released by radioactive decay is an important source of internal heat for planetary bodies, driving convection in Earth's mantle and core. In the early Solar System radioactive nuclides were more abundant, and species that are now extinct were then significant sources of energy. The abundance of ^{26}Al was sufficient to melt planetesimals a few hundred kilometres in size, and radioactive heating was the cause of thermal processing of meteorites.

radio astronomy Study of a wide range of objects from the coolest, most quiescent astronomical objects (namely the neutral hydrogen gas in the Galaxy) to some of the most energetic (such as PULSARS and QUASARS) in the radio region of the spectrum. Every type of astronomical object has been studied at radio wavelengths, including the Sun, planets, stars, nebulae, galaxies and the cosmic microwave background. The radio wavelength region covers the range from around 10 to 20 m (30 to 15 MHz), where the radio waves are absorbed and scattered by the ionosphere, and 350 μm (850 GHz), where atmospheric water vapour absorbs the radiation. The short-wavelength region from 350 μm to 10 mm is now considered the domain of SUBMILLIMETRE-WAVE ASTRONOMY and MILLIMETRE-WAVE ASTRONOMY, and the region between 1 mm and 30 mm (300 GHz and 1 GHz) is also known as the MICROWAVE

R

▲ **radio galaxy** Jets emerging from the active nucleus of Cygnus A produce two lobes of strong radio emission to either side of the galaxy, as seen in this image from the Very Large Array. Many radio galaxies show similar patterns of emission as material they eject impacts on the surrounding intergalactic medium.

region. Radio astronomers use frequency as often as wavelength to characterize the signal because of the historical roots of the area.

The initial discovery that radio waves from outside the Solar System were detectable on Earth was made in 1931 by Karl JANSKY, whose name is used for the unit of measurement of radio signal. The words 'radio astronomy' were first used in the late 1940s. The discovery was not thought to be of great significance to the development of astronomy until after World War II, when the new techniques that had been developed during the War were used to study these radio emissions and it was appreciated that non-thermal processes must be operating in the Sun and in more distant regions of the Universe. Originally (reflecting the nature of the equipment), radio emission was measured in decibels since it was 'signal strength'; now FLUX DENSITY as weak as micro-janskys can be detected. The 250-ft (76.2-m) Mark I RADIO TELESCOPE at JODRELL BANK is an early example of a large steerable telescope; it began work in 1957. The ARECIBO telescope in Puerto Rico (with a diameter of 305 m) is suspended across a sinkhole; it started operations in the early 1960s. However, the power of the radio telescope is seen to best advantage when several radio telescopes are used together, either through APERTURE SYNTHESIS, to improve the detail in maps, or through INTERFEROMETRY, to improve positional accuracy (*see also* MULTI-ELEMENT RADIO-LINKED INTERFEROMETER NETWORK; RADIO INTERFEROMETER; VERY LONG BASELINE INTERFEROMETRY).

Two early discoveries (in the 1950s) confirmed the importance of radio astronomy. One was the development by Soviet physicists of the theory of SYNCHROTRON RADIATION in the Milky Way. It was shown that the motion of relativistic electrons in the galactic magnetic field would produce radio emission such that the difficulty in the observed spectrum was removed. The other discovery concerned the nature of the localized (point-like) sources of emission. Although localized sources existed in the Milky Way (particularly SUPERNOVA REMNANTS such as the Crab Nebula), collaboration with the astronomers using the 200-inch (5-m) optical telescope on Palomar led to the surprising conclusion that the majority of the sources were distant extragalactic objects. These became known as RADIO GALAXIES. An important stage was reached in 1960 when, using the 200-inch telescope, Rudolph Minkowski identified one of these radio galaxies with a galaxy having a REDSHIFT of 0.46 – the most distant object then known, assessed to be 4.5×10^9 l.y. away, with a recessional velocity of 40% of the speed of light. Radio astronomy is able to detect radio galaxies to much larger distances than can be observed optically. The number of radio sources increases with distance, showing that the Universe is evolving.

Another discovery from radio astronomy, by Arno PENZIAS and Robert Wilson in 1965, namely that of the COSMIC MICROWAVE BACKGROUND, confirmed the theory that the Universe evolves with time. Recent work in this area has been undertaken with the COSMIC BACKGROUND EXPLORER (COBE) satellite, and a new satellite (PLANCK) will continue the study is greater detail.

Yet another important discovery by radio astronomers was that of pulsars, by the group at Cambridge (including Anthony HEWISH and Jocelyn BELL) in 1967. The initial discovery was of a source emitting a pulse of radio waves every 1.337 seconds with a precision better than one part in 10^7. Over 700 pulsars have been discovered, and although there are very few optical identifications it is widely accepted that they are rapidly rotating NEUTRON STARS – the collapsed very high-density remnants of stars that have undergone a SUPERNOVA explosion. The beam of radio emission, from the pole of the neutron star, swings rapidly across our line of sight, like the beam of light from a lighthouse. Pulsars have been detected with millisecond periods, in binary systems, and even with planets orbiting around them. The final stage of stellar evolution following the supernova explosion is the creation of a supernova remnant, which can be observed at radio wavelengths as the material from the supernova interacts with the interstellar medium.

Neutral hydrogen was known to be the principal component of the Galaxy; Hendrik van de HULST predicted that it should be possible to observe a spectral line from it in 1945. The line at 21 cm (TWENTY-ONE CENTIMETRE LINE) was detected in 1951. The motion of the neutral hydrogen gas can be studied using the 21-cm line, which gave the first unambiguous determination of the spiral structure of the Galaxy. Many other galaxies are close enough to be resolved and mapped in the radio part of the spectrum, and it has been found that the neutral gas extends farther from the galaxy centres than the visible light and stars indicate. The motions of the neutral hydrogen gas can be used to track the distribution of the mass in the Galaxy, showing that in most normal galaxies only about 10% of the mass of the galaxy is observed (the rest is DARK MATTER).

Several important molecules are observed in the radio region: OH (the hydroxyl radical) at 18 cm was the first, in 1963. Over 100 molecules have been detected from interstellar space in the Galaxy, from other galaxies, and from the envelopes around late-type stars in the Galaxy. The emission from molecules can be enhanced by the MASER process, and the masers in the outer envelopes of late-type stars can be accurately tracked showing the motions, and expansion, of the stellar envelope. These changes can be seen on short time-scales (months and years). Until molecular hydrogen could be directly detected, CO (carbon monoxide) was a very important molecule because it was co-located with molecular hydrogen in cold clouds and could be used to trace it. *See also* INTERSTELLAR MATTER

radio galaxy Galaxy that emits in the radio region of the ELECTROMAGNETIC SPECTRUM at rates around a million times stronger than the Milky Way. Many radio galaxies have optical counterparts, which are almost invariably ELLIPTICAL GALAXIES. The emission from a typical radio galaxy is concentrated in two huge radio LOBES lying well outside the galaxy and frequently located symmetrically about it. In appearance these lobes look as though they have been explosively ejected from the central galaxy. The lobe sizes for the biggest radio galaxies are around 16 million l.y., comparable with the size of a typical cluster of galaxies (our own Galaxy is around 65,000 l.y. across). These giant galaxies are the largest known objects in the Universe.

CYGNUS A was the first radio galaxy to be detected, the strongest source of radio emission outside the Galaxy. The lobes are only around 3 million years old, compared to perhaps 10 billion years old for the stars in the central galaxy. Both Cygnus A and CENTAURUS A are thought to be galaxies in collision. In Centaurus A the small spiral galaxy, engulfed by the giant elliptical galaxy, can be seen in the INFRARED, and the optical image shows a large dark dust lane across the galaxy. Centaurus A is also an X-ray source. The optical galaxy M87 coincides with the radio source Virgo A, a powerful radio galaxy. A JET of gas 8000 l.y. long can be seen, and the light is polarized, showing that the synchrotron mechanism is at work, and there is a strong magnetic field present.

The positions of the radio galaxies must be determined with very high precision, using VERY LONG BASELINE INTERFEROMETRY. This enables telescopes to image the faint optical counterparts. The structure of the galaxies, the lobes and jets can be mapped using APERTURE SYNTHESIS, so that details can be investigated using telescopes such as the VERY LARGE ARRAY. Radio astronomers have found that the extended radio lobes are almost invariably connected to the central galaxy by jets, which seem to originate from the galactic nucleus. High-frequency VLBI observations have revealed strong, point-like, active radio sources in the nuclei of many objects. It seems clear that the basic 'engine' that drives the explosion lies in the galaxy's nucleus, although it is unclear what the engine might be. Whilst the collision of two galaxies may contribute to the engine, INTERACTING GALAXIES are usually weak radio sources. Radio galaxies may have more in common with QUASARS and BL LACERTAE OBJECTS, which are as powerful radio emitters as radio galaxies (or slightly more powerful). Radio galaxies have more emission at lower frequencies: in the famous *THIRD CAMBRIDGE CATALOGUE* (3C) about 70% of sources are found to be radio galaxies, but in surveys made at higher frequency, such as the Parkes 2700 MHz survey, less than 50% are believed to be radio galaxies. Radio galaxies may be old quasars or failed quasars. There is still much to learn.

radio interferometer Two (or more) RADIO TELESCOPES separated by a distance greater than (or comparable to) their diameters that are looking at the same astronomical object; their signals are combined electronically. Radio interferometers were developed in order to permit radio telescopes to attain much higher resolution than was possible with a single telescope (usually more than 60″). Radio telescopes are used as interferometers to get more accurate positions for radio sources, allowing telescopes at other wavelengths (most commonly optical telescopes) to identify the source of the radio emission, and to map sources in greater detail to identify the places of most intense radio emission.

When the two signals from the two separate telescopes are combined correctly (*see* INTENSITY INTERFEROMETER) they will either reinforce or cancel each other, depending on the amount of delay between the two signals. If the two telescopes of the interferometer are arranged along an east–west line on the Earth, a radio source moving across the sky will produce a series of maxima and minima as the delays between the signals received by the two telescopes change. It can be difficult to decide whether the delay is zero or a whole number of wavelengths, but the ambiguities can often be removed by observing with the radio telescopes at different separations from each other (changing the amount of the delay). For this reason many radio interferometer telescopes are mounted on railway tracks, which permit the spacings to be changed easily.

LONG BASELINE INTERFEROMETRY and VERY LONG BASELINE INTERFEROMETRY do not use physical connections; instead the signals from each telescope are recorded on high-quality video tape recorders, with very accurate time signals (from atomic clocks or masers) recorded on a time-track along with the sky signals. The separate tape recorders have to be synchronized at playback to a very high accuracy.

In a normal radio interferometer with two telescopes, the voltage levels from the signals received at each telescope are not added together but multiplied. In this case the power output of the combined signal is proportional to the product of the individual voltages and the individual telescope dish diameters, so a small dish can be very effective as an interferometer when working with a second larger dish of large collecting area. Interferometers can reach resolving powers of better than a thousandth of an arcsecond.

The first radio interferometers were built in the early 1950s and were used to measure the sizes and positions of the strong sources then known. Objects such as the peculiar galaxy CYGNUS A were shown to be emitting vast quantities of radio energy from small regions. The identification of the strong radio source Taurus A with the CRAB NEBULA supernova remnant was only possible after an accurate position had been measured with the Dover Heights Sea Interferometer near Sydney, Australia. This instrument used only one aerial perched on the top of a cliff overlooking the sea; the second aerial was formed by the reflection of the first in the water.

radio scintillation Irregular rapid changes in the apparent intensity of radio sources, caused by variations in the electron density in the IONOSPHERE (mainly the F2 layer) and in the interplanetary gas. It occurs only with sources of less than 1″ diameter. *See also* SCINTAR

radio telescope Instrument used to collect and measure ELECTROMAGNETIC RADIATION emitted by astronomical bodies in the radio region of the spectrum. The radio region extends from around 10 mm to around 10–20 m. The short-wavelength region from 10 mm to 350 μm is now defined as the MILLIMETER-WAVE ASTRONOMY and SUBMILLIMETRE-WAVE ASTRONOMY range, but much of the equipment and many of the techniques are the same as those in the radio region. Almost all the types of object studied with optical telescopes have also been observed with radio telescopes. These include the Sun, planets, stars, gaseous nebulae and galaxies. Furthermore, RADIO ASTRONOMY has been responsible for the discovery of several new and unsuspected types of astronomical phenomena, such as QUASARS, PULSARS and the COSMIC MICROWAVE BACKGROUND.

The first radio telescope was built by Karl JANSKY in the early 1930s, and he recognized that the 'noise' he was detecting came from the Milky Way. Following World War II, there was a glut of cheap (or free) radar and communications equipment. Using this equipment, experimental telescopes were built, mainly in Australia and England, but there was interest in many countries including USA, Soviet Union, France and Japan. Since radio wavelengths are large, the paraboloid receiving dish does not have to be solid (as in an optical telescope): if the radio telescope is operating at wavelengths longer than around 20 cm, it can use mesh instead. There are several types of radio telescope in addition to the standard steerable telescope. Occasionally, radio dishes have been built on EQUATORIAL MOUNTINGS, similar to those used for most optical telescopes, but these have not proved popular. Radio telescopes are normally built with ALTAZIMUTH MOUNTINGS. That is, their basic movements are up-down and around parallel to the horizon. The alt–az mount, as it is known,

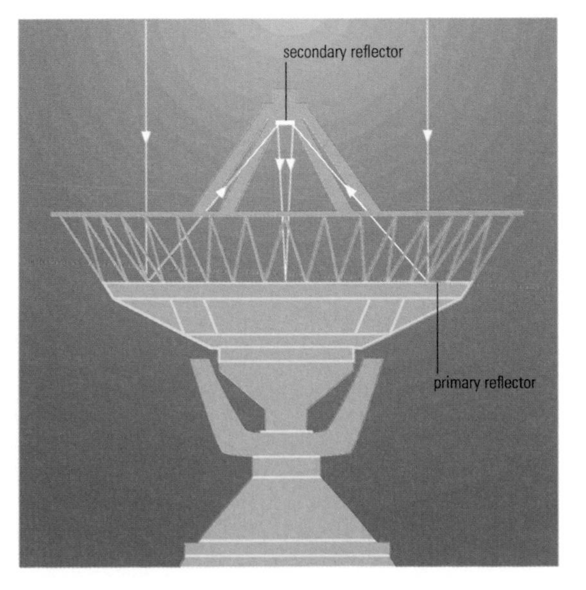

◀ **radio telescope** Just like an optical reflector, a radio telescope depends on collection of electromagnetic radiation by a large aperture, parabolic reflector, often in the form of a dish. Radio waves are brought to a focus on the detector. Most radio telescopes can be steered in altitude and azimuth.

R

presents many engineering advantages, particularly with the advent of powerful computer control systems. In fact, many new optical telescopes are being built with alt–az mounts. The largest steerable telescopes are the 100-m Effelsberg Radio Telescope (near Bonn) and the 100-m Green Bank Telescope (in Virginia), followed closely by one of the oldest, the 250-ft (76.2-m) JODRELL BANK telescope. The third largest steerable telescope is the 64-m (210-ft) PARKES RADIO TELESCOPE in Australia. The 300-m (1000-ft) telescope in ARECIBO, Puerto Rico, is suspended over a sinkhole and does not move, although tracking capability is provided by the secondary mirror. Even with these very large telescopes the resolving power is poor by optical standards (perhaps around 60″–100″ at typical wavelengths); the problem has been overcome by combining several radio telescopes using APERTURE SYNTHESIS to mimic a bigger telescope (up to the size of the Earth). Another approach is to use two (or more) telescopes as a RADIO INTERFEROMETER to improve the resolution. VERY LONG BASELINE INTERFEROMETRY using a few telescopes, very widely separated, can produce a resolving power of a few thousandths of an arcsecond.

A single radio telescope does not produce a picture directly (unlike an optical telescope); the dish must be scanned forwards and backwards to build it up, and the data then fed through a computer. The computer is also vital for controlling the telescope (which may weigh 1000 tonnes), pointing it accurately at the required position in the sky, moving continually in two coordinates (altitude and azimuth), and correcting for the inevitable deformation of a large dish. The accuracy required is often better than 10″, which becomes millimetres when translated into dish movements. Most radio telescopes are not located inside buildings, so the computers have to correct for weather too, such as high winds, cold and heat. Radio telescopes can also be adversely influenced by ice, rain and lightning.

radio waves ELECTROMAGNETIC RADIATION with wavelengths ranging from millimetres to hundreds of kilometres. Radio waves were first generated artificially by Heinrich Hertz (1857–94) in 1888. Radio astronomers measure radio waves in frequency units (cycles per second, called hertz, symbol Hz) as often as in wavelength units, for example the TWENTY-ONE CENTIMETRE LINE of neutral hydrogen and the four lines of OH at 1612, 1665, 1667 and 1720 MHz. To convert from one to the other is quite straightforward: frequency × wavelength = velocity of light ($f\lambda = c$).

radio window Range of wavelengths, from around a few millimetres to about 20 m, to which the terrestrial atmosphere is transparent. The window has several subdivisions, which have become separate areas of astronomy because of the techniques used for observing them: the submillimetre, the millimetre and the microwave. At short wavelengths, the radio waves are absorbed by water vapour in Earth's atmosphere, and at long wavelengths the radio waves are absorbed and scattered by the IONOSPHERE. *See also* ELECTROMAGNETIC SPECTRUM; MICROWAVES; MILLIMETRE-WAVE ASTRONOMY; SUBMILLIMETRE-WAVE ASTRONOMY

▶ **Ranger** Paving the way for the later Apollo programme, the Ranger missions provided early high-resolution pictures of potential lunar landing sites. This is the first US spacecraft image of the Moon, obtained by Ranger 7 on 1964 July 31, showing the crater Alphonsus at right centre.

radius vector Imaginary straight line between an orbiting celestial body and its primary, such as the line from the Sun to a planet. *See also* KEPLER'S LAWS

Ramsden eyepiece Basic telescope EYEPIECE consisting of two simple elements. While it has a wide field of view, the Ramsden eyepiece suffers from CHROMATIC ABERRATION and poor EYE RELIEF, so the KELLNER EYEPIECE is usually preferred. It is named after the English optician Jesse Ramsden (1735–1800).

Ranger First NASA space probes to investigate another Solar System body, in this case the Moon. Of the nine probes launched in 1961–65, only the last three were successful. The first success came with Ranger 7 in 1964 July, when 4316 television images of the surface at ever-increasing resolutions were returned before the spacecraft hit the region now known as Mare Cognitum. A MARE area was selected because of its relatively level and uncratered surface, two essential criteria for the future APOLLO PROGRAMME manned landings.

The final wide-angle frame covered an area about 1.6 km (1 mi) square, showing craters down to about 9 m (30 ft) in diameter. The best-resolution images showed craters as small as 1 m (3 ft) in an area about 30 by 49 m (100 by 160 ft).

Ranger 8 was targeted to a point in southwest Mare Tranquillitatis, while Ranger 9 headed for a more scientific target, the crater Alphonsus. The images showed that the lunar mare areas were free from small, deep craters and widespread rock fields, and that they would support the weight of a spacecraft. The Rangers also provided accurate values of the Moon's radius and mass ratio with respect to the Earth.

RAS Abbreviation of ROYAL ASTRONOMICAL SOCIETY

Rasalgethi The star α Herculis, an irregularly variable red giant with an estimated diameter about 400 times the Sun's, distance about 400 l.y. It is of spectral type M5 Ib or II and is so large that it pulsates in size, ranging between about mags. 2.7 and 4.0 with no set period. It is also a binary, with a wide bluish companion of mag. 5.4, easily visible in small telescopes, which orbits it in 3600 years or so. The name comes from the Arabic *ra's al-jathī*, meaning 'the kneeler's head'; the figure of Hercules is traditionally depicted in a kneeling position.

Rasalhague The star α Ophiuchi, visual mag. 2.08, distance 47 l.y., spectral type A5 V. Its name comes from the Arabic *ra's al-hawwā'*, meaning 'head of the serpent-bearer', since Ophiuchus was traditionally depicted holding Serpens.

RASC Abbreviation of ROYAL ASTRONOMICAL SOCIETY OF CANADA

RASNZ Abbreviation of ROYAL ASTRONOMICAL SOCIETY OF NEW ZEALAND

RATAN-600 *See* SPECIAL ASTROPHYSICAL OBSERVATORY

Rayet, Georges Antoine Pons *See* WOLF, CHARLES JOSEPH ÉTIENNE

Rayleigh scattering Scattering of light by particles smaller than the wavelength of the light. It is named after the British physicist Lord Rayleigh (1842–1919). The efficiency of the scattering increases rapidly as the wavelength decreases, thus leading to blue skies on Earth, since the blue light from the Sun is scattered much more than the red by molecules in the atmosphere.

rays Radial ALBEDO features that diverge from young craters on planetary bodies that lack atmospheres; these systems of straight or curved streaks look like splash marks. Rays are prominent on the Moon and Mercury; they are also seen on some satellites of the outer planets. Rays are believed to be surface deposits of finely comminuted EJECTA excavated from craters. Rays diverge from well-formed, young craters for typical distances of 10 diameters. They traverse relief without deviation or interruption, indicating ballistic emplacement. Most rays are brighter than the underlying surface, probably due to light scattering from small grains. Rays often contain small secondary craters, which are caused by the impact of large chunks of excavated material. It is not known why this material tends to go in specific directions rather than being uniformly distributed. Probably most craters have rays when first formed, but they are gradually erased by micrometeorite bombardment and by darkening from exposure to the SOLAR WIND. TYCHO, the most prominent lunar rayed crater, 87 km (54 mi) in diameter, has rays extending to 23 diameters; they are easily visible in a small telescope, particularly close to full moon. Tycho's age is estimated at about 100 million years, testifying to the slowness of erasure and thus the much greater ages of most other lunar craters.

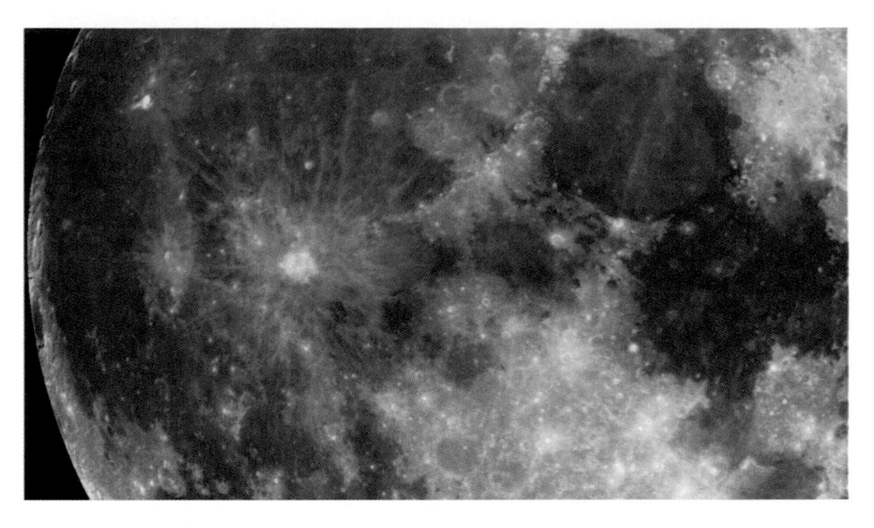

▲ **rays** This Galileo image of the Moon's northeast quadrant, was taken as the spacecraft sped past Earth en route to Jupiter. A bright ray is seen crossing the oval Mare Serenitatis at left.

R Coronae Borealis star (RCB) Star that exhibits sudden, unpredictable fades. R Coronae Borealis stars are high-luminosity stars of spectral types Bpe–R, carbon- and helium-rich but hydrogen deficient. The fades may be anywhere between 1 and 9 magnitudes; the resultant deep minima may last for a few weeks, months or for more than a year. These fades occur completely at random, without the least sign of any periodicity, and once a decline has commenced it is usually rapid, especially if the fading is of several magnitudes. The subsequent rise back to maximum is very much slower, and there are many fluctuations in brightness as the star rises. R Coronae Borealis stars spend most of their time at maximum, some then pulsating about 0.5 magnitude in a semi-periodic manner. Several show periods of around 35 to 50 days for these slow pulsations.

R Coronae Borealis stars are a small class of SUPERGIANTS, numbering about three dozen. The type star, R Coronae Borealis, was first found to be a variable in 1795. It is normally about 6th magnitude, and may remain at this maximum brightness for as long as ten years before suddenly beginning to fade. The minimum may be anywhere between 7th and 15th magnitude: there is no way of determining just how much it will fade. It remains at minimum for a few weeks or several years: again, there is no way of knowing for how long. Other stars in this group behave in a similar manner, although they are not as bright at maximum as R Coronae Borealis. RY Sagittarii when at maximum is about half a magnitude fainter than R Coronae Borealis, and the remaining members of the group are very much fainter.

In those stars for which adequate spectra are available, chemical abundances are estimated at 67% carbon, 27% hydrogen and 6% light metals.The greatest attention spectroscopically has been given to the two brightest examples, R Coronae Borealis and RY Sagittarii. Investigators agree that both stars are much more abundant in carbon than hydrogen and also that lithium is overabundant in both, with the C_2 absorption bands only weakly present. Other RCB stars have strong C_2 absorption bands and are presumably cooler.

During the initial decline the normal ABSORPTION spectrum is rapidly replaced by a rich EMISSION SPECTRUM. All RCB variables have an infrared excess. Variations in the infrared do not occur at the same time as changes in the visual, and changes in either may be unrelated to changes in the other.

A number of models have been proposed to account for the behaviour of RCB variables. In one, a gas cloud forms above the photosphere and moves outwards. As it does so it cools and condenses, forming small particles of graphite. These are very efficient absorbers of light and so would account for the sudden declines in brightness. The cloud would disperse, becoming at first patchy and transparent in places. Later, the temperature and pressure would become so low that the carbon particles would condense into soot, revealing the star once more.

Alternatively, particles form in the upper photosphere, cutting off the chromosphere from its source of excitation so that it gradually decays. A third theory suggests that a cloud is ejected from the star. The chromospheric spectrum then results from an eclipse by the dense cloud in the observer's line of sight. This spectrum is in many ways similar to that of the solar chromosphere during a total eclipse of the Sun. At first the cloud might be much smaller than the star, but as it moved away it would grow in size and, if it were centred over the star, cause an eclipse that would result in the deep declines that are observed.

The RCB variables provide a very good example of why professional astronomers welcome the assistance of amateurs. The professionals are interested in certain parts of the light-curves of these stars, especially the commencement of deep declines. The cost of modern instruments and the demands on them mean that they cannot be used night after night, often over periods of years, in the hope that a decline of an RCB star will be observed. Professionals therefore rely on amateurs to monitor these stars continuously and to advise them whenever a fade is detected. In this way the amateurs also produce the data needed to determine how RCB stars pulsate at maxima and the period in which they do so.

Reber, Grote (1911–) American pioneer of radio astronomy. In 1932, while working as a designer of radio sets, he heard of Karl JANSKY's discovery of cosmic radio emission and set about designing and building equipment to investigate the phenomenon for himself. Six years later, he completed a 9.57-m (31.4-ft) paraboloidal dish erected behind his house, and the ancillary instrumentation. Using a receiver working at a frequency of 160 MHz

▼ **R Coronae Borealis star** The light-curve of a typical R Coronae Borealis star shows long intervals spent at maximum brightness, punctuated by irregular episodes of abrupt dimming caused by condensation of carbon in the stellar atmosphere. Some fades are deeper than others, and the time taken to recover to maximum varies from one maximum to the next.

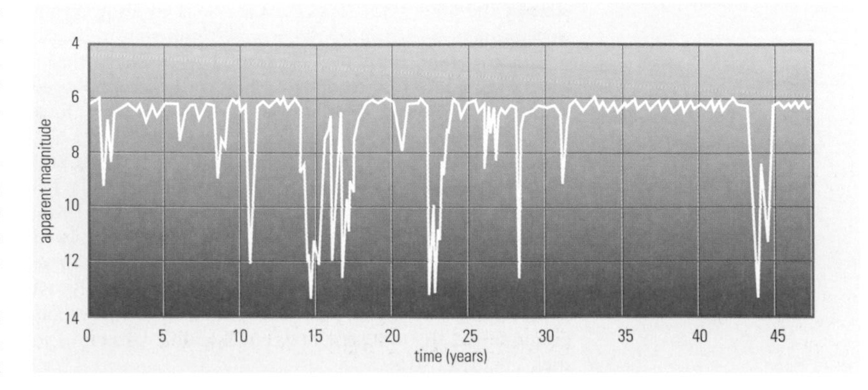

R

(wavelength 1.87 m), he succeeded in recording radio emission from the Milky Way.

From then until after World War II, when Stanley HEY and others began to follow up their wartime research, Reber was the world's only radio astronomer. He detected radio emission from many sources (then called 'radio stars') that did not correspond to visible stars, such as CASSIOPEIA A and CYGNUS A. He also found that the Sun and the Andromeda Galaxy emitted at radio wavelengths. In 1948, with the potential of radio astronomy rapidly coming to be appreciated, he tried unsuccessfully to raise funds for a 67-m (220-ft) US radio telescope; instead the world's first large dish was constructed at JODRELL BANK, UK, under the direction of Bernard LOVELL. Reber continued to map the radio sky, particularly at 1–2 MHz (150–300 m) with modest equipment of his own construction, moving to Tasmania in 1954.

reciprocity failure In astrophotography, pronounced diminution in the effective sensitivity or speed of a photographic emulsion with longer than optimum exposures. In bright conditions, image brightness is inversely proportional to exposure time (hence **reciprocity**). Thus, doubling the exposure time exactly compensates for a halving of image brightness. But the reciprocity fails with the very low image brightnesses encountered in astrophotography, and much longer exposures are necessary. One way of countering this effect is to subject the film to HYPERSENSITIZATION.

recombination Capture of an electron by a positive ion. In a PLANETARY NEBULA or DIFFUSE NEBULA, hydrogen atoms (as well as other atoms and ions) are ionized (or further ionized) by ultraviolet radiation from hot stars. Protons and heavier ions then capture the free electrons, which takes the ions to their next lower ionization states (hydrogen becoming neutral). Recombination to an atom or ion can take place on any energy level. As electrons move from the levels on to which they are captured to lower levels, they radiate EMISSION LINES.

Recta Rupes (Straight Wall) Lunar fault running from 20°S 8°W to 23°S 9°W. This fault is radial to the Mare IMBRIUM; it was probably formed by the shock wave from that impact and then covered by its EJECTA. Later, basalts flooded the region, placing stress on the fracture until it activated as a fault. The fault has a slope angle of approximately 40°, though in telescopes it appears much steeper. The mare ridges surrounding Recta Rupes are from the submerged walls of an ancient crater (*see* GHOST CRATER).

recurrent nova (NR) BINARY STAR system in which the secondary is a late type (G, K or M) giant filling its ROCHE LOBE and transferring material to a less massive WHITE DWARF primary. At intervals of years to tens of years this accreted material on the surface of the white dwarf explodes in a thermonuclear reaction, causing the brightness of the star to increase by 7 to 10 magnitudes for about 100 days in the visible and for double that at other wavelengths. These stars are brighter at minimum than an ordinary NOVA, but not because of the cycle length; the stars with longer cycles tend to show faster development and brighter maximum luminosity.

Recurrent novae may be regarded as ordinary novae that have had more than one outburst (of lesser amplitude than normal novae) or as DWARF NOVAE with very long intervals between outbursts. Typical recurrent novae are: T Coronae Borealis, which had outbursts in 1866 and 1946 (with two smaller outbursts in 1963 and 1975); RS Ophiuchi, with outbursts in 1898, 1933, 1958 and 1985; and T Pyxidis, which flared up in 1890, 1902, 1920, 1944 and 1966. At present only six confirmed recurrent novae are known, with a few additional suspected systems. Apart from RS Ophiuchi and T Pyxidis, which closely resemble one another, the systems show considerable variations in their characteristics.

reddening Phenomenon exhibited by radiation as it passes through the interstellar medium and is absorbed by gas or dust and re-radiated. Most of this re-radiation is at a longer wavelength than the original incident radiation. As such, visible light is re-radiated at a 'redder' wavelength than it originally had. Generically, radiation is described as reddening as it travels, and this description is even applied in the infrared.

red dwarf Star at the lower (cool) end of the MAIN SEQUENCE with spectral type K or M. Red dwarfs' surface temperatures are between 2500 and 5000 K and their masses are in the range 0.08 to 0.8 times that of the Sun. Red dwarfs are the most common type of star in our Galaxy, comprising at least 80% of the stellar population. Examples are BARNARD'S STAR and PROXIMA CENTAURI.

Red dwarfs have a similar composition to the Sun, with a thick outer convection zone compared to the radius of the star. They exhibit many solar-type phenomena, such as spots, a chromosphere, a corona and flares. The rotation rate of many red dwarfs has been determined from fluctuations in their light-curves due to the presence of large, cool starspots analogous to SUNSPOTS. The slower the rotation, the less active the star. Red dwarfs with rotation rates above about 3 to 5 km/s (2–3 mi/s) often show flare activity and are known as FLARE STARS.

All red dwarfs close enough to be studied in detail show a spectrum consistent with the existence of a chromosphere. X-ray emission suggests the existence of coronae; the more active the red dwarf, the higher the X-ray emission, suggesting hotter coronae. *See also* DWARF STAR

red giant Star that has finished burning hydrogen in its core and that is experiencing hydrogen shell burning. As a consequence, its atmosphere expands and its effective temperature falls to between 2000 and 4000 K, making it appear red in colour. Red giants have SPECTRAL TYPE K or M and lie in the upper right hand part of the HERTZSPRUNG–RUSSELL DIAGRAM. They have diameters 10 to 1000 times that of the Sun. ARCTURUS and ALDEBARAN are red giants.

Red giants are often VARIABLE STARS, with slowly pulsating surface layers. Because of the red giants' large radii, the gravitational effect falls considerably at their outer layers; they often lose substantial amounts of material from STELLAR WINDS, producing a PLANETARY NEBULA.

The Infrared Astronomy Satellite (IRAS) discovered many red giant stars embedded within extensive dust shells, which, radiating at a temperature of a few hundred Kelvin, are detectable only in the infrared.

Red Planet Common name for the planet MARS, owing to its striking red colour.

redshift (*z*) Lengthening of the wavelengths of ELECTROMAGNETIC RADIATION caused by the expansion of the Universe and curved SPACETIME. Wavelength shifts are commonly ascribed to the DOPPLER EFFECT, first described by Christiaan Doppler (1803–53) in 1842. Though the 'redshifts' that pertain to the Universe are not caused by the Doppler effect, it must be examined first.

The Doppler effect is encountered in all manner of wave motion, from water waves through sound waves to light. Its most familiar aspect is the drop in the pitch of sound from a passing automobile or aircraft. As a body approaches, the waves seem to be shifted to shorter wavelengths; if the body is receding, the wavelengths appear longer. The Doppler shift is readily detected in astronomical bodies by the shifts in the wavelengths of spectrum lines. An approaching body has its spectrum lines shifted towards longer wavelengths, in the vernacular of astronomy 'towards the blue' (a 'blue shift', no matter what the lines' actual colours). The lines of a receding object are shifted towards longer wavelengths, similarly 'towards the red' (a 'red shift'). The greater the speed along the line of sight (the RADIAL VELOCITY), the greater the wavelength shift. Under ordinary low-velocity circumstances (such that relativity is not required), the relative shift (change in wavelength divided by wavelength) equals the radial velocity (v) divided by the speed of light (c), or

$$(\lambda_{observed} - \lambda_0)/\lambda_0 = v/c$$

where $\lambda_{observed}$ and λ_0 are respectively the observed and rest (laboratory) wavelengths.

Half the stars show blue shifts in their spectra, the other half red shifts. The single-word term 'redshift', however, is reserved for galaxies. The 'redshift' (z) is simply the observed relative wavelength change:

$$z = (\lambda_{observed} - \lambda_0)/\lambda_0$$

which in Doppler notation is v/c, or $v = zc$, which works as long as the velocities and distances are – by the standards of cosmology – small.

In 1914 Vesto M. SLIPHER, at the Lowell Observatory in Arizona, began studying the spectra of objects then called 'spiral nebulae', but today known to be galaxies. He found that 11 of the 15 he examined displayed redshifts, thereby indicating a strong preference for recessional motion. Years later, Edwin HUBBLE at the Mount Wilson Observatory in California proved these objects to be (by present standards) relatively nearby galaxies, each rather like our own Galaxy. By 1929 Hubble had succeeded in proving that the redshift of a galaxy is directly proportional to its distance from Earth. This discovery, today called the HUBBLE LAW, can be written as $v = H_0 d = zc$, where d is the distance to a galaxy and H_0 is a term of proportionality called the HUBBLE CONSTANT. The value of H_0, which has been hotly contended, is currently set at around 70 km/s/Mpc; that is, the recession velocity of a galaxy increases by 70 km/s for every megaparsec away from the Earth. The Hubble law tells us that the Universe is expanding – its galaxies (really the clusters of galaxies) are getting farther apart.

If megaparsecs and kilometres are scaled to each other, H_0 has units of inverse seconds. The inverse of H_0 therefore has units of seconds, and thus gives an age for the Universe (discounting gravitational drag or Einstein's accelerative cosmological force) of around 13 billion years. The Hubble law is also reversed to find distances to remote galaxies simply by measuring their redshifts. If $v = zc = H_0 d$, then $d = zc/H_0$. 'Redshift distances' are being used to map great portions of the Universe.

Redshifts of the most distant galaxies are so great that they can be determined through colour alone. At maximum, z exceeds 5. The simple Doppler formula would suggest that such bodies are moving away at over five times the speed of light, which is impossible. The expansion of the Universe is properly described by the theory of GENERAL RELATIVITY, formulated by Albert EINSTEIN. According to this description, space expands with time. Galaxies are not moving through space, but are riding along with space as it grows larger. Redshifts of galaxies caused by the expansion of space are therefore not the result of the Doppler effect. When a galaxy radiates energy, its photons become trapped in the web of space along with everything else. As the photons approach us, space expands, and therefore the photons stretch and their wavelengths lengthen and shift to the red. The farther away the galaxy, the longer the flight time and the greater the stretch. Distant galaxies therefore have greater redshifts than nearby ones. Superimposed on those redshifts related to the expansion of the Universe are genuine Doppler shifts caused by the local motions of galaxies through space. Such Doppler effects can mask the effects of expansion and have historically made the determination of H_0 quite difficult.

The Doppler formula ($v = zc$) gives velocities when the redshifts are low, much less than 1. But when the redshift z approaches 1, zc no longer equals v, and a correct formula, given by the theory of general relativity, must be used. Unfortunately, there is no unique simple formula, because the relation depends on the structure of the Universe. Indeed, we can use redshift and distance measurements to learn about the structure of the Universe.

The redshift of a given body tells us how much the Universe has expanded since the light left that body. The fractional change in wavelength during the flight time of a photon is $\lambda_{observed}/\lambda_{emitted}$, where $\lambda_{emitted}$ is the 'rest' wavelength λ_0. This ratio equals the fractional degree to which the Universe has expanded, given by $R_{observed}/R_{emitted}$, where $R_{emitted}$ is the distance between two galaxies at the time of photon emission and $R_{observed}$ is the distance at the time it was received. The redshift z is therefore

$$(\lambda_{observed} - \lambda_0)/\lambda_0 = \lambda_{observed}/\lambda_0 - 1 = R_{observed}/R_{emitted} - 1$$

Thus $R_{observed}/R_{emitted}$, which could be thought of as a local 'radius' of the Universe, equals $z + 1$. That is, if the redshift z for a given galaxy is 0.5, the Universe has expanded by a factor of 1.5 since the light departed the galaxy on its way to us.

The most highly redshifted photons are those in the COSMIC MICROWAVE BACKGROUND radiation. These photons have been travelling towards us since matter and radiation decoupled from each other (that is, when the Universe went from being opaque to transparent) about 300,000 years after the Big Bang. They exhibit a redshift of $z = 1000$, meaning that their wavelengths have been stretched a thousand-fold during their 13-billion-year journey to Earth.

redshift survey Method of mapping the three-dimensional distribution of galaxies in space, with the aim of determining the LARGE-SCALE STRUCTURE of the Universe, by measuring the REDSHIFTS of a large, controlled sample of galaxies, either to a certain magnitude limit or in a specific region of the sky. A notable early achievement was the CfA Redshift Survey by Margaret GELLER and her team. Current redshift surveys, such as the 6dF (Six-degree Field) Galaxy Survey or the SLOAN DIGITAL SKY SURVEY, employ bundles of optical fibres, each precisely aimed at a different galaxy, to enable the spectra – and hence the redshifts – of several hundred galaxies to be obtained at one pointing of the telescope.

Rees, Martin John (1942–) English cosmologist and, from 1995, the fifteenth ASTRONOMER ROYAL, who has studied quasars, galaxy formation and clustering, and the cosmic microwave background radiation. His models of quasars were the first to show convincingly that they are supermassive black holes at the centres of distant galaxies. In the 1960s, his studies of quasar redshifts showed that their distribution increased with increasing distance, providing strong observational evidence for the Big Bang theory, and against the steady-state model.

reference stars Stars whose positions on the celestial sphere are known to a high degree of accuracy, allowing them to be used to measure the relative positions of other celestial objects. Those whose positions are most accurately known are classed as FUNDAMENTAL STARS, their positions recorded in a FUNDAMENTAL CATALOGUE.

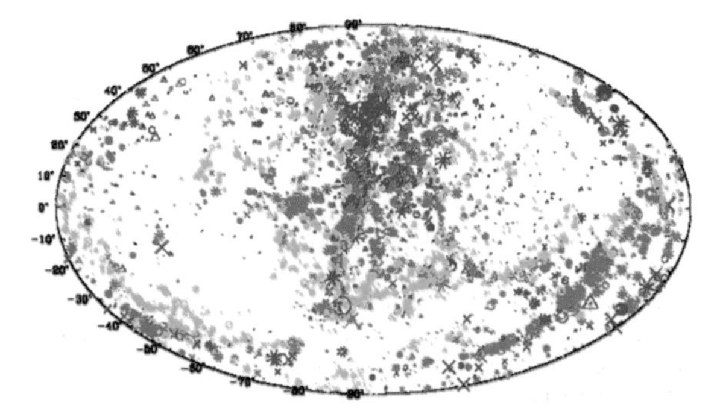

◀ **redshift survey** Combining data from several surveys, this plot shows redshifts for nearly 14,000 galaxies. Among the large-scale structure revealed is the Virgo Cluster, near the centre of the plot.

R

reflectance spectroscopy Technique that involves shining light on to a geological sample so that the interaction of the light with the atoms and crystal structure of the material produces a characteristic pattern of absorption and reflectance bands. This technique is used by planetary geologists to determine the mineralogical properties of the surfaces of planets and satellites without having actually to sample the rocks or subject them to geochemical analysis in the laboratory. *See also* SPECTROSCOPY

reflecting telescope (reflector) Telescope that collects and focuses light using mirrors. After an unsuccessful attempt by James GREGORY in 1663 to make an all-mirror telescope, the first workable reflecting telescope was demonstrated to the Royal Society, London, by Isaac NEWTON in 1668. The lenses of the time all suffered from CHROMATIC ABERRATION, and the idea was to replace refracting lenses with a curved mirror. With the reflecting surface on the front face of the mirror, there was no need for the material of which the mirror was made to be transparent, so materials other than glass could be used. The first reflectors experimented with metal mirrors, as these were easier to manufacture. An alloy made from 68 parts copper and 32 parts tin, called speculum metal, was routinely used until the late 19th century, and it was not until 200 years after Newton that Jean FOUCAULT made the first silver-on-glass mirror.

The NEWTONIAN TELESCOPE uses a parabolic mirror surface to reflect the light falling upon it back to a single point known as the PRIME FOCUS. By introducing a small flat secondary mirror at 45° just before this focus, the cone of light is diverted through a right angle to an EYEPIECE mounted on the outside of the telescope tube.

The magnification of the image is the FOCAL LENGTH of the mirror divided by the focal length of the eyepiece. With a mirror of 2000 mm focal length, for example, a 20-mm eyepiece will magnify 100 times. The FOCAL RATIO (focal length divided by APERTURE) of common types of amateur reflectors is usually around *f*/8 or *f*/6. This means, for example, that a typical *f*/8 200-mm (8-in.) reflector will have a tube about 1.6 m (5.2 ft) long.

In the 17th century, other mirror arrangements were devised. Some, like the GREGORIAN TELESCOPE, had such long focal lengths that they proved unwieldy and soon fell out of favour. The development in the late 17th century of the CASSEGRAIN TELESCOPE, in which light is reflected back from the secondary through a hole in the primary mirror, allowed long focal lengths to be accommodated in shorter, less cumbersome tube assemblies. In order not to have to perforate the primary, a small flat mirror can be introduced into the optical system to direct the light path outside the tube (sometimes along the declination axis). This system is known as the **Cassegrain/coudé** (*see* COUDÉ FOCUS).

Very large mirrors suffer from distortion as they expand and contract with changes in temperature, posing problems for the constructors of large professional reflectors. This problem was reduced in the 1930s by making telescope mirrors from the newly developed Pyrex, which has a very low coefficient of linear expansion. The recent introduction of quartz and ceramics means that modern mirrors hardly expand or contract at all.

Bernhard SCHMIDT and Dmitri MAKSUTOV designed and built the first corrector-plate telescopes. They were initially intended for photographic use, but can be adapted to provide portable instruments for visual observing. All the various systems incorporate a corrector plate in front of the main mirror. There are several optical designs for this type of telescope, each giving a wide field at the prime focus which enables photographs to be taken of large areas of the sky.

The most important and delicate parts of any reflecting telescope are the reflecting surfaces, and over the years many different materials have been tried in the quest for a durable COATING that will reflect the greatest possible amount of light. For many years silver was used, as a newly applied burnished finish will reflect 95% of visible light. Silver's tendency to oxidize and tarnish renders it less durable than other coats, but it can readily be renewed. Aluminium was used from the 1930s. It reflects 87% of visible light, but its application requires the mirror to be placed in vacuum during deposition, so ALUMINIZING is a job for professionals.

Other metals have been tried, but their hardness requires abrasives to be used to remove them before recoating. Rhodium is least affected by chemical and physical ageing but is very expensive and has less reflectivity than either silver or aluminium. Overcoating aluminium-coated mirrors has been experimented with to try to reduce oxidation and has proved successful in extending the life of the mirror surface. The length of time between aluminizations, however, will always depend on the care taken in looking after the telescope, the amount of use it gets and the air pollution at the telescope site. *See also* REFRACTING TELESCOPE; TILTED-COMPONENT TELESCOPE

reflection grating DIFFRACTION GRATING in which light is dispersed into a SPECTRUM by means of a series of closely spaced, equidistant parallel grooves ruled on to a metal surface. A typical spacing density for the grooves in an optical reflection grating for astronomy is about 1000 grooves per millimetre, and the size of the separation (about one micrometre) is not much bigger than the wavelength of light. Light of different wavelengths therefore reflects from the grating in different directions (DIFFRACTION) creating a high-quality spectrum. *See also* SPECTROSCOPE; TRANSMISSION GRATING

reflection nebula Interstellar gas and dust cloud that is seen because it scatters light from another source. That source is usually a nearby star. The efficiency of the scattering process increases rapidly as the wavelength of the light decreases, and so these nebulae often appear blue in colour. The nebulosity around the PLEIADES is a well-known example of a reflection nebula.

reflection variable (R) Recently designated type of EXTRINSIC VARIABLE; it is a close binary system, in which light from the hot primary is reflected (re-radiated) from the surface of the cooler secondary. The lightcurves of reflection variables have visual amplitudes of 0.5–1.0 magnitude and are essentially sinusoidal, with the maximum occurring when the hot star passes in front of the cool companion. Eclipses may occur in certain systems.

reflector *See* REFLECTING TELESCOPE

► **reflection nebula** A small part of the nebulosity surrounding the Pleiades star cluster can be seen in this Hubble Space Telescope image. The star Merope is just out of frame at top right.

refracting telescope (refractor) Telescope that utilizes the refraction of light through lenses to form images of distant objects. In its simplest form, a refracting telescope consists of two LENSES, an OBJECTIVE and an EYEPIECE. In practice, both objective and eyepiece are compound lenses, consisting of two or more components. The objective is a lens of large APERTURE and long FOCAL LENGTH that forms an image of a remote object in its focal plane. The eyepiece is a smaller lens of short focal length that magnifies the image. The magnification equals the focal length of the objective divided by that of the eyepiece.

The invention of the refracting telescope is usually credited to the Dutch optician Hans LIPPERSHEY, in 1608. The first astronomer to make serious regular telescopic observations was GALILEO in 1610, with telescopes of his own design and construction. The Galilean refractor had a convex ('positive') objective and a concave ('negative') eyepiece which was placed in front of the focal point of the objective. As a result it produced bright, erect images, but the field of view was small and the instruments were difficult to use.

Simple lenses and telescope systems suffer from a range of optical defects, or ABERRATIONS, notably chromatic aberration and spherical aberration. Seventeenth-century astronomers realized that spherical aberration and, to a lesser extent, chromatic aberration, could be reduced by making gently curved lenses of very long focal length. This approach resulted in some extraordinary instruments. For example, Johannes HEVELIUS constructed several telescopes up to 46 m (150 ft) in length, with wooden lattice tubes which flexed badly. Christiaan HUYGENS developed an alternative design – the **aerial telescope** – with the objective fitted in a short tube at the top of a mast, and connected only by a cord to an eyepiece at ground level.

In 1729 the English lawyer and amateur optician Chester Moor Hall (1703–71) realized that chromatic aberration could be substantially reduced by combining a positive lens made from ordinary crown glass with a negative lens made from flint glass – a harder material which contained lead and had a higher refractive index and a higher dispersion (a measure of the extent to which the different wavelengths are separated out by refraction). By careful selection of lens shapes, the chromatic aberration introduced by one lens could to a large extent be cancelled out by the other. A lens of this type is called achromatic, and the first one was built for Hall by the optician George Bass in 1733.

An ACHROMAT (two-lens objective) can bring only two wavelengths to a focus at exactly the same point, but the residual spread of focal points for the other wavelengths – the **secondary spectrum** – can be reduced by a factor of 30 or so compared with a single-lens objective. An apochromat is an objective – normally a triplet made from three types of glass – which brings three wavelengths to the same focus and reduces the secondary spectrum even further. In practice, very few astronomical refractors have anything other than two-element objectives. Eyepieces, too, must be designed to minimize optical aberrations.

With continuing improvements in materials, skills and optical technology, the 19th century became the heyday of the refractor. Although most refractors were conventional instruments, specialized designs were produced for specific purposes (*see* TRANSIT INSTRUMENT; HELIOMETER). Coudé refractors used mirrors placed after the objective to reflect light to a fixed observing position, the COUDÉ FOCUS. In another variant, light was reflected into a fixed refractor by a SIDEROSTAT or COELOSTAT.

The building of big refractors culminated in the 36-inch (0.9-m) instrument at LICK OBSERVATORY, and the 40-inch (1-m) refractor of the YERKES OBSERVATORY, installed in 1888 and 1897, respectively. The Yerkes instrument is highly unlikely ever to be surpassed. Compared with modern reflectors, large refractors have many disadvantages. Chromatic aberration cannot wholly be eliminated. As lenses become larger and thicker a significant amount of light is absorbed in them. Since a lens can be supported only around its edge (whereas a mirror can be supported across the whole of its rear surface), a large lens tends to flex under its own weight; the long tube itself tends to flex and needs to be housed in a large, expensive building. An achromatic doublet has four surfaces to be accurately shaped whereas a mirror has only one. All these factors combine to make a refractor very much more expensive than a reflector of the same aperture.

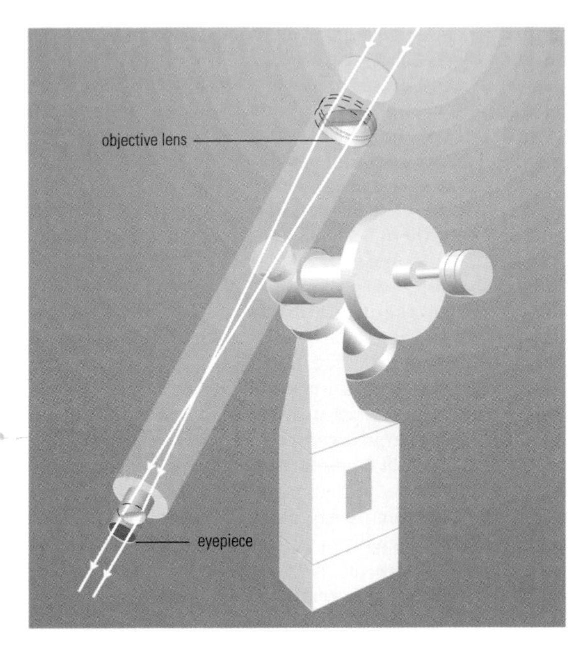

◄ **refracting telescope** The optical layout of a refractor is shown in this schematic cutaway. Light is collected by the main, objective lens at the front end of the telescope and brought to a focus. The observer views a magnified image of the focused light through a smaller lens, or set of lenses, making up the eyepiece. The telescope shown here is on a German equatorial mount.

Smaller refractors are widely used as guide instruments and still have a number of attractions for visual observers. Their optical components are less likely to go out of alignment than those of a reflector (*see* COLLIMATION). Refractors often seem to be less affected by temperature changes than reflectors, and the lack of obstructions in the light path together with the wholly enclosed tube (which cuts down air currents) is reckoned by devotees to contribute to better, steadier images on average, than those produced by reflectors of equal aperture. *See also* REFLECTING TELESCOPE

refraction, atmospheric *See* ATMOSPHERIC REFRACTION

refractor *See* REFRACTING TELESCOPE

Regiomontanus (1436–76) Name by which the German astronomer and mathematician Johannes Müller was known. Regiomontanus was one of the first publishers of astronomy literature, completing the translation of Ptolemy's *Almagest* (which was published posthumously in 1496) begun by his mentor Georg von PURBACH. This work was not just a translation but also a critique of the Ptolemaic system, and it encouraged Nicholas Copernicus to devise his heliocentric model. After his death, a letter by Regiomontanus was discovered which stated in part that 'the motion of the stars must vary a tiny bit on account of the motion of the Earth'; this is possibly the earliest exposition of the idea of stellar parallax.

Regiomontanus Irregularly shaped lunar crater (30°S 48°E), 129 × 105 km (80 × 65 mi) in diameter. This crater is from the earliest period of the Moon, and so has experienced considerable degradation from impact erosion. Indeed, the walls and rims to the south and west are nearly gone, along with any evidence of this crater's ejecta. A prominent central peak is offset to the north because of loss of the crater's original wall as a result of the impact that created PURBACH.

regolith Loose incoherent material of any origin on the surface of a planet or satellite. The term was first suggested for Earth's materials at the end of the 19th century, but it was not widely used until the 1960s, when it was applied to lunar surface material formed by multiple meteorite impacts. Lunar regolith consists mostly of unsorted fragments (from micrometres to metres across) of local bedrock, with an admixture of material brought ballistically from distant areas. Among the other components of regolith are the products of impact melting, such as pieces of glass (often containing unmelted mineral grains), and so-called regolith BRECCIA, in which small fragments of the bedrock and pieces of glass are cemented within a fine-grained matrix of the same composition. Lunar regolith is a few metres thick in lunar maria and thicker in lunar highlands. Regolith of Mercury and asteroids seems to be rather similar to lunar regolith. Regolith of Mars is thought to have been formed not only by meteorite bombardment but also by aeolian and, in the planet's early history, fluvial erosion and accumulation. Regolith of icy satellites is fragmented ice with an admixture of non-icy components, such as organic materials and possibly silicates.

regression of the nodes Slow retrograde (westwards) motion of the NODES of the Moon's orbit on the ecliptic, as a result of the gravitational attraction of the Sun. A full circuit takes 18.6 years, and it is the cause of the largest components of the NUTATION of the Earth's axis of rotation, which also has a period of 18.6 years. In fact, regression of the nodes is a common property of all planets and satellites. For the planets it is caused by PERTURBATIONS of other planets, and for satellites it is caused mainly by the oblateness of the planet. The direction of motion is opposite to the direction of motion of the satellite, so for retrograde satellites (for example Triton, and some artificial Earth satellites), the nodes move around the equatorial plane of the planet in the direct sense.

Regulus The star α Leonis, visual mag. 1.36, distance 77 l.y., spectral type B7 V. It is the faintest first-magnitude star. Regulus has a wide companion of mag. 7.7 visible in binoculars and small telescopes. The name comes from the Latin meaning 'little king'.

Reinmuth, Karl (1892–1979) German astronomer who worked at the Königstuhl Observatory, Heidelberg (1912–57) on the astrometry of asteroids. Reinmuth discovered 270 asteroids, including many Trojan asteroids, and Hermes, which in 1937 came within 800,000 km (500,000 mi) of Earth. He also discovered two short-period comets.

relativity *See* GENERAL RELATIVITY; SPECIAL RELATIVITY

relative sunspot number Daily index of sunspot activity, R, defined as $R = k\,(10g + f)$, where k is a factor based on the estimated efficiency of observer and telescope (usually 1), g is the number of groups of sunspots (irrespective of the number of spots each contains), and f is the total number of individual spots in all the groups. The relative sunspot number has also been called the Wolf number and the Zürich number, after the pioneering sunspot records begun in 1848 by Rudolf WOLF at the Zürich Federal Observatory. While empirical, the formula gives a good indication of overall solar activity.

▶ **regolith** This picture from Surveyor 5 shows the spacecraft's footpad on the lunar surface following its 1967 landing. These missions helped to establish that the regolith of the Moon had sufficient mechanical strength to support later manned Apollo landers.

relaxation time (R_T) Interval during which individual stars' orbits within a star cluster will be changed significantly by the gravitational perturbations of other stars within the cluster. As a result, some stars will then have orbits that take them out of the cluster, and 1% of the stars will so escape over the relaxation time. In about 40 R_T, the stars of the cluster will either escape or collapse down to the centre to form a black hole, and this is therefore the maximum lifetime of a star cluster. For a GALACTIC CLUSTER like the PLEIADES, the relaxation time is around 10 million years, for a GLOBULAR CLUSTER it is a few hundred million years, while for a galaxy like the Milky Way it is a few hundred thousand million years.

Renaissance astronomy Astronomy as practised in Renaissance Europe, from the recovery of authentic Greek astronomical texts in the mid-15th century, to the transition from 'classical' to Newtonian and telescopic astronomy in the mid-17th century. Renaissance astronomy grew out of the astronomical traditions of MEDIEVAL EUROPEAN ASTRONOMY, with its underlying concern with the calendar, tabular computation, and the cosmologies of Ptolemy and Aristotle. Its initial point of departure was the recovery of authentic Greek astronomical texts, to replace the corrupted Latin translations of the medieval universities, following the influx of Byzantine Greek scholars and books into Italy, especially after the fall of Constantinople to the Ottoman Turks in 1453.

In Rome in the 1460s, Cardinal Johannes Bessarion (1395–1472) encouraged astronomers such as Georg von PURBACH and REGIOMONTANUS to study Ptolemy in the original Greek, and to undertake a programme of observations. Indeed, the run of solar declination observations made by Bernhard Walther (1430–1504) at Nuremberg with a large set of Ptolemy's rulers between 1475 and 1504 effectively began observatory research in northern Europe. Reverential to the past as all of these astronomers were, they were active not only in producing accurate digests of *Almagest*, but also in comparing the Greek and Arab observations with positional measurements in their own day. Indeed, it was a concern with checking the ancients that initiated original, observation-based research in Renaissance Europe.

Problems with the retrograde motions of Mars, Jupiter and Saturn led Nicholas COPERNICUS, who had previously made original planetary position observations in Bologna and Rome, to devise his heliocentric system of the heavens in 1513, though his monumental *De revolutionibus* was not published until 1543. And, like a good classical scientist, Copernicus searched for heliocentric schemes amongst the ancient Greeks before publishing his own model. *De revolutionibus*, however, marked a decisive turning-point in astronomical history, because of the inevitable challenge put up by the heliocentric theory. If the Earth, against reason and common sense, both rotated on its axis and revolved around the Sun, then that motion should be detectable through the slight seasonal discrepancies displayed by astronomical bodies.

The need to settle the question of a moving or stationary Earth by observation gave rise to the great enterprise of Tycho BRAHE. Uraniborg, his observatory on Hven island, Denmark, was Renaissance Europe's greatest centre of scientific research, with its graduate students, craftsmen, laboratories, workshops and printing press. Yet even Tycho's superb instruments were unable to detect an annual PARALLAX. What they did produce, however, was a body of observational data from which Johannes KEPLER would derive his three laws of planetary motion based on elliptical orbits, which demanded the abandonment of the ancient pre-requisite of uniform circular motion.

Renaissance astronomy came to its greatest fruition in GALILEO's revolutionary telescopic observations after 1610. He used his observations of the lunar 'seas', Jupiter's four large satellites, sunspots and the stars of the Milky

Way to launch a full-scale assault upon the limitations of the ancients, in order to advance the Copernican heliocentric theory. Yet before Galileo forced the issue after 1616, the Catholic Church was not opposed to Copernicanism. By 1640, however, astronomy had moved away from being a purely 'classical' to an instrument-based observational science.

réseau Reference grid of dots or small crosses superimposed on an astronomical image to facilitate the measurement of positions.

residuals Differences between predicted and observed values of some quantity, such as those that arise in the analysis of astronomical observations. Residuals of the measured positions of Solar System bodies compared with predicted positions are used in the process of improving the predicted orbit. For example, the predicted orbit of a natural satellite is computed from a mathematical model, or theory, which consists of a number of periodic PERTURBATION terms affecting the orbital elements, caused by the Sun, other satellites and the oblateness of the planet. The theory contains a number of parameters, the numerical values of which are determined from observations and have to be re-determined occasionally as the time base of the observations is extended. Typical parameters are the six ORBITAL ELEMENTS, the masses of the planet and other perturbing satellites, coefficients defining the oblate gravity field of the planet, and the orientation of the planet's equator. Residuals of the observations from the theory are formed, using the current best set of parameters. The process of least squares is then used, which determines corrections to the parameters by minimizing the sum of the squares of the residuals. New residuals are formed using the corrected parameters. Mostly these residuals will be caused by the inevitable measurement errors that arise in making the observations. Ideally they should be scattered randomly, but there are always biases due to systematic errors of the observation method. There may also be signatures in the residuals as a result of inadequacies of the perturbation theory used, and this is the real incentive for the celestial mechanician – to produce a better theory, and perhaps to discover some new unexpected source of perturbation.

resolution Smallest detail visible in the image formed by an optical system. The resolution of a TELESCOPE and its instruments is limited by many factors. The over-riding one is the theoretical RESOLVING POWER of the telescope, which is determined by its APERTURE. Other limitations are the quality and design of the optics and the effects caused by the atmosphere. If the image is a direct representation of an object in the sky then the resolution will determine the level of spatial detail that can be discerned. If the image is produced by a SPECTROGRAPH then the resolution will determine the level of wavelength detail produced.

resolving power Ability of an optical system to distinguish objects close to one another, such as the two components of a binary star, or a single small object. Resolving power is measured in angular units. A telescope's theoretical resolving power in arcseconds is given by dividing 115.8 by its aperture in millimetres (or 4.56 by the aperture in inches). This measure is called DAWES' LIMIT. A 150-mm (6-in.) telescope should be able to resolve objects separated by 0″.8, but in practice ABERRATION, SEEING and imperfection in the observer's eye prevent this limit from being achieved. The unaided human eye can resolve objects separated by about 1′.

resonance Increased perturbing effect that occurs when an external force on a dynamical system has a frequency close to one of the natural frequencies of the system. In astronomy the prominent example is the large perturbing effect on an orbit as a result of a close COMMENSURABILITY of orbital period with another body. In recent work in CELESTIAL MECHANICS it has become common to use the word 'resonance' in place of the word 'commensurability'. Thus one would refer to 'the 2:1 resonance' rather than 'the resonance effect caused by the 2:1 commensurability'. Using this terminology, recent work in celestial mechanics commenced with efforts to explain the large number of resonances that occur between pairs of satellites. It is likely that TIDAL EVOLUTION of the orbits has caused one of the orbits to spiral outwards until a resonance was encountered. In many cases the satellite would pass straight through the resonance, but in some circumstances capture into the resonance can occur. The intricate detail of Saturn's rings revealed by the VOYAGER spacecraft, and the tenuous rings of Jupiter, Uranus and Neptune, have many features that have been explained by resonances with satellites (*see also* SHEPHERD MOON). The same variety of types of resonance occur as those between pairs of satellites, but because they are acting on a ring of particles rather than a single body their effects are rather different, causing clusters around a ring, and radial density variations. Several types of resonance have been given particular names taken from the field of galactic dynamics, including linblad, vertical and corotation resonances.

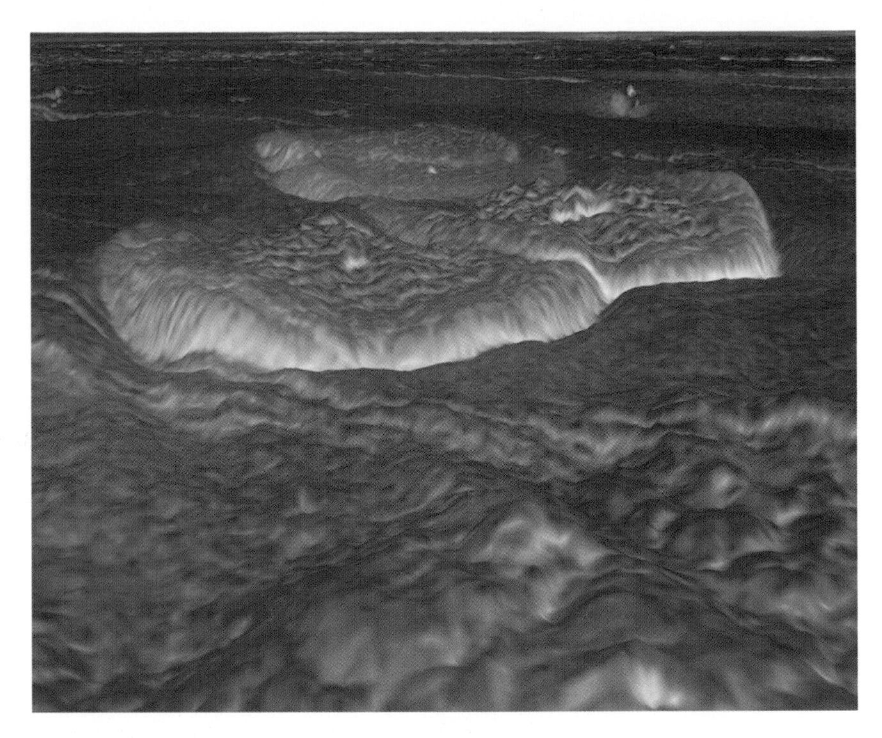

▲ **resurfacing** Alpha Regio on Venus, as imaged by radar from the Magellan orbiter. Venus has undergone extensive resurfacing in comparatively recent geological time as a result of volcanic activity in areas such as Alpha Regio.

rest mass Mass of a body when it is stationary. Strictly, it is the mass of a body measured when it is at rest in an inertial frame of reference. According to the theory of RELATIVITY, if a body has a finite rest mass, its mass increases as its speed increases and would become infinite if it could be made to travel at the speed of light. Photons (which travel at the speed of light) have zero rest mass.

resurfacing Renewal of a portion of a planetary surface by covering with fresh material, which is usually erupted from beneath the surface. Rocky bodies are resurfaced by the eruption of lava, whereas icy bodies are resurfaced by floods of water freezing on the surface. Resurfacing erases previously existing features and 'resets the clock' for accumulation of craters, by which the new surface may be dated.

retardation Change in moonrise times on successive nights. As the Moon orbits Earth, it moves eastwards across the sky, rising later (being 'retarded') each night.

R

RETICULUM (GEN. RETICULI, ABBR. RET)

Small southern constellation near the Large Magellanic Cloud, introduced by Lacaille in the 18th century to commemorate the reticle, a grid-like device in his telescope's eyepiece used for measuring star positions. α Ret, its brightest star, is visual mag. 3.33, distance 163 l.y., spectral type G7 or G8 III. ζ Ret is a wide double for binoculars or even the naked eye, consisting of two yellow dwarfs similar to the Sun, spectral types G1 V and G2 V, mags. 5.2 and 5.5, both 39 l.y. away.

Reticulum *See* feature article

retrograde asteroid ASTEROID with an orbital inclination of greater than 90°, so that it orbits the Sun in the opposite sense to the planets and most other Solar System bodies. By late 2001 four such asteroids were known, all with long orbital periods. These may be cometary nuclei that have exhausted their volatile constituents. *See also* DAMOCLES; LONG-PERIOD ASTEROID

retrograde motion (1) Orbital or rotational motion in the opposite direction to that of the Earth around the Sun, that is, clockwise when viewed from above the Sun's north pole. It is the opposite of DIRECT MOTION (prograde motion). Comet HALLEY, together with some of Jupiter's outermost satellites, moves in a retrograde motion. Venus, Uranus and Pluto have retrograde rotational motion.

retrograde motion (2) Temporary apparent east-to-west movement relative to the background stars of a superior planet before and after OPPOSITION. Because the planets orbit the Sun at different velocities, those closer to it travelling faster than those farther away, the Earth periodically catches up with and 'overtakes' a superior planet when it is at opposition. This has the effect of causing that planet to appear to stand still against the stellar background and to describe a loop (most prominently in the case of Mars) before continuing its normal apparent west–east movement. The path it traces out is known as a retrograde loop, and the places at which it appears to change direction are known as stationary points.

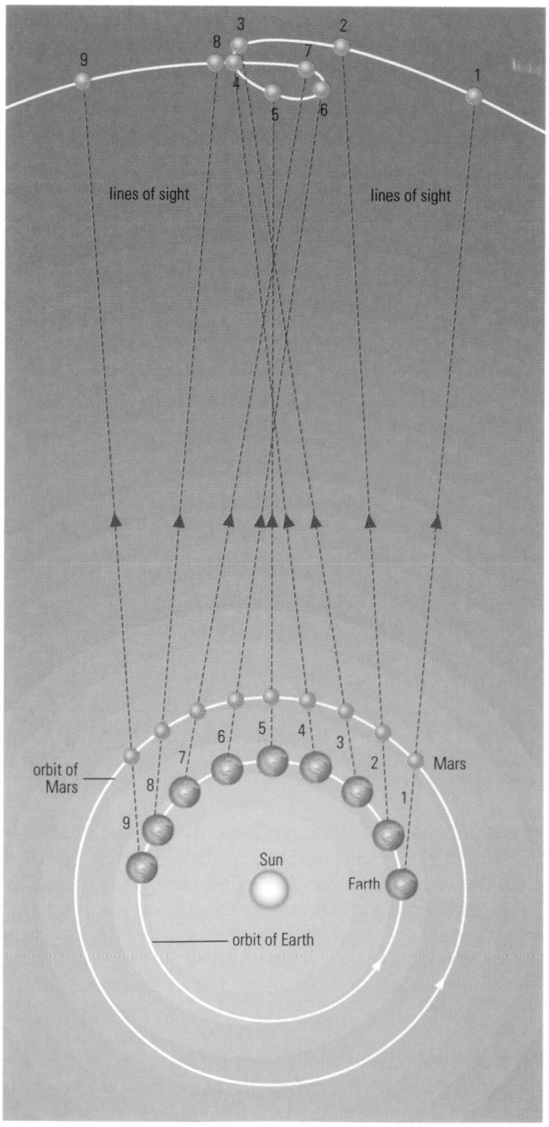

► **retrograde motion** (2) The cause of the apparent reversal of superior planets in their motion along the ecliptic close to the time of opposition is explained in this schematic diagram. Here, Earth (inner orbit) is seen catching up on Mars (outer orbit). Before opposition, Mars appears to move eastwards (top). As Earth, moving more rapidly on its smaller orbit, catches up, Mars' eastwards motion apparently slows, then reverses relative to the background stars as Earth overtakes around opposition. After opposition, the apparent eastwards motion of Mars resumes.

revolution Movement of a planet or other celestial body around its ORBIT, as distinct from the ROTATION of the body around its axis.

RFT Abbreviation of RICH-FIELD TELESCOPE

Rhaeticus (1514–74) Name by which German mathematician-astronomer Georg Joachim von Lauchen was known. He was an early supporter of the Copernican model of the Solar System, and held mathematics positions at the Universities of Wittenberg and Leipzig. He wrote *Narratio prima* (1540) as a popular exposition of the theories of Copernicus, whom he later persuaded to publish *DE REVOLUTIONIBUS ORBIUM COELESTIUM* with the patronage of Duke Albert of Prussia.

Rhea Second-largest satellite of SATURN; it was discovered in 1672 by G.D. CASSINI. Rhea's icy surface is very heavily scarred by impact craters, with no clear evidence of CRYOVOLCANISM or other resurfacing processes. It is thus one of the few large satellites of the outer planets to have a surface that resembles what was expected before the role that tidal heating can play in driving these processes came to be appreciated.

The VOYAGER image coverage of Rhea was patchy. Details as small as about 2 km (1 mi) were revealed in the north polar region, which is heavily cratered, whereas most images of the Saturn-facing hemisphere are only good enough to show details down to 20 km (12 mi). The resolution of the best images of the opposite hemisphere is even poorer. There appear to be a few fractures and some variations in crater density, but Rhea has little in common with its neighbouring satellite, DIONE, other than that observations by the Hubble Space Telescope have revealed ozone trapped within the ice on both bodies. The ozone is probably formed by the breakdown of water molecules in the same manner as on Jupiter's icy Galilean satellites. *See* data at SATURN

Rho Cassiopeiae Puzzling SEMIREGULAR VARIABLE star near β Cassiopeiae. Usually it is around mag. 5, with slight fluctuations and a periodicity of about 320 days. On rare occasions it has fallen to below mag. 6, most notably between 1945 and 1947. Its type is uncertain; the spectrum is F8.

Rho Ophiuchi Dark Cloud Complex group of NEBULAE on the Ophiuchus–Scorpius border. It is about 500 to 600 l.y. from the Earth and its apparent size on the sky extends over an area about 300 times larger than the Moon. Nearby nebulosities around α Scorpii (Antares) and σ Scorpii add to the complexity of the region. The cloud contains a large amount of irregularly distributed dust, which blocks the visible light of the stars within it and behind it. The apparently empty dark areas represent the regions where the absorption of light is the greatest (*see* DARK NEBULAE and PIPE NEBULA). Nevertheless, the cloud can be penetrated to the core at radio and infrared wavelengths, and this has revealed evidence for the existence of more than 60 young stars buried in the cloud and several compact HII REGIONS. It is estimated that about 10 per cent of the mass of the cloud has already condensed to form new stars. The thick parts of the cloud are very cold (around 10 K) and consist mainly of hydrogen in the molecular form (H_2), with the densest

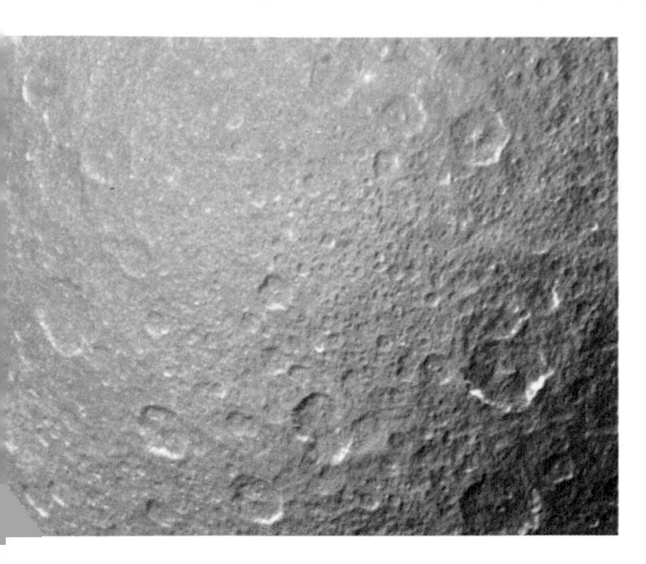

◀ **Rhea** This Voyager 1 image of Saturn's satellite Rhea was obtained from a distance of 128,000 km (79,500 mi) in 1980 November. Features as small as 2.5 km (1.5 mi) in diameter are visible on Rhea's heavily cratered surface.

regions containing more than 10^{10} molecules m^{-3} (*see* GIANT MOLECULAR CLOUD). The molecular hydrogen cannot be observed directly, but the presence of molecules such as formaldehyde, hydroxyl (OH) or carbon monoxide (CO) can be mapped through their absorption at radio wavelengths. The regions of strongest absorption correlate well with the visibly thick parts of the obscuring cloud, confirming that the dust and the molecules cohabit in the cloud. The cloud is remarkable because it contains so many REFLECTION NEBULAE, showing a range of colours from blue to red, and so few extended HII regions.

Riccioli, Giovanni Battista (1598–1671) Italian Jesuit astronomer and topographer whose *Almagestum novum* (1651) included a lunar map based on many telescopic observations by him and Francesco Maria Grimaldi (1618–63). The map named major craters after Ptolemy and his followers, while relegating the supporters of Copernicus to minor features. Riccioli also observed Jupiter's coloured belts, and may have been the first to see the ashen light on Venus.

Riccioli Lunar crater (3°S 75°W), 160 km (100 mi) in diameter. In its original state, Riccioli was a complex crater with a central peak and terraced walls. It then underwent intense bombardment, notably by the ejecta from the ORIENTALE basin-forming event. Later yet, lava flooded the inner region, and meteorite erosion further degraded the crater. Now Riccioli appears as a deeply degraded rim with a central lava plain. It is crossed by numerous rilles.

Richard of Wallingford (*c.*1292–1336) English churchman-astronomer, abbot of St Albans and a pioneer of mechanical clockwork. He designed a clock for the abbey whose dial was a planispheric astrolabe showing the daily rotation of the sky, and the positions of the Sun and Moon. The most sophisticated device of its kind, it was essentially a model of the medieval cosmos.

Richer, Jean (1630–96) French astronomer, explorer and surveyor whose work during a scientific expedition to Cayenne, French Guyana (1671–73), helped to establish the scale of the Solar System and the shape of the Earth. From Richer's observations of Mars from South America, Giovanni Domenico CASSINI derived a good approximation for the ASTRONOMICAL UNIT. Richer also found that a pendulum at Cayenne, near the equator, ran more slowly than at Paris, which meant that gravity at the equator was slightly weaker than at far northerly latitudes, a result later used by Isaac Newton and others to show that the Earth is flattened at its poles.

rich-field telescope (RFT) Low-power telescope that is equipped with a wide-angle eyepiece (such as a NAGLER EYEPIECE) designed to show a wide (2–5°) field of view at a relatively low magnification. Such an instrument is ideal for studying starfields and hunting for novae and comets.

Rigel The star β Orionis, seventh-brightest in the sky, with a visual mag. of 0.18. It is a blue supergiant of spectral type B8 Ia lying 773 l.y. away, and is 40,000 times as luminous than the Sun. Rigel has a companion of mag. 6.8, difficult to see in small telescopes because of the glare from the primary. The name comes from the Arabic *rijl*, meaning 'foot'.

right ascension (RA) Measure of angular distance along the CELESTIAL EQUATOR and one of the two coordinates of the EQUATORIAL COORDINATE system, the other being DECLINATION. Right ascension is the equivalent of LONGITUDE on the Earth and is measured eastwards from the FIRST POINT OF ARIES to where the HOUR CIRCLE of a celestial body intersects the celestial equator. It is occasionally expressed in degrees but is usually measured in hours, minutes and seconds of sidereal time and is equal to the interval between the TRANSIT or CULMINATION of the First Point of Aries and that of the celestial body. One hour of right ascension is equal to 15° of arc. *See also* CELESTIAL SPHERE

Rigil Kentaurus Popular name for the star ALPHA CENTAURI, sometimes abbreviated to *Rigil Kent*.

rille Linear or curvilinear surface depression on the Moon and other planetary bodies. Rilles have a variety of appearances and causes, and so are subdivided into the following major types: SINUOUS RILLES, arcuate rilles and floor fractures.

▼ **Rho Ophiuchi Dark Cloud** The region of the Milky Way just north of Antares (the orange-red star at bottom left) is laden with dust, some of which can be seen reflecting starlight in this image. Running across this region at top left is the obscuring dusty material of the Rho Ophiuchi Dark Cloud. It cuts off the light from the rich starfields beyond. Our view in this direction is towards the centre of the Galaxy.

Sinuous rilles are characterized as meandering channels of relatively small width, with sloping sides; they terminate in MARE regions.

Arcuate rilles have flat floors between steep sided walls (*see* GRABEN). They occur in parallel sets that are roughly concentric with a basin ring. Examples of arcuate rilles are seen around Mare Serenitatis and Mare Humorum.

Floor fractures are characterized by radial or concentric fractures in crater floors. Such craters occur along mare boundaries, suggesting that the lava that filled the basin also tracked up faults beneath these craters and collected beneath their floors, fracturing them into plates.

ring, planetary Band of particles orbiting a planet inside its ROCHE LIMIT, where tidal forces prevent them from accreting into a satellite. All four of the large outer planets are surrounded by ring systems, which have complex and varied structures.

JUPITER has a narrow, tenuous ring of small particles, probably of silicate composition, at about 1.8 planetary radii. It is surrounded by a halo of charged grains of micrometre size, which are levitated by the planet's magnetic field. The outer 'gossamer' ring has a very low density and extends out to 2.9 radii.

SATURN has three broad main rings, named C, B and A, in order of distance outwards from the planet. They extend from 1.2 to 2.3 planetary radii. There is also a faint inner D ring, which may extend down to the top of the planet's atmosphere. Despite their great width, these rings are less than 100 m (330 ft) thick. They consist of particles of sizes from about 1 cm to 10 m (0.3–33 ft), which are composed mostly of water ice. The main rings have radial variations in density and several gaps, such as the CASSINI DIVISION, which are produced by satellite perturbations. Saturn also has a narrow F ring confined between two small satellites, PANDORA and PROMETHEUS. The tenuous E and G rings consist mainly of micrometre-sized grains; the E ring extends out to about 8 planetary radii.

URANUS has 11 narrow rings of widths 5 to 100 km (3–60 mi) in the region 1.45 to 2 radii. They are separated by wide gaps containing dust and believed to be occupied by unseen small satellites. NEPTUNE has two broad and four narrow dusty rings between 1.65 and 2.4 radii; the outermost contains denser regions known as RING ARCS. The rings of Uranus and Neptune consist of dark material, probably of carbon-rich composition; both sets of rings contain bodies of metre to kilometre size.

Much of our knowledge of these ring systems comes from the VOYAGER spacecraft. Their varied configurations are due primarily to gravitational perturbations by satellites in their vicinity. Saturn's rings may either be primordial or the result of tidal disruption of a captured large cometary body or CENTAUR. The other ring systems, which are much less massive, are probably the result of disruption of small satellites by cometary impacts. The fine dust in these systems is constantly renewed by meteoroid bombardment of the larger ring particles.

▲ **ring arcs** Material in the tenuous rings around Neptune is distributed in clumps, rather than as a uniform sheet. Voyager 2 obtained this image of denser concentrations (ring arcs) in Neptune's outer Adams ring in 1989 August.

ring arcs Longitudinal structures within the outermost of Neptune's rings over which the ring's brightness varies by about a factor of three. They extend for between 1° and 10°. Five such regions were discovered by the VOYAGER 2 spacecraft. As orbital motions of ring particles would quickly even out such features, they must be maintained by some mechanism, most likely resonant perturbations by one or more small satellites. *See also* RESONANCE

ring galaxy GALAXY, clearly distinct from a spiral or elliptical, in which a ring of stars surrounds the nucleus like the rim of a wheel. Ring galaxies are thought to arise as a result of a DENSITY WAVE produced when a small galaxy passes through a larger one. Ring galaxies are also distinct from ringed galaxies, in which normal stellar dynamics (often in the presence of a bar) have channelled stars into a prominent ring about the galaxy's centre. Collisional rings often have an off-centre nucleus, or none at all. *See also* CARTWHEEL GALAXY

▼ **ring galaxy** NGC 4560A in Centaurus, as shown in this Very Large Telescope (VLT) image, appears to consist of a lenticular galaxy surrounded by a ring of dust and stars. The structure is almost certainly the result of a collision between two galaxies.

Ring Nebula (M57, NGC 6720) Bright PLANETARY NEBULA located midway between β and γ Lyrae (RA $18^h 53^m.6$ dec. +33°02′). The nebula has an elliptical profile, with a diameter across the major axis of 71″ and a darker central region. Although the Ring Nebula, at overall magnitude +8.8, is an easy object for small telescopes, even large instruments struggle to reveal the magnitude +15.3 central star. The nebula is toroidal in shape and is seen almost end-on. It was formed about 5500 years ago and lies at a distance of 4100 l.y. The Ring Nebula was discovered by the French astronomer Antoine Darquier (1718–1802) in 1779 January.

Ring-Tail Galaxy *See* ANTENNAE

Ritchey, George Willis (1864–1945) American optical designer and craftsman, co-inventor of the RITCHEY–CHRÉTIEN TELESCOPE design. He served under George Ellery HALE as chief optician and then head of instrument construction at Yerkes Observatory (1899–1904) and Mount Wilson Observatory (1905–09). Ritchey designed and built both the optics and mounting for Mount Wilson's 60-inch (1.5-m) reflector, and figured the optics for the 100-inch (2.5-m) HOOKER TELESCOPE. He used these instruments to take photographs of nebulae and galaxies that showed more detail than could be observed visually, and concluded that spiral galaxies are similar to our own Milky Way. In 1917 he made some of the first distance estimates to M31, using novae and supernovae as 'standard candles'. In 1923 Ritchey emigrated to France, becoming director of Paris Observatory's astrophotographic laboratory (1924–30). With the Frenchman Henri Chrétien (1879–1956), he designed a type of telescope, now named after them, that minimizes spherical aberration and coma.

Ritchey–Chrétien telescope Modified form of CASSEGRAIN TELESCOPE, designed originally by George Willis RITCHEY and Henri Chrétien. The primary and secondary mirrors are HYPERBOLOIDAL, although there are variants with near-hyperboloidal, PARABOLOIDAL or ELLIP-

R

◀ **Ring Nebula** The planetary nebula M57, the Ring Nebula in Lyra, has been produced by ejection of material from the hot white dwarf star at its centre.

SOIDAL components. The system is free from COMA over a wide field, though there is some ASTIGMATISM and FIELD CURVATURE. *See also* VERY LARGE TELESCOPE

Rittenhouse, David (1732–96) American surveyor, instrument-maker and astronomer. Rittenhouse made many precise clocks and surveying instruments, including the vernier compass, of which he was the purported inventor. To astronomy he introduced the use of spider-web crosshairs and gratings. For the 1769 transit of Venus he constructed a transit telescope, quadrant and pendulum clock. Rittenhouse also made ORRERIES and other instruments to demonstrate astronomical phenomena.

Roche, Édouard Albert (1820–83) French mathematician, professor of pure mathematics at Montpellier, who studied the shapes of rotating fluid masses, deriving the ROCHE LIMIT for planetary satellites. He found that for a parent and satellite of equal density the equilibrium point is about 2.5 times the radius of the parent; a satellite venturing any closer than this will be broken apart by tidal gravitational forces.

Roche limit Minimum distance from a planet at which a satellite can remain intact, without being torn apart by gravitational forces. It was noted in 1848 by Édouard ROCHE that a satellite orbiting close to its planet is subjected to great stresses because the nearer parts of its body try to orbit faster than the more distant parts. If close enough, these stresses will exceed the mechanical strength of the satellite, and it will be ripped apart. Also, at this distance a ring of particles will never be able to form into a satellite by ACCRETION. The Roche limit is usually quoted as 2.5 times the planet's radius, though it is now recognised that it depends strongly on the internal strength of the satellite, and also slightly on the densities of the planet and satellite. For a body in hydrostatic equilibrium (that is, little internal strength) the limit is about 2.46 radii, and for a small rocky satellite it is about 1.44 radii. The ring systems of Jupiter, Saturn, Uranus and Neptune all lie within their planet's hydrostatic Roche limit.

Roche lobe Surface that defines the maximum sizes of stars in a BINARY STAR system relative to their separation. If both components are well within their Roche lobes, the system is termed a DETACHED BINARY. If one component fills its Roche lobe, material escapes through the inner LAGRANGIAN POINT, L_1, on to its companion, and the system is termed a SEMIDETACHED BINARY. If both stars fill their Roche lobes, the system is a CONTACT BINARY. In contact binaries, material can escape entirely from the system through the outer Lagrangian points, L_2, creating common envelope binaries.

ROE Abbreviation of ROYAL OBSERVATORY, EDINBURGH

rocket astronomy Astronomical research carried out using instruments flown on suborbital rockets. Today, most astronomical observations are undertaken either from ground-based observatories or from orbiting satellites, but important discoveries have been made using instruments carried on balloons (*see* BALLOON ASTRONOMY) and suborbital sounding rockets. Sounding rockets provide the only means of making *in situ* measurements between the maximum altitude for balloons, about 50 km (30 mi), and the minimum altitude for satellites, about 160 km (100 mi). Larger rockets can be flown to altitudes of more than 1300 km (800 mi) and can carry payloads of up to 550 kg. Rocket-borne experiments are valuable for studies of the upper atmosphere and the Earth's radiation belts, and have also led to breakthroughs in our knowledge of the Sun, stars and galaxies.

During the first half of the 20th century, astronomers became increasingly aware that, although visible light from cosmic sources penetrates to the Earth's surface, most other wavelengths are absorbed or reflected by the atmosphere. The solution was to lift instruments above the blanket of air, but the enabling technology became available only after World War II, when captured German V-2 missiles were put to more peaceful uses.

On 1946 October 19, V-2 carried a spectrometer above the ozone layer to make the first detection of ultraviolet radiation from the Sun. Proof that X-rays were emitted by the Sun came in 1949, when Herbert FRIEDMAN flew Geiger counters on a modified V-2 launched from White Sands, New Mexico. Another early pioneer of rocket astronomy was James VAN ALLEN, who flew experiments from White Sands on the Aerobee rocket to observe cosmic rays before they collided with Earth's upper atmosphere.

The use of sounding rockets for scientific research received a boost during the INTERNATIONAL GEOPHYSICAL YEAR (IGY) of 1957–58, which coincided with a period of enhanced solar activity. Almost 300 suborbital rockets were launched during the IGY by the United States, and another 175 by the Soviet Union. By the late 1950s, X-ray observations from sounding rockets had shown that the temperature of the solar corona reaches several million degrees. Such observations revolutionized our understanding of nuclear reactions inside the Sun, but it was more difficult to capture the small amounts of short-wave radiation from more distant cosmic sources.

The initial breakthrough came when a rocket-borne spectrograph developed by Donald Morton (1933–) and Lyman SPITZER of Princeton University made the first detection of ultraviolet light from two stars in the constellation Scorpius. Within a short time, several hundred stars as faint as magnitude 6.5 had been observed in the ultraviolet.

The first detection of cosmic X-rays came on 1962 June 18, when a team led by Italian physicist Riccardo GIACCONI discovered a distant source in Scorpius, which they called SCORPIUS X-1, and a completely unexpected diffuse glow of X-rays known as the cosmic X-ray background. The following year, Friedman found a second source, Taurus X-1, which was later shown to be the X-ray counterpart of the CRAB NEBULA. Over the next 10 years, rocket surveys revealed more than 40 X-ray

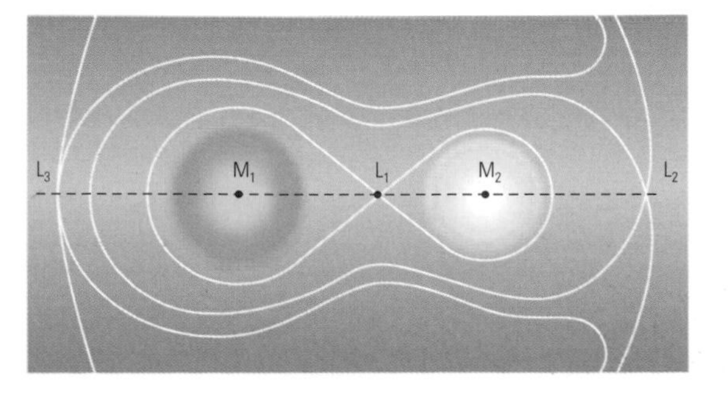

◀ **Roche lobe** In a binary star system, the stars (here marked M_1 and M_2) have their own gravitational sphere of influence, defined by the Roche lobe. Dependent on the evolutionary state and relative masses of the partners in a binary system, material from one star, or both, may fill the respective Roche lobes. Overspill of material from a distended, highly evolved star from its Roche lobe into that of a smaller more massive partner is an important driving mechanism in several forms of cataclysmic variable stars.

R

sources, two of which were extragalactic (the active galaxy M87 and the nearest quasar, 3C 273).

Advances in infrared astronomy suffered from the difficulties of developing instruments and operating them at very low temperatures. The first success was the project HISTAR to launch a 166-mm (6½-in.) telescope cooled by liquid helium. Seven flights were made from White Sands between 1972 April and December, with two more from Australia in 1974 to survey the southern sky. Although the total observing time was only about 30 minutes, the project generated a catalogue of more than 2000 celestial infrared sources.

Much of the research previously undertaken with sounding rockets has since been taken over by orbiting observatories, but sounding rockets are still widely used in the United States, Europe, Japan, India and Brazil. NASA currently uses 13 different suborbital launchers and conducts about 25 launches annually from sites at White Sands, Wallops Island and Poker Flat in Alaska, although many of these are devoted to microgravity experiments rather than astronomy.

Their rapid launch capability and recovery time (a typical flight lasts no more than half an hour) and relatively low cost also ensure that suborbital rockets still have an important role to play. Their quick response capability was demonstrated by the launch of six NASA rockets from Australia in 1987–88 to monitor SUPERNOVA 1987A.

Sounding rockets also serve as low-cost test-beds for new techniques, instrumentation and technology intended for future satellite missions. For example, the payloads flown to study Supernova 1987A included the first photon-counting CCD X-ray camera ever flown. NASA missions that have benefited from precursor flights include the Compton Gamma Ray Observatory, the Solar and Heliospheric Observatory (SOHO) and the Transition Region And Coronal Explorer (TRACE).

Current research areas for rocket astronomy include observations of the ionosphere, the aurora and processes in the magnetosphere, together with X-ray studies of the Sun and supernova remnants. These frequently take place as part of coordinated observational programmes involving ground-based instruments and satellite overpasses, for example during total solar eclipses.

Römer, Ole Christensen (1644–1710) Danish astronomer, the first to obtain a reasonable approximation for the value of the speed of light (*c*). In 1671 Römer, then professor of astronomy at the University of Copenhagen, was offered a post at PARIS OBSERVATORY by Jean Picard (1620–82), who was impressed with the accuracy of the Dane's measurements. At Paris, the director Giovanni Domenico CASSINI had instigated a broad programme of observations with the object of drawing up tables that could be used by navigators to find longitude at sea. It had been suggested that the periodic eclipses by Jupiter of the Galilean satellites would provide a suitable standard.

In 1675 Römer, checking eclipse times based on predictions by Cassini, found that they did not match the observed times: intervals between successive eclipses decreased as the orbital motions of the Earth and Jupiter brought them closer together, and increased as the two planets moved farther apart. Römer correctly deduced that the differences were caused by light taking a finite time to travel the intervening distance. The next year, he was able to announce a value for *c* equivalent to about 225,000 km/s (140,000 mi/s), three-quarters of the true value.

Römer returned to his homeland in 1681, becoming the Danish Astronomer Royal in Copenhagen, where he invented the transit telescope.

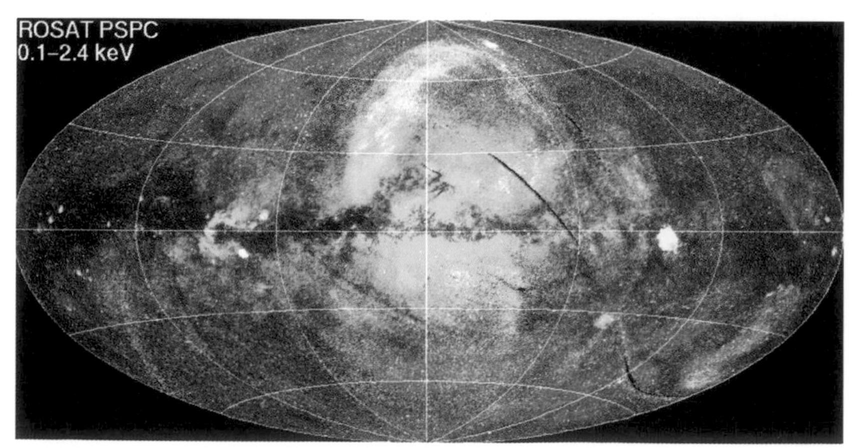

▼ **Rosat** This all-sky map of diffuse X-ray emission was obtained from Rosat. The brightest sources are supernova remnants and the central region of the Galaxy.

roof prism Optical component often used in compact binoculars to produce an image that is the right way up and the right way round as seen by the user. The OBJECTIVES and EYEPIECES in a pair of binoculars would produce inverted images if used alone, so another component such as a roof prism must be placed between them to cancel out this inversion. Unlike the PORRO PRISMS that they are beginning to replace where size is important, roof prisms permit the objectives and eyepieces to be in line, leading to a more compact shape.

Roque de los Muchachos Observatory (Observatorio del Roque de los Muchachos, ORM) Major optical astronomy facility on the island of La Palma in the Canary Islands at an elevation of 2400 m (7870 ft). It occupies some 2 sq km (0.8 sq mi) on the highest peak of the Caldera de Taburiente and is named after the 'Rock of the Companions' at the very highest point. The observatory belongs to the INSTITUTO DE ASTROFÍSICA DE CANARIAS (IAC), and was developed jointly by Spain, the United Kingdom, Denmark and Sweden. The site was established as an observatory in 1979 and inaugurated in 1985. Its clear skies, atmospheric stability and freedom from light pollution make it one of the northern hemisphere's best locations for optical astronomy.

The ORM's facilities include the 4.2-m (165-in.), 2.54-m (100-in.) and 1.0-m (39-in.) telescopes of the ISAAC NEWTON GROUP, the 3.5-m (138-in.) GALILEO NATIONAL TELESCOPE and the 2.6-m (102-in.) NORDIC OPTICAL TELESCOPE. Other instruments include the Carlsberg Automatic Meridian Circle for positional astronomy, the HEGRA gamma-ray facility and a solar tower operated by Sweden. In 2001 the 1.2-m (48-in.) Mercator Telescope of the Catholic University of Leuven, Belgium, was completed, and the 2.0-m (79-in.) LIVERPOOL TELESCOPE went into operation the following year. When it is completed in 2003, the largest conventional telescope on the site will be the 10.4-m (34-ft) GRAN TELESCOPIO CANARIAS, although a 17-m (56-ft) Cherenkov radiation telescope is also under construction.

Rosalind One of the small inner satellites of URANUS, discovered in 1986 by the VOYAGER 2 imaging team. Rosalind is about 58 km (36 mi) in size. It takes 0.558 days to circuit the planet, at a distance of 69,900 km (43,400 mi) from its centre, in a near-circular, near-equatorial orbit.

Rosat (Roentgen Satellite) Joint German/UK/US mission launched in 1990; it produced the first high-resolution, all-sky astronomical surveys at X-ray and extreme-ultraviolet (EUV) wavelengths. It detected about 150,000 X-ray sources and 600 EUV sources, and obtained more than 9000 observations of objects such as quasars, galaxy clusters, black holes, supernova remnants and protostars. Discoveries included X-ray emissions from comets. The mission ended on 1999 February 12.

Rosetta EUROPEAN SPACE AGENCY (ESA) spacecraft to be launched in 2003 May. It will become the first spacecraft to orbit a comet, Wirtanen, and to deploy a lander to make the first landing on the nucleus of the comet, in 2012. At the time Wirtanen will be in an active phase, close to the Sun. Rosetta will require a series of planetary gravity-assisted flybys to pick up enough speed to reach Wirtanen. It will reach Mars in 2005 May and then fly back towards a rendezvous with the Earth in 2005 October and 2007 October, before being directed towards its path for Wirtanen. En route, Rosetta will fly past two asteroids, 4979 Otawara and 140 Siwa (the largest asteroid to be explored by a spacecraft), at a distance of 1000 km (600 mi), in 2006 July and 2008 July respectively. After a journey of 5.3 billion km (3.3 billion mi), Rosetta will rendezvous with Wirtanen in 2011 November, eventually entering orbit around the comet the following May, while a small lander will fly 2 km (1.2 mi) to the surface of the nucleus, anchoring itself by a small harpoon.

Rosette Nebula (NGC 2237-2239) Large, visually rather faint EMISSION NEBULA in northern Monoceros (RA 06^h 32^m.3 dec. +05°03′). Covering an area of 80′ × 60′, the Rosette emits strongly in the wavelength of hydrogen-alpha and is well recorded on red-sensitive films. The Rosette lies 4900 l.y. away and has an actual diameter of 90 l.y. Embedded within it is the bright star cluster NGC 2244, the stars of which formed from the nebulosity about 500,000 years ago. Ultraviolet emission from NGC 2244's hot O-class stars has cleared a central cavity of about 30 l.y. in diameter in the nebula. Star formation is probably still going on here: detailed images reveal the presence of numerous Bok GLOBULES in the Rosette. Several sections of the outer ring have been assigned their own NGC numbers.

▲ **Rosette Nebula** Shining in HII emission excited by the young stars in the cluster at its centre, the Rosette Nebula in Monoceros is a stellar nursery. Difficult to see visually, the Rosette records well in long-duration photographs.

Ross, Frank Elmore (1874–1960) American astronomer whose work covered celestial mechanics, optical design and astrophotography. In 1905, while at the Carnegie Institution, he computed the orbit of Saturn's satellite Phoebe. As director of the International Latitude Observatory in Maryland (1905–15), he designed and built the first PHOTOGRAPHIC ZENITH TUBE (1911) for use in a programme to determine precisely the wobble of the Earth's rotational axis. Ross spent the next nine years at Eastman Kodak, where he designed wide-angle photographic lenses and developed highly sensitive photographic emulsions for imaging faint astronomical objects. He then moved to Yerkes Observatory (1924–39), where he carried out photographic surveys of our Galaxy that resulted in *Atlas of the Northern Milky Way* (1934–36) and *New Proper-Motion Stars* (1929–39), a catalogue of stars he discovered with high proper motions.

Rosse, Third Earl of (William Parsons) (1800–1867) Irish amateur astronomer who constructed large telescopes used to discover the spiral structure of galaxies. After serving as a member of parliament (1821–34), he determined to build reflecting telescopes to rival those of William HERSCHEL. His largest instrument, known as the Leviathan of Parsonstown, had an aperture of 72 inches (1.8 m). Working at special workshops built at his estate in Birr, Ireland, Parsons developed a new method of casting mirrors from a solid disk; he invented a ventilator that solved the common problem of the metal cracking during the cooling process. The Leviathan's primary mirror was cast in 1842; Parsons built a steam-driven machine to grind and polish its surface to the desired shape. The rest of the telescope was completed in 1845.

Parsons used his big reflectors to survey the star clusters and nebulous objects of the northern sky; they clearly showed the spiral arms of galaxies such as M51, the Whirlpool Galaxy, which he sketched. He discovered 15 spiral galaxies and gave the Crab Nebula (M1) its name. His son Laurence Parsons (1840–1908), the fourth earl, continued to work with the Leviathan, but after he died it was dismantled. Another son, Charles Algernon Parsons, was an engineer who continued Howard Grubb's telescope-making business under the name GRUBB, PARSONS & CO. *See also* BIRR CASTLE ASTRONOMY

Rossi, Bruno Benedetto (1905–93) Italian-American physicist regarded as the founder of high-energy astrophysics. After chairing the physics department (1932–38) at Padua, Italy, Rossi emigrated to Denmark, England and finally the USA, where he held the chair of physics at the Massachusetts Institute of Technology (1946–71). In the 1930s Rossi had shown that cosmic rays are highly energetic, positively charged particles that constantly bombard the Earth; in the late 1950s he made pioneering studies of the interplanetary medium, finding it to consist of highly ionized gases. In 1963 Rossi and his team discovered Scorpius X-1, the first known extraterrestrial source of X-rays besides the Sun, opening up the field of high-energy astronomy. The Rossi X-Ray Timing Explorer (RXTE) satellite, launched in 1995, was named after him.

rotating variable VARIABLE STAR that exhibits variations in light output as it rotates; the variations are a result of its ellipsoidal shape (with consequent changes in apparent surface area) or its non-uniform surface luminosity. The latter may arise through starspots or through magnetic effects, as in stars where the rotational axis does not coincide with the magnetic axis ('oblique rotators'), or through reflection. *See also* ALPHA^2 CANUM VENATICORUM STAR; BY DRACONIS STAR; ELLIPSOIDAL VARIABLE; REFLECTION VARIABLE

rotation Spinning motion of a body around an axis as opposed to orbital motion about another body. The time it takes a celestial body to complete one revolution about its rotational axis is a measure of the length of its day. Some degree of rotation seems to be a general property of all classes of celestial body.

rotation curve Plot of the way in which the average linear velocities through space of stars and nebulae within a GALAXY vary with their distances from the centre of the galaxy. If stars orbited in a manner similar to the planets in the Solar System, then the rotation curve would show a steady decrease with increasing distance from the centre. Instead, for many galaxies the velocity is more-or-less constant away from the central regions of the galaxy. This is probably due to large amounts of matter being present in the galaxy's halo and corona.

Rowland, Henry Augustus (1848–1901) American physicist who devised a method of ruling DIFFRACTION

GRATINGS more finely than previously, achieving nearly 1700 lines per millimetre (43,000 lines per inch). Using his own gratings, Rowland also made a detailed map of the solar spectrum that was named after him.

Royal Astronomical Society (RAS) Britain's main organization for professional astronomers (though amateurs are admitted), founded in 1820 and receiving its Royal Charter in 1831. The Society's aims are the encouragement and promotion of astronomy and geophysics, which it achieves by publishing the results of scientific research and holding regular meetings. Its main publications are *Monthly Notices* and *Geophysical Journal International*, both of which have international reputations. The RAS headquarters are at Burlington House, Piccadilly, London.

Royal Astronomical Society of Canada (RASC) Principal astronomical body in Canada, with a membership of around 4500 amateur and professional astronomers. Originally established in the mid-19th century, the organization was granted its Royal Charter in 1903. The Royal Astronomical Society of Canada (RASC) is based in Toronto, with 25 regional centres across the country. A bi-monthly *Journal* publishes observational reports and analyses, and the annual RASC *Observer's Handbook* is a useful guide to astronomical phenomena.

Royal Astronomical Society of New Zealand (RASNZ) New Zealand's main organization for amateur and professional astronomers, who work in close collaboration principally on variable star observations and their analysis. Founded in 1920, the RASNZ received its Royal Charter in 1967 and is based in Wellington. It has around 200 members drawn from the islands. Observational work is coordinated by special interest groups and sections, with reports published in a quarterly journal *Southern Stars*.

Royal Greenwich Observatory (RGO) Britain's former national astronomy institution, so named after its move from GREENWICH OBSERVATORY in 1948. The new location at Herstmonceux Castle, near Eastbourne, Sussex, offered much-improved observing conditions, and by 1958 the move was complete. A comprehensive suite of telescopes at the new site included the 0.91-m (36-in.) Yapp Reflector (which was formerly at Greenwich), the largest instrument at Herstmonceux until the 2.5-m (98-in.) ISAAC NEWTON TELESCOPE was completed in 1967. In the 1970s, the RGO played a major role in developing the ROQUE DE LOS MUCHACHOS OBSERVATORY, eventually the home of the ISAAC NEWTON GROUP OF TELESCOPES. With the shift in emphasis to overseas observing, there was no need for telescopes on British soil, and the RGO underwent a controversial and protracted move to Cambridge in 1990. Further controversy over the location of a UK ASTRONOMICAL TECHNOLOGY CENTRE led to the closure of the RGO in 1998 October, bringing to an end 323 years of distinguished astronomical achievement.

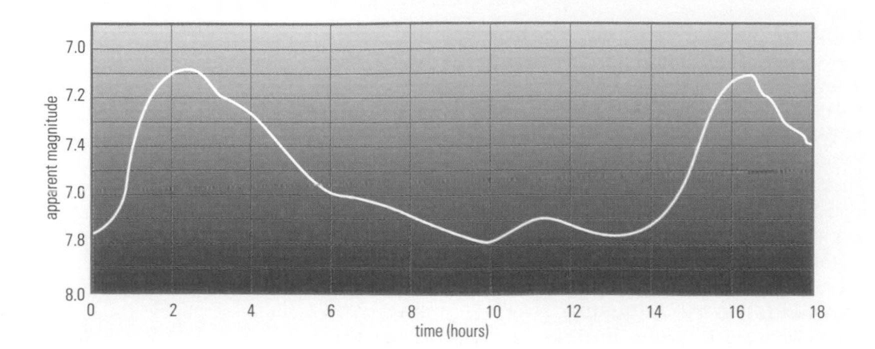

▼ **RR Lyrae variable** This is the light-curve of a typical RR Lyrae star. Like Cepheids, these stars show a rapid rise to peak light followed by a slower decline. Commonly found in globular clusters, RR Lyrae stars have much shorter periods than Cepheids.

Royal Observatory, Cape of Good Hope Originally a southern-hemisphere outstation of the GREENWICH OBSERVATORY, the observatory was proposed in 1821 and completed in 1828 near Cape Town, South Africa. During the 19th century, it generated most of the accurate star positions measured in the southern skies. Under David GILL the observatory's work was expanded to include the compilation of star catalogues from photographic surveys (*see* CAPE PHOTOGRAPHIC DURCHMUSTERUNG). At the beginning of 1972 the observatory broke its 140-year-long association with the ROYAL GREENWICH OBSERVATORY and became the SOUTH AFRICAN ASTRONOMICAL OBSERVATORY. The original observatory buildings serve as the new organization's headquarters.

Royal Observatory, Edinburgh (ROE) Scotland's national observatory from 1822 until the formation of the UK ASTRONOMICAL TECHNOLOGY CENTRE (ATC) in 1998. It began its life at Calton Hill, Edinburgh, where it rose to prominence under the direction of Charles Piazzi SMYTH, Astronomer Royal for Scotland 1846–1888. A move in 1896 to its present site at Blackford Hill, 3 km (2 mi) south of the city centre, allowed the observatory to escape the pollution of Edinburgh, and in 1928 a new 0.91-m (36-in.) reflector was installed. The ROE's outstation at Monte Porzio Catone, Italy, was opened in 1967, followed by the UNITED KINGDOM SCHMIDT TELESCOPE in 1973 and the UNITED KINGDOM INFRARED TELESCOPE in 1979. Growing expertise in astronomical technology eventually led to ROE being selected as the UK's ATC in 1998. With the transfer of its remaining astronomy support functions to the INSTITUTE FOR ASTRONOMY, UNIVERSITY OF EDINBURGH (with which the observatory had always had close ties), ROE formally ceased to exist. However, the name is retained for the site and the Visitor Centre.

r process Method of creating heavy stable nuclei within the interiors of stars by successive capture of neutrons. In the r process, neutrons are added rapidly before the nuclei has time to decay, a process that occurs in SUPERNOVAE. *See also* NUCLEOSYNTHESIS; P PROCESS; S PROCESS

RR Lyrae variable (RR) Type of pulsating VARIABLE STAR. RR Lyrae stars are radially pulsating GIANTS of spectral classes A–F, with periods that are generally fractions of a day. At visible wavelengths their amplitudes are 0.2–2 mag. These stars are often called short-period Cepheids or cluster variables. The latter name originated when Solon BAILEY, in 1895, discovered numerous examples in GLOBULAR CLUSTERS. It was soon found that some of the brightest globular clusters contained many of these stars, and within a few years hundreds had been found.

These rapidly pulsating stars have light-curves that differ from classical CEPHEID VARIABLES. Most rise very quickly to maximum in only a tenth or less of their total period. Their minima are comparatively prolonged, so that for a few hours their light remains constant. Their periods range from 1.2 to 0.2 days. In 1900 the first short-period Cepheid was found outside a globular cluster; it was discovered by Williamina Fleming, at Harvard, and was RR Lyrae, the type star of these variables. At first it was thought that this star must have escaped from a globular cluster, but as more and more such stars were discovered away from globular clusters the idea was dropped. There are now more than 6000 of these stars known (including subtypes), and roughly half are found in clusters. Most have periods between 9 and 17 hours, with many around 13 hours. In the Small Magellanic Cloud some RR Lyrae stars have periods as long as two days, but apart from these there is a fairly sharp cut-off in periods after 1.2 days.

RR Lyrae variables are giant stars; those in our Galaxy belong to the halo component. Some are known to have variable light-curve shapes as well as variable periods. If these changes are periodic, they indicate what is called the BLAZHKO EFFECT. The maximum expansion velocity of the stars' surface layers almost coincides with maximum light: in this respect they resemble the classical Cepheids. Some stars show two simultaneous operating pulsating modes – the fundamental and the first overtone; such stars are placed in a subtype of RR Lyrae variables designated RRB. Another subtype (RRAB) has stars with asymmetric light-curves, with steep rises and periods from 0.3 to 1.2 days. Their amplitudes range from 0.5 to 2 mags. The type star, RR Lyrae, belongs to the RRAB subtype. Stars of another subtype (RRC) have nearly symmetrical (sinusoidal) light-curves with periods of 0.2 to 0.5 days and amplitudes that do not exceed 0.8 mag. The absolute magnitude of RR Lyrae variables is about +0.5. They are too faint to be seen in any but the nearest external galaxies, such as the dwarf system in Sculptor. Those found in the Magellanic Clouds are at about the observable limit.

RR Lyrae variables may be used as distance indicators out to about 650,000 l.y. The method used depends on the star's motion in space. It is assumed that all the stars in a globular cluster are at about the same distance and therefore have the same velocities in space. The RADIAL VELOCITY is one-third of the total space velocity. With these assumptions, the approximate distance to a globular cluster is found and this, in turn, gives its absolute magnitude. From the absolute magnitudes of clusters the size of the Galaxy may be calculated, because the globular clusters form a halo around it. It is then possible to extend these measurements to the nearest external galaxies.

RS Canum Venaticorum star (RS) Type star for a distinctive set of binary stars in which one star interacts with the other to produce giant starspots, resulting in quasi-periodic variations in light with amplitudes of up to 0.2 mag. RS Canum Venaticorum stars are X-ray sources and generate radio emission from flares.

R star Warm CARBON STAR. Class R stars, recognized by bands of C_2, CN and CH, describe spectra that track oxygen-rich stars from class G4 to M1 (5000 to 3600 K). The class – along with the cooler N STARS – has been subsumed into C, with subclasses R0–R9 becoming roughly C0–C5. The classical carbon stars are on the second-ascent asymptotic giant branch of the HERTZSPRUNG–RUSSELL DIAGRAM. The warmer R stars may be helium-burning red giants.

Rubin, Vera Cooper (1928–) American astronomer who discovered the first persuasive observational evidence for the existence of dark matter in galaxies. Working at the Carnegie Institution from 1965, Rubin specialized in galaxy dynamics and clustering. In 1978 she and her colleagues observed that giant gas clouds in galaxies showed high rotational velocities that were consistent with a 'dark matter' model first proposed five years earlier by Jeremiah Ostriker (1937–) and Jim PEEBLES. She realized that only haloes of dark matter beyond the visible boundaries of these galaxies could keep the gas clouds orbiting. Rubin also discovered (1951) peculiar velocities of galaxies as they move through intergalactic space, suggesting that the Universe's LARGE-SCALE STRUCTURE is less uniform than previously thought.

runaway star Star moving with very high space velocity, typically hundreds of kilometres per second. They are usually of spectral class O or B. Some are thought to have been ejected from a CLOSE BINARY system when its companion exploded as a supernova. Some of the youngest are likely to have escaped from their birthplaces through the slingshot interactions of single and binary stars, and several can be traced to particular star-forming regions. For example, the massive stars μ Columbae, AE Aurigae and the binary ι Orionis all seem to have left the Orion Nebula about 2.5 million years ago. They may indeed have originally been members of two close binaries that underwent a close encounter, flinging two stars away in opposite directions and leaving the other two in a very eccentric mutual orbit. *See also* HIGH-VELOCITY STAR

◀ **runaway star** O and B stars ejected by supernova explosions from star-forming regions show high proper motions and may produce a bow shock in the interstellar medium ahead of their direction of travel. Shown here is HD 77581, companion to the supernova remnant Vela X-1.

Russell, Henry Norris (1877–1957) American astronomer, famous for first publishing what is now called the HERTZSPRUNG–RUSSELL DIAGRAM (HR diagram). After graduating from Princeton University, he received his PhD there (1899) for a new way of determining binary star orbits. From 1902 he worked with Arthur Robert Hinks (1873–1945) at Cambridge University, determining stellar distances (and hence absolute magnitudes, M) by parallactic measurements on photographic plates. Russell returned to Princeton, where he became professor of astronomy (1911–27) and observatory director (1912–47), and a major figure in US astrophysics.

Back at Princeton, he built on his work at Cambridge and discovered a correlation between M and the spectral types that were being assigned by Annie CANNON at Harvard. In 1913 he produced what was initially known as the **Russell diagram**, on which he plotted M against Harvard spectral type for over 300 stars. Immediately apparent were the MAIN SEQUENCE running from top left to bottom right, and giant stars across the top. (Unknown to Russell, Ejnar HERTZSPRUNG had done the same some years previously, but had published his results in an obscure journal and in a form less easy to interpret.) The Hertzsprung–Russell diagram is the starting-point for all modern theories of stellar evolution, though Russell first thought, erroneously, that a star began its life as a giant and evolved down the main sequence to become a red dwarf.

Russell continued to study binaries throughout his life, developing a method for estimating the sizes and orbits of eclipsing binaries. He also studied the solar spectrum, using the SAHA EQUATION, and in 1928–29 was able to confirm Cecilia PAYNE-GAPOSCHKIN's earlier finding that hydrogen is the major constituent of the Sun and other stars.

Russell–Vogt theorem *See* VOGT–RUSSELL THEOREM

Russian Aviation and Space Agency Organization controlling Russia's activities in space. The Agency's major goals are formally linked to the solution of social and economic problems at home as well as the implementation of Russia's international interests as the first spacefaring power. They include the provision of communications and broadcast facilities across Russia and CIS territories, environmental monitoring, fundamental research including planetary science and astrophysics, and manned spaceflight.

Rutherfurd, Lewis Morris (1816–92) Pioneer American astrophotographer who subsequently turned to spectroscopy, using improved diffraction gratings of his own design and making. Beginning in 1858, he used an 11.5-inch (290-mm) refractor by Henry Fitz (1808–63) to take some of the earliest photographs of the Sun, Moon and planets, and of stars as faint as 5th magnitude.

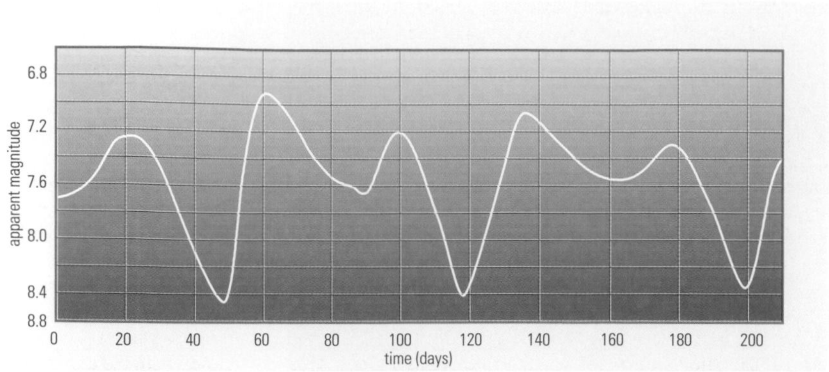

▲ **RV Tauri star** The light-curve of a typical RV Tauri star shows alternating deep and shallow minima. The deep minima result from different modes of pulsation in the star's outer layers coming into phase with each other. At other times, the pulsation mechanism may apparently switch off, leaving only the shallow minima.

In 1864 he built the first purely photographic telescope, designed to focus the violet wavelengths to which photographic emulsions were most sensitive. Rutherfurd obtained a solar spectrum that showed three times as many spectral lines as the best previous examples.

RV Tauri star (RV) Highly luminous, pulsating VARIABLE STAR. Members of this small group are mainly yellow or orange SUPERGIANT stars in spectral classes G and K, with some F stars. Examples are RV Tauri, R Sagittae and R Scuti. Their enormous, very extended atmospheres of gas emit infrared radiation. The light-curves of the RV Tauri stars are characterized by alternating deep and shallow minima, which may occasionally exchange places. Their periods range from about 30 to 145 days. Sometimes the light variations may become rather irregular, particularly for the stars having the longest periods. For this reason the RV Tauri stars are classified as SEMIREGULAR VARIABLES. They also have a variation in their COLOUR INDEX (an indication of the temperature of the star) that looks like the light-curve, but goes through its maximum shortly before the star reaches minimum brightness. There are two distinct photometric subtypes: one (RVA) exhibits the behaviour just described; in the second subtype (RVB) the characteristic variations are superimposed on a longer-period fluctuation, with an amplitude of about two magnitudes and a typical period of 600–1500 days. R Sagittae and RV Tauri itself are both RVB stars.

RW Aurigae variable Very young star, a member of a subclass of NEBULAR VARIABLE or T TAURI STARS. The RW Aurigae stars have quite large amplitudes and are almost all G-type DWARFS. Very few are brighter than 10th magnitude at maximum. They show rapid and extremely irregular variations in their light-curves, although some stars appear to have a pseudo-periodic wave superimposed upon the primary light variations. Many RW Aurigae stars are found associated with the T ORIONIS VARIABLES.

Ryle, Martin (1918–84) English astronomer, the twelfth ASTRONOMER ROYAL, who established Cambridge as a centre for radio astronomy, and pioneered radio interferometry and aperture synthesis. He joined Cambridge University's Cavendish Laboratory after working on radar during World War II, and became director of the MULLARD RADIO ASTRONOMY OBSERVATORY (MRAO) in 1957, and the university's first professor of radio astronomy two years later.

Ryle saw that the immediate task facing radio astronomers was to map the sky at radio wavelengths. He chose to do this by building the first RADIO INTERFEROMETERS, for which MRAO became famous. In 1950 he instigated the first of the Cambridge surveys, the most important of which was the third, completed in 1959, which yielded the *THIRD CAMBRIDGE CATALOGUE* ('3C') of some five hundred radio sources. He and his colleagues developed the technique of APERTURE SYNTHESIS in which an array of small instruments provide the sensitivity of a much larger single collector. The first such instrument, the One-Mile Telescope, was completed in 1963; its successor was the Five-Kilometre Telescope, since renamed the RYLE TELESCOPE.

The Cambridge surveys and radio telescopes advanced knowledge of PULSARS and QUASARS, and the number of distant radio sources lent support to the Big Bang theory. For his pioneering work in radio astronomy he shared the 1974 Nobel Prize for Physics with Antony HEWISH.

Ryle Telescope Eight-element, east–west radio interferometer operated by the MULLARD RADIO ASTRONOMY OBSERVATORY. The individual elements are steerable 13-m (43-ft) antennae, four of which are mounted on a 1.2-km (0.75-mi) rail track while the others are fixed at 1.2-km (0.75-mi) intervals.

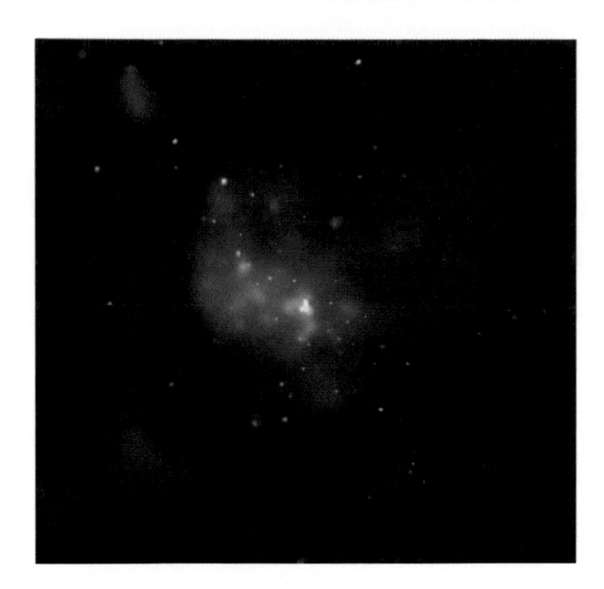

Safronov, Viktor Sergeyevich (1917–99) Russian planetary astronomer, the foremost theorist of COSMOGONY in the 20th century. His detailed, comprehensive and mathematically rigorous theory of the formation of planets from a circumstellar disk of gas and dust underlies present-day research. He recognized that the rate of collisions among planetesimals, which led to growth by ACCRETION, was controlled by their relative velocities; the velocities were governed by mutual gravitational perturbations, which were in turn controlled by the mass distribution. He also examined the formation of PLANETESIMALS by gravitational instability of a dust layer in the protoplanetary nebula.

Sagan, Carl Edward (1934–96) American astronomer known for his Solar System studies, especially of planetary atmospheres and surfaces, and investigations into life and its origin on Earth. After working at the Smithsonian Astrophysical Observatory he moved to Cornell in 1968, becoming director of its Laboratory for Planetary Studies and professor of astronomy and space science. In the early 1960s he showed that, since a greenhouse effect should be operating on VENUS, its surface temperature could be as high as 800 K. In 1963 Sagan and colleagues at NASA repeated the Miller–Urey experiment (*see* Harold UREY) using a different set-up and mixture of gases, and showed that not only amino acids but also sugars and ATP – the chemical used by cells to store energy – could have formed in the primordial atmosphere of the Earth.

Sagan's interest in the possibility of LIFE IN THE UNIVERSE involved him in SETI studies and led to his co-authorship of Iosif SHKLOVSKII's *Intelligent Life in the Universe* (1966). He also helped to compose messages carried on board the VOYAGER and some of the PIONEER probes that could convey information about humans and the Earth to any intelligent entity that might eventually find them. From the 1970s, Sagan was a successful popularizer of astronomy, with books on planetary science and other areas, including evolution, as well as the TV series *Cosmos*. In 1997, the year after his death, the MARS PATHFINDER lander was named the Carl Sagan Memorial Station in his honour.

Sagitta *See* feature article

Sagittarius *See* feature article

Sagittarius A (Sgr A) Complex centre of our Galaxy, and the most intense part of the Milky Way's radio emission. The actual centre is marked by a smaller radio source, Sgr A*, less than 4 AU across, pin-pointed using the Very Long Baseline Array of radio telescopes. Sgr A* may mark the presence of a massive BLACK HOLE. Very close to Sgr A* is an intense infrared source (IRS 16), which is a large cluster of newly formed hot B stars. In addition to the radio and infrared emission from the region, X-rays and gamma-rays have also been detected.

Detailed maps of Sgr A in radio emission, mostly made by the VERY LARGE ARRAY, show remarkable structures, loops, arcs and a series of narrow parallel filaments at right angles to the plane of the Milky Way, which radiate by the SYNCHROTRON mechanism. The filaments may indicate the presence of a strong magnetic field, but they could also be created by winds. Within 3 l.y. of the centre is a theta-shaped pattern. The ring of the theta is made of gas that has probably piled up there as a result of some outflow from the centre; alternatively it may be the tattered remnant of one or more nebulae that were shredded by moving in the gravitational pull of the central stars. The bar of the theta, also gaseous, is surprisingly hot and apparently has a different origin. There is a thin streamer of gas that may link the shell of material around the putative black hole with the nearest gas cloud. Sgr A remains a very mysterious object.

Sagittarius Arm Next spiral arm of our GALAXY after the ORION ARM, moving in towards the centre of the Galaxy. At its closest it is about 5000 l.y. from the Sun, and it contains the LAGOON, EAGLE and ETA CARINAE NEBULAE. It extends in the sky from Serpens to Centaurus, and may also continue into Carina to form

◄ **Sagittarius A** The bright point at the centre of this image of Sagittarius A is an X-ray flare from the area near the black hole at the centre of the Milky Way. It brightened dramatically in a few minutes and then declined over a period of three hours.

SAGITTA (GEN. SAGITTAE, ABBR. SGE)

Third-smallest constellation but a distinctive one, representing an arrow, between Vulpecula and Aquila. Its brightest star, γ Sge, is mag. 3.5. ζ Sge is a triple system with bluish-white and bluish components, mags. 5.6 and 9.0, separation 8″.3; the former has a very close companion, mag. 6.0. WZ SAGITTAE is a recurrent nova; FG SAGITTAE is an unusual variable which slowly brightened from about mag. 13.6 in 1890 to mag. 9.4 in 1967. The brightest deep-sky object is M71 (NGC 6838), an 8th-magnitude globular cluster.

SAGITTARIUS (GEN. SAGITTARII, ABBR. SGR)

Large, conspicuous southern zodiacal constellation, 'the Archer', usually depicted as a centaur aiming an arrow at the heart of neighbouring Scorpius. It lies between Ophiuchus and Capricornus, in a region rich in star clouds of the Milky Way, in the direction of the centre of the Galaxy. The brightest stars are ε Sgr (Kaus Australis), mag. 1.8, and σ Sgr (NUNKI), mag. 2.1. β Sgr (Arkab) is a wide, naked-eye double with bluish-white and pale yellow components, mags. 4.0 and 4.3; the former has a white companion, mag. 7.2, separation 28″.5. RY Sgr is an R CORONAE BOREALIS STAR which is usually around 6th magnitude but from time to time drops unpredictably to 14th magnitude. Eight of the constellation's brighter stars form an asterism known as the TEAPOT.

The brightest part of the Milky Way, just to the north of γ Sgr (Alnasl), is known as the Great Sagittarius Star Cloud, while another bright part, to the north of μ Sgr, is sometimes called the Small Sagittarius Star Cloud. Bright star clusters and nebulae in Sagittarius include: the open clusters M25 (IC 4725), M21 (NGC 6531) and M23 (NGC 6494); the 5th-magnitude globular cluster M22 (NGC 6656), the third brightest in the sky; and the LAGOON NEBULA (M8, NGC 6523), the OMEGA NEBULA (M17, NGC 6618) and the TRIFID NEBULA (M20, NGC 6514), all 6th-magnitude emission nebulae. Also in Sagittarius is the SAGITTARIUS DWARF GALAXY, a dwarf spheroidal galaxy in the Local Group, about 80,000 l.y. away, and the radio source SAGITTARIUS A.

BRIGHTEST STARS

	Name	RA h	RA m	dec. ° ′	Visual mag.	Absolute mag.	Spectral type	Distance (l.y.)
ε	Kaus Australis	18	24	−34 23	1.79	−1.4	B9.5	145
σ	Nunki	18	55	−26 18	2.05	−2.1	B2.5	224
ζ	Ascella	19	03	−29 53	2.60	0.4	A3	89
δ	Kaus Media	18	21	−29 50	2.72	−2.1	K3	306
λ	Kaus Borealis	18	28	−25 25	2.82	0.9	K1	77
π		19	10	−21 01	2.88	−2.8	F2	440
γ	Alnasl	18	06	−30 25	2.98	0.6	K0	96
η		18	18	−36 46	3.10	−0.2	M2	149
φ		18	46	−26 59	3.17	−1.1	B8.5	231
τ		19	07	−27 40	3.32	0.5	K1/K2	120

the CARINA ARM. It may best be traced from its TWENTY-ONE CENTIMETRE radio emissions from hydrogen and by the presence of young blue stars.

Sagittarius Dwarf Galaxy Disrupted companion to our GALAXY, discovered by detailed star counts and distance estimates. It is the closest known galaxy to the centre of our Galaxy, lying mostly behind the central bulge from our point of view so that its sparse stars are lost in the myriad foreground objects. It probably contains the globular cluster M54. The elongated shape revealed by the distribution of stars belonging to the Sagittarius Dwarf suggests that it is being tidally disintegrated by our Galaxy's gravity in the kind of event that may have happened many times during our Galaxy's formation and growth.

Saha, Meghnad (1894–1956) Indian nuclear physicist and astrophysicist who derived the formula (SAHA'S EQUATION) linking the degree of ionization in a gas to temperature and electron pressure that is vital to the interpretation of stellar spectra. He confirmed his theory of thermal ionization (1920) by studying the solar chromosphere and spectra of stars, especially novae.

Saha equation Equation, formulated by the Indian physicist Meghnad Saha (1893–1956), that determines how the relative number densities of the various levels of ionization (including the neutral atom) of an element change with temperature. It takes the form:

$$\frac{N_{I+1}}{N_I} = \frac{2U_{I+1}}{U_I}\left(\frac{2\pi\, m_e kT}{h^2}\right)^{3/2} e^{-\chi_I/kT}$$

where the subscript I or $I+1$ denotes the level of ionization, N is the number density, U is an atomic constant called the partition function, m_e is the electron mass, k is BOLTZMANN'S CONSTANT, T is the temperature, h is PLANCK'S CONSTANT and χ_I is the energy required to ionize the Ith ion to the $(I+1)$th ion.

St Andrews University Observatory Teaching and research facility of the University of St Andrews' School of Physics and Astronomy, 50 km (30 mi) north-east of Edinburgh. It is notable for its early work on large SCHMIDT–CASSEGRAIN TELESCOPES and the construction of the University's 0.94-m (37-in.) James Gregory Telescope, one of the largest optical telescopes in Britain.

Sakigake Space probe launched to Comet HALLEY in 1985 January by Japan's Institute of Space and Astronautical Science (ISAS). Japan's first deep-space mission, it investigated the interaction between the SOLAR WIND and the comet. Sakigake flew past the comet's sunward side on 1986 March 11 at a distance of 6.9 million km (4.3 million mi).

Sakurai's object V4334 in Sagittarius; the central object in a PLANETARY NEBULA discovered by the Japanese amateur astronomer Yukio Sakurai in 1996 February. He observed a rapid brightening of the 11th-magnitude object, which remained bright for about two years before dimming in mid-1998. The star had probably experienced a late HELIUM FLASH in 1994. The planetary nebula has since been observed to be expanding at a rate of 31 km/s (19 mi/s). It has an apparent angular diameter of 44″.

Sakurai's object is the third late helium flash star to be observed. The first was V605 Aql, which brightened and faded during 1919 to 1923, and the second was FG SAGITTAE, which reached its maximum brightness in 1970 and faded in 1992. It is presumed to have experienced its helium flash at the beginning of the 19th century. *See also* SHELL BURNING

Salpeter process *See* TRIPLE-α PROCESS

Sandage, Allan Rex (1926–) American astronomer who made the first optical identification of a quasar and for many years has worked to establish the distance scale of the Universe and a low value of the HUBBLE CONSTANT. Since 1952 Sandage has worked at the Mount Wilson and Palomar Observatories, initially as assistant to Edwin Hubble. In 1960 he and Thomas Arnold Matthews (1927–) used Palomar's 200-inch (5-m) HALE TELESCOPE to identify a faint, star-like object in the same position as the radio source 3C 48 listed in the *THIRD CAMBRIDGE CATALOGUE*; three years later, Maarten SCHMIDT found the immense redshift in the spectrum of the object, which turned out to be a quasar. In 1965 Sandage found the first radio-quiet quasar.

Sandage also worked on the COSMOLOGICAL DISTANCE SCALE, calibrating the various 'standard candles' in order to establish the distances of galaxies. With Martin SCHWARZSCHILD, he studied the evolution of globular clusters in order to determine their ages. The ages of their oldest stars were not compatible with a relatively young, small Universe, as favoured by Gerard DE VAUCOULEURS. In 1976 Sandage and Gustav Andreas Tammann (1932–) gave a value for the Hubble constant of H_0 = 50 km/s/Mpc, half that given by De Vaucouleurs. In 1979 Sandage proposed that one of the main tasks of the Hubble Space Telescope (HST) should be to strive towards fixing the value of H_0. Since its launch, Sandage and others have used the HST to examine CEPHEID VARIABLES and supernovae in distant galaxies, deriving in the late 1990s a value of H_0 = 60 km/s/Mpc.

Saiph The star κ Orionis, visual mag. 2.07, distance 722 l.y., spectral type B0.5 Ia. The name comes from the Arabic *saif*, meaning 'sword', but is wrongly applied as this star actually marks the right leg of Orion.

Salpeter, Edwin Ernest (1924–) Austrian-American astrophysicist, who emigrated first to Australia, then to England and the USA, and specialized in stellar evolution. He spent most of his career at Cornell University, where he worked with Hans BETHE applying relativity theory to atomic physics. He was the first to explain how highly evolved stars convert hydrogen to helium to carbon by the triple-α reaction (known also as the Salpeter reaction). By relating the observed abundances of stars of various luminosities to his models of stellar evolution, Salpeter was able to derive the initial mass function (known also as the Salpeter function), which predicts the rate at which these stars will form in the Milky Way based upon their masses.

SALT Abbreviation of SOUTHERN AFRICAN LARGE TELESCOPE

Salyut space stations Name given to seven Soviet SPACE STATIONS. Two different design bureaux were involved in their development, and each bureau produced its own versions in 1971–76, one series for 'civilian' use (Salyuts 1, 2 and 4) and another 'military' version known as 'Almaz' (Salyuts 3 and 5). Both versions weighed about 18.5 tonnes.

Crews of two or three cosmonauts were ferried up to Salyut by the SOYUZ spacecraft. The first five Salyuts were equipped with only one docking port. Commencing with Salyut 6, a second docking port was made available. This allowed automatic Progress and heavy Cosmos ferry craft to carry supplies and fuel to the stations, and enabled semi-permanent occupation of the stations by long-duration crews. The last of the series, Salyut 7, re-entered the atmosphere in 1991 February.

Various astronomical experiments were carried on the stations, including instruments to observe the Sun, X-ray and infrared telescopes (Salyut 4) and a submillimetre telescope (Salyut 6).

SAMPEX Abbreviation of SOLAR ANOMALOUS AND MAGNETOSPHERIC PARTICLE EXPLORER

S Andromedae First extragalactic SUPERNOVA to be detected. Discovered near the centre of the ANDROMEDA GALAXY (M31) in 1885, the star attained an apparent

S

magnitude of +6.5. Supernovae were then not recognized as a distinct class of objects. The assumption that the star was an ordinary NOVA led to a distance estimate for M31 of about 8000 l.y., well within our own Galaxy. Later it was realized that M31 was a galaxy similar to our own and that S Andromedae, shining temporarily with a sixth of the total light of the galaxy, was vastly more brilliant than a typical nova.

SAO Abbreviation of SMITHSONIAN ASTROPHYSICAL OBSERVATORY and of SPECIAL ASTROPHYSICAL OBSERVATORY

SARA Abbreviation of SOCIETY OF AMATEUR RADIO ASTRONOMERS

satellite Smaller body orbiting a larger one. At the most massive end of the scale, this can be taken to refer to small galaxies orbiting a larger primary as, for example, M32 and NGC 205 in relation to the Andromeda Galaxy. Likewise, in this sense, planets are satellites of their star, though they are rarely spoken of in this way. Instead, the term is most commonly applied to a body in orbit about a planet. Every planet in the Solar System, except Mercury and Venus, has at least one natural satellite (as opposed to ARTIFICIAL SATELLITE) in orbit about it. The discovery of planetary satellites was important because measurement of the period and dimensions of a satellite's orbit enables the mass of its planet (or, strictly, the combined mass of planet and satellite) to be determined.

Written without a capital 'M', the term 'moon' is synonymous with 'natural satellite', whereas the Moon is the name of the Earth's only such satellite. The MOON and the Earth have been companions since the final stages of the Earth's formation, 4.5 billion years ago, and the Moon is believed to have formed from the debris of a giant collision (*see* GIANT IMPACTOR THEORY). A similar process could account for the origin of Pluto's only known satellite, CHARON, which is the most massive satellite relative to its planet. In contrast Mars has two small irregularly shaped satellites, PHOBOS and DEIMOS, which are asteroids captured by Mars probably within the past billion years.

Each of the four GIANT PLANETS has an extensive satellite system. At the outer fringes of the planetary RING system are small, bright, icy moonlets in circular orbits; they may be fragments of larger satellites destroyed by collisions or tidal forces. Slightly nearer the planet come several satellites, such as Jupiter's GALILEAN SATELLITES, that are large enough (greater than about 400 km/250 mi in diameter) for their own gravity to pull them into a spheroidal shape. These too are in virtually circular orbits and are believed to have grown by ACCRETION within circumplanetary disks while the planets were forming. This far from the Sun, the SOLAR NEBULA was cold enough to allow the direct condensation of ice, so the large satellites are mostly icy bodies with deeply buried rocky interiors. At Jupiter, the ice is composed just of water, but with increasing distance from the Sun more volatile species are present, such as methane, ammonia and nitrogen.

Neptune, exceptionally, has only one large satellite, TRITON. It is in a RETROGRADE orbit and is presumed to be a captured EDGEWORTH–KUIPER BELT object. Any family of large satellites previously possessed by Neptune is likely to have been lost during the capture process.

Two of Saturn's large satellites, TETHYS and DIONE, are accompanied in their orbits by small (less than 30 km/20 mi in diameter) irregularly shaped satellites that occupy stable LAGRANGIAN POINTS and bear the same geometric and dynamic relation to them as do the TROJAN ASTEROIDS to Jupiter. Beyond their large satellites, the giant planets have numerous smaller satellites, mostly in inclined, eccentric, and in some cases retrograde, orbits. These satellites are dark irregular objects, possibly coated by carbonaceous material, and are probably captured asteroids or comet nuclei.

TIDAL FRICTION has caused the rotation period of most satellites to become synchronous with their orbital period, so that like our own Moon they always keep the same face towards their planet (*see* SYNCHRONOUS ROTATION). Slowing down of a large, originally rapidly spinning satellite in this way could cause considerable internal TIDAL HEATING, which would have allowed internal DIFFERENTIATION to take place, with the formation of a rocky CORE (and possibly an iron-rich inner core) beneath an icy MANTLE. Tides are also important in causing satellite orbital ECCENTRICITY to decrease until, in the absence of other perturbations, the orbit would become circular. However, the large satellites of Jupiter, Saturn and Uranus are massive enough to have a significant gravitational influence on one another. The orbits of adjacent satellites have often evolved into a state of orbital RESONANCE, meaning for example that one satellite has twice the orbital period of the next.

When in a state of orbital resonance, the influence of other satellites prevents an orbit from becoming exactly circular. At times the varying forces on the TIDAL BULGES raised on a large satellite by its planet may then be sufficient to act as a significant source of tidal heating, leading to volcanism and/or fracturing of the surface. This activity occurs at the present time within Jupiter's satellites IO and EUROPA, and it clearly affected several other large satellites in the past. Without tidal heating, satellites would simply have ancient surfaces heavily scarred by a 4-billion-year history of impact cratering. One of the greatest revelations of the VOYAGER missions, which were the first to explore the satellites of the outer planets in any detail, was the amazing variety of landscapes they possess.

Some asteroids are now known to have their own tiny satellites. The first, named DACTYL, was discovered on images of the asteroid IDA sent back by the GALILEO spacecraft in 1993. The second, an object 18 km (11 mi) across, was discovered telescopically in orbit around the asteroid EUGENIA (diameter 215 km/134 mi) in 1998.

satellite laser ranger (SLR) Specialized telescope that uses time-of-flight measurements of short pulses of laser light to determine the distances to satellites in Earth orbit. The principle is the same as that of radar, except that light, rather than radio waves, is used. Pulses of laser light are beamed from the telescope at dedicated satellites covered in retro-reflectors. These work like 'cat's eyes' and reflect the light back in the direction from which it came. By timing how long it takes the light to travel to and from the satellite, the distance can be computed to an accuracy of a few centimetres. Because the satellites are in very stable and well-known orbits, the results provide information about the rotation of the Earth, the geometry and deformation of its surface, and variations in its gravitational field.

SLR stations are located all over the globe and the results from each one make it possible to measure very accurately the distance between points on the Earth's surface, making it a good technique for studying plate tectonics. Satellite laser rangers have replaced the PHOTOGRAPHIC ZENITH TUBE (PZT) as the primary means of measuring variations in the Earth's rotation.

Saturn Sixth planet in the Solar System and the second largest; it was the most distant planet known to man before the development of the telescope. The telescopic appearance of the planet is dominated by the majestic system of rings, which were probably first seen by Galileo in 1610, even though he did not recognize their nature (he believed Saturn to be a triple planet). The rings lie in the plane of the planet's equator and are tilted by 27° with respect to its orbit. Consequently, the faces of the rings will be alternately inclined towards the Sun and then the Earth by up to 27°. At intervals of approximately 15 years, the rings become edge on to the Earth and are virtually invisible to the observer. This situation occurred in 1995–96. The rings are extremely reflective, so that they can add significantly to the total brightness of the planet. The distance from one edge of the rings to the other is more than 275,000 km (171,000 mi), which is nearly

► **Saturn** In this high-resolution image of Saturn from the Hubble Space Telescope, the belts in the atmosphere are clearly visible. The open angle of the rings means that some of the finer details can be seen.

SATURN: DATA	
Globe	
Diameter (equatorial)	120,000 km
Diameter (polar)	107,100 km
Density	0.69 g/cm^3
Mass (Earth = 1)	95.2
Volume (Earth = 1)	764
Sidereal period of axial rotation (equatorial)	$10^h\ 14^m$
Escape velocity	35.5 km/s
Albedo	0.47
Inclination of equator to orbit	26° 43′
Temperature at cloud-tops	95 K
Surface gravity (Earth = 1)	1.19
Orbit	
Semimajor axis	9.539 AU = 1427 × 10^6 km
Eccentricity	0.056
Inclination to ecliptic	2° 29′
Sidereal period of revolution	29.46y
Mean orbital velocity	9.65 km/s
Satellites	30

three-quarters the distance of the Earth to the Moon. Saturn is now known to have at least 30 satellites.

Saturn is composed of hydrogen and helium, but they are not in solar proportions. There is a helium depletion in Saturn where the mass fraction is only 11%, compared with the solar abundance of 27%. This significant difference, when compared with JUPITER, may be due to the differing internal structures and current stages of evolution of the two planets. It is thought that the internal temperatures are too low for helium to be uniformly mixed with hydrogen throughout the deep interior. Instead, the helium may be condensing at the top of this region, where the gravitational energy is then turned into heat. This process may have started 2000 million years ago when the temperatures first dropped to the helium condensation point. For Jupiter, this situation can only have been reached recently. The other primary constituents of the atmosphere are ammonia, methane, acetylene, ethane, phosphine and water vapour.

Saturn has a similar visible appearance to Jupiter, with alternating light and dark cloud bands, known as zones and belts, respectively. Saturn's clouds are more subtle and yellowish and, therefore, less colourful than those found on Jupiter. Contrast is also muted by an overlying haze layer. Consequently, the cloud features and associated spots, although varied in nature and colour, are less prominent on Saturn. Several stable ovals of various colours (white, brown and red) have been observed in Saturn's atmosphere. The white clouds are composed of ammonia particles, and the other colours are generated through dynamical and photochemical actions and reactions in the atmosphere. Three brown spots, situated at 42°N during the VOYAGER encounters, were seen to behave in a similar fashion to the Jovian white ovals. The largest features seen were a reddish cloud 10,000 × 6000 km (6000 × 4000 mi) at 72°N and a red spot 5000 × 3000 km (3000 × 2000 mi) at 55°S. These features demonstrate the non-uniqueness of the Jovian Great Red Spot (GRS) and other cloud ovals. A major characteristic of the Saturnian mid-latitude weather systems is a jet stream, which produces alternate high- and low-pressure systems in the same way as the terrestrial phenomenon. Anti-cyclonically rotating cloud systems, like the Jovian GRS and trains of vortices, familiar in the Earth's atmosphere, are also seen. The Saturnian weather systems, like those of Jupiter, are strongly zonal. At the equator the cloud top winds reach 500 m/s (1600 ft/s), which is equivalent to three-quarters of the speed of sound at this level. Although Saturn has a strong internal heat source, the weather systems of Saturn (and of Jupiter and the Earth) are driven by the transport of energy from small-scale features into the main zonal flow.

Probably the most dramatic cloud features witnessed in the Saturnian atmosphere are the Great White Spots, which break out at roughly 30-year intervals coincident with the planet's northern hemisphere midsummer, as in 1876, 1903, 1933, 1960 and 1990.

The interior of Saturn is thought to consist of an Earth-sized iron-rich core of ammonia, methane and water, which is enclosed by about 21,000 km (13,000 mi) of liquid metallic hydrogen, above which extends the liquid molecular hydrogen and the extensive cloud layers. It is in the metallic hydrogen region that the magnetic field is created by the dynamo action from the rapidly rotating

► **Saturn** The Hubble Space Telescope's STIS (Space Telescope Imaging Spectrograph) instrument imaged Saturn's aurorare in 1994. They only glow in ultraviolet light and so are only detectable from above the Earth's atmosphere.

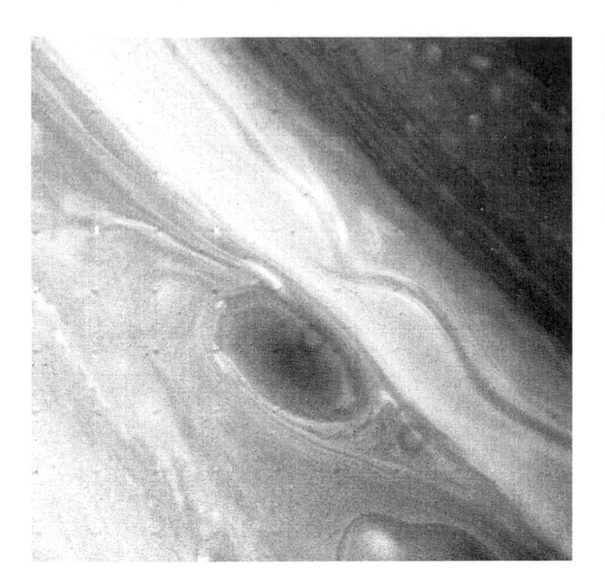

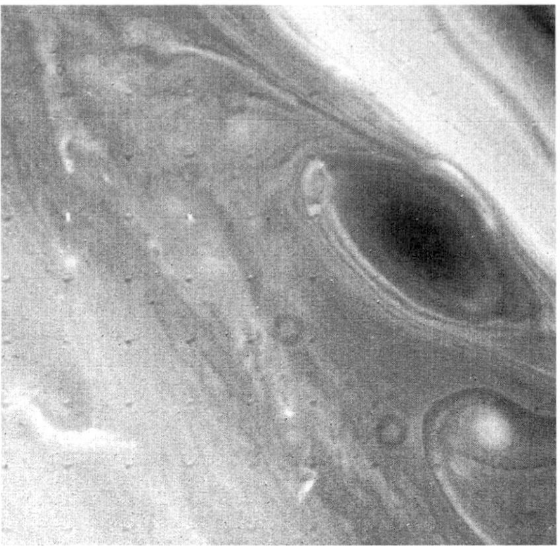

◀ **Saturn** The rapid changes in Saturn's atmosphere can be seen in this pair of Voyager 2 images taken just over ten hours apart. The smallest features visible are more than 50 km (30 mi) across.

planet. The first detection of the Saturnian magnetic field was made from the PIONEER 11 spacecraft in 1979. At the cloud tops the equatorial field has a strength of 0.21 G, compared with the Earth's value of 0.31 G. The magnetic axis is within 1° of the axis of rotation; it is therefore the least tilted field in the Solar System.

The MAGNETOSPHERE of Saturn is intermediate between those of the Earth and Jupiter in terms of both extent and population of the trapped charged particles. The average distance of the BOW SHOCK is 1.8 million km (1.1 million mi), while the MAGNETOPAUSE itself lies much closer to the planet, at 500,000 km (300,000 mi). These distances are, of course, extremely variable, since their precise positions will depend on the temporal behaviour of the SOLAR WIND. The largest satellite, TITAN, is situated near the magnetopause boundary and regularly crosses this division. The magnetosphere is divided into several definite regions. At about 400,000 km (250,000 mi) there is a torus of ionized hydrogen and oxygen atoms; the plasma's ions and electrons spiral up and down the magnetic field lines, contributing to the local field. Beyond the inner torus, there is a region of plasma, extending out to 1 million km (600,000 mi), produced by material coming partly from Saturn's outer atmosphere and partly from Titan. The magnetotail has a diameter of about 80 R_S (Saturnian radii). There is a strong interaction between the charged particle environment and the embedded satellites and rings that surround Saturn. All these bodies absorb the charged particles and have the effect of sweeping a clear path through the region where they are located. There is also auroral activity in the polar regions, where the charged particles cascade into the upper atmosphere. The Saturnian aurorae are about two to five times brighter than the equivalent terrestrial phenomena.

Saturn is a powerful radio source, emitting broad band emissions in the ran ge from about 20 KHz to about

SATURN: SATELLITES AND RINGS

Rings ring or gap	distance from centre (thousand km)	width (km)
D	67–74.5	7500
C (Crepe)	74.5–92	17500
Maxwell Gap	88	270
B	92–117.5	25500
Cassini Division	117.5–122.2	4700
A	122.2–136.8	14600
Encke Division	133.6	325
Keeler Gap	136.5	35
F	140.2	30–500
G	165.8–173.8	8000
E	180–480	300000

Satellites	diameter (km)	distance from centre of planet (thousand km)	orbital period (days)	mean opposition magnitude
Pan	20	134	0.58	19
Atlas	36 × 34 × 28	138	0.60	18.0
Prometheus	150 × 90 × 70	139	0.61	15.8
Pandora	110 × 85 × 60	142	0.63	16.5
Epimetheus	140 × 120 × 105	151	0.69	15.7
Janus	200 × 180 × 150	151	0.69	14.5
Mimas	398	186	0.94	12.9
Enceladus	498	238	1.37	11.7
Tethys	1058	295	1.89	10.2
Telesto	30 × 26 × 16	295	1.89	18.5
Calypso	30 × 16 × 16	295	1.89	18.7
Dione	1120	377	2.74	10.4
Helene	32	377	2.74	18.4
Rhea	1528	527	4.52	9.7
Titan	5150	1221	15.95	8.3
Hyperion	370 × 280 × 225	1481	21.28	14.2
Iapetus	1440	3561	79.33	10.2–11.9
Phoebe	230 × 210	12,940	550.5 *R*	16.4

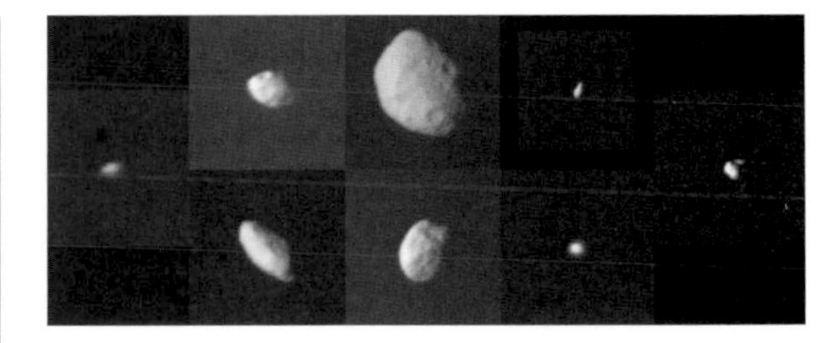

▲ **Saturn** These Voyager images of some of Saturn's smaller satellites are in order of distance from the planet and at their correct relative sizes. The paired satellites orbit at the same distance from the giant planet.

▼ **Saturn** Most of Saturn's satellites orbit the planet directly, but the outer satellites Iapetus and Phoebe have retrograde orbits and are in a different plane from the others, which may imply that they are captured asteroids.

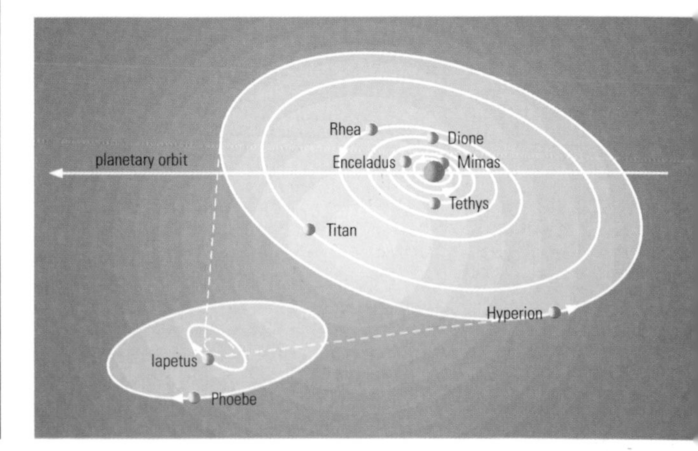

S

▲ **Saturn rocket** Many of NASA's missions have been launched on Saturn rockets, including all of the Apollo trips to the Moon. Here, a Saturn 1B rocket is carrying the third and final crew to Skylab in 1973.

1 MHz. The maximum intensity occurs between 100 and 500 KHz with a period of $10^h\,39^m.4$, which corresponds to the System III rotation period. The IONOSPHERE, residing above the neutral mesosphere, contains mainly ionized hydrogen, which is strongly controlled by the solar wind.

The interaction between the charged particle environment and the rings is unique at Saturn, where radial features or spokes have been seen in the B RING. These features are confined to the central portion of the B ring and appear to correspond to a location where only small, micrometre-size particles are present. The ring particles become electrically charged and appear as a torus above the plane of the rings. Each spoke, which is generally wedge-shaped and about 6000 km (3700 mi) long, has a lifetime of a few planetary rotations. The detection of electrostatic discharges (lightning) at radio wavelengths suggests that this mechanism is related to the formation of the spokes.

Saturn Nebula (NGC 7009) PLANETARY NEBULA in western Aquarius (RA $21^h\,04^m.2$ dec. $-11°22'$). At an estimated distance of 160 l.y., it is one of the closest to our Solar System. Discovered by William HERSCHEL in 1782, the Saturn Nebula takes its popular name – given by Lord Rosse in the 1850s – from the extensions, or ansae, seen to either side, which give it a resemblance to Saturn when its rings are presented close to edge-on. The nebula has overall magnitude +8.0 and covers an area of $0'.4 \times 1'.6$. The central star is of magnitude +11.5.

Saturn rocket Rocket developed by NASA in the early 1960s for manned spaceflight. The Saturn V, the largest, most powerful rocket ever built, was used for the APOLLO missions to the Moon. Its thirteenth and last flight was to insert the SKYLAB space station into Earth orbit in 1973. The final Saturn IB launch took place in 1975 during the Apollo–Soyuz Test Programme.

saros Cycle of 18 years 10.3 days after which the Sun, Moon and the NODES of the Moon's orbit return to almost the same relative positions. It is a result of the REGRESSION OF THE NODES. The saros was known to the ancient Babylonian astronomers; it was used to predict eclipses, since any eclipse is usually followed by a similar one 18 years 10.3 days later. Slight variations from cycle to cycle mean that the eclipses are not identical. For example, the total SOLAR ECLIPSE of 1999 August 11 was potentially visible along a track from south-west England to France, Germany, Turkey, Iran and India: the next one in the cycle, on 2017 August 21, will be visible from the United States.

scale height Increase in height over which a physical quantity, for example atmospheric density or pressure, declines by a factor of e (e = 2.71828). For example, at an altitude of one scale height, the density of an atmosphere is $1/e$ (= 0.37) of the value at its base. In the case of the Earth's troposphere, the scale height is approximately 8.5 km (5.3 mi).

scarp Steep slope of some extent along the margin of a plateau, terrace or bench. Linear scarps on planets and satellites often represent the surface expression of FAULTS; sinuous scarps are usually the result of fluvial (water, lava) erosion. The inner slopes of volcanic craters and calderas are often scarps. In planetary nomenclature, such features are identified by the term Rupes.

scattering Deflection of ELECTROMAGNETIC RADIATION by particles. Where the particles are very much larger than the wavelength, scattering consists of a mixture of reflection and DIFFRACTION, and is largely independent of wavelength. Where the particle sizes are very much smaller than the wavelength, the amount of scattering is inversely proportional to the fourth power of wavelength and so blue light is scattered about ten times more efficiently than red; this is known as RAYLEIGH SCATTERING. The blueness of the sky is due to Rayleigh scattering of sunlight by atoms and molecules in the atmosphere. The setting sun appears red because far more blue than red light is scattered out of the line of sight and thus a greater proportion of red light penetrates to the observer. Scattering is described as elastic if no change in wavelength is produced, and inelastic if the wavelength is increased or decreased by the encounter. Interstellar dust produces both scattering and ABSORPTION of starlight; the combined effect is known as EXTINCTION. The amount of extinction is inversely proportional to the wavelength.

Scheat The star β Pegasi, distance 199 l.y., one of the four stars that make up the SQUARE OF PEGASUS. It is a red giant, spectral type M2 II or III, and varies irregularly between mags. 2.3 and 2.7. Its name comes from the Arabic *saq*, meaning 'leg'.

Scheiner, Christoph(er) (1575–1650) German Jesuit scholar and astronomer who, independently of GALILEO, discovered sunspots (1611). He initially tried to explain these spots, not as features on the Sun's surface, which would imply that the Sun was 'imperfect', but as the shadows of small intra-Mercurian planets. Galileo, who claimed that his own sunspot observations predated those of Scheiner, countered by showing that sunspots are features of the solar disk and do not move independently of it. Scheiner was the first to make systematic observations of sunspots (1611–25). Now acknowledging that sunspots move across the solar disk, he demonstrated from plots of their paths that the Sun's axis of rotation is inclined by 7° 30′ to the ecliptic. He invented the first specialized solar telescope, which he called a *heliotropii telioscopici*, or 'helioscope'. This telescope was significant for

S

two reasons: it was the first to have an equatorial mounting, and the first to project the image of the solar disk on to a plane surface beyond the eyepiece.

Schiaparelli, Giovanni Virginio (1835–1910) Italian astronomer whose mapping of Mars gave rise to the idea of Martian canals, and who discovered the relations between comets and meteor showers. Schiaparelli joined Brera Observatory in 1860, becoming director two years later and remained there for the rest of his working life.

His studies of cometary tails in the 1860s, and his researches into previously observed comets, persuaded him that the particles that cause meteors are shed from cometary tails and follow elliptical or parabolic orbits around the Sun, identical with the orbits of comets. Schiaparelli originated the convention of naming showers after the constellation containing their radiant.

In 1877 Brera acquired new equipment in time for the opposition of Mars that year. Schiaparelli prepared a map of the surface of Mars, introducing a nomenclature for the various features that remained standard until the planet was mapped by space probes. His adoption of Angelo SECCHI's term *canali* (*see* CANALS, MARTIAN), to describe vague linear markings was the cause of much subsequent controversy; his reporting two years later of the apparent doubling of some canals (he called it 'gemination') further fuelled the debate over their origin.

Schiaparelli prepared maps of Venus and Mercury, believing – mistakenly – that they had SYNCHRONOUS ROTATION. He also made important studies of double stars, and discovered the asteroid (69) Hesperia.

Schickard Very much degraded lunar crater (44°S 54°W), 202 km (125 mi) in diameter, with rim components reaching 2800 m (9500 ft) above the floor. Because of foreshortening, this crater appears oblong, though its actual shape is more nearly circular. Long after its formation, Schickard was flooded by lunar basalts, giving the floor a dark patchy appearance.

Schlesinger, Frank (1871–1943) American astronomer who directed Allegheny Observatory, Pittsburgh (1905–20) and Yale University Observatory (1920–41). He used astrophotography to determine stellar parallaxes, supplanting the old visual methods. Schlesinger compiled several extensive 'zone catalogues' of precise positions and other data for more than 150,000 stars, and the first edition of the *BRIGHT STAR CATALOGUE*.

Schmidt, Bernhard Voldemar (1879–1935) Estonian optician of Swedish–German parentage who invented the SCHMIDT CAMERA. He made lenses and mirrors for several European observatories, including 300-mm (12-in.) and 600-mm (24-in.) mirrors for the University of Prague, and he refigured a 500-mm (20-in.) Steinheil objective later used by Ejnar Hertzsprung to measure very close double stars. In 1930 he made a telescope with a 440-mm ($17\frac{1}{2}$-in.) primary mirror and a 360-mm (14-in.) corrector plate which he figured while it was being deformed under vacuum – when the vacuum was released, the plate assumed the proper shape to eliminate coma and other aberrations of the primary mirror.

Schmidt, Maarten (1929–) Dutch-American astronomer, the first to realize the great distance of quasars when he identified highly redshifted lines in their unusual spectra. Schmidt left Leiden Observatory in 1959 for the CALIFORNIA INSTITUTE OF TECHNOLOGY, and became director of the Hale Observatories (Mount Wilson and Palomar) in 1978.

Following the discovery of the faint optical counterpart of the radio source 3C 48 in 1960 by Allan SANDAGE and Thomas Matthews (1927–), it was found to have a curious spectrum. These and similar objects, including 3C 273, were named 'quasi-stellar objects', or QUASARS. Schmidt showed that broad emission lines in the spectrum of 3C 273 were immensely redshifted, and that if this were a Doppler effect, the quasar had to be a billion light-years away. Subsequently, all quasars were found to have highly redshifted spectra.

By the late 1960s, enough quasars had been found for Schmidt to use them as a cosmological test for the then rival BIG BANG and STEADY-STATE THEORIES. According to the latter, the large-scale structure of the Universe should be the same at all places and all times. However, the number of quasars was found to increase with distance, supporting the Big Bang theory.

Schmidt camera Type of reflecting TELESCOPE incorporating a spherical primary mirror and a specially shaped glass correcting plate to achieve a wide field of view combined with fast focal ratio.

The conventional paraboloidal primary mirrors of reflecting telescopes only form a perfect point image as long as the incoming parallel light is exactly aligned with the axis. Departure from this causes steadily increasing degradation of the images due to coma, which is noticeable at 'off-axis' angles as small as a few arcminutes.

The Schmidt camera, which was invented by Bernhard Schmidt in 1930, overcomes this problem by the use of a spherical rather than a paraboloidal, primary mirror. This eliminates the problem of coma but introduces SPHERICAL ABBERATION. To correct this, a thin glass plate, known as a 'Schmidt corrector' and with a very shallow profile worked into one or both of its surfaces, is placed at the centre of curvature. The profile is complex in shape and has to be precisely computed to match and counteract the spherical aberration of the mirror, thus producing perfect images over a wide field of view. Because the image is formed on a curved surface, the photographic film or plate must be deformed by clamping it in a suitably shaped holder.

Schmidt cameras are used photographically to record detailed images of very large areas of sky. This makes them ideal for observing extensive objects such as comets or certain nebulæ, or for recording large numbers of more compact sources, such as stars or distant galaxies, for statistical studies. Perhaps their most important role is as survey telescopes used for identifying unusual or specific kinds of object, which can then be investigated in more detail with larger, conventional (narrow-field) instruments. Hybrid designs incorporating secondary mirrors (Schmidt–Cassegrains) or additional correctors (BAKER–SCHMIDTS) have also been developed. *See also* RITCHEY–CHRÉTIEN TELESCOPE

Schmidt–Cassegrain telescope (SCT) Short-focus telescope combining features of the SCHMIDT CAMERA and the CASSEGRAIN TELESCOPE. The SCT has the Schmidt's spherical primary mirror and specially figured corrector

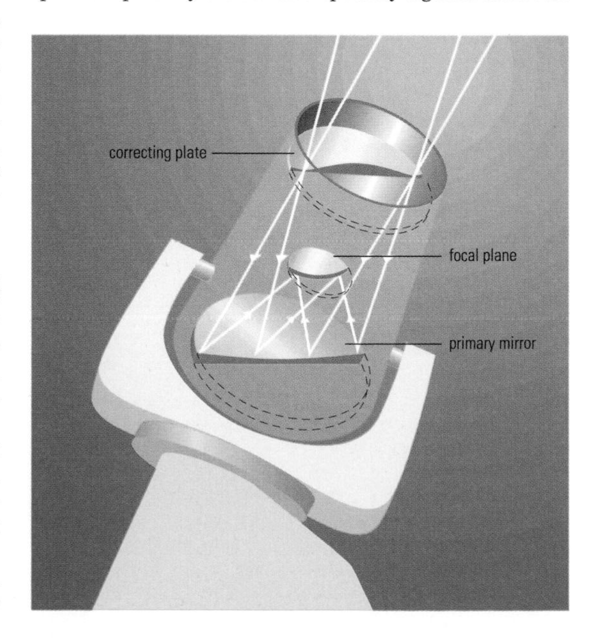

◀ **Schmidt camera** This type of reflecting telescope gives a wide field of view. They are used only for astrophotography and are useful in survey work.

plate, but light is reflected back down the tube by a convex secondary mirror mounted behind the plate and through a hole in the primary to a Cassegrain focus. This makes for a highly compact and portable telescope that has become very popular with amateur astronomers.

Schönberg–Chandrasekhar limit Upper limit on the mass of the helium core of a MAIN-SEQUENCE star. A star evolves off the main sequence when it has finished burning hydrogen in its core. This occurs when the star has used about 10% of its total mass in hydrogen burning. The limit can be larger if electron degeneracy becomes important within the core; this can occur in lower-mass stars where the core has a higher density. The limit is named after Mario Schönberg (1916–) and Subrahmanyan CHANDRASEKHAR.

Schramm, David Norman (1945–97) American theoretical astrophysicist who pursued links between particle physics, astrophysics and cosmology. In 1977 he was part of a team that constructed a model using the cosmic abundances of light elements to show that there could be no more than four families of elementary particles, a prediction confirmed in the late 1980s by particle accelerator experiments. Calculations he made suggested that the matter observable by telescopes accounts for only a small fraction of the Universe's total substance – the rest must be some form of dark matter. Schramm developed cosmological models that incorporated quarks to explain the Universe's LARGE-SCALE STRUCTURE.

Schröter, Johann Hieronymus (1745–1816) German amateur astronomer and chief magistrate of Lilienthal, near Bremen, where he erected his observatory. Schröter's 19-inch (480-mm) Newtonian reflector was for a time the largest in continental Europe. He made detailed observations of the topographic features of the Moon and planets. His two-volume *Selenographische Fragmente* (1791, 1802) contained descriptions, drawings and height measurements for hundreds of lunar craters and mountains. His 1785 October observations of small, dark spots on Jupiter that soon vanished may have been the earliest record of a comet impacting that planet, predating the SHOEMAKER–LEVY 9 impacts by over two centuries. In 1787 he discovered the solar GRANULATION and also the 'light bridge' features of sunspot umbrae. *See also* CELESTIAL POLICE

SCORPIUS (GEN. SCORPII, ABBR. SCO)

Striking southern zodiacal constellation between Ophiuchus and Ara, and one of the few that even remotely resembles the object after which it was named – the mythological scorpion that killed the hunter Orion. Its brightest star, ANTARES (which marks the heart of the scorpion), is a red giant irregular variable (range 0.9–1.2) and close binary; its bluish-white companion, mag. 5.4, separation 2″.9, appears pale green in comparison to its primary. Next brightest at mag. 1.6 is λ Sco (SHAULA). β^1 Sco (known as Acrab or Graffias) is a double star with bluish-white components, mags. 2.6 and 4.9, separation 13″.9, while ν Sco consists of two bluish-white stars, mags. 4.1 and 6.8, separation 41″, each of which has a fainter, close companion. U Sco is a recurrent nova which is usually around 18th magnitude but which flared up to 9th magnitude in 1863, 1906, 1936 and 1979. Bright star clusters and nebulae in Scorpius include the naked-eye open clusters NGC 6231, M7 (NGC 6475) and the BUTTERFLY CLUSTER (M6, NGC 6405); the globular clusters M4 (NGC 6121), 6th magnitude, and M80 (NGC 6093), 7th magnitude; and the Bug Nebula (NGC 6302), a 10th-magnitude planetary nebula. Also in Scorpius is SCORPIUS X-, the brightest X-ray source in the sky.

BRIGHTEST STARS

	Name	RA h m	dec. ° ′	Visual mag.	Absolute mag.	Spectral type	Distance (l.y.)
α	Antares	16 29	−26 26	0.9–1.2	−5.4–−5.1	K0	604
λ	Shaula	17 34	−37 06	1.62	−5.0	B1.5	703
θ	Girtab	17 37	−43 00	1.86	−2.7	F1	272
δ	Dschubba	16 00	−22 37	2.29	−3.2	B0.2	402
ϵ		16 50	−34 18	2.29	0.8	K2	65
κ		17 42	−39 02	2.39	−3.4	B1.5	464
β^1	Acrab	16 05	−19 48	2.56	−3.5	B0.5	530
υ		17 31	−37 18	2.70	−3.3	B2	519
τ		16 36	−28 13	2.82	−2.8	B0	430
π		15 59	−26 07	2.89	−2.9	B1	459
σ		16 21	−25 36	2.90	−3.9	B1	735
ι^1		17 48	−40 08	2.99	−5.7	F3	1792
μ^2		16 52	−38 03	2.9–3.2	−4.1–−3.8	B1.5	822
G		17 50	−37 03	3.19	0.2	K0/K1	127
η		17 12	−43 14	3.32	−1.6	F3	72

Schröter effect *See* PHASE ANOMALY

Schröteri Vallis (Schröter's Valley) Enormous lunar valley in the ARISTARCHUS uplift. It appears to be a vast lava flow structure, which emptied out into Mare IMBRIUM. Embedded within the rille is another much smaller SINUOUS RILLE. The valley begins in an elongated depression called the Cobra Head, which was probably the source vent for the basalts.

Schwabe, (Samuel) Heinrich (1789–1875) German pharmacist and amateur astronomer who discovered (1843) the approximately 11-year solar cycle. In 1825 he began to search for an intra-Mercurian planet. He realized that he needed to keep a careful record of sunspots, as any planet crossing the solar disk would closely resemble one of these features. Acquiring a Fraunhofer refractor, he continued recording sunspots, and 17 years later he noticed a periodicity in their number. Schwabe's 11-year cycle was later confirmed by the discovery (1852) that the Sun's magnetic field also varies over the same period. Schwabe made one of the earliest drawings of Jupiter's GREAT RED SPOT (1831).

Schwarzschild, Karl (1873–1916) German astronomer remembered chiefly for his theoretical work, especially for finding the first solution to the equations of Einstein's general theory of relativity. In the five years leading up to World War I, Schwarzschild directed both Göttingen and Potsdam Observatories; in 1914 he volunteered for military service, and died two years later of a skin disease contracted on the Eastern Front.

Schwarzschild designed and improved a number of instruments and techniques for use in spectroscopy and astrophotography. He pioneered photometric methods for measuring stellar magnitudes, demonstrating that there is a difference (the COLOUR INDEX) between a star's magnitude measured from a photographic plate and the visually estimated value.

He was one of the first, in the mid-1900s, to examine RADIATIVE TRANSFER in stars, showing that radiation and gravitation are in equilibrium. His most famous work was done in the final year of his life. As early as 1900, he had suggested that the geometry of space could be non-Euclidean. In 1916, shortly after the theory of general relativity was published, Schwarzschild worked out the first exact solution of Einstein's field equations, from which he developed the concepts of what are now called the SCHWARZSCHILD RADIUS and the BLACK HOLE. His son, Martin SCHWARZSCHILD, became an astrophysicist.

Schwarzschild, Martin (1912–97) German-American astrophysicist, the son of Karl SCHWARZSCHILD, who made major contributions to stellar structure and evolution. After taking his doctoral degree at Göttingen University (1935), he emigrated to the USA, joining Princeton in 1951 and remaining there for the rest of his life. He made detailed studies of the structure of the Sun and other nearby stars, comparing their masses, temperatures and other properties. Schwarzschild's research placed upper and lower limits on the masses of the stars, demonstrating that objects smaller than 10 Jupiter masses lack sufficient mass to initiate hydrogen fusion, while stars of more than 65 solar masses are rare. He calculated Z_{He}, the total mass density of elements heavier than helium, showing that this quantity varies with the age of a particular star. His work on stellar evolution mainly concerned 'pulsation theory', which explains the light-curves of variable stars in terms of expansions and contractions of these objects.

Schwarzschild radius Radius that a body must exceed if light from its surface is to reach an outside observer. If an object collapses below this radius, its escape velocity rises to above the speed of light and the object becomes a BLACK HOLE. For a black hole, the Schwarzschild radius is equal to the radius of the EVENT HORIZON.

The Schwarzschild radius is proportional to the mass of a body. For a non-rotating body of mass M and no charge, the Schwarzschild radius, R_s, equals $2GM/c^2$, where G is the gravitational constant and c is the speed of light. For a body the size of the Sun, the Schwarzschild radius is around 3 km (2 mi); for the Earth it is about 9 mm (0.4 in.).

Schwassmann–Wachmann 1, Comet 29P/ One of three periodic comets discovered by Arnold Schwassmann (1870–1964) and Arno Wachmann (1902–90) at Hamburg Observatory. Comet Schwassmann–Wachmann 1 was found photographically on 1927 November 15. The comet has a more or less circular orbit, just beyond that of Jupiter, at a distance of 6 AU from the Sun. It can be observed for much of the year. Normally around magnitude +17 to +18, it is prone to unexplained, irregular outbursts, on occasion reaching magnitude +10.

scintar Source that shows the effects of interplanetary RADIO SCINTILLATION and is therefore point-like rather than extended. This property allows QUASARS and radio stars to be distinguished from RADIO GALAXIES and nebulae. Stars and planets can be distinguished from each other because Earth's atmospheric variations cause the point-like stars to twinkle and the planets (that have a visible disk) to remain steady; a similar twinkling occurs with radio sources but from a different cause. Radio waves are refracted by the plasma clouds that move outwards from the Sun through the Solar System. The point-like radio sources are affected by the clouds, and so scintillate. This means the sources are either intrinsically very small, or very distant (and so appear small). The scintillation method was used to discover quasars, with the 4.5-acre radio telescope in Cambridge. To see the twinkling it was necessary to make the radio receiver respond very quickly to changes in radio intensity. This made it possible for Jocelyn BELL to discover PULSARS during the course of the scintillation measurements.

scintillation Twinkling of stars and, to a lesser extent, of planets as a result of the uneven refraction of light in areas of different density in the Earth's atmosphere. To the naked eye, scintillation appears as a change in brightness and colour, while in the telescope it may also make the star appear to make rapid slight movements. It is greatest at low altitudes, when the object's light shines through a greater amount of the atmosphere. Most of the effect is caused below 9000 m (30,000 ft).

The planets twinkle less than stars because they present small disks rather than points of light and the whole disk is unlikely to be affected simultaneously. In a telescope, however, the disk will appear blurred; this effect is usually referred to as bad SEEING. Scintillation depends on the weather, and is usually greatest at about noon. Modern observatories are sited where the effect occurs least. *See also* RADIO SCINTILLATION; SCINTAR

Sciama, Dennis William (1926–99) English cosmologist who applied general relativity to problems as far-ranging as black hole thermodynamics and Mach's principle, which holds that the appearance and structure of the 'local' or nearby Universe is affected by matter located at cosmological distances. Originally a proponent of the STEADY-STATE THEORY of cosmology, Sciama re-focused his research towards Einstein–de Sitter models of a BIG BANG universe when observational evidence showed that the former theory was seriously flawed.

SCULPTOR (GEN. SCULPTORIS, ABBR. SCL)

Inconspicuous southern constellation between Cetus and Phoenix. It was named Apparatus Sculptoris (the Sculptor's Workshop) by Lacaille in the 18th century. Its brightest star, α Scl, is mag. 4.3; ε Scl is a binary with pale yellow and yellow components, mags. 5.4 and 8.8, separation 4″.9, period about 1200 years. Deep-sky objects include NGC 288, an 8th-magnitude globular cluster; NGC 253 and NGC 55, 8th-magnitude edge-on spiral galaxies; the Sculptor Dwarf Galaxy, a dwarf spheroidal galaxy in the LOCAL GROUP, about 280,000 l.y. away; and the CARTWHEEL GALAXY.

Scorpius *See* feature article

Scorpius Centaurus association Group of about 45 O and B-type stars that are all about 500 l.y. away from us. The stars are found across the two constellations; they form the nearest OB ASSOCIATION to the Solar System.

Scorpius X-1 First cosmic X-ray source to be discovered (in 1962), and the brightest known apart from certain X-RAY TRANSIENTS. It is a low-mass X-RAY BINARY with an orbital period of 19.2 hours. One member is a NEUTRON STAR and the other member is unknown, except that it has an ULTRAVIOLET EXCESS and optical flaring, and it supplies material to the ACCRETION DISK around the neutron star.

Scotch mount Simple camera mount, based in two boards joined by a hinge, that can be driven by hand or a small motor to follow the stars' sidereal motion, allowing moderate-length wide-angle exposures without trailing. The hinge is aligned to the celestial pole. By turning a screw of known pitch at a defined rate, the upper board – to which the camera is attached – can be turned on the hinge away from the lower to mimic the westwards tracking of an EQUATORIAL MOUNTING. For exposures up to about 20 minutes, this gives satisfactory non-trailed star images. The mount takes its name from the Scottish roots of George Youngson Haig (1928–) who designed and popularized it in the early 1970s; it is also sometimes known as the **Haig mount**, while American amateurs know it as the **barn-door mount**.

SCT Abbreviation of SCHMIDT–CASSEGRAIN TELESCOPE

Sculptor *See* feature article

Scutum *See* feature article

S Doradus star (SDOR, Hubble–Sandage variable, luminous blue variable) Hot, extremely luminous ERUPTIVE VARIABLE. S Doradus stars are SUPERGIANTS of spectral class Op–Fp, with masses that exceed 30 solar masses and luminosities generally at least one million times that of the Sun. They are among the brightest stars in their parent galaxies. Their variations have magnitudes of between 1 and 7, although greater ranges have been noted, and the variations occur over periods of tens of years. Although often irregular, some stars show cyclic activity. Many are associated with expanding envelopes.

SCUTUM (GEN. SCUTI, ABBR. SCT)

Small, inconspicuous southern constellation to the south-west of Aquila. It was named Scutum Sobiescianum (Sobieski's Shield) by Johannes Hevelius in 1684, to honour King Jan Sobieski III of Poland. Its brightest star, α Sct, is mag. 3.9. R Sct is an RV Tauri star, varying between 4.2 and 8.6 with a period of about 147 days; δ Sct, a pulsating variable (range 4.6–4.8, period 0.19 day), is the prototype DELTA SCUTI STAR. The brightest deep-sky object is the WILD DUCK CLUSTER (M11, NGC 6705).

The mass-loss rate changes with time and the corresponding changes in the envelope contribute to the observed visual brightness changes. The ejection of matter is probably connected with the rapid evolution of a star whose luminosity is near the upper limit for stable stars. ETA CARINAE is generally included in this class, as is P CYGNI. *See also* GAMMA CASSIOPEIAE STAR

SDSS Abbreviation of SLOAN DIGITAL SKY SURVEY

Sea Launch US-led international company that markets launches to equatorial geostationary transfer orbit (GTO) from an offshore platform positioned on the equator in the mid-Pacific Ocean. It uses a modified Ukrainian ZENIT 2 booster with a Russian upper stage. A launch from the equator saves energy, which can be converted into added payload weight. Launches to GTO from higher or lower latitudes require more propellant for 'dog-leg' manoeuvres to reach an equatorial orbit, so payload capability is lost. The Sea Launch Zenit 3SL flies from the Odyssey platform, a converted semi-submersible oil rig positioned close to a launch control ship, *Sea Commander*. The vessels are based at San Diego, California. Sea Launch offers flights of payloads weighing 6 tonnes to GTO. Sea Launch has flown five times, four times successfully.

Seares, Frederick Hanley (1873–1964) American astronomer who standardized the stellar magnitude system and successfully mapped the Milky Way's structure. After directing the Laws Observatory of the University of Missouri (1901–09), he joined the staff of Mount Wilson Observatory, where he spent the remainder of his career (1909–45). In 1922, the IAU adopted his NORTH POLAR SEQUENCE of fundamental stellar magnitudes as the basis of its photographic and photovisual systems of stellar photometry. Collaborating with Frank E. ROSS, Seares compiled an extremely accurate catalogue of 2271 northern circumpolar stars brighter than 9th magnitude. In his many research projects, he often collaborated with his wife, **Mary (Cross) Joyner Seares**.

season Part of a cyclical variation in the climatic conditions on the surface of a planetary body over the course of a single revolution around the Sun, brought about by the inclination of the body's axis of rotation to the plane of its orbit.

The Earth's axis of rotation is inclined at 23°27′ so that, during the course of a year, the northern and southern hemispheres alternately receive longer or shorter periods of daylight. When the northern hemisphere is tilted towards the Sun it experiences summer, while the southern hemisphere experiences winter. Six months later, when the Earth is on the other side of its orbit around the Sun, the situation is reversed; the southern hemisphere enjoying summer whilst it is winter in the north. Between the two extremes lie spring and autumn.

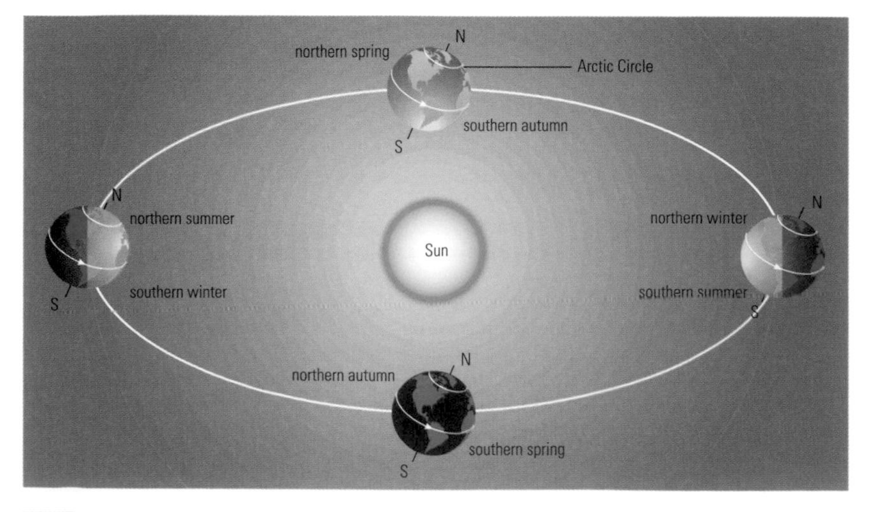

▼ **season** Because the Earth's axis is inclined relative to its plane of orbit, the Earth experiences seasons. The side that is tilted towards the Sun gets more light and has longer days and so experiences summer.

The beginning and end of each season is defined by the position of the Sun with respect to the ecliptic. At the time of the VERNAL EQUINOX, on March 21, it is said to be the first day of spring in the northern hemisphere. Spring lasts until the SUMMER SOLSTICE on June 21, which, although marking the longest day, is deemed to be the first day of summer. The end of summer and the start of autumn is marked by the AUTUMNAL EQUINOX on September 23, with the WINTER SOLSTICE on 22 December, the shortest day, defining the end of autumn and the beginning of winter. These seasons are reversed for the southern hemisphere.

Mars, the rotational axis of which is inclined at 25°11′, also displays seasonal variations, particularly to its POLAR CAPS. On Neptune's moon Triton, the polar cap migrates from pole to pole during the course of the planet's 165-year orbital period.

Secchi, (Pietro) Angelo (1818–78) Italian Jesuit astronomer, director of the Collegio Romano Observatory (1849–78) and a pioneer of solar physics and stellar spectroscopy. Although forced to take refuge at Georgetown (Washington, D.C.) Observatory in 1847, when the Jesuits were driven into exile, Secchi spent most of his career at the Collegio Romano. His observations of 10th-magnitude stars shining through the tail of Biela's Comet (1852) led him to conclude that cometary tails were gaseous, not solid. Secchi was one of the first (1859) to compile a complete photographic lunar atlas. He organized several expeditions to view solar eclipses and another, to Maddapur, India, to observe the 1874 transit of Venus.

Secchi proved that PROMINENCES, which he classified as quiescent or eruptive, were a physical part of the Sun, and he was the first to describe SPICULES. Simultaneously with William HUGGINS, Secchi was the first astronomer to make systematic observations of stellar spectra. From his analysis of the characteristics of the spectra of over 4000 stars, Secchi devised the first spectral classification scheme, which formed the basis of the system developed by Edward C. PICKERING and his staff at Harvard.

secondary (mirror) Small mirror in a REFLECTING TELESCOPE that diverts the converging beam of light from the PRIMARY towards the EYEPIECE.

secondary body Any of the smaller members of a system of celestial bodies that orbit around the largest, the primary body. The planets are secondary bodies in their orbits around the Sun, but each planet is the primary body when considering its own satellite system, in which the satellites are the secondary bodies.

secondary crater CRATER formed by the impact of pieces of EJECTA from a primary impact crater. Secondary craters are typically shallower than primary ones. They are abundant on bodies that lack an atmosphere. In rare cases, however, secondary craters have been observed even on Venus, the very dense atmosphere of which usually brakes the ejecta flight. Secondary craters are typically between one and two orders of magnitude smaller than the primary. They form radial and concentric chains and clusters around the primary crater.

second contact Moment during a SOLAR ECLIPSE when the leading (easterly) limb of the Moon appears to touch the easterly limb of the Sun. During a TOTAL ECLIPSE, this marks the arrival of TOTALITY. Second contact in a LUNAR ECLIPSE defines the moment when the Moon's trailing (westerly) limb enters Earth's UMBRA; in a total lunar eclipse, this is the time when totality begins.

second of arc (symbol ″) Unit of angular measure also known as an arcsecond; it is equivalent to 1/60 of a MINUTE OF ARC and 1/360 of a degree. The unit is widely used in astronomy, particularly as a measure of angular separation or diameter of celestial bodies. The RESOLVING POWER of a telescope is also usually expressed in arcseconds.

secular acceleration Continuous, non-periodic rate of change of the orbital motion of a body. For example, in the Earth–Moon system, frictional effects in the TIDES raised on the Earth by the Moon cause a slight lag in the Earth's response to the tidal force, with the result that the TIDAL BULGE does not lie precisely along the Earth–Moon line. The resulting slight asymmetry in the gravitational attraction between the Earth and Moon causes a transfer of angular momentum and energy from the Earth's rotation into the Moon's orbit. This slows the Earth's rotation, increasing the length of the day by 0.000014 seconds a year, and causes the Moon to recede at about 3.8 cm (1.5 in.) a year. A similar effect occurs for the Martian satellite Phobos, but because it is below SYNCHRONOUS ORBIT height, it is spiralling into the planet instead of receding. The radius of Phobos' orbit is decreasing by about 1.9 cm/yr (0.7 in./yr) and it is expected that it will enter the ROCHE LIMIT and break up in about 38 million years time. *See also* TIDAL EVOLUTION

secular parallax Angular displacement of a celestial body over time caused by the Sun's motion through space relative to the LOCAL STANDARD OF REST. Secular parallax provides us with a method of measuring the distance to nearby groups of stars. Relative to nearby stars, the Sun is travelling at around 20 km/s (12 mi/s) in the direction of the constellation Hercules. The apparent displacement of a star arising from this solar motion is called its parallactic motion and by measuring the parallactic motion of a homogeneous group of stars, it is possible to obtain the secular parallax of the group. This method has been applied to estimate the average distances of groups of variable stars, such as RR LYRAE and CEPHEIDS, which are beyond the range of TRIGONOMETRIC PARALLAX.

secular variable Star that is suspected to have increased or decreased markedly and presumably permanently since ancient times. Thus PTOLEMY and other early observers ranked MEGREZ (δ Ursae Majoris) as the equal of the other stars in the Plough pattern, whereas it is now obviously fainter; DENEBOLA, in Leo, was ranked of the first magnitude, but is now below the second, and so on. On the other hand RASALHAGUE (α Ophiuchi) was given as magnitude 3 and is now 2. On the whole it seems very unlikely that any of these changes are real: more probably they are errors of observation or – even more plausibly – translation or interpretation.

segmented mirror Very large telescope mirror made up of a number of smaller mirrors that fit together to form one continuous optical surface. A segmented mirror offers the advantages of being much thinner and lighter in weight than a solid mirror of the same size. The 10-metre primary mirrors of the twin Keck telescopes on Hawaii are each made up of 36 hexagonal pieces, each about two metres in diameter. During observing, a computer-controlled system of sensors and actuators adjusts the position of each segment relative to its neighbours to an accuracy of four nanometres. *See also* ACTIVE OPTICS; SPIN CASTING

seeing Effect of the Earth's atmosphere on the quality of images produced by optical telescopes. The term is also used to express the quality of observing conditions and is usually expressed in arcseconds (″). It corresponds to the angular diameter of a star image produced at the focal plane of a telescope; the smaller that image is the better. Seeing refers to the spatial effect on the image and should not be confused with transparency, which is a measure of how much (or how little) the atmosphere attenuates the light. Ironically, clear nights when the stars are bright can suffer from poor seeing, and the seeing can be very good on nights when the stars are dimmed by slightly hazy conditions.

Meaningful comparisons of seeing measurements can only be made when consistent methods of defining and measuring the diameter of the star image are used. Seeing also varies with time, so the timescale over which measurements are made is also significant.

Where important decisions rely on consistent seeing measurements, such as choosing a site for an observatory, differential image motion monitoring is often used. Two or more images of the same star are formed by parallel paths through a single telescope, and the relative motion of these images is converted into an assessment of the seeing. Prior to quantitative measurements of seeing the ANTONIADI SCALE was used.

The effect of seeing is caused by light passing through air of differing temperature and therefore differing refractive index. These differences are mostly due to air masses mixing in the upper TROPOSPHERE, but the local effect of heat sources in the telescope building and heat rising from the surrounding ground can be significant. Astronomers go to great lengths to choose sites with good inherent seeing, where the airflow is smooth with little turbulent mixing above the site. Increasingly, they pay great attention to the design of telescope buildings to minimize their effect on seeing. On world-class sites 'good seeing' would be better than half an arcsecond, whereas one-arcsecond seeing would be considered very good on a site in the United Kingdom.

Avoiding atmospheric seeing is one of the main reasons for placing optical telescopes such the HUBBLE SPACE TELESCOPE into orbit. The detail revealed in Hubble pictures clearly illustrates what can be achieved when seeing is removed and the quality of the image depends only on the size and quality of the optics.

◀ **segmented mirror** The mirror of the Gran Telescopio Canarias is made up of 36 hexagonal segments measuring 936 mm (36.85 in) on each side. The thin segments, one of which is seen here being polished, combine to form a large, flexible surface that would not be achievable with a single mirror.

seismology Study of the shockwaves produced by earthquakes and other disturbances, such as meteoritic impacts and explosions, that propagate through planetary bodies. The strength and characteristics of the waves are measured by instruments called seismographs. The principal types are P (pressure) and S (shear) waves, which travel through rock at different speeds, depending on its density and mechanical properties. Measurements of such waves provide information on the internal structure of a planet, such as its division into CRUST, MANTLE and CORE. Seismographs left by the APOLLO astronauts provided valuable information about the Moon's interior.

selected areas Set of 262 small, uniformly distributed regions of sky in which the magnitudes, spectral and luminosity classes of stars have been accurately measured to provide standard comparison data and statistics on the distribution of stars.

Selene Japan's Selenological and Engineering Explorer, which is scheduled to be launched in about 2003–2004 to orbit the Moon and demonstrate a lunar landing. Selene's primary objective is to investigate lunar origins and evolution and to develop technology for future missions. The spacecraft, comprising a mission module and a propulsion module with a small data relay satellite attached, will eventually be manoeuvred from its initial

SERPENS (GEN. SERPENTIS, ABBR. SER)

Rather inconspicuous constellation, representing a huge snake coiled around the body of Ophiuchus, the serpent-bearer. The constellation is unique in consisting of two separate parts: Serpens Caput (the Serpent's Head), between Böotes and Hercules, and Serpens Cauda (the Serpent's Tail), between Ophiuchus and Scutum. The brightest star in Serpens, Unukalhai, is mag. 2.6. Alya (θ Ser) is a wide double with white components, mags. 4.6 and 5.0, separation 22″; δ Ser is a close binary, also with white components, mags. 4.2 and 5.3, separation 4″.0. Deep-sky objects include M5 (NGC 5904), a 6th-magnitude globular cluster, and the EAGLE NEBULA (IC 4703), an emission nebula containing M16 (NGC 6611), an open cluster of more than 60 stars fainter than 8th magnitude.

BRIGHTEST STARS

	Name	RA h m	dec. °	Visual mag.	Absolute mag.	Spectral type	Distance (l.y.)
α	Unukalhai	15 44	+06 26	2.63	0.9	K2	73
η		18 21	−02 54	3.23	1.8	K0	62

elliptical orbit into a circular polar orbit of 100 km (60 mi). During the transition, the data relay satellite will be deployed. It will relay the Doppler ranging signal between the orbiter and ground station to measure the far-side gravitational field. The orbiter will map the Moon for a year, after which the propulsion module will separate and will be deorbited. It will make a landing and send radio signals from the landing site for differential VERY LARGE BASELINE INTERFEROMETRY observation.

semidetached binary BINARY STAR system in which only one star fills its ROCHE LOBE. Material from the lobe-filling star escapes through the inner LAGRANGIAN POINT and accretes on to the companion, usually via an ACCRETION DISK. *See also* CLOSE BINARY; DETACHED BINARY

semimajor axis (symbol a) Half of the longest axis of an ELLIPSE. For an object in an elliptical orbit, it is the mean distance from the primary body (that is half way between the nearest distance and the farthest distance). It is one of the ORBITAL ELEMENTS, and is related to the mean motion n (the average rate of angular motion around the primary) by the relation $n^2a^3 = G(M + m)$, where G is the GRAVITATIONAL CONSTANT, and M and m are the masses of the primary and the object.

semiregular variable (SR) Pulsating giant or supergiant VARIABLE STAR, generally of late spectral type (M, C, S or Me, Ce, Se), with period 20 to 2000 days or more, and amplitude from a few hundredths to several magnitudes. There are various subtypes. Some (designated SRA) differ little from MIRA STARS except in having amplitudes less than 2.5 mag., but the shapes of their light-curves vary. Other stars (subtype SRB) show a definite periodicity, interrupted at times by irregularities, or intervals of constant light. A small subtype (SRC) comprises supergiants of late spectral type with amplitudes of no more than 1 magnitude and periods from 30 to several thousand days. MU CEPHEI is an interesting example of this subtype. There are also stars classified as semiregulars but with earlier spectral types (F, G and K), sometimes with emission lines. These have been assigned to subtype SRD. *See also* RV TAURI STAR

separation Angular distance, measured in seconds of arc, between two celestial bodies, particularly the members of a visual binary or multiple star system. Separation is one measure of the relative positions of the components of a binary system, the other being POSITION ANGLE.

sensitivity Ability of a detector or instrument to measure a faint signal from an astronomical object, making the detector many times more sensitive than the naked eye. The term can also be used to describe the region of the electromagnetic spectrum where the instrument or telescope works best, for example the Jodrell Bank telescope is sensitive to radio waves, whereas the Hubble Space Telescope is not sensitive to radio waves.

Serenitatis, Mare (Sea of Serenity) Lunar lava plain, roughly 700 km (430 mi) in diameter, located in the north-east quadrant of the Moon. This area was struck by an enormous object, producing a multi-ring impact basin, the inner region of which was later flooded by lava. The inner ring is marked by a series of circular mare (wrinkle) ridges. Apollo 17 landed in the Taurus–Littrow Valley, in the outer ring of Serenitatis, to sample a dark-mantling deposit.

Serpens *See* feature article

Service Module *See* APOLLO PROGRAMME

Setebos One of the several small outer satellites of URANUS; it was discovered in 1999 by J.J. Kavelaars and others. Setebos is about 20 km (12 mi) in size. It takes 2273 days to circuit the planet, at an average distance of 17.88 million km (11.11 million mi). It has a RETROGRADE orbit (inclination near 158°) with a substantial eccentricity (0.551). *See also* CALIBAN

SETI Acronym for Search for Extraterrestrial Intelligence, the umbrella term for all endeavours, especially those using radio telescopes, to find signs of intelligent life elsewhere in the Universe. SETI assumes that life exists on some EXTRASOLAR PLANETS; a small percentage of that life is sufficiently advanced to have developed civilizations and technological capabilities; such civilizations are either unintentionally emitting or deliberately transmitting signals that are detectable across intervening space; and such signals if received will be identifiably of artificial origin. Almost 50 years of SETI research has produced no evidence for extraterrestrial intelligence (ETI).

A vast number of stars must first be checked for the presence of suitable planetary systems, very few of which will contain planets potentially suitable for complex life, and very few of those may be expected to host technologically advanced civilizations. Natural cosmic radio sources produce 'noise' spread across a wide frequency range. Radio SETI experiments look for narrowband signals, with a frequency spread of just a few hertz, characteristic of a purpose-built transmitter – these are the easiest signals to detect as their energy is concentrated in a small region of the radio spectrum.

The earliest serious attempt to detect ETI signals was **Project Ozma** (the name is from Frank Baum's *The Wizard of Oz*). In 1960 Frank DRAKE and others used a 26-m (85-ft) antenna at Green Bank to search at the 21-cm hydrogen line. Ozma was unsuccessful and was abandoned after a few months. Since then a number of other projects, mostly privately funded, have scanned the heavens in various regions of the radio spectrum.

Project Phoenix targets 1000 carefully selected stars within 200 l.y. of the Sun. A 28 million channel receiver provides high sensitivity to weak signals. Observations began at Parkes, NSW, Australia in 1995 February, and at Green Bank in 1996 September; observations are also made for two 3-week sessions each year at Arecibo. By mid-1999, Phoenix had examined about half its targeted stars.

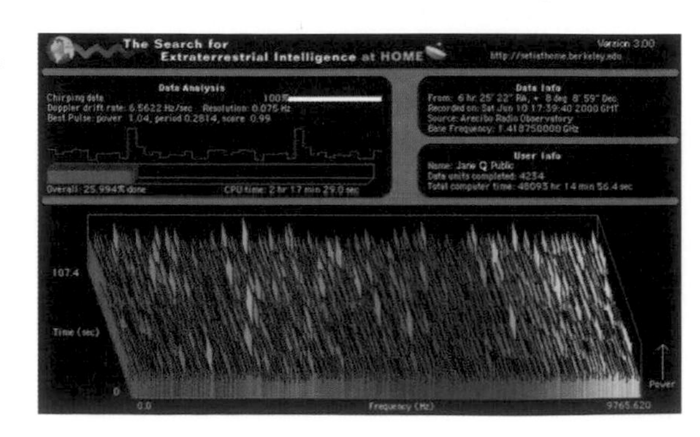

▶ **SETI** Shown here is the SETI@home project screensaver. More than 890,000 years of computer-processing has been performed by the 3.5 million participants in the project.

S

In the **Serendip** project, spectrum analysers developed and operated by the University of California (UC), Berkeley, have been used on various radio telescopes. Serendip I, a 100 channel per second analyser, operated at Hat Creek Observatory in 1979. The most recent instrument, Serendip IV, was installed at Arecibo in 1997 June and examines 168 million channels every 1.7 seconds.

A 1971 report from NASA's Ames Research Center proposed an array of between 1000 and 2500 100-m (330-ft) antennae to synthesize a single 10-km (6-mi) diameter dish to search for ETI signals from up to 1000 l.y. away. **Project Cyclops**, as this was called, would have cost about $20 billion at 2001 prices and the project was stillborn, but the idea of an integrated multi-dish array has resurfaced in the ALLEN TELESCOPE ARRAY (ATA). Its ability to scan many areas of the sky at once, with more channels than in previous searches and for 24 hours a day, will give the ATA a much greater capability than Project Phoenix.

Analysis of the vast amounts of data generated in SETI research has low priority on large astronomical computers. The UC Berkeley SETI team realized if a large number of small computers – PCs – could work simultaneously on different parts of the analysis, the job could be done in a relatively short time. By mid-2001, the **SETI@home** project had 3.5 million computers in homes, offices, schools and colleges linked to the Berkeley through the Internet and processing data during idle time.

Until recently, SETI was restricted to radio astronomy techniques. However, **optical SETI** (OSETI) aims to detect pulsed lasers, infrared messages or other artificial optical signals. Theoretical studies began in the early 1960s, since when developments in optical technology have made OSETI capable of detecting ET signals. Charles Townes (1915–) suggested in 1962 that extraterrestrial civilizations might use lasers for interstellar communication, for which the directionality of laser beams makes them ideal. He calculated that a 10-kW laser directed through a 5-m (200-in.) space telescope would appear brighter than the Sun, and would therefore be detectable by an ETI on a target planet, using a similar telescope, 100 l.y. away. Attempts to search for such signals using small telescopes and fast photon counters are already under way, and more sensitive searches are planned.

Seven Sisters *See* PLEIADES

Sextans *See* feature article

sextant Astronomical instrument containing a 60° arc (from Latin *sextans*, meaning one-sixth of a circle). In the 1570s Tycho BRAHE developed large wooden sextants of 1.8 m (6 ft) radius mounted on a ball joint, whereby two astronomers working together could measure horizontal or right ascension angles between pairs of stars, for celestial map-making. The term 'sextant', however, is more commonly used nowadays to describe a precision hand-held instrument used by navigators for measuring celestial angles. The modern navigator's sextant is based on the original 1731 design by John Hadley (1682–1743) in which the observation of a stellar image reflected from a small built-in mirror doubles the instrument's amplitude from 60° to 120°; it replaced the mariner's ASTROLABE. Eighteenth-century precision engineering, as embodied in the dividing engine made in 1775 by Jesse Ramsden (1735–1800), soon brought the sextant to technical maturity, enabling it to be used to find the longitude at sea from the position of the Moon.

Seyfert, Carl Keenan (1911–60) American astronomer best known for his investigations of what are now known as SEYFERT GALAXIES. He spent his career at McDonald Observatory (1936–40), Mount Wilson Observatory (1940–46) and Vanderbilt University (1946–60). He studied the spectra and distribution of stars in the Milky Way and developed new equipment and techniques for photoelectric photometry and wide-field astronomical photography. His most famous research, published in 1943, concerned galaxies with inconspicuous spiral arms and very bright nuclei, the class of active galaxy now named after him.

SEXTANS (GEN. SEXTANTIS, ABBR. SEX)

Small, insignificant equatorial constellation, representing a sextant, between Leo and Hydra. It was named Sextans Uraniae (Urania's Sextant) by Johannes Hevelius in 1687 to commemorate the astronomical instrument. Its brightest star, α Sex, is mag. 4.5. The brightest deep-sky object in Sextans is the Spindle Galaxy (NGC 3115), a 9th-magnitude lenticular galaxy.

Seyfert galaxy Class of ACTIVE GALACTIC NUCLEUS, defined as showing a starlike nucleus and EMISSION LINES that are of high ionization and large velocity width. The class was first recognized by Carl SEYFERT. When examined carefully, about 5% of bright galaxies have Seyfert nuclei, which are found mostly in spirals of early type (Hubble classes S0, Sa, Sb). There are two subtypes: type 1, in which the broad BALMER LINES of hydrogen are much broader than other emission lines, spanning thousands of km/s; and type 2 in which all emission-line widths are comparable (usually a thousand km/s or less). Type 1 Seyferts have spectra very much like QUASARS. As in the case of quasars and RADIO GALAXIES, there is a unified model in which type 2 Seyfert nuclei are type I objects seen through an obscuring torus. This model is supported in many cases by polarization measurements and by the detection of cones of highly ionized material to each side of the (sometimes invisible) nucleus itself. These cones are not generally well aligned with the galaxy's overall rotation axis. Some Seyferts show small-scale radio structure resembling distorted JETS, although their interstellar gas seems to prevent the jets from propagating very far.

Like quasars, Seyfert nuclei are variable, and the response of the surrounding gas to changes in the core brightness has been used to map the size and structure of the gas seen in emission lines. Much of the gas seen in the broad emission lines must be situated only a light-day or so from the core. This small size combined with the large linewidths, indicating that a very deep gravitational field is needed to keep the gas from escaping, are points giving rise to the popular picture of a massive BLACK HOLE powering Seyfert galaxies.

The best-known Seyfert galaxies are M77 (NGC 1068), the prototype for type 2, and NGC 4151, the standard type 1. Such galaxies as M81 also show weak Seyfert activity when examined closely.

shadow bands Phenomenon seen immediately before and after totality at a total SOLAR ECLIPSE, consisting of rapidly moving ripples of light and dark passing over the ground. Attempts to photograph them have failed, but these shadow bands have been recorded on video when seen crossing a white screen. It is believed that they result

◄ **Seyfert galaxy** M77 (NGC 1068) is the prototype type 2 Seyfert galaxy. It has equally broad hydrogen emission and forbidden lines in its spectrum, indicating that there is high-velocity gas in its inner regions. Visually, Seyfert galaxies are characterized by their bright, small nuclei.

from small-scale changes in the refractive index of the air along the eclipse track, with the brighter bands being images of the tiny remaining visible sliver of the solar disk exposed within a few moments of totality. Observers who witnessed a strong display of shadow bands at the 1998 February total solar eclipse in the Caribbean likened the effect to the play of sunlight on the bottom of a water-filled swimming pool.

Shakerley, Jeremy (1626–*c.*1655) English astronomer. By 1649 he had discovered the writings of the recently deceased Jeremiah HORROCKS, William GASCOIGNE and William CRABTREE, and was a fervent supporter of the Copernican system. Between 1649 and 1653 he published three books attacking astrology. He clearly considered Horrocks his scientific role model, and his *Anatomy of Urania Practica* (1649) was the first occasion that Horrocks' work was acknowledged in print. Shakerley was making telescopic observations of comets and eclipses after 1650, and of the 1651 transit of Mercury.

Shapley, Harlow (1885–1972) American astronomer who first defined the nature and extent of the Milky Way galaxy. His early years were spent at Mount Wilson Observatory (1914–21), where he became the first (1918) accurately to measure the distances to globular clusters, by calibrating Henrietta LEAVITT's period–luminosity law for Cepheid and RR Lyrae variable stars in the clusters. The globular clusters are concentrated near the central bulge of the Milky Way, so these studies enabled Shapley to identify a region in Sagittarius at galactic longitude 325° as the galactic centre. He calculated its distance as 50,000 l.y., which, though about 60% greater than the distance obtained if the effects of interstellar reddening are taken into account, still implied that the Galaxy was much larger than previously thought. Shapley defended his idea of a large 'metagalaxy' in the famous 1920 GREAT DEBATE with Heber Curtis.

From 1921 to 1951 Shapley directed HARVARD COLLEGE OBSERVATORY, modernizing the facilities and equipment and establishing the BOYDEN OBSERVATORY in South Africa. Shapley catalogued many thousands of galaxies beyond the Milky Way, publishing, with Adelaide Ames (1900–1932), the first comprehensive catalogue of these objects – *A Survey of the External Galaxies Brighter Than the Thirteenth Magnitude* (1932). He was one of the first to recognize that galaxies are not distributed uniformly in the Universe, but arranged in groups and clusters. In 1938 Shapley discovered the Fornax and Sculptor dwarf galaxies.

► **Shepard, Alan Bartlett, Jr** In 1961 Alan Shepard piloted Freedom 7 to the edge of space. Here he is seen in his space suit during preflight testing.

Shaula The star λ Scorpii, visual mag. 1.62, distance 703 l.y., spectral type B1.5 IV. It lies in the tail of the scorpion, and its name is the Arabic word for 'stinger'.

Shedir The star α Cassiopeiae, visual mag. 2.24, distance 229 l.y., spectral type K0 III. Small telescopes show a wide, unrelated companion of mag. 8.9. Its name, which is also spelled *Schedar* or *Schedar*, comes from the Arabic *sadr*, meaning 'breast'.

Sheliak Name of the star BETA LYRAE.

shell burning Nuclear burning of elements in thin spherical shells in the atmosphere of stars. Shell burning occurs when nuclear FUSION has been exhausted in the core, whereupon the core collapses producing thermal energy, which raises the temperature of the material surrounding the core sufficiently for nuclear fusion to occur there. The shell burning gradually moves outwards, causing the star to expand, until all the nuclear fuel has been exhausted in the envelope.

Hydrogen shell burning occurs in stars as they evolve off the MAIN SEQUENCE to become RED GIANTS. More massive stars will burn other elements in their cores and have other episodes of shell burning. *See also* STELLAR EVOLUTION

shell galaxy ELLIPTICAL GALAXY that shows faint, thin shells of stars on either side of the nucleus. Attention was first called to these galaxies in 1979, when the photographic image-enhancement techniques of David Malin (1941–) showed these subtle features. They appear to result when a DISK GALAXY merges with an elliptical, and the coherent stellar motions from the disk lead to its stars wrapping across the centre of the elliptical, producing narrow shells where they reverse direction. As many as 28 concentric shells have been seen in a single galaxy. Related shell-like features have also been found in S0 and a few Sa galaxies, and their occurrence has been used as evidence of advanced mergers or galactic CANNIBALISM.

shell star MAIN-SEQUENCE star that is surrounded by a shell of gas, generally of spectral type Be. Examples are γ Cassiopeiae, BU Tauri and 48 Librae.

The spectrum of a shell star typically shows a rapidly rotating underlying B star, with broad absorption lines superposed with emission lines and additional sharp absorption features. The star is rotating rapidly, losing mass to an expanding shell that is pulled into a disk around the equator. The sharp absorption features are caused by the disk. Be-shell stars are believed to be the same as 'classical' BE STARS viewed along the equator.

Other Be stars are in CLOSE BINARY systems, with the material surrounding them being accreted from their companion stars. Herbig Be objects are believed to be a more massive version of T TAURI stars.

Shen Gua (or Guo or Kua) (*c.*1031–*c.*1095) Chinese civil and military administrator, mathematician and astronomer. He devised a model of the Solar System using new mathematical techniques to explain the planets' retrograde movements, and developed a theory of the Moon's complex orbit; his ability to predict celestial events ingratiated him with the emperor's government. Shen designed and built astronomical instruments, including an improved GNOMON and a highly accurate ARMILLARY SPHERE. His other major achievement was to reform the calendar.

Shen Zhou Chinese 8-tonne spacecraft, which, in about 2003–2004, should carry the first Chinese astronauts into orbit. The craft made two unmanned LONG MARCH 2F-boosted orbital flights in 1999 and 2001, the descent capsule being safely returned to Earth. Shen Zhou is

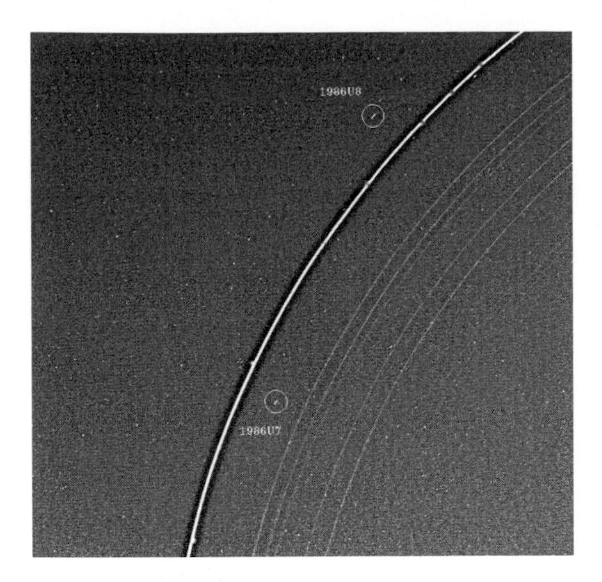

◀ **shepherd moon** Cordelia (1986 U7) and Ophelia (1986 U8) act as shepherd moons to Uranus' Epsilon ring. Their gravitational influences keep the particles within the ring in place.

based on the design of the Russian Soyuz U spacecraft but its forward docking module is cylindrical rather than spherical and can be left in orbit as an unmanned orbital facility after the crew capsule, which can carry two 'taikonauts', has returned to Earth. The rear service module of Shen Zhou is equipped with a manoeuvring engine and solar panels. After some manned test flights, China plans to dock two Shen Zhou spacecraft together in orbit, forming an interim space station.

Shepard, Alan Bartlett, Jr (1923–98) American astronaut who made the first suborbital Mercury space flight in 1961. He was selected in 1959 as one of the original Mercury astronauts, and on 1961 May 5 he piloted the *Freedom 7* spacecraft to an altitude of 188 km (117 mi) before returning to Earth. Although grounded for medical reasons (1964–69), Shepard later commanded the Apollo 14 mission (1971), becoming the fifth man to walk on the Moon.

shepherd moon Minor moon, the gravitational influence of which holds the particles of a planetary ring in place, preventing them from dispersing. Often found in pairs, the shepherd moons share the orbit of the particles in the ring system, an example being PROMETHEUS and PANDORA, which stabilize Saturn's F RING. The moon closer to Saturn moves slightly faster, having the effect of speeding up any particles which stray, while the outer moon moves more slowly and drags any stray particles back into the ring. Two shepherd moons, CORDELIA and OPHELIA, are found in the Uranian system, situated on either side of the planet's Epsilon ring.

Shergotty METEORITE that fell in Bihar, India, in 1865 August; a single stone of approximately 5 kg was recovered. Shergotty is the type specimen for the Shergottites, one of the main subdivisions of MARTIAN METEORITES (SNCs). Shergottites are subdivided into the basaltic and lherzolitic shergottites. The former are fine-grained pyroxene–plagioclase rocks, which formed as flows at or close to the Martian surface. The latter are more coarse-grained olivine–pyroxene plutonic rocks.

Shklovskii, Iosif Samuilovich (1916–85) Ukrainian astrophysicist who in 1953 explained that the continuum radio and X-ray emission from the Crab Nebula was SYNCHROTRON RADIATION and correctly predicted that OH would radiate at microwave frequencies. He discovered many X-ray binary systems, and improved the calculation of distances to the planetary nebulae associated with white dwarfs. Shklovskii also showed that the temperature of the solar corona is about 1 million K, and how magnetic fields outline its structure. He also was an early proponent of SETI studies, and his *Intelligent Life in the Universe* (1966), co-written with Carl SAGAN, is a classic text.

shock Important process in many branches of astronomy and space science. Though the simple shock produced by a supersonic aircraft is familiar, the processes involved in astrophysical shocks are less well understood. The general description of a shock is that there is some boundary, the shock surface, and that as matter crosses the boundary there is a rather sudden change in the state of the matter. This change is an irreversible change in thermodynamic terms. The usual form of shock is that there is conversion of kinetic or flow energy into random thermal energy. The temperature is higher on one side of the shock.

Examples of shocks are found in the SOLAR WIND, including the BOW SHOCK caused by planetary MAGNETOSPHERES. Shocks are also probably involved in solar FLARES. In the interstellar medium, for the clouds of gas and dust that form the region between the stars as well as the more spectacular nebulae, shocks form important boundaries. They are thought to be involved in various early stages of star formation from these clouds and are certainly involved in the late stages of STELLAR EVOLUTION such as SUPERNOVAE. On a larger scale it is thought that shock processes occur in RADIO GALAXIES.

Shoemaker, Eugene Merle (1928–97) American geologist and astronomer who extended geological principles to the Moon and other bodies, advanced the impact theory of cratering, and pioneered the study of NEAR-EARTH OBJECTS. In 1937 he joined the UNITED STATES GEOLOGICAL SURVEY (USGS), with which he remained associated until his death. In 1951 he married future collaborator, **Carolyn (Jean) Spellmann Shoemaker** (1929–). The Shoemakers' visit to METEOR CRATER the next year convinced Gene that terrestrial and lunar craters resulted from asteroidal impacts, which he followed up with a study of craters formed in the deserts of Nevada during US nuclear bomb tests.

In the 1960s, Shoemaker organized the geological experiments on NASA's manned and unmanned lunar missions, and in 1965 was appointed chief scientist of the USGS's new Center of Astrogeology. In 1969 his interest turned from impact craters to the objects that produced them. Initially with Eleanor Helin and later with Carolyn Shoemaker, he began a search for near-Earth objects, using PALOMAR OBSERVATORY's 0.46-m (18-in.) Schmidt telescope. The programme resulted in the discovery, from 1973, of many NEAR-EARTH ASTEROIDS and, from 1983, the first of many comets now bearing the Shoemaker name, the most famous of which was Comet SHOEMAKER–LEVY 9. It established Palomar as the leading discovery site for asteroids of all kinds.

Shoemaker–Levy 9, Comet Comet discovered on 1993 March 25 by Eugene and Carolyn Shoemaker and David Levy (1948–) on a photographic patrol plate exposed at Mount Palomar, San Diego, USA. The

◀ **Shoemaker–Levy 9, Comet** The scars left in Jupiter's atmosphere by the impacts of Comet Shoemaker–Levy 9 were probably gases brought up from deeper within the planet's atmosphere by the explosions. They appear dark red because they absorb light at different wavelengths from the high-level clouds.

S

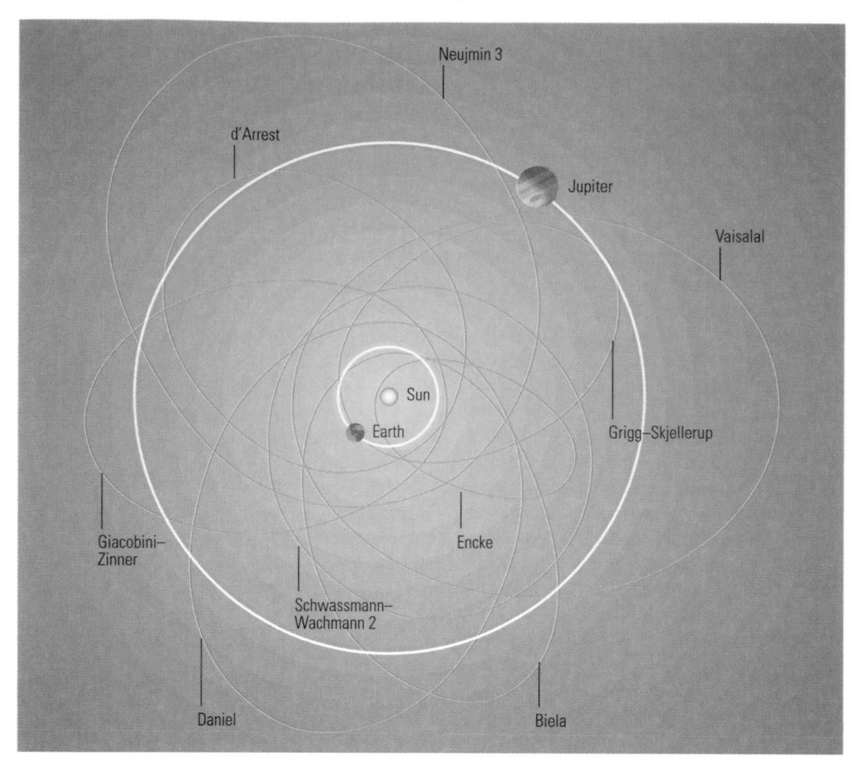

▲ **short-period comet** Most short-period comets are thought to have originated in the Edgeworth–Kuiper belt and to have been perturbed into the inner Solar System by the gravitational influence of the gas giants. Encke has an orbital period of only 3.3 years, Grigg–Skjellerup 5.1 years and Giacobini–Zinner 6.5 years. Biela's Comet suffered the probable fate of all short-period comets: on one orbit, its nucleus was seen to have broken into at least two pieces and it never reappeared.

discovery image appeared unusually elongated, and further investigation revealed that the comet had, in fact, been broken into a number of fragments. Analysis of the comet's motion showed it to be in a two-year orbit around Jupiter, into which it had probably been captured as long ago as 1929. At a close (21,000 km/13,000 mi) perijove in 1992 July, tidal stress had disrupted the comet's fragile nucleus into at least 21 sub-kilometre-sized fragments.

Calculations showed that these fragments would impact on to Jupiter over the course of the week of 1994 July 16–22. The fragments were named alphabetically in order of anticipated impact, and as they continued along their terminal orbit they became spread out into a 'string of pearls' imaged from ground-based observatories and the Hubble Space Telescope. Ahead of 1994 July, there was much speculation as to what effect, if any, the impacts would have on the giant planet.

Intensive observing programmes were established worldwide. The impacts themselves occurred just beyond the jovian limb, out of sight from Earth (but visible from the Galileo spacecraft at that time en route to Jupiter). The planet's rapid rotation would carry the impact sites into view from Earth within about an hour of each event.

Observers were surprised by the scale and violence of the impacts. Cameras on Galileo revealed the entry fireballs as bright flashes, and infrared observations from the terrestrial viewpoint showed huge plumes of material thrown high above the jovian cloud-decks following the impacts. The energies involved were most graphically shown by the easily visible Earth-sized dark spots that marked each impact site on Jupiter. Over the course of the week of the comet's demise a series of dark impact scars peppered Jupiter's clouds at 44°S latitude, with impacts coming, on average, about seven hours apart. The impact scars remained visible for some weeks, eventually merging into a new dark belt on Jupiter, persisting for the next 18 months.

Searches through historical observations of Jupiter have produced only flimsy candidates for previous similar impacts during the telescopic era since the early 17th century. It has been suggested that such events may occur once in a thousand years.

The Shoemaker–Levy 9 impacts again emphasized the fragility of comet nuclei (many of which appear to have been captured as distant, small satellites by both Jupiter and Saturn), and the catastrophic energy scales involved in cosmic collisions.

shooting star Popular, non-scientific description of a METEOR.

Short, James (1710–68) Scottish optician who worked in London, the first to give telescopic mirrors a true parabolic figure. Short realized that a mirror having a parabolic, instead of a spherical, surface would be freer from spherical aberration. He also made the first successful telescope based on the design of Laurent Cassegrain (1629–93) (*see* CASSEGRAIN TELESCOPE). This instrument, completed around 1740 and nicknamed 'Dumpy', had a diameter of 6 inches (150 mm) and a focal length of 24 inches (600 mm).

short-period comet COMET for which the interval between successive perihelion returns is less than 200 years. Abut 150 such comets are currently known. Many have become faint due to continued depletion of their volatile materials. Comet 1P/HALLEY is perhaps the best-known, bright example.

Sickle (of Leo) Distinctive pattern (asterism) of stars forming the head of LEO the lion, shaped like a sickle or reversed question mark. REGULUS is at the southern end of the Sickle.

sidereal Pertaining to the stars or measured relative to the stars; the term is used particularly to denote astronomical measurements of time with respect to the stellar background. The SIDEREAL DAY, which lasts 23^h 56^m $4^s.091$ of MEAN SOLAR TIME, is the time taken for the Earth to complete a single revolution on its axis, relative to the stars rather than to the Sun, which is the basis for a civil day. Astronomers use sidereal clocks, which run at a slightly different rate from conventional timepieces, to determine whether a celestial object can be observed at any given time.

sidereal day Time interval between two successive transits across an observer's meridian of the FIRST POINT OF ARIES or of any given star. The sidereal DAY is a measure of the period of rotation of the Earth, relative to the background stars, and is equivalent to 23^h 56^m $4^s.091$ of MEAN SOLAR TIME.

sidereal month Time taken for the Moon to complete a single revolution around the Earth, measured relative to a fixed star; it is equivalent to 27.32166 days of MEAN SOLAR TIME. *See also* ANOMALISTIC MONTH; DRACONIC MONTH; MONTH; SYNODIC MONTH; TROPICAL MONTH

sidereal period Orbital period of a planet or other celestial body around its primary, as measured relative to the stellar background. The sidereal period of a planet provides a true measure of its year as opposed to the SYNODIC PERIOD, which is a measure of the time taken for the body to complete a single orbit as observed from the Earth.

sidereal time Time measured by the rotation of the Earth relative to the stars, rather than to the Sun. The Earth rotates on its axis once a DAY, at the same time moving in its orbit around the Sun, which it takes a YEAR to complete. This means that, relative to the stars, it completes one extra rotation during the course of a year, resulting in a SIDEREAL DAY being approximately 3^m 56^s shorter than a mean solar day (*see* MEAN SOLAR TIME).

Sidereal time provides an indication as to whether a celestial object is observable at any given time, since objects cross the local meridian at a local sidereal time equal to their RIGHT ASCENSION. Astronomers therefore use clocks that show sidereal time and that run at a slightly different rate from ordinary clocks.

The zero point for measuring sidereal time is the FIRST POINT OF ARIES, the position on the celestial sphere where the ecliptic crosses the celestial equator. When this transits the observer's meridian, it is 0^h sidereal time. Sidereal and solar times coincide once a year near the VERNAL

EQUINOX, on or about March 21, when the Sun lies in the direction of the First Point of Aries.

sidereal year Time taken for the Earth to complete a single revolution of the Sun, measured relative to the fixed stars; it is equivalent to 365.25636 mean solar days. Because of the effects of PRECESSION, a sidereal year is 20 minutes longer than a TROPICAL YEAR. *See also* ANOMALISTIC YEAR

siderite Obsolete name for IRON METEORITE

siderolite Obsolete name for STONY-IRON METEORITE

siderostat Flat, altazimuth-mounted mirror driven to counteract the rotation of the Earth and continuously to direct the light from the Sun on to a fixed, focusing mirror. From here it can be directed towards instruments such as a SPECTROHELIOSCOPE. Unlike the similar HELIOSTAT, which is equatorially mounted, the mirror of the siderostat has to be driven about two axes simultaneously in order to follow the Sun across the sky. The benefit of this arrangement is that the beam is directed horizontally, rather than in a direction parallel to the Earth's rotation axis, allowing greater flexibility in the positioning of the fixed mirror. As in the heliostat, the final image slowly rotates. *See also* COELOSTAT

Siding Spring Observatory Australia's major optical observatory, 20 km (12 mi) west of Coonabarabran in New South Wales. The observatory is at an elevation of 1150 m (3770 ft) on Siding Spring Mountain, one of the higher ridges of the Warrumbungle Range. It was developed in the early 1960s as an outstation of MOUNT STROMLO OBSERVATORY and remains the property of the Australian National University (ANU). The first group of telescopes had apertures of 0.41, 0.61 and 1.02 m (16, 24 and 40 in.). In 1984 the ANU added a 2.3-m (90-in.) telescope of advanced design known as the Advanced Technology Telescope (ATT), which for the first time incorporated in a single instrument a thin mirror, an altazimuth mounting and a rotating building. The ANU also operates the 0.5-m (20-in.) Uppsala Schmidt Telescope on behalf of NASA for near-Earth object searches.

Siding Spring also hosts facilities belonging to other organisations, including the two telescopes of the ANGLO-AUSTRALIAN OBSERVATORY (the ANGLO-AUSTRALIAN TELESCOPE and the UNITED KINGDOM SCHMIDT TELESCOPE, respectively the largest of their kind in Australia), the Automated Patrol Telescope of the University of New South Wales and the southern FAULKES TELESCOPE. Siding Spring's modest elevation and easterly location do not provide the best atmospheric stability, but it does enjoy particularly dark skies and it remains Australia's premier site for optical astronomy.

Sigma Octantis Closest naked-eye star to the south celestial pole, visual mag. 5.45, distance 270 l.y., spectral type F0 III. Currently it lies about 1° from the pole, and the distance is increasing due to precession.

signal-to-noise ratio Measure of signal clarity. All measurements (M) are the sum of the signal (S) and noise (N), so when the signal-to-noise ratio (R) is calculated the noise must be removed from the measurement first (to give the actual signal) before dividing by the noise: $R = S/N = (M - N)/N$. If the background noise is very high, the signal is overwhelmed by it, and the signal-to-noise ratio is very low (much less than 1). Technology continually strives to lower the noise in the detector in order to raise the signal-to-noise ratio.

Simms, William *See* COOKE, TROUGHTON & SIMMS

singularity Point in space or spacetime at which the current laws of physics make non-real predictions for the values of some quantities. Thus at the centre of a BLACK HOLE, the density, the force of gravity and the curvature of spacetime are all predicted to be infinite. It is likely that radical changes to the laws of physics or new laws will be required to deal correctly with singularities. This may however not be necessary if the idea of cosmic censorship holds. Cosmic censorship requires that all singularities be hidden from the rest of the Universe by EVENT HORIZONS. If correct, then the difficulties presented by a singularity will never have to be dealt with in the real world (*see also* NAKED SINGULARITY). It is possible that the Universe as a whole originated in a singularity at the start of the BIG BANG.

Sinope One of JUPITER's outer moons, discovered in 1914 by Seth Nicholson (1891–1963). It is about 36 km (22 mi) in size. Sinope takes 758 days to orbit Jupiter, at an average distance of 23.94 million km (14.88 million mi), in an orbit of eccentricity 0.250. It has a path (inclination 158°) in common with other members of its group. *See also* ANANKE

sinuous rille Winding channel that typically emanates from distinct, circular or elongated regions of collapse. Sinuous rilles are known on the the Moon, Mars and Venus. They are usually a few hundred metres to 1 km deep, 1 to 3 km (0.6–2 mi) wide and up to 100 km (60 mi) long. Their morphology and association with volcanic plains and constructs suggests that they originated from thermal erosion by flowing lava. APOLLO 15 landed close to Hadley Rille, one of the largest sinuous rilles of the Moon. Astronauts took detailed photographs of the rille's inner slopes and collected samples of BASALTS from its edge.

Sirius The star α Canis Majoris, the brightest star in the sky, at visual mag. −1.44. At a distance of only 8.6 l.y., it is also one of the closest stars to us. Sirius is popularly known as the Dog Star, since it lies in the constellation of the greater dog. It is a main-sequence star of spectral type A0 with a luminosity 22 times that of the Sun. The HIPPARCOS satellite detected signs of variability, but the range and cause are unknown. Sirius has a spectrum that shows enhanced absorption lines due to heavy metals, and is classified as a metallic-line A star, or AM STAR. Accompanying it is a white dwarf known as **Sirius B**, discovered in 1862 by the American astronomer Alvan G. CLARK. Sirius B is of visual mag. 8.44 and has just 0.005% of the Sun's luminosity. It orbits Sirius every 50 years but is too close to be seen separately with small telescopes. Sirius B was originally the more massive of the two stars and evolved more quickly, transferring some of its gas to Sirius A and hence giving rise to the unusual spectrum. The name Sirius comes from the Greek word *seirios*, meaning 'scorching'.

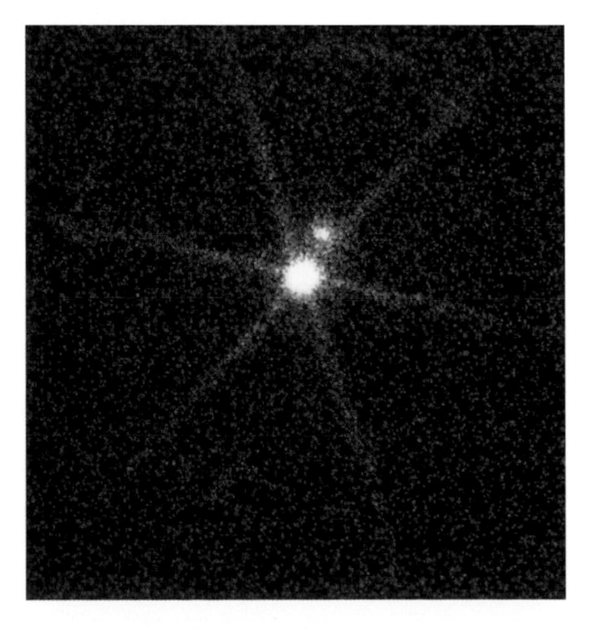

◀ **Sirius** Shown here are Sirius A and B viewed at X-ray wavelengths. Although Sirius A outshines it in visible light, Sirius B is far brighter at these low-energy wavelengths. This is because it is far hotter – its surface temperature is some 25,000 K. The spike pattern is an optical effect of the detector. Sirius A is not in fact hot enough to be seen at X-ray wavelengths and it is thought that some ultraviolet radiation from it leaked on to the detector to give the fainter component.

Sirrah Alternative name for the star α Andromedae. *See* ALPHERATZ

SIRTF *See* SPACE INFRARED TELESCOPE FACILITY

Sisyphus APOLLO ASTEROID discovered in 1972; number 1866. It is the largest known Apollo, being about 10 km (6 mi) in size. *See* table at NEAR-EARTH ASTEROID

Sitterly, Charlotte (Emma) Moore (1898–1990) American astronomer, an expert on the solar spectrum and standard spectral identifications. Working with Henry Norris RUSSELL at Princeton University, she studied binary star systems and the masses of their components. At Mount Wilson Observatory she made a detailed study of the solar spectrum. Later (1945–90), at the National Bureau of Standards and Naval Research Laboratory, Sitterly did much to standardize the nomenclature of solar spectral lines and general spectral line multiplets, including data for the near-ultraviolet obtained by rocket-launched UV instruments.

Sixty-one Cygni (61 Cygni) First star to have its parallax measured, by the German astronomer Friedrich Wilhelm BESSEL in 1838. It also has the greatest annual proper motion of any naked-eye star: 5″.2. 61 Cygni is a binary star, consisting of two orange dwarfs divisible in small telescopes or even binoculars, mags. 5.20 and 6.05 (4.79 combined), spectral types K5 V and K7 V, with an orbital period of 659 years. The stars are so widely spaced that they have measurably different distances from Earth – 11.36 and 11.43 l.y. respectively.

Skalnaté Pleso Observatory Observatory of the Astronomical Institute of the Slovak Academy of Sciences, located at an elevation of 1780 m (5850 ft) at Tatranska Lomnica in the eastern Tatras Mountains. The main instrument is a modern 0.6-m (24-in.) reflector with equipment for photoelectric stellar photometry. The observatory was founded by Antonin Bečvář.

Skjellerup–Maristany, Comet (C/1927 X1) Bright long-period comet discovered by John Francis Skjellerup (1875–1952) at Melbourne, Australia, on 1927 December 3, and independently by Maristany, at La Plata, Argentina, on December 6. At perihelion (0.18 AU) on December 18, the comet was visible in daylight only 5° from the Sun; peak brightness was estimated at magnitude −6. At the end of 1927 December, the comet was visible in dark skies, with a tail approaching 40° in length. The orbit is elliptical with a period of 36,500 years.

Skylab First US SPACE STATION. Skylab was developed from the third stage of the SATURN V rocket. The 75-tonne station was launched into a near circular orbit 433 km (270 mi) above the Earth's surface on 1973 May 14.

Skylab consisted of four sections, the largest being the orbital workshop, 14.7 m (48 ft) long and 6.6 m (22 ft) in diameter. This section also contained the living quarters. An airlock module contained equipment for the control of the station and also the hatch for space walks. A multiple docking facility contained an Apollo docking port at one end and a second, rescue port. Six telescopes for monitoring the Sun were powered by a windmill-shaped array of four solar panels.

Just after launch, the station's micrometeorite shield deployed prematurely and broke away, destroying one solar panel and damaging the other. After being briefed on the damage, the first crew was launched on May 25 and, after some difficulties, managed to erect a sunshield and cut free the jammed solar panel. The astronauts Charles Conrad (1930–99), Joseph Kerwin (1932–) and Paul Weitz (1932–) returned to Earth on June 22, after a record stay of just over 28 days. In spite of the early technical problems, the Apollo Telescope Mount was operated for 88% of the planned time, making almost 30,000 exposures. Nearly 10,000 images of the Earth were also obtained.

On July 28, a second crew, comprising Alan Bean (1932–), Owen Garriott (1930–) and Jack Lousma (1936–), was launched to the space station. The astronauts installed a new and more efficient sunshield and also installed a micrometeorite detector on the telescope mount truss. During the mission they took 25,000 photographs of the Sun and studied in detail more than 100 FLARES, including a major disturbance on August 21. The crew also returned with 16,800 pictures for use in crop surveys, land-use planning and searches for natural resources. They returned to Earth on September 25 after 59.5 days in space, then a record.

Gerald Carr (1932–), Edward Gibson (1936–) and William Pogue (1930–) entered Skylab on November 16. They spent much of their time repairing and replacing failed equipment. Nevertheless, they spent nearly two and a half times longer than planned observing the Sun and Comet KOHOUTEK, and more than three times the allotted time on materials-processing experiments. They returned to Earth after a record 84 days on 1974 February 4.

During the three missions over 120,000 photographs of the Sun were taken and 72 km (45 mi) of magnetic tape were used to record data from on-board experiments. Five years later, Skylab re-entered the atmosphere over the Indian Ocean, scattering some debris over Australia. Fortunately no one was injured.

Slipher, Earl Carl (1883–1964) American planetary astronomer (younger brother of Vesto SLIPHER) who spent virtually his entire career at LOWELL OBSERVATORY (1906–64). Slipher was noted for the huge number of high-quality photographs, especially of Mars, that he took over a 50-year period. He confirmed (1937) the blue clearing phenomenon, when the Martian atmosphere, normally opaque to blue and violet light, suddenly becomes transparent at those wavelengths, allowing surface features to be seen clearly. In 1954 Slipher discovered Mars' 'W-clouds', which formed over the Tharsis region. Slipher was among the last of the leading planetary astronomers to hold to the belief that the CANALS of Mars were real physical features.

Slipher, Vesto Melvin (1875–1969) American astronomer (elder brother of Earl SLIPHER) who discovered the recession of the galaxies. He spent his entire career at LOWELL OBSERVATORY, which he directed from 1916 to 1952, and where he supervised Percival Lowell's search for Pluto, discovered by Clyde TOMBAUGH in 1930. He used a large Brashear spectrograph attached to Lowell's 24-inch (0.61-m) Clark refractor accurately to measure the rotation periods of Venus, Mars, Jupiter, Saturn and Uranus. His spectrograms of Jupiter showed bands of methane and ammonia that were later confirmed by Rupert WILDT and others.

Slipher became the first (1912) to recognize the redshifts of lines in the spectrum of the Andromeda Galaxy (M31) and 36 other galaxies, a discovery later used by Edwin HUBBLE to determine the scale of the Universe. Obtaining radial velocities for galaxies required very long exposures with the Lowell spectrograph – as long as 60 hours – and meticulous calibration of the equipment. He announced a radial velocity of 300 km/s for M31, a result that was initially met with disbelief, as it implied that this was an 'island universe' far beyond the Milky Way (*see* GREAT DEBATE). Slipher discovered that other spiral galaxies were receding at high speeds and that they rotated, too; in 1916 he announced a rotation rate for M31 and later determined similar rates for other galaxies.

He also investigated the nature of the interstellar medium, being the first to confirm the existence of interstellar calcium and sodium atoms. Slipher discovered reflection nebulae, explaining their luminescence. Expecting that the bluish nebulosity around Merope and other stars of the Pleiades would show a spectrum similar to the Orion Nebula and other bright diffuse nebulae, Slipher instead found that the Pleiades' nebulosity had spectra corresponding to the hot, young stars in the cluster, and correctly surmised that the Merope Nebula must shine by reflected starlight.

Sloan Digital Sky Survey (SDSS) Large-scale SURVEY, the first to be made with purely electronic detectors, thus giving unprecedented photometric quality in five colours. It began in 1998, aimed at covering one-quarter of the sky. Redshifts are also being measured, and it is hoped that the resulting REDSHIFT SURVEY will plot the distribution of galaxies in a volume of space over 30 times as large as in any previous survey. The SDSS has a dedicated 2.5-m (100-in.) reflector at APACHE POINT OBSERVATORY. It is named after Alfred Pritchard Sloan, Jr (1875–1966), whose foundation financed the project.

slow motions Manual controls that allow movement of a telescope around the axes of its mounting and allow a celestial target to be followed. Adjustments are normally effected by turning a hand-wheel or fine screw. For a telescope on an EQUATORIAL MOUNTING, only the right ascension axis requires a slow motion control. An instrument on an ALTAZIMUTH MOUNTING will require controls for both altitude and azimuth.

slow nova (NB) NOVA that takes at least 150 days to decline 3 magnitudes below maximum light. Such a decline excludes any dip and recovery such as that exhibited by the slow nova DQ Herculis.

SLR *See* SATELLITE LASER RANGER

Small Astronomy Satellites (SAS) Name given to three NASA spacecraft launched in the 1970s for X-ray and gamma-ray observations. SAS-1 was also known as UHURU. SAS-2, launched in 1972, carried a gamma-ray telescope. This was followed in 1975 by SAS-3, which discovered a type of X-ray star known as a 'rapid burster'. *See also* GAMMA-RAY ASTRONOMY

Small Magellanic Cloud (SMC, Nubecula Minor) Small GALAXY that is the second nearest external galaxy to us. It is about 190,000 l.y. away and has about 2% of the mass of our GALAXY. It extends over about 3° of the sky within the constellation of Tucana (RA 01^h 05^m dec. $-72°$), where it is easily visible to the naked eye, appearing like a detached portion of the MILKY WAY. The SMC, along with the LARGE MAGELLANIC CLOUD (LMC), is named after Ferdinand Magellan, who observed them during his voyage around the world in 1519, but they were known before that date. It is an irregular galaxy with a HUBBLE CLASSIFICATION of Irr I. The SMC's physical width is about 15,000 l.y., but it is highly elongated with the long axis along the line of sight. Its depth is therefore some 60,000 l.y. The curious, long, twisted structure that has been found for the galaxy (and that may in fact divide the cloud into two distinct portions) suggests that strong tidal forces have distorted the SMC. This may have occurred during a close passage some 200 million years ago between the SMC and our Galaxy. This same close passage may have produced the Magellanic Stream, one of the largest HIGH-VELOCITY CLOUDS known. Although the SMC contains less gas and dust than the LMC, like the latter it also contains many HII and star-forming regions. The SMC orbits our Galaxy roughly at right angles to the plane of the Milky Way, and it may at some time in the future be disrupted and captured by our Galaxy.

SMC *See* SMALL MAGELLANIC CLOUD

Smithsonian Astrophysical Observatory (SAO) Major US research institute, founded in 1890 as a bureau of the Smithsonian Institution primarily for studies of the Sun. It moved from Washington, D.C. to Cambridge, Massachusetts in 1955 and established itself as a pioneering centre for astrophysics and space science. In 1973 its close ties with the HARVARD COLLEGE OBSERVATORY were cemented by the creation of the HARVARD–SMITHSONIAN CENTER FOR ASTROPHYSICS (CfA), and the SAO became part of one of the largest and most diverse astrophysical institutions in the world. Today, in collaboration with CfA and a number of other institutions, the SAO runs several major facilities including the FRED L. WHIPPLE OBSERVATORY, the MMT OBSERVATORY and the SUBMILLIMETER ARRAY. The SAO retains its headquarters in Cambridge, Mass.

Smithsonian Astrophysical Observatory Star Catalog (SAO) Catalogue giving the positions and proper motions for 258,997 stars to 9th magnitude, published in 1966. It was the first large 'synthetic' catalogue, created on a computer by combining data from several large astrometric catalogues.

Smoot, George Fitzgerald III (1945–) American observational cosmologist who led the COSMIC BACKGROUND EXPLORER (COBE) team that in 1992 discovered minute variations in the cosmic microwave background radiation. Working at Lawrence Berkeley National Laboratory since 1970, Smoot has designed and led several missions to map the afterglow of the Big Bang, using instruments known as Differential Microwave Radiometers (DMRs) designed to measure minute fluctuations in the cosmic microwave background. Three ultra-sensitive DMRs launched on COBE detected temperature variations of just ±30 millionths of a degree. These findings, announced by Smoot's team on 1992 April 23, also supported the theories of INFLATION and COLD DARK MATTER.

Smyth, Charles Piazzi (1819–1900) Scottish astronomer, the second Astronomer Royal for Scotland (1846–88), whose 1856 expedition to Tenerife, in the Canary Islands, proved the value of high-altitude observatories. He was born in Naples, Italy, the son of noted British amateur astronomer William Henry SMYTH, and his middle name honours his father's friend Giuseppe PIAZZI. Because the Scottish observatory at Calton Hill was so poorly funded, he did much of his research abroad.

His 1856 voyage to Tenerife was undertaken to compare observations of the planets and stars made at sea level

▼ **Skylab** Three American crews spent a total of more than 170 days in Skylab during 1973 and 1974. Despite technical setbacks, the science achieved exceeded expectations.

S

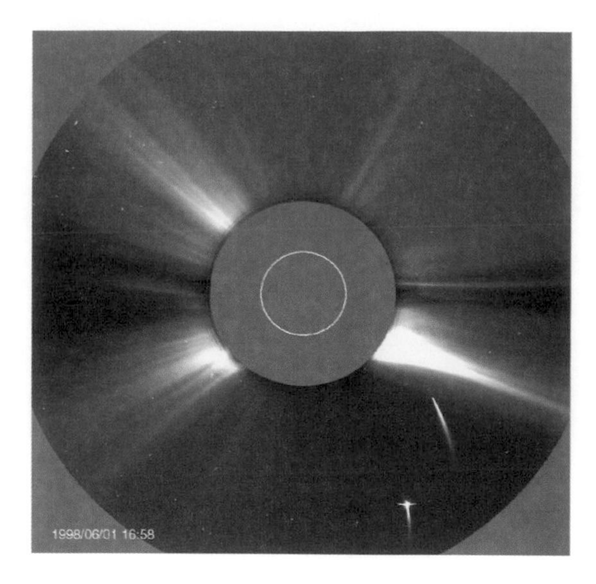

► **SOHO, Comets** The Solar and Heliospheric Observatory's LASCO camera 3 saw these two comets approaching the Sun in 1998. Shortly after they were occulted by the disc used to block out the Sun's photosphere, two coronal mass ejections occurred, leading to the conclusion that they had both impacted on the Sun. (The bar across the lower comet is an optical effect.)

from England with similar observations made from a temporary observatory located near the summit of a volcano, at 3718 m (12,200 ft). From this altitude, bad weather was less of a problem than at sea level; and the thinner air resulted in better seeing, which enabled him to detect fine detail on the Moon and planets and to resolve very close double stars. This expedition also marked the beginnings of infrared astronomy, as Smyth was able to make delicate measurements of the Moon's heat radiation that would normally have been absorbed by the lower levels of Earth's atmosphere, which are opaque to the infrared.

Smyth, William Henry (1788–1865) English naval officer, hydrographer and amateur astronomer, famous for his *Cycle of Celestial Objects* (1844). Inspired by his work assisting Giuseppe Piazzi in preparing a star catalogue, Smyth built a well-equipped private observatory at Bedford, near London, in 1830. Its main instrument was a 5.6-inch (142-mm) Tulley–Dollond refractor equipped with the first weighted clock drive designed by Richard Sheepshanks (1794–1855). Smyth published a decade of observations as the *Cycle*, which included as its second volume *The Bedford Catalogue*, a listing of 850 double stars, star clusters and nebulae. This work, aimed principally at amateur astronomers, was the first comprehensive guidebook to observing celestial objects beyond the Solar System.

Smythii, Mare (Smyth's Sea) Lunar lava plain located in the far eastern limb of the Moon. It formed from a multi-ring impact basin, the centre of which was later flooded with lava. Quite irregular in shape, Mare Smythii is best observed during a favourable LIBRATION.

SNC meteorite *See* MARTIAN METEORITE

Snow Telescope Horizontal solar telescope of MOUNT WILSON OBSERVATORY, and the oldest telescope there. It was moved to Mount Wilson in 1904 on loan from YERKES OBSERVATORY, to which it never returned. Named after its benefactor, Helen Snow, it consists of a COELOSTAT feeding a concave mirror of 0.61 m (24 in.) aperture and 18 m (59 ft) focal length. By 1910 the Snow Telescope had been superseded by the 60-ft (18-m) and 150-ft (45-m) Solar Towers, which are still used for solar research. The Snow Telescope itself is now used for educational programmes.

SNR *See* SUPERNOVA REMNANT

SOAR Abbreviation of SOUTHERN ASTROPHYSICAL RESEARCH TELESCOPE

Society of Amateur Radio Astronomers (SARA) Organization that was formed in the early 1980s to coordinate what is probably the least practised aspect of amateur astronomy – radio observations of Solar System and cosmic radio sources. SARA encourages the continual surveillance of wide areas of the sky in search of new or unusual radio emissions. It publishes a monthly *Journal*; the membership also includes professional radio astronomers.

SOFIA Abbreviation of STRATOSPHERIC OBSERVATORY FOR INFRARED ASTRONOMY

software, astronomical Modern professional astronomy relies extensively on software – the programmed sequences of instructions that control modern digital computers. Astronomers use computers networked through the Internet and World Wide Web to plan observations. Some observatories use intelligent scheduling programs to determine when observations will be made. During an observation, a telescope is controlled by programs that correct for telescope deformation and keep the telescope tracking the target object (*see* ACTIVE OPTICS, ADAPTIVE OPTICS). The instrumentation generates electronic data (perhaps passing it first through a processing sequence that removes any instrument-specific effects) in a standard format that can be handled by one of the standard astronomical data reduction systems. If the observation forms part of a large-scale survey, the data may be automatically processed to determine basic characteristics such as object type and redshift, and stored in an Internet-accessible DATABASE.

The astronomer may wish to examine the data in detail, displaying it in different ways, experimenting with IMAGE PROCESSING techniques, probing for the secrets it contains. The Flexible Image Transport System (FITS), developed by astronomers to encode definitions of image data and the data themselves, is a platform-independent system now widely used for interchanging data between observatories that has also become widely used outside astronomy.

Astronomical data reduction is a specialist process requiring specialist software, usually written by programmers employed at observatories or other astronomical institutions such as universities. Telescope and instrument control software is generally even more specialized, since each instrument has its own particular characteristics, so the software is usually written by those building the instrument or telescope. The software required for a new instrument now often represents a significant fraction of the total cost.

The main astronomical data reduction systems are NOAO's Image Reduction and Analysis Facility (IRAF) and ESO's Munich Image Data Analysis System (MIDAS) in the optical region, and NRAO's Astronomical Image Processing System (AIPS, and its successor, AIPS++) in the radio region. These can be seen as software component frameworks, with a structure that allows new components to be added easily and provides most of the basic services (for example data file access, user interaction) that the new component needs. Data acquisition systems can be structured in the same way, although, being real-time systems, they are significantly more complex. Most major observatories have developed their own data acquisition frameworks, but there are only a few such systems, notably the ROYAL GREENWICH OBSERVATORY's ADAM (Astronomical Data Acquisition Monitor) and its descendant, DRAMA, developed at the Anglo-Australian Observatory and first used on the Two-degree Field (2dF) project (*see* ANGLO-AUSTRALIAN TELESCOPE).

Astronomical software is not restricted to large research institutions. Commercial packages are available for amateur telescope control (*see* DRIVE; GO TO TELESCOPE). Image-processing techniques previously used only by professionals are now available to amateur observers wishing to use their personal computers to enhance images obtained with CCD cameras or to eliminate the effects of LIGHT POLLUTION. 'Virtual planetarium' programs can bring the Universe to one's home.

SOHO Abbreviation of SOLAR AND HELIOSPHERIC OBSERVATORY

SOHO, Comets Numerous comets, making close perihelion approaches to the Sun, discovered by the LASCO C2 and C3 coronagraphs aboard the SOHO spacecraft. As of late 2001, almost 380 had been found, most of them very small objects on KREUTZ SUNGRAZER orbits. Nearly all of these comets were tiny, with nuclei perhaps only a few tens of metres in diameter. Very few SOHO comets have been observed to survive perihelion passage. One notable exception was C/1998 J1 (SOHO), discovered on 1998 May 3 as a magnitude 0 object in the coronagraphs' field of view. The comet, which was not a Kreutz sungrazer, faded rapidly after perihelion, but was reasonably well seen in small telescopes from southern hemisphere locations.

C/1996 B2 HYAKUTAKE was visible in the SOHO coronagraph field at its 1996 May 1 perihelion, and 2P/ENCKE has also been observed close to the Sun using this spacecraft. The SOHO comet discoveries follow on from several made using the SOLWIND coronagraph aboard the US military P78-1 satellite in the late 1970s and early 1980s.

Solar and Heliospheric Observatory (SOHO) Part of the European Space Agency's INTERNATIONAL SOLAR TERRESTRIAL PHYSICS programme, together with the CLUSTER programmes, contributing to an international effort involving many spacecraft from Europe, the USA, Japan and other countries. The Solar and Heliospheric Observatory is a 1.8-tonne spacecraft; it was launched by an Atlas booster in 1995 December. Its mission is to make continuous observations of the solar photosphere, corona and solar wind to investigate the processes that form and heat the CORONA, maintain it and give rise to the expanding SOLAR WIND. It is also investigating the internal structure of the Sun. It carries 11 instruments, including a range of spectrometers and particle analysers. SOHO is situated at the L_1 Lagrangian point between the Sun and the Earth (1.5 million km/0.9 million mi from the Earth), where it points to the Sun continuously, with the Earth behind the spacecraft.

Solar Anomalous and Magnetospheric Particle Explorer (SAMPEX) NASA satellite launched in 1992 July to observe the activity of the Sun and to investigate the origin and transport of galactic cosmic rays. SAMPEX studies the changes in the effect of different geomagnetic influences on the spacecraft, allowing studies to be made of the ionization and composition of irregular, or anomalous, components of cosmic rays having the same properties irrespective of the direction from which they come. The craft also observes precipitating magnetospheric electrons that interact with the atmosphere and investigates the isotropic composition of particles originating in solar flares.

Solar B Japanese-led international spacecraft, involving the USA and the UK, to be launched in 2005 to make detailed observations of the Sun. It will study the way in which magnetic fields emerge through different layers of the Sun's atmosphere, creating the violent disturbances that affect the Earth. Solar B will be placed into a Sun-synchronous polar orbit around the Earth, in which the craft's optical telescope, X-ray telescope and ultraviolet imaging spectrometer will remain in continuous sunlight for nine months each year.

solar constant Total amount of radiant solar energy reaching the top of the Earth's atmosphere at a mean Sun–Earth distance of one ASTRONOMICAL UNIT. The mean value of the solar constant from 1978 to 1998 was 1366.2±1.0 watts per square metre. Variations in the solar constant were first reliably determined in the 1980s when suitable instruments were placed aboard satellites such as the SOLAR MAXIMUM MISSION. The solar constant rises and falls in step with the SOLAR CYCLE, but with a total change of just 0.1% from 1978 to 1998. During the two cycle minima of this period, the mean value was 1365.6 watts/m^2. SUNSPOTS crossing the visible solar disk briefly decrease the solar constant by a few tenths of one percent for a few days: such decreases are outweighed by the brightness increase due to FACULAE and PLAGES at times of high solar activity. Observations of sunlike stars suggest that larger-scale variations can occur. A 0.25% reduction in the value of the solar constant is capable of explaining the estimated drop of about 0.5 K in global mean temperature during the MAUNDER MINIMUM.

solar cycle Cyclical variation in solar activity with a period of about 11 years between maxima (or minima). The solar cycle is characterized by waxing and waning of various forms of solar activity, such as ACTIVE REGIONS, CORONAL MASS EJECTIONS, FLARES, the SOLAR CONSTANT and SUNSPOTS. The rise to maximum activity is usually much more rapid than the subsequent decline.

Over the course of an approximate 11-year cycle, sunspots vary both in number and latitude. The conventional onset for the start of a solar cycle is the time when the smoothed number of sunspots (the 12-month moving average) has decreased to its minimum value. At the commencement of a new cycle sunspots erupt around latitudes of 40 degrees north and south. Sunspots and their associated active regions are found in belts or zones at both sides of the equator with latitudes that move closer to the equator over the course of the cycle, finishing at around 5 degrees north and south. This pattern can be demonstrated graphically as a BUTTERFLY DIAGRAM and is known as SPÖRER'S LAW. The magnetic polarity of sunspots varies over a 22-year cycle, described by Hale's law. The leading (westernmost) spots of sunspot groups in the northern hemisphere generally share the same polarity, while the following (easternmost) spots have the opposite polarity. In the southern hemisphere, the leading and following spots also exhibit opposite polarities, but their magnetic configuration is the reverse of that in the northern hemisphere. All of the spots' magnetic polarities reverse each roughly 11-year solar activity cycle, so the complete cycle of the solar magnetic field is around 22 years. Reversal of field polarity can occasionally happen earlier for one hemisphere than the other, meaning that there may be intervals of up to a year when northern and southern sunspots actually have the same magnetic configuration.

The solar cycle may be maintained by a DYNAMO EFFECT, driven by differential rotation and convection. Longer-term variations, with periods of the order of a thousand years may also occur, and there have also been episodes such as the MAUNDER MINIMUM when the normal workings of the cycle appear to have been suspended. Observations of chromospheric activity in other stars suggests that similar activity cycles are typical of stars with convective zones.

solar day Mean time interval between two successive noons. *See also* DAY; EQUATION OF TIME; MEAN SUN

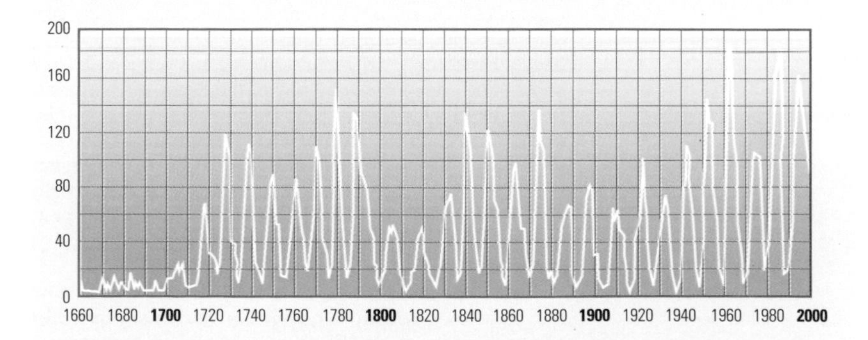

▼ **solar cycle** The number of sunspots observed peaks roughly every 11 years, although the peak intensity can vary widely. During the second half of the 17th century there were very few spots and this period is known as the Maunder Minimum.

S

▲ **solar eclipse** The solar corona varies in intensity during the course of the solar cycle and this can be seen most easily during total solar eclipses. The eclipse of 1981 July 11 revealed a particularly beautiful and complex corona.

solar eclipse Eclipse of the Sun. It occurs when the Moon's shadow falls on Earth; for this to happen the Moon must be in line between Sun and Earth. In fact, it is an OCCULTATION of the Sun by the Moon and the purist will call it an eclipse of the Earth, since an eclipse is by definition the passage of a celestial body through the shadow of another.

A solar eclipse can occur only at new moon (when the Moon is in conjunction with the Sun), but not at every new moon because of the 5°.15 inclination of the Moon's orbit to the plane of the ecliptic. An eclipse will result only when the new moon roughly coincides with a NODE (the intersection between the orbit of the Moon and the Earth). This coincidence need not be exact, an eclipse can occur up to 18.75 days before or after the alignment, thus creating the 'eclipse season'. Two eclipses can occur in every eclipse season because the synodic month (29.5 days) is less than the eclipse season (37.5 days).

The nodes shift gradually westwards along the ecliptic, so the Moon reaches the opposite node less than six months later and realignment with the original node takes place after 346.62 days, representing the 'ECLIPSE YEAR'.

Four solar eclipses can occur in one year, but since the calendar year is greater than the eclipse year a fifth eclipse is possible in one calendar year on rare occasions, and then only in January or December. The maximum of five solar eclipses will next happen again in 2206. At least two solar eclipses of some kind must occur every year. A LUNAR ECLIPSE precedes or follows a solar eclipse by about two weeks because the same conditions prevail for the Moon before or after that interval.

It has been known since Babylonian times that the nodes regain their original positions after 18 years 10.3 days, the SAROS period. It lasts 223 synodic months ($29.5306 \times 223 = 6585.32$ days). This cycle closely corresponds with 19 eclipse years ($346.62 \times 19 = 6585.78$ days), hence eclipses recur after such cycles and form series. The added 0.32 of a day of the Saros is responsible for the westwards shift of subsequent eclipses by one-third of the Earth's circumference (120° longitude). They also shift 2° to 3° north or south due to the 0.46 day difference between 19 eclipse years and the saros. This eventually causes the series to end by passing one or the other pole. Each series comprises some 70 eclipses over a period of about 1262 years.

The apparent sizes of Sun and Moon as seen from Earth are very similar, but subject to variation due to the elliptical orbits of the Earth and Moon. A total eclipse occurs when the Moon appears larger than the Sun and the shadow cone reaches the Earth. When the Moon is at its largest (at perigee) and the Sun at its smallest (Earth at aphelion) a long total eclipse occurs; the maximum duration is 7 minutes 31 seconds, and this can happen only if the shadow cone reaches the Earth near the equator around local noon.

An ANNULAR ECLIPSE results in the opposite situation, when the shadow cone fails to reach the Earth. A ring of Sun will surround the Moon at mid-eclipse. An annular-total eclipse occurs if the apparent size of Sun and Moon are the same; it will be annular along most of the path, but in the middle of the eclipse path the shadow cone will just reach the Earth and in this location a very short totality is seen. A PARTIAL ECLIPSE results if only part of the Sun is occulted by the Moon. A partial eclipse is seen over a wide area on Earth where the Moon's penumbra reaches the Earth's surface. A total eclipse can be seen only on the narrow path caused by the shadow-cone sweeping over the Earth's surface from west to east with a velocity of some 3200 km/hr (2000 mph), the maximum width of this path is 270 km (170 mi). An observer situated outside this path will see a partial eclipse.

Various phenomena can be seen at a total eclipse. The mystery of a total eclipse is enhanced because the uninitiated have no warning of the impending spectacle, since the Moon cannot be seen approaching the Sun: the partial phase passes unnoticed unless the observer knows to look for it. It is dangerous to look at the Sun at any time, especially with optical aids. The only safe way to observe the partial phase is to project the image of the Sun on to a white surface. Only during totality is it perfectly safe to look directly at the occulted Sun and the CORONA.

The eclipse begins when the east limb of the Moon appears in the same line of sight as the opposite limb of the Sun and seems to encroach upon the Sun. This is the FIRST CONTACT. This moment goes by quite unnoticed. The projected image of the Sun will show a small notch some 10 seconds after first contact, this increasing in size as the Moon travels across the face of the Sun during the next hour or so. The light reduction is imperceptible at first and the temperature drops very little until the last five minutes of the partial phase. Then the real drama begins: the sky becomes darker, often with an eerie greenish tinge, quite unlike the darkening caused by clouds. Far on the

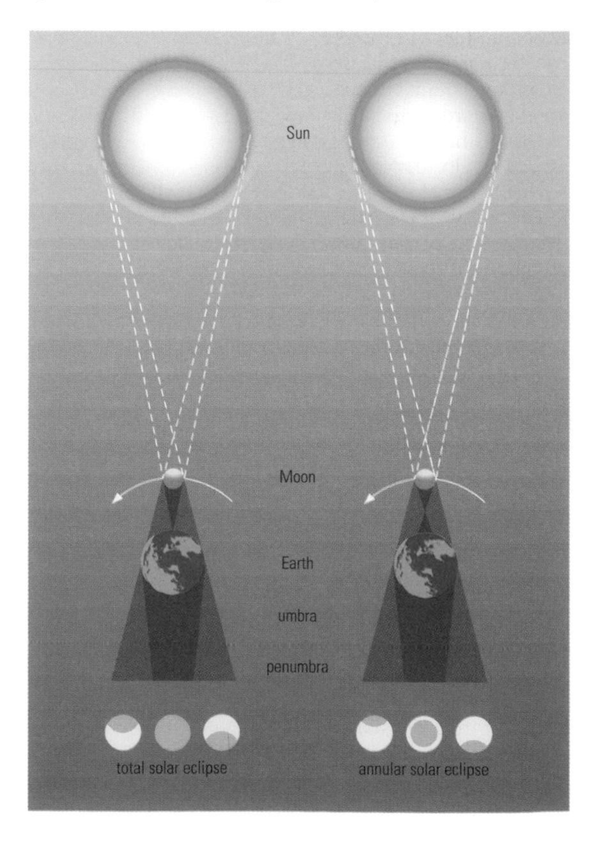

► **solar eclipse** In a total solar eclipse (left) the Sun's photosphere is completely obscured by the Moon's disk in the area covered by the cone of shadow. In an annular eclipse (right) the Moon is slightly farther away and so the cone of shadow does not reach the Earth.

western horizon an ominous cloud-like darkening appears to be increasing in size; this is the approaching shadow of the Moon. At the same time curious moving ripples of dark and light bands appear on any white smooth surface – a strange atmospheric phenomenon known as SHADOW BANDS. During the last few seconds of the partial phase light fails rapidly, it becomes noticeably cooler, birds settle down to roost, some flower petals close, and the wind tends to drop. As the last rays of sunlight fade, a dramatic change of the scene occurs: darkness descends on the countryside. The last sliver of the Sun is broken up by the Moon's irregular limb, forming BAILY'S BEADS, and as the last bead disappears SECOND CONTACT has occurred: totality has begun. The observing site on the path of totality is engulfed in darkness, illuminated only by the beautiful pearly white corona surrounding the pitch black Moon. The resulting brightness on the Earth's surface varies from eclipse to eclipse; it is comparable to that of the full moon. In contrast to the slow progress of the partial phase, events around second contact progress with incredible rapidity.

A few seconds after the disappearance of the last Baily's bead the pink CHROMOSPHERE (shining in the light of HYDROGEN-ALPHA emission) becomes visible on the eastern edge of the Moon's dark disk, only to be covered quickly as the Moon advances. PROMINENCES of various shapes and sizes are seen during totality as pink flame-like projections. Large prominences remain visible throughout totality, while smaller ones appear and disappear as the advancing Moon uncovers or covers them; initially, those on the eastern limb are best seen, while towards the end of totality, prominences on the western limb become uncovered. The corona is the most striking feature of the total eclipse. The bright inner corona contains elegantly shaped arches, loops and helmet-like structures tapering off into the fainter streamers of the outer corona for a distance of several solar diameters. These various forms are created by the solar magnetic field. The shape of the corona varies with the 11-year solar cycle. At solar minimum, the corona is drawn out into long equatorial streamers, extending to enormous distances east and west, whilst the poles are studded with shorter plume-like jets. At solar maximum, the corona surrounds the Sun more evenly, the whole circumference of the Sun being surrounded by medium-sized streamers of intricate structure.

It pays to take the eye off the features surrounding the Moon and look at the sky, where planets and bright stars can be seen with the now dark-adapted eyes. The surrounding landscape shows a 360° orange glow – similar to sunset – bordering the shadow of the Moon.

Brightening at the Moon's western limb heralds the end of totality. As THIRD CONTACT occurs, the first rays of sunlight from the brilliant photosphere, shining through a lunar valley, give rise to the famous DIAMOND-RING EFFECT. The corona and the brightest planets may be discernible to the still dark-adapted eyes for some 10–20 seconds, but the main spectacle is almost over. Events now happen in reverse: as the sky brightens one can see the shadow of the Moon receding towards the eastern horizon, and shadow bands reappear; the temperature gradually rises, cocks crow as in the early morning, and day-time activity resumes again after the short interruption. The projected image of the Sun shows the gradual uncovering of the solar disk by the advancing Moon. The partial phase lasts another hour or so until the last notch on the solar disk dwindles and finally disappears: the Moon has parted from the Sun: fourth contact has occurred and the eclipse is over.

Much information can be gained at a total eclipse. The corona, the outer atmosphere of the Sun, can be studied both visually and spectroscopically only during a total or annular eclipse. However, Bernard LYOT'S CORONAGRAPH creates an artificial eclipse and allows limited study of the inner corona from Earth. Spacecraft-borne coronagraphs have proven particularly useful. Prominences, also first seen during an eclipse, can now be studied in more detail with the spectroscope or the interference filter.

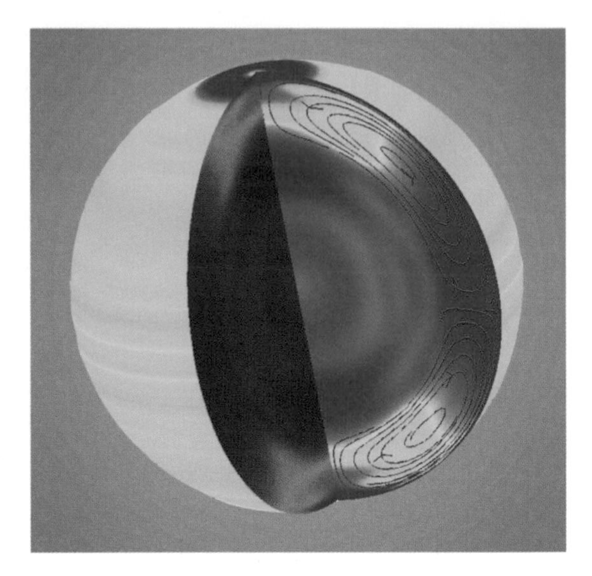

◀ **solar interior** Not only does the Sun rotate faster at the equator than at the poles but also belts of plasma have been observed moving at different speeds. These belts drift down towards the equator during the solar cycle and, as sunspots appear to form on the boundaries, may be a major factor in the changes in the Sun's magnetic field. There are also deeper currents within the convective zone, which travel from the equator towards the pole.

Timing the four contacts is still important today. The results may show a slight discrepancy between observed and predicted times, due principally to perturbations of the lunar orbit and irregularities of Earth's rotation.

The FLASH SPECTRUM can be photographed at second and third contacts, when the dark absorption bands of the photosphere change to the bright EMISSION LINES of the lower chromosphere, thus giving information on the intensities of the various spectral lines. The shadow bands occur as a result of effects in Earth's atmosphere, and have been successfully recorded using video equipment.

Einstein's theory of GENERAL RELATIVITY was first tested during the total solar eclipse of 1919 May 29. Starlight was proved to be deflected by the Sun's gravitational field. Ionospheric studies found that the ultraviolet radiation cut-off during totality resulted in an alteration of the electrical conductivity of the upper atmosphere.

solar interior Sun's internal structure. It comprises three principal regions: a central CORE, above which is found the radiative zone, and finally the convective zone, from which material rises to the visible surface of the PHOTOSPHERE.

All of the Sun's energy is produced by NUCLEAR REACTIONS fusing hydrogen into helium in the dense high temperature core, which extends for about one quarter of the solar radius (174,000 km/108,000 mi) from the centre. The core accounts for only 1.6% of the Sun's volume, but about half its mass. At its centre, the core has a temperature of 15.6 million K and a density of 148,000 kg/m^3.

The radiative zone surrounds the core out to 71.3% of the Sun's radius, 496,000 km (308,000 mi) from the solar centre. Core radiation, initially in the form of gamma rays, is continuously absorbed and re-emitted at lower temperatures (and longer wavelengths) as it travels out through the radiative zone. Recent computations indicate that it takes about 170,000 years, on average, for the radiation to work its way out from the Sun's core through the radiative zone to the convective zone.

At the bottom of the convective zone, the temperature has become cool enough, at about one million K, to allow some heavy nuclei to capture electrons. Their light-absorbing ability (opacity) obstructs the outflowing radiation and causes the PLASMA to become hotter than it would otherwise be. Because of its low density, the hot plasma rises, carrying energy through the convective zone from bottom to top in about 10 days. On reaching the visible solar disk, the hot material cools by radiating sunlight into space and then sinks back down to become reheated and rise again. These churning convection cells create a GRANULATION pattern in white-light images of the PHOTOSPHERE, which marks the top of the convective zone.

Turbulent motions in the convective zone excite sound waves that echo and resonate through the Sun. When these sound waves strike and rebound from the photosphere, they cause the gas there to rise and fall with a period of about five minutes. Observations of these oscillations with instruments on the SOLAR AND HELIOSPHERIC OBSERVATORY (SOHO), and from the GLOBAL OSCILLATION NETWORK GROUP (GONG), have been used to examine sound waves with different paths inside the Sun, determining its internal structure and dynamics with the techniques of HELIOSEISMOLOGY.

The dominant factor affecting each sound is its speed, which in turn depends on the temperature and composition of the solar regions through which it passes. The sound waves move faster through higher-temperature gas. Helioseismologists determine the difference between the observed sound speed and that calculated from the STANDARD MODEL of solar structure. Relatively small differences between the theoretical calculations and the observed sound speed are used to fine-tune the model and establish the Sun's radial variation in temperature, density and composition.

A small but definite change in sound speed has been detected at the lower boundary of the convective zone, pinpointing its radius. After suitable refinements to the Standard Solar Model, the measured and predicted sound velocities do not differ from each other by more than 0.2%, from 0.95 solar radii down to 0.05 radii from the centre. This places the central temperature of the Sun very close to the 15.6 million K of the model, confirming predictions of the expected amounts of SOLAR NEUTRINOS.

Measurements of sound wave frequencies to infer rotational and other motions inside the Sun have shown that the DIFFERENTIAL ROTATION, discovered by observations of SUNSPOTS, persists to just below the convective zone. The equatorial regions in this zone spin rapidly, while the regions near the poles rotate with slower speed. At deeper levels the rotation rate remains independent of latitude, becoming uniform from pole to equator to pole. The Sun's magnetism is probably generated at the interface between the deep interior, which rotates at one speed, and the overlying gas, which spins faster in the equatorial middle. Relative motions between neighbouring layers of electrified gas at this deep level probably help to amplify and generate the Sun's magnetic field by the DYNAMO EFFECT.

Material in the Sun's interior flows in ways other than rotation. Broad zonal bands sweep around the equatorial regions at different speeds. The velocity of the faster zonal flows is about 5 m/s (16 ft/s) higher than gases to either side, but this difference is about 400 times slower than the mean velocity of rotation. A single zonal band is more than 65,000 km (40,000 mi) wide and 20,000 km (12,000 mi) deep. These zonal bands gradually drift from high latitudes towards the equator during the 11-year SOLAR CYCLE, moving in step with a similar motion of sunspots.

Both the sunspots and the zonal bands are moving against another steady flow from the equator to the poles, which has a speed of about 20 m/s (66 ft/s). This flow penetrates to a depth of at least 25,000 km (15,500 mi). Researchers suspect that a return flow towards the equator exists at deeper levels, but detailed motions have not yet been observed at this depth.

solar mass (symbol $M_\odot$) Mass of the Sun, used as a benchmark against which the masses of other stars are compared. One solar mass is equivalent to 1.99×10^{30} kg and accounts for over 99% of the total mass of the Solar System. Other stellar masses range from less than 0.2 $M_\odot$ up to 100 $M_\odot$ or more.

Solar Maximum Mission (SMM) NASA satellite launched in 1980 February to study the Sun during a period of maximum activity at the peak of the SOLAR CYCLE. It failed after nine months, but repairs were successfully done by a Space Shuttle crew in 1984. The Solar Maximum Mission satellite re-entered Earth's atmosphere in 1989. One of its important results was the discovery of the variability of the SOLAR CONSTANT.

solar nebula PROTOPLANETARY DISK that surrounded the early Sun and produced the planets of our Solar System. It had the same composition as the Sun, mostly hydrogen and helium, with about 1% of heavier elements. In order to contain enough heavy elements to produce the planets, the total mass of the solar nebula must have been at least a few per cent of the Sun's mass; it may have exceeded a tenth of a solar mass if planetary formation was inefficient. Most of that mass, in the form of gas and dust, dissipated after the planets had formed. *See also* COSMOGONY

solar neutrinos Electron neutrinos (*see* NEUTRINO ASTRONOMY) produced in vast quantities by NUCLEAR REACTIONS in the Sun's energy-generating core. Every second, an estimated 3×10^{15} solar electron neutrinos enter each square metre of the Earth's Sun-facing side and pass out through the other side unimpeded. The neutrino reaction rate with matter is so small that a special Solar Neutrino Unit (SNU) is used to specify the flux. One SNU is equal to one neutrino interaction per second for every 10^{36} atoms. To measure this flux, neutrino detectors must contain large amounts of material and be placed deep underground to filter out confusing energetic particles generated by COSMIC RAYS.

The observed flux of electron neutrinos is compared with the number predicted by the STANDARD MODEL of the Sun. Typically, only one-third to one-half the predicted number were found in the pioneering experiments, a discrepancy known as the solar neutrino problem. For example, the HOMESTAKE MINE experiment obtained an average flux of 2.55 ± 0.25 SNU compared with the predicted 8.0 ± 1.0 SNU. Other detectors, including Kamiokande, Sage and Gallex, gave similar findings.

The solar neutrino problem might have been solved if there was a mistake in our models of the SOLAR INTERIOR. However, HELIOSEISMOLOGY observations accurately confirm the predicted core temperature of 15.6 million K, indicating that the neutrino problem could not be solved with plausible variations in the standard model.

In one alternative explanation, the electron neutrinos produced by nuclear reactions in the Sun's core can change into muon or tau neutrinos on their way out of the

▼ **solar particles** Charged particles from a solar flare were detected by SOHO's LASCO camera just 3 minutes after the flare erupted from the Sun's surface. The resulting full-halo coronal mass ejection can be seen spreading out from the camera's occulting disc.

S

Sun, thereby becoming invisible to these detectors, which responded only to electron neutrinos. This metamorphosis is known as neutrino oscillation, since a neutrino oscillates back and forth between types as it travels through space and time. The neutrino transformation that occurs when travelling through matter, such as the Sun or Earth, is named the MSW effect after the surname initials of the scientists who developed the theory in the 1970s and 80s.

Experiments at the underground Sudbury Neutrino Observatory in Canada confirm the previously observed deficit in solar neutrinos to be a result of neutrino oscillation. Containing a 1000-tonne reservoir of 'heavy water' (D_2O) in which the hydrogen has been replaced with a heavy isotope, deuterium, the Sudbury detector is, uniquely, capable of distinguishing between neutrino types. Results published in 2001 showed the expected numbers of neutrinos, as predicted by the standard model of the solar interior.

solar neutrino unit (SNU) Measurement of the rate of flow of neutrinos per unit area (neutrino flux). One SNU equals one neutrino-induced event per 10^{36} target atoms in a neutrino detector. *See also* SOLAR NEUTRINOS

solar parallax Apparent shift in position of the Sun as measured using the diameter of the Earth as a baseline. PARALLAX is an important astronomical tool when determining distances to celestial objects. The apparent change in position of an object, when viewed from opposite ends of a chosen baseline of known length, allows its distance to be computed using simple trigonometry. For objects within the Solar System, the diameter of the Earth is used, which gives the DIURNAL PARALLAX (or geocentric parallax). For more distant objects, the Earth's orbit around the Sun provides a more appropriate baseline. Solar parallax is defined as being the angular size of the Earth's equatorial radius as measured from a distance of one astronomical unit (1 AU). This method of determining the distance to the Sun has been very important in establishing the scale of the Solar System.

solar particles Energetic particles that are expelled by the Sun into interplanetary space. Electrons, protons and other ions are being continuously blown in all directions into interplanetary space by the perpetual SOLAR WIND. CORONAL MASS EJECTIONS and FLARES hurl more energetic charged particles into space, producing powerful gusts in the solar wind that are important in solar–terrestrial interaction. Nuclear reactions in the SOLAR INTERIOR produce SOLAR NEUTRINOS, which are emitted from the Sun's energy-generating core in all directions at nearly the speed of light.

Solar Radiation and Climate Explorer (SORCE) NASA satellite to be launched in 2002 to provide total irradiance measurements and the full spectral irradiance measurements in ultraviolet, visible and near-infrared wavelengths required for climate studies. The spacecraft is a combination of two original spacecraft merged into one.

Solar System Collective name given to the Sun and all the celestial bodies within its gravitational influence. This includes the nine planets and their accompanying satellites and ring systems, asteroids, comets, meteoroids and tenuous interplanetary dust and gas. The point at which the Solar System ends is called the heliopause, estimated to lie at a distance of 100 AU from the Sun. *See also* HELIOSPHERE

solar telescope Fixed telescope, usually located within a tower, and dedicated to making observations of the Sun. Observing the Sun is not as straightforward as making observations of the stars or other celestial objects because of the extreme heating effect of its light. The Earth absorbs solar radiation, producing a layer of hot, turbulent air near ground-level, causing images formed by mirrors at this height to be blurred and unsteady. To overcome this, the light-gathering equipment is placed high above ground level at the top of a tall tower, sometimes known as a solar tower.

▲ **solar telescope** The New Swedish Solar Telescope on La Palma is operated by the Institute for Solar Physics of the Royal Swedish Academy of Sciences. As well as studying the Sun, it will be used for high-resolution observations of Mercury and other objects in the Solar System.

Light rays from the Sun are collected by an instrument such as a COELOSTAT or HELIOSTAT. These track the apparent motion of the Sun across the sky and use plain mirrors to direct its light down into the tower towards a curved, focusing mirror located at the bottom. From here, the light beam is reflected back up the tower and, via one or more flat mirrors, into a room where large instruments such as high-dispersion SPECTROHELIOGRAPHS may be located.

The use of a tower allows long focal lengths to be obtained, which are necessary in order to form an image of the Sun with easily distinguishable detail. The tower, which is usually painted white in order to reduce the amount of solar radiation it absorbs, also protects the internal mirrors from wind and vibration. Its interior is often cooled or evacuated in order to further reduce distortion of the image by moving currents of warm air.

solar wind Solar material flowing into interplanetary space. The Sun's atmosphere is expanding radially outwards in all directions at supersonic speeds of hundreds of kilometres per second, filling interplanetary space with charged particles and MAGNETIC FIELDS. This solar wind is made up of an equal number of electrons and protons, with lesser amounts of heavier ions. It carries solar material out into interstellar space at a rate of almost a million tonnes each second, flowing past the planets, which essentially orbit within the Sun's outer atmosphere. The solar wind carves out a huge bubble in space with the Sun at its centre, known as the HELIOSPHERE, extending out to about 100 AU from the Sun.

The existence of the solar wind was suggested from observations of COMET ion tails by the German astronomer Ludwig Biermann (1907–86) in the 1950s. A comet's ion tail always points away from the Sun. While the RADIATION PRESSURE of sunlight is sufficient to push comets' curved dust tails away from the Sun, it is not enough to create their ion tails. Biermann proposed that electrically charged particles pour out from the Sun at all times and in all directions, accelerating the ions to high speeds and pushing them radially away from the Sun in straight ion tails.

In 1958 Eugene Parker (1927–) of the University of Chicago showed how a flow might work, dubbing it the solar wind. It would naturally result from expansion of the Sun's CORONA. At a critical distance of a few solar radii, the corona's thermal energy overcomes the gravitational

▼ **solar wind** This composite image of the Sun shows the solar disk and corona with a plot of the solar wind speed (in km/s). Combining observations in this way helps solar physicists to understand the way in which different phenomena on the Sun relate.

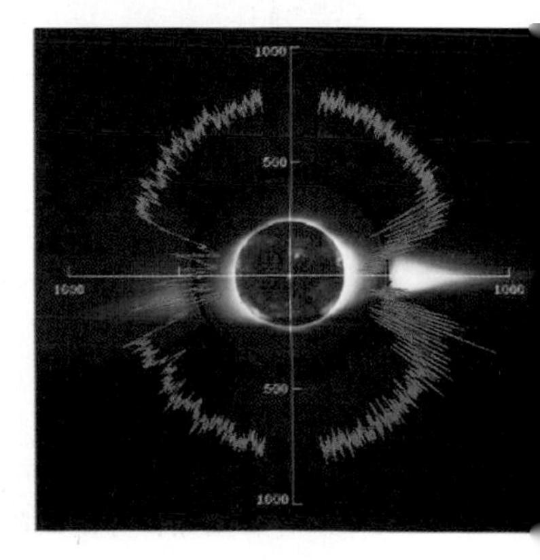

S

attraction of the Sun, allowing coronal plasma to expand supersonically into interplanetary space. Parker also demonstrated how the Sun's magnetic fields would be pulled into interplanetary space, acquiring a spiral shape as a result of the combined effects of radial solar wind flow and the Sun's rotation.

Spacecraft have been making *in situ* measurements of the solar wind since 1959. The average solar wind density near the Earth was shown in 1962–63 by Mariner 2 to be 5×10^6 particles per cubic metre. Such a low density close to Earth's orbit is a natural consequence of the wind's expansion into an ever-greater volume. The Mariner 2 data also indicated that the solar wind has a slow and a fast component. The slow component travels at a mean speed of about 400 km/s (250 mi/s) and emanates from coronal streamers close to the solar equator; the fast component travels at a mean speed of about 7500 km/s (4700 mi/s) and originates from the coronal holes. Subsequent spacecraft showed that the interplanetary magnetic field has a strength of approximately 0.00006 Gauss (6×10^{-9} Tesla) at Earth's orbit distance from the Sun.

When fast solar wind streams interact with slow streams, they produce Co-rotating Interaction Regions (CIRs), in which forward and reverse shocks are generated. Intense magnetic fields are also produced, and solar particles can be accelerated to high energies by CIRs. In 1994–95, near a minimum in the SOLAR CYCLE, the ULYSSES spacecraft made measurements of the solar wind over the full range of heliographic latitudes and at a distance comparable to that of the Earth. Ulysses' velocity data conclusively prove that a uniform, fast, low-density wind pours out at high latitudes near the solar poles, and that a gusty, higher-density, slow wind emanates from the Sun's equatorial regions at solar minimum.

Simultaneous observations with SOHO and YOHKOH pinpointed the sources of the solar wind on the Sun. The slow wind is associated with the narrow stalks of coronal streamers, at least during minimum in the solar activity cycle. Much of the high-speed wind escapes from polar coronal holes. Instruments on these spacecraft also showed that the high-speed wind is accelerated very close to the Sun (within just a few solar radii), and that the slow component attains full speed about ten times farther away.

The SOHO observations also indicated that the fast wind in polar coronal holes emanates from the boundaries of the magnetic network seen in the CHROMOSPHERE, and that heavier particles move faster than light particles in polar coronal holes. Magnetic waves might provide this preferential acceleration. For example, oxygen ions have agitation speeds 60 times greater than those of protons in coronal holes. Ulysses has detected magnetic fluctuations, attributed to ALFVÉN WAVES, far above the Sun's poles; they may block COSMIC RAYS.

The solar wind plays an important role in shaping Earth's MAGNETOSPHERE and the magnetospheres of the other planets. Particles and magnetic fields carried by the solar wind drive MAGNETIC STORMS.

▼ **Sombrero Galaxy** Seen almost edge-on, just enough of of the disk of M104 (NGC 4594) is visible to see that it does have a spiral structure. The central bulge is unusually large and surrounded by the glow of old, yellow stars.

Solis Lacus *See* SOLIS PLANUM

Solis Planum (formerly Solis Lacus) Dark oval feature on MARS (25°.5S 86°.5W). Set against the bright neighbouring Syria and Thaumasia deserts, Solis Planum is nicknamed the 'Eye of Mars'. It is highly variable in albedo and extent because of the frequent incidence of dust storms, which excavate and redistribute the lighter dusty surface deposits.

solstice Time when the Sun is at its greatest declination, 23°.5N or 23°.5S, marking the northern and southern limits of its annual path along the ECLIPTIC. In the northern hemisphere the SUMMER SOLSTICE occurs around June 21, when the Sun reaches its highest altitude in the sky and is overhead at the TROPIC OF CANCER. This marks the longest day of the year, the period of maximum daylight. The WINTER SOLSTICE occurs around December 22, when the Sun reaches its lowest altitude, being overhead at the TROPIC OF CAPRICORN. This is the point of the shortest day, when daylight hours are at a minimum. The solstices are reversed in the southern hemisphere.

solstitial colure GREAT CIRCLE, or HOUR ANGLE, on the CELESTIAL SPHERE that passes through both the north and south celestial poles, and intersects the celestial equator at the RIGHT ASCENSION of the summer and winter SOLSTICES. *See also* EQUINOCTIAL COLURE

Sombrero Galaxy (M104, NGC 4594) Spiral galaxy in the constellation Virgo (RA $12^h\ 40^m.0$ dec. $-11°37'$), seen almost edge-on. It has a prominent lane of dark material, obscuring part of the nuclear bulge. The galaxy was discovered by Pierre MÉCHAIN in 1781. The Sombrero Galaxy is a member of the Virgo–Coma supercluster, at a distance of 65 million l.y. It has apparent dimensions of $7'.1 \times 4'.4$, and magnitude +8.0. The actual diameter is probably in excess of 135,000 l.y.

SORCE *See* SOLAR RADIATION AND CLIMATE EXPLORER

source count Number of sources, such as stars or galaxies, per unit area and/or per unit flux/brightness interval. The source count is often used to determine the rate of change. It can demonstrate, for example, that there is a dark cloud obscuring faint stars like the Coalsack Nebula or the Horsehead Nebula, or that there are more galaxies at large distances than nearby.

South African Astronomical Observatory (SAAO) Major South African optical astronomy facility founded with UK participation in 1972. Its headquarters are at the old ROYAL OBSERVATORY, CAPE OF GOOD HOPE, but the telescopes are located near Sutherland in the semi-desert of the Karoo, about 320 km (200 mi) north-east of Cape Town, at an elevation of 1760 m (5775 ft). The main instruments are a 1.88-m (74-in.) reflector, moved from the Radcliffe Observatory, Pretoria, in 1976, and a 1.02-m (40-in.) reflector, originally erected at the Cape Observatory in 1964 and known as the Queen Elizabeth Telescope. By 2004 the observatory will be home to the multi-national 9.2-m (30-ft) SOUTHERN AFRICAN LARGE TELESCOPE (SALT).

South Atlantic anomaly Region located off the east coast of South America in which the magnetic field on the Earth's surface is at its weakest. Since inner RADIATION BELT particles circling the Earth follow contours of constant magnetic field strength, they move to lower altitudes at this point above the South Atlantic. Within this region, energetic particles, and COSMIC RAYS in general, have the greatest chance of precipitating into the atmosphere or adversely affecting low-Earth-orbiting satellites. A major fraction of particle loss from the inner radiation belt to the atmosphere occurs within the South Atlantic anomaly.

Southern African Large Telescope (SALT) Major optical telescope under construction at the SOUTH AFRICAN ASTRONOMICAL OBSERVATORY at Sutherland, South Africa, expected to become operational in 2004. The telescope is being built in partnership with Poland, New Zealand, and universities in Germany and the United States. It is a close copy of the HOBBY–EBERLY TELESCOPE (HET), with a similar 11-m (36-ft) hexagonally segmented mirror fixed in a 'tilted ARECIBO' configuration. However, SALT will use an improved version of the HET's Spherical Aberration Corrector to allow more of the mirror's surface to be used at any one time, giving it a slightly larger effective aperture than the HET's 9.2 m (30 ft). Like the HET, it will have both imaging and spectroscopic capabilities. SALT will be the largest single-mirror telescope in the southern hemisphere.

Southern Astrophysical Research Telescope (SOAR) Optical telescope of 4.2-m (165-in.) aperture located at Cerro Pachón, close to the southern telescope of the GEMINI OBSERVATORY and on the same ridge system as the CERRO TOLOLO INTER-AMERICAN OBSERVATORY. The telescope was built by the Southern Observatory for Astrophysical Research (SOAR), a consortium consisting of Brazil, the US NATIONAL OPTICAL ASTRONOMY OBSERVATORY and two US universities. It became operational in 2002.

Southern Cross Popular name for the constellation CRUX.

southern lights (aurora australis) *See* AURORA

South, James (1785–1867) English surgeon, amateur astronomer and observer of double stars, remembered for his 'telescope war' with the telescope-maker Edward Troughton (*see* COOKE, TROUGHTON & SIMMS). With John HERSCHEL he co-authored a standard catalogue of 380 double stars, later discovering another 160 pairs. Realizing that further discoveries required a grander telescope, in 1829 South bought an 11¾-inch (300-mm) objective lens – then the world's largest – by the French optician Robert Cauchoix (1776–1845). To build an equatorial mount he contracted with his then-friend Troughton. But Troughton, who by 1832 was ageing and had never built such a large mount, did not complete the job to South's satisfaction, as the mount suffered from excessive vibration. South refused to pay Troughton, and years of costly litigation followed. When South was ordered to pay for the work he destroyed the mount and sold it as scrap metal.

South Pole–Aitken Basin Lunar basin (55°S 180°W). It is the largest (2500 km/1600 mi diameter) and deepest (12 km/7 mi) impact structure on the Moon. Most of this basin is on the farside of the Moon, but the outer ring does come across the south pole region on to the nearside. The South Pole–Aitken Basin formed when an enormous object struck the Moon. This basin is from the oldest period of the Moon, and so is deeply degraded. Scattered mare regions of limited area occur inside the basin.

South Tropical Disturbance Major feature on JUPITER. It was in the same latitude as the GREAT RED SPOT, which it periodically passed. Its mean rotation period was $9^h\ 55^m\ 27^s.6$. Seen from 1901 to 1940, it has not subsequently been observed and has presumably disappeared permanently.

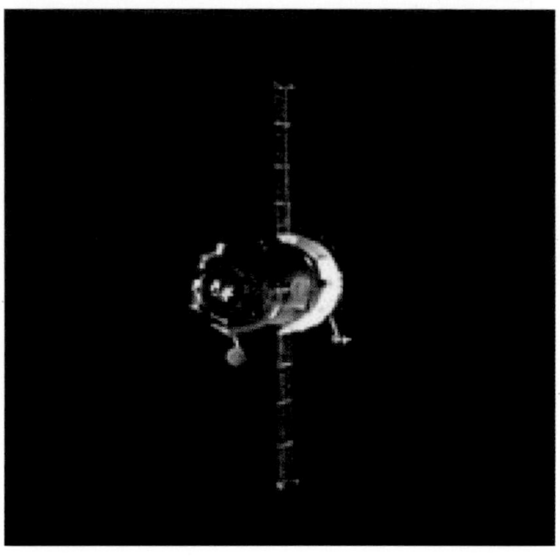

◀ **Soyuz TM** In addition to ferrying crews to and from Mir and the International Space Station, Soyuz modules have been used to transport equipment and satellites. The unmanned version is called Progress and it was one of these that collided with Mir.

Soyuz TM Russian-crewed spacecraft that is used as a ferry to and from SPACE STATIONS. It is based on the original Soyuz craft, which was first flown on a manned spaceflight in 1967, upgrades to which have been made during the craft's distinguished career. The Soyuz TM first flew in 1986 and there were 37 manned launches of the original Soyuz, 14 of the Soyuz T and, to the end of 2001, 33 of the Soyuz TM. The spacecraft, which weighs approximately 7 tonnes, is launched on a booster of the same name. It carries a maximum of three crew, who are housed in the flight cabin–descent module in the centre of the spacecraft. It has at its rear a service module with an in-orbit manoeuvring engine and two solar panels, and at the front it has an orbital module with a docking system at the forward end. Although earlier Soyuz modules flew independent flights, the TMs remained docked to the MIR space station for months, providing immediate availability for a return to Earth. A Soyuz TM is always docked to the INTERNATIONAL SPACE STATION (ISS) to provide emergency evacuation for the early three-person crews. Later, the ISS may have two Soyuz TMs available at all times for the six crew.

Space Infrared Telescope Facility (SIRTF) Last of the NATIONAL AERONAUTICS AND SPACE ADMINISTRATION (NASA) Great Observatory series spacecraft, after the HUBBLE SPACE TELESCOPE (1990), the COMPTON GAMMA RAY OBSERVATORY (1991) and the CHANDRA X-RAY OBSERVATORY (1999). To be launched in 2002, it is expected to be named after a notable infrared astronomer. SIRTF is much smaller than originally planned and will be placed into an Earth-trailing solar orbit, providing excellent sky visibility and a benign thermal environment. It will continue the work conducted by the INFRARED ASTRONOMICAL SATELLITE, COSMIC BACKGROUND EXPLORER and the INFRARED SPACE OBSERVATORY. Its liquid-helium-cooled telescope is expected to operate for about three years before the cryogenic coolant is depleted.

Spacelab Manned laboratory developed by the EUROPEAN SPACE AGENCY to fly scientists and experiments into space in the cargo bay of the SPACE SHUTTLE. There were two basic elements: a pressurized laboratory, where scientists worked in a 'shirt-sleeve' environment; and one or more external pallets, which carried instruments that were directly exposed to space. Various combinations of these elements were possible. Several astronomy missions were flown during the 22 Spacelab flights between 1983 and 1998.

space probe Space vehicle launched to investigate the Moon, the planets and their satellites, comets and asteroids, or to study interplanetary space.

S

Space Shuttle Probably the most famous space vehicle today. The USA's Space Shuttle had flown over 100 missions by the end of 2001, 20 years since its maiden flight. The Shuttle has made space travel almost look routine, albeit not quite as regular an occurrence as was hoped when the programme began life in 1972 with President Richard Nixon's signature. At the time 50 missions a year were envisaged, rather than the seven to eight actually flown.

The Space Shuttle comprises three main elements. The orbiter is equipped with three main engines called the SSMEs, which are fuelled by liquid oxygen and liquid hydrogen fed from a large brown external tank attached to the belly of the orbiter. The external tank (ET) is the only part of the Shuttle that is expendable, being abandoned to re-enter the Earth's atmosphere when the orbiter has reached initial orbit. Attached to either side of the ET are two solid rocket boosters (SRBs), which supplement the SSMEs during the first 2 minutes of flight and which are recovered from the sea after each flight. Most parts of the SRBs are used again on future Shuttle flights. The Shuttle's payloads are carried primarily in the payload bay, which is 18.3 m (60 ft) long and 4.6 m (15 ft) wide. Additional cargo and experiments, as well as equipment and consumables for the crew, are located in the mid-deck, which is under the flight deck. The mid-deck also acts as the wardroom, kitchen and temporary gym and is equipped with a toilet. The maximum payload capability of 24,947 kg has not yet been carried on any Shuttle mission. The payload capability depends on the orbital inclination taken by a mission. Each degree higher than 28° takes 226 kg off the payload weight. Missions from the Kennedy Space Center can fly up the eastern seaboard of the USA into a 57° orbit – on one reconnaissance satellite deployment mission, this was extended to 62°.

Six Shuttle orbiters have been built, starting with the *Enterprise* in 1977. This was not a spaceworthy craft and was used for atmospheric glide tests before the space missions could be attempted. The first orbiter to make an orbital spaceflight was *Columbia* in 1981 and the second, *Challenger*, in 1983. This was followed by *Discovery* in 1984 and *Atlantis* in 1985. After *Challenger* was lost in 1986, a replacement orbiter, *Endeavour*, was built , which flew for the first time in 1992. The orbiter is 37.24 m (122.2 ft) long with a wingspan of 23.79 m (78.06 ft) and is 17.27 m (56.67 ft) high from the undercarriage to the top of the vertical stabilizer. The Space Shuttle on the launch pad and fully loaded for lift-off weighs 2,041,186 kg, of which the orbiter accounts for about 113,399 kg. The SRBs are 45.46 m (149.15 ft) and 3.70 m (12.14 ft) in diameter. The ET is 47 m (154 ft) long and 8.38 m (27.5 ft) wide. The total length of the whole stack from the tip of the ET to the tail of the SRBs is 56.14 m (184.2 ft).

Most flights of the Space Shuttle carry the Remote Manipulator System (RMS), which is a sophisticated robot arm controlled from the flight deck by specialist RMS operator astronauts, who use a hand controller, a computer, TV cameras on the RMS and the view out of the rear window of the flight deck. The RMS is used to deploy and retrieve payloads and as a mobile carrier for space-walking astronauts who stand on a foot restraint at the end of the arm. The 15.24-m (50-ft) long RMS is fitted with a 'shoulder', 'elbow', 'wrist' and 'hand', to enable it to be moved into all kinds of positions. The RMS has proved invaluable to the Shuttle programme and its technology is being utilized for the INTERNATIONAL SPACE STATION.

At the end of a mission, after the retrofire, the Shuttle begins its transition from a spaceship to a glider as it starts re-entry at about 121,920 m (400,000 ft), approximately 8000 km (5000 mi) from the landing site at a speed of about Mach 25. There are six types of thermal material to protect parts of the orbiter that experience varied heat levels, the wing leading edges and underside bearing the brunt of the re-entry heating (1530 K), while the upper part of the fuselage is less exposed to heat. The approach to the landing site, usually at the Kennedy Space Center, is made at an angle seven times steeper and a speed 20 times faster than an airliner, and at touchdown the speed is over 320 km/h (200 mph).

▶ **Space Shuttle** Columbia was first launched on 1981 April 12 and, after a major refit, was brought back into service in 2002 March for the third Hubble servicing mission.

space station Large orbiting structure with substantial living and working accommodation which is designed to be permanently or intermittently manned.

Space stations have many potential roles. These include experimental and observational work in the pure sciences (astronomy, physics, life sciences); environmental monitoring; research and development work in applied sciences; industrial activity utilizing the space environment (virtual zero gravity, ultra-high vacuum) for materials processing; providing a servicing base for satellites and free-flying structures; and providing a base for further constructional work in space. There are also various potential military applications, such as surveillance, servicing of satellites and anti-missile activity. Space stations may eventually provide a base for the launching and return of interplanetary exploration missions, manned or unmanned.

The first space station to be placed in orbit was the Soviet SALYUT 1 in 1971. The lone US competition came in 1973 with the launch of SKYLAB. A major advance came with the introduction of the second generation of Soviet space stations and the assembly of separate units in orbit. The first demonstration of this technique was the linking of the COSMOS 1267 module to Salyut 6 in 1981. However, the real breakthrough came in 1986 February, when the Soviet Union launched the core module of its MIR ('Peace') station. Mir was equipped with four radial docking ports in addition to the usual access at the front and rear. Five additional modules and a US-made docking unit were attached to the station in 1987–96.

Even before Mir re-entered the atmosphere in 2001 March, construction of an INTERNATIONAL SPACE STATION had begun. Under the leadership of the USA, 16 nations came together to contribute to the largest space structure ever placed in orbit. Assembly is scheduled to be completed by 2006.

Space Telescope Science Institute (STScI) Astronomical research centre responsible for operating the HUBBLE SPACE TELESCOPE. Located at the Homewood campus of the Johns Hopkins University in Baltimore, Maryland, the STScI was founded in 1980 and is operated by AURA under contract to NASA. Besides its service role to the astronomical community, it is also concerned with the further development of the HST, and the pursuit of new space- and ground-based initiatives.

spacetime Four-dimensional CONTINUUM made up of three spatial dimensions and one time dimension. The laws of physics proposed by Isaac Newton in the late 17th century were based on the supposition that three-dimensional space was fundamentally different from time. In fact, time played a role of only increasing at a constant rate everywhere in the Universe. In SPECIAL RELATIVITY, time lost its special role and became another dimension, not quite equivalent to the other three, but could not be assumed to tick off at the same rate for every observer. Thus, relativity transformed our picture of the Universe from three spatial dimensions to a four-dimensional spacetime continuum. GENERAL RELATIVITY attributes gravity to the warping or bending of spacetime due to the presence of mass and energy.

space velocity True speed and direction of a star through space, relative to the Sun. From observations of a star's position it is possible to measure its PROPER MOTION – its movement across our line of sight – and from that to calculate its TANGENTIAL VELOCITY. Observations of the Doppler shift in its spectrum reveal how fast it is travelling away from us, in other words its RADIAL VELOCITY. Because the two lie at right angles to one another, simple trigonometry can be used to calculate their resultant, which gives a measure of the star's actual speed and direction through space.

Spacewatch ASTEROID and COMET search programme based at the University of Arizona, USA, employing telescopes at Kitt Peak. Spacewatch, founded by Tom Gehrels (1925–), was the first NEAR-EARTH OBJECT project to make use of charge-coupled devices. Many comets discovered in this project have been given the name Spacewatch.

space weather Responses of Earth's local space environment to variations on the Sun, such as CORONAL MASS EJECTIONS, FLARES, solar particle events and associated interplanetary shocks. Monitoring and predicting these variations is becoming more widespread because many modern technological systems are increasingly susceptible to these effects. For example, these severe solar disturbances propagate out to the Earth and drive GEOMAGNETIC STORMS in the MAGNETOSPHERE. These in turn cause dramatic increases in the particle populations in the VAN ALLEN BELTS and in the electrical currents flowing in the magnetosphere and IONOSPHERE. Navigation, communication and weather satellites orbiting the Earth may be damaged or destroyed if they pass through these enhanced radiation environments. Hazardous radiation levels may also be experienced by astronauts, and in extreme cases by passengers in aircraft flying high over the poles of the Earth.

Communication between spacecraft and the ground may be affected by disturbances in the ionosphere. The ionospheric drag on low-Earth-orbit satellites may increase, thus reducing the satellites' operational lifetime. Ionospheric currents generate magnetic fields on the ground, which may disrupt the operations of large power-distribution grids. For example, such magnetic fields were the cause of a 9-hour blackout affecting 9 million people in Quebec, Canada, in 1989 March. These fields can also introduce significant errors in geomagnetic surveys used in the commercial exploration of natural resources, such as drilling for oil.

Space weather effects are particularly prevalent during the maximum of the SOLAR CYCLE, although they may occur at any time. Various agencies around the world are engaged in routine monitoring of the Sun, the solar wind and the terrestrial magnetosphere and attempt to predict which events will have major impact on the Earth environment. Such predictions allow appropriate protective measures, such as powering down susceptible systems, to be implemented before damage occurs.

spallation NUCLEAR REACTION wherein a nucleus is struck by a particle whose energy is greater than 50 MeV. Usually this is a COSMIC RAY impact with the nucleus of an atom in the interstellar medium. It leads to fragmentation of the nucleus and the production of light elements (*see* ASTROCHEMISTRY and NUCLEOSYNTHESIS).

Spallation is also the ejection of debris from the back of a surface, or around the point of impact, for an object hit at high speed by a macroscopic particle, such as micrometeoroid impacts on spacecraft.

Special Astrophysical Observatory (SAO) Research institution of the Russian Academy of Sciences, established in 1966, and the only Russian centre for ground-based astronomy. The observatory is located near the town of Zelenchukskaya in the North Caucasus. Its two facilities are the BTA (Bolshoi Teleskop Azimutal'ny, or Large Altazimuth Telescope), completed in 1976, and – 20 km (12 mi) to the south-east – the RATAN-600 (for 600-m Radio Telescope Antenna), opened in 1977. The BTA is at an elevation of 2100 m (6890 ft) and, with its 6.05-m (238-in.) mirror, was for 15 years the largest optical telescope in the world. The RATAN-600 is similarly generously proportioned, with a 0.6-km (0.4-mi) diameter multi-element ring antenna that can be pointed by tilting the individual panels of the ring. The SAO carries out research over the full range of modern astrophysics, and has branches in St Petersburg and Moscow.

special relativity Theory of mechanics proposed by Albert EINSTEIN in 1905 that correctly describes the motions of objects moving near the speed of light. Einstein proposed a version of mechanics based on the LORENTZ TRANSFORMATIONS. Einstein's theory returns results identical to Newton's laws for small velocities relative to the speed of light, but predicts substantial differences for objects moving close to the speed of light. A primary postulate for special relativity is that we live in a four-dimensional SPACETIME continuum, and that time is relative to the reference frame of the observer. A second postulate is that observers always measure the speed of light at the same value: lightspeed is an invariant quantity in relativity.

Lorentz transformations predict that time runs slower in the reference frame of the object as it moves relative to a stationary observer: time slows down with increasing velocity. These propositions also show that in the reference frame of a stationary observer, the object in motion is physically smaller in the direction of motion, being smaller when closer to the speed of light.

Finally, special relativity predicts that mass increases with velocity. These important considerations led to the famous equation $E = mc^2$, spelling out the equivalence of mass and energy. The mass (m) multiplied by the speed of light (c) squared yields a tremendous amount of energy (E), since the speed of light is huge (roughly 300,000 km/s or 186,000 mi/s). This discovery allowed astronomers to explain energy generation in the Sun as caused by NUCLEAR REACTIONS converting hydrogen to helium. Every aspect of special relativity has been confirmed to a high degree of accuracy numerous times.

speckle interferometry Technique for improving the resolution of astronomical images by removing the blurring caused by turbulence in the Earth's atmosphere. When the light from a star passes through the atmosphere, it is refracted in random directions over periods of tens of milliseconds by constantly moving pockets of air of varying density and temperature. This refraction produces multiple images of the star that appear to dance around (twinkle) many times a second and be blurred. The effect is known as SEEING.

The light refracted from the individual atmospheric cells reaches different parts of the telescope's optics at different times and from different directions, resulting in either constructive INTERFERENCE (peaks of separate waves coincide) or destructive interference (peaks and troughs coincide). This produces a series of bright and dark regions, giving a highly contrasted mottled or grainy appearance. The bright regions are known as speckles, several hundred of which may be seen, and the speckle pattern represents the random distribution of atmospheric irregularities. Because the atmosphere is turbulent, this pattern changes quite rapidly with time so that images of stars taken with exposures of about a second or more become smeared into what is known as the 'seeing disk'. For short time-exposures, however, the atmospheric motion is frozen and the full speckle pattern is revealed.

Speckle interferometry involves the use of a detector such as a CCD to take a sequence of short-exposure (10–15 milliseconds) 'snapshots' of an object, freezing its apparent motion. Each snapshot produces an instantaneously distorted image and the sequence can then be combined by computer image-processing, electronically removing the atmospheric effects, to produce a distortion-free image.

The technique is not limited to stellar observations. In recent years, speckle interferometry has been successfully applied to various solar phenomena and has also been used to map surface features on Saturn's moon Titan.

spectral classification System of classifying the spectra of stars (*see* SPECTRUM). In the mid-19th century, when astronomers began to observe the brighter stars with SPECTROSCOPES, they discovered a rich variety of spectra that required some system of classification. Of several original schemes, that by Father Angelo SECCHI was the most widely adopted. His system contained five types that were based on both the stars' colours and the details within their spectra: Type I contained blue-white stars with strong hydrogen ABSORPTION LINES (for example, Sirius and Vega); Type II referred to yellow and orange stars with numerous metallic spectrum lines (the Sun, Capella, Arcturus); Type III contained orange-red stars with metallic lines and bands, now known to be caused by titanium oxide (TiO), that shaded to the blue (Betelgeuse, Antares); Type IV included deep-red stars that had dark carbon bands shaded to the red; Type V was reserved for stars with bright EMISSION LINES.

As spectroscopes improved and photography became available, better spectroscopic detail required a more comprehensive description. A system devised between 1890 and 1901 by Harvard's E.C. PICKERING and his assistants (Williamina FLEMING, Antonia MAURY and Annie J. CANNON) originally used letters from A to Q based primarily on the strengths of the hydrogen lines such that A–D belonged to Secchi's Type I, E–L to Type II, M to Type III, carbon-line N to Type IV, and O–Q to Type V.

With improved observations, several letters were dropped as unnecessary, while the work of Maury and Cannon showed better continuity of all the lines if B were to come before A and O before B. The final result was the classic spectral sequence OBAFGKM. Cannon then added for each class simple numerical subtypes from 0 to 9, such that B0 to B9 is followed by A0 to A9 and so on (the modern system beginning at O3). Cannon's classification of about 225,000 stars was published between 1918 and 1924 in the *HENRY DRAPER CATALOGUE*. (A later extension increased the count to 359,082.)

Based in part on the presence of emission lines, M was originally subdivided into Ma to Md and O from Oa to Oe, but these were eventually made numeric as well ('e' and 'f' are now descriptive terms for emission lines). R STARS, which are the warmer versions of class N CARBON STARS, were added in 1908. S STARS, which are intermediate carbon stars with bands of zirconium oxide, were added in 1924.

The system was quickly seen to correlate with colour that progressed from blue to red, and thus with temperature, class O to M ranging from 50,000 to 2000 K. The development of physical theory demonstrated that the spectral sequence O to M represents an ionization and excitation sequence, not one of chemical composition. R, N and S, however, are different, containing various enhancements of carbon. Though some carbon stars are DWARF STARS (with their own classification schemes), all the R, N and S stars are GIANT STARS. Temperatures in the classes R and N (now combined into class C) characterize the track on the Hertzsprung–Russell diagram of ordinary stars evolving from mid-G to late M, whereas class S temperatures track only class M.

Improved infrared technology in the last decade of the 20th century began to turn up stars that were not classifiable on the original scheme. In 1999 class L (*see* L STARS) was added to account for deep-red (really infrared) stars in which the strong TiO bands of class M (*see* M STARS) were replaced by powerful absorptions of hydrides and the alkali metals, the temperatures falling from 2000 to 1500 K. T STARS, at the end, near 1000 K, are defined by methane bands. Class L is a mixture of RED DWARFS and BROWN DWARFS, while T is reserved for low-mass brown dwarfs.

Various comments are commonly added to the basic spectral types to describe details. Dwarfs and giants were originally prefixed by 'd' and 'g' (the Sun, for example, is a dG2 star, Arcturus is gK1). A separate classification by Maury used a, b and c to describe line widths. Sharp-lined stars, class c, were later seen to be SUPERGIANTS. The descriptive is still in use: Lc stars, for example, are irregular supergiant variables. Also, 'e', 'm', 'p', 'wk', respectively, stand for 'emission', 'metallic', 'peculiar' and 'weak-line' (Sirius is an A1m star).

The HARVARD SYSTEM was not adequate to deal with the differences between dwarfs, giants and supergiants, that is, stars of different luminosities at a given temperature. In 1943 William MORGAN, Philip Childs Keenan (1908–2000) and Edith Marie Kellman (1911–) redefined the spectral types and introduced a two-dimensional classification system. The MORGAN–KEENAN CLASSIFICATION (MK or MKK) scheme retained the decimally subdivided OBAFGKM temperature sequence. To each of the classes is added a Roman numeral that describes the luminosity of the star, I to V standing for supergiant to dwarf, the Sun (a G2 dwarf) is thus classified as G2 V. Keenan later added Arabic '0' to define super-supergiants (hypergiants). The MK system gives precise details of which ratios of spectrum line strengths are to be used to determine the spectral types, and it provides photographic examples of bright stars that act as classification standards. Summaries of both Harvard and MK classes are given in the accompanying tables.

For accurate work, finer subdivisions are necessary, such as Iab (intermediate between Ia and Ib) and IIa or

GENERAL SPECTRAL CHARACTERISTICS

Class	Colour	Temp.[1]	
O	blue-white	32,000–50,000	ionized helium; neutral helium and weak hydrogen lines are also visible
B	blue-white	10,000–30,000	neutral helium, hydrogen lines strengthening toward the later subtypes
A	white	7200–9500	very strong hydrogen decreasing towards later subtypes as ionized calcium increases
F	yellow-white	6000–7000	ionized calcium continuing to increase in strength and hydrogen weakening; lines of neutral elements strengthen
G	yellow	5300–5900	ionized calcium very strong, hydrogen weaker; neutral metal lines, particularly of iron and calcium prominent
K	orange	4000–5200	strong neutral metallic lines; molecular bands of CH and CN become prominent
M	orange-red	2000–3900	strong absorption bands of titanium oxide and large numbers of metallic lines
L[2]	red-infrared	1500–2000	metallic hydrides and alkali metals prominent
T[2]	infrared	1000	methane bands prominent
S[3]	red	2000–3600	intermediate carbon content; zirconium oxide bands replace titanium oxide
R[3]	orange-red	3600–5000	warm carbon stars with prominent C_2, CN and CH
N[3]	deep red	2000–3500	cool carbon stars with strong C_2, CN and CH
C[3]		2000–5000	combined classes R and N

1 Dwarf temperature scale except for R, N, S and C
2 Added in 1999; no MK standards exist
3 Giants only

MK LUMINOSITY CLASSES	
Class	Star Type
0[1]	extremely luminous super-supergiants, or hypergiants; present in only small numbers in the Magellanic Clouds and the Galaxy
Ia	luminous supergiants
Ib	supergiants of lower luminosity
II	bright giants
III	ordinary giants
IV	subgiants
V	dwarfs (main-sequence stars)
VI[2]	subdwarfs
VII[2]	white dwarfs

[1] Arabic 0, standard but added later
[2] Non-standard and not part of the MK system, but occasionally seen

IIb (respectively on the bright or faint side of class II). The decimal divisions can also be more finely divided, B0.5 falling between B0 and B1. Intermediate luminosity classes are expressed by hyphenated Roman numerals, 09 IV–V falling between subgiant and giant.

The MK system is applicable only to stars of normal (that is, solar) chemical composition. Roman numeral VI for metal-poor SUBDWARFS is therefore inconsistent. WHITE DWARFS were originally classified on a pseudo-Harvard scheme, with D preceding the type: DA white dwarfs have strong hydrogen lines; DB and DO strong helium lines; and DC only a continuum. These types bear more relation to odd chemical compositions than they do to temperature. DA now refers to white dwarfs with hydrogen envelopes, DB or non-DA to the others (those with helium-rich atmospheres). A more complex scheme is now available. For stars that contain spectral abnormalities, an MK type may be assigned together with an indication of the strength of the peculiarity; for example K0III-CN3 shows that the star has anomalously strong bands of CN, and K2II-Ba5 indicates that the K2 giant is an extreme BARIUM STAR. Carbon content is also added to the spectral class, for example C2,4, where the latter number indicates band strength.

To be properly classified on the MK system, the spectrum of a star must be photographed in the blue-violet with a SPECTROGRAPH of prescribed type and with an appropriate classification dispersion. Higher dispersions can reveal lines that are not seen under low-dispersion conditions and can lead to mis-classification. It is common, however, to calibrate a different system of absorption lines in a different part of the spectrum (red, ultraviolet) against the MK standards. Such calibration is also necessary for classifications achieved with modern CCD (charge-coupled device) detectors, whose data are rendered graphically rather than photographically. A variety of quantitative and computer-aided schemes are also available. For consistency, however, all need to be calibrated against the original MK standards.

spectral type Series of divisions into which stars are classified according to the appearance of their SPECTRUM. A star's spectral type is ultimately based on temperature, luminosity and to some degree chemical composition. *See* SPECTRAL CLASSIFICATION

spectrogram Recording of a SPECTRUM. Spectrograms were originally recorded on photographic plates. With the electronic revolution in astronomy, in particular with the advent of charge-coupled devices (CCDs), spectrograms are now recorded digitally for graphical and computer-aided display. Photographic spectrograms are difficult to calibrate in terms of light intensity, while the electronic (digital) versions easily lend themselves to analysis.

spectrograph Device that takes and records SPECTROGRAMS. Spectrographs are the permanent-recording versions of the original spectroscopes through which astronomers made direct visual examinations of spectra. The first spectrographs used prisms as dispersing elements and recorded the spectra on photographic plates.

Modern spectrographs are large and complex pieces of equipment. They are often positioned at a stationary focus of a telescope, such as the COUDÉ or NASMYTH FOCUS. Light from a star or other celestial object enters a narrow slit and passes through a collimator, producing a parallel beam. This is directed on to a DIFFRACTION GRATING, a piece of reflective material ruled with thousands of parallel lines, to disperse the light. The resultant spectrum is then imaged on to a detector such as a CCD. The use of a narrow entrance slit means that only an image of this, rather than the star image, which will be distorted by atmospheric effects, is recorded. This prevents adjacent wavelengths being smeared out and keeps the spectrum sharp.

Spectrographs contain different types of diffraction gratings to produce either high-, medium- or low-resolution spectra, depending on the required application. An Echelle spectrograph uses an ECHELLE GRATING, which produces spectra with a high degree of resolution over a narrow band of wavelengths. The grooves of an echelle grating are stepped and their spacing relatively wide, producing a number of high-resolution overlapping spectra, which are then separated by a second, lower-dispersion grating. An astronomer might use a variety of different spectrographs to study an object. For example, if he or she were attempting to identify the optical counterpart of a known X-ray source, a low-resolution spectrograph might be used to obtain rapidly the spectra of all the possible candidates. From this survey, the X-ray object could be identified by its spectral properties, at which point an intermediate-resolution instrument would be used to obtain more detailed information, isolating features of interest. If even greater detail were required, a high-resolution instrument would be brought into use. Modern, large telescopes are equipped with a variety of such instruments for this very purpose.

An examination of some of the spectroscopic instruments available for astronomers to use on the twin Keck telescopes reveals the diversity of today's equipment. ESI, an Echellette Spectrograph and Imager, is used for obtaining high-resolution spectra of very faint galaxies and quasars from the blue to the infrared in a single exposure. It can also be configured to record lower-resolution spectra of several objects simultaneously. HIRES, as its name suggests, is a high-resolution instrument that operates in the 0.3–1.1 μm range and can measure the precise intensity of thousands of separate wavelengths. LRIS is a low-resolution imaging spectrograph that can take spectra of up to 30 objects simultaneously, making it suitable for studying stellar populations of distant galaxies as well as galactic clusters and quasars. LWS is a long-wavelength spectrograph, designed for the study of planetary nebulae, protostellar objects and galactic cores. Finally, NIRSPEC, the near-infrared spectrometer, is an echelle spectrograph that operates in the 1–5 μm range and is designed for the study of very high redshift radio galaxies and the motions of stars near the galactic centre. It can also be used to study active galactic nuclei, interstellar chemistry and stellar physics. *See also* SPECTROSCOPY; SPECTROMETER

spectroheliograph Instrument used for imaging the Sun in light of a particular wavelength (monochromatically). In principle, the spectroheliograph operates like a SPECTROHELIOSCOPE with a photographic plate or digital detector placed very closely behind the exit slit. An image of the Sun is projected from the telescope on to the narrow entrance slit, which is used to select the part of the solar disk or CHROMOSPHERE to be observed. A DIFFRACTION GRATING disperses the light into a SPECTRUM, and a second slit is then used to select the exact wavelength of light required, usually corresponding to one of the Sun's main elements, such as hydrogen (which produces the spectral line Hα) or calcium (K). A whole image of the Sun, known as a spectroheliogram and showing the distribution of that element, can be obtained by scanning the

S

entire solar disk with the entrance slit. Increasingly fine detail can be resolved as the slit widths are reduced, but this lessens the final image brightness so that the speed of the scan has to be reduced proportionately.

The spectroheliograph was invented in the 1890s independently by George HALE and Henri DESLANDRES. As with the spectrohelioscope, the apparatus is usually so bulky that it is mounted with its entrance slit at the focus of a fixed telescope served by a HELIOSTAT or COELOSTAT and second mirror.

spectrohelioscope High-dispersion SPECTROSCOPE that provides a visual image of the Sun in monochromatic light. The spectrohelioscope is the visual equivalent of the SPECTROHELIOGRAPH, in that an image of the solar disk is projected from a telescope into a narrow entrance slit. A DIFFRACTION GRATING disperses the light into a SPECTRUM, and the exact wavelength to be observed is set using a second slit. The observer can thus select the element within the Sun and the depth in the solar atmosphere that they wish to view. By rapidly scanning the solar disk and its prominences with a repetition rate exceeding ten per second, by persistence of vision the observer sees a stationary monochromatic image. Scanning is commonly achieved by vibrating the two slits synchronously at high frequency, but some instruments have fixed slits that are scanned optically.

This apparatus is usually so bulky that it is mounted with its entrance slit at the focus of a fixed telescope served by a HELIOSTAT or COELOSTAT and second mirror. The focal ratio of the spectrohelioscope must be the same as the fixed telescope so that, although the optics are invariably folded to put the entrance and exit slits adjacent to each other, the complete system is longer than twice the focal length of the fixed telescope.

The wide range of possible settings and its suitability for the measurement of radial velocity of violent solar disturbances make the spectrohelioscope a more versatile monochromatic instrument than optical filters. The detail seen increases as the slit width is reduced, but this necessitates a high scanning rate to ensure adequate image brightness. Wider slit widths show limb prominences only but disk detail is not visible.

spectrometer Instrument used for observing and measuring features of a SPECTRUM, such as the spectral lines, by direct observation. The spectrometer incorporates a SPECTROSCOPE, to split light from a celestial object into its component wavelengths, producing a spectrum, together with a device such as a FABRY–PEROT INTERFEROMETER, to measure accurately the positions and intensities of the emission and absorption lines within the spectrum. *See also* SPECTROSCOPY

spectrophotometer Instrument used for measuring the relative intensities of ABSORPTION and EMISSION LINES in different parts of a SPECTRUM. The line profile of the spectrum that the spectrophotometer produces is a graph of how intensity varies with wavelength. This measure of the amount of radiation of a particular wavelength that has been absorbed or emitted by a celestial object is an indication of the relative concentration of chemical elements present within the object. The spectrophotometer can be used for the study of spectra at ultraviolet and infrared as well as visible wavelengths.

spectroscope Instrument that splits electromagnetic radiation into its constituent wavelengths, producing a characteristic SPECTRUM. The spectrum can be observed visually or recorded in some form, in which case the instrument is known as a SPECTROGRAPH. Since all astronomical data is now recorded for future analysis, the term spectrograph is almost universally used to describe this sort of instrument.

spectroscopic binary BINARY STAR that is too close for its components to be resolved optically; its orbital motion is deduced from periodic shifts in its spectral lines, indicating variable radial velocity. A spectroscopic binary is likely to be a CLOSE BINARY with a short orbital period and relatively high radial velocity. Binaries cannot be detected spectroscopically if their orbit is perpendicular to the Earth.

If the components have similar brightness, both sets of SPECTRAL LINES are observed, and the system is called a double-lined spectroscopic binary. A composite spectrum binary has a spectrum that consists of two sets of lines from stars of dissimilar spectral type. Systems with only one set of spectral lines observable are known as single-lined spectroscopic binaries.

The RADIAL VELOCITY of the star (or stars) is determined by measuring the DOPPLER SHIFT of the spectral lines. In double-lined spectroscopic binaries, the relative velocities (v) of the two components can be determined and hence the relative masses (M), using:

$$\frac{v_2}{v_1} = \frac{M_1}{M_2}$$

The individual masses can be determined if the inclination of the orbit is known, for example when the system is also an ECLIPSING BINARY.

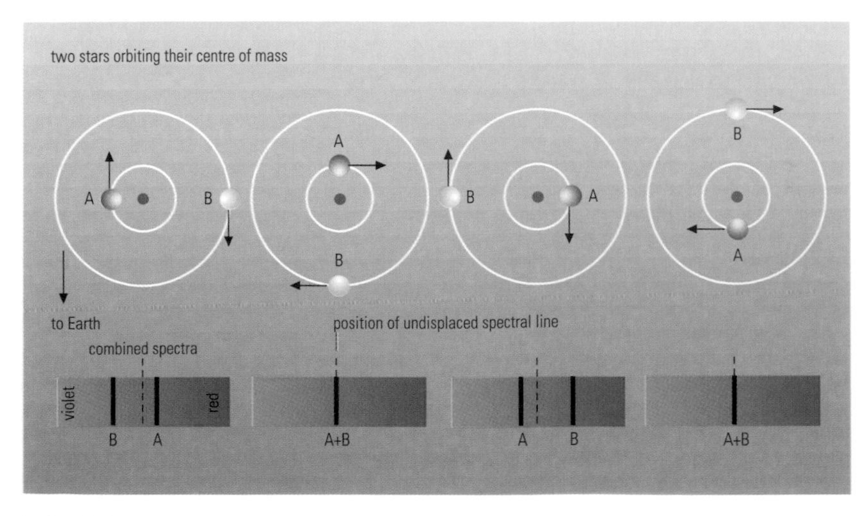

▼ spectroscopic binary Redshift is used to ascertain the relative motions of stars in close binary pairs. When a star is moving away, its spectrum is shifted to the red, whereas the spectrum of a star moving towards us will be blueshifted. When both stars are moving at right angles to our line of sight their spectra are not displaced.

spectroscopic parallax Method of obtaining the distances of stars from their SPECTRAL TYPES (*see* SPECTRAL CLASSIFICATION) and apparent magnitudes. Analysis of the SPECTRUM reveals the full spectral class in the MORGAN–KEENAN CLASSIFICATION system. These classes have all been calibrated in terms of absolute visual magnitude through the measurement of the distances of stars by TRIGONOMETRIC PARALLAX and by main-sequence cluster fitting. The spectral class of any star thus gives its absolute visual magnitude. Comparison with the apparent visual magnitude gives the distance modulus and therefore, through the magnitude equation, the distance itself.

The method is dependent on an assessment of the degree of interstellar EXTINCTION to the star. This can be found from the degree of REDDENING through comparison of the actual colour with that inferred from the spectral class. The chief unknown is the ratio of total to selective extinction (that is, the relation between reddening and total absorption), which can vary from place to place depending on the nature of the absorbing dust grains.

spectroscopy Practice of obtaining, and the study of, the SPECTRUM of an astronomical object. Astronomical spectroscopy is the key technique by which the physical properties of astronomical bodies are revealed. Spectra are obtained with a SPECTROGRAPH, and recordings are made of the distribution of light relative to wavelength for the object concerned.

Current instruments allow the precise measurement of the following: the flux distribution (energy per unit area per unit time per unit wavelength) of the continuum; the fluxes, wavelengths and shapes of EMISSION LINES; and the wavelengths, strengths (the amounts of energy extracted from the continuum) and the detailed profiles (exact distribution of radiation with wavelength) of ABSORPTION LINES.

Reduction of spectrographic data to actual radiative fluxes requires calibration of the sensitivity of the detector, and also requires calibration with standard astronomical sources to correct for wavelength-dependent absorption by the Earth's atmosphere (if the telescope is ground based) and the telescope/spectrograph optics. The determination of precise wavelengths for RADIAL VELOCITIES requires observation of wavelength standards. In older spectrographs, these were iron arcs set within the spectrograph. The iron spectrum was displayed photographically on either side of the spectrum of the source. In modern spectrographs, the standards are gas emission tubes whose spectra are digitally compared with those of the source.

The data then allow accurate physical analysis of astronomical sources. For example, the energy distribution of the continuum reveals the nature of the source, whether thermal (produced by a hot gas) or non-thermal (perhaps synchrotron emission from a supernova remnant). Emission line fluxes and wavelengths of nebulae, galaxies, quasars and a variety of other sources allow the determination of temperatures, densities, velocities (both radial velocities and those related to internal motions) and chemical compositions. Absorption line strengths and profiles allow the deduction of temperatures, densities, surface gravities, magnetic fields, rotation speeds and chemical compositions of stars.

These principles extend to all wavelength domains. Gamma-ray, X-ray and ultraviolet spectra are observed by space-borne telescopes and spectrographs. Infrared as well as visual spectra are observed with both ground-based and orbiting spectrographs. Radio telescopes are also fitted with spectrographs that use electronic techniques for the observation of emission, absorption and continuous spectra from a great variety of radio sources, from molecular clouds to galaxies.

spectrum Distribution of ELECTROMAGNETIC RADIATION with wavelength; a 'map' of brightness plotted against wavelength (or frequency). A continuous spectrum is an unbroken distribution over a broad range of wavelengths. At visible wavelengths, for example, white light may be split into a continuous band of colours ranging from red to violet. An EMISSION SPECTRUM comprises light emitted at particular wavelengths only (EMISSION LINES), and an absorption spectrum consists of a series of dark ABSORPTION LINES superimposed on a continuous spectrum.

The spectrum of a star normally consists of a continuous spectrum together with dark absorption lines, though in some cases emission lines are also present. Emission line spectra are typical of luminous nebulae. Astronomical spectra extend to all wavelengths, from the gamma-ray domain through X-ray, ultraviolet, optical (visual) and infrared, to radio. High-energy gamma-ray and X-ray spectra are usually plotted against energy (kiloelectron volts, keV, or megaelectron volts, MeV) rather than wavelength. Wavelength in nanometres (1 nm = 10^{-9} m) or angstroms (1 Å = 10^{-8} m) is used for ultraviolet and optical, whereas infrared observations use micrometres (1 μm = 10^{-6} m). Radio spectra, on the other hand, use frequency units (cycles per second, or hertz, kilohertz, megahertz or gigahertz, expressed as Hz, kHz, MHz, GHz respectively).

The term 'spectrum' is also applied to a distribution of particle energies: that is, an energy spectrum is a plot of the numbers of particles with particular energies against the range of possible energies.

spectrum variable Star showing low amplitude light variations, but having a spectrum with ABSORPTION LINES that vary in strength by larger amounts, sometimes in a clearly periodic manner. The class of spectrum variables

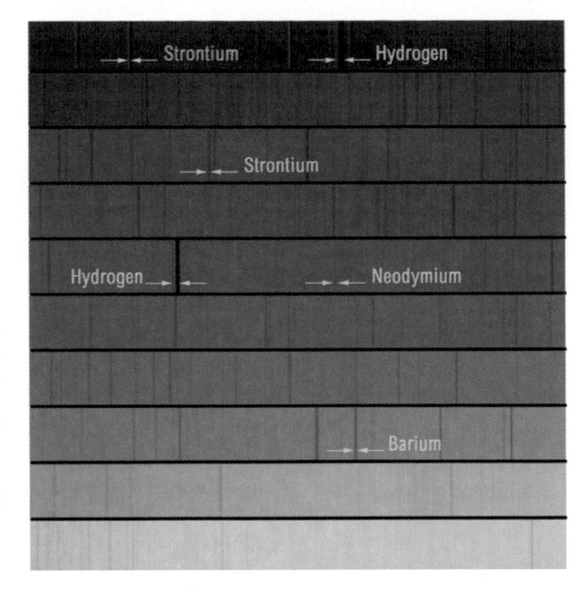

◀ **spectrum** All elements absorb or emit radiation at particular wavelengths. This partial spectrum shows a few of the absorption lines in the the atmosphere of a star.

includes peculiar stars of spectral type A (Ap) and early F (Fp). The brightest and best studied example is the Ap star ALPHA^2 CANUM VENATICORUM. In this star, variations in spectrum, brightness and magnetic field are all cyclical, with a period of 5.46939 days, equivalent to its rotation period. Models derived from observations suggest that spectrum variable stars are objects in which quantities of particular metallic elements are concentrated in patches in the stellar atmosphere; absorption lines due to specific elements are most prominent when these patches are presented in line of sight by the star's rotation. Some spectral type B stars also show such variations.

speculum metal *See* REFLECTING TELESCOPE

spherical aberration Imperfections in an image produced by lenses or mirrors having spherical surfaces. When light parallel to the optical axis is focused by a spherical lens or mirror, the light close to the axis will be focused at a different point from light farther from the axis, thus producing an imperfect image. Spherical aberration is the main ABERRATION affecting simple optical systems; it can be reduced by the use of aspheric optics such as PARABOLOIDS.

spherical astronomy Largely obsolete term describing the mathematical calculations necessary to determine the positions and distances of celestial objects on the celestial sphere. The sky can be thought of as the inside of a sphere surrounding the Earth; to locate positions on the sphere requires the precise measurement of the angular displacement between the object in question and a set of reference points. Distances to nearby objects may be determined using TRIGONOMETRIC PARALLAX. *See also* ANNUAL PARALLAX; ASTROMETRY; CELESTIAL COORDINATES; PARALLAX

spheroid Shape formed by rotating an ELLIPSE about one of its axes. If the major axis is chosen, the spheroid is prolate; if the minor is chosen, it is oblate. A self-gravitating, rotating planetary body is distorted by centrifugal force into an oblate spheroid, with equatorial radius greater than polar. The shape, called a Maclaurin spheroid, depends on the body's density and rotation rate.

Spica The star α Virginis, visual mag. 0.98 (but slightly variable), distance 262 l.y., spectral type B1 V. It is a spectroscopic binary with a period of 4 days that owes its light variation of just under 0.1 mag. not to eclipses but to tidal distortion in the shape of the orbiting components. The name is Latin for 'ear of grain'.

spicules Narrow, predominantly vertical, short-lived jets of solar material, extending from the CHROMOSPHERE into the CORONA. They have a spiky appearance. Spicules are seen in spectrograms or SPECTROHELIOGRAMS, especially those taken in the red wing of the HYDROGEN-ALPHA LINE. Spicules change rapidly, having a lifetime of five to fifteen minutes and velocities of about 25 km/s (16 mi/s). Typically, spicules are about one kilometre thick, and thousands of kilometres long. They are not distributed uniformly on the Sun, being concentrated along the cell boundaries of the SUPERGRANULATION pattern.

spider Thin, usually wire or metal support holding the SECONDARY mirror in the tube of a REFLECTING TELESCOPE. The support may commonly have three or four 'legs', DIFFRACTION effects from which give rise to the cross patterns often seen in astronomical images.

spin casting Technique for manufacturing large parabolic telescope mirrors by spinning molten glass in a rotating furnace and allowing centrifugal forces to produce the required curvature in the surface. Astronomical mirrors are traditionally cast flat, and large quantities of glass are then ground out and discarded to create the correct shape, in a time-consuming and costly process. Spin casting involves heating blocks of borosilicate glass in a honeycomb mirror mould to around 1453 K. As the furnace rotates, the surface of the molten glass assumes a paraboloidal shape. It is then allowed to cool and solidify whilst still spinning.

spiral galaxy GALAXY that has a thin disk of stars, gas and dust, in which a more or less continuous spiral pattern appears. Most spiral galaxies also contain a central spherical bulge of old stars. The spiral pattern may be maintained as a DENSITY WAVE moving through the disk, or by differential rotation shearing star-forming regions into locally tilted segments. Thus spirals range from grand design patterns, with between two and four arms traceable through complete turns around the galaxy, to flocculent galaxies, in which only small, discontinuous pieces of the spiral pattern exist. Most spiral galaxies have at least a weak bar, an elongation of the nuclear bulge, in the plane of the disk. Spirals are classified into various subtypes in the HUBBLE CLASSIFICATION. Most maintain active star formation and have a significant reservoir of interstellar gas to fuel additional generations of stars.

▼ **spiral galaxy** Spiral galaxies vary in shape and structure, but all have a nucleus containing older stars surrounded by a disk where star formation is occurring in spiral arms.

Spitzer, Lyman Jr (1914–97) American astrophysicist who studied stellar dynamics and interstellar matter, and originated the concept of a telescope in Earth orbit. After helping to develop sonar during World War II, he succeeded Henry Norris RUSSELL as head of the astronomy department and observatory director at Princeton University, where he spent the rest of his working life.

From the late 1930s, Spitzer investigated interstellar matter (ISM), establishing it as a major research field. Observing that elliptical galaxies contained old stars and lacked ISM, whereas spirals contained younger stars and were rich in ISM, he concluded that stars formed from condensing clouds of interstellar gas and dust. He studied the effect of heating and cooling and of interstellar magnetic fields on the ISM.

This interest in magnetic fields led Spitzer to another line of enquiry, the possibility of power generation by nuclear fusion, and the founding, with Hannes ALFVÉN and others, of the study of plasma physics. The plasma of which stars are made is held together by the star's gravitation, but to contain plasma in terrestrial laboratories would require powerful magnetic fields. A figure-of-eight chamber constructed for this purpose at Princeton's Plasma Physics Laboratory was the forerunner of today's experimental fusion reactors.

In 1990 Spitzer witnessed the launch of the Hubble Space Telescope. He had proposed an orbiting telescope as early as 1946, since when he had fostered the idea with continual political encouragement and technical advice. He also directed the development of the COPERNICUS orbiting observatory.

spokes Features discovered in Saturn's B RING by VOYAGER. They take the form of radial shadows that are visible projected on the rings when viewed at low angles. They are thought to be the shadows of particles held above the plane of the rings by electrostatic forces. Since the discovery of the spokes, a reanalysis of drawings has shown that shadings have been recorded for over three centuries when conditions have been favourable; these are thought, in part, to represent observations of the spokes.

sporadic meteor Completely random METEOR that does not belong to any recognized METEOR SHOWER, being produced instead by the atmospheric entry of debris from the general dust-cloud filling the inner Solar System. Most sporadics originate from long-defunct meteor streams, which have been dispersed by gravitational perturbations and solar radiation effects. Sporadic activity varies over the course of the year, being highest in the autumn months. Rates also show a diurnal variation, and are highest just before dawn. Depending on the time of night and time of year, between three and 12 sporadic meteors may be seen per hour.

Spörer, Gustav Friedrich Wilhelm (1822–95) German solar astronomer who co-discovered the Sun's differential rotation and first noticed the pattern of sunspot drift called SPÖRER'S LAW. He joined Potsdam Observatory in 1874, working with Johann ENCKE on the motions and distribution of sunspots. Spörer discovered, independently of Richard CARRINGTON, the Sun's differential rotation: spots near the Sun's equator rotate faster than spots nearer its poles. By carefully plotting sunspot positions throughout the approximately 11-year solar cycle, Spörer discovered the patterns of sunspot migration collectively known as Spörer's law. His study (1887) of historical records revealed a scarcity of sunspots between 1645 and 1715, a period now known as the Maunder minimum after E. Walter MAUNDER, who publicized it.

Spörer's Law Appearance of sunspots at lower latitudes over the course of the SOLAR CYCLE. It is named after Gustav SPÖRER, who first studied it in detail. It is also depicted graphically in the BUTTERFLY DIAGRAM. A new cycle starts with sunspots at high latitudes of around 40° north and south of the equator. As the cycle progresses, the spots appear at decreasing latitudes until some occur within 5° north or south of the equator at solar minimum. Before minimum has been reached, high-latitude spots of the new cycle can appear, causing an overlap of the cycles.

spring equinox Alternative name for VERNAL EQUINOX

spring tides High-level high TIDES that occur when the Sun and Moon lie in the same or opposite directions (that is at new moon or full moon).

s process Method of creating heavy stable nuclei within the interiors of stars by successive capture of neutrons. In the s process, neutrons are added slowly to nuclei, which have time to decay before the next neutron is added. It is the process that occurs in the interiors of S-type red giants. *See also* NUCLEOSYNTHESIS; P PROCESS; R PROCESS

Sputnik (Fellow Traveller) Name given to the first ten satellites launched by the Soviet Union. Sputnik 1, launched on 1957 October 4 was the first ARTIFICIAL SATELLITE. It weighed 83.6 kg and orbited the Earth in 96 minutes. Sputnik 2 (launched 1957 November 3) carried the dog Laika, the first living creature in space. Sputnik 3 (1958 May 15) was a highly successful scientific satellite. Sputniks 4, 5 and 6 tested the VOSTOK re-entry capsules, with Sputnik 5 successfully returning the dogs Belka and Strelka from space. Sputniks 7 and 8 were failed Venus missions. Sputniks 9 and 10 were also Vostok test flights with dogs on board.

Square Kilometre Array (SKA) International project to design and ultimately build a centimetre-wavelength radio telescope with an effective collecting area of 1 sq km (0.4 sq mi). It will synthesize an aperture with a diameter of approximately 1000 km (625 mi). Such a telescope would probe the gaseous content of the very early Universe, complementing planned facilities in other wavebands such as the NEXT-GENERATION SPACE TELESCOPE. A consortium representing the United States, Europe, Australia, Canada and Asia is planning to select a site having minimal interference from radio communications by 2005.

Square of Pegasus A large asterism, over 15° across, formed by the stars α, β and γ Pegasi, and α Andromedae; this latter star was in former times regarded as common to Andromeda and Pegasus, when it had the alternative designation δ Pegasi. The Square is high in the sky on northern autumn evenings. Despite its considerable size, there are remarkably few naked-eye stars within it.

SS Cygni star *See* U GEMINORUM STAR

SS433 Very unusual BINARY STAR system that lies at the centre of the supernova remnant W50 in AQUILA. SS433 was discovered in 1976 by the Ariel 5 X-ray satellite. It has strong EMISSION LINES from hydrogen and helium, and it was classified as object 433 in a catalogue of emission-line stars published (1977) by Bruce Stephenson and Nicholas Sanduleak (1933–90) of Case Western Reserve University. The emission lines have broad and narrow components; the broad lines vary slightly with a period of 13.1 days, while the narrow lines vary significantly in wavelength over a period of 164 days.

It is generally accepted that SS433 is a binary system composed of a hot, massive star (of 10 to 20 solar masses) and a NEUTRON STAR (about 1.5 to 3 solar masses), which orbit each other every 13.1 days. The neutron star is probably the stellar remnant from the SUPERNOVA that produced W50. A stellar wind from the hot star produces the 'stationary' emission features. Matter is transferred from the massive star to the neutron star via an ACCRETION DISK. Not all the material reaches the surface of the neutron star, however, and some is ejected at high speed, probably by RADIATION PRESSURE, via two finely collimated jets. The jets sweep around the sky every 164 days, producing the peculiar drifting spectral features that appear to make SS433 unique in astronomy.

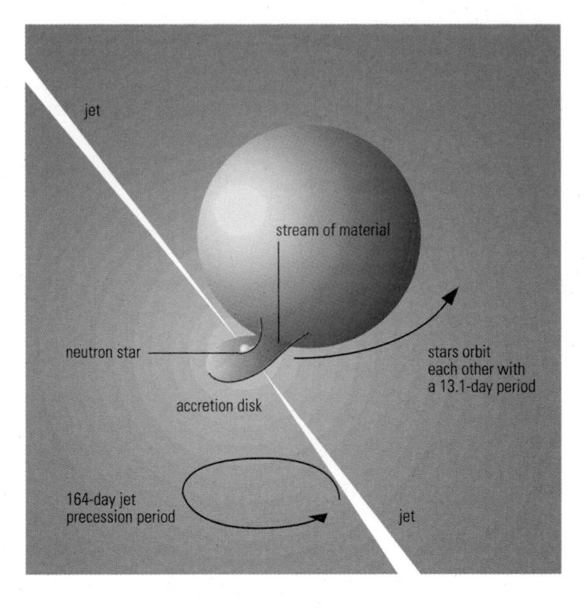

◀ **SS433** The unusual binary pair SS433 is thought to consist of a neutron star and a red giant. Material from the red giant is being accreted by the neutron star and some of it is being expelled in twin jets. The jets are observed to precess in a 164-day period probably because of gravitational interactions between the two stars.

S star Member of a class of GIANT STARS with the same temperature range as M STARS, but with absorption bands of zirconium oxide (ZrO) rather than titanium oxide (TiO). The overabundance of zirconium, as well as carbon and many heavy elements, is a result of convective mixing that brings by-products of nuclear reactions (principally those of hydrogen fusion, helium fusion and slow neutron capture) up to the stellar surfaces.

S stars are the evolutionary intermediaries between M giants and CARBON STARS. As M giants within a particular (but uncertain) mass range dredge carbon from below, and in the process become carbon stars, they pass through the S star state. While the atmospheres of M giants have more oxygen than carbon and carbon stars more carbon than oxygen, S stars have approximately equal amounts of the two. The carbon readily combines with the oxygen. What little free oxygen is left combines more readily with the enhanced zirconium (a slow neutron-capture product) to make zirconium oxide rather than titanium oxide. M stars with weak ZrO bands are assigned the intermediate spectral type MS, while SC stars are intermediate between classic S stars and carbon stars.

S stars often show hydrogen emission lines in their spectra and commonly pulsate as MIRA STARS, the brightest being CHI CYGNI, which can peak at third magnitude.

Stadius Lunar GHOST CRATER (11°N 14°W), 70 km (44 mi) in diameter. It originally appeared as a complex crater. After its formation, however, lunar basalts flooded the region to a depth greater than the height of Stadius' rim. Only a few of the highest rim peaks are visible above the lava. Numerous secondary craters from the EJECTA of COPERNICUS cover the region.

standard epoch Particular date and time chosen as a reference point against which to measure astronomical data in order to remove the effects of PRECESSION, PROPER MOTION and gravitational PERTURBATION. The standard epoch currently in use is called 2000.0 and all celestial coordinates published in star catalogues are quoted as being correct on 2000 January 1. It is customary to change the standard epoch every 50 years.

standard model (1) Mathematical description of the SOLAR INTERIOR, based on a spherical Sun; it specifies the current variation with radius of density, temperature, luminosity and pressure. The Sun is presumed, for this purpose, to have condensed from a cloud of primordial gas composed primarily of hydrogen and helium. Theoretical models use appropriate equations describing nuclear energy generation by hydrogen burning (*see* PROTON–PROTON REACTION) in the central core of the Sun, hydrostatic equilibrium that balances the outward force of gas pressure and the inward force of gravity (that is, the Sun is in a steady state and perfect gas laws can be used), energy transport by radiative diffusion and convection, and an opacity determined from atomic physical calculations. Computers are used to integrate these models over the 4.6-billion year lifetime of the Sun. A standard model is arrived at that best describes the Sun's observed luminosity, size and mass. Theoretical fluxes of SOLAR NEUTRINOS are also calculated from the model. The standard model is consistent with HELIOSEISMOLOGY measurements of the Sun's internal temperatures.

NEAREST STARS

Star	RA	dec.	Visual mag.	Spectral type	Absolute mag.	Distance
	h m	° ′				l.y.
Sun	– –		–26.78	G2 V	4.82	–
Proxima Centauri	14 30	–62 41	11.01v	M5 V	15.45	4.22
α Centauri A	14 40	–60 50	–0.01	G2 V	4.34	4.39
α Centauri B			1.35	K1 V	5.70	4.39
Barnard's Star	17 58	+04 41	9.54	M4 V	13.24	5.94
Wolf 359	10 56	+07 01	13.46v	M6.5 Ve	16.57	7.8
Lalande 21185	11 03	+35 58	7.49	M2 V	10.46	8.31
Sirius A	06 45	–16 43	–1.44	A0 V	1.45	8.60
Sirius B			8.44	DA2	11.34	8.60
UV Ceti A	01 39	–17 56	12.56v	M5.5 Ve	15.42	8.7
UV Ceti B			12.96v	M5.5 Ve	15.81	8.7
Ross 154	18 50	–23 50	10.37	M3.5 Ve	13.00	9.69
Ross 248	23 42	+44 09	12.27	M5.5 Ve	14.77	10.3
ε Eridani	03 33	–09 27	3.72	K2 V	6.18	10.50
HD 217987	23 06	–35 51	7.35	M2 V	9.76	10.73
Ross 128	11 48	+00 48	11.12v	M4.5 V	13.50	10.89
L789–6	22 39	–15 17	12.32	M5 V	14.63	11.2
61 Cygni A	21 07	+38 45	5.20v	K5 V	7.49	11.36
Procyon A	07 39	+05 13	0.40	F5 IV–V	2.68	11.41
Procyon B			10.7	DF	13.00	11.41
61 Cygni B	21 07	+38 45	6.05v	K7 V	8.33	11.42
HD 173740	18 43	+59 38	9.70	K5 V	11.97	11.47
HD 173739	18 43	+59 38	8.94	K5 V	11.18	11.64
GX Andromedae A	00 18	+44 01	8.09v	M1 V	10.33	11.64
GX Andromedae B			11.10	M4 V	13.35	11.64
G51–15	08 30	+26 48	14.81	M6.5 Ve	17.01	11.8
ε Indi	22 03	–56 47	4.69	K5 Ve	6.89	11.83
τ Ceti	01 44	–15 56	3.49	G8 V	5.68	11.90
L372–58	03 36	–44 30	13.03	M4.5 V	15.21	11.9

BRIGHTEST STARS

Star		RA		dec.		Visual	Absolute	Spectral	Distance
		h	m	°	′	mag.	mag.	type	(l.y.)
Sirius	α Canis Marjoris	06	45	–16	43	–1.44	1.45	A0m	8.6
Canopus	α Carinae	06	24	–52	41	–0.62	–5.53	F0	313
Rigil Kent	α Centauri	14	40	–60	50	–0.28	4.07	G2 + K1	4.4
Arcturus	α Boötes	14	16	+19	11	–0.05	–0.3	K2	37
Vega	α Lyrae	18	37	+38	47	0.03	0.6	A0	25
Capella	α Aurigae	05	17	+46	00	0.08	–0.5	G2 + G6	42
Rigel	β Orionis	05	15	–08	12	0.18	–6.69	B8	773
Procyon	α Canis Minoris	07	39	+05	14	0.40	2.68	F5	11.4
Achernar	α Eridani	01	38	–57	14	0.45	–2.8	B3	144
Betelgeuse	α Orionis	05	55	+07	24	0.45(v)	–5.14	M2	427
Hadar	β Centauri	14	04	–60	23	0.61	–5.42	B1	525
Altair	α Aquilae	19	51	+08	52	0.76	2.2	A7	17
Acrux	α Crucis	12	27	–63	06	0.77	–4.19	B0.5 + B1	321
Aldebaran	α Tauri	04	36	+16	31	0.87	–0.6	K5	65
Antares	α Scorpii	16	29	–26	26	0.9–1.2	–5.4––5.1	K0	604
Spica	α Virginis	13	25	–11	10	0.98	–3.5	B1	262
Pollux	β Geminorum	07	45	+28	02	1.16	1.1	K0	34
Fomalhaut	α Piscis Austrini	22	58	–29	37	1.16	2.0	A3 V	25
Deneb	α Cygni	20	41	+45	17	1.25	–8.73	A2	3230
Becrux	β Crucis	12	48	–59	41	1.25	–3.92	B0.5	353
Regulus	α Leonis	10	08	+11	58	1.36	–0.5	B7	77
Adhara	ε Canis Marjoris	06	59	–28	58	1.50	–4.10	B2	431
Castor	α Geminorum	07	35	+31	53	1.58	0.6	A1 + A2	52
Gacrux	γ Crucis	12	31	–57	07	1.59	–0.56	M4	88
Shaula	λ Scorpii	17	34	–37	06	1.62	–5.0	B1.5	703
Bellatrix	γ Orionis	05	25	+06	21	1.64	–2.72	B2	243
Alnath	β Tauri	05	26	+28	36	1.65	–1.4	B7	131
Miaplacidus	β Carinae	09	13	–69	42	1.67	–0.99	A1	111
Alnilam	ε Orionis	05	36	–01	12	1.69	–6.38	B0	1300

standard model (2) Reigning model of ELEMENTARY PARTICLE physics; it successfully combines the weak and strong forces into a consistent quantized theory. The standard model utilizes quarks to build protons, neutrons and other subatomic particles.

standard time Legal time kept in any community. This is likely to be based on its local TIME ZONE, but it may differ slightly for reasons of convenience. Many countries also observe daylight saving time, advancing their clocks by one hour during the summer months. The standard time in the UK is UNIVERSAL TIME (UT), popularly known as GREENWICH MEAN TIME (GMT). In summer, the standard time is UT $+1^{h}$ (British Summer Time).

star Gaseous body that emits radiation generated within itself by nuclear FUSION. Mainly composed of hydrogen and helium, a star is a fine balancing act between the force produced within itself and the gravitational force. Many stars belong to binary or multiple systems.

A star begins its life when the temperature and pressure within a collapsing cloud of interstellar material cause nuclear reactions to start. Below about 0.08 solar mass, conditions do not reach the point where nuclear fusion can occur (*see* BROWN DWARF). Above about 150 solar masses, a star becomes unstable. The mass of a star has a great influence on its evolution (*see* STELLAR MASS). A star's evolution is dictated throughout its life by the balancing act between gravitation and the energy produced by the nuclear reactions within the star. As a star evolves, its chemical composition changes. Stars belonging to CLOSE BINARY systems often have their evolution altered by interactions with their companion.

The composition of stars is initially mainly hydrogen and helium, with a proportion of heavier elements, depending on the material from which it is forming. Stars produce energy within their core by nuclear fusion (*see* STELLAR INTERIORS). The energy released has to force its way out through opaque gas, and at every point within the star the gravitational pull of the parts within exactly balances the outward thrust and the star is in hydrostatic equilibrium (*see* STELLAR ATMOSPHERE).

The internal structure of a star cannot be observed directly. The equations of stellar structure, on which computer models of the conditions within stars depend, are based on known laws of physics. Physical data such as mass, luminosity, radius, chemical composition and EFFECTIVE TEMPERATURE can be measured for some stars, and the observed stellar conditions are satisfactorily predicted by the equations of stellar structure.

Stars can be broadly classified by their mass and stage of evolution and also by their effective temperature. A star's SPECTRAL TYPE is an indication of its temperature, but it also gives information about its chemical composition. Stars are also classified by the percentage of heavy elements in their composition (*see* POPULATIONS, STELLAR). The percentage of heavy elements affects a star's evolution.

Many stars are observed to fluctuate in brightness (*see* VARIABLE STARS). These fluctuations can be due to a star belonging to an ECLIPSING BINARY system or a close binary system, or they may be a result of instabilities caused by its evolutionary phase (*see* T TAURI STARS, MIRA STARS).

starburst galaxy GALAXY that is undergoing a strong, temporary increase in its rate of star formation. Such activity may span the whole galaxy, but is often confined to the nucleus. Starbursts may appear as ultraviolet-bright galaxies or may be identified by strong optical emission lines or excess radio emission, but the largest number to date have been found from far-infrared

◀ **starburst galaxy** The bright pink areas in this irregular galaxy are areas of starbirth and the blue areas are newly emerged young stars. Many starburst galaxies are triggered by gravitational interactions between galaxies.

emission (as by the Infrared Astronomical Satellite). This is because many starbursts are quite dusty, and when the dust absorbs the ultraviolet and visible light from the massive stars, it radiates in the far-infrared (typically at 0.03–0.1 mm). Starbursts can be triggered by tidal interactions and mergers between galaxies, and the energy output of the massive stars and supernovae can be strong enough to drive a wind sweeping gas completely out of the galaxy (and terminating the starburst). Starburst galaxies are useful guides for the conditions expected in genuinely young galaxies in the early Universe, when star formation was more intense than is usually found in present-day galaxies.

star clouds Parts of the MILKY WAY, especially in Sagittarius and its surrounding constellations, where the light from millions of faint stars combines to give the appearance of irregular faintly glowing clouds. The star distribution is actually fairly uniform and the irregularities result from the stars' obscuration in some regions by DARK NEBULAE such as the COALSACK and the PIPE NEBULA.

star cluster Group of stars that are physically associated. *See* ASSOCIATION, STELLAR; EMBEDDED CLUSTER; GLOBULAR CLUSTER; MOVING GROUP; OB ASSOCIATION; OPEN CLUSTER; T ASSOCIATION

star diagonal Flat mirror or prism enabling an object to be viewed at right angles to the direction in which a telescope is pointing. The image is reversed left-to-right, but is the correct way up. Star diagonals are popular with users of REFRACTING TELESCOPES or SCHMIDT–CASSEGRAIN TELESCOPES, in which viewing of objects at high altitudes might otherwise prove uncomfortable.

Stardust NASA Discovery programme spacecraft, launched in 1999 to rendezvous with the comet Wild 2 in 2004 January and to return to Earth in 2006 January with samples of dust from the comet's coma. Stardust, which will also return with samples of INTERPLANETARY DUST, is equipped with two collectors, on either side of a mast, consisting of ultra-low-density silica aerogel material. One hundred particles of interplanetary dust between 0.1 and 1 μm and 1000 particles of comet dust larger than 15 μm will be returned to Earth. The collector will be deposited into a capsule, which will enter the Earth's atmosphere as Stardust makes a return to its home planet, to be recovered in Utah. The samples will be analysed at the Planetary Material Curatorial Facility at the NASA Johnson Space Center in Houston, Texas.

Star of Bethlehem Celestial portent of the birth of Jesus Christ, mentioned in the biblical book of Matthew. The precise nature of the Star of Bethlehem has been of great interest to Christians and astronomers alike. Many books and articles have been written on the subject, some authors regarding it as a mythical or miraculous event and others supposing that it was a real astronomical phenomenon that ordinary mortals could have seen.

The first problem encountered when trying to relate the star to a real astronomical event is that the precise year of the nativity is uncertain, though it is believed to have been between 7 BC and 1 BC. A census of allegiance to Augustus Caesar, which may have been the census mentioned in the Bible, took place in 3 BC. Numerous astronomical events such as planetary conjunctions and conjunctions of planets with stars occurred during those years; several comets and a nova were also recorded. Although there is no general agreement on the precise events that guided the Magi on their path to Bethlehem, one theory is that successive conjunctions of Jupiter and Venus in 3 BC August and 2 BC June were the 'stars' that led them westwards. Around 2 BC December 25, the planet Jupiter seemed to stand still among the stars because of its change from (apparent) retrograde to prograde motion. Seen from Jerusalem, it would have appeared to be above Bethlehem.

star streaming Grouping of the motions of individual stars, relative to their neighbours, around two opposite preferred directions in space. The effect was discovered by Jacobus KAPTEYN, who made statistical studies showing that, in general, the peculiar motions of stars (that is relative to the average motion of neighbouring stars) appear to be in one of two directions in space, rather than moving in random directions. The phenomenon is caused by the rotation of the Galaxy and many later studies of distances and spatial arrangements of the stars in the Milky Way arose from this discovery.

star trailing Extension of the images of stars into arcs caused by the rotation of the Earth, and hence the movement of stars across the sky, on a photograph taken with a stationary camera or telescope. Images of stars near either celestial pole form tight arcs, while at the equator star trails

▼ **star trailing** Because of the Earth's rotation, the light from stars will form trails on unguided astronomical images. This image is of star trails around the southern celestial pole.

S

are straight lines. Lengthening the exposure time produces longer trails. Trailing is avoided by driving the camera to track the stars (*see* TELESCOPE DRIVE).

stationary point Point at which the motion of a SUPERIOR PLANET, as observed from Earth, appears to change direction relative to the background stars due to the Earth catching it up in its orbit and overtaking it. *See also* RETROGRADE MOTION

steady-state theory Theory, originated by Hermann BONDI, Thomas GOLD, Fred HOYLE, Jayant NARLIKAR and others, designed to satisfy the PERFECT COSMOLOGICAL PRINCIPLE, which states that the local properties of the Universe – when averaged over some suitable distance – are the same from whatever point in space and time they are determined. This principle is consistent with the model derived by Willem DE SITTER from Einstein's field equations. The steady-state universe is expanding, but a constant density of matter is maintained by invoking CONTINUOUS CREATION of matter at all places and all times by a so-called C-field. It avoids the problem, associated with the BIG BANG THEORY, of an initial SINGULARITY – a fixed starting point and, apparently, some instrument of creation.

For nearly two decades, steady state and Big Bang were rival theories. There were two main reasons for the demise of the steady-state theory: the discovery of the COSMIC MICROWAVE BACKGROUND in 1965, and the realization that galaxies have evolved over time, neither of which it could readily explain. Its lasting importance was the stimulus it provided to astrophysics and cosmology, leading, for example, to the first detailed work on NUCLEOSYNTHESIS. Ironically, the steady-state universe strongly resembles some versions of INFLATION, the model proposed to account for the initial rapid expansion that followed the Big Bang.

Stebbins, Joel (1878–1966) American astronomer who pioneered photoelectric photometry. Stebbins spent most of his career (1918–48) at the University of Wisconsin's Washburn Observatory. In 1910 he attached a primitive photocell to a telescope and discovered the secondary minimum in the light curve of the eclipsing binary star Algol. Stebbins devised methods for reducing the huge amounts of data generated by his photoelectric instruments, which he used to study 'regular' stars, various types of variable stars, the solar corona, globular clusters and galaxies. With Albert Edward Whitford (1905–?1983) he developed increasingly sensitive photoelectric cells that eventually supplanted photography as a means of measuring star magnitudes. The superiority of the photoelectric method (photographic emulsions are usually overly sensitive to either blue or red light) allowed Stebbins and his colleagues to quantify the 'reddening' effects of interstellar matter and to adjust stellar magnitudes accordingly.

Stefan–Boltzmann constant (symbol σ) Constant relating the energy emitted by a BLACK BODY to its temperature (*see* STEFAN–BOLTZMANN LAW). It has the value 5.6697×10^{-8} W m^{-2} K^{-4}.

Stefan–Boltzmann law Law giving the total radiant energy (flux) emitted by a BLACK BODY (*see also* RADIATION). It was formulated by the Austrian physicist Joseph Stefan (1853–93) and by Ludwig BOLTZMANN. It takes the form $F = \sigma\mu^2T^4$, where F is the flux, σ is the STEFAN–BOLTZMANN CONSTANT, μ is the refractive index of the medium and T is the temperature.

stellar association *See* ASSOCIATION, STELLAR

stellar atmosphere Atmosphere of a STAR; it consists of the layers from which radiation is directly observed. Stellar atmospheres comprise plasmas composed of many types of particles, including atoms, ions, free electrons, molecules and sometimes dust grains. The inner boundary of a stellar atmosphere is with the stellar interior, about which no direct information is available, while the outer boundary is taken to be where it merges with the interstellar medium.

Although the atmosphere of a star can extend a long way out into space, the majority of radiation emitted from it comes from a relatively thin layer called the PHOTOSPHERE. Generally, a stellar photosphere is about one thousandth of the star's radius, although hotter stars have relatively thicker photospheres. The photosphere is the innermost part of a stellar atmosphere.

The outer layer of a stellar atmosphere is the STELLAR WIND, a region where the outflow velocities are comparable to or larger than the local speed of sound. In a few stars there is a further, more remote layer: some very young stars may still be enveloped by the material from which they formed.

The structure of the atmosphere above the photosphere differs markedly between cool and hot stars. Cool stars, defined as stars with an EFFECTIVE TEMPERATURE of less than about 7500 K, have been observed to have a CHROMOSPHERE and CORONA similar to that of the Sun. The atmospheres of cool stars are very complex and are characterized by increasing temperature with altitude in their outer layers. The mechanism causing this temperature increase is not fully understood, but heat conduction becomes an important process in regions of low density such as in stellar coronae. Energy may also be transported by MAGNETOHYDRODYNAMIC or acoustic waves, which are responsible for heating the corona non-radiatively. These waves are generated by strong, geometrically complex magnetic fields, which in turn are created by a dynamo effect in the convection zone beneath the photosphere. Coronal heating is believed to be the source of the X-rays detected from these stars.

The high coronal temperature in cool MAIN-SEQUENCE stars results in a large gas pressure, which induces a weak pressure-driven stellar wind. Fast, but very tenuous, this wind is so optically thin throughout that it cannot be detected spectroscopically; it is often not included in the atmosphere of such stars. In evolved cool stars, however, a much denser wind is observed. These stars have low surface gravity so the mass-loss rate of a coronal wind is larger. In addition, other mechanisms, such as RADIATION PRESSURE on dust grains formed in the outer layers or acceleration in pulsating layers, become efficient drivers of the stellar wind.

Strong, fast winds typify the atmospheres of early-type main-sequence stars (O and early B-type) and O, B and A-type SUPERGIANTS. These fast, supersonic winds are driven by radiation pressure generated by absorption of photons. The X-ray emission detected in O and early B-type stars is attributed to shock-heated material in the wind and not to the corona. The winds of red supergiants are slow.

stellar diameters Stars range in size from having diameters of a few tens of kilometres (NEUTRON STARS) to several hundreds of millions of kilometres (SUPERGIANTS). Generally the sizes of stars are expressed in terms of the solar radius ($R_\odot$).

The diameter or radius of a star, along with its mass, LUMINOSITY and EFFECTIVE TEMPERATURE, helps to determine its position on the HERTZSPRUNG–RUSSELL DIAGRAM. When a star's radius is very large or very small, its effect on the luminosity can dominate over the effect due to temperature. A star's diameter can be measured directly if its angular diameter and distance from Earth can be measured. Measuring the angular diameters of stars directly, however, is difficult because of the large distances involved. The diameters of some stars can been measured when they are occulted by the Moon.

Optical INTERFEROMETRY can also be used, and as the resolution of interferometers improves, the diameters of smaller stars can be measured. Another technique used is SPECKLE INTERFEROMETRY. The diameters of some stars can be obtained indirectly, for example if they are members of ECLIPSING BINARY systems. If the effective temperature (T_{eff}) can be measured from its spectrum, and a value of its luminosity (L) determined from its apparent brightness and distance, then a value of the star's radius (R) can be inferred from $L = 4\pi R^2 \sigma T^2_{eff}$, where σ is the STEFAN–BOLTZMANN CONSTANT.

Some stellar diameters have been obtained when a star is involved in MICROLENSING events.

stellar evolution Development of a STAR over its lifetime. The theory of stellar evolution revolves around mathematical models of stellar interiors and current laws of physics. It gives a fairly detailed account of the evolution of any particular star, and it can be checked by observation of stars at each of the predicted phases. The evolutionary stages of stars are represented on the HERTZSPRUNG–RUSSELL DIAGRAM (HR diagram). Timescales for evolution lie between hundreds of thousands, and thousands of millions of years; hence the evolution of a particular star cannot be directly observed, although very occasionally changes in a star over the course of a human lifetime can be observed as it passes through a rapid phase, as with FG SAGITTAE. The evolution of a star in a close binary system can be affected when mass transfer occurs from its companion, because changing the star's mass alters conditions in its interior.

Stars form from a dense interstellar cloud known as a GIANT MOLECULAR CLOUD. A large cloud is able to collapse under its own gravity and will subsequently fragment into several hundred smaller clouds, each of which contracts further and heats up to become a PROTOSTAR. Protostars are very red and can be observed at infrared wavelengths; they are found in regions of our Galaxy where there is an abundance of gas and dust. The progress of a PRE-MAIN-SEQUENCE STAR on the HR diagram is represented by the HAYASHI and HENYEY tracks.

A protostar will continue to collapse under gravity until its centre is hot enough (about 15 million degrees K) for nuclear FUSION reactions to commence. As it collapses, its outer layers fall in to take up a normal star-like form and it becomes bluer in colour. T TAURI STARS are low-mass stars in this final phase.

Once nuclear fusion has commenced in the core of a star, the star adopts a stable structure, with its gravity being balanced by the heat from its centre (the photons created in the nuclear fusion reactions diffuse outwards, exerting RADIATION PRESSURE). The controlled nuclear fusion reaction is the fusion of hydrogen atom nuclei (protons) to form helium nuclei. This period is known as the MAIN-SEQUENCE phase of a star's life; it is the longest phase in the life of a star, and consequently at any particular moment most stars are in the main-sequence period of their existence. The Sun has a main-sequence lifetime of some ten thousand million years, of which only a half has expired. Less massive stars are redder and fainter than the Sun and stay on the main sequence longer; more massive stars are bluer and brighter and have a shorter lifetime. As a star ages, the composition of its core changes from mainly hydrogen to mainly helium, causing it to become very slightly brighter and bluer; on the main sequence its position moves slightly upwards and to the right of its zero-age position (*see* ZERO-AGE MAIN SEQUENCE).

The internal structure of a main-sequence star depends on its mass. Massive stars have a convective core and a radiative mantle (*see* CONVECTION and RADIATIVE TRANSFER). Low-mass stars, including the Sun, have a radiative core and a convective mantle. The stirring that occurs in the core of a massive star causes it to have a uniform composition within that core, which affects the details of the next phase of evolution.

Eventually, the hydrogen supply in the core of the star runs out. With the central energy source removed, the core will collapse under gravity, and heat itself up further, until hydrogen fusion is able to take place in a spherical shell surrounding the core (hydrogen SHELL BURNING). As this change occurs, the outer layers of the star expand considerably, and the star becomes a RED GIANT, or in the case of the most massive stars, a red SUPERGIANT. The speed of evolution to the red giant phase depends on the mass. A low-mass star changes gradually to become a red giant as the hydrogen exhaustion spreads outward from its centre. A high-mass star evolves quickly to become a red giant because its whole convective core runs out of hydrogen at the same time. Giant stars are sufficiently cool to have a spectral type of K or M.

While it is a red giant, the star's central temperature will reach 100 million K, and the fusion of helium to carbon will commence in the core. In the case of low-mass stars, the onset of helium fusion is sudden (the HELIUM FLASH), and the star reduces its radius to become a bluer but fainter HORIZONTAL BRANCH STAR as observed in GLOBULAR CLUSTERS; it may subsequently return to being a red giant. In high-mass stars, the onset of helium fusion occurs more gradually, and the star remains a red giant.

The evolution of a star beyond the helium fusion phase is more difficult to calculate with certainty. When the helium fuel in the core has been converted into carbon, the core of the star will again collapse and heat up. In the case of a low-mass star, the central temperature will not rise high enough to initiate carbon fusion, and the outer layers of the star will contract, then gradually cool down, so that the star becomes a WHITE DWARF. It is likely that before a white dwarf cools appreciably, it throws off its outer layers in the form of a PLANETARY NEBULA. White dwarf stars are plentiful, but they are difficult to observe because of their faintness. It is believed that white dwarfs cool forever, becoming BLACK DWARFS – mere cinders of stars.

In the case of a high-mass star, the contraction of the carbon core will lead to further episodes of nuclear fusion, during which the star will remain very bright. Many of these stars fluctuate in brightness as MIRA STARS. At this stage the internal chemistry can change, or gas of a different chemical mix deep within can rise to the surface by convection. These stars may then appear as CARBON STARS (C stars). Most of these objects are extremely red and thus rather faint.

Once a star reaches the point where no more nuclear fusion can occur, it will collapse to become a white dwarf, unless the core exceeds the CHANDRASEKHAR LIMIT of 1.4 solar masses, in which case further contraction occurs, producing a NEUTRON STAR. When the core is composed of elements close to iron in the periodic table, no further fusion is possible. At this point, it is most likely that the core will collapse explosively, and the star will throw off its outer layers in a SUPERNOVA explosion, becoming for a few weeks bright enough to outshine the galaxy in which it is situated. The dead core of the massive star will remain as a neutron star or, if the core is massive enough, it will collapse further to become a stellar BLACK HOLE.

stellar interferometer OPTICAL INTERFEROMETER used to measure the angular diameter of stars or the angular separation of the components of double stars. An interferometer operates by splitting and re-combining a beam of light using mirrors in order to examine the INTERFERENCE fringes produced by the re-combined beam. Because a star image is a disk, not a point source, it is possible to adjust the separation of the two mirrors, d, until the fringes disappear, bright fringes from one side of the stellar disk coinciding with dark fringes from the other. At this point, $d = 1.22\ \lambda/\theta$, where λ is the wavelength of light and θ the angle subtended by the stellar disk.

stellar interior Internal structure of a STAR. Knowledge of stellar interiors is gained by observing external properties and formulating a stellar model using known laws of physics to explain the properties observed. The equations that are used to construct models of stellar interiors are known as the equations of stellar structure.

The determination of the density, temperature, composition and luminosity at any point within a star requires the solutions to the conservation laws of energy, mass and momentum at that point. Different conditions exist in stars of different mass and at different stages of evolution, the main differences being the form of energy transport and the composition.

CONVECTION and RADIATIVE TRANSFER are the principal means of energy transport in stellar interiors. For MAIN-SEQUENCE stars below about 1.5 solar masses, the temperature gradient within the core is too low for convection to occur. Energy transfer within these stars is primarily by radiative transfer, with convection occurring in shallow regions near the surface known as convective shells. More massive main-sequence stars have convective cores.

Most stars at a stable point in their evolution are assumed to be in hydrostatic equilibrium, whereby the forces at any point within a star balance and the internal pressures counteract the gravitational pull. For main-sequence stars the internal pressure is a combination of gas and radiation pressure. For evolved stars like WHITE DWARFS and NEUTRON STARS the internal pressure is produced by the DEGENERATE MATTER in their cores.

stellar mass The mass of a STAR is a very important factor in its evolution and determines its basic structure and lifetime. Stellar masses are generally expressed in terms of the Sun's mass ($M_{\odot}$).

Stellar masses range from 0.08 solar mass to about 150 solar masses. Below 0.08 solar mass, temperatures at the core do not become sufficiently high for nuclear FUSION to begin. Stellar-like objects with mass less than 0.08 solar mass are called BROWN DWARFS.

Upper theoretical limitations suggest that stars with masses above about 150 solar masses are unstable because of the effect of RADIATION PRESSURE. The most massive MAIN-SEQUENCE stars observed are within the R136 cluster in the 30 Doradus Nebula of the LARGE MAGELLANIC CLOUD. These stars are around 155 solar masses; they show signs of high mass loss similar to that observed in WOLF–RAYET STARS. Stellar masses can be determined directly from BINARY SYSTEMS. They can also be determined indirectly, using stellar models, if a star's LUMINOSITY, EFFECTIVE TEMPERATURE and distance are known.

According to the VOGT–RUSSELL THEOREM, a star's properties and evolution are determined by its initial mass and chemical composition. Since the composition of stars varies relatively little, it is the initial mass that dictates basic structure and evolution. Higher-mass stars have shorter lifespans because the higher temperatures at their cores enable them to burn their fuel quicker and thus evolve faster. *See also* INITIAL MASS FUNCTION; MASS–LUMINOSITY RELATION; STELLAR EVOLUTION

▼ **stellar wind** The Bubble Nebula is caused by the strong stellar wind from the star at the bottom of the image. When young stars undergo a phase called the T-Tauri phase their extremely strong stellar winds blow away much of the nebula in which they were formed.

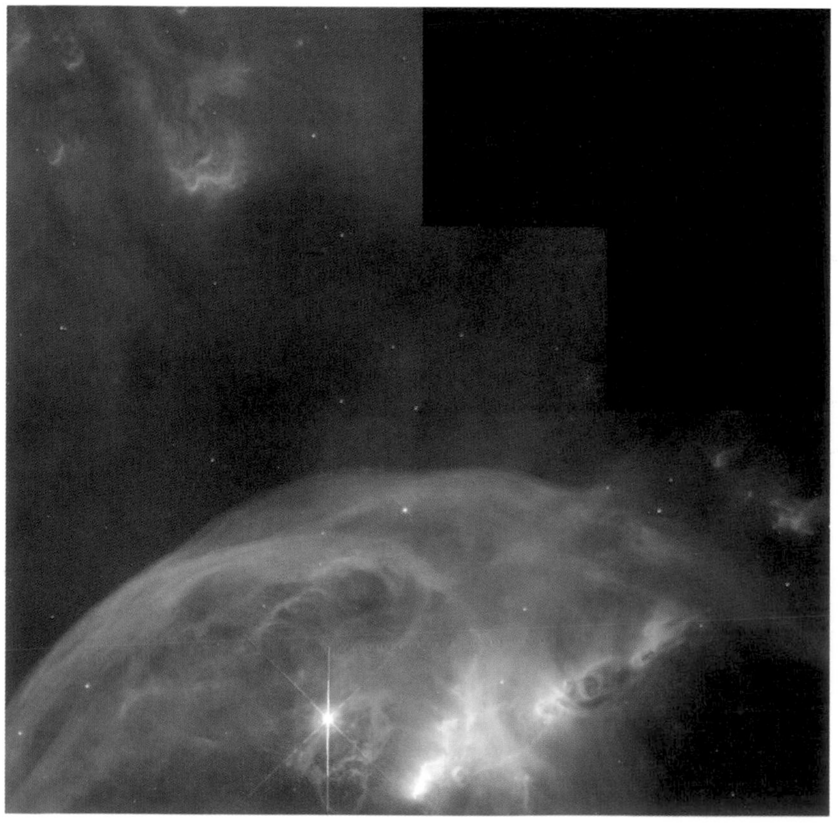

stellar nomenclature System for assigning designations to stars. The brightest stars have proper names originally given to them by Greek, Roman or Arab astronomers, for example Sirius, Capella and Aldebaran. In addition to any proper name, stars down to around 5th magnitude within each constellation are identified by a letter or number followed by the name of the constellation in the genitive (possessive) form (*see* constellations table, pages 94–95). The brightest stars are indicated by Greek letters, as in α (Alpha) Lyrae or ε (Epsilon) Eridani. These letters are known as **Bayer letters** because they were assigned by Johann BAYER in his star atlas *Uranometria* of 1603, usually in order of brightness. Stars that lack Bayer letters are prefixed by a number, as in 61 Cygni. These are known as **Flamsteed numbers** because they were assigned to stars charted in John FLAMSTEED's *Historia coelestis britannica*, in order of right ascension (RA). Capital and lower-case roman letters are also used, particularly for stars in the southern hemisphere. VARIABLE STARS have their own system of nomenclature.

Fainter stars (and deep-sky objects) are identified according to their listing in any of a number of other catalogues, by the catalogue's abbreviation followed by its designation in the catalogue. The designation scheme can be a single sequence of numbers (often in order of RA), as in HD 74156 or NGC 6543, or the numbering can be based on zones, as in BD +31 643. Modern designations tend to use truncated coordinates: for example, PSR 0531+21 is the designation of a pulsar having its 1950.0 coordinates RA $05^h\ 31^m$, dec. +21°; while in PSR J0534+2200, the 'J' indicates truncated 2000.0 coordinates. A list of the various designation letters is in a *Dictionary of Nomenclature of Celestial Objects* maintained under the auspices of the International Astronomical Union (IAU).

There are commercial organizations from which star names can be bought. Although this is not illegal, and individuals might like to name a star after a loved one, such star names have absolutely no status and are not recognized by the IAU.

stellar wind Stream of charged particles, mostly PROTONS and ELECTRONS, that is continually emitted from the surface of a STAR, including the Sun (*see* SOLAR WIND).

Young stars evolving towards the MAIN SEQUENCE have powerful stellar winds, sometimes up to a thousand times stronger than the SOLAR WIND. These winds crash into the surrounding gas clouds and ionize them, producing expanding shock waves in the interstellar medium. While on the main sequence, high-mass stars lose a significant fraction of their mass to the interstellar medium. A star of around 120 solar masses at birth could lose as much as 50 solar masses of material during its main-sequence lifetime. A star of around 60 solar masses at birth could lose around 12 solar masses. Old stars evolving off the main sequence to become RED GIANTS also have strong stellar winds.

Stephano One of the several small outer satellites of URANUS; it was discovered in 1999 by J.J. Kavelaars and others. Stephano is about 20 km (12 mi) in size. It takes 674 days to circuit the planet, at an average distance of 7.98 million km (4.96 million mi). It has a RETROGRADE orbit (inclination near 144°) with a moderate eccentricity (0.228). *See also* CALIBAN

Stephan's Quintet (NGC 7317, NGC 7318 A and B, NGC 7319, NGC 7320) Compact group of galaxies in the constellation Pegasus (RA $22^h\ 36^m.0$ dec. +33°58′); the group was discovered in 1877 by the French astronomer Édouard Stephan (1837–1923). All the constituent galaxies are about 13th magnitude and appear within an area of sky

S

only 3′.5 across. Four of the galaxies lie at a distance of 270 million l.y., and two – NGC 7318 A and B – are interacting with each other. The fifth galaxy, NGC 7320, is very much closer; it is not a true member of the group but simply a relatively nearby galaxy in the line of sight.

step method *See* ARGELANDER STEP METHOD; POGSON STEP METHOD

stereo comparator Type of COMPARATOR that enables two photographs of the same area of sky, taken at different times, to be viewed simultaneously to reveal objects that have changed position or brightness. The instrument has two optical paths so that both photographs can be viewed together with binocular vision. Discordant images then appear to stand out from the plane of the picture.

Steward Observatory Major US astronomical institution operated by University of Arizona, Tucson. It incorporates the University's Department of Astronomy. During the 1980s, the observatory pioneered novel techniques for manufacturing large telescope mirrors, and its Mirror Laboratory has now produced optics for some of the world's biggest telescopes. Today, the instruments operated with the participation of the Steward Observatory include the 6.5-m (256-in.) telescope of the MMT OBSERVATORY, the two 6.5-m MAGELLAN TELESCOPES and the 10-m (33-ft) HEINRICH HERTZ TELESCOPE. The twin 8.4-m (331-in.) instruments that comprise the LARGE BINOCULAR TELESCOPE are due to become operational in 2004. The Steward Observatory also runs the 2.3-m (90-in.) Bok Telescope at KITT PEAK NATIONAL OBSERVATORY.

Stingray Nebula (Hen-1357) Youngest-known PLANETARY NEBULA, with estimated age 200 years. Located in the southern constellation of Ara (RA 17^h $16^m.4$ dec. $-59°29'$), the Stingray is 18,000 l.y. away and has an apparent diameter of less than one arcsecond. Its actual diameter is about 130 times that of the Solar System.

Stjerneborg Second observatory constructed by Tycho BRAHE in about1584; the name means 'Castle of the Stars'. Its observing chambers were mainly below ground level to provide stability and were covered by rotating roofs equipped with openable shutters. *See also* URANIBORG

Stockholm Observatory Observatory founded by the Royal Swedish Academy of Sciences in Stockholm in 1748. In 1931 it moved to Saltsjöbaden, 20 km (12 mi) south-east of Stockholm. New instruments were added, including a 1.0-m (39-in.) reflector. In 2001, with its scientists using overseas facilities such as the EUROPEAN SOUTHERN OBSERVATORY and the NORDIC OPTICAL TELESCOPE, the observatory moved again to become part of the Stockholm Centre for Physics, Astronomy and Biotechnology near the city centre. The telescopes remain at Saltsjöbaden, and are used for public outreach.

Stöfler Lunar crater (41°S 6°E), 145 km (90 mi) in diameter. It is an ancient crater, as can be seen by its walls, which are deeply incised from later impacts, and its rim, which has been rounded by impact erosion. It has central peaks. Stöfler sits astride several even older crater units. Under high Sun illumination, bright rays from TYCHO are visible on its floor.

stony-iron meteorite METEORITE with approximately equal proportions of silicate minerals and iron–nickel metal. Stony-irons are subdivided into two big groups, MESOSIDERITES and PALLASITES, which have very different origins and histories.

stony meteorites METEORITES that are made from the same elements as terrestrial rocks, dominantly silicon, oxygen, iron, magnesium, calcium and aluminium.

Like terrestrial rocks, stony meteorites are assemblages of minerals, such as pyroxene, olivine and plagioclase, but unlike terrestrial rocks they also contain metal and sulphides. Stony meteorites are subdivided into two big groups, CHONDRITES and ACHONDRITES. The former have not melted since aggregation of their original components, and thus they are unfractionated with respect to the Sun. The latter formed from melts on their various parent bodies.

◀ **Stephan's Quintet** The gravitational interactions between the member galaxies of Stephan's Quintet have spawned massive star formation. The stars in the bright regions are between 2 million and 1 billion years old.

Strasbourg Astronomical Observatory A laboratory of the Université Louis Pasteur at Strasbourg, France. Besides teaching and research activities, it hosts the Strasbourg Astronomical Data Centre (CENTRE DE DONNÉES ASTRONOMIQUES DE STRASBOURG).

stratosphere Layer in the Earth's ATMOSPHERE extending above the tropopause (the upper limit of the TROPOSPHERE) as far as the MESOSPHERE. There is an atmospheric temperature minimum at the tropopause, above which the temperature at first remains steady with increasing height. Above 20 km (12 mi), it begins to increase, reaching a a maximum of about 273 K at an altitude of 50 km (31 mi), which marks the stratopause, the top of the stratosphere. The heating is primarily through the absorption of solar ultraviolet radiation by ozone. The ozone layer (sometimes called the ozonosphere) lies at an altitude of between about 15 and 50 km (9 and 31 mi) and is essentially identical with the stratosphere. The highest concentration of ozone occurs at a height of about 20–25 km (12–16 mi).

Stratospheric Observatory for Infrared Astronomy (SOFIA) Boeing 747 aircraft modified to accommodate a 2.5-m (100-in.) reflecting telescope – the largest airborne telescope in the world – for INFRARED ASTRONOMY. The aircraft is being modified by a US–German team and will fly in 2002 from its base at NASA's Ames Research Center in California.

strewnfield Area over which material from a specific impact event is scattered. It is most often applied to TEKTITES and to METEORITE shower falls. Strewnfields are usually elliptical in shape and indicate the direction from which the meteorite came. Following atmospheric disruption, larger meteorites are slowed down by frictional heating less than smaller ones, and so are carried further distances; smaller meteorites are deposited at the incoming direction of the strewnfield, larger ones at the far end.

Strömgren, Bengt George Daniel (1908–87) Danish astrophysicist, born in Sweden, who explained the nature of ionized hydrogen (HII) gas clouds that surround hot stars. He succeeded his father, **Svante Elis Strömgren** (1870–1947), as director of Copenhagen Observatory (1940), and later worked at Yerkes and McDonald Observatories and Princeton University's Institute for Advanced Study. Strömgren became particularly interested in the astrophysics of the Orion Nebula and Milky Way's other HII regions. He developed theoretical models of what were later called STRÖMGREN SPHERES, which he used to estimate the size and density of a hydrogen cloud by measuring the luminosity of its exciting star. These models allowed Strömgren and others, especially William W. MORGAN, to map the spiral structure of the Milky Way.

Strömgren sphere Volume of ionized gas around a hot star (*see also* HII REGION). The region will only be spherical when the interstellar gas in which it forms is of uniform density. A Strömgren sphere has a definite size that is reached when the number of ultraviolet photons emitted by the star exactly balances the number of ionizations occurring in the gas. The radii of Strömgren spheres were first calculated by Bengt STRÖMGREN; they range from 1 to 100 l.y. for O5 stars (within high and low density HI REGIONS respectively), and from 0.1 to 15 l.y. for B0 stars.

Struve family German–Russian family that produced astronomers of note across four generations. As a youth, (Friedrich Georg) Wilhelm von STRUVE (1793–1864) left Germany for Latvia, later moving to Russia to establish PULKOVO OBSERVATORY. One of his sons was **Otto (Wilhelm) Struve** (1819–1905), known to historians as Otto I Struve. Like his father, whom he succeeded as director at Pulkovo in 1862, Otto studied double stars, discovering around 500 of them. He determined an accurate value for the rate of precession, taking into account the motion of the Sun with respect to nearby stars, which he measured. **(Karl) Hermann Struve** (1854–1920), who became director of Berlin Observatory (1904), and **(Gustav Wilhelm) Ludwig Struve** (1858–1920), director of Kharkhov Observatory from 1894, were sons of Otto, while **Georg Struve** (1886–1933) was a son of Hermann; they all studied the Solar System. Otto Ludwig STRUVE (1897–1963), a son of Ludwig, emigrated to the USA in 1921 and became a naturalized American. He is known to historians as Otto II Struve.

Struve, (Friedrich Georg) Wilhelm von (1793–1864) German astronomer, the first major observer and cataloguer of double stars, and one of the first to measure a stellar parallax. He founded the most distinguished of astronomical dynasties – the STRUVE FAMILY. In 1808 he left his homeland to escape conscription into the occupying Napoleonic army, settling in Dorpat (modern Tartu, in Estonia), becoming director of Dorpat Observatory in 1817. From 1835 he helped to establish PULKOVO OBSERVATORY, which he directed until his retirement in 1862.

In 1824 Struve commenced an extensive survey of double and multiple stars with the 9½-inch (240-mm) *f*/18 Dorpat Refractor, made by Joseph FRAUNHOFER and equipped with a clock drive and a filar micrometer – the largest and best telescope of its day. His measurements of position angles and separations, comparable with modern ones in their precision, were published in a catalogue (1827) and two later supplements, which between them listed 3112 doubles, of which 2343 were discovered by Struve himself. The designations from these catalogues, prefixed by Σ (the Greek capital S) are still in use. The collected results were published in *Stellarum duplicium et multiplicium Mensurae Micrometricae per magnum Fraunhoferi tubum in Specula Dorpatensi* (1837).

Between 1835 and 1838 Struve made careful measurements of Vega, seeking to detect a parallax. He completed his analysis shortly before Friedrich Wilhelm BESSEL had done the same for the star 61 Cygni, but Bessel was the first to publish his result. Struve obtained a parallax for Vega of 0″.262, equivalent to a distance of 12.5 l.y., half the present-day value. His other achievements included the finding, in 1846, that starlight is absorbed in the galactic plane, which he correctly attributed to the presence of interstellar matter.

Struve, Otto Ludwig (1897–1963) Russian-American astrophysicist, great-grandson of Wilhelm Struve, who applied spectroscopy to the study of binary and variable stars, stellar rotation and interstellar matter. Born in Russia, his work at Kharkov Observatory, where his father, Ludwig Struve, was director, was halted by the turmoil following World War I, and in 1921 he moved to Yerkes Observatory, becoming a naturalized American in 1927. He was director of the observatory and professor of astrophysics at the University of Chicago (1932–47), and later became the first director of McDonald Observatory (1939–50) and of the National Radio Astronomy Observatory (1960–62). Struve's most important contribution (1929) was to show, with Boris Petrovich Gerasimovich (1889–1937), that interstellar matter pervades the whole Galaxy, and is not localized as had been thought. In 1937 he detected interstellar hydrogen.

STS Abbreviation of Space Transportation System – the SPACE SHUTTLE

STScI Abbreviation of SPACE TELESCOPE SCIENCE INSTITUTE

Subaru Telescope Major optical/infrared telescope with an aperture of 8.2 m (323 in.) operated by the NATIONAL ASTRONOMICAL OBSERVATORY OF JAPAN at MAUNA KEA OBSERVATORY, Hawaii. The telescope's name is Japanese for 'Pleiades'. Construction began in 1991, and the telescope became operational in 2000 with a sea-level base facility in nearby Hilo. The Subaru is notable for its ACTIVE OPTICS, sophisticated drive systems and cylindrical dome, which helps to suppress local atmospheric turbulence. It also has four foci and an auto-exchanger system for instruments at its prime focus, giving it a reputation as the Rolls-Royce of 8–10-metre telescopes.

subdwarf DWARF STAR of class F, G or K with low metal abundance; it appears below the normal MAIN SEQUENCE in the HERTZSPRUNG–RUSSELL DIAGRAM or colour–magnitude diagram. The term is misleading because subdwarfs are not below the standard main sequence but to the left of it. The low metal content yields spectra that make subdwarfs look too early for their temperatures; it decreases the atmospheric opacities to make them seem too hot and blue for their masses. Subdwarfs tend towards high velocities and are considered to be Population II halo stars.

subdwarf (sd) Any member of a group of stars that are less luminous by one to two magnitudes than MAIN-SEQUENCE stars of the same SPECTRAL TYPE; subdwarfs are of luminosity class VI. Mainly of spectral type F, G and K, subdwarfs are generally old halo Population II stars with low metal content. They lie below the main sequence on the HERTZSPRUNG–RUSSELL DIAGRAM. *See also* POPULATIONS, STELLAR

subgiant Star that has the same SPECTRAL TYPE as GIANT STARS, mainly G and K, but lower luminosity; subgiants are of luminosity class IV. They are stars evolving off the MAIN SEQUENCE in the process of becoming giant stars.

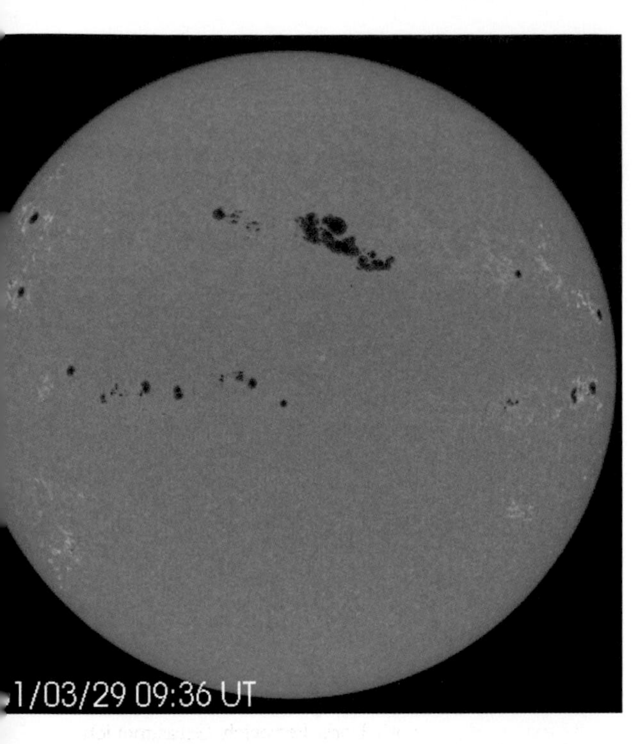

◀ **Sun** The Sun is a typical main-sequence dwarf star. Although it is at a stable part of its life cycle, it is not unchanging and undergoes a roughly 11-year cycle, during which its magnetic field 'winds up' and then declines. Sunspots are the most obvious result of these changes.

subluminous star Star that is less luminous than a MAIN-SEQUENCE star of the same temperature. Subluminous stars are usually old, evolved, metal-poor, Population II stars. Examples are WHITE DWARFS, SUBDWARFS and HIGH-VELOCITY STARS.

Submillimeter Array (SMA) Joint project by the SMITHSONIAN ASTROPHYSICAL OBSERVATORY and the Academia Sinica Institute of Astronomy and Astrophysics (Taiwan) to build an array of eight 6-m (20-ft) submillimetre telescopes at MAUNA KEA OBSERVATORY, Hawaii. It became operational in 2002.

Submillimeter Telescope Observatory (SMTO) *See* HEINRICH HERTZ TELESCOPE

submillimetre-wave astronomy Study of extraterrestrial objects that emit in the region of the ELECTROMAGNETIC SPECTRUM between 350 μm and 1 mm. It was originally regarded as part of the radio region or the infrared region – part of the radio region because it was the shortest non-optical wavelength that could be observed from the ground. Alternatively, it can be regarded as very far-infrared because the instruments used to detect submillimetre radiation are very similar to those used for INFRARED ASTRONOMY. The instruments must be cooled with liquid helium, and the telescopes must be at high, very dry sites because atmospheric water vapour easily absorbs submillimetre radiation. Submillimetre-wave astronomy focuses on molecular clouds, star-forming regions, cold dust and molecules. Molecules and cold dust are seen in planetary nebulae and galaxies, and many other types of astronomical object. A far-infrared and submillimetre satellite (the HERSCHEL SPACE OBSERVATORY) will be launched in 2007.

There are several submillimetre observatories, but the JAMES CLERK MAXWELL TELESCOPE (JCMT) on Hawaii has made a significant impact in the area of submillimetre astronomy, particularly with the array instrument SCUBA. High-redshift galaxies in the HUBBLE DEEP FIELD have been found to emit strongly in the submillimetre. Cold dust disks around young and main-sequence stars (such as Vega and ϵ Eridani) have been resolved by JCMT. Spectral emission lines from higher energy levels of molecules can be detected in the submillimetre (and compared with those in the millimetre wave region), so that the warmer and denser regions of star-forming clouds can be observed, probing for the initial stages of the collapse, both for high-mass and low-mass stars. The submillimetre can be used to map cold dust and molecules in the direction of the centre of our Galaxy.

subsolar point Point on the surface of a body in the Solar System, particularly the Earth, at which the Sun would be at the ZENITH at any given moment. Similar points are also defined for other pairs of bodies.

substellar object Body with mass above that of about ten Jupiter masses but below that of a BROWN DWARF. The more massive a substellar object, the more closely its characteristics resemble those of a brown dwarf. *See also* EXTRASOLAR PLANETS

Suisei Second of two Japanese space probes sent to Comet HALLEY. Suisei was launched in 1985 August and flew past the comet at a distance of 151,000 km (94,000 mi) on 1986 March 8. It was used to investigate the growth and decay of the comet's hydrogen corona and the interaction of the SOLAR WIND with the cometary ionosphere.

summer solstice Moment when the Sun reaches its greatest declination and highest altitude in the sky. In the northern hemisphere this occurs around June 21 when the Sun's declination is 23°.5N, marking the northern limit of its annual path along the ECLIPTIC. At this point it is overhead at the TROPIC OF CANCER and the hours of daylight are at a maximum, the day of the solstice also marking the longest day of the year. In the southern hemisphere the summer solstice occurs around December 22, when the Sun is overhead at the TROPIC OF CAPRICORN. *See also* WINTER SOLSTICE

Summer Triangle Popular name for the large and prominent triangle formed by the first-magnitude stars ALTAIR (in Aquila), DENEB (in Cygnus) and VEGA (in Lyra). It is overhead on summer nights in northern temperate latitudes, but remains visible well into the northern autumn.

Sun MAIN-SEQUENCE star of spectral type G2; it is the central body of the SOLAR SYSTEM around which all the planets, asteroids, comets and meteoroids revolve in their orbits. The Sun's light and heat are essential for life on Earth.

The Sun's energy source is the nuclear FUSION of hydrogen into helium, taking place in the CORE, which occupies about 25% of the solar radius. SOLAR NEUTRINOS produced by these NUCLEAR REACTIONS are detected at Earth. The structure and dynamics of the SOLAR INTERIOR are studied by HELIOSEISMOLOGY. A radiative zone surrounds the core, and a convective zone occupies the outer 28.7% of the solar radius. The visible disk of the Sun is called the PHOTOSPHERE. ACTIVE REGIONS may manifest as PORES, SUNSPOTS and FACULAE in the photosphere. These

SOLAR: DATA

Globe		
Mean distance, AU	1.4959787×10^8 km	
Light travel time Sun to Earth	499.004782 seconds	
Solar parallax	8.194148″	
Radius $R_\odot$	6.955×10^5 km (= 109 Earth radii)	
Volume	1.412×10^{24} km^3 (1.3 million Earths)	
Mass $M_\odot$	1.989×10^{30} kg (332,946 Earth masses)	
Escape velocity at photosphere	6178 km/s	
Mean density	109 kg/m^3	
Solar constant $f_\odot$	1366 W/m^2	
Luminosity $L_\odot$	2.854×10^{26} W	
Apparent magnitude m_v	−26.74	
Absolute magnitude M_v	+4.83	
Principal chemical constituents		
	(by number atoms)	**(by mass of fraction)**
Hydrogen	92.1%	70.68 %
Helium	7.8%	27.43%
All others	0.1%	1.89%
Age	4.566 billion years	
Density (centre)	151,300 kg/m^3	

S

regions are associated with strong MAGNETIC FIELDS from 2000 to 4000 Gauss. The white-light photosphere shows fine-scale mottling known as GRANULATION, and a larger-scale SUPER GRANULATION; both are caused by convection. DIFFERENTIAL ROTATION is seen in the photosphere, which has an effective temperature of 5780 K.

The inner solar atmosphere consists of the thin CHROMOSPHERE, lying just above the photosphere. SPECTROHELIOGRAMS or SPECTROGRAMS of the chromosphere reveal structures known as PROMINENCES, SPICULES, FIBRILS, PLAGES and FLOCCULI. The Sun's outer atmosphere forms the million-degree CORONA. Observed at X-ray wavelengths, the corona shows features such as CORONAL HOLES and coronal loops.

All forms of activity vary in a roughly 11-year SOLAR CYCLE, including CORONAL MASS EJECTIONS, FLARES and the occurrence of active regions and their associated sunspots. The total flux of solar radiation incident on the Earth (the SOLAR CONSTANT) also varies in step with this cycle; longer-term changes in the solar constant may underpin episodes of climatic change, such as the MAUNDER MINIMUM. Solar activity may vary on millennial timescales. Intense X-ray radiation from flares affects Earth's IONOSPHERE, whilst energetic particles released from them may be a hazard to astronauts and artificial satellites. Coronal mass ejections influence SPACE WEATHER and can lead to magnetic storms, accompanied by intensified displays of the AURORA.

sundial Simple instrument of great antiquity for determining the time of day from a shadow cast by the sun. It usually consists of a flat plate graduated in hours and minutes, and a style or gnomon which casts the shadow. Many variations of this simple pattern exist. Without correction, the dial shows the apparent solar time. However, an accurate sundial may be constructed for a particular location. A simple date-dependent correction for the non-circularity of the Earth's orbit can be applied to obtain the time correct to about a minute.

▼ **sunspot** The magnetic field around the region of a sunspot prevents the normal flow of plasma. The spot is cooler than the surrounding regions and so appears dark.

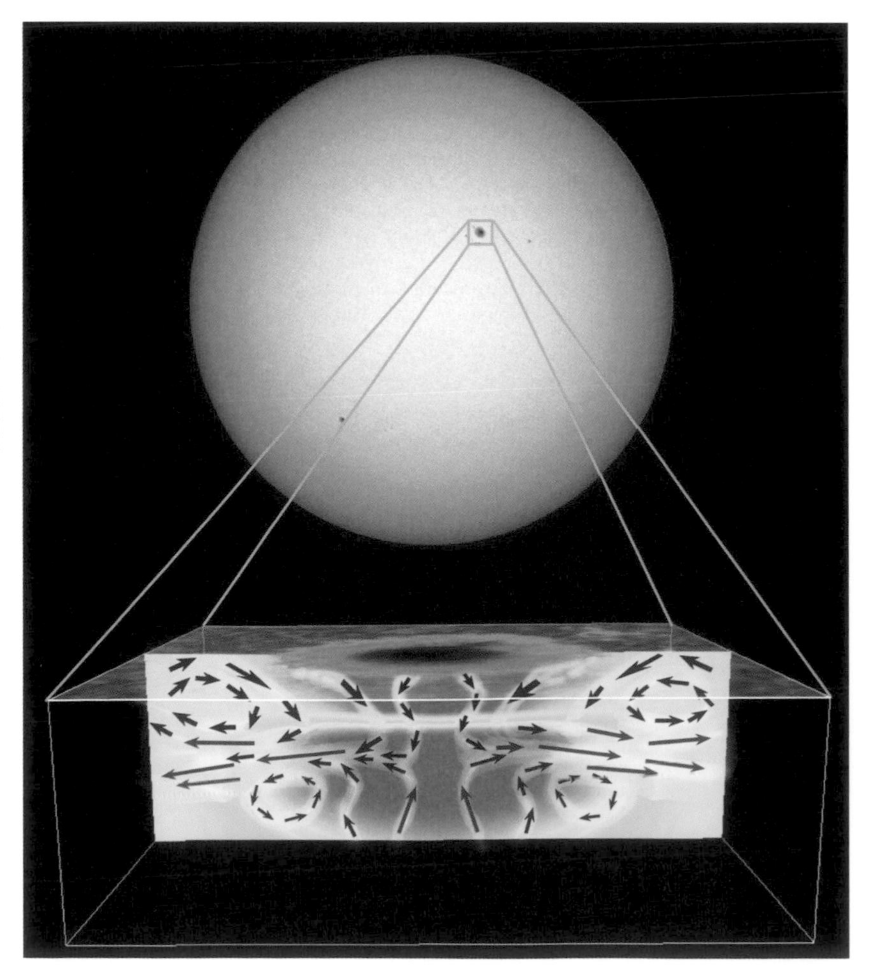

sundog Popular name for a PARHELION

sungrazer COMET that at perihelion passes very close to the Sun. *See also* KREUTZ SUNGRAZER

sunspot Dark, temporary concentration of strong magnetic fields detected in the white light of the Sun's PHOTOSPHERE. Sunspots are the most visible manifestations of ACTIVE REGIONS, which also include features such as FACULAE and PLAGES. The central magnetic field of a sunspot is vertical, and typically has a strength of 2000 to 4000 Gauss. Most sunspots have a central dark umbra and a lighter, grey surrounding region, called the penumbra, although either feature can exist without the other. A sunspot is cooler than the surrounding material and therefore appears darker. The effective temperature of the sunspot umbra is about 4000 K, compared with 5780 K in the neighbouring photosphere. The penumbra consists of linear bright and dark elements known as filaments, which extend radially from the umbra if the spot is more or less circular; the penumbra in complex spot groups may be more irregular.

Sunspots vary in size from PORES about 1000 km (600 mi) in diameter to about 1 million km (600,000 mi). The largest spots become visible to the protected naked eye (under no circumstances should an observer look directly at the Sun without using an approved, safe filter). The duration of sunspots varies from a few hours to a few weeks, or months for the very biggest. The number and location of sunspots vary over the SOLAR CYCLE. The number of sunspots is described by the RELATIVE SUNSPOT NUMBER, and their locations change over the cycle in accordance with SPÖRER'S LAW, shown graphically as the BUTTERFLY DIAGRAM. Heliographic latitudes and longitudes are measured relative to a standard reference frame defined by the CARRINGTON ROTATIONS. Sunspots usually occur in pairs or groups of opposite magnetic polarity that move in unison across the face of the Sun as it rotates. *See also* EVERSHED EFFECT; WILSON EFFECT

supercluster Extensive grouping of galaxies which may include multiple clusters as well as surrounding smaller groups and extensive regions of enhanced galaxy density in the forms of sheets or filaments. We are in the Local Supercluster, centered on the VIRGO CLUSTER about 60 million l.y. away, and occupying a flattened volume roughly perpendicular to the plane of the Milky Way. As is typical, the Local Supercluster is still growing because its dynamical influence has time to be felt over larger and larger regions, so that eventually the LOCAL GROUP of galaxies is expected to fall into the supercluster centre. Other well-known superclusters include the Shapley Supercluster, which is related to the GREAT ATTRACTOR; the Perseus–Pisces Supercluster, which includes seven rich clusters plus outlying filaments, together spanning more than 40° in our sky and about 1 billion l.y. in space; and the Coma/Abell 1367 Supercluster, which is part of the so-called Great Wall of galaxies seen almost edge-on about 300 million l.y. away.

The mass of a supercluster can be estimated from non-Hubble distance estimates of outlying objects on its near and far borders, which may reveal the characteristic pattern of infall under the supercluster's influence. At their outskirts, superclusters blend into the overall LARGE-SCALE STRUCTURE of clumps, walls and filaments of galaxies which spans the observable Universe, on scales of up to 1 billion l.y. Their existence is important for cosmology, since enough time must have elapsed for them to have collected such a massive concentration of material, and the mass distribution in the early Universe must have been uneven enough to start the process of accumulation.

supergiant Largest and most luminous star known. Supergiants occur with SPECTRAL TYPES from O to M. Red (M type) supergiants have the largest radius, of the order of 1000 times that of the Sun. Betelgeuse is a type M supergiant; Rigel is a type B. As there is a an upper

S

limit to the ABSOLUTE BOLOMETRIC MAGNITUDE of red supergiants, they can be used as distance indicators. Supergiants lie above the MAIN SEQUENCE and GIANT region on the HERTZSPRUNG–RUSSELL DIAGRAM. Instabilities created by RADIATION PRESSURE mean that supergiants are often variable. *See also* CEPHEID VARIABLES; ETA CARINAE; S DORADUS STAR

supergranulation Large convective cells seen in the solar PHOTOSPHERE. They have dimensions of about 30,000 km (19,000 mi), lifetimes of about 20 hours, and internal plasma motion velocities of about 1 km/s (0.6 mi/s). Unlike the GRANULATION, the supergranulation is not visible in the white light of the photosphere and is instead detected by the DOPPLER EFFECT. The dominant flow in the cells is horizontal and outwards from the centre, but there is a weak upward flow at the cell centre and a downward flow at the cell boundaries. The cells are outlined by the magnetic network in the CHROMOSPHERE, the boundaries of which contain concentrations of magnetic fields and SPICULES.

superior conjunction Point in the orbit of one of the INFERIOR PLANETS at which it lies on the far side of the Sun, as seen from Earth, and the three bodies are in alignment. CONJUNCTION occurs when two bodies have the same celestial longitude as viewed from the Earth. For an inferior planet, this can also occur when it is on the same side of the Sun as the Earth. At this point it is said to be at INFERIOR CONJUNCTION. *See* diagram at CONJUNCTION

superior planet Term that is used to describe any planet with an orbit that lies beyond that of the Earth; the superior planets are Mars, Jupiter, Saturn, Uranus, Neptune and Pluto.

superluminal expansion Apparent faster-than-light effect. Special relativity asserts that massive particles can never reach or exceed the speed of light. Yet, radio lobes seen in distant radio galaxies and quasars have been measured to be separating from the central source at velocities exceeding the speed of light. This faster-than-light or 'superluminal' expansion is merely an effect of the finite speed of light that occurs in things that are moving close to the speed of light and very close to the line of sight. Time dilation effects coupled with the time for light to travel between the reference object (the quasar or galaxy core) and the emitting object (the knot in the jet that is moving close to the speed of light) can cause the distant observer to overestimate the speed by a large fraction.

supermassive black holes BLACK HOLES that are thought to reside in the nuclei of galaxies and quasars and to contain 10^6 to 10^9 times the mass of the Sun. These black holes are a million to a billion times more massive than black holes that result from stellar evolution. It is not clear how they form in galactic centres, but the evidence for their existence is very strong. The direct evidence for the existence of supermassive black holes comes from observations of high-velocity gas clouds orbiting the centres of radio galaxies. The gravitational potential necessary to bind these clouds in orbit requires 10^6 to 10^8 solar masses contained within a very small region of space. The mass density required assures that the supermassive object would immediately collapse to a black hole if it were not one already. Indirect evidence for supermassive black holes is also found in observations of quasars, which suggest that a large amount of energy is generated within a volume the size of our Solar System. The only energy generation mechanism efficient enough to accomplish this is the conversion of gravitational energy into light by a supermassive black hole.

supermassive star Very rare star with mass of the order of 100 solar masses and above. It is not known if there is an upper limit for STELLAR MASSES, but some estimates suggest an upper limit of 440 solar masses. The most massive MAIN-SEQUENCE stars observed are within the R136 cluster in the 30 Doradus Nebula of the LARGE MAGELLANIC CLOUD. They are around 155 solar masses. *See also* ETA CARINAE; S DORADUS STAR

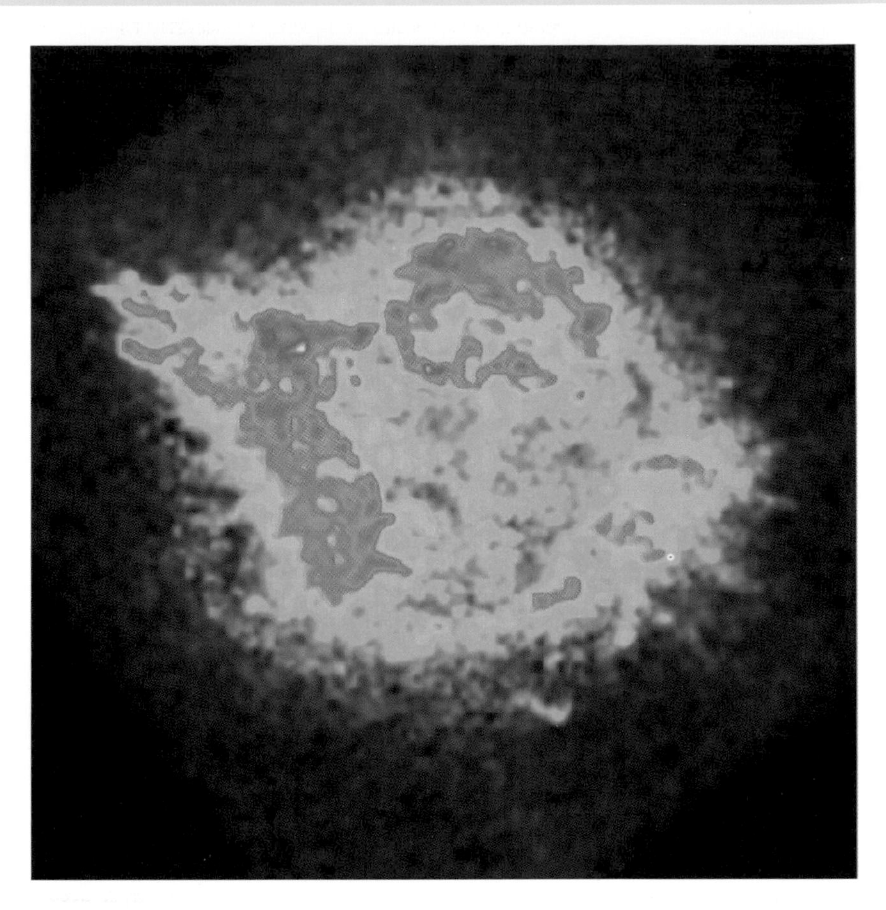

▲ **supernova** Elements heavier than iron cannot be formed in normal stellar fusion, and supernovae are thought to be the mechanism that produces them. Lighter elements are also created during supernovae – this image shows the presence of silicon in the supernova remnant Cassiopeia A.

supernova Stellar explosion that involves the disruption of virtually an entire STAR. Supernovae are classified by their spectra into two broad types, Type I and Type II, according to the presence or absence of hydrogen in their spectra. Type I supernovae have no hydrogen in their spectra, whereas Type II do. It is now known that the cause of supernovae is not determined by the presence or lack of hydrogen in the spectra, thus further categorization has been made.

There are two basic models for the cause of supernovae. The 'core collapse supernovae' are massive stars that have exhausted the nuclear fuel in their cores. However, because the mass of the core reaches beyond the CHANDRASEKHAR LIMIT further core collapse occurs until neutron degeneracy pressure sets in, and the outer atmosphere is thrown off as a result of shock waves. Other supernovae are thought to occur in CLOSE BINARY systems in which a WHITE DWARF is sent over the Chandrasekhar limit by MASS TRANSFER from its companion.

Type Ia supernovae appear in all types of galaxies but are less frequent in the spiral arms of spiral galaxies. They have elements such as magnesium, silicon, sulphur and calcium in their spectra near maximum light, and iron later on. The light-curve of a Type Ia supernova shows an initial rise over about two weeks, then a more gradual decay over timescales of months. A Type Ia supernova is thought to be the explosion, as a result of mass transfer, of an old, low-mass, long-lived star in a binary system. As Type Ia supernovae are so bright, they have been used to estimate distances to faraway galaxies.

Type II supernovae do not appear in elliptical galaxies, but instead occur mostly in the spiral arms of spiral galaxies or sometimes in irregular galaxies. They show ordinary stellar abundances in their spectra. The light-curve of a Type II supernova rises to a peak in a week or so, remains constant for about a month, then drops suddenly over a few weeks, returning to obscurity over a timescale of months. A Type II supernova is believed to be the result of

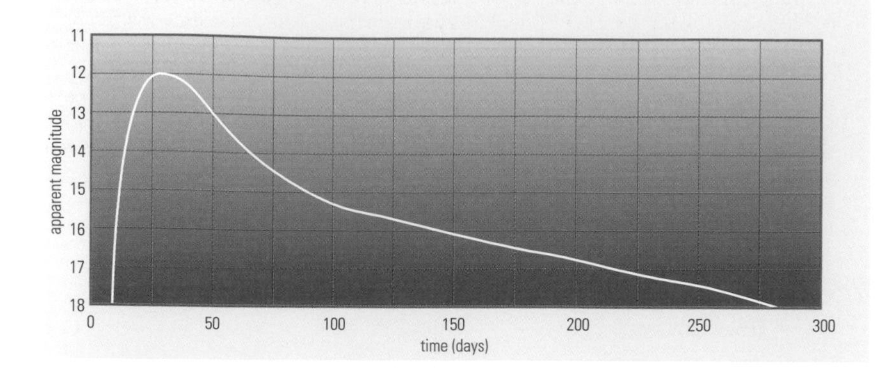

▲ **supernova** The light-curve of a supernova shows a sudden brightening as the outer atmosphere is thrown off after the stellar core has collapsed. This peak is followed by a steady decline for about two months, after which this fading slows further.

an explosion in the core of a RED GIANT star with a massive extended envelope.

Types Ib and Ic supernovae appear to explode only in the spiral arms of spiral galaxies. Both show evidence of oxygen, magnesium and calcium in their spectra after peak brightness. In addition, Type Ib supernovae show evidence of helium at times near maximum light. The light-curves of both Ib and Ic are similar to Type Ia but dimmer at maximum light. They are often strong radio sources, whereas no radio emissions have yet been detected from Type Ia. Types Ib and Ic supernovae are thought to be produced by explosions of the cores of massive stars that have been stripped of their hydrogen, and, in the case of Type Ic, of their helium as well.

Supernovae produce NEUTRON STARS, and many have been observed at the centre of the remains of the disintegrated stellar envelope. The nebula created by the supernova is termed a SUPERNOVA REMNANT. Supernova remnants with PULSARS at their centres are termed PLERIONS.

Supernovae are relatively rare, discovered at the rate of about one every century in an average galaxy. They are discovered relatively infrequently in edge-on spiral galaxies because their light is dimmed by dust. Only five have been discovered in the Milky Way in the last millennium, with SUPERNOVA SN 1987A occurring in the close companion galaxy, the LARGE MAGELLANIC CLOUD, in 1987.

Recent technological advances, especially the CCD, becoming available to amateur astronomers have meant a dramatic increase in the number of supernovae discoveries in other galaxies in recent years. Computer-controlled telescopes equipped with CCDs scan a large number of remote galaxies each clear night, and comparisons made with previous images reveal any supernovae explosions.

The supernova of 1054 was identified by Edwin HUBBLE as the progenitor of the Crab Nebula. Like the supernovae of 1006 and 1181, it was recorded by oriental astronomers as they scanned the sky for celestial portents. Chinese, Korean, Japanese, Arabic and European astronomers contributed to records of these supernovae: the 1054 supernova was probably depicted in Native American art.

The supernova of 1572 was carefully observed by Tycho BRAHE. He recorded data on its unchanging position and its stellar magnitude as it faded day by day. Brahe demonstrated that the supernova, which was circumpolar from Denmark, had no PARALLAX as the Earth rotated. This placed the star well beyond the Moon. Its lack of motion over the 18 months that it was visible meant that it was beyond Saturn, the most distant planet then known. This discovery placed the supernovae among the 'fixed stars' and proved that they were subject to the same laws of change as terrestrial phenomena.

The supernova of 1604 is known as Kepler's star, although Johannes Kepler was not the first to see it. Both Kepler's and Tycho's supernovae inspired a wealth of comment by many writers of the 16th and 17th centuries, including Richard Corbet, Henry More, John Donne, Edmund Spenser and John Dryden.

There is evidence that a supernova exploded in Cassiopeia in around 1680. An expanding gaseous remnant of an exploded star is observed as a powerful emitter of radio radiation; it is known as CASSIOPEIA A. No optical outburst was recorded, but the star may have ejected a great deal of its outer layers before exploding, or been relativity small and compact.

supernova remnant (SNR) What is left after a SUPERNOVA explosion has occurred. For some supernovae, there are two components to the supernova remnant – the central NEUTRON STAR, PULSAR or BLACK HOLE and the expanding cloud of gas. Only the last of these, however, is included in the usual meaning of the term SNR.

In appearance SNRs may resemble some HII REGIONS, but they often have more of a filamentary, shell-like morphology than the latter. The radio emissions from the two types of object allow them to be differentiated unequivocally, since recombination lines appear in the radio spectra of HII regions but not in those from SNRs. Furthermore, SYNCHROTRON RADIATION produces much of the emission from SNRs and this is strongly polarized, whereas the emission from HII regions is unpolarized.

Given that they are created by an explosion, it is clear that SNRs must be expanding. The initial expansion velocities can be 10,000 to 20,000 km/s (6000–12,000 mi/s), though this reduces to a few hundred km/s with time. Nonetheless after a century or two, the nebula's physical size is measured in light-years. The CRAB NEBULA for example is about 13 l.y. across, the Cygnus Loop about 100 l.y. and the LOCAL BUBBLE, which may have resulted from a much older supernova explosion, is some 300 l.y. in diameter. The kinetic temperature of the material initially exceeds 10^6 K, but this is reduced as energy is radiated away. The particle densities range from 10^6 m^{-3} to 10^{12} m^{-3}, with 10^9 m^{-3} being fairly typical for the visible portions of an SNR. The masses of the nebulae range from a fraction to a few times the solar mass.

The development of an SNR occurs in several stages. At first the hot gases exploding outwards at around 10,000 km/s (6000 mi/s) have densities so much higher than that of the surrounding interstellar medium that the expansion is essentially into a vacuum. After about a century, when the SNR is around 1 l.y. or so in size, the interstellar material swept up by the expanding remnant reaches densities at which it starts to impede the expansion. This creates a turbulent shock region, with synchrotron emission coming from the relativistic electrons. The expansion velocity will then slowly decrease until at about 100 km/s (60 mi/s) optical line emission from the heavier elements becomes significant. The last identifiable stage may be as a hot low-density cavity blown into the interstellar medium. Finally, after some few hundred thousand years, the remnant will merge with the interstellar medium.

Most SNRs have the appearance of a thin spherical shell, with much fine structure. Type Ia supernovae, such as that seen by Tycho BRAHE in 1572, lead to shells that are brighter towards their outer edges at radio and X-ray wavelengths, with only Balmer emission lines detected in the visible. Type II supernovae, such as the one that produced Cassiopeia A, result in shells containing numerous rapidly moving knots of material that are very bright at both radio and X-ray wavelengths, with thermal emission dominating the X-rays. A few SNRs, however, like the Crab Nebula, are filled-in to their centres. These are given the name PLERIONS, and their difference probably arises from the presence of a central

SUPERNOVAE

Year	Constellation	Duration	Mag.
1006	Lupus	Two years	much brighter than Venus
1054	Taurus	24 months	−3.5
1181	Cassiopeia	6 months	−1
1572	Cassiopeia	483 days	−4.0
1604	Ophiuchus	365 days	−2.6
1987	Large Magellanic Cloud	?	2.3

S

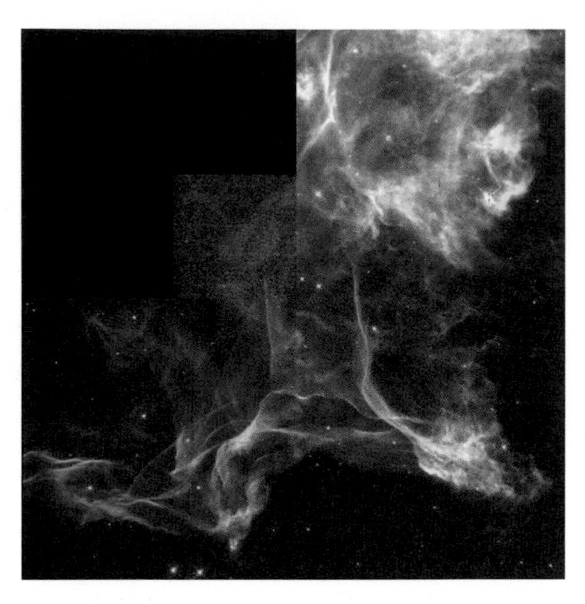

◀ **supernova remnant** In a supernova explosion, the outer atmosphere of a star is flung off and continues to expand away from the progenitor star. The Cygnus Loop is the remnant of an explosion 30,000 years ago.

pulsar that can continue to supply high-energy electrons to the centre of the nebula. An SNR should be detectable for a period of a few hundred thousand years after the explosion. Since the estimated rate of supernovae in our GALAXY is around ten per century, there should be several thousand SNRs in our Galaxy. Only about a hundred are actually known, however, and most of these are detected by their radio emissions. The disparity is probably a result of SNRs' low surface brightnesses and the obscuration of the more distant examples.

The radio emission from SNRs is mostly produced by synchrotron radiation. Initially, the electrons producing this radiation result from the supernova explosion. Later, except in plerions, the electrons probably gain their energies whilst passing through the shock fronts where the expanding SNR meets the interstellar medium. A few SNRs are observable outside the radio and X-ray regions. In particular, the Crab Nebula emits detectable radiation from the infrared to the gamma-ray region. Most of this again appears to be due to synchrotron radiation, as shown by the high degree of optical polarization of the nebula. Some of the visible light, however, results from RECOMBINATION after the gas has been ionized by the ultraviolet synchrotron emission. The X-ray emission from SNRs results from synchrotron radiation and from electron–ion collisions in the hot gas.

SNRs affect their locality in several ways. Since Type II supernovae originate from short-lived massive stars, they will often occur whilst the star is still inside a dense gas cloud. The pressure of the expanding SNR may then initiate further star formation. The energy of the SNR adds to the energy and turbulence of the interstellar medium. It also enriches it in elements heavier than helium, which were synthesized in the supernova and its preceding star. Finally, as particles are accelerated during interactions between the turbulent high-velocity gases and the magnetic fields of an SNR, COSMIC RAYS may be produced.

Supernova SN 1987A Bright SUPERNOVA that flared up in the Large Magellanic Cloud in 1987 February, reaching naked-eye visibility. It was the first supernova to do so since Kepler's star of 1604. Supernova SN 1987 A was a Type II supernova, and a flood of neutrinos was detected from its eruption, suggesting that a NEUTRON STAR was formed. No subsequent evidence of a neutron star has been found, but it may be obscured in some way.

The precursor star (Sanduleak $-69°202$) of the supernova explosion was a compact blue SUPERGIANT, rather than the red giant or supergiant that current theories predict should give rise to a Type II supernova. It has been suggested that the star may have had a binary companion that was consumed as the progenitor became a red giant, increasing both its size and temperature. The set of three nested rings that have subsequently been observed may be related to the fact that the system was a binary system, but this remains unclear.

superstring theory Theory of ELEMENTARY PARTICLES that asserts that the most fundamental forms of matter and energy are not particles but minute vibrating strings. The shape, topology and frequency of vibrations of these strings determine the physical properties of the particles. The scale of these strings is very small, of the order of 10^{-33} cm. This in itself is a problem since they are the smallest possible quanta and thus can never be observed. However, if one assumes these strings do exist, and they do make up elementary particles and fields, then one can construct a model that could possibly describe everything in the universe – a THEORY OF EVERYTHING (TOE).

supersymmetry Principle that attempts to explain the different strengths of the FUNDAMENTAL FORCES in the STANDARD MODEL. This model accurately describes the three fundamental forces and provides a grand unification. If the interaction strengths of the three GUT forces (*see* GRAND UNIFIED THEORY) are extrapolated to high energies, they all come together around 10^{16} GeV. The interaction strength of each force needs to be specified independently in the standard model; this problem is called the hierarchy problem. Supersymmetry attempts to explain the hierarchy problem in terms of a symmetric principle that assigns a symmetric partner to every known elementary particle and also relates fermions to bosons.

surface gravity Acceleration, g, due to the GRAVITATIONAL FORCE experienced by a freely falling object close to the surface of a massive body. It is given by:

$$g = \frac{GM}{R^2}$$

where G is the gravitational constant and M and R are the mass and radius respectively of the massive body. For the Earth, g has a value of 9.81 m/s^2.

survey, astronomical Systematic large-scale observation made with one or a small number of dedicated instruments to gather reference data for the whole sky or a significant fraction of it. Examples are the *PALOMAR OBSERVATORY SKY SURVEY* and the SLOAN DIGITAL SKY SURVEY, both in the visible region, the TWO MICRON ALL SKY SURVEY in the infrared, and the NVSS (NRAO VLA Sky Survey) at radio wavelengths. REDSHIFT

▼ **Supernova SN1987A** Although not in the Milky Way galaxy, SN1987A was visible with the naked eye.

S

► **Surveyor** The Surveyor lunar probes provided useful information for the development of the Apollo programme. This image is of Surveyor 1's shadow on the lunar surface.

SURVEYS are aimed at systematic observations of the recessional velocities of very large numbers of galaxies.

Surveyor Series of seven soft-landing lunar spacecraft launched by NASA between 1966 May and 1968 January in preparation for the Apollo landings. Five of them were successful in sending back high-resolution images and data on the nature of the lunar surface. Each Surveyor had a steerable television camera with filters. Other payloads included a combined surface-sampler/trench-digger and a simple soil composition analyser. Among the wealth of data returned were panoramic and close-up images, data on the bearing strength, optical and thermal properties of the surface, the content of magnetic material and its major-element chemistry.

Su Song (or Su Sung) (1020–1101) Chinese astronomer, noted for his clock-driven astronomical instruments. In 1090 Su Song finished his two-storey clock tower at Kaifeng. Inside was a clock-driven celestial globe, while moving figures indicating the time were visible from the outside. Surmounting the tower was an armillary sphere fitted with a sighting tube. Globe and sphere automatically followed the rotation of the skies, and were the first astronomical instruments to do so, anticipating Western automatic clock-driven instruments by six centuries.

SU Ursae Majoris star (UGSU) Subtype of CATACLYSMIC VARIABLE known generically as a DWARF NOVA (commonly called U GEMINORUM STAR). Their behaviour resembles that of the classical U Geminorum/SS Cygni subtype (UGSS) in that there are short and long maxima. In the SU Ursae Majoris stars, however, the latter are known as supermaxima, because they are half to one magnitude brighter than normal outbursts, and remain at maximum for 10 to 20 days (as against 4 days for a normal outburst). For almost all maxima, stars of this type rise to near maximum brightness in 24 hours or less. Some have a short pause at an intermediate magnitude on the rise for an hour or two. They also have periodic oscillations, termed superhumps, superimposed on the light-curve with amplitudes of 0.2 to 0.5 mag. Their period is about 3% longer than the orbital period.

Swan Nebula *See* OMEGA NEBULA

Swasey, Ambrose *See* WARNER & SWASEY CO.

Swedish ESO–Submillimetre Telescope (SEST) Radio telescope of 15-m (49-ft) aperture operating in the frequency range 70–365 GHz. Built in 1987, it is located at LA SILLA OBSERVATORY and operated in a partnership between the EUROPEAN SOUTHERN OBSERVATORY (ESO) and ONSALA SPACE OBSERVATORY, which is responsible for receivers and computer software. Hardware and mechanical components are maintained by ESO.

Swedish Extremely Large Telescope (SELT) Proposal for a 50-m (164-ft) optical telescope made by astronomers at LUND OBSERVATORY and elsewhere during the late 1990s. The telescope would require multi-conjugate ADAPTIVE OPTICS to achieve its potential resolution, and would use the segmented-mirror technology pioneered at the W.M. KECK OBSERVATORY. An alternative is to spin-cast the mirror as a monolithic structure at the observing site. By 2002 it seems likely that the SELT proposal would be subsumed into the OVERWHELMINGLY LARGE TELESCOPE project.

Swift, Lewis (1820–1913) American astronomer who, between 1855 and 1901, discovered more than 1200 star clusters, nebulae and galaxies, ranking him third behind only the English astronomers William HERSCHEL and John HERSCHEL. Swift also found 13 comets, including 109P/SWIFT–TUTTLE, which produces the annual PERSEID meteor shower. He served as director of the H.H. Warner Observatory in Rochester, New York and the T.S. Lowe Observatory on Echo Mountain, near Pasadena, California.

Swift–Tuttle, Comet 109P/ Short-period comet discovered by Lewis Swift on 1862 July 16, and independently three days later by Horace Tuttle (1837–1923). The comet became a naked-eye object, at mag. +2.0 in early September, when the tail reached a length of 25–30°. The last observations in the discovery apparition were made at the end of October. Uncertainties over the orbital period remained until the next return – ten years later than some predictions – in 1992.

Comet 109P/Swift–Tuttle returned to perihelion on 1992 December 12, reaching mag. +5.0. As at the previous return, the comet's NUCLEUS showed considerable jet activity, a source of NON-GRAVITATIONAL FORCE, which partly explains the earlier difficulty in deriving an orbit. The period is now taken to be close to 135 years, and the comet is, as originally proposed by Brian Marsden (1937–) of the Harvard Smithsonian Center for Astrophysics, identical with Comet Kegler of 1737. Its next return in 2126 will see 109P/Swift–Tuttle make a close approach to Earth. The comet is the parent of the PERSEID

► **Swift–Tuttle, Comet** The comet's most recent return was in 1992 and was not particularly spectacular. It will be much closer to Earth at its next approach in 2126.

meteor shower, which showed enhanced activity around the time of its 1992 perihelion.

Sword Handle *See* DOUBLE CLUSTER

SX Phoenicis Pulsating VARIABLE STAR, 6°.5 west of α Phoenicis; it exhibits periods of 79 and 62 minutes, with a beat period of 278 minutes. The visual range is from mag. 7.1 to 7.5. Its distance is about 140 l.y. Its behaviour closely resembles that of a DELTA SCUTI STAR, and it has been taken as the prototype (designated SXPHE) for subdwarf stars that belong to globular clusters and the older regions of the Galaxy and have multiple periods.

Sycorax One of the several small outer satellites of URANUS; it was discovered in 1997 by Philip Nicholson (1951–) and others. Sycorax is about 120 km (75 mi) in size. It takes 1289 days to circuit the planet, at an average distance of 12.18 million km (7.57 million mi). It has a retrograde orbit (inclination near 159°) with a substantial eccentricity (0.523). *See also* CALIBAN

symbiotic star (variable) BINARY STAR that exhibits the spectral characteristics of two grossly different temperature regimes. Typically there are absorption bands such as those found in giants and supergiants of late spectral classes around 3000 K, together with emission from a hot star of about 20,000 K. In addition, emission from hot, surrounding nebulosity is often detectable. Symbiotic stars exhibit a wide range of characteristics. They may all be classed as CATACLYSMIC VARIABLES, in which gas from the cool star is accreted by the smaller companion, giving rise to outbursts. Z ANDROMEDAE STARS resemble DWARF NOVAE, but with a giant (rather than main-sequence) secondary. In certain systems, such as RR Telescopii, there is a high rate of MASS TRANSFER, producing systems with extremely extended outbursts. Such systems were formerly known as 'very slow novae' but are now more commonly called 'symbiotic novae'. Some systems – most notably R Aquarii – appear to consist of a pulsating MIRA STAR transferring mass to an ACCRETION DISK and thence to a condensed object that is ejecting material in a pair of luminous jets. This is very similar to the configuration found in the strange object SS 433, where the condensed component appears to be a neutron star, with an accretion disk and exceptionally strong polar jets.

synchronous orbit Orbit in which a satellite's period of REVOLUTION is the same as the planet's period of axial ROTATION. For Earth, the radius of the orbit is 42,164 km (26,200 mi). The satellite appears to hover over one point on the equator, neither rising nor setting. The main significance for natural satellites is that the direction of TIDAL EVOLUTION is inwards or outwards according to whether the satellite is below or above the synchronous orbit. Many artificial satellites of the Earth have been placed in such orbits, but the only natural example among the planets is Pluto's satellite, CHARON. It is an ultimate state resulting from tidal evolution; it requires that the secondary is a substantial fraction of the mass of the primary, in order that it can absorb the spin angular momentum of the primary. Eventually the Moon will reach this state. The Earth's rotation is slowing as the Moon recedes from the Earth (*see also* SECULAR ACCELERATION), and this will continue until the day has lengthened to be the same as the month (but this will take about 10 billion years).

synchronous rotation Coincidence of the rotational period of a celestial body with its orbital period. This has the effect of causing a planetary moon, for example, always to present the same face towards the planet it orbits, as is the case with the Earth and the Moon. The phenomenon, also known as captured rotation, is caused by tidal friction. *See also* TIDE

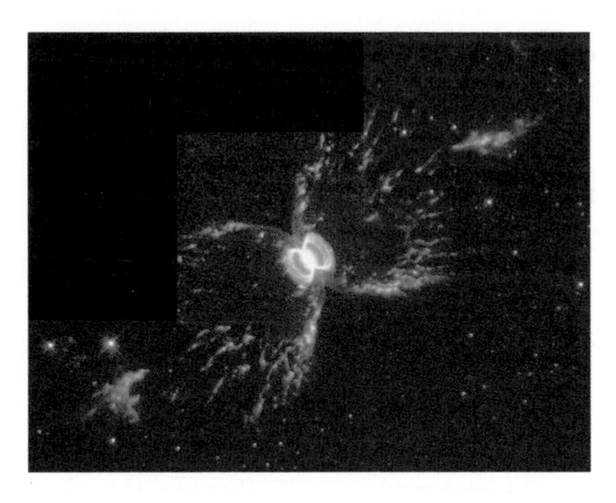

◄ **symbiotic star** This peculiar structure is thought to be the result of at least three successive nova explosions on a red giant/white dwarf pair. The hourglass structure in the centre results from the most recent explosion, while the insect-like structure is the remnant of the previous episode.

synchrotron radiation ELECTROMAGNETIC RADIATION emitted by charged particles (usually electrons) moving in magnetic fields at large fractions of the speed of light. The charged particles follow helical paths along magnetic lines of force and emit radiation because of the accelerations to which they are subjected. The process is closely related to FREE–FREE RADIATION. The wavelength of the emitted radiation decreases as the energy of the electrons increases, and is strongly polarized. Synchrotron radiation produces much of the emission from the CRAB NEBULA, PULSARS and RADIO GALAXIES.

synodic Strictly, a term that means 'pertaining to two successive conjunctions of celestial bodies'; it is more commonly used to describe a single cycle of any phase of a celestial body viewed from a given point. For example, a synodic month is the period between successive new or full moons.

synodic month (lunar month) Period between successive new or full moons. This is the same duration as one LUNATION and is equivalent to 29.53059 days of MEAN SOLAR TIME. *See also* ANOMALISTC MONTH; DRACONIC MONTH; MONTH; SIDEREAL MONTH; TROPICAL MONTH

synodic period Interval between successive OPPOSITIONS, or CONJUNCTIONS, of a planetary body. It is a measure of that body's orbital period as observed from the Earth, as opposed to its SIDEREAL PERIOD, which provides a true measure.

Syrtis Major Planitia Most conspicuous feature on MARS; it is centred near 9°.5N 290°.5W. It is dark, wedge-shaped and an easy telescopic object. Formerly known as the Syrtis Major, it was first drawn by Christiaan HUYGENS as long ago as 1659. It is now known to be a plateau rather than a vegetation-filled depression, as was once thought.

Systems I and II Grouping by rotation period of the clouds of JUPITER, which do not rotate uniformly. The groups are: System I, $9^h\ 50^m\ 30^s$, found in equatorial regions; and System II, $9^h\ 55^m\ 41^s$, found above latitude 10°N or S. A third, System III, defined by radio emissions, has a rotation period of $9^h\ 55^m\ 29^s.37$.

syzygy Approximate alignment of the Sun, Earth and Moon, or the Sun, Earth and another planet. Syzygy thus occurs when the Moon is either new or full or, for a planet, at CONJUNCTION and OPPOSITION. The term is also used to describe the alignment of any three celestial bodies.

S

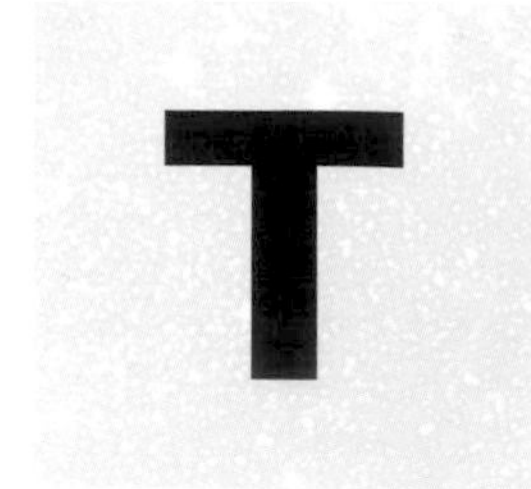

tachyon Hypothetical elementary particle that has a velocity greater than the speed of light. According to relativity, the tachyon, if it exists, would be travelling backwards in time. No such particle has ever been detected, but some elementary particle theories allow for their existence.

Taeduk Radio Astronomy Observatory (TRAO) Radio observatory founded in 1986 in the science town of Taeduk, Taejon City, Korea. The facility is part of the Korea Astronomy Observatory, and operates a 14-m (46-ft) millimetre-wave telescope enclosed in a radome. The telescope was equipped with new high-performance receivers in 1995.

TAI *See* INTERNATIONAL ATOMIC TIME

tail Extended structure that is directed anti-sunwards from the NUCLEUS of an active COMET. Bright comets usually show two tails – a dust tail driven away by the pressure of solar radiation, and an ion tail dragged 'downwind' by the interplanetary magnetic field in the SOLAR WIND. Comets' tails are very tenuous, and background stars are usually readily visible through them. *See also* ANTITAIL

Taiwan Oscillation Network (TON) Ground-based network to measure solar intensity oscillations to study the internal structure of the Sun. Solar images in the light of the calcium K line are taken at the rate of one per minute with identical telescopes in Tenerife, Huairou (near Beijing), BIG BEAR SOLAR OBSERVATORY and Uzbekistan. The network has been gathering data since 1994, and two extra observing stations are planned in order to improve longitude coverage.

▼ **Tarantula Nebula** This emission nebula in the Large Magellanic C loud is glowing with the radiation from hot stars buried within it. Its pinkish colour is caused by ionized hydrogen (HII).

tangential velocity (transverse velocity) Velocity of a star measured across the line of sight of the observer. Observations of a star's gradual change in position over a long period of time give a measure of its annual PROPER MOTION; if its distance in PARSECS is also known, its tangential velocity can then be calculated. Measurements of the Doppler shift in its spectrum provide its RADIAL VELOCITY (velocity along the line of sight) and from the two, the star's true velocity (SPACE VELOCITY) can be computed.

Tarantula Nebula (NGC 2070, 30 Doradus) Emission nebula in the LARGE MAGELLANIC CLOUD (RA $05^h\ 39^m$ dec. $-69°57'$); it is illuminated by an embedded cluster of recently formed hot O and B stars. The Tarantula Nebula, also known as the Loop Nebula, takes its name from 'spidery' extensions of nebulosity on its outer edge. Its apparent angular size of 30′ corresponds to a true diameter of about 900 l.y.

Tarazed The star γ Aquilae, visual mag. 2.72, distance 460 l.y., spectral type K3 II. With Alshain (β Aquilae), mag. 3.71, it forms the 'wings' either side of the 'eagle star', Altair. The names Alshain and Tarazed come from a Persian expression referring to a beam or balance.

T association Region of recent and active star formation in which the dominant visible population comprises low-mass, roughly solar, stars. These stars, in their pre-MAIN-SEQUENCE phase of life, are styled T TAURI STARS. They are named after the prototype in the TAURUS DARK CLOUDS complex, which, at a distance of 500 light-years, is one of the nearest stellar nurseries to the Sun.

Low-mass stars form throughout the dark clouds of T associations, and EMBEDDED CLUSTERS can be observed at infrared wavelengths. Occasionally, high-mass young stars are also found in T associations, but only within the darkest, densest cores of clouds.

The importance of T associations is that their members afford a series of snapshots of our young Sun through its infancy. If we could order these different stars correctly, we would have constructed an evolutionary sequence for our Sun. It might also be clear how and when the planets formed, and over what period.

The T Tauri stars themselves are irregular optical variables. In the nearby Taurus–Auriga T association the stars are of the 11th to 19th visual magnitude. Infrared brightness characterizes T association stars, because they are still surrounded by dusty obscuring circumstellar material, which will eventually be accreted by the central star. Of topical interest is the manner in which the already formed stars of a T association interact with their parent gas clouds. Vigorous winds terminate the protostellar ACCRETION phase of T Tauri stars (perhaps after only a million years) and stir up the ambient medium of the T association through BIPOLAR FLOWS. Such flows are likely to act against the ready formation of new low-mass stars in their immediate vicinities.

Taurids METEOR SHOWER active between October 15 and November 25, producing low rates even during its broad peak in the opening week of November. At best, observed rates may reach 5 meteors/hr, from radiants near the Hyades and Pleiades. The shower is produced by debris from Comet 2P/ENCKE. Material in the Taurid meteor stream has become very spread out. Taurids are slow (30 km/s or 19 mi/s) and a reasonable proportion are bright. The stream is encountered again during June, when it produces the daytime BETA TAURIDS.

Taurus dark clouds Large complex of gas and dust clouds in Taurus forming a GIANT MOLECULAR CLOUD. They are about 500 l.y. away. The clouds are the site of continuing star formation and of many T TAURI STARS.

Taurus *See* feature article

Tautenburg Schmidt Telescope *See* KARL SCHWARZSCHILD OBSERVATORY

Taylor column Relatively stagnant column that forms over an obstacle immersed in a rotating fluid. Such columns were first studied in the laboratory by the British physicist Geoffrey Taylor (1886–1975). In the 1960s a Taylor column was suggested as the explanation of Jupiter's GREAT RED SPOT. This theory has since been discounted.

Taylor, Joseph *See* HULSE, RUSSELL ALAN

T corona Thermal emission from dust near the Sun, detected at INFRARED wavelengths. The dust producing the T corona also scatters sunlight to produce the F CORONA. *See also* CORONA

T Coronae Borealis *See* BLAZE STAR

TCT Abbreviation of TILTED-COMPONENT TELESCOPE

Teapot (of Sagittarius) Distinctive shape formed by eight stars in Sagittarius: γ, ε and δ Sagittarii form the spout of the teapot, λ Sagittarii is the tip of its lid, while φ, ζ, τ and σ Sagittarii outline its handle.

Tebbutt, Comet (C/1861 J1) Long-period comet discovered on 1861 May 13 by the Australian astronomer John Tebbutt (1834–1916). The comet reached perihelion 0.82 AU from the Sun on June 12, and became a bright, magnitude 0 object as it moved into northern hemisphere skies. Earth passed through the comet's tail on June 30 at the time of closest approach (0.13 AU), the first recorded instance of such an encounter. The tail reached a maximum extent of about 100°. Comet Tebbutt has an orbital period of 409 years.

technetium star Star, generally a GIANT STAR on the asymptotic giant branch of the HERTZSPRUNG–RUSSELL DIAGRAM, that shows technetium in its spectra. Technetium is an unstable radioactive element. It does not occur naturally on Earth but was the first element to be created artificially. As it has a HALF-LIFE of 2.1×10^5 years, shorter than the life of the stars in which it is observed, it must be created in the star. It is believed to be created during SHELL BURNING.

tectonics Deformation of the LITHOSPHERE of a planetary body. On Earth, the dominant mechanism for producing large-scale topographic features is 'plate tectonics'. Large sections of the lithosphere, thousands of kilometres in extent, are shifted by convective motions in the underlying MANTLE. Differential motions of these 'plates', moving at a few centimetres per year, result in distortions. Plates sliding past each other are bounded by strike-slip faults; these faults are sources of earthquakes, when accumulated stress causes rupture of the lithosphere. Head-on collisions at thrust faults cause compression and uplift, producing mountain ranges, such as the Himalayas at the conjunction of the Indian and Asiatic plates. Stretching of continental plates produces normal faults and linear depressed features (GRABENS). Oceanic crust is subducted into the MANTLE at trenches; it is renewed by lava rising at mid-ocean ridges where plates are pulling apart.

Plate tectonics is responsible for continental drift; over Earth's history the continents have been rearranged like pieces of a gigantic puzzle. Other planets have different styles of tectonics. Mercury's numerous scarps or ridges appear to be the result of compression; they apparently were produced by global shrinkage of the planet as its core cooled and solidified. Venus' large circular features (coronae) may have been produced by upward or downward motions of magma below the lithosphere. Mars has numerous grabens associated with its large volcanoes; the huge canyon VALLES MARINERIS may be a rift valley, although there is little evidence for plate tectonics. Jupiter's satellites Europa and Ganymede show evidence of motions of sections of their icy crusts.

TAURUS (GEN. TAURI, ABBR. TAU)

Conspicuous northern zodiacal constellation, between Aries and Orion, representing the bull disguise assumed by Zeus when he carried away Princess Europa in Greek mythology. Its brightest star, ALDEBARAN (which marks the bull's right eye), is a red giant irregular variable (range 0.75–0.95) that lies to the top-left of a V-shaped cluster of stars, the HYADES. ALNATH (or Elnath), the constellation's second-brightest star, used to be designated γ Aurigae. ALCYONE, mag. 2.9, is the brightest member of another naked-eye cluster, the PLEIADES. θ^2 and θ^1 Tau form a wide naked-eye double, mags. 3.4 and 3.8, separation over 5′, the former being the brightest star in the Hyades. Two interesting variables are the irregular variable T Tau (range 9.3–13.5 – *see* T TAURI STAR), which is embedded in and illuminates Hind's Variable Nebula (NGC 1555), a small reflection nebula that shows changes in both brightness and extent; and the highly luminous pulsating variable RV Tau (range 9.8–13.3, period about 79 days, superimposed on a cycle of about 1224 days – *see* RV TAURI STAR). Taurus also contains the CRAB NEBULA (M1, NGC 1952), at 8th magnitude, the brightest supernova remnant in the sky, whose centre is marked by the CRAB PULSAR (known at radio and X-ray wavelengths as Taurus A and Taurus X-1). The Taurid meteor shower radiates from Taurus.

BRIGHTEST STARS

	Name	RA h m	dec. ° ′	Visual mag.	Absolute mag.	Spectral type	Distance (l.y.)
α	Aldebaran	04 36	+16 31	0.87	−0.6	K5	65
β	Alnath	05 26	+28 36	1.65	−1.4	B7	131
η	Alcyone	03 47	+24 06	2.85	−2.4	B7	368
ζ		05 38	+21 09	2.97	−2.6	B4	417
θ^2		04 29	+15 52	3.40	0.1	A7	149

tektite Object of natural glass found scattered across four main areas, known as STREWNFIELDS, of the Earth's

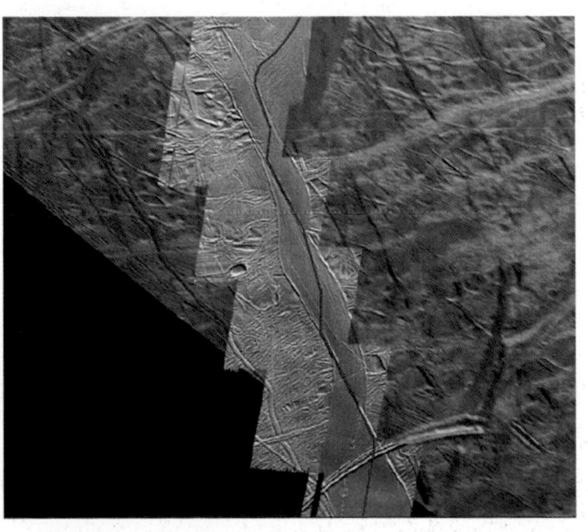

◀ **tectonics** On the left is Earth's San Andreas fault and on the right, the strike-slip fault Astypalaea Linea on Jupiter's satellite Europa. The icy surface of Europa appears to fracture in blocks as it moves over the underlying ocean.

TELESCOPIUM (GEN. TELESCOPII, ABBR. TEL)

Small, inconspicuous southern constellation, representing a telescope, between Ara and Indus. It was introduced by Lacaille in the 18th century. Its brightest star, α Tel, is mag. 3.5. RR Tel is a NOVA-LIKE VARIABLE, and the only such star to have been recognized as variable before it began its slow (1600-day), unprecedented rise to 7th magnitude in 1944.

surface. The Australasian strewnfield stretches from Tasmania, across Australia to Sumatra and the Philippines and to the Southeast Asian mainland. Australian tektites, known as australites, are 830,000 years old, compared with 690,000 years old for the remainder of the strewnfield, thus two groups may be represented. The parent crater for this strewnfield is unknown. The Ivory Coast tektites are 1.3 million years old. They are related to the impact crater of similar age at Lake Bosumtwi, Ghana. Moldavites are tektites found in the Czech Republic. Their age, of 14.7 million years, links them with the Ries impact structure in south Germany. The North American tektites, found in Texas and Georgia (and one only from Martha's Vineyard, Massachusetts), are 34 million years old. They are now linked to the Chesapeake Bay impact structure. Micrometeorites related to the Australasian, Ivory Coast and North American tektites occur in marine sediments.

Tektites are unrelated to volcanism. Their usual shapes, such as teardrop, disc or dumbbell, suggest that they were once fluid, but they may be corroded. The Southeast Asian tektites are the largest, with masses of up to 15 kg; the Muong-Nong type have laminar flow structures. Many of the south-east Australian tektites have features indicative of high-velocity atmospheric flight, such as a 'button' shape, atmospheric ablation having melted the front part of an original sphere, causing melt to flow backwards to form a flange.

Ivory Coast and Australasian tektites are opaque, unless very thin. Moldavites are green and North American tektites brown. All are silica-rich, containing inclusions of lechatelierite (pure silica glass) and other residual minerals. A lunar origin was suggested for tektites, but the return of lunar samples reinforced the terrestrial chemical and isotopic signatures. Tektites are terrestrial material that has been subjected to extraterrestrial impact. Following the impact, molten or vaporized ejected material coagulated and re-entered the atmosphere to land as tektites.

▼ **terminator** Because features near the day–night boundary on the Moon are lit obliquely, more details can be observed than at other times. This Galileo orbiter image of the Moon reveals central peaks within craters close to the terminator, while little structure is seen in other areas.

tele-compressor Converging lens inserted into the light path of a telescope to reduce the focal length; it produces a wider field of view and a faster focal ratio for photographic purposes.

tele-extender Extension tube containing an optical device and used in conjunction with an EYEPIECE to increase the effective focal length of a telescope.

telescope Device that augments the ability of the eye to observe distant or faint objects, particularly celestial objects. The Egyptians were making glass around 3000 BC, and the Greeks had a good understanding of the refraction and reflection of light by glass lenses in 300 BC, but it was nearly two thousand years before the telescope made its appearance. In 1608 the Dutch spectacle-maker Hans LIPPERSHEY applied for a patent for his 'instrument for seeing at a distance', the first REFRACTING TELESCOPE. While Lippershey is generally acknowledged to be the inventor of the telescope, there were enough competing claims that his patent application was denied. There is also some evidence that devices resembling the telescope were around in the previous century (*see* DIGGES).

It was in May 1609 that Galileo first heard about the telescope while travelling in Venice. He proceeded to build one within 24 hours of returning home. His design, the GALILEAN TELESCOPE, enabled him to observe four moons of Jupiter, mountains and valleys on the Moon, sunspots and the phases of Venus. However, it suffered from severe CHROMATIC ABERRATION and SPHERICAL ABERRATION.

Around 1616, Niccolò Zucchi (1586–1670), a professor of mathematics from Rome, proposed a solution to these problems by replacing the objective lens with a mirror, creating the first REFLECTING TELESCOPE, but got poor results from the example he constructed and abandoned the effort. Johannes KEPLER, René Descartes (1596–1650), Marin Mersenne (1588–1648), James GREGORY and Robert HOOKE all developed alternative designs, but the technologies of the day were unable to produce workable instruments from their specifications.

By the late 1650s, the only solution to the aberrations that plagued the refractors of the day was to increase the FOCAL LENGTH. Chromatic and spherical aberration are both reduced when the focal length is increased, and so the telescopes grew. By 1670 almost unusable *f*/300 monsters stretching 67 m (220 ft) from objective to eyepiece were being built.

In 1671 Isaac Newton began to experiment with the possibility of an achromatic refractor. Due to a flaw in his experimental procedure, Newton concluded that an achromatic refractor was not possible, and began to research Zucchi's reflecting designs. The result was the NEWTONIAN TELESCOPE, although it would be almost 300 years before the design gained the popularity it enjoys today. Appearing at about the same time was the CASSEGRAIN TELESCOPE, which would eventually become the most popular design for large professional instruments, despite Newton's intense criticism of the design. *See also* DALL–KIRKHAM TELESCOPE; DOBSONIAN TELESCOPE; HERSCHELIAN TELESCOPE; MAKSUTOV TELESCOPE; RITCHEY–CHRÉTIEN TELESCOPE; SCHMIDT–CASSEGRAIN TELESCOPE; TILTED-COMPONENT TELESCOPE.

Telescopium *See* feature article

Telesto Small satellite of SATURN, discovered in 1980 by Brad Smith and others in images returned by the VOYAGER missions. Telesto is irregular in shape, measuring about 30 × 26 × 16 km (19 × 16 × 10 mi). *See also* CALYPSO

telluric line SPECTRUM line produced by the Earth's atmosphere superimposed on celestial spectra. FRAUNHOFER LINES A and B (at 759.4 and 686.7 nm) are telluric absorptions from oxygen (O_2), while Fraunhofer a (716.5 nm) is from water. Telluric carbon dioxide and water absorptions block much of the infrared; telluric ozone blocks the ultraviolet. Telluric absorption lines strengthen as the body sets, allowing them to be identified. Telluric lines can in part be avoided by Doppler shifts caused by the relative motion of the Earth and the body being observed. Telluric auroral emission lines also contaminate astronomical spectra.

Tempel, (Ernst) Wilhelm (Leberecht) (1821–89) German observational astronomer, a prolific discoverer of comets. He found his first comet (C/1859 G1) from Italy, using a 4-inch (100-mm) Steinheil refractor. Over the next 18 years he discovered twelve more of these objects, five asteroids and several nebulae, including the reflection nebula around Merope in the PLEIADES. Most of Tempel's comet discoveries were made from Marseilles Observatory; he also worked as Giovanni SCHIAPARELLI's assistant at Brera Observatory (Milan) and directed ARCETRI OBSERVATORY (Florence).

Tempel-1, Comet 9P/ Short-period comet discovered on 1867 April 3 by Wilhelm Tempel. It is always faint and was lost from 1879 until 1967. The orbit was modified by passages close to Jupiter in 1881, 1941 and 1953. Currently, the perihelion distance is 1.49 AU, and the period is 5.51 years.

Tempel–Tuttle, Comet 55P/ Short-period comet discovered on 1865 December 19 by Wilhelm Tempel, Marseilles, France, and independently on 1866 January 5 by Horace Tuttle (1837–1923), Harvard, USA. The comet currently has a period of 33.22 years, and the orbit is retrograde, being inclined to the ecliptic by 162°.5. During its discovery apparition, the comet reached peak magnitude +3, and showed a short tail. Although the orbit was established at this return, Comet 55P/Tempel–Tuttle was not observed again until 1965. At one of its past returns, in 1366, it made the third-closest approach (0.0229 AU) of a comet to Earth in historical times. The most recent return, with perihelion on 1998 February 28, was quite favourable, with the comet reaching peak magnitude +6.5. Comet 55P/Tempel–Tuttle is the parent of the LEONIDS, which have produced meteor storms close to several of the comet's past returns (most recently in 1966, 1999 and 2001); they show substantially enhanced activity for about five years to either side of perihelion.

Tempel-2, Comet 10P/ Short-period comet discovered on 1873 July 3 by Wilhelm Tempel. Its current period is 5.47 years. IRAS observations in 1983 showed 10P/Tempel-2 to have an associated dust trail more than 0.67 AU (100,000,000 km or 62,000,000 mi) long.

temperature Property of an object that determines whether heat energy will flow into it or out of it when it is in contact with another object. The direction of flow of heat energy is from the object at the higher temperature to that at the lower. If there is no energy flow then the two objects are in THERMAL EQUILIBRIUM.

Temperature is quantified through various physical phenomena. The most generally applicable measure is the EFFECTIVE TEMPERATURE, which is based on the temperature of an equivalent BLACK BODY. Other temperatures include: the RADIATION TEMPERATURE, which is based on the black body emission at a single wavelength; the colour temperature, based upon the COLOUR INDEX; the thermal or kinetic temperature, based upon the thermal motions of the atoms, ions, electrons and so on; the excitation temperature, based upon the levels of EXCITATION of the electrons within an atom or ion; and the ionization temperature, based upon the levels of IONIZATION of an element. The value obtained for the temperature may vary depending on the property upon which it is based. Thus the effective temperature of the solar CORONA is about

◀ **Tethys** Ithaca Chasma can be seen to the right in this image of Saturn's satellite Tethys. Younger craters are superimposed on the giant trough, indicating that it is an ancient feature.

100 K, while the kinetic temperature of the electrons within it is about 1,000,000 K.

temperature scale Graduated scale of degrees for measuring TEMPERATURE. The principal temperature scales are the ABSOLUTE (or Kelvin) scale and the CELSIUS (or Centigrade) scale. The unit of temperature on the absolute scale, the kelvin (K), is equal in magnitude to one degree on the Celsius scale (°C). The zero points of the two scales differ however, with 0°C equal to 273.15 K. To convert from one scale to the other, 273.15 has to be added to the temperature on the Celsius scale or subtracted from the temperature on the absolute scale. All physical formulae require the temperature to be expressed on the absolute scale.

Tempe Terra Region of MARS east of ACIDALIA PLANITIA (40°.0N 71°.0W). It is associated with the long 'ditch' Tempe Fossa.

terminator Boundary between the illuminated and non-illuminated hemispheres of a planet or satellite; in other words, the dividing line between day and night. From the Earth, the terminator is only visible on those bodies that exhibit PHASES, that is Mercury, Venus, Mars and the Moon. Because of its relative proximity, it is possible to see that the mountainous surface of the Moon causes its terminator to appear jagged and irregular, rather than as a smooth line.

terracing Formation of terraces – relatively flat, horizontal or gently inclined surfaces, bounded by a steeper ascending slope on one side and by a steeper descending slope on the opposite side. Terracing is typical on the inward-looking slopes of large impact CRATERS.

terrestrial dynamical time (TDT) *See* DYNAMICAL TIME

terrestrial planet Term used to describe the inner, rocky planets of the Solar System (Mercury, Venus, Earth and Mars), which have similar characteristics in terms of size and density. These contrast with the largely gaseous giant or JOVIAN planets. (Pluto, having more in common with cometary planetesimals, is not generally included in this group.) The differences between the terrestrial and Jovian planets are thought to have been caused by their respective distances from the Sun during the early formation of the Solar System.

Tethys Fifth-largest SATELLITE of SATURN. It was discovered by G.D. CASSINI in 1684, the year in which he also discovered its outer neighbour, DIONE. Tethys shares its orbit with the tiny irregular satellites TELESTO and CALYPSO, which travel in stable LAGRANGIAN POINTS 60°

ahead of it and 60° behind it, respectively. Tethys is similar in size to Dione and is heavily cratered. In one region the plains between large ancient craters have been flooded by CRYOVOLCANISM up to the crater rims. This happened a long time ago, because there has been time for many younger (and smaller) craters to have formed on the plains. Tethys is remarkable in having one enormous crater, named ODYSSEUS, the diameter (440 km/270 mi) of which is 40% of the diameter of the satellite itself. The impact that created this crater was almost violent enough to break Tethys into fragments.

A giant trough up to 100 km (60 mi) wide and 3 km (2 mi) deep, named Ithaca Chasma, encircles the globe 90° away from Odysseus. The trough may be a fracture formed in response to the impact that created the crater. The only other substantial satellite to have a crater of this magnitude with respect to the size of the whole body is MIMAS. However, whereas the giant crater on Mimas has retained its pristine shape, the topography of Odysseus has become subdued, probably because the LITHOSPHERE of Tethys was too weak to support it. *See* data at SATURN

Thalassa Small inner satellite of NEPTUNE, discovered in 1989 by the VOYAGER 2 imaging team. Thalassa is about 80 km (50 mi) in size. It takes 0.312 days to circuit the planet, at a distance of 50,100 km (31,100 mi) from its centre, in a near-circular, near-equatorial orbit.

Thales of Miletus (*c.*625–*c.*547 BC) Greek polymath, born in what is now modern Turkey, famous for his teaching that everything in the Universe sprang from a primordial mass of water. According to Herodotus (*c.*484–425 BC), Thales predicted the total eclipse of the Sun that occurred in 585 BC, though this may be apocryphal. Thales travelled to Egypt and Babylon, bringing back to Greece the astronomical knowledge of those civilizations. He defined the constellation Ursa Minor, which, because of its inclusion of the North Pole Star, he used for navigational purposes.

Tharsis Montes (Tharsis Ridge) Main volcanic region of MARS. Tharsis is a complex area with two rises. In the northern part stands the volcano ALBA PATERA. On the main Tharsis Ridge, in the southern half of Tharsis, stand three aligned volcanoes, ARSIA MONS, PAVONIS MONS and ASCRAEUS MONS, each separated from the next by about 700 km (430 mi).

The Astronomer (TA) Monthly UK publication providing rapid dissemination of raw reports from observers around the world, established as *The Casual Astronomer* in 1964. TA also performs a relatively informal co-ordinating role for various PATROLS, notably for novae and supernovae, alongside the BRITISH ASTRONOMICAL ASSOCIATION.

Thebe One of the inner moons of JUPITER, discovered in 1979–80 by Stephen Synnott in images obtained by the VOYAGER project. Thebe is irregular in shape, measuring about 110 × 90 km (70 × 60 mi). It orbits Jupiter at an average distance of 221,900 km (137,900 mi), giving it a period of 0.675 days. Its orbit is of low eccentricity (0.018) and is inclined to the Jovian equator by only 1°.1.

Theophilus Lunar crater (12°S 26°E), 101 km (63 mi) in diameter. It is a complex crater, with terraced sidewalls and central peaks. Theophilus is just old enough for most of the bright rays to have been eroded away, though a few are still visible. Its continuous and discontinuous EJECTA are easily observed at low Sun angles. Theophilius overlies part of the older crater CYRILLUS.

theory of everything (TOE) Theory that is an attempt to unify gravity with the other three known FUNDAMENTAL FORCES. The GRAND UNIFIED THEORY (GUT) unifies the strong, weak and electromagnetic forces into a consistent quantized theory, but does not allow for the quantization of gravity. A TOE would also unify all three GUT forces with gravity in some type of all-encompassing theory. The SUPERSTRING THEORY is one such theory that attempts to understand all four forces and all of the known ELEMENTARY PARTICLES in one grand theory of everything. *See also* QUANTUM THEORY OF GRAVITY

thermal equilibrium State in which all parts of a system have the same TEMPERATURE and there is no flow of heat energy from one part of the system to another or to or from the surroundings.

thermal radiation RADIATION that arises as a consequence of the thermal energy in a body. For astronomical sources it is usually either BLACK BODY RADIATION or FREE–FREE RADIATION. Non-thermal radiation sources include SYNCHROTRON RADIATION and MASER emission. If these latter sources are mistakenly interpreted as thermal radiation, then the inferred temperature may be many orders of magnitude too high.

thermosphere Outermost layer of the Earth's ATMOSPHERE. It lies above the mesopause, at a height of 86–100 km (53–62 mi), which is the upper boundary of the MESOSPHERE. Within the thermosphere, the temperature rises with height as the Sun's far-ultraviolet and X-ray radiation is absorbed by the oxygen and nitrogen of the thin air. This process produces ionized atoms and molecules and is responsible for the formation of the layers in the IONOSPHERE, at altitudes of between 60 and 500 km (40 and 310 mi). AURORAE and METEORS also occur in the thermosphere. The temperature climbs to about 780 K at 150 km (90 mi) and 1600 K at 500 km (310 mi). However, the atmospheric density is extremely low, decreasing from seven-millionths of its sea-level value at the mesopause to only about one-million-millionth (10^{-12}) at 500 km (310 mi) altitude. This density varies with solar activity, increasing when the Sun is active, when the additional heating of the uppermost layers causes the atmosphere to expand. The region above 500 km (310 mi) is often known as the EXOSPHERE.

Third Cambridge Catalogue (3C) Third in a series of four catalogues of radio sources compiled from surveys carried out from the Cavendish Laboratory, Cambridge, during the 1950s and 1960s by a team that included Antony HEWISH, Martin RYLE and Francis GRAHAM-SMITH. The Third Cambridge Catalogue Survey was completed in 1959, using collectors of greater area and higher sensitivity than its two predecessors; it became the definitive listing of radio sources. Sources in this catalogue are still known by their '3C' designations, for example 3C 273, the first object to be identified as a quasar.

third contact During a SOLAR ECLIPSE, the moment when the Moon's trailing (westerly) limb appears to touch the Sun's more westerly limb. In a TOTAL ECLIPSE, this marks the end of TOTALITY and is usually quickly followed by the appearance of the DIAMOND RING EFFECT and BAILY'S BEADS, as the Sun's westerly edge is once more revealed.

At a LUNAR ECLIPSE, third contact is the time when the Moon's leading, easterly limb arrives at the eastern edge of Earth's UMBRA, re-emerging into sunlight at totality's end.

Thomson, William *See* KELVIN, LORD (WILLIAM THOMSON)

three-body problem Fundamental problem in CELESTIAL MECHANICS, to determine the motions of three bodies under the influence of their mutual gravitational attractions. There is no exact general solution, only solutions for some special cases, such as the LAGRANGIAN POINTS. It is the formulation that is used as the basis for most theoretical and perturbational studies in celestial mechanics. There has been a vast amount of work to investigate the types of motion that occur close to a COMMENSURABILITY, and all of this takes the three-body

T

problem as the basis, usually with various simplifications in order to isolate the crux of the problem. For example, when analysing the motion of an asteroid close to a commensurability with Jupiter, one might use the plane circular restricted three-body problem. This considers Jupiter to be in a circular orbit, unperturbed by other planets, and the asteroid to be of zero mass and to move in the plane of Jupiter's orbit.

The task of determining analytically the PERTURBATIONS of the orbit of a planet or satellite caused by other bodies is invariably reduced to a series of three-body problems. For example, what is referred to as 'the main problem' of lunar theory is to determine the orbit of the Moon around the Earth perturbed by the Sun, with all three bodies regarded as point masses. Perturbations by other planets, and effects of the oblateness of the Earth and the Moon are added later. Even so, it is a substantial problem. The lunar theory constructed by Ernest BROWN occupied him for most of his life; it contains about 1020 periodic terms from the main problem, with a further 580 terms from planetary perturbations. *See also* MANY-BODY PROBLEM

3C Abbreviation of *THIRD CAMBRIDGE CATALOGUE*

3C 273 (1226+023) First-identified and optically brightest quasar, at z = 0.158 in the direction of Virgo. The optical object was originally identified by timing the disappearance of the radio emission during a lunar occultation, as observed with the Parkes radio telescope. This object has by far the brightest optical and X-ray jet among quasars.

3 kiloparsec arm Spiral arm lying about 3.5 kpc from the centre of our GALAXY; it is visible as 21 cm neutral hydrogen radio emission (*see* TWENTY-ONE CENTIMETRE LINE). The emission lies within about 20° longitude from the galactic centre; it is expanding outwards at about 50 km/s (30 mi/s).

Thuban The star α Draconis, visual mag. 3.67, distance 309 l.y., spectral type A0 III. It was the North Pole Star around 2800 BC. The name is from the Arabic *thu'bān*, meaning 'serpent'.

tidal bulge Deformation of a celestial body in orbit around another as a result of the higher gravitational attraction on the nearside and lower attraction on the farside of the body, compared with the attraction at the centre of the body. The resulting distortion lies along the line joining the two bodies, with bulges both towards and away from the attracting body (*see diagram at* TIDES). The effect on a fluid ocean is easily understood, but even solid bodies are not completely rigid and will deform because of tidal forces. The Sun raises tidal bulges on the planets, and planets raise tidal bulges on the satellites orbiting them. Satellites also raise tidal bulges on their planets. In the case of the Earth and Moon, the tides raised on the Earth are large, because the Moon is a fairly large fraction (1/81) of the mass of the Earth. However even a tiny satellite such as Phobos raises tides on its planet, Mars. These tides are small, and have no significant effect on Mars, but the attraction of this bulge acting back on Phobos causes very significant TIDAL EVOLUTION of the orbit of Phobos.

tidal evolution Mechanism that can cause changes to the spin rates of planets and satellites, and to the orbital radii of satellites. The rates of change are very slow, but the accumulated effect can be very significant over the age of the Solar System. The attraction of one body on another raises a TIDAL BULGE on that body. TIDAL FRICTION causes a slight delay in the rise and fall of the bulge, with the result that it does not quite lie along the line between the two bodies: it will either lead or lag, depending on whether the attracting body is orbiting faster or slower than the body is rotating. Thus the gravitational attraction between a satellite and its planet is asymmetrical, with the force having a component at right angles to the line between the bodies; this force exerts a torque. In the case of a satellite orbiting slower than the planet is rotating (that is, it is above the SYNCHRONOUS ORBIT), the nearer bulge exerts a larger torque than the farther bulge, and so there is a net torque acting in the direction of motion of the satellite, and an equal and opposite torque acting on the rotation of the planet. This causes the satellite to spiral outwards, and slows the rotation of the planet. A satellite below the synchronous orbit would spiral inwards, and the planet would rotate faster.

Tidal evolution has had many effects on the Solar System. Tides raised by the Sun on planets can despin the planets; this is a probable explanation for the present slow rotation rates of Mercury and Venus. Tides raised by satellites on planets have had numerous effects: causing the Moon to recede and the Earth's rotation to slow down; causing Phobos to spiral into Mars; and causing satellites of Jupiter and Saturn to spiral outwards and become trapped in COMMENSURABILITIES. Tides raised by planets on satellites have resulted in most satellites being in SYNCHRONOUS ROTATION with their planet; they have caused intense heating within some satellites, notably IO.

tidal friction Dissipation of a small fraction of the energy contained in a TIDAL BULGE as it rises and falls. In the case of the Earth the main dissipation of tidal energy arises from friction at the bottom of shallow seas as the tides sweep through. In bodies without oceans the dissipation mechanisms are not so obvious, and estimates of the rate of dissipation are very uncertain. In a few cases the dissipation of tidal energy within a body can cause significant heating of the body, such as for Jupiter's satellite IO. More often the most significant effect of the tidal friction is to cause a slight delay in the rise and fall of the tidal bulge, with the result that it does not quite lie along the line between the two bodies. This slight asymmetry causes a torque between the two bodies, which can cause TIDAL EVOLUTION.

tidal heating Heat produced by the tidal force exerted on a satellite by its primary; this force can distort the satellite. The tidal force varies with distance, and if the satellite's orbit is eccentric the satellite will change shape slightly during each revolution. The work done by this

◀ **tidal heating** Jupiter's satellite Io is the best example in the Solar System of the effects of tidal heating. Without the effects of Jupiter's gravity Io would be a dead world. Instead it is dotted with active volcanoes through which it releases energy.

T

flexing is dissipated as heat; the energy is derived from its orbital energy. Tidal heating drives VOLCANISM on Jupiter's satellite IO. It probably maintains a liquid layer beneath the icy crust of EUROPA.

Tidbinbilla Tracking Station *See* DEEP SPACE NETWORK

tides Distortions induced in a celestial body by the gravitational attraction of one or more others. The force a body experiences is greatest on the side nearest the attracting body, and least on the side farthest away, causing it to elongate slightly towards and away from the attracting body, acquiring a TIDAL BULGE on each side.

On the Earth tides are raised by the Moon and by the Sun, the effect of the Moon being about three times that of the Sun. As the Earth rotates, different parts of the surface move into the tidal bulge region, causing two high tides and two low tides in just over a day. The tides most obviously affect the fluid ocean, but even the solid crust, which is supported by a fluid mantle, is able to flex, experiencing tides of up to 25-cm (*c.*10-in.) amplitude. When the Sun and the Moon are exerting a pull in the same direction (as at new moon or full moon) their effects are additive and the high tides are higher (spring tides). When the pull of the Sun is at right angles to that of the Moon (as at first or last quarter) the high tides are lower (neap tides).

TIDAL FRICTION caused by the tidal ebb and flow of water over the ocean floor causes a slight delay in the response of the oceans to the tidal attracting forces, with the result that the tidal bulge does not lie precisely along the line to the attracting body. This asymmetry causes a torque between the Earth and the Moon (*see also* TIDAL EVOLUTION), which transfers angular momentum and energy from the Earth's rotation to the orbit of the Moon. This results in a slowing of the Earth's rotation and thus an increase in the length of the day (*see also* LEAP SECOND), and in the SECULAR ACCELERATION of the Moon, causing it to recede from the Earth at about 3.8 cm/yr (1.5 in./yr). A similar mechanism is probably responsible for evolution of the orbits of many of the natural satellites of the Solar System.

The tidal forces between the Earth and the Moon also cause a tidal bulge on the Moon. A similar effect of the transfer of angular momentum from the Moon's rotation to its orbit has resulted in the present state of SYNCHRONOUS ROTATION, with the same face of the Moon always towards the Earth. Similarly most of the satellites of the major planets are in synchronous rotation with their primary. For the Jovian satellite IO the tidal effect is very large, due to its relative closeness and the large mass of Jupiter. Also Io has a fairly large orbital eccentricity (0.0043), and this causes the tidal bulge on Io to rise and fall and oscillate around a mean position as the distance and direction of Jupiter vary in each orbital revolution. There is a dynamical process that would normally react on the orbit and reduce the eccentricity to eliminate this effect, but it cannot operate for Io, as its eccentricity is actually caused by its 2:1 COMMENSURABILITY with Europa. The result is that the continued dissipation of tidal energy has caused internal heating and is responsible for the violent volcanic activity on Io.

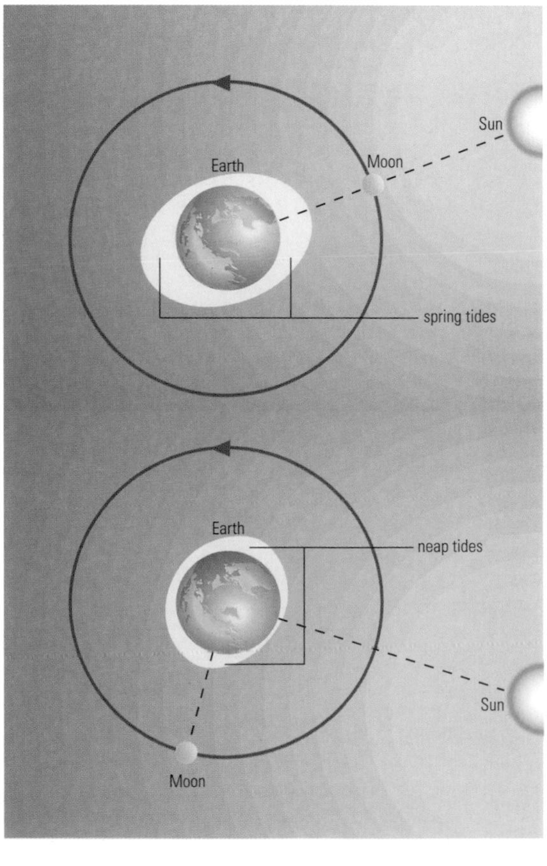

► **tides** The gravitational effects of the Sun and Moon combine to distort the oceans on Earth. When the Sun and Moon are lying in the same direction, high tides are at their highest and the difference between high and low tides is at its greatest magnitude. When the Sun and Moon are pulling in different directions, neap tides result – high tides are not particularly high, and the difference between high and low tide is less.

tilted-component telescope (TCT) REFLECTING TELESCOPE that avoids obstructing the optical path by tilting or warping some or all of the optical elements. The first attempt at a tilted-element telescope was the **brachyt**, developed around 1900. This was essentially a CASSEGRAIN TELESCOPE cut in half, and it had off-axis mirrors that were virtually impossible to grind. An improvement was the **neo-brachyt** in 1953; it had normal mirrors but required a warping harness to distort the secondary into the proper curve. The **schiefspiegler**, invented in 1940 by Anton Kutter (1903–85), was much easier to make, having a convex and concave mirror of the same curvature, so that the secondary could be made from the tool that ground the primary. The schiefspiegler was limited to long focal ratios in the *f*/20 to *f*/30 range. Kutter later refined this design into the **tri-shiefspielgler** in 1965. In 1957 Arthur S. Leonard invented the **Yolo**, a re-interpretation of the neo-Brachyt design with better performance, but still requiring the warping harness. Leonard's later design, the **Solano** in 1971, added a third mirror and eliminated the harness. (Both designs are named after Californian counties.)

The culmination of TCT designs is the **Stevick–Paul**. Invented in 1993 by amateur David Stevick (1944–) and based on the design for a corrector assembly by French optician Maurice Paul from 1935, the Stevick–Paul has a tilted focal plane but negligible other distortions and can be built to reasonably fast (low) focal ratios.

time Property of the Universe determined by the observed order of events and in which effects follow causes. Newtonian or classical physics allowed the existence of an absolute time framework, whereby different observers could agree on the time of occurrence of an event by making an allowance for the travel times of light over different distances. Such an absolute time framework permitted the simultaneous occurrence of events taking place at different points in space. The concept of simultaneity breaks down under relativistic physics. With RELATIVITY, different observers measure time elapsing at different rates depending on their relative velocities, accelerations and strengths of their local gravitational fields (or SPACETIME distortions). Nonetheless, even under relativity events always follow causes, and a sequence of events occurring at a single point in space is seen by all observers to occur in the same order. Only if information could be passed from observer to observer faster than the speed of light in a vacuum, could this causality basis of time be upset, thus allowing travel backwards in time.

Time, along with the three dimensions of space, becomes spacetime as conceived by relativity. The separation of events in spacetime is invariant for all observers and is given by:

$$s^2 = x^2 + y^2 + z^2 - c^2t^2$$

where x, y and z are the spatial coordinates, t is the time, and c the velocity of light in a vacuum. Newtonian gravitational forces are then replaced by particles following geodesic tracks through a spacetime that has been distorted from Euclidean geometry by the presence of masses.

The measurement of time is the subject of ordinary observation and experience and is based upon sequences of events occurring at regular intervals. Annual and diurnal changes led in pre-history to the concept of the year and the day. Less clearly, lunar cycles may have led to the month. The development of clocks of increasing accuracy and reliability began with the sundial and nocturne, and proceeded through the development of the water clock, pendulum, balance wheel, quartz crystal to the atomic clock of today.

The current unit of time is the second, which has been defined since 1967 as 'the duration of 9,192,631,770 periods of the radiation corresponding to the transition between the two hyperfine levels of the fundamental state of the atom caesium 133'. Within the limits of measurement, this definition was identical to the one in use previously, which was based upon the Earth's orbital motion. This 'ephemeris second' was defined as 1/31,556,925.9747 of the TROPICAL YEAR for 1900 January 0^d 12^h. INTERNATIONAL ATOMIC TIME (TAI) is now used as the starting point for all other types of time scale.

The time scale used for civil purposes is UNIVERSAL TIME (UT), otherwise known as GREENWICH MEAN TIME (GMT). It is defined as the HOUR ANGLE of the MEAN SUN plus 12 hours (so that the start of the civil day is midnight, not midday). UT is based upon the Earth's rotation, not its orbital motion, and it is therefore affected by changes in the rotation of the Earth. The UT corrected for the CHANDLER WOBBLE is called UT1, and is the basis of civil timekeeping. It is kept to within 0.9 seconds of TAI by the occasional insertion or removal of a LEAP SECOND. The difference between the time given by a sundial (solar time) and UT is called the EQUATION OF TIME.

The SOLAR DAY is 86,400 atomic seconds long. The Earth's actual rotation period is given by the SIDEREAL DAY, and this is 86,164.1 atomic seconds (23^h 56^m $4^s.1$) long. *See also* CALENDAR; JULIAN DATE; TIME ZONES.

time dilation effect Effect whereby a clock on a moving object runs slower than a similar clock stationary with respect to the observer (*see* SPECIAL RELATIVITY). The dilation is by a factor of:

$$\sqrt{1-\frac{v^2}{c^2}}$$

where v is the object's velocity and c the velocity of light. Thus an astronaut visiting a planet 25 l.y. away at a constant speed of 0.999c (ignoring acceleration and deceleration times) would age by one year, while 25 years would elapse on Earth. GENERAL RELATIVITY predicts gravitational time dilation, whereby clocks run slowly in strong gravitational fields. Both kinds of time dilation have been confirmed experimentally.

timekeeping Our concept of the passage of TIME is based on the regular rising and setting of the Sun each day, in other words the rotation of the Earth, which provides a ready-made framework by which to regulate events in our lives. The Earth, however, is not a good timekeeper; its rotation rate is irregular and slowing because of the effects of tidal braking. This did not become apparent, however, until the early 20th century, when the free pendulum clock was developed, which was accurate to within a second a year. Subsequent quartz crystal-controlled clocks and atomic clocks confirmed the small, but significant, variations in the Earth's rate of rotation.

The first crude methods of accounting for the passage of time involved measuring the length of the shadow cast by a stick placed upright in the ground. Early mechanical clocks appeared during the 15th and 16th centuries and it was then that the system of splitting the day up into hours of equal length came into regular use. The invention of the pendulum clock in the 17th century provided a more accurate means of timekeeping. During the 18th century clocks were perfected that were accurate enough to be used at sea to help determine longitude and also for scientific purposes. But even with these man-made methods of accounting for time, the basic way of measuring its passing was still the rotation of the Earth, and clocks were adjusted accordingly to keep in step with it.

Until 1955 the scientific standard of time, the second, was based on the Earth's period of rotation and was defined as 1/86,400 of the mean solar day. The current standard, the SI second, was defined in 1967 as being the duration of 9,192,631,770 periods of the radiation corresponding to the transition between two hyperfine levels of the ground state of the caesium-133 atom (*see* ATOMIC TIME). The internationally adopted timescale INTERNATIONAL ATOMIC TIME (TAI) is formed by intercomparing the results from many different atomic clocks around the world.

In the UK, atomic clocks at the National Physical Laboratory (NPL) at Teddington keep the country's time accurate to within one second in three million years. These clocks are used to disseminate time in a variety of ways, including via the MSF time signal, which is broadcast from the Rugby Radio Station, and the familiar 'six pips'. For civil time-keeping purposes a uniform TIMESCALE is used; it is formed by atomic clocks, but kept in step with the rotation of the Earth through the periodic introduction of LEAP SECONDS. For convenience, we also advance our clocks in summer by one hour in order to make better use of the increased daylight hours.

Our need for, and use of, accurate time is expanding. For everyday life, today's watches and clocks are sufficient, but for other applications, more accurate methods are needed. Telecommunications systems rely on being able to measure precisely small intervals of time in order to ensure that data transmitted down telephone lines is sent at the correct rate. This has become ever more imperative with the all-pervasive Internet and the increasing use of electronic means of communication such as e-mails. Navigation systems, used aboard shipping and aircraft (and even becoming widely available for road vehicles), use the Global Positioning System (GPS) of satellites which have atomic clocks on board. These satellites broadcast timing signals that enable ground positions to be established to within a few tens of metres.

The quest for ever-more accurate methods of timekeeping goes on. Research is currently underway into the development of clocks using 'ion traps', in which ions held in an electromagnetic field are cooled by laser beam almost to absolute zero, keeping them stationary. It is anticipated that clocks using this technology will be around 1000 times more accurate than the current generation of atomic clocks. That means that they would lose no more than a single second over the entire lifetime of the Universe. *See also* GREENWICH MEAN TIME (GMT)

timescale Specified system of measuring the passage of TIME based on a given frame of reference. There are two kinds of timescale: those based on the rotational or orbital period of the Earth, and those that measure continuous units of time defined by atomic processes.

The best-known timescale in the world is GREENWICH MEAN TIME (GMT), which is based on the MEAN SOLAR TIME at the Greenwich MERIDIAN; it is officially now known as UNIVERSAL TIME (UT). UT is a non-uniform timescale, derived by accurately and continually measuring the rotation of the Earth. Raw data from these measurements are used to produce a timescale called UT0, which is then corrected for the shift in longitude caused by the slight wandering of the Earth's geographical poles (*see* POLAR MOTION). This corrected timescale is designated UT1.

Since the rotation of the Earth is non-uniform, being subjected to tidal braking, any timescale based on this rotation will also be non-uniform. For scientific purposes, therefore, a continuous, uniform timescale is required and this is provided by INTERNATIONAL ATOMIC TIME (TAI), which is based on the SI second (*see* ATOMIC TIME). TAI has run continuously since 0^h 0^m 0^s GMT on

T

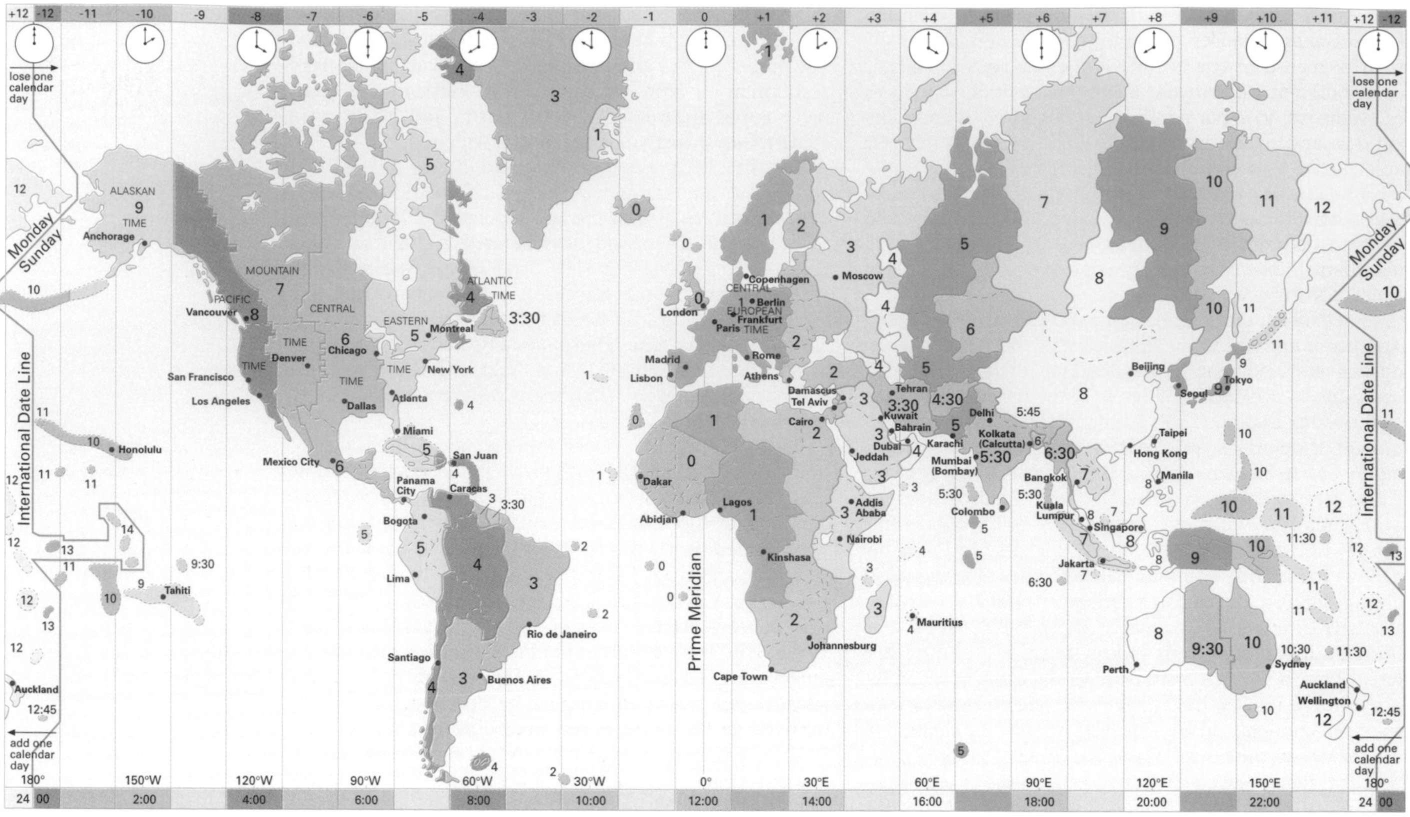

time zone

10 Hours slow or fast of UT or Coordinated Universal Time

Zones using UT (GMT)

Zones slow of UT (GMT)

International boundaries

Zones fast of UT (GMT)

Half-hour zones

Time zone boundaries

International Date Line

Actual Solar Time when time at Greenwich is 12:00 (noon)

1958 January 1, and it is used as a standard against which other clocks can be measured.

In calculations of the predicted positions of celestial bodies, which involve gravitational equations of motion, a timescale called DYNAMICAL TIME, which is also based on the SI second, is used. In 1984 this replaced EPHEMERIS TIME, which was previously used for such calculations and was based on the Earth's orbital motion around the Sun.

For practical, everyday use we require a civil timescale that is both accurate and uniform but that is also linked to the rotation of the Earth. COORDINATED UNIVERSAL TIME (UTC) fulfils this purpose, being derived from atomic clocks but kept in step with the rotation of the Earth through the periodic introduction of LEAP SECONDS. Because of this, TAI and UTC run at the same uniform rate but differ by an integral number of seconds.

The International Bureau of Weights and Measures (BIPM) in Paris co-ordinates international timescales. In the United Kingdom, responsibility for providing a national time service lies with the National Physical Laboratory (NPL) at Teddington.

time zone Geographical division of the Earth's surface, 15° of longitude wide, where civil time is deemed to be the same throughout. The globe is divided into 24 such zones, each one an hour different from its neighbours. The zone centred on the local MERIDIAN at the GREENWICH OBSERVATORY, and where GREENWICH MEAN TIME or UNIVERSAL TIME is kept, was adopted in 1884 as the reference point from which all other zones should be measured. Those east of Greenwich are ahead by an integral number of hours whilst those to the west are behind. At the 180° longitude line, the transition between the two is marked by the INTERNATIONAL DATE LINE.

Tinsley, Beatrice Muriel Hill (1941–81) New Zealand astrophysicist who revolutionized the study of the extragalactic distance scale by taking the evolutionary stage of galaxies into account when measuring distances to clusters of galaxies. Her PhD thesis at the University of Texas incorporated models of star formation and evolution, nucleosynthesis of the heavier elements, and the interstellar medium, opening up a new field of astrophysical cosmology. Tinsley showed that a galaxy's star formation history determines its colour and other properties. At Yale University (1975–1981) she extended this research, comparing UBV colours of 'normal' galaxies with those of 'peculiar' galaxies disturbed by collisions, concluding that 'normal' spirals have lower star-formation rates and older average stellar ages.

Titan By far the largest SATELLITE of SATURN; it was found in 1655 by Christiaan HUYGENS. Titan was the first planetary satellite to be discovered since Jupiter's GALILEAN SATELLITES in 1610. It is also the only satellite to have a substantial atmosphere, which was discovered in 1944 by Gerard KUIPER.

Titan is the first outer planet satellite to have been targeted by a landing probe: the HUYGENS lander is due to detach from the CASSINI Saturn orbiter for a parachute descent to Titan's surface in 2004 November. Cassini will map Titan's surface by imaging radar. Huygens carries optical imagers for use during descent as well as a variety of analytical experiments to examine the nature of the atmosphere and surface.

VOYAGER images of Titan obtained in 1980 and 1981 merely showed an opaque orange atmosphere overlain by a blue haze layer. The HUBBLE SPACE TELESCOPE, however, has been able to see through the atmosphere to reveal an apparently stable pattern of ALBEDO markings.

Titan has a similar density to Jupiter's largest satellite, GANYMEDE, and is almost as big. Indeed, some old tables add the 200 km (120 mi) thickness of the opaque part of Titan's atmosphere to its radius, thereby counting Titan as the largest planetary satellite of all. Titan's density of 1.88 g/cm^3 argues for it being largely an icy body, but it is not known to what extent it has become internally differentiated (*see* DIFFERENTIATION) to form a rocky or iron-rich core.

Most of Titan's nitrogen- and methane-rich atmosphere is probably inherited from gases scavenged directly from the SOLAR NEBULA and the circum-Saturn gas and dust cloud within which Titan grew. However, it is likely to have been augmented by subsequent degassing from Titan's interior. The reason why the young Titan was able to scavenge and retain an atmosphere, whereas Ganymede and CALLISTO could not, is that Titan grew in a colder environment, being farther from the Sun and receiving less warmth from its planet. Saturn's other satellites are much smaller, so their gravity is insufficient for them to have clung on to any gas.

Titan's atmosphere shields its surface from radiation, so water-ice does not get broken down or liberated as a vapour, and the surface temperature of 90 K is far too cold for ice to evaporate naturally. Titan's atmosphere is, therefore, both dry and free of oxygen. As a result, it contains compounds now extremely rare in the atmospheres of the TERRESTRIAL PLANETS, but which may have been abundant there originally. Compounds detected spectroscopically include hydrocarbons, such as ethane (C_2H_6), ethene (C_2H_4), ethyne (C_2H_2) and propane (C_3H_8), and nitrogen compounds, such as hydrogen cyanide (HCN), cyanogen (C_2N_2) and cyanoacetylene (HC_3N). Many of these gases condense in Titan's atmosphere, contributing to the smog that obscures the surface. The atmosphere's orange colour is probably caused by these simple hydrocarbon molecules linking into longer chains by means of reactions triggered by solar ultraviolet radiation.

Ethane could play a fascinating role in Titan's meteorology, because condensed droplets of ethane should fall to the surface and collect there as liquid. If so, Titan's cold and dark nature is made even more miserable by a continual ethane drizzle. Winds blow at 100 m/s (330 ft/s) in Titan's upper atmosphere, but there is currently no knowledge of the near-surface wind pattern.

The strength of reflections from ground-based radar shows that Titan is not covered by a global ocean of ethane as was once believed. However, there could be widespread seas and lakes from which ethane could evaporate back into the atmosphere in a cycle analogous to Earth's water cycle. If it does rain ethane on Titan, there may be rivers as well as seas and lakes. Add to this an intricate weathering cycle, involving hydrocarbons mixed with or coating the surface ice, and the possibility of transport and reworking of surface material by the wind, and it becomes clear that Titan's landforms could turn out to be strikingly Earth-like. *See* data at SATURN

Titan IV Major US military satellite launcher. It began life as the Titan II Intercontinental Ballistic Missile, which still provides the core stage of the vehicle. The Titan IV is the latest and final upgrade of the Titan launch family. It comprises a Titan II core stage with new upper stages and two powerful strap-on solid rocket boosters. It is based on the design of the Titan 3C. The Titan IV first flew in 1989 and is available in six models of Titan IVA and IVB, without an upper stage or with IUS (Inertial Upper Stage) or Centaur upper stages. They are capable of placing 22 tonnes into low Earth orbit and up to 5.8 tonnes directly into geostationary orbit. The vehicle carries military optical and radar reconnaissance, missile early warning and communications satellites. By the end of 2001 Titan IV had made 26 successful and five failed launches. It will be retired from service in about 2003–2004, to be replaced by the heavier version of the DELTA IV.

Titania Largest SATELLITE of URANUS. It was discovered in 1787 by William HERSCHEL, who found its neighbour OBERON in the same year. The best VOYAGER image of Titania is only good enough to show details down to about 7 km (4 mi) across. This indicates a surface that is more heavily cratered than ARIEL but less cratered than UMBRIEL or Oberon. It is traversed by fault scarps, 2–5 km (1–3 mi) high and up to 1500 km (900 mi) long. These faults may have the same origin as those on Ariel, but there are no visible signs that Titania's faulted valleys have been flooded by CRYOVOLCANISM. Another indication that Titania has been active more recently than Umbriel and Oberon is that its largest craters have unusually low rims, as if the icy LITHOSPHERE lacked the strength to support such large features when they were formed. *See* data at URANUS

Titius–Bode law *See* BODE'S LAW

Titov, Gherman Stepanovich (1935–) Soviet cosmonaut who piloted the Vostok 2 spacecraft in 1961 August, so becoming the second man to orbit the Earth (and the first to suffer from 'space sickness'). He completed 17 orbits in a flight lasting more than 24 hours.

TNG Abbreviation of Telescopio Nazionale Galileo. *See* GALILEO NATIONAL TELESCOPE

TNO Abbreviation of TRANS-NEPTUNIAN OBJECT

TOE Abbreviation of THEORY OF EVERYTHING

Tokyo Astronomical Observatory *See* NATIONAL ASTRONOMICAL OBSERVATORY OF JAPAN

Toliman Alternative name for the star ALPHA CENTAURI

Tombaugh, Clyde William (1906–97) American astronomer who discovered Pluto in 1930. Tombaugh was hired by LOWELL OBSERVATORY in 1929 to photograph the sky with the 330-mm (13-in.) astrographic telescope and to scan the photographs with a BLINK COMPARATOR to search for a trans-Neptunian planet. On 1930 February 18, he found a 17th-magnitude object near the star δ Geminorum that appeared to have moved when two photographic plates taken six nights apart were compared; Percival LOWELL announced the discovery of Pluto on 1930 March 13. Tombaugh continued his photographic survey for another 15 years, examining millions of star images over 65% of the sky, discovering a comet, numerous asteroids, a nova (TV Corvi) and several star clusters. After World War II, he designed and built telescopic cameras for tracking rockets launched from White Sands, New Mexico.

TOMS Abbreviation of TOTAL OZONE MAPPING SPECTROMETER

topocentric Pertaining to observations made, or coordinates measured, from a point on the surface of

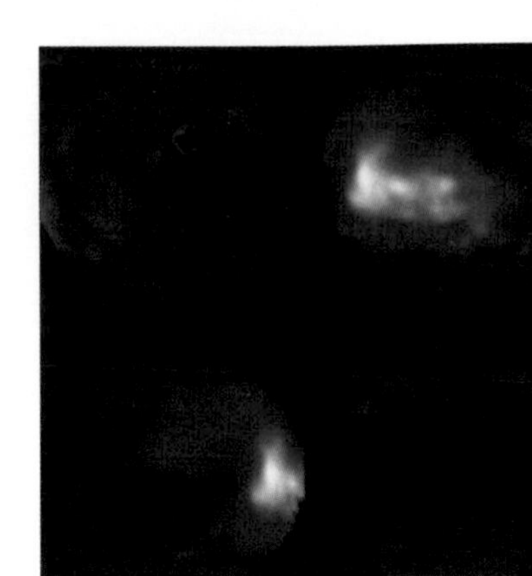

▲ **Titan** The Hubble Space Telescope observed Titan in the near-infrared in order to penetrate the atmospheric haze. Titan always keeps one hemisphere facing its parent planet, Saturn, much as the same side of the Moon always faces Earth.

◀ **Titan IV** As well as carrying satellites into space, the Titan IV has the ability to boost probes into space. On 1997 October 15, this Titan IVB/Centaur launched the Cassini Huygens probe on its journey towards Titan.

T

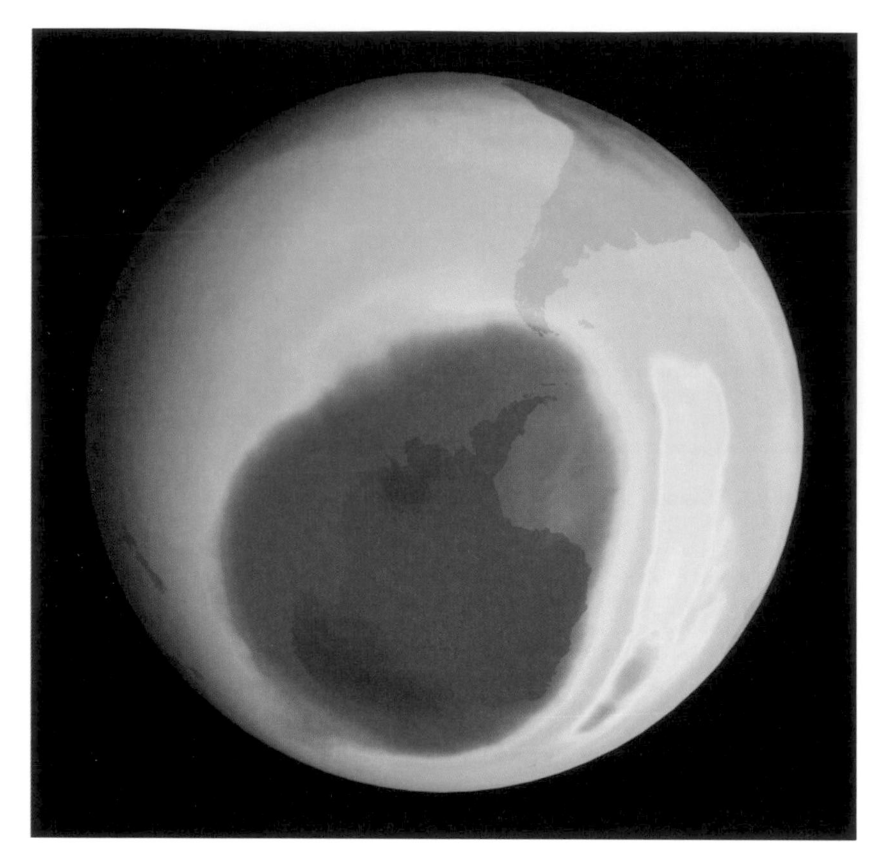

▲ **Total Ozone Mapping Spectrometer** In 2000 September, TOMS observed an enormous 'hole' in the ozone layer over the Antarctic (blue). In effect, it covered the whole continent (the outline of the Antarctic Peninsula can be seen towards the top right of the hole) and reached as far north as Tierra del Fuego.

the Earth. This is in contrast to geocentric coordinates or observations, which are measured relative to the centre of the Earth.

T Orionis variable Irregular ERUPTIVE VARIABLE that exhibits Algol-like fades. Like other Orion variables, such stars are found associated with bright or dark diffuse nebulae.

Toro APOLLO ASTEROID discovered in 1948; number 1685. Toro is remarkable because its orbital period is resonant with both Venus and the Earth. It is also one of the largest Apollos, being about 6 km (4 mi) in size. *See* table at NEAR-EARTH ASTEROIDS

Torun Radio Astronomy Observatory (TRAO) Part of the Torun Centre for Astronomy, an educational and research facility of the Nicolaus Copernicus University, Torun, Poland. The main instrument is a modern 32-m (105-ft) radio telescope at Piwnice, 15 km (9 mi) north of Torun. It is used extensively for very long baseline interferometry (VLBI), pulsar timing and spectroscopy. TRAO has been a member of the EUROPEAN VLBI NETWORK since 1998.

▼ **Toutatis** Radar images of the asteroid Toutatis show that its spin is highly complex, as it rotates around four axes. [S. Ostro (JPL/NASA)]

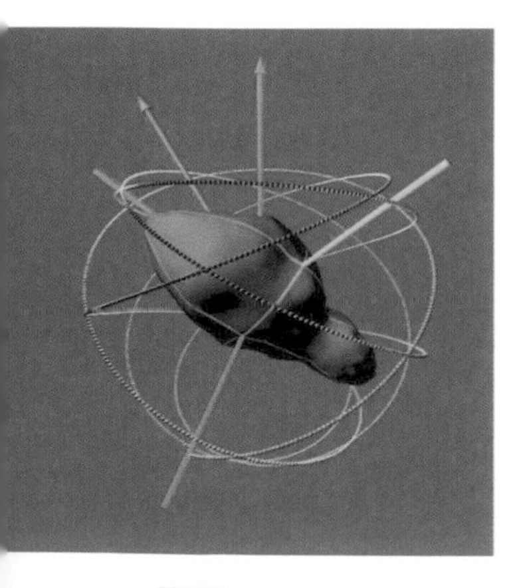

total eclipse SOLAR ECLIPSE in which the Moon completely covers the bright disk of the Sun; also a LUNAR ECLIPSE during which the Moon becomes completely immersed in the UMBRA of Earth's shadow.

totality During a total SOLAR ECLIPSE, the interval between SECOND CONTACT and THIRD CONTACT, when the Moon completely obscures the Sun's bright disk; totality at such an eclipse can last up to a maximum of 7 minutes 31 seconds. At a total LUNAR ECLIPSE, totality is the interval between second and third contacts, when the Moon is completely immersed in Earth's dark UMBRA, lasting up to a maximum of 1 hour 47 minutes.

Total Ozone Mapping Spectrometer (TOMS) Satellite dedicated to mapping global atmospheric ozone levels, the key to understanding ozone depletion. It was launched in 1996 July. The spacecraft complements other craft and instruments on board meteorological satellites, keeping a close watch on the status of the hole in the ozone layer.

Toutatis APOLLO ASTEROID discovered in 1989; number 4179. Toutatis makes repeated close approaches to the Earth (to within 0.006 AU) because of its small orbital inclination (only half a degree). Radar imaging has shown it to be irregularly shaped, roughly 4.6 × 2.4 × 1.9 km (2.9 × 1.5 × 1.2 mi) in size. These images suggest that Toutatis is actually made up of two distinct rocky components in contact with each other and held together by self-gravity. Toutatis is also unusual in that rather than undergoing simple rotation about a principal axis, it actually tumbles as it moves along its orbit. *See* table at NEAR-EARTH ASTEROIDS

TRACE Abbreviation of TRANSITION REGION AND CORONAL EXPLORER

train Brief 'afterglow', resulting from ionization in the upper atmosphere, that lingers after the extinction of a METEOR. In most cases, the train – if present at all – lasts only a matter of a few seconds: those of less than a second's duration are described as wakes. Trains are most commonly seen associated with bright meteors, and with meteors from cometary streams that impact on the upper atmosphere at high velocities: the PERSEIDS, ORIONIDS and LEONIDS are each noted for their large proportion of train-producing events. Longer-duration trains may drift and undergo distortion in high-atmosphere winds, blowing at up to 400 km/hr (250 mph).

Tranquillitatis, Mare (Sea of Tranquillity) Lunar lava plain, roughly 470 km (290 mi) in diameter, located in the central and eastern region of the Moon. This mare is irregular in shape. The dark coloration is the result of flooding with lunar basalts. Other volcanic products include mare ridges and domes (small shield volcanoes). Arcuate RILLES occur near the southern and eastern edges. A nearly circular mare ridge system, named Lamont, marks a buried crater.

transfer orbit Path followed by a satellite or spacecraft in moving from one ORBIT to another, for example from low Earth orbit to a higher one, or from the Earth to another planet.

transient lunar phenomenon *See* LUNAR TRANSIENT PHENOMENON

transit (1) Passage of a celestial body across an observer's MERIDIAN. As the Earth turns on its axis, so the CELESTIAL SPHERE appears to rotate, completing one circuit in a SIDEREAL DAY. All celestial objects reach their greatest altitude at the point they cross, or transit, the observer's meridian (the north–south line that connects points of the same longitude). Transits of celestial bodies have always been important in measuring their position (using an instrument called a TRANSIT CIRCLE) and for the determination of time. *See also* TIMEKEEPING

transit (2) Passage of a celestial body, particularly one of the INFERIOR PLANETS, directly between the Sun and the Earth, crossing the Sun's disk.

Since their orbits are slightly inclined to the ecliptic, MERCURY and VENUS usually pass north or south of the Sun at INFERIOR CONJUNCTION. If CONJUNCTION occurs when the orbit of the inferior planet crosses the ecliptic (*see* NODE), then it is observed from Earth as a small, dark spot moving from east to west across the Sun's disk. This is known as a transit. There are four phases during a transit: two at the start (ingress) and two at the finish (egress). (i) First exterior contact occurs when the planet first appears to touch the Sun's edge or limb; (ii) first internal contact is the point at which the planet is fully upon the Sun's disk but still contiguous with its limb; (iii) second internal contact occurs when the planet touches the oppo-

◀ **transit** The 1992 November 15 transit of Mercury across the limb of the Sun. The tiny planet was dwarfed by many of the sunspots visible.

site limb of the Sun, having crossed its disk; and (iv) second external contact, the moment when the planet's trailing limb finally clears the Sun's disk. In the past, transit observations were used as a method of determining the ASTRONOMICAL UNIT.

Transits of Mercury can only occur during May and November, the former being when the planet is near APHELION and the latter when it is close to PERIHELION. November events are twice as numerous as those in May but last for a shorter time; neither is visible to the naked eye. The interval between successive events varies from three to 13 years, with the next due in 2003 and 2006, and again in 2016 and 2019.

Transits of Venus are very rare but occur in pairs, eight years apart, during June or December. The next events are due in 2004 and 2012, but then not again until 2117 and 2125. Unlike those of Mercury, transits of Venus are visible to the naked eye (although under no circumstances should an observer look directly at the Sun without using an approved, safe filter).

transit (3) Passage of a planetary satellite across the disk of the parent planet. The four main moons of Jupiter can frequently be observed transiting its disk. The inner moons of Saturn transit its disk at times close to the rings' edge-on presentation towards Earth.

transit (4) Passage of an atmospheric or surface feature of a celestial body across its CENTRAL MERIDIAN as it rotates. Observations and timings of these transits provide a means of measuring the body's period of rotation. This information can also be used to determine cloud motions relative to a standard reference frame for features on GAS GIANTS (principally Jupiter). *See also* SYSTEMS I AND II

transit instrument (transit circle, meridian circle) TELESCOPE mounted so that it can only move about a fixed horizontal axis in a north–south plane. It can observe objects as they cross (transit) the MERIDIAN in order accurately to determine their positions.

The simple transit instrument consists of a telescope fixed at right angles to a horizontal axis and free to rotate upon pivots on two fixed piers. The telescope can thus be moved up and down from horizon to horizon via the zenith, but not from side to side, so that its optical axis will always be in the plane of the observer's meridian. A star or other celestial object is observed using cross-wires in the eyepiece of the transit instrument as it crosses the meridian and the exact time is recorded. When related to a fundamental reference frame, this time provides the star's RIGHT ASCENSION or CELESTIAL LONGITUDE. Its ALTITUDE can be measured directly at the telescope from precisely graduated circles mounted on the horizontal axis and used to determine the star's DECLINATION or CELESTIAL LATITUDE.

An example of a modern transit instrument is the Carlsberg Meridian Circle at the ROQUE DE LOS MUCHACHOS OBSERVATORY on La Palma in the Canary Islands. This fully automated instrument is equipped with a CCD camera and is capable of accurately measuring the positions of between 100,000 and 200,000 stars a night down to magnitude 17. Observations from it over a number of years have enabled the positions, PROPER MOTIONS and MAGNITUDES of some 180,000 stars down to magnitude 15 to be precisely determined. In addition, it has made 25,000 measurements of the positions and magnitudes of 180 Solar System objects including the planet Pluto and many asteroids. It was from these observations that a more precise orbit for Pluto was determined, thereby removing speculation about the existence of a 'PLANET X'. *See also* ASTROMETRY

transition region Thin region in the solar atmosphere, less than 100 km (60 mi) thick, between the CHROMOSPHERE and CORONA. It is characterized by a large rise in temperature from about 20,000 to 1,000,000 K. The plasma density decreases as the temperature increases, keeping the pressure constant.

Transition Region and Coronal Explorer (TRACE) Satellite launched in 1998 April to perform the first US mission dedicated to solar science since the SOLAR MAXIMUM MISSION launched in 1980. TRACE takes high-resolution images of the TRANSITION REGION between the Sun's chromosphere and the corona to obtain measurements of the temperature regimes, to complement data being collected by other spacecraft, including the SOLAR HELIOSPHERIC OBSERVATORY. Scientists are also interested in the transition region because many of the physical phenomena that are found, such as plasma heating, arise throughout astrophysics.

transmission grating DIFFRACTION GRATING made from transparent material, on the surface of which are a series of closely spaced, equidistant parallel grooves. A typical spacing density for the grooves is about 1000 per millimetre, and the size of the separation (about one micrometre) is not much bigger than the wavelength of light. Light of different wavelengths is transmitted through the grating in different directions (diffraction) creating a high-quality SPECTRUM. *See also* REFLECTION GRATING; SPECTROSCOPE

trans-Neptunian object (TNO) Any of a large population of sub-planetary size Solar System bodies that orbit the Sun in generally low-inclination, low-eccentricity orbits beyond 30 AU, the position of Neptune. The first TNO was discovered in 1992, although the existence of such objects had been predicted several decades earlier (*see* EDGEWORTH–KUIPER BELT). Through to late 2001 more than 500 TNOs had been discovered. Based on their observed brightnesses, these objects are mostly between 100 and 500 km (60–300 mi) in size, although some may be rather larger, an example being VARUNA. The low end of the size range represents the magnitude

▼ **trans-Neptunian object** These are the discovery images of TNO 1993 SC against a background of stars and galaxies. These small bodies are formed of a mixture of ice and rock. If perturbed into the inner Solar System, they will become comets.

TRIANGULUM (GEN. TRIANGULI, ABBR. TRI)

Small but distinctive northern constellation, between Andromeda and Aries. Its three brightest stars, β, α (Mothallah) and γ Tri, mags. 3.0, 3.4 and 4.0, form a narrow isosceles triangle with α at the apex. 6 Tri is a double star, with yellow and pale yellow components, mags. 5.3 and 6.8, separation 3″.9. The brightest deep-sky object is the 6th-magnitude PINWHEEL GALAXY (or Triangulum Galaxy; M33, NGC 598).

BRIGHTEST STARS

	Name	RA h m	dec. ° ′	Visual mag.	Absolute mag.	Spectral type	Distance (l.y.)
β		02 10	+34 59	3.00	0.1	A5	124
α	Mothallah	01 53	+29 35	3.42	2.0	F6	64

TRIANGULUM AUSTRALE (GEN. TRIANGULI AUSTRALIS, ABBR. TRA)

Small but distinctive southern constellation whose three brightest stars form an almost equilateral triangle, between Norma and Apus. It was introduced by Keyser and de Houtman at the end of the 16th century. Its brightest star, Atria, is mag. 1.9. The constellation's brightest deep-sky object is NGC 6025, an open cluster of about 60 stars fainter than 7th magnitude.

BRIGHTEST STARS

	Name	RA h m	dec. ° ′	Visual mag.	Absolute mag.	Spectral type	Distance (l.y.)
α	Atria	16 49	−69 02	1.91	−3.6	K2	415
β		15 55	−63 26	2.83	2.4	F2	40
γ		15 19	−68 41	2.87	−0.9	A1	183

limit attainable with ground-based telescopes, rather than a real absence of smaller TNOs: sky sampling with the Hubble Space Telescope has indicated that TNOs in the size range 1–10 km (0.6–6 mi) certainly exist.

Formally classed as ASTEROIDS (minor planets), the TNOs are probably more similar to COMETS in nature than to main-belt asteroids, being composed largely of ices rather than rock and metal.

Many of the TNOs occupy orbits with periods that are precise multiples of that of Neptune. This relationship leads to stability, as in the case of the orbital periods of Uranus, Neptune and Pluto, which are in the ratio 6:3:2. The PLUTINOS are TNOs residing in the 3:2 resonance with Neptune, as does Pluto. Some astronomers regard the plutinos as being a separate population to the rest of the TNOs, in the same way as the Hilda asteroids are not considered to be part of the main belt.

▼ **Trifid Nebula** The dark lanes in this emission nebula are in fact dust, which obscures the nebula behind. This HII region is pinkish rather than red because of the light of the hot blue stars within it.

Although most TNOs are in near-circular orbits, about 10% of the currently known population have moderate to large eccentricities, with perihelia at 30–45 AU and aphelia at 60–200 AU. These bodies, termed scattered disk objects, have perhaps been pushed into higher-eccentricity orbits by non-collisional encounters with other TNOs. Some such objects may have been scattered into orbits with perihelia inside the orbit of Neptune, whereupon strong perturbations by that and, successively, the other giant planets may pass the objects inwards. The CENTAURS and perhaps some of the LONG-PERIOD ASTEROIDS are thought to have originated in this way.

Trapezium Popular name of the multiple star θ^1 Orionis, which lies within the Orion Nebula and illuminates it. The four brightest components, of mags. 5.1, 6.7, 6.7 and 8.0, form a trapezium shape easily seen with a small telescope. The two faintest stars are spectroscopic binaries with ranges of 0.8 and 0.5 mags. and periods of 65.4 and 6.5 days, respectively. Larger apertures reveal two additional stars in the group, of 11th magnitude. Like the Orion Nebula itself, the Trapezium lies about 1500 l.y. away.

Treptow Observatory Public observatory in the south-eastern Berlin suburb of Treptow. It is formally known as the Archenhold Observatory in honour of its first director, Friedrich Simon Archenhold (1861–1939). It was built as a temporary structure for the Berlin Trade Exhibition of 1896, and housed a refracting telescope with an aperture of 0.70 m (27 in.) and the extraordinarily large focal length of 21 m (69 ft). A permanent building was erected in 1910, and the telescope is still on public exhibition.

Triana NATIONAL AERONAUTICS AND SPACE ADMINISTRATION (NASA) satellite to be stationed in an orbit between the Earth and the Sun; it will return a continuous stream of real-time images of the full disk of the Earth to be viewed live via the Internet, giving people a daily view of their home planet in space. The concept has been expanded to include additional science relevant to monitoring the Earth's environment. Triana may be launched in 2003–2004.

Triangulum *See* feature article

Triangulum Australe *See* feature article

Triangulum Galaxy (M33, NGC 598) *See* PINWHEEL GALAXY

Triesnecker Lunar crater (4°N 4°E), 23 km (14 mi) in diameter. It is a complex crater, having side-wall terracing and central peaks. Triesnecker is a relatively recent crater, with a bright ray system best seen at high Sun illumination. From the north-east to the south-east of this crater is an extensive set of RILLES; these are likely to have been caused by subsurface stresses, and so are TECTONIC in origin.

Trifid Nebula (M20, NGC 6514) Emission and reflection nebula in the constellation Sagittarius (RA $18^h 02^m.3$ dec. −23°02′). Long-exposure photographs show the reddish emission nebula, illuminated by an embedded O-class star, to be split by three dark lanes, from which the object takes its popular name. The presence of Bok GLOBULES indicates that star formation is still ongoing. Bluish (dust) reflection nebulosity lies nearby. The nebula has an overall magnitude +6.3 and angular diameter 20′. It lies 6700 l.y. away.

trigonometric parallax Method of determining the distance to a nearby celestial object by making observations from either end of a baseline of known length. Stellar distances are measured using the Earth's orbit around the Sun as a baseline, giving the ANNUAL PARALLAX (heliocentric PARALLAX). For objects within

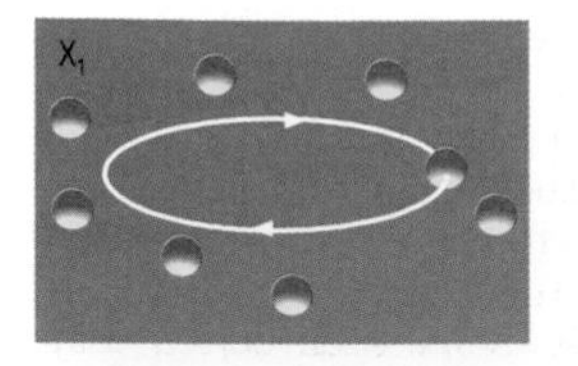

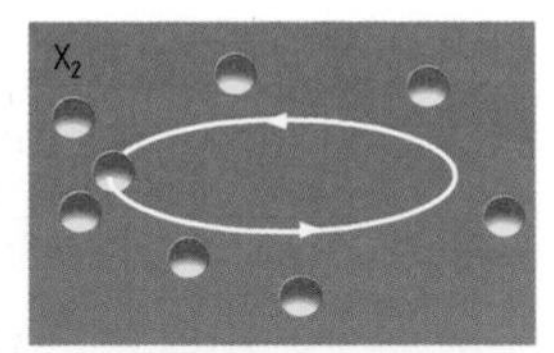

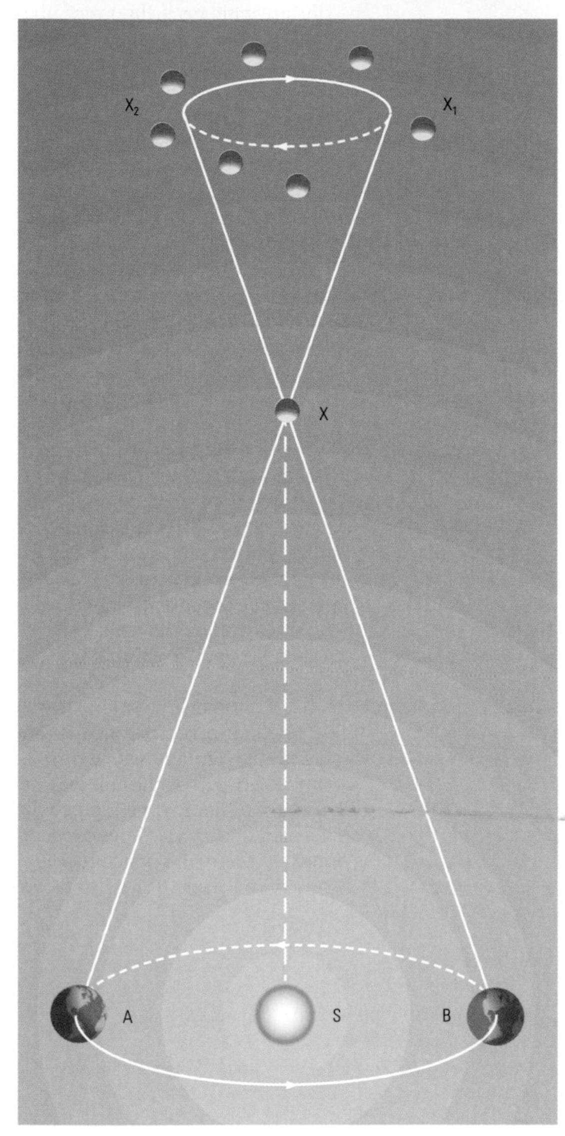

◀ trigonometric parallax If a nearby object (X) is observed from opposite sides of the Earth's orbit (A and B), it will appear in different positions against the stellar background (X_1 and X_2). As the distance between A and B is known, the distance to X can be calculated.

the Solar System the radius of the Earth is used as a baseline, producing a measure of DIURNAL PARALLAX (geocentric parallax).

As the Earth revolves around the Sun the directions to nearby stars change relative to the distant, background stars. When observed six months apart from opposite sides of the Earth's orbit, the position of a nearby star will appear to have shifted by an angle $\Delta\theta$. Half of this angle, π, gives the annual parallax, which is equal to the angle subtended at the observed star by the semimajor axis, or radius, of the Earth's orbit, which is one astronomical unit (1 AU). Therefore,

$$d = \frac{1\ \text{AU}}{\tan \pi}$$

where d is the distance between the Sun and the star.

The reciprocal of annual parallax in arcseconds (″) gives the distance to the star in PARSECS (pc), a parsec being the distance at which a star would have a parallax of one arcsecond, so $d = 1/\pi$ pc. The star Proxima Centauri has the largest known parallax, at 0″.772, which is equivalent to a distance of 1.3 pc (4.22 l.y.).

In practice, the parallax of a star is measured from observations taken about six months apart. Allowance is made for the PROPER MOTION of the star and the possibility that some of the background stars may also have significant parallaxes. The measurement is deemed 'absolute' when it has been corrected for the parallaxes of the background stars. The astrometric satellite HIPPARCOS, which operated between 1989 and 1993, extended the range of accurate stellar parallax distances by roughly 10 times, at the same time increasing the number of stars with good parallaxes by a much greater factor.

triple-α process NUCLEAR REACTION that converts helium to carbon by the near simultaneous combination of three α-particles (helium-4 nuclei).

The next stage after the production of helium-4 from hydrogen inside stars is not the production of beryllium-8, because that decays back to two α-particles in just 2×10^{-16} s, instead it is the triple-α reaction, producing carbon-12. The triple-α reaction involves two α-particles combining and then a third being added before the beryllium-8 can decay. It requires a temperature of about 10^8 K before it will start because the nuclear reactions involved are slightly endothermic (requiring the supply of energy) and therefore need the input of thermal energy to get them to occur. The triple-α reaction overall is exothermic: more energy is released by the process than is used in starting it, because the carbon-12 nucleus that is the result of the reaction is in an excited state and emits a high-energy gamma ray. Additional helium captures can produce oxygen, neon and a number of heavier elements (*see* NUCLEOSYNTHESIS). The reaction is believed to be the dominant energy-generating process in RED GIANTS.

triplet LENS assembly made up of three separate lenses that may or may not be cemented together but that work effectively as a single lens. Triplets are used to correct the ABERRATIONS that a single lens would produce. One example is the OBJECTIVE of an APOCHROMAT; it normally consists of a triplet designed to reduce CHROMATIC ABERRATION to a minimum. An ORTHOSCOPIC EYEPIECE contains a triplet lens.

Triton Only major SATELLITE of NEPTUNE; it has a diameter of 2706 km (1682 mi). Triton was discovered in 1846 by William LASSELL only a few weeks after the planet itself had been found. Unique among large planetary satellites, Triton orbits its planet in a RETROGRADE direction, with an orbit inclined at an angle of 157° to Neptune's equator.

Triton has an unusually bright icy surface with an albedo of about 0.8. Its density of 2.1 g/cm³ is exceeded among planetary satellites only by IO and EUROPA. Triton's strange orbit suggests that it did not share its origin with Neptune. It most likely began as a large EDGEWORTH–KUIPER BELT

▼ Triton Voyager 2 imaged Triton in 1989, and this mosaic shows some of the varied terrains. The broad area at the bottom of the picture is the southern polar ice cap, which probably consists mainly of nitrogen ice.

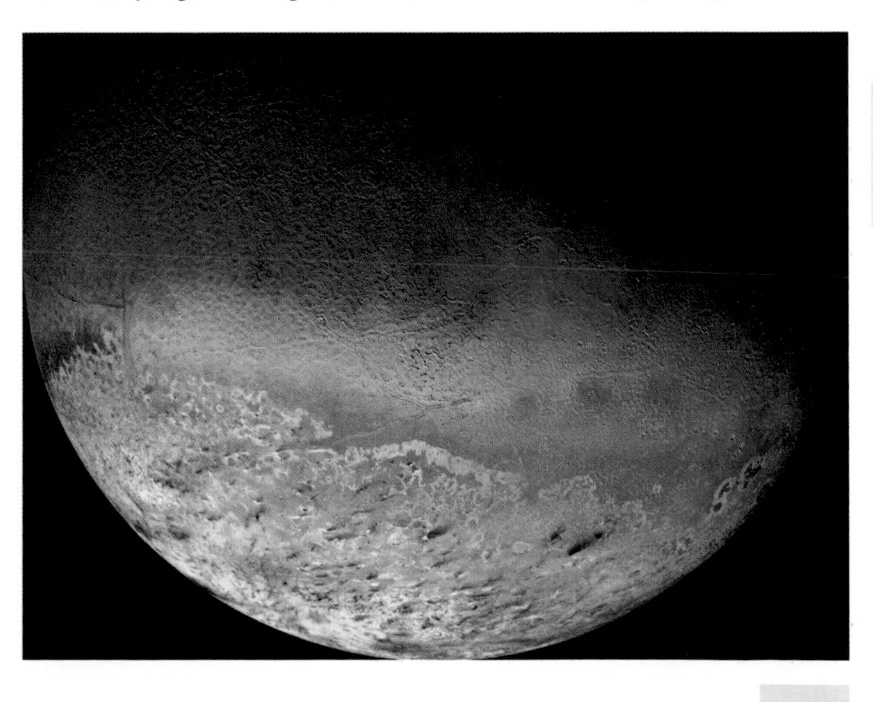

► **Trojan asteroid** The asteroids known as the Trojans orbit at the Lagrangian points in front of or behind Jupiter. They oscillate around these points over a period of 150–200 years.

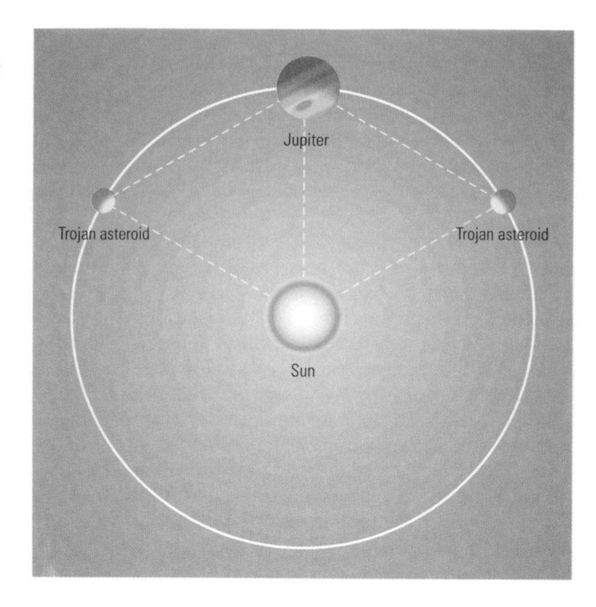

object, and was later, perhaps about a billion years ago, captured into orbit around Neptune. While Triton was being captured and its orbit made circular, any previous large satellites of Neptune would have been destroyed by mutual collisions or ejected from the Neptune system. It is possible that NEREID and PROTEUS are collisional remnants dating from this time.

Triton must have suffered greatly in the aftermath of capture. Even if not involved in any collisions, there must have been much TIDAL HEATING during the circularization of its orbit and while TIDAL FRICTION was slowing its rotation until it became SYNCHRONOUS with its orbital period.

Triton's surface is remarkably varied, as revealed on close-up images sent back by VOYAGER 2 in 1989. There are few impact craters, attesting to the young age of the surface. There are many signs of CRYOVOLCANISM, including long ridges, where viscous melts have oozed out, and a large tract, named 'cantaloupe terrain', that is pockmarked by 30-km-diameter (20 mi) dimples. The surface temperature of 38 K makes Triton the coldest body yet visited by a space probe. Spectroscopic observations have demonstrated the presence on the surface of ices formed from methane, carbon dioxide, carbon monoxide and nitrogen. Ammonia, which is almost impossible to detect, is probably present too. The scale of the topographic features demonstrates strength consistent with water ice, and presumably this lies beneath a thin veneer of the more exotic ices. Triton's density suggests that a large rocky core occupies about two-thirds of its radius, and there may be a smaller iron-rich core in the middle.

Triton has a complex 688-year-long cycle of seasons, resulting from a combination of Neptune's 164-year orbit and Triton's orbital inclination, which precesses (*see* PRECESSION) about Neptune's axis of rotation. At the time of the Voyager 2 flyby it was high summer in Triton's southern hemisphere. There was a large and apparently shrinking south polar cap of nitrogen ice, but the north polar region was in darkness. Triton has an extremely tenuous atmosphere composed mostly of nitrogen. Its surface pressure doubled between 1989 and 1997, when it reached about 30 millionths of the Earth's atmospheric pressure. Triton's atmosphere is probably derived from the polar caps, and it must vary with the seasons as the sunlit polar cap shrinks by sublimation and the opposite cap grows by condensation. Voyager 2 saw several jets of gas bursting through the south polar cap and flinging columns of sooty particles to heights of about 8 km (5 mi), where they were deflected sideways by high altitude winds. These jets are probably solar-powered geysers rather than true cryovolcanic features powered by heat from within. However, Triton ranks with Io as one of the only two satellites on which present-day geological processes have been observed in action. *See* data at NEPTUNE

Trojan asteroid ASTEROID that has an orbital period very close to that of Jupiter and is positioned in one of two groups surrounding the LAGRANGIAN POINTS located 60° in front of (L_4) and behind (L_5) the planet. After numbering, these objects are generally given names associated with the heroes of Homer's *Iliad*, the leading set being the Greek warriors and the followers being the defenders of Troy. By 2001 more than a thousand such Trojan asteroids had been discovered, 62% of them at L_4 and 38% at L_5.

Unlike most other planet-crossing objects, the Trojan asteroids are dynamically stable, being protected from close approaches to Jupiter by their resonant orbits. It is believed that most of them have occupied their present locations since they formed from the SOLAR NEBULA at the same time as the rest of the solid bodies in the Solar System. Trojan asteroids associated with Saturn's orbit have not been found, presumably because Jupiter's gravitational perturbations render unstable the corresponding Lagrangian points for Saturn. There are, however, several MARS-TROJAN ASTEROIDS known.

The first Trojan asteroid to be recognized was ACHILLES, in 1906; it was closely followed by PATROCLUS and HEKTOR. It has recently been realised that a Trojan asteroid re-discovered as 1999 RM11 had been previously observed in 1904.

tropical month Time taken for the Moon to complete a single revolution around the Earth, measured relative to the FIRST POINT OF ARIES; it is equivalent to 27.32158 days of MEAN SOLAR TIME. *See also* ANOMALISTIC MONTH; DRACONIC MONTH; MONTH; SIDEREAL MONTH, SYNODIC MONTH

tropical year Time taken for the Earth to complete one revolution of the Sun, measured relative to the vernal equinox; it is equivalent to 365.24219 mean solar days. Because of the effects of PRECESSION, which cause the FIRST POINT OF ARIES to drift westwards along the ecliptic (in the opposite direction to the Sun), a tropical year is 20 minutes shorter than a SIDEREAL YEAR. A tropical year is also sometimes known as a solar year. *See also* ANOMALISTIC YEAR

Tropic of Cancer Line of LATITUDE on the Earth at 23°.5N; it marks the most northerly DECLINATION reached by the Sun during its annual path along the ECLIPTIC. This occurs around June 21, the time of the SUMMER SOLSTICE in the northern hemisphere. At this point the Sun reaches its highest altitude in the sky, being directly overhead at the Tropic of Cancer; it also marks the longest day of the year, the period of maximum daylight in the northern hemisphere. *See also* TROPIC OF CAPRICORN

Tropic of Capricorn Line of LATITUDE on the Earth at 23°.5S; it marks the most southerly DECLINATION reached by the Sun during its annual path along the ECLIPTIC. This occurs around 22 December, the time of the WINTER SOLSTICE in the northern hemisphere. At this point the Sun reaches its lowest altitude in the sky, being directly overhead at the Tropic of Capricorn; it also marks the shortest day of the year, the period of minimum daylight in the northern hemisphere. *See also* TROPIC OF CANCER

tropopause Name for the temperature minimum that marks the boundary between the TROPOSPHERE and the overlying STRATOSPHERE. It is defined as the height at which the lapse rate (the decline in temperature with height) becomes 2 K per kilometre or less. The altitude of the tropopause varies from 5–8 km (3–5 mi) at the poles, where it is indistinct in winter, to 14–18 km (9–11 mi) at the equator. Although considerable fluctuations occur, the temperature is generally about 218 K at the polar tropopause and 193 K at the equatorial tropopause.

troposphere Lowest layer in the Earth's ATMOSPHERE; it extends from the surface to the tropopause. The troposphere contains three-quarters of the atmosphere by mass and has the majority of clouds and weather systems. It is heated by infrared radiation and convection from the ground. Within the troposphere, the temperature generally falls with increasing height, reaching the first atmospheric minimum at the TROPOPAUSE, where the atmospheric density decreases to approximately one-quarter of its sea-level value.

Troughton, Edward *See* COOKE, TROUGHTON & SIMMS

true anomaly *See* ANOMALY

Trumpler, Robert Julius (1886–1956) Swiss-American observational astronomer who spent the main part of his career at Lick Observatory (1918–38) making detailed studies of star clusters and pioneer investigations of the interstellar medium. In 1930 he showed that the Milky Way's globular clusters were reddened and dimmed, by 0.2 magnitudes for every 1000 l.y., by interstellar dust. He called this effect interstellar extinction. Trumpler carefully computed the sizes of, distances to and distribution of many clusters, discovering that their member stars belong to different 'populations' that reflect their average age. He was a pioneer in measuring the radial velocities of stars in clusters, especially for the massive O stars

Trumpler classification Classification of OPEN CLUSTERS using three criteria. The criteria are the number of stars in the cluster, the amount of stellar concentration towards the centre of the cluster, and the range of brightnesses of the member stars. It was devised by Robert TRUMPLER.

Tsiolkovskii, Konstantin Eduardovich (1857–1935) Russian rocket pioneer, now justly regarded as the 'father of astronautics'. Between 1903 and 1929 he wrote many seminal papers on fundamental spaceflight technology, such as liquid propellants and multi-stage launch vehicles. Many of his ideas were collected in *Investigations of Outer Space* (1911) and *Aims of Astronautics* (1914).

Tsiolkovsky Lunar crater (29°S 129°E). It is conspicuously dark as the only large (180 km/113 mi diameter) lava-filled crater on the lunar farside.

T star New spectral class formally added to the cool end of the SPECTRAL CLASSIFICATION system (as were L STARS) in 1999, making the sequence OBAFGKMLT. The class was needed because new infrared technologies were able to find stars that radiate almost entirely in the infrared. T stars, at the extreme cool end of the modern spectral sequence, are characterized principally by methane (CH_4) in their spectra. Absorptions by water and alkali metals, specifically neutral caesium, are also present. The first T star found, Gliese 229 B (companion to the class M1 red dwarf Gliese 229 A), has a temperature of only 1000 K and a mass estimated to be between 0.03 and 0.055 solar mass, making it also the first-known BROWN DWARF. Indeed, all the T stars are brown dwarfs (and are sometimes called 'methane brown dwarfs'). Dwarfs that fuse hydrogen on the full proton–proton chain have higher masses and thus higher temperatures; they start their upward mass progression in class L. Evolved stars (giants and supergiants) cannot become as cool as class T. Little is understood about T stars. The class is not yet subdivided, and the full range of temperatures and masses is unknown. Chromospheric emission has been seen in spite of the low masses, fully convective natures and presumably low magnetic field strengths of the stars

T Tauri star Very young star still settling on to the MAIN SEQUENCE, typified by the VARIABLE STAR T Tauri. T Tauri stars are found to the right of the main sequence on a HERTZSPRUNG–RUSSELL DIAGRAM, where HAYASHI TRACKS trace the evolution of PRE-MAIN-SEQUENCE STARS.

Characteristics of the T Tauri stars include: emission

◄ **T-Tauri star** These very young stars are still contracting and are all irregular variables. They are extremely luminous and have strong bipolar outflows with speeds of several hundred kilometres per second.

T

▲ **Tucana** One of the deep-sky objects found in the southern constellation Tucana is the Small Magellanic Cloud. This companion galaxy to the Milky Way is at a distance of about 200,000 l.y.

lines in their spectra indicating an extended atmosphere of gas; high velocities in outflowing or infalling gas as the stars adjust to the recent onset of NUCLEOSYNTHESIS; association with dark clouds where star formation occurs; and a high proportion of lithium, which is radioactive and quickly becomes depleted in most stars. Many T Tauri stars show evidence of a strong, continuous flow of material from the surface of the star, known as the T Tauri wind.

Other characteristics of these stars include the proximity of HERBIG–HARO OBJECTS, the presence of jets of gas, and perhaps the existence of an even younger companion star. T Tauri itself lies at the centre of a Herbig–Haro object. T Tauri stars are usually less massive than the Sun. Heavier stars either pass through the T Tauri phase while still obscured in their parent clouds or have a different appearance at the same stage of life. *See also* T ASSOCIATION

Tucana *See* feature article

Tully–Fisher relation Relationship between the widths of the 21-cm radio emission line of hydrogen (*see* TWENTY-ONE CENTIMETRE LINE) emitted by giant gas clouds in spiral galaxies to the optical brightnesses of the galaxies. The larger the galaxy, the faster the rotation and the broader the emission line width, and also the brighter the galaxy the more stars it has. This relationship can be used to measure distances to galaxies out to about 100 Mpc.

Tunguska event Very bright FIREBALL that was visible early in the morning of 1908 June 30 over the remote Tunguska region of Siberia. Contemporary reports record that explosions like thunder were heard. The first expedition to visit the area was in 1927, to search for the remnants of what was presumed to have been a meteorite impact. A crater was not found; instead there was an area of devastation *c.*60 km (*c.*37 mi) across. In the centre, a belt of dead trees, which appeared to have been blasted in an intense forest fire, surrounded an approximately circular area of swamp. Up to 30 km (19 mi) farther out from the centre, trees had been flattened. The alignment of the trees, pointing radially outwards from the swamp, suggested that they had been felled by a shock wave. Only small particles of meteoritic material have been found in peat and in tree resin. It is assumed that an impactor, possibly *c.*50 m (*c.*160 ft) across, exploded 6–8 km (4–5 mi) up in the atmosphere, and that radiation and shock wave from the explosion burnt and blew down the trees. It is still not certain whether the impactor was a comet or an asteroid. The energy of the impact has been estimated as *c.*10^{17} J, which is equivalent to 50 million tonnes of TNT.

tuning-fork diagram Diagram of GALAXY types, originated by Edwin HUBBLE. The 'handle' of the fork consists of elliptical galaxies (E), numbered according to their degree of elongation. The two prongs of the fork are made up of ordinary (S) and barred (SB) spirals, designated a, b, or c according to how tightly wound the spiral arms are. S0 or lenticular galaxies occur at the branch of the fork; they have both nonbarred and barred varieties and share stellar populations with elliptical galaxies. The classification was originally interpreted as a sequence of evolution, with ellipticals developing into spirals, but this is now known to be incorrect.

Turner, Herbert Hall (1861–1930) English astronomer who greatly advanced the precise determination and cataloguing of star positions. Turner was Savilian Professor of Astronomy at Oxford University (1894–1930), and a leader of the international *CARTE DU CIEL* photographic sky survey, which occupied him from 1887 to the end of his life. To expedite this massive undertaking, which required the Oxford astronomers to measure over 400,000 star images, Turner invented a measuring engine which replaced the usual micrometer screw with an eyepiece scale, greatly speeding up data analysis. He was the first to make use of the coelostat to observe solar eclipses, obtaining some of the first detailed spectra of the inner corona.

turnoff point Point on the HERTZSPRUNG–RUSSELL DIAGRAM (HR diagram) where a star evolves off the MAIN SEQUENCE. The turnoff point depends on the initial STELLAR MASS: both luminosity and temperature of the turnoff point increase with stellar mass. Stellar age also depends on the initial stellar mass: the more massive the star, the quicker it burns its core hydrogen and the shorter time it spends on the main sequence. Plotting stars belonging to a GLOBULAR CLUSTER on an HR diagram gives a distinct turnoff point which can be used to determine the age of the cluster.

Tuttle, Comet 8P/ Short-period comet originally discovered by Pierre MÉCHAIN on 1790 January 8. The comet was re-discovered on 1858 January 5 by Horace TUTTLE, and over the next two months sufficient observations were obtained to confirm its identity with the 1790 object. It was successfully recovered in October 1871. The orbital period is 13.51 years, and 8P/Tuttle was last seen in 1993. The comet is the parent of the URSID meteor stream.

twenty-one centimetre line EMISSION LINE of neutral hydrogen in interstellar space, resulting from the energy difference between two sublevels of the lowest energy state of hydrogen. The line at 21 cm (1420 MHz) lies in the RADIO WAVE region of the ELECTROMAGNETIC SPECTRUM. If a hydrogen atom is in the upper sublevel, its electron and proton have their spins aligned in the same direction. There is a small probability that the

TUCANA (GEN. TUCANAE, ABBR. TUC)

Small, rather inconspicuous southern constellation, representing a toucan, between Indus and Hydrus. It was introduced by Keyser and de Houtman at the end of the 16th century. Its brightest star, α Tuc, is mag. 2.9. β Tuc is a multiple star with bluish-white components of mags. 4.4 and 4.5 (separation 27″) and mag. 5.2 (10′ to the south-west). Tucana's deep-sky objects include the SMALL MAGELLANIC CLOUD and the 4th-magnitude globular cluster FORTY-SEVEN TUCANAE (NGC 104).

T

spinning electron will flip over so that the spins are in opposite directions, and this change emits a photon at 21 cm. The existence of the 21 cm line was predicted by Hendrik van de HULST in 1944, and was discovered by him and others in 1951. Neutral atomic hydrogen will be in its lowest possible energy state in interstellar space, and it is this that can be mapped using the 21 cm line, to show the distribution of hydrogen gas throughout the Galaxy. Radio waves penetrate the interstellar dust with ease, and so the 21 cm line can be observed from everywhere in the Galaxy, including the centre and the far side of the Galaxy. The velocity of the gas can be determined, so that the spiral arm structure can be traced (once a model is used to relate how each part of the Galaxy rotates about the centre).

twilight Period before sunrise and after sunset when the illumination of the sky gradually increases and decreases respectively. The effect is caused by the scattering of sunlight by dust particles and air molecules in the Earth's atmosphere and its duration is greatest at higher latitudes, being dependent on the angle of the Sun's apparent path with respect to the horizon. Three definitions of twilight exist. **Civil twilight** begins or ends when the centre of the Sun's disk is 6° below the horizon. This is regarded as the point at which normal daytime activities are no longer possible. **Nautical twilight** begins or ends when the centre of the Sun's disk is 12° below the horizon and the marine horizon is no longer visible. **Astronomical twilight** begins or ends when the centre of the Sun's disk is 18° below the horizon; it is the time when faint stars can be seen with the naked eye.

twinkling *See* SCINTILLATION

2MASS Abbreviation of TWO MICRON ALL SKY SURVEY

Two Micron All Sky Survey (2MASS) Survey of the entire sky at near-infrared wavelengths, begun in 1998 and due for completion in late 2002. It uses two dedicated automated 1.3-m (51-in.) telescopes, one at Mount Hopkins Observatory and one at Cerro Tololo Inter-American Observatory, Chile. The objectives of 2MASS include producing detailed views of the Milky Way unobscured by dust, and an all-sky catalogue of over a million galaxies.

Tycho Lunar crater (43°S 11°W), 87 km (54 mi) in diameter. Tycho is a complex crater with side-wall terracing and central peaks. This recent crater has a bright ray pattern, extending out for hundreds of kilometres. Extensive secondary impact fields are visible in the surrounding terrain. The EJECTA appear dark, probably from minerals dredged up in the cratering process.

Tycho Brahe *See* BRAHE, TYCHO

◀ **Tunguska event** This photograph of flattened trees and blasted trees, taken in 1927, shows the devastation that was caused by the aerial explosion in Siberia in 1908. The effects on a city would have been terrible.

Tycho Catalogue (TYC) *See HIPPARCOS CATALOGUE*

Tychonian system GEOCENTRIC THEORY of the Universe suggested by Tycho BRAHE in 1583. After the publication of Copernicus' heliocentric theory in 1543, astronomers were divided between the PTOLEMAIC SYSTEM and the new COPERNICAN SYSTEM. While the Earth-centred Universe seemed to accord with common sense and the geocentric physics of ARISTOTLE, placing the Sun at the centre accounted straightforwardly for the planets' retrograde motions and other phenomena. Tycho, a great admirer of Copernicus, realized that the question could be decided only by acquiring better observational data, and he developed instruments with which angular measurements of unprecedented accuracy could be made.

However, Tycho could detect no PARALLAX or any other sign of the Earth's motion in space. He therefore devised a cosmology that kept the Earth immobile at the centre of the Universe, but with the Moon and Sun revolving around it, and all the other planets revolving around the Sun. This was an ingenious compromise: his solar centre of planetary motion incorporated Copernican explanations of the retrograde motions, while his fixed Earth explained why heavy terrestrial bodies fall downwards. The Tychonian system soon gained many adherents. Although rejected by Galileo, it could still explain such Galilean phenomena as the phases of Venus. Not until after 1670 did the weight of observational evidence swing decisively from the Tychonian system towards the Copernican.

Tycho's Star First SUPERNOVA whose position and changing brightness were accurately estimated, with important observations having been made by Tycho BRAHE. It appeared in Cassiopeia in 1572 and was visible for 18 months. It was historically significant for the development of astronomy because it demonstrated that changes could (and did) occur in the heavens.

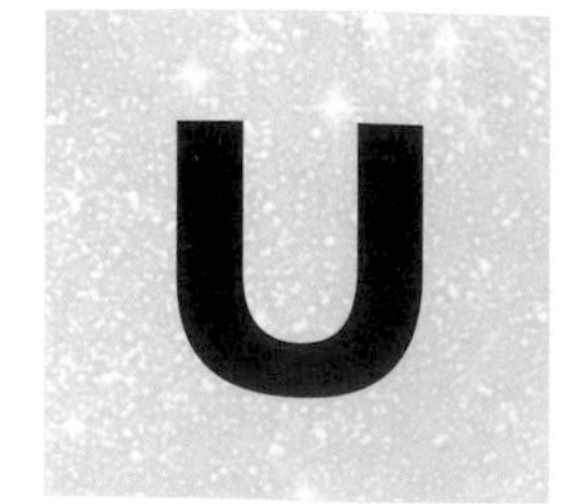

UARS Abbreviation of UPPER ATMOSPHERE RESEARCH SATELLITE

UBV System System of colour indices using differences of magnitudes at three selected spectral bands. The bands are (U) ultraviolet (360 nm), (B) blue (440 nm) and (V) visual (550 nm). It is now extended into the red (R), near-infrared (I) and further (J, K and L). It is occasionally irreverently referred to as the 'Alphabet Soup System'. *See also* PHOTOMETER

U Cephei Algol-type eclipsing VARIABLE STAR, which varies between magnitudes 6.6 and 9.8. The primary minimum lasts just under 10 hours and occurs every 2.493 days. The period of the star is slowly increasing, and also shows some irregular, spontaneous variations. The system consists of two stars moving in a nearly circular orbit. The smaller of the two stars is brighter and more massive; it is of spectral type B8V and has a mass 4.7 times that of the Sun. The larger component is a G-type giant star, with a mass only 1.9 times that of the Sun. A stream of gas is flowing from the larger, less dense star. This slows the rotation of the large star and speeds up the smaller, more massive companion. The surfaces of the two stars are only about 10 million km (6 million mi) apart, just about the same as the diameter of the giant star.

U Geminorum star (USGS) Subtype of VARIABLE STAR; they are also known as SS Cygni stars, after the brightest member, and as DWARF NOVAE, along with the Z CAMELOPARDALIS STARS and SU URSAE MAJORIS STARS. U Geminorum stars are named after the prototype, discovered by John Russell Hind (1823–95) in 1855. Almost 400 are now known. They are characterized by rapid outbursts of between 2 to 6 magnitudes, followed by a slower return to minimum, where they remain until the next outburst.

Periods range from 10 days to several years and are fairly constant when averaged out over many cycles, but there may be considerable variations in both period and brightness from one cycle to the next. Members of the SU Ursae Majoris subclass have both normal maxima and 'supermaxima', which last about five times as long and are twice as bright. Another group, with periods between 50 and 80 days, sometimes show anomalous maxima in which the rise takes several days, instead of the usual day or so.

Almost all U Geminorum stars show rapid irregular flickering of around mag. 0.5 at minimum. Because they are all binaries, some also show eclipses at minimum.

All are close binaries, comprising a subgiant or dwarf star of type K to M, which has filled its ROCHE LOBE, and a WHITE DWARF surrounded by an ACCRETION DISK of infalling matter. Variations in the flow of matter and/or variations in the atmospheres are thought to cause the outbursts. The longer the period of the star, the greater the variation, which suggests a relationship with the RECURRENT NOVAE, though the physical basis for this, if any, is unclear. There have been suggestions that U Geminorum stars evolve from W URSAE MAJORIS variables but this remains speculative.

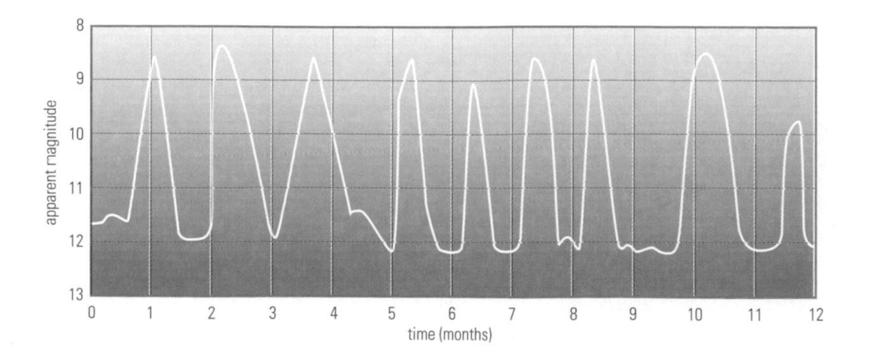

▼ **U Geminorum star** The light-curve of a typical U Geminorum star shows repeated outbursts at intervals ranging from about ten days to several years. During outburst, the brightness climbs rapidly, with a slower decline. Outbursts occur as a result of mass transfer to an accretion disk around the smaller, more massive star in these close binary systems.

At minimum the spectra of U Geminorum stars show wide emission lines, which during the brightening merge gradually into the continuous background, reappearing at maximum as absorption lines. At least two possible U Geminorum stars have been found in globular clusters (M5 and M30).

Uhuru (SAS-1) NASA Small Astronomy Satellite launched in 1970 December to carry out the first X-ray sky survey. It discovered neutron stars in binary systems and extended X-ray sources in clusters of galaxies. Uhuru means 'freedom' in Swahili. *See also* X-RAY ASTRONOMY

UKIRT Abbreviation of UNITED KINGDOM INFRARED TELESCOPE

UKST Abbreviation of UNITED KINGDOM SCHMIDT TELESCOPE

ultraviolet astronomy Study of astronomical objects at wavelengths in the ELECTROMAGNETIC SPECTRUM usually between 90 and 310 nm. The regions from 380 to 310 nm (near-ultraviolet) and from 90 to 10 nm (extreme-ultraviolet) are also included in this area. The interstellar medium becomes opaque at 91.2 nm, the limit to the Lyman series of the hydrogen atom, but the gas is patchy and inhomogeneous so astronomical objects can be detected in the extreme-ultraviolet range. Many stars and galaxies emit in the ultraviolet and the main strength of ultraviolet astronomy has been SPECTROSCOPY, especially that provided by the INTERNATIONAL ULTRAVIOLET EXPLORER (IUE) satellite, with over 104,000 spectra. Ultraviolet astronomy probes the physics and chemistry of hot gas around stars, nebulae and the interstellar medium, since the spectra provide information about motion, rotation and winds.

The ozone high in the STRATOSPHERE of Earth's atmosphere absorbs electromagnetic radiation at wavelengths shorter than 300 nm, making ground-based observations of the ultraviolet impossible. Early experiments (down to 200 nm) were possible from balloon platforms, which could reach altitudes of 45 km (28 mi). The first satellite experiments involved studying the ultraviolet spectrum of the Sun, culminating in an American series of satellites, the Orbiting Solar Observatories (OSO). Other early satellites included the American COPERNICUS satellite, the European TD-1 and the Dutch ANS satellites. The INTERNATIONAL ULTRAVIOLET EXPLORER (IUE) was launched in 1978 and operated until 1996. A conventional aluminium-coated mirror cannot be used for ultraviolet astronomy because aluminium oxide absorbs ultraviolet radiation at wavelengths shorter than 160 nm. Alternatives such as magnesium fluoride or lithium fluoride are often used. When the EXTREME ULTRAVIOLET EXPLORER (EUVE) satellite, operating between 1 and 100 nm, was launched in 1992, it detected hundreds of sources, hot stars and even some extragalactic sources. Several other satellites have been launched, including the FAR ULTRAVIOLET SPECTROSCOPIC EXPLORER (FUSE) and the extreme-ultraviolet (EUV) camera on ROSAT; the HUBBLE SPACE TELESCOPE (HST) has ultraviolet capabilities because it is sensitive in the wavelength range 110 to 2500 nm. There have also been ultraviolet astronomy experiments aboard the SPACE SHUTTLE, including the Hopkins Ultraviolet Telescope (HUT).

Ultraviolet observations contribute to all branches of astronomy. Most atoms and ions found between the stars have their strongest resonance ABSORPTION LINES in the ultraviolet. Observations in this wavelength region have, therefore, been significant for studies of the composition and motions of the interstellar gas. Discoveries include the way in which some elements are depleted from the gas, with the atoms sticking on to interstellar grains, and the existence in the interstellar gas of highly ionized regions, which may be 'fossil' supernova remnants. An extensive ionized halo to our Galaxy has also been discovered.

In the Solar System, ultraviolet astronomy led to the discovery of new constituents in planetary atmospheres. HUT and EUVE examined the Venus EUV (extreme-ultraviolet) dayglow emission, and FUSE has discovered molecular hydrogen (H_2) at 107 nm in the Martian atmosphere. Striking results have come from the observations of comets. IUE observed comets Halley and Iras–Araki–Alcock. The data revealed the way in which the gases spread out from the cometary nucleus and are gradually dissociated and ionized by the solar radiation. These studies also reveal details of the composition of the cometary material itself. Such studies are important because comets may be samples of material unchanged since the formation of the Solar System.

Ultraviolet astronomy has also been important for the study of stars. Many hot stars show strong stellar winds, and binary stars transfer material from one star to the other. In both cases EMISSION LINES in the ultraviolet region provide information on the nature of the flows and their effect on the evolution of the stars concerned. In extreme cases the massive, hot star is losing a significant amount of material in its wind, perhaps more than 10^{-6} solar masses/year, and this wind makes a noticeable impact on the surrounding interstellar medium. In addition, the emission spectra of gas nebulae excited by the radiation from hot stars pose interesting spectroscopic problems and lead to accurate measurements of abundances of materials, like carbon, which have few strong spectral lines at optical wavelengths. The ROSAT EUV camera showed the local bubble of high-temperature (around 10,000 K) gas around the Sun, with the region of lowest absorption by the gas in the direction of Canis Major, where tunnel-like structures permeate the bubble wall. Hot white dwarfs emit strongly in the ultraviolet and EUVE has catalogued over 400 of them. When white dwarfs are members of binary star systems, the radiation from their hot accretion disks is also very bright in the ultraviolet. Ultraviolet astronomy has also shown that coronae and transition regions are absent for stars cooler than a certain limit: this is thought to reflect a difference in the details of the stellar wind flows from these stars.

Ultraviolet astronomy has made important contributions to extragalactic studies. First, it was shown that in many ways the ultraviolet spectra of nearby active galaxies resemble very closely those of the quasars whose ultraviolet spectra are seen from the ground by virtue of the redshift. Secondly, the stellar content of galaxies has been studied in the ultraviolet, where the galaxy spectrum is sensitive to the presence of hot stars in the galaxy. But perhaps one of the most spectacular results in this field is the work on the variations of the ultraviolet spectrum of the active galaxy NGC 4151, which led to an estimate of 1000 million solar masses for the weight of the black hole in the nucleus of the galaxy.

ultraviolet excess Excess signal in the NEAR-ULTRAVIOLET and blue (originally measured in the Johnson UBV photometric system) when compared to normal stars (often defined by the Hyades cluster). It showed whether a star or sample of stars lacked heavy elements (metals), so they could be defined as 'metal-poor'. If there were fewer heavy elements, there were fewer absorption lines in the near-ultraviolet, hence the excess emission.

ultraviolet radiation Wavelength range from around 380 nm beyond the visible violet down to 90 nm, the limit of the LYMAN SERIES of hydrogen lines, with the region from around 90 to 10 nm called the extreme-ultraviolet. The region from 380 to 310 nm (known as the NEAR-ULTRAVIOLET) can be observed from the ground, and beyond this observations must be made from satellites or balloons. Ultraviolet radiation is strongly absorbed by glass, so lenses are made from quartz or fluorite.

Ulugh Beg (1393–1449) Islamic Turkish ruler of Maverannakhr (now Uzbekistan) and astronomer, whose name means 'great prince'; his real name was Muhammad Taragi ibn Shah-Rukh ibn-Timur. In 1420 Beg built a *madrasa* (university) equipped with a three-story observatory at Samarkand. Its instruments included the Fakhri 60° sextant – the largest transit instrument ever built, made of marble and standing 40 m (130 ft) tall. The Fakhri was used by Beg and his assistants to measure the altitudes of stars as they crossed the meridian to an accuracy of a few arcseconds and to calculate the obliquity of the ecliptic very precisely. Beg produced a catalogue of 992 stars, the *Zīj-i Gurgānī* (1437), a revision of the *ALMAGEST*, and compiled important tables of planetary positions.

▲ **ultraviolet astronomy** Hot, newly formed stars emit strongly in the ultraviolet. Star formation in the face-on galaxy NGC 4214 is revealed in the lower half of this Hubble Space Telescope ultraviolet image.

Ulysses First spacecraft to orbit the Sun around its poles, out of the ecliptic plane in which the planets orbit. Ulysses was launched aboard the Space Shuttle in 1990 October and passed over the south solar pole in 1994 September. This EUROPEAN SPACE AGENCY craft was one of two being planned for an International Solar Polar Mission but the US probe was cancelled. Ulysses itself was delayed from 1983 because of upper stage issues and the *Challenger* accident in 1986.

Departing from Earth orbit under the boost of its upper stage, Ulysses became the fastest spacecraft in history, travelling at 15.4 km/s (9.6 mi/s). The craft used a Jupiter gravity assist flyby in 1992 February, which

▼ **Ulysses** The Ulysses spacecraft is shown here undergoing final tests at Cape Canaveral prior to launch in 1990.

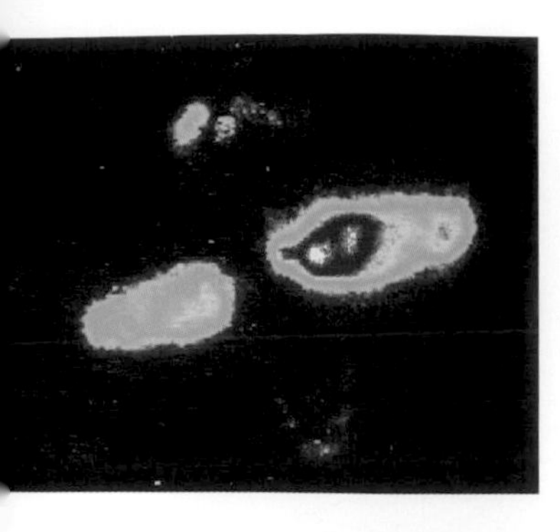

▲ **United Kingdom Infrared Telescope** HII emission from the Egg Nebula can be seen in this UKIRT image. A torus of emission is apparent in the plane of rotation.

swung it out of the ecliptic plane. Equipped with a suite of esoteric solar study experiments, Ulysses flew over the north solar pole in 1995 July and repeated a pass over the south pole in 2001 January. Further passes over the poles were scheduled, starting with a north polar pass in 2001 October, although the Ulysses budget limits operations until 2001 December. The spacecraft is capable, new budgets permitting, of continuing its unique exploration, with further passes in 2006 and 2007.

umbra (1) Dark central cone of the shadow cast by a planet or satellite. *See also* PENUMBRA

umbra (2) Darkest, coolest, usually central region of a SUNSPOT, with a typical temperature around 4000 K. The umbra is commonly surrounded by a grey, less cool penumbra.

Umbriel Icy SATELLITE of URANUS. It was discovered in 1851 by William LASSELL, who found its neighbour ARIEL in the same year. Umbriel is Ariel's near-twin in size and is only slightly less dense; it is curious, therefore, that it should appear so different. Similarities would perhaps be revealed if images were available of the unseen two-thirds of each globe.

As its name suggests, Umbriel is darker than the other large satellites of Uranus, having an albedo of 0.19. The imaged part of the globe is much more heavily cratered, and therefore older, than the cratered terrain on Ariel. Its dark colour could be a result of exceptional radiation-darkening of methane with age or a particularly high carbon content. The only bright material visible on Umbriel is the floor of the 150-km-diameter (90-mi) crater Wunda and in the central peak of the 110-km-diameter (70-mi) crater Vuver, both of which have albedos of about 0.5. *See* data at URANUS

United Kingdom Infrared Telescope (UKIRT) Telescope of 3.8 m (150 in.) aperture located at MAUNA KEA OBSERVATORY in Hawaii, one of the first large telescopes designed specifically to work in the infrared. When it was opened in 1979, it was operated as an outstation of the ROYAL OBSERVATORY, EDINBURGH, in collaboration with the Netherlands and the University of Hawaii. A sea-level headquarters was built in nearby Hilo. Today, UKIRT is the responsibility of the JOINT ASTRONOMY CENTRE in Hilo. Instrumentation developed for the telescope at Edinburgh during the 1980s pioneered novel techniques of imaging and spectroscopy in the infrared. It gave this branch of astronomy similar capabilities to optical astronomy. A wide-field infrared camera allows large-area surveys to be carried out.

United Kingdom Schmidt Telescope (UKST) Large SCHMIDT TELESCOPE inaugurated in 1973 at SIDING SPRING OBSERVATORY in central New South Wales, Australia. With its 1.2-m (48-in.) corrector plate and 1.8-m (72-in.) spherical mirror, the instrument was modelled on the OSCHIN SCHMIDT TELESCOPE at Palomar Observatory. The UKST was an outstation of the ROYAL OBSERVATORY, EDINBURGH until 1988, when its close relationship with the nearby ANGLO-AUSTRALIAN TELESCOPE was formalized and it became part of the ANGLO-AUSTRALIAN OBSERVATORY. Its principal task during its first two decades was to conduct photographic surveys of the entire southern sky in various wavebands, which were then digitized at Edinburgh. In the early 1980s, experiments with fibre optics led to new modes of operation for the telescope, culminating in 2001 with the deployment of a robotic 150-object spectroscopy system known as 6dF.

▼ **unsharp masking** Before processing (left) the fine detail in this image of part of the Lagoon Nebula (M8, NGC 6253) is slightly blurred. After unsharp masking (right) fine details, such as small stars, are less blurred.

United States Geological Survey (USGS) Sole science agency for the US Department of the Interior. Its Flagstaff Field Center was established in Arizona in 1963, originally for lunar studies and training astronauts for the APOLLO PROGRAMME in geology. Since then, Flagstaff's activities have expanded to include the study of the origin and evolution of the terrestrial planets and the major satellites of the Solar System, the development of techniques of remote sensing and image processing to aid in the interpretation of planetary surfaces, and to produce topographic and geological maps of planetary bodies.

United States Naval Observatory (USNO) One of the oldest scientific institutions in the United States, founded in 1830 to maintain the US Navy's chronometers and navigational instruments. It moved to its present location in Washington, D.C. in 1893. A 0.66-m (26-in.) refractor, built by ALVAN CLARK & SONS in 1873 (and for ten years the world's largest refractor), is still used at Washington for SPECKLE INTERFEROMETRY. At the main observing site at Flagstaff, Arizona (established in 1955), USNO operates a 1.55-m (61-in.) astrometric reflector, used to take the photographs that led to the discovery of Pluto's moon, CHARON, in 1978. It also operates the Navy Prototype Optical Interferometer at the Anderson Mesa site of LOWELL OBSERVATORY.

Unity First US module of the INTERNATIONAL SPACE STATION to be launched, in 1988. Unity is a node to provide connections to other parts of the ISS.

Universal Time (UT) Non-uniform TIMESCALE based on the rate of rotation of the Earth. In the past this was determined by making accurate observations of star positions using an instrument known as a PHOTOGRAPHIC ZENITH TUBE (PZT), but today it utilizes a technique called satellite laser ranging. Pulses of laser light are reflected off artificial satellites in known Earth orbits, allowing the position of the laser telescope relative to the satellite, and hence the rotation of the Earth, to be measured to a high degree of accuracy (*see* SATELLITE LASER RANGER).

This measure of the Earth's rate of rotation provides a timescale designated UT0. This timescale is dependent upon the location the observations were made from and has to be corrected for the shift in longitude caused by the slight wandering of the Earth's geographical poles (POLAR MOTION). The corrected timescale is called UT1 and it is this that is generally implied when the term UT is used.

COORDINATED UNIVERSAL TIME (UTC) is a uniform ATOMIC TIME scale, based on caesium-beam atomic clocks; it is the basis of civil time. Because the Earth's rate of rotation is gradually slowing due to the effects of tidal braking, UTC and UT1 are gradually diverging. In order to keep them linked, and thus ensure that time as kept by a clock is in step with that as measured by the position of the Sun in the sky, a LEAP SECOND is periodically introduced to UTC to keep it within 0.9 second of UT1. *See also* GREENWICH MEAN TIME (GMT); INTERNATIONAL ATOMIC TIME (TAI); TIMEKEEPING

Universe Everything we can detect, see, feel, know, or that has ever had any effect on our region of space. It encompasses all of SPACETIME, not just out to the visible horizon, and includes all particles, fields and interactions.

URANUS: DATA	
Globe	
Diameter (equatorial)	51,118 km
Diameter (polar)	49,947 km
Density	1.29 g/cm³
Mass (Earth = 1)	14.53
Volume (Earth = 1)	62.18
Sidereal period of axial rotation (equatorial)	$17^h\ 14^m$ (retrograde)
Escape velocity	21.3 km/s
Albedo	0.50
Inclination of equator to orbit	97° 52′
Temperature at cloud-tops	55 K
Surface gravity (Earth = 1)	0.79
Orbit	
Semimajor axis	19.22 AU = 2871 × 10^6 km
Eccentricity	0.046
Inclination to ecliptic	0° 46′
Sidereal period of revolution	84.01y
Mean orbital velocity	6.81 km/s
Satellites	21

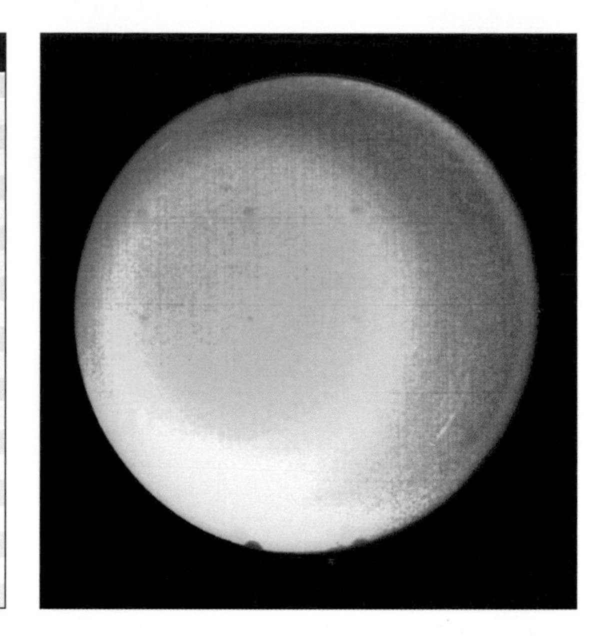

◀ **Uranus** Cloud features on Uranus are fairly elusive. This heavily processed Voyager 2 image from 1986 shows some structure in the planet's atmosphere at lower right.

Our Universe is commonly thought to have only three dimensions of space and one of time, but some theories postulate that there are 'rolled-up' dimensions through which some quantum fields work. These rolled up dimensions also constitute a part of our Universe, but we do not see them as a normal dimension, rather we feel their effects as forces or fields. Another common misconception is that spacetime and matter must be expanding out into 'something' or that spacetime warps in another dimension by mass. Since this dimension or direction is not within our Universe, it can never be seen or experienced by definition; thus its characteristics are irrelevant. The singularity in a BLACK HOLE, although consisting of infinitely warped spacetime, remains part of our Universe although a signal cannot escape. However, some ideas in GENERAL RELATIVITY claim that black holes, if combined with a black hole in another universe, can form a bridge or WORMHOLE to another universe under special conditions.

Additionally, some interpretations of quantum theory state that every time a wave function is collapsed, in other words something is measured, a new universe is created. In the case of tossing a coin, if it comes up heads it is in one universe, but there is another universe created where it came out tails. Most quantum theorists do not take seriously this multi-universe idea of the so-called Copenhagen interpretation of quantum physics, but theoretically there is no reason why it cannot be true.

University of Arizona Department of Astronomy *See* STEWARD OBSERVATORY

unsharp masking Photographic technique used for revealing fine detail in astronomical images. The process involves making a positive contact copy of the negative image on a glass photographic plate using low-contrast film. The film is placed in contact with the back of the original plate and a diffuse light source used, producing a blurred copy of the large-scale structure from the original plate. When replaced in contact with the back of the glass plate, the blurred image of the unsharp mask cancels out the large scale structure on the original, revealing the fine detail.

Unsöld, Albrecht Otto Johannes (1905–95) German astrophysicist who, from the 1920s, used quantum physics to improve the modelling of stellar atmospheres. He was thus able to explain the behaviour of the hydrogen atoms that produce the Balmer lines in the solar spectrum (1929). For this research, Unsöld obtained high-resolution spectroheliograms with Mount Wilson's 45-m (150-ft) tower telescope; subsequently, he obtained high-resolution spectra of other stars as a visiting scientist at Yerkes and McDonald Observatories. His long-term investigations of the BO star θ Scorpii yielded the first highly detailed analysis of a stellar atmosphere to discuss the phenomena of spectral line shifts and pressure broadening.

Upper Atmosphere Research Satellite (UARS) First spacecraft of the NASA Mission to Planet Earth programme; it was at the time the largest ever flown for atmospheric research. It was deployed into Earth orbit by the Space Shuttle in 1991 September. It investigates the mechanisms controlling the structure and variability of the upper atmosphere to help create a comprehensive database in order better to understand the depletion of ozone in the stratosphere. UARS helped confirm that man-made chlorofluorocarbons are responsible for the depletion of ozone. The spacecraft was equipped with ten instruments, some of which have failed.

upper culmination Passage of a CIRCUMPOLAR STAR across the MERIDIAN between the POLE and the ZENITH. At this point the star's HOUR ANGLE is exactly 0h. *See also* CULMINATION

Uppsala Astronomical Observatory Observatory of Uppsala University, Sweden, with origins in the early 17th century. The present building in central Uppsala dates from 1853, when it replaced the old Celsius Observatory of 1741. In 2000 the Astronomy Department vacated the 1853 observatory building and moved to a new laboratory in the city, having merged with the Department of Space Physics. An outstation at Kvistaberg Observatory, 50 km (30 mi) south of Uppsala, has a 1.0-m Schmidt telescope; the 0.5-m (20-in) Schmidt of the former Uppsala Southern Station at SIDING SPRING OBSERVATORY, Australia, is now operated on behalf of NASA for near-Earth object searches.

Uraniborg First observatory constructed on the Danish island of Hven by Tycho BRAHE during 1576-80; the name means 'Castle of the Heavens'. It included accommodation, a library, an instrument workshop and even a printing press. *See also* STJERNEBORG

Uranus Seventh planet in the Solar System, the first to be discovered telescopically. In 1781 William HERSCHEL, observing from Bath, was examining the small stars in the neighbourhood of H Geminorum when he found a star 'visibly larger than the rest', which he suspected to be a comet. As no one had previously discovered a planet this view is understandable. Uranus is the third largest of the major planets. It is more than four times the size of the Earth and 14 times more massive.

▼ **Uranus** The extreme axial tilt of Uranus is compared with that of Earth.

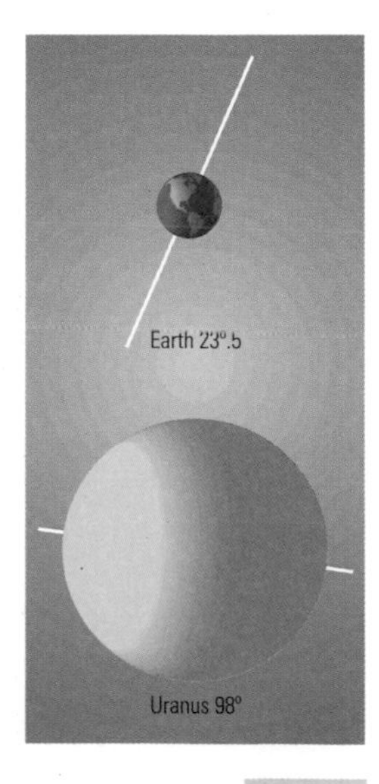

▶ **Uranus** The rings of Uranus were discovered during observations from the Kuiper Airborne Observatory of a stellar occultation by the planet in 1977. A series of dips in the star's brightness prior to its disappearance behind the planet indicated the presence of the rings. Similar dips were seen after the star re-emerged from behind the planet, confirming that the occulting material was in rings extending around Uranus.

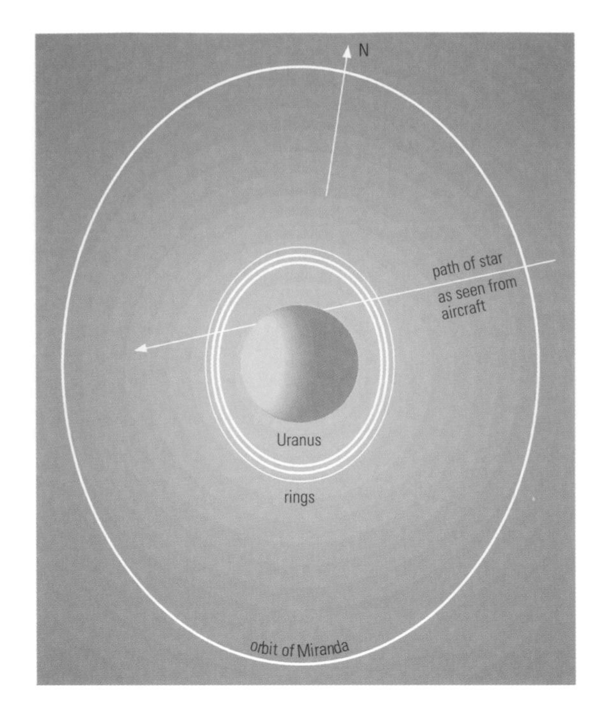

Uranus is denser than the larger gas giants, Jupiter and Saturn. It is composed primarily of hydrogen and helium, in approximately solar proportions, with small amounts of methane, ammonia and their photochemical products. Uranus has a greenish visual appearance, due to the absorption of sunlight in the blue/green portions of the spectrum by methane gas in its atmosphere. Uranus is also thought to contain heavier materials, such as oxygen, nitrogen, carbon, silicon and iron. Beneath the extensive layers of clouds, there is thought to be an ocean of superheated water, forming a 'mantle' overlying the Earth-sized rocky core. It has been suggested that this extensive ocean may have originated from ACCRETION of billions of cometary planetesimals, primarily composed of aqueous materials and found in large numbers in the outer parts of the Solar System during its early formation. The numerous collisions and high pressure of the region where the water is currently located accounts for its super-heated state.

Uranus' axis of rotation is tilted by 97°52′ with respect to its orbit. Consequently, it is unique in our Solar System since its axis of rotation is close to its orbital plane. In its motion around the Sun, each pole is presented towards the Sun and the Earth. During the encounter by the VOYAGER 2 spacecraft in 1986 January, the south pole of the planet was directed towards the oncoming probe and the Earth. The evolution of this axial tilt is probably a result of collisions with planetesimals early during the planet's formation – a major collision after the planet had grown to a large size would have disrupted it. A possible candidate for these collisions is a swarm of comets, an event that would also be consistent with the deep oceans of superheated water in the interior of Uranus. This unusual interior structure may also account for Uranus not possessing a significant internal heat source, in contrast to the other jovian planets, Jupiter, Saturn and Neptune.

▼ **Uranus** Features in the atmosphere of Uranus are less pronounced at visible wavelengths, as seen in the 1997 Hubble Space Telescope image at left. The right hand image, in red wavelengths where absorption by methane can be detected, does reveal banding structure, resembling that seen in the clouds on the other gas giant planets.

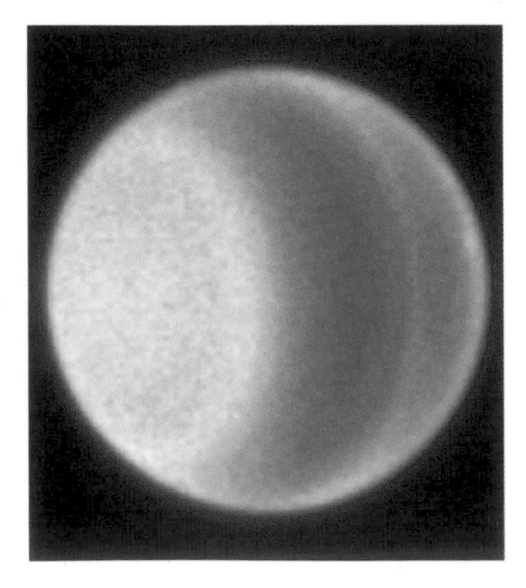

Since Uranus is so distant from the Earth, the Voyager 2 spacecraft encounter provided a dramatic increase in the understanding of it. The planetary magnetic field is extraordinary, since it is inclined at 60° to the rotational axis of the planet. The dipole field has a strength of 0.25 G, which compares with field strengths of 0.21 G at Saturn and 0.31 G at the Earth. Uranus' is the most inclined magnetic field in the Solar System, with a correspondingly large difference from the planetary rotational axis. It is possible that the magnetic field may be in some stage of reversing its polarity, which is currently south-seeking towards the Earth. The planet has an extensive MAGNETOSPHERE, which stretches for more than 18 R_U and contains a hot plasma environment with temperatures reaching more than 10,000 K. There is a complicated interaction between this hostile region, containing the trapped charged particles, and the embedded rings, satellites and the upper atmosphere of the planet. The magnetosphere sweeps away tiny particles from the rings and affects the chemistry of the satellites by heating their surfaces by particle bombardment. On the sunlit side of the planet there is an intense electroglow, while on the dark side auroral activity is detected through these magnetospheric/atmospheric interactions, which are also related to a huge glow in the atmosphere, spreading outwards for about 2 R_U (50,000 km/31,000 mi).

Uranus, like Jupiter and Saturn, has a banded appearance, which is barely visible from Earth. The cloud motions reveal a predominantly zonal circulation, where the winds are blowing in a east–west direction rather than from north to south. This circulation resembles the flow on Jupiter and Saturn and, to a lesser extent, the motions on every planet in the Solar System. The motions are zonal in spite of the substantial differences in their solar heating distributions, confirming that it is the rotation of the planet that organizes the weather systems. Unlike all the other planets, however, the polar regions are slightly warmer than the equatorial regions of the planet. Extensive haze layers, created photochemically and composed of acetylene and ethane particles, obscure the major weather systems on Uranus. During the Voyager 2 encounter, several discrete clouds composed of methane particles were seen in the southern mid-latitudes, at the 1-2 bar pressure level. Some of these clouds resemble the convective systems seen on Earth. The temperature at the cloud-top levels is approximately 64 K. There is a small seasonal variation of about 5 K in the cloud-top temperatures, but this will have a negligible effect on the meteorological systems since the Uranus atmosphere responds sluggishly to solar radiation, like a very deep ocean. Although the upper clouds are composed of methane, layers of ammonia, ammonia polymers and water clouds may exist deeper in the atmosphere.

Uranus is now known to possess at least 21 satellites. The five main satellites, MIRANDA, ARIEL, UMBRIEL, TITANIA and OBERON, were discovered during Earth-based telescopic observations. These satellites have densities in the range 1.26–1.5 g/cm^3 and are probably, therefore, composed mainly of water ice and rock, with small amounts of methane clathrate. However, they do appear to have varied surfaces. Oberon, Titania and Umbriel are the least geologically active of the satellite family, with Umbriel the darkest and most inactive of this set. The tectonic activity increases from Ariel to Miranda, which both show evidence of surfaces with fracture patterns, scarps, valleys and layered terrains. Miranda has the appearance of a body reassembled following a major collision at some earlier stage in its existence, with features resembling the grooves of Ganymede, the canyons of Mars and the compressional faults of Mercury, all on this tiny body. The 10 new satellites all reside in the region between Miranda and the rings, with two of them acting as shepherds for the outer Epsilon ring. These satellites range in size from about 15–170 km (9–110 mi) in diameter. All are surprisingly dark, with albedos of only about 0.05. It is possible that the darkness of all the satellites and the rings is caused by their bombardment by charged particles in the magnetosphere, which alters their surface heating and chemistry.

Stellar occultation observations in 1977 provided the first evidence of a system of nine rings around Uranus. Two further rings were found during the Voyager 2 encounter. This ring system extends from 37,000–51,000 km (23,000–32,000 mi) from Uranus' centre, although there appear to be hundreds of very thin rings and ring arcs around the planet. The Epsilon ring seems to be composed of 1 m (3 ft) boulders rather than the micrometre-size particles typical of the rings of Jupiter and Saturn. It is perhaps surprising that only one pair of shepherding satellites have been found so far, namely Cordelia and Ophelia, which are adjacent to the Epsilon ring. It is likely that smaller satellites, currently below the level of detection, do exist, shepherding the other rings.

ureilites Enigmatic group of ACHONDRITE meteorites. Ureilites are carbon-rich igneous rocks, the origin of which is uncertain. They have been described either as partial melt residues or as igneous cumulates. Ureilites show a wide range in oxygen isotopic compositions, and so have not achieved isotopic equilibration on a single parent body.

Urey, Harold Clayton (1893–1981) American chemist who founded the field of COSMOCHEMISTRY, explaining the origin and abundance of the chemical elements in the Solar System and the Universe at large. Urey worked at Johns Hopkins University (1924–29), Columbia University (1929–45), the University of Chicago (1945–58) and the University of California at San Diego (1958–72). His studies of the isotope ^{18}O enabled him to explain the origin of the chemical elements found on Earth, by comparing their terrestrial and solar abundances. It was under his direction that Stanley Lloyd Miller (1930–) performed what is called the **Miller–Urey experiment**. Miller passed a strong electric discharge (to simulate lightning) through a

▲ **Uranus** This false-colour image was generated from Hubble Space Telescope observations of Uranus in 1998. The planet's north pole is presented towards the observer. Banding and bright clouds can be seen, and enhanced contrast views of the ring system and inner satellites have been added.

URANUS SATELLITES AND RINGS

Rings

Ring or gap	Distance from centre (thousand km)	Width (km)
1986U2R	37–39.5	2500
6	41.8	1–3
5	42.2	2–3
4	42.6	2–3
Alpha	44.7	7–12
Beta	45.7	7–12
Eta	47.2	0–2
Gamma	47.6	1–4
Delta	48.3	3–9
Lambda	50	1–2
Epsilon	51.1	20–100

Satellites

	Diameter (km)	Distance from centre of planet (thousand km)	Orbital period (days)	Mean opposition magnitude
Cordelia	26	50	0.34	24.1
Ophelia	32	54	0.38	23.8
Bianca	44	59	0.43	23.0
Cressida	66	62	0.46	22.2
Desdemona	58	63	0.47	22.5
Juliet	84	64	0.49	21.5
Portia	108	66	0.51	21.0
Rosalind	58	70	0.56	22.5
Belinda	68	75	0.62	22.1
Puck	154	86	0.76	20.2
Miranda	480 × 468 × 466	129	1.41	16.3
Ariel	1158	191	2.52	14.2
Umbriel	1170	266	4.14	14.8
Titania	1580	436	8.71	13.7
Oberon	1520	583	13.46	13.9
Caliban	60	7230	579 *R*	22
Stephano	20	7980	674 *R*	24
Sycorax	120	12,180	1289 *R*	20
Prospero	20	16,670	2037 *R*	24
Setebos	20	17,880	2273 *R*	24

▼ **Uranus** The orbits of Uranus' larger satellites are shown to scale.

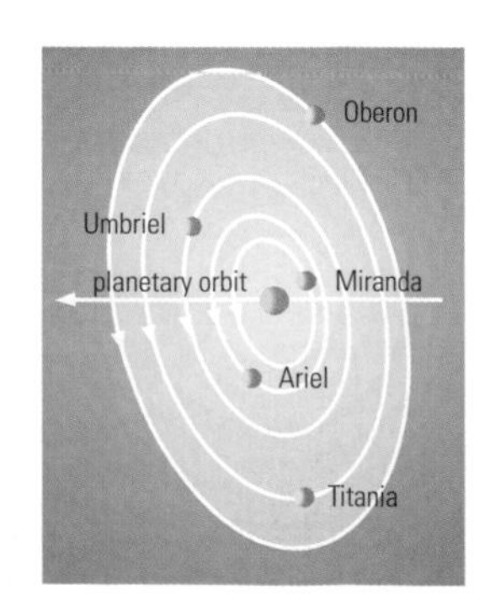

▶ **Ursa Major** This wide-field view of Ursa Major is dominated by the seven bright stars of the Plough. The constellation sprawls below and to the right of the Plough, with fainter stars marking out the Great Bear's paws and head.

URSA MAJOR (GEN. URSAE MAJORIS, ABBR. UMA)

Third-largest constellation and one of the best known, 'the Great Bear' lies in the northern sky between Draco and Leo. Many ancient northern cultures pictured a bear in this part of the sky. In Greek mythology these stars represent either Callisto, who fell victim to Juno's passion, was turned into a bear and was then placed by Zeus in the sky for safety; or Adrastea, a Cretan nymph who raised the infant Zeus and whom he placed among the stars in gratitude. It is easily recognized by the asterism of the PLOUGH (Big Dipper), the five central stars of which are part of the URSA MAJOR MOVING CLUSTER.

Ursa Major's brightest star, ALIOTH, is mag. 1.8. MIZAR is a quadruple system, whose brightest components are both white, mags. 2.3 and 3.9, separation 14″.4, which forms a wide optical double with ALCOR, mag. 4.0, separation 11′.8. Xi UMa is another multiple system, consisting of a binary with yellow components, mags. 4.3 and 4.8, separation 0″.9–3″.1, period 59.8 years, both of which are themselves spectroscopic binaries. Other interesting stars include Lalande 21185, mag. 7.5, which at just 8.3 l.y. away is the fourth-closest star to the Sun; and Groombridge 1830, mag. 6.4, which has the third-largest proper motion of any star.

Deep-sky objects in Ursa Major include the OWL NEBULA (M97, NGC 3587), an 11th-magnitude planetary nebula; M81 (NGC 3031), a 7th-magnitude spiral galaxy; M82 (NGC 3034), an 8th-magnitude irregular STARBURST GALAXY, probably disrupted by a near collision with M81 about 40 million years ago; and M101 (NGC 5457), an 8th-magnitude face-on spiral galaxy.

BRIGHTEST STARS

	Name	RA h m	dec. ° ′	Visual mag.	Absolute mag.	Spectral type	Distance (l.y.)
ε	Alioth	12 54	+55 58	1.76	−0.2	A0	81
α	Dubhe	11 04	+61 45	1.81	−1.1	F7	124
η	Alkaid	13 48	+49 19	1.85	−0.6	B3	101
ζ	Mizar	13 24	+54 56	2.23	0.3	A2	78
β	Merak	11 02	+56 23	2.34	0.4	A1	79
γ	Phad	11 54	+53 42	2.41	0.4	A0	84
ψ		11 10	+44 30	3.00	−0.3	K1	147
μ	Tania Australis	10 22	+41 30	3.06	−1.4	M0	249
ι	Talitha	08 59	+48 03	3.12	2.3	A7	48
θ		09 33	+51 41	3.17	2.5	F6	44
δ	Megrez	12 15	+57 02	3.32	1.3	A3	81
ο	Muscida	08 30	+60 43	3.35	−0.4	G4	184
λ	Tania Borealis	10 17	+42 55	3.45	0.4	A2	134
ν	Alula Borealis	11 18	+33 06	3.49	−2.1	K3	421

URSA MINOR (GEN. URSAE MINORIS, ABBR. UMI)

Small, rather inconspicuous northern circumpolar constellation, 'the Little Bear' lies between Draco, which surrounds it on three sides, and the north celestial pole, which is at present close to its brightest star, POLARIS, mag. 2.0. In Greek mythology it represents Ida, a Cretan nymph who helped her sister, Adrastea, raise the infant Zeus and whom he placed among the stars alongside her sister (the nearby constellation of Ursa Major) in gratitude. Ursa Minor's seven brightest stars form an asterism similar in shape to the Plough sometimes called the Little Dipper. β and γ UMi (KOCHAB and Pherkad) are known as the 'Guardians of the Pole'.

BRIGHTEST STARS

	Name	RA h m	dec. ° ′	Visual mag.	Absolute mag.	Spectral type	Distance (l.y.)
α	Polaris	02 32	+89 16	1.97	−3.6	F7	431
β	Kochab	14 51	+74 09	2.07	−0.9	K4	126
γ	Pherkad	15 21	+71 50	3.00	−2.8	A3	480

mixture of gases believed to have been present in the Earth's primordial atmosphere. The products included various organic molecules, including two amino acids – substances essential to life.

Ursa Major *See* feature article

Ursa Major Moving Cluster Nearest OPEN CLUSTER to the Earth; it is at a distance of 75 l.y. The URSA MAJOR Moving Cluster includes five of the seven stars of the well-known asterism the Plough (Big Dipper). The star at the tip of the tail, η UMa, and the pointer that lies nearer to the pole, α UMa, are the non-members. Over long periods of time the motion of these two stars will distort the present star pattern. The other five stars share the same basic motion through the Galaxy, forming a MOVING CLUSTER, and have about the same distance from us. They resemble a very open cluster of a few stars.

When the motions of other bright stars through the Galaxy are plotted, it is found that more show the same motion as the Ursa Major stars. Sirius is one such star, and others include δ Leonis, β Aurigae, β Eridani and α Coronae Borealis. In total about 100 members of this MOVING GROUP are known. Many are widely scattered across the sky from Ursa Major. Their separation in space is moderately large, certainly larger than in a normal cluster. Our Sun is taking us through the outskirts of the group, which is why its member stars can appear on all sides of the sky.

Ursa Minor *See* feature article

Ursid meteors METEOR SHOWER active between December 19 and 24 from a RADIANT close to β Ursae Minoris. Activity peaks around December 22–23, usually at zenithal hourly rate no greater than 10. However, some noteworthy displays have been seen. The 1945 return, which first brought the shower to prominence, gave observed rates of 100 meteors/hr. Lesser enhancements occurred in 1982 and 1986, and it is possible that further good Ursid displays have been missed. Ursid meteors are fairly swift. The stream parent is Comet 8P/Tuttle.

USGS Abbreviation of UNITED STATES GEOLOGICAL SURVEY

USNO Abbreviation of UNITED STATES NAVAL OBSERVATORY

UT Abbreviation of UNIVERSAL TIME

Utopia Planitia Vast plain on MARS, 3200 km (2000 mi) in diameter, centred near 50°.0N 242°.0W. It was the landing site of VIKING 2 in 1976.

UV Ceti Active red dwarf FLARE STAR that exhibits large flares every few hours UV Ceti is a 12th-magnitude binary, both components spectral type M5.5 V. At 8.7 l.y. distant, it is one of the Sun's closest neighbours. *See* table of nearest STARS

UX Ursae Majoris star Eclipsing VARIABLE STAR with Algol-like characteristics and extremely short period. Unlike normal ALGOL STARS, members of this small group show rapid and completely random brightness variations outside of eclipse, up to as much as 1.5 magnitudes in the case of RW Trianguli. During eclipse this star has an asymmetric light-curve, often with 'shoulders' before and after minimum. The maximum brightness of UX Ursae Majoris is subject to variations of amplitude 0.01 to 0.2 magnitude over periods of one to 20 minutes. The average maximum and minimum brightnesses also vary, and there are oscillations in the period. UX Ursae Majoris stars have a bright gaseous cloud surrounding the smaller, brighter component of the system, which ejects gas erratically. The larger, less luminous companion may lose mass to the smaller star.

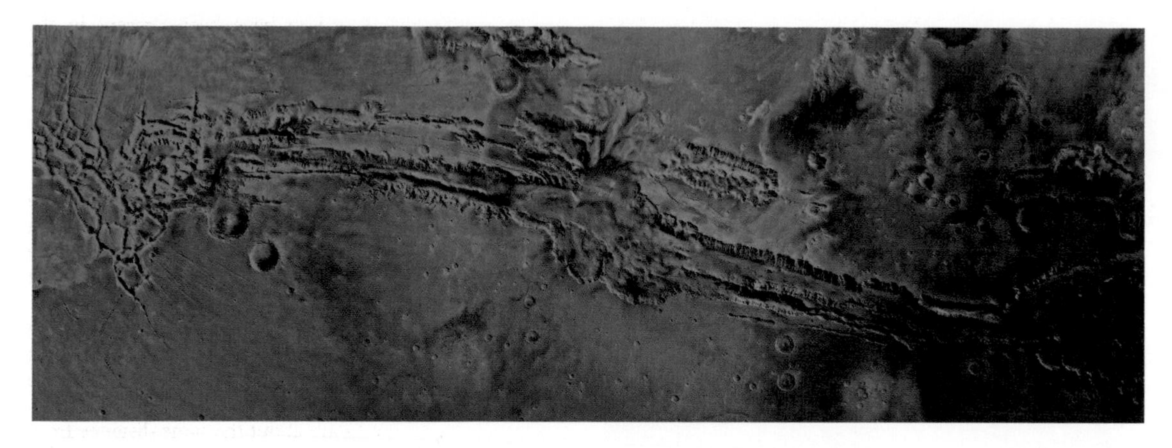

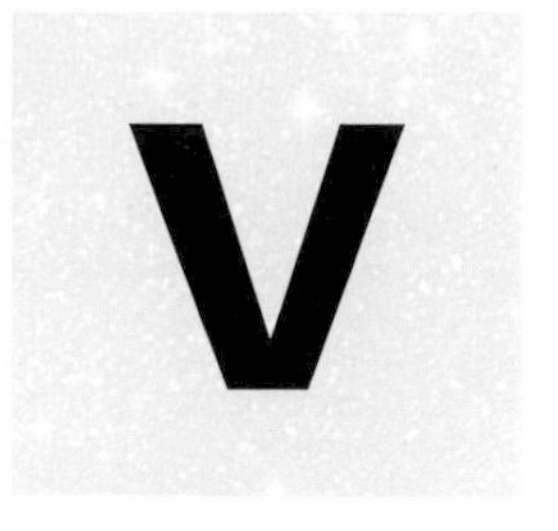

◀ **Valles Marineris** This composite Viking image shows the huge, sprawling canyon system of the Valles Marineris (Mariner Valley) on Mars. Ancient dry river channels run northwards from the chaotic terrain at the east (right).

vacuum Volume of space containing a low pressure gas. A perfect vacuum would contain no particles at all, but this is unreachable in practice. On the Earth, a soft vacuum contains gas at a pressure of 0.01 Pa, a hard vacuum, gas at 10^{-6} Pa. The interstellar medium, away from interstellar gas clouds, contains gas at a pressure of around 10^{-12} to 10^{-14} Pa.

Valles Marineris Series of enormous, interconnected canyons on MARS, situated between 30°W and 110°W, just south of the Martian equator. These canyons, produced by extensive faulting of the crust, extend from the crest of THARSIS MONTES to the vicinity of CHRYSE PLANITIA, a distance of 4500 km (2800 mi). In the west the canyon system emerges from Noctis Labyrinthus then trends approximately west–east. The main section of the system comprises a number of parallel, straight-sided chasms, beginning in the west with Ius Chasma and Tithonium Chasma. The central section sees both the widest and deepest expression: here three vast troughs, each 200 km (120 mi) wide and up to 7 km (4 mi) deep, slice through the plains. At the eastern end the canyons pass into chaotic terrain. Ophir Chasma and Candor Chasma are situated to the north of the central canyon complex. The west and east ends of Ophir Chasma are both rather blunted and the trend of the canyon walls is continued laterally by lines of crater chains, indicating an underlying structural control. Spacecraft images reveal the existence not only of erosional debris but also of rhythmically layered deposits, which, by analogy with terrestrial rocks, would be expected to have been laid down in quiet conditions beneath a cover of water. Such deposits are seen generally to form flat-topped hills (mesas) or irregular blocks within the troughs. The mesas and canyon walls are finely stratified by both light and dark materials on a scale of a few metres, indicating a complex geological history. Very recent evidence points to the deposition of the layered strata prior to the large-scale faulting that gave rise to the canyon system.

Valhalla Largest multiple-ringed impact BASIN on Jupiter's satellite CALLISTO. At 4000 km (2500 mi) in diameter, it is in fact the largest such structure in the entire Solar System. A 600-km-wide (400-mi) bright circular patch representing the impact site is surrounded by a series of concentric fractures.

Van Allen belts RADIATION BELT regions in the terrestrial MAGNETOSPHERE in which charged particles are trapped as they spiral along the magnetic field and bounce up and down between reflection points located towards the magnetic poles. These regions were discovered in 1958 by James Van Allen on analysis of Geiger counter data from the EXPLORER 1 satellite. The inner region of the Van Allen belts is located between 1000 and 5000 km (600 and 3000 mi) above the equator; it contains protons and electrons, which are either captured from the SOLAR WIND or originate from collisions between upper atmosphere atoms and high-energy COSMIC RAYS. The outer belt region is located between 15,000 and 25,000 km (9000 and 16,000 mi) above the equator, but it curves downwards towards the magnetic poles. This region contains mainly electrons from the solar wind. The Van Allen belts are a potential hazard to Earth-orbiting spacecraft, since the radiation levels are high enough to have an adverse effect on the electronic subsystems and on-board instrumentation, especially following a GEOMAGNETIC STORM. Similar radiation belt regions of trapped, charged particles have been discovered around Jupiter, Saturn, Uranus and Neptune.

Van Allen, James Alfred (1914–) American physicist who discovered (1958) the radiation belts surrounding Earth. He has spent most of his career (1951–) at the University of Iowa, investigating Solar System radiation and magnetic fields. After World War II, he used captured German V-2 rockets to measure cosmic rays, inventing the 'rockoon' method of reaching far greater altitudes than previously possible by launching rockets first carried to high altitude by balloons. An organizer of the INTERNATIONAL GEOPHYSICAL YEAR, he analysed data from the early Explorer spacecraft to map two torus-shaped regions of radiation above the Earth's equator now named the VAN ALLEN BELTS. More recently, he and his colleagues have made detailed studies of SOLAR–TERRESTRIAL INTERACTIONS. In all, Van Allen has served as principal investigator for 24 different space missions.

Van Biesbroeck, George (1880–1974) Belgian-American astronomer (born Georges Achille van Biesbroeck), an observer of double stars at Yerkes Observatory (1917–64). He moved to the United States when Belgium was invaded in World War I. At Yerkes, he used the 40-inch (1-m) telescope to carefully re-observe 1200 double stars identified at Lick Observatory by William Joseph Hussey (1862–1926). During his long career, Van Biesbroeck preferred computing and refining orbits of known pairs to seeking new doubles. He discovered three new comets (1925, 1936, 1954) and the faint companion of the star BD +4° 4048, with an absolute

▼ **Van Allen belts** Earth is surrounded by radiation belts, which contain energetic particles trapped by the magnetic field. The inner belt dips low over the South Atlantic, where its particle population can present a hazard to satellites.

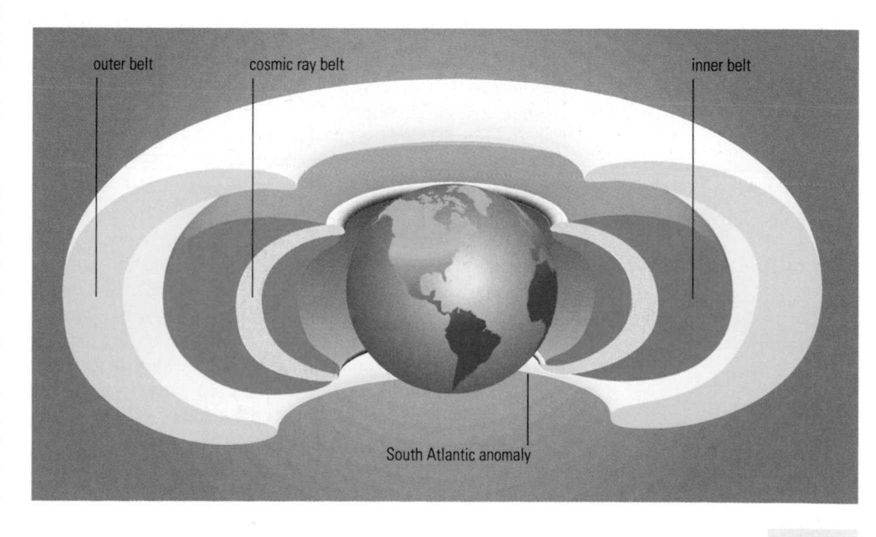

▲ **Veil Nebula** The delicate wisps of the Veil Nebula in Cygnus are the remnant of a supernova explosion that occurred at least 30,000 years ago. Material like this, thrown into interstellar space by supernovae, is enriched in heavy elements which are recycled in future generations of stars.

magnitude of only +19. Named **Van Biesbroeck's Star**, this was for many years the least luminous star known.

Vandenberg Air Force Base US military launch facility about 240 km (150 mi) north-west of Los Angeles, operated by the US Air Force Space Command's 30th Space Wing. It began as a US Army training centre in 1941 but from 1957 was developed as the nation's first space and ballistic missile operations and training base under the Air Force. In 1958 it was renamed to honour General Hoyt S. Vandenberg. The base supports west-coast launches for the NATIONAL AERONAUTICS AND SPACE ADMINISTRATION (NASA) and commercial contractors as well as the Air Force and the Department of Defense. All US spacecraft destined for near-polar orbit are launched from Vandenberg.

Van Maanen, Adriaan (1884–1946) Dutch-American astronomer who conducted astrometric and spectroscopic studies at Mount Wilson Observatory (1912–46). There, he used the 60- and 100-inch (1.5- and 2.5-m) telescopes to make a comprehensive survey of parallaxes for 500 starfields. He discovered **Van Maanen's Star**, a nearby 12th-magnitude white dwarf. He also made many measurements of proper motions, among them many Cepheid variables, thus helping to determine the scale of the Galaxy. This led Van Maanen to extend his earlier research on star streaming, carried out with Jacobus KAPTEYN, to determine which stars belong to Double Cluster in Perseus, the Pleiades and Hyades, and other clusters. This work greatly extended our knowledge of the density and luminosity of stars in the solar neighbourhood.

variable star Star whose brightness changes with time. These variations in light may be the result of some inherent feature in a star or its atmosphere, or of some geometrical alignment. The former are termed INTRINSIC VARIABLES; the latter EXTRINSIC VARIABLES. There are about 35,000 known variable stars and at least 15,000 objects that are suspected to vary in light, but have not yet been confirmed. These numbers do not include variables in GLOBULAR CLUSTERS or those in other galaxies. A few variable stars are known by their Greek letters or proper names, for example α Orionis (BETELGEUSE). The system of naming variable stars is to assign the letter R to the first discovered in a constellation; the second becomes S and so on down to Z. Then comes RR to RZ, SS to SZ, and so on. This sequence ends at ZZ, after which comes AA to AZ, BB to BZ and so on down to QZ, but the letter J is never used. After QZ the next variable in any constellation would be the 335th: it becomes V335, and each subsequent variable in that constellation gets a higher number. The letters and numbers are followed by the name of the constellation.

Early discoveries of variables were made visually and a few are still discovered by this method. However, most discoveries result from conventional or CCD photography. Two images taken of the same region of the sky may be placed in a blink or stereo COMPARATOR, which ensures that any difference in a star is soon noticed. CCD images may be compared electronically with a reference image of the same area of sky. The plates, films or images must be taken some time apart and with the same instrument and, if possible, under the same sky conditions.

Variable-star work is a field in which a large contribution is made by amateur astronomers, working in co-operation with central organizations such as the AMERICAN ASSOCIATION OF VARIABLE STAR OBSERVERS (AAVSO), the BRITISH ASTRONOMICAL ASSOCIATION (BAA) or the ROYAL ASTRONOMICAL SOCIETY OF NEW ZEALAND (RASNZ). There are so many variable stars that the professional astronomers can only observe a few. They wish to select those stars that pose special problems, or to observe them at certain phases of their light variations. As a result, they rely on amateurs to provide data that will enable them to select the stars that interest them and to provide light-curves and predictions for the more regular variables so that they can select the phases that they wish to observe. The amateur astronomers are thus able to make a very valuable contribution by consistently following these stars.

The intrinsic variables are divided into many classes, each of which has a number of divisions depending on the way in which the stars change in brightness or because of some other property inherent in them. Broadly speaking, there are seven classes. The first are called ERUPTIVE VARIABLES because of violent processes taking place in their chromospheres and coronae. The resultant flares are often accompanied by shell events, as matter is carried off by stellar winds, and by interaction with the interstellar matter that surrounds the star. Typical examples are the various types of Orion variables. The second class are the pulsating variables, which have fairly regular expansion and contraction of their surface layers. MIRA STARS and the many types of CEPHEIDS belong to this class. The third class consists of rotating stars, with light variations resulting from their axial rotation or from starspots or some feature of their atmospheres caused by a magnetic field. This class includes both intrinsic stars and extrinsic variables. The fourth class comprises the explosive or CATACLYSMIC VARIABLES such as NOVAE, SUPERNOVAE and DWARF NOVAE. The fifth class comprises the ECLIPSING BINARY systems such as the Algols, which are really extrinsic variables. The sixth class consists of X-ray sources in which the variability is in the X-ray radiation. Finally, there is a seventh class, which consists of unique objects that cannot be assigned to any other type. Included in this class are the BL LACERTAE OBJECTS and optically varying QUASARS, which are, of course, extragalactic objects.

The observation of many of the foregoing stars has undergone a change in recent years, with the availability of photoelectric PHOTOMETERS and their associated equipment. It is now possible to obtain precise three-colour data on a wide range of stars. Similarly, the use of CCD equipment has enabled amateurs to reach fainter magnitudes than were possible with purely visual methods. All such observations have been welcomed by the professionals because there are so many variable stars to be observed: in 1786 there were only 12; by 1866 this had grown to 119; by 1907 the number had increased to 1425; the advent of photographic methods of discovery saw numbers grow rapidly thereafter until by 1941 there were 8445 known variable stars. Since then the number has continued to increase rapidly so that today there are tens of thousands of these stars.

To observe variable stars, both professionals and amateurs need detailed maps of small areas of the sky in which the variable is situated and on which both the variable and the surrounding stars are clearly identified, preferably with a sequence of comparison stars of constant brightness. This is often impossible for the simple reason that the magnitudes of the surrounding stars have not been accurately determined. This problem may be overcome in several ways, one of which is to select what appear to be suitable comparison stars from photographic plates and to assign letters to them as symbols. This then enables visual observations to be made pending the determination of accurate magnitudes later. For making a visual estimate of a variable star, the observer compares it with two stars of known, unchanging brightness. There are three principal methods – the ARGELANDER STEP METHOD, the FRACTIONAL METHOD and the POGSON STEP METHOD.

The main sources of the necessary charts are the Variable Star Sections of the BAA and RASNZ, and the AAVSO. The importance of variable stars is shown by the fact that about one-third of the astronomical literature is concerned with them in some way.

variation Term with several meanings in different contexts. **Lunar variation** is the second-largest periodic PERTURBATION of the Moon's longitude, caused by the Sun, displacing it from its mean position by ± 39′29″.9 with a period of 14.765 days (half of the synodic month). The perturbation, discovered by Tycho Brahe at around 1580, was the first major advance in the knowledge of the Moon's orbit since the discovery of the evection by Ptolemy in about AD 140. The perturbation varies as the sine of twice the longitude difference of the Sun and Moon, and hence it is zero at new and full moons. Thus it does not affect the times of occurrence of eclipses, which allowed it to escape detection for 14 centuries. **Annual variation** is the annual rate of change of the coordinates of a star due to the combined effects of PRECESSION and PROPER MOTION. **Secular variation** is the rate of change per century of the annual precessional change in a star's coordinates. **Magnetic variation** is the angle between the directions of magnetic north and true north at any particular location on the Earth's surface; its value changes with time.

Varuna Large TRANS-NEPTUNIAN OBJECT catalogued as asteroid number 20000. It was discovered late in the year 2000, although images of it on photographic plates dating back to 1955 have been identified. Varuna takes 285 years to orbit the Sun, in a low-eccentricity orbit with perihelion at 40.9 AU and aphelion at 45.7 AU. Varuna is particularly notable because its absolute magnitude (H=3.7) indicates that it may be among the largest known asteroids. If its albedo is as low as 0.05, then its equivalent diameter would be about 1100 km (680 mi), which is larger than CERES. If its albedo is 0.25, however, its diameter would be only 500 km (300 mi).

Vastitas Borealis Northern circumpolar plain of MARS.

Vatican Advanced Technology Telescope (VATT) Research facility consisting of the 1.8-m (72-in.) Alice P. Lennon Telescope and the Thomas J. Bannan Astrophysics Facility located at the MOUNT GRAHAM INTERNATIONAL OBSERVATORY. It is operated by the Vatican Observatory Research Group, and observing time is shared with the UNIVERSITY OF ARIZONA DEPARTMENT OF ASTRONOMY. The telescope has an unusually deep primary mirror whose focal length is equal to its diameter. It was made at the University of Arizona Mirror Laboratory to test new spin-casting and optical polishing techniques that have since been used for mirrors up to 8.4 m (330 in.) in diameter (in the LARGE BINOCULAR TELESCOPE). The VATT saw first light in 1993.

Vega The star α Lyrae, the fifth-brightest in the sky, visual mag. 0.03 (although HIPPARCOS found it to be DELTA SCUTI STAR – a variable with fluctuations of a few hundredths of a magnitude). It is a blue-white main-sequence star, spectral type A0 V and 50 times as luminous as the Sun, at a distance of 25 l.y. The Infrared Astronomical Satellite detected a disk of gas and dust surrounding Vega, possibly a PROTOPLANETARY DISK. The star's name comes from of the word *wāqi'*, part of an Arabic phrase meaning 'swooping eagle'.

Vega 1 and 2 Two space probes launched by the Soviet Union in 1984 December. Seven months later, on their way to Halley's Comet, they released landers on to the surface and two balloons into the atmosphere of VENUS. The landers sent back data for just under an hour, while the balloons transmitted for 46.5 hours as they drifted around the planet. Their closest approaches to the comet took place on 1986 March 6 and 9 respectively.

Although the Vegas were used as pathfinders for the GIOTTO spacecraft, they also obtained valuable images of the comet's nucleus and other data on its coma. The flyby distances were 8890 km (5520 mi) and 8030 km (4990 mi) on the sunward side of the nucleus. The probes carried identical payloads of two cameras, infrared spectrometers, gas and dust mass spectrometers, dust impact detectors and plasma analysers. The name Vega was derived from Venera and Galley (the Russian pronunciation of Halley).

Veil Nebula Faint SUPERNOVA REMNANT in Cygnus; it is spread over an area about 3° wide, just south of ε Cyg. Also known as the Cygnus Loop, the Veil Nebula has three principal sections, each with separate NGC designations: NGC 6960 at the west (RA 20^h 45^m.7 dec. +30°43′); NGC 6974 and 6979 in the centre (RA 20^h 50^m.9 dec. +32°00′); and NGC 6992 and 6995 at the east (RA 20^h 56^m.8 dec. +31°28′). While quite prominent on long-exposure images, the filaments of the Veil Nebula are rather difficult to observe visually without

VELA (GEN. VELORUM, ABBR. VEL)

Southern constellation between Pyxis and Carina, part of the old Greek figure of ARGO NAVIS, the ship of the Argonauts; Vela represents the ship's sails. Its brightest star, Regor, is an optical double (separation 41″) with bluish-white components, one the brightest WOLF–RAYET STARS in the sky (range 1.81–1.87), the other mag. 4.3. δ and κ Vel, together with Iota and ε Carinae, form an asterism known as the FALSE CROSS, because it can be confused with Crux (the Southern Cross). Deep-sky objects include IC 2391, a large, bright open cluster; NGC 3201, a 7th-magnitude globular cluster; the Eight-Burst Nebula (NGC 3132), a 9th-magnitude planetary nebula; and the supernova remnants the GUM NEBULA and the much more recent VELA SUPERNOVA REMNANT, at the heart of which lies the VELA PULSAR.

BRIGHTEST STARS

	Name	RA h m	dec. ° ′	Visual mag.	Absolute mag.	Spectral type	Distance (l.y.)
γ	Regor	08 10	−47 20	1.75	−5.3	WC8 + O9	841
δ		08 45	−54 43	1.93	0.0	A1	80
λ	Suhail	09 08	−43 26	2.23	−4.0	K4	573
κ	Markeb	09 22	−55 01	2.47	−3.6	B2	539
μ		10 47	−49 25	2.69	−0.1	G5	116
N		09 31	−57 02	3.16	−1.2	K5	238

V

▲ **Vela Supernova Remnant** Tendrils of material were ejected into interstellar space by the explosion of a supernova in the southern constellation Vela about a thousand years ago. The green line across the photograph is the path of a satellite that traversed the field of view during the exposure of the green-sensitive plate.

the aid of special OIII filters. The supernova that gave rise to the structure is thought to have occurred 30,000 to 40,000 years ago. The nebulosity is expanding at about 100 km/s (60 mi/s). It lies 2500 l.y. away.

Vela *See* feature article, page 427

Vela Pulsar Strong, short-period radio PULSAR, discovered in 1968, and identified with a SUPERNOVA REMNANT. It has a period of 89 milliseconds and in 1977 an optical counterpart was discovered that flashes with the same period. The Vela Pulsar occasionally increases its spin rate abruptly; this is called a GLITCH.

Vela Supernova Remnant SUPERNOVA REMNANT (SNR) in Vela (RA 08^h 34^m dec. $-45°45'$). It is some 200 l.y. in diameter and about 1600 l.y. away. It is the result of the same supernova that produced the VELA PULSAR. From the spin-down rate of the latter, the explosion can be dated to about 1000 years ago. The Vela SNR is within the GUM NEBULA. Adjacent to it in the sky, but actually four times more distant, is the Puppis SNR.

velocity Rate of change of position of a body in a specified direction. Velocity has both magnitude ('speed') and direction, and so is a vector quantity.

velocity–distance relation *See* HUBBLE LAW

velocity of light *See* LIGHT, VELOCITY OF

Vendelinus Eroded, ancient lunar crater (16°S 62°E), 165 km (103 mi) in diameter. The lava-flooded floor contains numerous elongated secondary craters from the EJECTA of LANGRENUS to the north. Vendelinus appears oblong due to foreshortening, but it is actually more circular in shape.

Venera Series of Soviet VENUS probes launched 1961–83. Venera 4 was the first spacecraft to send back data during descent through the Venusian atmosphere. The first successful landing on Venus was accomplished by Venera 7 in 1970. Venera 9, the first spacecraft to enter orbit around Venus, also made history by deploying a lander that returned the first picture from the rocky surface. Pictures were also returned from landers on Veneras 10, 13 and 14. The landers collected atmospheric data during descent and, on later missions, used a gamma-ray detector and instruments to detect electrical (lightning) discharges. Veneras 15 and 16, the last in the series, were orbiters equipped with radar to map the surface between 30°N and the north pole at a resolution of 1–2 km (0.6–1.2 mi).

Venus Second planet in the Solar System from the Sun and, apart from the Sun and the Moon, the brightest object in the sky. Venus is one of the TERRESTRIAL PLANETS, similar in nature to the Earth but slightly smaller. Like the Earth it has a substantial atmosphere. Since its orbit lies inside that of the Earth, Venus never strays farther from the Sun than 47°. Accordingly, it can only be observed either in the east as a morning object or in the west as an evening star. As a result, like the Moon, it goes through a sequence of phases and at its brightest is actually a thin crescent close to inferior conjunction. Galileo in the 17th century helped to affirm Copernicus' heliocentric theory through his discovery of the phases of Venus.

All that can be seen optically on Venus is the upper deck of a uniform, unbroken layer of yellowish cloud, the top of which may be at an altitude of about 100 km (62 mi) above the planet's surface. The cloud reflects 76% of the incident sunlight, and Venus has the highest known planetary albedo (0.76). This contrasts with the Earth, which, on average, has only 50% cloud cover and an albedo of about 0.37. Venus' atmosphere is huge, more than 90 times more massive than that of the Earth and is composed primarily of carbon dioxide (CO_2). Traces of hydrochloric and hydrofluoric acids, carbon monoxide, nitrogen, water vapour, hydrogen sulphide, carbonyl sulphide, sulphur dioxide (SO_2), argon, krypton and xenon have also been detected. Venus' atmosphere extends to an altitude of 250 km (155 mi) above the surface, but it possesses only a thermosphere and troposphere in its layered structure. On the day side of Venus, there is a terrestrial-type thermosphere with temperatures increasing from about 180 K at 100 km (62 mi) to about 300 K in the exosphere. The thermosphere does not exist on the night side of the planet, where the temperature falls from about 180 K at 100 km (62 mi) to 100 K at 150 km (93 mi). The transition from day-side to night-side temperatures across the terminator is very abrupt.

The atmosphere is extremely variable in the neighbourhood of the cloud tops in the altitude range of 75–100 km (47–62 mi); diurnal fluctuations of as much as 25 K have been observed at the 95 km (59 mi) level. Below these variable haze layers are the ubiquitous clouds, which occupy a substantial portion of the

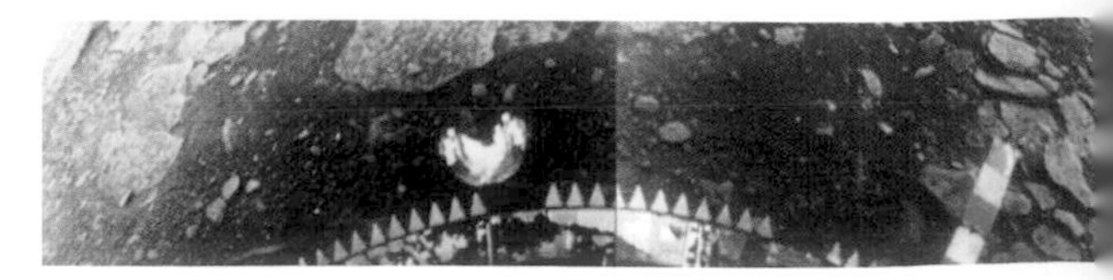

► **Venera** The Russian Venera 13 lander survived the extreme conditions on Venus' surface long enough to send back this mosaic of images, covering a 170° panorama, on 1982 March 1. Flat slabs of basalt-like rock can be seen together with loose soil.

V

VENUS: DATA	
Globe	
physical characteristics	
diameter	12,104 km
density	5.24 g/cm^3
mass (Earth = 1)	0.8149
volume (Earth = 1)	0.8568
escape velocity	10.40 km/s
inclination of equator to orbit	177°.3
sidereal period of axial rotation	243.0 days (retrograde)
surface gravity (Earth = 1)	0.903
surface temperature	750 K
albedo	0.76
mean visual magnitude (greatest elongation)	−4.4
orbital constants	
semimajor axis	0.723 AU
eccentricity	0.007
inclination to ecliptic	3°.395
sidereal period of revolution	224.701 days
mean synodic period	583.92 days
mean orbital velocity	35.02 km/s
mean longitude of perihelion	131°.571
satellites	none

▲ **Venus** Centred on longitude 180°, this is a composite view of Venus prepared from Magellan Imaging Radar results in 1991.

troposphere. There are three distinct cloud layers in the region of 45–60 km (28–37 mi), with differing particle sizes and concentrations. Sulphuric acid droplets make up the main composition of the clouds of Venus. The droplets are formed from the reaction between H_2O and SO_2 high in the atmosphere, aided by ultraviolet radiation from the Sun. The sulphuric acid clouds form a layer of varying concentration extending from 40–75 km (25–47 mi), with rain occurring in the lower layers.

The temperature increases steadily from the cloud tops at about 300 K to the surface at 750 K, where there is virtually no wind. Beneath the clouds and on the surface of Venus, the temperatures are the same everywhere. About 90% of the volume of the entire atmosphere lies between the surface and an altitude of 28 km (17 mi). At this level the atmosphere resembles a massive ocean, dense and sluggish in its response to the very weak solar heating. Only 2% of the incident sunlight actually reaches the surface of the planet. The surface pressure on Venus is 90 times greater than that on Earth and the surface temperature is the highest known in the Solar System. This crushing CO_2 environment at the surface of Venus is the result of a runaway GREENHOUSE EFFECT. The basic make-up of the Earth and Venus is very similar, but there is now as much CO_2 in the atmosphere of Venus as we find in the limestone rocks of the Earth. Because Venus is nearer to the Sun, it receives twice the sunlight than is incident on the Earth, and thus its surface has rapidly heated up by the greenhouse mechanism to its current state. This effect cannot be achieved by the CO_2 alone. The small traces of H_2O and SO_2 are also essential for the efficient greenhouse effect. Venus' surface temperature will not increase further since the atmosphere and surface are in chemical equilibrium.

The weather systems of Venus, at the level of the cloud tops, are strange. Although the planet itself is rotating very slowly, the equatorial clouds have a rotation period of four days, indicating wind speeds of 100 m/s (330 ft/s). Consequently, the cloud tops are moving in a retrograde direction 60 times faster than the surface of the planet. Almost all the solar energy is absorbed in the cloud tops, and this provides the main driving mechanism for the super-rotation of the atmosphere. The cloud tops of Titan and the upper atmosphere of the Earth are the only other regions known to super-rotate.

The surface of Venus is surprisingly varied, which may suggest that the initial geological developments took place before the massive atmosphere evolved into its current state. Images from the Soviet VENERA landers show a stony desert landscape, with outcrops and patches of dark material suggesting some chemical erosion. The subsequent radioactive analyses of Venus' soil suggest the composition is similar to BASALT, but with an unusually high concentration of potassium. Some of the basaltic materials are similar to those found on the terrestrial seabed. About 70% of the surface is covered by huge rolling plains, 20% by depressional regions and the remaining 10% by highlands, which are concentrated in two main areas. A more detailed understanding of the nature and global distribution of the planet's landforms has resulted from studies of the improved-resolution MAGELLAN imagery. Features disclosed by this mission include aeolian dunes, wind-streaks, channels, lava flow lobes, impact craters and outflow material associated with impact ejecta.

Each of the major highlands or continent-like units (Terra) is separated from the others by low lying plains or basins (Planitia). The most extensive is Atalanta Planitia in the northern hemisphere. Atalanta Planitia is about 1.4 km (0.9 mi) below the mean planetary radius, and is about the size of the Gulf of Mexico. Rising from the plains are smaller uplands (Regio), some of which are joined by deep, elongated, steep-sided depressions (Chasma). There are some craters about 25–48 km

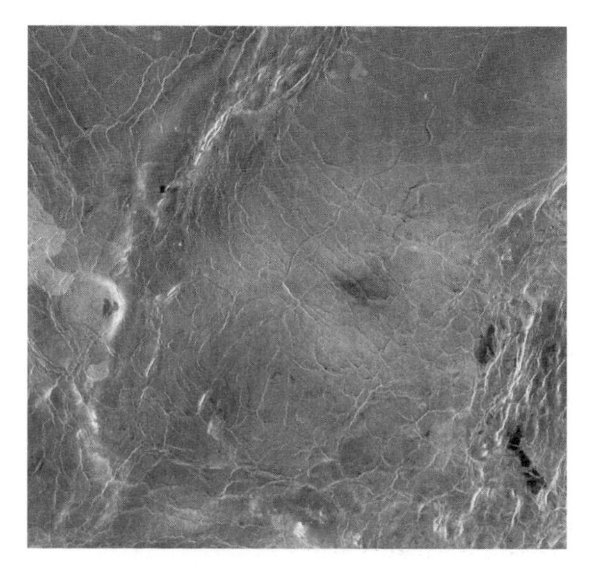

◀ **Venus** This Magellan Imaging Radar mosaic shows a 600 km (360 mi) section of the longest channel on Venus. Details as small as 120 m (400 ft) can be resolved. The channel has a total length of 7000 km (4200 mi).

V

SELECTED SURFACE FEATURES OF VENUS

Name	Latitude	Longitude	Description
Aino Planitia	40°.5S	94°.5E	extensive plain to south of Aphrodite Terra
Akna Montes	68°.9N	318°.2E	high mountain range on western border of the plateau Lakshmi Planum
ALPHA REGIO	25°.5S	0°.3E	isolated highland massif
APHRODITE TERRA	5°.8S	104°.8E	most extensive upland region on the planet stretching for some 10,000 km (6200 mi)
Atalanta Planitia	45°.6N	165°.8E	major region of low elevation, 2050 km (1270 mi) greatest extent
Atla Regio	9°.2N	200°.1E	chain of high mountains, forms east section of Aphrodite Terra
BETA REGIO	25°.3N	282°.8E	one of the four main upland areas
Devana Chasma	16°.0N	285°.0E	comparable with East African Rift; some 2500 km (1550 mi) long and 90 km (56 mi) wide; it descends to 2.5 km (1.6 mi) below datum at its deepest point
Diana Chasma	14°.8S	154°.8E	lowest point on surface, 2 km (1.2 mi) below datum
ISHTAR TERRA	70°.4N	27°.5E	massive northern uplands of continent size
Lada Terra	62°.5S	20°.0E	most recently discovered highland
Lakshmi Planum	68°.6N	339°.3E	striking plateau covering an area roughly twice that of the Tibetan Plateau
Maxwell Mons	65°.2N	3°.3E	highest mountains on the planet, occupying eastern end of Ishtar Terra
Phoebe Regio	6°.0S	282°.8E	complex ridged terrain
Rhea Mons	32°.4N	282°.2E	large volcanic structure
Sappho Patera	7°.0N	15°.0E	radar bright feature, probably volcanic in nature
Theia Mons	22°.7N	281°.0E	large volcanic structure in Beta Regio, evidence of collapse into a 3 km (2 mi) deep summit caldera

(16–30 mi) in diameter, suggesting that impacts occurred before the atmosphere reached its current density. The highland areas include mountains (Mons or Montes), plateaux or high plains (Planum) and volcanic areas important in the evolution of the planet and its atmosphere. The major highlands are ISHTAR TERRA, in high northern latitudes, and APHRODITE TERRA, which spans the equator. A third, the most recently discovered, is Lada Terra, revealed by Magellan, which is located largely south of latitude 50°S. Lada Terra is bordered on the north by lowland plains, and is joined to ALPHA REGIO, some 3500 km (2200 mi) to the north, by a complex of deep rifts. BETA REGIO, another highland area, contains two large shield volcanoes, Rhea Mons and Theia Mons. These tower some 4 km (2.5 mi) above the surface on a faultline that extends north–south. They are similar to the Hawaiian volcanoes and may well be active. Indeed, there has been a noticeable reduction in the measured amount of atmospheric SO_2 during the past 10 years, which could be explained by past volcanic eruptions and the associated atmospheric adjustment. Electrical activity beneath the clouds in the form of lightning has been inferred from measurements made during the descent of Venera 11 and 12. Again this could be caused by volcanic activity.

Venus does not have a magnetic field in spite of possessing a large nickel–iron core. This is due to the very slow rotation of the planet, which is unable to generate a field by dynamo action. However, a very weak magnetic field is induced in the planet's ionosphere through the interaction of the SOLAR WIND; the resulting BOW SHOCK region acts as a buffer between the interplanetary medium and the atmosphere of Venus.

Venus was once described as the twin of the Earth. The reality is otherwise. Our knowledge and understanding, derived from ground-based measurement and numerous spacecraft missions, involving flybys, orbiters, atmospheric probes and the Soviet/French floating balloons from the VEGA mission en route to the HALLEY comet encounter in 1986, reveals a hostile world, utterly different in atmospheric composition, meteorology and surface condition from its terrestrial counterparts.

▼ Very Large Telescope Shown here are the four lined 8 m (312 in.) instruments of the Very Large Telescope (VLT) at the European Southern Observatory, Cerro Paranal in the Atacama Desert. Working together as an interferometer, these instruments will deliver resolution equivalent to that of a single 16 m (624 in.) telescope.

vernal equinox (spring equinox) Moment at which the Sun appears to cross the CELESTIAL EQUATOR from south to north, on or near March 21 each year, at the FIRST POINT OF ARIES. At this time, the Sun is directly overhead at the equator and rises and sets due east and due west respectively on that day, the hours of daylight and darkness being equal in length. The name is also used as an alternative for the First Point of Aries, indicating the point on the celestial sphere where the ECLIPTIC and the celestial equator intersect at the Sun's ASCENDING NODE, and the zero point for measuring CELESTIAL LONGITUDE. *See also* AUTUMNAL EQUINOX; EQUINOX

Very Large Array (VLA) One of the world's premier radio observatories, located at an elevation of 2120 m (6970 ft) on the plains of San Augustin about 80 km (50 mi) west of Socorro, New Mexico, and completed in 1980. The VLA is operated by the NATIONAL RADIO ASTRONOMY OBSERVATORY, a facility of the National Science Foundation. The telescope consists of 27 antennae arranged in a huge 'Y' shape with a maximum extent of 36 km (22 mi). Each antenna has an aperture of 25 m (82 ft), weighs 230 tonnes and can be moved along a track to adjust the spacing of the array. The signals are combined electronically, and four different configurations are used – from compact to far-flung. At its highest resolution (with the antennae at their maximum spacing), the VLA has the same resolution as a single dish 36 km (22 mi) across. This corresponds to 0″.04 at the maximum operating frequency of 43 GHz. The instrument has the sensitivity of a single telescope 130 m (427 ft) in diameter.

Very Large Telescope (VLT) Largest optical telescope in the world, consisting of four 8.2-m (27-ft) reflectors located at PARANAL OBSERVATORY and operated by the EUROPEAN SOUTHERN OBSERVATORY (ESO). The four separate instruments, known as Unit Telescopes, are named Antu (completed in 1998), Kueyen (1999), Melipal (2000) and Yepun (2001). The names are those of the Sun, the Moon, the Southern Cross, and Venus, respectively, in the local Maopuche language. They can be operated either independently or together; in combination they have an effective aperture of 16.4 m (54 ft). When operated in a mode called VLTI, they are used with three 1.8-m (72-in) outrigger telescopes to carry out optical INTERFEROMETRY. The telescopes of the VLT have mirrors made of a single piece of glass–ceramic material as opposed to the segmented mirrors at the W.M. KECK OBSERVATORY. Unique among the new instruments (not least because it was built in a non-ESO nation) is the OzPoz robotic spectrograph feed supplied by the ANGLO-AUSTRALIAN OBSERVATORY in 2002. ESO is also investing in multi-conjugate ADAPTIVE OPTICS for the VLT, partly to demonstrate the viability of the OVERWHELMINGLY LARGE TELESCOPE.

Very Long Baseline Array (VLBA) Series of ten radio antennae spread across the United States and its overseas territories, from St Croix in the Virgin Islands to MAUNA KEA OBSERVATORY, Hawaii. Their signals are recorded separately on tape and later combined at the Operations Centre in Socorro, New Mexico. Each antenna is 25 m (82 ft) in diameter, and the maximum separation of the array is 8600 km (5400 mi), giving it an angular resolution of 0″.001. Rather than supplanting the VERY LARGE ARRAY, it is complementary to it, providing astronomers with a finer scale of resolution. Construction of the VLBA began in 1986, and the first observation with the complete array was made in 1993. It is operated by the NATIONAL RADIO ASTRONOMY OBSERVATORY with funding from the National Science Foundation.

very long baseline interferometry (VLBI) Combining of RADIO TELESCOPES separated by thousands of kilometres to act as a RADIO INTERFEROMETER in order to observe the same astronomical object with a resolving power of 0″.0002. The radio telescopes for VLBI are located around the world, many in the United States, Europe and Australia, plus several in other countries. Several radio telescopes in Europe can also combine together to form the European VLBI Network (EVN). VLBI can study structure in quasar jets, and it showed the superluminal motions in the jet of the quasar 3C 273 and the active galaxy M87. Knowledge and understanding of radio sources has been revolutionized by VLBI. Intercontinental VLBI has also used measurements of a known quasar to track the positions of the telescopes, monitoring the steady drift of the continents and the wandering of the poles.

Very Small Array (VSA) Ground-based instrument designed to map the cosmic microwave background radiation on angular scales of about 1°. It has 14 antennae with apertures of 0.32 m (13 in.) on a single tip-table mounting located at an elevation of 2385 m (7820 ft) at the Observatorio del Teide of the INSTITUTO DE ASTROFÍSICA DE CANARIAS (IAC) in Tenerife. The project is a collaboration between the MULLARD RADIO ASTRONOMY OBSERVATORY, JODRELL BANK OBSERVATORY and the IAC.

Vesta Large MAIN-BELT ASTEROID; number 4. Its size, albedo and location, near the inner edge of the main belt, mean that Vesta is the brightest of all such asteroids and may just attain naked-eye visibility. Images obtained with the HUBBLE SPACE TELESCOPE show Vesta to be roughly spherical, with a very large depression around its southern pole, which may be a vast impact crater. Like PALLAS, Vesta has a high average density (*c*.3.9 g/cm^3), which may be due to a metallic interior similar to nickel–iron meteorites, despite its surface reflectivity indicating a stony composition. Its mass is about 3.0 × 10^{20} kg (0.41% the mass of the Moon). Several smaller asteroids have similar orbits to Vesta (*see* HIRAYAMA FAMILY) and this group, formed from the fragmentation of an earlier body, is thought to be the source of the METEORITES known as eucrites, howardites and diogenites.

video astronomy Use of video cameras for astronomical imaging. The large image scales produced by a CCD offer dramatic detailed views of the Sun, Moon, planets and brighter planetary satellites. Unlike the long integration times possible with cooled CCD imagers, conventional video cameras are limited to short exposure times comprised of two consecutive fields combined to produce a stream of continuous pictures at 25 or 30 frames per second. This fast frame ability makes video an excellent tool for accurately timing occultation events. The fast exposure time of continuous video also provides an effective way to capture an unpredictable moment of steady seeing. A standard 3-hour videotape can capture over 270,000 individual images, and frame-by-frame playback makes for easy identification of the best images recorded, which can be selected and combined by IMAGE PROCESSING.

Throughout the 1990s, developments in highly light sensitive CCD image sensors have paved the way for convenient astronomical imaging with video cameras. In 1998 a video camera was used on the historic 60-inch (1.5-m) telescope at Mount Wilson by a Boston University team imaging notoriously difficult Mercury. They obtained unprecedented ground-based images of a hemisphere not previously imaged during the Mariner 10 flyby.

Video cameras are also widely used to record fireball and meteor showers. In 1999 members of the INTERNATIONAL OCCULTATION TIMING ASSOCIATION captured video evidence of meteoric impact events on the Moon during the November Leonid meteor shower. To study the radiant of an extremely faint meteor shower, a wide-angle lens is coupled to the input window of an image intensifier, and a video camera then records the image at the output window of the intensifier. Image intensifiers can be used for real-time viewing and recording of many deep-sky objects.

To produce useful images, detail is important and the number of TV lines a camera produces governs the resolution of the video image. Many amateurs use modified low-light security cameras that output between 400 and 600 lines. The output signal can be recorded directly to videotape or a computer using a video capture interface. Analogue S-VHS video machines are capable of recording up to 400 lines, while digital recorders and computer capture devices are capable of more. To preserve spatial resolution sacrificed in most single-chip colour cameras and maintain optimal signal-to-noise ratios, monochrome cameras produce the best results when used for tri-colour imaging of planets.

As with conventional cooled CCD cameras, video images transferred to computer can be further processed using image-processing software. Since a video camera is an uncooled integrating device, individual frames suffer from some amount of thermal dark noise generated by the camera's electronics. Stacking or layering several selected individual images can produce a smoother, more aesthetically pleasing final picture with improved depth of detail.

▲ **video astronomy** These tri-colour images of Jupiter were caught in 1999 October (left) and 2000 December (right). Using a monochrome video camera, consecutive exposures were made in red, green and blue light.

vignetting Partial shadowing of an optical image caused by obstructions within the light beam or by optical components too small to pass all the beam. The effect of vignetting is to make some parts of the image dimmer than they should be, but it is not always detected by the user. It is often present in cheap binoculars and can be detected by looking at the bright spot visible in the eyepiece when viewed from a distance. This spot should be circular and evenly illuminated. If it is not, then some vignetting, usually caused by the prisms, is present. The effect is sometimes present when taking photographs through a telescope. It may not be noticed on a photograph of the sky but a terrestrial shot will often be brighter in the middle and darker at the corners.

► **Viking** Compiled from images obtained by the Viking orbiters, this composite shows the Amphitrites Patera region of Mars. Radial ridges extend northwards for about 400 km (250 mi) from this old volcano.

Viking Mission to Mars consisting of two orbiter-lander spacecraft. Viking 1 was launched by a Titan-Centaur booster on 1975 September 9, and injected into Mars orbit on 1976 June 19. The first month of orbiter operations consisted of locating and verifying a safe landing site for the lander spacecraft. The lander spacecraft separated from the orbiter and landed on the surface on 1976 July 20. Viking 2 was launched on 1975 August 20, injected into Mars orbit on 1976 August 7, and its lander spacecraft touched down on 1976 September 3.

Because Mars is the most hospitable of the planets, and because of the long history of debate regarding the possibility of life on Mars (*see* LIFE IN THE UNIVERSE; CANALS, MARTIAN), the landers carried experiments to search for life. All the experiments operated successfully, and no indications of life were found at either landing site (though some have criticized the design of the experiments, claiming the outcome was ambiguous), but this does not preclude life at other locations. Meteorological instruments on the landers regularly reported the weather and monitored surface changes and seasonal effects.

Each orbiter carried two vidicon cameras and each lander carried two facsimile cameras. The orbiters mapped the entire surface of Mars at 150–300 m (500–1000 ft) resolution and selected areas at resolutions down to 8 m (26 ft). In all, they returned more than 55,000 images, including high-resolution pictures of the satellites Phobos and Deimos.

Virgo *See* feature article

Virgo cluster Rich concentration of galaxies in the direction of the constellation Virgo, although some lie across the border in Coma Berenices. The centre of the cluster is about 65 million l.y. away, and is marked by the giant elliptical galaxy M87, also known as the radio source Virgo A, which is ejecting a strong jet of gas. In all there are 16 Messier objects in the cluster; the total number of galaxies in the cluster is about 3000.

virial theorem Theorem concerning the conservation of energy equations that relates the kinetic energy of gas or stars to the gravitational potential energy:

$$E = P_G + \mathrm{KE}$$

where E is the total energy, P_G is the gravitational potential energy, and KE is the kinetic energy or energy of motion. It is useful in determining the mass of a system based on its gravitational field and the orbital velocities of stars or gas clouds orbiting around it.

Visible–Infrared Survey Telescope for Astronomy *See* VISTA

visible spectrum Wavelength band that is visible to the human eye. The visible spectrum, sometimes called the 'optical spectrum', extends approximately from 400 to 700 nm. It is subdivided (from long to short wave) by the major visual colours red, orange, yellow, green, indigo and violet. It is flanked by the 'near-ultraviolet' and the 'near-infrared'. The visible spectrum was the first to be explored, and it contains the FRAUNHOFER LINES as well as myriad other ABSORPTION LINES and EMISSION LINES. *See also* ELECTROMAGNETIC SPECTRUM

VISTA (acronym for 'Visible–Infrared Survey Telescope for Astronomy') British project to build a southern-hemisphere 4-m (157-in.) wide-field survey telescope to support the current generation of 8–10-metre class telescopes. The telescope will be located at Paranal Observatory and will be shared with the EUROPEAN SOUTHERN OBSERVATORY (ESO) as part of the UK's entry agreement with ESO. It will be operational by 2004. Originally, the telescope was planned to have wide-field imagers in both the visible and infrared wavebands, but the visible-light function will now be carried out by ESO's VST (*see* PARANAL OBSERVATORY), so VISTA will be an infrared-only telescope.

visual binary Gravitationally bound BINARY STAR system in which both components can be resolved, sometimes without a telescope. The brighter star is termed the primary, the fainter is termed the companion. When the separation is too large for orbital motion to be observed, they are called common proper motion stars.

If orbital motion can be observed, the orbit of the companion with respect to the primary can be determined. If the distance to the pair is known, the total mass of the binary can be determined from Kepler's third law (*see* KEPLER'S LAWS). The individual masses of the components can be determined if the absolute orbit of each star can be measured.

visual magnitude Apparent brightness of an astronomical body as seen by the eye, whose maximum sensitivity is at a wavelength of 550 nm. Such magnitudes are now determined photographically or photoelectrically, using appropriate filters, and are called PHOTOVISUAL MAGNITUDES.

VLA Abbreviation of VERY LARGE ARRAY

VLBA Abbreviation of VERY LONG BASELINE ARRAY

VLBI Abbreviation of VERY LONG BASELINE INTERFEROMETRY

VLT Abbreviation of VERY LARGE TELESCOPE

Vogel, Hermann Carl (1841–1907) German astronomer who pioneered the use of spectroscopy to determine stellar diameters and masses, leading to the

VIRGO (GEN. VIRGINIS, ABBR. VIR)

Second-largest constellation and one of the signs of the zodiac, which has been associated with a succession of female deities since Babylonian times. It lies between Leo and Libra and is not particularly prominent, except for its brightest star, SPICA (or Azimech), mag. 1.0. Porrima is a fine binary with yellow components, both mag. 3.6, separation 2″.7, period 168.7 years. There are no bright star clusters or nebulae in Virgo, but there are numerous galaxies, many of which are members of the VIRGO CLUSTER. These include the 8th-magnitude M49 (NGC 4472) and the 9th-magnitude M87 (NGC 4486), both giant elliptical galaxies (the latter containing the radio source Virgo A); and the SOMBRERO GALAXY (M104, NGC 4594), which is not a member of the cluster. The constellation also contains 3C 273, the first quasar to have been discovered (in 1963), mag. 12.9.

BRIGHTEST STARS

	Name	RA h m	dec. ° ′	Visual mag.	Absolute mag.	Spectral type	Distance (l.y.)
α	Spica	13 25	−11 10	0.98	−3.5	B1	262
γ	Porrima	12 42	−01 27	2.74	2.4	F0	39
ε	Vindemiatrix	13 02	+10 58	2.85	0.4	G8	102
ζ		13 35	−00 36	3.38	1.6	A3	73
δ		12 56	+03 24	3.39	−0.6	M3	202

discovery of the first SPECTROSCOPIC BINARIES. Using the 'reversion spectroscope' invented by Friedrich ZÖLLNER, Vogel was the first to calculate the Sun's rotational velocity by measuring Doppler shifts in spectra of its opposite limbs; later he applied the technique to other stars, finding that Algol and Spica were each a spectroscopic binary consisting of stars of nearly equal mass but very different luminosities. In 1876, he documented changes in the spectrum of Nova Cygni as it faded, the first time this had been accomplished. As director of Potsdam Astrophysical Observatory (1881–1907), he supervised Potsdam's participation in the *CARTE DU CIEL* sky-mapping project.

Vogt–Russell theorem Theorem proposed independently in 1926 by the German physicist Heinrich Vogt (1890–1968) and H.N. RUSSELL stating that if a star's mass and chemical composition are known, then all its other properties can be determined by the laws of physics. Since initial chemical composition varies relatively slightly between stars, it is thus principally a star's initial mass that determines its basic structure and evolution.

voids Large areas in the Universe where there are apparently few, if any, galaxies. This is in contrast to areas where large clusters of galaxies reside. *See also* LARGE-SCALE STRUCTURE

Volans *See* feature article

volatile Element or compound that evaporates at a relatively low temperature. All of the noble gases and other constituents of planetary atmospheres, such as hydrogen, nitrogen, methane, ammonia and carbon dioxide, are volatiles. The most significant volatile is water. Sulphur and its compounds are also classed as volatiles. Materials with higher temperatures of evaporation, such as most metals and silicates, are called refractory.

volcanism Eruption of molten material at the surface of a planetary body. MAGMA is usually less dense than the surrounding solid rock, and it tends to rise through any cracks or zones of weakness. Planetary crusts consist of a variety of minerals, so heating these materials does not necessarily lead to a unique melting temperature. Some minerals may melt while others, still solid, may be carried along in a fluidized medium. VOLATILES, including water, carbon dioxide and sulphur compounds, dissolve in molten silicates at high pressures; they are released at lower pressure near the surface. The mixture's volatile component, viscosity and yield strength will vary between planets and between different rock types on a given planet; thus there are different types of eruption, the extremes being 'quiescent' and 'violent'. On Earth, large volumes of basaltic lava are erupted relatively quiescently from calderas (as in Hawaii), producing gently sloping shield volcanoes, and from fissures (such as in Iceland), producing lava plains. Steeper cones are produced by explosive pyroclastic eruptions of silica-rich material in which the gas pressure shatters the lava and rock into fountains of fragments.

Earth displays a wide variety of volcanic landforms. The volcanic materials reach the surface through conduits and fissures associated with the active margins of tectonic plates and mid-oceanic ridges. Other planets have a variety of volcanic landforms. Mercury appears to show some volcanic features. On the Moon, lavas erupted from fractures, filling the large impact basins (maria) with basaltic magma; some small volcanic domes and lava tubes are also present. Venus appears to have been resurfaced by lava flows of great extent. Mars has several extinct shield volcanoes; OLYMPUS MONS, with a height of 24 km (15 mi) and a diameter greater than 700 km (430 mi), is the largest volcanic edifice known. Apart from Earth, only IO is known to have active volcanoes. Subject to TIDAL HEATING produced by Jupiter's gravitational pull, Io has numerous volcanoes in nearly constant state of eruption. The products of these eruptions, which are in both pyroclastic and flow form, are unique in their high sulphur content.

In the early Solar System, impacts during ACCRETION produced heat; much of the Moon was probably melted in this manner. Larger planets were melted by conversion of gravitational energy into heat during DIFFERENTIATION. Heating is also produced by natural RADIOACTIVITY in planetary interiors. *See also* CRYOVOLCANISM

VOLANS (GEN. VOLANTIS, ABBR. VOL)

Small, inconspicuous southern constellation, representing a flying fish, south of Carina. It was introduced by Keyser and de Houtman at the end of the 16th century. Its brightest star, β Vol, is mag. 3.8. γ Vol is a binary with yellow and pale yellow components, mags. 3.8 and 5.7, separation 14″.1.

Von Kármán, Theodor (1881–1963) Hungarian-American aeronautical engineer and 'father of supersonic flight'. He was one of the first to apply higher mathematics to the new fields of aeronautics and astronautics. In 1930 he became head of the Guggenheim Aeronautical Laboratory of the California Institute of Technology, developing jet and rocket engine designs. Von Kármán was one of the founders of the JET PROPULSION LABORATORY, responsible for most of NASA's major unmanned space probes.

Voskhod Two modified VOSTOK spacecraft that were used as a stopgap before the introduction of the SOYUZ. Voskhod 1 (1974 October 12–13) was the first Soviet spacecraft to carry three men. During Voskhod 2, Alexei LEONOV made the first space walk on 1965 March 18.

Vostok Series of six Soviet manned spacecraft. In Vostok 1, Yuri GAGARIN became the first man to orbit the Earth on 1961 April 12. Valentina Tereshkova (1937–), the first woman in space, flew in Vostok 6, 1963 June 16–19. The world solo spaceflight record ($4^d\ 23^h\ 6^m$) is held by Valeri Bykovsky (1934–) in Vostok 5.

Voyager Two NATIONAL AERONAUTICS AND SPACE ADMINISTRATION (NASA) spacecraft; they were intended to visit only Jupiter and Saturn, but Voyager 2 went on to complete a 'GRAND TOUR' of the four giant outer planets. The two spacecraft were identical except for the more powerful Radioisotope Thermoelectric

◄ **volcanism** The ASTER instrument aboard NASA's Terra satellite obtained this image of Mount Etna in eruption during 2001 July. Lava flows (red and yellow) can be seen advancing on the volcano's southern flank.

V

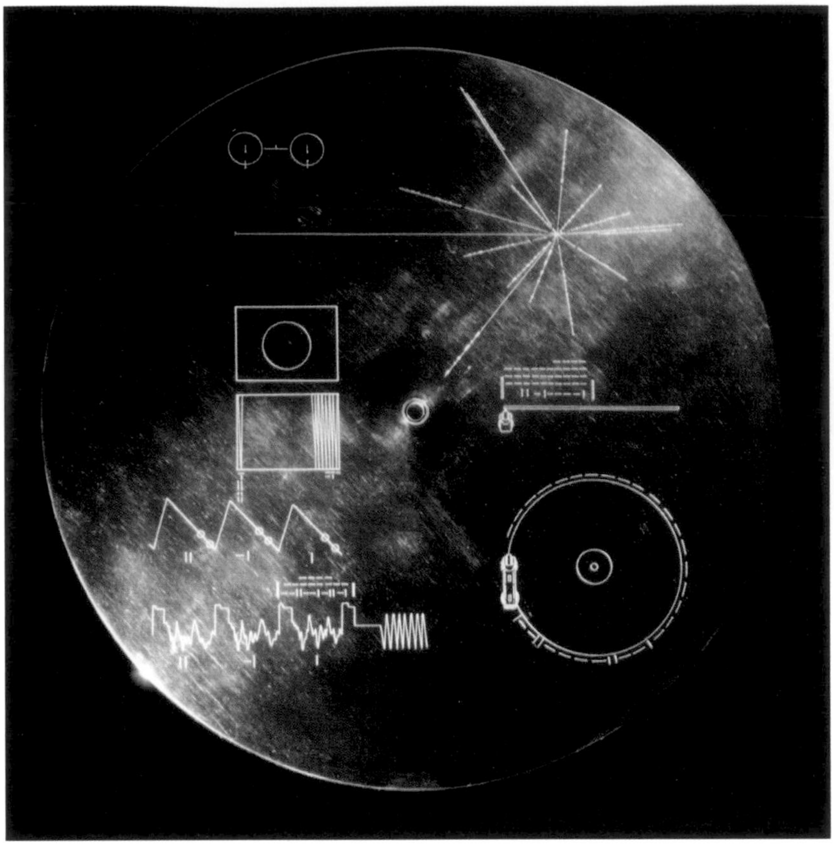

▲ **Voyager** The two Voyager spacecraft, now heading into interstellar space, each carry a copy of a gold-plated record of 'Sounds of Earth'. The records are protected from micrometeorite bombardments by a gold aluminium case, engraved with information showing Earth's location.

Generator (RTG) on Voyager 2, which it was hoped might rendezvous with Uranus and Neptune as well as Jupiter and Saturn. The launch weight of the vehicle and propulsion rocket was in each case 2016 kg, of which the Voyager itself accounted for 792 kg. Ten instruments were carried on each spacecraft.

Voyager 2 was the first to be launched from Cape Canaveral, on 1977 August 20. Voyager 1 followed on September 5. However, Voyager 1 was travelling on a faster, shorter trajectory and overtook its twin during the crossing of the ASTEROID BELT. On 1979 March 5 Voyager 1 passed Jupiter at a distance of 350,000 km (217,500 mi) and sent back the best images ever obtained of the planet and its moons. Its discoveries included volcanic eruptions on IO, linear features on EUROPA and the existence of a dark, dusty ring.

Voyager 1 then went on to a rendezvous with SATURN on 1980 November 12, at a minimum distance of 124,200 km (77,000 mi). Once again, it sent back high-resolution images of the planet and the satellites TITAN, RHEA, DIONE and MIMAS. The complex nature of the rings was fully revealed, and new data were obtained about the magnetic field, radiation belts and other phenomena. Voyager 1's encounter with Titan bent the spacecraft's path northward, so that it continued on an orbit out of the ECLIPTIC plane at an angle of 35°. Had Titan not been satisfactorily imaged, then Voyager 2 would have been targeted to carry out a survey, so losing the opportunity to rendezvous with either of the outer giant planets.

Voyager 2 passed Jupiter on 1979 July 9 at a distance of 645,000 km (400,000 mi). The results fully complemented those of its predecessor, and showed marked changes in the Jovian atmosphere and the volcanic activity on Io. A small satellite, now known as ADRASTEA, was found close to the rings. Voyager 2 then went on to Saturn, passing the planet on 1981 August 25 at a distance of 101,300 km (63,000 mi). Quite detailed images of the satellites IAPETUS, HYPERION, TETHYS and ENCELADUS were obtained. However, after closest approach to Saturn, the scan platform that carried the camera apparently jammed, so some vital information was lost. The problem solved itself, but subsequently the platform was manoeuvred only at reduced rate.

After the gravity assist at Saturn, Voyager 2 was directed towards URANUS. The closest approach took place on 1986 January 24, at a distance of 107,000 km (67,000 mi). This part of the mission was particularly important, because comparatively little had been known about Uranus. Many discoveries were made, including ten new moons (including a number of small 'shepherd' satellites near the rings), the weird terrain of MIRANDA and the planet's remarkable, tilted magnetic field.

Voyager 2's 'Grand Tour' was completed with the flyby of NEPTUNE and its large moon TRITON on 1989 August 24. As the spacecraft passed within 4900 km (3000 mi) of Neptune's cloud tops, Voyager 2 obtained remarkable images of giant storm systems on the planet. It also found six new moons and evidence of geyser activity on Triton's icy surface.

Voyager 2's trajectory was radically altered during the Neptune encounter, sending it below the ecliptic plane at an angle of 48°. Both Voyagers are now well beyond the orbit of Pluto and heading out of the Solar System in opposite directions. During this extended Voyager Interstellar Mission, it is hoped that they will determine the location of the heliopause, the boundary of the HELIOSPHERE. With the remote chance that the Voyagers will one day be picked up by some alien race in another solar system, each probe carries a 31-cm (12-in.) gold-plated record containing 'Sounds of Earth', together with information upon how to play it with the cartridge and needle provided.

Vulcan Planet once thought to circle the Sun inside the orbit of Mercury. It was invoked by Urbain LE VERRIER in 1859 to explain the faster than predicted advance of the heliocentric longitude of the perihelion of Mercury. To Le Verrier this implied either the existence of a body roughly the size of Mercury but at half its distance from the Sun, or a similarly placed ring of smaller bodies. Following his announcement in September of that year, Le Verrier had news from a French amateur astronomer, Edmond Lescarbault of Orgres, who reported that in March he had seen an unidentified dark object transit a segment of the Sun. Le Verrier hurried to Orgres and satisfied himself the claim was authentic. He named the object Vulcan and calculated its orbit. More sightings were reported but all turned out to be cases of mistaken identity. Interest revived in 1878, following the observation of mysterious objects near the Sun at the total eclipse of July 29. But the claim degenerated into confusion and controversy, and it was ultimately discarded. There were other false alarms in 1882 and 1900, but in 1915 after a radical reappraisal, which altered our understanding of Newtonian gravitational principles, Albert EINSTEIN neatly resolved the problem with his general theory of relativity.

While Le Verrier's Vulcan has slipped from currency, questions remain. For instance theoreticians have revived the idea of an inner ring of asteroids, or 'Vulcanoids', and have defined a number of possible orbits. The discovery by SOHO of numerous small comets close to the Sun gives credence to the idea. With a better appreciation of Earth-crossing asteroids and interplanetary objects in general, it is now easier to understand why these 19th-century astronomers believed they had observed Vulcan.

Vulpecula *See* feature article

VULPECULA (GEN. VULPECULAE, ABBR. VUL)

Small, inconspicuous northern constellation, representing a fox, south of Cygnus. It was named Vulpecula cum Ansere (the Fox and Goose) by Johannes Hevelius in 1687. Its brightest star, α Vul, is mag. 4.4. The brightest deep-sky objects are the COATHANGER (Cr 399, also known as Brocchi's Cluster), an open cluster of about a dozen stars between mags. 6 and 8, and the DUMBBELL NEBULA (M27, NGC 6853), an 8th-magnitude planetary nebula.

walled plain Older term, still used by some lunar observers, for large lunar CRATERS.

Walter Ancient lunar crater (33°S 1°E), 129 km (80 mi) in diameter. Its walls are deeply incised from later impacts, to the point that they have been eroded away to the north and south. The ejecta have also been removed by impact erosion. While the floor appears to contain peaks to the north-east, these are mostly crater rims that have merged because of the proximity of the impacts.

Wargentin Lunar crater (50°S 60°W), 89 km (55 mi) in diameter. It was originally a complex crater with central peaks, wall terracing and an ejecta blanket. After its formation, however, lunar basalts flooded the interior, covering the central peaks and the inner wall structures. Indeed, the lava level is higher than the surrounding lunar surface. The ejecta blanket has been eroded away by meteorite erosion.

Warner & Swasey Co. Machine-tool company based in Cleveland, Ohio, that made several large telescopes for major observatories in the late 1800s and early 1900s. **Worcester Reed Warner** (1846–1929) was a machinist from Massachusetts who had an interest in astronomy; in 1880 he opened a factory in Chicago with **Ambrose Swasey** (1846–1937), a mechanical engineer from New Hampshire. After a year, they relocated their operations to Cleveland, where there were more skilled machinists, and the business prospered.

Warner & Swasey's first telescope was a 9½-inch (240-mm) refractor sold to Beloit College, Wisconsin, with a lens by ALVAN CLARK & SONS. Buoyed by this success, the firm began making larger German equatorial mountings. Large, sturdy mountings of this type were necessary to support the large refracting telescopes favoured by major observatories in the late 19th century. Warner & Swasey were the first telescope-makers to approach telescope design as a total engineering problem, and their highly accurate equatorial drives gained them renown. One of their innovations was to replace the tangle of ropes and chains used to control the long tube assemblies with a more elegant and efficient system of rods and gear trains controlled from the eyepiece end of the telescope.

By 1886 Warner & Swasey had completed LICK OBSERVATORY'S 36-inch (0.9-m) refracting telescope which, with another Clark lens, was then the world's largest. They topped this achievement with the 40-inch (1-m) refractor installed at YERKES OBSERVATORY (1893). From 1885 to 1915 John BRASHEAR supplied most of the lenses for the firm's telescopes.

Water Jar Y-shaped asterism in Aquarius, formed by γ, ζ, η and π Aquarii. It represents the jar from which Aquarius is visualized to be pouring water.

wavelength (symbol λ) Linear distance along a wave between successive maxima or minima or any other points of equal phase. *See also* ELECTROMAGNETIC RADIATION; ELECTROMAGNETIC SPECTRUM; FREQUENCY

wave mechanics Equations formulated by the French physicist Louis Victor de Broglie (1892–1987) and the Austrian physicist Erwin Schrödinger (1887–1961) to describe the wave structure of atomic particles. Wave mechanics is completely contained within the more general structure of QUANTUM MECHANICS.

weakly interacting massive particle (WIMP) Particle postulated to make up the missing mass in galaxies and clusters of galaxies. These hypothetical particles would rarely interact with ordinary matter, hence the description 'weakly interacting', and would have masses 10 to 10,000 times the mass of a proton. These types of particles are suggested by certain theories of ELEMENTARY PARTICLES but have never been experimentally detected.

Webb Society International society of amateur and professional astronomers who specialize in observing double stars and deep-sky objects (DSOs), founded in 1967 and named in honour of the 19th-century amateur observer the Reverend T.W. WEBB. The Webb Society publishes a quarterly journal, *Deep Sky Observer*, and specialist observing guides for various categories of DSO.

Webb, Thomas William (1807–85) English clergyman whose *Celestial Objects for Common Telescopes* (1859) did much to popularize amateur astronomy in Britain. His book was based on 25 years of observing planets and variable and binary stars with a 3.7-inch refractor, though after 1864 he began to popularize the new silver-on-glass reflecting telescopes of George Henry With (1827–1904). Webb was pre-eminently an observer, not concerned with making scientific discoveries, just wishing to enjoy the beauty of the night sky. He inspired many to do likewise, and is therefore one of the founders of modern amateur astronomy.

wedge Adjustable, angled platform placed under a FORK MOUNTING to adjust it to the local latitude. Many modern SCHMIDT–CASSEGRAIN TELESCOPES are equipped with computer-controlled drives allowing them to be used without a wedge for visual purposes, or with a wedge for ASTROPHOTOGRAPHY. The wedge is required to create a true EQUATORIAL MOUNTING and eliminate FIELD ROTATION.

Weizsäcker, Carl Friedrich von (1912–) German theoretical physicist and astrophysicist known for his theories of stellar energy generation and the formation of the Solar System. In 1938 he proposed, independently of Hans BETHE, that stars generate energy in their extremely hot cores via the successive fusion of nuclei of hydrogen, helium and carbon. Weizsäcker calculated how much energy each reaction would produce. In 1944 he proposed a theory of the formation of the Solar System, a variant of Pierre-Simon de LAPLACE'S NEBULAR HYPOTHESIS. In Weizsäcker's COSMOGONY, vortices develop in a disk of condensing gas and dust surrounding the primordial Sun, triggering the formation of bodies by accretion.

West, Comet (C/1975 V1) Long-period comet, the brightest of the 1970s, discovered photographically by Richard West (1941–) in 1975 November. Comet West became a bright morning object in 1976 February and March. Perihelion, 0.20 AU from the Sun, was reached on

◀ **West, Comet** A brilliant object in pre-dawn skies of early 1976, Comet West showed a strong, complex, yellow dust tail. Also seen here, in this March 10 photograph, is the bluish ion tail at left.

1976 February 25. The nucleus broke into four fragments soon afterwards. Comet West reached a peak brightness of mag. −1, and showed a prominent fan-shaped dust tail 25–30° long. The orbital period is of the order of 500,000 years, and it is likely that the individual nuclear fragments will each make their returns well separated from the others.

Westerbork Synthesis Radio Telescope (WSRT) Major radio telescope array near the village of Westerbork in the north-east of the Netherlands, operated by the Netherlands Foundation for Research in Astronomy. Completed in 1970, it consists of 14 antennae 25 m (82 ft) in diameter arranged along a 2.7-km (1.7-mi) east–west baseline. Ten of the antennae are fixed in position, while the remaining four can be moved along rail tracks to synthesize a telescope up to 2.7 km in aperture.

Wezen The star δ Canis Majoris, visual mag. 1.84. It is a white supergiant of spectral type F8 Ia with a luminosity around 50,000 times that of the Sun, lying some 1800 l.y. away. The name comes from the Arabic *wazn*, meaning 'weight', the significance of which is unknown.

Whipple, Fred Lawrence (1906–) American planetary scientist who devised the 'dirty snowball' picture of cometary structure. He spent most of his scientific life at Harvard, where he directed the SMITHSONIAN ASTROPHYSICAL OBSERVATORY (1955–73). In the 1930s, his photographic studies of meteor paths provided evidence that meteor showers occur when swarms of debris left behind by comets intersect the Earth's orbit. Whipple's comet model (1949) described them as large aggregates of ices and silicate debris; when the comet nears the Sun and is heated, sublimating ices are ejected in jets along with silicates, generating the comet's dust tail. He showed that these jets, confirmed by 1986 spacecraft studies of Comet HALLEY, can affect the comet's orbital motions. The FRED L. WHIPPLE OBSERVATORY is named in his honour.

▼ **Whirlpool Galaxy** The magnificent Whirlpool Galaxy (M51, NGC 5194) takes its name from its pronounced spiral structure. The small companion galaxy at left, NGC 5195, appears to be interacting with the Whirlpool.

Whirlpool Galaxy (M51, NGC 5194) Face-on spiral galaxy in the constellation Canes Venatici (RA $13^h\ 29^m.9$ dec. +47°12′), with apparent magnitude +8.4 and angular diameter 11′ × 7′. Discovered by Charles MESSIER in 1773, the Whirlpool Galaxy takes its name from drawings and descriptions made in 1845 by Lord Rosse, who saw its prominent spiral arms through his 72-inch (1.8-m) reflector. M51 appears to be interacting with a smaller companion galaxy, NGC 5195. It is the most prominent member of a small group that lies 37 million l.y. away.

whistler Effect that occurs when a plasma disturbance, caused by a lightning stroke, travels out along lines of magnetic force of the Earth's magnetic field and is reflected back to its point of origin. The disturbance may be picked up electromagnetically and when converted to sound gives the characteristic drawn-out descending pitch of the whistler. This is a dispersion effect caused by the greater velocity of the higher-frequency components of the disturbance. The reflections from ionized meteor trails (*see* RADAR ASTRONOMY) can be observed as a 'whistler' in a receiver tuned to a distant transmitting station. Here the whistler is an interference beat between the ground wave from the transmitter and the wave reflected from the ionized meteor trail.

white dwarf Evolved STAR with mass similar to the Sun but with a radius only about as large as that of the Earth. White dwarfs are composed of DEGENERATE MATTER and are supported by electron degeneracy pressure. The density of the matter from which they are made is enormously greater than that of any terrestrial material: one cubic centimetre of white dwarf matter weighs approximately one tonne.

The first white dwarf to be discovered was the companion of Sirius, now called SIRIUS B. In 1844 Friedrich BESSEL discovered irregularities in the motion of Sirius, and concluded that it must have a companion, the pair forming a BINARY SYSTEM with a period of about 50 years. In order for its gravity to swing Sirius around in its orbit, it was clear that the companion must have a comparable mass (quantitative analyses using Kepler's third law for the binary orbit giving a mass between 75% and 95% of the Sun's mass), and yet it must be very faint to have remained undetected by direct observation. In 1862 ALVAN CLARK and his son Alvan Graham Clark discovered a very faint star close to Sirius, which was subsequently identified as the elusive companion. As the distance to the system was known, this allowed the companion's luminosity to be estimated as only about 1/360 of that of the Sun.

In 1914 W.S. ADAMS made the surprising discovery that Sirius B had the spectrum of a 'white' star, with a surface temperature about two and a half to three times that of the Sun. The total radiation from the surface of a hot body rises as the fourth power of its temperature, so that each square centimetre of Sirius B's surface should radiate between thirty and eighty times as much as the Sun. The only way to reconcile this with the very low total luminosity inferred above was to conclude that the radius of Sirius B was far smaller than that of any star known at the time. The result remained uncertain until in 1925 Adams succeeded in detecting the gravitational redshifts of several absorption lines in the spectrum of Sirius B. This general relativistic effect depends on the ratio of the star's mass to its radius. As the mass was already known from the properties of the orbit, Adams was able to find the radius of the star in an independent way, giving an estimate consistent with his earlier result.

It was very quickly realized that their enormous matter densities make white dwarfs fundamentally different in nature from 'normal' stars. In a star like the Sun, it is the thermal pressure of the gas and radiation of the star that prevents it simply collapsing under its own weight. As early as 1926, the British astronomer Ralph Howard Fowler (1889–1944) showed that this cannot be the case for white dwarfs. Instead, the pressure required to hold up these stars against their own gravity is provided by a fundamental quantum-mechanical effect which had been discovered only months before by US physicist Enrico Fermi (1901–54) and British physicist Paul Dirac (1902–84).

As in most stars, the matter inside a white dwarf is almost completely ionized, that is, the electrons have all been ripped off their parent atoms (here because of the frequent collisions between atoms) leaving the matter as a mixture of free electrons and nuclei. What Fermi and Dirac had discovered is that there is a fundamental limit to how close together two electrons can be pushed. Electrons resist being in the same place with the same speeds: push them closer together and they will react by moving faster. This motion amounts to a pressure, called degeneracy pressure: it differs from ordinary thermal pressure in depending only on the density of the constituent particles, rather than on the product of the density and the gas temperature.

The larger the mass of a star supported by degeneracy pressure, the faster the electrons inside it must move, until ultimately, for stars of masses only slightly greater than that of the Sun, they are moving with speeds close to that of light. The theory of RELATIVITY forbids motions faster than light, so that relativistic corrections must be made to the degeneracy pressure for such stars. The pressure is then found to increase rather less rapidly with the density.

In 1931 Subrahmanyan CHANDRASEKHAR computed the structure of a white dwarf with these relativistic effects included. He made the discovery that a white dwarf cannot have a mass more than about 1.4 times the Sun's mass (*see* CHANDRASEKHAR LIMIT). For larger masses, gravity would always overwhelm the pressure forces and the star would collapse under its own weight. Stars with initial masses ranging from about 0.08 to about 8 solar masses (over 95% of the stars of our Galaxy) will eventually end up as white dwarfs.

All stars lose energy into space by radiation. In response to this constant drain of energy, the central parts of the star gently sink closer together during huge expanses of time, becoming denser and more tightly bound by their mutual gravitational attraction. If the star is supported in the normal way by thermal pressure, it has to become hotter in order to supply the increased pressure demanded by the stronger gravity. Thus in response to the loss of energy by radiation from its surface, the star has actually had to become hotter. This means that radiation losses from its surface will continue, driving the evolution to more tightly bound and hotter configurations ever onwards: the star cannot simply cool down and stop evolving. In the white dwarf state, however, we have seen that the supporting degeneracy pressure is independent of the temperature of the stellar material: the star can now radiate all its thermal energy into space without any consequences for its internal structure, and the evolution of the star ends with it quietly cooling down to a cold, dark, inert configuration known as a BLACK DWARF.

In practice the star will by then consist entirely of helium and heavier elements, its original hydrogen having been transmuted in the course of its earlier evolution. The existence of the Chandrasekhar limiting mass means that this endpoint is not available to all stars, but only to those whose mass is below the limit when they reach the white dwarf state. It is now known that more massive stars must end their evolution either as NEUTRON STARS or BLACK HOLES. Neutron stars are also supported by degeneracy pressure, but of neutrons rather than electrons; for them, too, there is a maximum mass, although its precise value is less certain than for white dwarfs.

Many white dwarfs are now known, and their observed properties are in good accord with theoretical predictions. Much interest now centres on white dwarfs in binary systems, particularly those that are close enough for gas to be pulled off the companion star on to the white dwarf. Such systems are called CATACLYSMIC VARIABLES and include NOVAE and DWARF NOVAE. Evidence also suggests that a very old, faint population of white dwarfs exists in the halo of our Galaxy.

◀ **white dwarf** This close-up view from the Hubble Space Telescope shows part of the globular cluster M4 in Scorpius. While most of the objects seen here are old, red stars, several white dwarfs (circled in blue) have also been found.

white hole Inverse evaporating BLACK HOLE. Although nothing can escape a black hole's gravity, a black hole actually has a temperature and particles can tunnel out of it. The temperature and emission of particles and energy by a black hole depend on the inverse of the temperature of the black hole; low-mass black holes emit particles at a higher rate than high-mass black holes. A white hole is the final moments of evaporation of a black hole when information and matter emerge from the hole in an unpredictable fashion.

Widmanstätten pattern *See* IRON METEORITE

Wien's law Law, formulated in 1893 by the German physicist Wilhelm Wien (1864–1928), relating the wavelength of the peak emission from a BLACK BODY to its temperature. It takes the form:

$$\lambda_{max} = \frac{0.0028979}{T}$$

where λ_{max} is the wavelength of the peak emission, and T is the temperature. As a consequence of the operation of Wien's law, cool stars appear reddish in colour, medium temperature stars like the Sun are yellow-white, while hot stars appear blue-white.

CLASSIFICATION OF WHITE DWARFS

Type	Spectral Characteristics
DA	spectrum dominated by Balmer series of hydrogen
DO	spectrum dominated by ionized helium
DB	spectrum dominated by neutral helium
DC	featureless spectrum
DQ	spectrum shows presence of carbon, generally in molecular form
DZ	lines of heavy elements other than carbon, for example calcium, magnesium or iron
DBQ or DBZ	weak carbon or heavy elements together with dominant neutral helium
PG1159	ionized helium as well as highly ionized carbon and nitrogen features

▲ **William Herschel Telescope** Located on La Palma, the William Herschel Telescope is one of the Isaac Newton Group of instruments. The telescope, a 4.2 m (163 in.) reflector, is the largest in Europe.

Wild Duck Cluster (M11, NGC 6705) Rich open star cluster in the constellation Scutum (RA 18^h 51^m.1 dec. −06°16′); it was originally discovered by Gottfried Kirch (*see* KIRCH FAMILY) in 1681. The popular name comes from a description by the 19th-century English astronomer W.H. SMYTH of the V-shaped cluster as resembling a flight of ducks. The cluster is compact, with apparent diameter 13′. Its stars are fairly uniform in magnitude, with the exception of a single brighter member near the eastern tip. M11 has overall magnitude +5.8. It lies 6200 l.y. away, and has an estimated age of 250 million years.

Wildt, Rupert (1905–76) German-American astrophysicist and planetary scientist known for his accurate models of the giant planets and stellar atmospheres. He was the first (1931) to correctly ascribe absorption lines in the spectra of Jupiter and Saturn to methane and ammonia in their atmospheres. In Wildt's models of the gas giants, a small rocky core was surrounded by ice and shrouded in a deep, hydrogen-rich atmosphere. In 1939 he explained the 'missing opacity' of the solar atmosphere by showing that hydrogen ions slow the radiative flow of energy from the interior, making the outer layers more transparent.

Wilhelm Förster Observatory Public observatory in Berlin, dating back to 1889. Its first astronomer, Friedrich Simon Archenhold (1861–1939), was also director of the nearby TREPTOW OBSERVATORY. Originally called the Urania Observatory, the institution was renamed in 1953 and moved to its present building in 1963. Its telescopes include the 0.31-m (12-in.) Bamberg Refractor of 1889 and a 0.75-m (30-in.) reflector added in the late 1980s.

Wilkins, John (1614–72) English churchman and writer who helped to found the Royal Society. He was probably the first writer to explore the ideas of Galileo and Kepler for an English readership. In *Discovery of a World in the Moone* (1638) he speculated on the Moon's structure and discussed the design of a machine that could fly there.

William Herschel Telescope (WHT) Optical telescope with a mirror 4.2 m (165 in.) in diameter, the largest of the three Anglo-Dutch telescopes of the ISAAC NEWTON GROUP at ROQUE DE LOS MUCHACHOS OBSERVATORY in La Palma. Opened in 1986, the WHT was the last telescope to be built by the firm of GRUBB, PARSONS & CO., which closed down once it was delivered. The WHT's mounting allows it to be switched between PRIME, CASSEGRAIN and NASMYTH FOCI, and there is a comprehensive instrument suite taking advantage of this versatility. High-dispersion and multi-object spectroscopy can be carried out with the Utrecht Echelle Spectrograph and the WYFFOS spectrograph respectively, while there is also a prime-focus imaging camera.

Wilson, Alexander (1714–86) Scottish scientist who made pioneering solar observations. Over many years, he carefully observed sunspots and their shapes. In 1769 he noticed the changed appearance in a spot's penumbra when a spot is near the limb, now known as the WILSON EFFECT, which he attributed to the fact that sunspots are depressions in the solar surface. Wilson published his cosmological theories in *Thoughts on General Gravitation* (1770), proposing that the entire Universe rotates about a discrete central point.

Wilson, Robert Woodrow *See* PENZIAS, ARNOS ALLAN

Wilson effect Roughly circular SUNSPOT that appears elliptical when seen foreshortened near the LIMB. The umbra appears displaced towards the Sun's centre, while the penumbra nearer the limb appears wider than that towards the solar centre, suggesting that sunspots are bowl-shaped depressions. The phenomenon was discovered by the Scottish astronomer Alexander WILSON in the 18th century.

Wilson–Harrington Small inner Solar System body that has been given dual COMET and ASTEROID designations. It was first observed in 1949 as an object showing weak comet-like outgassing. When it was rediscovered in 1979, however, no cometary activity was detected. It has been classified both as periodic comet 107P/Wilson–Harrington and also as an APOLLO ASTEROID (4015) Wilson–Harrington. *See also* CHIRON; ELST-PIZARRO; table at NEAR-EARTH ASTEROID

WIMP Abbreviation of WEAKLY INTERACTING MASSIVE PARTICLE

Wind Satellite in the INTERNATIONAL SOLAR TERRESTRIAL PHYSICS (ISTP) programme launched in 1994 November. Wind and the POLAR satellite are NASA's contribution to the ISTP project. Wind investigates the sources, acceleration mechanisms and propagation processes of energetic particles and the SOLAR WIND. It will help to complete observations of the plasma, energetic particle and magnetic field influences on the magnetosphere and ionosphere. Wind was placed into an L_1 halo orbit between the Earth and Sun via a lunar gravity assist swing-by. It is equipped with a suite of ten instruments provided by the USA, Russia and France. The Konus instrument was the first instrument from the former Soviet Union to fly on a US spacecraft.

winonaite Subgroup of the ACHONDRITE meteorites. Winonaites are related to the silicate inclusions in certain IRON METEORITES and may derive from the same parent body. They are also described as primitive achondrites.

winter solstice Moment when the Sun reaches its greatest declination and lowest altitude in the sky. In the northern hemisphere this occurs around December 22 when the Sun's declination is 23°.5S, marking the southern limit of its annual path along the ecliptic. At this point it is overhead at the TROPIC OF CAPRICORN and the hours of daylight are at a minimum, the day of the solstice also marking the shortest day of the year. In the southern hemisphere the winter solstice occurs around June 21, when the Sun is overhead at the TROPIC OF CANCER. *See also* SUMMER SOLSTICE

Winthrop, John (1714–79) Colonial American scientist, professor of mathematics at Harvard University (1738–79), where he made long-term meteorological and sunspot observations; he gave Harvard its first telescope (1672). Winthrop correctly predicted the return of Comet HALLEY in 1759 and made detailed observations of the apparition. He observed the transits of Venus in 1761 and 1769, organizing a successful expedition to Newfoundland for the earlier event.

WIYN Telescope Highly innovative 3.5-m (138-in.) optical telescope at KITT PEAK NATIONAL OBSERVATORY, completed in 1994. It is owned and operated by the WIYN Consortium, which consists of the University of Wisconsin, Indiana University, Yale University and the NATIONAL OPTICAL ASTRONOMY OBSERVATORY. The telescope combines a spin-cast mirror (made at the STEWARD OBSERVATORY Mirror Laboratory) with a compact, lightweight structure and a low-turbulence enclosure. The WIYN telescope has only one-eighth of the moving mass of the nearby (and similarly sized) MAYALL TELESCOPE, built in 1973.

W.M. Keck Observatory Major optical observatory operating the two largest single-mirror telescopes in the world. The identical 10-m (394-in.) instruments are located at an elevation of 4145 m (13,796 ft) at MAUNA KEA OBSERVATORY, Hawaii, and the facility has its sea-level headquarters at nearby Kamuela. Built with funding from the W.M. Keck Foundation, the observatory is operated jointly by the CALIFORNIA INSTITUTE OF TECHNOLOGY, the University of California and (since 1996) NASA.

Construction of the first telescope, Keck I, began in 1986, pioneering the use of segmented-mirror technology in large telescopes. The mirror is composed of 36 hexagonal segments, each 1.8 m (72 in.) wide and 75 mm (3 in.) thick, under computer control to maintain a continuous reflective surface. The technique has since been widely adopted in, for example, the HOBBY–EBERLY TELESCOPE.

Keck I began science operations in 1993 May, and was followed in 1996 October by Keck II. A large and varied suite of instruments has been developed for the telescopes, most notably an ADAPTIVE OPTICS system that permits the Kecks' imaging cameras and spectrographs to achieve maximum efficiency. Also available is a mode that allows the two telescopes to be combined along with several 2.0-m (79-in.) outrigger telescopes, allowing high-resolution optical INTERFEROMETRY to be carried out.

Wolf, Charles Joseph Étienne (1827–1918) and **Rayet, Georges Antoine Pons** (1839–1906) French astronomers who discovered a new class of high-temperature stars, now known as WOLF–RAYET STARS. Rayet worked at the Paris and Bordeaux observatories and was first director (1879) of the observatory at Floriac, France. Wolf also worked at Paris, and it was there, while examining spectra of the nova that appeared in 1866 May, that he and Rayet noticed wide, bright emission lines (later identified with helium, carbon and oxygen) that they later found (1867) also in the spectra of three hot, luminous stars in Cygnus.

Wolf diagram Plot of the number of stars within a given brightness range against APPARENT MAGNITUDE. The number of stars normally increases by about 250% for each decrease in the brightness by one stellar magnitude. However, if there is a DARK NEBULA along the line of sight, then the observed curve will deviate from the theoretical one at some point. The apparent magnitude where the curves deviate may be used to estimate the distance of the nebula. For thin nebulae, the physical depth of the nebula may also be found from the magnitude at which the observed curve resumes its theoretical trend. Maximilian WOLF first plotted the diagram.

Wolf, (Johann) Rudolf (1816–93) Swiss solar astronomer who independently discovered the 11-year solar activity cycle by logging the numbers of sunspots that appear over this period of time. Director of the Bern Observatory (1847–55) and founder (1864) of the Zürich Observatory, Wolf related sunspot activity to geomagnetic storms on Earth. In 1849 he devised a way of accurately counting the number of sunspots and groups, formerly known as the **Zürich** or **Wolf number** (*see* RELATIVE SUNSPOT NUMBER).

Wolf, Maximilian Franz Joseph Cornelius ('Max') (1863–1932) German astronomer, who was a pioneer of astrophotography, discovering 232 minor planets and dark nebulae in the Milky Way. He was the long-time (1893–1932) director of Heidelberg's Königstuhl Observatory. Among Wolf's asteroid discoveries were (323) Brucia, the first found photographically, in 1898, and (588) Achilles, the first TROJAN ASTEROID identified, in 1906. His star counts of his wide-field Milky Way photographs enabled him to map huge dark clouds of obscuring gas and dust. He discovered that planetary nebulae are gaseous shells. Before it was understood that the 'spiral nebulae' were other galaxies, Wolf obtained many spectra of these objects, showing (1908) that they were composed of stars. He observed and catalogued over 6000 nebulae and galaxies, including many members of the Coma and Perseus galaxy clusters; the majority of these objects were newly discovered.

Wolf number *See* RELATIVE SUNSPOT NUMBER

Wolf–Rayet star (WR star) Hot star that is experiencing high mass loss. The spectra of Wolf–Rayet stars show broad emission lines, making them easy to identify in spectroscopic observations, even at large distances.

First recognized by the French astronomers Charles WOLF and George Rayet in 1867, Wolf–Rayet stars are divided into three broad spectroscopic classes – WN, WC and WO. WN stars show EMISSION LINES predominantly of helium and nitrogen, although emission lines of carbon, silicon and hydrogen can be observed in some spectra. The spectra of WC stars are dominated by carbon and helium emission lines with no hydrogen and nitrogen. The more rare WO stars are similar to WC stars, with oxygen lines and lines of other elements showing high ionization.

Wolf–Rayet stars have surface temperatures between 25,000 and 50,000 K, luminosities between 100,000 and one million times that of the Sun, and masses from 10 to 50 solar masses. The very high luminosities generate a radiation pressure that drives a wind of gas up to a rate of 10 solar masses in a million years. The average maximum velocity of material ejected from WR stars is 800 to over 3000 km/s (500–1900 mi/s).

Wolf–Rayet stars demonstrate quite dramatically the effect that mass loss can have on massive stars. Their original hydrogen-rich envelopes have been stripped away to expose regions in which the products of NUCLEAR

▼ **Wolf–Rayet star** The extremely hot star HD 56925 has ejected this nebulosity, catalogued as NGC 2359. Wolf–Rayet stars like HD 56925 show enormous rates of mass loss via stellar winds, and rapidly become unstable.

W

REACTIONS are present. The type of WR star that emerges after the mass loss depends on how far evolution had progressed before the hydrogen envelope was lost.

Because of their spatial association with O STARS, and their peculiar surface abundances, WR stars are believed to be descended from O stars. Around 50% of WR stars are found in BINARY SYSTEMS. Many central stars of PLANETARY NEBULAE are WR stars. About 220 WR stars are known in our Galaxy, but this is a lower limit because most WR stars are hidden from view by dust. One WR star is easily visible to the unaided eye: the 2nd-magnitude southern-hemisphere star γ Velorum is a WC star with a SPECTROSCOPIC BINARY companion of spectral type O7.

Wollaston, William Hyde (1766–1828) English chemist and physicist who made some of the earliest observations of the solar spectrum. In 1801 he noticed five dark (absorption) lines in the solar spectrum, but misinterpreted them as boundaries between different colours. In 1814 Joseph von FRAUNHOFER confirmed the existence of these features, and in 1859 Gustav KIRCHHOFF correctly explained their nature. Wollaston was one of the first to observe ultraviolet light. He also invented the Wollaston prism, which is used in some POLARIMETERS.

Woolley, Richard van der Riet (1906–86) English astrophysicist who greatly advanced the development of astronomy in Australia. Woolley directed MOUNT STROMLO OBSERVATORY (1939–55), acquired a 1.88-m (74-inch) reflecting telescope, at the time the southern hemisphere's largest. As England's tenth ASTRONOMER ROYAL, he directed the Royal Greenwich Observatory (1956–71). Woolley's contributions spanned studies of the solar photosphere, chromosphere and corona; the measurement of stellar radial velocities; investigations of RR Lyrae variables and the stellar populations in globular clusters; and the determination of large-scale galactic motions.

wormhole Connection of two black holes either from two locations in our Universe, or one from our Universe and one from another, separate universe. These black holes are connected by their SINGULARITIES and theoretically can provide a bridge or shortcut from one location to another, or from one universe to another. Wormholes are also called Einstein–Rosen bridges after the first physicists to consider the mathematical possibility of such connections as suggested by GENERAL RELATIVITY.

Wright, Thomas ('Thomas Wright of Durham') (1711–86) English philosopher, instrument-maker, teacher and popularizer of astronomy. In *An Original Theory, or New Hypothesis of the Universe* (1750), Wright described the Milky Way as a flat disk of stars with the Solar System at its periphery, an essentially correct model. He also correctly described Saturn's rings as being composed of particulate matter.

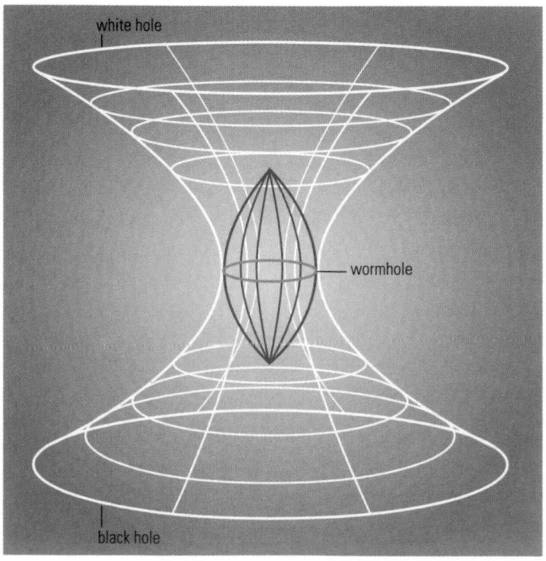

► **wormhole** This is a mathematical representation of the distortion of spacetime produced by a Schwarzschild wormhole connecting two universes via a black hole and a white hole. While predicted by Einstein's equations, such wormholes cannot exist in reality, since the occurrence of white holes is forbidden by the second law of thermodynamics.

wrinkle ridge Long, sinuous, narrow ridge. Wrinkle ridges are abundant on the plains of the Moon, Mercury and Mars; they are interpreted to be compressional tectonic structures. Wrinkle ridges are typically not more than a few hundred metres high, a few kilometres wide and tens to hundreds of kilometres long. On Earth wrinkle ridges are rare because of surface erosion.

W Serpentis Eclipsing VARIABLE STAR with extremely unusual light variations, between magnitudes of about 8.9 and 10.3. The generally accepted period is 14.153 days. There appears to be an overlying longer period of about 270 days. In addition to the deep minimum, there are also two shallow minima and three maxima. The system consists of two stars of roughly the same size, but of different brightness, rotating inside a common envelope. From one of the components (of spectral class A or F) ejection of material is taking place. The system evolves rapidly and the period is increasing. One of the two stars itself may be an INTRINSIC VARIABLE.

W Ursae Majoris star (EW) One of the three main subtypes of ECLIPSING BINARY. W Ursae Majoris stars exhibit continuously varying light-curves, with primary and secondary minima of identical or almost identical depth. They differ from BETA LYRAE STARS in being smaller, less luminous stars of nearly identical brightness. The stellar cores are surrounded by a common convective envelope. W Ursae Majoris stars are CONTACT BINARIES. They are thought to be possible precursors of DWARF NOVAE and other CATACLYSMIC VARIABLES.

W Virginis star (CW) Type of pulsating VARIABLE STAR superficially similar to a CEPHEID VARIABLE. On the HERTZSPRUNG–RUSSELL DIAGRAM they typically lie in the spectral class range G0 to M0 and the absolute luminosity range $M_v = -1$ to -4; they are thus giant stars. Their masses could be as low as 0.5 solar mass, suggesting that they have evolved from low-mass MAIN-SEQUENCE stars. They lie on the INSTABILITY STRIP on the HR diagram, which goes from Delta Scuti stars on the main sequence through the RR Lyrae stars and Cepheids to the irregular variables and Mira stars; therefore the cause of their variation is thought to be the same, that is, a layer of ionized helium within the star.

W Virginis stars may be distinguished from classical Cepheids by the less-regular shape of their light-curves and the double-peaked nature of their maxima. Their periods range from approximately one day to 100 days and their PERIOD–LUMINOSITY LAW is distinctly different from that of the Cepheids. They do not form a homogeneous group: some belong to Population I and some to Population II. Occasionally W Virginis stars show small period changes. RU Camelopardalis, however, is a remarkable case: its variations suddenly stopped in 1964 for about three years. The spectral type of RU Camelopardalis (KO to R2 with excess carbon) is later than usual for W Virginis stars, but the star is otherwise unremarkable. It is not understood how pulsations can stop, or start, in such a short time.

Because these stars are brighter than the RR Lyraes, but fainter than the Cepheids, and have a well-defined period–luminosity relationship, they may be used to estimate distances for galactic and extragalactic objects.

WZ Sagittae Star, originally classified as a RECURRENT NOVA, that has many similarities with the U GEMINORUM DWARF NOVA class; it may be a link between the two types. Four outbursts have been observed, in 1913, 1946, 1978 and 2001. It is also an eclipsing system, with a period of 81.5 minutes. The light-curve at eclipse resembles that of a W URSAE MAJORIS STAR. WZ Sagittae is normally around magnitude 15.5 at minimum, but has reached 7.0 during an outburst. Possible masses of the components are 0.59 and 0.03 solar mass.

XMM Abbreviation of X-Ray Multi-Mirror spacecraft, now called NEWTON.

X-ray astronomy Study of energetic processes and extreme physical conditions – high temperatures, winds and accretion disks, neutron stars and black holes. X-rays are generated by several environments: hot gas with a temperature greater than 1,000,000 K will emit radiation through atomic lines and thermal bremsstrahlung; high-energy electrons scattered by low-energy electrons produce X-rays (inverse COMPTON EFFECT); electrons in strong magnetic fields emit SYNCHROTRON RADIATION; and hot stars emit thermal BLACK BODY RADIATION. X-rays are also generated when material such as hydrogen ignites in NUCLEAR REACTIONS. The X-ray wavelength region is around 0.04 to 2.5 nm, but this is more usually expressed as from 0.5 to 30 keV, since X-rays are often measured in energy terms as kilo-electron-volts (keV). Hard X-rays are usually regarded as those in the range 10 keV and above, and soft X-rays lower than 10 keV. Many types of astronomical object emit X-rays, from the Sun, with a CORONA of around 2,000,000 K, to BINARY STARS containing a compact object (*see* X-RAY BINARY), to galaxies with a supermassive BLACK HOLE at the centre. However, X-rays from astronomical objects cannot be detected from the ground: 10 cm of air at ground-level, or three sheets of paper, will stop 90% of 3-keV X-rays.

X-ray astronomy began in 1949 with the discovery that the Sun emitted X-rays, and the subsequent study of its general properties, by Herbert FRIEDMAN, Burnight and co-workers, using a Geiger counter on a V2 and other sounding rockets. This work, together with that using the UK Ariel I satellite, built and launched by NASA in 1962, established that high-temperature (about 10^7 K) plasma was created in the corona during solar FLARES. Solar X-ray studies continued through the 1960s and 1970s with several NASA satellites, including the Orbiting Solar Observatory and the SOLAR MAXIMUM MISSION. The Sun is not a strong emitter of X-rays, so it was a major surprise when Riccardo GIACCONI and colleagues discovered a strong X-ray source in the constellation of Scorpius (SCORPIUS X-1) in 1962 using a sounding rocket. By 1970, after some dozens of sounding rocket flights that gave a total exposure of only a few hours, about 50 individual X-ray sources had been discovered, including the CRAB NEBULA, the galaxy CENTAURUS A, the quasar 3C 273, and a diffuse X-ray emission component (the X-ray background), which appeared to be isotropic at energies above 2 keV.

The launching of the UHURU satellite by NASA in 1970 December marked the beginning of a new era for X-ray astronomy. It carried out an all-sky survey resulting in the discovery of more than 400 discrete X-ray sources. More satellites followed, including the UK Ariel V satellite and the NASA COPERNICUS satellite. In 1978 NASA launched the EINSTEIN OBSERVATORY satellite, which carried an X-RAY TELESCOPE with an angular resolution of a few arcseconds. This satellite ushered in a new era in X-ray astronomy. Consisting of a large nested grazing-incidence X-ray telescope, it responded to photons in the 0.1–4 keV energy range. The combination of good angular resolution and high sensitivity led to the discovery of several thousand new X-ray sources during the 2½-year operational life of the satellite. Recent X-ray satellites include EXOSAT launched in 1983, ROSAT in 1990, and CHANDRA and NEWTON, both launched in 1999.

There are now more than 150,000 X-ray sources known. X-rays have even been detected from the Moon (reflected solar coronal X-rays) and from comets. X-ray spectra show highly ionized lines of oxygen, iron and other heavy elements, often with so many electrons stripped from the atom that they mimic the hydrogen spectral series. It has been found that many types of ordinary star emit X-rays. BROWN DWARFS have been detected with X-rays from their coronae. A population of WHITE DWARFS has been found to be continuously emitting X-rays because material is being accreted from the companion on to the surface of the white dwarf and burnt, with temperatures of several hundred thousand degrees – a unique situation of steady nuclear burning at the surface of a compact star.

X-ray astronomy established that NEUTRON STARS were present in binary star systems; CYGNUS X-1 is one of the best candidates for a system containing a BLACK HOLE. It has revolutionized our understanding of the end of a star's life and of compact objects. The surface of a PULSAR in Vela has been detected in X-rays, as have some other young pulsars. The surface temperature of the neutron star in the Vela Pulsar is around 1,000,000 K. Binary star systems containing a compact object show a variety of phenomena, depending on the mass and geometry of the members. The presence of an ACCRETION DISK in the system means that the X-ray emission can be intermittent (*see* X-RAY BURSTERS; X-RAY NOVAE). The strange X-ray binary SS433, with its relativistic JETS, was found to emit X-rays. Some 200 SUPERNOVA REMNANTS have been found by their extended X-ray emission, which comes from very hot gas behind the shock front. SUPERNOVA 1987A was observed by ROSAT several times; X-rays were first detected in 1992, becoming steadily stronger.

When the ANDROMEDA GALAXY was observed by ROSAT, around 550 sources were detected, about the same number as the bright X-ray sources in our own Galaxy. The LARGE MAGELLANIC CLOUD (LMC) has a source (LMC X-3) that is very similar to Cygnus X-1, with a possible black hole as the compact object in the massive X-ray binary star system. Many galaxies emit X-rays (*see* X-RAY GALAXY); clusters of galaxies (for example Coma, Virgo, Perseus) emit X-rays from hot intergalactic gas. Observations have shown that the Sun is embedded in a bubble of hot gas around 160 l.y. in diameter, with a temperature of over 1,000,000 K – perhaps heated by a long-past supernova explosion. In Cygnus there is a superbubble of hot (1,000,000 K) gas, with a density of about 0.001 particles/cm^3. The superbubble is over 1000 l.y. across, possibly the largest structure in the Galaxy, and this may be the result of a cluster of stars becoming supernovae.

X-ray binary Close BINARY STAR system emitting X-rays, either continuously or at intervals. X-ray binaries usually

▼ **X-ray astronomy** This Chandra X-ray Observatory image shows a young supernova remnant in Centaurus. A rapidly expanding shell of gas 36 l.y. in diameter, G292.0+1.8 contains large amounts of oxygen, neon, magnesium, sulphur and silicon. It has a pulsar at its centre.

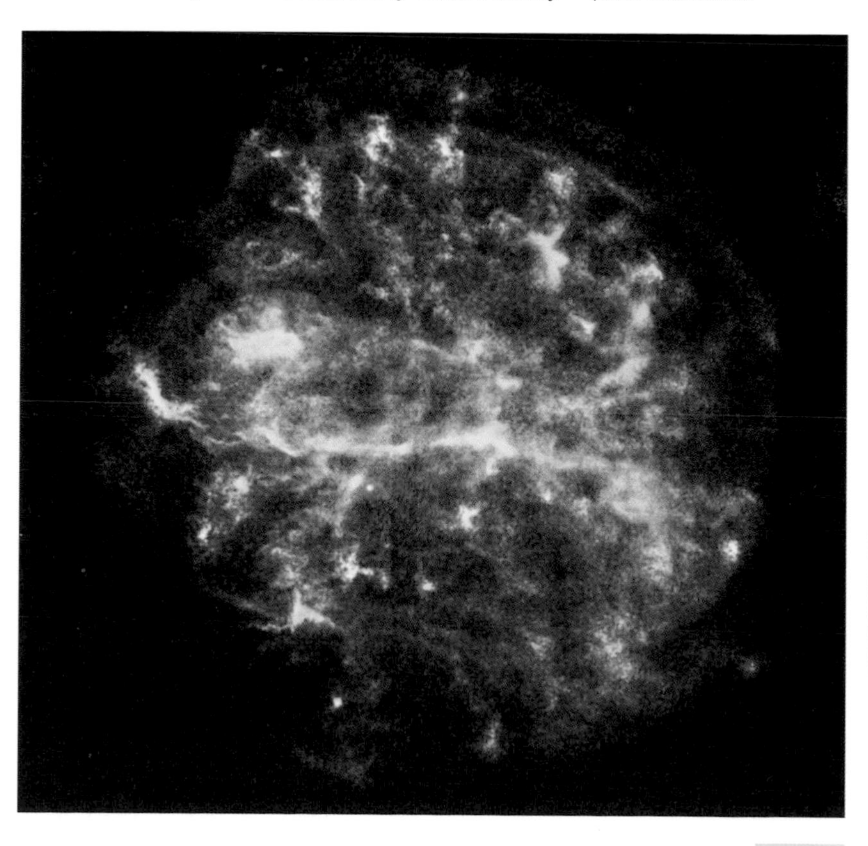

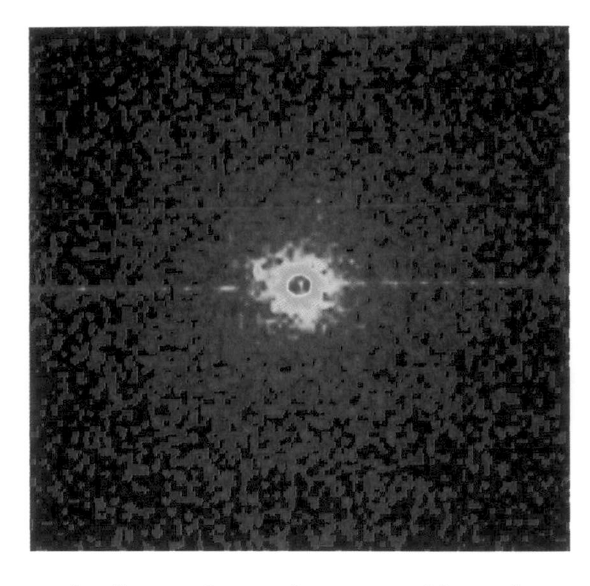

▶ **X-ray binary** The strong X-ray source Cygnus X-3 is a close binary system in which matter from a normal star is being drawn into a neutron star or black hole. This Chandra X-ray Observatory image shows a halo of emission resulting from scattering by interstellar dust in the line of sight, allowing the distance to Cygnus X-3 to be estimated at 30,000 l.y.

consist of a normal star and a compact object such as a WHITE DWARF, NEUTRON STAR or BLACK HOLE. The UHURU satellite originally showed that some X-ray stars underwent eclipses, revealing that they were close binary systems (for example HERCULES X-1). There are several types of X-ray binary system, with properties that depend upon the nature of the compact object. There are massive (high-mass) X-ray binaries and low-mass X-ray binaries (low-mass implies stars around the mass of the Sun). Low-mass X-ray binaries containing a neutron star tend to be X-RAY BURSTERS, for example SCORPIUS X-1 and Hercules X-1. Low mass X-ray binaries with a white dwarf as the compact object are CATACLYSMIC VARIABLES. When the companion star shedding the material is a massive star (for CYGNUS X-1 the mass is more than 30 solar masses), the star can often be detected optically. The strong stellar wind transfers material directly on to the compact object. X-ray pulsars are usually found in binary systems with high-mass companions. SS433 is an unusual massive X-ray binary, with a possible black hole as the compact object; it shows relativistic JETS.

X-ray burster Source with intense flashes, or bursts, of X-rays. The bursts have rise times of around 1 second and can last as little as 10 seconds, yet the energy released is equivalent to a week's output from the Sun (or more). A few X-ray bursts have been detected at optical wavelengths. X-ray bursters are BINARY STAR systems in which one star is a NEUTRON STAR. The neutron star is surrounded by a hydrogen-rich ACCRETION DISK from a low-mass companion star (around one solar mass). The hydrogen flows from the accretion disk on to the surface of the neutron star and is burnt to helium (*see* NUCLEAR REACTIONS). When the amount of helium on the surface of the neutron star exceeds a critical amount, the helium suddenly ignites, causing a burst of X-rays. The temperature can rise to 30 million K at the peak of the burst and 10 seconds later fall back to 15 million K. Almost 30 X-ray bursters have been found in GLOBULAR CLUSTERS; for example, the first was found in NGC 6624 (Omega Centauri has five, M22 has four). They are generally found close to the centre of the cluster, where the star density is very high. These X-ray bursters are also thought to be binary stars, with a neutron star accreting material from a low-mass companion star.

▼ **X-ray galaxy** A strong source of radio waves, Centaurus A (NGC 5128) is seen optically to have a dark lane crossing its centre, believed to result from a merger between two galaxies – a giant elliptical and a spiral – several hundred million years ago. X-ray observations reveal further activity, in the form of a jet extending from the nucleus towards top left in this image.

X-ray galaxy Galaxy that emits powerfully in the X-ray part of the ELECTROMAGNETIC SPECTRUM, indicating that it probably has a large or supermassive black hole at its centre, with an ACCRETION DISK. These galaxies include active galaxies (such as NGC 4151 and Centaurus A), SEYFERT GALAXIES, QUASARS and BL LACERTAE OBJECTS, many of which are also radio sources. X-ray galaxies often appear to be merging or colliding systems.

X-ray nova BINARY STAR system that occasionally becomes bright at X-ray wavelengths. One member of the binary is a WHITE DWARF and the other is a low-mass giant (usually around one solar mass). Material from the giant star falls on to the white dwarf, and eventually there is so much material on the surface that it ignites (*see* NUCLEAR REACTIONS) causing the X-ray emission, which will fade away over weeks or months. It is assumed that this will recur, but the timescale may be decades or centuries. DWARF NOVAE have ACCRETION DISKS, so the material falls on to the white dwarf from the disk rather than from the companion star. Although the outburst is less dramatic (hence the adjective 'dwarf'), the outburst can occur frequently, even as often as every week or two.

X-ray star Star that is a powerful source of X-rays. The detection of X-rays from a star usually means the star is a CLOSE BINARY system, with one member being a compact object, either a WHITE DWARF, NEUTRON STAR or BLACK HOLE (*see* X-RAY BINARY). However, even the Sun emits X-rays, from its CORONA, as do other stars. Massive hot O stars (such as CYGNUS X-3) emit X-rays because of their high temperatures and high mass-loss in their stellar winds.

X-ray telescope Instrument for imaging X-ray sources. All X-ray telescopes are satellite-borne because X-rays do not penetrate the Earth's atmosphere. Unlike less energetic electromagnetic radiation, X-rays cannot be focused by reflection from a conventional concave mirror (as in a reflecting telescope) because, for all but the shallowest angles of incidence, they penetrate the surface. In grazing incidence X-ray telescopes the focusing element is a pair of coaxial surfaces, one PARABOLOIDAL and the other HYPERBOLOIDAL, from which incoming X-rays are reflected at a very low 'grazing' angle towards a focus. The detecting element is often a CCD adapted for X-ray wavelengths. An alternative instrument is the microchannel plate detector, in which the incident radiation falls on a plate made up of many fine tubes, rather like a short, wide fibre-optic bundle. The plate is charged, so that the radiation generates electrons which are accelerated down the tubes and together form an image that can be read off. X-ray telescopes have been flown on missions such as YOHKOH, the CHANDRA X-RAY OBSERVATORY and NEWTON.

X-ray transient X-ray outburst from a BINARY STAR system; it is similar to an X-RAY NOVA outburst but lasts for a longer period. The X-ray source will brighten suddenly and fade away after a few weeks (or occasionally months). More than 30 objects have been found to show X-ray transients. Some NEUTRON STARS have hot companion stars and some have cool (K or M), low-mass companions. The transient indicates that mass transfer is erratic for some (unknown) reason. *See also* X-RAY BURSTER

Yagi antenna Basic form of ANTENNA, used in RADIO TELESCOPES and for work on the IONOSPHERE. It consists of several parallel elements mounted on a straight member; it is the familiar form of a television aerial. A Yagi antenna often forms the basis of cheap arrays used for APERTURE SYNTHESIS. It was developed by the Japanese engineer Hietsugu Yagi (1886–1976).

year Time taken for the Earth to complete a single revolution around the Sun. Several different types of year are defined, with differing lengths, according to which point of reference is chosen. An ANOMALISTIC YEAR is one revolution relative to PERIHELION; it is equivalent to 365.25964 mean solar days. An ECLIPSE YEAR is one revolution relative to the same node of the Moon's orbit; it is equivalent to 346.62003 mean solar days. Nineteen eclipse years are 6585.78 days, almost exactly the same as a SAROS. A SIDEREAL YEAR is one revolution relative to the fixed stars; it equivalent to 365.25636 mean solar days. A TROPICAL YEAR or solar year is one revolution relative to the EQUINOXES; it equivalent to 365.24219 mean solar days. For convenience, the civil or CALENDAR year is set at a whole number of days, usually 365 but 366 in a LEAP YEAR.

Yerkes Observatory Observatory of the University of Chicago, located at an elevation of 334 m (1050 ft) at Williams Bay, Wisconsin, near Chicago. It is famous for its 1.02-m (40-in.) refracting telescope, built in 1897 with optics figured by ALVAN CLARK & SONS and a mounting constructed by the WARNER & SWASEY COMPANY, and still the largest refractor in the world. The observatory owes its foundation to George Ellery HALE, who persuaded the streetcar magnate Charles Tyson Yerkes (1837–1905) to finance the great telescope. The observatory also has 1.02-m (40-in.) and 0.6-m (24-in) reflectors and a number of smaller instruments. Gerard KUIPER used the 40-inch to discover MIRANDA in 1948 and NEREID the year after. Today, though LIGHT POLLUTION limits observational astronomy from the William Bay site, Yerkes is still a major research institution and its astronomers use such facilities as the Astrophysics Research Consortium's 3.5-m (138-in.) telescope at Apache Point, New Mexico, the HUBBLE SPACE TELESCOPE and the telescopes of the W.M. KECK OBSERVATORY.

'Y-feature' Characteristic atmospheric feature of VENUS. Although this planet is virtually featureless in visible light, ultraviolet photographs reveal a banded structure that resolves into a characteristic shape, a horizontal letter Y. This feature rotates around the planet in a period of only four to five days, implying wind speeds of 100 m/s (330 ft/s) in the upper atmosphere. The solid surface rotates at only about 4 m/s (13 ft/s).

Yohkoh (Sunbeam) Solar X-ray satellite, a collaboration between the Japanese Institute of Space and Astronautical Science (ISAS), NASA and the UK. Launched from the Kagoshima Space Centre in 1991 August, the spacecraft has observed the solar atmosphere continuously for more than an entire cycle of solar activity.

Yohkoh's main scientific objective is to observe the energetic phenomena taking place on the Sun, especially X-ray and gamma-ray emissions from solar FLARES. It carries four instruments that detect these energetic emissions – two spectrometers, a soft X-ray telescope (SXT) and a hard X-ray telescope (HXT).

The observations of spectral lines provide information about the temperature and density of the hot plasma in the Sun's atmosphere, and about motions of the plasma along the line of sight. The SXT images X-rays in the range 0.25–4.0 keV and can resolve features down to 2″.5 in size. Flare images can be obtained every 2 seconds. Smaller images with a single filter can be obtained as frequently as once every 0.5 seconds. The HXT observes hard X-rays in four energy bands and can resolve structures with angular sizes of about 5″. These images can also be obtained once every 0.5 seconds.

Yohkoh played an important role in improving understanding of processes in the solar atmosphere. Of particular interest were the origin of solar flares, the link between such flares and coronal mass ejections, studies of coronal holes and the evolution of magnetic loops.

The Yohkoh spacecraft is in a slightly elliptical low Earth orbit, with an altitude ranging from approximately 570 to 730 km (355 to 455 mi); 65 to 75 minutes of each 90-minute orbit are spent in sunlight.

Young, Charles Augustus (1834–1908) American solar astronomer at Dartmouth (1866–77) and Princeton (1877–1905) who made early spectroscopic studies of the solar corona, proving its gaseous nature. He obtained the first FLASH SPECTRA of the solar chromosphere during a total eclipse in 1870 and compiled a catalogue of bright solar spectral lines to measure the Sun's rotational velocity. Young obtained the first successful photographs of solar prominences in visible light.

▼ **Yohkoh** This Yohkoh image of the active Sun at X-ray wavelengths shows hot coronal plasma in magnetic loops above active regions. Cooler, dark regions above the poles are described as coronal holes; these extend to lower solar latitudes at sunspot minimum.

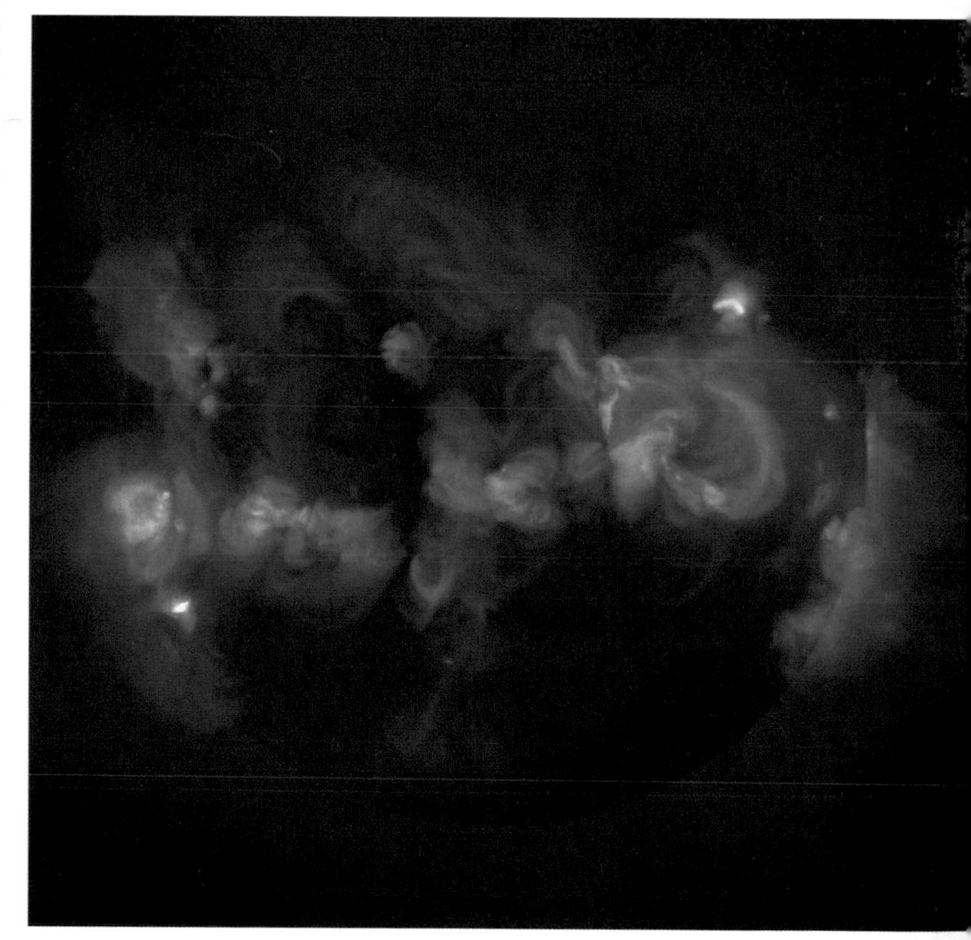

Z

z Symbol for REDSHIFT

Zach, Franz Xaver von (1754–1832) Hungarian astronomer known for his part in the discovery of Ceres. In 1786 he began building the Seeberg Observatory near Gotha, Germany; three years later he began searching for the hypothetical planet between Mars and Jupiter predicted by BODE'S LAW. Concluding that a more organized search by the world's most skilled observers would be necessary to find the 'missing' planet, in 1800 he convened a meeting at the private observatory of Johann SCHRÖTER, one of the CELESTIAL POLICE. On 1801 January 1, Giuseppe PIAZZI discovered CERES, orbiting between Mars and Jupiter. The asteroid was subsequently lost, but Zach recovered it using orbital calculations by Carl Friedrich GAUSS.

ZAMS Abbreviation of ZERO-AGE MAIN SEQUENCE

Z Andromedae star (ZAND) Type of CATACLYSMIC VARIABLE, included in the larger SYMBIOTIC STAR grouping. Z Andromedae star systems resemble DWARF NOVAE, but instead of a main-sequence secondary, they contain a red giant or supergiant together with a hot white dwarf. The components are close together, and MASS TRANSFER appears to occur either in the form of a stream of material or as an enhanced stellar wind, depending on the system. The material is accreted by the white dwarf, either directly or through an ACCRETION DISK, giving rise to occasional outbursts. Z Andromedae itself has a magnitude range of 8.3 to 12.4.

zap crater Most commonly, a very small crater created by micrometeorite impact. It is a usually lined with glasses and surrounded by fractures.

Zarya Russian Functional Energy module for the INTERNATIONAL SPACE STATION (ISS). It was the first ISS component to be launched, in 1988.

Z Camelopardalis star (UGZ) ERUPTIVE VARIABLE star, a subtype of DWARF NOVA. Z Camelopardalis stars differ from the common dwarf nova class (*see* U GEMINORUM STAR) in that they experience occasional 'standstills', remaining more or less constant in brightness for a long period. These standstills always seem to begin during a decline from maximum. When the standstill ends, the star drops to minimum brightness, and then resumes its 'normal' behaviour. Both the occurrence of the standstills and their duration, ranging from a few days to many months, are quite unpredictable. Z Camelopardalis itself is the brightest member of the class. It may reach magnitude 10.2 at brightest and falls to about 14.5 at minimum. The usual interval between outbursts is roughly 22 days.

Zeeman effect Splitting of a spectral line into several components by the presence of a magnetic field. Where the components are unresolved, a broadened line is seen. The effect allows the measurement of the strengths of magnetic fields on the Sun, the stars and even in the interstellar medium.

Zel'dovich, Yakov Borisovich (1914–87) Russian astrophysicist and cosmologist, born in Minsk, modern Belorussia, who originated the 'pancake model' of LARGE-SCALE STRUCTURE in the Universe. With Rashid Alievich Sunyaev (1943–) he described the **Sunyaev–Zel'dovich effect**, an apparent reduction in the temperature of the cosmic background radiation as it passes through hot ionized gas between members of galaxy clusters. In the early 1970s, Zel'dovich and others developed a model of the early Universe in which huge discrete lumps of primordial matter collapsed asymmetrically under their own weight as they cooled, forming thin 'pancakes'. The model correctly predicts the arrangement of galaxies in sheets and voids revealed by REDSHIFT SURVEYS.

Zelenchukskaya Astrophysical Observatory *See* SPECIAL ASTROPHYSICAL OBSERVATORY

Zenit Heavyweight former Soviet Union satellite launcher, first flown in 1985, which can place payloads weighing 13 tonnes into low Earth orbit. The two-stage Zenit, built in the Ukraine, had 28 successful and 9 failed launches to 2001. It mainly carried Soviet electronic intelligence and Earth observation satellites. The booster also provides the basis of the SEA LAUNCH commercial satellite launcher.

zenith Point on the CELESTIAL SPHERE directly overhead an observer and 90° from the horizon. This is known as the astronomical zenith. Because the Earth is not a sphere, the geocentric zenith is defined as a line joining the centre of the Earth to the observer. The point 180° away from the zenith, directly beneath an observer, is the NADIR.

zenith distance Angular distance from the ZENITH to a celestial body, measured along a GREAT CIRCLE. It is usually expressed as a topocentric measure, from the observer's position on the Earth's surface, but sometimes geocentric, as measured by a hypothetical observer at the centre of the Earth. The zenith distance is equal to 90° minus the altitude of the body above the horizon.

zenithal hourly rate (ZHR) Useful index of METEOR SHOWER activity, allowing comparison of observations made at different times and under different sky conditions. Zenithal hourly rate is determined by allowing for the altitude (a) of the shower RADIANT during observations, the stellar limiting magnitude (LM), and the POPULATION INDEX (r, indicative of the proportions of faint meteors that might be lost to background skyglow in the shower under study). The derived ZHR corresponds to the expected number of meteors seen in a perfectly transparent sky ($LM = +6.5$) with the radiant overhead; it is calculated by multiplying the observed hourly count by $1/\sin a \times r^{6.5-LM}$.

zero-age main sequence (ZAMS) MAIN SEQUENCE as defined by stars of zero age, which is the point when they have achieved a stable state, with core temperatures sufficiently high for nuclear fusion to begin. As the star evolves and changes hydrogen into helium, its chemical composition changes and it shifts to the right of its zero-age position on the HERTZSPRUNG–RUSSELL DIAGRAM.

zero gravity Apparent absence of GRAVITATIONAL FORCES within a free-falling system. A body in a 'zero gravity' or free-fall state experiences no sensation of weight, hence the 'weightlessness' of astronauts, since the spacecraft is continuously free falling towards the Earth while its transverse motion ensures that it gets no closer. The term 'zero gravity' does not imply a total absence of gravity, rather it refers to an absence of any detectable gravitational forces. Although gravity becomes very weak at large distances from massive bodies, it nowhere declines absolutely to zero. *See also* ACCELERATION OF FREE FALL

Zeta Aurigae ECLIPSING BINARY with a period of 972 days; the visual range is from magnitude 3.7 to 4.2. It is the faintest of the three 'Kids' near Capella. The Zeta Aurigae system consists of a hot B-type star and a supergiant companion of type K; the distance is about 790 l.y. During the partial phase of the eclipse of the hot star, its light shines through the outer, rarefied layers of the supergiant, and there are complicated and informative spectroscopic effects. The eclipse of the B-type star is total for 38 days; this is preceded and followed by partial stages lasting for 32 days each.

Zhang Heng (AD 78–139) Chinese scientist, the first in China to build a rotating celestial globe and an armillary sphere with horizon and meridian rings. With these and other simple instruments, he observed and catalogued 320

bright stars, and estimated the total number of naked-eye stars as 11,520. He concluded that 'the sky is large and the Earth small' – a radical concept at the time. Zhang understood that the Earth and Moon are spherical, lunar eclipses are caused by the Earth's shadow falling upon the Moon, and the Moon shines by reflected sunlight.

ZHR Abbreviation for ZENITHAL HOURLY RATE

Zijin Shan Observatory *See* PURPLE MOUNTAIN OBSERVATORY

zodiac Belt of CONSTELLATIONS, roughly 8° on either side of the ECLIPTIC, through which the Sun, Moon and planets (except Pluto) appear to pass. The zodiac includes the 12 familiar constellations – Aries, Taurus, Gemini, Cancer, Leo, Virgo, Libra, Scorpio, Sagittarius, Capricornus, Aquarius and Pisces, which are of varying sizes. In ASTROLOGY, however, the zodiac is divided into 12 equal signs, each 30° long, but because of the effects of PRECESSION and re-definitions of constellation boundaries they no longer coincide with the constellations of the same name. Precession has also caused the ecliptic to now pass through the constellation Ophiuchus, while the zodiac now also includes parts of Cetus, Orion and Sextans.

zodiacal band Very faint band of light that extends along the ECLIPTIC and joins the ZODIACAL LIGHT to the GEGENSCHEIN. Its brightness is variable and it can only be observed under conditions of extreme clarity when no Moon is present. Like the zodiacal light, it is caused by the scattering of sunlight towards the Earth by a cloud of dust particles surrounding the Sun (*see* INTERPLANETARY DUST). The band is faintest at around 135° from the Sun and brightens towards the cone of the zodiacal light and towards the gegenschein.

zodiacal catalogue Catalogue of stars in a narrow zone straddling the ECLIPTIC (the zodiac), through which most of the planets and asteroids move. OCCULTATIONS of stars in this region by the Moon, planets or smaller members of the Solar System permit highly accurate position measurements of the occulting body, and may also reveal the binary nature of an occulted star, or even allow an estimation of its radius. Zodiacal catalogues therefore give highly accurate positions of stars. James Robertson's *Catalogue of 3539 Zodiacal Stars* brighter than 9th magnitude was published in 1940; the largest zodiacal catalogue (USNO-SA2.0) lists about 50 million stars.

zodiacal dust *See* INTERPLANETARY DUST

zodiacal light Faint, diffuse conical skyglow seen extending along the ecliptic soon after twilight ends at sunset, or before dawn begins to brighten the sky ahead of sunrise. The zodiacal light is comparable in brightness to the Milky Way, and it is best seen from temperate latitudes in the spring evening sky about 90 minutes after sunset, or in the autumn morning sky about 90 minutes before sunrise. At these times, the ecliptic is steeply inclined relative to the western or eastern horizon respectively. From lower latitudes – between the tropics and the equator – viewing conditions for the zodiacal light are favourable throughout the year. Transparent skies and the absence of moonlight (even a crescent moon can swamp it) are essential for successful observation of the zodiacal light.

Broadest at its base, the zodiacal light extends some 60–90° along the ecliptic from the Sun, and it is produced by the scattering of sunlight from myriad small (1–300 μm diameter) particles lying in the plane of the planets' orbits. This material, originating from emissions by comets close to perihelion and collisions in the asteroid belt, forms a vast zodiacal dust complex, which permeates the inner Solar System out to the orbit of Jupiter, 5 AU from the Sun. Spacecraft measurements show that the dust is very much less abundant beyond Jupiter.

The zodiacal light is joined around the ecliptic to the GEGENSCHEIN by narrow faint extensions known as the zodiacal bands. Variations in the intensity of the zodiacal light are thought to occur, with maximum brightness being found at sunspot minimum when interplanetary space is pervaded by fast-flowing particle streams from coronal holes.

The zodiacal dust complex contains a mass of material estimated to be equivalent to that of a typical comet nucleus; without continual replenishment from active comets, the complex and its associated zodiacal light would probably disappear within about 10,000 years.

Zöllner, (Johann Karl) Friedrich (1834–82) German inventor of astronomical instruments and pioneer astrophysicist. In the late 1850s and early 1860s, he perfected the astronomical PHOTOMETER, with which the relative brightnesses of stars are measured accurately by comparing them with an artificial star produced by a petroleum lamp. The Potsdam Observatory used this instrument to compile the first photometric star catalogue, the *Photometrische Durchmusterung des nördlichen Himmels*. He also invented the 'reversion spectroscope' – based on the same principles as the HELIOMETER – which Hermann Carl VOGEL used to calculate the Sun's rotation period. In theoretical astrophysics, Zöllner introduced the idea that a star's temperature determines its spectral characteristics, and that these attributes are both related to the star's evolutionary stage.

Zöllner photometer Visual PHOTOMETER that uses a fixed and a rotating polarizing element to vary the apparent brightness of an artificial star until it is the same as a real star seen in the same field of view. The amount of rotation of the polarizer can be calibrated to give the APPARENT MAGNITUDE of the real star.

◄ **zodiacal light** Produced by scattering of sunlight from interplanetary dust, the zodiacal light appears as a conical glow extending along the ecliptic in the late-evening or pre-dawn sky. In this photograph, the planet Venus is visible at lower right.

Zond Eight unmanned Soviet spacecraft launched in 1964–70. Zond 1, 2 and 4 returned no data. Zond 3 photographed the farside of the Moon in 1965. Zond 5, 6, 7 and 8 went around the Moon and returned to Earth as part of preparations for a crewed circumlunar mission.

zone of avoidance Region of the sky near the plane of the Milky Way, where dust absorption and the high concentration of stars make it difficult to locate other galaxies optically. It typically spans 10° on either side of the galactic plane. Some kinds of galaxies in the zone of avoidance can now be detected by their radio, infrared or X-ray emissions, which are less vulnerable to dust and gas absorption. Surveys of such galaxies are important for tracing LARGE-SCALE STRUCTURE, since some important nearby SUPERCLUSTERS and the GREAT ATTRACTOR either cross the zone of avoidance or lie mostly within it.

Zürich number *See* RELATIVE SUNSPOT NUMBER

Zvezda Russian Service Module for the INTERNATIONAL SPACE STATION (ISS).

Zwicky, Fritz (1898–1974) Swiss physicist and astrophysicist, born in Bulgaria, known chiefly for his observational and theoretical work on supernovae and his cataloguing of clusters of galaxies. He left Switzerland for the United States in 1925, joining the CALIFORNIA INSTITUTE OF TECHNOLOGY in 1927, where he was appointed professor of astrophysics in 1942. Although Zwicky lived in America for nearly fifty years, he retained his Swiss citizenship.

In 1934 he and Walter BAADE coined the term 'supernova'. They had noted that ordinary novae in the Andromeda Galaxy (M31) reached a maximum average apparent magnitude of 17. However, the 'nova' observed in 1885 in that galaxy, designated S Andromedae, had reached 7th magnitude, and now that the great distance to M31 was appreciated, it was clear that this was a phenomenon of a different order from ordinary novae. Zwicky and Baade suggested that a supernova is a cataclysmic stellar explosion that leaves behind a neutron star, and that the CRAB NEBULA was a supernova remnant (the latter confirmed in 1968 with the discovery of the CRAB PULSAR). From 1936, using the newly developed Schmidt camera, Zwicky discovered and examined many supernovae in other galaxies.

In 1933 Zwicky inferred the presence of dark matter by observing that outlying members of the COMA CLUSTER were moving more rapidly than could be explained by the calculated mass of the cluster, and four years later he suggested that dark matter could be investigated via gravitational lensing by intervening galaxies. His extensive studies of galaxies culminated in the six-volume *Catalogue of Galaxies and Clusters of Galaxies* (1961–68), listing 30,000 galaxies and 10,000 clusters, compiled with colleagues and completed just before his death.

ZZ Ceti star (ZZ) White dwarf VARIABLE STAR, with low amplitude (0.001–0.2 mag.). The variations arise from non-radial pulsations, and generally multiple periods are present simultaneously in each star. Periods range from about 30 seconds to 25 minutes. There are three recognized subtypes based upon the presence of hydrogen, helium, or helium and carbon absorption lines in the spectra. Approximately 50 examples are currently known.

STAR MAPS

CREATED BY WIL TIRION

Maps 1 & 2: Northern Stars to dec. +30°

Map 3: Equatorial Stars, RA 18 to 0^h, dec. +60° to –60°

Map 4: Equatorial Stars, RA 12 to 18^h, dec. +60° to –60°

Map 5: Equatorial Stars, RA 6 to 12^h, dec. +60° to –60°

Map 6: Equatorial Stars, RA 0 to 6^h, dec. +60° to –60°

Maps 7 & 8: Southern Stars to dec. –30°

This complex of hot stars and nebulosity is designated Chamaeleon I.
It lies in the constellation of the same name.

These star maps were specially created for this edition by Wil Tirion. Between them they cover the whole of the sky, with some overlap. Maps 1 and 2, on these pages, represent the northern stars down to declination +30°. Maps 3 to 6 cover a broad strip of the sky extending 60° either side of the celestial equator, suitably oriented for northern hemisphere observers. (Southern-hemisphere observers should simply turn the book through 180°.) Maps 7 and 8 show the southern stars, to declination –30°.

All 88 constellations are shown on these maps. (An index showing on which map(s) each constellation appears will be found alongside maps 7 and 8.) The

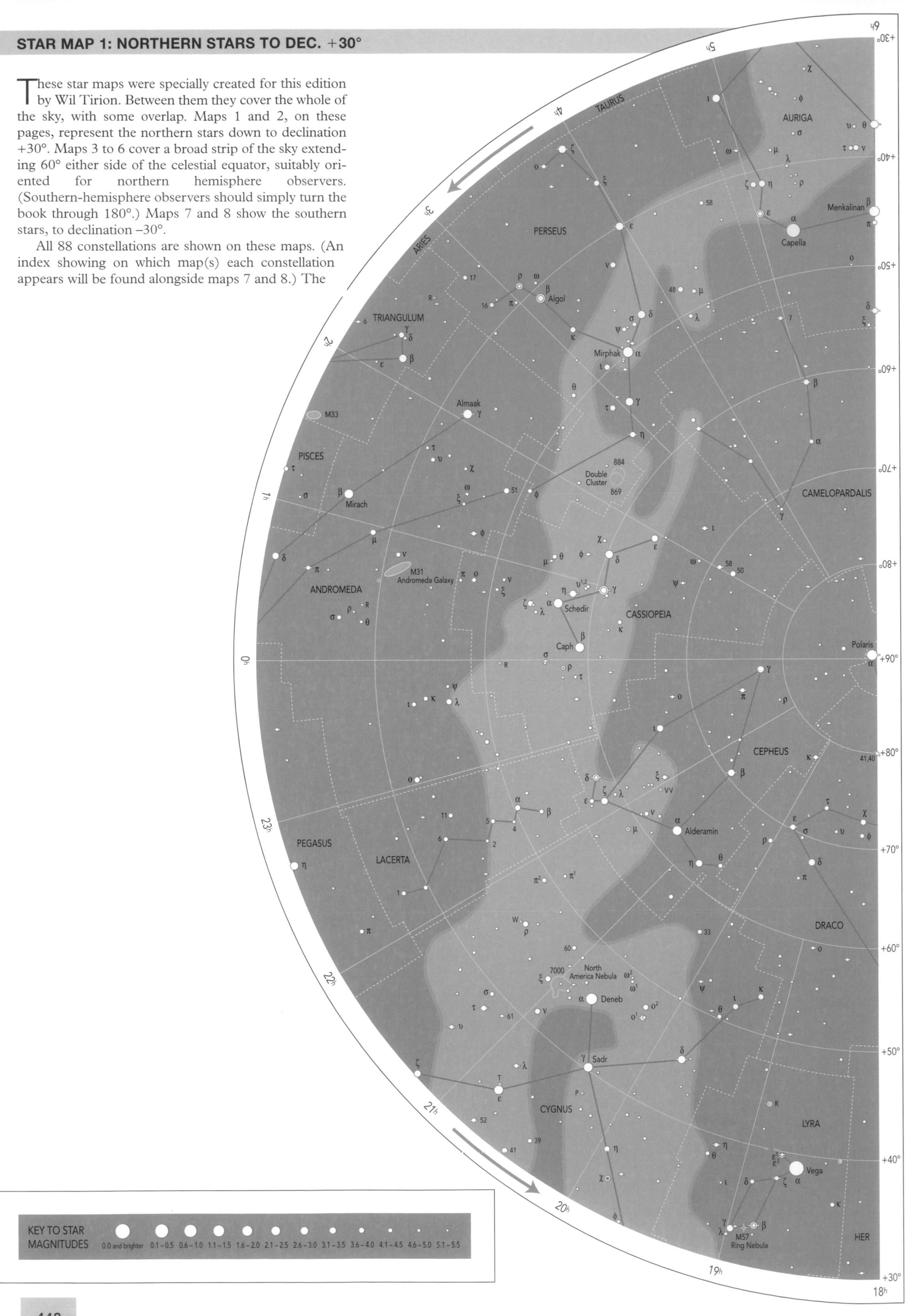

constellation names are in capital letters, while star names are lower case with an initial capital. A few other features, for example prominent asterisms such as the Square of Pegasus, are also named. The Milky Way is shown in a lighter blue than the main background. The ecliptic (the Sun's path against the background stars) is shown as a dashed red line. The Moon and the planets will always be found near the ecliptic. Stars to magnitude 5.5 are shown (in half-magnitude steps), together with major deep-sky objects. A key to star magnitudes and a key to map symbols accompany each chart. The brightest stars are shown in their true colours.

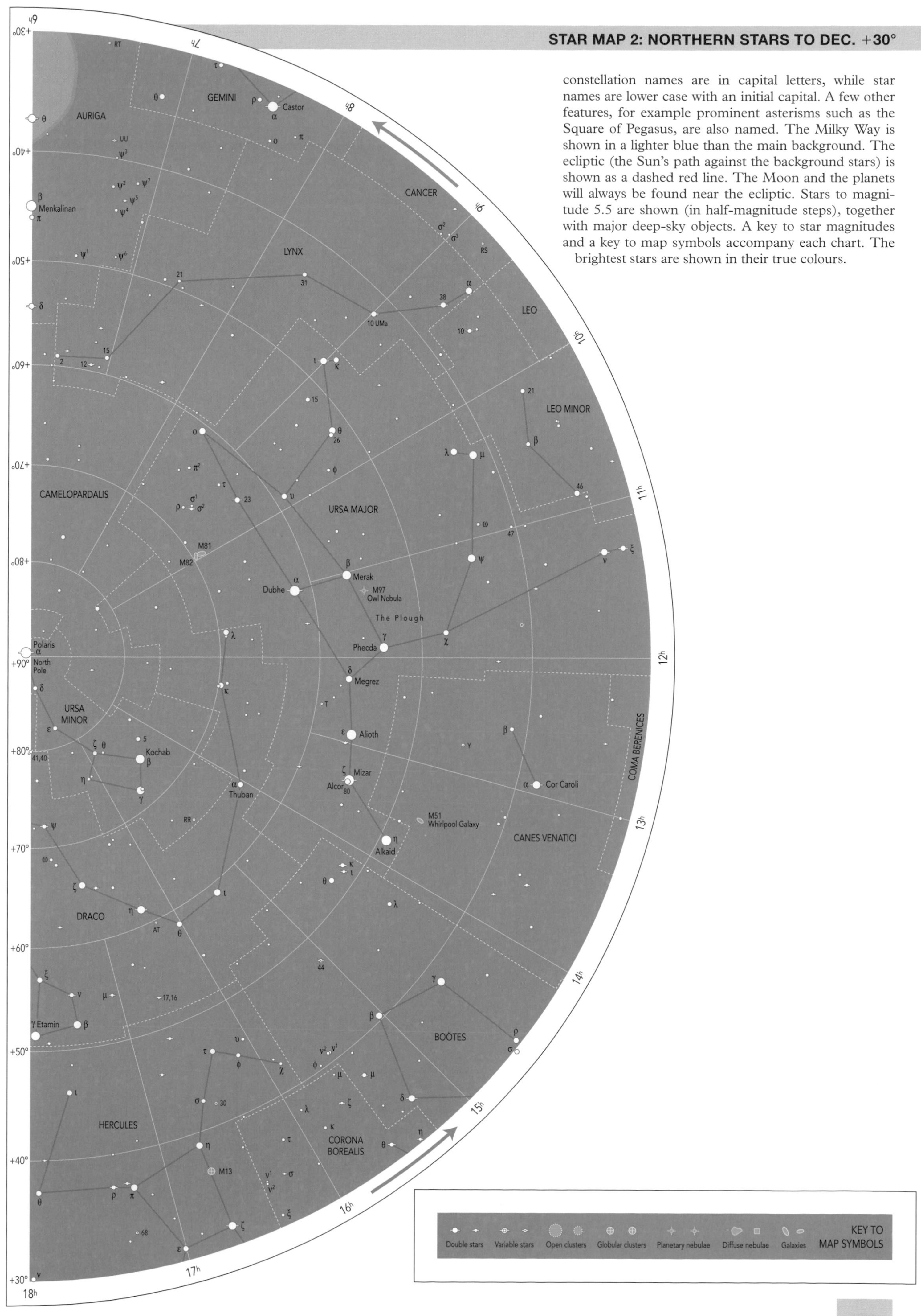

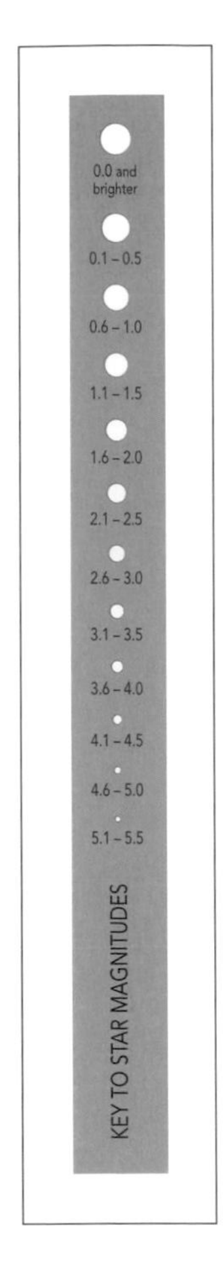
0.0 and brighter
0.1 – 0.5
0.6 – 1.0
1.1 – 1.5
1.6 – 2.0
2.1 – 2.5
2.6 – 3.0
3.1 – 3.5
3.6 – 4.0
4.1 – 4.5
4.6 – 5.0
5.1 – 5.5
KEY TO STAR MAGNITUDES

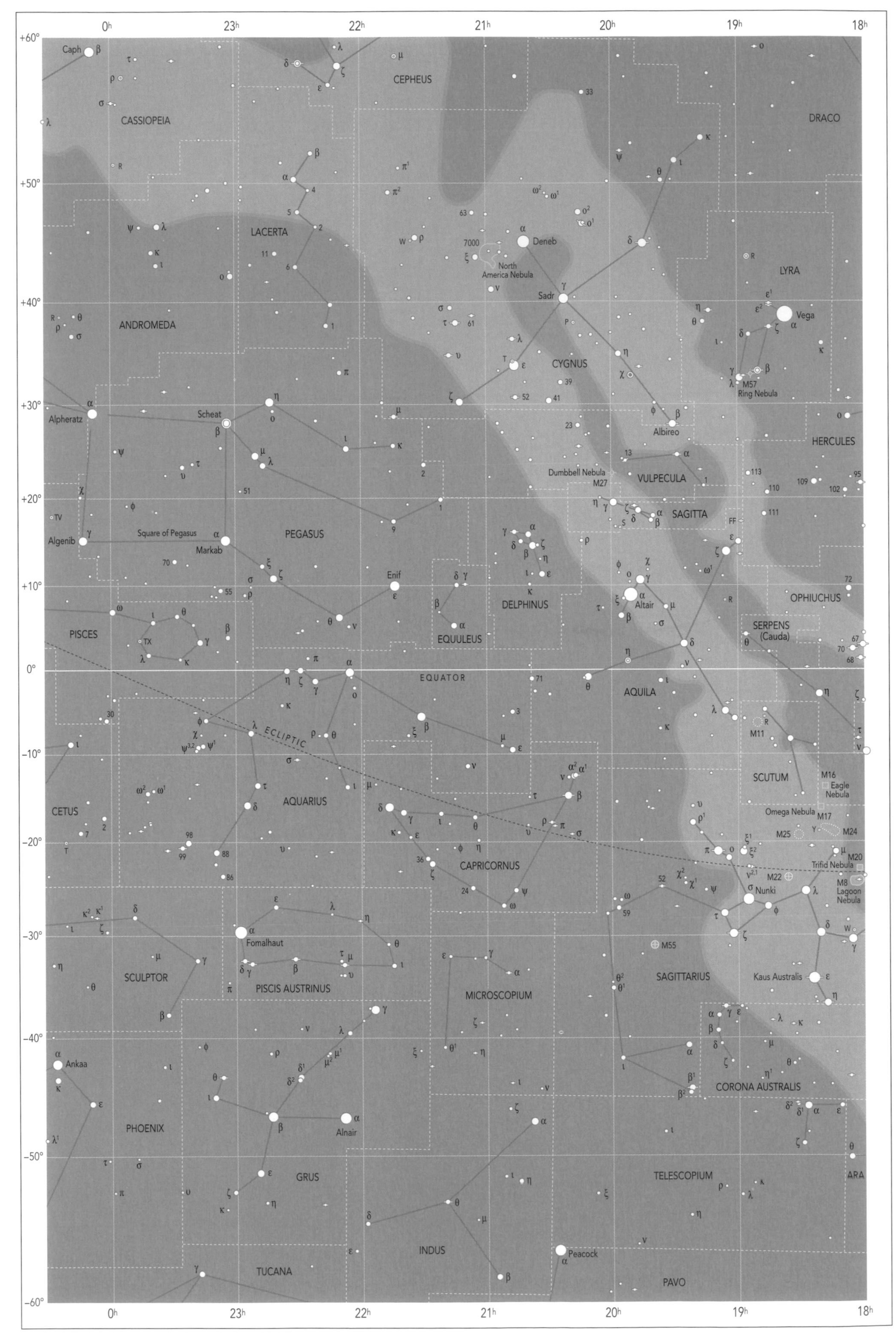
0h
23h
22h
21h
20h
19h
18h
+60°
+50°
+40°
+30°
+20°
+10°
0°
-10°
-20°
-30°
-40°
-50°
-60°
CASSIOPEIA
CEPHEUS
DRACO
LACERTA
LYRA
ANDROMEDA
CYGNUS
HERCULES
VULPECULA
SAGITTA
PEGASUS
DELPHINUS
OPHIUCHUS
SERPENS (Cauda)
EQUULEUS
PISCES
AQUILA
EQUATOR
ECLIPTIC
SCUTUM
CETUS
AQUARIUS
CAPRICORNUS
SAGITTARIUS
SCULPTOR
PISCIS AUSTRINUS
MICROSCOPIUM
CORONA AUSTRALIS
PHOENIX
GRUS
TELESCOPIUM
ARA
INDUS
TUCANA
PAVO
Caph
Deneb
North America Nebula
Sadr
Vega
M57 Ring Nebula
Albireo
Alpheratz
Scheat
Dumbbell Nebula M27
Square of Pegasus
Algenib
Markab
Enif
Altair
M11
M16 Eagle Nebula
Omega Nebula M17
M25
M24
M20 Trifid Nebula
M22
M8 Lagoon Nebula
Nunki
Fomalhaut
M55
Kaus Australis
Ankaa
Alnair
Peacock

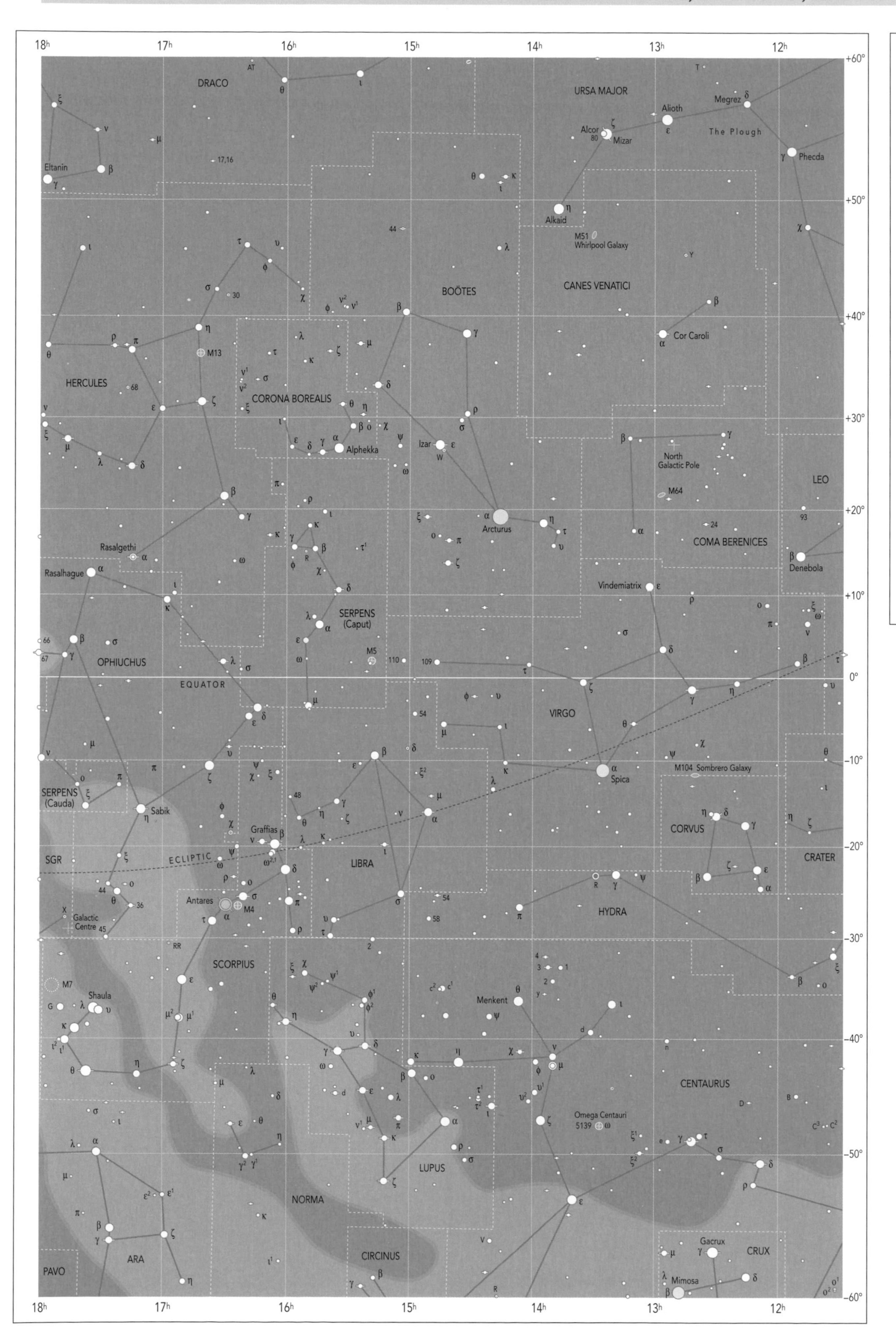

DRACO
URSA MAJOR
Alioth
Megrez
Alcor
Mizar
The Plough
Phecda
Eltanin
Alkaid
M51 Whirlpool Galaxy
BOÖTES
CANES VENATICI
Cor Caroli
HERCULES
M13
CORONA BOREALIS
Alphekka
Izar
North Galactic Pole
M64
LEO
Arcturus
COMA BERENICES
Denebola
Rasalgethi
Rasalhague
Vindemiatrix
SERPENS (Caput)
M5
OPHIUCHUS
EQUATOR
VIRGO
Spica
M104 Sombrero Galaxy
SERPENS (Cauda)
Sabik
Graffias
CORVUS
ECLIPTIC
LIBRA
SGR
CRATER
Antares
M4
HYDRA
Galactic Centre
SCORPIUS
M7
Shaula
Menkent
CENTAURUS
Omega Centauri
5139
LUPUS
NORMA
ARA
CIRCINUS
Gacrux
CRUX
Mimosa
PAVO

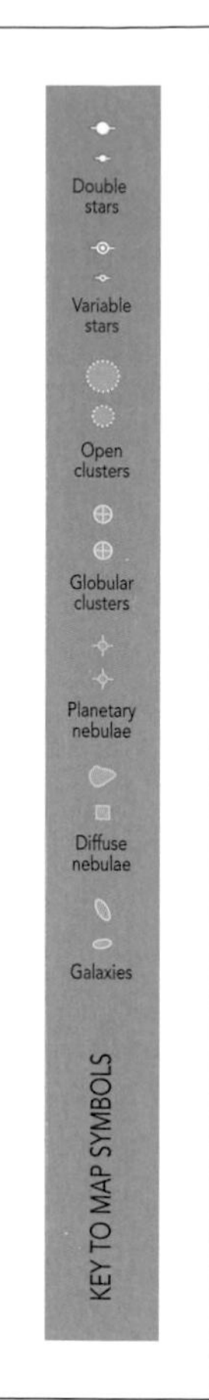

KEY TO MAP SYMBOLS
Double stars
Variable stars
Open clusters
Globular clusters
Planetary nebulae
Diffuse nebulae
Galaxies

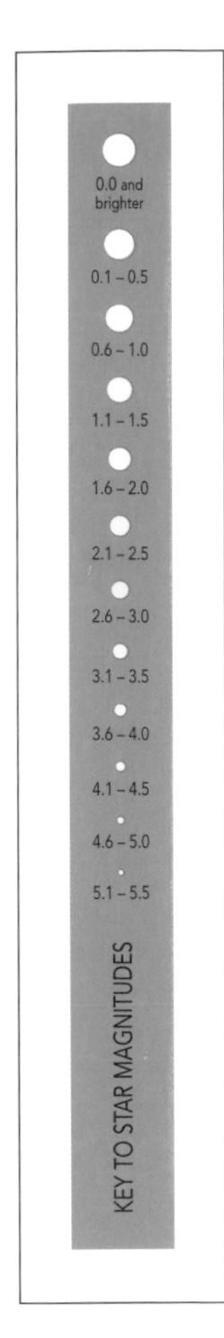
0.0 and brighter
0.1 – 0.5
0.6 – 1.0
1.1 – 1.5
1.6 – 2.0
2.1 – 2.5
2.6 – 3.0
3.1 – 3.5
3.6 – 4.0
4.1 – 4.5
4.6 – 5.0
5.1 – 5.5
KEY TO STAR MAGNITUDES

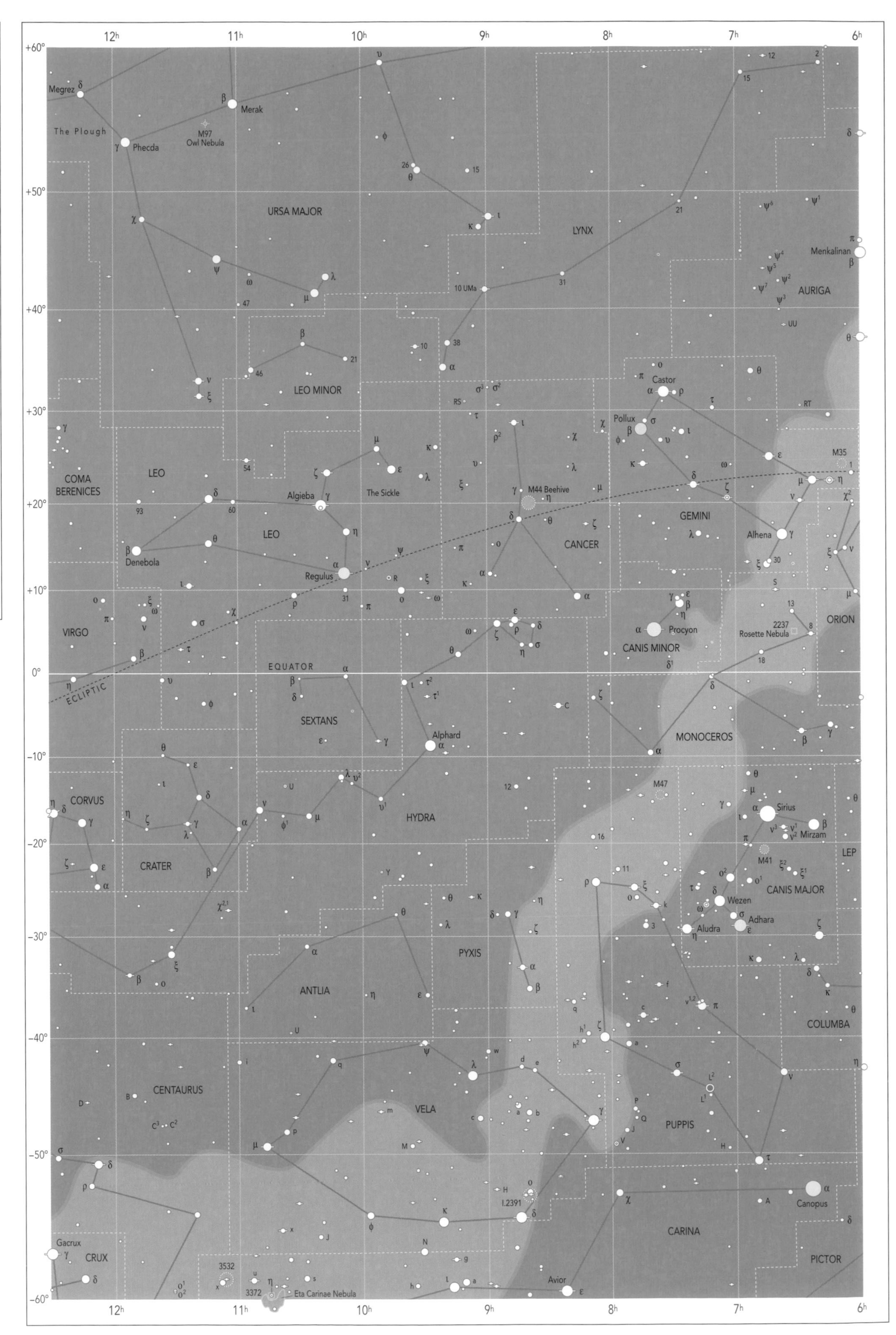
12h
11h
10h
9h
8h
7h
6h
+60°
+50°
+40°
+30°
+20°
+10°
0°
−10°
−20°
−30°
−40°
−50°
−60°
Megrez
Merak
The Plough
Phecda
M97 Owl Nebula
URSA MAJOR
LYNX
Menkalinan
AURIGA
LEO MINOR
Castor
Pollux
COMA BERENICES
LEO
M35
Algieba
The Sickle
M44 Beehive
GEMINI
Alhena
Denebola
CANCER
Regulus
VIRGO
Procyon
CANIS MINOR
2237 Rosette Nebula
ORION
EQUATOR
ECLIPTIC
SEXTANS
MONOCEROS
Alphard
CORVUS
HYDRA
M47
Sirius
Mirzam
CRATER
M41
LEP
CANIS MAJOR
Wezen
Adhara
Aludra
PYXIS
ANTLIA
COLUMBA
CENTAURUS
VELA
PUPPIS
Canopus
I.2391
CARINA
Gacrux
CRUX
3532
3372
Eta Carinae Nebula
Avior
PICTOR

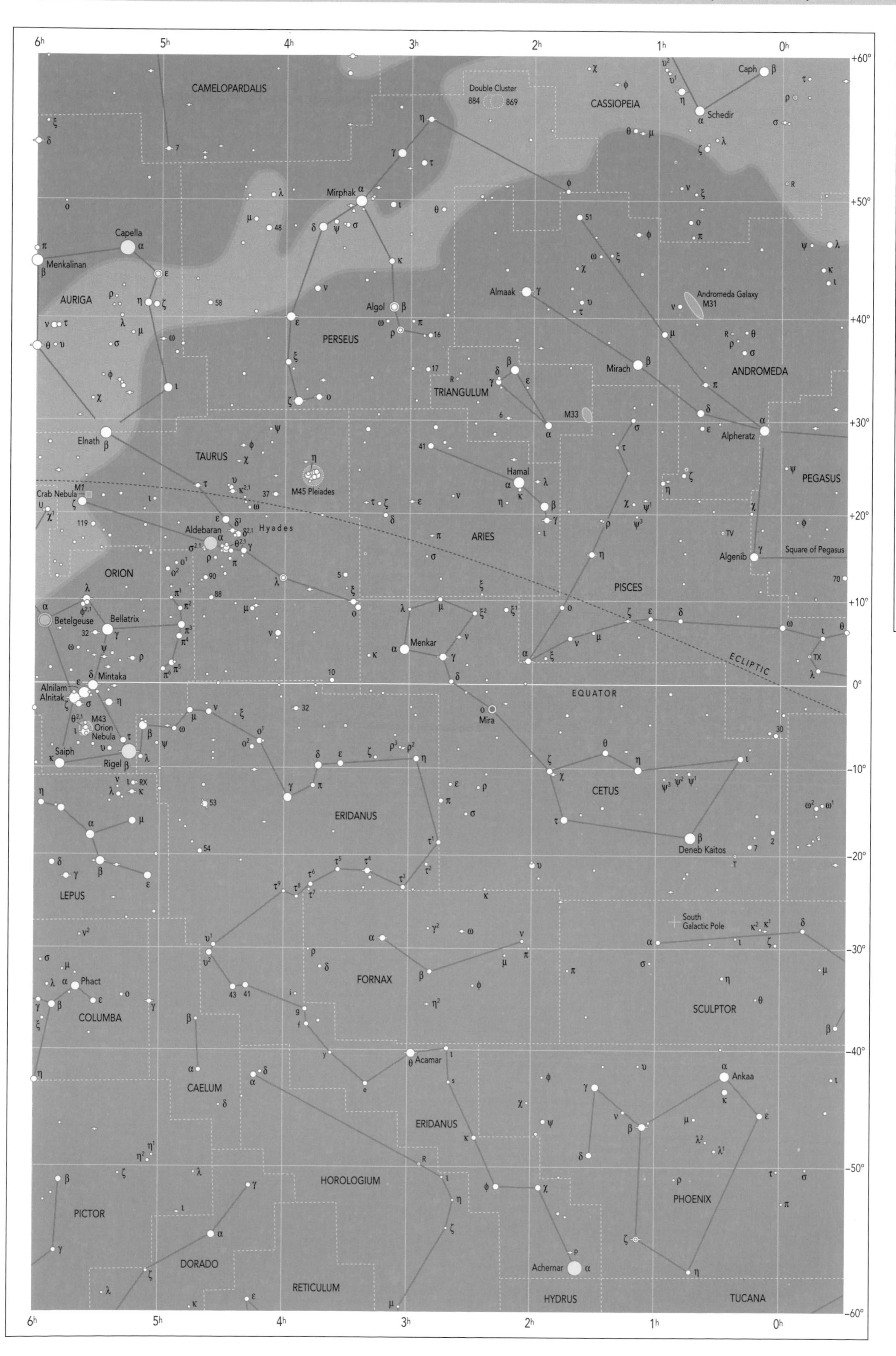
6h
5h
4h
3h
2h
1h
0h
+60°
+50°
+40°
+30°
+20°
+10°
0°
-10°
-20°
-30°
-40°
-50°
-60°
CAMELOPARDALIS
Double Cluster
884
869
CASSIOPEIA
Caph
Schedir
Mirphak
Capella
Menkalinan
AURIGA
Algol
PERSEUS
Almaak
Andromeda Galaxy
M31
Mirach
ANDROMEDA
TRIANGULUM
M33
Alpheratz
Elnath
TAURUS
M45 Pleiades
Hamal
PEGASUS
Crab Nebula
M1
Aldebaran
Hyades
ARIES
Algenib
Square of Pegasus
ORION
PISCES
Betelgeuse
Bellatrix
Menkar
ECLIPTIC
Mintaka
Alnilam
Alnitak
EQUATOR
Mira
M43
Orion Nebula
Saiph
Rigel
CETUS
ERIDANUS
Deneb Kaitos
LEPUS
South Galactic Pole
Phact
FORNAX
COLUMBA
SCULPTOR
Acamar
CAELUM
Ankaa
ERIDANUS
HOROLOGIUM
PHOENIX
PICTOR
DORADO
Achernar
RETICULUM
HYDRUS
TUCANA

Double stars
Variable stars
Open clusters
Globular clusters
Planetary nebulae
Diffuse nebulae
Galaxies
KEY TO MAP SYMBOLS

STAR MAP 7: SOUTHERN STARS TO DEC. −30°

The constellations and their names are approved by the International Astronomical Union.

Constellation	Map
Andromeda	1, 3, 6
Antlia	5
Apus	8
Aquarius	3
Aquila	3
Ara	3, 4, 8
Aries	6
Auriga	1, 2, 5, 6
Boötes	4
Caelum	6
Camelopardalis	1, 2
Cancer	5
Canes Venatici	4
Canis Major	5
Canis Minor	5
Capricornus	3
Carina	8
Cassiopeia	1
Centaurus	4, 5, 8
Cepheus	1
Cetus	6
Chamaeleon	8
Circinus	4, 8
Columba	5, 6
Coma Berenices	4
Corona Australis	3, 7
Corona Borealis	4
Corvus	4
Crater	5
Crux	8
Cygnus	3
Delphinus	3
Dorado	6, 7
Draco	3
Eridanus	6
Fornax	6
Gemini	5
Grus	3
Hercules	3, 4
Horologium	6, 7
Hydra	4, 5
Hydrus	7

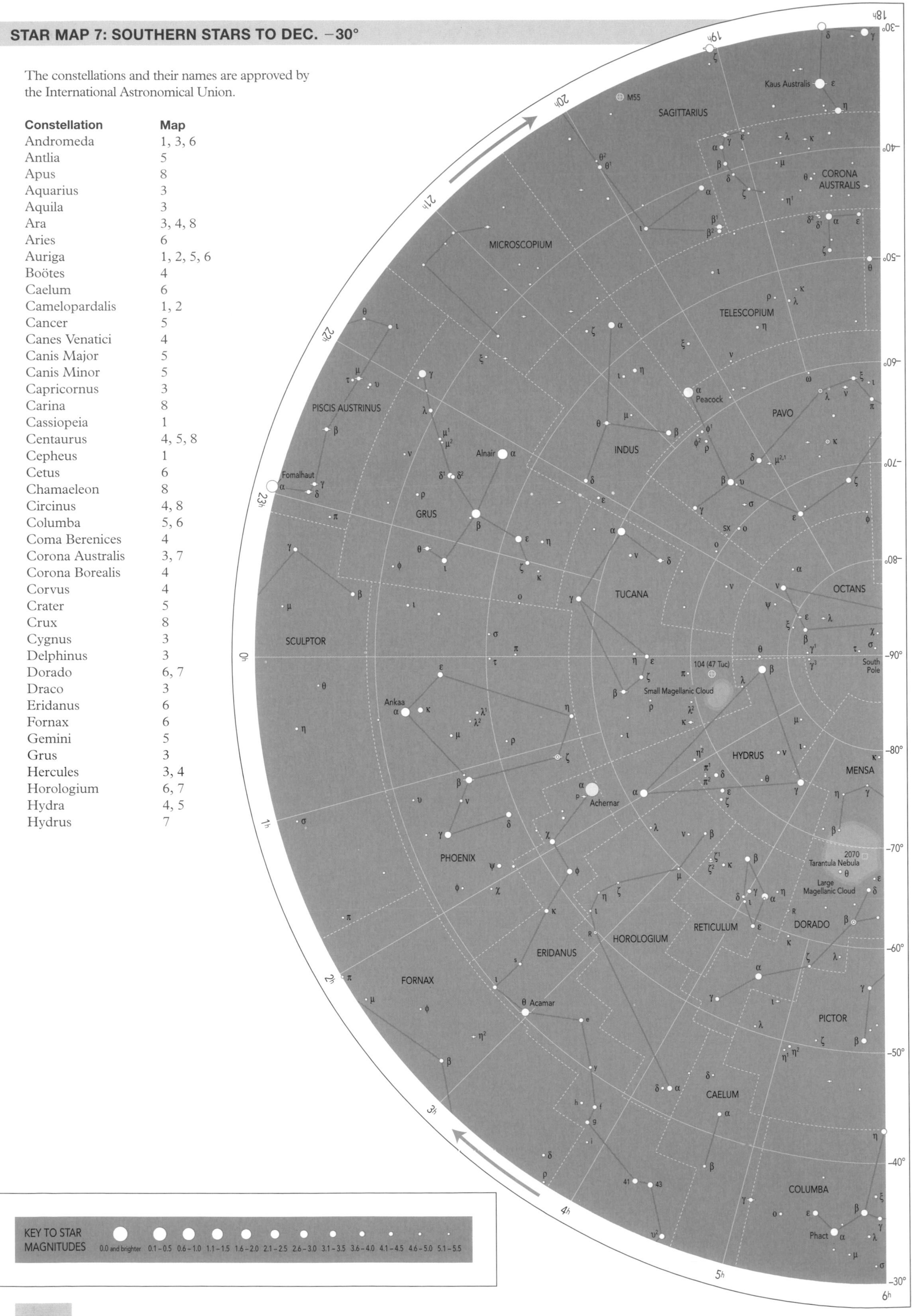

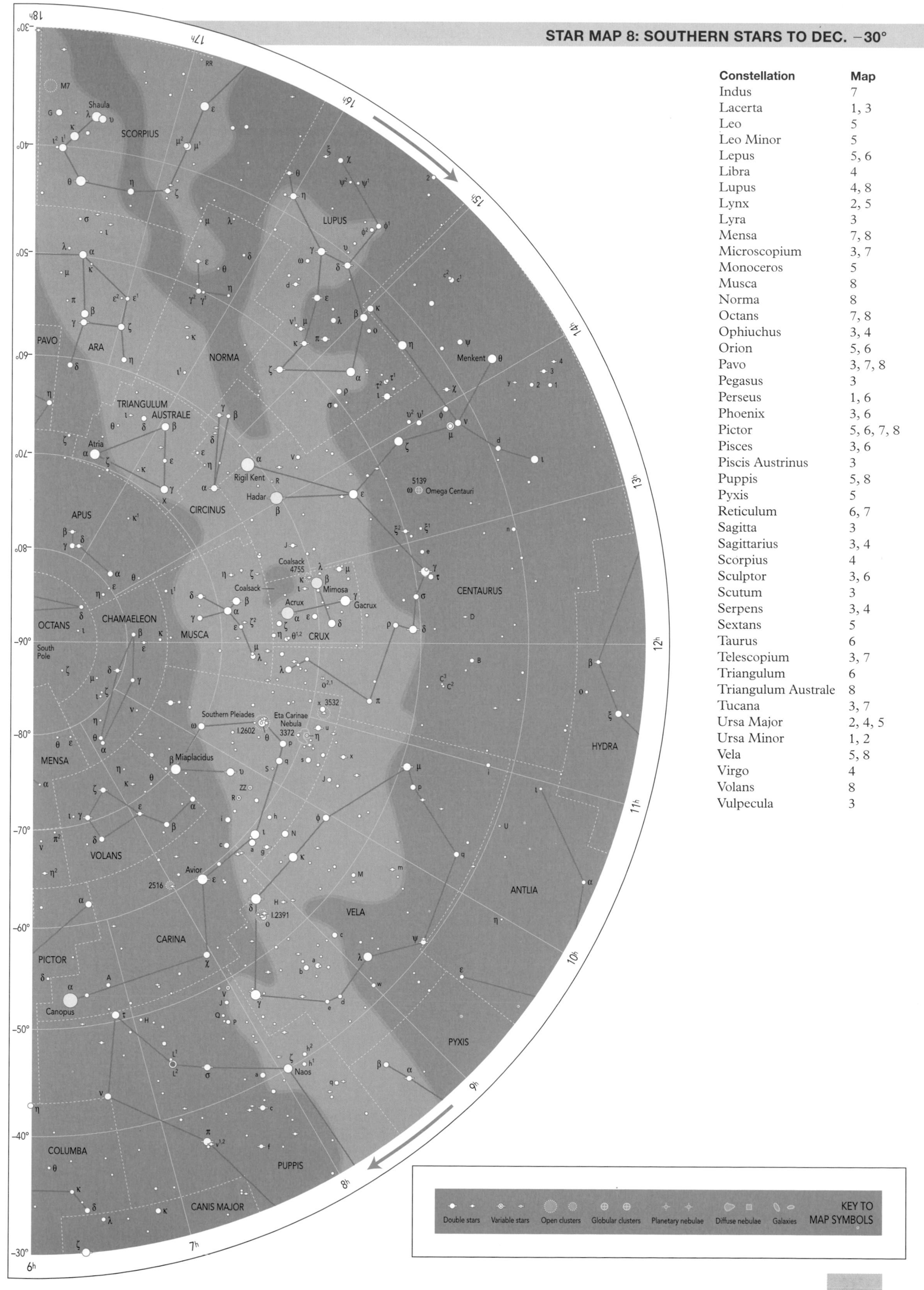

Constellation	Map
Indus	7
Lacerta	1, 3
Leo	5
Leo Minor	5
Lepus	5, 6
Libra	4
Lupus	4, 8
Lynx	2, 5
Lyra	3
Mensa	7, 8
Microscopium	3, 7
Monoceros	5
Musca	8
Norma	8
Octans	7, 8
Ophiuchus	3, 4
Orion	5, 6
Pavo	3, 7, 8
Pegasus	3
Perseus	1, 6
Phoenix	3, 6
Pictor	5, 6, 7, 8
Pisces	3, 6
Piscis Austrinus	3
Puppis	5, 8
Pyxis	5
Reticulum	6, 7
Sagitta	3
Sagittarius	3, 4
Scorpius	4
Sculptor	3, 6
Scutum	3
Serpens	3, 4
Sextans	5
Taurus	6
Telescopium	3, 7
Triangulum	6
Triangulum Australe	8
Tucana	3, 7
Ursa Major	2, 4, 5
Ursa Minor	1, 2
Vela	5, 8
Virgo	4
Volans	8
Vulpecula	3

ACKNOWLEDGEMENTS

Abbreviations
AMANDA – Antarctic Muon and Neutrino Detector Array
AUI – Associated Universities, Inc.
AURA – Association of Universities for Research in Astronomy, Inc.
Caltech – California Institute of Technology
CfA – Harvard-Smithsonian Center for Astrophysics
CNES – Centre National d'Études Spatiales, France
CXC – Chandra X-ray Observatory Center, Harvard-Smithsonian Center for Astrophysics
DLR – Deutschen Zentrum für Luft- und Raumfahrt
DMI – David Malin Images
DMSP – Defense Meteorological Satellite Program
ERSDAC – Earth Remote Sensing Data Analysis Center, Tokyo
ESA – European Space Agency
ESO – European Southern Observatory
FTS – Fourier Transform Spectrometer (Kitt Peak)
GSFC – Goddard Space Flight Center
IAC – Instituto de Astrofísica de Canarias
ING – Isaac Newton Group of Telescopes
IPAC – Infrared Processing and Analysis Center of California Institute of Technology
ISAS – Institute of Space and Astronautical Science, Japan
JAROS – Japan Resources Observation System Organization
JHU – Johns Hopkins University
JPL – Jet Propulsion Laboratory
JSC – Johnson Space Center
KSC – Kennedy Space Center
LaRC – Langley Research Center
MISR – Multi-angle Imaging SpectroRadiometer
MIT – Massachusetts Institute of Technology
MITI – Ministry of International Trade and Industry
MPE – Max-Planck-Institut für extraterrestrische Physik, Garching
NASA – National Aeronautics and Space Administration
NCSSM – North Carolina School of Science and Mathematics
NSO – National Solar Observatory
NOAA – National Oceanic and Atmospheric Administration
NOAO – National Optical Astronomy Observatory
NRAO – National Radio Astronomy Observatory
NSF – National Science Foundation
RGO – Royal Greenwich Observatory
SAO – Smithsonian Astrophysical Observatory
SOHO – Solar and Heliospheric Observatory. SOHO is a project of international cooperation between ESA and NASA
STScI – Space Telescope Science Institute
ST-ECF – Space Telescope European Coordinating Facility
SwRI – Southwest Research Institute, Boulder, Colorado
TRACE – Transition Region and Coronal Explorer, Lockheed Martin Solar and Astrophysics Laboratories
USAF – US Air Force
USGS – US Geological Survey
WIYN – Univ. Wisconsin, Indiana Univ., Yale Univ. and NOAO

t – top, *b* – bottom, *l* – left, *r* – right, *c* – centre

Endpapers T. A. Rector/B. A. Wolpa/NOAO/AURA/NSF **Half-title** ESO **Opposite title** ESO **Opposite Foreword** ESO **Opposite page 1** NASA/JSC **3** Nigel Sharp/NOAO/AURA/NSF **4*l*** NASA/JPL/Malin Space Science Systems **4*br*** NOAO/AURA/NSF **5*t*** ESO **5*b*** NASA/GSFC **6** NASA **7*t*** Science Photo Library **7*b*** NASA/JPL **8** NASA **10** NASA/JPL **13*t*** NASA/JPL/Caltech **13*b*** NASA/JPL/Cornell Univ. **14** Science Photo Library **16** T. A. Rector/B. A. Wolpa/NOAO/AURA/NSF **18** Sir Patrick Moore Collection **19** AMANDA Collaboration **21** NASA/JPL **22*t*** NASA **22*b*** NASA/David R. Scott **25** National Astronomy and Ionosphere Center/Cornell Univ./NSF **26** NASA/JPL **27*t*** NASA/JPL **27*b*** NASA/JPL/USGS **28*t*** NASA/USAF **28*b*** Chuck Claver, Nigel Sharp (NOAO)/WIYN/NOAO/NSF (Courtesy WIYN Consortium, Inc. All Rights Reserved) **32** J. and M. Tichá, Klet Observatory, Czech Republic **34** David Parker/Science Photo Library **39** NASA **43** © Akira Fujii/DMI **45** © Anglo-Australian Observatory, Photograph by David Malin **46** S. Ostro (JPL/NASA) **47** The Boomerang Collaboration **48** ESO **49*t*** All Rights Reserved Beagle 2. (http://www.beagle2.com) **49*b*** Robin Scagell/Science Photo Library **50** © Akira Fujii/DMI **51** ESO **52** NASA/JPL **53*t*** A. Dupree (CfA)/NASA/STScI **53*b*** Big Bear Solar Observatory **56** Nik Szymanek (Univ. Herts) based on data in the ING archive **57** NRAO/NSF **58** Bruce Balick and Jason Alexander (Univ. Washington)/Arsen Hajian (US Naval Obs.)/Yervant Terzian (Cornell Univ.)/Mario Perinotto (Univ. Florence)/Patrizio Patriarchi (Arcetri Observatory)/NASA **59** Rex Saffer (Villanova Univ.)/Dave Zurek (STScI) **62*t*** NASA/JPL **62*b*** T. Nakajima and S. Kulkarni (Caltech)/S. Durrance and D. Golimowski (JHU) **63** NOAO/AURA/NSF **65** NASA/JPL **68*t*** NASA/JPL **68*b*** NASA/JPL **70** Carl Grillmair (Caltech)/NASA/STScI **71** © Natural History Museum, London **72** © Anglo-Australian Observatory, Photograph by David Malin **73** NASA/JPL/Univ. Arizona **74** © Akira Fujii/DMI **75*t*** NASA/CXC/SAO **75*b*** S. Ostro (JPL/NASA) **79** NOAO/AURA/NSF **82** ESO **83** NASA/MIT/F. Baganoff et al **84** NASA/JPL/USGS **86** NASA/JPL/USGS **87** ESA **88** H. Weaver (JHU)/NASA **89** NOAO/AURA/NSF **91** © Anglo-Australian Observatory, Photograph by David Malin **92** © Anglo-Australian Observatory, Photograph by David Malin **97** John Caldwell (York Univ. Ontario)/Alex Storrs (STScI)/NASA **98*t*** Courtesy of the TRACE consortium, Stanford-Lockheed Institute for Space Research/NASA **98*b*** NASA/JPL **99** Courtesy SOHO/LASCO Consortium **100** DMR/COBE/NASA/Four-Year Sky Map **101** DMR/COBE/NASA/Two-Year Sky Map **103** P. Garnavich (CfA) et al/STScI/NASA **104** ESO **105** ESO **106** NASA/JPL **107*t*** ©Akira Fujii/DMI **107*b*** NASA/JPL/Brown Univ. **109*t*** NASA/JPL **109*b*** © Anglo-Australian Observatory, Photograph by David Malin **110** Courtesy of the Deep Impact Mission/NASA (http://deepimpact.jpl.nasa.gov) **111** NASA **112** NASA/JPL **113** ESO **114** © Akira Fujii/DMI **116** ESO **118** © Akira Fujii/DMI **119** Bill Schoening/NOAO/AURA/NSF **121** ESO **122** NASA/JSC **125** A. Cochran (Univ. Texas)/STScI/NASA **126** L. J. King (Univ. Manchester)/STScI/NASA **127** NASA/JPL/USGS **129** NOAO/AURA/NSF **130** © Anglo-Australian Observatory/Royal Observatory, Edinburgh, Photograph by David Malin **131*t*** NASA/JPL/USGS **131*b*** NASA/JPL **134** NASA/JHU Applied Physics Laboratory **135*t*** A. Fruchter and the ERO Team (STScI)/NASA **135*b*** ©Anglo-Australian Observatory/Royal Observatory, Edinburgh, Photograph by David Malin **136** Univ. Arizona/DLR/NASA/JPL **137** ESO **138** NASA **141*t*** Courtesy SOHO/MDI Consortium **141*b*** Dr S. R. McCandliss, Dr K. R. Sembach and the FUSE team **142** SEC/NOAA/US Department of Commerce **145** Courtesy SOHO EIT Consortium **147** © Anglo-Australian Observatory, Photograph by David Malin **148** Nigel Sharp/NOAO/NSO/Kitt Peak FTS/AURA/NSF **149** NASA/JPL **150** © Akira Fujii/DMI **152** NASA/JPL/DLR **153** NASA/JPL/Univ. Arizona **154** ESO **155** NASA/JPL/Brown Univ. **156** NASA/JPL/USGS **157** Gemini Observatory/NSF/Univ. Hawaii Institute for Astronomy **159** S. Ostro (JPL/NASA) **160** Nigel Sharp/NOAO/AURA/NSF **161** ESA/MPE **162** NOAO/AURA/NSF **163*t*** Michael Rich, Kenneth Mighell and James D. Neill (Columbia Univ.)/Wendy Freedman (Carnegie Observatories)/NASA **163*b*** © Anglo-Australian Observatory/Royal Observatory, Edinburgh, Photograph by David Malin **164** S. Ostro (JPL/NASA) **165*t*** G. Scharmer/Swedish Vacuum Solar Telescope **165*b*** A. Fruchter and the ERO Team (STScI, ST-ECF)/NASA **166** ESO **167** NASA/JPL/Cornell Univ. **170** © Akira Fujii/DMI **173** (x-ray) NASA/UMass/D. Wang et al (optical) NASA/HST/D. Wang **174** Courtesy SOHO/MDI Consortium **175** © Anglo-Australian Observatory, Photograph by David Malin **176** NASA/JPL/USGS **177** ESO **178** Sheila Terry/Science Photo Library **181** NASA/JPL **182** ESA **183** Sir Patrick Moore Collection/Ludolf Meyer **184** © Anglo-Australian Observatory, Photograph by David Malin **185** Raghvendra Sahai and John Trauger (JPL)/WFPC2 science team/NASA **187** R. Williams (STScI)/NASA **188** NASA/JSC **189** NASA/JPL **190** © Akira Fujii/DMI **193*tl*** NASA/JPL/USGS **193*tr*** NASA/JPL **193*b*** Roger Lynds/NOAO/AURA/NSF **194*l*** Johan Knapen/Nik Szymanek (Univ. Herts) **194*r*** Johan Knapen/Nik Szymanek (Univ. Herts) based on data in the ING Archive **195** NASA/GSFC/LaRC/JPL/MISR Team **196** IPAC/JPL **197** IPAC/JPL **198** ESA/ISOCAM/ISOGAL Team **199** NASA/Hubble Heritage Team (STScI/AURA) **200** NASA/STS-108 Crew **201** ESA **204** NASA/JPL/Lunar and Planetary Laboratory **206*t*** IPAC **206*b*** Natural History Museum, London **207** Nik Szymanek (Univ. Herts) **209** NASA/JPL **210** Ian Morison/Jodrell Bank Observatory **212** NASA/JPL/Univ. Arizona **213*t*** Reta Beebe/Amy Simon (New Mexico State Univ.)/STScI/NASA, **213*b*** NASA/ESA/John Clarke (Univ. Michigan) **214** NASA/JPL **215** © Anglo-Australian Observatory, Photograph by David Malin **216** NASA **218** © Anglo-Australian Observatory, Photograph by David Malin **219*t*** NOAO/AURA/NSF **219*b*** S. Ostro (JPL/NASA) **221** A. Caulet (ST-ECF, ESA)/NASA **222*t*** NASA/JPL **222*b*** NASA/LaRC **223** NASA/JPL **228** Alcatel Space Industries **230** NASA/GSFC/DMSP **231** NASA/Harold Weaver (JHU)/HST Comet LINEAR Investigation Team **236*t*** NASA/JSC **236*b*** NASA HQ **237** NASA/JSC **239** NOAO/AURA/NSF **241** © Anglo-Australian Observatory/Royal Observatory, Edinburgh, Photograph by David Malin **244** ESO **245** NASA/JPL **246*t*** NASA/JPL/USGS **246*b*** NASA/JPL/Malin Space Science Systems **247** NASA/JPL/Malin Space Science Systems **249*t*** NASA/JPL/GSFC **249*b*** NASA/JPL **250** ESA/ESO/MACHO Project Team **251*t*** NASA/JHU Applied Physics Team **251*b*** Richard Wainscoat/Gemini Observatory/AURA/NSF **254** NASA/JPL **255** NASA/JPL **256** Daniel Bramich (ING)/Nik Szymanek (Univ. Herts) **257** Alan Fitzsimmons **262** © Anglo-Australian Observatory, Photograph by David Malin **263** NASA/JPL **264*t*** NASA/JSC **264*b*** © Akira Fujii/DMI **265** NASA/JPL/USGS **268** NASA/JPL/USGS **269** NASA/JSC **274** Alistair Gunn, Jodrell Bank Observatory **276** Scott Manley and Duncan Steel **278*t*** NASA/JPL **278*b*** Lawrence Sromovsky (Univ. Wisconsin-Madison)/STScI/NASA **279*l*** NASA/JPL **279*r*** NASA/JPL **280** NASA/JPL **281*t*** Institute for Cosmic Ray Research, Univ. Tokyo **281*b*** NASA/NCSSM/C. Olbert et al **284** Robin Scagell/Galaxy Picture Library **285*t*** Ian King **285*b*** Mike Shara, Bob Williams and David Zurek (STScI)/NASA **288*t*** NASA/UMass/D. Wang et al **288*b*** NASA/JPL/USGS **289** NASA/JPL **290** NASA/JPL/Arizona State Univ. **291** NASA/JPL/USGS **292*t*** © Anglo-Australian Observatory, Photograph by David Malin **292*b*** Todd Boroson/NOAO/AURA/NSF **295** © Akira Fujii/DMI **296** Mark McCaughrean/ESO **297** NOAO/AURA/NSF **298** C. M. Lowne **299** Dr Seth Shostak/Science Photo Library **302*t*** NASA/JPL/USGS **302*b*** John Bally (Univ. Colorado)/NOAO/AURA/NSF **303*t*** Mir 27 Crew © CNES **303*b*** Courtesy SOHO/MDI Consortium **305** NASA/JPL/Malin Space Science Systems **308** Bill Schoening/NOAO/AURA/NSF **310** NASA/ESA/Hubble Heritage Team (STScI/AURA) **313** © Anglo-Australian Observatory/Royal Observatory, Edinburgh, Photograph by David Malin **314*t*** Dr R. Albrecht ESA/ESO/ST-ECF/NASA **314*b*** A. Stern (SwRI), B. Buie (Lowell Observatory)/ESA/NASA **316** ESO **320** Courtesy SOHO/EIT Consortium **322** NASA/JPL **323** C. R. O'Dell/Rice Univ./NASA/STScI **324** NASA/JPL **325** NASA/CXC/SAO **327** John Bahcall, Institute for Advanced Study, Princeton/Mike Disney (Univ. Wales)/NASA **328*t*** Dr John Hutchings/Dominion Astrophysical Observatory/STScI **328*b*** Gregory L. Slater/Gary A. Lindford/NASA/ISAS/Lockheed-Martin Solar and Astrophysics Laboratory of Japan/Univ. Tokyo **329** NASA/JPL **332** NRAO/AUI **334** NASA/JPL **335** NASA/JPL/USGS **337** Prof. J. Huchra/Harvard-Smithsonian Center for Astrophysics **338** NASA/Hubble Heritage Team (STScI/AURA) **340** NASA/JPL **341** NASA/JPL **343*t*** NASA/JPL **343*b*** © Anglo-Australian Observatory/Royal Observatory, Edinburgh, Photograph by David Malin **344*t*** NASA/JPL **344*b*** ESO **345** Bill Schoening/NOAO/AURA/NSF **346** MPE **347** © Anglo-Australian Observatory/Royal Observatory, Edinburgh, Photograph by David Malin **349** ESO **351** NASA/MIT/F. Baganoff et al **354*t*** NASA/Hubble Heritage Team (STScI/AURA)/R. G. French (Wellesley College)/J. Cuzzi and J. Lissauer (NASA/Ames Reseach Center)/L. Dones (SwRI) **354*b*** J. T. Trauger (JPL)/NASA **355*t*** NASA/JPL **355*cr*** NASA/JPL **356** NASA/KSC **361** © Gran Telescopio CANARIAS (GTC), Instituto de Astrofísica de Canarias **362** SETI@home/Univ. California, Berkeley **363** NOAO/AURA/NSF **364** NASA/KSC **365*t*** NASA/JPL **365*b*** Hubble Space Telescope Comet Team/NASA **368** CXC/NASA/SAO **369** NASA **370** Courtesy SOHO/LASCO Consortium **372** © Akira Fujii/DMI **373** Courtesy SOHO/MDI Consortium **374** Courtesy SOHO/LASCO Consortium **375*t*** Rolf Kever/The Institute for Solar Physics of the Royal Swedish Academy of Sciences **375*b*** ESA **376** © Anglo-Australian Observatory, Photograph by David Malin **377** Genesis Photo Library **378** NASA/KSC **383** NOAO/AURA/NSF **384** NOAO/AURA/NSF **387*t*** © Anglo-Australian Observatory, Photograph by David Malin **387*b*** © Anglo-Australian Observatory, Photograph by David Malin **390** NASA/Donald Walter (S. Carolina State Univ.)/ Paul Scowen and Brian Moore (Arizona State Univ.) **391** NASA/Jayanne English (Univ. Manitoba)/Sally Hunsberger, Sarah Gallagher and Jane Charlton (Pennsylvania State Univ.)/Zolt Levay (STScI) **393** Courtesy SOHO/MDI Consortium **394** Courtesy SOHO/MDI Consortium **395** NASA/GSFC/U. Hwang et al **396*t*** Jeff Hester (Arizona State Univ.)/NASA **396*b*** © Anglo-Australian Observatory, Photograph by David Malin **398*t*** NASA/JPL **398*b*** © Akira Fujii/DMI **399** Romano Corradi (IAC)/Mario Livio (STScI)/Ulisse Munari (Osservatorio Astronomico di Padova)/Hugo Schwarz (Nordic Optical Telescope)/NASA **400** © Anglo-Australian Observatory, Photograph by David Malin **401** NASA/JPL/Univ. Arizona **402** NASA/JPL **403** NASA/JPL **405** NASA/JPL/Lunar and Planetary Observatory **409*t*** STScI **409*b*** NASA/JPL **410*t*** NASA/GSFC **410*b*** S. Ostro (JPL/NASA) **411*t*** Bill Livingston/NOAO/AURA/NSF **411*b*** Alan Fitzsimmons **412** Robert Dalby/Nik Szymanek (Univ. Herts) **413** NASA/JPL/USGS **415** ESO **416** © Anglo-Australian Observatory/Royal Observatory, Edinburgh, Photograph by David Malin **417** Courtesy Tunguska Page of Bologna Univ. (http://www.th.bo.infn.it/tunguska) **419*t*** NASA/Hubble Heritage Team (STScI) **419*b*** NASA/KSC **420*t*** Courtesy of the UK Infrared Telescope, Mauna Kea Observatory, Hawaii **420*bl*** Nik Szymanek (Univ. Herts) **420*br*** Nik Szymanek (Univ. Herts) **421** NASA/JPL **422** Heidi Hammel (MIT)/NASA/STScI **423** Kenneth Seidelmann, US Naval Observatory/NASA/STScI **424** © Akira Fujii/DMI **425** NASA/JPL/USGS **426** © IAC/RGO/Malin, Photograph by David Malin **428*t*** © Anglo Australian Observatory, Photograph by David Malin **428*b*** Novosti **429*t*** NASA/JPL **429*b*** NASA/JPL **430** ESO **431*l*** Steve Massey **431*r*** Steve Massey **432** NASA/JPL/USGS **433** NASA/GSFC/MITI/ERSDAC/JAROS and US/Japan ASTER **434** NASA/JPL **435** © Akira Fujii/DMI **436** Javier Méndez (ING)/Nik Szymanek (Univ. Herts) **437** Harvey Richer (Univ. British Columbia)/STScI/NASA **438** Nik Szymanek (Univ. Herts) **439** © Anglo-Australian Observatory, Photograph by David Malin **441** NASA/CXC/Rutgers/J. Hughes et al **442*t*** NASA/SRON/MPE **442*b*** (X-ray) NASA/CXC/SAO (optical) AURA/NOAO/NSF **443** ISAS **445** Domenic Cantin **447** ESO

Illustration acknowledgements
Artworks prepared by Philip's/Raymond Turvey
Stefan Chabluk **40**
Cartography by Philip's **122** and **408**
Star Maps created by Wil Tirion **448–55**